ENCYCLOPEDIA OF COMPUTER SCIENCE AND TECHNOLOGY

REVISED EDITION

HARRY HENDERSON

Facts On File
An imprint of Infobase Publishing

In memory of my brother,
Bruce Henderson,
who gave me my first opportunity to explore
personal computing almost 30 years ago.

ENCYCLOPEDIA OF COMPUTER SCIENCE AND TECHNOLOGY, Revised Edition

Facts On File, Inc.
An imprint of Infobase Publishing
132 West 31st Street
New York NY 10001

Library of Congress Cataloging-in-Publication Data

Henderson, Harry, 1951–
Encyclopedia of computer science and technology / Harry Henderson.—Rev. ed.
p. cm.
Includes bibliographical references and index.
ISBN-13: 978-0-8160-6382-6
ISBN-10: 0-8160-6382-6
1. Computer science—Encyclopedias. 2. Computers—Encyclopedias. I. Title.
QA76.15.H43 2008
004.03—dc22   2008029156

Text design by Erika K. Arroyo
Cover design by Salvatore Luongo
Illustrations by Sholto Ainslie
Photo research by Tobi Zausner, Ph.D.

Printed in the United States of America

MP Hermitage 10 9 8 7 6 5 4 3

This book is printed on acid-free paper and contains
30 percent postconsumer recycled content.

CONTENTS

ACKNOWLEDGMENTS

I wish to acknowledge with gratitude the patient and thorough management of this project by my editor, Frank K. Darmstadt. I can scarcely count the times he has given me encouragement and nudges as needed. I also wish to thank Tobi Zausner, Ph.D., for her ability and efficiency in obtaining many of the photos for this book.

INTRODUCTION TO THE REVISED EDITION

Chances are that you use at least one computer or computer-related device on a daily basis. Some are obvious: for example, the personal computer on your desk or at your school, the laptop, the PDA that may be in your briefcase. Other devices may be a bit less obvious: the "smart" cell phone, the iPod, a digital camera, and other essentially specialized computers, communications systems, and data storage systems. Finally, there are the "hidden" computers found in so many of today's consumer products—such as the ones that provide stability control, braking assistance, and navigation in newer cars.

Computers not only seem to be everywhere, but also are part of so many activities of daily life. They bring together willing sellers and buyers on eBay, allow you to buy a book with a click on the Amazon.com Web site, and of course put a vast library of information (of varying quality) at your fingertips via the World Wide Web. Behind the scenes, inventory and payroll systems keep businesses running, track shipments, and more problematically, keep track of where people go and what they buy. Indeed, the infrastructure of modern society, from water treatment plants to power grids to air-traffic control, depends on complex software and systems.

Modern science would be inconceivable without computers to gather data and run models and simulations. Whether bringing back pictures of the surface of Mars or detailed images to guide brain surgeons, computers have greatly extended our knowledge of the world around us and our ability to turn ideas into engineering reality.

The revised edition of the *Facts On File Encyclopedia of Computer Science and Technology* provides overviews and important facts about these and dozens of other applications of computer technology. There are also many entries dealing with the fundamental concepts underlying computer design and programming, the Internet, and other topics such as the economic and social impacts of the information society.

The book's philosophy is that because computer technology is now inextricably woven into our everyday lives, anyone seeking to understand its impact must not only know how the bits flow, but also how the industry works and where it may be going in the years to come.

NEW AND ENHANCED COVERAGE

The need for a revised edition of this encyclopedia becomes clear when one considers the new products, technologies, and issues that have appeared in just a few years. (Consider that at the start of the 2000 decade, Ajax was still only a cleaning product and *blog* was not even a word.)

The revised edition includes almost 180 new entries, including new programming languages (such as C# and Ruby), software development and Web design technologies (such as the aforementioned Ajax, and Web services), and expanded coverage of Linux and other open-source software. There are also entries for key companies in software, hardware, and Web commerce and services.

Many other new entries reflect new ways of using information technology and important social issues that arise from such use, including the following:

- blogging and newer forms of online communication that are influencing journalism and political campaigns

- other ways for users to create and share content, such as file-sharing networks and YouTube

- new ways to share and access information, such as the popular Wikipedia

- the ongoing debate over who should pay for Internet access, and whether service providers or governments should be able to control the Web's content

- the impact of surveillance and data mining on privacy and civil liberties

- threats to data security, ranging from identity thieves and "phishers" to stalkers and potential "cyberterrorists"

- the benefits and risks of social networking sites (such as MySpace)

- the impact of new technology on women and minorities, young people, the disabled, and other groups

Other entries feature new or emerging technology, such as

- portable media devices (the iPod and its coming successors)

- home media centers and the gradual coming of the long-promised "smart house"

- navigation and mapping systems (and their integration with e-commerce)

- how computers are changing the way cars, appliances, and even telephones work

- "Web 2.0"—and beyond

Finally, we look at the farther reaches of the imagination, considering such topics as

- nanotechnology

- quantum computing

- science fiction and computing

- philosophical and spiritual aspects of computing

- the ultimate "technological singularity"

In addition to the many new entries, all existing entries have been carefully reviewed and updated to include the latest facts and trends.

GETTING THE MOST OUT OF THIS BOOK

This encyclopedia can be used in several ways: for example, you can look up specific entries by referring from topics in the index, or simply by browsing. The nearly 600 entries in this book are intended to read like "mini-essays," giving not just the bare definition of a topic, but also developing its significance for the use of computers and its relationship to other topics. Related topics are indicated by SMALL CAPITAL LETTERS. At the end of each entry is a list of books, articles, and/or Web sites for further exploration of the topic.

Every effort has been made to make the writing accessible to a wide range of readers: high school and college students, computer science students, working computer professionals, and adults who wish to be better informed about computer-related topics and issues.

The appendices provide further information for reference and exploration. They include a chronology of significant events in computing; a listing of achievements in computing as recognized in major awards; an additional bibliography to supplement that given with the entries; and finally, brief descriptions and contact information for some important organizations in the computer field.

This book can also be useful to obtain an overview of particular areas in computing by reading groups of related entries. The following listing groups the entries by category.

AI and Robotics

artificial intelligence
artificial life
Bayesian analysis
Breazeal, Cynthia
Brooks, Rodney
cellular automata
chess and computers
cognitive science
computer vision
Dreyfus, Hubert L.
Engelberger, Joseph
expert systems
Feigenbaum, Edward
fuzzy logic
genetic algorithms
handwriting recognition
iRobot Corporation
knowledge representation
Kurzweil, Raymond C.
Lanier, Jaron
Maes, Pattie
McCarthy, John
Minsky, Marvin Lee
MIT Media Lab
natural language processing
neural interfaces
neural network
Papert, Seymour
pattern recognition
robotics
singularity, technological
software agent
speech recognition and synthesis
telepresence
Weizenbaum, Joseph

Business and E-Commerce Applications

Amazon.com
America Online (AOL)
application service provider (ASP)
application software
application suite
auctions, online
auditing in data processing
banking and computers
Bezos, Jeffrey P.
Brin, Sergey
business applications of computers
Craigslist
customer relationship management (CRM)
decision support system
desktop publishing (DTP)

enterprise computing
Google
groupware
home office
management information system (MIS)
middleware
office automation
Omidyar, Pierre
online advertising
online investing
online job searching and recruiting
optical character recognition (OCR)
Page, Larry
PDF (Portable Document Format)
personal health information management
personal information manager (PIM)
presentation software
project management software
smart card
spreadsheet
supply chain management
systems analyst
telecommuting
text editor
transaction processing
trust and reputation systems
word processing
Yahoo!

Computer Architecture

addressing
arithmetic logic unit (ALU)
bits and bytes
buffering
bus
cache
computer engineering
concurrent programming
cooperative processing
Cray, Seymour
device driver
distributed computing
embedded system
grid computing
parallel port
reduced instruction set computer (RISC)
serial port
supercomputer
USB (Universal Serial Bus)

Computer Industry

Adobe Systems
Advanced Micro Devices (AMD)
Amdahl, Gene Myron
Apple Corporation
Bell, C. Gordon
Bell Laboratories
benchmark

certification of computer professionals
Cisco Systems
compatibility and portability
computer industry
Dell, Inc.
education in the computer field
employment in the computer field
entrepreneurs in computing
Gates, William III (Bill)
Grove, Andrew
IBM
Intel Corporation
journalism and the computer industry
marketing of software
Microsoft Corporation
Moore, Gordon E.
Motorola Corporation
research laboratories in computing
standards in computing
Sun Microsystems
Wozniak, Steven

Computer Science Fundamentals

Church, Alonzo
computer science
computability and complexity
cybernetics
hexadecimal system
information theory
mathematics of computing
measurement units used in computing
Turing, Alan Mathison
von Neumann, John
Wiener, Norbert

Computer Security and Risks

authentication
backup and archive systems
biometrics
computer crime and security
computer forensics
computer virus
copy protection
counterterrorism and computers
cyberstalking and harassment
cyberterrorism
Diffie, Bailey Whitfield
disaster planning and recovery
encryption
fault tolerance
firewall
hackers and hacking
identity theft
information warfare
Mitnick, Kevin D.
online frauds and scams
phishing and spoofing
RFID (radio frequency identification)

graphics formats
graphics tablet
image processing
media center, home
multimedia
music and video distribution, online
music and video players, digital
music, computer
online gambling
online games
photography, digital
podcasting
PostScript
RSS (real simple syndication)
RTF (Rich Text Format)
sound file formats
streaming (video or audio)
Sutherland, Ivan Edward
video editing, digital
YouTube

Hardware Components

CD-ROM and DVD-ROM
flash drive
flat-panel display
floppy disk
hard disk
keyboard
monitor
motherboard
networked storage
optical computing
printers
punched cards and paper tape
RAID (redundant array of inexpensive disks)
scanner
tape drives

Internet and World Wide Web

active server pages (ASP)
Ajax (Asynchronous JavaScript and XML)
Andreessen, Marc
Berners-Lee, Tim
blogs and blogging
bulletin board systems (BBS)
Bush, Vannevar
cascading style sheets (CSS)
Cerf, Vinton G.
certificate, digital
CGI (common gateway interface)
chat, online
chatterbots
conferencing systems
content management
cookies
Cunningham, Howard (Ward)
cyberspace and cyber culture
digital cash (e-commerce)
digital convergence

domain name system (DNS)
eBay
e-books and digital libraries
e-commerce
e-mail
file-sharing and P2P networks
flash and smart mob
HTML, DHTML, and XHTML
hypertext and hypermedia
Internet
Internet applications programming
Internet cafes and "hot spots"
Internet organization and governance
Internet radio
Internet service provider (ISP)
Kleinrock, Leonard
Licklider, J. C. R.
mashups
Netiquette
netnews and newsgroups
online research
online services
portal
Rheingold, Howard
search engine
semantic Web
social networking
TCP/IP
texting and instant messaging
user-created content
videoconferencing
virtual community
Wales, Jimmy
Web 2.0 and beyond
Web browser
Web cam
Web filter
Webmaster
Web page design
Web server
Web services
wikis and Wikipedia
World Wide Web
XML

Operating Systems

demon
emulation
file
input/output (I/O)
job control language
kernel
Linux
memory
memory management
message passing
microsoft windows
MS-DOS
multiprocessing

multitasking
operating system
OS X
system administrator
regular expression
Ritchie, Dennis
shell
Stallman, Richard
Torvalds, Linus
UNIX

Other Applications

bioinformatics
cars and computing
computer-aided design and manufacturing (CAD/CAM)
computer-aided instruction (CAI)
distance education
education and computers
financial software
geographical information systems (GIS)
journalism and computers
language translation software
law enforcement and computers
legal software
libraries and computing
linguistics and computing
map information and navigation systems
mathematics software
medical applications of computers
military applications of computers
scientific computing applications
smart buildings and homes
social sciences and computing
space exploration and computers
statistics and computing
typography, computerized
workstation

Personal Computer Components

BIOS (Basic Input-Output System)
boot sequence
chip
chipset
clock speed
CPU (central processing unit)
green PC
IBM PC
laptop computer
microprocessor
personal computer (PC)
PDA (personal digital assistant)
plug and play
smartphone
tablet PC

Program Language Concepts

authoring systems
automatic programming
assembler

Backus-Naur Form (BNF)
compiler
encapsulation
finite state machine
flag
functional languages
interpreter
loop
modeling languages
nonprocedural languages
ontologies and data models
operators and expressions
parsing
pointers and indirection
procedures and functions
programming languages
queue
random number generation
real-time processing
recursion
scheduling and prioritization
scripting languages
Stroustrup, Bjarne
template
Wirth, Niklaus

Programming Languages

Ada
Algol
APL
awk
BASIC
C
C#
C++
Cobol
Eiffel
Forth
FORTRAN
Java
JavaScript
LISP
LOGO
Lua
Pascal
Perl
PHP
PL/1
Prolog
Python
RPG
Ruby
Simula
Tcl
Smalltalk
VBScript

Social, Political, and Legal Issues

anonymity and the Internet
censorship and the Internet

computer literacy
cyberlaw
developing nations and computing
digital divide
disabled persons and computing
e-government
electronic voting systems
globalization and the computer industry
government funding of computer research
identity in the online world
intellectual property and computing
Lessig, Lawerence
net neutrality
philosophical and spiritual aspects of computing
political activism and the Internet
popular culture and computing
privacy in the digital age
science fiction and computing
senior citizens and computing
service-oriented architecture (SOA)
social impact of computing
Stoll, Clifford
technology policy
women and minorities in computing
young people and computing

Software Development and Engineering

applet
application program interface (API)
bugs and debugging
CASE (computer-aided software engineering)
design patterns
Dijkstra, Edsger
documentation of program code
documentation, user
document model
DOM (document Object Model)
error handling
flowchart
Hopper, Grace Murray
information design

internationalization and localization
library, program
macro
Microsoft .NET
object-oriented programming (OOP)
open source movement
plug-in
programming as a profession
programming environment
pseudocode
quality assurance, software
reverse engineering
shareware
Simonyi, Charles
simulation
software engineering
structured programming
systems programming
virtualization

User Interface and Support

digital dashboard
Engelbart, Doug
ergonomics of computing
haptic interface
help systems
installation of software
Jobs, Steven Paul
Kay, Alan
Macintosh
mouse
Negroponte, Nicholas
psychology of computing
technical support
technical writing
touchscreen
Turkle, Sherry
ser groups
user interface
virtual reality
wearable computers

A

abstract data type   *See* DATA ABSTRACTION.

active server pages (ASP)

Many users think of Web pages as being like pages in a book, stored intact on the server, ready to be flipped through with the mouse. Increasingly, however, Web pages are dynamic—they do not actually exist until the user requests them, and their content is determined largely by what the user requests. This demand for greater interactivity and customization of Web content tends to fall first on the server (see CLIENT-SERVER COMPUTING and WEB SERVER) and on "server side" programs to provide such functions as database access. One major platform for developing Web services is Microsoft's Active Server Pages (ASP).

In ASP programmers work with built-in objects that represent basic Web page functions. The RecordSet object can provide access to a variety of databases; the Response object can be invoked to display text in response to a user action; and the Session object provides variables that can be used to store information about previous user actions such as adding items to a shopping cart (see also COOKIES).

Control of the behavior of the objects within the Web page and session was originally handled by code written in a scripting language such as VBScript and embedded within the HTML text (see HTML and VBSCRIPT). However, ASP .NET, based on Microsoft's latest Windows class libraries (see MICROSOFT .NET) and introduced in 2002, allows Web services to be written in full-fledged programming languages such as Visual Basic .NET and C#, although in-page scripting can still be used. This can provide several advantages: access to software development tools and methodologies available for established programming languages, better separation of program code from the "presentational" (formatting) elements of HTML, and the speed and security associated with compiled code. ASP .NET also emphasizes the increasingly prevalent Extensible Markup Language (see XML) for organizing data and sending those data between objects using Simple Object Access Protocol (see SOAP).

Although ASP .NET was designed to be used with Microsoft's Internet Information Server (IIS) under Windows, the open-source Mono project (sponsored by Novell) implements a growing subset of the .NET classes for use on UNIX and Linux platforms using a C# compiler with appropriate user interface, graphics, and database libraries.

An alternative (or complementary) approach that has become popular in recent years reduces the load on the Web server by avoiding having to resend an entire Web page when only a small part actually needs to be changed. See AJAX (asynchronous JavaScript and XML).

Further Reading
Bellinaso, Marco. *ASP .NET 2.0 Website Programming: Problem—Design—Solution.* Indianapolis: Wiley Publishing, 2006.
Liberty, Jesse, and Dan Hurwitz. *Programming ASP .NET.* 3rd ed. Sebastapol, Calif.: O'Reilly, 2005.
McClure, Wallace B., et al. *Beginning Ajax with ASP .NET.* Indianapolis: Wiley Publishing, 2006.
Mono Project. Available online. URL: http://www.mono-project.com/Main_Page. Accessed April 10, 2007.

Ada

Starting in the 1960s, the U.S. Department of Defense (DOD) began to confront the growing unmanageability of its software development efforts. Whenever a new application such as a communications controller (see EMBEDDED SYSTEM) was developed, it typically had its own specialized programming language. With more than 2,000 such languages in use, it had become increasingly costly and difficult to maintain and upgrade such a wide variety of incompatible systems. In 1977, a DOD working group began to formally solicit proposals for a new general-purpose programming language that could be used for all applications ranging from weapons control and guidance systems to barcode scanners for inventory management. The winning language proposal eventually became known as Ada, named for 19th-century computer pioneer Ada Lovelace see also (BABBAGE, CHARLES). After a series of reviews and revisions of specifications, the American National Standards Institute officially standardized Ada in 1983, and this first version of the language is sometimes called Ada-83.

LANGUAGE FEATURES

In designing Ada, the developers adopted basic language elements based on emerging principles (see STRUCTURED PROGRAMMING) that had been implemented in languages developed during the 1960s and 1970s (see ALGOL and PASCAL). These elements include well-defined control structures (see BRANCHING STATEMENTS and LOOP) and the avoidance of the haphazard jump or "goto" directive.

Ada combines standard structured language features (including control structures and the use of subprograms) with user-definable data type "packages" similar to the classes used later in C++ and other languages (see CLASS and OBJECT-ORIENTED PROGRAMMING). As shown in this simple example, an Ada program has a general form similar to that used in Pascal. (Note that words in boldface type are language keywords.)

```
with Ada.Text_IO; use Ada.Text_IO;
procedure Get_Name is
Name : String (1..80);
Length : Integer;

begin
Put ("What is your first name?");
Get_Line (Name, Length);
New_Line;
Put ("Nice to meet you,");
Put (Name (1..Length));
end Get_Name;
```

The first line of the program specifies what "packages" will be used. Packages are structures that combine data types and associated functions, such as those needed for getting and displaying text. The Ada.Text.IO package, for example, has a specification that includes the following:

```
package Text_IO is
type File_Type is limited private;
type File_Mode is (In_File, Out_File, Append_File);
```

```
procedure Create (File : in out File_Type;
Mode : in File_Mode := Out_File;
Name : in String := "");
procedure Close (File : in out File_Type);
procedure Put_Line (File : in File_Type;
Item : in String);
procedure Put_Line (Item : in String);
end Text_IO;
```

The package specification begins by setting up a data type for files, and then defines functions for creating and closing a file and for putting text in files. As with C++ classes, more specialized packages can be derived from more general ones.

In the main program **Begin** starts the actual data processing, which in this case involves displaying a message using the Put function from the Ada.Text.IO function and getting the user response with Get_Line, then using Put again to display the text just entered.

Ada is particularly well suited to large, complex software projects because the use of packages hides and protects the details of implementing and working with a data type. A programmer whose program uses a package is restricted to using the visible interface, which specifies what parameters are to be used with each function. Ada compilers are carefully validated to ensure that they meet the exact specifications for the processing of various types of data (see DATA TYPES), and the language is "strongly typed," meaning that types must be explicitly declared, unlike the case with C, where subtle bugs can be introduced when types are automatically converted to make them compatible.

Because of its application to embedded systems and real-time operations, Ada includes a number of features designed to create efficient object (machine) code, and the language also makes provision for easy incorporation of routines written in assembly or other high-level languages. The latest official version, Ada 95, also emphasizes support for parallel programming (see MULTIPROCESSING). The future of Ada is unclear, however, because the Department of Defense no longer requires use of the language in government contracts.

Ada development has continued, particularly in areas including expanded object-oriented features (including support for interfaces with multiple inheritance); improved handling of strings, other data types, and files; and refinements in real-time processing and numeric processing.

Further Reading

"Ada 95 Lovelace Tutorial." Available online. URL: http://www.adahome.com/Tutorials/Lovelace/lovelace.htm. Accessed April 18, 2008.

Ada 95 On-line Reference Manual (hypertext) Available online. URL: http://www.adahome.com/Resources/refs/rm95.html. Accessed April 18, 2008.

Barnes, John. *Programming in Ada 2005 with CD.* New York: Pearson Education, 2006.

Dale, Nell, and John W. McCormick. *Ada Plus Data Structures: An Object-Oriented Approach.* 2nd ed. Sudbury, Mass.: Jones and Bartlett, 2006.

addressing

In order for computers to manipulate data, they must be able to store and retrieve it on demand. This requires a way to specify the location and extent of a data item in memory. These locations are represented by sequential numbers, or addresses.

Physically, a modern RAM (random access memory) can be visualized as a grid of address lines that crisscross with data lines. Each line carries one bit of the address, and together, they specify a particular location in memory (see MEMORY). Thus a machine with 32 address lines can handle up to 32 bits, or 4 gigabytes (billions of bytes) worth of addresses. However the amount of memory that can be addressed can be extended through indirect addressing, where the data stored at an address is itself the address of another location where the actual data can be found. This allows a limited amount of fast memory to be used to point to data stored in auxiliary memory or mass storage thus extending addressing to the space on a hard disk drive.

Some of the data stored in memory contains the actual program instructions to be executed. As the processor executes program instructions, an instruction pointer accesses the location of the next instruction. An instruction can also specify that if a certain condition is met the processor will jump over intervening locations to fetch the next instruction. This implements such control structures as branching statements and loops.

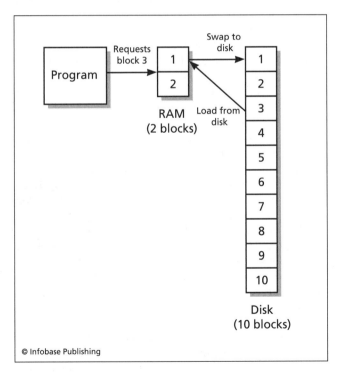

Virtual memory uses indirect addressing. When a program requests data from memory, the address is looked up in a table that keeps track of each block's actual location. If the block is not in RAM, one or more blocks in RAM are copied to the swap file on disk, and the needed blocks are copied from disk into the vacated area in RAM.

ADDRESSING IN PROGRAMS

A variable name in a program language actually references an address (or often, a range of successive addresses, since most data items require more than one byte of storage). For example, if a program includes the declaration

```
Int Old_Total, New_Total;
```

when the program is compiled, storage for the variables Old_Total and New_Total is set aside at the next available addresses. A statement such as

```
New_Total = 0;
```

is compiled as an instruction to store the value 0 in the address represented by New_Total. When the program later performs a calculation such as:

```
New_Total = Old_Total + 1;
```

the data is retrieved from the memory location designated by Old_Total and stored in a register in the CPU, where 1 is added to it, and the result is stored in the memory location designated by New_Total.

Although programmers don't have to work directly with address locations, programs can also use a special type of variable to hold and manipulate memory addresses for more efficient access to data (see POINTERS AND INDIRECTION).

Further Reading
"Computer Architecture Tutorial." Available online. URL: http://www.cs.iastate.edu/~prabhu/Tutorial/title.html. Accessed April 10, 2007.
Murdocca, Miles J., and Vincent P. Heuring. *Principles of Computer Architecture.* Upper Saddle River, N.J.: Prentice Hall, 2000.

Adobe Systems

Adobe Systems (NASDAQ symbol ADBE) is best known for products relating to the formatting, printing, and display of documents. Founded in 1982 by John Warnock and Charles Geschke, the company is named for a creek near one of their homes.

Adobe's first major product was a language that describes the font sizes, styles, and other formatting needed to print pages in near-typeset quality (see POSTSCRIPT). This was a significant contribution to the development of software for document creation (see DESKTOP PUBLISHING), particularly on the Apple Macintosh, starting in the later 1980s. Building on this foundation, Adobe developed high-quality digital fonts (called Type 1). However, Apple's TrueType fonts proved to be superior in scaling to different sizes and in the precise control over the pixels used to display them. With the licensing of TrueType to Microsoft for use in Windows, TrueType fonts took over the desktop, although Adobe Type 1 remained popular in commercial typesetting applications. Finally, in the late 1990s Adobe, together with Microsoft, established a new font format called OpenType, and by 2003 Adobe had converted all of its Type 1 fonts to the new format.

Adobe's Portable Document Format (see PDF) has become a ubiquitous standard for displaying print documents. Adobe greatly contributed to this development by making a free Adobe Acrobat PDF reader available for download.

IMAGE PROCESSING SOFTWARE

In the mid-1980s Adobe's founders realized that they could further exploit the knowledge of graphics rendition that they had gained in developing their fonts. They began to create software that would make these capabilities available to illustrators and artists as well as desktop publishers. Their first such product was Adobe Illustrator for the Macintosh, a vector-based drawing program that built upon the graphics capabilities of their PostScript language.

In 1989 Adobe introduced Adobe Photoshop for the Macintosh. With its tremendous variety of features, the program soon became a standard tool for graphic artists. However, Adobe seemed to have difficulty at first in anticipating the growth of desktop publishing and graphic arts on the Microsoft Windows platform. Much of that market was seized by competitors such as Aldus PageMaker and QuarkXPress. By the mid-1990s, however, Adobe, fueled by the continuing revenue from its PostScript technology, had acquired both Aldus and Frame Technologies, maker of the popular FrameMaker document design program. Meanwhile PhotoShop continued to develop on both the Macintosh and Windows platforms, aided by its ability to accept add-ons from hundreds of third-party developers (see PLUG-INS).

MULTIMEDIA AND THE WEB

Adobe made a significant expansion beyond document processing into multimedia with its acquisition of Macromedia (with its popular Flash animation software) in 2005 at a cost of about $3.4 billion. The company has integrated Macromedia's Flash and Dreamweaver Web-design software into its Creative Suite 3 (CS3). Another recent Adobe product that targets Web-based publishing is Digital Editions, which integrated the existing Dreamweaver and Flash software into a powerful but easy-to-use tool for delivering text content and multimedia to Web browsers. Buoyed by these developments, Adobe earned nearly $2 billion in revenue in 2005, about $2.5 billion in 2006, and $3.16 billion in 2007.

Today Adobe has over 6,600 employees, with its headquarters in San Jose and offices in Seattle and San Francisco as well as Bangalore, India; Ottawa, Canada; and other locations. In recent years the company has been regarded as a superior place to work, being ranked by *Fortune* magazine as the fifth best in America in 2003 and sixth best in 2004.

Further Reading

"Adobe Advances on Stronger Profit." *Business Week Online,* December 18, 2006. Available online. URL: http://www.businessweek.com/investor/content/dec2006/pi20061215_986588.htm. Accessed April 10, 2007.

Adobe Systems Incorporated home page. Available online. URL: http://www.adobe.com. Accessed April 10, 2007.

"Happy Birthday Acrobat: Adobe's Acrobat Turns 10 Years Old." *Print Media* 18 (July–August 2003): 21.

Advanced Micro Devices (AMD)

Sunnyvale, California-based Advanced Micro Devices, Inc., (NYSE symbol AMD) is a major competitor in the market for integrated circuits, particularly the processors that are at the heart of today's desktop and laptop computers (see MICROPROCESSOR). The company was founded in 1969 by a group of executives who had left Fairchild Semiconductor. In 1975 the company began to produce both RAM (memory) chips and a clone of the Intel 8080 microprocessor.

When IBM adopted the Intel 8080 for its first personal computer in 1982 (see INTEL CORPORATION and IBM PC), it required that there be a second source for the chip. Intel therefore signed an agreement with AMD to allow the latter to manufacture the Intel 9806 and 8088 processors. AMD also produced the 80286, the second generation of PC-compatible processors, but when Intel developed the 80386 it canceled the agreement with AMD.

A lengthy legal dispute ensued, with the California Supreme Court finally siding with AMD in 1991. However, as disputes continued over the use by AMD of "microcode" (internal programming) from Intel chips, AMD eventually used a "clean room" process to independently create functionally equivalent code (see REVERSE ENGINEERING). However, the speed with which new generations of chips was being produced rendered this approach impracticable by the mid-1980s, and Intel and AMD concluded a (largely secret) agreement allowing AMD to use Intel code and providing for cross-licensing of patents.

In the early and mid-1990s AMD had trouble keeping up with Intel's new Pentium line, but the AMD K6 (introduced in 1997) was widely viewed as a superior implementation of the microcode in the Intel Pentium—and it was "pin compatible," making it easy for manufacturers to include it on their motherboards.

Today AMD remains second in market share to Intel. AMD's Athlon, Opteron, Turion, and Sempron processors are comparable to corresponding Intel Pentium processors, and the two companies compete fiercely as each introduces new architectural features to provide greater speed or processing capacity.

In the early 2000s AMD seized the opportunity to beat Intel to market with chips that could double the data bandwidth from 32 bits to 64 bits. The new specification standard, called AMD64, was adopted for upcoming operating systems by Microsoft, Sun Microsystems, and the developers of Linux and UNIX kernels. AMD has also matched Intel in the latest generation of dual-core chips that essentially provide two processors on one chip. Meanwhile, AMD strengthened its position in the high-end server market when, in May 2006, Dell Computer announced that it would market servers containing AMD Opteron processors. In 2006 AMD also moved into the graphics-processing field by merging with ATI, a leading maker of video cards, at a cost of $5.4 billion. Meanwhile AMD also continues to be a leading maker of flash memory, closely collaborating with Japan's Fujitsu Corporation (see FLASH DRIVE). In 2008 AMD continued its aggressive pursuit of market share, announcing a variety of products, including a quad-core Opteron chip that it expects to catch up to if not surpass similar chips from Intel.

Further Reading
AMD Web site. Available online. URL: http://www.amd.com/us-en/. Accessed April 10, 2007.
Rodengen, Jeffrey L. *The Spirit of AMD: Advanced Micro Devices.* Ft. Lauderdale, Fla.: Write Stuff Enterprises, 1998.
Tom's Hardware [CPU articles and charts]. Available online. URL: http://www.tomshardware.com/find_by_topic/cpu.html. Accessed April 10, 2007.

advertising, online   *See* ONLINE ADVERTISING.

agent software   *See* SOFTWARE AGENT.

AI   *See* ARTIFICIAL INTELLIGENCE.

Aiken, Howard
(1900–1973)
American
Electrical Engineer

Howard Hathaway Aiken was a pioneer in the development of automatic calculating machines. Born on March 8, 1900, in Hoboken, New Jersey, he grew up in Indianapolis, Indiana, where he pursued his interest in electrical engineering by working at a utility company while in high school. He earned a B.A. in electrical engineering in 1923 at the University of Wisconsin.

By 1935, Aiken was involved in theoretical work on electrical conduction that required laborious calculation. Inspired by work a hundred years earlier (see BABBAGE, CHARLES), Aiken began to investigate the possibility of building a large-scale, programmable, automatic computing device (see CALCULATOR). As a doctoral student at Harvard, Aiken aroused interest in his project, particularly from Thomas Watson, Sr., head of International Business Machines (IBM). In 1939, IBM agreed to underwrite the building of Aiken's first calculator, the Automatic Sequence Controlled Calculator, which became known as the Harvard Mark I.

MARK I AND ITS PROGENY
Like Babbage, Aiken aimed for a general-purpose programmable machine rather than an assembly of special-purpose arithmetic units. Unlike Babbage, Aiken had access to a variety of tested, reliable components, including card punches, readers, and electric typewriters from IBM and the mechanical electromagnetic relays used for automatic switching in the telephone industry. His machine used decimal numbers (23 digits and a sign) rather than the binary numbers of the majority of later computers. Sixty registers held whatever constant data numbers were needed to solve a particular problem. The operator turned a rotary dial to enter each digit of each number. Variable data and program instructions were entered via punched paper tape. Calculations had to be broken down into specific instructions similar to those in later low-level programming languages such as "store this number in this register" or "add this number to the number in that register" (see ASSEMBLER). The results (usually tables of mathematical function values) could be printed by an electric typewriter or output on punched cards. Huge (about 8 feet [2.4 m] high by 51 feet [15.5 m] long), slow, but reliable, the Mark I worked on a variety of problems during World War II, ranging from equations used in lens design and radar to the designing of the implosive core of an atomic bomb.

Aiken completed an improved model, the Mark II, in 1947. The Mark III of 1950 and Mark IV of 1952, however, were electronic rather than electromechanical, replacing relays with vacuum tubes.

Compared to later computers such as the ENIAC and UNIVAC, the sequential calculator, as its name suggests, could only perform operations in the order specified. Any looping had to be done by physically creating a repetitive tape of instructions. (After all, the program as a whole was not stored in any sort of memory, and so previous instructions could not be reaccessed.) Although Aiken's machines soon slipped out of the mainstream of computer development, they did include the modern feature of parallel processing, because different calculation units could work on different instructions at the same time. Further, Aiken recognized the value of maintaining a library of frequently needed routines that could be reused in new programs—another fundamental of modern software engineering.

Aiken's work demonstrated the value of large-scale automatic computation and the use of reliable, available technology. Computer pioneers from around the world came to Aiken's Harvard computation lab to debate many issues that would become staples of the new discipline of computer science. The recipient of many awards including the Edison Medal of the IEEE and the Franklin Institute's John Price Award, Howard Aiken died on March 14, 1973, in St. Louis, Missouri.

Further Reading
Cohen, I. B. *Howard Aiken: Portrait of a Computer Pioneer.* Cambridge, Mass.: MIT Press, 1999.
Cohen, I. B., R. V. D. Campbell, and G. Welch, eds. *Makin' Numbers: Howard Aiken and the Computer.* Cambridge, Mass.: MIT Press, 1999.

Ajax (Asynchronous JavaScript and XML)
With the tremendous growth in Web usage comes a challenge to deliver Web-page content more efficiently and with greater flexibility. This is desirable to serve adequately the many users who still rely on relatively low-speed dial-up Internet connections and to reduce the demand on Web servers. Ajax (asynchronous JavaScript and XML) takes advantage of several emerging Web-development technologies to allow Web pages to interact with users while keeping the amount of data to be transmitted to a minimum.

In keeping with modern Web-design principles, the organization of the Web page is managed by coding in XHTML, a dialect of HTML that uses the stricter rules and grammar of the data-description markup language XML (see HTML, DHTML, AND XHTML and XML). Alternatively, data can be stored directly in XML. A structure called the DOM (Document Object Model; see DOM) is used to request data from the server, which is accessed through an object called httpRequest. The "presentational" information (regarding such matters as fonts, font sizes and styles, justification of paragraphs, and so on) is generally incorporated in an associated cascading style sheet (see CASCADING STYLE SHEETS). Behavior such as the presentation and processing of forms or user controls is usually handled by a scripting language (for example, see JAVASCRIPT). Ajax techniques tie these forms of processing together so that only the part of the Web page affected by current user activity needs to be updated. Only a small amount of data needs to be received from the server, while most of the HTML code needed to update the page is generated on the client side—that is, in the Web browser. Besides making Web pages more flexible and interactive, Ajax also makes it much easier to develop more elaborate applications, even delivering fully functional applications such as word processing and spreadsheets over the Web (see APPLICATION SERVICE PROVIDER).

Some critics of Ajax have decried its reliance on JavaScript, arguing that the language has a hard-to-use syntax similar to the C language and poorly implements objects (see OBJECT-ORIENTED PROGRAMMING). There is also a need to standardize behavior across the popular Web browsers. Nevertheless, Ajax has rapidly caught on in the Web development community, filling bookstore shelves with books on applying Ajax techniques to a variety of other languages (see, for example, PHP).

Ajax can be simplified by providing a framework of objects and methods that the programmer can use to set up and manage the connections between server and browser. Some frameworks simply provide a set of data structures and functions (see APPLICATION PROGRAM INTERFACE), while others include Ajax-enabled user interface components such as buttons or window tabs. Ajax frameworks also vary in

how much of the processing is done on the server and how much is done on the client (browser) side. Ajax frameworks are most commonly used with JavaScript, but also exist for Java (Google Web Toolkit), PHP, C++, and Python as well as other scripting languages. An interesting example is Flapjax, a project developed by researchers at Brown University. Flapjax is a complete high-level programming language that uses the same syntax as the popular JavaScript but hides the messy details of sharing and updating data between client and server.

DRAWBACKS AND CHALLENGES

By their very nature, Ajax-delivered pages behave differently from conventional Web pages. Because the updated page is not downloaded as such from the server, the browser cannot record it in its "history" and allow the user to click the "back" button to return to a previous page. Mechanisms for counting the number of page views can also fail. As a workaround, programmers have sometimes created "invisible" pages that are used to make the desired history entries. Another problem is that since content manipulated using Ajax is not stored in discrete pages with identifiable URLs, conventional search engines cannot read and index it, so a copy of the data must be provided on a conventional page for indexing. The extent to which XML should be used in place of more compact data representations is also a concern for many developers. Finally, accessibility tools (see DISABLED PERSONS AND COMPUTERS) often do not work with Ajax-delivered content, so an alternative form must often be provided to comply with accessibility guidelines or regulations.

Despite these concerns, Ajax is in widespread use and can be seen in action in many popular Web sites, including Google Maps and the photo-sharing site Flickr.com.

Further Reading
Ajaxian [news and resources for Ajax developers]. Available online. URL: http://ajaxian.com/. Accessed April 10, 2007.
Crane, David, Eric Pascarello, and Darren James. *Ajax in Action.* Greenwich, Conn.: Manning Publications, 2006.
"Google Web Toolkit: Build AJAX Apps in the Java Language." Available online. URL: http://code.google.com/webtoolkit/. Accessed April 10, 2007.
Holzner, Steve. *Ajax for Dummies.* Hoboken, N.J.: Wiley, 2006.
Jacobs, Sas. *Beginning XML with DOM and Ajax: From Novice to Professional.* Berkeley, Calif.: Apress, 2006.

Algol

The 1950s and early 1960s saw the emergence of two high-level computer languages into widespread use. The first was designed to be an efficient language for performing scientific calculations (see FORTRAN). The second was designed for business applications, with an emphasis on data processing (see COBOL). However many programs continued to be coded in low-level languages (see ASSEMBLER) designed to take advantages of the hardware features of particular machines.

In order to be able to easily express and share methods of calculation (see ALGORITHM), leading programmers

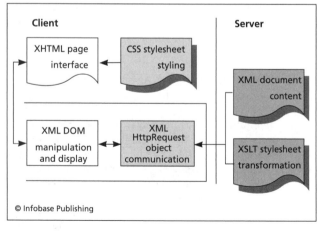

Ajax is a way to quickly and efficiently update dynamic Web pages—formatting is separate from content, making it easy to revise the latter.

algorithm   7

began to seek a "universal" programming language that was not designed for a particular application or hardware platform. By 1957, the German GAMM (Gesellschaft für angewandte Mathematik und Mechanik) and the American ACM (Association for Computing Machinery) had joined forces to develop the specifications for such a language. The result became known as the Zurich Report or Algol-58, and it was refined into the first widespread implementation of the language, Algol-60.

LANGUAGE FEATURES

Algol is a block-structured, procedural language. Each variable is declared to belong to one of a small number of kinds of data including integer, real number (see DATA TYPES), or a series of values of either type (see ARRAY). While the number of types is limited and there is no facility for defining new types, the compiler's type checking (making sure a data item matches the variable's declared type) introduced a level of security not found in most earlier languages.

An Algol program can contain a number of separate procedures or incorporate externally defined procedures (see LIBRARY, PROGRAM), and the variables with the same name in different procedure blocks do not interfere with one another. A procedure can call itself (see RECURSION). Standard control structures (see BRANCHING STATEMENTS and LOOP) were provided.

The following simple Algol program stores the numbers from 1 to 10 in an array while adding them up, then prints the total:

```
begin
  integer array ints[1:10];
  integer counter, total;
  total := 0;
  for counter :=1 step 1 until counter > 10
  do
    begin
    ints [counter] := counter;
    total := total + ints[counter];
  end;
  printstring "The total is:";
  printint (total);
  end
```

ALGOL'S LEGACY

The revision that became known as Algol-68 expanded the variety of data types (including the addition of boolean, or true/false values) and added user-defined types and "structs" (records containing fields of different types of data). Pointers (references to values) were also implemented, and flexibility was added to the parameters that could be passed to and from procedures.

Although Algol was used as a production language in some computer centers (particularly in Europe), its relative complexity and unfamiliarity impeded its acceptance, as did the widespread corporate backing for the rival languages FORTRAN and especially COBOL. Algol achieved its greatest success in two respects: for a time it became the language of choice for describing new algorithms for

computer scientists, and its structural features would be adopted in the new procedural languages that emerged in the 1970s (see PASCAL and C).

Further Reading

"Algol 68 Home Page." URL: http://www.algol68.org. Accessed April 10, 2007.

Backus, J. W., and others. "Revised Report on the Algorithmic Language Algol 60." Originally published in *Numerische Mathematik*, the *Communications of the ACM*, and the *Journal of the British Computer Society*. Available online. URL: http://www.masswerk. at/algol60/report.htm. Accessed April 10, 2007.

algorithm

When people think of computers, they usually think of silicon chips and circuit boards. Moving from relays to vacuum tubes to transistors to integrated circuits has vastly increased the power and speed of computers, but the essential idea behind the work computers do remains the algorithm. An algorithm is a reliable, definable procedure for solving a problem. The idea of the algorithm goes back to the beginnings of mathematics and elementary school students are usually taught a variety of algorithms. For example, the procedure for long division by successive division, subtraction, and attaching the next digit is an algorithm. Since a bona fide algorithm is guaranteed to work given the specified type of data and the rote following of a series of steps, the algorithmic approach is naturally suited to mechanical computation.

ALGORITHMS IN COMPUTER SCIENCE

Just as a cook learns both general techniques such as how to sauté or how to reduce a sauce and a repertoire of specific recipes, a student of computer science learns both general problem-solving principles and the details of common algorithms. These include a variety of algorithms for organizing data (see SORTING AND SEARCHING), for numeric problems (such as generating random numbers or finding primes), and for the manipulation of data structures (see LIST PROCESSING and QUEUE).

A working programmer faced with a new task first tries to think of familiar algorithms that might be applicable to the current problem, perhaps with some adaptation. For example, since a variety of well-tested and well-understood sorting algorithms have been developed, a programmer is likely to apply an existing algorithm to a sorting problem rather than attempt to come up with something entirely new. Indeed, for most widely used programming languages there are packages of modules or procedures that implement commonly needed data structures and algorithms (see LIBRARY, PROGRAM).

If a problem requires the development of a new algorithm, the designer will first attempt to determine whether the problem can, at least in theory, be solved (see COMPUTABILITY AND COMPLEXITY). Some kinds of problems have been shown to have no guaranteed answer. If a new algorithm seems feasible, principles found to be effective in the past will be employed, such as breaking complex problems down into component parts or building up from the simplest case to generate a solution (see RECURSION). For example, the merge-sort algorithm divides the data to be sorted into successively smaller portions until they are sorted, and then merges the sorted portions back together.

Another important aspect of algorithm design is choosing an appropriate way to organize the data (see DATA STRUCTURES). For example, a sorting algorithm that uses a branching (tree) structure would probably use a data structure that implements the nodes of a tree and the operations for adding, deleting, or moving them (see CLASS).

Once the new algorithm has been outlined (see PSEUDO-CODE), it is often desirable to demonstrate that it will work for any suitable data. Mathematical techniques such as the finding and proving of loop invariants (where a true assertion remains true after the loop terminates) can be used to demonstrate the correctness of the implementation of the algorithm.

PRACTICAL CONSIDERATIONS

It is not enough that an algorithm be reliable and correct, it must also be accurate and efficient enough for its intended use. A numerical algorithm that accumulates too much error through rounding or truncation of intermediate results may not be accurate enough for a scientific application. An algorithm that works by successive approximation or convergence on an answer may require too many iterations even for today's fast computers, or may consume too much of other computing resources such as memory. On the other hand, as computers become more and more powerful and processors are combined to create more powerful supercomputers (see SUPERCOMPUTER and CONCURRENT PROGRAMMING), algorithms that were previously considered impracticable might be reconsidered. Code profiling (analysis of which program statements are being executed the most frequently) and techniques for creating more efficient code can help in some cases. It is also necessary to keep in mind special cases where an otherwise efficient algorithm becomes much less efficient (for example, a tree sort may work well for random data but will become badly unbalanced and slow when dealing with data that is already sorted or mostly sorted).

Sometimes an exact solution cannot be mathematically guaranteed or would take too much time and resources to calculate, but an approximate solution is acceptable. A so-called "greedy algorithm" can proceed in stages, testing at each stage whether the solution is "good enough." Another approach is to use an algorithm that can produce a reasonable if not optimal solution. For example, if a group of tasks must be apportioned among several people (or computers) so that all tasks are completed in the shortest possible time, the time needed to find an exact solution rises exponentially with the number of workers and tasks. But an algorithm that first sorts the tasks by decreasing length and then distributes them among the workers by "dealing" them one at a time like cards at a bridge table will, as demonstrated by Ron Graham, give an allocation guaranteed to be within 4/3 of the optimal result—quite suitable for most applications. (A procedure that can produce a practical,

though not perfect solution is actually not an algorithm but a heuristic.)

An interesting approach to optimizing the solution to a problem is allowing a number of separate programs to "compete," with those showing the best performance surviving and exchanging pieces of code ("genetic material") with other successful programs (see GENETIC ALGORITHMS). This of course mimics evolution by natural selection in the biological world.

Further Reading

Berlinksi, David. *The Advent of the Algorithm: The Idea That Rules the World.* New York: Harcourt, 2000.

Cormen, T. H., C. E. Leiserson, R. L. Rivest, and Clifford Stein. *Introduction to Algorithms.* 2nd ed. Cambridge, Mass.: MIT Press, 2001.

Knuth, Donald E. *The Art of Computer Programming.* Vol. 1: *Fundamental Algorithms.* 3rd ed. Reading, Mass.: Addison-Wesley, 1997. Vol. 2: *Seminumerical Algorithms.* 3rd ed. Reading, Mass.: Addison-Wesley, 1997. Vol. 3: *Searching and Sorting.* 2nd ed. Reading, Mass.: Addison-Wesley, 1998.

ALU   *See* ARITHMETIC LOGIC UNIT.

Amazon.com

Beginning modestly in 1995 as an online bookstore, Amazon.com became one of the first success stories of the early Internet economy (see also E-COMMERCE).

Named for the world's largest river, Amazon.com was the brainchild of entrepreneur Jeffrey Bezos (see BEZOS, JEFFREY P.). Like a number of other entrepreneurs of the early 1990s, Bezos had been searching for a way to market to the growing number of people who were going online. He soon decided that books were a good first product, since they were popular, nonperishable, relatively compact, and easy to ship.

Several million books are in print at any one time, with about 275,000 titles or editions added in 2007 in the United States alone. Traditional "brick and mortar" (physical) bookstores might carry a few thousand titles up to perhaps 200,000 for the largest chains. Bookstores in turn stock their shelves mainly through major book distributors that serve as intermediaries between publishers and the public.

For an online bookstore such as Amazon.com, however, the number of titles that can be made available is limited only by the amount of warehouse space the store is willing to maintain—and no intermediary between publisher and bookseller is needed. From the start, Amazon.com's business model has capitalized on this potential for variety and the ability to serve almost any niche interest. Over the years the company's offerings have expanded beyond books to 34 different categories of merchandise, including software, music, video, electronics, apparel, home furnishings, and even nonperishable gourmet food and groceries. (Amazon.com also entered the online auction market, but remains a distant runner-up to market leader eBay).

EXPANSION AND PROFITABILITY

Because of its desire to build a very diverse product line, Amazon.com, unusually for a business startup, did not expect to become profitable for about five years. The growing revenues were largely poured back into expansion. In the heated atmosphere of the Internet boom of the late 1990s, many other Internet-based businesses echoed that philosophy, and many went out of business following the bursting of the so-called dot-com bubble of the early 2000s. Some analysts questioned whether even the hugely popular Amazon.com would ever be able to convert its business volume into an operating profit. However, the company achieved its first profitable year in 2003 (with a modest $35 million surplus). Since then growth has remained steady and generally impressive: In 2005, Amazon.com earned $8.49 billion revenues with a net income of $359 million. By then the company had about 12,000 employees and had been added to the S&P 500 stock index.

In 2006 the company maintained its strategy of investing in innovation rather than focusing on short-term profits. Its latest initiatives include selling digital versions of books (e-books) and magazine articles, new arrangements to sell video content, and even a venture into moviemaking. By year end, annual revenue had increased to $10.7 billion.

In November 2007 Amazon announced the Kindle, a book reader (see E-BOOKS AND DIGITAL LIBRARIES) with a sharp "paper-like" display. In addition to books, the Kindle can also subscribe to and download magazines, content from newspaper Web sites, and even blogs.

As part of its expansion strategy, Amazon.com has acquired other online bookstore sites including Borders.com and Waldenbooks.com. The company has also expanded geographically with retail operations in Canada, the United Kingdom, France, Germany, Japan, and China.

Amazon.com has kept a tight rein on its operations even while continually expanding. The company's leading market position enables it to get favorable terms from publishers and manufacturers. A high degree of warehouse automation and an efficient procurement system keep stock moving quickly rather than taking up space on the shelves.

INFORMATION-BASED STRATEGIES

Amazon.com has skillfully taken advantage of information technology to expand its capabilities and offerings. Examples of such efforts include new search mechanisms, cultivation of customer relationships, and the development of new ways for users to sell their own goods.

Amazon's "Search Inside the Book" feature is a good example of leveraging search technology to take advantage of having a growing amount of text online. If the publisher of a book cooperates, its actual text is made available for online searching. (The amount of text that can be displayed is limited to prevent users from being able to read entire books for free.) Further, one can see a list of books citing (or being cited by) the current book, providing yet another way to explore connections between ideas as used by different authors. Obviously for Amazon.com, the ultimate reason for offering all these useful features is that more potential customers may be able to find and purchase books on even the most obscure topics.

Amazon.com's use of information about customers' buying histories is based on the idea that the more one knows about what customers have wanted in the past, the more effectively they can be marketed to in the future through customizing their view of the site. Users receive automatically generated recommendations for books or other items based on their previous purchases (see also CUSTOMER RELATIONSHIP MANAGEMENT). There is even a "plog" or customized Web log that offers postings related to the user's interests and allows the user to respond.

There are other ways in which Amazon.com tries to involve users actively in the marketing process. For example, users are encouraged to review books and other products and to create lists that can be shared with other users. The inclusion of both user and professional reviews in turn makes it easier for prospective purchasers to determine whether a given book or other item is suitable. Authors are given the opportunity through "Amazon Connect" to provide additional information about their books. Finally, in late 2005 Amazon replaced an earlier "discussion board" facility with a wiki system that allows purchasers to create or edit an information page for any product (see WIKIS AND WIKIPEDIA).

The company's third major means of expansion is to facilitate small businesses and even individual users in the marketing of their own goods. Amazon Marketplace, a service launched in 2001, allows users to sell a variety of items, with no fees charged unless the item is sold. There are also many provisions for merchants to set up online "storefronts" and take advantage of online payment and other services.

Another aspect of Amazon's marketing is its referral network. Amazon's "associates" are independent businesses that provide links from their own sites to products on Amazon. For example, a seller of crafts supplies might include on its site links to books on crafting on the Amazon site. In return, the referring business receives a commission from Amazon.com.

Although often admired for its successful business plan, Amazon.com has received criticism from several quarters. Some users have found the company's customer service (which is handled almost entirely by e-mail) to be unresponsive. Meanwhile local and specialized bookstores, already suffering in recent years from the competition of large chains such as Borders and Barnes and Noble, have seen in Amazon.com another potent threat to the survival of their business. (The company's size and economic power have elicited occasional comparisons with Wal-Mart.) Finally, Amazon.com has been criticized by some labor advocates for paying low wages and threatening to terminate workers who sought to unionize.

Further Reading

Amazon.com Web site. Available online. URL: http://www.amazon. com. Accessed August 28, 2007.
Daisey, Mike. *21 Dog Years: Doing Time @ Amazon.com*. New York: The Free Press, 2002.
Marcus, James. *Amazonia*. New York: New Press, 2005.

Shanahan, Francis. *Amazon.com Mashups.* New York: Wrox/Wiley, 2007.

Spector, Robert. *Amazon.com: Get Big Fast: Inside the Revolutionary Business Model That Changed the World.* New York: Harper-Business, 2002.

Amdahl, Gene Myron
(1922–   )
American
Inventor, Entrepreneur

Gene Amdahl played a major role in designing and developing the mainframe computer that dominated data processing through the 1970s (see MAINFRAME). Amdahl was born on November 16, 1922, in Flandreau, South Dakota. After having his education interrupted by World War II, Amdahl received a B.S. from South Dakota State University in 1948 and a Ph.D. in physics at the University of Wisconsin in 1952.

As a graduate student Amdahl had realized that further progress in physics and other sciences required better, faster tools for computing. At the time there were only a few computers, and the best approach to getting access to significant computing power seemed to be to design one's own machine. Amdahl designed a computer called the WISC (Wisconsin Integrally Synchronized Computer). This computer used a sophisticated procedure to break calculations into parts that could be carried out on separate processors, making it one of the earliest examples of the parallel computing techniques found in today's computer architectures.

DESIGNER FOR IBM
In 1952 Amdahl went to work for IBM, which had committed itself to dominating the new data processing industry. Amdahl worked with the team that eventually designed the IBM 704. The 704 improved upon the 701, the company's first successful mainframe, by adding many new internal programming instructions, including the ability to perform floating point calculations (involving numbers that have decimal points). The machine also included a fast, high-capacity magnetic core memory that let the machine retrieve data more quickly during calculations. In November 1953 Amdahl became the chief project engineer for the 704 and then helped design the IBM 709, which was designed especially for scientific applications.

When IBM proposed extending the technology by building a powerful new scientific computer called STRETCH, Amdahl eagerly applied to head the new project. However, he ended up on the losing side of a corporate power struggle, and did not receive the post. He left IBM at the end of 1955.

In 1960 Amdahl rejoined IBM, where he was soon involved in several design projects. The one with the most lasting importance was the IBM System/360, which would become the most ubiquitous and successful mainframe computer of all time. In this project Amdahl further refined his ideas about making a computer's central processing unit more efficient. He designed logic circuits that enabled the processor to analyze the instructions waiting to be executed (the "pipeline") and determine which instructions could be executed immediately and which would have to wait for the results of other instructions. He also used a cache, or special memory area, in which the instructions that would be needed next could be stored ahead of time so they could be retrieved immediately when needed. Today's desktop PCs use these same ideas to get the most out of their chips' capabilities.

Amdahl also made important contributions to the further development of parallel processing. Amdahl created a formula called Amdahl's law that basically says that the advantage gained from using more processors gradually declines as more processor are added. The amount of improvement is also proportional to how much of the calculation can be broken down into parts that can be run in parallel. As a result, some kinds of programs can run much faster with several processors being used simultaneously, while other programs may show little improvement.

In the mid-1960s Amdahl helped establish IBM's Advanced Computing Systems Laboratory in Menlo Park, California, which he directed. However, he became increasingly frustrated with what he thought was IBM's too rigid approach to designing and marketing computers. He decided to leave IBM again and, this time, challenge it in the marketplace.

CREATOR OF "CLONES"
Amdahl resolved to make computers that were more powerful than IBM's machines, but that would be "plug compatible" with them, allowing them to use existing hardware and software. To gain an edge over the computer giant, Amdahl was able to take advantage of the early developments in integrated electronics to put more circuits on a chip without making the chips too small, and thus too crowded for placing the transistors.

Thanks to the use of larger scale circuit integration, Amdahl could sell machines with superior technology to that of the IBM 360 or even the new IBM 370, and at a lower price. IBM responded belatedly to the competition, making more compact and faster processors, but Amdahl met each new IBM product with a faster, cheaper alternative. However, IBM also countered by using a sales technique that opponents called FUD (fear, uncertainty, and doubt). IBM salespersons promised customers that IBM would soon be coming out with much more powerful and economical alternatives to Amdahl's machines. As a result, many would-be customers were persuaded to postpone purchasing decisions and stay with IBM. Amdahl Corporation began to falter, and Gene Amdahl gradually sold his stock and left the company in 1980.

Amdahl then tried to repeat his success by starting a new company called Trilogy. The company promised to build much faster and cheaper computers than those offered by IBM or Amdahl. He believed he could accomplish this by using the new, very-large-scale integrated silicon wafer technology in which circuits were deposited in layers on a single chip rather than being distributed on separate chips on a printed circuit board. But the problem of dealing with the electrical characteristics of such dense circuitry, as well as some design errors, somewhat crippled the new computer design. Amdahl was forced to repeatedly delay the introduction of the new machine, and Trilogy failed in the marketplace.

Amdahl's achievements could not be overshadowed by the failures of his later career. He has received many industry awards, including Data Processing Man of the Year by the Data Processing Management Association (1976), the Harry Goode Memorial Award from the American Federation of Information Processing Societies, and the SIGDA Pioneering Achievement Award (2007).

Further Reading
"Gene Amdahl." Available online. URL: http://www.thocp.net/biographies/amdahl_gene.htm. Accessed April 10, 2007.
Slater, Robert. *Portraits in Silicon.* Cambridge, Mass.: MIT Press, 1987.

America Online (AOL)

For millions of PC users in the 1990s, "going online" meant connecting to America Online. However, this once dominant service provider has had difficulty adapting to the changing world of the Internet.

By the mid-1980s a growing number of PC users were starting to go online, mainly dialing up small bulletin board services. Generally these were run by individuals from their homes, offering a forum for discussion and a way for users to upload and download games and other free software and shareware (see BULLETIN BOARD SYSTEMS). However, some entrepreneurs saw the possibility of creating a commercial information service that would be interesting and useful enough that users would pay a monthly subscription fee for access. Perhaps the first such enterprise to be successful was Quantum Computer Services, founded by Jim Kimsey in 1985 and soon joined by another young entrepreneur, Steve Case. Their strategy was to team up with personal computer makers such as Commodore, Apple, and IBM to provide special online services for their users.

In 1989 Quantum Link changed its name to America Online (AOL). In 1991 Steve Case became CEO, taking over from the retiring Kimsey. Case's approach to marketing AOL was to aim the service at novice PC users who had trouble mastering arcane DOS (disk operating system) commands and interacting with text-based bulletin boards and primitive terminal programs. As an alternative, AOL provided a complete software package that managed the user's connection, presented "friendly" graphics, and offered point-and-click access to features.

Chat rooms and discussion boards were also expanded and offered in a variety of formats for casual and more formal use. Gaming, too, was a major emphasis of the early AOL, with some of the first online multiplayer fantasy role-playing games such as a version of Dungeons and Dragons called *Neverwinter Nights* (see ONLINE GAMES). A third popular application has been instant messaging (IM), including a feature that allowed users to set up "buddy lists" of their friends and keep track of when they were online (see also TEXTING AND INSTANT MESSAGING).

INTERNET CHALLENGE
By 1996 the World Wide Web was becoming popular (see WORLD WIDE WEB). Rather than signing up with a proprietary service such as AOL, users could simply get an account with a lower-cost direct-connection service (see INTERNET SERVICE PROVIDER) and then use a Web browser such as Netscape to access information and services. AOL was slow in adapting to the growing use of the Internet. At first, the service provided only limited access to the Web (and only through its proprietary software). Gradually, however, AOL offered a more seamless Web experience, allowing users to run their own browsers and other software together with the proprietary interface. Also, responding to competition, AOL replaced its hourly rates with a flat monthly fee ($19.95 at first).

Overall, AOL increasingly struggled with trying to fulfill two distinct roles: Internet access provider and content provider. By the late 1990s AOL's monthly rates were higher than those of "no frills" access providers such as NetZero. AOL tried to compensate for this by offering integration of services (such as e-mail, chat, and instant messaging) and news and other content not available on the open Internet.

AOL also tried to shore up its user base with aggressive marketing to users who wanted to go online but were not sure how to do so. Especially during the late 1990s, AOL was able to swell its user rolls to nearly 30 million, largely by providing millions of free CDs (such as in magazine inserts) that included a setup program and up to a month of free service. But while it was easy to get started with AOL, some users began to complain that the service would keep billing them even after they had repeatedly attempted to cancel it. Meanwhile, AOL users got little respect from the more sophisticated inhabitants of cyberspace, who often complained that the clueless "newbies" were cluttering newsgroups and chat rooms.

In 2000 AOL and Time Warner merged. At the time, the deal was hailed as one of the greatest mergers in corporate

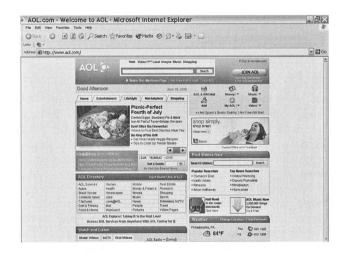

America Online (AOL) was a major online portal in the 1990s, but has faced challenges adapting to the modern world of the Web. (SCREEN IMAGE CREDIT: AOL)

history, bringing together one of the foremost Internet companies with one of the biggest traditional media companies. The hope was that the new $350 billion company would be able to leverage its huge subscriber base and rich media resources to dominate the online world.

FROM SERVICE TO CONTENT PROVIDER

By the 2000s, however, an increasing number of people were switching from dial-up to high-speed broadband Internet access (see BROADBAND) rather than subscribing to services such as AOL simply to get online. This trend and the overall decline in the Internet economy early in the decade (the "dot-bust") contributed to a record loss of $99 billion for the combined company in 2002. In a shakeup, Time-Warner dropped "AOL" from its name, and Steve Case was replaced as executive chairman. The company increasingly began to shift its focus to providing content and services that would attract people who were already online, with revenue coming from advertising instead of subscriptions.

In October 2006 the AOL division of Time-Warner (which by then had dropped the full name America Online) announced that it would provide a new interface and software optimized for broadband users. AOL's OpenRide desktop presents users with multiple windows for e-mail, instant messaging, Web browsing, and media (video and music), with other free services available as well. These offerings are designed to compete in a marketplace where the company faces stiff competition from other major Internet presences who have been using the advertising-based model for years (see YAHOO! and GOOGLE).

Further Reading

AOL Web site. Available online. URL: http://www.aol.com. Accessed August 28, 2007.

Kaufeld, John. *AOL for Dummies.* 2nd ed. Hoboken, N.J.: Wiley, 2004.

Klein, Alec. *Stealing Time: Steve Case, Jerry Levin, and the Collapse of AOL Time Warner.* New York: Simon & Schuster, 2003.

Mehta, Stephanie N. "Can AOL Keep Pace?" *Fortune,* August 21, 2006, p. 29.

Swisher, Kara. *AOL.COM: How Steve Case Beat Bill Gates, Nailed the Netheads, and Made Millions in the War for the Web.* New York: Times Books, 1998.

analog and digital

The word *analog* (derived from Greek words meaning "by ratio") denotes a phenomenon that is continuously variable, such as a sound wave. The word *digital,* on the other hand, implies a discrete, exactly countable value that can be represented as a series of digits (numbers). Sound recording provides familiar examples of both approaches. Recording a phonograph record involves electromechanically transferring a physical signal (the sound wave) into an "analogous" physical representation (the continuously varying peaks and dips in the record's surface). Recording a CD, on the other hand, involves sampling (measuring) the sound level at thousands of discrete instances and storing the results in a physical representation of a numeric format that can in turn be used to drive the playback device.

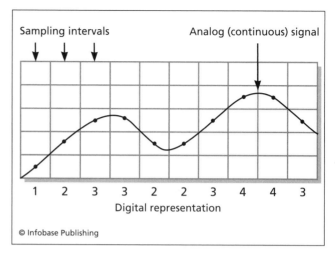

Most natural phenomena such as light or sound intensity are analog values that vary continuously. To convert such measurements to a digital representation, "snapshots" or sample readings must be taken at regular intervals. Sampling more frequently gives a more accurate representation of the original analog data, but at a cost in memory and processor resources.

Virtually all modern computers depend on the manipulation of discrete signals in one of two states denoted by the numbers 1 and 0. Whether the 1 indicates the presence of an electrical charge, a voltage level, a magnetic state, a pulse of light, or some other phenomenon, at a given point there is either "something" (1) or "nothing" (0). This is the most natural way to represent a series of such states.

Digital representation has several advantages over analog. Since computer circuits based on binary logic can be driven to perform calculations electronically at ever-increasing speeds, even problems where an analog computer better modeled nature can now be done more efficiently with digital machines (see ANALOG COMPUTER). Data stored in digitized form is not subject to the gradual wear or distortion of the medium that plagues analog representations such as the phonograph record. Perhaps most important, because digital representations are at base simply numbers, an infinite variety of digital representations can be stored in files and manipulated, regardless of whether they started as pictures, music, or text (see DIGITAL CONVERGENCE).

CONVERTING BETWEEN ANALOG AND DIGITAL REPRESENTATIONS

Because digital devices (particularly computers) are the mechanism of choice for working with representations of text, graphics, and sound, a variety of devices are used to digitize analog inputs so the data can be stored and manipulated. Conceptually, each digitizing device can be thought of as having three parts: a component that scans the input and generates an analog signal, a circuit that converts the analog signal from the input to a digital format, and a component that stores the resulting digital data for later use. For example, in the ubiquitous flatbed scanner a moving head reads varying light levels on the paper and converts them to a varying level of current (see SCANNER). This analog signal is in turn converted into a digital reading by an analog-to-digital converter, which creates numeric information that represents discrete spots (pixels) representing either levels of gray or of particular colors. This information is then written to disk using the formats supported by the operating system and the software that will manipulate them.

Further Reading

Chalmers, David J. "Analog vs. Digital Computation." Available online. URL: http://www.u.arizona.edu/~chalmers/notes/analog.html. Accessed April 10, 2007.

Hoeschele, David F. *Analog-to-Digital and Digital-to-Analog Conversion Techniques.* 2nd ed. New York: Wiley-Interscience, 1994.

analog computer

Most natural phenomena are analog rather than digital in nature (see ANALOG AND DIGITAL). But just as mathematical laws can describe relationships in nature, these relationships in turn can be used to construct a model in which natural forces generate mathematical solutions. This is the key insight that leads to the analog computer.

The simplest analog computers use physical components that model geometric ratios. The earliest known analog computing device is the Antikythera Mechanism. Constructed by an unknown scientist on the island of Rhodes around 87 B.C., this device used a precisely crafted differential gear mechanism to mechanically calculate the interval between new moons (the synodic month). (Interestingly, the differential gear would not be rediscovered until 1877.)

Another analog computer, the slide rule, became the constant companion of scientists, engineers, and students until it was replaced by electronic calculators in the 1970s. Invented in simple form in the 17th century, the slide rule's movable parts are marked in logarithmic proportions, allowing for quick multiplication, division, the extraction of square roots, and sometimes the calculation of trigonometric functions.

The next insight involved building analog devices that set up dynamic relationships between mechanical movements. In the late 19th century two British scientists, James Thomson and his brother Sir William Thomson (later Lord Kelvin) developed the mechanical integrator, a device that could solve differential equations. An important new principle used in this device is the closed feedback loop, where the output of the integrator is fed back as a new set of inputs. This allowed for the gradual summation or integration of an equation's variables. In 1931, VANNEVAR BUSH completed a more complex machine that he called a "differential analyzer." Consisting of six mechanical integrators using specially shaped wheels, disks, and servomechanisms, the differential analyzer could solve equations in up to six independent variables. As the usefulness and applicability of the device became known, it was quickly replicated in various forms in scientific, engineering, and military institutions.

These early forms of analog computer are based on fixed geometrical ratios. However, most phenomena that scientists and engineers are concerned with, such as aerodynamics, fluid dynamics, or the flow of electrons in a circuit, involve a mathematical relationship between forces where the output changes smoothly as the inputs are changed. The "dynamic" analog computer of the mid-20th century took advantage of such force relationships to construct devices where input forces represent variables in the equation, and

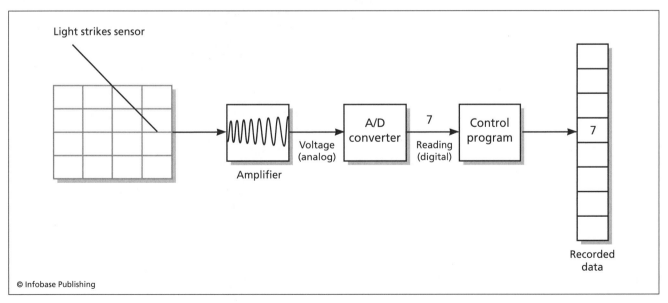

© Infobase Publishing

Converting analog data to digital involves several steps. A sensor (such as the CCD, or charge-coupled device in a digital camera) creates a varying electrical current. An amplifier can strengthen this signal to make it easier to process, and filters can eliminate spurious spikes or "noise." The "conditioned" signal is then fed to the analog-to-digital (A/D) converter, which produces numeric data that is usually stored in a memory buffer from which it can be processed and stored by the controlling program.

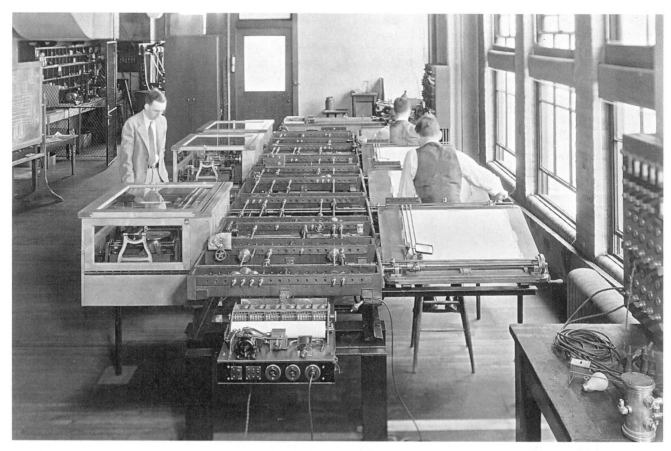

Completed in 1931, Vannevar Bush's Differential Analyzer was a triumph of analog computing. The device could solve equations with up to six independent values. (MIT MUSEUM)

nature itself "solves" the equation by producing a resulting output force.

In the 1930s, the growing use of electronic circuits encouraged the use of the flow of electrons rather than mechanical force as a source for analog computation. The key circuit is called an operational amplifier. It generates a highly amplified output signal of opposite polarity to the input, over a wide range of frequencies. By using components such as potentiometers and feedback capacitors, an analog computer can be programmed to set up a circuit in which the laws of electronics manipulate the input voltages in the same way the equation to be solved manipulates its variables. The results of the calculation are then read as a series of voltage values in the final output.

Starting in the 1950s, a number of companies marketed large electronic analog computers that contained many separate computing units that could be harnessed together to provide "real time" calculations in which the results could be generated at the same rate as the actual phenomena being simulated. In the early 1960s, NASA set up training simulations for astronauts using analog real-time simulations that were still beyond the capability of digital computers.

Gradually, however, the use of faster processors and larger amounts of memory enabled the digital computer to surpass its analog counterpart even in the scientific programming and simulations arena. In the 1970s, some hybrid machines combined the easy programmability of a digital "front end" with analog computation, but by the end of that decade the digital computer had rendered analog computers obsolete.

Further Reading

"Analog Computers." Computer Museum, University of Amsterdam. Available online. URL: http://www.science.uva.n/museum/AnalogComputers.html. Accessed April 18, 2007.

Hoeschele, David F., Jr. *Analog-to-Digital and Digital-to-Analog Conversion Techniques.* 2nd ed. New York: John Wiley, 1994.

Vassos, Basil H., and Galen Ewing, eds. *Analog and Computer Electronics for Scientists.* 4th ed. New York: John Wiley, 1993.

Andreessen, Marc

(1971–)
American
Entrepreneur, Programmer

Marc Andreessen brought the World Wide Web and its wealth of information, graphics, and services to the desktop, setting the stage for the first "e-commerce" revolution of the later 1990s. As founder of Netscape, Andreessen also created the first big "dot-com," or company doing business on the Internet.

Born on July 9, 1971, in New Lisbon, Wisconsin, Andreessen grew up as part of a generation that would become familiar with personal computers, computer games, and graphics. By seventh grade Andreessen had his own PC and was programming furiously. He then studied computer science at the University of Illinois at Urbana-Champaign, where his focus on computing was complemented by a wide-ranging interest in music, history, literature, and business.

By the early 1990s the World Wide Web (see WORLD WIDE WEB and BERNERS-LEE, TIM) was poised to change the way information and services were delivered to users. However, early Web pages generally consisted only of linked pages of text, without point-and-click navigation or the graphics and interactive features that adorn Web pages today.

Andreessen learned about the World Wide Web shortly after Berners-Lee introduced it in 1991. Andreessen thought it had great potential, but also believed that there needed to be better ways for ordinary people to access the new

Marc Andreessen, Chairman of Loudcloud, Inc., speaks at Fortune magazine's "Leadership in Turbulent Times" conference on November 8, 2001, in New York City. (PHOTO BY MARIO TAMA/GETTY IMAGES)

medium. In 1993, Andreessen, together with colleague Eric Bina and other helpers at the National Center for Supercomputing Applications (NCSA), set to work on what became known as the Mosaic Web browser. Since their work was paid for by the government, Mosaic was offered free to users over the Internet. Mosaic could show pictures as well as text, and users could follow Web links simply by clicking on them with the mouse. The user-friendly program became immensely popular, with more than 10 million users by 1995.

After earning a B.S. in computer science, Andreessen left Mosaic, having battled with its managers over the future of Web-browsing software. He then met Jim Clark, an older entrepreneur who had been CEO of Silicon Graphics. They founded Netscape Corporation in 1994, using $4 million seed capital provided by Clark.

Andreessen recruited many of his former colleagues at NCSA to help him write a new Web browser, which became known as Netscape Navigator. Navigator was faster and more graphically attractive than Mosaic. Most important, Netscape added a secure encrypted facility that people could use to send their credit card numbers to online merchants. This was part of a two-pronged strategy: First, attract the lion's share of Web users to the new browser, and then sell businesses the software they would need to create effective Web pages for selling products and services to users.

By the end of 1994 Navigator had gained 70 percent of the Web browser market. *Time* magazine named the browser one of the 10 best products of the year, and Netscape was soon selling custom software to companies that wanted a presence on the Web. The e-commerce boom of the later 1990s had begun, and Marc Andreessen was one of its brightest stars. When Netscape offered its stock to the public in summer 1995, the company gained a total worth of $2.3 billion, more than that of many traditional blue-chip industrial companies. Andreessen's own shares were worth $55 million.

BATTLE WITH MICROSOFT

Microsoft (see MICROSOFT and GATES, BILL) had been slow to recognize the growing importance of the Web, but by the mid-1990s Gates had decided that the software giant had to have a comprehensive "Internet strategy." In particular, the company had to win control of the browser market so users would not turn to "platform independent" software that could deliver not only information but applications, without requiring the use of Windows at all.

Microsoft responded by creating its own Web browser, called Internet Explorer. Although technical reviewers generally considered the Microsoft product to be inferior to Netscape, it gradually improved. Most significantly, Microsoft included Explorer with its new Windows 95 operating system. This "bundling" meant that PC makers and consumers had little interest in paying for Navigator when they already had a "free" browser from Microsoft. In response to this move, Netscape and other Microsoft competitors helped promote the antitrust case against Microsoft that would result in 2001 in some of the company's practices being declared an unlawful use of monopoly power.

Andreessen tried to respond to Microsoft by focusing on the added value of his software for Web servers while making Navigator "open source," meaning that anyone was allowed to access and modify the program's code (see OPEN SOURCE). He hoped that a vigorous community of programmers might help keep Navigator technically superior to Internet Explorer. However, Netscape's revenues began to decline steadily. In 1999 America Online (AOL) bought the company, seeking to add its technical assets and Webcenter online portal to its own offerings (see AMERICA ONLINE).

After a brief stint with AOL as its "principal technical visionary," Andreessen decided to start his own company, called LoudCloud. The company provided Web-site development, management, and custom software (including e-commerce "shopping basket" systems) for corporations that had large, complex Web sites. However, the company was not successful; Andreessen sold its Web-site-management component to Texas-based Electronic Data Systems (EDS) while retaining its software division under the new name Opsware. In 2007 Andreessen scored another coup, selling Opsware to Hewlett-Packard (HP) for $1.6 billion.

In 2007 Andreessen launched Ning, a company that offers users the ability to add blogs, discussion forums, and other features to their Web sites, but faced established competitors such as MySpace (see also SOCIAL NETWORKING). In July 2008 Andresseen joined the board of Facebook.

While the future of his recent ventures remains uncertain, Marc Andreessen's place as one of the key pioneers of the Web and e-commerce revolution is assured. His inventiveness, technical insight, and business acumen made him a model for a new generation of Internet entrepreneurs. Andreessen was named one of the Top 50 People under the Age of 40 by *Time* magazine (1994) and has received the Computerworld/Smithsonian Award for Leadership (1995) and the W. Wallace McDowell Award of the IEEE Computer Society (1997).

Further Reading

Clark, Jim. *Netscape Time: The Making of the Billion-Dollar Startup That Took on Microsoft.* New York: St. Martin's Press, 1999.

Guynn, Jessica. "Andreessen Betting Name on New Ning." *San Francisco Chronicle,* February 27, 2006, p. D1, D4.

Payment, Simone. *Marc Andreessen and Jim Clark: The Founders of Netscape.* New York: Rosen Pub. Group, 2006.

Quittner, Joshua, and Michelle Slatala. *Speeding the Net: The Inside Story of Netscape and How It Challenged Microsoft.* New York: Atlantic Monthly Press, 1998.

animation, computer

Ever since the first hand-drawn cartoon features entertained moviegoers in the 1930s, animation has been an important part of the popular culture. Traditional animation uses a series of hand-drawn frames that, when shown in rapid succession, create the illusion of lifelike movement.

COMPUTER ANIMATION TECHNIQUES

The simplest form of computer animation (illustrated in games such as *Pong*) involves drawing an object, then erasing it and redrawing it in a different location. A somewhat more sophisticated approach can create motion in a scene by displaying a series of pre-drawn images called *sprites*—for example, there could be a series of sprites showing a sword-wielding troll in different positions.

Since there are only a few intermediate images, the use of sprites doesn't convey truly lifelike motion. Modern animation uses a modern version of the traditional drawn animation technique. The drawings are "keyframes" that capture significant movements by the characters. The keyframes are later filled in with transitional frames in a process called *tweening.* Since it is possible to create algorithms that describe the optimal in-between frames, the advent of sufficiently powerful computers has made computer animation both possible and desirable. Today computer animation is used not only for cartoons but also for video games and movies. The most striking use of this technique is morphing, where the creation of plausible intermediate images between two strikingly different faces creates the illusion of one face being transformed into the other.

Algorithms that can realistically animate people, animals, and other complex objects require the ability to create a model that includes the parts of the object that can move separately (such as a person's arms and legs). Because the movement of one part of the model often affects the positions of other parts, a treelike structure is often used to describe these relationships. (For example, an elbow moves an arm, the arm in turn moves the hand, which in turn moves the fingers). Alternatively, live actors performing a repertoire of actions or poses can be digitized using wearable sensors and then combined to portray situations, such as in a video game.

Less complex objects (such as clouds or rainfall) can be treated in a simpler way, as a collection of "particles" that move together following basic laws of motion and gravity. Of course when different models come into contact (for example, a person walking in the rain), the interaction between the two must also be taken into consideration.

While realism is always desirable, there is inevitably a tradeoff between the resources available. Computationally intensive physics models might portray a very realistic spray of water using a high-end graphics workstation, but simplified models have to be used for a program that runs on a game console or desktop PC. The key variables are the frame rate (higher is smoother) and the display resolution. The amount of available video memory is also a consideration: many desktop PCs sold today have 256MB or more of video memory.

APPLICATIONS

Computer animation is used extensively in many feature films, such as for creating realistic dinosaurs (*Jurassic Park*) or buglike aliens (*Starship Troopers*). Computer games combine animation techniques with other techniques (see COMPUTER GRAPHICS) to provide smooth action within a vivid 3D landscape. Simpler forms of animation are now a staple of Web site design, often written in Java or with the aid of animation scripting programs such as Adobe Flash.

The intensive effort that goes into contemporary computer animation suggests that the ability to fascinate the human eye that allowed Walt Disney to build an empire is just as compelling today.

Further Reading

"3-D Animation Workshop." Available online. URL: http://www.webreference.com/3d/indexa.html. Accessed April 12, 2007.
Comet, Michael B. "Character Animation: Principles and Practice." Available online. URL: http://www.comet-cartoons.com/toons/3ddocs/charanim. Accessed April 12, 2007.
Hamlin, J. Scott. *Effective Web Animation: Advanced Techniques for the Web*. Reading, Mass.: Addison-Wesley, 1999.
O'Rourke, Michael. *Principles of Three-Dimensional Computer Animation: Modeling, Rendering, and Animating with 3D Computer Graphics*. New York: Norton, 1998.
Parent, Rick. *Computer Animation: Algorithms and Techniques*. San Francisco: Morgan Kaufmann, 2002.
Shupe, Richard, and Robert Hoekman. *Flash 8: Projects for Learning Animation and Interactivity*. Sebastapol, Calif.: O'Reilly Media, 2006.

anonymity and the Internet

Anonymity, or the ability to communicate without disclosing a verifiable identity, is a consequence of the way most Internet-based e-mail, chat, or news services were designed (see E-MAIL, CHAT, TEXTING AND INSTANT MESSAGING, and NETNEWS AND NEWGROUPS). This does not mean that messages do not have names attached. Rather, the names can be arbitrarily chosen or pseudonymous, whether reflecting development of an online persona *or* the desire to avoid having to take responsibility for unwanted communications (see SPAM).

ADVANTAGES

If a person uses a fixed Internet address (see TCP/IP), it may be possible to eventually discover the person's location and even identity. However, messages can be sent through anonymous remailing services where the originating address is removed. Web browsing can also be done "at arm's length" through a proxy server. Such means of anonymity can arguably serve important values, such as allowing persons living under repressive governments (or who belong to minority groups) to express themselves more freely precisely because they cannot be identified. However, such techniques require some sophistication on the part of the user. With ordinary users using their service provider accounts directly, governments (notably China) have simply demanded that the user's identity be turned over when a crime is alleged.

Pseudonymity (the ability to choose names separate from one's primary identity) in such venues as chat rooms or online games can also allow people to experiment with different identities or roles, perhaps getting a taste of how members of a different gender or ethnic group are perceived (see IDENTITY IN THE ONLINE WORLD).

Anonymity can also help protect privacy, especially in commercial transactions. For example, purchasing something with cash normally requires no disclosure of the purchaser's identity, address, or other personal information.

Various systems can use secure encryption to create a cash equivalent in the online world that assures the merchant of valid payment without disclosing unnecessary information about the purchaser (see DIGITAL CASH). There are also facilities that allow for essentially anonymous Web browsing, preventing the aggregation or tracking of information (see COOKIES).

PROBLEMS

The principal problem with anonymity is that it can allow the user to engage in socially undesirable or even criminal activity with less fear of being held accountable. The combination of anonymity (or the use of a pseudonym) and the lack of physical presence seems to embolden some people to engage in insult or "flaming," where they might be inhibited in an ordinary social setting. A few services (notably The WELL) insist that the real identity of all participants be available even if postings use a pseudonym.

Spam or deceptive e-mail (see PHISHING AND SPOOFING) takes advantage both of anonymity (making it hard for authorities to trace) and pseudonymity (the ability to disguise the site by mimicking a legitimate business). Anonymity makes downloading or sharing files easier (see FILE-SHARING AND P2P NETWORKS), but also makes it harder for owners of videos, music, or other content to pursue copyright violations. Because of the prevalence of fraud and other criminal activity on the Internet, there have been calls to restrict the ability of online users to remain anonymous, and some nations such as South Korea have enacted legislation to that effect. However, civil libertarians and privacy advocates believe that the impact on freedom and privacy outweighs any benefits for security and law enforcement.

The database of Web-site registrants (called Whois) provides contact information intended to ensure that someone will be responsible for a given site and be willing to cooperate to fix technical or administrative problems. At present, Whois information is publicly available. However, the Internet Corporation for Assigned Names and Numbers (ICANN) is considering making the contact information available only to persons who can show a legitimate need.

Further Reading

Lessig, Lawrence. *Code: Version 2.0*. New York: Basic Books, 2006.
Rogers, Michael. "Let's See Some ID, Please: The End of Anonymity on the Internet?" *The Practical Futurist* (MSNBC), December 13, 2005. Available online. URL: http://www.msnbc.msn.com/ID/10441443/. Accessed April 10, 2007.
Wallace, Jonathan D. "Nameless in Cyberspace: Anonymity on the Internet." CATO Institute Briefing Papers, no. 54, December 8, 1999. Available online. URL: http://www.cato.org/pubs/briefs/bp54.pdf. Accessed April 10, 2007.

AOL   *See* AMERICA ONLINE.

API   *See* APPLICATIONS PROGRAM INTERFACE.

APL (a programming language)

This programming language was developed by Harvard (later IBM) researcher Kenneth E. Iverson in the early 1960s as a way to express mathematical functions clearly and consistently for computer use. The power of the language to compactly express mathematical functions attracted a growing number of users, and APL soon became a full general-purpose computing language.

Like many versions of BASIC, APL is an interpreted language, meaning that the programmer's input is evaluated "on the fly," allowing for interactive response (see INTERPRETER). Unlike BASIC or FORTRAN, however, APL has direct and powerful support for all the important mathematical functions involving arrays or matrices (see ARRAY).

APL has over 100 built-in operators, called "primitives." With just one or two operators the programmer can perform complex tasks such as extracting numeric or trigonometric functions, sorting numbers, or rearranging arrays and matrices. (Indeed, APL's greatest power is in its ability to manipulate matrices directly without resorting to explicit loops or the calling of external library functions.)

To give a very simple example, the following line of APL code:

$$X [\Delta X]$$

sorts the array X. In most programming languages this would have to be done by coding a sorting algorithm in a dozen or so lines of code using nested loops and temporary variables.

However, APL has also been found by many programmers to have significant drawbacks. Because the language uses Greek letters to stand for many operators, it requires the use of a special type font that was generally not available on non-IBM systems. A dialect called J has been devised to use only standard ASCII characters, as well as both simplifying and expanding the language. Many programmers find mathematical expressions in APL to be cryptic, making programs hard to maintain or revise. Nevertheless, APL Special Interest Groups in the major computing societies testify to continuing interest in the language.

Further Reading

ACM Special Interest Group for APL and J Languages. Available online. URL: http://www.acm.org/sigapl/. Accessed April 12, 2007.

"APL Frequently Asked Questions." Available from various sites including URL: http://home.earthlink.net/~swsirlin/apl.faq.html. Accessed May 8, 2007.

Gilman, Leonard, and Allen J. Rose. *APL: An Interactive Approach.* 3rd ed. (reprint). Malabar, Fla.: Krieger, 1992.

"Why APL?" Available online. URL: http://www.acm.org/sigapl/whyapl.htm. Accessed.

Apple Corporation

Since the beginning of personal computing, Apple has had an impact out of proportion to its relatively modest market share. In a world generally dominated by IBM PC-compatible machines and the Microsoft DOS and Windows operating systems, Apple's distinctive Macintosh computers and more recent media products have carved out distinctive market spaces.

Headquartered in Cupertino, California, Apple was cofounded in 1976 by Steve Jobs, Steve Wozniak, and Ronald Wayne (the latter sold his interest shortly after incorporation). (See JOBS, STEVE, and WOZNIAK, STEVEN.) Their first product, the Apple I computer, was demonstrated to fellow microcomputer enthusiasts at the Homebrew Computer Club. Although it aroused considerable interest, the hand-built Apple I was sold without a power supply, keyboard, case, or display. (Today it is an increasingly valuable "antique.")

Apple's true entry into the personal computing market came in 1977 with the Apple II. Although it was more expensive than its main rivals from Radio Shack and Commodore, the Apple II was sleek, well constructed, and featured built-in color graphics. The motherboard included several slots into which add-on boards (such as for printer interfaces) could be inserted. Besides being attractive to hobbyists, however, the Apple II began to be taken seriously as a business machine when the first popular spreadsheet program, VisiCalc, was written for it.

By 1981 more than 2 million Apple IIs (in several variations) had been sold, but IBM then came out with the IBM PC. The IBM machine had more memory and a somewhat more powerful processor, but its real advantage was the access IBM had to the purchasing managers of corporate America. The IBM PC and "clone" machines from other companies such as Compaq quickly displaced Apple as market leader.

THE MACINTOSH

By the early 1980s Steve Jobs had turned his attention to designing a radically new personal computer. Using technology that Jobs had observed at the Xerox Palo Alto Research Center (PARC), the new machine would have a fully graphical interface with icons and menus and the ability to select items with a mouse. The first such machine, the Apple Lisa, came out in 1983. The machine cost almost $10,000, however, and proved a commercial failure.

In 1984, however, Apple launched a much less expensive version (see MACINTOSH). Viewers of the 1984 Super Bowl saw a remarkable Apple commercial in which a female figure runs through a group of corporate drones (representing IBM) and smashes a screen. The "Mac" sold reasonably well, particularly as it was given more processing power and memory and was accompanied by new software that could take advantage of its capabilities. In particular, the Mac came to dominate the desktop publishing market, thanks to Adobe's PageMaker program.

In the 1990s Apple diversified the Macintosh line with a portable version (the PowerBook) that largely set the standard for the modern laptop computer. By then Apple had acquired a reputation for stylish design and superior ease of use. However, the development of the rather similar Windows operating system by Microsoft (see MICROSOFT WINDOWS) as well as constantly dropping prices for IBM-compatible hardware put increasing pressure on Apple and kept its market share limited. (Apple's legal challenge to

Microsoft alleging misappropriation of intellectual property proved to be a protracted and costly failure.)

Apple's many Macintosh variants of the later 1990s proved confusing to consumers, and sales appeared to bog down. The company was accused of trying to rely on an increasingly nonexistent advantage, keeping prices high, and failing to innovate.

However, in 1997 Steve Jobs, who had been forced out of the company in an earlier dispute, returned to the company and brought with him some new ideas. In hardware there was the iMac, a sleek all-in-one system with an unmistakable appearance that restored Apple to profitability in 1998. On the software side, Apple introduced new video-editing software for home users and a thoroughly redesigned UNIX-based operating system (see OS X). In general, the new incarnation of the Macintosh was promoted as the ideal companion for a media-hungry generation.

CONSUMER ELECTRONICS

Apple's biggest splash in the new century, however, came not in personal computing, but in the consumer electronics sector. Introduced in 2001, the Apple iPod has been phenomenally successful, with 100 million units sold by 2006. The portable music player can hold thousands of songs and easily fit into a pocket (see also MUSIC AND VIDEO PLAYERS, DIGITAL). Further, it was accompanied by an easy-to-use interface and an online music store (iTunes). (By early 2006, more than a billion songs had been purchased and downloaded from the service.) Although other types of portable MP3 players exist, it is the iPod that defined the genre (see also PODCASTING). Later versions of the iPod include the ability to play videos.

In 2005 Apple announced news that startled and perhaps dismayed many long-time users. The company announced that future Macintoshes would use the same Intel chips employed by Windows-based ("Wintel") machines like the IBM PC and its descendants. The more powerful machines would use dual processors (Intel Core Duo). Further, in 2006 Apple released Boot Camp, a software package that allows Intel-based Macs to run Windows XP. Jobs's new strategy seems to be to combine what he believed to be a superior operating system and industrial design with industry-standard processors, offering the best user experience and a very competitive cost. Apple's earnings continued strong into the second half of 2006.

In early 2007 Jobs electrified the crowd at the Macworld Expo by announcing that Apple was going to "reinvent the phone." The product, called iPhone, is essentially a combination of a video iPod and a full-featured Internet-enabled cell phone (see SMARTPHONE). Marketed by Apple and AT&T (with the latter providing the phone service), the iPhone costs about twice as much as an iPod but includes a higher-resolution 3.5-in. (diagonal) screen and a 2 megapixel digital camera. The phone can connect to other devices (see BLUETOOTH) and access Internet services such as Google Maps. The user controls the device with a new interface called Multitouch.

Apple also introduced another new media product, the Apple TV (formerly the iTV), allowing music, photos, and video to be streamed wirelessly from a computer to an existing TV set. Apple reaffirmed its media-centered plans by announcing that the company's name would be changed from Apple Computer Corporation to simply Apple Corporation.

In the last quarter of 2006 Apple earned a record-breaking $1 billion in profit, bolstered mainly by very strong sales of iPods and continuing good sales of Macintosh computers.

Apple had strong Macintosh sales performance in the latter part of 2007. The company has suggested that its popular iPods and iPhones may be leading consumers to consider buying a Mac for their next personal computer.

Meanwhile, however, Apple has had to deal with questions about its backdating of stock options, a practice by which about 200 companies have, in effect, enabled executives to purchase their stock at an artificially low price. Apple has cleared Jobs of culpability in an internal investigation, and in April 2007 the Securities and Exchange Commission announced that it would not take action against the company.

Further Reading
Carlton, Jim. *Apple: The Inside Story of Intrigue, Egomania and Business Blunders.* New York: Random House, 1997.
Deutschman, Alan. *The Second Coming of Steve Jobs.* New York: Broadway Books, 2000.
Hertzfeld, Andy. *Revolution in the Valley.* Sebastapol, Calif.: O'Reilly, 2005.
Kunkel, Paul. *AppleDesign: The Work of the Apple Industrial Design Group.* New York: Graphis, 1997.
Levy, Steven. *Insanely Great: The Life and Times of Macintosh, The Computer that Changed Everything.* New York: Penguin Books, 2000.
Linzmayer, Owen W. *Apple Confidential 2.0: The Definitive History of the World's Most Colorful Company.* 2nd ed. San Francisco, Calif.: No Starch Press, 2004.

applet

An applet is a small program that uses the resources of a larger program and usually provides customization or additional features. The term first appeared in the early 1990s in connection with Apple's AppleScript scripting language for the Macintosh operating system. Today Java applets represent the most widespread use of this idea in Web development (see JAVA).

Java applets are compiled to an intermediate representation called bytecode, and generally are run in a Web browser (see WEB BROWSER). Applets thus represent one of several alternatives for interacting with users of Web pages beyond what can be accomplished using simple text markup (see HTML; for other approaches see JAVASCRIPT, PHP, SCRIPTING LANGUAGES, and AJAX).

An applet can be invoked by inserting a reference to its program code in the text of the Web page, using the HTML applet element or the now-preferred object element. Although the distinction between applets and scripting code (such as in PHP) is somewhat vague, applets usually run in their own window or otherwise provide their own interface, while scripting code is generally used to tailor the behavior of separately created objects. Applets are also rather like plug-ins, but the latter are generally used to provide a particular capability (such as the ability to read or play a particular kind of media file), and have a standardized facility for their installation and management (see PLUG-IN).

Some common uses for applets include animations of scientific or programming concepts for Web pages supporting class curricula and for games designed to be played using Web browsers. Animation tools such as Flash and Shockwave are often used for creating graphic applets.

To prevent badly or maliciously written applets from affecting user files, applets such as Java applets are generally run within a restricted or "sandbox" environment where, for example, they are not allowed to write or change files on disk.

Further Reading
"Java Applets." Available online. URL: http://en.wikibooks.org/wiki/Java_Programming/Applets. Accessed April 10, 2007.
McGuffin, Michael. "Java Applet Tutorial." Available online. URL: http://www.realapplets.com/tutorial/. Accessed April 10, 2007.

application program interface (API)

In order for an application program to function, it must interact with the computer system in a variety of ways, such as reading information from disk files, sending data to the printer, and displaying text and graphics on the monitor screen (see USER INTERFACE). The program may need to find out whether a device is available or whether it can have access to an additional portion of memory. In order to provide these and many other services, an operating system such as Microsoft Windows includes an extensive application program interface (API). The API basically consists of a variety of functions or procedures that an application program can call upon, as well as data structures, constants, and various definitions needed to describe system resources.

Applications programs use the API by including calls to routines in a program library (see LIBRARY, PROGRAM and PROCEDURES AND FUNCTIONS). In Windows, "dynamic link libraries" (DLLs) are used. For example, this simple function puts a message box on the screen:

```
MessageBox (0, "Program Initialization Failed!",
"Error!", MB_ICONEXCLAMATION | MB_OK | MB_
SYSTEMMODAL);
```

In practice, the API for a major operating system such as Windows contains hundreds of functions, data structures, and definitions. In order to simplify learning to access the necessary functions and to promote the writing of readable code, compiler developers such as Microsoft and Borland have devised frameworks of C++ classes that package related functions together. For example, in the Microsoft Foundation Classes (MFC), a program generally begins by deriving a class representing the application's basic characteristics from the MFC class CWinApp. When the program wants to display a window, it derives it from the CWnd class, which has the functions common to all windows, dialog boxes, and controls. From CWnd is derived the specialized class

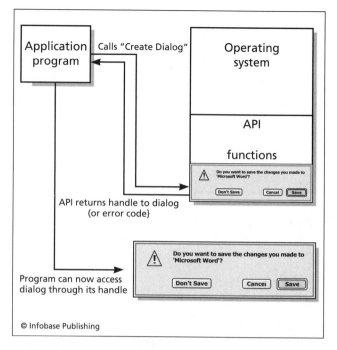

Modern software uses API calls to obtain interface objects such as dialog boxes from the operating system. Here the application calls the CreateDialog API function. The operating system returns a pointer (called a handle) that the application can now use to access and manipulate the dialog.

for each type of window: for example, CFrameWnd implements a typical main application window, while CDialog would be used for a dialog box. Thus in a framework such as MFC or Borland's OWL, the object-oriented concept of encapsulation is used to bundle together objects and their functions, while the concept of inheritance is used to relate the generic object (such as a window) to specialized versions that have added functionality (see OBJECT-ORIENTED PROGRAMMING and ENCAPSULATION INHERITANCE).

In recent years Microsoft has greatly extended the reach of its Windows API by providing many higher level functions (including user interface items, network communications, and data access) previously requiring separate software components or program libraries (see MICROSOFT.NET).

Programmers using languages such as Visual Basic can take advantage of a further level of abstraction. Here the various kinds of windows, dialogs, and other controls are provided as building blocks that the developer can insert into a form designed on the screen, and then settings can be made and code written as appropriate to control the behavior of the objects when the program runs. While the programmer will not have as much direct control or flexibility, avoiding the need to master the API means that useful programs can be written more quickly.

Further Reading
"DevCentral Tutorials: MFC and Win32." Available online. URL: http://devcentral.iftech.com/learning/tutorials/submfc.asp. Accessed April 12, 2007.

Petzold, Charles. *Programming Windows: the Definitive Guide to the Win32 API.* 5th ed. Redmond, Wash.: Microsoft Press, 1999.
"Windows API Guide." Available online. URL: http://www.vbapi.com/. Accessed April 12, 2007.

application service provider (ASP)

Traditionally, software applications such as office suites are sold as packages that are installed and reside on the user's computer. Starting in the mid-1990s, however, the idea of offering users access to software from a central repository attracted considerable interest. An application service provider (ASP) essentially rents access to software.

Renting software rather than purchasing it outright has several advantages. Since the software resides on the provider's server, there is no need to update numerous desktop installations every time a new version of the software (or a "patch" to fix some problem) is released. The need to ship physical CDs or DVDs is also eliminated, as is the risk of software piracy (unauthorized copying). Users may be able to more efficiently budget their software expenses, since they will not have to come up with large periodic expenses for upgrades. The software provider, in turn, also receives a steady income stream rather than "surges" around the time of each new software release.

For traditional software manufacturers, the main concern is determining whether the revenue obtained by providing its software as a service (directly or through a third party) is greater than what would have been obtained by selling the software to the same market. (It is also possible to take a hybrid approach, where software is still sold, but users are offered additional features online. Microsoft has experimented with this approach with its Microsoft Office Live and other products.)

Renting software also has potential disadvantages. The user is dependent on the reliability of the provider's servers and networking facilities. If the provider's service is down, then the user's work flow and even access to critical data may be interrupted. Further, sensitive data that resides on a provider's system may be at risk from hackers or industrial spies. Finally, the user may not have as much control over the deployment and integration of software as would be provided by outright purchase.

The ASP market was a hot topic in the late 1990s, and some pundits predicted that the ASP model would eventually supplant the traditional retail channel for mainstream software. This did not happen, and more than a thousand ASPs were among the casualties of the "dot-com crash" of the early 2000s. However, ASP activity has been steadier if less spectacular in niche markets, where it offers more economical access to expensive specialized software for applications such as customer relationship management, supply chain management, and e-commerce related services—for example, Salesforce.com. The growing importance of such "software as a service" business models can be seen in recent offerings from traditional software companies such as SAS. By 2004, worldwide spending for "on demand" software had exceeded $4 billion, and Gartner Research has predicted that in the second half of the decade about a third of all software will be obtained as a service rather than purchased.

WEB-BASED APPLICATIONS AND FREE SOFTWARE

By that time a new type of application service provider had become increasingly important. Rather than seeking to gain revenue by selling online access to software, this new kind of ASP provides the software for free. A striking example is Google Pack, a free software suite offered by the search giant (see GOOGLE). Google Pack includes a variety of applications, including a photo organizer and search and mapping tools developed by Google, as well as third-party programs such as the Mozilla Firefox Web browser, RealPlayer media player, the Skype Internet phone service (see VOIP), and antivirus and antispyware programs. The software is integrated into the user's Windows desktop, providing fast index and retrieval of files from the hard drive. (Critics have raised concerns about the potential violation of privacy or misuse of data, especially with regard to a "share across computers" feature that stores data about user files on Google's servers.) America Online has also begun to provide free access to software that was formerly available only to paid subscribers.

This use of free software as a way to attract users to advertising-based sites and services could pose a major threat to companies such as Microsoft that rely on software as their main source of revenue. In 2006 Google unveiled a Google Docs & Spreadsheets, a program that allows users to create and share word-processing documents and spreadsheets over the Web. Such offerings, together with free open-source software such as Open Office.org, may force traditional software companies to find a new model for their own offerings.

Microsoft in turn has launched Office Live, a service designed to provide small offices with a Web presence and productivity tools. The free "basic" level of the service is advertising supported, and expanded versions are available for a modest monthly fee. The program also has features that are integrated with Office 2007, thus suggesting an attempt to use free or low-cost online services to add value to the existing stand-alone product line.

By 2008 the term *cloud computing* had become a popular way to describe software provided from a central Internet site that could be accessed by the user through any form of computer and connection. An advantage touted for this approach is that the user need not be concerned with where data is stored or the need to make backups, which are handled seamlessly.

Further Reading

Chen, Anne. "Office Live Makes Online Presence Known." *eWeek,* November 2, 2006. Available online. URL: http://www.eweek.com/article2/0,1759,2050580,00.asp. Accessed May 22, 2007.

Focacci, Luisa, Robert J. Mockler, and Marc E. Gartenfeld. *Application Service Providers in Business.* New York: Haworth, 2005.

Garretson, Rob. "The ASP Reincarnation: The Application Service Provider Name Dies Out, but the Concept Lives on among Second-Generation Companies Offering Software as a service." *Network World,* August 29, 2005. Available online.

URL: http://www.networkworld.com/research/2005/082905-asp.html. Accessed May 22, 2007.

"Google Spreadsheets: The Soccer Mom's Excel." *eWeek*, June 6, 2006. Available online. URL: http://www.eweek.com/article2/0,1759,1972740,00.asp. Accessed May 22, 2007.

Schwartz, Ephraim. "Applications: SaaS Breaks Down the Wall: Hosted Applications Continue to Remove Enterprise Objections." *Infoworld*, January 1, 2007. Available online. URL: http://www.infoworld.com/article/07/01/01/01FEtoyapps_1.html. Accessed May 22, 2007.

application software

Application software consists of programs that enable computers to perform useful tasks, as opposed to programs that are concerned with the operation of the computer itself (see OPERATING SYSTEM and SYSTEMS PROGRAMMING). To most users, applications programs *are* the computer: They determine how the user will accomplish tasks.

The following table gives a selection of representative applications:

DEVELOPING AND DISTRIBUTING APPLICATIONS

Applications can be divided into three categories based on how they are developed and distributed. Commercial applications such as word processors, spreadsheets, and general-purpose Database Management Systems (DBMS) are developed by companies specializing in such software and distributed to a variety of businesses and individual users (see WORD PROCESSING, SPREADSHEET, and DATABASE MANAGEMENT SYSTEM). Niche or specialized applications (such as hospital billing systems) are designed for and marketed to a particular industry (see MEDICAL APPLICATIONS OF COMPUTERS). These programs tend to be much more expensive and usually include extensive technical support. Finally, in-house applications are developed by programmers within a business or other institution for their own use. Examples might include employee training aids or a Web-based product catalog (although such applications could also be developed using commercial software such as multimedia or database development tools).

While each application area has its own needs and priorities, the discipline of software development (see SOFTWARE ENGINEERING and PROGRAMMING ENVIRONMENT) is generally applicable to all major products. Software developers try to improve speed of development as well as program reliability by using software development tools that simplify the writing and testing of computer code, as well as the manipulation of graphics, sound, and other resources used by the program. An applications developer must also have a good understanding of the features and limitations of the relevant operating system. The developer of commercial software must work closely with the marketing department to work out issues of feature selection, timing of releases, and anticipation of trends in software use (see MARKETING OF SOFTWARE).

Further Reading

"Business Software Buyer's Guide." Available online. URL: http://businessweek.buyerzone.com/software/business_software/buyers_guide1.html. Accessed April 12, 2007.

ZDnet Buyer's Guide to Computer Applications. Available online. URL: http://www.zdnet.com/computershopper/edit/howto-buy/. Accessed April 12, 2007

GENERAL AREA	APPLICATIONS	EXAMPLES
Business Operations	payroll, accounts receivable, inventory, marketing	specialized business software, general spreadsheets and databases
Education	school management, curriculum reinforcement, reference aids, curriculum expansion or supplementation, training	attendance and grade book management, drill-and-practice software for reading or arithmetic, CD or online encyclopedias, educational games or simulations, collaborative and Web-based learning, corporate training programs
Engineering	design and manufacturing	computer-aided design (CAD), computer-aided manufacturing (CAM)
Entertainment	games, music, and video	desktop and console games, online games, digitized music distribution (MP3 files), streaming video (including movies)
Government	administration, law enforcement, military	tax collection, criminal records and field support for police, legal citation databases, combat information and weapons control systems
Health Care	hospital administration, health care delivery	hospital information and billing systems, medical records management, medical imaging, computer-assisted treatment or surgery
Internet and World Wide Web	web browser, search tools, e-commerce	browser and plug-in software for video and audio, search engines, e-commerce support and secure transactions
Libraries	circulation, cataloging, reference	automated book check-in systems, cataloging databases, CD or online bibliographic and full-text databases
Office Operations	e-mail, document creation	e-mail clients, word processing, desktop publishing
Science	statistics, modeling, data analysis	mathematical and statistical software, modeling of molecules, gene typing, weather forecasting

application suite

An application suite is a set of programs designed to be used together and marketed as a single package. For example, a typical office suite might include word processing, spreadsheet, database, personal information manager, and e-mail programs.

While an operating system such as Microsoft Windows provides basic capabilities to move text and graphics from one application to another (such as by cutting and pasting), an application suite such as Microsoft Office makes it easier to, for example, launch a Web browser from a link within a word processing document or embed a spreadsheet in the document. In addition to this "interoperability," an application suite generally offers a consistent set of commands and features across the different applications, speeding up the learning process. The use of the applications in one package from one vendor simplifies technical support and upgrading. (The development of comparable applications suites for Linux is likely to increase that operating system's acceptance on the desktop.)

Applications suites have some potential disadvantages as compared to buying a separate program for each application. The user is not necessarily getting the best program in each application area, and he or she is also forced to pay for functionality that may not be needed or desired. Due to their size and complexity, software suites may not run well on older computers. Despite these problems, software suites sell very well and are ubiquitous in today's office.

(For a growing challenge to the traditional standalone software suite, see APPLICATION SERVICE PROVIDER.)

Further Reading

Villarosa, Joseph. "How Suite It Is: One-Stop Shopping for Software Can Save You Both Time and Money." Available online. *Forbes* magazine online. URL: http://www.forbes.com/buyers/070.htm. Accessed April 12, 2007.

arithmetic logic unit (ALU)

The arithmetic logic unit is the part of a computer system that actually performs calculations and logical comparisons on data. It is part of the central processing unit (CPU), and in practice there may be separate and multiple arithmetic and logic units (see CPU).

The ALU works by first retrieving a code that represents the operation to be performed (such as ADD). The code also specifies the location from which the data is to be retrieved and to which the results of the operation are to be stored. (For example, addition of the data from memory to a number already stored in a special accumulator register within the CPU, with the result to be stored back into the accumulator.) The operation code can also include a specification of the format of the data to be used (such as fixed or floating-point numbers)—the operation and format are often combined into the same code.

In addition to arithmetic operations, the ALU can also carry out logical comparisons, such as bitwise operations that compare corresponding bits in two data words, corresponding to Boolean operators such as AND, OR, and XOR (see BITWISE OPERATIONS and BOOLEAN OPERATORS).

The data or operand specified in the operation code is retrieved as words of memory that represent numeric data, or indirectly, character data (see MEMORY, NUMERIC DATA, and CHARACTERS AND STRINGS). Once the operation is performed, the result is stored (typically in a register in the CPU). Special codes are also stored in registers to indicate characteristics of the result (such as whether it is positive, negative, or zero). Other special conditions called exceptions indicate a problem with the processing. Common exceptions include overflow, where the result fills more bits than are available in the register, loss of precision (because there isn't room to store the necessary number of decimal places), or an attempt to divide by zero. Exceptions are typically indicated by setting a flag in the machine status register (see FLAG).

THE BIG PICTURE

Detailed knowledge of the structure and operation of the ALU is not needed by most programmers. Programmers who need to directly control the manipulation of data in the ALU and CPU write programs in assembly language (see ASSEMBLER) that specify the sequence of operations to be performed. Generally only the lowest-level operations involving the physical interface to hardware devices require this level of detail (see DEVICE DRIVER). Modern compilers can produce optimized machine code that is almost as efficient as directly-coded assembler. However, understanding the architecture of the ALU and CPU for a particular chip can help predict its advantages or disadvantages for various kinds of operations.

Further Reading

Kleitz, William. *Digital and Microprocessor Fundamentals: Theory and Applications*. 4th ed. Upper Saddle River, N.J.: Prentice Hall, 2002.

Stokes, Jon. "Understanding the Microprocessor." *Ars Technica.* Available online. URL: http://arstechnica.com/paedia/c/cpu/part-1/cpu1-1.html. Accessed May 22, 2007.

array

An array stores a group of similar data items in consecutive order. Each item is an *element* of the array, and it can be retrieved using a *subscript* that specifies the item's location relative to the first item. Thus in the C language, the statement

```
int Scores (10);
```

sets up an array called *Scores,* consisting of 10 integer values. The statement

```
Scores [5] = 93;
```

stores the value 93 in array element number 5. One subtlety, however, is that in languages such as C, the first element of the array is [0], so [5] represents not the fifth but the sixth element in *Scores.* (Many versions of BASIC allow for setting either 0 or 1 as the first element of arrays.)

In languages such as C that have pointers, an equivalent way to access an array is to declare a pointer and store the address of the first element in it (see POINTERS AND INDIRECTION):

```
int * ptr;
ptr = &Scores [0];
```

(See POINTERS AND INDIRECTION.)

Arrays are useful because they allow a program to work easily with a group of data items without having to use separately named variables. Typically, a program uses a loop to traverse an array, performing the same operation on each element in order (see LOOP). For example, to print the current contents of the *Scores* array, a C program could do the following:

```
int index;
for (index = 0; i < 10; i++)

    printf ("Scores [%d] = %d \n", index,
        Scores [index]);
```

This program might print a table like this:

```
Scores [0] = 22
Scores [1] = 28
Scores [2] = 36
```

and so on. Using a pointer, a similar loop would increment the pointer to step to each element in turn.

An array with a single subscript is said to have one dimension. Such arrays are often used for simple data lists, strings of characters, or vectors. Most languages also support multidimensional arrays. For example, a two-dimensional array can represent X and Y coordinates, as on a screen display. Thus the number 16 stored at Colors[10][40] might represent the color of the point at X=10, Y=40 on a 640 by 480 display. A matrix is also a two-dimensional array, and languages such as APL provide built-in support for mathematical operations on such arrays. A four-dimensional array might hold four test scores for each person.

Some languages such as FORTRAN 90 allow for defining "slices" of an array. For example, in a 3 × 3 matrix, the expression MAT(2:3, 1:3) references two 1 × 3 "slices" of the matrix array. Pascal allows defining a subrange, or portion of the subscripts of an array.

ASSOCIATIVE ARRAYS

It can be useful to explicitly associate pairs of data items within an array. In an *associative array* each data element has an associated element called a *key*. Rather than using subscripts, data elements are retrieved by passing the key to a hashing routine (see HASHING). In the Perl language, for example, an array of student names and scores might be set up like this:

%Scores = ("Henderson" => 86, "Johnson" => 87, "Jackson" => 92);

The score for Johnson could later be retrieved using the reference:

$Scores ("Johnson")

Associative arrays are handy in that they facilitate look-up tables or can serve as small databases. However, expanding the array beyond its initial allocation requires rehashing all the existing elements.

PROGRAMMING ISSUES

To avoid error, any reference to an array must be within its declared *bounds*. For example, in the earlier example, Scores[9] is the last element, and a reference to Scores[10] would be out of bounds. Attempting to reference an out-of-bounds value gives an error message in some languages such as Pascal, but in others such as standard C and C++, it simply retrieves whatever happens to be in that location in memory.

Another issue involves the allocation of memory for the array. In a *static* array, such as that used in FORTRAN 77, the necessary storage is allocated before the program runs, and the amount of memory cannot be changed. Static arrays use memory efficiently and reduce overhead, but are inflexible, since the programmer has to declare an array based on the largest number of data items the program might be called upon to handle. A *dynamic* array, however, can use a flexible structure to allocate memory (see HEAP). The program can change the size of the array at any time while it is running. C and C++ programs can create dynamic arrays and allocate memory using special functions (malloc and free in C) or operators (new and delete in C++).

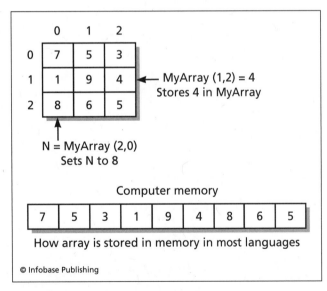

How array is stored in memory in most languages

© Infobase Publishing

A two-dimensional array can be visualized as a grid, with the array subscripts indicating the row and column in which a particular value is stored. Here the value 4 is stored at the location (1,2), while the value at (2,0), which is 8, is assigned to N. As shown, the actual computer memory is a one dimensional line of successive locations. In most computer languages the array is stored row by row.

In the early days of microcomputer programming, arrays tended to be used as an all-purpose data structure for storing information read from files. Today, since there are more structured and flexible ways to store and retrieve such data, arrays are now mainly used for small sets of data (such as look-up tables).

Further Reading

Jensen, Ted. "A Tutorial on Pointers and Arrays in C." Available online. URL: http://pw2.netcom.com/~tjensen/ptr/pointers. htm. Accessed April 12, 2007.

Sebesta, Robert W. *Concepts of Programming Languages*. 8th ed. Boston: Addison-Wesley, 2008.

art and the computer

While the artistic and technical temperaments are often viewed as opposites, the techniques of artists have always shown an intimate awareness of technology, including the physical characteristics of the artist's tools and media. The development of computer technology capable of generating, manipulating, displaying, or printing images has offered a variety of new tools for existing artistic traditions, as well as entirely new media and approaches.

Computer art began as an offshoot of research into image processing or the simulation of visual phenomena, such as by researchers at Bell Labs in Murray Hill, New Jersey, during the 1960s. One of these researchers, A. Michael Noll, applied computers to the study of art history by simulating techniques used by painters Piet Mondrian and Bridget Riley in order to gain a better understanding of them. In addition to exploring existing realms of art, experimenters began to create a new genre of art, based on the ideas of Max Bense, who coined the terms "artificial art" and "generative esthetics." Artists such as Manfred Mohr studied computer science because they felt the computer could provide the tools for an esthetic strongly influenced by mathematics and natural science. For example, Mohr's *P-159/A* (1973) used mathematical algorithms and a plotting device to create a minimalistic yet rich composition of lines. Other artists working in the minimalist, neoconstructivist, and conceptual art traditions found the computer to be a compelling tool for exploring the boundaries of form.

By the 1980s, the development of personal computers made digital image manipulation available to a much wider group of people interested in artistic expression, including the more conventional realms of representational art and photography. Programs such as Adobe Photoshop blend art and photography, making it possible to combine images from many sources and apply a variety of transformations to them. The use of computer graphics algorithms make realistic lighting, shadow, and fog effects possible to a much greater degree than their approximation in traditional media. Fractals can create landscapes of infinite texture and complexity. The computer has thus become a standard tool for both "serious" and commercial artists.

Artificial intelligence researchers have developed programs that mimic the creativity of human artists. For example, a program called Aaron developed by Harold Cohen

Air, *created by Lisa Yount with the popular image-editing program Adobe Photoshop, is part of a group of photocollages honoring the ancient elements of earth, air, water, and fire. The "wings" in the center are actually the two halves of a mussel shell.* (LISA YOUNT)

can adapt and extend existing styles of drawing and painting. Works by Aaron now hang in some of the world's most distinguished art museums.

An impressive display of the "state of the computer art" could be seen at a digital art exhibition that debuted in Boston at the SIGGRAPH 2006 conference. More than 150 artists and researchers from 16 countries exhibited work and discussed its implications. Particularly interesting were dynamic works that interacted with visitors and the environment, often blurring the distinction between digital arts and robotics. In the future, sculptures may change with the season, time of day, or the presence of people in the room, and portraits may show moods or even converse with viewers.

IMPLICATIONS AND PROSPECTS

While traditional artistic styles and genres can be reproduced with the aid of a computer, the computer has the potential to change the basic paradigms of the visual arts. The representation of all elements in a composition in digital form makes art fluid in a way that cannot be matched by traditional media, where the artist is limited in the ability to rework a painting or sculpture. Further, there is no hard-and-fast boundary between still image and animation, and the creation of art works that change interactively in response to their viewer becomes feasible. Sound, too, can be integrated with visual representation, in a way far more sophisticated than that pioneered in the 1960s with "color organs" or laser shows. Indeed, the use of virtual reality technology makes it possible to create art that can be experienced "from the inside," fully immersively (see VIRTUAL REALITY). The use of the Internet opens the possibility of huge collaborative works being shaped by participants around the world.

The growth of computer art has not been without misgivings. Many artists continue to feel that the intimate physical relationship between artist, paint, and canvas cannot be matched by what is after all only an arrangement of light on a flat screen. However, the profound influence of the computer on contemporary art is undeniable.

Further Reading

Computer-Generated Visual Arts (Yahoo). Available online. URL: http://dir.yahoo.com/Arts/Visual_Arts/Computer_Generated/. Accessed April 13, 2007.

Ashford, Janet. *Arts and Crafts Computer: Using Your Computer as an Artist's Tool.* Berkeley, Calif.: Peachpit Press, 2001.

Kurzweil Cyber Art Technologies homepage. Available online. URL: http://www.kurzweilcyberart.com/index.html. Accessed May 22, 2007.

Popper, Frank. *Art of the Electronic Age.* New York: Thames & Hudson, 1997.

Rush, Michael. *New Media in Late 20th-Century Art.* New York: Thames & Hudson, 1999.

SIGGRAPH 2006 Art Gallery. "Intersections." Available online. URL: http://www.siggraph.org/s2006/main.php?f=conference&p=art. Accessed May 22, 2007.

artificial intelligence

The development of the modern digital computer following World War II led naturally to the consideration of the ultimate capabilities of what were soon dubbed "thinking machines" or "giant brains." The ability to perform calculations flawlessly and at superhuman speeds led some observers to believe that it was only a matter of time before the intelligence of computers would surpass human levels. This belief would be reinforced over the years by the development of computer programs that could play chess with increasing skill, culminating in the match victory of IBM's Deep Blue over world champion Garry Kasparov in 1997. (See CHESS AND COMPUTERS.)

However, the quest for artificial intelligence would face a number of enduring challenges, the first of which is a lack of agreement on the meaning of the term *intelligence,* particularly in relation to such seemingly different entities as humans and machines. While chess skill is considered a sign of intelligence in humans, the game is deterministic in that optimum moves can be calculated systematically, limited only by the processing capacity of the computer. Human chess masters use a combination of pattern recognition, general principles, and selective calculation to come up with their moves. In what sense could a chess-playing computer that mechanically evaluates millions of positions be said to "think" in the way humans do? Similarly, computers can be provided with sets of rules that can be used to manipulate virtual building blocks, carry on conversations, and even write poetry. While all these activities can be perceived by a human observer as being intelligent and even creative, nothing can truly be said about what the computer might be said to be experiencing.

In 1950, computer pioneer Alan M. Turing suggested a more productive approach to evaluating claims of artificial intelligence in what became known as the Turing test (see TURING, ALAN). Basically, the test involves having a human interact with an "entity" under conditions where he or she does not know whether the entity is a computer or another human being. If the human observer, after engaging in teletyped "conversation" cannot reliably determine the identity of the other party, the computer can be said to have passed the Turing test. The idea behind this approach is that rather than attempting to precisely and exhaustively define intelligence, we will engage human experience and intuition about what intelligent behavior is like. If a computer can successfully imitate such behavior, then it at least may become problematic to say that it is *not* intelligent.

Computer programs have been able to pass the Turing test to a limited extent. For example, a program called ELIZA written by Joseph Weizenbaum can carry out what appears to be a responsive conversation on themes chosen by the interlocutor. It does so by rephrasing statements or providing generalizations in the way that a nondirective psychotherapist might. But while ELIZA and similar programs have sometimes been able to fool human interlocutors, an in-depth probing by the humans has always managed to uncover the mechanical nature of the response.

Although passing the Turing test could be considered evidence for intelligence, the question of whether a computer might have consciousness (or awareness of self) in the sense that humans experience it might be impossible to answer. In practice, researchers have had to confine themselves to producing (or simulating) intelligent *behavior,* and they have had considerable success in a variety of areas.

TOP-DOWN APPROACHES

The broad question of a strategy for developing artificial intelligence crystallized at a conference held in 1956 at Dartmouth College. Four researchers can be said to be founders of the field: Marvin Minsky (founder of the AI Laboratory at MIT), John McCarthy (at MIT and later, Stanford), and Herbert Simon and Allen Newell (developers of a mathematical problem-solving program called Logic Theorist at the Rand Corporation, who later founded the AI Laboratory at Carnegie Mellon University). The 1950s and 1960s were a time of rapid gains and high optimism about the future of AI (see MINSKY, MARVIN and MCCARTHY, JOHN).

Most early attempts at AI involved trying to specify rules that, together with properly organized data, can enable the machine to draw logical conclusions. In a *production system* the machine has information about "states" (situations) plus rules for moving from one state to another—and ultimately, to the "goal state." A properly implemented production system cannot only solve problems, it can give an explanation of its reasoning in the form of a chain of rules that were applied.

The program SHRDLU, developed by Marvin Minsky's team at MIT, demonstrated that within a simplified "microworld" of geometric shapes a program can solve problems and learn new facts about the world. Minsky later developed a more generalized approach called "frames" to provide the computer with an organized database of knowledge about the world comparable to that which a human child assimilates through daily life. Thus, a program with the appropriate frames can act as though it understands a story about two people in a restaurant because it "knows" basic facts such as that people go to a restaurant to eat, the meal is cooked for them, someone pays for the meal, and so on.

While promising, the frames approach seemed to founder because of the sheer number of facts and relationships needed for a comprehensive understanding of the world. During the 1970s and 1980s, however, expert systems were developed that could carry out complex tasks such as determining the appropriate treatment for infections (MYCIN) and analysis of molecules (DENDRAL). Expert systems combined rules of inference with specialized databases of facts and relationships. Expert systems have thus been able to encapsulate the knowledge of human experts and make it available in the field (see EXPERT SYSTEMS and KNOWLEDGE REPRESENTATION).

The most elaborate version of the frames approach has been a project called Cyc (short for "encyclopedia"), developed by Douglas Lenat. This project is now in its third decade and has codified millions of assertions about the world, grouping them into semantic networks that represent dozens of broad areas of human knowledge. If successful, the Cyc database could be applied in many different domains, including such applications as automatic analysis and summary of news stories.

BOTTOM-UP APPROACHES

Several "bottom-up" approaches to AI were developed in an attempt to create machines that could learn in a more humanlike way. The one that has gained the most practical success is the neural network, which attempts to emulate the operation of the neurons in the human brain. Researchers believe that in the human brain perceptions or the acquisition of knowledge leads to the reinforcement of particular neurons and neural paths, improving the brain's ability to perform tasks. In the artificial neural network a large number of independent processors attempt to perform a task. Those that succeed are reinforced or "weighted," while those that fail may be negatively weighted. This leads to a gradual improvement in the overall ability of the system to perform a task such as sorting numbers or recognizing patterns (see NEURAL NETWORK).

Since the 1950s, some researchers have suggested that computer programs or robots be designed to interact with their environment and learn from it in the way that human infants do. Rodney Brooks and Cynthia Breazeal at MIT have created robots with a layered architecture that includes motor, sensory, representational, and decision-making elements. Each level reacts to its inputs and sends information to the next higher level. The robot Cog and its descendant Kismet often behaved in unexpected ways, generating complex responses that are emergent rather than specifically programmed.

The approach characterized as "artificial life" adds a genetic component in which the successful components pass on program code "genes" to their offspring. Thus, the power of evolution through natural selection is simulated, leading to the emergence of more effective systems (see ARTIFICIAL LIFE and GENETIC ALGORITHMS).

In general the top-down approaches have been more successful in performing specialized tasks, but the bottom-up approaches may have greater general application, as well as leading to cross-fertilization between the fields of artificial intelligence, cognitive psychology, and research into human brain function.

APPLICATION AREAS

While powerful artificial intelligence is not yet ubiquitous in everyday computing, AI principles are being successfully used in a number of application areas. These areas, which are all covered separately in this book, include

- devising ways of capturing and representing knowledge, making it accessible to systems for diagnosis and analysis in fields such as medicine and chemistry (see KNOWLEDGE REPRESENTATION and EXPERT SYSTEMS)

- creating systems that can converse in ordinary language for querying databases, responding to customer service calls, or other routine interactions (see NATURAL LANGUAGE PROCESSING)

- enabling robots to not only see but also "understand" objects in a scene and their relationships (see COMPUTER VISION and ROBOTICS)

- improving systems for voice and face recognition, as well as sophisticated data mining and analysis (see SPEECH RECOGNITION AND SYNTHESIS, BIOMETRICS, and DATA MINING)

- developing software that can operate autonomously, carrying out assignments such as searching for and evaluating competing offerings of merchandise (see SOFTWARE AGENT)

PROSPECTS

The field of AI has been characterized by successive waves of interest in various approaches, and ambitious projects have often failed. However, expert systems and, to a lesser extent, neural networks have become the basis for viable products. Robotics and computer vision offer a significant potential payoff in industrial and military applications. The creation of software agents to help users navigate the complexity of the Internet is now of great commercial interest. The growth of AI has turned out to be a steeper and more complex path than originally anticipated. One view suggests steady progress. Another, shared by science fiction writers such as Vernor Vinge, suggests a breakthrough, perhaps arising from artificial life research, might someday create a true—but truly alien—intelligence (see SINGULARITY, TECHNOLOGICAL).

Further Reading

American Association for Artificial Intelligence. "Welcome to AI Topics." Available online. URL: http://www.aaai.org/Pathfinder/html/welcome.html. Accessed April 13, 2007.

"An Introduction to the Science of Artificial Intelligence." Available online. URL: http://library.thinkquest.org/2705/. Accessed April 13, 2007.

Feigenbaum, E. A. and J. Feldman, eds. *Computers and Thought.* New York: McGraw-Hill, 1963.

Henderson, Harry. *Artificial Intelligence: Mirrors for the Mind.* New York: Facts On File, 2007.

Jain, Sanjay, et al. *Systems that Learn: An Introduction to Learning Theory.* 2nd ed. Cambridge, Mass: MIT Press, 1999.

Kurzweil, Ray. *The Age of Spiritual Machines: When Computers Exceed Human Intelligence.* New York: Viking, 1999.

McCorduck, Pamela. *Machines Who Think.* 25th Anniversary update. Notick, Mass.: A. K. Peters, 2004.

Shapiro, Stuart C. *Encyclopedia of Artificial Intelligence.* 2nd ed. New York: Wiley, 1992.

artificial life (AL)

This is an emerging field that attempts to simulate the behavior of living things in the realm of computers and robotics. The field overlaps artificial intelligence (AI) since intelligent behavior is an aspect of living things. The design of a self-reproducing mechanism by John von Neumann in the mid-1960s was the first model of artificial life (see VON NEUMANN, JOHN). The field was expanded by the development of cellular automata as typified in John Conway's Game of Life in the 1970s, which demonstrated how simple components interacting according to a few specific rules could generate complex emergent patterns. A program by Craig Reynolds uses this principle to model the flocking behavior of simulated birds, called "boids" (see CELLULAR AUTOMATA).

The development of genetic algorithms by John Holland added selection and evolution to the act of reproduction. This approach typically involves the setting up of numerous small programs with slightly varying code, and having them attempt a task such as sorting data or recognizing patterns. Those programs that prove most "fit" at accomplishing the task are allowed to survive and reproduce. In the act of reproduction, biological mechanisms such as genetic mutation and crossover are allowed to intervene (see GENETIC ALGORITHMS). A rather similar approach is found in the neural network, where those nodes that succeed better at the task are given greater "weight" in creating a composite solution to the problem (see NEURAL NETWORK).

A more challenging but interesting approach to AL is to create actual robotic "organisms" that navigate in the physical rather than the virtual world. Roboticist Hans Moravec of the Stanford AI Laboratory and other researchers have built robots that can deal with unexpected obstacles by improvisation, much as people do, thanks to layers of software that process perceptions, fit them to a model of the world, and make plans based on goals. But such robots, built as full-blown designs, share few of the characteristics of artificial life. As with AI, the bottom-up approach offers a different strategy that has been called "fast, cheap, and out of control"—the production of numerous small, simple, insectlike robots that have only simple behaviors, but are potentially capable of interacting in surprising ways. If a meaningful genetic and reproductive mechanism can be included in such robots, the result would be much closer to true artificial life (see ROBOTICS).

The philosophical implications arising from the possible development of true artificial life are similar to those involved with "strong AI." Human beings are used to viewing themselves as the pinnacle of a hierarchy of intelligence and creativity. However, artificial life with the capability of rapid evolution might quickly outstrip human capabilities, perhaps leading to a world like that portrayed by science fiction writer Gregory Benford, where flesh-and-blood humans become a marginalized remnant population.

Further Reading

"ALife Online 2.0." Available online. URL: http://alife.org/. Accessed April 13, 2007.

"Karl Sims Retrospective." Available online. URL: http://www.biota.org/ksims/. Accessed April 13, 2007.

Langton, Christopher G., ed. *Artificial Life: an Overview.* Cambridge, Mass.: MIT Press, 1995.

Levy, Stephen. *Artificial Life: the Quest for a New Creation.* New York: Pantheon Books, 1992.

Tierra homepage. Available online. URL: http://www.his.atr.jp/çray/tierra. Accessed.

ASP   *See* APPLICATION SERVICE PROVIDER.

assembler

All computers at bottom consist of circuits that can perform a repertoire of mathematical or logical operations. The earliest computers were programmed by setting switches for operations and manually entering numbers in working storage, or memory. A major advance in the flexibility of computers came with the idea of stored programs, where a set of instructions could be read in and held in the machine in the same way as other data. These instructions were in machine language, consisting of numbers representing instructions (operations to be performed) and other numbers representing the address of data to be manipulated (or an address containing the address of the data, called indirect addressing—see ADDRESSING). Operations include basic arithmetic (such as addition), the movement of data between storage (memory) and special processor locations called registers, and the movement of data from an input device (such as a card reader) and an output device (such as a printer).

Writing programs in machine code is obviously a tedious and error-prone process, since each operation must be specified using a particular numeric instruction code together with the actual addresses of the data to be used. It soon became clear, however, that the computer could itself be used to keep track of binary codes and actual addresses,

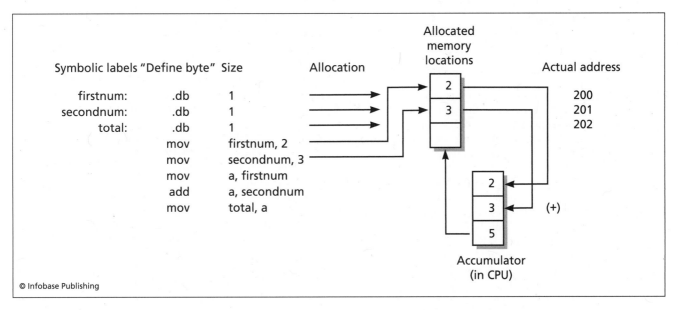

Symbolic labels	"Define byte"	Size	Allocation	Allocated memory locations	Actual address
firstnum:	.db	1	→	2	200
secondnum:	.db	1	→	3	201
total:	.db	1	→		202
	mov	firstnum, 2			
	mov	secondnum, 3			
	mov	a, firstnum		2	
	add	a, secondnum		3	(+)
	mov	total, a		5	

Accumulator
(in CPU)

© Infobase Publishing

In this assembly language example, the "define byte" (.db) directive is used to assign one memory byte to each of the symbolic names (variables) firstnum, secondnum, and total. The two mov commands then load 2 and 3 into firstnum and secondnum, respectively. Firstnum is then loaded into the processor's accumulator (a), and secondnum is then added to it. Finally, the sum is moved into the memory location labeled total.

allowing the programmer to use more human-friendly names for instructions and data variables. The program that translates between symbolic language and machine language is the assembler.

With a symbolic assembler, the programmer can give names to data locations. Thus, instead of saying (and having to remember) that the quantity Total will be in location &H100, the program can simply define a two-byte chunk of memory and call it Total:

```
Total DB
```

The assembler will take care of assigning a physical memory location and, when instructed, retrieving or storing the data in it.

Most assemblers also have macro capability. This means that the programmer can write a set of instructions (a procedure) and give it a name. Whenever that name is used in the program, the assembler will replace it with the actual code for the procedure and plug in whatever variables are specified as operands (see MACRO).

APPLICATIONS

In the mainframe world of the 1950s, the development of assembly languages represented an important first step toward symbolic programming; higher-level languages such as FORTRAN and COBOL were developed so that programmers could express instructions in language that was more like mathematics and English respectively. High-level languages offered greater ease of programming and source code that was easier to understand (and thus to maintain). Gradually, assembly language was reserved for systems programming and other situations where efficiency or the need

to access some particular hardware capability required the exact specification of processing (see SYSTEMS PROGRAMMING and DEVICE DRIVER).

During the 1970s and early 1980s, the same evolution took place in microcomputing. The first microcomputers typically had only a small amount of memory (perhaps 8–64K), not enough to compile significant programs in a high-level language (with the partial exception of some versions of BASIC). Applications such as graphics and games in particular were written in assembly language for speed. As available memory soared into the hundreds of kilobytes and then megabytes, however, high level languages such as C and C++ became practicable, and assembly language began to be relegated to systems programming, including device drivers and other programs that had to interact directly with the hardware.

While many people learning programming today receive little or no exposure to assembly language, some understanding of this detailed level of programming is still useful because it illustrates fundamentals of computer architecture and operation.

Further Reading

Abel, Peter. *IBM PC Assembly Language and Programming.* 5th ed. Upper Saddle River, N.J.: Prentice Hall, 2001.

Duntemann, Jeff. *Assembly Language Step by Step: Programming with DOS and Linux.* 2nd ed. New York: Wiley, 2000.

Miller, Karen. *An Assembly Language Introduction to Computer Architecture Using the Intel Pentium.* New York: Oxford University Press, 1999.

asynchronous JavaScript and XML   *See* AJAX.

Atanasoff, John Vincent
(1903–1995)
American
Computer Engineer

John V. Atanasoff is considered by many historians to be the inventor of the modern electronic computer. He was born October 4, 1903, in Hamilton, New York. As a young man, Atanasoff showed considerable interest in and a talent for electronics. His academic background (B.S. in electrical engineering, Florida State University, 1925; M.S. in mathematics, Iowa State College, 1926; and Ph.D. in experimental physics, University of Wisconsin, 1930) well equipped him for the design of computing devices. He taught mathematics and physics at Iowa State until 1942, and during that time, he conceived the idea of a fully electronic calculating machine that would use vacuum tubes for its arithmetic circuits and would store binary numbers on a rotating drum memory that used high and low charges on capacitors. Atanasoff and his assistant Clifford E. Berry built a suc-

According to a federal court it was John Atanasoff, not John Mauchly and Presper Eckert, who built the first digital computer. At any rate the "ABC" or Atanasoff-Berry computer represented a pioneering achievement in the use of binary logic circuits for computation. (IOWA STATE UNIVERSITY)

cessful computer called ABC (Atanasoff-Berry computer) using this design in 1942. (By that time he had taken a wartime research position at the Naval Ordnance Laboratory in Washington, D.C.)

The ABC was a special-purpose machine designed for solving up to 29 simultaneous linear equations using an algorithm based on Gaussian elimination to eliminate a specified variable from a pair of equations. Because of inherent unreliability in the system that punched cards to hold the many intermediate results needed in such calculations, the system was limited in practice to solving sets of five or fewer equations.

Despite its limitations, the ABC's design proved the feasibility of fully electronic computing, and similar vacuum tube switching and regenerative memory circuits were soon adopted in designing the ENIAC and EDVAC, which unlike the ABC, were general-purpose electronic computers. Equally important was Atanasoff's use of capacitors to store data in memory electronically: The descendent of his capacitors can be found in the DRAM chips in today's computers.

When Atanasoff returned to Iowa State in 1948, he discovered that the ABC computer had been dismantled to make room for another project. Only a single memory drum and a logic unit survived. Iowa State granted him a full professorship and the chairmanship of the physics department, but he never returned to that institution. Instead, he founded the Ordnance Engineering Corporation in 1952, which grew to a 100-person workforce before he sold the firm to Aerojet General in 1956. He then served as a vice president at Aerojet until 1961.

Atanasoff then semi-retired, devoting his time to a variety of technical interests (he had more than 30 patents to his name by the time of his death). However, when Sperry Univac (owner of Eckert and Mauchly's computer patents) began demanding license fees from competitors in the mid-1960s, the head lawyer for one of these competitors, Honeywell, found out about Atanasoff's work on the ABC and enlisted his aid as a witness in an attempt to overturn the patents. After prolonged litigation, Judge Earl Richard Larson ruled in 1973 that the two commercial computing pioneers had learned key ideas from Atanasoff's apparatus and writings and that their patent was invalid because of this "prior art."

Atanasoff received numerous awards for his work for the Navy on acoustics and for his pioneering computer work. These awards included the IEEE Computer Pioneer Award (1984) and the National Medal of Technology (1990). In addition, he had both a hall at Iowa State University and an asteroid (3546-Atanasoff) named in his honor. John Atanasoff died on June 15, 1995, in Monrovia, Maryland.

Further Reading

Burks, A. R., and A. W. Burks. *The First Electronic Computer: the Atanasoff Story.* Ann Arbor, Mich: University of Michigan Press, 1988.

Lee, J. A. N. *Computer Pioneers.* Los Alamitos, Calif.: IEEE Computer Society Press, 1995.

"Reconstruction of the Atanasoff-Berry Computer (ABC)." Available online. URL: http://www.scl.ameslab.gov/ABC/ABC. html. Accessed April 13, 2007.

auctions, online

By the late 1990s, millions of computer users had discovered a new way to buy and sell an immense variety of items ranging from traditional collectibles to the exotic (such as a working German Enigma encoding machine).

Since its founding in 1995, leading auction site eBay has grown to 78 million users in mid-2006, with revenue of about $7.6 billion in 2007 (see EBAY). (Two other e-commerce giants, Amazon.com and Yahoo!, also entered the online auction market, but with much more modest results.)

PROCEDURES

Online auctions differ from traditional auctions in several ways. Traditional auction firms generally charge the seller and buyer a commission of around 10 percent of the sale or "hammer" price. Online auctions charge the buyer nothing, and the seller typically pays a fee of about 3–5 percent of the amount realized. Online auctions can charge much lower fees because unlike traditional auctions, there is no live auctioneer, no catalogs to produce, and little administration, since all payments pass from buyer to seller directly.

An online auction is like a mail bid auction in that bids can be posted at any time during the several days a typical auction runs. A buyer specifies a maximum bid and if he or she becomes the current high bidder, the high bid is adjusted to a small increment over the next highest bid. As with a "live" auction, however, bidders can revise their bids as many times as they wish until the close of the auction. An important difference between online and traditional live auctions is that a traditional auction ends as soon as no one is willing to top the current high bid. With an online auction, the bidding ends at the posted ending time. This has led to a tactic known as "sniping," where some bidders submit a bid just over the current high bid just before the auction ends, such that the previous high bidder has no time to respond.

FUTURE AND IMPLICATIONS

Online auctions have become very popular, and an increasing number of people run small businesses by selling items through auctions. The markets for traditional collectibles such as coins and stamps have been considerably affected by online auctions. Knowledgeable buyers can often obtain items for considerably less than a dealer would charge, or sell items for more than a dealer would pay. However, many items are overpriced compared to the normal market, and faked or ill-described items can be a significant problem. Attempts to hold the auction service legally responsible for such items are met with the response that the auction service is simply a facilitator for the seller and buyer and does not play the role of traditional auctioneers who catalog items and provide some assurance of authenticity. If courts or regulators should decide that online auctions must bear this responsibility, the cost of using the service may rise or the variety of items that can be offered may be restricted.

Further Reading

Cohen, Adam. *The Perfect Store: Inside eBay.* Boston: Little, Brown, 2002.

Encell, Steve, and Si Dunn. *The Everything Online Auctions Book: All You Need to Buy and Sell with Success—on eBay and Beyond.* Avon, Mass.: Adams Publishing Group, 2006.

Kovel, Ralph M., and Terry H. Kovel. *Kovels' Bid, Buy, and Sell Online: Basic Auction Information and Tricks of the Trade.* New York: Three Rivers Press, 2001.

auditing in data processing

The tremendous increase in the importance and extent of information systems for all aspects of commerce and industry has made it imperative that businesses be able to ensure the accuracy and integrity of their accounting systems and corporate databases. Errors can result in loss of revenue and even exposure to legal liability.

Auditing involves the analysis of the security and accuracy of software and the procedures for using it. For example, sample data can be extracted using automated scripts or other software tools and examined to determine whether correct and complete information is being entered into the system, and whether the reports on which management relies for decision making are accurate. Auditing is also needed to confirm that data reported to regulatory agencies meets legal requirements.

In addition to confirming the reliability of software and procedures, auditors must necessarily also be concerned with issues of security, since attacks or fraud involving computer systems can threaten their integrity or reliability (see COMPUTER CRIME AND SECURITY). The safeguarding of customer privacy has also become a sensitive concern (see PRIVACY IN THE DIGITAL AGE). To address such issues, the auditor must have a working knowledge of basic psychology and human relations, particularly as they affect large organizations.

Auditors recommend changes to procedures and practices to minimize the vulnerability of the system to both human and natural threats. The issues of backup and archiving and disaster recovery must also be addressed (see BACKUP AND ARCHIVE SYSTEMS). As part accountant and part systems analyst, the information systems auditor represents a bridging of traditional practices and rapidly changing technology.

Further Reading

Cannon, David L., Timothy S. Bergmann, and Brady Pamplin. *CISA: Certified Information Systems Auditor Study Guide.* Indianapolis: Wiley Publishing, 2006.

Champlain, Jack. *Auditing Information Systems.* Hoboken, N.J.: Wiley, 2003.

Information Systems Audit and Control Association. Available online. URL: http://www.isaca.org/. Accessed May 22, 2007.

Pathak, Jagdish. *Information Systems Auditing: An Evolving Agenda.* New York: Springer-Verlag, 2005.

authentication

This process by which two parties in a communication or transaction can assure each other of their identity is a fundamental requirement for any transaction not involving cash, such as the use of checks or credit or debit cards. (In practice, for many transactions, authentication is "one way"—the seller needs to know the identity of the buyer or at least have some way of verifying the payment, but the buyer need not confirm the identity of the seller—except, perhaps in order to assure proper recourse if something turns out to be wrong with the item purchased.)

Traditionally, authentication involves paper-based identification (such as driver's licenses) and the making and matching of signatures. Since such identification is relatively easy to fake, there has been growing interest in the use of characteristics such as voice, facial measurements, or the patterns of veins in the retina that can be matched uniquely to individuals (see BIOMETRICS). Biometrics, however, requires the physical presence of the person before a suitable device, so it is primarily used for guarding entry into high-security areas.

AUTHENTICATION IN ONLINE SYSTEMS

Since many transactions today involve automated systems rather than face-to-face dealings, authentication systems generally involve the sharing of information unique to the parties. The PIN used with ATM cards is a common example: It protects against the physical diversion of the card by requiring information likely known only to the legitimate owner. In e-commerce, there is the additional problem of safeguarding sensitive information such as credit card numbers from electronic eavesdroppers or intruders. Here a system is used by which information is encrypted before it is transmitted over the Internet. Encryption can also be used to verify identity through a digital signature, where a message is transformed using a "one-way function" such that it is highly unlikely that a message from any other sender would have the same encrypted form (see ENCRYPTION). The most widespread system is public key cryptography, where each person has a public key (known to all interested parties) and a private key that is kept secret. Because of the mathematical relationship between these two keys, the reader of a message can verify the identity of the sender or creator.

The choice of technology or protocol for authentication depends on the importance of the transaction, the vulnerability of information that needs to be protected, and the consequences of failing to protect it. A Web site that is providing access to a free service in exchange for information about users will probably not require authentication beyond perhaps a simple user/password pair. An online store, on the other hand, needs to provide a secure transaction environment both to prevent losses and to reassure potential customers that shopping online does not pose an unacceptable risk.

Authentication ultimately depends on a combination of technological and social systems. For example, cryptographic keys or "digital certificates" can be deposited with a trusted third party such that a user has reason to believe that a business is who it says it is.

Further Reading

Ratha, Nalini, and Ruud Bolie, eds. *Automatic Fingerprint Recognition Systems.* New York: Springer-Verlag, 2004.
Smith, Richard E., and Paul Reid. *User Authentication Systems and Role-Based Security.* Upper Saddle River, N.J.: Pearson Custom Publishing, 2004.

Tung, Brian. *Kerberos: A Network Authentication System.* Reading, Mass.: Addison-Wesley, 1999.

authoring systems

Multimedia presentations such as computer-based-training (CBT) modules are widely used in the corporate and educational arenas. Programming such a presentation in a high-level language such as C++ (or even Visual Basic) involves writing code for the detailed arrangement and control of graphics, animation, sound, and user interaction. Authoring systems offer an alternative way to develop presentations or courses. The developer specifies the sequence of graphics, sound, and other events, and the authoring system generates a finished program based on those specifications.

Authoring systems can use a variety of models for organizing presentations. Some use a scripting language that specifies the objects to be used and the actions to be performed (see SCRIPTING LANGUAGES). A scripting language uses many of the same features as a high-level programming language, including the definition of variables and the use of control structures (decision statements and loops). Programs such as the once ubiquitous Hypercard (for the Macintosh) and Asymetrix Toolbook for Windows organize presentations into segments called "cards," with instructions fleshed out in a scripting language.

As an alternative, many modern authoring systems such as Discovery Systems' CourseBuilder use a graphical approach to organizing a presentation. The various objects (such as graphics) to be used are represented by icons, and the icons are connected with "flow lines" that describe the sequence of actions, serving the same purpose as control structures in programming languages. This "iconic" type of authoring system is easiest for less experienced programmers to use and makes the creation of small presentations fast and easy. Such systems may become more difficult to use for lengthy presentations (due to the number of symbols and connectors involved), and speed of the finished program can be a problem. Other popular models for organizing presentations include the "timeline" of Macromedia Flash, which breaks the presentation into "movies" and specifies actions for each frame, as well as providing multiple layers to facilitate animation. With the migration of many presentations to the Internet, the ability of authoring systems to generate HTML (or DHTML) code is also important.

Further Reading

Makedon, Fillia, and Samuel A. Rebelsky, ed. *Electronic Multimedia Publishing: Enabling Technologies and Authoring Issues.* Boston: Kluwer Academic, 1998.
"Multimedia Authoring Systems FAQ." Available online. URL: http://fags.cs.uu.nl/na-dir/multimedia/authoring-systems/part1.html. Accessed August 8, 2007.
Murray, T. "Authoring Intelligent Tutoring Systems: An analysis of the state of the art." *International J. of Artificial Intelligence in Education* 10 (1999): 98–129.
Wilhelm, Jeffrey D., Paul Friedman, and Julie Erickson. *Hyperlearning: where Projects, Inquiry, and Technology Meet.* York, Me.: Stenhouse, 1998.

automatic programming

From the beginning of the computer age, computer scientists have grappled with the fact that writing programs in any computer language, even relatively high-level ones such as FORTRAN or C, requires painstaking attention to detail. While language developers have responded to this challenge by trying to create more "programmer friendly" languages such as COBOL with its English-like syntax, another approach is to use the capabilities of the computer to automate the task of programming itself. It is true that any high-level language compiler does this to some extent (by translating program statements into the underlying machine instructions), but the more ambitious task is to create a system where the programmer would specify the problem and the system would generate the high-level language code. In other words, the task of programming, which had already been abstracted from the machine code level to the assembler level and from that level to the high-level language, would be abstracted a step further.

During the 1950s, researchers began to apply artificial intelligence principles to automate the solving of mathematical problems (see ARTIFICIAL INTELLIGENCE). For example, in the 1950s Anthony Hoare introduced the definition of preconditions and postconditions to specify the states of the machine as it proceeds toward an end state (the solution of the problem). The program Logic Theorist demonstrated that a computer could use a formal logical calculus to solve problems from a set of conditions or axioms. Techniques such as deductive synthesis (reasoning from a set of programmed principles to a solution) and transformation (step-by-step rules for converting statements in a specification language into the target programming language) allowed for the creation of automated programming systems, primarily in mathematical and scientific fields (see also PROLOG).

The development of the expert system (combining a knowledge base and inference rules) offered yet another route toward automated programming (see EXPERT SYSTEMS). Herbert Simon's 1963 Heuristic Compiler was an early demonstration of this approach.

APPLICATIONS

Since many business applications are relatively simple in logical structure, practical automatic principles have been used in developing application generators that can create, for example, a database management system given a description of the data structures and the required reports. While some systems output code in a language such as C, others generate scripts to be run by the database management software itself (for example, Microsoft Access).

To simplify the understanding and specification of problems, a visual interface is often used for setting up the application requirements. Onscreen objects can represent items such as data files and records, and arrows or other connecting links can be dragged to indicate data relationships.

The line between automated program generators and modern software development environments is blurry. A programming environment such as Visual Basic encapsulates a great deal of functionality in objects called *controls*, which can represent menus, lists, buttons, text input boxes, and other features of the Windows interface, as well as other functionalities (such as a Web browser). The Visual Basic programmer can design an application by assembling the appropriate interface objects and processing tools, set properties (characteristics), and write whatever additional code is necessary. While not completely automating programming, much of the same effect can be achieved.

Further Reading

Andrews, James H. *Logic Programming: Operational Semantics and Proof Theory.* New York: Cambridge University Press, 1992.

"Automatic Programming Server." Available online. URL: http://www.cs.utexas.edu/users/novak/cgi/apserver.cgi. Accessed April 14, 2007.

"Programming and Problem Solving by Connecting Diagrams." Available online. URL: http://www.cs.utexas.edu/users/novak/cgi/vipdemo.cgi. Accessed April 14, 2007.

Tahid, Walid, ed. *Semantics, Applications and Implementation of Program Generation.* New York: Springer-Verlag, 2000.

awk

This is a scripting language developed under the UNIX operating system (see SCRIPTING LANGUAGES) by Alfred V. Aho, Brian W. Kernighan, and Peter J. Weinberger in 1977. (The name is an acronym from their last initials.) The language builds upon many of the pattern matching utilities of the operating system and is designed primarily for the extraction and reporting of data from files. A number of variants of awk have been developed for other operating systems such as DOS.

As with other scripting languages, an awk program consists of a series of commands read from a file by the awk interpreter. For example the following UNIX command line:

```
awk -f MyProgram > Report
```

reads awk statements from the file MyProgram into the awk interpreter and sends the program's output to the file Report.

LANGUAGE FEATURES

An awk statement consists of a *pattern* to match and an action to be taken with the result (although the pattern can be omitted if not needed). Here are some examples:

```
{print $1} # prints the first field of every
       # line of input (since no pattern
       # is specified)
/debit/ {print $2} # print the second field of
       # every line that contains the
       # word "debit"
if ( Code == 2 ) # if Code equals 2,
print $3 # print third field
         # of each line
```

Pattern matching uses a variety of regular expressions familiar to UNIX users. Actions can be specified using a limited but adequate assortment of control structures similar to those found in C. There are also built-in variables (including counters for the number of lines and fields), arithmetic functions, useful string functions for extracting text from fields, and arithmetic and relational operators. Formatting of output can be accomplished through the versatile (but somewhat cryptic) print function familiar to C programmers.

Awk became popular for extracting reports from data files and simple databases on UNIX systems. For more sophisticated applications it has been supplanted by Perl, which offers a larger repertoire of database-oriented features (see PERL).

Further Reading

Aho, Alfred V., Brian Kernighan, and Peter J. Weinberger. *The Awk Programming Language*. Reading, Mass.: Addison-Wesley, 1998.

Goebel, Greg. "An Awk Primer." Available online. URL: http://www.vectorsite.net/tsawk.html. Accessed May 22, 2007.

B

Babbage, Charles
(1791–1871)
British
Mathematician, Inventor

Charles Babbage made wide-ranging applications of mathematics to a variety of fields including economics, social statistics, and the operation of railroads and lighthouses. Babbage is best known, however, for having conceptualized the key elements of the general-purpose computer about a century before the dawn of electronic digital computing.

As a student at Trinity College, Cambridge, Babbage was already making contributions to the reform of calculus, championing new European methods over the Newtonian approach still clung to by British mathematicians. But Babbage's interests were shifting from the theoretical to the practical. Britain's growing industrialization as well as its worldwide interests increasingly demanded accurate numeric tables for navigation, actuarial statistics, interest rates, and engineering parameters. All tables had to be hand-calculated, a long process that inevitably introduced numerous errors. Babbage began to consider the possibility that the same mechanization that was revolutionizing industries such as weaving could be turned to the automatic calculation of numeric tables.

Starting in 1820, Babbage began to build a mechanical calculator called the difference engine. This machine used series of gears to accumulate additions and subtractions (using the "method of differences") to generate tables. His small demonstration model worked well, so Babbage undertook the full-scale "Difference Engine

Number One," a machine that would have about 25,000 moving parts and would be able to calculate up to 20 decimal places. Unfortunately, Babbage was unable, despite financial support from the British government, to overcome the difficulties inherent in creating a mechanical device of such complexity with the available machining technology.

Undaunted, Babbage turned in the 1830s to a new design that he called the Analytical Engine. Unlike the Difference Engine, the new machine was to be programmable using instructions read in from a series of punch cards (as in the Jacquard loom). A second set of cards would contain the variables, which would be loaded into the "store"—a series of wheels corresponding to memory in a modern computer. Under control of the instruction cards, numbers could be moved between the store and the "mill" (corresponding to a modern CPU) and the results of calculations could be sent to a printing device.

Collaborating with Ada Lovelace (who translated his lecture transcripts by L. F. Menebrea) Babbage wrote a series of papers and notes that explained the workings of the proposed machine, including a series of "diagrams" (programs) for performing various sorts of calculations.

Building the Analytical Engine would have been a far more ambitious task than the special-purpose Difference Engine, and Babbage made little progress in the actual construction of the device. Although Babbage's ideas would remain obscure for nearly a century, he would then be recognized as having designed most of the key elements of the modern computer: the central processor, memory, instructions, and data organization. Only in the lack of a capability

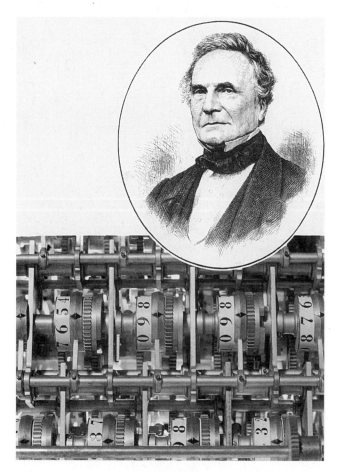

If it had been built, Charles Babbage's Analytical Engine, although mechanical rather than electrical, would have had most of the essential features of modern computers. These included input, (via punched cards), a processor, a memory (store), and a printer. A reproduction of part of the early Difference Engine is shown here. (PHOTO RESEARCHERS, INC.)

to manipulate memory addresses did the design fall short of a modern computer.

Further Reading

"The Analytical Engine: the First Computer." Available online. URL: http://www.fourmilab.ch/babbage/. Accessed April 20, 2007.

Babbage, Henry Prevost, ed. *Babbage's Calculating Engines: A Collection of Papers.* With a new introduction by Allan G. Bromley. Los Angeles: Tomash, 1982.

Campbell-Kelly, M., ed. *The Works of Charles Babbage.* 11 vols. London: Picerking and Chatto, 1989.

"Who Was Charles Babbage?" Charles Babbage Institute. Available online. URL: http://www.cbi.umn.edu/exhibits/cb.html. Accessed April 20, 2007.

Swade, Doron D. "Redeeming Charles Babbage's Mechanical Computer." *Scientific American,* February 1993.

backup and archive systems

The need to create backup copies of data has become increasingly important as dependence on computers has grown and the economic value of data has increased. Potential threats to data include bugs in the operating system or software applications, malicious acts such as the introduction of computer viruses, theft, hardware failure (such as in hard disk drives), power outages, fire, and natural disasters such as earthquakes and floods.

A variety of general principles must be considered in devising an overall strategy for creating and maintaining backups:

Reliability: Is there assurance that the data is stored accurately on the backup medium, and will automatic backups run reliably as scheduled? Can the data be accurately retrieved and restored if necessary?

Physical storage: Is the backed-up data stored securely and organized in a way to make it easy to retrieve particular disks or tapes? Is the data stored at the site where it is to be used, or off-site (guarding against fire or other disaster striking the workplace).

The daughter of poet Lord Byron, Lady Ada Lovelace (1815–52) acquired mathematical training usually denied to her gender. When she met Charles Babbage and learned about his computer design, she translated his work and wrote the world's first computer programs. (PHOTO RESEARCHERS, INC. / Science Photo Library)

Ease of Use: To the extent backups must be set up or initiated by human operators, is the system easy to understand and use with minimal training? Ease of use both promotes reliability (because users will be more likely to perform the backups), and saves money in training costs.

Economy: How does a given system compare to others in terms of the cost of the devices, software, media (such as tapes or cartridges), training, and administration?

The market for storage and backup software and services has grown rapidly in the mid-2000s, driven in part by a new awareness of the need of corporations to protect their vital data assets from natural disasters or possible terrorist attacks (see CYBERTERRORISM and DISASTER PLANNING AND RECOVERY). In many corporations the amount of data that needs to be backed up or archived grows at a rate of 50 percent per year or more.

CHOICE OF METHODS

The actual choice of hardware, software, and media depends considerably on how much data must be backed up (and how often) as well as whether the data is being generated on individual PCs or being stored at a central location. (See FILE SERVER, DATA WAREHOUSE.)

Backups for individual PCs can be accomplished using the backup software that comes with various versions of Microsoft Windows or through third-party software.

In addition to traditional tapes, the media used include CDs or DVDs (for very small backups), tiny USB "flash drives" (generally up to a few gigabytes of data), cartridge drives (up to 70 gigabytes or more), or even compact external USB hard drives that can store hundreds of gigabytes. (see CD AND DVD ROM, FLASH DRIVE, HARD DRIVE, TAPE DRIVE, and USB.)

In addition to backing up documents or other data generated by users, the operating system and applications software is often backed up to preserve configuration information that would otherwise be lost if the program were reinstalled. There are utilities for Microsoft Windows and other operating systems that simplify the backing up of configuration information by identifying and backing up only those files (such as the Windows Registry) that contain information particular to the installation.

The widespread use of local area networks makes it easier to back up data automatically from individual PCs and to store data at a central location (see LOCAL AREA NETWORK and FILE SERVER). However, having all data eggs in one basket increases the importance of building reliability and redundancy into the storage system, including the use of RAID (multiple disk arrays), "mirrored" disk drives, and uninterruptible power supplies (UPS). Despite such measures, the potential risk in centralized storage has led to advocacy of a "replication" system, preferably at the operating system level, that would automatically create backup copies of any given object at multiple locations on the network.

Another alternative of growing interest is the use of the Internet to provide remote (off-site) backup services.

By 2005 Gartner Research was reporting that about 94 percent of corporate IT managers surveyed were using or considering the use of "managed backup" services. IDC has estimated that the worldwide market for online backup services would reach $715 million by 2011. Online backup offers ease of use (the backups can be run automatically, and the service is particularly handy for laptop computer users on the road) and the security of off-site storage, but raise questions of privacy and security of sensitive information, particularly if encryption is not built into the process. Online data storage is also provided to individual users by a variety of service providers such as Google. Application Service Providers (ASPs) have a natural entry into the online storage market since they already host the applications their users use to create data (see APPLICATION SERVICE PROVIDER).

A practice that still persists in some mainframe installations is the tape library, which maintains an archive of data on tape that can be retrieved and mounted as needed.

ARCHIVING

Although using much of the same technology as making backups, archiving of data is different in its objectives and needs. An archive is a store of data that is no longer needed for routine current use, but must be retrievable upon demand, such as the production of bank records or e-mail as part of a legal process. (Data may also be archived for historical or other research purposes.) Since archives may have to be maintained for many years (even indefinitely), the ability of the medium (such as tape) to maintain data in readable condition becomes an important consideration. Besides physical deterioration, the obsolescence of file formats can also render archived data unusable.

MANAGEMENT CONSIDERATIONS

If backups must be initiated by individual users, the users must be trained in the use of the backup system and motivated to make backups, a task that is easy to put off to another time. Even if the backup is fully automated, sample backup disks or tapes should be checked periodically to make sure that data could be restored from them. Backup practices should be coordinated with disaster recovery and security policies.

Further Reading
Backup Review. Available online. URL: http://www.backupreview. info/index.php. Accessed April 22, 2007.
Jacobi, Jon L. "Online Backup Services Come of Age." *PC World Online,* July 28, 2005. Available online. URL: http://www. pcworld.com/article/id,121970-page,1-c,utilities/article.html. Accessed April 22, 2007.
Jackson, William. "Modern Relics: NIST and Others Work on How to Preserve Data for Later Use." Available online. URL: http:// www.gcn.com/print/25_16/41069-1.html. Accessed April 22, 2007.
Storage Search. Available online. URL: http://www.storagesearch. com/. Accessed April 22, 2007.
Preston, W. Curtis. *Backup & Recovery.* Sebastapol, Calif.: O'Reilly Media, 2006.

Backus-Naur form

As the emerging discipline of computer science struggled with the need to precisely define the rules for new programming languages, the Backus-Naur form (BNF) was devised as a notation for describing the precise grammar of a computer language. BNF represents the unification of separate work by John W. Backus and Peter Naur in 1958, when they were trying to write a specification for the Algol language.

A series of BNF statements defines the syntax of a language by specifying the combinations of symbols that constitute valid statements in the language.

Thus in a hypothetical language a program can be defined as follows:

```
<program> ::= program
    <declaration_sequence>
  begin
    <statements_sequence>
  end;
```

Here the symbol ::= means "is defined as" and items in brackets <> are *metavariables* that represent placeholders for valid symbols. For example, <declaration_sequence> can consist of a number of different statements defined elsewhere.

Statements in square brackets [] indicate optional elements. Thus the If statement found in most programming languages is often defined as:

```
<if_statement> ::= if >boolean_expression> then
    <statement_sequence>
  [ else
<statement_sequence> ]
  end if ;
```

This can be read as "an *If* statement consists of a boolean_expression (something that evaluates to "true" or "false") followed by one or more statements, followed by an optional else that in turn is followed by one or more statements, followed by the keywords *end if*." Of course each item in angle brackets must be further defined—for example, a Boolean_expression.

Curly brackets {} specify an item that can be repeated one or more times. For example, in the definition

```
<identifier> ::= <letter> { <letter> | <digit> }
```

An identifier is defined as a letter followed by one or more instances of either a letter or a digit.

An extended version of BNF (EBNF) offers operators that make definitions more concise yet easier to read. The preceding definition in EBNF would be:

Identifier = Letter

```
{Letter | Digit}
```

EBNF statements are sometimes depicted visually in railroad diagrams, so called because the lines and arrows indicating the relationship of symbols resemble railroad tracks. The definition of <identifier> expressed in a railroad diagram is depicted in the above figure.

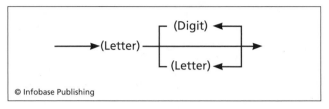

© Infobase Publishing

This "railroad diagram" indicates that an identifier must begin with a letter, which can be followed by a digit or another letter. The tracks curving back indicate that an element can appear more than once.

BNF and EBNF are useful because they can provide unambiguous definitions of the syntax of any computer language that is not context-dependent (which is to say, nearly all of them). It can thus serve as a reference for introduction of new languages (such as scripting languages) and for developers of parsers for compilers.

Further Reading

Garshol, Lars Marius. "BNF and EBNF: What Are They and How Do They Work?" Available online. URL: http://www.garshol.priv.no/download/text/bnf.html. Accessed April 23, 2007.
Jensen, K., N. Wirth et al. *Pascal User Manual and Report: ISO Pascal Standard.* New York: Springer-Verlag, 1985.
Sebesta, Robert W. *Concepts of Programming Languages.* 9th ed. Boston: Addison-Wesley, 2008.

bandwidth

In its original sense, bandwidth refers to the range of frequencies that a communications medium can effectively transmit. (At either end of the bandwidth, the transmission becomes too attenuated to be received reliably.) For a standard voice telephone, the bandwidth is about 3kHz.

In digital networks, bandwidth is used in a rather different sense to mean the amount of data that can be transmitted in a given time—what is more accurately described as the information transfer rate. A common measurement is Mb/sec (megabits per second). For example, a fast Ethernet network may have a bandwidth of 100 Mb/sec while a home phone-line network might have a bandwidth of from 1 to 10 Mb/sec and a DSL or cable modem runs at about 1 Mb/sec. (By comparison, a typical dial-up modem connection has a bandwidth of about 28–56 kb/sec, roughly 20 times slower than even a slow home network.)

The importance of bandwidth for the Internet is that it determines the feasibility of delivering new media such as sound (MP3), streaming video, and digital movies over the network, and thus the viability of business models based on such products. The growth of high-capacity access to the Internet (see BROADBAND) is changing the way people use the network.

Further Reading

Benedetto, Sergio, and Ezio Biglieri. *Principles of Digital Transmission: With Wireless Applications.* New York: Springer, 1999.
Smith, David R. *Digital Transmission Systems.* 3rd ed. New York: Kluwer Academic Publishers, 2003.

banking and computers

Beginning in the 1950s, banks undertook extensive automation of operations, starting with electronic funds transfer (EFT) systems. Check clearing (the sending of checks for payment to the bank on which they are drawn) was facilitated by the development of magnetic ink character recognition (MICR) that allowed checks to be automatically sorted and tabulated. Today an automated clearing house (ACH) network processes checks and other payments through regional clearinghouses.

Starting in the 1960s, the use of credit cards became an increasingly popular alternative to checks, and they were soon joined by automatic teller machine (ATM) networks and the use of debit cards (cards for transferring funds from a checking account at the point of sale).

Direct deposit of payroll and benefit checks has also been promoted for its safety and convenience. Credit card, ATM, and debit card systems rely upon large data processing facilities operated by the issuing financial institution. Because of the serious consequences of system failure both in immediate financial loss and customer goodwill, these fund transfer systems must achieve a high level of reliability and security. Reliability is promoted through the use of fault-tolerant hardware (such as redundant systems that can take over for one another in the event of a problem). The funds transfer messages must be provided a high level of security against eavesdropping or tampering through the use of algorithms such as the long-established DES (Data Encryption Standard)—see ENCRYPTION. Designers of EFT systems also face the challenge of providing a legally acceptable paper trail. Electronic signatures are increasingly accepted as an alternative to written signatures for authorizing fund transfers.

ONLINE BANKING

The new frontier of electronic banking is the online bank, where customers can access many banking functions via the Internet, including balance queries, transfers, automatic payments, and loan applications. For the consumer, online banking offers greater convenience and access to information than even the ATM, albeit without the ability to obtain cash.

From the bank's point of view, online banking offers a new way to reach and serve customers while relieving the strain on the ATM hardware and network. However, use of the Internet increases vulnerability to hackers and raises issues of privacy and the handling of personal information similar to those found in other e-commerce venues (see COMPUTER CRIME AND SECURITY and PRIVACY IN THE DIGITAL AGE). In 2006 a Pew Center survey found that 43 percent of Internet users were banking online—a total of about 63 million American adults. Other surveys have found about a third of Internet users now pay bills online. There are also a relatively small but growing number of Internet-only banks, many of which are affiliated with traditional banks. A particularly attractive feature of online banking is the ability to integrate bank services with popular personal finance software such as Quicken.

As impressive as it has been, the growth in online banking may have been inhibited by a perceived lack of security. A 2006 Gartner Research survey reported that nearly half of adults surveyed said that concerns over the potential for information theft and computer attacks had affected their use of online services such as banking and e-commerce transactions. Gartner translates this to an estimated 33 million U.S. adults who do not bank online because of such concerns. (Banks are frequently impersonated in deceptive emails and Web sites—see PHISHING AND SPOOFING.)

In response, government regulations (FFIEC or Federal Financial Institutions Examination Council) guidelines issued in October 2005 required banks by the end of 2006 to provide detailed risk assessments and mitigation plans for dealing with data breaches. Large banks spent about $15 million each on this process in 2006. Much greater expenses are likely as banks find themselves compelled to purchase and install more-secure user authentication software. They face the multiple challenge of securing their systems while reassuring their users and not forcing them to go through complicated, hard-to-remember log-in procedures.

Credit card issuers are also starting to turn to the Internet to provide additional services. According to the comScore service 524 million credit card bills were paid online in 2006. By 2007 about 70 percent of all credit card holders had logged on to their accounts at least once. Many customers have responded to incentives to discontinue receiving paper statements.

Further Reading

Fox, Susannah, and Jean Beier. "Online Banking 2006: Surfing to the Bank." Pew Internet & American Life Project, June 14, 2006. Available online. URL: http://www.pewinternet.org/pdfs/PIP_Online_Banking_2006.pdf. Accessed April 23, 2007.

Macklin, Ben. "Trust Has Value in E-Commerce," November 30, 2006. Available online. URL: http://www.emarketer.com/Article.aspx?1004323. Accessed April 23, 2007.

BASIC

The BASIC (Beginner's All-purpose Symbolic Instruction Code) language was developed by J. Kemeny and T. Kurtz at Dartmouth College in 1964. At the time, the college was equipped with a time-shared computer system linked to terminals throughout the campus, an innovation at a time when most computers were programmed from a single location using batches of punch cards. John G. Kemeny and Thomas Kurtz wanted to take advantage of the interactivity of their system by providing an easy-to-learn computer language that could compile and respond immediately to commands typed at the keyboard. This was in sharp contrast to the major languages of the time, such as COBOL, Algol, and FORTRAN in which programs had to be completely written before they could be tested.

Unlike the older languages used with punch cards, BASIC programs did not have to have their keywords typed in specified columns. Rather, statements could be typed like English sentences, but without punctuation and with a casual attitude toward spacing. In general, the syntax for decision and control structures is simpler than other languages. For example, a for loop counting from 1 to 10 in C looks like this:

```
for (i = 1; i <= 10; i++)
  printf("%d", i);
```
The same loop in BASIC reads as follows:

```
for i = 1 to 10
  print i
next i
```

BASIC AND MICROCOMPUTERS

During the 1960s and 1970s BASIC was used on a growing number of time-sharing computers. The language's simplicity and ease of use made it useful for writing short utility programs and for teaching basic principles of computing, particularly to noncomputer science majors. When the first personal computers became widely available in the early 1980s, they typically had memory capacities of 8KB–64KB, not enough to run the editor, compiler, and other utilities needed for a language such as C. However, a simple interpreter version of BASIC could be put on a read-only memory (ROM) chip, as was done with the Apple II, the early IBM PC, and dozens of other microcomputers. More advanced versions of BASIC (including compilers) could be loaded from tape (the first sales by a young entrepreneur named Bill Gates consisted of such products).

As a consequence of the adopting of BASIC for a variety of microcomputers, numerous dialects of the language came into existence. Commands for generating simple graphics and for manipulating memory and hardware directly (PEEK and POKE) made many BASIC programs platform specific.

Gradually, as microcomputers gained in memory capacity and processing power, languages such as Pascal (especially with the integrated development environment created at the University of California at San Diego) and C (from the UNIX community) began to supplant BASIC for the development of more complex microcomputer software.

CRITIQUE AND PROSPECTS

Most versions of BASIC used line numbers (a legacy of the early text editors that worked on a line-by-line basis) and a Goto statement could be used to make program control jump to a given line. While the language had simple subroutines (reached by a Gosub statement), it lacked the ability to explicitly pass variables to a procedure as in Pascal and C. Indeed, all variables were global, meaning that they could be accessed from anywhere in the program, leading to the danger of their values being unintentionally changed.

As interest in the principles of structured programming grew (see STRUCTURED PROGRAMMING), BASIC's structural shortcomings made it poorly regarded among computer scientists, who preferred Pascal as a teaching language and C for systems programming. In 1984, BASIC's original developers responded to what they saw as the problems of "street Basic" by introducing True BASIC, a modern, well-structured version of the language, and the 1988 ANSI BASIC standard incorporated similar features. These efforts had only limited impact. However, Microsoft introduced new BASIC

development systems (Quick BASIC in the 1980s and Visual Basic in the 1990s) that also featured improved control structures and data types and that dispensed with the need for cumbersome line numbers. Visual Basic in particular has achieved considerable success, offering a combination of the interactivity of traditional BASIC and access to powerful pre-packaged "controls" that provide menus, dialog boxes, and other features of the Windows user interface. Recent versions of Visual Basic have become increasingly object-oriented, using classes similar to those in C++.

While BASIC in its newer forms continues to have a significant following, it can be argued that what was most distinctive about the original BASIC (the quick, interactive approach to programming) is no longer much in evidence. The writing of short utility programs is now more likely to be undertaken in any of a variety of scripting languages.

Further Reading
Brin, David. "Why Johnny Can't Code," September 14, 2006. Available online. URL: http://www.salon.com/tech/feature/2006/09/14/basic/print.html. Accessed April 24, 2007.
Kemeny, J. G., and Thomas E. Kurtz. *Back to Basic: The History, Corruption, and Future of the Language.* Reading, Mass.: Addison-Wesley, 1985.
Lomax, Paul, and Ron Petrusha. *VB and VBA in a Nutshell: The Languages.* Sebastopol, Calif.: O'Reilly, 1998.
Neuberg, Matt. *REALbasic: The Definitive Guide.* 2nd ed. Sebastapol, Calif.: O'Reilly, 2001.
Sempf, Bill. *Visual Basic 2008 for Dummies.* Hoboken, N.J.: Wiley, 2008.

basic input/output system   *See* BIOS.

Bayesian analysis

Formerly obscure topics in mathematics have a way of suddenly becoming relevant in the information age. For example, the true/false algebraic logic invented by George Boole in the 19th century turned out to perfectly map the operation of electronic on/off in computer circuits.

The Reverend Thomas Bayes (1701?–1761) was another formerly obscure British mathematician who discovered a completely different way of looking at probability. Classical probability assumes that one can make no prior assumptions about the events to be tested. That is, when throwing a die, one does not base the probability that it will come up with a six on the results of any prior throws. Of course that approach is correct in that probability of a six is always 1 in 6 (as long as the dice are honest).

In some situations, however, what has already happened does influence the probability of a future event. Consider a blackjack player who wants to know the probability that the next card drawn will be a face card. If the deck has been properly shuffled, that probability starts out as 12/52 (or 3/13), since there are 12 face cards in the deck of 52 cards.

But suppose that, of the six cards dealt to three players in the first hand, two are face cards. When the dealer deals the next hand, the probability that any card will be a face card has changed. There are now two fewer face cards (12 - 2 = 10) and four fewer non-face cards (40 - 4 = 36), so the probability that a given card is a face card becomes 10/36 or 5/18.

While this is pretty straightforward, in many situations one cannot easily calculate the shifting probabilities. What Bayes discovered was a more general formula:

$$P(T|E) = \frac{P(E|T) * P(T)}{P(E)}$$

In this formula T is a theory or hypothesis about a future event. E represents a new piece of evidence that tends to support or oppose the hypothesis. P(T) is an estimate of the probability that T is true, before considering the evidence represented by E. The question then becomes: If E is true, what happens to the estimate of the probability that T is true? This is called a conditional probability, represented by the left side of the equation, P(T|E), which is read "the probability of T, given E." The right side of Bayes's equation considers the reverse probability—that E will be true if T turns out to be true. This is represented by P(E|T), multiplied by the prior probability of T and divided by the independent probability of E.

PRACTICAL APPLICATIONS

In the real world one generally has imperfect knowledge about the future, and probabilities are seldom as clear cut as those available to the card counter at the blackjack table. However, Bayes's formula makes it possible to continually adjust or "tune" estimates based upon the accumulating evidence. One of the most common applications of Bayesian analysis is in e-mail filters (see SPAM). Bayesian spam filters work by having the user identify a sample of messages as either spam or not spam. The filter then looks for patterns in the spam and non-spam messages and calculates probabilities that a future message containing those patterns will be spam. The filter then blocks future messages that are (above some specified threshold) probably spam. While it is not perfect and does require work on the part of the user, this technique has been quite effective in blocking spam.

A Bayesian algorithm's effectiveness can be expressed in terms of its rate of false positives (in the spam example, this would be the percentage of messages that have been mistakenly classified as spam). If the rate of "true positives" is too low, the algorithm is not effective enough. However, if the rate of false positives is too high, the negative effects (blocking wanted e-mail) might outweigh the positive ones (blocking unwanted spam).

Further Reading

Kantor, Andrew. "Bayesian Spam Filters Use Math that Works Like Magic." *USA Today* online, September 17, 2004. Available online. URL: http://www.usatoday.com/tech/columnist/andrewkantor/2004-09-17-kantor_x.htm. Accessed March 15, 2007.
Lee, Peter M. *Bayesian Statistics: An Introduction.* 3rd ed. New York: Wiley, 2004.
Sivia, D. S. *Data Analysis: A Bayesian Tutorial.* 2nd ed. New York: Oxford University Press, 2006.
"Thomas Bayes, 1702–1761." St. Andrews University Mac Tutor. Available online. URL: http://www-history.mcs.st-andrews.ac.uk/Mathematicians/Bayes.html. Accessed March 15, 2007.

BBS   *See* BULLETIN BOARD SYSTEM.

Bell, C. Gordon
(1934–)
American
Engineer, Computer Designer

Chester Gordon Bell (also known as Gordon Bennet Bell) was born August 19, 1934, in Kirksville, Missouri. As a young boy Bell worked in his father's electrical contracting business, learning to repair appliances and wire circuits. This work led naturally to an interest in electronics, and Bell studied electrical engineering at MIT, earning a B.S. in 1956 and an M.S. in 1957. After graduation and a year spent as a Fulbright Scholar in Australia, Bell worked in the MIT Speech Computation Laboratory (see SPEECH RECOGNITION AND SYNTHESIS). In 1960 he was invited to join the Digital Equipment Corporation (DEC) by founders Ken Olsen and Harlan Anderson.

Bell was a key architect of DEC's revolutionary PDP series (see MINICOMPUTER), particularly as designer of the input/output (I/O) hardware in the PDP-1 and the multi-tasking PDP-6. Bell left DEC to teach computer science at Carnegie Mellon University (1966–72), but then returned to DEC until his retirement in 1983 following a heart attack. During this time Bell developed a deployment plan for the new VAX series minicomputers, which were data-processing workhorses in many organizations during the 1970s and 1980s.

As a close observer of the computer industry, Bell formulated "Bell's Law of Computer Classes" in 1972. It basically states that as new technologies (such as the microprocessor) emerge, they result about once a decade in the emergence of new "classes" or computing platforms, each being generally cheaper and being perceived as a distinct product with new applications. Within a given class, price tends to hold constant while performance increases. Examples thus far include mainframes, minicomputers, personal computers and workstations, networks, cluster or grid computing, and today's ubiquitously connected wireless, portable devices. Bell has indeed suggested that the trend to ubiquitous computing will continue (see UBIQUITOUS COMPUTING and WEARABLE COMPUTERS).

After retirement Bell soon became active again. He founded Encore Computer, a company that specialized in multiprocessor computers, and later was a founding member of Ardent Computer as well as participating in the establishment of the Microelectronics and Computer Technology Corporation, a consortium that attempted to be America's answer to a surging competitive threat from Japanese companies. Bell was also active in debates over technology policy, playing an instrumental role as an assistant director in the National Science Foundation's computing initiatives and the early adoption of the Internet. In 1987 Bell established the Gordon Bell Prize for achievements in parallel processing.

Bell began the 1990s in a new role, helping Microsoft develop a research group, where he was still working as of 2008. Here Bell has developed what amounts to a new paradigm for managing personal data, a project called MyLifeBits. Its main idea is that pictures, e-mails, documents, and other materials that are important to a person's life and work should be organized according to their chronological and other relationships so they can be retrieved naturally and virtually automatically, eschewing the often arbitrary conventions of traditional file systems and interfaces. In 1992 Bell presciently told a *Computer World* interviewer that "twenty-five years from now . . . computers will be exactly like telephones. They are probably going to be communicating all the time."

Bell also retains a strong interest in the history of computing. He cofounded the Computer History Museum in Boston in 1979 and was also a founder of its successor, the Computer History Museum in Mountain View, California.

Bell is a distinguished member of the American Academy of Arts and Sciences, American Association for the Advancement of Science, the Association for Computing Machinery (ACM), and the Institute of Electrical and Electronic Engineering (IEEE). His awards include the IEEE Von Neumann Medal, the AEA Inventor Award, and the National Medal of Technology (1991).

Further Reading

Gordon Bell Home Page. Microsoft Bay Area Research Center. Available online. URL: http://research.microsoft.com/users/gbell/. Accessed April 30, 2007.

Slater, Robert. *Portraits in Silicon.* Boston: MIT Press, 1987.

"Vax Man: Gordon Bell." *Computerworld,* June 22, 1992, p. 13. Available online. URL: http://research.microsoft.com/~gbell//CGB%20Files/Computerworld%20Vax%20Ma n%20920622%20c.pdf. Accessed April 30, 2007.

Bell Laboratories

Bell Telephone Laboratories was established in 1925 in Murray Hill, New Jersey: It was intended to take over the research arm of the Western Electric division of American Telephone and Telegraph (AT&T) and was jointly administered by the two companies. The organization's principal task was to design and develop telephone switching equipment, but there was also research in facsimile (fax) transmission and television.

The research that would have the greatest impact, however, would come from a relative handful of Bell scientists who were given resources to undertake fundamental research. In the 1930s Bell scientist Karl Jansky, investigating interference with long-range radio transmissions, discovered that radio waves were arriving from space, leading to the development of radio astronomy. Other Bell Labs developments of the 1930s and 1940s included the vocoder, an early electronic speech synthesizer, and the photovoltaic cell, with its potential application to solar power systems.

Several Bell Labs technologies would have a direct impact on the computer field. The transistor, developed by Bell Labs researchers John Bardeen, Walter Brattain, and William Shockley, would make a new generation of more compact and reliable computers possible. Information theory (see INFORMATION THEORY and SHANNON, CLAUDE) would revolutionize telecommunications, signal processing, and data transfer. Work on the laser in the 1960s would eventually lead to the compact disc (see CD-ROM AND DVD-ROM). Other hardware contributions include the charge-coupled device (CCD) that would revolutionize astronomical and digital photography and fiber-optic cables for high-volume data communications.

In software engineering the most important achievements of Bell researchers were the development of the C programming language and the UNIX operating system in the early 1970s (see C; RITCHIE, DENNIS; and UNIX). The elegant design of the modular UNIX system is still admired today, and versions of UNIX and Linux power many servers and networks.

NEW CORPORATE DIRECTION

Perhaps ironically, AT&T's near monopolistic position in the telecommunications industry both provided substantial revenue for fundamental research and shielded the lab from competitive pressure and the need to tie research to the development of commercial products. As a result, Bell Labs arguably became the most important private research institution in the 20th century. By the end of the 1980s, however, court decisions had reshaped the landscape of the communications field, and Bell Labs became a victim of the company's change from monopolist to competitor.

In 1996 AT&T divested Bell Labs along with its main equipment manufacturing facilities into a new company, Lucent Technologies. A smaller group of researchers were retained and reorganized as AT&T Laboratories. As the 2000s began these researchers made new achievements, including tiny transistors whose size is measured in atoms, optical data routing (see OPTICAL COMPUTING) and nanotechnology, DNA-based computing (see MOLECULAR COMPUTING), and other esoteric but potentially momentous fields.

In recent years, however, the organization has largely changed its focus from long-term research in fundamental topics to the search for projects that can be quickly turned into commercial products—in essence the requirement that the Labs become a profit center. The merger of Lucent and another communications giant, Alcatel, in 2006 has led to renewed concerns that consolidation and even tighter integration of the Labs with corporate goals might come at the expense of the kind of research culture that has inspired the Labs' greatest breakthroughs.

Further Reading

Alcatel-Lucent Bell Laboratories. Available online. URL: http://www.alcatel-lucent.com. Accessed May 2, 2007.

Bell Labs Technical Journal. Available online. URL: http://www3.interscience.wiley.com. Accessed May 2, 2007.

Gehani, Narain. *Bell Labs: Life in the Crown Jewel.* Summit, N.J.: Silicon Press, 2003.

benchmark

A benchmark is a tool used to evaluate or compare the performance of computer software or systems. Typically, this involves the design of a program (or suite of programs) that performs a series of operations that mimic "real world" activities. For example, computer processors (CPUs) can be given calculations in floating-point arithmetic, yielding a result in "flops" (floating point operations per second). Similarly, several different C-language compilers can be given the same files of source code and rated according to how quickly they produce the executable code, as well as the code's compactness, speed, or efficiency.

Some examples of computer industry benchmarks include:

- Dhrystone and Whetstone for integer and floating point arithmetic, respectively

- MIPS (millions of instructions per second) and MFLOPS (millions of floating point instructions per second) for microprocessors

- FPS (frames per second) for various types of graphics

- 3DMark for three-dimensional graphics

- test suites using Linpack and LAPACK for supercomputers

The devising of appropriate benchmarks is important because they can help prospective purchasers decide which competing CPU, program development tool, database system, or Web server to buy. Often the aspects of systems that are highlighted in advertising are not those that are most relevant to determining their actual utility. For example, CPUs are often compared according to clock speed, but a chip with a superior architecture and algorithm for handling instructions might actually outperform chips with faster clock speeds. By putting chips through their paces using the same arithmetic, data transfer, or graphics instructions, the benchmark provides a more valid comparison.

The most relevant benchmarks tend to focus on re-creating real-world use. Thus database systems can be compared in their speed of retrieval or update of data records. Real-world benchmarks also help guard against manufacturers "tweaking" their systems to create artificially high benchmark results. Nevertheless, benchmarks cannot be used mechanically. While a given industry may have an "industry standard" benchmark, and a given product may be the highest performer using that test, the user must consider how well that benchmark reflects the actual work for which the system or program is being purchased. Performance, however well benchmarked, is usually only one key consideration, with environment (such as network connections), reliability, security, ease of use, and of course cost being other considerations.

Further Reading
comp.benchmarks (USENET newsgroup).
Jones, Capers. *Software Assessments, Benchmarks, and Best Practices*. Boston: Addison-Wesley, 2000.

Netlib [repository for mathematical benchmarking software]. Available online. URL: http://www.netlib.org/. Accessed May 10, 2007.

Berners-Lee, Tim
(1955–)
British
Computer Scientist

A graduate of Oxford University, Tim Berners-Lee created what would become the World Wide Web in 1989 while working at CERN, the giant European physics research institute. At CERN, he struggled with organizing the dozens of incompatible computer systems and software that had been brought to the labs by thousands of scientists from around the world. With existing systems each requiring a specialized access procedure, researchers had little hope of finding out what their colleagues were doing or of learning about existing software tools that might solve their problems.

Berners-Lee's solution was to bypass traditional database systems and to consider text on all systems as "pages" that would each have a unique address, a universal document identifier (later known as a uniform resource locator, or URL). He and his assistants used existing ideas of hypertext to link words and phrases on one page to another page (see HYPERTEXT AND HYPERMEDIA), and adapted existing hypertext editing software for the NeXT computer to create the first World Wide Web pages, a server to provide access to the pages and a simple browser, a program that could be used to read pages and follow the links as the reader desired (see WEB SERVER and WEB BROWSER). But while existing hypertext systems were confined to browsing a single file or at most, the contents of a single computer system, Berners-Lee's World Wide Web used the emerging Internet to provide nearly universal access.

Between 1990 and 1993, word of the Web spread throughout the academic community as Web software was written for more computer platforms (see WORLD WIDE WEB). As demand grew for a body to standardize and shape the evolution of the Web, Berners-Lee founded the World Wide Web Consortium (W3C) in 1994 and continues as its director. Together with his colleagues, he has struggled to maintain a coherent vision of the Web in the face of tremendous growth and commercialization, the involvement of huge corporations with conflicting agendas, and contentious issues of censorship and privacy. His general approach has been to develop tools that would empower the user to make the ultimate decision about the information he or she would see or divulge.

Berners-Lee now works as a senior researcher at the Massachusetts Institute of Technology Computer Science and Artificial Intelligence Laboratory. In his original vision for the Web, users would create Web pages as easily as they could read them, using software no more complicated than a word processor. While there are programs today that hide the details of HTML coding and allow easier Web page creation, Berners-Lee feels the Web must become even easier to use if it is to be a truly interactive, open-ended knowledge system. He is also interested in developing software that can take better advantage of the rich variety of information on the Web, creating a "semantic" Web of meaningful connections that would allow for logical analysis and permit human beings and machines not merely to connect, but to actively collaborate (see SEMANTIC WEB and XML).

In the debate over a possible tiered Internet service (see INTERNET ACCESS POLICY) Berners-Lee has spoken out for "net neutrality," the idea that priority given to material passing over the Internet should not depend on its content or origin. He describes equal treatment to be a fundamental democratic principle, given the primacy of the Net today.

Berners-Lee has garnered numerous awards and honorary degrees. In 1997 he was made an Officer of the British Empire, and in 2001 he became a Fellow of the British Royal Society. Berners-Lee also received the Japan Prize in 2002 and in that same year shared the Asturias Award with fellow Internet pioneers Lawrence Roberts, Robert Kahn, and Vinton Cerf. In 2007 Berners-Lee received the Charles Stark Draper Prize of the U.S. National Academy of Engineering.

Further Reading

Berners-Lee, Tim. Home page with biography and links: Available online. URL: http://www.w3.org/People/Berners-Lee/. Accessed April 20, 2007.
———. Papers on Web design issues. Available online. URL: http://www.w3.org/DesignIssues/.
———. "Proposal for the World Wide Web, 1989." Available online. URL: http://www.w3.org/History/1989/proposal.html.
Berners-Lee, Tim, and Mark Fischetti. *Weaving the Web*. San Francisco: HarperSanFrancisco, 1999.
Henderson, Harry. *Pioneers of the Internet*. San Diego, Calif.: Lucent Books, 2002.
Markoff, John. "'Neutrality' Is New Challenge for Internet Pioneer." *New York Times,* September 27, 2006. Available online. URL: http://www.nytimes.com/2006/09/27/technology/circuits/27neut.html. Accessed April 25, 2007.

Bezos, Jeffrey P.

(1964–)
American
Entrepreneur

With its ability to display extensive information and interact with users, the World Wide Web of the mid-1990s clearly had commercial possibilities. But it was far from clear how traditional merchandising could be adapted to the online world, and how the strengths of the new medium could be translated into business advantages. In creating Amazon.com, "the world's largest bookstore," Jeff Bezos would show how the Web could be used to deliver books and other merchandise to millions of consumers.

Jeff Bezos was born on January 12, 1964, and grew up in Miami, Florida. He would be remembered as an intense, strong-willed boy who was fascinated by gadgets but also liked to play football and other sports. His uncle, Preston Gise, a manager for the Atomic Energy Commission, encouraged young Bezos's interest in technology by giving him electronic equipment to dismantle and explore. Bezos

Jeff Bezos, founder and CEO of Amazon.com, poses for a portrait in the Internet retailer's distribution center. (© JACK KURTZ/THE IMAGE WORKS)

also liked science fiction and became an enthusiastic advocate for space colonization.

Bezos entered Princeton University in 1982. At first he majored in physics, but later switched to electrical engineering, graduating in 1986 with highest honors. By then Bezos had become interested in business software applications, particularly financial networks. At the age of only 23, he led a project at Fitel, a financial communications network, managing 12 programmers and commuting each week between the company's New York and London offices.

As a vice president at Bankers Trust, a major Wall Street firm in the late 1980s, Bezos became very enthusiastic about the use of computer networking and interactive software for providing timely information for managers and investors. However, he found that the "old line" Wall Street firms resisted his efforts and declined to invest in these new uses of information technology.

In 1990, however, Bezos was working at the D.E. Shaw Company and his employer asked him to research the commercial potential of the Internet, which was starting to grow (even though the World Wide Web would not reach most consumers for another five years). Bezos ranked the top 20 possible products for Internet sales. They included computer software, office supplies, clothing, music—and books.

Analyzing the publishing industry, Bezos identified ways in which he believed it was inefficient. Even large bookstores could stock only a small portion of the available titles, while on the other hand many books that were in stock stayed on the shelves for months, tying up money and space. Bezos believed that by combining a single huge warehouse with an extensive tracking database, an online ordering system, and fast shipping, he could satisfy many more customers while keeping costs low.

Bezos pitched his idea to D.E. Shaw. When the company declined to invest in the venture, Bezos decided to put his promising corporate career on hold and start his own online business. By then it was the mid-1990s and the World Wide Web was just starting to become popular, thanks to the new graphical Web browsers such as Netscape.

Looking for a place to set up shop, Bezos decided on Seattle, partly because the state of Washington had a relatively small population (the only customers who would have to pay sales tax) yet had a growing pool of technically trained workers, thanks to the growth of Microsoft and other companies in the area. After several false starts he decided to call his store Amazon, deciding that the name of the Earth's biggest river would be suited to earth's biggest bookstore. Amazon's first headquarters was a converted garage.

Bezos soon decided that the existing software for mail-order businesses was too limited and set a gifted programmer named Shel Kaphan to work creating a custom program that could keep track not only of each book in stock, but how long it would take to get more copies from the publisher or book distributor.

By mid-1995 Amazon.com was ready go online from a new Seattle office using $145,553 contributed by Bezos's mother from the family trust. As word about the store spread through Internet chat rooms and a listing on Yahoo!, the orders began to pour in and Bezos had to struggle to keep up. Despite the flood of orders, the business was losing money as expenses piled up even more quickly.

Bezos went to Silicon Valley in search of venture capital. Bezos's previous experience as a Wall Street "star," together with his self-confidence, enabled him to raise $1 million. Bezos believed that momentum was the key to long-term success. The company's motto became "get big fast." Revenue was poured back into the business, expanding sales into other product lines such as music, video, electronics, and software. The other key element of Bezos's growth strategy was to take advantage of the vast database that Amazon was accumulating—not only information about books and other products, but about what products a given individual or type of customer was buying. Once a customer has bought something from Amazon, he or she is greeted by name and given recommendations for additional purchases based upon what items other customers who had bought that item had also purchased. Customers are encouraged to write reviews of books and other items so that each customer gets the sense of being part of a virtual peer group.

By 1997, the year of its first public stock offering, Amazon seemed to be growing at an impressive rate. A year later the stock was worth almost $100 a share, and by 1999 Jeff Bezos's personal wealth neared $7.5 billion. Bezos and Amazon proved to be one of the few Internet businesses to weather the "dot-bust" collapse of 2000 and 2001. In 2003 Amazon.com chalked up its first annual profit, and the company's stock prices tripled during that time.

Bezos gained a reputation as a very demanding CEO, insisting on recruiting top talent, then demanding that projects set bold goals and complete them ahead of schedule. The pressure resulted in high turnover of top executives, but Bezos has also been quick to encourage and reward initiative. (The company's "Just Do It" program encourages managers to start projects without asking permission of their superiors.)

Aside from Amazon.com, Bezos has maintained his interest in space travel. In 2002 he founded a company called Blue Origin, whose spaceship project has remained shrouded in secrecy. However, in January 2007 the company released video of the first successful (albeit brief) test of a prototype suborbital passenger craft.

Bezos has written a new chapter in the history of retailing, making him a 21st-century counterpart to such pioneers as Woolworth and Montgomery Ward. *Time* magazine acknowledged this by making him its 1999 Person of the Year, while *Internet Magazine* put Bezos on its list of the 10 persons who have most influenced the development of the Internet.

Further Reading
Blue Origin website. Available online. URL: http://public.blueorigin.com/index.html. Accessed April 10, 2007.
Byers, Ann. *Jeff Bezos: The Founder of Amazon.com*. New York: Rosen Publishing Group, 2006.
Marcus, James. *Amazonia: Five Years at the Epicenter of the Dot. Com Juggernaut*. New York: New Press, 2004.
Spector, Robert. *Amazon.com: Get Big Fast: Inside the Revolutionary Business Model that Changed the World*. New York: Harper Business, 2000.

binding

Designers of program compilers are faced with the question of when to translate a statement written in the source language into final instructions in machine language (see also ASSEMBLER). This can happen at different times depending on the nature of the statement and the decision of the compiler designer.

Many programming languages use formal data types (such as integer, floating point, double, string, and so on) that result in allocation of an exact amount of storage space to hold the data (see DATA TYPES). A statement that declares a variable with such a type can be effectively bound immediately (that is, a final machine code statement can be generated). This is also called compile-time binding.

However, there are a variety of statements for which binding must be deferred until more information becomes available. For example, it is common for programmers to use libraries of precompiled routines. A statement that calls such a routine cannot be turned immediately into machine language because the compiler doesn't know the actual address where the routine will be embedded in the final compiled program. (That address will be determined by a program called a linker that links the object code from the source program to the library routines called upon by that code.)

Another aspect of binding arises when there is more than one object in a program with the same name. In languages such as C or Pascal that use a nested block structure, lexical binding can determine that a name refers to the closest declaration of that name—that is, the smallest scope that contains that name (see VARIABLE). In a few languages such as Lisp, however, the reference for a name depends on how (or for what) the function is being called, so binding can be done only at run time.

BINDING AND OBJECT-ORIENTED LANGUAGES
The use of polymorphism in object-oriented languages such as C++ raises a similar issue. Here there can be a base class

and a hierarchy of derived classes. A function in the base class can be declared to be virtual, and versions of the same function can be declared in the derived classes. In this case a statement containing a pointer to the function in the base class cannot be bound until run time, because only then will it be known which version of the virtual function is being called. However, compilers for object-oriented languages can be written so they do early binding on statements for which it is safe (such as those involving static data types), but do dynamic binding when necessary.

From the point of view of efficiency, early binding is better because memory can be allocated efficiently. Dynamic binding provides greater flexibility, however, and facilitates debugging—for example, because the name of a variable is normally lost once it is bound and the machine code is generated.

Further Reading
Aho, Alfred V., et al. *Compilers: Principles, Techniques, and Tools.* 2nd ed. Reading, Mass.: Addison-Wesley, 2006.
Scott, Michael L. *Programming Language Pragmatics.* 2nd ed. San Francisco: Morgan Kaufmann, 2005.

bioinformatics
Broadly speaking, bioinformatics (and the related field of computational biology) is the application of mathematical and information-science techniques to biology. This undertaking is inherently difficult because a living organism represents such a complex interaction of chemical processes. Understanding any one process in isolation gives little understanding of the role it plays in physiology. Similarly, as more has been learned about the genome of humans and other organisms, it has become increasingly clear that the "programs" represented by gene sequences are "interpreted" through complex interactions of genes and the environment. Given this complexity, the great strides that have been made in genetics and the detailed study of metabolic and other biological processes would have been impossible without advances in computing and computer science.

APPLICATION TO GENETICS
Since information in the form of DNA sequences is the heart of genetics, information science plays a key role in understanding its significance and expression. The sequences of genes that determine the makeup and behavior of organisms can be represented and manipulated as strings of symbols using, for example, indexing and search algorithms. It is thus natural that the advent of powerful computer workstations and automated lab equipment would lead to the automation of gene sequencing (determining the order of nucleotides), comparing or determining the relationship between corresponding sequences, and identifying and annotating regions of interest. The completion of the sequencing of the human genome well ahead of schedule was thus a triumph of computer science as much as biology. Today the systematic search for genetic and metabolic interactions has been greatly sped up by the use of microarrays, silicon chips with grids of tiny holes that each contain a

A scientist observes an experiment performed by robotic equipment. (ANDREI TCHERNOV/ISTOCKPHOTO)

specified material that can be automatically tested for reaction to a given sample.

EVOLUTIONARY BIOLOGY
The ability to compare genes and to account for the effects of mutation has also established evolutionary biology on a firm foundation. Given a good estimate of the mutation rate (a "molecular clock") in mitochondrial DNA, the chronology of species and common ancestors can be determined with considerable accuracy using statistical methods and appropriate data structures (see TREE). The results of such research have cast intriguing if sometimes controversial light on such issues in paleontology as the relationship between early modern humans and Neanderthals. Computational genetics can also measure the biodiversity of a present-day ecosystem and predict the likely future of particular species in it.

FROM GENES TO PROTEINS
Gene sequences are only half of many problems in biology. Computational techniques are also being increasingly applied to the analysis and simulation of the many intricate

chemical steps that link genetic information to expression in the form of a particular protein and its three-dimensional structure in the process known as protein folding. Already molecular simulations and predictive techniques are being used to determine which of thousands of possible molecular configurations might have promising pharmaceutical applications. The development of better algorithms and more powerful computing architectures for such analysis can further speed up research, avoid wasteful "dead ends," and bring effective treatments for cancer and other serious diseases to market sooner. Recently, the unlikely platform of a Sony PlayStation 3 and its powerful new processor has been harnessed to turn gamers' idle time to the processing of protein-folding data in the Folding@Home project.

SIMULATION

A variety of other types of biological computer simulation have been employed. Examples include the chemical components (metabolites and enzymes) that are responsible for metabolic activity in organisms, the structure of the nervous system and the brain (see NEURAL NETWORK), and the interaction of multiple predators and food sources in an ecosystem. Simulations can also incorporate algorithms first devised by artificial intelligence researchers (see GENETIC ALGORITHMS and ARTIFICIAL LIFE). Simulations are combined with sophisticated graphics to enable researchers to visualize structure. Such visualization can

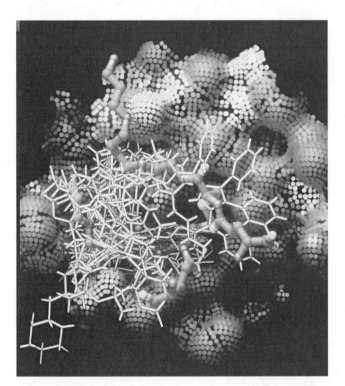

Computers can create detailed representations that give scientists unprecedented ability to visualize nature's most intricate structures. This is a computer model of trypanathione Reductase, a protein crystal. (NASA PHOTO; MARSHALL SPACE FLIGHT CENTER IMAGE EXCHANGE)

provide insight and encourage intuitive "leaps" that might be missed when working only with formulas. Visualization algorithms developed for biomedical research can also be applied to the development of advanced MRI and other scans for use in diagnosis and therapy.

A FRUITFUL RELATIONSHIP

Bioinformatics has been one of the "hottest" areas in computing in recent years, often following trends in the broader "biotech" sector. This challenging field involves such diverse subjects as genetics, biochemistry, physiology, mathematics (structural and statistical), database analysis and search techniques (see DATA MINING), simulation, modeling, graphics, and image analysis. Major projects often involve close cooperation between bioinformatics specialists and other researchers. Many computer scientists may find it profitable to study biology just as biologists will need to learn about and master the latest software tools. Researchers must also consider how the availability of ever-increasing computing power might make previously impossible projects feasible (see SUPERCOMPUTER and GRID COMPUTING). (The National Institutes of Health (NIH) currently funds seven biomedical computation centers, including the National Center for Physics-based Simulation of Biological Structures at Stanford University.)

The relationship between biology and computer science seems destined to be even more fruitful in coming years. As software tools allow researchers to probe ever more deeply into biological processes and to bridge the gap between physics, biochemistry, and the emergent behavior of living organisms, understanding of those processes may in turn inspire the creation of new architectures and algorithms in areas such as artificial intelligence and robotics.

Further Reading

Bader, David A. "Computational Biology and High-Performance Computing." *Communications of the ACM* 47, 11 (2004): 34–41.

Brent, Roger, and Jehoshua Bruck. "Can Computers Help to Explain Biology?" *Nature* 440 (March 23, 2006): 416.

Campbell, A. Malcolm, and Laurie J. Heyer. *Discovering Genomics, Proteomics, and Bioinformatics.* 2nd ed. San Francisco: Benjamin Cummings, 2006.

Claverie, Jean-Michel, and Cedric Notredame. *Bioinformatics for Dummies.* 2nd ed. Indianapolis: Wiley, 2006.

Cohen, Jacques. "Computer Science and Bioinformatics." *Communications of the ACM* 48 (2005): 72–78.

"Just the Facts: A Basic Introduction to the Science Underlying NCBI Resources: Bioinformatics." National Center for Biotechnology Information. Available online. URL: http://www.ncbi.nlm.nih.gov/About/primer/bioinformatics.html. Accessed April 24, 2007.

biometrics

The earliest use of biometrics was probably the development by Alphonse Bertillon in 1882 of anthropometry, a system of classification by physical measurements and description. While this was soon supplanted by the discovery that fingerprints could serve as an easier to use means of unique identification of persons, the need for a less invasive means of physical identification has led to the development of a variety of biometric scanners that take Bertillon's ideas to a much more detailed level.

TECHNOLOGIES

In general, biometric scanning involves four steps: the capture of an image using a camera or other device, the extraction of key features from the image, the creation of a template that uniquely characterizes the person being scanned, and the matching of the template to stored templates in order to identify the person.

There are several possible targets for biometric scanning, including the following areas:

FACIAL SCANNING

Facial scanning uses cameras and image analysis software that looks at areas of the human face that change little during the course of life and are not easily alterable, such as the upper outline of the eye sockets and the shape of the cheekbones. Researchers at MIT developed a series of about 125 grayscale images called eigenfaces from which features can be combined to characterize any given face. The template resulting from a scan can be compared with the one on file for the claimed identity, and coefficients expressing the degree of similarity are calculated. Variance above a specified level results in the person being rejected. Facial scanning is often viewed as less intrusive than the use of fingerprints, and it can also be applied to surveillance images.

FINGER SCANNING

Finger scanning involves the imaging and automatic analysis of the pattern of ridges on one or more fingertips. Unlike traditional fingerprinting, the actual fingerprint is not saved, but only enough key features are retained to provide a unique identification. This information can be stored in a database and also compared with full fingerprints stored in existing databases (such as that maintained by the Federal Bureau of Investigation). Finger scanning can meet with resistance because of its similarity to fingerprinting and the association of the latter with criminality.

HAND GEOMETRY

This technique measures several characteristics of the hand, including the height of and distance between the fingers and the shape of the knuckles. The person being scanned places the hand on the scanner's surface, aligning the fingers to five pegs. Hand-scanning is reasonably accurate in verifying an individual compared to the template on file, but not accurate enough to identify a scan from an unknown person.

IRIS AND RETINA SCANNING

These techniques take advantage of many unique individual characteristics of these parts of the eye. The scanned characteristics are turned into a numeric code similar to a bar code. Retina scanning can be uncomfortable because it involves shining a bright light into the back of the eye, and has generally been used only in high-security installations.

However, iris scanning involves the front of the eye and is much less intrusive, and the person being scanned needs only to look into a camera.

VOICE SCANNING

Voice scanning and verification systems create a "voiceprint" from a speech sample and compare it to the voice of the person being verified. It is a quick and nonintrusive technique that is particularly useful for remote transactions such as telephone access to banking information.

BEHAVIORAL BIOMETRICS

Biometrics are essentially invariant patterns, and these can be found in behavior as well as in physical features. One of the most promising techniques (recently patented) analyzes the pace or rhythm of a person's typing on a keyboard and generates a unique numeric code. A similar approach might be applicable to mouse usage.

APPLICATIONS OF BIOMETRICS

Due to the expense of the equipment and the time involved in scanning, biometrics were originally used primarily in verifying identity for people entering high-security installations. However, the development of faster and less intrusive techniques, combined with the growing need to verify users of banking (ATM) and other networks has led to a growing

A portable iris recognition scanner being demonstrated at the Biometrics 2004 exhibition and conference in London. (IAN WALDIE/ GETTY IMAGES)

interest in biometrics. For example, a pilot program in the United Kingdom has used iris scanning to replace the PIN (personal identification number) as a means of verifying ATM users.

The general advantage of biometrics is that it does not rely on cards or other artifacts that can be stolen or otherwise transferred from one person to another, and in turn, a person needing to identify him or herself doesn't have to worry about forgetting or losing a card. However, while workers at high-security installations can simply be required to submit to biometric scans, citizens and consumers have more choice about whether to accept techniques they may view as uncomfortable, intrusive, or threatening to privacy.

Recent heightened concern about the stealing of personal identification and financial information (see IDENTITY THEFT) may promote greater acceptance of biometric techniques. For example, a built-in fingerprint reader (already provided on some laptop computers) could be used to secure access to the hard drive or transmitted to authenticate an online banking customer.

Of course every security measure has the potential for circumvention or misuse. Concerns about the stealing and criminal use of biometric data (particularly online) might be addressed by a system created by Emin Martinian of the Mitsubishi Electric Research Laboratories in Cambridge, Massachusetts. The algorithm creates a unique code based on a person's fingerprint data. The data itself is not stored, and the code cannot be used to re-create it, but only to match against the actual finger.

The growing use of biometrics by government agencies (such as in passports and border crossings) is of concern to privacy advocates and civil libertarians. When combined with surveillance cameras and central databases, biometrics (such as face analysis and recognition) could aid police in catching criminals or terrorists, but could also be used to strip the anonymity from political protesters. The technology is thus double-edged, with the potential both to enhance the security of personal information and to increase the effectiveness of surveillance.

Further Reading

Ashborn, Julian D. M. *Biometrics: Advanced Identity Verification, the Complete Guide.* New York: Springer-Verlag, 2000.

"Biometrics Overview." Available online. URL: http://www.biometricgroup.com/a_bio1/_technology/research_a_technology.htm. Accessed April 20, 2007.

Biometrics Research Homepage at Michigan State University. Available online. URL: http://biometrics.cse.msu.edu/. Accessed April 24, 2007.

"Biometrics: Who's Watching You?" Electronic Frontier Foundation. Available online. URL: http://www.eff.org/Privacy/Surveillance/biometrics/. Accessed April 24, 2007.

Harreld, Heather. "Biometrics Points to Greater Security." *Federal Computer Week,* July 22, 1999. Available online. URL: http://www.cnn.com/TECH/computing/9907/22/biometrics.idg/index.html.

Jain, Anil, Ruud Bolle, and Sharath Pankanti. *Biometrics: Personal Identification in Networked Society.* Norwell, Mass.: Kluwer Academic Publishers, 1999.

Woodward, John D., Nicholas M. Orlans, and Peter T. Higgins. *Biometrics: Identity Assurance in the Information Age.* New York: McGraw-Hill, 2002.

BIOS (Basic Input-Output System)

With any computer system a fundamental design problem is how to provide for the basic communication between the processor (see CPU) and the devices used to obtain or display data, such as the video screen, keyboard, and parallel and serial ports.

In personal computers, the BIOS (Basic Input-Output System) solves this problem by providing a set of routines for direct control of key system hardware such as disk drives, the keyboard, video interface, and serial and parallel ports. In PCs based on the IBM PC architecture, the BIOS is divided into two components. The fixed code is stored on a PROM (programmable read-only memory) chip commonly called the "ROM BIOS" or "BIOS chip." This code handles interrupts (requests for attention) from the peripheral devices (which can include their own specialized BIOS chips). During the boot sequence the BIOS code runs the POST (power-on self test) and queries various devices to make sure they are functional. (At this time the PC's screen will display a message giving the BIOS manufacturer, model, and other information.) Once DOS is running, routines in the operating system kernel can access the hardware by making calls to the BIOS routines. In turn, application programs can call the operating system, which passes requests on to the BIOS routines.

The BIOS scheme has some flexibility in that part of the BIOS is stored in system files (in IBM PCs, IO.SYS and IBMIO.COM). Since this code is stored in files, it can be upgraded with each new version of DOS. In addition, separate device drivers can be loaded from files during system startup as directed by DEVICE commands in CONFIG.SYS, a text file containing various system settings.

For further flexibility in dealing with evolving device capabilities, PCs also began to include CMOS (complementary metal oxide semiconductor) chips that allow for the storage of additional parameters, such as for the configuration of memory and disk drive layouts.

In modern PCs the BIOS setup screen also allows users to specify the order of devices to be used for loading system startup code. This, for example, might allow a potentially corrupted hard drive to be bypassed in favor of a bootable CD or DVD with disk repair tools. Another scenario would allow users to boot from a USB memory stick (see FLASH DRIVE) and use a preferred operating system and working files without disturbing the PC's main setup.

The data on these chips is maintained by a small onboard battery so settings are not lost when the main system power is turned off.

Additionally, modern PC BIOS chips use "flash memory" (EEPROM or "electrically erasable programmable read-only memory") to store the code. These chips can be "flashed" or reprogrammed with newer versions of the BIOS, enabling the support of newer devices without having to replace any chips.

BEYOND THE BIOS

While the BIOS scheme was adequate for the earliest PCs, it suffered from a lack of flexibility and extensibility. The routines were generic and thus could not support all the functions of newer devices. Because BIOS routines for such tasks as graphics tended to be slow, applications programmers often bypassed the BIOS and dealt with devices directly or created device drivers specific to a particular model of device. This made the life of the PC user more complicated because programs (particularly games) may not work with some video cards, for example, or at least required an updated device driver.

While both the main BIOS and the auxiliary BIOS chips on devices such as video cards are still essential to the operation of the PC, modern operating systems, such as Microsoft Windows and applications written for them, generally do not use BIOS routines and employ high performance device drivers instead. (By the mid-1990s BIOSes included built-in support for "Plug and Play," a system for automatically loading device drivers as needed. Thus, the BIOS is now usually of concern only if there is a hardware failure or incompatibility.)

Further Reading
"System BIOS Function and Operation." Available online. URL: http://www.pcguide.com/ref/mbsys/bios/func.htm. Accessed April 20, 2007.

bitmapped image

A bitmap is a series of bits (within a series of bytes in memory) in which the bits represent the pixels in an image. In a monochrome bitmap, each pixel can be represented by one bit, with a 1 indicating that the pixel is "on." For grayscale or color images several bits must be used to store the information for each pixel. The pixel value bits are usually preceded by a data structure that describes various characteristics of the image.

For example, in the Microsoft Windows BMP format, the file for an image begins with a BITMAPFILEHEADER that includes a file type, size, and layout. This is followed by a BITMAPINFOHEADER that gives information about the image itself (dimensions, type of compression, and color format). Next comes a color table that describes each

color found in the image in terms of its RGB (red, green, blue) components. Finally comes the consecutive bytes representing the bits in each line of the image, starting from lower left and proceeding to the upper right.

The actual number of bits representing each pixel depends on the dimensions of the bitmap and the number of colors being used. For example, if the bitmap has a maximum of 256 colors, each pixel value must use one byte to store the index that "points" to that color in the color table. However, an alternative format stores the actual RGB values of each pixel in three consecutive bytes (24 bits), thus allowing for a maximum of 24 (16,777,216) colors (see COLOR IN COMPUTING).

SHORTCOMINGS AND ALTERNATIVES

The relationship between number of possible colors and amount of storage needed for the bitmap means that the more realistic the colors, the more space is needed to store an image of a given size, and generally, the more slowly the bitmap can be displayed. Various techniques have been used to shrink the required space by taking advantage of redundant information resulting from the fact that most images have areas of the same color (see DATA COMPRESSION).

Vector graphics offer an alternative to bitmaps, particularly for images that can be constructed from a series of lines. Instead of storing the pixels of a complete image, vector graphics provides a series of vectors (directions and lengths) plus the necessary color information. This can make for a much smaller image, as well as making it easy to scale the image to any size by multiplying the vectors by some constant.

Further Reading
Artymiak, Jacek. *Dynamic Bitmap Graphics with PHP and Gd. 2nd ed.* Lublin, Poland: devGuide.net, 2007.
"Microsoft Windows Bitmap File Format Summary." FileFormat-Info. Available online. URL: http://www.fileformat.info/format/bmp/egff.htm. Accessed May 10, 2007.
Slaybaugh, Matt. *Professional Web Graphics.* Boston: Course Technology, 2001.

bits and bytes

Computer users soon become familiar with the use of bits (or more commonly bytes) as a measurement of the capacity of computer memory (RAM) and storage devices such as disk drives. They also speak of such things as "16-bit color," referring to the number of different colors that can be specified and generated by a video display.

In the digital world a bit is the smallest discernable piece of information, representing one of two possible states (indicated by the presence or absence of something such as an electrical charge or magnetism, or by one of two voltage levels). Bit is actually short for "binary digit," and a bit corresponds to one digit or place in a binary (base 2) number. Thus an 8-bit value of

11010101

corresponds, from right to left, to $(1 * 2^0) + (0 * 2^1) + (1 * 2^2) + (0 * 2^3) + (1 * 2^4) + (0 * 2^5) + (1 * 2^6) + (1 * 2^7)$, or 213 in terms of the familiar decimal system.

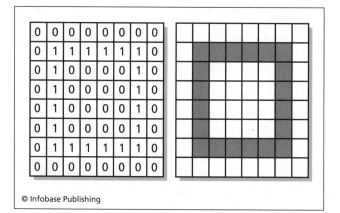

© Infobase Publishing

In a monochrome bitmapped image, a 1 is used to represent a pixel that is turned on, while the empty pixels are represented by zeroes. Color bitmaps must use many more bits per pixel to store color numbers.

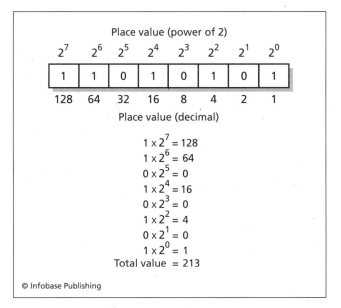

Place value (power of 2)

2^7 2^6 2^5 2^4 2^3 2^2 2^1 2^0

| 1 | 1 | 0 | 1 | 0 | 1 | 0 | 1 |

128 64 32 16 8 4 2 1

Place value (decimal)

$1 \times 2^7 = 128$
$1 \times 2^6 = 64$
$0 \times 2^5 = 0$
$1 \times 2^4 = 16$
$0 \times 2^3 = 0$
$1 \times 2^2 = 4$
$0 \times 2^1 = 0$
$1 \times 2^0 = 1$
Total value = 213

© Infobase Publishing

One byte in memory can store an 8-bit binary number. Just as each place to the left in a decimal number represents the next higher power of 10, the places in the byte increase as powers of 2. Here the places with 1 in them add up to a total decimal value of 213.

With regard to computer architectures the number of bits is particularly relevant to three areas: (1) The size of the basic "chunk" of data or instructions that can be fetched, processed, or stored by the central processing unit (CPU); (2) The "width" of the data bus over which data is sent between the CPU and other devices—given the same processor speed, a 32-bit bus can transfer twice as much data in a given time as a 16-bit bus; and (3) The width of the address bus (now generally 32 bits), which determines how many memory locations can be addressed, and thus the maximum amount of directly usable RAM.

The first PCs used 8-bit or 16-bit processors, while today's PC processors and operating systems often use 32-bits at a time, with 64-bit processors now entering the market. Besides the "width" of data transfer, the number of bits can also be used to specify the range of available values. For example, the range of colors that can be displayed by a video card is often expressed as 16 bit (65,536 colors) or 32 bit (16,777,777,216 colors, because only 24 of the bits are used for color information).

Since multiple bits are often needed to specify meaningful information, memory or storage capacity is often expressed in terms of bytes. A byte is 8 bits or binary digits, which amounts to a range of from 0 to 255 in terms of decimal (base 10) numbers. A byte is thus enough to store a small integer or a character code in the standard ASCII character set (see CHARACTER). Common multiples of a byte are a kilobyte (thousand bytes), megabyte (million bytes), gigabyte (billion bytes), and occasionally terabyte (trillion bytes). The actual numbers represented by these designations are actually somewhat larger, as indicated in the accompanying table.

Further Reading
"How Bits and Bytes work." Available online. URL: http://www.howstuffworks.com/bytes.htm. Accessed April 22, 2007.

bitwise operations

Since each bit of a number (see BITS AND BYTES) can hold a truth value (1 = true, 0 = false), it is possible to use individual bits to specify particular conditions in a system, and to compare individual pairs of bits using special operators that are available in many programming languages.

Bitwise operators consist of logical operators and shift operators. The logical operators, like Boolean operators in general (see BOOLEAN OPERATORS), perform logical comparisons. However, as the name suggests, bitwise logical operators do a bit-for-bit comparison rather than comparing the overall value of the bytes. They compare the corresponding bits in two bytes (called source bits) and write result bits based on the type of comparison.

The AND operator compares corresponding bits and sets the bit in the result to one if both are one. Otherwise, it sets it to zero.

Example: 10110010 AND 10101011 = 10100010

The OR operator compares corresponding bits and sets the bit in the result to one if either or both of the bits are ones.

Example: 10110110 OR 10010011 = 10110111

The XOR ("exclusive OR") operator works like OR except that it sets the result bit to one only if either (not both) of the source bits are ones.

Example: 10110110 XOR 10010011 = 00100101

The COMPLEMENT operator switches all the bits to their opposites (ones for zeroes and zeroes for ones).

Example: COMPLEMENT 11100101 = 00011010

MEASUREMENT	NUMBER OF BYTES		EXAMPLES OF USE
byte	1		small integer, character
kilobyte	2^{10}	1,024	RAM (PCs in the 1980s)
megabyte	2^{20}	1,048,576	hard drive (PCs to mid-1990s)
			RAM (modern PCs)
gigabyte	2^{30}	1,073,741,824	hard drive (modern PCs)
			RAM (latest PCs)
terabyte	2^{40}	1,099,511,627,776	large drive arrays

The shift operators simply shift all the bits left (LEFT SHIFT) or right (RIGHT SHIFT) by the number of places specified after the operator. Thus

00001011 LEFT SHIFT 2 = 00101100

and

00001011 RIGHT SHIFT 2 = 00000010 (bits that shift off the end of the byte simply "drop off" and are replaced with zeroes).

While we have used words in our general description of these operators, actual programming languages often use special symbols that vary somewhat with the language. The operators used in the C language are typical:

& AND

| OR

^ Exclusive OR

~ Complement

>> Right Shift

<< Left Shift

MASKING

There are a number of programming tasks where the contents of individual bits must be read or manipulated. Operating systems and network protocols often have data structures where several separate pieces of information are stored in a single byte in order to save space. (For example, in IBM architecture PC's interrupts are often enabled or disabled by setting particular bits in a mask register.) Operations using bitmapped images can also involve bit manipulation.

Suppose the right three bits of a byte contain a desired piece of information. The byte is ANDed with a prepared byte called a mask in which the desired bits are set to one and the rest of the bits are zero: in this case it would be 00000111. Thus if the byte contains 11010110:

      11010110 AND 00000111 = 000000110

The result contains only the value of the right three bits. Similarly, if one wants to set a particular bit to zero, one simply ANDs the byte with a byte that has a zero in that position and ones in the rest of the byte. Thus to "zero out" the second bit from the left in 11010110:

      11010110 AND 10111111 = 10010110

Further Reading

"Bitwise Operators in C and C++." Available online. URL: http://www.cprogramming.com/tutorial/bitwise_operators.html. Accessed September 17, 2007.

"Java Lesson 7: Bitwise Operations with Good Examples." Available online. URL: http://www.javafaq.nu/java-article402.html. Accessed September 17, 2007.

"Logic and Bitwise Operators in PHP." Available online. URL: http://theopensourcery.com/phplogic.htm. Accessed September 17, 2007.

blogs and blogging

As the 20th century drew to a close, a new form of personal self-expression began to appear on the Web. Called "Web logs" but soon universally shortened to *blogs,* this new type of online journal caught on rapidly, being adopted not only by Web-savvy designers and writers, but by millions of ordinary users wanting to express opinions on the news of the day, critique music or restaurants, analyze technological developments, or just keep relatives informed about family doings. (By 2006 the Pew Internet and American Life project was reporting that about 16 percent of the American population—around half of all Internet users—was writing or at least reading blogs.) Additionally, today's blogs can have institutional as well as personal roles. They have created a new form of journalism that challenges the mainstream media, have kept researchers in touch with new developments, and have provided a new way for corporations to communicate with customers or prospective investors.

FORMATS AND SOFTWARE

The "classic" blog resembles a diary or journal. The writer simply adds a new entry either on a regular basis such as daily or weekly, or when there is something new to be said or responded to. Indeed, what makes blogs different from traditional journals is two things: linkage and interactivity. When a "blogger" writes about something such as a news story, he or she almost always includes a Web link that can take the reader directly to the source in question. The interactivity comes in readers having the opportunity to click a button and write their own response—either to the original journal entry or to someone's earlier response.

In order for blogging to become ubiquitous, there needed to be software that anyone could use without knowing anything about Web design or HTML coding. Most commonly, the software is hosted on a Web site, and users only need a Web browser to create and manage their blogs. One of the first popular blogging applications was developed in the late 1990s by David Winer of Userland Software. Google's Blogger.com is another popular choice. Many blogging applications are free and open source, such as Drupal, Mephisto, and WordPress (which can be used stand-alone or as a hosted service). Today anyone can start and maintain a blog with just a few clicks.

As blogs proliferated, the value of a search engine devoted specifically to finding blogs and blog entries became evident. While a general search engine can find blog entries that match keywords, the results generally do not show the context or the necessary links to follow the threads of discussion. In addition to such services as Bloglines, general search engines such as Google include options for searching the burgeoning "blogosphere."

As with many other Web developments, what began as primarily a textual medium soon embraced multimedia. The availability of inexpensive cameras makes it easy for bloggers to engage in "video blogging." Anyone who wants to see these videos regularly can "subscribe" and have them downloaded automatically to their PC or portable player (see PODCASTING).

Blogging can also be seen as part of a larger trend toward Web users taking an active role in expressing and sharing opinion and resources (see USER-CREATED CONTENT, FILE-SHARING AND P2P NETWORKS, and YOUTUBE).

SOCIAL AND ECONOMIC IMPACT

Blogs first emerged in popular consciousness as a new way in which people caught in the midst of a tragedy such as the September 11, 2001, attacks could reassure friends about their safety while describing often harrowing accounts. The Iraq war that began in 2003 was the first war to be blogged on a large scale. Like their journalistic counterparts, bloggers, whether American or Iraqi, were "embedded" in the often-violent heart of the protracted conflict, but they were also effectively beyond the control of government or military authorities. (See also POLITICAL ACTIVISM AND THE INTERNET.)

Blogs are also being used widely in business. Within a company, a blog can highlight ongoing activities and relevant resources that might otherwise be overlooked in a large corporate network. Software developers can also report on the progress of bug fixes or enhancements and solicit comments from end users. There has been some concern, however, that corporate blogs are not sufficiently supervised to prevent the dissemination of sensitive information or the posting of libelous or inflammatory material. (For the collaborative creation of large bodies of structured knowledge, see WIKIS AND WIKIPEDIA.)

Blogs have provided an outlet where other means of expression are unavailable because of war (as in Iraq), disaster (Hurricane Katrina), or government censorship—although China in particular has hired hundreds of censors to remove offending postings as well as requiring blog providers such as MSN to police their content (see CENSORSHIP AND THE INTERNET).

Further Reading

Blogger. Available online. URL: http://www.bloger.com. Accessed September 2, 2007.
Bloglines. Available online. URL: http://www.bloglines.com. Accessed April 10, 2007.
Blood, Rebecca. *The Weblog Handbook: Practical Advice on Creating and Maintaining Your Blog.* Cambridge, Mass.: Perseus, 2002.
Burden, Matthew Currier. *The Blog of War: Front-Line Dispatches from Soliders in Iraq and Afghanistan.* New York: Simon & Schuster, 2006.
Dedman, Jay. *Videoblogging.* New York: Wiley, 2006.
Farber, Dan. "Reflections on the First Decade of Blogging." February 25. 2007. Available online. URL: http://blogs.zdnet.com/BTL/?p=4541&tag=nl.e539. Accessed April 10, 2007.
Hasin, Hayder. *WordPress Complete: Set Up, Customize, and Market Your Blog.* Birmingham, U.K.: Packt Publishing, 2006.
Radio Userland. Available online. URL: http://radio.userland.com. Accessed September 2, 2007.
Rebecca's Pocket. Available online. URL: http://www.rebeccablood.net/. Accessed April 10, 2007.
Technorati. Available online. URL: http://www.technorati.com. Accessed April 10, 2007.
WordPress. Available online. URL: http:// www.wordpress.com. Accessed April 10, 2007.

Bluetooth

Loosely named after a 10th-century Danish king, Bluetooth is a wireless data communications and networking system designed for relatively short-range operation (generally within the same room, although it can be used over longer distances up to several hundred feet [tens of meters]). The radio transmission is in the 2.4-GHz band and is typically low power, making it suitable for battery-powered devices such as laptops.

APPLICATIONS

Bluetooth was originally developed by Ericsson Corporation to provide a wireless connection for mobile telephone headsets. Today it is often used to "sync" (update data) between a PDA such as a Blackberry or Palm (see PDA) with a Bluetooth-equipped laptop or desktop. Many cell phones are also equipped with Bluetooth, allowing them to be dialed from a PDA, although the growing use of phones that combine telephony and PDA functions is making this scenario less common (see SMARTPHONE). Bluetooth can also be used for wireless keyboards, mice, or printers.

It is possible to connect PDAs or PCs to the Internet and local area networks using a Bluetooth wireless access point (WAP) attached to a router, but faster and longer range Wifi (802.11) wireless connections are much more widely used for this application (see WIFI).

Bluetooth connections between devices are specified using profiles. Profiles have been developed for common kinds of devices, specifying how data is formatted and exchanged. For example, there are profiles for controlling telephones, printers and faxes, digital cameras, and audio devices. Most modern operating systems (including Windows Mobile, Linux, Palm OS, and Mac OS X) include support for basic Bluetooth profiles. Functions fundamental to all Bluetooth operations are found in Bluetooth Core Specifications (version 2.1 as of August 2007). Planned future enhancements include accommodation for ultra-wide band (UWB) radio technology, allowing for data transfer up to 480 megabits per second. At the same time, Bluetooth is extending the ultra-low-power modes that are particularly important for wearable or implanted medical devices.

Further Reading

"Bluetooth." Wikipedia. Available online. URL: http://en.wikipedia.org/wiki/Bluetooth. Accessed July 20, 2007.
Bluetooth Special Interest Group. Available online. URL: http://www.bluetooth.com/bluetooth/. Accessed July 20, 2007.
Layton, Julia, and Curt Franklin. "How Bluetooth Works." Available online. URL: http://www.howstuffworks.com/bluetooth.htm. Accessed September 3, 2007.

Boolean operators

In 1847, British mathematician George Boole proposed a system of algebra that could be used to manipulate propositions, that is, assertions that could be either true or false. In his system, called propositional calculus or Boolean Algebra, propositions can be combined using the "and" and "or" operators (called Boolean operators), yielding a new proposition that is also either true or false. For example:

"A cat is an animal" AND "The sun is a star" is true because both of the component propositions are true.

"A square has four sides" AND "The Earth is flat" is false because only one of the component propositions is true.

However "A square has four sides" OR "The Earth is flat" is true, because *at least one* of the component propositions is true.

A chart called a truth table can be used to summarize the AND and OR operations. Here 1 means true and 0 means false, and you read across from the side and down from the top to see the result of each combination.

AND TABLE		
	0	1
0	0	0
1	0	1

OR TABLE		
	0	1
0	0	1
1	1	1

A variant of the OR operator is the "exclusive OR," sometimes called "XOR" operator. The XOR operator yields a result of true (1) if *only one* of the component propositions is true:

XOR TABLE		
	0	1
0	0	1
1	1	0

Additionally, there is a NOT operator that simply reverses the truth value of a proposition. That is, NOT 1 is 0 and NOT 0 is 1.

APPLICATIONS

Note the correspondence between the two values of Boolean logic and the binary number system in which each digit can have only the values of 1 or 0. Electronic digital computers are possible because circuits can be designed to follow the rules of Boolean logic, and logical operations can be harnessed to perform arithmetic calculations.

Besides being essential to computer design, Boolean operations are also used to manipulate individual bits in memory (see BITWISE OPERATIONS), storing and extracting information needed for device control and other purposes. Computer programs also use Boolean logic to make decisions using branching statements such as If and loop statements such as While. For example, the Basic loop

```
While (Not Eof()) OR (Line = 50)
  Read (Line$)
  Print (Line$)
  Line = Line + 1

Endwhile
```

will read and print lines from the previously opened file until *either* the Eof (end of file) function returns a value of True or the value of Line reaches 50. (In some programming languages different symbols are used for the operators. In C, for example, AND is &&, OR is ||, and NOT is !.)

Users of databases and Web search engines are also familiar with the use of Boolean statements for defining search criteria. In many search engines, the search phrase "computer science" AND "graduate" will match sites that have both the phrase "computer science" and the word "graduate," while sites that have only one or the other will either not be listed or will be listed after those that have both (see SEARCH ENGINE).

Further Reading
University at Albany Libraries. "Boolean Searching on the Internet." Available online. URL: http://www.albany.edu/library/internet/boolean.html.
Whitesitt, J. E. *Boolean Algebra and Its Applications*. New York: Dover, 1995.

boot sequence

All computers are faced with the problem that they need instructions in order to be able to read in the instructions they need to operate. The usual solution to this conundrum is to store a small program called a "loader" in a ROM (read-only memory) chip. When the computer is switched on, this chip is activated and runs the loader. The loader program has the instructions needed to be able to access the disk containing the full operating system. This process is called booting (short for "bootstrapping").

BOOTING A PC

While the details of the boot sequence vary with the hardware and operating system used, a look at the booting of a "Wintel" machine (IBM architecture PC running DOS and Microsoft Windows) can serve as a practical example.

When the power is turned on, a chip called the BIOS (basic input-output system) begins to execute a small program (see BIOS). The first thing it does is to run a routine called the POST (power-on self test) that sends a query over the system bus (see BUS) to each of the key devices (memory, keyboard, video display, and so on) for a response that indicates it is functioning properly. If an error is detected, the system generates a series of beeps, the number of which indicates the area where the problem was found, and then halts.

Assuming the test runs successfully (sometimes indicated by a single beep), the BIOS program then queries the devices to see if they have their own BIOS chips, and if so, executes their programs to initialize the devices, such as the video card and disk controllers. At this point, since the video display is available, informational and error messages can be displayed as appropriate. The BIOS also sets various parameters such as the organization of the disk drive, using information stored in a CMOS chip. (There is generally a way the user can access and change these information screens, such as when installing additional memory chips.)

The BIOS now looks for a disk drive that is bootable—that is, that contains files with the code needed to load the operating system. This is generally a hard drive, but could be a floppy disk or even a CD-ROM or USB device. (The order in which devices are checked can be configured.) On a hard drive, the code needed to start the operating system is found in a "master boot record."

The booting of the operating system (DOS) involves the determination of the disk structure and file system and the loading of the operating system kernel (found in files called IO.SYS and MSDOS.SYS), and a command interpreter (COMMAND.COM). The latter can then read the contents of the files AUTOEXEC.BAT and CONFIG.SYS, which specify system parameters, device drivers, and other programs to be loaded into memory at startup. If the system is to run Microsoft Windows, that more elaborate operating system will then take over, building upon or replacing the foundation of DOS.

Further Reading
PC Guide. "System Boot Sequence." Available online. URL: http://www.pcguide.com/_ref/mbsys/bios/bootSequence-c.html. Accessed April 10, 2008.

branching statements

The simplest calculating machines (see CALCULATOR) could only execute a series of calculations in an unalterable sequence. Part of the transition from calculator to full computer is the ability to choose different paths of execution according to particular values—in some sense, to make decisions.

Branching statements (also called decision statements or selection statements) give programs the ability to choose one or more different paths of execution depending on the results of a logical test. The general form for a branching statement in most programming languages is

```
if (Boolean expression)

statement

else statement
```

For example, a blackjack game written in C might have a statement that reads:

```
if ((Card_Count + Value(This_Card)) > 21)
  printf ("You're busted!");
```

Here the Boolean expression in parenthesis following the if keyword is evaluated. If it is true, then the following statement (beginning with printf) is executed. (The Boolean expression can be any combination of expressions, function calls, or even assignment statements, as long as they evaluate to true or false—see also BOOLEAN OPERATORS.)

The else clause allows the specification of an alternative statement to be executed if the Boolean expression is *not* true. The preceding example could be expanded to:

```
if (Card_Count + Value (This_Card) > 21)
  printf ("You're busted!");
else
  printf("Do you want another card?");
```

In most languages if statements can be nested so that a second if statement is executed only if the first one is true. For example:

```
if (Turn > Max_Turns)
  {
  if (Winner() )
    PrintScore();
  }
```

Here the first if test determines whether the maximum number of turns in the game has been exceeded. If it has, the second if statement is executed, and the Winner() function is called to determine whether there is a winner. If there is a winner, the PrintScore() function is called. This example also illustrates the general rule in most languages that wherever a single statement can be used a block of statements can also be used. (The block is delimited by braces in the C family of languages, while Pascal uses Begin . . . End.)

The switch or case statement found in many languages is a variant of the if statement that allows for easy testing of several possible values of a condition. One could write:

```
if (Category = = "A")
  AStuff();

else if (Category = = "B")
  BStuff();

else if (Category = = "C")
  CStuff();

else
  printf "(None of the above\n");
```

However, C, Pascal, and many other languages provide a more convenient multiway branching statement (called switch in C and case in Pascal). Using a switch statement, the preceding test can be rewritten in C as:

```
switch (Category) {
case "A":
  AStuff();
  break;
case "B":
  BStuff();
  break;
case "C"
  CStuff();
  break;
default:
  printf ("None of the above\n");
}
```

(Here the break statements are needed to prevent execution from continuing on through the other alternatives when only one branch should be followed.)

Further Reading
Sebesta, Robert W. *Concepts of Programming Languages*. 8th ed. Boston: Addison-Wesley, 2008.

Breazeal, Cynthia

(1968–)
American
Roboticist

Born in Albuquerque, New Mexico, in 1968, Cynthia Breazeal (pronounced like "Brazil") grew up in California. Her father was a mathematician and her mother was a computer scientist at the Lawrence Livermore Laboratory. When she was only eight, Breazeal saw the 1970s film *Star Wars* and became intrigued with the "droids."

Besides robots, as a student the young Breazeal was also fascinated by medicine and astronomy. When she attended the University of California at Santa Barbara, Breazeal considered a future career in NASA. UC also had a robotics center, and Breazeal encountered there the possibility of building planetary robot rovers.

After getting her undergraduate degree in electrical and computer engineering, Breazeal applied for graduate school to the Massachusetts Institute of Technology. The MIT robotics lab, headed by Rodney Brooks, was developing a new generation of small, agile robotic rovers based in part on observing how insects moved. Breazeal's work on two such robots, named Attila and Hannibal, helped prove the feasibility of mobile robots for planetary exploration while furnishing her a topic for her master's thesis.

Besides its implications for space research, Breazeal's work with Attila and Hannibal demonstrated the feasibility of building robots that were controlled by hundreds of small, interacting programs that detected and responded to

MIT researcher Cynthia Breazeal, shown here with her robot "Leonardo," specializes in "sociable" robots that can interact and learn much like human children. (SAM OGDEN / PHOTO RESEARCHERS, INC.)

specified conditions or "states." It gave concrete reality to Brooks's and Breazeal's belief that robots, like living organisms, grew by building more complex behaviors on top of simpler ones, rather than depending on some single top-down design.

Brooks then announced that he was starting a new project: to make a robot that could interact with people in much the same way people encounter one another socially. The result of the efforts of Brooks, Breazeal, and their colleagues was the creation of a robot called Cog. Cog attempted to replicate the sense perceptions and reasoning skills of a human infant. Cog had eyes that focused like those of a person. Like an infant, Cog could pick up on what people nearby were doing, and what they were focused on.

Breazeal had done much of the work in designing Cog's stereovision system. She and another graduate student also programmed many of the interacting feedback routines that allowed Cog to develop its often-intriguing behavior. Cog could focus on and track moving objects and sound sources. Eventually, the robot gained the kind of hand-eye coordination that enabled it to throw and catch a ball and even play rhythms on a snare drum.

For her doctoral research, Breazeal decided to design a robot unlike the 6-foot, 5-inch (1.96 m) Cog; one that instead would be more child-sized and childlike. She named the new robot Kismet, from the Turkish word for fate or fortune. Kismet looks a bit like the alien from the film *ET: The Extra-Terrestrial*. The robot is essentially a head without arms or legs. With big eyes (including exaggerated eyebrows), pink ears that can twist, and bendable surgical tubing for lips that can "smile," Kismet has a "body language" that conveys a kind of brush-stroked essence of response and emotion. Kismet has a variety of hardware and software features that support its interaction with humans.

Like Cog, Kismet's camera "eyes" function much like the human eye. However, the vision system is more sophisticated than that in the earlier robot. Kismet looks for colorful objects, which are considered to be toys, for potential play activities. An even higher priority is given to potential playmates, which are recognized by certain facial features, such as eyes, as well as the presence of flesh tones. Kismet does not actually understand the words spoken to it; however, it perceives the intonation and rhythms of human speech and identifies them as corresponding to emotional states. If a visitor addresses Kismet with tones of friendly praise (as perhaps one might a baby, or a dog), the robot moves to a "happy" emotional state. On the other hand, a harsh, scolding tone moves Kismet toward an "unhappy" condition.

Kismet's "emotions" are not just simple indicators of what state the software decides the robot should be in, based on cues it picks up from humans. Rather, the robot has been so carefully "tuned" in its feedback systems that it establishes a remarkably natural rhythm of vocalization and visual interaction. Kismet reacts to the human, which in turn elicits further human responses.

Kismet's successor is called Leonardo. Unlike Kismet, Leonardo has a full torso with arms and legs and looks rather like a furry little *Star Wars* alien. With the aid of artificial skin and an array of 32 separate motors, Leonardo's facial expressions are much more humanlike than Kismet's. Body language now includes shrugs. The robot can learn new concepts and tasks both by interacting with a human teacher and by imitating what it sees people do, starting with facial expressions and simple games.

Breazeal's group at MIT is currently investigating ways in which computers can use "body language" to communicate with users and even encourage better posture. "RoCo" is a computer whose movable "head" is a monitor screen. Using a camera, RoCo can sense the user's posture and emotional state.

Breazeal has also created "responsive" robots in new forms, and for venues beyond the laboratory. In 2003 the Cooper-Hewitt National Design Museum in New York hosted a "cyberfloral installation" designed by Breazeal. It featured "flowers" of metal and silicone that exhibit behaviors such as swaying and glowing in bright colors when a person's hand comes near.

Besides earning her a master's degree (1993) and doctoral degree (2000) from MIT, Breazeal's work has brought her considerable acclaim and numerous appearances in the media. She has been widely recognized as being a significant young inventor or innovator, such as by *Time* magazine and the *Boston Business Forward*. Breazeal is one of 100 "young innovators" featured in MIT's *Technology Review*.

Further Reading

Bar-Cohen, Yoseph, and Cynthia Breazeal. *Biologically Inspired Intelligent Robots*. Bellingham, Wash.: SPIE Press, 2003.
Biever, Celeste. "Robots Like Us: They Can Sense Human Moods." *San Francisco Chronicle,* May 6, 2007. Available online. URL: http://www.sfgate.com/cgi-bin/article.cgi?f=/c/a/2007/05/06/ING9GPK9U51.DTL. Accessed May 7, 2007.
Breazeal, Cynthia. *Designing Sociable Robots*. Cambridge, Mass.: MIT Press, 2002.
Brooks, Rodney. *Flesh and Machines: How Robots Will Change Us*. New York: Pantheon Books, 2002.
Dreifus, Claudia. "A Passion to Build a Better Robot, One with Social Skills and a Smile." *New York Times,* June 10, 2003, p. F3.
Henderson, Harry. *Modern Robotics: Building Versatile Machines*. New York: Chelsea House, 2006.
Robotic Life Group (MIT Media Lab). Available online. URL: http://robotic.media.mit.edu/. Accessed May 1, 2007.

Brin, Sergey

(1973–)
Russian-American
Entrepreneur

Cofounder and current president of technology at Google, Sergey Brin has turned the needs of millions of Web users to find information online into a gigantic and pervasive enterprise.

Brin was born in Moscow, Russia, on August 21, 1973 to a Jewish family (his father, Michael, was a mathematician and economist). However, the family immigrated to the United States in 1979, settling in Maryland. Brin's father supplemented his education, particularly in math-

ematics. Brin graduated with honors from the University of Maryland in 1993, earning a bachelor's degree in computer science and mathematics. Brin then went to Stanford, receiving his master's degree in computer science in 1995. Along the way to his Ph.D., however, Brin was "sidetracked" by his growing interest in the Internet and World Wide Web, particularly in techniques for searching for and identifying data (see also DATA MINING).

SEARCH ENGINES AND GOOGLE

The year 1995 was pivotal for Brin because he met fellow graduate student Larry Page (see PAGE, LARRY). Page shared Brin's interests in the Web, and they collaborated on a seminal paper titled "The Anatomy of a Large-Scale Hypertextual Web Search Engine." This work (including the key "PageRank" algorithm) would form the basis for the world's most widely used search engine (see GOOGLE and SEARCH ENGINE).

In 1998 Brin took a leave of absence from the Ph.D. program. The fall of that year Brin and Page launched Google. The search engine was much more useful and accurate than existing competitors, and received a Technical Excellence Award from *PC* magazine in 1999. Google soon appeared near the top of many analysts' lists of "companies to watch." In 2004 the company went public, and Brin's personal net worth is now estimated to be more than $16 billion. (Brin and Page remain closely involved with Google, promoting innovation such as the aggregation and presentation of information including images and maps.)

Besides Google, Brin's diverse interests include moviemaking (he was an executive producer of the film *Broken Arrow*) and innovative transportation (he is an investor in Tesla Motors, makers of long-range electric vehicles). In 2005 Brin was named as one of *Time* magazine's 100 most influential people. In 2007 Brin was named by *PC World* as number one on their list of the 50 most important people on the Web.

Further Reading

Brin, Sergey, and Lawrence Page. "The Anatomy of a Large-Scale Hypertextual Web Search Engine." Available online. URL: http://infolab.stanford.edu/~backrub/google.html. Accessed September 3, 2007.
"The Founders of Google." NPR *Fresh Air* interview, October 14, 2003 [audio]. Available online. URL: http://www.npr.org/templates/story/story.php?storyId=1465274. Accessed September 3, 2007.
Sergey Brin's Home Page. Available online. URL: http://infolab.stanford.edu/~sergey/. Accessed September 3, 2007.
"Sergey Brin Speaks with UC Berkeley Class" [video]. Available online. URL: http://video.google.com/videoplay?docid=7582902000166025817. Accessed September 3, 2007.

broadband

Technically, broadband refers to the carrying of multiple communications channels in a single wire or cable. In the broader sense used here, broadband refers to high-speed data transmission over the Internet using a variety of technologies (see DATA COMMUNICATIONS and TELECOMMU-

NICATIONS). This can be distinguished from the relatively slow (56 Kbps or slower) dial-up phone connections used by most home, school, and small business users until the late 1990s. A quantitative change in speed results in a qualitative change in the experience of the Web, making continuous multimedia (video and sound) transmissions possible.

BROADBAND TECHNOLOGIES

The earliest broadband technology to be developed consists of dedicated point-to-point telephone lines designated T1, T2, and T3, with speeds of 1.5, 6.3, and 44.7 Mbps respectively. These lines provide multiple data and voice channels, but cost thousands of dollars a month, making them practicable only for large companies or institutions.

Two other types of phone line access offer relatively high speed at relatively low cost. The earliest, ISDN (Integrated Services Digital Network) in typical consumer form offers two 64 Kbps channels that can be combined for 128 Kbps. (Special services can combine more channels, such as a 6 channel 384 Kbps configuration for videoconferencing.) The user's PC is connected via a digital adapter rather than the usual analog-to-digital modem.

The most common telephone-based broadband system today is the digital subscriber line (see DSL). Unlike ISDN, DSL uses existing phone lines. A typical DSL speed today is 1–2 Mbps, though higher speed services up to about 5 Mbps are now being offered. The main drawback of DSL is that the transmission rate falls off with the distance from the telephone company's central office, with a maximum distance of about 18,000 feet (5,486.4 m).

The primary alternative for most consumers uses existing television cables (see CABLE MODEM). Cable is generally a bit faster (1.5–3 Mbps) than DSL, with premium service of up to 8 Mbps or so available in certain areas. However, cable speed slows down as more users are added to a given circuit. With both DSL and cable upload speeds (the rate at which data can be sent from the user to an Internet site) are generally fixed at a fraction of download speed (often about 128 kbps). While this "throttling" of upload speed does not matter much for routine Web surfing, the growing number of applications that involve users uploading videos or other media for sharing over the Internet (see USER-CREATED CONTENT) has led to some pressure for higher upload speeds.

ULTRA BROADBAND

Rather surprisingly, the country that brought the world the Internet has fallen well behind many other industrialized nations in broadband speed. In Japan, DSL speeds up to 40 Mbps are available, and at less cost than in the United States. South Korea also offers "ultra broadband" speeds of 20 Mbps or more. American providers, on the other hand, have tended to focus on expanding their networks and competing for market share rather than investing in higher speed technologies. However, this situation is beginning to improve as American providers ramp up their investment in fiber networks (see FIBER OPTICS). For example, in 2005 Verizon introduced Fios, a fiber-based DSL service that can reach speeds up to 15 Mbps. However, installing fiber networks is expensive, and as of 2007 it was available in only about 10 percent of the U.S. market.

Cable and phone companies typically offer Internet and TV as a package—many are now including long-distance phone service (and even mobile phone service) in a "triple play" package. (For long-distance phone carried via Internet, see VOIP).

WIRELESS BROADBAND

The first wireless Internet access was provided by a wireless access point (WAP), typically connected to a wired Internet router. This is still the most common scenario in homes and public "hot spots" (see also INTERNET CAFÉS AND "HOT SPOTS"). However, with many people spending much of their time with mobile devices (see LAPTOP, PDA, and SMARTPHONE), the need for always-accessible wireless connectivity at broadband speeds has been growing. The largest U.S. service, Nextlink, offered wireless broadband in 37 markets in 2007 (including many large and mid-sized cities) at speeds starting at 1.5 Mbps. An alternative is offered by cell phone companies such as Verizon and Sprint, which "piggy back" on the existing infrastructure of cell phone towers. However, the speed of this "3G" service is slower, from 384 kbps up to 2 Mbps.

Yet another alternative beginning to appear is WiMAX, a technology that is conceptually similar to Wifi but has much greater range because its "hot spots" can be many miles in diameter. WiMAX offers the possibility of covering entire urban areas with broadband service, although questions about its economic viability have slowed implementation as of 2008.

Satellite Internet services have the advantage of being available over a wide area. The disadvantage is that there is about a quarter-second delay for the signal to travel from a geostationary satellite at an altitude of 22,300 km. (Lower-altitude satellites can be used to reduce this delay, but then more satellites are needed to provide continuous coverage.)

ADOPTION AND APPLICATIONS

By mid-2007, 53 percent of adult Americans had a broadband connection at home. This amounts to 72 percent of home Internet users. (About 61 percent of broadband connections used cable and about 37 percent DSL.)

With dial-up connections declining to less than 25 percent, Web services are increasingly designed with the expectation that users will have broadband connections. This, however, has the implication that users such as rural residents and the inner-city poor may be subjected to a "second class" Web experience (see also DIGITAL DIVIDE). Meanwhile, as with connection speed, many other countries now surpass the United States in the percentage of broadband users.

Broadband Internet access is virtually a necessity for many of the most innovative and compelling of today's Internet applications. These include downloading media (see PODCASTING, STREAMING, and MUSIC AND VIDEO DISTRIBUTION, ONLINE), uploading photos or videos to sites such as Flickr and YouTube, using the Internet as a substitute for a traditional phone line (see VOIP), and even gaming (see ONLINE GAMES). Broadband is thus helping drive the integration of many forms of media (see DIGITAL CONVERGENCE) and the continuous connectivity that an increasing number of people seem to be relying on (see UBIQUITOUS COMPUTING).

Further Reading

Bates, Regis. *Broadband Telecommunications Handbook.* 2nd ed. New York: McGraw-Hill, 2002.
Bertolucci, Jeff. "Broadband Expands." *PC World* (August 2007): 77–90.
Cybertelecom. "Statistics: Broadband." Available online. URL: http://www.cybertelecom.org/data/broadband.htm. Accessed July 17, 2007.
Gaskin, James E. *Broadband Bible.* New York: Wiley, 2004.
Hellberg, Chris, Dylan Greene, and Truman Boyes. *Broadband Network Architectures: Designing and Deploying Triple-Play Services.* Upper Saddle River, N.J.: Prentice Hall, 2007.

Brooks, Rodney
(1954–   )
Australian, American
Roboticist

Rodney Brooks's ideas about robots have found their way into everything from vacuum cleaners to Martian rovers. Today, as director of the Artificial Intelligence Laboratory at the Massachusetts Institute of Technology, Brooks has extended his exploration of robot behavior into new approaches to artificial intelligence.

Brooks was born in Adelaide, Australia, in 1954. As a boy he was fascinated with computers, but it was still the mainframe era, and he had no access to them. Brooks decided to build his own logic circuits from discarded electronics modules from the defense laboratory where his father worked. Brooks also came across a book by Grey Walter, inventor of the "cybernetic tortoise" in the late 1940s. He tried to build his own and came up with "Norman," a robot that could track light sources while avoiding obstacles. In 1968, when young Brooks saw the movie *2001: A Space Odyssey,* he was fascinated by the artificial intelligence of its most tragic character, the computer HAL 9000 (see ARTIFICIAL INTELLIGENCE and ROBOTICS).

Brooks majored in mathematics at Flinders University in South Australia, where he designed a computer language and development system for artificial intelligence projects. He also explored various AI applications such as theorem solving, language processing, and games. He was then able to go to Stanford University in Palo Alto, California, in 1977 as a research assistant.

While working for his Ph.D. in computer science, awarded in 1981, Brooks met John McCarthy, one of the "elder statesmen" of AI in the Stanford Artificial Intelligence Lab (SAIL). He also joined in the innovative projects being conducted by researchers such as Hans Moravec, who were revamping the rolling robot called the Stanford Cart and teaching it to navigate around obstacles.

In 1984 Brooks moved to the Massachusetts Institute of Technology. For his Ph.D. research project, Brooks and his fellow graduate students equipped a robot with a ring of sonars (adopted from a camera rangefinder) plus two cameras. The cylindrical robot was about the size of R2D2 and was connected by cable to a minicomputer. However, the calculations needed to enable a robot to identify objects as they appear at different angles were so intensive that the robot could take hours to find its way across a room.

Brooks decided to take a lesson from biological evolution. He realized that as organisms evolved into more complex forms, they could not start from scratch each time they added new features. Rather, new connections (and ways of processing them) would be added to the existing structure. For his next robot, called Allen, Brooks built three "layers" of circuits that would control the machine's behavior. The simplest layer was for avoiding obstacles: If a sonar signal said that something was too close, the robot would change direction to avoid a collision. The next layer generated a random path so the robot could "explore" its surroundings freely. Finally, the third layer was programmed to identify specified sorts of "interesting" objects. If it found one, the robot would head in that direction.

Each of these layers or behaviors was much simpler than the complex calculations and mapping done by a traditional AI robot. Nevertheless, the layers worked together in interesting ways. The result would be that the robot could explore a room, avoiding both fixed and moving obstacles, and appear to "purposefully" search for things.

In the late 1980s, working with Grinell More and a new researcher, Colin Angle, Brooks built an insectlike robot called Genghis. Unlike Allen's three layers of behavior, Genghis had 51 separate, simultaneously running computer programs. These programs, called "augmented finite state machines," each kept track of a particular state or condition, such as the position of one of the six legs. It is the interaction of these small programs that creates the robot's ability to scramble around while keeping its balance. Finally, three special programs looked for signals from the infrared sensors, locked onto any source found, and walked in its direction.

Brooks's new layered architecture for "embodied" robots offered new possibilities for autonomous robot explorers. Brooks's 1989 paper, "Fast, Cheap, and Out of Control: A Robot Invasion of the Solar System," envisaged flocks of tiny robot rovers spreading across the Martian surface, exploring areas too risky when one has only one or two very expensive robots. The design of the *Sojourner* Mars rover and its successors, *Spirit* and *Opportunity,* would partially embody the design principles developed by Brooks and his colleagues.

In the early 1990s Brooks and his colleagues began designing Cog, a robot that would embody human eye movement and other behaviors. Cog's eyes are mounted on gimbals so they can easily turn to track objects, aided by the movement of the robot's head and neck (it has no legs). Cog also has "ears"—microphones that can help it find the source of a sound. The quest for more humanlike robots continued in the late 1990s with the development of

Kismet, a robot that includes dynamically changing "emotions." Brooks's student Cynthia Breazeal would build her own research career on Kismet and what she calls "sociable robots" (see BREAZEAL, CYNTHIA).

By 1990, Brooks wanted to apply his ideas of behavior-based robotics to building marketable robots that could perform basic but useful tasks, and he enlisted two of his most innovative and hard-working students, Colin Angle and Helen Greiner (see IROBOT CORPORATION). The company is best known for the Roomba robotic vacuum cleaner. Brooks remains the company's chief technical officer.

Meanwhile Brooks has an assured place as one of the key innovators in modern robotics research. He is a Founding Fellow of the American Association for Artificial Intelligence and a Fellow of the American Association for the Advancement of Science. Brooks received the 1991 Computers and Thought Award of the International Joint Conference on Artificial Intelligence. He has participated in numerous distinguished lecture series and has served as an editor for many important journals in the field, including the *International Journal of Computer Vision.*

Further Reading

Brockman, John. "Beyond Computation." Edge 2000. Available online. URL: http://www.edge.org/3rd_culture/brooks_beyond/beyond_index.html. Accessed May 3, 2007.
———. "The Deep Question: A Talk with Rodney Brooks." Edge 29 (November 19, 1997). Available online. URL: http://www.edge.org/documents/archive/edge29.html. Accessed May 3, 2007.
Brooks, Rodney. *Flesh and Machines: How Robots Will Change Us.* New York: Pantheon Books, 2002.
Computer Science and Artificial Intelligence Laboratory (CSAIL), MIT. Available online. URL: http://www.csail.mit.edu/index.php. Accessed May 3, 2007.
Henderson, Harry. *Modern Robotics: Building Versatile Machines.* New York: Chelsea House, 2006.
O'Connell, Sanjida. "Cog—Is It More than a Machine?" *London Times* (May 6, 2002): 10. Rodney Brooks [homepage]. CSAIL. Available online. URL: http://people.csail.mit.edu/brooks/. Accessed May 3, 2007.
"Rodney Brooks—The Past and Future of Behavior Based Robotics" [Podcast]. Available online. URL: http://lis.epfl.ch/resources/podcast/mp3/TalkingRobots-RodneyBrooks.mp3. Accessed May 3, 2007.

buffering

Computer designers must deal with the way different parts of a computer system process data at different speeds. For example, text or graphical data can be stored in main memory (RAM) much more quickly than it can be sent to a printer, and in turn data can be sent to the printer faster than the printer is able to print the data. The solution to this problem is the use of a buffer (sometimes called a spool), or memory area set aside for the temporary storage of data. Buffers are also typically used to store data to be displayed (video buffer), to collect data to be transmitted to (or received from) a modem, for transmitting audio or video content (see STREAMING) and for many other devices (see INPUT/OUTPUT). Buffers can also be used for data that must be reorganized in some way before it can be further processed. For example, character data is stored in a communications buffer so it can be serialized for transmission.

BUFFERING TECHNIQUES

The two common arrangements for buffering data are the pooled buffer and the circular buffer. In the pool buffer, multiple buffers are allocated, with the buffer size being equal to the size of one data record. As each data record is received, it is copied to a free buffer from the pool. When it is time to remove data from the buffer for processing, data is read from the buffers in the order in which it had been stored (first in, first out, or FIFO). As a buffer is read, it is marked as free so it can be used for more incoming data.

In the circular buffer there is only a single buffer, large enough to hold a number of data records. The buffer is set up as a queue (see QUEUE) to which incoming data records are written and from which they are read as needed for processing. Because the queue is circular, there is no "first" or "last" record. Rather, two pointers (called In and Out) are maintained. As data is stored in the buffer, the In pointer is incremented. As data is read back from the buffer, the Out pointer is incremented. If either pointer reaches around back to the beginning, it begins to wrap around. The software managing the buffer must make sure that if the In pointer goes past the Out pointer, then the Out pointer must not go past In. Similarly, if Out goes past In, then In must not go past Out.

The fact that programmers sometimes fail to check for buffer overflows has resulted in a seemingly endless series of security vulnerabilities, such as in earlier versions of the UNIX sendmail program. In one technique, attackers can use a too-long value to write data, or worse, commands into the areas that control the program's execution, possibly taking over the program (see also COMPUTER CRIME AND SECURITY).

Buffering is conceptually related to a variety of other techniques for managing data. A disk cache is essentially a special buffer that stores additional data read from a disk in anticipation that the consuming program may soon request it. A processor cache stores instructions and data in anticipation of the needs of the CPU. Streaming of multimedia (video or sound) buffers a portion of the content so it can be played smoothly while additional content is being received from the source.

Depending on the application, the buffer can be a part of the system's main memory (RAM) or it can be a separate memory chip or chips onboard the printer or other device. Decreasing prices for RAM have led to increases in the typical size of buffers. Moving data from main memory to a peripheral buffer also facilitates the multitasking feature found in most modern operating systems, by allowing applications to buffer their output and continue processing.

Further Reading

Buffer Overflow [articles]. Available online. URL: http://doc.bughunter.net/buffer-overflow/. Accessed May 23, 2007.
Grover, Sandeep. "Buffer Overflow Attacks and Their Countermeasures." *Linux Journal,* March 3, 2003. Available online. URL: http://www.linuxjournal.com/article/6701. Accessed May 23, 2007.

bugs and debugging

In general terms a bug is an error in a computer program that leads to unexpected and unwanted behavior. (Lore has it that the first "bug" was a burnt moth found in the relays of the early Mark I computer in the 1940s; however, as early as 1878 Thomas Edison had referred to "bugs" in the design of his new inventions.)

Computer bugs can be divided into two categories: syntax errors and logic errors. A syntax error results from failing to follow a language's rules for constructing statements, or from using the wrong symbol. For example, each statement in the C language must end with a semicolon. This sort of syntax error is easily detected and reported by modern compilers, so fixing it is trivial.

A logic error, on the other hand, is a syntactically valid statement that does not do what was intended. For example, if a C programmer writes:

```
if Total = 100
```

instead of

```
if Total == 100
```

the programmer may have intended to test the value of Total to see if it is 100, but the first statement actually *assigns* the value of 100 to Total. That's because a single equals sign in C is the assignment operator; testing for equality requires the double equals sign. Further, the error will result in the if statement always being true, because the truth value of an assignment is the value assigned (100 in this case) and any nonzero value is considered to be "true" (see BRANCHING STATEMENTS).

Loops and pointers are frequent sources of logical errors (see LOOP and POINTERS AND INDIRECTION). The boundary condition of a loop can be incorrectly specified (for example, < 10 when < = 10 is wanted). If a loop and a pointer or index variable are being used to retrieve data from an array, pointing beyond the end of the array will retrieve whatever data happens to be stored out there.

Errors can also be caused in the conversion of data of different types (see DATA TYPES). For example, in many language implementations the compiler will automatically convert an integer value to floating point if it is to be assigned to a floating point variable. However, while an integer can retain at least nine decimal digits of precision, a float may only be able to guarantee seven. The result could be a loss of precision sufficient to render the program's results unreliable, particularly for scientific purposes.

DEBUGGING TECHNIQUES

The process of debugging (identifying and fixing bugs) is aided by the debugging features integrated into most modern programming environments. Some typical features include the ability to set a *breakpoint* or place in the code where the running program should halt so the values of key variables can be examined. A *watch* can be set on specified certain variables so their changing values will be displayed as the program executes. A *trace* highlights the source code to show what statements are being executed as the program runs. (It can also be set to follow execution into and through any procedures or subroutines called by the main code.)

During the process of software development, debugging will usually proceed hand in hand with software testing. Indeed, the line between the two can be blurry. Essentially, debugging deals with fixing problems so that the program is doing what it intends to do, while testing determines whether the program's performance adequately meets the needs and objectives of the end user.

Further Reading

Agans, David J. *Debugging: The Nine Indispensable Rules for Finding Even the Most Elusive Software and Hardware Problems.* New York: AMACOM, 2002.

Robbins, John. *Debugging Applications.* Redmond, Wash.: Microsoft Press, 2000.

Rosenberg, Jonathan B. *How Debuggers Work: Algorithms, Data Structures, and Architecture.* New York: Wiley, 1996.

bulletin board systems (BBS)

An electronic bulletin board is a computer application that lets users access a computer (usually with a modem and phone line) and read or post messages on a variety of topics. The messages are often organized by topic, resulting in *threads* of postings, responses, and responses to the responses. In addition to the message service, many bulletin boards provide files that users can download, such as games and other programs, text documents, pictures, or sound files. Some bulletin boards expect users to upload files to contribute to the board in return for the privilege of downloading material.

The earliest form of bulletin board appeared in the late 1960s in government installations and a few universities participating in the Defense Department's ARPANET (the ancestor to the Internet). As more universities came online in the early 1970s, the Netnews (or Usenet) system offered a way to use UNIX file-transfer programs to store messages in topical newsgroups (see NETNEWS AND NEWSGROUPS). The news system automatically propagated messages (in the form of a "news feed") from the site where they were originally posted to regional nodes, and from there throughout the network.

By the early 1980s, a significant number of personal computer users were connecting modems to their PCs. Bulletin board software was developed to allow an operator (called a "sysop") to maintain a bulletin board on his or her PC. Users (one or a few at a time) could dial a phone number to connect to the bulletin board. In 1984, programmer Tom Jennings developed the Fido BBS software, which allowed participating bulletin boards to propagate postings in a way roughly similar to the distribution of UNIX Netnews messages.

DECLINE OF THE BBS

In the 1990s, two major developments led to a drastic decline in the number of bulletin boards. The growth of major services such as America Online and CompuServe (see ONLINE SERVICES) offered users a friendlier user interface, a comprehensive selection of forums and file downloads, and richer content than bulletin boards with their character-based interface and primitive graphics. An even greater impact resulted from the development of the World Wide Web and Web browsing software, which offered access to a worldwide smorgasbord of services in which each Web home page had the potential of serving as a virtual bulletin board and resource center (see WORLD WIDE WEB and WEB BROWSER). As the 1990s progressed, increasingly rich multimedia content became available over the Internet in the form of streaming video, themed "channels," and the sharing of music and other media files.

Traditional bulletin boards are now found mostly in remote and underdeveloped areas (where they can provide users who have only basic phone service and perhaps obsolescent PCs with an e-mail gateway to the Internet). However the BBS contributed much to the grassroots online culture, providing a combination of expansive reach and a virtual small-town atmosphere (see also VIRTUAL COMMUNITY). Venues such as The Well (see CONFERENCING SYSTEMS) retain much of the "feel" of the traditional bulletin board system.

Further Reading
"The BBS Corner." Available online. URL: http://www.dmine.com/bbscorner/. Accessed August 14, 2007.

Byrant, Alan D. *Growing and Maintaining a Successful BBS: The Sysop's Handbook.* Reading, Mass.: Addison-Wesley, 1995.

The BBS History Library. Available online. URL: http://www.bbshistory.org/. Accessed May 23, 2007.

O'Hara, Robert. *Commodork: Sordid Tales from a BBS Junkie.* Morrisville, N.C.: Lulu.com, 2006.

Sanchez, Julian. "The Prehistory of Cyberspace: How BBSes Paved the Way for the Web." *Reason* 37 (December 1, 2005): 61 ff. Available online. URL: http://www.reason.com/news/show/36324.html. Accessed May 23, 2007.

bus

A computer bus is a pathway for data to flow between the central processing unit (CPU), main memory (RAM), and various devices such as the keyboard, video, disk drives, and communications ports. Connecting a device to the bus allows it to communicate with the CPU and other components without there having to be a separate set of wires for each device. The bus thus provides for flexibility and simplicity in computer architecture.

Mainframe computers and large minicomputers typically have proprietary buses that provide a wide multipath connection that allows for data transfer rates from about 3 MB/s to 10 MB/s or more. This is in keeping with the use of mainframes to process large amounts of data at high speeds (see MAINFRAME).

MICROCOMPUTER BUSES
The bus played a key role in the development of the modern desktop computer in the later 1970s and 1980s. In the microcomputer, the bus is fitted with connectors called expansion slots, into which any expansion card that meets connection specifications can be inserted. Thus the S-100 bus made it possible for microcomputer pioneers to build

a variety of systems with cards to expand the memory and add serial and parallel ports, disk controllers, and other devices. (The Apple II had a similar expansion capability.) In 1981, when IBM announced its first PC, it also defined an 8-bit expansion bus that became known as the ISA (Industry Standard Architecture) as other companies rushed to "clone" IBM's hardware.

In the mid-1980s, IBM advanced the industry with the AT (Advanced Technology) machine, which had the 16-bit Intel 80286 chip and an expanded bus that could transmit data at up to 2 MB/s. The clone manufacturers soon matched and exceeded these specifications, however. IBM responded by trying both to improve the microcomputer bus and to define a proprietary standard that it could control via licensing. The result was called the Micro-Channel Architecture (MCA), which increased data throughput to 20 MB/s with full 32-bit capability. This bus had other advanced features such as a direct connection to the video system (Video Graphics Array) and the ability to configure cards in software rather than having to set physical switches. In addition, cards could now incorporate their own processors and memory in a way similar to that of their powerful mainframe counterparts (this is called *bus mastering*). Despite these advantages, however, the proprietary nature of the MCA and the fact that computers using this bus could not use any of the hundreds of ISA cards led to a limited market share for the new systems.

Instead of paying IBM and adopting the new standard, nine major clone manufacturers joined to develop the EISA (Extended ISA) bus. EISA was also a 32-bit bus, but its maximum transfer rate of 33 MB/s made it considerably faster than the MCA. It was tailored to the new Intel 80386 and 80486 processors, which supported the synchronous transfer of data in rapid bursts. The EISA matched and exceeded the MCA's abilities (including bus mastering and no-switch configuration), but it also retained the ability to use older ISA expansion cards. The EISA soon became the industry standard as the Pentium family of processors were introduced.

However, the endless hunger for more data-transfer capability caused by the new graphics-oriented operating systems such as Microsoft Windows led to the development

A Standard ISA bus PC expansion card. This "open architecture" allowed dozens of companies to create hundreds of add-on devices for IBM-compatible personal computers.

of *local buses*. A local bus is connected to the processor's memory bus (which typically runs at half the processor's external speed rather than the much slower system bus speed), a considerable advantage in moving data (such as graphics) from main memory to the video card.

Two of these buses, the VESA (or VL) bus and the PCI bus came into widespread use in higher-end machines, with the PCI becoming dominant. The PCI bus runs at 33 MHz and supports features such as Plug and Play (the ability to automatically configure a device, supported in Windows 98 and later) and Hot Plug (the ability to connect or reconnect devices while the PC is running). The PCI retains compatibility with older 8-bit and 16-bit ISA expansion cards. At the end of the 1990s, PC makers were starting to introduce even faster buses such as the AGP (accelerated graphics port), which runs at 66 MHz.

Two important auxiliary buses are designed for the connection of peripheral devices to the main PC bus. The older SCSI (Small Computer Systems Interface) was announced in 1986 (with the expanded SCSI-2 in 1994). SCSI is primarily used to connect disk drives and other mass storage devices (such as CD-ROMs), though it can be used for scanners and other devices as well. SCSI-2 can transfer data at 20 MB/s over a 16-bit path, and SCSI-3 (still in development) will offer a variety of high-speed capabilities. SCSI was adopted as the standard peripheral interface for many models of Apple Macintosh computers as well as UNIX workstations. On IBM architecture PCs SCSI is generally used for servers that require large amounts of mass storage. Multiple devices can be connected in series (or "chained").

The newer USB (Universal Serial Bus) is relatively slow (12 MB/s) but convenient because a simple plug can be inserted directly into a USB socket on the system board or the socket can be connected to a USB hub to which several devices can be connected. In 2002, USB 2.0 entered the marketplace. It offers 480 MB/s data transfer speed. (See USB.)

It is uncertain whether the next advance will be the adoption of a 64-bit PCI bus or the development of an entirely different bus architecture. The latter is attractive as a way to get past certain inherent bottlenecks in the PCI design, but the desire for downward compatibility with the huge number of existing ISA, EISA, and PCI devices is also very strong.

Further Reading
PC Guide. "System Buses." Available online. URL: http://www.pcguide.com/ref/mbsys/_buses/index.htm. Accessed May 23, 2007.

Bush, Vannevar
(1890–1974)
American
Engineer and Inventor

Vannevar Bush, grandson of two sea captains and son of a clergyman, was born in Everett, Massachusetts, just outside of Boston. Bush earned his B.S. and M.S. degrees in engineering at Tufts University, and received a joint doctorate from Harvard and MIT in 1916. He went on to full professorship at MIT and became dean of its Engineering School in 1932.

Bush combined an interest in mathematics with the design of mechanical devices to automate calculations. During his undergraduate years he invented an automatic surveying machine using two bicycle wheels and a recording instrument. His most important invention was the differential analyzer, a special type of computer that used combinations of rotating shafts and cams to incrementally add or subtract the differences needed to arrive at a solution to the equation (see also ANALOG COMPUTER). His improved model (Rockefeller Differential Analyzer, or RDA2) replaced the shafts and gears with an electrically-driven system, but the actual integrators were still mechanical. Several of these machines were built in time for World War II, when they served for such purposes as calculating tables of ballistic trajectories for artillery.

Later, Bush turned his attention to problems of information processing. Together with John H. Howard (also of MIT), he invented the Rapid Selector, a device that could retrieve specific information from a roll of microfilm by scanning for special binary codes on the edges of the film. His most far-reaching idea, however, was what he called the "Memex"—a device that would link or associate pieces of information with one another in a way similar to the associations made in the human brain. Bush visualized this as a desktop workstation that would enable its user to explore the world's information resources by following links, the basic principle of what would later become known as hypertext (see HYPERTEXT AND HYPERMEDIA).

In his later years, Bush wrote books that became influential as scientists struggled to create large-scale research teams and to define their roles and responsibilities in the cold war era. He played the key role in establishing the National Science Foundation in 1950, and served on its advisory board from 1953 to 1956. He then became CEO of the drug company Merck (1955–1962) as well as serving as chairman (and then honorary chairman) of the MIT Corporation (1957–1974).

Bush would receive numerous honorary degrees and awards that testified to the broad range of his interests and achievements not only in electrical and mechanical engineering, but also in social science. In 1964, he received the National Medal of Science. Bush died on June 28, 1974, in Belmont, Massachusetts.

Further Reading
Bush, Vannevar. *Pieces of the Action*. New York: William Morrow, 1970.
———. *Science: The Endless Frontier*. Washington, D.C.: U.S. Government Printing Office, 1945.
Nyce, J. M., and P. Kahn. *From Memex to Hypertext: Vannevar Bush and the Mind's Machine*. Boston: Academic Press, 1991. [Includes two essays by Bush: "As We May Think" and "Memex II."]
Zachary, G. Pascal. *Endless Frontier: Vannevar Bush, Engineer of the American Century*. Cambridge, Mass.: MIT Press, 1999.

business applications of computers
Efficient and timely data processing is essential for businesses of all sizes from corner shop to multinational corporation.

Business applications can be divided into the broad categories of Administration, Accounting, Office, Production, and Marketing and Sales.

Administrative applications deal with the organization and management of business operations. This includes personnel-related matters (recruiting, maintenance of personnel records, payroll, pension plans, and the provision of other benefits such as health care). It also includes management information or decision support systems, communications (from simple e-mail to teleconferencing), and the administration of the data processing systems themselves.

The Accounting category includes databases of accounts receivable (money owed to the firm) and payable (such as bills from vendors). While this software is decidedly unglamorous, in a large corporation small inefficiencies can add up to significant costs or lost revenue. (For example, paying a bill before it is due deprives the firm of the "float" or interest that can be earned on the money, while paying a bill too late can lead to a loss of discounts or the addition of penalties.) A variety of reports must be regularly generated so management can spot such problems and so taxes and regulatory requirements can be met.

The Office category involves the production and tracking of documents (letters and reports) as required for the day-to-day operation of the business. Word processing, desktop publishing, presentation and other software can be used for this purpose (see APPLICATION SUITE, WORD PROCESSING, SPREADSHEET, and PRESENTATION SOFTWARE).

Production is a catchall term for the actual product or service that the business provides. For a manufacturing business this may require specialized design and manufacturing programs (see COMPUTER-AIDED DESIGN AND MANUFACTURING CAD/CAM) as well as software for tracking and scheduling the completion of tasks. For a business that markets already produced goods the primary applications will be in the areas of transportation (tracking the shipping of goods [see also SUPPLY CHAIN MANAGEMENT]), inventory and warehousing, and distribution. Service businesses will need to establish accounts for customers and keep track of the services performed (on an hourly basis or otherwise).

Marketing and Sales includes market research, advertising, and other programs designed to make the public aware of and favorably disposed to the product or service (see CUSTOMER RELATIONSHIP MANAGEMENT). Once people come to the store to buy something, the actual retail transaction must be provided for, including the point-of-sale terminal (formerly "cash register") with its interface to the store inventory system and the verification of credit cards or other forms of payment.

CHANGING ROLE OF COMPUTERS

Computer support for business functions can be provided in several forms. During the 1950s and 1960s (the era of mainframe dominance), only the largest firms had their own computer facilities. Many medium- to small-sized businesses contracted with agencies called service bureaus to provide computer processing for such functions as payroll processing. Service bureaus and in-house data processing facilities often developed their own software (typically using the COBOL language).

The development of the minicomputer (and in the 1980s, the desktop microcomputer) allowed more businesses to undertake their own data processing, in the expectation (not always fulfilled) that they would be able both to save money and to create systems better tailored to their needs. Areas such as payroll and accounts payable/receivable generally still relied upon specialized software packages. However, the growing availability of powerful database software (such as dBase and its descendants) as well as spreadsheet programs enabled businesses to maintain and report on a variety of information.

During the 1980s, the daily life of the office began to change in marked ways. The specialized word processing machines gave way to programs such as WordStar, Word-Perfect, and Microsoft Word running on desktop computers. Advanced word processing and desktop publishing software moved more of the control of the appearance of documents into the hands of office personnel. The local area network (LAN) made it possible to share resources (such as the new laser printers and databases on a powerful file server PC) as well as providing for communication in the form of e-mail.

As the Internet and the World Wide Web came into prominence in the later 1990s, another revolution was soon under way. Every significant organization is now expected to have its own Web site or sites. These Web pages serve a Janus-like function. On the one hand, they present the organization's face to the world, providing announcements, advertising, catalogs, and the capability for online purchasing (e-commerce). On the other hand, many organizations now put their databases and other records on Web sites (in secured private networks) so that employees can readily access and update them. The growth in mobile computing and readily available Internet connections (including wireless services) increasingly enables traveling business-persons to effectively take the office and its resources with them on the road.

Further Reading
Bodnar, George H,. and William S. Hopwood. *Accounting Information Systems*. Upper Saddle River, N.J.: Prentice Hall, 2000.

Cortada, James W. *21st Century Business: Managing and Working in the New Digital Economy*. Upper Saddle River, N.J.: Prentice Hall, 2000.

O'Brien, James A. *Introduction to Information Systems*. New York: McGraw-Hill, 2000.

C

The C programming language was developed in the early 1970s by Dennis Ritchie, who based it on the earlier languages BCPL and B. C was first used on DEC PDP-11 computers running the newly developed UNIX operating system, where the language provided a high-level alternative to the use of PDP Assembly language for development of the many utilities that give UNIX its flexibility. Since the 1980s, C and its descendent, C++, have become the most widely used programming languages.

LANGUAGE FEATURES

Like the earlier Algol and the somewhat later Pascal, C is a procedural language that reflects the philosophy of programming that was gradually taking shape during the 1970s (see STRUCTURED PROGRAMMING). In general, C's approach can be described as providing the necessary features for real world computing in a compact and efficient form. The language provides the basic control structures such as if and switch (see BRANCHING STATEMENTS) and while, do, and for (see LOOP). The built-in data types provide for integers (int, short, and long), floating-point numbers (float and double), and characters (char). An array of any type can be declared, and a string is implemented as an array of char (see DATA TYPES and CHARACTERS AND STRINGS).

Pointers (references to memory locations) are used for a variety of purposes, such as for storing and retrieving data in an array (see POINTERS AND INDIRECTION). While the use of pointers can be a bit difficult for beginners to under-

stand, it reflects C's emphasis as a systems programming language that can "get close to the hardware" in manipulating memory.

Data of different types can be combined into a record type called a struct. Thus, for example:

```
struct Employee_Record {
  char [10] First_Name;
  char [1] Middle_Initial;
  char [20] Last_Name;
  int Employee_Number;
} ;
```

(There is also a union, which is a struct where the same structure can contain one of two different data items.)

The standard mathematical and logical comparison operators are available. There are a couple of quirks: the equals comparison operator is = =, while a single equal sign = is an assignment operator. This can create a pitfall for the wary, since the condition

```
if (Total = 10)
  printf ("Finished!");
```

always prints Finished, since the assignment Total = 10 returns a value of 10 (which not being zero, is "true" and satisfies the if condition).

C also features an increment ++ and decrement - - operator, which is convenient for the common operation of raising or lowering a variable by one in a counting loop. In C the following statements are equivalent:

```
Total = Total + 1;
Total += 1;
Total ++;
```

Unlike Pascal's two separate kinds of procedures (func, or function, which returns a value, and proc, or procedure, which does not), C has only functions. Arguments are passed to functions by value, but can be passed by reference by using a pointer. (See PROCEDURES AND FUNCTIONS.)

SAMPLE PROGRAM

The following is a brief example program:

```
#include <stdio.h>
float Average (void);
main () {
printf ("The average is: %f", Average() );
}
float Average (void) {
int NumbersRead = 0;
int Number;
int Total = 0;
while (scanf("%d\n", &Number) == 1)
   {
     Total = Total + Number;
     NumbersRead = NumbersRead + 1;
   }
return (Total / NumbersRead);
}
}
```

Statements at the beginning of the program that begin with # are preprocessor directives. These make changes to the source code before it is compiled. The #include directive adds the specified source file to the program. Unlike many other languages, the C language itself does not include many basic functions, such as input/output (I/O) statements. Instead, these are provided in standard libraries. (The purpose of this arrangement is to keep the language itself simple and portable while keeping the implementation of functions likely to vary on different platforms separate.) The stdio.h file here is a "header file" that defines the I/O functions, such as printf() (which prints formatted data) and scanf() (which reads data into the program and formats it).

The next part of the program declares any functions that will be defined and used in the program (in this case, there is only one function, Average). The function declaration begins with the type of data that will be returned by the function to the calling statement (a floating point value in this case). After the function name comes declarations for any parameters that are to be passed to the function by the caller. Since the Average function will get its data from user input rather than the calling statement, the value (void) is used as the parameter.

Following the declaration of Average comes the main() function. Every C program must have a main function. Main is the function that runs when the program begins to execute. Typically, main will call a number of other functions to perform the necessary tasks. Here main calls Average

within the printf statement, which will print the average as returned by that function. (Calling functions within other statements is an example of C's concise syntax.)

Finally, the Average function is defined. It uses a loop to read in the data numbers, which are totaled and then divided to get the average, which is sent back to the calling statement by the return statement.

A programmer could create this program on a UNIX system by typing the code into a source file (test.c in this case) using a text editor such as vi. A C compiler (gcc in this case) is then given the source code. The source code is compiled, and linked, creating the executable program file a.out. Typing that name at the command prompt runs the program, which asks for and averages the numbers.

```
% gcc test.c
% a.out
5
7
9
.
```

The average is: 7.000000

SUCCESS AND CHANGE

In the three decades after its first appearance, C became one of the most successful programming languages in history. In addition to becoming the language of choice for most UNIX programming, as microcomputers became capable of running high-level languages, C became the language of choice for developing MS-DOS, Windows, and Macintosh programs. The application programming interface (API) for Windows, for example, consists of hundreds of C functions, structures, and definitions (see APPLICATION PROGRAMMING INTERFACE and MICROSOFT WINDOWS).

However, C has not been without its critics among computer scientists. Besides containing idioms that can encourage cryptic coding, the original version of C (as defined in Kernighan and Ritchie's *The C Programming Language*) did not check function parameters to make sure they matched the data types expected in the function definitions. This problem led to a large number of hard-to-catch bugs. However, the development of ANSI standard C with its stricter requirements, as well as type checking built into compilers has considerably ameliorated this problem. At about the same time, C++ became available as an object-oriented extension and partial rectification of C. While C++ and Java have considerably supplanted C for developing new programs, C programmers have a relatively easy learning path to the newer languages and the extensive legacy of C code will remain useful for years to come.

Further Reading

Kernighan, B. W., and D. M. Ritchie. *The C Programming Language,* 2nd ed. Upper Saddle River, N.J.: Prentice-Hall, 1988.
Prata, Stephen. *C Primer Plus.* 5th ed. Indianapolis: SAMS, 2004.
Ritchie, D. M. "The Development of the C Language," in *History of Programming Languages II,* ed. T. J. Bergin and R. G. Gibson, 678–698. Reading, Mass.: Addison-Wesley, 1995.

C#

Introduced in 2002, C# (pronounced "C sharp") is a programming language similar to C++ and Java but simplified in several respects and tailored for use with Microsoft's latest programming platform (see MICROSOFT.NET). C# is a general-purpose language and is thoroughly object-oriented—all functions must be declared as members of a class or "struct," and even fundamental data types are derived from the System.Object class (see CLASS and OBJECT-ORIENTED PROGRAMMING).

Compared with C++, C# is stricter about the use and conversion of data types, not allowing most implicit conversions (such as from an enumeration type to the corresponding integer—see DATA STRUCTURES). Unlike C++, C# does not permit multiple inheritance (where a type can be derived from two or more base types), thereby avoiding an added layer of complexity in class relationships in large software projects. (However, a similar effect can be obtained by declaring multiple "interfaces" or specified ways of accessing the same class.)

Unlike Java (but like C++), C# includes pointers (and a safer version called "delegates"), enumerations (enum types), structs (treated as lightweight classes), and overloading (multiple definitions for operators). The latest version of the language, C# 3.0 (introduced in 2007), provides additional features for list processing and functional programming (see FUNCTIONAL LANGUAGES).

The canonical "Hello World" program looks like this in C#:

```
using System;
// A "Hello World!" program in C#
namespace HelloWorld
{
    class Hello
    {
        static void Main()
        {
        System.Console.WriteLine("Hello World!");
    }
}
}
```

Essentially all program structures must be part of a class. The first statement brings in the System class, from which are derived basic interface methods. A program can have one or more namespaces, which are used to organize classes and other structures to avoid ambiguity. This program has only one class (Hello), which includes a Main function (every program must have one and only one). This function calls the Console member of the System class, and in turn uses the WriteLine method to display the text.

C++ AND MICROSOFT DEVELOPMENT

C# is part of a family of languages (including C++, J# [an equivalent version of Java], and Visual Basic.NET). All these languages compile to a common intermediate language (IL). The common class framework, Microsoft.NET, has replaced earlier frameworks for Windows programming and, increasingly, for modern Web development (see also AJAX).

Although it has been primarily associated with Microsoft development and Windows, the Mono and Dot GNU projects provide C# and an implementation of the Common Language Infrastructure, and many (but not all) of the .NET libraries for the Linux/UNIX environment.

Further Reading
"The C# Language." MSDN. Available online. URL: http://msdn2. microsoft.com/en-us/vcsharp/aa336809.aspx. Accessed April 28. 2007.
Davis, Stephen Randy. *C# for Dummies*. New York: Hungry Minds, 2002.
Hejlsberg, Andres, Scott Wiltamuth, and Peter Golde. *The C# Programming Language*. 2nd ed. Upper Saddle River, N.J.: Addison-Wesley, 2006.

C++

The C++ language was designed by Bjarne Stroustrup at AT&T's Bell Labs in Murray Hill, New Jersey, starting in 1979. By that time the C language had become well established as a powerful tool for systems programming (see C). However Stroustrup (and others) believed that C's limited data structures and function mechanism were proving inadequate to express the relationships found in increasingly large software packages involving many objects with complex relationships.

Consider the example of a simple object: a stack onto which numbers can be "pushed" or from which they can be "popped" (see STACK). In C, a stack would have to be implemented as a struct to hold the stack data and stack pointer, and a group of separately declared functions that could access the stack data structure in order to, for example "push" a number onto the stack or "pop" the top number from it. In such a scheme there is no direct, enforceable relationship between the object's data and functions. This means, among other things, that parts of a program could be dependent on the internal structure of the object, or could directly access and change such internal data. In a large software project with many programmers working on the code, this invites chaos.

An alternative paradigm already existed (see OBJECT-ORIENTED PROGRAMMING) embodied in a few new languages (see SIMULA and SMALLTALK). These languages allow for the structuring of data and functions together in the form of objects (or classes). Unlike a C struct, a class can contain both the data necessary for describing an object and the functions needed for manipulating it (see CLASS). A class "encapsulates" and protects its private data, and communicates with the rest of the program only through calls to its defined functions.

Further in object-oriented languages, the principle of inheritance could be used to proceed from the most general, abstract object to particular versions suited for specific tasks, with each object retaining the general capabilities and revising or adding to them. Thus, a "generic" list foundation class could be used as the basis for deriving a variety of more specialized lists (such as a doubly-linked list).

While attracted to the advantages of the object-oriented approach, Stroustrup also wanted to preserve the C language's ability to precisely control machine behavior needed for systems programming. He thus decided to build a new language on C's familiar syntax and features with object-oriented extensions. Stroustrup wrote the first version, called "C with Classes" as his Ph.D. thesis at Cambridge University in England. This gradually evolved into C++ through the early 1980s.

C++ FEATURES

The fundamental building block of C++ is the class. A class is used to create objects of its type. Each object contains a set of data and can carry out specified functions when called upon by the program. For example, the following class defines an array of integers and declares some functions for working with the array. Typically, it would be put in a header file (such as stack.h):

```
const int Max_size=20; // maximum elements
   in Stack

class Stack { // Declare the Stack class
public: // These functions are available
   outside
Stack(); // Constructor to create Stack
   objects
void push (int); // push int on Stack
int pop(); // remove top element
private: // This data can only be used in
   class
int index;
int Data[Max_size];
};
```

Next, the member functions of the Stack class are defined. The definitions can be put in a source file Stack.cpp:

```
#include "Stack.h" // bring in the declarations
Stack::Stack() { index=0;} // set zero for
   new stack
void Stack::push (int item) { // put a num-
   ber on stack
Data[index++] = item;
}
int Stack::pop(){ // remove top number
return Data [index-];
}
```

Now a second source file (Stacktest.cpp) can be written. It includes a main() function that creates a Stack object and tests some of the class functions:

```
#include "Stack.cpp" // include the Stack
   class
#include <iostream.h> // include standard I/O
   library
main() {
Stack S; // Create a Stack object called S
int index;
```

```
for (index = 1; index <= 5; index++)
   S.push(index); // put numbers 1-5 on stack
for (index = 1; index <=5; index++)
   cout < S.pop(); // print the stack
}
```

The stack implementation is completely separate from any program code that uses stack objects. Thus, a programmer could revise the stack class (perhaps using an improved algorithm or generalizing it to work with different data types). As long as the required parameters for the member functions aren't changed, programs that use stack objects won't need to be changed.

In addition to classes and inheritance, C++ has some other important features. The data types for function parameters can be fully defined, and types checked automatically (although programmers can bypass this type checking if they really want or need to). New operators can be added to a class by defining special operator functions, and the same operator can be given different meanings when working with different data types. (This is called overloading.) Thus, the + operator can be defined with a String class to combine (concatenate) two strings. The operator will still mean "addition" when used with numeric data.

An abstract object (one with no actual implementation) can be used as the basis for virtual functions. These functions can be redefined in each derived object so that whenever an object of that type is encountered the compiler will automatically search "downward" from the base class and find the correct derived class function.

Later versions of C++ include a related concept called templates. A template is an abstract specification that can be used to generate class definitions for data types passed to it (see TEMPLATE). Thus, a list template could be passed a vector and a 2D array and it will create a list class definition for each of these types. Templates are generally used when there is no hierarchical inheritance relationship between the types (in that case the virtual base class is a better approach).

C++ provides object-oriented alternatives to the standard libraries. For example, input/output uses a stream model, and I/O operators can be overloaded so they'll work with new classes. There is also an improved error-handling mechanism using appropriate objects.

GROWTH OF C++

During the late 1980s and 1990s, C++ became a very popular language for a variety of applications ranging from systems programming to business applications and games. The growth of the language coincided with the development of more powerful desktop computers and the release of inexpensive, easy-to-use but powerful development environments from Microsoft, Borland, and others. Since these compilers could also handle traditional C code, programmers could "port" existing code and use the object-oriented techniques of C++ as they mastered them. By the late 1990s, however, C++, although still dominant in many areas, was being challenged by Java, a language that simplified some of the more complex features of C++ and that was designed particularly for writing software to run on Web servers and browsers (see JAVA). For an alternative approach to creating a somewhat more "streamlined" C-type language, see C#.

Further Reading

"C++ Archive." Available online. URL: http://www.austinlinks. com/CPlusPlus/. Accessed May 24, 2007.

"Complete C++ Language Tutorial." Available online. URL: http://www.cplusplus.com/_doc/tutorial/. Accessed May 24, 2007.

Prata, Stephen. *C++ Primer Plus.* 5th ed. Indianapolis: SAMS, 2004.

Stroustrup, Bjarne. "A History of C++: 1979–1991." In *History of Programming Languages II,* edited by Thomas J. Bergin, Jr., and Richard G. Gibson, Jr. New York: ACM Press; Reading, Mass.: Addison-Wesley, 1996, 699–755.

———. *The C++ Programming Language.* Special 3rd ed. Reading, Mass.: Addison-Wesley, 2000.

cable modem

One of the most popular ways to connect people to the Internet takes advantage of the cable TV infrastructure that already exists in most communities. (For another pervasive alternative, using telephone lines, see DSL.)

Cable systems offer high-speed access (see BROADBAND) up to about 6 megabits/second (Mb/s), at least 20 times faster than an ordinary telephone modem and generally suitable for receiving today's multimedia offerings, including streaming video. (Upload speeds are usually "throttled" to 384 kb/s or fewer.)

In a typical installation, a splitter is used to separate the signal used for cable TV from the one used for data transmission. The data cable is then connected to the modem. The modem can then either be connected directly to a computer using a standard Ethernet "Cat 5" cable, or connected to a switch (or more commonly, a router) that will in turn provide the Internet connection to computers on the local network. (If the cable modem is connected directly to a computer, additional security against intrusions should also be provided. See FIREWALL.)

A typical cable TV system has from 60 channels to several hundred, most of which are used for TV programming. A few channels are dedicated to providing Internet service. Users in a given division of the cable network (typically a small neighborhood) thus share a fixed pool of bandwidth, which can reduce speed at times of high usage. The cable system can adjust by reallocating channels from TV to data or by adding new channels.

DOCSIS (Data Over Cable Service Interface Specification) is the industry standard for cable modems in North America.

MARKETING CONSIDERATIONS

As of 2007 there were about 30 million households in North America with cable Internet service. Monthly service fees are $40–$60, though cable providers generally try to bundle their cable TV and Internet services. Increasingly they are also offering telephone service over the cable network, using voice over Internet protocol (see VOIP).

In turn, telephone companies compete with cable companies by offering DSL Internet access. Although "traditional" DSL is generally somewhat slower than cable modems, Verizon in 2005 announced a new, much faster fiber-based form of DSL called FiOS, with speeds of up to 15 Mb/s (see also FIBER OPTICS). And just as cable companies can now offer phone service over the Internet, phone companies can offer video content, potentially competing with cable TV services. (Verizon has announced its own Internet-based television network, IPTV.) In general there is likely to be increased competition and more (if sometimes perplexing) choices for consumers.

Further Reading

Cable Industry Insider. Available online. URL: http://www.light-reading.com/cdn/. Accessed May 10, 2007.

Cable Modem Information Network. Available online. URL: http://www.cable-modem.net/. Accessed May 10, 2007.

Dominick, Joseph R., Barry L. Sherman, and Fritz J. Messere. *Broadcasting, Cable, the Internet and Beyond: An Introduction to Electronic Media.* 6th ed. New York: McGraw-Hill, 2007.

Dutta-Roy, Amitava. *Cable Modem: Technology and Applications.* New York: Wiley-Interscience, 2007.

cache

A basic problem in computer design is how to optimize the fetching of instructions or data so that it will be ready when the processor (CPU) needs it. One common solution is to use a cache. A cache is an area of relatively fast-access memory into which data can be stored in anticipation of its being needed for processing. Caches are used mainly in two contexts: the processor cache and the disk cache.

CPU CACHE

The use of a processor cache is advantageous because instructions and data can be fetched more quickly from the cache (static memory chips next to or within the CPU) than they can be retrieved from the main memory (usually dynamic RAM). An algorithm analyzes the instructions currently being executed by the processor and tries to anticipate what instructions and data are likely to be needed in the near future. (For example, if the instructions call for a possible branch to one of two sets of instructions, the cache will load the set that has been used most often or most recently. Since many programs loop over and over again through the same instructions until some condition is met, the cache's prediction will be right most of the time.)

These predicted instructions and data are transferred from main memory to the cache while the processor is still executing the earlier instructions. If the cache's prediction was correct, when it is time to fetch these instructions and data they are already waiting in the high-speed cache memory. The result is an effective increase in the CPU's speed despite there being no increase in clock rate (the rate at which the processor can cycle through instructions).

The effectiveness of a processor cache depends on two things: the mix of instructions and data being processed and the location of the cache memory. If a program uses long sequences of repetitive instructions and/or data, caching will noticeably speed it up. A cache located within the CPU itself (called an L1 cache) is faster (albeit more expensive) than an L2 cache, which is a separate set of chips on the motherboard.

Changes made to data by the CPU are normally written back to the cache, not to main memory, until the cache is full. In multiprocessor systems, however, designers of processor caches must deal with the issue of cache coherency. If, for example, several processors are executing parts of the same code and are using a shared main memory to communicate, one processor may change the value of a variable in memory but not write it back immediately (since its cache is not yet full). Meanwhile, another processor may load the old value from the cache, unaware that it has been changed. This can be prevented by using special hardware that can detect such changes and automatically "write through" the new value to the memory. The processors, having received a hardware or software "signal" that data has been changed, can be directed to reread it.

DISK CACHE

A disk cache uses the same general principle as a processor cache. Here, however, it is RAM (either a part of main memory or separate memory on the disk drive) that is the faster medium and the disk drive itself that is slower. When an application starts to request data from the disk, the cache reads one or more complete blocks or sectors of data from the disk rather than just the data record being requested. Then, if the application continues to request sequential data records, these can be read from the high-speed memory on the cache rather than from the disk drive. It follows that disk caching is most effective when an application, for example, loads a database file that is stored sequentially on the disk.

Similarly, when a program writes data to the disk, the data can be accumulated in the cache and written back to the drive in whole blocks. While this increases efficiency, if a power outage or other problem erases or corrupts the cache contents, the cache will no longer be in synch with the drive. This can cause corruption in a database.

Microsoft's Windows Vista introduced an ingenious type of cache at the system level. The "ReadyBoost" features allows many inexpensive USB flash drives to be used automatically as disk caches to store recently used data that had been paged out of main RAM memory.

NETWORK CACHE

Caching techniques can be used in other ways. For example, most Web browsers are set to store recently read pages on disk so that if the user directs the browser to go back to such a page it can be read from disk rather than having to be retransmitted over the Internet (generally a slower process). Web servers and ISPs (such as cable services) can also cache popular pages so they can be served up quickly.

Further Reading

Nottingham, Mark. "Caching Tutorial for Web Authors and Webmasters." Available online. URL: http://www.wdvl.com/Internet/Cache/index._html. Accessed May 24, 2007.
"System Cache." Available online. URL: http://www.pcguide.com/ref/mbsys/cache/. Accessed April 14, 2008.
Peir, J.-K., W. Hsu, and A. J. Smith. "Implementation Issues in Modern Cache Memories." *IEEE Transactions on Computers*, 48, 2 (1998): 100–110.

calculator

The use of physical objects to assist in performing calculations begins in prehistory with such practices as counting with pebbles or making what appears to be counting marks on pieces of bone. Nor should such simple manipulations be despised: In somewhat more sophisticated form it yielded the abacus, whose operators regularly outperformed mechanical calculators until the advent of electronics.

Generally, however, the term *calculator* is used to refer to a device that is able to store a number, add it to another number, and mechanically produce the result, taking care of any carried digits. In 1623, astronomer Johannes Kepler commissioned such a machine from Wilhelm Schickard. The machine combined a set of "Napier's bones" (slides marked with logarithmic intervals, the ancestor of the slide rule) and a register consisting of a set of toothed wheels that could be rotated to displays the digits 0 to 9, automatically carrying one place to the left. This ingenious machine was destroyed in a fire before it could be delivered to Kepler.

In 1642, French philosopher and mathematician Blaise Pascal invented an improved mechanical calculator. Its mechanism used a carry mechanism with a weight that would drop when a carry was reached, pulling the next wheel into position. This avoided having to use excessive force to carry a digit through several places. Pascal produced a number of his machines and tried to market them to accountants, but they never really caught on.

Schikard's and Pascal's calculators could only add, but in 1674 German mathematician Gottfried Wilhelm Leibniz invented a calculator that could work with all the digits of a number at once, rather than carrying from digit to digit. It worked by allowing a variable number of gear teeth to be engaged in each digit wheel. The operator could, for example, set the wheels to a number such as 215, and then turn a crank three times to multiply it by three, giving a result of 645. This mechanism, gradually improved, would remain fundamental to mechanical calculators for the next three centuries.

The first calculator efficient enough for general business use was invented by an American, Dorr E. Felt, in 1886. His machine, called a Comptometer, used the energy transmitted through the number-setting mechanism to perform the addition, considerably speeding up the calculating process. Improved machines by William Burroughs and others would replace the arm of the operator with an electric motor and provide a printing tape for automatically recording input numbers and results.

ELECTRONIC CALCULATORS

The final stage in the development of the calculator would be characterized by the use of electronics to replace mechanical (or electromechanical) action. The use of logic circuits to perform calculations electronically was first seen in the giant computers of the late 1940s, but this was obviously impractical for desktop office use. By the late 1960s, however, transistorized calculators comparable in size to mechanical desktop calculators came into use. By the 1970s, the use of integrated circuits made it possible to shrink the calculator down to palm-size and smaller. These calculators use a microprocessor with a set of "microinstructions" that enable them to perform a repertoire of operations ranging from basic arithmetic to trigonometric, statistical, or business-related functions.

The most advanced calculators are programmable by their user, who can enter a series of steps (including perhaps decisions and branching) as a stored program, and then apply it to data as needed. At this point the calculator can be best thought of as a small, somewhat limited computer. However, even these limits are constantly stretched: During the 1990s it became common for students to use graphing calculators to plot equations. Calculator use is now generally accepted in schools and even in the taking of the Scholastic Aptitude Test (SAT). However, some educators are concerned that overdependence on calculators may be depriving students of basic numeracy, including the ability to estimate the magnitude of results.

Further Reading

Aspray, W., ed. *Computing Before Computers*. Ames: Iowa State University Press, 1989.

The Old Calculator Museum. Links to Interesting Calculator-Related Sites. Available online. URL: http://www.oldcalculatormuseum.com/links.html. Accessed May 25, 2007.

cars and computing

Development of automotive technology has tended to be incremental rather than revolutionary. The core "hardware" such as the engine and drive train has changed little over several decades, other than the replacement of carburetors with fuel injection systems, and some improvements in areas such as brake design. On the other hand there have been significant improvements in safety features such as seat belts, air bags, and improved crash absorption barriers.

In recent years, however, the incorporation of computers in automobile design (see also EMBEDDED SYSTEM) has led to a number of significant advances in areas such as fuel efficiency, traction/stability, crash response, and driver information and navigation. Put simply, cars are becoming "smarter" and are making driving easier and safer.

Hybrid cars (such as gas/electric systems) depend on computers to sense how the car is being driven and when to augment electric power with the gas engine, as well as controlling the feeding of power back into the batteries (as in regenerative braking). In all cars, a general-purpose computing platform (such as one that has been developed by Microsoft) can keep drivers up to date on everything from road conditions to regular maintenance reminders. Many purchasers of higher-end vehicles are purchasing services such as OnStar that provide a variety of communication, navigation, and security and safety features. An example of

the latter includes the automatic sending of a signal when air bags are deployed. An operator then tries to determine if assistance is needed, and contacts local dispatchers. Drivers who lock themselves out accidentally can also have their cars unlocked remotely.

Another promising approach is to build systems that can monitor the driver's condition or behavior. For example, by analyzing images of the driver's eyes, facial features, and posture (such as slumping), the car may be able to tell when the driver has a high probability of being impaired (sleepy, drunk, or sick) and take appropriate action. (Of course many drivers may object to having their car "watch" them all the time.)

ULTIMATE SMART CARS

Much future progress in car computing will depend on creating integrated networking between vehicles and the road. An advanced navigation system could take advantage of real-time information being transmitted by the surrounding vehicles. For example, a stalled car would transmit warning messages to other drivers about the impending obstacle. Vehicles that sense an oil slick, ice, or other road hazard could also "mark" the location so it can be avoided by subsequent drivers. Data about the speed and spacing of traffic could provide real-time information about traffic jams, possibly routing vehicles into alternative lanes or other roads to reduce congestion and travel time (see MAPPING AND NAVIGATION SYSTEMS).

For many futurists, the ultimate "smart car" is one that can drive itself with little or no input from its human occupant. Such cars (with appropriate infrastructure) could eliminate most accidents, use roads more efficiently, and maintain mobility for a rapidly aging population. Such events as the annual DARPA automated vehicle challenge show considerable progress being made: Automated cars are already driving cross-country, with the human driver or follow-on vehicle serving only as a safety backup. In 2005 for the first time some competitors actually made it across the finish line. "Stanley," a robotic Volkswagen Touareg designed by Stanford University, won the race over an arduous 131-mile

This Mercedes Benz has an integrated navigation system—a feature appearing increasingly in other higher-end cars. (© WOLF-GANG MEIER / VISUM / THE IMAGE WORKS)

Mojave Desert course, navigating by means of a camera, laser range finders, and radar. In 2007 the contest entered a more difficult arena, where the robot vehicles had to deal with simulated urban traffic, negotiate intersections and traffic circles, and merge with traffic, all while obeying traffic laws.

Meanwhile efforts continue for developing a practical automated system that could be used for everyday driving. A "tethered" system using magnetic or radio frequency guides embedded in the road would reduce the complexity of the on-board navigation system, but would probably require dedicated roads. A "free" system linked only wirelessly would be much more flexible, but would require the ability to visualize and assess a constantly changing environment and, if necessary, make split-second decisions to avoid accidents. Such systems may also feature extensive automatic communication, where cars can provide each other with information about road conditions as well as their intended maneuvers.

The biggest obstacles to implementation of a fully automated highway system may be human rather than technical: the cost of the infrastructure, the need to convince the public the system is safe and reliable, and concerns about potential legal liability.

Ironically, just as information technology is making cars safer, such activities as cell phone use, text messaging, and use of in-car entertainment systems seem to be making drivers more distracted. Whether cars will get smart fast enough to compensate for increasingly inattentive drivers remains an open question.

Further Reading

DARPA Grand Challenge. Available online. URL: http://www.darpa.mil/grandchallenge/index.asp. Accessed May 18, 2007.

Edwards, John. "Robotic Cars Get Street Smart." *Electronic Design* 55 (June 29, 2007): 89 ff.

Shladover, Steven E. "What if Cars Could Drive Themselves?" Available online. URL: http://faculty.washington.edu/jbs/itrans/ahspath.htm. Accessed May 18, 2007.

Whelan, Richard. *Smart Highways, Smart Cars.* Boston: Artech House, 1995.

cascading style sheets (CSS)

Most word processor users are familiar with the use of styles in formatting text. Using a built-in style or defining one's own, particular characteristics can be assigned to the structural parts of a document, such as headings, lead and body paragraphs, quotations, references, and so on. There are several advantages to using styles. Once a style is associated with an element, the formatting attached to that style can automatically be applied to all instances of the element. If the writer decides that, for example, level two headings should be in italics rather than normal font, a simple change to the "head2" style will change all level two headings to italics.

Cascading style sheets (CSS) extend this idea to the creation of Web pages. The style sheet defines the structural elements of the document and applies the desired formatting. Instead of the main text of the document being filled with formatting directives (see HTML), a style sheet is associated with the document. When a compatible Web browser loads the page, it also loads the associated style sheet and

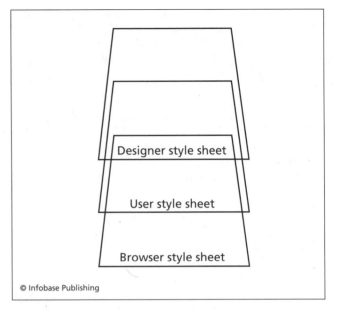

Cascading style sheets enable the appearance and formatting of a Web page to be handled separately from the page contents. Specifications provided in one sheet can be inherited or modified by other sheets.

uses it to determine how the page will be displayed. In other words, the structure of the document is separated from the details of its presentation. This not only makes it easier to change styles (as with word processing), but it also means that different style sheets can be used to tailor the document to different viewing situations (for example, viewing in a browser on a handheld PDA).

CSS uses a standard "box model" for laying out the presentation of a page. From outside in, the areas are defined as outer edge, margin, border, padding, inner edge, and the content area. Styles are applied in an order that depends on the relationship of the affected elements. For example, a style defined for the text body will be inherited by the paragraph, which can then redefine one or more of its elements. Similarly, an emphasis style used within a sentence might override the paragraph style in turn. It is this flowing of definitions down through the hierarchy of styles that creates the "cascading" part of CSS.

As CSS developed further, separate specifications have been provided for different media that can be included in a Web page: speech (to be read by a speech synthesizer), Braille (for a tactile Braille system), Emboss (for Braille printing), Handheld (for PDAs and other devices with limited display space), Print, Projection (for computer projection or transparencies), Screen, Tty (teletype-like displays with fixed-width characters), and TV.

Further Reading

"CSS From the Ground Up." Web Page Design. Available online. URL: http://www.wpdfd.com/editorial/basics/index.html. Accessed May 19, 2007.

Lie, Hakon Wium, and Bert Ros. *Cascading Style Sheets: Designing for the Web.* 3rd ed. Addison-Wesley Professional, 2005.

Meyer, Eric A. *CSS: The Definitive Guide.* 3rd ed. Sebastapol, Calif.: O'Reilly, 2007.

"Zen Garden: The Beauty of CSS Design." Available online. URL: http://www.csszengarden.com. Accessed May 19, 2007.

CASE (computer-aided software engineering)

During the late 1950s and 1960s, software rapidly grew more complex—especially operating system software and large business applications. With the typical program consisting of many components being developed by different programmers, it became difficult both to see the "big picture" and to maintain consistent procedures for transferring data from one program module to another. As computer scientists worked to develop sounder principles (see STRUCTURED PROGRAMMING) it also occurred to them that the power of the computer to automate procedures could be used to create tools for facilitating program design and managing the resulting complexity. CASE, or computer-aided software engineering, is a catchall phrase that covers a variety of such tools involved with all phases of development.

DESIGN TOOLS

The earliest design tool was the flowchart, often drawn with the aid of a template that could be used to trace the symbols on paper (see FLOWCHART). With its symbols for the flow of execution through branching and looping, the flowchart provides a good tool for visualizing how a program is intended to work. However large and complex programs often result in a sea of flowcharts that are hard to relate to one another and to the program as a whole. Starting in the 1960s, the creation of programs for manipulating flow symbols made it easier both to design flowcharts and to visualize them in varying levels of detail.

Another early tool for program design is pseudocode, a language that is at a higher level of abstraction than the target programming language, but that can be refined by adding details until the actual program source code has been specified (see PSEUDOCODE). This is analogous to a writer outlining the main topics of an essay and then refining them into subtopics and supporting details. Attempts were made to create a well-defined pseudocode that could be automatically parsed and transformed into compilable language statements, but they met with only limited success.

During the 1980s and 1990s, the graphics capabilities of desktop computers made it attractive to use a visual rather than linguistic approach to program design. Symbols (sometimes called "widgets") represent program functions such as reading data from a file or creating various kinds of charts. A program can be designed by connecting the widgets with "pipes" representing data flow and by setting various characteristics or properties.

CASE principles can also be seen in mainstream programming environments such as Microsoft's Visual Basic and Visual C++, Borland's Delphi and Turbo C++, and others (see also PROGRAMMING ENVIRONMENT). The design approach begins with setting up forms and placing objects (controls) that represent both user interface items (such as menus, lists, and text boxes) and internal processing (such as databases and Web browsers). However these environments do not in themselves provide the ability of full CASE tools to manage complex projects with many components.

ANALYSIS TOOLS

Once a program has been designed and implementation is under way, CASE tools can help the programmers maintain consistency across their various modules. One such tool (now rather venerable) is the data dictionary, which is a database whose records contain information about the definition of data items and a list of program components that use each item (see DATA DICTIONARY). When the definition of a data item is changed, the data dictionary can provide a list of affected components. Database technology is also applied to software design in the creation of a database of objects within a particular program, which can be used to provide more extensive information during debugging.

INTEGRATION AND TRENDS

A typical CASE environment integrates a variety of tools to facilitate the flow of software development. This process may begin with design using visual flowcharting,

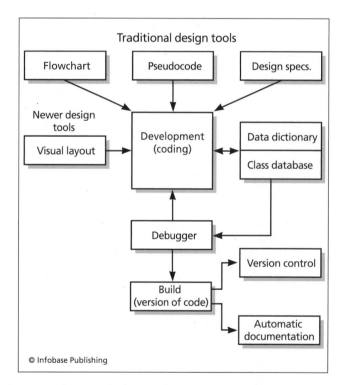

© Infobase Publishing

Many tools are used today to aid the complex endeavor of software engineering. Design tools include the traditional flowchart, pseudocode, and design specifications document. Additionally, many systems today use interactive, visual layout tools. During the coding and debugging phase, a data dictionary and/or class database can be used to describe and verify relationships and characteristics of objects in the program. Once the code is "built," a version control system keeps track of what was changed, and various automatic documentation features can be used to obtain listings of classes, functions, and other program elements.

"rapid prototyping," or other design tools. Once the overall design is settled, the developer proceeds to the detailed specification of objects used by the program and perhaps creates a data dictionary or other databases with information about program objects. During the coding process, source control or versioning facilities help log and keep track of the changes to code and the succession of new versions ("builds"). While testing the program, an integrated debugger (see BUGS AND DEBUGGING) can use information from the program components database to help pinpoint errors. As the code is finished, other tools can automatically generate documentation and other supporting materials (see TECHNICAL WRITING and DOCUMENTATION OF PROGRAM CODE).

Just as some early proponents of the English-like COBOL language proclaimed that professional programmers would no longer be needed for generating business applications, CASE tools have often been hyped as a panacea for all the ills of the software development cycle. Rather than causing the demise of the programmer, however, CASE tools have played an important role in keeping software development viable.

In recent years, tools for managing or debugging code have been supplemented with tools to aid the design process itself (see MODELING LANGUAGES). There are also tools to aid in refactoring, or the process of reorganizing and clarifying code to make it easier to maintain.

In a broader sense, CASE can also include tools for managing the programming team and its efforts. Even social networking tools (see BLOGS AND BLOGGING and WIKIS AND WIKIPEDIA) can play a part in keeping programmers in touch with issues and concerns relating to many different aspects of a project.

Further Reading

Carnegie Mellon Software Engineering Institute. "What Is a CASE Environment?" Available online. URL: http://www.sei.cmu.edu/legacy/case/case_whatis.html. Accessed May 18, 2007.

CASE Tool Index. Available online. URL: http://www.cs.queensu.ca/Software-Engineering/tools.html. Accessed May 18, 2007.

Stahl, Thomas, and Markus Voelter. *Model-Driven Software Development: Technology, Engineering, Management.* New York: Wiley, 2006.

CD-ROM and DVD-ROM

CD-ROM (compact disk read-only memory) is an optical data storage system that uses a disk coated with a thin layer of metal. In writing data, a laser etches billions of tiny pits in the metal. The data is encoded in the pattern of pits and spaces between them (called "lands"). Unlike the case with a magnetic hard or floppy disk, the data is written in a single spiral track that begins at the center of the disk. The CD-ROM drive uses another laser to read the encoded data (which is read from the other side as "bumps" rather than pits). The drive slows down as the detector (reading head) moves toward the outer edge of the disk. This maintains a constant linear velocity and allows for all sectors to be the same size. This system was adapted from the one used for

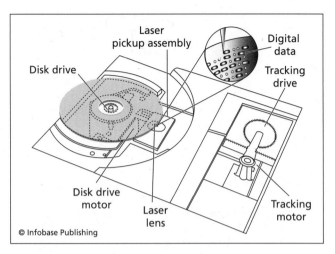

Schematic of the components of a CD drive. The tracking drive and tracking motor move the laser pickup assembly across the spinning disk drive to position it to the correct track. The laser beam hits the disk surface, reflecting differently from the pits and flat areas (lands). This pattern of differences encodes the data as ones and zeros.

the audio CDs that largely supplanted phonograph records during the 1980s.

A CD can hold about 650 MB of data. By the early 1990s, the CD had become inexpensive and ubiquitous, and it has now largely replaced the floppy disk as the medium of software distribution. The relatively large capacity meant that one CD could replace multiple floppies for a distribution of products such as Microsoft Windows or Word, and it also made it practical to give users access to the entire text of encyclopedias and other reference works. Further, the CD was essential for the delivery of multimedia (graphics, video, and sound) to the desktop, since such applications require far more storage than is available on 1.44-MB floppy disks. CD drives declined in price from several hundred dollars to about $50, while their speeds have increased by a factor of 30 or more, allowing them to keep up with games and other software that needs to read data quickly from the disk.

RECORDABLE CDS

In the late 1990s, a new consumer technology enabled users to create their own CDs with data or audio tracks. The cheapest kind, CD-R (Compact Disk Recordable) uses a layer of a dyed material and a thin gold layer to reflect the laser beam. Data is recorded by a laser beam hitting the dye layer in precise locations and marking it (in one of several ways, depending on technology). The lengths of marked ("striped") track and unmarked track together encode the data.

A more versatile alternative is the CD-RW (Compact Disk, Readable/Writeable), which can be recorded on, erased, and re-recorded many times. These disks have a layer made from a mixture of such materials as silver, antimony, and rare earths such as indium and tellurium. The mixture forms many tiny crystals. To record data, an infrared laser beam is directed at pinpoint spots on the layer. The heat from the beam melts the crystals in the target spot into an amorphous mass. Because the amorphous state has lower reflectivity than the original crystals, the reading laser can distinguish the marked "pits" from the surrounding lands. Because of a special property of the material, a beam with a heat level lower than the recording beam can reheat the amorphous material to a point at which it will, upon cooling, revert to its original crystal form. This permits repeated erasing and re-recording.

DVD-ROM

The DVD (alternatively, Digital Video Disc or Digital Versatile Disc) is similar to a CD, but uses laser light with a shorter wavelength. This means that the size of the pits and lands will be considerably smaller, which in turns means that much more data can be stored on the same size disk. A DVD disk typically stores up to 4.7 GB of data, equivalent to about six CDs. This capacity can be doubled by using both sides of the disk.

The high capacity of DVD-ROMs (and their recordable equivalent, DVD-RAMs) makes them useful for storing feature-length movies or videos, very large games and multimedia programs, or large illustrated encyclopedias. The development of high-definition television (HDTV) standards spurred the introduction of higher capacity DVD formats. The competition between Sony's Blu-Ray and HD-DVD (backed by Toshiba and Microsoft, among others) was resolved by 2008 in favor of the former. Blu-Ray offers high capacity (25GB for single layer discs, 50GB for dual layer).

Further Reading

About.com "Home Recording: Burning CDs." Available online. URL: http://homerecording.about.com/cs/burningcds/. Accessed May 10, 2007.
Taylor, Jim. *DVD Demystified*. 3rd ed. New York: McGraw-Hill, 2006.
White, Ron, and Timothy Edward Downs. *How Computers Work*. 8th ed. Indianapolis: Que, 2005.

cellular automata

In the 1970s, British mathematician John H. Conway invented a pastime called the Game of Life, which was popularized in Martin Gardner's column in *Scientific American*. In this game (better termed a simulation), each cell in a grid "lived" or "died" according to the following rules:

1. A living cell remains alive if it has either two or three living neighbors.
2. A dead cell becomes alive if it has three living neighbors.
3. A living cell dies if it has other than two or three living neighbors.

Investigators created hundreds of starting patterns of living cells and simulated how they changed as the rules were repeatedly applied. (Each application of the rules to the cells in the grid is called a *generation*.) They found, for

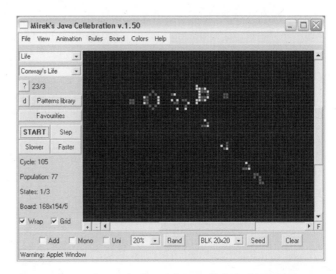

A screen from a Game of Life simulator called Mirek's Celebration. (This version runs as a Web browser–accessible Java applet.) This and other programs make it easy to experiment with a variety of Life patterns and track them across hundreds of "generations."

example, that a simple pattern of three living cells in a row "blinked" or switched back and forth between a horizontal and vertical orientation. Other patterns, called "glider guns" ejected smaller patterns (gliders or spaceships) that traveled across the grid.

The Game of Life is an instance of the general class called cellular automata. Each cell operates like a tiny computer that takes as input the states of its neighbors and produces its own state as the output. (See also FINITE STATE MACHINE.) The cells can be arranged in one (linear), two (grid), or three dimensions, and a great variety of sets of rules can be applied to them, ranging from simple variants of Life to exotic rules that can take into account how long a cell has been alive, or subject it to various "environmental" influences.

APPLICATIONS

Cellular automata theory has been applied to a variety of fields that deal with the complex interrelationships of components, including biology (microbe growth and population dynamics in general), ecology (including forestry), and animal behavior, such as the flight of birds. (The cues that a bird identifies in its neighbors are like the input conditions for a cell in a cellular automaton. The "output" would be the bird's flight behavior.)

The ability of cellular automatons to generate a rich complexity from simple components and rules mimics the development of life from simple components, and thus cellular automation is an important tool in the creation and study of ARTIFICIAL LIFE. This can be furthered by combining a set of cellular automation rules with a GENETIC ALGORITHM, including a mechanism for inheritance of characteristics. Cellular automation principles can also be applied to engineering in areas such as pattern or image recognition.

In 2002, computer scientist and mathematician Stephen Wolfram (developer of the *Mathematica* program) published a book titled *A New Kind of Science* that undertakes the modest project of explaining the fundamental structure and behavior of the universe using the principles of cellular automation. Time will tell whether this turns out to be simply an idiosyncratic (albeit interesting) approach or a generally useful paradigm.

Further Reading
Gutowitz, Howard, ed. *Cellular Automata.* Cambridge, Mass.: MIT Press, 1991.
"Patterns, Programs, and Links for Conway's Game of Life." Available online. URL: http://www.radicaleye.com/lifepage/. Accessed May 28, 2007.
Wojtowicz, Mirek. "Welcome to Mirek's Celebration.: 1D and 2D Cellular Automation Explorer." Available online. URL: http://www.mirwoj.opus.chelm.pl/ca/. Accessed May 28, 2007.
Wolfram, S. *A New Kind of Science.* Champaign, Ill.: Wolfram Media, 2002.
———. *Theory and Applications of Cellular Automata.* Singapore: World Scientific, 1986.

censorship and the Internet

Governments have always to varying degrees concerned themselves with the content of public media. The growing use of the Internet for expressive activities (see BLOGS AND BLOGGING and JOURNALISM AND COMPUTERS) has prompted authoritarian governments such as that of China to attempt to block "objectionable" material both through filtering techniques (see WEB FILTER) and through pressure on service providers. Further, users identified as creators of banned content may be subjected to prosecution. However because of the Internet's decentralized structure and the ability of users to operate relatively anonymously, Internet censorship tends to be only partially effective (see ANONYMITY AND THE INTERNET).

In the democratic West, Internet censorship generally applies to only a few forms of content. Attempts to criminalize the online provision of pornography to minors in the 1996 Communications Decency Act have generally been overturned by the courts as excessively infringing on the right of adults to access such content. However, a succession of bills seeking to require schools and libraries to install Web-filtering software culminated in the Children's Internet Protection Act, which was upheld by the U.S. Supreme Court in 2003.

Another area of potential censorship involves the rights of bloggers and other nontraditional journalists to post or link to documents that might be involved with a legal case.

Although the term "censorship" is sometimes limited to government action under criminal law, there are other ways in which Internet content may be restricted. For example, content providers seek to protect their work from unauthorized copying or distribution (see INTELLECTUAL PROPERTY AND COMPUTING). Civil sanctions can be brought to bear on violators of copyright or in cases of libel. However, as with other forms of censorlike activity on the Internet, the targeted behavior can be curtailed only to a limited extent.

CENSORSHIP IN CHINA
China has played a central role in the debate over censorship. The rapidly growing Chinese economy offers seemingly unlimited market potential for Internet-based businesses and sellers of software and hardware. However the Chinese government's desire to closely control the spread of "subversive" ideas has brought it into collision with the liberal ideas shared by many of the Internet's most important developers.

Human rights organizations such as Amnesty International have criticized online service providers such as Yahoo, Google, and Microsoft for providing the Internet addresses of users who have then been arrested. The companies have been accused of putting the potential profits of China's huge market ahead of ensuring free access to information. Generally, the companies say they have no choice but to comply with all local laws and legal demands for information about users. However, critics charge that the technology companies have often gone well beyond mere compliance to the provision of sophisticated filtering software for Web sites, blogs, and online chat and discussion groups.

The actual extent of censorship in China seems to vary considerably, depending on shifting political considerations. The nation's increasingly sophisticated users often find ways around the censorship, such as through using "proxy servers" that are inside the "Great Firewall" but can connect to the outside Internet. (Encrypted protocols such as VPN [virtual private networks] and SSH [secure shell] can also be used, because their content is not detected by monitoring and filtering software.)

Although generally not as highly organized, Internet censorship can also be found in countries such as Burma (Myanmar), North Korea, Iran, and Syria and to a lesser extent in South Korea and Saudi Arabia.

While Internet censorship can be viewed as being ultimately a political problem, technical realities limit its effectiveness, and curtailing the free exchange of information and open-ended communication that the Net affords is likely to have economic costs as well.

Further Reading
Amnesty International. Available online. URL: http://www.amnestyusa.org. Accessed May 22, 2007.
Axelrod-Contrada, Joan. *Reno v. ACLU: Internet Censorship.* New York: Benchmark Books, 2006.
Chase, Michael. *You've Got Dissent! Chinese Dissident Use of the Internet and Beijing's Counter-Strategies.* Santa Monica, Calif.: RAND Corporation, 2002.
Herumin, Wendy. *Censorship on the Internet: From Filters to Freedom of Speech.* Berkeley Heights, N.J.: Enslow, 2004.
Reporters without Borders. Handbook for Bloggers and Cyber-Dissidents. Available online. URL: http://www.rsf.org/rubrique.php3?id_rubrique=542. Accessed May 8, 2007.
Ringmar, Erik. *A Blogger's Manifesto: Free Speech and Censorship in the Age of the Internet.* London: Anthem Press, 2007.

While some parents and many schools use filtering software to block Web sites considered to be inappropriate for children, another approach is to provide a site with "child friendly" material and links. (IMAGE COURTESY OF THE ESTATE OF KEITH HARING, WWW.HARINGKIDS.COM)

central processing unit *See* CPU.

Cerf, Vinton D.

(1943–)
American
Computer Scientist

Vinton (Vint) Cerf is a key pioneer in the development of the packet-switched networking technology that is the basis for the Internet. In high school, Cerf distinguished himself from his classmates by wearing a jacket and a tie and carrying a large brown briefcase, which he later described as "maybe a nerd's way of being different." He has a lifelong love for fantasy and science fiction, both of which explore difference. Finally, Cerf was set apart by being hearing-impaired as a result of a birth defect. He would overcome this handicap through a combination of hearing aids and communications strategies. And while he was fascinated by chemistry and rocketry, it would be communications, math, and computer science that would form his lifelong interest.

After graduating from Stanford in 1965 with a B.S. in mathematics, Cerf worked at IBM as an engineer on its time-sharing systems, while broadening his background in computer science. At UCLA he earned on M.S. and then a Ph.D. in computer science while working on technology that could link one computer to another. Soon he was working with Len Kleinrock's Network Measurement Center to plan the ARPA network, a government-sponsored computer link. In designing software to simulate a network that as yet existed only on paper, Cerf and his colleagues had to explore the issues of network load, response time, queuing, and routing, which would prove fundamental for the real-world networks to come.

By the summer of 1968, four universities and research sites (UCLA, UC Santa Barbara, the University of Utah, and SRI) as well as the firm BBN (Bolt Beranek and Newman) were trying to develop a network. At the time, a custom combination of hardware and software had to be devised to connect each center's computer to the other. The hardware, a refrigerator-sized interface called an IMP, was still in development.

By 1970, the tiny four-node network was in operation, cobbled together with software that allowed a user on one machine to log in to another. This was a far cry from a system that would allow any computer to seamlessly communicate with another, however. What was needed on the software end was a universal, consistent language—a protocol—that any computer could use to communicate with any other computer on the network.

In a remarkable display of cooperation, Cerf and his colleagues in the Network Working Group set out to design such a system. The fundamental idea of the protocol is that data to be transmitted would be turned into a stream of "packets." Each packet would have addressing information that would enable it to be routed across the network and then reassembled back into proper sequence at the destination. Just as the Post Office doesn't need to know what's in a letter to deliver it, the network doesn't need to know whether the data it is handling is e-mail, a news article, or something else entirely. The message could be assembled and handed over to a program that would know what to do with it.

With the development of what eventually became TCP/IP (Transmission Control Protocol/Internet Protocol) Vint Cerf and Bob Kahn essentially became the fathers of the Internet we know today (see TCP/IP). As the online world began to grow in the 1980s, Cerf worked with MCI in the development of its electronic mail system, and then set up systems to coordinate Internet researchers.

In later years, Cerf undertook new initiatives in the development of the Internet. He was a key founder and the first president of the Internet Society in 1992, serving in that post until 1995 and then as chairman of the board, 1998–1999. This group seeks to plan for expansion and change as the Internet becomes a worldwide phenomenon. Cerf's interest in science fiction came full circle in 1998 when he joined an effort at the Jet Propulsion Laboratory (JPL) in Pasadena, California. There they are designing an "interplanetary Internet" that would allow a full network connection between robot space probes, astronauts, and eventual colonists on Mars and elsewhere in the solar system.

In 2005 Cerf joined Google as its "chief Internet evangelist," where he has the opportunity to apply his imagination to network applications and access policies. Cerf also served as chairman of the board of the Internet Corporation for Assigned Names and Numbers (ICANN), a position that he left in 2007.

Cerf has received numerous honors, including the IEEE Kobayashi Award (1992), International Telecommunications Union Silver Medal (1995), and the National Medal of Technology (1997). In 2005 Cerf (along with Robert Kahn) was awarded the Presidential Medal of Freedom, the nation's highest civilian award.

Further Reading
"Cerf's Up." Personal Perspectives. Available online. URL: http://global.mci.com/ca/resources/cerfs_up/personal_perspective/. Accessed May 28, 2007.
Hafner, Katie and Matthew Lyon. *Where Wizards Stay Up Late: the Origins of the Internet.* New York: Simon & Schuster, 1996.
Henderson, Harry. *Pioneers of the Internet.* San Diego, Calif.: Lucent Books, 2002.

certificate, digital

The ability to use public key encryption over the Internet makes it possible to send sensitive information (such as credit card numbers) to a Web site without electronic eavesdroppers being able to decode it and use it for criminal purposes (see ENCRYPTION and COMPUTER CRIME AND SECURITY). Any user can send information by using a person or organization's public key, and only the owner of the public key will be able to decode that information.

However, the user still needs assurance that a site actually belongs to the company that it says it does, rather than being an imposter. This assurance can be provided by a trusted third party certification authority (CA), such as VeriSign, Inc. The CA verifies the identity of the appli-

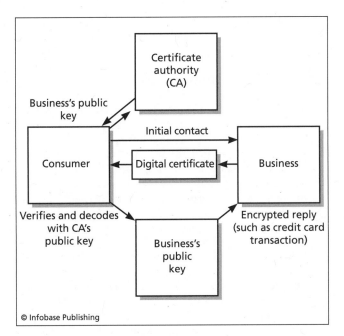

Digital certification relies upon public key cryptography and the existence of a trusted third party, the Certificate Authority (CA). First a business properly identifies itself to the CA and receives a digital certificate. A consumer can obtain a copy of the business's digital certificate and use it to obtain the business's public key from the CA. The consumer can now send encrypted information (such as a credit card number) to the business.

cant and then provides the company with a digital certificate, which is actually the company's public key encrypted together with a key used by the CA and a text message. (This is sometimes called a digital signature.) When a user queries the Web site, the user's browser uses the CA's public key to decrypt the certificate holder's public key. That public key is used in turn to decrypt the accompanying message. If the message text matches, this proves that the certificate is valid (unless the CA's private key has somehow been compromised).

The supporting technology for digital certification is included in a standard called Secure Sockets Layer (SSL), which is a protocol for sending encrypted data across the Internet. SSL is supported by leading browsers such as Microsoft Internet Explorer and Netscape. As a result, digital certification is usually transparent to the user, unless the user is notified that a certificate cannot be verified.

Digital certificates are often attached to software such as browser plug-ins so the user can verify before installation that the software actually originates with its manufacturer and has not been tampered with (such as by introduction of a virus).

The use of digital certification is expanding. For example, VeriSign and the federal General Services Administration (GSA) have begun an initiative called ACES (Access Certificates for Electronic Services) that will allow citizens a secure means to send information (such as loan applications) and to view benefits records. The IRS has a pilot program for accepting tax returns that are digitally certified and signed.

Further Reading
Altreya, Mohan, et al. *Digital Signatures*. Berkeley, Calif.: Osborne/McGraw-Hill, 2002.
Brands, Stefan A. *Rethinking Public Key Infrastructures and Digital Certificates*. Cambridge, Mass.: MIT Press, 2000.
Feghhi, Jalal, and Peter Williams. *Digital Certificates: Applied Internet Security*. Reading, Mass.: Addison-Wesley, 1998.

certification of computer professionals

Unlike medicine, the law, or even civil engineering, the computer-related fields do not have legally required certification. Given society's critical dependence on computer software and hardware for areas such as infrastructure management and medical applications, there have been persistent attempts to require certification or licensing of software engineers. However, the fluid nature of the information science field would make it difficult to decide which application areas should have entry restrictions.

At present, a variety of academic degrees, professional affiliations, and industry certificates may be considered in evaluating a candidate for a position in the computing field.

ACADEMIC AND PROFESSIONAL CREDENTIALS
The field of computer science has the usual levels of academic credentials (baccalaureate, master's, and doctoral degrees), and these are often considered prerequisites for an academic position or for industry positions that involve research or development in areas such as ROBOTICS or ARTIFICIAL INTELLIGENCE. For business-oriented IT positions, a bachelor's degree in computer science or information systems may be required or preferred, and candidates who also have a business-oriented degree (such as an MBA) may be in a stronger position. However, degrees are generally viewed only as a minimum qualification (or "filter") before evaluating experience in the specific application or platform in question. While not a certification, membership in the major professional organizations such as the Association for Computing Machinery (ACM) and Institute for Electrical and Electronic Engineers (IEEE) can be viewed as part of professional status. Through special interest groups and forums, these organizations provide computer professionals with a good way to track emerging technical developments or to broaden their knowledge.

In the early years of computing and again, in the microcomputer industry of the 1980s, programming experience and ability were valued more highly than academic credentials. (Bill Gates, for example, had no formal college training in computer science.) In general, degree or certification requirements tend to be imposed as a sector of the information industry becomes well defined and established in the corporate world. For example, as local area networks came into widespread use in the 1980s, certifications were developed by Microsoft, Novell, and others. In turn, colleges and trade schools can train technicians, using the certificate examinations to establish a curriculum, and numerous books and packaged training courses have been marketed.

In a newly emerging sector there is less emphasis on credentials (which are often not yet established) and more emphasis on being able to demonstrate knowledge through having actually developed successful applications. Thus, in the late 1990s, a high demand for Web page design and programming emerged, and a good portfolio was more important than the holding of some sort of certificate. However as e-commerce and the Web became firmly established in the corporate world, the cycle is beginning to repeat itself as certification for webmastering and e-commerce applications is developed.

INDUSTRY CERTIFICATIONS

Several major industry certifications have achieved widespread acceptance.

Since 1973, the Institute for Certification of Computing Professionals (ICCP) has offered certification based on general programming and related skills rather than mastery of particular platforms or products. The Associate Computing Professional (ACP) certificate is offered to persons who have a basic general knowledge of information processing and who have mastered one major programming language. The more advanced Certified Computing Professional (CCP) certificate requires several years of documented experience in areas such as programming or information systems management. Both certificates also require passing an examination.

A major trade group, the Computing Technology Industry Association (CompTIA) offers the A+ Certificate for computer technicians. It is based on passing a Core Service Technician exam focusing on general hardware-related skills and a DOS/Windows Service Technician exam that emphasizes knowledge of the operating system. The exams are updated regularly based on required job skills as assessed through industry practices.

Networking vendor Novell offers the Certified NetWare Engineer (CNE) certificate indicating mastery of the installation, configuration, and maintenance of its networking products or its GroupWise messaging system. The Certified NetWare Administrator (CNA) certificate emphasizes system administration.

Microsoft offers a variety of certificates in its networking and applications development products. The best known is the Microsoft Certified System Engineer (MCSE) certificate. It is based on a series of required and elective exams that cover the installation, management, configuration, and maintenance of Windows 2000 and other Microsoft networks.

A number of other vendors including Cisco Systems and Oracle offer certification in their products. Given the ever-changing marketplace, it is likely that most computer professionals will acquire multiple certificates as their career advances.

Further Reading

CompTIA Certification Page. Available online. URL: http://www.comptia.org/. Accessed May 28, 2007.

Institute for Certification of Computing Professionals. Available online. URL: http://www.iccp.org. Accessed May 28, 2007.

"MCSE Guide." Available online. URL: http://www.mcseguide.com/

Novell Education Page. Available online. URL: http://www.novell.com/training/certinfo/howdoi.htm. Accessed May 28, 2007.

CGI (common gateway interface)

By itself, a Web page coded in HTML is simply a "static" display that does not interact with the user (other than for the selection of links). (See HTML, DHTML, and XHTM.) Many Web services, including online databases and e-commerce transactions, require that the user be able to interact with the server. For example, an online shopper may need to browse or search a catalog of CD titles, select one or more for purchase, and then complete the transaction by providing credit card and other information. These functions are provided by "gateway programs" on the server that can access databases or other facilities.

One way to provide interaction with (and through) a Web page is to use the CGI (common gateway interface). CGI is a facility that allows Web browsers and other client programs to link to and run programs stored on a Web site. The stored programs, called scripts, can be written in various languages such as JavaScript or PHP (see SCRIPTING LANGUAGES) and placed in a cgi-bin folder on the Web server.

The CGI script is referenced by an HTML hyperlink on the Web page, such as

```
<A HREF="http://www.MyServer.com/cgi-bin/
MyScript">MyScript </A>
```

Or more commonly, it is included in an HTML form that the user fills in, then clicks the Submit button. In either case, the script executes. The script can then process the information the user provided on the form, and return information to the user's Web browser in the form

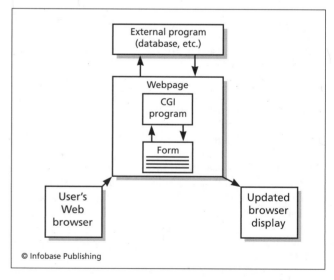

© Infobase Publishing

CGI or Common Gateway Interface allows a program linked to a Web page to obtain data from databases and use it to generate forms to be shown on users' Web browsers. For example, a CGI program can link a Web user to a "shopping cart" and inventory system for online purchases.

of an HTML document. The script can perform additional functions such as logging the user's query for marketing purposes.

The complexity of Web features and the heavy load on servers have prompted a number of strategies for serving dynamic content more efficiently. Traditionally, each time a CGI request is passed to the URL for a script, the appropriate language interpreter must be loaded and initialized. However, modern Web servers such as Apache have built-in modules for commonly used scripting languages such as PHP, Perl, Python, and Ruby. This allows the Web server to run the script directly without the overhead of starting a new interpreter process.

A more fundamental shift in implementation is the development of methods to tie together DHTML and XML with a document model and scripting languages to allow for dynamic changes in page content without having to reload the page (see AJAX).

Note: the acronym CGI can also stand for "computer-generated imagery" (see COMPUTER GRAPHICS).

Further Reading

"A Guide to HTML and CGI Scripts." Available online. URL: http://snowwhite.it.brighton.ac.uk/~mas/mas/courses/html/html.html. Accessed May 30, 2007.

Hamilton, Jacqueline D. *CGI Programming 101*. Houston, Tex.: CGI101.com, 2000. (First six chapters are available free online at URL: http://www.cgi101.com/book/.) Accessed August 12, 2007.

"The Most Simple Intro to CGI." Available online. URL: http://bignosebird.com/prcgi.shtml. Accessed August 12, 2007.

characters and strings

While the attention of the first computer designers focused mainly on numeric calculations, it was clear that much of the data that business people and others would want to manipulate with the new machines would be textual in nature. Billing records, for example, would have to include customer names and addresses, not just balance totals.

The "natural" representation of data in a computer is as a series of two-state (binary) values, interpreted as binary numbers. The solution for representing text (letters of the alphabet, punctuation marks, and other special symbols) is to assign a numeric value to each text symbol. The result is a character code, such as ASCII (American Standard Code for Information Interchange), which is the scheme used most widely today. (Another system, EBCDIC (Extended Binary-Coded Decimal Interchange Code) was used during the heyday of IBM mainframes, but is seldom used today.)

The seven-bit ASCII system is compact (using one byte of memory to store each character), and was quite suitable for early microcomputers that required only the basic English alphabet, punctuation, and a few control characters (such as carriage return). In an attempt to use characters to provide simple graphics capabilities, an "extended ASCII" was developed for use on IBM-compatible PCs. This used eight bits, increasing the number of characters available from 128 to 256. However, the use of bit-mapped graphics in Windows and other operating systems made this version of ASCII unnecessary. Instead, the ANSI (American National Standards Institute) eight-bit character set used the additional character positions to store a variety of special symbols (such as fractions and the copyright symbol) and various accent marks used in European languages.

TABLE OF 7-BIT ASCII CHARACTER CODES

The following are control (nonprinting) characters:

0	Null (nothing)
7	Bell (rings on an old teletype; beeps on most PCs)
8	Backspace
9	Tab
10	Line feed (goes to next line without changing column position)
13	Carriage return (positions to beginning of next line)
26	End of file
27	[Esc] (Escape key)

The characters with codes from 32 to 127 produce printable characters.

32	[space]	64	@	96	`
33	!	65	A	97	a
34	"	66	B	98	b
35	#	67	C	99	c
36	$	68	D	100	d
37	%	69	E	101	e
38	&	70	F	102	f
39	'	71	G	103	g
40	(	72	H	104	h
41	)	73	I	105	i
42	*	74	J	106	j
43	+	75	K	107	k
44	'	76	L	108	l
45	-	77	M	109	m
46	.	78	N	110	n
47	/	79	O	111	o
48	0	80	P	112	p
49	1	81	Q	113	q
50	2	82	R	114	r
51	3	83	S	115	s
52	4	84	T	116	t
53	5	85	U	117	u
54	6	86	V	118	v
55	7	87	W	119	w
56	8	88	X	120	x
57	9	89	Y	121	y
58	:	90	Z	122	z
59	;	91	[	123	{
60	<	92	\	124		
61	=	93	]	125	}
62	>	94	^	126	~
63	?	95	_	127	[delete]

As computer use became more widespread internationally, even 256 characters proved to be inadequate. A new standard called Unicode can accommodate all of the world's alphabetic languages including Arabic, Hebrew, and Japanese (Kana Unicode schemes can also be used to encode ideographic languages (such as Chinese) and languages such as Korean that use syllabic components. At present each ideograph has its own character code, but Unicode 3.0 includes a scheme for describing ideographs through their component parts (radicals). Most modern operating systems use Unicode exclusively for character representation. However, support in software such as Web browsers is far from complete, though steadily improving. Unicode also includes many sets of internationally used symbols such as those used in mathematics and science. In order to accommodate this wealth of characters, Unicode uses 16 bits to store each character, allowing for 65,535 different characters at the expense of requiring twice the memory storage.

PROGRAMMING WITH STRINGS

Before considering how characters are actually manipulated in the computer, it is important to realize that what the binary value such as 1000001 (decimal 65) stored in a byte of memory actually represents depends on the context given to it by the program accessing that location. If the program declares an integer variable, then the data is numeric. If the program declares a character (char) value, then the data will be interpreted as an uppercase "A" (in the ASCII system).

Most character data used by programs actually represents words, sentences, or longer pieces of text. Multiple characters are represented as a *string*. For example, in traditional BASIC the statement:

```
NAME$ = "Homer Simpson"
```

declares a string variable called NAME$ (the $ is a suffix indicating a string) and sets its value to the character string "Homer Simpson." (The quotation marks are not actually stored with the characters.)

Some languages (such as BASIC) store a string in memory by first storing the number of characters in the string, followed by the characters, with one in each byte of memory. In the family of languages that includes C, however, there is no string type as such. Instead, a string is stored as an array of char. Thus, in C the preceding example might look like this:

```
char Name [20] = "Homer Simpson";
```

This declares Name as an array of up to 20 characters, and initializes it to the string literal "Homer Simpson."

An alternative (and equivalent) form is:

```
char * Name = "Homer Simpson";
```

Here Name is a pointer that returns the memory location where the data begins. The string of characters "Homer Simpson" is stored starting at that location.

Unlike the case with BASIC, in the C languages, the number of characters is not stored at the beginning of the data. Rather, a special "null" character is stored to mark the end of the string.

Programs can test strings for equality or even for greater than or less than. However, programmers must be careful to understand the collating sequence, or the order given to characters in a character set such as ASCII. For example the test

```
If State = "CA"
```

will fail if the current value of State is "ca." The lowercase characters have different numeric values than their uppercase counterparts (and indeed must, if the two are to be distinguished). Similarly, the expression:

```
"Zebra" < "aardvark"
```

is true because uppercase Z comes before lowercase "a" in the collating sequence.

Programming languages differ considerably in their facilities for manipulating strings. BASIC includes built-in functions for determining the length of a string (LEN) and for extracting portions of a string (substrings). For example given the string Test consisting of the text "Test Data," the expression Right$ (Test, 4) would return "data."

Following their generally minimalist philosophy, the C and C++ languages contains no string facilities. Rather, they are provided as part of the standard library, which can be included in programs as needed. In the following little program:

```
#include <iostream.h>
#include <string.h>
void main ()
{
char String1[20];
char String2[20];
strcpy (String1, "Homer");
strcpy (String2, "Simpson");
//Concatenate string2 to the end of string1
strcat (String1, String2);
cout String1 <<endl;
}
```

Here the strcpy function is used to initialize the two strings, and then the strcat (string concatenate) function is used to combine the two strings and store the result back in string1, which is then sent to the output.

As an alternative, one can take advantage of the object orientation of C++ and define a string class. The addition operator (+) can then be extended, or "overloaded" so that it will concatenate strings. Then, the preceding program, instead of using the strcat function, can use the more natural syntax:

```
cout << String1 + String2
```

to display the combined strings.

STRING-ORIENTED LANGUAGES

Sophisticated string processing (such as parsing and pattern matching) tends to be awkward to express in traditional number-oriented programming languages. Several languages have been designed especially for manipulating textual data. Snobol, designed in the early 1960s, is best known for its sophisticated pattern-matching and pattern processing capabilities. A similar language, Icon, is widely used for specialized string-processing tasks today. Many programmers working with textual data in the UNIX environment have found that the awk and Perl languages are easier to use than C for extracting and manipulating data fields. (See AWK and PERL.)

Further Reading
Gillam, Richard. *Unicode Demystified: A Practical Programmer's Guide to the Encoding Standard.* Reading, Mass.: Addison-Wesley, 2002.
Korpela, Jukka. *Unicode Explained.* Sebastapol, Calif.: O'Reilly, 2006.
A Tutorial on Character Code Issues. Available online. URL: http://www.cs.tut.fi/~jkorpela/chars.html. Accessed May 31, 2007.
Unicode Consortium. *Unicode Standard, Version 5.0.* 5th ed. Reading, Mass.: Addison-Wesley, 2006.

chat, online

In general terms, to "chat" is to communicate in real time by typing messages to other online users who can immediately type messages in reply. It is this conversational immediacy that distinguishes chat services from conferencing systems or bulletin boards.

COMMERCIAL SERVICES
Many PC users have become acquainted with chatting through participating in "chat rooms" operated by online services such as AMERICA ONLINE (AOL). A chat room is a "virtual space" in which people meet either to socialize generally or to discuss particular topics. At their best, chat rooms can develop into true communities whose participants develop long-term friendships and provide one another with information and emotional support (see VIRTUAL COMMUNITY).

However, the essentially anonymous character of chat (where participants often use "handles" rather than real names) that facilitates freedom of expression can also provide a cover for mischief or even crime. Chat rooms have acquired a rather lurid reputation in the eyes of the general public. There has been considerable public concern about children becoming involved in inappropriate sexual conversation. This has been fueled by media stories (sometimes exaggerated) about children being recruited into face-to-face meetings with pedophiles. AOL and other online services have tried to reduce such activity by restricting online sex chat to adults, but there is no reliable mechanism for a service to verify its user's age. A chat room can also be supervised by a host or moderator who tries to prevent "flaming" (insults) or other behavior that the online service considers to be inappropriate.

DISTRIBUTED SERVICES
For people who find commercial online services to be too expensive or confining, there are alternatives available for just the cost of an Internet connection. The popular Internet Relay Chat (IRC) was developed in Finland by Jarkko Oikarinen in the late 1980s. Using one of the freely available client programs, users connect to an IRC server, which in turn is connected to one of dozens of IRC networks. Users can create their own chat rooms (called channels). There are thousands of IRC channels with participants all over the world. To participate, a user simply joins a channel and sees all messages currently being posted by other users of the channel. In turn, the user's messages are posted for all to see. While IRC uses only text, there are now enhanced chat systems (often written in Java to work with a Web browser) that add graphics and other features.

There are many other technologies that can be used for conversing via the Internet. Some chat services (such as Cu-SeeMe) enable participants to transmit their images (see VIDEOCONFERENCING and WEB CAM). Voice can also be transmitted over an Internet connection (see VOIP). For a very pervasive form of "ad hoc" textual communication, see TEXTING AND INSTANT MESSAGING.

Further Reading
McDonald, Wayne. *Chat Rooms in Wonderland.* Frederick, Md.: PublishAmerica, 2005.
Ploch, Nicolas. "A Short IRC Primer." Available online. URL: http://www.irchelp.org/irchelp/ircprimer.html. Accessed June 1, 2007.
Wasuki, Dennis D. *Self-Games and Body-Play: Personhood in Online Chat and Cybersex.* Bern: Peter Lang, 2003.
Weverka, Peter. *Mastering ICQ: the Official Guide.* Dulles, Va.: ICQ Press, 2001.

chatterbots

The famous Turing test (see TURING, ALAN M.) proposes that if a human is unable to reliably distinguish messages from a computer from those of another person, the computer program involved can at least be provisionally declared to be "intelligent." The advent of textual communication via the Internet (see TEXTING AND INSTANT MESSAGING) has afforded a variety of ways to attempt to meet this challenge. Programs that mimic human conversational styles have come to be known as "chatterbots."

The prototypical chatterbot was ELIZA, developed by Joseph Weizenbaum in the mid-1960s (see WEIZENBAUM, JOSEPH). ELIZA mimicked a form of nondirective psychotherapy in which the therapist echoes or plays off of the client's statements as a form of gentle encouragement and validation. Thus if one types, "My father didn't really like me," ELIZA might reply, "Tell me more about your father." Although primitive, ELIZA once inadvertently fooled an executive into thinking he was exchanging messages with Weizenbaum. Other classic chatterbots include Parry, designed to mimic a paranoid, and the story-generating Racter.

The first online chatterbot, Julia, was created by Michael Mauldin in 1994. (Mauldin also coined the term *chatterbot.*) More sophisticated chatterbots include Jabberwocky, which includes learning algorithms, and ELLA, which applies linguistic principles to produce more realistic speech (see NATURAL LANGUAGE PROCESSING). On a negative note, some chatterbots have been designed to enter chat rooms and spew advertising (spambots) or even to elicit personal information (see IDENTITY THEFT and PHISHING AND SPOOFING).

One of the best known and most versatile modern chatterbots is ALICE (Artificial Linguistic Internet Computer Entity), developed by Richard Wallace. ALICE does much more than simply echo statements; it includes an elaborate parser and sentence generator and specialized scripts for initiating or responding to conversational gambits. ALICE can even tell lies and spread "gossip" that it learns from other users.

In the following dialog ALICE seems to be able to respond appropriately:

Human: Are you intelligent?

ALICE: I am highly intelligent. I have many intellectual functions.

Human: How do you know you are intelligent?

ALICE: By introspection.

However the illusion breaks down as the matter is pressed further and the bot uses a "deflection" strategy:

Human: What were you thinking while you said that?

ALICE: I was minding my own business.

Human: How do you feel about being interrogated in this way?

ALICE: The opposite of not being interrogated in this way.

Still, with more than 25,000 conversation templates stored in XML files that are modifiable by the user, ALICE has repeatedly won the Loebner Contest, a modern version of the Turing test.

Although many are created as programming exercises or for amusement, chatterbots embody principles that are important in artificial intelligence research, including natural language processing and machine learning (see ARTIFICIAL INTELLIGENCE). Techniques first developed with chatterbots can contribute to the creation of programs designed to provide answers to users' questions or other forms of assistance (see SOFTWARE AGENT).

Further Reading

A.L.I.C.E. Artificial Intelligence Foundation. Available online. URL: http://www.alicebot.org/. Accessed April 27, 2007.
Chatterbot Central (The Simon Laven Page). Available online. URL: http://www.simonlaven.com/. Accessed April 27, 2007.
Loebner Prize. Available online. URL: http://www.loebner.net/Prizef/loebner-prize.html. Accessed April 27, 2007.

chess and computers

With simple rules but endless permutations, chess has fascinated millions of players for hundreds of years. When mechanical automatons became fashionable in the 18th century, onlookers were intrigued by "the Turk," a chess-playing automaton. While the Turk was eventually shown to be a hoax (a human player was hidden inside), the development of the electronic digital computer in the mid-20th century provided the opportunity to create a true automatic chess player.

In 1950 Claude Shannon outlined the two basic strategies that would be used by future chess-playing programs. The "brute force" strategy would examine the possible moves for the computer chess player, the possible replies of the opponent to each move, the possible next moves by the computer, and so on for as many half moves or "plies" as possible. The moves would be evaluated by a "minimax" algorithm that would find the move that best improves the computer's position despite the opponent's best play.

The fundamental problem with the brute force is the "combinatorial explosion": Looking ahead just three moves (six plies) would involve evaluating more than 700,000,000 positions. This was impractical given the limited computing power available in the 1950s. Shannon realized this and decided that a successful chess program would have to incorporate principles of chess strategy that would enable it to quickly recognize and discard moves that did not show a likelihood of gaining material or improving the position (such as by increasing control of center squares). As a result of this "pruning" approach, only the more promising initial moves would result in the program looking ahead—but those moves could be analyzed much more deeply.

The challenge of the pruning approach is the need to identify the principles of good play and codify them in such a way that the program can use them reliably. Progress was slow at first—programs of the 1950s and 1960s could scarcely challenge an experienced amateur human player, let alone a master. A typical program would play a mixture of reasonable moves, odd-looking but justifiable moves, and moves that showed the chess version of "nearsightedness." By the 1970s, however, computing power was rapidly increasing, and a new generation of programs such as Chess 4.0 from Northwestern University abandoned most pruning techniques in favor of brute-force searches that could now extend further ahead. In practice, each programmer chose a particular balance between brute force and pruning-selection

In the 18th century the Turk, a mechanical chess player, astonished onlookers. Although the original Turk was a fraud (a small human player was hidden inside), the modern computer chess program Fritz 9 pays its homage by simulating its predecessor. (FRITZ 9, CHESSBASE GMBH, WWW.CHESSBASE.COM)

techniques. An ever-increasing search base could be combined with evaluation of particularly important positional features (such as the possibility of creating a "passed pawn" that could be promoted to a queen).

By the end of the 1970s, International Master David Levy was still beating the best chess programs of the time (defeating Chess 4.7 in 1978). A decade later, however, Levy was defeated in 1989 by Deep Thought, a program that ran on a specially designed computer that could examine hundreds of millions of positions per move. That same year World Champion Garry Kasparov decisively defeated the machine. In 1996, however, the successor program Deep Blue (sponsored by IBM) shocked the chess world by beating Kasparov in the first game of their match. Kasparov went on to win the match, but the following year an updated version of Deep Blue defeated Kasparov 3 1/2–2 1/2. A computer had arguably become the strongest chess player in the world. As a practical matter, the match brought IBM invaluable publicity as a world leader in supercomputing.

CHESS AND AI

The earliest computer chess theorists such as Claude Shannon and Alan Turing saw the game as one potential way to demonstrate true machine intelligence. Ironically, by the time computers had truly mastered chess, the artificial intelligence (AI) community had concluded that mastering the game was largely irrelevant to their goals. AI pioneers Herbert Simon and John McCarthy have referred to chess as "the Drosophila of AI." By this they mean that, like the ubiquitous fruit flies in genetics research, chess became an easy way to measure computer prowess. But what was it measuring? The dominant brute-force approach was more a measure of computing power than the application of such AI techniques as pattern recognition. (There is, however, still some interest in writing chess programs that "think" more like a human player.) In recent years there has been some interest in programming computers to play the Asian board game Go, where positional and structural elements play a greater role than in chess. However, even the latest generation of Go programs seem to be relying more on a statistical approach than a deep conceptual analysis.

Further Reading

Computer History Museum. "Mastering the Game: A History of Computer Chess." Available online. URL: http://www.computerhistory.org/chess/. Accessed April 28, 2007.

Hsu, Feng-Hsiung. *Behind Deep Blue: Building the Computer That Defeated the World Chess Champion.* Princeton, N.J.: Princeton University Press, 2004.

Levy, David, and Monty Newborn. *How Computers Play Chess.* New York: Computer Science Press, 1991.

Shannon, Claude E. "Programming a Computer for Playing Chess." *Philosophical Magazine* 41 (1950): 314. Available from Computer History Museum. Available online. URL: http://archive.computerhistory.org. Accessed April 27, 2007.

chip

As early as the 1930s, researchers had begun to investigate the electrical properties of materials such as silicon and germanium. Such materials, dubbed "semiconductors," were neither a good conductor of electricity (such as copper) nor a good insulator (such as rubber). In 1939, one researcher, William Shockley, wrote in his notebook "It has today occurred to me that an amplifier using semiconductors rather than vacuum [tubes] is in principle possible." In other words, if the conductivity of a semiconductor could be made to vary in a controlled way, it could serve as an electronic "valve" in the same way that a vacuum tube can be used to amplify a current or to serve as an electronic switch.

The needs of the ensuing wartime years made it evident that a solid-state electronic device would bring many advantages over the vacuum tube: compactness, lower power usage, higher reliability. Increasingly complex electronic equipment, ranging from military fire control systems to the first digital computers, further underscored the inadequacy of the vacuum tube.

In 1947, William Shockley, along with John Bardeen and Walter Brattain, invented the transistor, a solid-state electronic device that could replace the vacuum tube for most low-power applications, including the binary switching that is at the heart of the electronic digital computer. But as the computer industry strove to pack more processing power into a manageable volume, the transistor itself began to appear bulky.

Starting in 1958, two researchers, Jack Kilby of Texas Instruments and Robert Noyce of Fairchild Semiconductor, independently arrived at the next stage of electronic miniaturization: the integrated circuit (IC). The basic idea of the IC is to make semiconductor resistors, capacitors, and diodes, combine them with transistors, and assemble them into complete, compact solid-state circuits. Kilby did this by embedding the components on a single piece of germanium called a substrate. However, this method required the painstaking and expensive hand-soldering of the tiny gold wires connecting the components. Noyce soon came up with a superior method: Using a lithographic process, he was able to print the pattern of wires for the circuit onto a board containing a silicon substrate. The components could then be easily connected to the circuit. Thus was born the ubiquitous PCB (printed circuit board). This technology would make the minicomputer (a machine that was roughly refrigerator-sized rather than room-sized) possible during the 1960s and 1970s. Besides the PCBs being quite reliable compared to hand-soldered connections, a failed board could be easily "swapped out" for a replacement, simplifying maintenance.

FROM IC TO CHIP

The next step to the truly integrated circuit was to form the individual devices onto a single ceramic substrate (much smaller than the printed circuit board) and encapsulate them in a protective polymer coating. The device then functioned as a single unit, with input and output leads to connect it to a larger circuit. However, the speed of this "hybrid IC" is limited by the relatively large distance between components. The modern IC that we now call the "computer chip" is a monolithic IC. Here the devices, rather than being attached to the silicon substrate, are formed by altering the substrate itself with tiny amounts of impurities (a process called "doping"). This creates regions with an excess of electrons (n-type, for negative) or a deficit (p-type for positive). The junction between a p and an n region functions as a diode. More complex arrangements of p and n regions form transistors. Layers of transistors and other devices can be formed on top of one another, resulting in a highly compact integrated circuit. Today this is generally done using optical lithography techniques, although as the separation between components approaches 100 nm (nanometers, or billionths of a meter) it becomes limited by the wavelength of the light used.

In computers, the IC chip is used for two primary functions: logic (the processor) and memory. The microprocessors of the 1970s were measured in thousands of transistor equivalents, while chips such as the Pentium and Athlon being marketed by the late 1990s are measured in tens of millions of transistors (see MICROPROCESSOR). Meanwhile, memory chips have increased in capacity from the 4K and 16K common around 1980 to 256 MB and more. In what became known as "Moore's law," Gordon Moore has observed that the number of transistors per chip has doubled roughly every 18 months.

FUTURE TECHNOLOGIES

Although Moore's law has proven to be surprisingly resilient, new technologies will be required to maintain the pace of progress.

In January 2007, Intel and IBM separately announced a process for making transistors out of the exotic metal hafnium. It turns out that hafnium is much better than the traditional silicon at preventing power leakage (and resulting inefficiency) through layers that are only about five atoms thick. Hafnium transistors can also be packed more closely together and/or run at a higher speed.

Another approach is to find new ways to connect the transistors so they can be placed closer together, allowing signals to travel more quickly and thus provide faster operation. Hewlett-Packard (HP) is developing a way to place the connections on layers above the transistors themselves, thus reducing the space between components. The scheme uses two layers of conducting material separated by a layer of insulating material that can be made to conduct by having a current applied to it. Although promising, the approach faces difficulties in making the wires (only about 100 atoms thick) reliable enough for applications such as computer memory or microprocessors.

Ultimately, direct fabrication at the atomic level (see NANOTECHNOLOGY) will allow for the maximum density and efficiency of computer chips.

Further Reading

Baker, R. Jacob, Harry W. Li, and David E. Boyce. *CMOS Circuit Design, Layout and Simulation.* New York: IEEE Press, 1998.
Saint, Christopher and Judy Saint. *IC Layout Basics.* New York: McGraw-Hill, 2001.
Semiconductor Industry Association. Available online. URL: http://www.sia-online.org/home.cfm. Accessed August 13, 2007.

Thompson, J. M. T., ed. *Visions of the Future: Physics and Electronics.* New York: Cambridge University Press, 2001.

chipset

In personal computers a chipset is a group of integrated circuits that together perform a particular function. System purchasers generally think in terms of the processor itself (such as a Pentium III, Pentium IV, or competitive chips from AMD or Cyrix). However they are really buying a *system chipset* that includes the microprocessor itself (see MICROPROCESSOR) and often a memory cache (which may be part of the microprocessor or a separate chip—see CACHE) as well as the chips that control the memory bus (which connects the processor to the main memory on the motherboard—see BUS.) The overall performance of the system depends not just on the processor's architecture (including data width, instruction set, and use of instruction pipelines) but also on the type and size of the cache memory, the memory bus (RDRAM or "Rambus" and SDRAM) and the speed with which the processor can move data to and from memory.

In addition to the system chipset, other chipsets on the motherboard are used to support functions such as graphics (the AGP, or Advanced Graphics Port, for example), drive connection (EIDE controller), communication with external devices (see PARALLEL PORT, SERIAL PORT, and USB), and connections to expansion cards (the PCI bus).

At the end of the 1990s, the PC marketplace had chipsets based on two competing architectures. Intel, which originally developed an architecture called Socket 7, has switched to the more complex Slot-1 architecture, which is most effective for multiprocessor operation but offers the advantage of including a separate bus for accessing the cache memory. Meanwhile, Intel's main competitor, AMD, has enhanced the Socket 7 into "Super Socket 7" and is offering faster bus speeds. On the horizon may be completely new architecture. In choosing a system, consumers are locked into their choice because the microprocessor pin sockets used for each chipset architecture are different.

Further Reading

Intel. "Desktop Chipsets." Available online. URL: http://www.intel.com/products/desktop/chipsets/. Accessed June 6, 2007.
"Motherboards." Available online. URL: http://www.motherboards.org/index.html. Accessed June 6, 2007.
Walrath, Josh. "Chipsets Today and Tomorrow." ExtremeTech. Available online. URL: http://www.extremetech.com/article2/0,1697,1845493,00.asp. Accessed June 6, 2007.

Church, Alonzo

(1903–1995)
American
Mathematician

Born in Washington, D.C., mathematician and logician Alonzo Church made seminal contributions to the fundamental theory of computation. Church was mentored by noted geometer Oswald Veblen and graduated from Princeton with an A.B. in mathematics in 1924. Veblen encouraged Church to devote his graduate thesis to the investigation of the fundamental problem of computability. At the time, mathematician David Hilbert and his followers were attempting to create a formal way to express mathematical propositions.

In 1927, Church received his Ph.D. from Princeton for a dissertation on the axiom of choice in set theory. During the 1930s, Church developed the lambda calculus, which provided rules for substituting bound variables in generating mathematical functions. The Church thesis (also called the Church-Turing thesis, because Alan Turing [see TURING, ALAN] approached the same conclusion from a different angle) stated that every calculable function in number theory could be defined in lambda calculus and was also computable in Turing's sense (see COMPUTABILITY AND COMPLEXITY). This provided the theoretical confidence that given appropriate technology, computers could tackle a variety of problems reliably. At the same time, another of Church's achievements, the Church theorem, proved that there were theorems that could not be proven by any computer.

Church's lambda calculus became important for the design and verification of computer languages, and the LISP language in particular was based on lambda expressions. Computer scientists working with problems in list processing and the use of recursion also have owed much to Church's pioneering work.

Church taught at Princeton for many years. In 1961, he received the title of Professor of Mathematics and Philosophy. In 1967, he took the same position at UCLA, where he was active until 1990. He received numerous honorary degrees, and in 1990 an international symposium was held in his honor at the State University of New York at Buffalo.

Further Reading

Barendregt, H. "The Impact of the Lambda Calculus in Logic and Computer Science." *The Bulletin of Symbolic Logic* 3: 181–215.
Church, Alonzo. *Introduction to Mathematical Logic.* Princeton, N.J.: Princeton University Press, 1956.
Copeland, Jack. "The Church-Turing Thesis." AlanTuring.net. Available online. URL: http://www.alanturing.net/turing_archive/pages/Reference%20Articles/The%20Turing-Church%20Thesis.html. Accessed June 6, 2007.
Davis, M. *The Undecidable: Basic Papers on Undecidable Propositions, Unsolvable Problems, and Computable Functions.* Hackett, N.Y.: Raven Press, 1965.

Cisco Systems

Cisco Systems (NASDAQ symbol: CSCO) builds much of the physical infrastructure of the Internet—the routers and switches that direct the streams of data between Web servers and millions of users, as well as specialized networking, security, and storage devices.

Cisco was founded in 1984 by Leo Bosack and Sandy Lerner, a married couple who worked in computer operations at Stanford University. (The name "Cisco" is from "San Francisco," and the company's logo is a stylized version of the Golden Gate Bridge.)

The company focused on networking at a time when that sector of the computer industry was still rather small. They were able to build one of the first routers that could link otherwise incompatible computers over the Internet. (Eventually, when the protocol was standardized (see TCP/IP), routers could focus on the burgeoning traffic in IP packets.)

As the market for basic hardware became relatively saturated, Cisco began to emphasize the development of more intelligent "application aware" routing solutions as well as equipment geared for distributed processing (see GRID COMPUTING).

Cisco grew along with the Internet/Web boom of the late 1990s. In 2000 Cisco was for a time the most valuable company in the world, with a market capitalization of more than half a trillion dollars. (Today that has shrunk to a "mere" $180 billion or so—still one of the world's most valuable companies.)

THE "LAST MILE"

In the telecommunications industry, "the last mile" refers to the connections and equipment that actually bring content to users' homes and businesses. One source of Cisco's continued growth in the 2000 decade is the way it has addressed the consumer sector through strategic acquisitions. In 2003, Cisco acquired Linksys, maker of home Internet routers and wireless access points. In 2005, Scientific Atlanta—maker of cable modems, digital cable boxes, and other consumer equipment—also became a Cisco company.

The company has also entered the area of Internet telephony (see VOIP) by teaming up with Skype to build a cordless phone that can connect to a computer to make phone calls over the Internet.

Moving from hardware into software, Cisco in 2007 purchased Utah Street Networks, a San Francisco–based maker of software to link online communities (see also SOCIAL NETWORKING) and operator of the Tribe.net Web site. Around the same time, Cisco made a much larger buy, acquiring WebEx, maker of online collaboration software, for $3.2 billion.

In 2007 Cisco had revenue of $35 billion, with more than 63,000 employees.

Further Reading

Burrows, Peter. "Microsoft and Cisco: Product Promises: The Tech Giants' New Spirit of Cooperation Is Promising, but CEOs Ballmer and Chambers Say Making the Alliance Work Will Be Difficult." *Business Week Online,* August 20, 2007. Available online. URL: http://www.businessweek.com/technology/content/aug2007/tc20070820_282297.htm?chan=search. Accessed September 3, 2007.
Cisco Corporation Web site. Available online. URL: http://www.cisco.com/. Accessed September 3, 2007.
Paulson, E. *Inside Cisco: The Real Story of Sustained M&A Growth.* New York: Wiley, 2001.
Stauffer, Davide. *Nothing but Net: Business the Cisco Way.* Milford, Conn.: Capstone Publishing, 2000.
Velte, Toby J., and Anthony T. Velte. *Cisco: A Beginner's Guide.* 4th ed. New York: McGraw-Hill, 2007.
Waters, John K. *John Chambers and the Cisco Way: Navigating through Volatility.* New York: Wiley, 2002.

class

A class is a data type that combines both a data structure and methods for manipulating the data. For example, a string class might consist of an array to hold the characters in the string and methods to compare strings, combine strings, or extract portions of a string (see CHARACTERS AND STRINGS).

As with other data types, once a class is declared, objects (sometimes called instances) of the class can be created and used. This way of structuring programs is called object-oriented programming because the class object is the basic building block (see OBJECT-ORIENTED PROGRAMMING).

Object-oriented programming and classes provide several advantages over traditional block-structured languages. In a traditional BASIC or even Pascal program, there is no particular connection between the data structure and the procedures or functions that manipulate it. In a large program one programmer might change the data structure without alerting other programmers whose code assumes the original structure. On the other hand, someone might write a procedure that directly manipulates the internal data rather than using the methods already provided. Either transgression can lead to hard-to-find bugs.

With a class, however, data and procedures are bound together, or encapsulated. This means that the data in a class object can be manipulated only by using one of the methods provided by the class. If the person in charge of maintaining the class decides to provide an improved implementation of the data structure, as long as the data parameters expected by the class methods do not change, code that uses the class objects will continue to function properly.

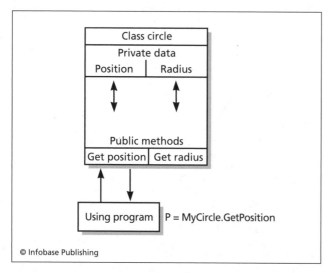

© Infobase Publishing

A class encapsulates (or hides) its internal information from the rest of the program. When the program calls MyCircle.GetPosition, the GetPosition member function of the MyCircle Circle class object retrieves the private Position data and sends it back to the calling statement, where it is assigned to the variable P. Private data cannot be directly accessed or changed by an outside caller.

Most languages that use classes also allow for inheritance, or the ability to create a new class that derives data and methods from a "parent" class and then modifies or extends them. For example, a class that provides support for 3D graphics could be derived from an existing class for 2D graphics by adding data items such as a third (Z) coordinate and replacing a method such as "line" with a version that works with three coordinates instead of two.

In designing classes, it is important to identify the essential features of the physical situation you are trying to model. The most general characteristics can be put in the "base class" and the more specialized characteristics would be added in the inherited (derived) classes.

CLASSES AND C++

Classes first appeared in the Simula 67 language, which introduced the terms class and object (see SIMULA). As the name suggests, the language was used mainly for simulation and modeling, but its object-oriented ideas would prove influential. The Smalltalk language developed at Xerox PARC in the 1970s ran on the Alto computer, which pioneered the graphic user interface that would become popular with the Macintosh in the 1980s. Smalltalk used classes to build a seamless and extensible operating system and environment (see SMALLTALK).

However it was Bjarne Stroustrup's C++ language that brought classes into the programming mainstream (see C++). C++ essentially builds its classes by extending the C struct so that it contains both methods (class functions) and data. An access mechanism allows class variables to be designated as completely accessible (public), which is rare, accessible only by derived classes (protected), or accessible only within the class itself (private). The creation of a new object of the class is specified by a constructor function, which typically allocates memory for the object and sets initial default values. The corresponding destructor function frees up the memory when the object no longer exists.

C++ allows for multiple inheritance, meaning that a class can be derived from more than one parent or base class. The language also provides two powerful mechanisms for extending functionality. The first, called virtual functions, allows a base class and its derived classes to have functions based on the same *interface*. For example, a base graphics class might have virtual line, circle, setcolor, and other functions that would be implemented in derived classes for 3D objects, 3D solid objects, and so on. When the program calls a method in a virtual class, the compiler automatically searches the class's "family tree" until it finds the class that corresponds to the actual data type of the object.

A template specifies how to create a class definition based on the type of data to be used by the class. In other words, where a regular procedure takes and manipulates data parameters and returns data, a template takes data parameters and returns a definition of a class for working with that data (see TEMPLATE).

Other languages of the 1980s and later have embraced classes. Examples include descendants of the Algol family of languages (see PASCAL, ADA, C++'s close cousin—JAVA), and Microsoft's Visual Basic. (There is even a version of COBOL with classes.)

The use of class frameworks, such as the Microsoft Foundation Classes (MFC), the C++ STL (Standard Template Library) and various Java implementations, has provided a superior way to organize the complexities of data access and operating system functions.

Further Reading
Sebesta, Robert W. *Concepts of Programming Languages*. 8th ed. Boston: Addison-Wesley, 2007.
Stroustrup, Bjarne. *The C++ Programming Languages*. Special 3rd ed. Reading, Mass.: Addison-Wesley, 2000.

clean room   *See* REVERSE ENGINEERING.

client-server computing

It is often more efficient to have a large, relatively expensive computer provide an application or service to users on many smaller, inexpensive computers that are linked to it by a network connection. The term server can apply to both the application providing the service and the machine running it. The program or machine that receives the service is called the client.

A familiar example is browsing the Web. The user runs a Web browser, which is a client program. The browser connects to the Web server that hosts the desired Web site. Another example is a corporate server that runs a database. Users' client programs connect to the database over a local area network (LAN). Many retail transactions are also handled using a client-server arrangement. Thus, when a travel or theater booking agent sells a ticket, the agent's client program running on a PC or terminal connects to the server containing the database that keeps track of what seats are available (see TERMINAL).

There are several advantages to using the client-server model. Having most of the processing done by one or more servers means that these powerful and more costly machines can be used to the greatest efficiency. If more processing capacity is needed, more servers can be brought online without having to revamp the whole system. Users, on the other hand, only need PCs (or terminals) that are powerful enough to run the smaller client program to connect to the server.

Keeping the data in a central location helps ensure its integrity: If a database is on a server, transactions can be committed in an orderly way to ensure that, for example, the same ticket isn't sold to two people. A client-server model also offers flexibility to users. Any client program that meets the standards supported by the server can be used to make a connection. (The marketplace generally decides which clients will be supported: for example most Web sites today support both Microsoft Internet Explorer and Firefox, although they may cater to some features unique to one or the other and other browsers will also work to some extent.)

Client-server computing does have potential disadvantages. If there is only one server, a failure of the server (whether from a hardware failure, a bug, or a hacker attack) brings the whole system to a halt, since the client has no ability to complete transactions on its own. The clients' access to the server is also dependent on the network that connects them. A network failure or traffic bottleneck will also prevent the client from getting any work done.

EXTENDING THE MODEL

One way used in larger organizations to improve the efficiency of the client-server model is to introduce an intermediary between the client and the server. The intermediary program can cache frequently requested data so it can be supplied immediately rather than having to be retrieved from the server (see CACHE). The intermediary can also act as a "traffic cop" to route client requests to the server that currently has the least load or the fastest network access.

Another design consideration is the distribution of processing between the client and the server. At one extreme is the "thin client," where the client machine may only display forms and transmit information to and display information from the server. A POS (point of sale) terminal typifies this approach. On the other hand, a "fat client" running on a full-featured desktop PC may perform functions such as verifying the completeness and validity of data before sending it to the server, or use information from the server to generate graphics (this is typical with online games, where limiting the amount of information that must be sent over the network can be crucial to speed).

The ultimate extension of the client-server model is "distributed object computing." This is an application of object-oriented programming principles to the organization of the resources needed for data processing. In this model each object (such as a database table) is accessible throughout the network by all other objects, regardless of their physical location. This scheme provides the ultimate in flexibility, because objects can be moved freely among physical machines in order to even out the load. One popular implementation of distributed object computing is CORBA (Common Object Request Broker Architecture—see CORBA). For Windows-based programs, Microsoft has developed the DCOM (Distributed Component Object Model), which allows controls (that is, objects with functional interfaces) written using ActiveX to communicate with each other in a networked environment. (For example, an Excel spreadsheet in an ActiveX control can be embedded in a Word document, and instructed to update itself regularly by obtaining data from a Microsoft Access database table on another machine.) The Microsoft.NET initiative is also geared toward creating applications that can fluidly interoperate over the Internet (see MICROSOFT .NET).

Further Reading
Fox, Dan. *Building Distributed Applications with Visual Basic .NET*. Indianapolis: Sams, 2002.
Goodyear, Mark, ed. *Enterprise System Architectures*. Grand Rapids, Mich.: CRC Press, 1999.
Graham, Steve [and others]. *Building Web Services with Java: Making Sense of XML, SOAP, WSDL and UDDI*. 2nd ed. Indianapolis: Sams, 2004.
"Network Design Manual: Client-Server Fundamentals." Available online. URL: http://www.networkcomputing.com/netdesign/1005part1a.html. Accessed January 25, 2008.

Sinclair, Joseph T., and Mark S. Merkow. *Thin Clients Clearly Explained*. San Francisco: Morgan Kaufmann, 2000.

clock speed

The transfer of data within the microprocessor and between the microprocessor and memory must be synchronized to ensure that the data needed to execute each instruction is available when the flow of execution has reached an appropriate point. This synchronization is accomplished by moving data in intervals that correspond to the pulses of the system clock (a quartz crystal). This is done by sending control signals that tell the components of the processor and memory when to send or wait for data. Thus, if the microprocessor is the heart of the computer, the clock is the heart's pacemaker. Because most devices cannot run at the same pace as the processor, circuits in various parts of the motherboard create secondary control signals that run at various ratios of the actual system clock speed.

The following table shows the speed of various system components in relation to the system clock rate. Although the example uses a 600-MHz clock, the ratios will generally hold for faster processors.

DEVICE	SPEED	RELATIONSHIP
Processor	600	System bus * 4.5
System (Memory) Bus	133	(depends on multiplier)
Level 2 Cache	300	Processor / 2
AGP	66	System bus / 2
PCI bus	33	System bus / 4

Microprocessors are rated according to the frequency (that is, number of pulses per second) of their associated clock. For example, a 1.2-GHz Pentium IV processor has 1.2 billion (giga-) pulses per second. It follows that all other things being equal, the higher a processor's clock frequency, the more instructions it can process per second. An alternative way to rate processors is according to the number of a standard type of instruction that it can process per second, hence MIPS (millions of instructions per second).

The relationship between clock speed and processor performance is not as simple as the preceding might imply, however. Each processor is designed with circuits that can move data at a certain rate. In some cases a processor can be run at a higher clock rate than specified (this is called overclocking), but then reliability comes into question. Also, the actual processing power of a processor depends on many other factors. If a processor implements instructions in its microcode that are more efficient for handling certain operations (such as floating point math or graphics rendering), applications that depend on these operations may run faster on one processor than on another, even if the two processors run at the same clock speed. The speed of the system bus (which connects the processor to the RAM memory) also affects the speed at which data can be fetched, processed, and stored. A processor with a clock speed of 733 MHz should perform better on a motherboard with a bus speed of 133 MHz than on one with a bus speed of only 100 MHz.

Speed is "sexy" in marketing terms, so the major chip manufacturers always tout their fastest chips. However, the difference in speed between, for example, a 2.2-GHz version of a processor and a 2.0-GHz version may be unnoticeable to the user of all but the most processor-intensive applications (such as image processing). Indeed, if the system with the slower chip has a faster bus, faster memory (such as RDRAM), or a larger processor cache (see CACHE) it may well outperform the one with a faster chip.

Another reason for caution in interpreting clock speed is that many recent PCs have two or even four processors (see MULTIPROCESSING). Performance in such systems is likely to depend at least as much on optimization of the operating system and applications as on any multiple of raw clock speed. This trend to multicore CPUs is also seen as an alternative to any substantial increase in processor speed, because higher speeds bring increasing concerns about heat and power usage.

In PCs the term "clock" can also refer to the battery-powered "real-time" clock that provides a timing interval that can be accessed by the operating system and applications.

Further Reading

Clock speed resources. TechRepublic. Available online. URL: http://search.techrepublic.com.com/search/clock+speed. html?t=11& s=0&o=0. Accessed June 6, 2007.
"Understanding System Memory and CPU Speeds." Available online. URL: http://www.directron.com/fsbguide.html. Accessed June 6, 2007.
"What Is CPU Overclocking?" Available online. URL: http://www. webopedia.com/DidYouKnow/Computer_Science/2005/over clocking.asp. Accessed June 6, 2007.

COBOL

Common Business-Oriented Language was developed under the impetus of a 1959 Department of Defense initiative to create a common language for developing business applications that centered on the processing of data from files. (The military, after all, was a "business" whose inventory control and accounting needs dwarfed those of all but the largest corporations.) At the time, the principal business-oriented language for mainframe computers was FLOW-MATIC, a language developed by Grace Hopper's team at Remington-Rand UNIVAC and limited to that company's computers (see HOPPER, GRACE MURRAY). The first COBOL compilers became available in 1960, and the American National Standards Institute (ANSI) issued a standard specification for the language in 1968. Expanded standards were issued in 1974 and 1985 (COBOL-74 and COBOL-85) with a new standard issued in 2002.

The committee that outlined the language that would become COBOL focused on making program statements resemble declarative English sentences rather than the mathematical expressions used by FORTRAN for scientific programming. COBOL's designers hoped that accountants, managers, and other business professionals could quickly master the language, reducing if not removing the need for professional programmers. (This theme of "programming without programmers" would recur with regard to other languages such as RPG, BASIC, and various database systems, always with limited success.)

PROGRAM STRUCTURE

A COBOL program as a whole resembles a business form in that it is divided into specific sections called divisions, each with required and optional items.

The Identification division simply identifies the programmer and gives some information about the program:

```
IDENTIFICATION DIVISION.
PROGRAM-ID WEEKLY REPORT.
AUTHOR JAMES BRADLEY.
DATE-WRITTEN DECEMBER 10, 2000.
DATE-COMPILED DECEMBER 12, 2000.
REMARKS THIS IS AN EXAMPLE PROGRAM.
```

The Environment division contains specifications about the environment (hardware) for which the program will be compiled. In some cases (for example, microcomputer versions of COBOL) it may not be needed. In other cases, it might simply have a Configuration section that specifies the machine to be used:

```
ENVIRONMENT DIVISION.
CONFIGURATION SECTION.
SOURCE-COMPUTER IBM-370.
OBJECT-COMPUTER IBM-370.
```

(The reason for the separate source and object computers is that programs were sometimes compiled on one computer for use on another, often smaller, one.)

In some cases, the Environment Division must also include an Input-Output section that specifies devices and files that will be used by the program. For example:

```
INPUT-OUTPUT SECTION.
FILE-CONTROL.
    SELECT STUDENT-FILE ASSIGN TO READER
    SELECT STUDENT-LISTING ASSIGN TO LOCAL-
    PRINTER
```

The Data division gives a description of the data records and other items that will be processed by the program. It is roughly comparable to the declarations of variables in languages such as Pascal, C, or BASIC. Since COBOL focuses on the processing of file records and the formatting of reports, it tends to have fewer data types than many other languages, but it makes it easier to describe the kinds of data structures commonly used in business applications. For example, it is easy to describe records that have fields and subfields by using level numbers to indicate the relationship:

```
DATA DIVISION.
FILE SECTION.
FD INFILE
    LABEL RECORDS ARE OMITTED.

01 STUDENT-DATA.
```

```
02 STUDENT-ID PIC 999999.
02 STUDENT-NAME.
   03 LAST-NAME PIC X(15).
   03 INITIAL PIC X.
   03 FIRST-NAME PIC X(10).
02 GPA PIC 9.99
```

The "PIC" or picture clause specifies the type of data (using 9's and a decimal point for numbers and X for text) and the length. In addition to specifying the input records, the Data division often includes items that specify the format of the lines of output that are to be printed.

The Procedure division provides the statements that perform the actual data manipulation. Procedures can be organized as subroutines (roughly equivalent to procedures or functions on other languages). Some sample procedure statements are:

```
READ STUDENT-DATA INTO STUDENT-WORK-RECORD
  AT END MOVE 'E' TO PROC-FLAG-ST
    GO TO EXIT-PRINT
ADD 1 TO TOTAL-STUDENT-RECORDS
```

Mathematical expressions can be computed using a Compute statement:

```
COMPUTE GPA = TOTAL-GRADES / CLASSES
```

Branching (if) statements are available, and looping is provided by the Perform statement, for example:

```
PERFORM 100-PRINT-LINE
  UNTIL LINES-FL IS EQUAL TO 'E'
```

(As with older versions of BASIC, subroutines are numbered.)

IMPACT AND PROSPECTS

From the 1960s through the 1980s, COBOL became the workhorse language for business applications for mainframe and mid-size computers, and it is still widely used today. (The concerns about possible problems at the end of the century often involved older programs written in COBOL, see Y2K PROBLEM.) The main line of programming language evolution bypassed COBOL and went through Algol (a contemporary of COBOL) and on into Pascal, C, and other block-structured languages (see also STRUCTURED PROGRAMMING).

Some modern versions of COBOL have incorporated later developments in structured programming (such as modularization) and even object-oriented design. COBOL has also shown considerable versatility in accommodating modern development frameworks, including Microsoft.NET as well as processing now-ubiquitous XML data. Nevertheless, usage of COBOL continues to decline slowly as developers increasingly turn to languages such as C++, scripting languages, or database development systems.

Further Reading

Bivar de Oliveria, Rui. *The Power of COBOL: For Systems Developers of the 21st Century*. Charleston, S.C.: BookSurge, 2006.
COBOL Portal. Available online. URL: http://www.cobolportal. com. Accessed June 8, 2007.

Murcah, Mike, Anne Prince, and Raul Menendez. *Murach's Mainframe COBOL*. Fresno, Calif.: Murach and Associates, 2004.

Sammet, J. E. "The Early History of COBOL," in *History of Programming Languages*. Wexelblat, R. L., ed., 199–276. New York: Academic Press, 1985.

codec

Short for "coder/decoder," a codec is essentially an algorithm for encoding (and compressing) a stream of data for transmission, and then decoding and decompressing it at the receiving end. Usually the data involved represents audio or video content (see STREAMING). Typically the data is being downloaded from a Web site to be played on a personal computer or portable player (see MULTIMEDIA and MUSIC AND VIDEO PLAYERS, DIGITAL).

A codec is described as "lossy" if some of the original information is lost in the compression process. It then becomes a question of whether the loss in quality is perceived by the user as significant. A codec that preserves all the information needed to re-create the original file is "lossless." For most purposes, the much greater size of the lossless version of a file is not worth the (often imperceptible) increase in quality or fidelity.

A codec is usually used in connection with a "container format" that specifies how the encoded data is to be stored in a file. Often a container can hold more than one data stream and even more than one kind of media (such as video and audio). When one refers to a Windows WAV file, for example, one is actually referring to a container.

Most of the popular codecs and file formats are proprietary, which creates something of a dilemma for users who prefer open-source solutions. However, while most Linux distributions do not include support for formats such as MP3 out of the box, distributions such as Ubuntu are now making it easier for users to choose nonsupported proprietary codecs if desired.

The preceding table lists some codecs likely to be encountered by program developers and consumers.

CODEC	CONTAINER DESCRIPTION
AAC	advanced audio coding; developed as a successor to MP3 and especially used by Apple (iTunes, iPod, iPhone, etc.)
AIFF	audio interchange file format; audio container format for transferring content between applications
ALAC	Apple lossless audio codec
AVI	audio video interleave; video and movies container format
FLACC	free lossless audio codec; music, open source, lossless
MP3	actually MPEG-3, probably the most common music codec
MPEG	Moving Picture Experts Group; video, movies, audio (four layers MPEG-1 through MPEG-4)
Ogg Vorbis	music, open source (often used on Linux systems)
Quick Time	Apple multimedia
Real Audio and and RealVideo	developed by RealNetworks for many platforms
RIFF	resource interchange file format; container format
Vorbis	free, open-source audio codec (often used in Linux)
WAV	Windows audio format (usually uncompressed)
WMA	Windows media audio
WMV	Windows media video

Further Reading

Audio Files. Available online. URL: http://www.fileinfo.net/filetypes/audio. Accessed September 3, 2007.

Harte, Lawrence. *Introduction to MPEG*. Fuquay Varina, N.C.: Althos Publishing, 2006.

Rathbone, Andy. *MP3 for Dummies*. 2nd ed. New York: Hungry Minds, 2001.

Richardson, Iain E. G. *Video Codec Design: Developing Image and Video Compression Systems*. New York: Wiley, 2002.

Roberts-Breslin, Jan. *Making Media: Foundations of Sound and Image Production*. Boston: Focal Press, 2003.

Thurott, Paul. *PC Magazine Windows XP Digital Media Solutions*. Indianapolis: Wiley, 2005.

Video Files. Available online. URL: http://www.fileinfo.net/filetypes/video. Accessed September 3. 2007.

cognitive science

Cognitive science is the study of mental processes such as reasoning, memory, and the processing of perception. It is necessarily an interdisciplinary approach that includes fields such as psychology, linguistics, and neurology. The importance of the computer to cognitive science is that it offers a potential nonhuman model for a thinking entity. The attempts at artificial intelligence over the past 50 years have used the insights of cognitive science to help devise artificial means of reasoning and perception. At the same time, the models created by computer scientists (such as the neural network and Marvin Minsky's idea of "multiple intelligent agents") have in turn been applied to the study of human cognition (see MINSKY, MARVIN LEE and NEURAL NETWORK).

Since the late 19th century, technological metaphors have been used to describe the human mind. The neurons and synapses of the brain were compared to the multitude of switches in a telephone company central office. The invention of digital computers seemed to offer an even more compelling correspondence between neurons and their electrochemical states and the binary state of a vacuum tube or transistor. It is only a small further step to assert that human mental processes can be reduced in principle to computation, albeit a very complex tapestry of computation. Various schools of popular psychology and personal improvement have offered simplistic images of the human mind suffering from "bad programming" that can be debugged or manipulated through various processes. The simulation of some forms of reasoning and language construction by AI programs certainly suggests that there are fruitful analogies between human and machine cognition, but construction of a detailed model that would be applicable to both human and artificial intelligences seemed almost as distant in the science fictional year of 2001 as it was when Alan Turing and other AI pioneers first considered such questions in the early 1950s (see TURING, ALAN MATHISON).

SYMBOLISTS AND CONNECTIONISTS

Unlike standard computer memory cells, neurons can have hundreds of potential connections (and thus states). If a human being is a computer, it must be to a considerable extent an analog computer, with input in the form of levels of various chemicals and electrical impulses. Yet in the 1980s, Allen Newell and Herbert Simon suggested that the "output" of human mental experience can be effectively mapped as relationships between symbols (words, images, and so forth) that correspond to physical states (this is called the Physical Symbol System Hypothesis). If so, then such a symbol system would be "computable" in the Turing-Church sense (see COMPUTABILITY AND COMPLEXITY). Working from the computer end, AI researchers have created a variety of programs that seem to "understand" restricted universes of discourse such as a table with variously shaped blocks upon it or "story frames" based upon common human activities such as eating in a restaurant. Thus, symbol manipulators can at least appear to be intelligent.

The "connectionists," however, argue that it is not symbolic representations that are significant, but the structure within the mind that generates them. By designing neural networks (or distributed processor networks) the connectionists have been able to create systems that produce apparently intelligent behavior (such as pattern recognition) without any reference to symbolic representation.

Critiques have also come from philosophers. Herbert Dreyfus has pointed out that computers lack the body, senses, and social milieu that shape human thought. That machines can generate symbolic representations according to some sort of programmed rules doesn't make the machine truly intelligent, at least not in the way experienced by human beings. John Searle responded to the famous Turing test (which states that if a human being can't distinguish a computer's conversation from a human's, the computer is arguably intelligent). Searle's "Chinese Room" imagines a room in which an English-speaking person who knows no Chinese is equipped with a program that lets him manipulate Chinese words in such a way that a Chinese observer would think he knows Chinese. Similarly, Searle argues, the computer might act "intelligently," but it doesn't really understand what it is doing.

Advances in cognitive science will both influence and depend on developments in brain research (especially the connection between physical states and cognition) and in artificial intelligence.

Further Reading
Bechtel, William, and Adele Abrahamson. *Connectionism and the Mind: Parallel Processing, Dynamics, and Evolution in Networks.* 2nd ed. Cambridge, Mass.: Blackwell, 2000.
"Cognitive Science." Stanford Encyclopedia of Philosophy. Available online. URL: http://plato.stanford.edu/entries/cognitive-science/. Accessed June 10, 2007.
Horgan, Terence, and John Tienson. *Connectionism and the Philosophy of Psychology.* Cambridge, Mass.: MIT Press, 1996.
Sobel, Carolyn. *Cognitive Science: An Interdisciplinary Approach.* New York: McGraw-Hill, 2001.
Thagard, Paul. *Mind: Introduction to Cognitive Science.* 2nd ed. Cambridge, Mass.: MIT Press, 2005.

color in computing

With the exception of a few experimental systems, color graphics first became widely available only with the beginnings of desktop computers in the late 1970s. The first microcomputers were able to display only a few colors (some, indeed, displayed only monochrome or grayscale). Today's PC video hardware has the potential to display millions of colors, though of course the human eye cannot directly distinguish colors that are too close together. There are several important schemes that are used to define a "color space"—that is, a range of values that can be associated with physical colors.

RGB

One of the simplest color systems displays colors as varying intensities of red, green, and blue. This corresponds to the electronics of a standard color computer monitor, which uses three electron guns that bombard red, green, and blue phosphors on the screen. A typical RGB color scheme uses 8 bits to store each of the red, green, and blue components for each pixel, for a total of 24 bits (16,777,216 colors). The 32-bit color system provides the same number of colors but includes 8 bits for *alpha,* or the level of transparency. The number of bits per pixel is also called the bit depth or color depth.

CMYK

CMYK stands for cyan, magenta, yellow, and black. This four component color system is standard for most types of color printing, since black is an ink color in printing but is simply the absence of color in video. One of the more difficult tasks to be performed by DESKTOP PUBLISHING software is to properly match a given RGB screen color to the corresponding CMYK print color. Recent versions of Microsoft Windows and the Macintosh operating system include a CMS (color matching system) to support color matching.

PALETTES

Although most color schemes now support thousands or millions of colors, it would be wasteful and inefficient to use three or four bytes to store the color of each pixel in memory. After all, any given application is likely to need only a few dozen colors. The solution is to set up a *palette,* which is a table of (usually 256) color values currently in use by the program. (A palette is also sometimes called a CLUT, or color lookup table.) The color of each pixel can then be stored as an index to the corresponding value in the palette.

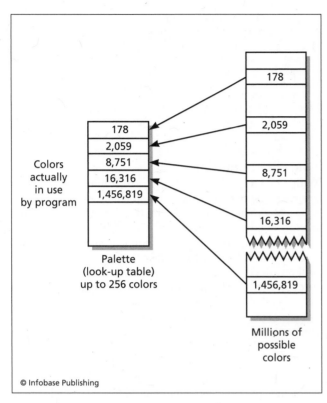

A color lookup table (CLUT) or palette can be used to store the colors actually being used by an image. Here up to 256 colors can be selected out of millions of possibilities.

The user of a paint program can select a palette from the full range of colors available from the operating system. Many color graphics image formats such as GIF (graphic interchange format) store a palette of the colors used by the image. When converting an image that has more colors that the palette can hold, various algorithms can be used to choose a palette that preserves as much of the color range as possible.

Further Reading
"Color" Webopedia. Available online. URL: http://www.webopedia. com/Graphics/Color/. Accessed June 10, 2007.

Drew, John, and Sarah Meyer. *Color Management: A Comprehensive Guide for Graphic Designers.* East Sussex, U.K.: RotoVision, 2005.

"Introduction to Color and Color Management Systems." Apple Computer Developer Connection. Available online. URL: http://developer.apple.com/documentation/mac/ACI/ACI46. html. Accessed June 10, 2007.

Koren, Norman. "Color Management and Color Science: Introduction." Available online. URL: http://www.normankoren.com/ color_management.html. Accessed June 10, 2007.

COM (common object model)   *See* MICROSOFT .NET.

common gateway interface   *See* CGI.

common object request broker architecture
See CORBA.

compatibility and portability
The computers of the 1940s were each hand built and unique. When the first commercial models were developed, such as the UNIVAC and the first IBM mainframes, the question of compatibility was born. Broadly speaking, compatibility is the degree to which a program or hardware device designed for one system can work with or run on another.

The designers of high-level languages usually intend that a source program written using the proper language syntax will compile and run on any system for which a compiler is available. However, there are many factors that can destroy compatibility. For example, if one machine stores the bytes of a numeric value from least significant to most significant while another does it in the opposite order, program code that depends on directly referencing memory locations will give the wrong results on one machine or another. Similarly, standard data sizes such as "integer" might be 16 bits on one system and 32 bits on another.

Language designers can minimize such problems by separating hardware-related issues from the language itself, as is the case with C and C++. A program is then linked with standard libraries implemented for each hardware or operating system environment.

Manufacturers often design newer models of their computers so they are "upwardly compatible" with existing models. This means that a program written for the smaller machine should run correctly on the new, larger one. This is of obvious benefit to users who do not want to have to rewrite their software every time they upgrade their machine. Often, however, such systems are not "downwardly compatible"—a program written for the new, larger machine may rely on features or architectural characteristics that are not available on the older, smaller machines. Sometimes a "compatibility mode" can be specified for a compiler or operating system. This restricts the use of features to those available on the older system.

Compatibility is also important with regard to software. Generally speaking, a newer version of a program such as a word processor will be able to read files that were originally created by a previous version, although this may not be true for more than a few versions back. However, files saved from the newest version may well be incompatible with older versions, because they contain formatting or other information that is not understandable by the earlier version. Sometimes an intermediate format (for example, see RTF, or Rich Text Format) can be used to transfer files between otherwise incompatible systems.

Compatibility between vendors can be an important competitive issue. If a developer wants to enter a market where one or two products are viewed as industry standards, the new product will have to be compatible with at least most files created by the dominant products. A technically superior product can thus be a market disaster if it is not compatible with the industry standard. In areas (such as graphics file formats) where there are many alternatives in widespread use, most programs will support multiple formats.

PORTABILITY

Portability is the ability to adapt software or hardware to a wide variety of platforms (that is, computer systems or operating systems). Developers want their products to be portable so they can adapt to an often rapidly changing marketplace. A typical strategy for portability is to choose a language that is in widespread use and a compiler that is certified as meeting the ANSI or other standard for the language. The program should be written in such a way that it makes as few assumptions as possible about hardware-dependent matters such as how data is stored in memory. It is also sometimes possible to use standard frameworks that provide the same functions in several different operating systems such as Windows, Macintosh, and UNIX.

However, there is a tradeoff: The more "generic" a program is made in order to be portable, the less optimized it will be for any given hardware or operating environment. The program will also not be able to take advantage of the special features of a given operating system, which may put it at a competitive disadvantage compared to the "native version" of a program. (This is particularly true with Windows, given that operating system's dominance in personal computing.)

The Internet has in general been a force for portability. The Java language, in particular, is designed to be platform-independent. A Java program is compiled into an intermediate language called byte code, which is interpreted or compiled by a "virtual machine" program running on each platform. Thus, the same Java program should run in a browser under Windows, Macintosh, or UNIX (see JAVA).

Further Reading

Hakuta, Mitsuari, and Masato Ohminami. "A Study of Software Portability Evaluation." *Journal of Systems and Software* 38 (August 1997): 145–154.
Robinson, John. "Delivering on Standards: Balancing Portability and Performance." Available online. URL: http://ipdps.cc.gatech.edu/1999/papers/it2.pdf. Accessed August 11, 2007.
"Software Portability Home Page." Available online. URL: http://www.cs.wvu.edu/~jdm/research/portability/home.html. Accessed June 11, 2007.

compiler

A compiler is a program that takes as input a program written in a source language and produces as output an equivalent program written in another (target) language. Usually the input program is in a high-level language such as C++ and the output is in assembly language for the target machine (see ASSEMBLER).

Compilers are useful because programming directly in low-level machine instructions (as had to be done with the first computers) is tedious and prone to errors. Use of assembly language helps somewhat by allowing substitution of symbols (variable names) for memory locations and the use of mnemonic names for operations (such as "add"

for addition, rather than some binary instruction code). An assembler is essentially a compiler that needs to make only relatively simple translations, because assembly language is still at a relatively low level.

Moving to higher-level languages with relatively English-like statements makes programming easier and makes programs easier to read and maintain. However, the task of translating high-level statements to machine-level code becomes a more complex multistep process.

THE COMPILATION PROCESS

Compilers are traditionally thought of as having a "front end" that analyzes the source code (high-level language statements) and a "back end" that generates the appropriate low-level code. The front end processing begins with *lexical analysis*. The compiler scans the source program looking for matches to valid tokens as defined by the language. A token is any word or symbol that has meaning in the language,

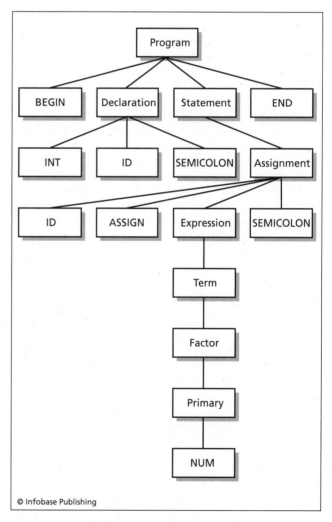

© Infobase Publishing

A parse tree showing how an assignment statement in Pascal can be broken down into its component parts. Here ID stands for a variable name, or identifier. An expression can be broken all the way down to a single number.

such as a keyword (reserved word) such as if or while. Next, the tokens are *parsed* or grouped according to the rules of the language. The result of parsing is a "parse tree" that resolves statements into their component parts. For example, an assignment statement may be parsed into an identifier, an assignment operator (such as =), and a value to be assigned. The value in turn may be an arithmetic expression that consists of operators and operands.

Parsing can be done either "bottom up" (finding the individual components of the statement and then linking them together) or "top down" (identifying the type of statement and then breaking it down into its component parts). A set of grammatical rules specifies how each construct (such as an arithmetic expression) can be broken into (or built up from) its component parts.

The next step is *semantic analysis*. During this phase the parsed statements are analyzed further to make sure they don't violate language rules. For example, most languages require that variables must be declared before they are referenced by the program. Many languages also have rules for which data types may be converted to other types when the two types are used in the same operation.

The result of front-end processing is an *intermediate representation* somewhere between the source statements and machine-level statements. The intermediate representation is then passed to the back end.

CODE GENERATION AND OPTIMIZATION

The process of code generation usually involves multiple passes that gradually substitute machine-specific code and data for the information in the parse tree. An important consideration in modern compilers is *optimization,* which is the process of substituting equivalent (but more efficient) constructs for the original output of the front end. For example, an optimizer can replace an arithmetic expression with its value so that it need not be repeatedly calculated while the program is running. It can also "hoist out" an invariant expression from a loop so that it is calculated only once before the loop begins. On a larger scale, optimization can also improve the communication between different parts (procedures) of the program.

The compiler must attempt to "prove" that the change it is making in the program will never cause the program to operate incorrectly. It can do this, for example, by tracing the possible paths of execution through the program (such as through branching and loops) and verifying that each possible path yields the correct result. A compiler that is too "aggressive" in making assumptions can produce subtle program errors. (Many compilers allow the user to control the level of optimization, and whether to optimize for speed or for compactness of program size.) During development, a compiler is often set to include special debugging code in the output. This code preserves potentially important information that can help the debugging facility better identify program bugs. After the program is working correctly, it will be recompiled without the debugging code.

The final code generation is usually accomplished by using templates that match each intermediate construction with a construction in the target (usually assembly)

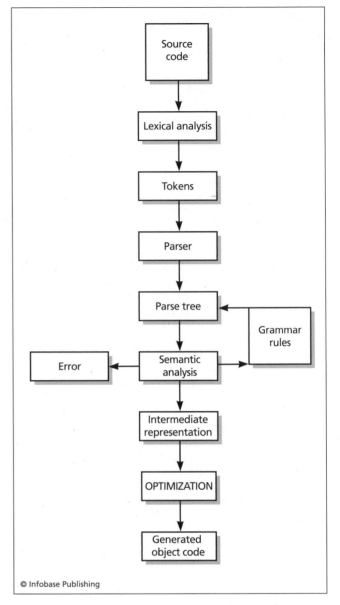

Compilation is a multistep process. Lexical analysis breaks statements down into tokens, which are then parsed and subjected to semantic analysis. The resulting intermediate representation can be optimized before the final object code is generated.

language, plugging items in as specified by the template. Often a final step, called *peephole optimization,* examines the assembly code and identifies redundancies or, if possible, replaces a memory reference so that a faster machine register is used instead.

In most applications the assembly code produced by the compiler is linked to code from other source files. For example, in a C++ applications class definitions and code that use objects from the classes may be compiled separately. Also most languages (such as C and C++) have operating system-specific libraries that contain commonly used support functions.

As an alternative to bringing the external code into the final application file, code can be "dynamically linked" to libraries that will be accessed only while the program is being run. This eliminates the waste that would occur if several running applications are all using the same standard library code (see LIBRARY, PROGRAM).

In mainframes compilers were usually invoked as part of a batch file using some form of JCL (job control language). With operating systems such as UNIX and MS-DOS a program called *make* is typically used with a file that specifies the compiler, linker, and other options to be used to compile the program. Modern visually oriented development environments (such as those provided by products such as Visual C++) allow options to be set via menus or simply by selecting from a variety of typical configurations.

Compiler design has become a highly complex field. A modern compiler is developed using a variety of tools (including packaged parsers and lexical analyzers), and involves a large team of programmers. Nevertheless, the principles of compiler design are emphasized in the general computer science curriculum because when a student understands even a simplified compiler in detail, he or she has become acquainted both with important ideas (such as language grammar, parsing, and optimization) and with many levels of understanding computer architecture.

Further Reading

Aho, Alfred V., Ravi Sethi, and Jeffrey D. Ullman. *Compiler Design: Principles, Techniques, and Tools.* 2nd ed. Reading, Mass.: Addison-Wesley, 2006.

"Compiler Connection: A Resource for Compiler Developers and Those Who Use Their Products and Services." Available online. URL: http://www.compilerconnection.com/. Accessed August 12, 2007.

Grune, Dick, et al. *Modern Compiler Design.* New York: Wiley, 2000.

component object model (Microsoft) *See* MICROSOFT .NET.

computability and complexity

Interestingly, one of the important discoveries of 20th-century mathematics is that certain kinds of problems were not computable. The Turing machine and Alonzo Church's lambda calculus provided equivalent models that could be used to determine what was computable (see TURING, ALAN MATHISON, and CHURCH, ALONZO). Thus far, the equivalence between the Turing machine and actual computers has held. That is, any decision problem (a problem with a "yes" or "no" answer) that can in theory be solved with a Turing machine can in theory be solved by any actual computer. Conversely, if a problem can't be solved by a Turing machine, it cannot be solved by a computer, no matter how powerful.

THE HALTING PROBLEM

The Halting problem is a classic example of an undecidable problem (or proposition). The problem is this: Given any computer program, can you determine whether the program will halt (end) given any input? There are specific programs that can be shown to halt on particular inputs. For example, this program:

```
If Input = 99 then end.
```

will obviously halt on an input of 99. But to decide whether a determination can be made for any program for any input, it is only necessary to construct a logical paradox. Assume that there is a program P that halts if and only if it receives input D. (Further assume that the program can print something to let you know that it has halted.)

Since the input can be anything, you can let it be a copy of the program itself. The question then becomes: Will the program halt if it is given a copy of itself? Create a procedure (or subroutine) called HaltTest, and define it as:

```
If P halts then print "Halted"
   else print "Didn't Halt."
```

Now create another program called Main. It calls HaltTest and is programmed to do the opposite of what HaltTest indicates.

```
If HaltTest (Main) prints "Yes" then loop
forever else halt;
```

But what happens when Main is run? It calls HaltTest, giving itself (Main) as input. If HaltTest halts, then Main loops forever. But if HaltTest doesn't halt, then Main halts. But this means that Main halts if it doesn't halt, and doesn't halt if it halts. This paradox shows that whether Main halts is undecidable.

The undecidability of the Halting problem has some interesting implications. For example, it means that there is no way a computer can reliably determine that a program does not contain an infinite loop. Also, because a mathematical function $f(x)$ is equivalent to a computer program with input x, similar proofs by contradiction can be written to show that it can't be decided whether a program will halt on all inputs (which is equivalent to $f(x)$ being defined for all x.) Nor can it be decided whether two different programs (or mathematical functions) are equivalent for all x.

It is important to realize that a program (or function) being undecidable in all cases doesn't necessarily mean that it can't be decided for some cases (or inputs). Indeed, the answer of the Halting Problem for any given input can be determined by feeding that particular input to the program, which will either halt or run forever.

COMPLEXITY

If a problem turns out to be computable, we then enter the realm of complexity—the analysis of how much computation will be required (see ALGORITHM). Sometimes a designer can devise a significantly faster algorithm for a given problem (such as finding prime factors or sorting). However, other problems appear to have complexity based on an exponential expression, meaning that they become more complex much more rapidly as the input increases. An example is the Traveling Salesman Problem, which is to find the most efficient route for a person traveling to a number of cities to visit each of the cities.

Mathematicians therefore categorize the complexity of problems as P (solvable in a polynomial period of time), EXP (requiring an exponential time), or an intermediate class NP, which means "nondeterministic polynomial." An NP problem is one that can be solved in polynomial time if one is able to guess (and then verify) the answer. The Traveling Salesman Problem is believed to be in the NP class.

While abstruse, the study of computability and complexity has important implications for practical applications. For example, determining the complexity of a cryptographic algorithm can help determine whether the resulting encryption is strong enough to withstand the efforts of a feasible attacker.

Further Reading

Boolos, George S., John P. Burgess, and Richard C. Jeffrey. *Computability and Logic.* 4th ed. New York: Cambridge University Press, 2002.

Jones, Neil D. *Computability and Complexity: From a Programming Perspective.* Cambridge, Mass.: MIT Press, 1997.

Sipser, Michael. *Introduction to the Theory of Computation.* 2nd ed. Boston: Thomson Course Technology, 2006.

computer-aided design and manufacturing (CAD/CAM)

The use of computers in the design and manufacturing of products revolutionized industry in the last quarter of the 20th century. Although computer-aided design (CAD) and computer-aided manufacturing (CAM) are different areas of activity, they are now so closely integrated that they are often discussed together as CAD/CAM.

COMPUTER-AIDED DESIGN

In 1950, science fiction writer Robert Heinlein had his future inventor create "Drafting Dan," an automated drafting system that would enable designers to turn their ideas into manufacturing plans in a fraction of the time required for the hand preparation of schematics and parts lists. By the 1960s, engineers had developed the first computer-assisted design programs, running on terminals attached to mainframe computers.

The activity of a CAD workstation centers on the creation of geometrical models (first 2D, then 3D). With the aid of models, a virtual representation of the product being designed can be built up. With its knowledge of geometrical and physical relationships, routines in the CAD system can perform not only measurement of dimensions and mass but also structural analysis. (In some cases CAD can be interfaced with systems that provide full-blown simulation of the effects of stresses, heat, and other factors.)

The growth of desktop computing power in the 1980s and 1990s moved CAD from the mainframe to the high-end workstation (such as those built by Sun Microsystems) and even to high-end personal computers. The growing processing power also meant that the geometric models could become more sophisticated, including solid models with realistically rendered surfaces rather than just wireframes. The model of surfaces can include such factors as reflectivity, friction, or even aerodynamic characteristics. In designing a product (or a subsystem of a product), engineers can now use simulation software to determine how well a group of parts in a complex assembly (such as a car's steering mechanism) will perform. The ability to get detailed data in real time means that the CAD operator can work in a feedback loop in which the design is incrementally refined until the required parameters are met.

This growing modeling capability has been combined with the use of detailed databases containing the standard parts used in a particular industry or application. Libraries of templates allow the designer to "plug in" standard assemblies of parts and then modify them. The databases can also be used with algorithms that can assist the designer in optimizing the design for some desired characteristic, such as strength, light weight, or lower cost. Recent systems even have the capability to set "strategic" design goals for a whole family of products and to identify particular optimizations that would help each part or subsystem achieve those goals.

COMPUTER-AIDED MANUFACTURING

The automated fabricating of products on the factory floor originally developed independently of computer-aided design. Numerically controlled machine tools and lathes can be programmed using specialized languages such as APT (Automatically Programmed Tool) or more recently, through a system that uses a graphical interface. Advances in pattern recognition and other artificial intelligence techniques have been used to improve the ability of the automatic tool to identify particular features (such as holes into which bolts are to be inserted) and to properly orient surfaces. At some point the programmability and flexibility of the system with regard to its ability to manipulate the environment gives it the characteristics of a robot (see ROBOTICS).

INTEGRATION OF CAD AND CAM

As CAD systems became more capable, it soon became evident that there could be substantial benefits to be gained from integrating the design and manufacturing process.

The CAD software can also output detailed parts and assembly specifications that can be fed into the CAM process. In turn, manufacturing considerations can be applied to the selection of parts during the design process.

The integration of design, simulation, and manufacturing continues. The goal is to give the engineer a seamless way to "tweak" a design and have a number of simulation modules automatically depict the effects of the design change. In essence, the designer or engineer would be working in a virtual world that accurately reflects the physical constraints that the product will face in the real world.

The automation of the design and manufacturing process has been mainly responsible for the increasing productivity of modern factories. Factories using traditional methods in producing complex products such as automobiles or consumer electronics have generally had to refit for CAD/CAM in order to remain competitive. Low-skill but relatively high-paying factory jobs characteristic of the earlier industrial era have given way to smaller numbers of more technical jobs. This has meant a greater emphasis on education and specialized training for the industrial workforce.

Further Reading
Amirouche, Farid M. *Principles of Computer Aided Design and Manufacturing*. 2nd ed. Upper Saddle River, N.J.: Prentice Hall, 2003.
CADLAB (MIT). Available online. URL: http://cadlab.mit.edu/. Accessed June 12, 2007.
"Computer-Aided Design" [outline and knowledge base]. Compinfo.ws. Available online. URL: http://www.compinfo-center.com/cad/cad.htm. Accessed June 12, 2007.
Duggal, Vijay. *CADD Primer: A General Guide to Computer Aided Design and Drafting*. New York: Mailmax Publications, 2000.

computer-aided instruction (CAI)

Also called computer based training (CBT), computer-aided instruction (CAI) is the use of computer programs to provide instruction or training. (See EDUCATION AND COMPUTERS for a more comprehensive discussion of the use of computers for teaching and learning.)

The American reaction to Soviet space achievements led to many attempts to modernize the educational system. While the high cost and limited capabilities of 1950s computing technology allowed only for theoretical research by IBM and some universities, by the 1960s more powerful solid-state computers were starting to make what were then called "teaching machines" practicable. The first large-scale initiative was the PLATO teaching system designed by the Computer-based Educational Research Laboratory at the University of Illinois, Urbana. PLATO used a large timesharing system to provide educational software to about a thousand users at terminals throughout the university. PLATO pioneered the use of graphics and what would later be called multimedia, and was eventually marketed by Control Data Corporation, a leading manufacturer of high-end mainframe computers. Stanford University also began a large-scale initiative to deliver computerized instruction.

The early CAI systems required expensive hardware, however, and generally could be sustained only by research funding or where they met the growing training needs of the military, the aerospace industry, or other specialized users. However, the advent of the personal computer in the late 1970s provided both a new technology for delivering educational software and a potential market. With its color graphics and astute marketing the Apple II had became a staple of classrooms by the mid-1980s, when its successor, the Macintosh, brought more advanced graphics (see MACINTOSH) and a program called Hypercard that made it easy for educators to create simple interactive presentations (see HYPERTEXT AND HYPERMEDIA). The Intel-based IBM PC and its "clones" also gained a foothold in the classroom, and Microsoft Windows brought a graphical interface similar to that on the Macintosh.

APPLICATIONS

The simplest (and probably least interesting) form of CAI is often called "drill and practice" programs. Such programs (usually found in the elementary grades) repetitively present math problems, reading vocabulary, or other exercises and test the user's understanding. (Teaching keyboard skills to young students is another common application.) In an attempt to hold the student's interest, many such programs provide a gamelike atmosphere and offer periodic rewards or reinforcement for success.

More sophisticated programs allow the student more creative scope, such as by letting the student program and test virtual "robots" as a means of mastering a programming language. Many computer games, while not designed explicitly for instruction, provide simulations that exercise thinking and planning skills (see COMPUTER GAMES). (For example, the strategy game *Civilization* incorporates concepts such as resource management, labor allocation, and a balanced economy.) Even more sophisticated programs use advanced programming (see ARTIFICIAL INTELLIGENCE) to interact with students in ways similar to those used by human teachers. For example, a program called Cognitive Tutor, now used in many schools, can recognize different "styles" of learning and approaches to solving, for example, an algebra problem. The program can also identify a student's specific weaknesses and tailor practice and supplemental instruction accordingly. These programs can teach and reinforce reasoning skills rather than just imparting specific knowledge.

Industry remains a large market for computer-based training. A variety of CBT packages are available for introducing and teaching programming languages such as C++ and Java as well as for preparing students to earn industry certificates such as the A+ certificate for computer technicians.

TRENDS

Two continuing trends in CAI are the growing use of graphics and multimedia, including video or movies, and the increasing delivery of training via the Internet. Some training software can be accessed directly over the Internet through a Web browser, without requiring special software on the user's PC. Increasingly, even products delivered on CD and run from the user's PC include links to supplemental material on the Web.

Further Reading
Horton, William. *E-Learning by Design*. San Francisco: Pfeiffer, 2006.
Ko, Suasan Schor and Steve Rossen. *Teaching On-line: A Practical Guide*. 2nd ed. Boston: Houghton Mifflin, 2003.
Rosenberg, Marc J. *E-Learning: Strategies for Delivering Knowledge in the Digital Age*. New York: McGraw-Hill, 2001.
Viadero, Debra. "New Breed of Digital Tutors Yielding Learning Gains." *Education Week*, April 2, 2007. Available online. URL: http://www.edweek.org/ew/articles/2007/04/02/31intelligent.h26.html. Accessed June 13, 2007.
Watkins, Ryan, and Michael Corry. *E-Learning Companion: A Student's Guide to Online Success*. Boston: Houghton Mifflin, 2004.
Web-Based Training Information Center. Available online. URL: http://wbtic.com. Accessed June 12, 2007.

computer crime and security

The growing economic value of information, products, and services accessible through computer systems has attracted increased attention from opportunistic criminals. In particular, the many potential vulnerabilities of online systems and the Internet have made computer crime attractive and pose significant challenges to professionals whose task it is to secure such systems.

The motivations of persons who use computer systems in unauthorized ways vary. Some hackers primarily seek detailed knowledge of systems, while others (often teenagers) seek "bragging rights." Other intruders have the more traditional criminal motive of gaining access to information such as credit card numbers and personal identities that can be used to make unauthorized purchases (see IDENTITY THEFT). Computer access can also be used to intimidate (see CYBERSTALKING AND HARASSMENT), as well as for extortion, espionage, sabotage, or terrorism (see CYBERTERRORISM). Attacking and defending information infrastructure is now a vital part of military and homeland security planning (see INFORMATION WARFARE).

According to the federal Internet Crime Complaint Center, in 2006 the most commonly reported computer-related crime was auction-related fraud (44.9 percent), followed by nondelivery of goods (19 percent)—these no doubt reflect the high volume of auction and e-commerce transactions. Various forms of financial fraud (including identity theft) make up most of the rest.

The new emphasis on the terrorist threat following September 11, 2001, has included some additional attention to cyberterrorism, or the attack on computers controlling key infrastructure (including banks, water and power systems, air traffic control, and so on). So far ideologically inspired attacks on computer systems have mainly amounted to simple electronic vandalism of Web sites. Internal systems belonging to federal agencies and the military tend to be relatively protected and isolated from direct contact with the Internet. However, the possibility of a crippling attack or electronic hijacking cannot be ruled out. Commercial systems may be more vulnerable to denial-of-service attacks (see below) that cause economic losses by preventing consumers from accessing services.

FORMS OF ATTACK

Surveillance-based attacks involve scanning Internet traffic for purposes of espionage or obtaining valuable information. Not only businesses but also the growing number of Internet users with "always-on" Internet connections (see BROADBAND) are vulnerable to "packet-sniffing" software that exploits vulnerabilities in the networking software or operating system. The main line of defense against such attacks is the software or hardware firewall, which both "hides" the addresses of the main computer or network and identifies and blocks packets associated with the common forms of attack (see FIREWALL).

In the realm of harassment or sabotage, a "denial of service" (DOS) attack can flood the target system with packets that request acknowledgment (an essential feature of network operation). This can tie up the system so that a Web server, for example, can no longer respond to user requests, making the page inaccessible. More sophisticated DOS attacks can be launched by first using viruses to insert programs in a number of computers (a so-called botnet), and then instructing the programs to simultaneously launch attacks from a variety of locations.

Computer viruses can also be used to randomly vandalize computers, impeding operation or destroying data (see COMPUTER VIRUS). But a virus can also be surreptitiously inserted as a "Trojan horse" into a computer's operating system where it can intercept passwords and other information, sending them to the person who planted the virus. Viruses were originally spread through infected floppy disks (often "bootleg" copies of software). Today, however, the Internet is the main route of access, with viruses embedded in e-mail attachments. This is possible because many e-mail and other programs have the ability to execute programs (scripts) that they receive. The main defense against viruses is regular use of antivirus software, turning off scripting capabilities unless absolutely necessary, and making a policy of not opening unknown or suspicious-looking e-mail attachments as well as messages that pretend to be from reputable banks or other agencies (see PHISHING AND SPOOFING).

COMPUTER SECURITY

Because there are a variety of vulnerabilities of computer systems and of corresponding types of attacks, computer security is a multifaceted discipline. The vulnerability of computer systems is not solely technical in nature. Sometimes the weakest link in a system is the human link. Hackers are often adept at a technique they call "social engineering." This involves tricking computer operators into giving out sensitive information (such as passwords) by masquerading as a colleague or someone else who might have a legitimate need for the information.

Since computer crimes and attacks can take so many forms, the best defense is layered or in depth. It includes appropriate software (firewalls and antivirus programs, and network monitoring programs for larger installations). It emphasizes proper training of personnel, ranging from security investigators to clerical users. Finally, if information is compromised, the use of strong encryption can make it much less likely to be usable (see ENCRYPTION).

While the flexibility and speed of the Internet can aid attackers, it can also facilitate defense. Emergency response networks and major vendors of antivirus software can quickly disseminate protective code or "patches" that close vulnerabilities in operating systems or applications.

The growing concern about vulnerability to computer intrusion and information theft has also been reflected in attempts to make operating systems inherently more secure. The introduction of new security features in Microsoft Windows Vista has received mixed reviews. Some features, such as User Account Control, make it harder for viruses or other automated attacks to access critical system resources, but also annoy users by constant requests for permission to carry out common tasks. This reflects a fundamental truth: Security features that make everyday computing more tedious tend to be turned off or bypassed by users.

Once a computer-based crime is detected, a systematic approach to evidence gathering and investigation is required (see COMPUTER FORENSICS). This is because evidence in computer crimes tends to be technical, intangible, and transient, and thus difficult to explain properly to judges and juries.

Individual consumers can reduce their vulnerability by ensuring that they do not give out personal information without verifying both the requester and the need for the data. Use of secure Web sites for credit card transactions has become standard. Generally speaking, vulnerability to computer crime is inversely proportional to the degree of privacy individuals have with regard to their personal information (see PRIVACY IN THE DIGITAL AGE). Public concern about privacy and security has led to recent laws and initiatives aimed at disclosure of organizations' privacy policy and limiting the redistribution of information once collected.

Further Reading
Balkin, J. M. *Cybercrime: Digital Cops in a Networked Environment.* New York: New York University Press, 2007.
CERT Coordination Center, Carnegie-Mellon University. Available online. URL: http://www.cert.org. Accessed August 12, 2007.
Easttom, Chuck. *Computer Security Fundamentals.* Upper Saddle River, N.J.: Prentice Hall, 2005.
McQuade, Sam C. *Understanding and Managing Cybercrime.* Boston: Allyn & Bacon, 2005.
Mitnick, Kevin, and William L. Simon. *The Art of Intrusion.* New York: Wiley, 2005.

computer engineering

Computer engineering involves the design and implementation of all aspects of computer systems. It is the practical complement to computer science, which focuses on the study of the theory of the organization and processing of information (see COMPUTER SCIENCE). Because hardware requires software (particularly operating systems) in order to be useful, computer engineering overlaps into software design, although the latter is usually considered to be a separate field (see SOFTWARE ENGINEERING).

To get an idea of the scope of computer engineering, consider the range of components commonly found in today's desktop computers:

PROCESSOR

The design of the microprocessor includes the number and width of registers, method of instruction processing (pipelining), the chipset (functions to be integral to the package with the microprocessor), the amount of cache, the connection to memory bus, the use of multiple processors, the order in which data will be moved and stored in memory (low or high-order byte first?), and the clock speed. (See MICROPROCESSOR, CHIPSET, CACHE, BUS, MULTIPROCESSING, MEMORY, and CLOCK SPEED.)

MEMORY

The design of memory includes the type (static or dynamic) and configuration of RAM, the maximum addressable memory, and the use of parity for error detection (see MEMORY, ADDRESSING, and ERROR CORRECTION). Besides random-access memory, other types of memory include ROM (read-only memory) and CMOS (rewritable persistent memory).

MOTHERBOARD

The motherboard is the platform and data transfer infrastructure for the computer system. It includes the main data bus and secondary buses (such as for high-speed connection between the processor and video subsystem—see BUS). The designer must also decide which components will be integral to the motherboard, and which provided as add-ons through ports of various kinds.

PERIPHERAL DEVICES

Peripheral devices include fixed and removable disk drives; CD and DVD-ROM drives, tape drives, scanners, printers, and modems.

DEVICE CONTROL

Each peripheral device must have an interface circuit that receives commands from the CPU and returns data (see GRAPHICS CARD).

INPUT/OUTPUT AND PORTS

A variety of standards exist for connecting external devices to the motherboard (see PARALLEL PORT, SERIAL PORT, and USB). Designers of devices in turn must decide which connections to support.

There are also a variety of input devices to be handled, including the keyboard, mouse, joystick, track pad, graphics tablet, and so on.

Of course this discussion isn't limited to the desktop PC; similar or analogous components are also used in larger computers (see MAINFRAME, MINICOMPUTER, and WORKSTATION).

NETWORKING

Networking adds another layer of complexity in controlling the transfer of data between different computer systems, using various typologies and transport mechanisms (such as Ethernet); interfaces to connect computers to the network; routers, hubs, and switches (see NETWORK).

OTHER CONSIDERATIONS

In designing all the subsystems of the modern computer and network, computer engineers must consider a variety of factors and tradeoffs. Hardware devices must be designed with a form factor (size and shape) that will fit efficiently into a crowded computer case. For devices that require their own source of power, the capacity of the available power supply and the likely presence of other power-consuming devices must be taken into account. Processors and other circuits generate heat, which must be dissipated. (In an increasingly energy-conscious world the reduction of energy consumption, such as through standby or "hibernation" modes, is also an important consideration—see GREEN PC.) Heat and other forms of stress affect reliability. And in terms of how a device processes input data or commands, the applicable standards must be met. Finally, cost is always an issue.

Moving beyond hardware to operating system (OS) design, computer engineers must deal with many additional questions, including the file system, how the OS will communicate with devices (or device drivers), and how applications will obtain data from the OS (such as the contents of input buffers). Today's operating systems include hundreds of system functions. Since the 1980s, the provision of all the objects needed for a standard user interface (such as windows, menus, and dialog boxes) has been considered to be part of the OS design. Finally, the building of security features into both hardware and operating systems has become an integral part of computer engineering (see, for example, BIOMETRICS and ENCRYPTION).

TRENDS

In the early days of mainframe computing (and again at the beginning of microcomputing) many distinctive system architectures entered the market in rapid succession. For example, the Apple II (1977), IBM PC (1981), and Apple Macintosh (1984) (see IBM PC and MACINTOSH). Because architectures are now so complex (and so much has been invested in legacy hardware and software), wholly new architectures seldom emerge today. Because of the complexity and cost involved in creating system architectures, development tends to be incremental, such as adding PCI card slots to the IBM PC architecture while retaining older ISA slots, or replacing IDE controllers with EIDE.

The growing emphasis on networks in general and the Internet in particular has probably diverted some effort and resources from the design of stand-alone PCs to network and telecommunications engineering. At the same time, new categories of personal computing devices have emerged over the years, including the suitcase-size "transportable" PC, the laptop, the book-sized notebook PC, the handheld PDA (personal digital assistant), as well as network-oriented PCs and "appliances." (See PORTABLE COMPUTERS and SMARTPHONE.)

As computing capabilities are built into more traditional devices (ranging from cars to home entertainment centers), computer engineering has increasingly overlapped other fields of engineering and design. This often means thinking of devices in nontraditional ways: a car that is able to plan travel, for example, or a microwave that can keep track of nutritional information as it prepares food (see EMBEDDED SYSTEM). The computer engineer must consider not only the required functionality but the way the user will access the functions (see USER INTERFACE).

Further Reading

IEEE Computer Society. Available online. URL: http://www.computer.org

Patterson, D. A. and J. L. Hennessy. *Computer Organization and Design.* 3rd ed. San Francisco: Morgan Kaufmann, 2004.

"PC Guide." Available online. URL: www.pcguide.com. Accessed June 18, 2007.

Stokes, John. *Inside the Machine: An Illustrated Introduction to Microprocessors and Computer Architecture.* San Francisco: No Starch Press, 2007.

computer forensics

Computer forensics is the process of uncovering, documenting, analyzing, and preserving criminal evidence that has been stored on (or created using) a computer system. (For the use of computers by police, see LAW ENFORCEMENT AND COMPUTERS.)

In general, computer forensics involves both adherence to legal evidentiary standards and the use of sophisticated technical tools. The legal standards require practices similar to those used in obtaining other types of criminal evidence (observing expectations of privacy, knowing when a warrant is needed to search and seize evidence, and so on).

Once there is a go-ahead for a search, the first step is to document the layout and nature of the equipment (generally by photographing it) and to identify both devices that might be problematic or notes or other materials that might reveal passwords for encrypted data.

If the system is running it may be viewed or scanned to determine what applications are running and what network connections may be active. However, this has to be done as unobtrusively as possible, since some machines can detect physical intrusions.

Step by step, the forensic technician must document each software program or other tool used, and why it is justified (such as the possibility that simply shutting down the system might lead to loss of data in RAM). There are a variety of such tools, particularly for UNIX/Linux environments, some of which have been ported to Windows. (In some cases a Linux "live" CD might be booted and used to explore a Windows file system.)

The next step is to collect the evidence from storage media in such a way as to ensure that it is accurately and completely preserved. A running machine must generally first be shut down in such a way as to prevent triggering any "trip wire" or intrusion-detection or self-destruct mechanism that may have been installed.

As a practical matter, once the system has been properly shut down or immobilized, it is usually taken to the forensic laboratory for extraction, copying, and documenting of the evidence (such as files on a hard drive or other storage device).

Once the data has been collected, each file or document must be analyzed to determine if it is relevant to the criminal investigation and what key information it contains. For example, e-mail headers may be analyzed to determine the source and routing of the message.

SOME TYPICAL CASES

Computer-based evidence may be relevant for almost any type of crime, but certain kinds of crimes are more likely to involve computer forensics. These include:

- financial crimes, such as embezzlement

- corporate crimes such as insider trading, where e-mails may reveal who knew what and when

- data or identity theft, including online scams or phishing

- stalking or harassment, particularly involving chat rooms or social networks

- child pornography, particularly distribution of images

In recent years many law enforcement agencies have become aware of the importance of proper investigation and treatment of evidence in our digital society, and demand for trained computer forensic specialists is expected to increase.

Further Reading

Britz, Marjie T. *Computer Forensics and Cyber Crime: An Introduction.* Upper Saddle River, N.J.: Prentice Hall, 2003.

Carrier, Brian. *File System Forensic Analysis.* Upper Saddle River, N.J.: Addison-Wesley Professional, 2005.

"Searching and Seizing Computers and Obtaining Electronic Evidence in Criminal Investigations." U.S. Dept. of Justice, July 2002. Available online. URL: http://www.usdoj.gov/criminal/cybercrime/s&smanual2002.htm. Accessed September 3, 2007.

Steel, Chad. *Windows Forensics: The Field Guide for Conducting Corporate Computer Investigations.* Indianapolis: Wiley, 2006.

Vacca, John R. *Computer Forensics: Computer Crime Scene Investigation.* 2nd ed. Hingham, Mass.: Charles River Media, 2005.

computer games

Today, playing games is one of the most popular computing activities. In the early days of computing, games offered a way to test AI techniques (see ARTIFICIAL INTELLIGENCE). Games have also encouraged the development of more sophisticated graphics (see COMPUTER GRAPHICS) and ways of interacting with the machine (see USER INTERFACE).

GAMES AND AI

Although modern computer games may draw upon several genres, several recognizably distinct types of games have been developed over the past half century or so. The first were computer versions of existing board games. "Deterministic" games (where there is no element of chance) such as tick-tack-toe and, more important, checkers and chess offered a challenge to the first computer scientists who were seeking to learn how to make machines perform tasks that are usually attributed to human intelligence. For example, Alan Turing and Claude Shannon both developed chess-playing programs, although Turing's came at a time when computers were still too primitive to handle the volume of calculations required, and was thus carried out by hand. By the time a computer program (Deep Blue) had defeated the world champion in 1997, the AI field had long since left the game behind (see CHESS AND COMPUTERS).

SIMULATION GAMES

Military planners had devised war games since the 19th century, but the complexity of modern warfare (including logistics as well as tactics) cried out for the help of the computer. By 1955 the U.S. military was running large-scale global cold war simulations pitting NATO against the USSR and the Eastern bloc. Unlike deterministic games such as chess, war games generally use complex rules to capture

A scene from the computer strategy game Civilization. Some games specialize in realistic physical simulation, while others (such as this one) embody sophisticated economic and strategic considerations.

the many interacting factors such as the morale, experience, and firepower of a military unit or the performance of an air defense system against different types of targets. The results will be more or less realistic depending on how many factors are properly accounted for—often only later combat experience will tell.

The use of game theory (the mathematics of competitive situations) and economics also proved to be fruitful areas for the use of computer simulations. In 1959 Carnegie Tech (later Carnegie-Melon University) introduced a simulation called "The Management Game." Until the 1980s, however, lack of inexpensive computing power kept sophisticated simulations limited to large institutions such as the military, government, universities, and major corporations.

Today simulation games are popular in both the educational and consumer markets. They include flight simulators, a variety of sports including baseball, football, soccer, and golf, and games in which the player strives to build a 19th-century railroad empire or run a modern city. Indeed some games, such as the popular kingdom-building simulator *Civilization* or the complex *Sim City,* while marketed primarily as entertainment, can easily fit into a social studies curriculum.

ARCADE AND GRAPHIC GAMES

Starting in the 1960s, CRT (television-like) displays gave the new minicomputers the means to display simple graphics. In 1962, an intrepid band of game hackers at MIT created *Spacewar,* the first interactive graphic game and the forerunner of the arcade boom of the 1970s. When the first home computers from Apple, Commodore, Atari, and IBM hit the market in the late 1970s and early 1980s, they included rudimentary (but often colorful) graphics capabilities. Many amateur programmers used the computers' built-in BASIC language to create games such as lunar lander simulators and *Star Trek*–style space battles. Around the same time, the home game cartridge machine was introduced by Atari and other companies, while the arcade game Pac-Man became a phenomenal success in 1980 (see GAME CONSOLES).

ROLE-PLAYING, REAL-TIME, AND SOCIAL WORLDS

Around the time of the first home computers, a noncomputer game called *Dungeons and Dragons* became extremely popular. "D&D" and other role-playing games allowed players to create and portray characters, with elaborate rules being used to resolve events such as battles. Role-playing games soon began to appear on PCs—early examples include the *Wizardry* and *Ultima* series. Meanwhile, text-based adventure games were becoming popular on early computer networks, particularly at universities. These evolved into MUDs (Multi-User Dungeons) where players' characters could interact with each other. Eventually many of these programs went beyond their adventuring roots to create a variety of social worlds in a sort of text-based virtual reality.

By the 1990s, the typical PC had a special circuit (see GRAPHICS CARD) capable of displaying millions of colors, together with video memory (now 256 MB or more) that could hold the complex images needed for high-resolution animation. Computer game graphics have become increasingly complex (see COMPUTER GRAPHICS), including realistic textures, shading and light, smooth animation, and special effects rivaling Hollywood. (Compare, for example, early wireframe graphics in games such as the *Wizardry* of 1980 with games such as *Diablo II* and *Warcraft* with animated characters moving in a richly textured world.)

The way players interact with the game world has also significantly changed. The first computer games tended to be divided into turn-based strategy and role-playing games and real-time arcade-style "shoot 'em ups." Today, however, most games, regardless of genre, run as RTS (real-time simulations) in which players must interact continuously with the game situation.

By the late 1990s gaming was no longer a solitary pursuit. The Internet made it possible to offer game worlds in which thousands of players could participate simultaneously (see ONLINE GAMES). Games such as *Everquest* and *Asheron's Call* have thousands of devoted players who spend many hours developing their characters' skills, while open-ended worlds such as *Second Life* seem to no longer be games at all, but a virtual, parallel universe with a full range of social interaction. However, the increased realism of modern games has also heightened the controversy about in-game violence and other antisocial behavior, as in the *Grand Theft Auto* series. (Although there is a rating system for games similar to that for movies, its effectiveness in keeping adult-themed games out of the hands of young children seems to be limited.)

GAME DEVELOPMENT

The emphasis on state-of-the-art animation and graphics and multiplayer design has changed the way game development is done. The earliest home computer games were typically the product of a single designer's vision, such as Chris Crawford's *Balance of Power* and Richard Garriott ("Lord British") in the *Ultima* series. Today, however, commercially competitive games are the product of teams that include graphics, animation, and sound specialists, actors and voice talent, and other specialists in addition to the game designers. While earlier games might be compared to books with single authors, modern game developers often compare their industry to the movie industry with its dominant studios. And, as with the movie industry, critics have argued that the high cost of development and of access to the market has led to much imitation of successful titles and less innovation.

On the other hand, a variety of modern programming environments (such as Visual Basic or even Macromedia Flash) make it easy for young programmers to get a taste of game programming, and for amateur programmers to create games that can be distributed via the Internet (see SHAREWARE AND FREEWARE). Although computer science programs have been slow to recognize the attraction and value of game programming, a variety of academic programs are now emerging. These range from computer arts, graphics, and animation programs to a full-fledged four-year degree program in game design at the University of California, Santa Cruz. This program includes not only courses in game design and programming, but also courses on the game business and even ethics.

Further Reading

Aronson, Sean. "School Fills Need for Game Designers." *Medianews,* June 18, 2007. Available online. URL: http://www. insidebayarea.com/sanmateocountytimes/localnews/ci_ 6168502. Accessed June 20, 2007.

Chaplain, Heather, and Aaron Ruby. *Smartbomb: The Quest for Art, Entertainment, and Big Bucks in the Videogame Revolution.* Chapel Hill, N.C.: Algonquin Books, 2005.

Crawford, Chris. *Chris Crawford on Game Design.* Indianapolis: New Riders, 2003.

———. *Chris Crawford on Interactive Storytelling.* Berkeley, Calif.: New Riders, 2005.

Game Developer. [magazine] Available online. URL: http://www. gdmag.com/homepage.htm. Accessed June 23, 2007.

Howland, Geoff. "How Do I Make Games? A Path to Game Development." Available online. URL: http://www.gamedev.net/ reference/design/features/makegames/. Accessed June 23, 2007.

Moore, Michael E., and Jennifer Sward. *Introduction to the Game Industry.* Upper Saddle River, N.J.: Prentice Hall, 2006.

computer graphics

Most early mainframe business computers produced output only in the form of punched cards, paper tape, or text printouts. However, system designers realized that some kinds of data were particularly amenable to a graphical representation. In the early 1950s, the first systems using the cathode ray tube (CRT) for graphics output found specialized application. For example, the MIT Whirlwind and the Air Force's SAGE air defense system used a CRT to display information such as the location and heading of radar targets. By 1960, the new relatively inexpensive minicomputers such as the DEC PDP series were being connected to CRTs by experimenters, who among other things created Spacewar, the first interactive video game.

By the late 1970s, the microcomputers from Apple, Radio Shack, Commodore, and others either included CRT monitors or had adapters that allowed them to be hooked up to regular television sets. These machines generally came with a version of the BASIC language that included commands for plotting lines and points and filling enclosed figures with color. While crude by modern standards, these graphics capabilities meant that spreadsheet programs could provide charts while games and simulations could show moving, interacting objects. Desktop computers that showed pictures on television-like screens seemed less forbidding than giant machines spitting out reams of printed paper (see GRAPHICS CARD).

Research at the Xerox PARC laboratory in the 1970s demonstrated the advantages of a graphical user interface based on visual objects, including menus, windows, dialog boxes, and icons (see USER INTERFACE). The Apple Macintosh, introduced in 1984, was the first commercially viable computer in which everything displayed on the screen (including text) consisted of bitmapped graphics. Microsoft's similar Windows operating environment became dominant on IBM architecture PCs during the 1990s. Today Apple, Microsoft, and UNIX-based operating systems include extensive graphics functions. Game and multimedia developers can call upon such facilities as Apple QuickDraw and Microsoft DirectX to create high resolution, realistic graphics (see also GAME CONSOLE).

BASIC GRAPHICS PRINCIPLES

The most basic capabilities needed for computer graphics are the ability to control the display of pixels (picture elements) on the screen and a way to specify the location of the spots to be displayed. A CRT screen is essentially a grid of pixels that correspond to phosphors (or groups of colored phosphors) that can be lit up by the electron beam(s). The first IBM PCs, for example, often displayed graphics on a 320 (horizontal) by 200 (vertical) grid, with 4 available colors.

A memory buffer is set up whose bytes correspond to the video display. (A simple monochrome display needs only one bit per pixel, but color displays must use additional space to store the color for each pixel.) A screen image is set up by writing the data bytes to the buffer, which then is sent to the video system. The video system uses the data to control the display device so the corresponding pixels are shown (in the case of a CRT, this means lighting up the "on" pixels with the electron gun[s]).

In most cases screen locations are defined in coordinates where point 0,0 is the upper left corner of the screen. The coordinates of the lower right corner depend on the screen resolution, At 320 by 200, the lower right corner would be 319,199.

For example, many versions of BASIC use statements such as the following:

```
PSET 50,50 ' draws a dot at X=50, Y=50
LINE (100,50)-(150,100), B ' draw square
   with UL
' corner at 100,50 and LR
' corner at 150,100
```

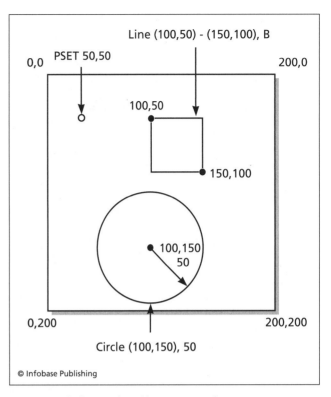

Some example figures plotted by BASIC graphics statements using screen coordinates.

```
CIRCLE (100,150), 50, 4 ' draw a circle of
   radius 50
' with center at 200,200 and
' color 4 (red)
```

Languages such as C, C++ and Java don't have built-in graphics commands, but functions can be provided in program libraries (see LIBRARY, PROGRAM). They would be used much like the BASIC commands given above.

More commonly, however, programmers use language-independent graphics platforms (see API). With Windows, this usually means DirectX, which includes Direct2D for 3D graphics, as well as a variety of multimedia libraries for sound, user interfacing, and networking. A competitor that is particularly popular in the Mac and UNIX/Linux worlds is OpenGL (Open Graphics Library). Both DirectX and OpenGL run on a wide variety of supported hardware.

GRAPHICS MODELS AND ENGINES

Modern applications (such as drawing programs and games) go well beyond simple two-dimensional objects. Indeed, multimedia developers typically use graphics engines designed to work with C++ or Java. A graphics engine provides a way to define and model 2D and 3D polygons. (Curves can be constructed by specifying "control points" for bicubic curves.)

Complex objects can be built up by specifying hierarchies (for example, a human figure might consist of a head, neck, upper torso, arms, hands, lower torso, legs, feet, and so on). By creating a hierarchy of arm, hand, fingers a transformation (scaling or rotation) of one object can be propagated to its dependent objects (see ANIMATION). In many cases graphics are created from real-world objects that have been digitally photographed or scanned, and then manipulated (see IMAGE PROCESSING).

In most scenes the relationships between graphical objects are also important. Modern graphics modeling programs use a virtual "camera" to indicate the position and angle from which the graphics are to be viewed. In rendering the scene, the Painter's Algorithm can be used to sort objects and draw closer surfaces on top of farther ones, as a painter might paint over the background. Alternatively, the Z-buffer algorithm stores depth information for each pixel to determine which ones are drawn. This technique requires less calculation (because surfaces don't need to be sorted), but more memory, since the depth of each pixel must be stored.

Within a scene, the effects of light (and its absence, shadows) must be realistically rendered. A simple technique can be used to calculate an overall light level for an object based on its angle in relation to the light source, plus a factor to account for ambient and diffuse light in the environment. The *Gouraud shading* technique can be used to smooth out the artifacts caused by the simple flat shading method. Another technique, *Phong shading,* can more realistically reproduce highlights (the sharp image of a light source being reflected within a surface). But the most realistic lighting effects are provided through ray tracing, which involves tracing how representative vectors (representing rays of light) reflect from or refract through various surfaces. However, ray tracing is also the most computationally intensive lighting technique.

Several techniques can be used to give objects more realistic surfaces. *Texture mapping* can be used to "paint" a realistic texture (perhaps scanned from a real-world object) onto a surface. For example, pieces in a chess game could be given a realistic wood grain or marble texture. This can be further refined through *bump mapping,* which calculates variations in the texture at each point based on light reflections.

APPLICATIONS AND TRADEOFFS

The most graphics-intensive applications today are games, multimedia programs, and scientific visualization or modeling applications. Because of the impact graphics have on users' perception of games and multimedia programs, developers spend a high proportion of their resources on graphics. Critics often complain that this is at the expense of core program functions. The software in turn places a high demand on user hardware: The contemporary "multimedia-ready" PC has a video card that includes special "video accelerator" hardware to speed up the display of graphics data and a video memory buffer of 256 MB or more.

Complex 3D graphics with lighting, shading, and textures may have to be displayed at a relatively low resolution (such as 640 × 480) because of the limitations of the main processor (which performs necessary calculations) and the video card. However as processor speed and memory capac-

ity continue to increase, many computer graphics now rival video and even film in realistic detail.

Further Reading

ACMSIGGRAPH. [graphics special interest group] Available online. URL: http://www.siggraph.org/. Accessed June 24, 2007.

Computer Graphics World. [magazine] Available online. URL: http://www.cgw.com/ME2/Default.asp. Accessed June 24, 2007.

Govil-Pai, Shalini. *Principles of Computer Graphics: Theory and Practice Using OpenGL and Maya.* New York: Springer, 2005.

Jones, Wendy. *Beginning DirectX 10 Game Programming.* Boston: Thomson Course Technology PTR, 2007.

computer industry

The U.S. computer industry began with the marketing of the Univac, designed by J. Presper Eckert and John Mauchly in the early 1950s. The first computers were made one at a time and only as ordered, and the market for the huge, expensive machines was thought to be limited to government agencies and the largest corporations. However, astute marketing by Sperry-Univac, Burroughs, and particularly, International Business Machines (see IBM) convinced a growing number of companies that modern data processing facilities would be essential for managing their growing and increasingly complex business (see MAINFRAME).

The mainframe market was controlled by a handful of vendors who typically provided the complete computer system (including peripherals such as printers) and a long-term service contract. (Eventually, third-party vendors began to make compatible peripherals.) Companies that could not afford their own computers began to contract with service bureaus for their data processing needs, such as payroll processing.

By the 1960s, transistorized circuitry was replacing the vacuum tube, and somewhat smaller machines became practicable (see MINICOMPUTER). While these computers were the size of a desk, not a desktop, models such as Digital Equipment Corporation's PDP series and competition from companies such as Data General provided computing power for engineers and scientists to use in factories and laboratories. During the 1970s, the dedicated word processing machine marketed by the Wang Corporation began the digital transformation of the office. By the end of that decade, the first general-purpose desktop microcomputers were marketed. The Apple II made a modest inroad into business, fueled by VisiCalc, the first spreadsheet program.

This new market attracted the attention of IBM, viewed by many microcomputer enthusiasts as a dinosaurlike relic of the mainframe age. Uncharacteristically, IBM management gave the developers of their personal computer (PC) project free rein, and the result was the IBM PC introduced in 1981. The machine had two major advantages. One was the IBM name itself, which was comforting to executives contemplating a bewildering new technology. The other was that IBM (again, uncharacteristically) had followed Apple's lead in designing their PC with an "open architecture," meaning that third-party manufacturers could market a variety of expansion cards to increase the machine's capabilities. By 1990, about 10 million PCs worth about $80 billion were being sold annually (see IBM PC).

Although IBM tried to prevent other manufacturers from "cloning" the IBM chipset itself, it was unable to prevent companies such as Compaq from creating "IBM compatible" PCs that often surpassed the capabilities of the IBM models. (IBM introduced its microchannel architecture in the late 1980s in an unsuccessful attempt to regain proprietary advantage.) By the 1990s the IBM-compatible PCs (sometimes called "Wintel," for the Microsoft Windows operating system and Intel-compatible processor) had become an industry standard and a commodity manufactured and marketed by everything from the big name brands such as Dell and Gateway down to the corner computer store's backroom operation.

The announcement of Apple's Macintosh computer in 1984 made a vivid impression on the public (see MACINTOSH). With its fully graphical user interface, mouse, drawing program, and fonts, it seemed light-years ahead of the text-based IBM PCs. However, the Mac's slow speed, relatively high price, and closed architecture limited its penetration into the business market. The Mac did attract an enthusiastic minority of consumer users and achieved a lasting niche presence in education and among graphics and video professionals. Gradually, as Microsoft's graphical Windows operating system improved in the early 1990s, the Mac's advantages over the IBM-compatible machines diminished.

During the 1990s, desktop computers came with a series of increasingly powerful series of Pentium processors, matched by offerings from AMD and Cyrix. Multimedia (including high-end graphics and sound capabilities) became a standard feature, particularly on consumer PCs. Increasingly, the business PC was being connected to a local area network, and both business and consumer PCs included modems or broadband access to online services and the Internet. The need to manage network files and services (such as Web servers) led to the development of server PCs featuring high-capacity mass storage. At the same time, high-end PCs also challenged the graphics workstations made by companies such as Sun. The traditional minicomputer and high performance workstation category began to melt away. By 2002, an estimated 600 million personal computers were in use worldwide, with about half of them in homes.

The personal computer also grew smaller. The suitcase-sized "luggable" computers of the 1980s gave way to a range of laptop, notebook-sized, and palm-sized computers. Today wireless networking technology allows users of diminutive machines to access the full resources of the World Wide Web and local networks.

The idea of "appliance computing" has also been a recurrent theme among industry pundits. Proponents argue that there are still many people who feel intimidated by a standard computer interface but have become comfortable with other consumer electronic products such as televisions, CD players, or microwaves. If computer functions could be built into such devices, people might use them comfortably. For example, WebTV is a box that allows the user to surf the Web from the same armchair where he or she watches TV, using controls little more complicated than those found on a regular TV remote. Kitchen appliances might be transformed, with the microwave providing recipes and the refrigerator keeping an inventory and automatically ordering from the grocery store. However, as with the fully automated "wired home," featured in Sunday newspaper supplements, the appliance computer has remained difficult to market to consumers (see SMART BUILDINGS AND HOMES).

THE SOFTWARE INDUSTRY

Hardware is useless without software. Since the operating system (OS) is the software that enables all other software to access the computer, the OS market is a key part of the computer industry. Through a historical accident, a young programmer-entrepreneur named Bill (WILLIAM) GATES and his MICROSOFT Corporation received the contract to develop the operating system for the first IBM PC. Microsoft bought and adapted an existing operating system to create MS-DOS (also called PC-DOS). Until the end of the 1980s, DOS was the dominant operating system for IBM-compatible PCs (see MS-DOS). In the early 1990s, Microsoft introduced Windows 3.0, the first successful version of its graphical operating environment (see MICROSOFT WINDOWS). The dominance of Windows became so complete that a federal antitrust case against Microsoft resulted in the company having to provide competitors greater access to the operating system.

The source of emerging challenges to Windows comes not from another desktop vendor but from the Internet, where Java offers the potential of delivering applications through the user's Web browser, regardless of whether that user is running Windows, the Macintosh OS, or Linux, a variant of UNIX that has been embraced by many enthusiasts. However, Java applications and Linux still represent only a tiny fraction of the market share held by Windows (see JAVA and LINUX).

The 1990s saw considerable consolidation in the office software arena. Microsoft's Office software suite overwhelmed once formidable competitors such as WordPerfect and Corel. Packages such as Microsoft Office create their own mini-industries where developers create templates and add-ins. However, the widespread use of high-speed Internet access (see BROADBAND) has made it practicable to offer many office software functions online, providing workers with convenient access from any location. The most significant offering here has been Google Apps, which includes calendar and communications features as well as Google Docs & Spreadsheets. In turn, Microsoft has been prompted to offer added-value online features to Microsoft Office.

Outside the office there is considerably more competition in the software industry. Today's consumers can choose from a wide variety of software that fills utility or other niche needs, including shareware ("try before you buy") offerings. In educational software and games some once-major innovators have been bought out or consolidated, but there is no one dominant company. Thousands of specialized software packages serve scientific, manufacturing, and business needs. While the general public is unaware of such programs, they make up much of the strength of the software industry.

OTHER PRODUCTS AND SERVICES

By the 2000s there were many new niches in the computer industry landscape. Powerful dedicated game machines such as the Microsoft Xbox 360 and the Sony PlayStation 3 make for a vigorous software industry that potentially goes beyond games (see GAME CONSOLES). Portable media players such as Apple's iPod are ubiquitous (see MUSIC AND VIDEO PLAYERS, DIGITAL). The personal digital assistant (see PDA) and the cell phone have largely merged and morphed (see SMARTPHONE), capable of running a variety of software including e-mail, Web browsing, games, and music. Meanwhile, digital cameras have virtually replaced film for all but the most high-end and specialized applications (see PHOTOGRAPHY, DIGITAL). The convergence and proliferation of all of these devices is continuing at a rapid pace, and competition is fierce.

The services sector of the computer industry lacks the visibility of new hardware products, but provides most of the industry's employment and much of its economic impact. In addition to the hundreds of thousands of programmers who provide business-related, consumer, and specialized software, there are the legions of help desk employees, computer and network technicians, creators of software development tools, writers of technical books and training products, industry investment analysts, reporters, and many others whose livelihood depends on the computer industry.

INTERNATIONAL COMPUTING

The computing industry came of age mainly in the United States. By the 1960s IBM had extended its dominant position to Britain and Europe despite the efforts of indigenous companies and government initiatives. Japan was considerably more successful in developing a competitive electronics and computer industry under the long-term guidance of MITI (Ministry of International Trade and Industry). The Japanese became dominant in industrial robotics and strong in consumer electronics, including game machines (Sony), digital cameras (Sony and Fujitsu), and laptop computers (Toshiba). They have been less successful in desktop computers, Internet-related technology, and commercial software. China has become an increasingly important player in the components and peripherals industry. The growing importance of Asia in the international computer industry is also underscored by the large number of programmers, engineers, and support personnel being trained in India (see GLOBALISM AND THE COMPUTER INDUSTRY).

Major Internet industry players such as Google and Yahoo! as well as hardware giant Dell have become heavily involved in the Chinese market, which boasted about 100 million users in 2006, second only to the United States.

A number of initiatives are helping spread computing even in the limited economies of many countries in Africa,

Asia, and Latin America (see DEVELOPING NATIONS AND COMPUTING). While illicit copying has hindered the marketing of commercial software in many countries, the alternative model of open-source software and very inexpensive laptops (the One Laptop Per Child initiative) may offer a viable path to the true globalization of computing.

EMERGING TRENDS

As the 2000 decade has progressed, a number of trends continue to reshape the computer industry. These include:

- The recovery from the "bust" years of 2001–3 was followed by more modest but significant growth, with rapid growth in particular sectors such as mobile devices, Web applications (see WEB 2.0), and security.

- Desktop PC sales were strong through 2005 (about 200 million that year) but now appear to be stagnating (in the United States at least) in favor of laptops, smaller portable computers, and smart phones.

- Although a new generation of multicore processors and the resource-hungry Microsoft Windows Vista operating system may eventually speed up the replacement of older PCs, businesses have been tending to keep slightly obsolescent machines and operating systems longer.

- Free or lower-cost alternative software and operating systems (see OPEN SOURCE and LINUX) are attracting considerable publicity, but it is unclear how much penetration they will achieve in the mainstream home and small-business computing sectors.

- Besides cost consciousness and other priorities (such as networking and security), the trend toward Web-based applications may be shifting sales away from hardware and traditional operating systems and software suites. (See APPLICATION SERVICE PROVIDER.)

- Outsourcing of many IT functions is continuing, including network administration, managed backup and storage, and even security. Meanwhile, there has been concern about lack of sufficient U.S. graduates in computer science and engineering.

While the computer hardware, software, and service industries are likely to continue growing vigorously, the boundaries between sectors and applications are blurring, making it harder to consider the industry as a whole as opposed to specific sectors and applications (see E-COMMERCE).

Further Reading

Chandler, Alfred D., Jr. *Inventing the Electronic Century: The Epic Story of the Consumer Electronics and Computer Industries.* New ed. Cambridge, Mass.: Harvard University Press, 2005.
Computer Industry Almanac. Available online. URL: http://www.c-i-a.com/. Accessed June 24, 2007.
Computerworld. Available online. URL: http://www.computerworld.com/. Accessed June 24, 2007.
Infoworld. Available online. URL: http://www.infoworld.com/. Accessed June 24, 2007.
International Data Corporation. Available online. URL: http://www.idc.com/. Accessed June 24, 2007.

PC World. Available online. URL: http://www.pcworld.com. Accessed June 24, 2007.

Plunkett's Info Tech, Computers & Software Industry. Available online. URL: http://www.plunkettresearch.com/Industries/InfoTechComputersSoftware/tabid/152/Default.aspx. Accessed June 24, 2007.

Yost, Jeffrey R. *The Computer Industry*. Westport, Conn.: Greenwood Press, 2005.

ZDNet. Available online. URL: http://www.zdnet.com/. Accessed June 24, 2007.

computer literacy

As computers became integral to business, industry, trades, and professions, educators and parents became increasingly concerned that young people acquire a basic understanding of computers and master the related skills. The term *computer literacy* suggested that computer skills were now as important as the traditional skills of reading, writing, and arithmetic. However, there has been disagreement about the emphasis for a computer literacy curriculum. Some educators, such as Seymour Papert, computer scientist and inventor of the Logo language, believe that students can and should understand the concepts underlying computing, and be able to write and appreciate a variety of computer programs (see LOGO). By gaining an understanding of what computers can (and cannot) do, students will be able to think critically about how to appropriately use the machines, rather than simply mastering route skills. Indeed, by gaining a good grasp of general principles, the student should be able to easily master specific skills.

An opposing view emphasizes the practical skills that most people (who will not become programmers) will need in everyday life and work. This sort of curriculum focuses on learning how to identify the parts of a computer and their functions, how to run popular applications such as word processors, spreadsheets, and databases, how to connect to the Internet and use its services, and so on. Computer literacy can also be broadened to include understanding the impact that computers are having on daily life and social issues that arise from computer use (such as security, privacy, and inequality).

Today computer literacy is an important part of every elementary and high school curriculum. Most students in middle-class or higher income brackets now have access to computers at home, and many thus gain considerable computer literacy outside of school. In addition, adult education and vocational schools often emphasize computer skills as a route to employment or career advancement. People also have the opportunity to learn on their own through books and videos.

The approach to computer literacy will vary with the background and resources of a given community. For example, programs for young people in developed countries can take advantage of the fact that many young people already have considerable experience with using computers, including related devices such as game consoles and music/video players. On the other hand, a program targeted at a poor or minority community must cope with the likelihood that many members of the community have had little opportunity to interact with computers (see DIGITAL DIVIDE). Programs for poor and developing countries may have to focus first on providing the basic infrastructure, as in the One Laptop Per Child Program.

Further Reading
Gookin, Dan. *PCs for Dummies.* 10th ed. New York: Wiley, 2005.

Jan's Illustrated Computer Literacy 101. Available online. URL: http://www.jegsworks.com/Lessons/index.html. Accessed June 24, 2007.

Parsons, June, and Dan Oja. *Practical Computer Literacy and Skills.* Boston: Thomson Course Technology, 2004.

White, Ron. *How Computers Work.* 8th ed. Indianapolis: Que, 2005.

computer science

Most generally, computer science is the study of methods for organizing and processing data in computers. The fundamental questions of concern to computer scientists range from foundations of theory to strategies for practical implementation.

FUNDAMENTAL THEORY

- What problems are susceptible to solving through an automated procedure? (See COMPUTABILITY AND COMPLEXITY.)

- Given that a problem is solvable, can it be solved without too much expenditure of time or computing resources?

- Can a step-by-step procedure be devised for solving a given problem? (See ALGORITHM.) How do different procedures (such as for sorting data) compare in efficiency and reliability? (See SORTING AND SEARCHING.)

- What methods of organizing data are most useful? (See DATA STRUCTURES.) What are the advantages and drawbacks of particular forms of organization? (See ARRAY, LIST PROCESSING, and QUEUE.)

- Which structures are best for representing the data needed for a given application? What is the best way to relate data to the procedures needed to manipulate it? (See ENCAPSULATION, CLASS, and PROCEDURES AND FUNCTIONS.)

THE TOOLS OF COMPUTING

- How can programs be structured so they are easier to read and maintain? (See STRUCTURED PROGRAMMING and OBJECT-ORIENTED PROGRAMMING.)

- Can programmers keep up with growth of operating systems and application programs that have millions of lines of code? (See SOFTWARE ENGINEERING and QUALITY ASSURANCE, SOFTWARE.)

- How can multiple simultaneous tasks (or even multiple processors) be coordinated to bring greater computing power to bear on problems? (See MULTITASKING and MULTIPROCESSING.)

- What is the best way to design an operating system, including the arrangement of different layers of the operating system such as the hardware-specific drivers, kernel (essential functions), and interfaces (shells or visual environments)? (See OPERATING SYSTEM, KERNEL, DEVICE DRIVER, and SHELL.)

- What should be emphasized in designing a programming language? How does one specify the grammar of statements the declaration and handling of data types, and the mechanism for handling functions or procedures? (See BACKUS-NAUR FORM, DATA TYPES, and PROCEDURES AND FUNCTIONS.)

- What considerations should be emphasized in designing a compiler for a given language? (See COMPILER.)

- How should a network be organized, and what protocols should be used for transferring data? (See NETWORK, INTERNET, DATA COMMUNICATIONS, TELE-COMMUNICATIONS, and TCP/IP.)

SPECIFIC APPLICATION AREAS

The general principles and tools must then be applied to a variety of application areas including:

- text processing (see WORD PROCESSOR, TEXT EDITOR, and FONT)

- graphics (see COMPUTER GRAPHICS and IMAGE PROCESSING)

- database management, including file structures and file access (such as indexing and hashing), and database architecture (relational databases) (see DATABASE MANAGEMENT SYSTEM, SQL, and XML).

- business data processing issues, including the design of MIS (management information systems) and decision support systems

- web applications, including commercial applications (see E-COMMERCE), multimedia, database access, integration of Web services (see BIOINFORMATICS SERVICE-ORIENTED ARCHITECTURE, MASHUPS, and WEB 2.0 AND BEYOND), and appropriate programming techniques (see AJAX and SCRIPTING LANGUAGES.)

- scientific programming issues, including data acquisition, maintaining accuracy in calculations, and creating visualizations driven by the data (see DATA ACQUISITION, NUMERIC DATA and SCIENTIFIC COMPUTING APPLICATIONS.)

- user interface design (designing the interaction between human beings and the operating system or application) (see USER INTERFACE)

- the broad area of artificial intelligence, which affects ways of representing information and modeling reasoning processes (see ARTIFICIAL INTELLIGENCE, NEURAL NETWORK, EXPERT SYSTEMS, and KNOWLEDGE REPRESENTATION.)

- robotics and control systems (an older term, "cybernetics," has also been used for this field) (see ROBOTICS AND CYBERNETICS.)

Clearly the concerns of computer science overlap a number of related fields. The design of computer hardware is often considered to be computer engineering, but designers of hardware must be familiar with the algorithms that will be used to operate it (see also COMPUTER ENGINEERING). Both artificial intelligence and user interface design are affected by cognitive science (or psychology), the study of human thought processes. Biology both inspires and is illuminated by artificial life simulations, genetic algorithms, and neural networks. The most abstract questions of information processing touch on the field of information science (or information theory).

HISTORY OF THE FIELD

The early computer pioneers such as Alan Turing, J. Presper Eckert, and John Mauchly brought backgrounds in mathematics or engineering (see TURING, ALAN; ECKERT, J. PRESPER; and MAUCHLY, JOHN). By the 1960s, however, a discipline and curriculum for computer science began to emerge. By the late 1990s more than 175 departments in American and Canadian universities offered a doctorate in computer science, with about a thousand new Ph.D.s being granted each year. However, in the following decade the number of students majoring in computer science declined by about 50 percent. (See EDUCATION IN THE COMPUTER FIELD for more details.)

The traditional computer science field emphasizes the theory of data representation, algorithms, and system architecture. In recent years a more practically oriented curriculum has emerged as an alternative. Under the titles of "Information Technology" or "Information Systems," this curriculum emphasizes application areas such as management information systems, database management, system administration, and Web development.

Further Reading
Association for Computing Machinery (ACM). Available online. URL: http://www.acm.org. Accessed June 24, 2007.
Biermann, Alan W. *Great Ideas of Computer Science with JAVA*. Cambridge, Mass.: MIT Press, 2001.
Computer Science [resources]. University of Albany Libraries. Available online. URL: http://library.albany.edu/subject/csci.htm. Accessed June 24, 2007.
Dale, Nell, and John Lewis. *Computer Science Illuminated*. 3rd ed. Sudbury, Mass.: Jones & Bartlett, 2006.
Hillis, Daniel W. *The Pattern on the Stone: The Simple Ideas That Make Computers Work*. New York: Basic Books, 1998.
IEEE Computer Society. Available online. URL: http://www.computer.org/portal/site/ieeecs/index.jsp. Accessed June 24, 2007.

computer virus

A computer virus is a program that is designed to copy itself into other programs. When the other programs are run, they carry out the virus's instructions, either instead of or in addition to their own. Since one of the primary tasks programmed into a virus is to reproduce itself, a virus program can spread rapidly. Viruses are generally programmed to seek out program files that are likely to be executed in the near future, such as those used by the operating system during the startup process. The result is a copy that can in turn generate an additional copy, and so on. (A virus disguised as an innocuous program is sometimes called a Trojan, short for "Trojan horse." A distinction is sometimes made between viruses and worms. A worm generally uses flaws in a networking system to send copies to other machines, without needing to insert code into a program.)

Appearing in the 1980s, the first computer viruses were generally spread by infecting programs on floppy disks, which were often passed between users. Today, viruses generally have instructions that enable them to gain access to network facilities (such as e-mail) to facilitate their spreading to other systems on a local network or on the Internet. The spread of viruses is complicated by the fact that operating systems (particularly Microsoft Windows) and applications (such as Microsoft Office) have the ability to run scripts or "macros" that are attached to documents. This facility can be useful for tasks such as sophisticated document formatting or form-handling, but it also means that viruses can attach themselves to scripts or macros and run whenever a document containing them is opened. Since modern e-mail programs have the ability to include documents as attachments to messages, this means that the unsuspecting recipient of a message can trigger a virus simply by opening a message attachment.

In today's Web-centric world, viruses are often spread using links in e-mail that either entices or frightens the reader into clicking on a link to a Web site, which can be made to closely resemble that of a legitimate institution such as a bank or e-commerce site (see PHISHING AND SPOOFING). Once connected to the site, the user's computer can be infected with a virus or with some other form of "malware" (see SPYWARE AND ADWARE). This route of infection is particularly dangerous because normal antivirus programs scan e-mail but not data being downloaded from a Web site, and firewalls are generally set to allow normal Web requests.

Once installed, a virus can be used for a variety of purposes according to the "payload" of instructions that are set to execute. Sensitive information such as credit card details can be stolen (see IDENTITY THEFT). Sometimes the infected computer can appear to be unaffected, but has had a stealthy "bot" (robot) program inserted. Thousands of bots can be linked into a "botnet" and later commanded to trigger large-scale "distributed denial of service" (DDOS) attacks to flood targeted Web sites with requests, crashing or disabling the site.

Viruses can be further disguised by programming them to remain dormant until a certain date, time, or other condition is reached. (Such a virus is sometimes called a *logic bomb*.) For example, a disgruntled programmer who is about to be dismissed might insert a virus that will wipe out payroll data at the beginning of the next month. A famous example of the time-triggered virus was the Michelangelo virus, so named because it was triggered to run on the artist's birthday, March 6, 1992. (See COMPUTER CRIME AND SECURITY.)

Viruses can be overtly destructive (such as by reformatting a computer's hard drive, wiping out its data). Other viruses can simply tie up system resources. The most infamous example of this was the "Internet Worm" introduced onto the network on November 2, 1988, by Robert Morris, Jr. This program was intended to reproduce slowly, planting its "segments" on networked computers by exploiting a flaw in the UNIX sendmail program. Unfortunately, Morris made an error that caused the worm to spread much more rapidly. Before the coordinated efforts of system administrators at affected sites came up with countermeasures, the worm had cost somewhere in the hundreds of thousands of dollars in lost computer and programmer time.

COUNTERMEASURES

The only certain defense of a computer system from viruses would be through abstaining from contact between it and any other computers, either directly through a network or indirectly through exchange of programs on floppy disks or other removable media. In today's highly networked world, this is usually impractical. A more practical defense is to install antivirus software. Antivirus programs work by comparing the contents of files (either those already on the disk or entering via the Internet) with "signatures" or patterns of data found in known viruses. More sophisticated antivirus programs include the ability to recognize program code that is similar to that found in known viruses or that attempts suspicious operations (such as attempts to reformat a disk or bypass the operating system and write directly to disk). If an antivirus program recognizes a virus, it warns the user and can be told to actually remove the virus. Because dozens of new viruses are identified each week, virus programs must be updated frequently with new virus signature files in order to remain effective. Many antivirus programs can update themselves by periodically linking to a Web site containing the update files.

Modern operating systems (such as Microsoft Windows Vista) have attempted to make it harder for unauthorized programs to access critical system files, such as by limiting default access permissions or prompting the user to approve various activities. Such operating systems also include an updating feature that can automatically download and install security "patches"—a vital task, as can be seen from the volume and variety of such updates that seem to appear every month. Indeed the use of "blended" threats (including more than one potential infection mechanism) and the development of new "exploits" for hundreds of different data file formats make system protection an ongoing challenge.

Reducing user temptation and enhancing user awareness is also important. Since unsolicited e-mail (see SPAM) is often a source of potentially malicious links and attachments, running a spam-blocking program can help protect the computer. There are also programs that can detect and block "phishing" messages and their related Web sites. Since none of these programs can completely keep up with the rapid appearance of new threats, caution and common sense on the part of the user remain an important last line of defense.

Further Reading

Antivirus Software Buying Guide. *PC World*. Available online. URL: http://www.pcworld.idg.com.au/index.php/id;316975074. Accessed June 24, 2007.

CERT Coordination Center. Available online. URL: http://www.cert.org. Accessed June 24, 2007.

Gregory, Peter H. *Computer Viruses for Dummies*. Indianapolis: Wiley, 2004.

Henderson, Harry. *Computer Viruses*. Detroit: Lucent Books/Thomson-Gale, 2006.

McAfee Corporation. Available online. URL: http://www.mcafee.com. Accessed June 24, 2007.

Symantec Corporation. Available online. URL: http://www.symantec.com. Accessed June 24, 2007.

computer vision

In the biological world, vision is the process of receiving light signals from the environment through the eyes and optic nerves, from which the brain can extract patterns that contain useful information (such as recognizing food or a potential predator). Computer vision (also known as machine vision) is the analogous process by which light is received by a sensor system (such as a digital camera). The light is then analyzed for meaningful patterns. Thus, a robot might be able to recognize the identity and positions of various parts on an assembly line.

Because computer vision involves pattern recognition, it is part of the discipline of artificial intelligence (see ARTIFICIAL INTELLIGENCE and PATTERN RECOGNITION). The challenge is not in getting information about a visual scene from the camera and turning it into digital information (a grid of pixels). Rather, it is the ability to recognize meaningful patterns in fragmented images, something human infants learn to do almost from birth when they encounter human faces.

One way to approach the problem is to constrain the kinds of images the computer (or robot) has to deal with. If you can guarantee that a robot's field of vision will contain only a few fixed objects (a hopper, perhaps, or a conveyer belt) plus one or more distinctively shaped parts, it is relatively easy to program the dimensions of the possible objects into the vision system so that the robot can identify objects by comparing them with stored templates. However, if the robot encounters an object it isn't prepared for, such as a stray bit of packing material, it will be unable to identify (or properly deal) with the object.

Vision is also complicated by the problem of parsing three-dimensional objects in the visual field. Seen head-on, the side of a cube appears to be a two-dimensional square. Seen at an angle, it appears to be a three-dimensional assemblage with some faces visible and some not. To interpret these and more complicated objects, the robot might be programmed with rules that help it infer that an object is really a cube, that all cubes have six equal sides, and so on. Another strategy is to give the robot more than one "eye" so that images can be compared, much as humans do unconsciously with binocular vision. Finally, the robot can be given the ability to move its head and eyes in order to find a viewpoint that yields more information about an ambiguous object.

Human infants, of course, are not born with a fully developed understanding of the types of objects in their world. They are always learning new ways to distinguish, for example, a stuffed teddy bear from a live dog. Robot vision systems, too, can be programmed to learn (or at least, refine their ability to recognize objects). A statistical technique can be used to "sample" objects in the environment and find which characteristics most reliably "predict" the true nature of an object. Characteristics can be resampled from different viewpoints to see which ones remain invariant (unchanged). For example, a cube will always have four edges on each face. Another approach is to use a neural network, where the visual information is processed by a grid of nodes that are reinforced to the extent they are successful in identifying features (such as edges).

APPLICATIONS

Computer vision is a problem of great theoretical interest because it engages so many questions about perception, the ability to build models of the world, and the ability to learn. The field also has considerable practical potential. Currently, most robots are fixed to stations on factory floors where they work with a limited number of objects (parts) in a highly constrained, stable environment. However "service robots" have been gradually developed to work in a much less constrained environment (such as carrying supplies down hospital corridors or even serving as mobile assistants to astronauts in the weightless environment of the International Space Station). These robots would benefit greatly by having robust vision systems so that they can, for example, recognize individual human faces or detect potentially dangerous situations.

Of course computer vision systems find many applications besides robotics. These include automatic quality control or inventory management systems, advanced medical imaging and computer-assisted surgery, as well as security, surveillance, and criminal investigation/forensics.

Further Reading

Davies, E. R. *Machine Vision: Theory, Algorithms, Practicalities*. 3rd ed. San Francisco: Morgan Kaufmann, 2004.

Henderson, Harry. *Modern Robotics: Building Versatile Machines*. New York: Chelsea House Publishers, 2006.

Hornberg, Alexander, ed. *Handbook of Machine Vision*. Weinheim, Germany: Wiley-VCH, 2006.

Machine Vision Online. Automated Imaging Association. Available online. URL: http://www.machinevisiononline.org/. Accessed June 24, 2007.

Shapiro, Linda G., and George Stockman. *Computer Vision*. Upper Saddle River, N.J.: Prentice Hall, 2001.

concurrent programming

Traditional computer programs do only one thing at a time. Execution begins at a specified point and proceeds according to decision statements or loops that control the processing. This means that a program generally cannot begin one step until a previous step ends.

Concurrent programming is the organization of programs so that two or more tasks can be executed at the same time. Each task is called a thread. Each thread is itself a traditional sequentially ordered program. One advantage of concurrent programming is that the processor can be used more efficiently. For example, instead of waiting for the user to enter some data, then performing calculations, then waiting for more data, a concurrent program can have a data-gathering thread and a data-processing thread. The data-processing thread can work on previously gathered data while the data-gathering thread waits for the user to enter more data. The same principle is used in multitasking operating systems such as UNIX or Microsoft Windows. If the system has only a single processor, the programs are allocated "slices" of processor time according to some scheme of priorities. The result is that while the processor can be executing only one task (program) at a time, for practical purposes it appears that all the programs are running simultaneously (see MULTITASKING).

Multiprocessing involves the use of more than one processor or processor "core." In such a system each task (or even each thread within a task) might be assigned its own processor. Multiprocessing is particularly useful for programs that involve intensive calculations, such as image processing or pattern recognition systems (see MULTIPROCESSING).

PROGRAMMING ISSUES

Regular programs written for operating systems such as Microsoft Windows generally require no special code to deal with the multitasking environment, because the operating system itself will handle the scheduling. (This is true with *preemptive multitasking,* which has generally supplanted an earlier scheme where programs were responsible for yielding control so the operating system could give another program a turn.)

Managing threads within a program, however, requires the use of programming languages that have special statements. Depending on the language, a thread might be started by a *fork* statement, or it might be coded in a way similar to a traditional subroutine or procedure. (The difference is that the main program continues to run while the procedure runs, rather than waiting for the procedure to return with the results of its processing.)

The coordination of threads is a key issue in concurrent programming. Most problems arise when two or more threads must use the same resource, such as a processor register (at the machine language level) or the contents of the same variable. Let's say two threads, A and B, have statements such as: Counter = Counter + 1. Thread A gets the value of Counter (let's say it's 10) and adds one to it. Meanwhile, thread B has also fetched the value 10 from Counter. Thread A now stores 11 back in counter. Thread B, now adds 1 and stores 11 back in Counter. The result is that Counter, which should be 12 after both threads have processed it, contains only 11. A situation where the result depends on which thread gets to execute first is called a *race condition.*

One way to prevent race conditions is to specify that code that deals with shared resources have the ability to "lock" the resource until it is finished. If thread A can lock the value of Counter, thread B cannot begin to work with it until thread A is finished and releases it. In hardware terms, this can be done on a single-processor system by disabling interrupts, which prevents any other thread from gaining access to the processor. In multiprocessor systems, an *interlock* mechanism allows one thread to lock a memory location so that it can't be accessed by any other thread. This coordination can be achieved in software through the use of a *semaphore,* a variable that can be used by two threads to signal when it is safe for the other to resume processing. In this scheme, of course, it is important that a thread not "forget" to release the semaphore, or execution of the blocked thread will halt indefinitely.

A more sophisticated method involves the use of message passing, where processes or threads can send a variety of messages to one another. A message can be used to pass data (when the two threads don't have access to a shared memory location). It can also be used to relinquish access to a resource that can only be used by one process at a time. Message-passing can be used to coordinate programs or threads running on a distributed system where different threads may not only be using different processors, but running on separate machines (a *cluster* computing facility).

Programming language support for concurrent programming originally came through devising new dialects of existing languages (such as Concurrent Pascal), building facilities into new languages (such as Modula-2), or creating program libraries for languages such as C and C++.

However, in recent years concurrent programming languages and techniques have been unable to keep up with the growth in multiprocessor computers and distributed computing (such as "clusters" of coordinated machines). With most new desktop PCs having two or more processing cores, there is a pressing need to develop new programs that can carry out tasks (such as image processing) using multiple streams of execution. Meanwhile, in very high-performance machines (see SUPERCOMPUTER), the Defense Advanced Research Projects Agency (DARPA) has been trying to work with manufacturers to develop languages to work better with computers that may have hundreds of processors as well as distributed systems or clusters. Such languages include Sun's Fortress, intended as a modern replacement for Fortran for scientific applications.

The new generation of concurrent languages tries to automate much of the allocation of processing, allowing programmers to focus on their algorithms rather than implementation issues. For example, program structures such as loops can be automatically "parallelized," such as by assigning them to separate cores.

Further Reading

Anthes, Gary. "Languages for Supercomputing Get 'Suped' Up." Computerworld.com, March 12, 2007. Available online. URL: http://www.computerworld.com/action/article.do?command=viewArticleBasic&articleId=283477&intsrc=hm_list. Accessed June 24, 2007.
Ben-Ari, M. *Principles of Concurrent & Distributed Programming.* 2nd ed. Reading, Mass.: Addison-Wesley, 2006.

Feldman, Michael. "Our Manycore Future." HPCWire. Available online. URL: http://www.hpcwire.com/hpc/1295541.html. Accessed June 24, 2007.

Lea, Douglas. *Concurrent Programming in Java: Design Principles and Patterns.* 3rd ed. Reading, Mass.: Addison-Wesley Professional, 2006.

Merritt, Rick. "Where Are the Programmers? Enrollment Wanes Just as Computer Scientists Grapple with Problem of Parallelism." *IEEE Times,* March 12, 2007. Available online. URL: http://www.eetimes.com/showArticle.jhtml?articleID=197801653. Accessed June 24, 2007.

Steele, Guy, and Jan-Willem Maessen. "Fortress Programming Language Tutorial Slides." Sun Microsystems. Available online. URL: http://research.sun.com/projects/plrg/PLDITutorialSlides9Jun2006.pdf. Accessed June 11, 2007.

conferencing systems

Conferencing systems are online communications facilities that allow users to log in and participate in discussions on a variety of topics. Although this is a rather amorphous category of software, some distinguishing characteristics can be identified. Conferencing is distinguished from chat or instant messaging systems because the messages are asynchronous (that is, one person at a time leaves a message, and there is no real-time interaction between participants). Unlike Netnews newsgroups, conferencing systems such as San Francisco Bay Area–based The Well tend to have users who are committed to long-term discussions in conferences (topical discussion areas) that tend to persist for weeks, months, or even years. Conferencing systems are often grouped under the umbrella term of Computer-Mediated Communications (CMC).

HISTORY

In the 1960s, researcher Murray Turoff at the Institute for Defense Analysis decided to adopt for computer use a discussion method called Delphi, developed at RAND corporation. This method was a collective process by which new ideas were discussed and voted on by a panel of experts. After he implemented Delphi as a system of messages passed via computer, he began to generalize his work into a more general method of facilitating online discussions. His Electronic Information Exchange System (EIES, pronounced "eyes") was designed to facilitate discussion within research communities of 10–50 members.

The emergence of topical online discussions can be seen in the development of the Usenet (or Netnews) newsgroups in the early 1980s, the development of communications or memo systems within large offices (particularly within the government), and the emergence of bulletin boards and online services for personal computer users. Most early news and bulletin board software had only rudimentary facilities for linking topics and responses. A more sophisticated approach to conferencing emerged within the PLATO educational computing network in the 1970s, in the form of Plato Notes. This system began as a simple way for users to leave messages or help requests in a text file, and evolved into a structure of "base notes" and linked response notes, a topic-and-response structure that became the general model for conferencing systems.

In the mid-1980s, the Well (Whole Earth 'Lectronic 'Link) began to provide online conferencing to anyone who subscribed. It used a text-based system called Picospan. With its improbable eclectic mix well salted with Grateful Dead fans and computer "nerds," the Well became a sort of petri dish for cultivating community (see VIRTUAL COMMUNITY). Long-term friendships (and feuds) and occasional romances have been nurtured by such conferencing systems.

TYPICAL STRUCTURE

A typical text-based conferencing system is divided into conferences, which are generally devoted to relatively broad subjects, such as UNIX, pop music, or politics. Each conference is further divided into topics, which usually reflect particular aspects of the general subject (such as a particular UNIX version, a pop music group, or a political issue). Most conferencing systems have a person or persons who act as a moderator (sometimes called a "host") who tries to encourage new users, keep discussions more or less on topic, and discourage personal attacks or vehement statements ("flames").

A user signs onto the system and "joins" one or more conferences. Each time the user visits a conference that he or she has joined, any topics (or responses in existing topics) that were posted since the last visit are presented. The user can read the postings and, if desired, enter a reply that becomes part of the thread of messages. (Users are also generally allowed to start new topics of their own.)

WEB-BASED CONFERENCING

Text-based systems such as Picospan are driven by the user entering command letters or words. While this paradigm is familiar to people who have experience with operating systems such as UNIX or MS-DOS, it can be more difficult for users who are used to the point-and-click approach of Windows programs and the World Wide Web. Many new conferencing systems use Web pages to present conference topics and messages, with buttons replacing text commands. (The Well continues to offer both the text-based Picospan and the Web-based Engaged.)

Although the Well and other conferencing systems such as The River continue in operation, conferencing systems have been largely supplanted by newer forms of online expression (see BLOGS AND BLOGGING, SOCIAL NETWORKING, and WIKIS AND WIKIPEDIA). (Note that "conferencing system" can also refer to video-based software such as Microsoft Live Meeting for facilitating meetings between geographically dispersed participants.)

Further Reading
Hafner, Katie. *The Well: A Story of Love, Death & Real Life in the Seminal Online Community.* New York: Carroll & Graf, 2001.
Rheingold, Howard. *The Virtual Community: Homesteading on the Electronic Frontier.* Rev. ed. Cambridge, Mass.: MIT Press, 2000.
Thurlow, Crispin, Laura Lengel, and Alice Tomic. *Computer Mediated Communication.* Thousand Oaks, Calif.: SAGE Publications, 2004.

Web Conferencing Review. Available online. URL: http://thinkofit. com/webconf/index.htm. Accessed June 25, 2007.

The Well. Available online. URL: http://www.well.com. Accessed June 25, 2007.

constants and literals

Constants and literals are ways of describing data that does not change while a program runs. For example, a statement in C such as

```
const float pi = 3.14159;
```

expresses a value that will be used in calculations, but not changed. Constants can be of any data type, including character strings as well as numbers. String constants are usually enclosed in single or double quotes:

```
char * Greeting = "Hello, World";
```

Actual strings and numerals found in programs are sometimes called *literals,* meaning that they are to be accepted exactly as given (literally) rather than standing for some other value. Thus 3.14159 and Hello, World as given above can be considered to be numeric and string literals respectively.

Because many languages consider a value of 1 as representing a "true" result for a branch or loop test, and 0 as representing "false," programs in languages such as C often include declarations such as:

```
const True = 1;
const False = 0;
```

This lets you later have a loop construction such as

```
while (True) {
' body of program
} ;
```

which is a more readable way to code an endless loop than:

```
while (1) {
' body of program
} ;
```

However languages such as Pascal and C++ have a special boolean data type (*bool* in C++) that allows for constants or variables that will have one of two values, **true** or **false.**

Some languages provide a way to set up an ordered group of constant values (see ENUMERATIONS AND SETS).

CONSTANTS VS. VARIABLES

The difference between a constant and a variable is that a variable represents a quantity that can change (and is often expected to). For example, in the statement

```
int Counter = 0;
```

Counter is set to a starting value of zero, but will presumably be increased as whatever is to be counted is counted.

Most compilers will issue an error message if they detect an attempt to change the value of a constant. Thus the sequence of statements:

```
const float Tax_Rate = 8.25;
Tax_Rate = Tax_Rate + Surtax;
```

would be illegal, since Tax_Rate was declared as a constant rather than as a variable.

Many compilers, as part of code optimization, can discover values or expressions that will remain constant throughout the life of the program, even if they include variables. Such constants can be "propagated" or substituted for variables. This can speed up execution because unlike a variable, a constant does not need to be retrieved from memory (see COMPILER).

Further Reading

Sebesta, Robert W. *Concepts of Programming Languages.* 8th ed. Reading, Mass.: Addison-Wesley, 2007.

content management

Content management is the process of creating, maintaining, and archiving data such as text and images to be used for a project such as a book, magazine, or Web site. Normally such projects involve a number of different people: content creators (such as writers or photographers), editors, reviewers, designers, and so on. A large project will often have many documents in various stages—early drafts, material approved for publication, existing publications in need of revision, older material ready to be archived, and so on.

The purpose of content management is to make sure every piece of a project has its status tracked, including who has worked on it and what has been done (or needs to be done). Because more than one person may want to work on a given piece at the same time, some form of "version control" (as with program code) must be used to either "lock" the material while one person is using it, or to merge their separate work into a new version of the document. Naturally there must also be a way for members of the team to communicate with each other in connection with specific parts of the project, and all members must be kept informed of key developments.

WORK FLOW

A key measurement of the effectiveness of a content management system (CMS) is how well it facilitates work flow, or the movement of documents through the production process. Work flow begins with the importing of material such as text documents or multimedia resources into the system. At this time the key users and their roles (such as editor or reviewer) are identified, and the system can then route the material to the next person automatically after each task is completed. Often messages are generated and sent to managers to keep them informed of progress or to alert them to problems.

Today Web sites are the most common large information-related projects, and managing them can be quite challenging. Usually multimedia material is included well beyond that found in printed projects, such as audio, video, animations, and information feeds (see RSS). Web sites, unlike most traditional publications, are under constant revision and review.

Once created, material will often be reused or repurposed for different projects. Thus an important part of most content management systems is the repository, which makes the material easily searchable and retrievable for later use. Material that is less likely to be used but still must be retained (such as for legal reasons) may be stored in a separate archive (see BACKUP AND ARCHIVE SYSTEMS). Note: The term *digital asset management* is also sometimes used for such systems.

SOFTWARE

Content management systems are usually built upon a framework or programming interface (see APPLICATION PROGRAM INTERFACE), often using languages such as Java, Perl, Python, or PHP. There are many products to choose from, including free and open-source alternatives.

An interesting alternative for some projects is to use a wiki as a content management system (see WIKIS AND WIKIPEDIA). Especially for textual content, wikis offer the advantage of already having revision tracking built in, and full-scale wikis such as MediaWiki have many additional features or plug-ins to aid in content management.

Further Reading
Boiko, Bob. *Content Management Bible.* 2nd ed. Indianapolis: Wiley, 2005.
CMS Matrix. Available online. URL: http://www.cmsmatrix.org/. Accessed September 4, 2007.
CMS Review. Available online. URL: http://www.cmsreview.com/. Accessed September 4, 2007.
Hackos, JoAnn T. *Content Management for Dynamic Web Delivery.* New York: Wiley, 2001.

cookies

Cookies are simply tiny text files that a Web server sends to the browser and retrieves each time the user accesses the Web site. The purpose is to maintain a sort of profile of the user containing such things as preferences as to how the user wants to view or use the site, shopping cart selections from previous sessions, and so on. In short, cookies enable a Web site to provide a more customized or personalized form of service and minimize the amount of repetitive data entry on the part of the user. (This type of cookie is called persistent, since it survives across sessions. There can also be temporary cookies that apply only to the current session.)

However, cookies also have benefits for the Web site owner. They can be used to track which pages or items the user has looked at in the past. This information can then be used (see DATA MINING) to create generic user profiles that can help with marketing or targeting advertising. In the case of some companies (notably Amazon.com) much more elaborate profiles associated with the cookie's identity can be used to create personalized recommendations, in effect continually directing targeted advertising at the user.

SECURITY AND PRIVACY CONCERNS

There are many popular misconceptions about cookies. Cookies contain only data, not executable code. This means they cannot function as worms or viruses or otherwise interact with the user's system. However, while cookies do not in themselves represent a security threat, they do have privacy implications. Although most profiles created using cookies are anonymous (containing no personal identifying data), an unscrupulous site could attach such data (such as addresses or credit card numbers entered by the user) to a profile and sell it for purposes ranging from spamming to identity theft.

Another risk comes from "third party" cookies such as are often included in advertisements (see ONLINE ADVERTISING). Potentially, these could be used to create a much more comprehensive profile of a user based on his or her actions on multiple Web sites.

Users do have some control over how cookies are stored. Most browsers allow the user to reject all cookies, accept or reject cookies from certain sites, or store cookies only temporarily. However, sites may in turn refuse services to users who do not accept cookies, and at any rate the user would see only a generic rather than a personalized view.

There has been a certain amount of government regulation of Web cookies. The U.S. government has strict rules for the use of cookies on federal Web sites. The European Union also has recommended (but not fully implemented) regulations that require that users be told how the stored data will be used and be given the opportunity to opt out.

Further Reading
Kuner, Christopher. *European Data Privacy Laws and Online Business.* New York: Oxford University Press, 2003.
Kymin, Jennifer. "What are HTTP Cookies?" Available online. URL: http://webdesign.about.com/cs/cookies/a/aa082498a.htm. Accessed September 4, 2007.
Levine, John R., Ray Everett-Church, and Gregg Stebben. *Internet Privacy for Dummies.* New York: Wiley, 2002.

cooperative processing

Historically there have been two basic ways to bring greater computer power to bear on a task. One is to build more powerful single computers (see SUPERCOMPUTER). The other is to link one or more computers or processors together and tightly coordinate them to process the data (see GRID COMPUTING). Both of these approaches require great expertise and considerable expense.

However, there is another quite interesting ad hoc approach to cooperative processing that first appeared with the SETI@Home project launched in 1999. The basic idea is to take advantage of the fact that millions of computer users are already connected via the Internet. The typical PC has many processing cycles to spare—idle time when the user is doing nothing and the operating system is doing very little.

A program like SETI@Home is designed to be downloaded to volunteer users. The program can run only when no other applications are being used (one way to ensure this is to make the program a screen saver), or it can run continuously but only use cycles not being requested by another program.

The data to be analyzed (signals from space in this case) is broken up into chunks or "work units" that are parceled out to the volunteers. When a given unit has been analyzed by the program on the user's machine, the results are sent back to the central server and a new work unit is sent.

Although no evidence of extraterrestrial intelligence had been found as of mid-2008, SETI@Home's more than 5 million participants have contributed more than 2 million years of CPU time, and can process at the collective rate of 256 TeraFLOPS (trillion floating point operations per second), comparable with the fastest single supercomputers.

There are currently a number of other cooperative distributed computing projects underway. Many of them are part of the Berkeley Open Infrastructure for Network Computing (BOINC), which includes SETI@Home, Proteins@home (protein folding), and the World Community Grid (humanitarian projects).

Ad hoc cooperative processing is not suitable for all types of projects. There must be a way to break the data into batches that can be separately processed. The project is also dependent on the number of volunteers and their degree of commitment.

Cooperative processing can be seen as part of a spectrum of emerging ways in which the line between producers and consumers of data is being blurred. Other examples include media-sharing services such as Gnutella (see FILE-SHARING AND P2P NETWORKS). Cooperative programs can also be used to gather information about software use and bugs from thousands of users to allow for faster debugging and optimization.

People can do more than passively share their computer's processors—they can add their own brains to the effort. Some of the most effective spam filters (see SPAM) use the "collective intelligence" of users by having them identify and mark spam messages, which can then be used by the software as a template for automatically rejecting similar messages. Another interesting application by the Carnegie Mellon Human Computation program uses a computer game where a pair of randomly selected volunteers assigns keywords to an image. For the players, the object of the game is to come up with matching keywords, thereby scoring points. However, the real work that is being accomplished is that thousands of previously uncategorized images are receiving appropriate keywords to enable them to be retrieved. In effect, the system is taking advantage of an image-recognition device that is far more capable than any computer algorithm—the human brain! (One might call this synergistic human–computer processing.)

Further Reading

Berkeley Open Infrastructure for Network Computing. Available online. URL: http://boinc.berkeley.edu/. Accessed September 4, 2007.

DeHon, Andre, et al. "Global Cooperative Computing." Available online. URL: http://www.ai.mit.edu/projects/iiip/colab/gcc-abstract.html. Accessed September 4, 2007.

Gomes, Lee. "Computer Scientists Pull a Tom Sawyer." *Wall Street Journal* (June 27, 2007): p. B1. Available online. URL: http://online.wsj.com/article/SB118288538741648871.html. Accessed September 4, 2007.

SETI@Home. Available online. URL: http://setiathome.berkeley.edu/. Accessed September 4, 2007.

Shankland, Stephen. "Cooperative Computing Finds Top Prime Number." ZDNet. Available online. URL: http://news.zdnet.com/2100-9584_22-5112827.html. Accessed September 4, 2007.

Taylor, Ian J. *From P2P to Web Services and Grids: Peers in a Client/Server World*. New York: Springer, 2004.

copy protection

Companies that produce software have had to cope with software that is expensive to develop, while the disks on which it is distributed are inexpensive to reproduce. The making and swapping of "pirated" copies of software is just about as old as the personal computer itself. Software piracy has taken a number of forms, ranging from teenaged hackers making extra copies of games to factories (often in Asia) that stamp out thousands of bogus copies of Windows operating systems and programs that would cost hundreds of dollars apiece if legitimate (see SOFTWARE PIRACY AND COUNTERFEITING).

To prevent such copying, software producers in the 1980s often recorded the programs on floppy disks in a special format that made them hard to copy successfully. One way to do this is to record key information on disk tracks that are not normally read by the operating system and thus not reproduced by an ordinary copy command. When such a program runs, it can use a special device control routine to read the "hidden" track. If it does not find the identifying information there, it knows the disk is not a legitimate copy.

Another way to do copy protection is by having the program look for a small hardware device called a "dongle" connected to the computer, usually to the parallel printer port. Since the dongle is distributed only with the legitimate program, it can serve as an effective form of copy protection. (Encryption can also be used to render copies unusable without the key.)

DECLINE OF COPY PROTECTION

Copy protection has a number of drawbacks. Because disk-based copy protection writes on nonstandard tracks, even legitimate programs may not work with certain models of disk or CD drive. And because the legitimate user is unable to make a backup copy of the disk, if it is damaged, the user will be unable to use the program. Dongles, on the other hand, can interfere with the operation of other devices connected to the port, and a user might be required to use multiple dongles for multiple programs.

During the 1990s, copy protection was generally phased out, except for some games. A variety of other strategies are used against software piracy. The Software Publishers Association (SPA) maintains a program in which disgruntled users can report unauthorized copying of software at their workplace. Companies that allow unauthorized copying of software can be sued for violating the terms of their software license. International trade negotiations can include provisions for cracking down on the massive "cloning" of major software packages abroad.

With modern software, "soft" copy protection generally still exists in the form of requiring the typing in of a serial number from the CD, often combined with online "activation" or "validation," as with Microsoft Windows and Office products. The online validation process can forestall the use of valid but duplicated serial numbers (see DIGITAL RIGHTS MANAGEMENT and SOFTWARE PIRACY AND COUNTERFEITING).

Hackers and cyber-libertarians have often argued that the problem of software piracy has been overrated, and that allowing the copying of software would enable more people who would not otherwise buy programs to try them out. Once someone likes the program, they might buy it not only for legitimacy of ownership, but in order to get access to the technical support and regular upgrades that are often required for complex business software packages. For less expensive software, an alternative channel (see SHAREWARE) allows for a "try before you buy" distribution of software.

Further Reading
Aldrich, John. "Implementing Simple Copy Protection: Technical Overview." Available online. URL: http://www.codeproject.com/win32/simplecopyprotection.asp?df=100&foru mid=425 0&exp=0&select=878288. Accessed September 6, 2007.
Gilmore, John. "What's Wrong with Copy Protection." February 16, 2001. Available online. URL: http://www.toad.com/gnu/whatswrong.html. Accessed September 6, 2007.
Wikipedia. "Copy Protection." Available online. URL: http://en.wikipedia.org/wiki/Copy_protection. Accessed September 6, 2007.

CORBA (Common Object Request Broker Architecture)

CORBA (Common Object Request Broker Architecture) is a standardized way to specify how different applications (on the same or different machines) can call upon the services of database objects (see DATABASE and OBJECT-ORIENTED PROGRAMMING). The CORBA standard is defined by the Object Management Group (OMG), a consortium of more than 700 companies or organizations, including the major players in distributed database technology.

STRUCTURE AND USAGE
Creating a CORBA application involves three basic steps. First, specifications are provided using an interface definition language (IDL) that specifies in generic terms what services an object will provide. An IDL compiler then creates a "skeleton" interface that the developer can fill in with actual code for a class for that object in a programming language (such as Java).

To use CORBA, a client application accesses an Object Request Broker (ORB), which is software that locates the referenced object on the network (thus the program does not need to know or keep track of specific locations). The ORB sends the request to the object, which processes it and returns the results, which are then sent back to the client application.

The intent of CORBA is to make objects implemented by different vendors fully interoperable (able to call one another using the same syntax). While CORBA 1.0 did not completely meet this goal, CORBA 2.0 explicitly provided for a protocol called IIOP (Internet Inter-ORB Protocol) that, if adhered to, does make brokers (ORBs) and objects interoperable across vendors and programming languages. CORBA 3 adds a new CORBA Component Model (CCM) and specifications that, among other things, provide for better negotiation with firewalls, a problem that had made CORBA hard to use in Web development.

CORBA SERVICES
In addition to the interfaces defined for particular objects, CORBA provides a number of services that apply to all objects. These services include creating, moving/copying, or removing objects; allowing more readable names for objects; concurrency and transaction control; setting properties for objects; and sending queries to objects.

A competing framework for distributed object computing is COM/DCOM (Common Object Model/Distributed Common Object Model), now supplanted by .NET (see MICROSOFT.NET). A simpler (though possibly less secure) way to connect programs running on different machines is to use the Simple Object Access Protocol (see SOAP).

Further Reading
Bolton, Fintan. *Pure CORBA*. Indianapolis: Sams, 2001.
"Introduction to CORBA" [Java implementation]. Available online. URL: http://java.sun.com/developer/onlineTraining/corba/corba.html. Accessed September 4, 2007.
McHale, Ciaran. "CORBA Explained Simply, 2007." Available online. URL: http://www.ciaranmchale.com/corba-explained-simply/. Accessed September 4, 2007.

counterterrorism and computers

Counterterrorism is the effort to detect, identify, and neutralize terrorist groups and prevent attacks. Not surprisingly, information technology plays a part in every phase of this effort—and sometimes even becomes part of the battlefield.

INTELLIGENCE AND SURVEILLANCE
The Web and other Internet services are an important part of the battle against terrorism, not least because terrorists themselves are beginning to use online tools effectively (see CYBERTERRORISM). The Internet inherently allows for considerable anonymity (see ANONYMITY AND THE INTERNET). However, any online activity leaves traces, however virtual, and surveillance, intelligence, and forensic techniques are being adapted to this new medium (see COMPUTER FORENSICS).

By putting so much material online, terrorists are exposing themselves to the increasingly sophisticated data mining and "semantic Web" tools that are being developed. These tools can, for example, identify material likely to be of interest (and summarize it) and even analyze the relationship between individuals or groups based on their writing or verbal communications. Of course such results must still be reviewed and acted upon by trained human analysts. Further, surveillance tools that are deployed too widely or indiscriminately are liable to raise privacy concerns.

In recent years the U.S. Department of Homeland Security has apparently been developing more sophisticated data-mining and pattern-recognition programs (see BIOMETRICS and DATA MINING). One is called ADVISE, or Analysis, Dissemination, Visualization, Insight, and Semantic Enhancement. This at least suggests an attempt not to simply find matches between e-mail, online postings, or other textual data, but to construct profiles of a person's activity and/or intentions, which could presumably then be compared with terrorist or criminal profiles.

Surveillance or wiretapping of specific individuals also raises legal issues, particularly with recent revelations of so-called warrantless wiretaps. Officials have claimed that there are relatively few such cases (perhaps fewer than 100 per year), but the Bush administration's claim that it did not need to follow Foreign Intelligence Surveillance Act (FISA) procedures raised considerable controversy, and a court decision forced the administration to seek affirmation of its powers by Congress.

Intelligence officials argue that existing FISA procedures are too cumbersome to deal with the Internet. Old-style wiretapping involved specific telephone instruments and lines, but on the Internet the routing of information is constantly changing, and a person may use several different devices and types of communication. Thus it is argued that the warrant must be broad enough to apply to the person, not a particular means of communication. It is also argued that the global nature of the network also means that distinctions about whether persons are inside or outside of the United States may no longer be as relevant.

Privacy and civil liberties advocates tend to agree that some updating of warrant procedures to deal with modern technology is necessary, but they point to secretiveness and lack of effective legal oversight resulting in a lack of accountability for government surveillance programs. This concern has also been fueled by a succession of revelations that surveillance programs are more extensive than previously thought. (This includes the involvement of telecommunications and Internet service providers and the use of FBI "national security letters"—essentially secret subpoenas.)

COORDINATING EFFORTS

Besides the gathering and analysis of intelligence, computer applications are used in the intelligence and counterterrorism community for many of the same functions found in any large enterprise. These applications include e-mail, personal information management, collaborative creation or review of documents, scheduling and project management, and so on.

Intelligence agencies are even adopting some popular emerging Web technologies. First came Intellipedia, a classified version of Wikipedia serving as a knowledge base for intelligence professionals (see WIKIS AND WIKIPEDIA). In late 2007 the director of national intelligence (DNI) launched A-Space, which includes Intellipedia, while adding other extensive databases, online office facilities (similar to Google Apps), and even blogs and a MySpace-like component (see SOCIAL NETWORKING).

Further Reading
Derosa, Mary. *Data Mining and Data Analysis for Counterterrorism.* Washington, D.C.: Center for Strategic & International Studies, 2004.
Hoover, J. Nicholas. "U.S. Spy Agencies Go Web 2.0 in Effort to Better Share Information." *InformationWeek,* August 23, 2007. Available online. URL: http://www.informationweek.com/story/showArticle.jhtml?articleID=201801990. Accessed September 10, 2007.
Lichtblau, Eric. "F.B.I. Data Mining Reached beyond Initial Targets." *New York Times,* September 9, 2007. Available online. URL: http://www.nytimes.com/2007/09/09/washington/09fbi.html. Accessed September 10, 2007.
Miller, Greg. "Spy Chief Reveals Details of Operations." *Los Angeles Times.* Available online. URL: http://www.latimes.com/news/nationworld/nation/la-na-intel23aug23,0,6229712.story?coll=la-home-center. Accessed September 10, 2007.
Mohammed, Arshad, and Sara Kehaulani Goo. "Government Increasingly Turning to Data Mining." *Washington Post,* June 15, 2006, p. D03. Available online. URL: http://www.whisperingwires.info/. Accessed September 10, 2007.
National Research Council. *Information Technology for Counterterrorism: Immediate Actions and Future Possibilities.* Washington, D.C.: National Academies Press, 2003.
Taipale, K. A. "Whispering Wires and Warrantless Wiretaps: Data Mining and Foreign Intelligence Surveillance." *Bulletin on Law & Security,* spring 2006. Available online. URL: http://www.whisperingwires.info/. Accessed September 10, 2007.

CPU

The CPU, or central processing unit, is the heart of a computer, the place where data is brought in from input devices, processed, and sent to output devices. (This article describes the CPU from the point of view of desktop microcomputers, where it is a single large silicon chip and supporting chips; see MAINFRAME for a discussion of that earlier architecture, MICROPROCESSOR for desktop and portable CPUs, and CHIP and CHIPSET for physical design of components.)

The CPU consists of two major parts. The arithmetic-logic unit performs arithmetic or logical operations on pairs of numbers brought in from memory and stored in special locations called registers (see ARITHMETIC LOGIC UNIT). For example, the CPU can add a value from main memory to a value stored in a register and store the result back into memory. In addition to addition, subtraction, multiplication, and division, the CPU can logically compare the individual bits in two values, performing such operations as AND, where the result is 1 only if both bits are ones, or OR, where the result is 1 if either bit is one. The power of a CPU is measured either in the number of clock cycles that drive it each second (see CLOCK SPEED) or the number of standard instructions it can execute in a second. For modern PCs, clock speeds range into the billions of cycles per second (gigahertz) and millions of instructions per second (most instructions take more than one cycle to be completed).

The other key part of the CPU is the control unit, which determines when (and which) instructions will be executed. Operations to be performed are specified by instruction values that are the lowest level representation of program code, sometimes called machine code. An index register is used to keep track of the current instruction. As instructions are processed, control signals can indicate special conditions,

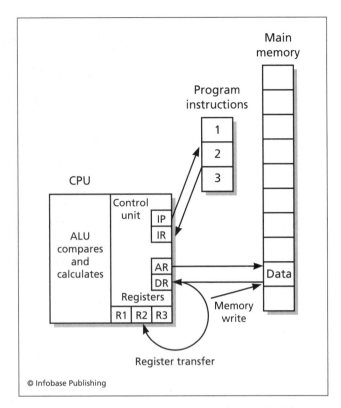

The CPU uses the Instruction Pointer (IP) to keep track of the address of the next instruction in memory, which is stored in the Instruction Register (IR). The Address Register (AR) and Data Register (DR) perform a similar function with program data. Data can also be moved between main memory and the CPU's registers, which are special fast-retrieval memory locations. Instructions are decoded by the control unit and passed to the arithmetic Logic Unit (ALU) for execution.

such as a result being negative. Based on the instructions and signals, the CPU can skip over some instructions, jumping to another location in the program.

The main memory or RAM (random access memory) contains both the program instructions and the data being used by the program, which in turn can be read from a disk or other medium or written back to storage. The effective speed of the system is derived not only from the clock speed but from the speed at which data travels over the system bus, a set of wires that each carry one data bit, as well as the operating speed of the memory chips themselves (see CLOCK SPEED and BUS).

The access of programs to the CPU is controlled in turn by the operating system. Modern operating systems share the CPU with several running programs, doling out execution time according to a scheduling algorithm that takes into account the possible special priority of some programs (see MULTITASKING).

Further Reading
Brain, Marshall. "How Microprocessors Work." Available online. URL: http://computer.howstuffworks.com/microprocessor.htm. Accessed June 30, 2007.

Mano, M. Morris, and Charles Kime. *Logic and Computer Design Fundamentals*. 4th ed. Upper Saddle River, N.J.: Prentice Hall, 2007.
Stokes, Jon. *Inside the Machine: An Illustrated Introduction to Microprocessors and Computer Architecture*. San Francisco: No Starch Press, 2006.

Craigslist

Some of the most successful Web services involve just one or two basic changes in a traditional business or social model. Online auctions, for example, came from the realization that the auctioneer and auction house could be eliminated and a platform provided by which people could buy from or sell to one another directly. (The platform, of course, does have to include such things as listing policies, payment methods, and feedback systems.)

Craigslist has done for the newspaper "personal" ad and laundromat bulletin board what eBay has done for auctions. It was founded in 1995 by Craig Newmark, a San Francisco Bay Area software developer who saw a need for an online forum for news about local events. The "list" part of Craigslist reflects its origin as an e-mail list.

News about the list spread rapidly in Newmark's milieu of well-connected professionals, and the volume of postings grew correspondingly large. Furthermore, many people began to post things other than event listings—including job openings, for which Newmark soon set up a separate category on the list. As the number and kinds of postings grew, the mailing list format became unwieldy, so Newmark and some volunteers put together a Web interface that users could use to browse the various categories. By 2000 Craigslist.org had become a full-time job for Newmark and nine employees

Craiglist's Web site is organized by community, including U.S. states and cities and a variety of other countries and international cities. Each local site is further divided into sections such as community activities (including people seeking or providing childcare or sharing rides), personal ads (seeking relationships), housing (mostly rentals), jobs, services, items for sale, and a variety of discussion forums.

As of 2007 Craigslist had 24 employees. The site is nearly completely free of charge, with revenue coming only from paid job listings and apartment broker listings in selected cities. The site's popularity has been impressive, with more than 5 billion page views, 10 million visitors, and over 10 million classified ads per month.

Newmark and CEO Jim Buckimaster have suggested that they have little interest in either turning Craigslist into a public company or "going commercial" and tapping what many observers consider to be much greater revenue potential.

PROBLEMS AND ISSUES
Craigslist's success has raised some issues. In 2004 eBay bought a 25 percent stake in the company, leading some supporters to worry about pressure to raise more revenue by carrying banner ads or charging for posting on the site. However, as of 2008 the site remains free to users.

As with eBay, Craigslist has to strike a balance between protecting users from criminal activity and exercising direct oversight with the attendant expenses and legal problems. For keeping out illegal ads (such as discriminatory housing or job offers or solicitations for prostitution), Craigslist has relied mainly on users to be "good citizens" and to "flag" offending ads for removal by the service. Nevertheless, police have reported use of Craigslist by prostitution rings and other organized criminals and identity thieves seeking personal information. A 2006 suit (subsequently dismissed) accused Craigslist of "allowing" discriminatory housing ads in Chicago. (Under federal law, Web sites are generally not liable for content posted by users, unless the site has edited content.)

Because Craigslist has been so successful, newspapers have complained that it has dried up much of their revenue from classified advertising, costing them an estimated $50–$65 million in 2004 in the Bay Area alone. This is particularly a concern of small local and independent newspapers for which ads may be their only source of revenue.

Craigslist has won a 2001 Webby award for Best Community Site, and was voted Best Local Web site in a 2003 Manhattan Reader's Poll.

Further Reading

Carney, Brian M. "Zen and the Art of Classified Advertising." *Wall Street Journal.com OpinionJournal,* June 17, 2006. Available online. URL: http://www.opinionjournal.com/editorial/feature.html?id=110008531. Accessed September 9, 2007.

Craigslist. Available online. URL: http://www.craigslist.org. Accessed September 9, 2007.

"Craigslist Hits Bay Area Classifieds Hard." *NewsInc* 16 (December 27, 2004): n.p. Available online. URL: http://www.newsinc.net/morgue/2004/NI041227.html. Accessed September 9, 2007.

Sentementes, Gus G. "Web Site Vice Stings." *Baltimoresun.com,* September 8, 2007. Available online. URL: http://www.baltimoresun.com/news/local/annearundel/bal-md.ar.prostitution08sep08,0,179750 1.story. Accessed September 9, 2007.

Cray, Seymour

(1925–1996)
American
Computer Engineer, Inventor

Seymour Cray was a computer designer who pioneered the development of high-performance computers that came to be called supercomputers. Cray was born in Chippewa Falls, Wisconsin. After serving in World War II as an army electrical technician, Cray went to the University of Minnesota and earned a B.S. in electrical engineering and then an M.S. in applied mathematics. (This combination is a common background for many of the designers who would have to combine mathematics and engineering principles to create the first computers.)

In 1951, he joined Engineering Research Associates (ERA), one of a handful of companies that sought to commercialize the digital computing technology that had been developed during and just after the war. Cray soon became

Seymour Cray is considered by many people to be the father of the supercomputer. His innovative Cray computers looked—and performed—like something out of science fiction. (CRAY RESEARCH)

known for his ability to grasp every aspect of computing from logic circuits to the infant discipline of software development. When ERA and its competitor, the Eckert-Mauchly Computer Company were bought by Remington Rand, Cray became the chief designer for the Univac, the first commercially successful computer. In 1957, however, Cray and two colleagues struck out on their own to form Control Data Corporation (CDC). Their CDC 1604 was one of the first computers to move from vacuum tubes to transistors. The CDC 6600 was considered by many to be technically superior to the IBM 360. However, by then IBM had become preeminent in the business computing market, while the CDC machines found favor with scientists.

By the late 1960s, Cray had persuaded CDC to provide him with production facilities within walking distance of his home in Chippewa Falls. There he designed the CDC 7600. This computer was hailed as the world's first supercomputer (see SUPERCOMPUTER). However CDC disagreed with Cray about the commercial feasibility of even more powerful computers. In 1972, Cray formed his own company, Cray Research, Inc. By then Cray's reputation as a computer architect was so great that investors flocked to buy stock in his company. His series of Cray supercomputers looked like sleek monoliths from a science fiction movie. The machines were the first supercomputers to use parallel processing, where tasks can be assigned to different processors to speed up throughput. While costing millions of dollars apiece, the Cray supercomputers made it possible to perform simulations in atomic physics, aerodynamics, and other fields that were far beyond the capabilities of earlier computers. However, the Cray Computer Corporation ran into financial problems and was bought by Silicon Graphics (SGI) in 1996.

Cray received many honors including the IEEE Computer Society Pioneer Award (1980) and the ACM/IEEE Eckert-Mauchly Award (1989). Cray died on October 5, 1996, in Colorado Springs, Colorado.

Further Reading

Bell, Gordon. "A Seymour Cray Perspective." Available online. URL: http://_research.microsoft.com/users/gbell/craytalk/. Accessed July 1, 2007.
Breckenridge, Charles W. "A Tribute to Seymour Cray." Available online. URL: http://www.cgl.ucsf.edu/home/tef/cray/tribute.html. Accessed July 1, 2007.
Murray, C. J. *The Supermen: the Story of Seymour Cray and the Technical Wizards behind the Supercomputer.* New York: John Wiley, 1997.
Smithsonian Institute. National Museum of American History. "Seymour Cray Interview." Available online. URL: http://americanhistory.si.edu/collections/comphist/cray.htm.

CRM   *See* CUSTOMER RELATIONSHIP MANAGEMENT.

CSS   *See* CASCADING STYLE SHEETS.

Cunningham, Howard (Ward)
(1949–   )
American
Software Developer

Today the first place many Web users look for information about a topic is Wikipedia, the vast and ever growing online collaborative encyclopedia. The type of software that makes Wikipedia (and thousands of other wikis) possible was invented by Howard G. Cunningham, better known as Ward Cunningham.

Born on May 26, 1949, Cunningham learned to program in high school. He then attended Purdue University, where he received a bachelor's degree in electrical engineering and computer science and then a master's in computer science. After graduation Cunningham worked as a researcher in microcomputer systems for Tektronix, where he encountered an intriguing style of programming (see SMALLTALK). In a later position at Wyatt Software, Cunningham became involved with larger-scale software projects and began to think about better ways to manage them.

In the early 1980s Cunningham encountered a book that looked at architecture in terms of the combining of intuitive patterns. Cunningham began to apply similar principles to the design of software (see also DESIGN PATTERNS). One result was the holding of the first conference on pattern languages at the University of Illinois at Urbana-Champaign in 1994.

Around that time, Cunningham was seeking a way for programmers to collaborate in working with design patterns. He had already encountered the power of linking (see HYPERTEXT) in HyperCard, developed by Apple for the Macintosh in the late 1980s. Because it was so easy to use, HyperCard encouraged many nonprofessional programmers (including teachers) to develop and share applications.

DEVELOPING THE WIKI

Using HyperCard, Cunningham built an application that allowed users to add free-form data to a database and link it to other entries by clicking a button. Users who tried it were fascinated by its potential. Cunningham then wanted to expand it so users could access it over networks. However, he was unable to develop a networked version of his HyperCard application.

One colleague suggested using the World Wide Web (see BERNERS-LEE, TIM and WORLD WIDE WEB). Cunningham implemented his free-form linking system as Web pages, and the result was something he at first thought of calling QuickWeb. He then remembered hearing the phrase *wiki wiki* or "quickly, quickly") in Hawaii, and he decided to call his system wikiwikiWeb. Today, it is just known as a wiki (see WIKIS AND WIKIPEDIA). This first wiki, called the Portland Pattern Repository, came online in 1995 and continues to operate today.

COLLABORATIVE SOFTWARE DEVELOPMENT

Cunningham worked for a few years on open-source projects at Microsoft. The giant software maker is not generally well regarded among open-source developers, though Cunningham has acknowledged its technical prowess. At any rate, Cunningham decided to move on. He served as director for community development at the Eclipse Foundation, which oversees development of Eclipse, a versatile and very popular open-source programming environment. In 2007 Cunningham left Eclipse to become chief technology officer (CTO) of AboutUs, a company founded to further develop wikis and collaborative communities.

Cunningham continues to be an enthusiastic proponent of open source. He argues that the most important advantage of open source is not lower cost, but the way it puts access to powerful tools into the hands of thousands of users and encourages them to develop new features and capabilities.

Cunningham's contributions to programming methods are also extensive, including the use of design patterns for "quick and agile" development and what became known as "extreme programming."

Further Reading

Cunningham, Ward. Home Page. Available online. URL: http://www.c2.com/cgi/wiki?WardCunningham. Accessed September 9, 2007.
Leuf, Bo, and Ward Cunningham. *The Wiki Way: Quick Collaboration on the Web.* Upper Saddle River, N.J.: Addison-Wesley Professional, 2001.
Siddalingaiah, Madhu. "Ward Cunningham Interview: Eclipse, Collaboration and Software Trends." Includes links to Cunningham's EclipseCon 2006 presentation. SQL Summit.

Available online [audio]. URL: http://www.sqlsummit.com/People/WCunningham.htm. Accessed September 9, 2007.

Taft, Darryl K. "Father of Wiki Speaks Out on Community and Collaborative Development." *eWeek,* March 20, 2006. Available online. URL: http://www.eweek.com/article2/0,1759,1939982,00.asp. Accessed September 9, 2007.

Wiki Wiki Web. Available online. URL: http://c2.com/cgi-bin/wiki?WikiWikiWeb. Accessed September 9, 2007.

customer relationship management (CRM)

In recent years there has been increasing emphasis, particularly in online business, on communicating with and "cultivating" customers as well as in systematically using information about transactions and customer behavior (see also E-COMMERCE). Collectively, these activities (and the software used to implement them) are often known as customer relationship management (CRM).

The basic data stream in CRM is a complete contact history for each customer, including not only purchases, but also product or customer support inquiries. The resulting database is used to ensure that with each new contact (such as through a call center), the person responding has access to all the information about previous contacts with the customer. Thus, for example, in the course of answering a query or solving a problem, the representative can review a list of which products the customer has purchased and suggest additional products that might help deal with the problem.

Besides dealing with customer-initiated contacts, CRM data can be very useful in designing marketing campaigns, advertising, promotions, and so on (see ONLINE ADVERTISING). The database can be analyzed to determine, for example, the likelihood that a customer who buys a digital camera might also buy a particular printer or memory card (see DATA MINING). Once this is known, a customer who is in the process of buying a camera might be offered a special price on a memory card during checkout. (For an example of extensive integration, mining, and use of CRM data, see AMAZON.COM.) For longer-term planning, "strategic CRM" can help a company decide on what types of products and markets to focus.

In addition to a database with extensive analysis and reporting facilities, a CRM system requires software that sales or support persons can use to access information in real time and update it with the results of the current call. Organizations can buy turnkey products or design their own CRM systems by selecting and integrating software components. However implemented, effective CRM requires that everyone in contact with a customer keep the ongoing cultivation of that relationship in mind, and search for ways to deliver more value than the competition.

Successful CRM also requires a balance between the desire to get as much information as possible and allaying customers' concerns. If the CRM software (or how it is used) slows down the resolution of support calls, ends up generating unwanted solicitations (particularly from third parties), or conveys a sense of disregard for privacy, it could damage customer relations and lead to loss of business and reputation.

Further Reading

Buttle, ·Francis. *Customer Relationship Management: Concepts and Tools.* Burlington, Mass.: Elsevier Butterworth-Heinemann, 2004.

CRM Today. Available online. URL: http://www.crm2day.com/. Accessed September 9, 2007.

Customer Relationship Management Association. Available online. URL: http://www.crmassociation.org/. Accessed September 9, 2007.

Kincaid, Judith W. *Customer Relationship Management: Getting It Right.* Upper Saddle River, N.J.: Prentice-Hall/HP Professional Books, 2002.

Kumar, V., and Werner J. Reinartz. *Customer Relationship Management: A Databased Approach.* New York: Wiley, 2005.

Prahalad, C. K., et al. *Harvard Business Review on Customer Relationship Management.* Cambridge, Mass.: Harvard Business School, 2001.

cyberlaw

Legal scholars and law schools have begun to use the term *cyberlaw* to refer to a variety of legal issues that are often involved in online interactions (see CYBERSPACE). While traditional legal fields such as contract law, property law, privacy, and jurisdiction do apply online, cyberlaw recognizes that certain common features of the digital world pose unique challenges.

The first question in any legal dispute is which court, if any, has jurisdiction. In the physical world there are well-demarcated spheres (in the United States) for federal, state, and municipal law. However, participants in an online transaction or other act may often be in different physical jurisdictions. Indeed, the World Wide Web's structure does not inherently follow physical boundaries, with linkage being largely semantic rather than geographical. Some Internet advocates such as John Perry Barlow have gone so far as to argue that the Web must develop its own laws and customs that reflect its technical and social nature—eventually forming its own social contract.

A more pragmatic approach is taken by Lawrence Lessig, who argues that a legal regime must evolve that takes into account the needs and concerns of both traditional physical jurisdictions and the new realm of cyberspace (see LESSIG, LAWRENCE).

DIVERSE ISSUES

In practice, when crimes or disputes occur online, political pressure or legal duty will impel federal and state officials to become involved. For example, users of file-sharing services are being sued for alleged violations of copyright law (see FILE-SHARING AND P2P NETWORKS and INTELLECTUAL PROPERTY AND COMPUTING). The question of whether the provider of an online service should be held responsible for violations by users must also be decided; in the United States, federal law has exempted providers from most legal liabilities. Matters can become even more complicated when people involved in a case are living in different countries. (Many countries have lax or no regulation of online activity, and activity prohibited in countries such as the United States can flourish there—see, for example, ONLINE GAMBLING.)

Many issues regarding freedom of speech and expression arise in the online world. Should a blogger be accorded the rights of a traditional journalist? Should an American company such as Google or Yahoo! be held responsible for turning dissidents over to Chinese authorities?

The growth of immersive and persistent online game worlds such as *Second Life* raises other difficult questions for cyberlaw (see IDENTITY IN THE ONLINE WORLD and ONLINE GAMES). Can promises (whether business contracts or even marriage proposals) made through online personas ("avatars") be binding? Who owns property (such as a house) created or purchased in the virtual world? What if someone steals or vandalizes the virtual property? Should a virtual world be treated as a kind of parallel jurisdiction and perhaps allowed to have its own legal system and courts, perhaps even a form of limited sovereignty? While these questions may seem far-fetched, they take on more urgency as millions of people begin to spend a significant part of their waking time in a virtual world and generate economic activity that can be denominated in real money. The resolution of these and other cyberlaw issues will both depend on and influence how the Internet itself is organized and governed (see INTERNET ORGANIZATION AND GOVERNANCE). For some organizations currently involved in trying to promote cyber rights and shape policy, see CYBERSPACE ADVOCACY GROUPS.

Further Reading

Barlow, John Perry. "A Declaration of the Independence of Cyberspace." Available online. URL: http://homes.eff.org/~barlow/Declaration-Final.html. Accessed September 9, 2007.

Electronic Frontier Foundation. "Internet Law Treatise." Available online. URL: http://ilt.eff.org/index.php/Table_of_Contents. Accessed April 23, 2008.

Gahtan, Alan. Cyberlaw Encyclopedia. Available online. URL: http://www.gahtan.com/cyberlaw/. Accessed September 9, 2007.

Ku, Raymond S., and Jacqueline D. Lipton. *Cyberspace Law: Cases and Materials.* 2nd ed. New York: Aspen Publishers, 2006.

Lessig, Lawrence. *Code: Version 2.0.* New York: Basic Books, 2006.

Zittrain, Jonathan L. *Jurisdiction (Internet Law Series).* New York: Foundation Press, 2005.

cybernetics

Cybernetics may not be familiar to many readers today, except as part of words like "cyberspace." The term was coined by mathematician Norbert Wiener (see WIENER, NORBERT) in his book about control and communication in animals and machines. The root comes from the Greek *kybernetes*, meaning steersman or governor.

Cybernetics looks at systems as a whole. A key concept is feedback, which allows a system to adjust itself in response to changes in the environment. A familiar example is a thermostat, which includes a switch that expands as the air heats, turning off the heater when the temperature reaches its indicated setting. Similarly, as the air cools the switch contracts and restarts the heater.

In addition to feedback, cybernetics looks at how information is communicated between the environment and a machine or organism, or between component parts. Cybernetics is also interested in structures that may be built up through feedback and communication—ultimately, in humans: the structures of self, identity, and consciousness.

Cybernetics is fundamental to the operation of robots (see ROBOTICS). Around the time of Wiener's book, Grey Walter built one of the earliest robots, a "cybernetic turtle" that could autonomously explore an environment, responding to changes in light.

In computers, any program that changes its behavior in response to new data might be called cybernetic. Cybernetics is relevant to a variety of fields in computer science that involve machine learning or reasoning (see ARTIFICIAL INTELLIGENCE, GENETIC PROGRAMMING, and NEURAL NETWORK).

During the 1950s and 1960s cybernetics concepts became quite influential and were applied to such diverse fields as neurology, cognitive science, psychology, philosophy, anthropology, sociology, and economics. However, the term *cybernetics* itself gradually fell out of favor, even though the concepts remain at the heart of systems thinking. For some writers such as Gregory Bateson and anthropologist Margaret Mead, the focus shifted to a "new cybernetics" or "second-order cybernetics" that studies the interaction of observers with phenomena and attempts to construct a model of the mind itself.

Further Reading

American Society for Cybernetics. Available online. URL: http://www.asc-cybernetics.org/. Accessed September 10, 2007.

Gasperi, Michael. "Grey Walter's Machina Speculatrix." Available online. URL: http://www.extremenxt.com/walter.htm. Accessed September 10, 2007.

Principia Cybernetica Web. Available online. URL: http://pespmc1.vub.ac.be/. Accessed September 10, 2007.

Wiener, Norbert. *Cybernetics: or Control and Communication in the Animal and the Machine.* 2nd ed. Cambridge, Mass.: MIT Press, 1961.

cyberspace advocacy groups

By the mid-1990s a number of issues were arising as the Internet and Web became an increasingly important factor in commerce and society (see CENSORSHIP AND THE INTERNET, INTELLECTUAL PROPERTY AND COMPUTING, and PRIVACY IN THE DIGITAL AGE). Often in response to proposed or enacted federal legislation, a number of advocates have organized groups to keep track of developments that they believe threaten the free exchange of information and expression, as well as opposing government surveillance and corporate practices believed to intrude on privacy.

Although there are dozens of groups advocating for the rights of Internet users, three groups have been particularly prominent and effective.

ELECTRONIC FRONTIER FOUNDATION

The Electronic Frontier Foundation (EFF) was founded in 1990 by Mitch Kapor, John Gilmore, and John Perry Barlow. Its immediate motivation was the federal search and seizure of computers belonging to Steve Jackson Games as part of an investigation into illegal distribution of proprietary documents. Although the game company was not involved in any crime, the seizure of its equipment and information threatened to put it out of business. Ultimately, Jackson prevailed in federal court, establishing that unconventional means of expression such as games were entitled to First Amendment protection. In another high-profile case, computer scientist Daniel Bernstein sued and won the right to publish encryption software and related papers, again extending First Amendment protections in the digital world.

The EFF has also been involved in the dispute between users of file-sharing services and the Recording Industry Institute of America (RIAA) over subpoenas of service providers seeking alleged illegal downloaders.

Most recently, the EFF has expanded its efforts further with regard to issues of government surveillance and the prosecution of computer crimes, such as collection and use of evidence.

CENTER FOR DEMOCRACY AND TECHNOLOGY

Founded in 1994, the Center for Democracy and Technology (CDT) somewhat overlaps the EFF in interests, but has a greater emphasis on the connections between online activities and the political process. The organization's first major battle involved the Computer Decency Act. While intended by its proponents to ban obscenity and particularly child pornography from the Internet, cyberspace-rights advocates saw the law as vague, poorly written, and likely to deny access to material that is constitutionally protected for adults—an argument that the Supreme Court ultimately accepted in *ACLU v. Reno* (1997).

More recently, the CDT has supported the free-speech rights of bloggers (see BLOGS AND BLOGGING), arguing that they should be accorded journalistic rights (see also JOURNALISM AND COMPUTERS). Besides issue advocacy, the organization's overall focus is on developing public policy that recognizes the unique features of cyberspace and promotes freedom of expression, protection of privacy, and widespread access to the Net (see also INTERNET ACCESS POLICY).

ELECTRONIC PRIVACY INFORMATION CENTER

Also founded in 1994, the Electronic Privacy Information Center (EPIC) is a Washington, D.C.–based public interest research center devoted to privacy and civil liberties issues. The group's electronic newsletter EPIC Alert provides a useful summary of ongoing developments, cases, and issues. The organization also publishes regularly updated compendiums on developments in open government/freedom of information, privacy and human rights, and privacy law.

(For online activists involved in general political issues and campaigns, see POLITICAL ACTIVISM AND THE INTERNET.)

Further Reading
Center for Democracy and Technology. Available online. URL: http://www.cdt.org/. Accessed September 10, 2007.
Center for Democracy and Technology. "CDT: A Decade of Internet Advocacy." Available online. URL: http://www.cdt.org/mission/2006aao.pdf. Accessed September 10, 2007.
Electronic Frontier Foundation. Available online. URL: http://www.eff.org/. Accessed September 10, 2007.
Electronic Privacy Information Center. Available online. URL: http://www.epic.org/. Accessed September 10, 2007.
Godwin, Mike. *Cyber Rights: Defending Free Speech in the Digital Age.* Revised ed. Cambridge, Mass.: MIT Press, 2003.
Privacy.org. Available online. URL: http://www.privacy.org/. Accessed September 10, 2007.

cyberspace and cyber culture

The term *cyberspace* first came to prominence when it appeared in *Neuromancer,* a 1984 novel by science fiction writer William Gibson. The word is a combination of "cyber" (meaning related to computers) and "space." As another SF writer, Bruce Sterling, wrote in *The Hacker Crackdown* (1993), cyberspace is "the place between the phones. The indefinite place *out there,* where the two of you, human beings, actually meet and communicate."

While the elite telegraphers of the 19th century and later telephone users first experienced the sense of disembodied electronic communication, it took the development of widespread computer terminals, personal computers, and connecting networks to create a sense of an ongoing place in which people meet and interact. The first "villages" in cyberspace came into being during the 1970s as research networks (ARPA), and the Usenet newsgroups of UNIX users began to carry messages and news postings. During the 1980s, many more settlements began to light up the map of cyberspace, ranging from cities (large online services such as The Source, BIX, and CompuServe) to thousands of villages (tiny bulletin board systems running on personal computers). (See ONLINE SERVICES and BULLETIN BOARD SYSTEMS.)

Wherever human beings build communities, they shape culture. The cyber culture that grew up in cyberspace has featured many diverse strands. Hackers (not originally a pejorative term) had their distinctive hangouts and lingo. Bulletin board cultures varied from the hacker hardcore to user groups that tried to assist beginners. On the nascent Internet multiplayer game worlds called MUDs (Multi-User Dungeons) and Muses used words to create richly detailed fantasy cyberspaces. Together with chat rooms and conferencing systems, they fostered virtual communities that, like physical communities, express a full range of human behavior (see BLOGS AND BLOGGING, CONFERENCING SYSTEMS, CHAT, SOCIAL NETWORKING, TEXTING AND INSTANT MESSAGING, and VIRTUAL COMMUNITY).

While cyber culture shares the characteristics of other human cultures, it also has unique characteristics that are dictated by the nature of the online, virtual medium. Since the online user reveals only what he or she chooses to reveal, identities can be fluid: playful or deceptive. While people are not physically vulnerable in cyberspace, they are certainly emotionally vulnerable. (Virtual eroticism, or "cyber sex" has even led to virtual rapes.) The issue of protecting privacy becomes important because sensitive personal information is constantly being exposed in order to carry on commerce (see IDENTITY IN THE ONLINE WORLD AND PRIVACY IN THE DIGITAL AGE.)

THE FUTURE OF CYBERSPACE

By the end of the 1990s, the face of cyberspace was no longer that of text screens but that of the World Wide Web with its graphical pages. Multiplayer games now often feature graphics and even real-time voice communication is possible. With ubiquitous digital cameras, the boundary between cyberspace and physical space has become fluid, with people able to enter into each other's physical environments in realistic ways. Meanwhile, the development of virtual reality techniques has made computer-generated worlds much more vivid and realistic (see VIRTUAL REALITY). As more people are linked continually to the network by broadband and wireless connections, cyberspace may eventually disappear as a separate reality, having merged with physical space.

Further Reading

Bell, David. *An Introduction to Cybercultures.* New York: Routledge, 2001.
Bell, David, and Barbara M. Kennedy, eds. *The Cybercultures Reader.* New York: Routledge, 2007.
Jenkins, Henry. *Convergence Culture: Where Old and New Media Collide.* New York: New York University Press, 2006.
Resource Center for Cyberculture Studies. Available online. URL: http://rccs.usfca.edu/default.asp. Accessed July 1, 2007.
Silver, David, and Adrienne Massari, eds. *Critical Cyberculture Studies.* New York: New York University Press, 2006.
Wired magazine. Available online. URL: http://www.wired.com. Accessed July 1, 2007.

cyberstalking and harassment

Cyberstalking and harassment or "cyber bullying" involve the use of online communications and facilities (such as instant messaging, chat rooms, e-mail, or Web sites) to stalk, harass, or otherwise abuse a person or group. These activities may be carried on entirely online or in connection with physical stalking or harassment.

Stalking and threatening a person has been a crime in the physical world for some time, and similar principles apply to online stalking. Generally, to be guilty of stalking, a person must repeatedly harass or threaten the victim, often following him or her and intruding or violating privacy.

Cyberstalkers take advantage of the fact that there is a great deal of information about many people online. (Indeed, the popularity of sites such as MySpace means that many users can unwittingly provide that information in well-organized, easy-to-access form—see SOCIAL NETWORKING.) The stalker can also use search engines to find e-mail or even physical addresses and phone numbers, or can join chat rooms used by the prospective victim.

Motives for stalking can range from sexual obsession to anger at some real or imagined slight, to more idiosyncratic reasons. As with physical stalking in an earlier generation, law enforcement agencies were often slow to acknowledge the potential seriousness of the crime or to develop effective ways to deal with it.

This began to change with the tragic and highly publicized case of Amy Boyer, who had been found online through a data broker, stalked, harassed, and ultimately murdered. In 1999 California became the first state to pass a law against cyberstalking, and in 2000 cyberstalking was made part of the federal Violence against Women Act.

CYBERBULLYING

Like traditional bullying in schools or other settings, cyberbullying involves harassment, sometimes organized, of people considered to be weak or different in some way. However, the ability to hide or disguise one's identity online (see ANONYMITY AND THE INTERNET) facilitates cyberbullying by making it harder for victims to identify and confront or report their tormentors. Media for cyberbullying include text and instant messaging, photos or videos, blogs, and increasingly, pages on social networking sites. Contents can include threats, racial or other slurs, and unwelcome sexual solicitations.

In March 2007 a number of organizations joined with the U.S. Department of Justice in a public service advertising campaign to educate young people about cyberbullying and what they can do to prevent it. Some schools are adopting anti-cyberbullying policies and programs.

Besides potentially serious psychological trauma to victims, cyberbullying can sometimes lead victims to lash out, and in extreme cases, cyberbullying may play a role in campus shootings.

Further Reading

Bocij, Paul. *Cyberstalking: Harassment in the Internet Age and How to Protect Your Family.* Westport, Conn.: Praeger, 2004.
Bolton, Jose, and Stan Graeve, eds. *No Room for Bullies: From the Classroom to Cyberspace: Teaching Respect, Stopping Abuse, and Rewarding Kindness.* Boys Town, Nebr.: Boys Town Press, 2005.
"Cyberbullying: Identifying the Causes and Consequences of Online Harassment" [news and resources]. Available online. URL: http://www.cyberbullying.us/. Accessed September 10, 2007.
Cyberbullying.org. Available online. URL: http://www.cyberbullying.org/. Accessed September 10, 2007.
Henderson, Harry. *Internet Predators (Library in a Book).* New York: Facts On File, 2005.
Widhalm, Shelley. "New Teen Bullies." *Washington Times,* September 10, 2007. Available online. URL: http://washingtontimes.com/article/20070910/METRO/109100038/1004. Accessed September 10, 2007.
Willard, Nancy E. *Cyberbullying and Cyberthreats: Responding to the Challenge of Online Social Aggression, Threats, and Distress.* Champaign, Ill.: Research Press, 2007.

cyberterrorism

Cyberterrorism can include several types of activities: the promotion of terrorist or militant groups on the Web (including propaganda and recruitment), the coordination or facilitation of terrorist activities, and actual attacks on Web sites or other information infrastructure.

TERRORISTS ON THE WEB

There is little doubt that terrorist groups are increasingly computer savvy and willing to use the technology to further their purposes. Many groups have Web sites that are used for propaganda and recruiting. (In 2007 a British court sentenced three men, calling them "cyber-jihadis" and saying they had used a Web site to urge Muslims to attack non-Muslims.) In fact extremist groups of many kinds (including neo-Nazis and other racial extremists) have long used Web sites to attract young followers through propaganda, music, and even games.

Other material posted by terrorist groups online includes bomb-making plans, lists of potential targets (possibly including maps or blueprints), and "tips" for penetrating defenses or evading detection. (A project called Dark Web at the University of Arizona searches for, compiles, and analyzes massive amounts of Web content generated by terrorist groups.)

ATTACKS ON WEB SITES

Attempts to jam or disrupt Web sites (such as denial of service attacks or DOS) have been made for a variety of reasons. At one end of the spectrum are individuals or small groups engaged in criminal activity (such as attempted extortion) or expressing political protest ("hacktivists"). At the other end are alleged online offensives by national governments (see INFORMATION WARFARE).

Although there have been no major disruptions as of mid-2008, terrorists (or sympathizers) have already conducted cyberattacks. One site has even offered a downloadable "electronic jihad" program that users can use to select from a list of targets to launch an automated DOS. While such sites are usually taken down after a few months, it is relatively easy to start another, especially because information provided for site registration is often not verified.

FIGHTING CYBERTERRORISM

Strategies and tactics to combat cyberterrorism involve both general antiterrorist intelligence and other techniques as well as those particularly adapted to the cyberspace arena (see COUNTERTERRORISM AND COMPUTERS, COMPUTER CRIME AND SECURITY, and COMPUTER FORENSICS).

The cyberterrorist threat also plays an important role in the effort to better protect vital infrastructure. Although attacks on banking and other financial computer systems have the potential to cause severe economic damage, much attention has focused on computer-based attacks that have the potential to directly injure or even kill people. Back in 2000, an individual hacker in Australia took over a pumping station and dumped more than 264,000 gallons of raw sewage into public lands and waterways. Although no humans were directly harmed, it is easy to see that such contamination in the drinking water supply could be deadly.

Regardless of the type of computer system, following best security practices can go a long way to "hardening" potential targets. Such practices include the use of robust firewalls and antivirus programs, regular security updates for the operating system and software, network monitoring and intrusion detection, sharing information about security threats, and training personnel to be aware of typical attacker techniques, including deception (social engineering). There needs to be a comprehensive protection plan for each facility that takes both physical and electronic security into account

ASSESSMENT

In recent years cyberterrorism has been a much publicized topic. Some critics believe that the threat of cyberterrorism has been overestimated—not because many computer systems are not vulnerable, but because the most vulnerable physical systems are generally not on the Internet and not easily accessible. It has also been argued that terrorists generally use simpler, more direct weapons (e.g., bombs) and aim to produce physically spectacular or terrifying results. Most cyberattacks would not seem to meet those criteria. On the other hand, a cyberattack might be launched in conjunction with physical attacks, either as a distraction or to make it harder for authorities to respond to the main attack.

Properly assessing risks and allocating resources will always be difficult, and will always be influenced by political and economic as well as technological factors.

Further Reading

Blane, John V., ed. *Cybercrime and Cyberterrorism: Current Issues.* New York: Novinka Books, 2003.

Brown, Lawrence V., ed. *Cyberterrorism and Computer Attacks.* New York: Novinka Books, 2006.

Chen, Hsinchun. "How Terrorists Use the Internet" [interview transcript]. *The Science Show,* March 31, 2007. Available online. URL: http://www.abc.net.au/rn/scienceshow/stories/2007/1885902.htm#transcript. Accessed September 7, 2007.

Colarik, Andrew M. *Cyber Terrorism: Political and Economic Implications.* Hershey, Penn.: Idea Group, 2006.

"Cyber-Terrorism: Propaganda or Probability?" About.com. Available online. URL: http://antivirus.about.com/library/weekly/aa090502a.htm. Accessed September 7, 2007.

Greenmeier, Larry. "Cyberterrorism: By Whatever Name, It's on the Increase." *InformationWeek,* July 7, 2007. Available online. URL: http://www.informationweek.com/story/showArticle.jhtml?articleID=200900812. Accessed July 12, 2007.

O'Day, Alan, ed. *Cyberterrorism.* Burlington, Vt.: Ashgate, 2004.

Wagner, Breanne. "Electronic Jihad: Experts Downplay Imminent Threat of Cyberterrorism. *National Defense* 92 (July 1, 2007): p. 34 ff.

data

Today the term *data* is associated in many peoples' minds mainly with computers. However, data (as in "given facts" or measurements) has been used as a term by scientists and scholars for centuries. Just as with a counting bead, a notch in a stick, or a handwritten tally, data as stored in a computer (or on digital media) is a *representation* of facts about the world. These facts might be temperature readings, customer addresses, dots in an image, the characteristics of a sound at a given instant, or any number of other things. But because computer data is not a fact but a representation of facts, its accuracy and usefulness depends not only on the accuracy of the original data, but on its *context* in the computer.

At bottom, computer data consists of binary states (represented numerically as ones or zeroes) stored using some physical characteristic such as an electrical or magnetic charge or a spot capable of absorbing or reflecting light. A string of ones and zeroes in a computer has no *inherent* meaning. Is the bit pattern 01000001 a number equivalent to 65 in the decimal system? Yes. Is it the capital letter "A"? It may be, if interpreted as an ASCII character code. Is it part of some larger number? Again, it may be, if the memory location containing this pattern is interpreted as part of a set of two, four, or more memory locations.

In order to be interpreted, data must be assigned a category such as integer, floating point (decimal), or character (see DATA TYPES). The programming language compiler uses the data type to determine how many memory locations make up that data item, and which bits in memory correspond to which bits in the actual number. Data items can be treated as a batch (see ARRAY) for convenience, or different kinds of data such as names, addresses, and Social Security numbers can be grouped together into records or structures that correspond to an entity of interest (such as a customer). In creating a structure within the program to represent the data, the programmer must be cognizant of its purpose and intended use.

The programming language and code statements define the context of data within the rules of the language. However, the *meaning* of data must ultimately be constructed by the human beings who use it. For example, whether a test score is good, bad, or indifferent is not a characteristic of the data itself, but is determined by the purposes of the test designer. This is why a distinction is often made between *data,* as raw numbers or characters, and *information* as data that has been placed in a meaningful context so that it can be useful and perhaps even enlightening to the user.

Further Reading

Bierman, Alan W. *Great Ideas in Computer Science: a Gentle Introduction.* 2nd ed. Cambridge, Mass.: MIT Press, 1997.
Hillis, Daniel W. *The Pattern on the Stone: the Simple Ideas that Make Computers Work.* New York: Basic Books, 1998.

data abstraction

Abstract data types are used to describe a "generic" type of data, specifying how the data is stored and what operations can be performed on it (see OBJECT-ORIENTED PROGRAMMING, LIST PROCESSING, STACK, and QUEUE).

For example, an abstract stack data type includes a structure for storing data (such as a list or array) and a set of operations, such as "pushing" an integer onto the stack and "popping" (removing) an integer from the stack. (For the process of combining data and operations into a single entity, see ENCAPSULATION.) Abstract data types can be implemented directly in object-oriented programming languages (see CLASS, C++, JAVA, and SMALLTALK).

One advantage of using abstract data types is that it separates a structure and functionality from its implementation. In designing the abstract stack type, for example, one can focus on what a stack does and its essential functions. One avoids becoming immediately bogged down with details, such as what sorts of data items can be placed on the stack, or exactly what mechanism will be used to keep track of the number of items currently stored. This approach also avoids "featuritis," the tendency to see how many possible functions or features one can add to the stack object. For example, while it might be useful to give a stack the ability to print out a list of its items, it is probably better to wait until one needs such a capability than to burden the basic stack idea with extra baggage that may make it more cumbersome or less efficient.

An abstract data type or its embodiment, a class, is not used directly by the program. Rather, it is used to create an entity (object) that is a particular instance of the abstract data type (for example, an actual stack that will be used to manipulate data). The data stored inside the object is not accessed directly, but through functions that the object receives from the abstract data type (such as the push and pop operations for a stack). (For more information about how such objects are used, see CLASS.)

Because the abstract data type is not directly used by the program, the implementation of how the data is stored or manipulated can be changed without affecting programs that use objects of that type. This *information hiding* is one of the chief benefits of object-oriented programming. Another advantage is inheritance, the ability to derive more specialized versions of the abstract data type or class. Thus, one can create a derived stack class that includes the printing function mentioned earlier.

Further Reading

Carrano, Frank M. *Data Abstraction and Problem Solving with C++: Walls and Mirrors.* 4th ed. Reading, Mass.: Addison-Wesley, 2004.

"Introduction to Data Abstraction." MIT Press. Available online. URL: http://mitpress.mit.edu/sicp/full-text/sicp/book/node27. html. Accessed July 3, 2007.

Koffman, Elliot B., and Paul A. T. Wolfgang. *Objects, Abstraction, Data Structures and Design Using Java Version 5.0.* New York: Wiley, 2004.

data acquisition

There are a variety of ways in which data (facts or measurements about the world) can be turned into a digital representation suitable for manipulation by a computer. For example, pressing a key on the keyboard sends a signal that is stored in a memory buffer using a value that represents the ASCII character code for the key pressed. Moving the mouse sends a stream of signals that are proportional to the rotation of the ball which in turn is calibrated into a series of coordinates and ultimately to a position on the screen where the cursor is to be moved. Digital cameras and scanners convert the varying light levels of what they "see" into a digital image.

Besides the devices that are familiar to most computer users, there are many specialized data acquisition devices (DAQs). Indeed, most instruments used in science and engineering to measure physical characteristics are now designed to convert their readings into digital form. (Sometimes the instrument includes a processor that provides a representation of the data, such as a waveform or graph. In other cases, the data is sent to a computer for processing and display.)

COMPONENTS OF A DATA ACQUISITION SYSTEM

The data acquisition system begins with a transducer, which is a device that converts a physical phenomenon (such as heat) into a proportional electrical signal. Transducers include devices such as thermistors, thermocouples, and pressure or strain gauges. The output of the transducer is then fed into a signal conditioning circuit. The purpose of signal conditioning is to make sure the signal fits into the range needed by the data processing device. Thus the

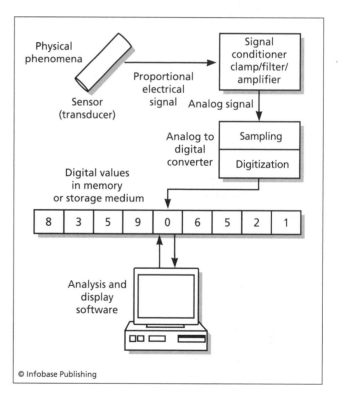

Data acquisition is the process of gathering real-time data from scientific instruments and making it available in digital form. Sensor signals are "conditioned" by filtering extraneous values, and are then sampled and digitized. Software can now provide elaborate graphic displays as well as alert scientists to unusual readings.

signal may be amplified or its voltage may be adjusted or scaled to the required level. Another function of signal conditioning is to isolate the incoming signal from the computer to which the acquisition device is connected. This is necessary both to protect the delicate computer circuits from possible "spikes" in the incoming signal and to prevent "noise" (extraneous electromagnetic signals created by the computer itself) from distorting the signal, and thus the ultimate measurements. Various sorts of filters can be added for this purpose.

The conditioned signal is fed as an analog input into the data acquisition device, which is often a board inserted into a personal computer. The purpose of the board is to sample the signal and turn it into a stream of digital data. The digital data is stored in a buffer (either on the board or in the computer's main memory). Software then takes over, analyzing the data and creating appropriate displays (such as digital readings, graphs, or warning signals) as configured by the user. If the data is being displayed in real time, the speed of the software, the operating system, and the computer's clock speed may become significant (see CLOCK SPEED).

PERFORMANCE CONSIDERATIONS

The sampling rate, or the number of times the signal is measured per second, is of fundamental importance. A higher sampling rate usually means a more accurate representation of the physical data (thus audio sampled at higher rates sounds more "natural"). The faster the sampling rate, the larger the amount of data to be processed and the greater the amount of computer resources needed. Thus, picking a sampling rate usually involves a tradeoff between accuracy and speed (for a real-time application, data must be processed fast enough so that whoever is using it can respond to it as it comes in).

Three internal factors determine the performance of a DAQ. The *resolution* is the number of bits available to quantify each measurement. Clearly the ability to measure thousands of voltage levels is useless if the resolution of a system is only 8 bits (256 possible values.) The *range* is the distance between the minimum and maximum voltage levels the DAQ can recognize. If a signal must be "squeezed" into too narrow a range, a corresponding amount of resolution will be lost. Finally, there is the *gain* or the ratio between changes in the measured quantity and changes in the signal strength.

APPLICATIONS

Data acquisition systems are essential to gathering and processing the detailed data required by scientific and engineering applications. The automated control of chemical or biochemical processes requires the ability of the control software to assess real-time physical data in order to make timely adjustments to such factors as temperature, pressure, and the presence of catalysts, inhibitors, or other components of the process. The highly automated systems used in modern aviation and increasingly, even in ground vehicles, depend on real-time data acquisition. It is not surprising, then, that data acquisition is one of the fastest-growing fields in computing.

Further Reading
Beyon, Jeffrey Y. *LabVIEW Programming, Data Acquisition and Analysis.* Upper Saddle River, N.J.: Prentice Hall, 2000.
"Data Acquisition (DAQ) Fundamentals." Available online. URL: http://zone.ni.com/devzone/cda/tut/p/id/3216. Accessed June 8, 2007.
James, Kevin. *PC Interfacing and Data Acquisition: Techniques for Measurement, Instrumentation and Control.* Boston: Newnes, 2000.

database administration

Database administration is the management of database systems (see DATABASE MANAGEMENT SYSTEM). Database administration can be divided into four broad areas: data security, data integrity, data accessibility, and system development.

DATA SECURITY

With regard to databases, ensuring data security includes the assignment and control of users' level of access to sensitive data and the use of monitoring tools to detect compromise, diversion, or unauthorized changes to database files (see DATA SECURITY). When data is proprietary, licensing agreements with both database vendors and content providers may also need to be enforced.

DATA INTEGRITY

Data integrity is related to data security, since the completeness and accuracy of data that has been compromised can no longer be guaranteed. However, data integrity also requires the development and testing of procedures for the entry and verification of data (input) as well as verifying the accuracy of reports (output). Database administrators may do some programming, but generally work with the programming staff in maintaining data integrity. Since most data in computers ultimately comes from human beings, the training of operators is also important.

Within the database structure itself, the links between data fields must be maintained (referential integrity) and a locking system must be employed to ensure that a new update is not processed while a pending one is incomplete (see TRANSACTION PROCESSING).

Internal procedures and external regulations may require that a database be periodically audited for accuracy. While this may be the province of a specially trained information processing auditor, it is often added to the duties of the database administrator. (See also AUDITING IN DATA PROCESSING.)

DATA ACCESSIBILITY

Accessibility has two aspects. First, the system must be reliable. Data must be available whenever needed by the organization, and in many applications such as e-commerce, this means 24 hours a day, 7 days a week (24/7). Reliability requires making the system as robust as possible, such as by "mirroring" the database on multiple servers (which in turn requires making sure updates are stored concurrently). Failure must also be planned for, which means the implementation of onsite and offsite backups and procedures for restoring data (see BACKUP AND ARCHIVE SYSTEMS).

SYSTEM DEVELOPMENT

An enterprise database is not a static entity. The demand for new views or applications of data requires the development and testing of new queries and reports. While this is normally done by the database programmers, the administrator may need to consider its impact on the operation of the system. The administrator also helps plan for the needs of a growing, changing, organization by designing or evaluating proposals for expanding the system, possibly moving it to new hardware or a new operating system or migrating the database applications to a new database management system (DBMS).

Because of the importance of database management to corporations, government, and other organizations, database administration became a "hot" employment area in the 1990s. Most database administrators specialize in a particular database platform, such as Oracle or Microsoft Access. The growing need to make databases accessible via the Internet has added a new range of challenges to the database administrator, including the management of servers, remote authentication of users, and the mastery of Java, Common Gateway Interface (CGI), and scripting languages in order to tie the database to the server and user (see JAVA, CGI, PERL, and XML).

Further Reading

About.com. Database Administration [links]. Available online. URL: http://databases.about.com/od/administration/Database_Administration.htm. Accessed July 8, 2007.

Alapati, Sam R. *Expert Oracle Database 10g Administration.* Berkeley, Calif.: Apress, 2005.

Mannino, Michael A. *Database Design, Application Development, and Administration.* 3rd ed. New York: McGraw-Hill, 2005.

Mullins, Craig S. *Database Administration: The Complete Guide to Practices and Procedures.* Reading, Mass: Addison-Wesley Professional, 2002.

MySQL AB. *MySQL Administrator's Guide and Language Reference.* Indianapolis: MySQL Press, 2005.

Wood, Dan, Chris Leiter, and Paul Turley. *Beginning SQL Server 2005 Administration.* Indianapolis: Wrox, 2006.

database management system (DBMS)

A database management system consists of a database (a collection of information, usually organized into records with component fields) and facilities for adding, updating, retrieving, manipulating, and reporting on data.

DATABASE STRUCTURE

In the early days of computing, a database generally consisted of a single file that was divided into data blocks that in turn consisted of records and fields within records. The COBOL language was (and is) particularly suited to reading, processing, and writing data in such files. This *flat file* database model is still used for many simple applications including "home data managers." However, for more complex applications where there are many files containing interrelated data, the flat file model proves inadequate.

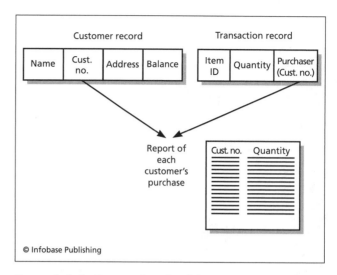

Because both the Customer Record and the Transaction Record include the Customer Number field, it is easy to pull information from both databases into a single report, such as a summary of purchases for each customer.

In 1970, computer scientist E. F. Codd proposed a *relational* model for data organization. In the relational model, data is not viewed as files containing records, but as a set of tables, where the columns represent fields and the rows individual entities (such as customers or transactions).

A field (column) that two tables have in common (called the *key*) can be used to link the two. For example, consider a table of customer information (name, customer number, address, current balance, and so on) and a table of transaction information (product number, quantity, customer number of purchaser, and so on).

To find all the items purchased by a particular customer, the relational database uses the common field (the customer account number) to *join* the two tables. A query can then select all records in the transaction file whose customer number field matches the current customer in the customer file. (Notice that the validity of a key field depends on its being unique: If each customer doesn't have one [and only one] customer number, any report of purchases will not be dependable.)

A procedure called *normalization* is often used to create a set of tables from a set of data files and records, such that no fields contain duplicate information. This is necessary in order to ensure that a piece of information can be updated and the update "propagated" to the entire database without missing any instances.

Relational databases usually also enforce *referential integrity*. This means preventing changes to the database from causing inconsistencies. For example, if table A and table B are linked and a record is deleted from table A, any links to that record from records in table B must be removed. Similarly, if a change is made in a linked field in a table, records in a linked table must be updated to reflect the change.

During the 1980s, the dBase relational database program became the most popular DBMS on personal computers. Microsoft Access is now popular on Windows systems,

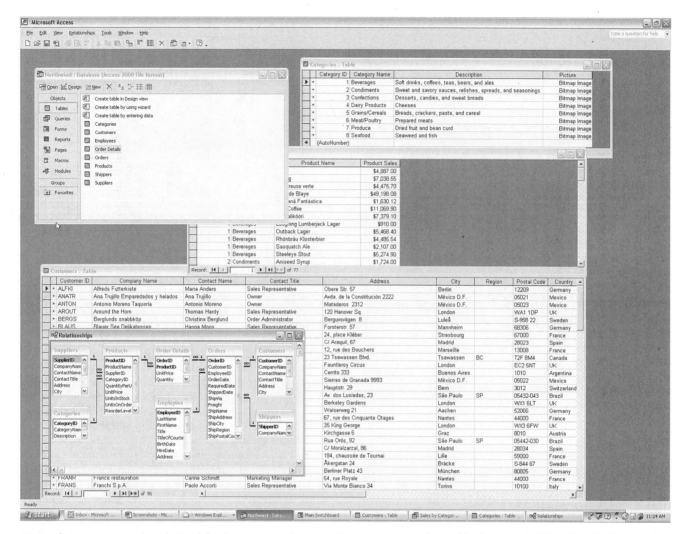

Microsoft Access is a popular relational database program for personal computers. It can be used for both simple ("flat file") databases and for complex databases with many interrelated files.

and Oracle is prominent in the UNIX world. Beginning in the 1980s, SQL (Structured Query Language) became a widely used standard for querying and manipulating data tables, and most DBMS implement SQL (see SQL).

TRENDS

The embracing of object-oriented programming principles starting in the 1980s has led to development of object-oriented database structures (see OBJECT-ORIENTED PROGRAMMING). In this approach tables, queries, views, and other components of the DBMS are treated as objects that present their functionality through interfaces (much in the way a class in an object-oriented program does). This approach can improve data integrity, flexibility (such as through the ability to define new operations), and the development of new capabilities derived from predecessor objects. Object models are also helpful in dealing with a networked world in which data tables are often stored on separate computers.

As important as changes in the architecture of databases have been, the impact of a changing environment has prob-

ably been even more significant. In particular, Web sites of all kinds are increasingly being driven by databases (such as for inventory and order processing for e-commerce). In turn, many databases of all sizes and types are now accessible and searchable via the Web. This has meant a new emphasis on rapid development of database programs, particularly using scripting languages, as well as fast and efficient Web-based database processing (see also AJAX). While the traditional high-end corporate database systems such as Oracle and SQL Server are still vital for the enterprise, open-source alternatives (particularly MySQL) are in widespread use for many applications including wikis and content-management systems. The use of flexibly structured data (see XML and SEMANTIC WEB) to link and transform databases has also expanded database concepts in the Web-centric world.

Further Reading

Allen, Christopher, Catherine Creary, and Simon Chatwin. *Introduction to Relational Databases*. Berkeley, Calif.: McGraw-Hill Osborne, 2003.

Hellerstein, Joseph S., and Michael Stonebreaker, eds. *Readings in Database Systems*. 4th ed. Cambridge, Mass.: MIT Press, 2005.

Hoffer, Jeffrey A., Mary Prescott, and Fred McFadden. *Modern Database Management*. 8th ed. Upper Saddle River, N.J.: Prentice Hall, 2006.

Powell, Gavin. *Beginning XML Databases*. Indianapolis: Wrox, 2006.

"Web Programming: Databases." Available online. URL: http://www.webreference.com/programming/databases.html. Accessed July 8, 2007.

Williams, Hugh E., and David Lane. *Web Database Applications with PHP and MySQL*. 2nd ed. Sebastapol, Calif.: O'Reilly Media, 2004.

data communications

Broadly speaking, data communications is the transfer of data between computers and their users. At its most abstract level, data communications requires two or more computers, a device to turn data into electronic signals (and back again), and a transmission medium. Telephone lines, fiber optic cable, network (Ethernet) cable, video cable, radio (wireless), or other kinds of links can be used. Finally, there must be software that can manage the flow of data.

Until recently, the modem was the main device used to connect personal computers to information services or

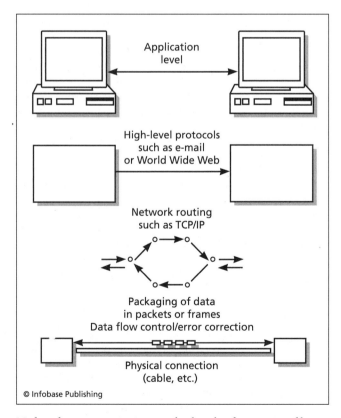

Modern data communications can be thought of as a series of layers, from the actual physical connection (such as a cable) at the "bottom" to the operations of software such as Web browsers or e-mail programs at the highest level.

networks (see MODEM). In general, data being sent over a communications link must be sent one bit at a time (this is called *serial* transmission, and is why an external modem is connected to a computer's serial port). However most phone cables and other links are *multiplexed,* meaning that they carry many channels (with many streams of data bits) at the same time.

To properly recognize data in a bit stream coming over a link, the transmission system must use some method of flow control and have some way to detect errors (see ERROR CORRECTION). Typically, the data is sent as groups or "frames" of bits. The frame includes a checksum that is verified by the receiver. If the expected and actual sums don't match, the recipient sends a "negative acknowledgment" message to the sender, which will retransmit the data. In the original system, the sender waited until the recipient acknowledged each frame before sending the next, but modern protocols allow the sender to keep sending while the frames being received are waiting to be checked.

The actual transmission of data over a line can be considered to be the lowest level of the data communications scheme. Above that is packaging of data as used and interpreted by software. Unless two computers are directly connected, the data is sent over a network, either a local area network (LAN) or a wide-area network such as the global Internet. A network consists of interconnected *nodes* that include switches or routers that direct data to its destination (see NETWORK). Networks such as the Internet use packet-switching: Data is sent as individual packets that contain a "chunk" of data, an address, and an indication of where the data fits within the message as a whole. The packets are routed at the routers using software that tries to find the fastest link to the destination. When the packets arrive at the destination, they are reassembled into the original message.

APPLICATIONS

Data communications are the basis both for networks and for the proper functioning of servers that provide services such as World Wide Web pages, electronic mail, online databases, and multimedia content (such as audio and streaming video). While Web page design and e-commerce are the "bright lights" that give cyberspace its character, data communications are like the plumbing without which computers cannot work together. The growing demand for data communications, particularly broadband services such as DSL and cable modems, translates into a steady demand for engineers and technicians specializing in the maintenance and growth of this infrastructure (see BROADBAND).

Besides keeping up with the exploding demand for more and faster data communications, the biggest challenge for data communications in the early 21st century is the integration of so many disparate methods of communications. A user may be using an ordinary phone line (19th-century technology) to connect to the Internet, while the phone company switches might be a mixture of 1970s or later technology. The same user might go to the workplace and use fast Ethernet cables over a local network, or connect to the Internet through DSL, an enhanced phone line. Traveling home, the user might use a personal digital assistant (PDA) with a wireless link to make a restaurant reservation (see WIRELESS COMPUTING). The user wants all these services to be seamless and essentially interchangeable, but today data communications is more like roads in the early days of the automobile—a few fast paved roads here and there, but many bumpy dirt paths.

Further Reading

Forouszan Behrouz. *Data Communications and Networking.* New York: McGraw Hill, 2006.

Stallings, William. *Data and Computer Communications.* 8th ed. Upper Saddle River, N.J.: Prentice Hall, 2006.

Strangio, Christopher E. "Data Communications Basics." Available online. URL: http://www.camiresearch.com/Data_Com_Basics/data_com_tutorial.html. Accessed July 8, 2007.

White, Curt. *Data Communications and Computer Networks: A Business User's Approach.* 4th ed. Boston: Course Technology, 2006.

data compression

The process of removing redundant information from data so that it takes up less space is called data compression. Besides saving disk space, compressing data such as e-mail attachments can make data communications faster.

Compression methods generally begin with the realization that not all characters are found in equal numbers in text. For example, in English, letters such as *e* and *s* are found much more frequently than letters such as *j* or *x*. By assigning the shortest bit codes to the most common characters and the longer codes to the least common characters, the number of bits needed to encode the text can be minimized.

Huffman coding, first developed in 1952, is an algorithm that uses a tree in which the pairs of the least probable (that is, least common) characters are linked, the next least probable linked, and so on until the tree is complete.

Another coding method, arithmetic coding, matches characters' probabilities to bits in such a way that the same bit can represent parts of more than one encoded character. This is even more efficient than Huffman coding, but the necessary calculations make the method somewhat slower to use.

Another approach to compression is to look for words (or more generally, character strings) that match those found in a dictionary file. The matching strings are replaced by numbers. Since a number is much shorter than a whole word or phrase, this compression method can greatly reduce the size of most text files. (It would not be suitable for files that contain numerical rather than text data, since such data, when interpreted as characters, would look like a random jumble.)

The Lempel-Ziv (LZ) compression method does not use an external dictionary. Instead, it scans the file itself for text strings. Whenever it finds a string that occurred earlier in the text, it replaces the later occurrences with an offset, or count of the number of bytes separating the

Original data:

they rode the waves
up and down and up again

1	the
2	up
3	and

Compressed data with index to dictionary

1y rode 1 waves
2 3 down 3 2 again

© Infobase Publishing

A basic approach to data compression is to look for recurring patterns and store them in a "dictionary." Each occurrence of the pattern can then be replaced by a brief reference to the dictionary entry. The resulting file may then be considerably smaller than the original.

occurrences. This means that not only common words but common prefixes and suffixes can be replaced by numbers. A variant of this scheme does not use offsets to the file itself, but compiles repeated strings into a dictionary and replaces them in the text with an index to their position in the dictionary.

Graphics files can often be greatly compressed by replacing large areas that represent the same color (such as a blue sky) with a number indicating the count of pixels with that value. However, some graphics file formats such as GIF are already compressed, so further compression will not shrink them much.

More exotic compression schemes for graphics can use fractals or other iterative mathematical functions to encode patterns in the data. Most such schemes are "lossy" in that some of the information (and thus image texture) is lost, but the loss may be acceptable for a given application. Lossy compression schemes are not used for binary (numeric data or program code) files because errors introduced in a program file are likely to affect the program's performance (if not "break" it completely). Though they may have less serious consequences, errors in text are also generally considered unacceptable.

TRENDS

There are a variety of compression programs used on UNIX systems, but variants of the Zip program are now the overwhelming favorite on Windows-based systems. Zip combines compression and archiving. Archiving, or the bundling together of many files into a single file, contributes a further reduction in file size. This is because files in most file systems must use a whole number of disk sectors, even if that means wasting most of a sector. Combining files into one file means that at most a bit less than one sector will be wasted.

Further Reading

Arimura, Mitsuharu. "Mitsuharu Arimura's Bookmarks on Source Coding/Data Compression." Available online. URL: http://www.hn.is.uec.ac.jp/~arimura/compression_links.html. Accessed July 8, 2007.

"Data Compression Reference Center." Available online. URL: http://www.rasip.fer.hr/_research/compress/index.html. Accessed July 8, 2007.

Saloman, David. *Data Compression: The Complete Reference.* 4th ed. New York: Springer-Verlag, 2006.

Sayood, Khalid. *Introduction to Data Compression.* 3rd ed. San Francisco: Morgan Kaufmann, 2005.

data conversion

The developer of each application program that writes data files must define a format for the data. The format must be able to preserve all the features that are supported by the program. For example, a word processing program will include special codes for font selection, typestyles (such as bold or italic), margin settings, and so on.

In most markets there are more than one vendor, so there is the potential for users to encounter the need to convert files such as word processing documents from one vendor's format to another. For example, a Microsoft Word user needing to send a document to a user who has WordPerfect, or the user may encounter another user who also has Microsoft Word, but a later version.

There are some ways in which vendors can relieve some of their users' file conversion issues (and thus potential customer dissatisfaction). Vendors often include facilities to read files created by their major rivals' products, and to save files back into those formats. This enables users to exchange files. Sometimes the converted document will look exactly like the original, but in some cases there is no equivalence between a feature (and thus a code) in one application and a feature in the other application. In that case the formatting or other feature may not carry over into the converted version, or may be only partially successful.

Vendors generally make a new version of an application *downwardly compatible* with previous versions (see also COMPATABILITY AND PORTABILITY). This means that the new version can read files created with the earlier versions. (After all, users would not be happy if none of their existing documents were accessible to their new software!) Similarly, there is usually a way to save a file from the later version in the format of an earlier version, though features added in the later version will not be available in the earlier format.

Another strategy for exchanging otherwise incompatible files is to find some third format that both applications can read. Thus Rich Text Format (RTF), a format that includes most generic document features, is supported by most modern word processors. A user can thus export a file as RTF and the user of a different program will be able to read it (see RTF). Similarly, many database and other programs can export files as a series of data values separated by commas (comma-delimited files), and the files can be then read by a different program and converted to its "native" format.

A variety of format conversion utilities are available as either commercial software or shareware. There are also businesses that specialize in data conversion. While their services can be expensive, using them may be the best way to convert large numbers of files, rather than having to individually load and save them. Data conversion services can also handle many "ancient" data files from the 1970s or even early 1980s whose formats are no longer supported by current software.

Further Reading

Heuser, Werner. "Data Conversion and Migration Tools." Available online. URL: http://dataconv.org/. Accessed July 8, 2007.

"Media Conversion: Online File Conversions." Available online. URL: http://www.iconv.com/. Accessed July 8, 2007.

data dictionary

A modern enterprise database system can contain hundreds of separate data items, each with important characteristics such as field types and lengths, rules for validating the data, and links to various databases that use that item (see DATABASE MANAGEMENT SYSTEM). There can also be many different *views* or ways of organizing subsets of the data, and stored procedures (program code modules) used to perform various data processing functions. A developer who is creating or modifying applications that deal with such a vast database will often need to check on the relationships between data elements, views, procedures, and other aspects of the system.

One fortunate characteristic of computer science is that many tools can be applied to themselves, often because the contents of a program is itself a collection of data. Thus, it is possible to create a database that keeps track of the elements of another database. Such a database is sometimes called a data dictionary. A data dictionary system can be developed in the same way as any other database, but many database development systems now contain built-in facilities for generating data dictionary entries as new data items are defined, and updating definitions as items are linked together and new views or stored procedures are defined. (A similar approach can be seen in some software development systems that create a database of objects defined within programs, in order to preserve information that can be useful during debugging.)

Data dictionaries are particularly important for creating data warehouses (see DATA WAREHOUSE), which are large collections of data items that are stored together with the procedures for manipulating and analyzing them.

Further Reading

Kreines, David. *Oracle Data Dictionary Pocket Reference.* Sebastapol, Calif.: O'Reilly Media, 2003.

Pelzer, Trudy. "MySQL 5.0 New Features: Data Dictionary." Available online. URL: http://dev.mysql.com/tech-resources/articles/mysql-datadictionary.html. Accessed July 8, 2007.

data glove   *See* HAPTICS.

data mining

The process of analyzing existing databases in order to find useful information is called data mining. Generally, a database, whether scientific or commercial, is designed for a particular purpose, such as recording scientific observations or keeping track of customers' account histories. However, data often has potential applications beyond those conceived by its collector.

Conceptually, data mining involves a process of refining data to extract meaningful patterns—usually with some new purpose in mind. First, a promising set or subset of the data is selected or sampled. Particular fields (variables) of interest are identified. Patterns are found using techniques such as regression analysis to find variables that are highly correlated to (or predicted by) other variables, or through clustering (finding the data records that are the most similar along the selected dimensions). Once the "refined" data is extracted, a representation or visualization (such as a report or graph) is used to express newly discovered information in a usable form.

Similar (if simpler) techniques are being used to target or personalize marketing, particularly to online customers. For example, online bookstores such as Amazon.com can find what other books have been most commonly bought by people buying a particular title. (In other words, identify a sort of reader profile.) If a new customer searches for that title, the list of correlated titles can be displayed, with an increased likelihood of triggering additional purchases. Businesses can also create customer profiles based on their longer-term purchasing patterns, and then either use them for targeted mailings or sell them to other businesses (see E-COMMERCE). In scientific applications, observations can be "mined" for clues to phenomena not directly related to the original observation. For example, changes in remote sensor data might be used to track the effects of climate or weather changes. Data-mining techniques can even be applied to the human genome (see BIOINFORMATICS).

TRENDS

Data mining of consumer-related information has emerged as an important application as the volume of e-commerce continues to grow, the amount of data generated by large systems (such as online bookstores and auction sites) increases, and the value of such information to marketers becomes established. However, the use of consumer data for purposes unrelated to the original purchase, often by companies that have no pre-existing business relationship to the consumer, can raise privacy issues. (Data is often rendered anonymous by removing personal identification information before it is mined, but regulations or other ways to assure privacy remain incomplete and uncertain.)

The most controversial applications of data mining are in the area of intelligence and homeland security. Because such applications are often shrouded in secrecy, the public and even lawmakers have difficulty in assessing their value and devising privacy safeguards. According to the Government Accountability Office, as of 2007 some 199 different data-mining programs were in use by at least 52 federal agencies. One of the most controversial is ADVISE (Analysis Dissemination, Visualization, Insight and Semantic Enhancement), developed by the Department of Homeland Security since 2003. The program purportedly can match and create profiles using government records and users'

Web sites and blogs. Privacy advocates and civil libertarians have raised concerns, and legislation has been introduced that would require that all federal agencies report their data-mining activities to Congress (see also COUNTERTERRORISM AND COMPUTERS and PRIVACY IN THE DIGITAL AGE.)

Further Reading
Clayton, Mark. "U.S. Plans Massive Data Sweep." *Christian Science Monitor,* February 9, 2006, n.p. Available online. URL: http://www.csmonitor.com/2006/0209/p01s02-uspo.html. Accessed July 8, 2007.
Dunham, Margaret H. *Data Mining: Introductory and Advanced Topics.* Upper Saddle River, N.J.: Prentice Hall, 2002.
Markov, Zdravko, and Daniel T. Larose. *Data Mining the Web: Uncovering Patterns in Web Content, Structure, and Usage.* Hoboken, N.J.: Wiley, 2007.
Tan, Pang-Ning, Michael Steinbach, and Vipin Kumar. *Introduction to Data Mining.* Upper Saddle River, N.J.: Pearson Education, 2006.

data security

In most institutional computing environments, access to program and data files is restricted to authorized persons. There are several mechanisms for restricting file access in a multiuser or networked system.

USER STATUS

Because of their differing responsibilities, users are often given differing restrictions on access. For example, there might be status levels ranging from root to administrator to "ordinary." A user with root status on a UNIX system is able to access any file or resource. Any program run by such a user inherits that status, and thus can access any resource. Generally, only the user(s) with ultimate responsibility for the technical functioning of the system should be given such access, because commands used by root users have the potential to wipe out all data on the system. A person with administrator status may be able to access the files of other users and to access certain system files (in order to change configurations), but will not be able to access certain core system files. Ordinary users typically have access only to the files they create themselves and to files designated as "public" by other users.

FILE PERMISSIONS

Files themselves can have permission status. In UNIX, there are separate statuses for the user, any group to which the user belongs, and "others." There are also three different activities that can be allowed or disallowed: reading, writing, and executing. For example, if a file's permissions are

User	Group	Other
rwx	rw-	r—

the user can read or write the file or (if it is a directory or program), execute it. Members of the same group can read or write, but not execute, while others can only read the file without being able to change it in any way. Operating systems such as Windows NT use a somewhat different structure and terminology, but also provide for varying user status and access to objects.

RECORD-LEVEL SECURITY

Security on the basis of whole directories or even files may be too "coarse" for many applications. In a particular database file, different users may be given access to different data fields. For example, a clerk may have read-only access to an employee's basic identification information, but not to the results of performance evaluations. An administrator may have both read and write access to the latter. Using some combination of database management and operating system level capabilities, the system will maintain lists of user accounts together with the objects (such as record types or fields) they can access, and the types of access (read only or read/write) that are permitted. Rather than assigning access capabilities separately for each user, they may be defined for a group of similar users, and then individual users can be assigned to the group.

OTHER SECURITY MEASURES

Security is also important at the program level. Because a badly written (or malicious) program might destroy important data or system files, most modern operating systems restrict programs in a number of ways. Generally, each program is allowed to access only such memory as it allocates itself, and is not able to change data in memory belonging to other running programs. Access to hardware devices can also be restricted: an operating system component may have the ability to access the innermost core of the operating system (where drivers interact directly with devices), while an ordinary applications program may be able to access devices only through facilities provided by the operating system.

There are a number of techniques that unauthorized intruders can use to try to compromise operating systems (see COMPUTER CRIME AND SECURITY). Access capabilities that are tied to user status are vulnerable if the user can get the login ID and password for the account. If the account has a high (administrator or root) status, then the intruder may be able to give viruses, Trojan horses, or other malicious programs the status they need in order to be able to penetrate the defenses of the operating system (see also COMPUTER VIRUS).

Files that have intrinsically sensitive or valuable data are often further protected by encoding them (see ENCRYPTION). Encryption means that even intruders who gain read access to the file will need either to crack the encryption (very difficult without considerable time and computer resources) or somehow obtain the key. Encryption does not prevent the deletion or copying of a file, however, just the understanding of its contents.

The dispersal of valuable or sensitive data (such as customers' social security numbers) across expanding networks increases the risk of "data breaches" where the privacy, financial security, and even identity of thousands of people are compromised (see also IDENTITY THEFT). In recent years, for example, there have been numerous cases where laptop computers containing thousands of sensitive records have been stolen from universities, financial institutions, or government agencies—in such cases there is often no way to know whether the thief will actually access the data. (Often affected individuals are notified that they may be at risk, and such prophylactic measures as credit monitoring are provided.) In response to public anxiety there has been pressure for federal or state legislation that would make companies responsible for breaches of their data and specify compensation or other recourse for affected customers. (Opponents of such laws cite government reports that find that most data breaches do not lead to identity theft, and that the regulations would increase the cost of millions of daily transactions.)

There is a continuing tradeoff between security and ease of use. From the security standpoint, it might be assumed that the more barriers or checkpoints that can be set up for verifying authorization, the safer the system will be. However, as security systems become more complex, it becomes more difficult to ensure that authorized users are not unduly inconvenienced. If users are sufficiently frustrated, they will be tempted to try to bypass security, such as by sharing IDs and passwords or making files they create "public."

Further Reading

Garretson, Cara. "The Do's and Don'ts of Data Breaches: How Security Professionals Can Lessen the Impact." *Network World,* June 18, 2007, p. 1.

Grant, Gross. "Gov't Report: Data Breaches Don't Often Result in ID Theft." *PC World,* July 6, 2007, n.p. Available online. URL: http://www.pcworld.com/article/id,134203-c,privacysecurity/article.html. Accessed July 8, 2007.

Killmeyer, Jan. *Information Security Architecture: An Integrated Approach to Security in the Organization.* 2nd ed. Boca Raton, Fla.: Auerbach Publications, 2006.

Rasch, Max. "Strict Liability for Data Breaches?" Available online. URL: http://www.securityfocus.com/columnists/387. Accessed July 8, 2007.

Tipton, Harold F., and Micki Krause. *Information Security Management Handbook.* 6th ed. Boca Raton, Fla.: Auerbach Publications, 2007.

data structures

A data structure is a way of organizing data for use in a computer program. There are three basic components to a data structure: a set of suitable basic data types, a way to organize or relate these data items to one another, and a set of operations, or ways to manipulate the data.

For example, the ARRAY is a data structure that can consist of just about any of the basic data types, although all data must be of the same type. The way the data is organized is by storing it in sequentially addressable locations. The operations include storing a data item (element) in the array and retrieving a data item from the array.

TYPES OF DATA STRUCTURES

The data structures commonly used in computer science include arrays (as discussed above) and various types of lists. The primary difference between an array and a list is that an array has no internal links between its elements, while a list has one or more pointers that link the elements. There are several types of specialized list. A tree is a list that has a root (an element with no predecessor), and each other element has a unique predecessor. The guarantee of a unique path to each tree node can make the operations of inserting or deleting an item faster. A STACK is a list that is accessible only at the top (or front). Any new item is inserted ("pushed") on top of the last item, and removing ("popping") an item always removes the item that was last inserted. This order of access is called LIFO (last in, first out). A list can also be organized in a first in, first out (FIFO) order. This type of list is called a QUEUE, and is useful in a situation where tasks must "wait their turn" for attention.

IMPLEMENTATION ISSUES
The implementation of any data structure depends on the syntax of the programming language to be used, the data types and features available in the language, and the algorithms chosen for the data operations that manipulate the structure. In traditional procedural languages such as C, the data storage part of a data structure is often specified in one part of the program, and the functions that operate on that structure are defined separately. (There is no mechanism in the language to link them.) In object-oriented languages such as C++, however, both the data storage declarations and the function declarations are part of the same entity, a CLASS. This means that the designer of the data structure has complete control over its implementation and use.

Together with algorithms, data structures make up the heart of computer science. While there can be numerous variations on the fundamental data structures, understanding the basic forms and being able to decide which one to use to implement a given algorithm is the best way to assure effective program design.

Further Reading
Drozdek, Adam. *Data Structures and Algorithms in C++*. 3rd ed. Boston: Course Technology, 2004.
Ford, William H., and William R. Topp. *Data Structures with Java*. Upper Saddle River, N.J.: Prentice Hall, 2004.
Lafore, Robert. *Data Structures & Algorithms in Java*. 2nd ed. Indianapolis: Sams, 2003.
Storer, J. A. *An Introduction to Data Structures and Algorithms*. New York: Springer, 2002.

data types
As far as the circuitry of a computer is concerned, there's only one kind of data—a series of bits (binary digits) filling a series of memory locations. How those bits are to be interpreted by the people using the computer is entirely arbitrary. The purpose of data types is to define useful concepts such as integer, floating-point number, or character in terms of how they are stored in computer memory.

Thus, most computer languages have a data type called integer, which represents a whole number that can be stored in 16 bits (two bytes) of memory. When a programmer writes a declaration such as:

```
int Counter;
```

in the C language, the compiler will create machine instructions that set aside two bytes of memory to hold the contents of the variable Counter. If a later statement says:

```
Counter = Counter + 1;
```

(or its equivalent, Counter++) the program's instructions are set up to fetch two bytes of memory to the processor's accumulator, add 1, and store the result back into the two memory bytes.

Similarly, the data type long represents four bytes (32 bits) worth of binary digits, while the data type float stores a floating-point number that can have a whole part and a decimal fraction part (see NUMERIC DATA). The char (character) type typically uses only a single byte (8 bits), which is enough to hold the basic ASCII character codes up to 255 (see CHARACTERS AND STRINGS).

The Bool (Boolean) data type represents a simple true or false (usually 1 or 0) value (see BOOLEAN OPERATORS).

STRUCTURED DATA TYPES
The preceding data types all hold single values. However, most modern languages allow for the construction of data types that can hold more than one piece of data. The ARRAY is the most basic structured data type; it represents a series of memory locations that hold data of one of the basic types. Thus, in Pascal an array of integer holds integers, each taking up two bytes of memory.

Many languages have composite data types that can hold data of several different basic types. For example, the struct in C or the record in Pascal can hold data such as a person's first and last name, three lines of address (all arrays of characters, or strings), an employee number (perhaps an integer or double), a Boolean field representing the presence or absence of some status, and so on. This kind of data type is also called a user-defined data type because programmers can define and use these types in almost the same ways as they use the language's built-in basic types.

What is the difference between data types and data structures? There is no hard-and-fast distinction. Generally, data structures such as lists, stacks, queues, and trees are more complex than simple data types, because they include data relationships and special functions (such as pushing or popping data on a stack). However, a list is the fundamental data type in list-processing languages such as Lisp, and string operators are built into languages such as Snobol. (See LIST PROCESSING, STACK, QUEUE, and TREE.)

Further, in many modern languages fundamental and structured data types are combined seamlessly into classes that combine data structures with the relevant operations (see CLASS and OBJECT-ORIENTED PROGRAMMING).

Further Reading
Prata, Stephen. *C Primer Plus*. 5th ed. Indianapolis: Sams, 2003.
———. *C++ Primer Plus*. 5th ed. Indianapolis: Sams, 2005.
"Type System." Wikipedia. Available online. URL: http://en.wikipedia.org/wiki/Type_system. Accessed July 8, 2007.
Watt, David A., and Deryck F. Brown. *Java Collections: An Introduction to Abstract Data Types, Data Structures, and Algorithms*. New York: Wiley, 2001.

data warehouse

Modern business organizations create and store a tremendous amount of data in the form of transactions that become database records. Increasingly, however, businesses are relying on their ability to use data that was collected for one purpose (such as sales, customer service, and inventory) for purposes of marketing research, planning, or decision support. For example, transaction data might be revisited with a view to identifying the common characteristics of the firm's best customers or determining the best way to market a particular type of product. In order to conduct such research or analysis, the data collected in the course of business must be stored in such a way that it is both accurate and flexible in terms of the number of different ways in which it can be queried. The idea of the data warehouse is to provide such a repository for data.

When data is used for particular purposes such as sales or inventory control, it is usually structured in records where certain fields (such as stock number or quantity) are routinely processed. It is not so easy to ask a different question such as "which customers who bought this product from us also bought this other product within six months of their first purchase?" One way to make it easier to query data in new ways is to store the data not in records but in arrays where, for example, one dimension might be product numbers and another categories of customers. This approach, called Online Analytical Processing (OLAP), makes it possible to extract a large variety of relationships without being limited by the original record structure.

IMPLEMENTATION

The key in designing a data warehouse is to provide a way that researchers using analytical tools (such as statistics programs) can access the raw data in the underlying database. Software using query languages such as SQL can serve as such a link. Thus, the researcher can define a query

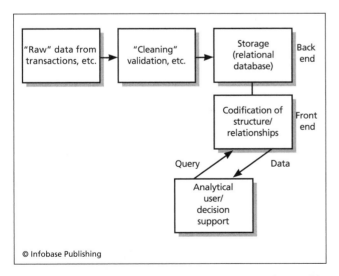

© Infobase Publishing

The general process of warehousing data. The data warehouse adds value to the data by further structuring it so relationships can be explored by analysts.

using the many dimensions of the data array, and the OLAP software (also called *middleware*) translates this query into the appropriate combination of queries against the underlying relational database.

The data warehouse is closely related to the concept of data mining. In fact, data mining can be viewed as the exploitation of the collection of views, queries, and other elements that can be generated using the data warehouse as the infrastructure (see DATA MINING).

Further Reading

Data Warehousing Information Center. Available online. URL: http://www.dwinfocenter.org/. Accessed July 8, 2007.
DM Review/dataWarehouse.com Available online. URL: http://www.datawarehouse.com/. Accessed July 8, 2007.
Inmon, W. H. *Building the Data Warehouse.* 4th ed. Indianapolis: Wiley, 2005.
Kimball, Ralph, and Margy Ross. *The Data Warehouse Toolkit: The Complete Guide to Dimensional Modeling.* 2nd ed. Indianapolis: Wiley, 2002.

DBMS   *See* DATABASE MANAGEMENT SYSTEM.

decision support system

A decision support system (DSS) is a computer application that focuses on providing access to or analysis of the key information needed to make decisions, particularly in business. (It can be thought of as a more narrowly focused approach to computer assistance to management—see MANAGEMENT INFORMATION SYSTEM.)

The development of DSS has several roots reaching back to the 1950s. This includes operational analysis and the theory of organizations and the development of the first interactive (rather than batch-processing) computer systems. Indeed, the SAGE automated air defense system developed starting in the 1950s could be described as a military DSS. The system presented real-time information (radar plots) and enabled the operator to select and focus on particular elements using a light pen. By the 1960s more-systematic research on DSS was underway and included the provocative idea of "human-computer symbiosis" for problem solving (see LICKLIDER, J. C. R.).

The "back end" of a DSS is one or more large databases (see DATA WAREHOUSE) that might be compiled from transaction records, statistics, online news services, or other sources. The "middle" of the DSS process includes the ability to analyze the data (online analytical processing, or OLAP; see also DATA MINING). Other elements that might be included in a DSS are rules-based systems (see EXPERT SYSTEM) and interactive models (see SIMULATION). These elements can help the user explore alternatives and "what if" scenarios.

The structure of a DSS is sometimes described as model driven (generally using a small amount of selected data), data driven (based on a large collection of historical data), knowledge driven (perhaps using an expert system), or communications driven (focusing on use of collaborative software—see GROUPWARE, as well as more recent developments) (see WIKIS AND WIKIPEDIA).

USER INTERFACE—THE "FRONT END"

All the data and tools in the world are of little use if the user cannot work with it effectively (see USER INTERFACE). Information or the results of queries or modeling must be displayed in a way that is easy to grasp and use. (A spreadsheet with nothing highlighted or marked would be a poor choice.) Graphical "widgets" such as dials, buttons, sliders, and so on can help the user see the results and decide what to look at next (see DIGITAL DASHBOARD).

Another key principle is that decision making in the modern world is as much a social as an individual process. Therefore a DSS should facilitate communication and collaboration (or interface with software that does so).

A variety of specialized DSSs have been developed for various fields. Examples include PROMIS (for medical decision making) and Carnegie Mellon's ZOG/KMS, which has been used in military and business settings.

Further Reading

Greenes, Robert A., ed. *Clinical Decision Support: The Road Ahead.* Orlando, Fla.: Academic Press, 2006.
Gupta, Jatinder N. D., Guisseppi A. Forgionne, and Manuel Mora T., eds. *Intelligent Decision-Making Support Systems.* New York: Springer, 2006.
Power, D. J. "A Brief History of Decision Support Systems." Version 4.0. Available online. URL: http://dssresources.com/history/dsshistory.html. Accessed September 10, 2007.
Turban, Efraim, et al. *Decision Support and Business Intelligence Systems.* 8th ed. Upper Saddle River, N.J.: Prentice-Hall, 2006.

Dell, Inc.

Dell Computer (NASDAQ: DELL) is one of the world's leading manufacturers and sellers of desktop and laptop computers (see PERSONAL COMPUTER). By 2008 Dell had more than 88,000 employees worldwide.

The company was founded by Michael Dell, a student at the University of Texas at Austin whose first company was PC's Limited, founded in 1984. Even at this early stage Dell successfully employed several practices that would come to typify the Dell strategy: Sell directly to customers (not through stores), build each machine to suit the customer's preferences, and be aggressive in competing on price.

In 1988 the growing company changed its name to Dell Computer Corporation. In the early 1990s Dell tried an alternative business model, selling through warehouse clubs and computer superstores. When that met with little success, Dell returned to the original formula. In 1999 Dell overtook Compaq to become the biggest computer retailer in America.

Generally, the Dell product line has aimed at two basic segments: business-oriented (OptiPlex desktops and Latitude laptops) and home/consumer (XPS desktops and Inspiron laptops, and in 2007, Inspiron desktops).

CHALLENGES AND DIVERSIFICATION

Around 2002, Dell, perhaps facing the growing commodity pricing of basic PCs, began to expand into computer peripherals (such as printers) and even home entertainment products (TVs and audio players). In 2003 the company changed its name to Dell, Inc. (dropping "Computer"). Dell also experienced an increase in international sales in 2005, while achieving a first place ranking in *Fortune* magazine as "most admired company." However, the company also made some missteps, losing $300 million because of faulty capacitors on some motherboards. Earnings continued to fall short of analysts' expectations, and in January 2007 Michael Dell returned as CEO after the resignation of Kevin B. Rollins, who had held the post since 2004.

Meanwhile, Dell has made further attempts at diversifying the product line. In 2006 the company began, for the first time, to introduce AMD (instead of Intel) processors in certain products, and in 2007 Dell responded to customer suggestions by announcing that some models could be ordered with Linux rather than Microsoft Windows installed. Also in 2007, Dell acquired Alienware, maker of high-performance gaming machines.

Dell has struggled to boost its sagging revenue as it lost ground to competitors, notably HP. Known primarily as a mail-order and online company, Dell has announced that it will also sell PCs through "big box" retailers such as Wal-Mart.

Dell continues to receive praise and criticism from various quarters. On the positive side, the company has been praised for its computer-recycling program by the National Recycling Coalition. Dell products also tend to score at or near the top in performance reviews by publications such as *PC Magazine.*

On the other hand, there have been complaints about Dell's technical support operation. Technicians apparently follow "scripts" very closely, making customers take systems apart and follow troubleshooting directions regardless of what the customer might already know or have done. The increasing "offshoring" of support has also led to complaints about language and communication problems.

Further Reading

Dell, Inc. Available online. URL: http://www.dell.com. Accessed September 10, 2007.
Dell, Michael, and Catherine Fredman. *Direct from Dell: Strategies that Revolutionized an Industry.* New York: HarperBusiness, 2006.
Holzner, Steven. *How Dell Does It: Using Speed and Innovation to Achieve Extraordinary Results.* New York: McGraw-Hill, 2006.

demon

The unusual computing term *demon* (sometimes spelled *daemon*) refers to a process (program) that runs in the background, checking for and responding to certain events. The utility of this concept is that it allows for automation of information processing without requiring that an operator initiate or manage the process.

For example, a print spooler demon looks for jobs that are queued for printing, and deals with the negotiations necessary to maintain the flow of data to that device. Another demon (called chron in UNIX systems) reads a file describing processes that are designated to run at particular dates or times. For example, it may launch a backup utility every morning at 1:00 A.M. E-mail also depends on the periodic operation of "mailer demons."

While the term *demon* originated in the UNIX culture, similar facilities exist in many operating systems. Even in the relatively primitive MS-DOS for IBM personal computers of the 1980s, the ability to load and retain small utility programs that could share the main memory with the currently running application allowed for a sort of demon that could spool output or await a special keypress. Microsoft Windows systems have many demon-like operating system components that can be glimpsed by pressing the Ctrl-Alt-Delete key combination.

The sense of autonomy implied in the term *demon* is in some ways similar to that found in *bots* or *software* agents that can automatically retrieve information on the Internet, or in the Web crawler, which relentlessly pursues, records, and indexes Web links for search engines. (See SOFTWARE AGENT and SEARCH ENGINE.)

Further Reading

Brock, Dean, and Bob Benites. *Mastering Tools, Taming Daemons: UNIX for the Wizard Apprentice.* Greenwich, Conn.: Manning Publications, 1995.

Stevens, W. Richard. *Advanced Programming in the UNIX Environment.* Upper Saddle River, N.J.: Addison-Wesley, 2005.

"UNIX Daemons in Perl." Available online. URL: http://www.webreference.com/perl/tutorial/9/. Accessed July 8, 2007.

Dertouzos, Michael L.
(1936–2001)
Greek-American
Computer Scientist, Futurist

Born in Athens, Greece, on November 5, 1936, Michael Dertouzos spent adventurous boyhood years accompanying his father (an admiral) in the Greek navy's destroyers and submarines. He became interested in Morse Code, shipboard machinery, and mathematics. At the age of 16 he read an article about Claude Shannon's work in information theory and a project at the Massachusetts Institute of Technology that sought to build a mechanical robot "mouse." He quickly decided that he wanted to come to America to study at MIT.

After the hardships of the World War II years intervened, Dertouzos received a Fulbright scholarship that placed him in the University of Arkansas, where he earned his bachelor's and master's degrees while working on acoustic-mechanical devices for the Baldwin Piano Company. He was then able to fulfill his boyhood dream by receiving his Ph.D. from MIT, then promptly joined the faculty. He was director of MIT's Laboratory for Computer Science (LCS) starting in 1974. The lab has been a hotbed of new ideas in computing, including computer time-sharing, Ethernet networking, and public-key cryptography. Dertouzos also embraced the growing Internet and serves as coordinator of the World Wide Web consortium, a group that seeks to create standards and plans for the growth of the network.

Combining theoretical interest with an entrepreneur's eye on market trends, Dertouzos started a small company called Computek in 1968. It made some of the first "smart terminals" that included their own processors.

In the 1980s, Dertouzos began to explore the relationship between developments and infrastructure in information processing and the emerging "information marketplace." However, the spectacular growth of the information industry has taken place against a backdrop of the decline of American manufacturing. Dertouzos's 1989 book, *Made In America,* suggested ways to revitalize American industry.

During the 1990s, Dertouzos brought MIT into closer relationship with the visionary designers who were creating and expanding the World Wide Web. When Tim Berners-Lee and other Web pioneers were struggling to create the World Wide Web consortium to guide the future of the new technology, Dertouzos provided extensive guidance to help them set their agenda and structure. (See WORLD WIDE WEB and BERNERS-LEE, TIM.)

Dertouzos was dissatisfied with operating systems such as Microsoft Windows and with popular applications programs. He believed that their designers made it unnecessarily difficult for users to perform tasks, and spent more time on adding fancy features than on improving the basic usability of their products. In 1999, Dertouzos and the MIT LCS announced a new project called Oxygen. Working in collaboration with the MIT Artificial Intelligence Laboratory, Oxygen was intended to make computers "as natural a part of our environment as the air we breathe."

As a futurist, Dertouzos tried to paint vivid pictures of possible future uses of computers in order to engage the general public in thinking about the potential of emerging technologies. His 1995 book, *What Will Be,* paints a vivid portrait of a near-future pervasively digital environment. His imaginative future is based on actual MIT research, such as the design of a "body net," a kind of wearable computer and sensor system that would allow people to not only keep in touch with information but also to communicate detailed information with other people similarly equipped. This digital world will also include "smart rooms" and a variety of robot assistants, particularly in the area of health care. However, this and his 2001 publication, *The Unfinished Revolution,* are not unalloyed celebrations of technological wizardry. Dertouzos has pointed out that there is a disconnect between technological visionaries who lack understanding of the daily realities of most peoples' lives, and humanists who do not understand the intricate interconnectedness (and thus social impact) of new technologies.

Dertouzos was given an IEEE Fellowship and awarded membership in the National Academy of Engineering, He died on August 27, 2001, after a long bout with heart disease. He was buried in Athens near the finish line for the Olympic marathon.

Further Reading

Dertouzos, Michael. L. *The Unfinished Revolution: How to Make Technology Work for Us—Instead of the Other Way Around.* New York: HarperCollins, 2002.

———. *What Will Be: How the New World of Information Will Change Our Lives.* New York: HarperCollins, 1997.

"Farewell to a Visionary of the Computer Age." *Business Week,* September 17, 2001, p. 101.

design patterns

Design patterns are an attempt to abstract and generalize what is learned in solving one problem so that it can be applied to future similar problems. The idea was first applied to architecture by Christopher Alexander in his book *A Pattern Language.* Alexander described a pattern as a description of situations in which a particular problem occurs, with a solution that takes into account the factors that are "invariant" (not changed by context). Guidance for applying the solution is also provided.

For example, a bus stop, a waiting room, and a line at a theme park are all places where people wait. A "place to wait" pattern would specify the problem to be solved (how to make waiting as pleasant as possible) and suggest solutions. Patterns can have different levels of abstraction or scales on which they apply (for example, an intimate theater and a stadium are both places of entertainment, but one is much larger than the other).

Patterns in turn are linked into a network called a pattern language. Thus when working with one pattern, the designer is guided to consider related patterns. For example, a pattern for a room might relate to patterns for seating or grouping the occupants.

PATTERNS IN SOFTWARE

The concept of patterns and pattern languages carries over well into software design. As with architectural patterns, a software pattern describes a problem and solution, along with relevant structures (see CLASS and OBJECT-ORIENTED PROGRAMMING). Note that patterns are not executable code; they are at a higher level (one might say abstract enough to be generalizable, specific enough to be applicable).

Software patterns can specify how objects are created and ways in which they function and interface with other objects. Patterns are generally documented using a common format; one example is provided in the book *Design Patterns.* This scheme has the following sections:

- name and classification

- intent or purpose

- alternative names

- problem—the kind of problem the pattern addresses, and conditions under which it can be used

- applicability—typical situations of use

- structure description—such as class or interaction diagrams

- participants—classes and objects involved in the pattern and the role each plays

- collaboration—how the objects interact with one another

- consequences—the expected results of using the pattern, and possible side effects or shortcomings

- implementation—explains a way to implement the pattern to solve the problem

- sample code—usually in a commonly used programming language

- known uses—actual working applications of the pattern

- related patterns—other patterns that are similar or related, with a description of how they differ

An example given in *Design Patterns* is the "publish-subscribe" pattern. This pattern describes how a number of objects (observers) can be dependent on a "subject." All observers are "subscribed" to the subject, so they are notified whenever any data in the subject changes. This pattern could be used, for example, to set up a system where different reports, spreadsheets, etc., need to be updated whenever notified by a controlling object that has received new data.

Some critics consider the use of patterns to be too abstract and inefficient. Since a pattern has to be re-implemented for each use, it has been argued that well-documented, reusable classes or objects would be more useful.

Proponents, however, argue that "design reuse" is more powerful than mere "object reuse." A pattern provides a whole "language" for talking about a problem and its proven solutions, and can help both the original designer and others understand and extend the design.

Further Reading

Alexander, Christopher. *A Pattern Language: Towns, Buildings, Construction.* New York: Oxford University Press, 1977.
"Design Patterns." IBM Research. Available online. URL: http://www.research.ibm.com/designpatterns/. Accessed September 10, 2007.
Freeman, Eric, and Elisabeth Freeman. *Head First Design Patterns.* Sebastapol, Calif.: O'Reilly, 2004.
Gamma, Erich, et al. *Design Patterns: Elements of Reusable Object-Oriented Software.* Upper Saddle River, N.J.: Addison-Wesley Professional, 1995.
Kurotsuchi, Brian T. "Welcome to the Wonderful World of Design Patterns." Available online. URL: http://www.csc.calpoly.edu/~dbutler/tutorials/winter96/patterns/. Accessed September 10, 2007.

desktop publishing (DTP)

Traditionally documents such as advertisements, brochures, and reports were prepared by combining typed or printed text with pasted-in illustrations (such as photographs and diagrams). This painstaking layout process was necessary in order to produce "camera-ready copy" from which a printing company could produce the final product.

Starting in the late 1980s, desktop computers became powerful enough to run software that could be used to create page layouts. In addition, display hardware gained a high enough resolution to allow for pages to be shown on the screen in much the same form as they would appear on the printed page. (This is known by the acronym WYSIWYG, or "what you see is what you get.") The final ingredient for the creation of desktop publishing was the advent of affordable laser or inkjet printers that could print near print quality text and high-resolution graphics (see PRINTERS).

This combination of technologies made it feasible for trained office personnel to create, design, and produce many documents in-house rather than having to send copy to a printing company. Adobe's PageMaker program soon became a standard for the desktop publishing industry, appearing first on the Apple Macintosh and later on systems running Microsoft Windows. (The Macintosh's support for fonts and WYSIWYG displays gave it a head start over the Windows PC in the DTP industry, and to this day many professionals prefer it.)

There is no hard-and-fast line between desktop publishing and the creation of text itself. Modern word processing software such as Microsoft Word includes a variety of features for selection and sizing of fonts, and the ability to define styles for creating headings, types of paragraphs, and so on (see WORD PROCESSING). Word and other programs also allow for the insertion and placement of graphics and tables, the division of text into columns, and other layout features. In general, however, word processing emphasizes the creation of text (often for long documents), while desktop publishing software emphasizes layout considerations and the fine-tuning of a document's appearance. Thus, while a word processor might allow the selection of a font in a given point size, a desktop publishing program allows for the exact specification of leading (space between lines) and kerning (the adjustment of space between characters). Most desktop publishing programs can import text that was originally created in a word processor. This is helpful because using desktop publishing software to create the original text can be tedious.

Desktop publishing is generally used for short documents such as ads, brochures, and reports. Material to be published as a book or magazine article is normally submitted by the author as a word processing document. The publisher's production staff then creates a print-ready version. Books and other long documents are generally produced using in-house computer typesetting facilities.

Today desktop publishing is part of a range of technologies used for the production of documents and presentations. Document designers also use drawing programs (such as Corel Draw) and photo manipulation programs (such as Adobe Photoshop) in preparing illustrations. Further, the growing use of the Web means that many documents must be displayable on Web pages as well as in print. Adobe's Portable Document Format (PDF) is one popular way of creating files that exactly portray printed text (see PDF).

Further Reading

Blattner, D., and N. Davis, eds. *The QuarkXPress Book: For Macintosh and Windows.* Berkeley, Calif.: Peachpit Press, 1998.
"Desktop Publishing News." Available online. URL: http://desktoppublishing.com/_news1.html. Accessed November 2007.
Parker, Roger C. *Web Design and Desktop Publishing for Dummies.* 2nd ed. New York: Hungry Minds, 1997.
"Resources for Desktop Publishers." Available online. URL: http://www.nlightning._com/dtpsbiblio.html. Accessed November 2007.
Shushan, R. and D. Wright, with L. Lewis. *Desktop Publishing by Design: Everyone's Guide to PageMaker6.* 4th ed. Redmond, Wash.: Microsoft Press, 1996.

developing nations and computing

Most writing about computer technology tends to focus on developments in technically advanced nations, such as the United States, European Union, and Japan. There is also growing coverage of the rapidly developing information economy in the world's two most populous nations, India and China. But what about the poorest or least developed nations, particularly those in Africa?

INFRASTRUCTURE

A common problem in developing countries is a lack of basic infrastructure to support electronic devices—phone lines, television cables, even a reliable power grid. (About two *billion* people on this planet still have no access to electricity!)

One way around this obstacle is to skip over the wired stage of development in favor of wireless connections, perhaps using battery or even solar power. The necessity for large government investments in infrastructure can then be avoided in favor of mobile, distributed, flexible access that can be gradually spread and scaled up. Already, in some of the poorest nations mobile phone use has been growing at an annual rate of 50 percent or more.

Once access to communications and data is provided, users can immediately start getting an economic return or otherwise improving their lives. Farmers, for example, can get weather reports and keep in touch with market prices. Of course online communications might also give farmers a tool for organizing themselves politically or economically (such as into co-ops). People start to get in touch with developments around the world that might affect them, and discover possible ways to a better life. However, authoritarian governments often resist such trends because they fear the development of well-connected democratic reform movements.

CLOSING THE GAP

Much of the barrier to developing countries joining the networked world is human rather than technological. Before people can learn to use computers, they need to be able to read. They also need some idea of what science and technology are about and why they are important for their economic well-being.

Beyond people learning to use computers to communicate, or in agriculture or commerce, a developing country needs to have enough people with the advanced skills needed for a self-sustaining information economy. These include technicians, support staff, teachers, engineers, programmers, and computer scientists.

One reason for the rapid growth of computing in India and especially China is that these countries, while still having millions of people living on subsistence, also have effective educational systems including advanced training. Their growing pool of skilled but relatively inexpensive workers in turn attracts foreign investment capital. In addition to China and India, other nations with strong electronics manufacturing industries include Singapore, Korea, Malaysia, Mexico, and Brazil.

The United Nations has developed the Technology Achievement Index (TAI) to measure the ability of a country to innovate, to effectively use new and existing technology, and to build a base of technically skilled workers.

ONE LAPTOP PER CHILD

While the conventional view of technological development stresses the importance of infrastructure and skills, some visionary educational activists are suggesting a way to "jump-start" the information economy in poor and developing countries. They note that despite the potential of wireless technology, adequate computing power for joining the world network has simply been too expensive for all but the elite in developing countries. (A $400 no-frills PC costs more than the annual per capita income of Haiti, for example.)

In response, MIT computer scientists (see MIT MEDIA LAB and NEGROPONTE, NICHOLAS) have started an initiative called One Laptop Per Child. Their machine (introduced as a prototype in 2005) includes the following features:

- very low power consumption (2–3 watts)

- lower and higher power modes (the latter, for example, can provide backlighting for the screen when an external power source is available)

- ability to use a variety of batteries or an external power source, including a hand-powered generator

- built-in wireless networking

- tough construction, including a water-resistant membrane keyboard

- flash memory instead of a hard drive or CD-ROM

- built-in color camera, microphone, and stereo speakers

- open-source Linux operating system and other software, including programming languages especially useful for learners

The computer is intended ultimately to cost no more than $100 per unit, and is to be distributed through participating governments. Countries that have made at least tentative commitments to the project as of 2007 include Argentina, Cambodia, Costa Rica, Dominican Republic, Egypt, Greece, Libya, Nigeria, Pakistan, Peru, Rwanda, Tunisia, Uruguay, and, in the United States, the states of Massachusetts and Maine.

The underlying philosophy of the project is based on "constructivist learning," the idea that children can learn powerful ideas through using suitable interactive systems (see LOGO and PAPERT, SEYMOUR). In a way it is intended to be a sort of lever to create a generation with the skills to function in the 21st-century information economy, without re-creating the cumbersome industrial-style educational systems of the previous 200 years.

Although, generally, some well received critics are concerned about the environmental impact of producing (and eventually disposing of) millions more computers, while others (including some officials in developing countries) believe the money for providing computers to children should be used instead for more urgent needs such as clean water, public health, and basic school supplies.

Whether using top-down or bottom-up approaches, the web of connection, communication, and information continues its rapid though uneven spread around the world. However, as new technologies continue to emerge in the developed world, the position of technological "have-nots" may worsen if effective education and access programs are not developed.

Further Reading

Desai, Meghnad, et al. "Measuring the Technological Achievement of Nations and the Capacity to Participate in the Network Age." *Journal of Human Development* 3 (2002): 95–122. Available online. URL: http://unpan1.un.org/intradoc/groups/public/documents/apcity/unpan014340.pdf. Accessed September 11, 2007.
One Laptop per Child. Available online. URL: http://laptop.org/vision/index.shtml. Accessed September 11, 2007.
Wilson, Ernest J., III. *The Information Revolution and Developing Countries*. Cambridge, Mass.: MIT Press, 2004.
Wireless Internet Institute. *The Wireless Internet Opportunity for Developing Countries*. Boston, Mass.: World Times, 2003.

device driver

A fundamental problem in computer design is the control of devices such as disk drives and printers. Each device is designed to respond to a particular set of control commands sent as patterns of binary values through the port to which the device is connected. For example, a printer will respond to a "new page" command by skipping lines to the end of the current page and moving the print head to the start of the next page, taking margin settings into account. The problem is this: When an applications program such as a word processor needs to print a document, how should the necessary commands be provided to the printer? If every application program has to include the appropriate set of commands for each device that might be in use, programs will be bloated and much development effort will be required for supporting devices rather than extending the functionality of the product itself. Instead, the manufacturers of printers and other devices such as scanners and graphics tablets typically provide a program called a driver. (A version of the driver is created for each major operating system in use.) The driver serves as the intermediary between the application, the operating system and the low-level device control system. It is sometimes useful to have drivers in the form of continually running programs that monitor the status of a device and wait for commands (see DEMON).

Modern operating systems such as Microsoft Windows typically take responsibility for services such as printing documents. When a printer is installed, its driver program is also installed in Windows. When the application program requests to print a document, Windows's print system accesses the driver. The driver turns the operating system's "generic" commands into the specific hardware control commands needed for the device.

While the use of drivers simplifies things for both program developers and users, there remains the need for users to occasionally update drivers because of an upgrade either in the operating system or in the support for device capabilities. Both Windows and the Macintosh operating system

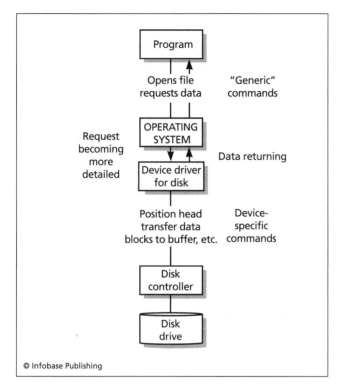

The device driver is the link between the operating system and the hardware that controls a specific device. Program requests are passed by the operating system to the device driver, which issues the detailed instructions needed by the device controller.

implement a feature called plug and play. This allows for a newly installed device to be automatically detected by the system and the appropriate driver loaded into the operating system (see PLUG AND PLAY). Other device management components enable the OS to keep track of the driver version associated with each device. Some of the newest operating systems include auto-update features that can search on the Web for the latest driver versions and download them.

The need to provide drivers for popular devices creates something of a barrier to the development of new operating systems. In a catch-22, device manufacturers are unlikely to support a new OS that lacks significant market share, while the lack of device support in turn will discourage users from adopting the new OS. (Users of the Linux operating system faced this problem. However, that system's open source and cooperative development system made it easier for enthusiasts to write and distribute drivers without waiting for manufacturers to do so.)

Further Reading
Mr. Driver: Device Drivers. Available online. URL: http://www.mrdriver.com. Accessed July 8, 2007.

Oney, Walter. *Programming the Microsoft Windows Driver Manual.* 2nd ed. Redmond, Wash.: Microsoft Press, 2005.

Rubini, Alessandro, and Jonathan Corbet. *Linux Device Drivers.* 3rd ed. Sebastapol, Calif.: O'Reilly, 2005.

Windows Driver Kit (WDK) Overview. Available online. URL: http://www.microsoft.com/whdc/devtools/wdk/default.mspx. Accessed July 8, 2007.

DHTML   *See* HTML, DHTML, and XHTML.

Diffie, Bailey Whitfield
(1944–   )
American
Mathematician, Computer Scientist

Bailey Whitfield Diffie created the system of public key cryptography that many computer users depend on today to protect their sensitive information (see ENCRYPTION).

Diffie was born on June 5, 1944, in the borough of Queens, New York City. As a youngster he read about secret codes and became fascinated. Although he was an indifferent high school student who barely qualified for graduation, Diffie scored so high on standardized tests that he won admission to the University of California, Berkeley, in 1962, where he studied mathematics for two years. However, in 1964 he transferred to the Massachusetts Institute of Technology (MIT) and obtained his B.S. in mathematics in 1965.

After graduation Diffie took a job at Mitre Corporation, a defense contractor, where he plunged into computer programming, helping create Mathlab, a program that allowed mathematicians to not merely calculate with a computer, but also to manipulate mathematical symbols to solve equations. (The program would eventually evolve into Macsyma, a software package used widely in the mathematical community—see MATHEMATICS SOFTWARE.)

By the early 1970s Diffie had moved to the West Coast, working at the Stanford Artificial Intelligence Laboratory (SAIL), where he met Lawrence Roberts, head of information processing research for ARPA, the Defense Department's research agency. Roberts's main project was the creation of the ARPAnet, the computer network that would later evolve into the Internet.

Roberts was interested in providing security for the new network, and (along with AI researcher John McCarthy) he helped revive Diffie's dormant interest in cryptography. By 1974 Diffie had learned that IBM was developing a more secure cipher system, the DES (Data Encryption Standard), under government supervision. However, Diffie soon became frustrated with the way the National Security Agency (NSA) doled out or withheld information on cryptography, making independent research in the field very difficult. Seeking to learn the state of the art, Diffie traveled widely, seeking out people who might have fresh thoughts on the subject.

Diffie found one such person in Martin Hellman, a Stanford professor who had also been struggling on his own to develop a better system of encryption. They decided to pool their ideas and efforts, and Diffie and Hellman came up with a new approach, which would become known as public key cryptography. It combined two important ideas that had already been discovered to an extent by other researchers. The first idea was the "trap-door function"—a mathematical operation that can be easily performed "forward" but that was very hard to work "backward." Diffie realized, however, that a trap-door function could be devised that could be worked backward easily *if* the person had the appropriate key.

The second idea was that of key exchange. In classical cryptography, there is a single key used for both encryption and decryption. In such a case it is absolutely vital to keep the key secret from any third party, so arrangements have to be made in advance to transmit and protect the key.

Diffie, however, was able to work out the theory for a system that generates pairs of mathematically interrelated keys: a private key and a public key. Each participant publishes his or her public key, but keeps the corresponding private key secret. If one wants to send an encrypted message to someone, one uses that person's public key (obtained from the electronic equivalent of a phone directory). The resulting message can only be decrypted by the intended recipient, who uses the corresponding secret, private key.

The public key system can also be used as a form of "digital signature" for verifying the authenticity of a message. Here a person creates a message encrypted with his or her private key. Since such a message can only be decrypted using the corresponding public key, any other person can use that key (together with a trusted third-party key service) to verify that the message really came from its purported author.

Diffie and Hellman's 1976 paper in the *IEEE Transactions on Information Theory* began boldly with the statement that "we stand today on the brink of a revolution in cryptography." This paper soon came to the attention of three researchers who would create a practical implementation called RSA (for Rivest, Shamir, and Adelman).

Through the 1980s Diffie, resisting urgent invitations from the NSA, served as manager of secure systems research for the phone company Northern Telecom, designing systems for managing security keys for packet-switched data communications systems (such as the Internet).

In 1991 Diffie was appointed Distinguished Engineer for Sun Microsystems, a position that has left him free to deal with cryptography-related public policy issues. The best known of these issues has been the Clipper Chip, a proposal that all new computers be fitted with a hardware encryption device that would include a "back door" that would allow the government to decrypt data. Along with many civil libertarians and privacy activists, Diffie did not believe users should have to trust largely unaccountable government agencies for the preservation of their privacy. Their opposition was strong enough to scuttle the Clipper Chip proposal by the end of the 1990s. Another proposal, using public key cryptography but having a third-party "key escrow" agency hold the keys for possible criminal investigation, also fared poorly. In 1998 Diffie and Susan Landau wrote *Privacy on the Line*, a book about the politics of surveillance and encryption. The book was revised and expanded in 2007.

Diffie has received a number of awards for both technical excellence and contributions to civil liberties. These include the IEEE Information Theory Society Best Paper Award (1979), the IEEE Donald Fink Award (1981), the Electronic Frontier Foundation Pioneer Award (1994), and even the National Computer Systems Security Award (1996), given by the NIST and NSA.

Further Reading
Diffie, Whitfield. "Interview with Whitfield Diffie on the Development of Public Key Cryptography." Conducted by Franco Furger; edited by Arnd Weber, 1992. Available online. URL: http://www.itas.fzk.de/mahp/weber/diffie.htm. Accessed September 12, 2007.
Diffie, Whitfield, and Susan Landau. *Privacy on the Line: the Politics of Wiretapping and Encryption.* Updated and expanded ed. Cambridge, Mass.: MIT Press, 2007.
Kahn, David. *The Codebreakers: The Story of Secret Writing.* Revised ed. New York: Scribner, 1996.
Levy, Steven. *Crypto: How the Code Rebels Beat the Government: Saving Privacy in the Digital Age.* New York: Viking Penguin, 2001.

digital cash

Also called digital money or e-cash, digital cash represents the attempt to create a method of payment for online transactions that is as easy to use as the familiar bills and coins in daily commerce (see E-COMMERCE). At present, credit cards are the principal means of making online payments. While using credit cards takes advantage of a well-established infrastructure, it has some disadvantages. From a security standpoint, each payment potentially exposes the payer to the possibility that the credit card number and possibly other identifying information will be diverted and used for fraudulent transactions and identity theft. While the use of secure (encrypted) online sites has reduced this risk, it cannot be eliminated entirely (see COMPUTER CRIME AND SECURITY). Credit cards are also impracticable for very small payments from cents to a few dollars (such as for access to magazine articles) because the fees charged by the credit card companies would be too high in relation to the value of the transaction.

One way to reduce security concerns is to make transactions that are anonymous (like cash) but guaranteed. Products such as DigiCash and CyberCash allow users to purchase increments of a cash equivalent using their credit cards or bank transfers, creating a "digital wallet." The user can then go to any Web site that accepts the digital cash and make a payment, which is deducted from the wallet. The merchant can verify the authenticity of the cash through its issuer. Since no credit card information is exchanged between consumer and merchant, there is no possibility of compromising it. The lack of wide acceptance and standards has thus far limited the usefulness of digital cash.

The need to pay for small transactions can be handled through micropayments systems. For example, users of a variety of online publications can establish accounts through a company called Qpass. When the user wants to read an article from the *New York Times,* for example, the fee for the article (typically $2–3) is charged against the user's Qpass account. The user receives one monthly credit card billing from Qpass, which settles accounts with the publications. Qpass, eCharge, and similar companies have had modest success. A similar (and quite successful) service is offered by companies such as PayPal and Billpoint, which allow winning auction bidders to send money from their credit card or bank account to the seller, who would not otherwise be equipped to accept credit cards. True micropayments would extend down to just a few cents.

"True" digital cash, allowing for anonymous payments and micropayments, has been slow to catch on. However, the successful digital cash system is likely to have the following characteristics:

- Protects the anonymity of the purchaser (no credit card information transmitted to the seller)

- Verifiable by the seller, perhaps by using one-time encryption keys

- The purchaser can create digital cash freely from credit cards or bank accounts

- Micropayments can be aggregated at a very low transaction cost

As use of digital cash becomes more widespread, it is likely that tax and law enforcement agencies will press for the inclusion of some way to penetrate the anonymity of transactions for audit or investigation purposes. They will be opposed by civil libertarians and privacy advocates. One likely compromise may be requiring that transaction information or encryption keys be deposited in some sort of escrow agency, subject to being divulged upon court order.

Further Reading

Kou, Weidong. *Payment Technologies for E-Commerce.* New York: Springer, 2003.
Lamb, Gregory M. "'Nickel and Diming' across the Internet." *Christian Science Monitor,* February 23, 2004, n.p. Available online. URL: http://www.csmonitor.com/2004/0223/p13s01-wmgn.html. Accessed July 8, 2007.
Orr, Bill. "Cashless Society, Ahoy! Suddenly Micropayments Are Hot Again." *ABA Banking Journal* 98 (March 2006): ff.
Rosenborg, Victoria. *PayPal for Dummies.* Hoboken, N.J.: Wiley, 2005.
Warwick, David R. "Violent Crime and Cash: The Connection; Privacy Concerns About Digital Cash May Be Overblown." *The Futurist,* May 1, 2007, p. 42.

digital convergence

Since the late 20th century, many forms of communication and information storage have been transformed from analog to digital representations (see ANALOG AND DIGITAL). For example, the phonograph record (an electromechanical analog format) gave way during the 1980s to a wholly digital format (see CD-ROM). Video, too, is now increasingly being stored in digital form (DVD or laser disks) rather than in the analog form of videotape. Voice telephony, which originally involved the conversion of sound to analogous electrical signals, is increasingly being digitized (as with many cell phones) and transmitted in packet form over the communications network.

The concept of digital convergence is an attempt to explore the implications of so many formerly disparate analog media now being available in digital form. All forms of digital media have key features in common. First, they are

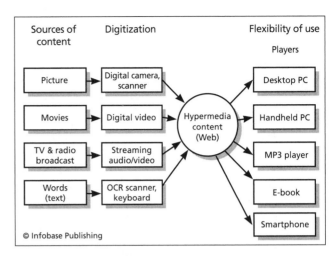

Digital convergence results from the fact that many formerly analog media (such as sound, film, and video) are now being acquired and processed digitally. Once in digital form, the content can be processed and played by a variety of software and used on many different platforms ranging from desktop computers to electronic books (e-books) and portable MP3 music players. Content can also be linked and organized using hypertext or hypermedia techniques, as on the Web.

essentially pure information (computer data). This means that regardless of whether the data originally represented still images from a camera, video, or film, the sound of a human voice, music, or some other form of expression, that data can be stored, manipulated, and retrieved under the control of computer algorithms. This makes it easier to create seamless multimedia presentations (see MULTIMEDIA and HYPERTEXT AND HYPERMEDIA). Services or products previously considered to be separate can be combined in new ways. For example, many radio stations now provide their programming in the form of "streaming audio" that can be played by such utilities as RealPlayer or Microsoft Windows Media Player (see STREAMING). Similarly, television news services such as CNN can offer selected excerpts of their coverage in the form of streaming video files. As more users gain access to broadband Internet connections (such as cable or DSL), it is gradually becoming feasible to deliver TV programs and even full-length feature films in digital format. By the middle of the decade, media delivery began to proliferate on new platforms that represent a further convergence of function. Many "smart phones" can play audio and video (see SMARTPHONE). In July 2007 Apple's iPhone entered the market, combining phone, media player, and Web browsing functions, and similar devices will no doubt follow (see also PDA).

EMERGING ISSUES

The merging of traditional media into a growing stream of digital content has created a number of difficult legal and social issues. Digital images or sounds from various sources can easily be combined, filtered, edited, or otherwise altered for a variety of purposes. As a result, the value of photographs as evidence may be gradually compromised.

The ownership and control of the intellectual property represented by music, video, and film has also been complicated by the combination of digitization and the pervasive Internet. For example, during 2000–2001 the legal battles involving Napster, a program that allows users to share music files pitted the rights of music producers and artists to control the distribution of their product against the technological capability of users to freely copy and distribute the material. While a variety of copy protection systems (both software and hardware-based) have been developed in an attempt to prevent unauthorized copying, historically such measures have had only limited effectiveness (see COPY PROTECTION, DIGITAL RIGHTS MANAGEMENT, and INTELLECTUAL PROPERTY AND COMPUTING).

Digital convergence also raises deeper philosophical issues. Musicians, artists, and scholars have frequently suggested that the process of digitization fails to capture subtleties of performance that might have been accessible in the original media. At the same time, the richness and immersive qualities of the new multimedia may be drawing people further away from the direct experience of the "real" analog world around them. Ultimately, the embodiment of digital convergence in the form of virtual reality likely to emerge in the early 21st century will pose questions as profound as those provoked by the invention of printing and the development of mass broadcast media (see VIRTUAL REALITY).

Further Reading
Covell, Andy. *Digital Convergence: How the Merging of Computers, Communications and Multimedia Is Transforming Our Lives.* Newport, R.I.: Aegis Publishing, 2000.
———. *Digital Convergence Phase 2: A Field Guide for Creator-Collaborators.* Champaign, Ill.: Stipes Publishing, 2004
Jenkins, Henry. *Convergence Culture: Where Old and New Media Collide.* New York: New York University Press, 2006.
Park, Sangin. *Strategies and Policies in Digital Convergence.* Hershey, Penn.: Information Science Reference, 2007.

digital dashboard

The dashboard of a car is designed to present vital real-time information to the driver, such as speed, fuel supply, and engine status. Ideally this information should be easy to grasp at a glance, allowing for prompt action when necessary. Conversely, unnecessary and potentially distracting information should be avoided, or at least relegated to an unobtrusive secondary display.

A digital dashboard is a computer display that uses similar concepts. Its goal is to provide an executive or manager with the key information that allows him or her to monitor the health of the enterprise and to take action when necessary. (A digital dashboard can also be part of a larger set of management tools—see DECISION SUPPORT SYSTEM.)

The screen display for a digital dashboard can use a variety of objects (see GRAPHICAL USER INTERFACE). These can include traditional charts (line, bar, or pie), color-coded maps, depictions of gauges, and a variety of other interface elements sometimes known as "widgets."

However information is depicted, the dashboard is designed to summarize the current status of business or other functions, identify trends, and warn the user when attention is required. For example, a dashboard might summarize production and shipping for each of a company's factories. Bars on a chart might be green when levels are within normal parameters, but turn red if, for example, production has fallen more than 20 percent below target goals. Dashboard displays can also be useful for graphically showing the degree to which project objectives are being met.

Digital dashboards can be custom built or obtained in forms specialized for various types of business. Typically the dashboard is hosted on the corporate Web server and is accessible through Web browsers—perhaps with an abbreviated version that can be viewed on PDAs and smart phones.

CRITIQUE

Today dashboards are in widespread use in many top corporations, from Microsoft to Home Depot. An oft-cited advantage of dashboard technology is that it keeps managers focused and provides for quick response in situations where time may be crucial. No longer is it necessary for the manager to track down key individuals and try to make sense of their reports over the phone.

Some critics, however, worry that dashboards may make management too "data driven." Those regular calls, after all, can form an important part of the relationship between an executive or manager and subordinates, as well as getting a sense of morale and possible personnel problems that may be affecting productivity. Overreliance on dashboards and "bottom line" numbers may also hurt the morale of salespeople and others who come to feel that they are being micromanaged. Further, the dashboard may omit important considerations that in turn are likely to receive less attention and support.

Further Reading
Ante, Spencer E., and Jena McGregor. "Giving the Boss the Big Picture: A 'Dashboard' Pulls Up Everything the CEO Needs to Run the Show." *BusinessWeek,* February 13, 2006. Available online. URL: http://www.businessweek.com/magazine/content/06_07/b3971083.htm. Accessed September 12, 2007.
Dashboard Examples. Available online. URL: http://www.enterprise-dashboard.com/. Accessed September 12, 2007.
Dashboard Insight. Available online. URL: http://www.dashboard insight.com/. Accessed September 12, 2007.
Eckerson, Wayne W. *Performance Dashboards: Measuring, Monitoring, and Managing Your Business.* New York: Wiley, 2006.
Few, Stephen. *Information Dashboard Design: The Effective Visual Communication of Data.* Sebastapol, Calif.: O'Reilly, 2006.

digital divide

The term *digital divide* was coined in the late 1990s amid growing concern that groups such as minorities, the elderly, and rural residents were not becoming computer literate and connecting to the Internet at the same rate as the young, educated, and relatively affluent.

Nearly a decade later this perception of a chasm has diminished somewhat. According to the Pew Internet & American Life project, as of 2006 about two-thirds (70 percent) of American adults were using the Internet, and the number has continued to increase, though more slowly (there is evidence of a "hard core" unconnected population). Groups that lagged in Internet usage included Americans 65 years or older (35 percent), African Americans (58 percent), and persons without at least a high school education (36 percent).

The digital divide is more severe if one looks at the world as a whole (see DEVELOPING NATIONS AND COMPUTING). Rapidly industrializing nations such as China and India are seeing considerable increases in the number of people with some form of computer and Internet access, though the numbers are still small in relation to the total population. In severely underdeveloped countries (such as many in Africa), connectivity may be improved by the "One Laptop per Child" project, which has designed a prototype computer designed to cost less than $100.

BROADBAND USE

Not all Internet access is equal. High-speed connections (see BROADBAND, CABLE MODEM, and DSL) encourage frequent Internet use throughout the day, and make it feasible to access and share rich media (images, videos, podcasts, and so on). According to the Pew Internet & American Life project, 47 percent of all adult Americans had a broadband Internet connection at home as of 2007. The rate of broadband adoption continues to lag for rural residents (31 percent) and African Americans (40 percent).

However, the broadband adoption rate for African Americans has been increasing rapidly (it was only 14 percent in early 2005). There are a number of factors that correlate with the likelihood that a person or community will have access to the Web. People in lower-income brackets are less likely to own PCs. Phone service may be less reliable (particularly in rural areas), and Internet access may require expensive toll charges. While schools and public libraries can offer an alternative venue for Internet access, inner-city schools have tended to lag behind in connecting to the Internet and in the ratio of networked computers to students. (The Net Day activities in the mid-1990s first publicized and sought to ameliorate this problem.)

Internet access also correlates to education. While persons lacking a college education are likely to be poorer than college graduates, they are also less likely to be working in jobs that include regular computer access. A deficiency in basic reading and keyboard skills can also serve as a barrier to participation in the online world (see also COMPUTER LITERACY). People over age 50 are also less likely to be online. They are more likely to have spent their career in noncomputerized jobs and may feel that they cannot master the new technology.

Targeted attempts to close the digital divide through providing more Internet access through schools and libraries are likely to continue to be successful. The marketplace itself is perhaps making the biggest contribution, since the price of an Internet-capable PC with a basic dial-up connection is now around $400 plus about $10/month.

Improvement in the teaching of general literacy as well as technical skills in the K-12 schools is necessary if the next generation is to be able to participate fully and equally in the online world.

Further Reading

Compaine, Benjamin M., ed. *The Digital Divide: Facing a Crisis or Creating a Myth?* Cambridge, Mass.: MIT Press, 2001.

Digital Divide Network. Available online. URL: http://www.digitaldivide.net/. Accessed July 9, 2007.

Norris, Pippa. *Digital Divide: Civic Engagement, Information Poverty, and the Internet Worldwide.* Cambridge, Mass.: Cambridge University Press, 2001.

Nulens, Gert. *The Digital Divide in Developing Countries: Towards an Information Society in Africa.* Brussels: VUB Brussels University Press, 2001.

Pew Internet & American Life Project. "Digital Divisions." Available online. URL: http://www.pewinternet.org/pdfs/PIP_Digital_Divisions_Oct_5_2005.pdf. Accessed July 9, 2007.

———. "Home Broadband Adoption 2007." Available online. URL: http://www.pewinternet.org/pdfs/PIP_Broadband%202007.pdf. Accessed July 9, 2007.

digital rights management (DRM)

By default, once information is digitized it is simply a pattern of bits that can be easily copied within the same or a different medium, using a variety of software or the built-in facilities of the operating system. Of course the development of tape-recording technology in the mid-20th century already made it possible to copy audio recordings, and the later development of videotape and the VCR did the same for video. However, while analog copying techniques lose some accuracy (or fidelity) with each generation of copying, digital files can be copied exactly each time. It is equally easy to e-mail, upload, or otherwise distribute audio or video files.

Legally, the creator of an original work can assert copyright—literally, the "right to copy" or to control when and how the work is distributed. Digital rights management (DRM) refers to a variety of technologies that can be used to enforce this right by making it at least difficult for the purchaser of one copy of a work to copy and distribute it in turn. (Similar technologies have also been used to prevent copying of software, which is, after all, just another pattern of bits—see COPY PROTECTION and SOFTWARE PIRACY AND COUNTERFEITING.)

DRM FOR FILM AND VIDEO

In the mid-1990s, movies on DVD were protected using the Content Scrambling System (CSS). This proprietary format was licensed only for certain hardware and operating systems, but in 1999 an activist programmer released DeCSS, a program that could decode protected discs and allow them to be played on operating systems such as Linux, which had not been licensed. A similar story unfolded in 2007 when hackers broke the Advanced Access Content System (AACS) that was used to protect the new high-definition HD DVD and Blu-Ray discs.

PROTECTING MUSIC

DRM has also been used on many audio CDs. Many consumers complained that their CD players (particularly when used with Windows PCs) were not compatible with the protected discs. A bigger controversy arose in 2005 when Sony began to use DRM technology that (without notification) installed a rootkit (a kind of "back door" to the operating system) that potentially left systems open to attack. Facing public outcry and several lawsuits, Sony withdrew the DRM, which, ironically, was rather ineffective at preventing copying. By 2007 music CD producers had concluded that DRM had more costs than benefits, and such protection is no longer found on audio CDs.

Music distributed online is often protected by DRM. However, some services such as Apple iTunes now offer the option of buying DRM-free music at a higher price. (Apple's Steve Jobs has called upon the online music industry to completely eliminate DRM.)

LEGAL AND OTHER ISSUES

Generally, the argument for DRM has been straightforward: If people can get something for free, they will not buy it. Content creators and publishers would go out of business. Organizations such as the Recording Institute Association of America (RIAA) have aggressively sued college students and others accused of sharing copyrighted music or video online and successfully forced the best known file-sharing service, Napster, to become a licensed music service (see FILE-SHARING AND P2P NETWORKS).

The principal legal means for enforcing DRM is the Digital Millennium Copyright Act (DMCA), passed in 1998. The law prohibits the production or dissemination of technology (software or hardware) that allows users to circumvent DRM. However, it has been difficult in practice to prevent the rapid dissemination of "cracks" for DRM over the Internet.

There are also a number of legal arguments against DRM. One is that it prevents certain actions allowed to consumers under copyright law, such as making a backup copy of media that one has purchased (see INTELLECTUAL PROPERTY AND COMPUTING). Also, because many DRM schemes work only with Windows or Macintosh machines, users of other operating systems (notably Linux) must "crack" DRM in order to be able to use the protected media. (Under the law, such action to promote "interoperability" is allowed, though not if the purpose is to facilitate illegal copying. But like most matters of intent, this can be hard to determine.)

There have also been First Amendment issues. Although the DMCA includes a "scholarly research" exception, some cryptography researchers have said that they have been inhibited from publishing analysis of DRM for fear of legal prosecution.

A number of activists and groups have opposed DRM, including open-source advocate Richard Stallman and the Electronic Frontier Foundation (see CYBERSPACE ADVOCACY GROUPS). One of their efforts has been promotion of the Free Software Foundation's General Public License (GPL3), which prohibits the use of DRM in products distributed under that open-source license.

Further Reading

Defective by Design: A Campaign of the Free Software Foundation. Available online. URL: http://defectivebydesign.org/. Accessed September 12, 2007.
Electronic Frontier Foundation. Available online. URL: http://www.eff.org. Accessed September 12, 2007.
May, Christopher. *Digital Rights Management: The Problem of Expanding Ownership Rights.* Oxford, U.K.: Chandos, 2007.
Motion Picture Association of America. Available online. URL: http://www.mpaa.org/. Accessed September 12, 2007.
Recording Industry Association of America. Available online. URL: http://www.riaa.org/. Accessed September 12, 2007.
Van Tassel, Joan. *Digital Rights Management: Protecting and Monetizing Content.* Burlington, Mass.: Focal Press, 2006.
Zeng, Wenjun, Heather Yu, and Ching-Yung Lin. *Multimedia Security Technologies for Digital Rights Management.* Burlington, Mass.: Academic Press, 2006.

Dijkstra, Edsger W.

(1930–2002)
Dutch
Computer Scientist

Edsger W. Dijkstra was born in Rotterdam, Netherlands, in 1930 into a scientific family (his mother was a mathematician and his father was a chemist). He received an intensive and diverse intellectual training, studying Greek, Latin, several modern languages, biology, mathematics, and chemistry. While majoring in physics at the University of Leiden in 1951, he attended a summer school at Cambridge that kindled what soon became a major interest in programming. He continued this pursuit at the Mathematical Center in Amsterdam in 1952 while finishing studies for his

Edsger Dijkstra's ideas about structured programming helped develop the field of software engineering, enabling programmers to organize and manage increasingly complex software projects. (DEPARTMENT OF COMPUTER SCIENCES, UT AUSTIN)

physics degree. At the time there were no degrees in computer science; indeed, programming did not yet exist as an academic discipline. Like most other computers of the time, the Mathematical Center's ARMAC was custom-built. With no high-level languages yet in use, programming required intimate familiarity with the machine's architecture and low-level instructions. Dijkstra soon found that he thrived in such an environment.

By 1956, Dijkstra had discovered an algorithm for finding the shortest path between two points. He applied the algorithm to the practical problem of designing electrical circuits that used as little wire as possible, and generalized it into a procedure for traversing treelike data structures.

During the 1960s, Dijkstra began to explore the problem of communication and resource-sharing within computers. He developed the idea of a semaphore. Like the railroad signaling device that allows only one train at a time to pass through a single section of track, the programming semaphore provides *mutual exclusion,* ensuring that two processes don't try to access the same memory or other resource at the same time.

Another problem Dijkstra tackled involved the sequencing of several processes that are accessing the same resources. He found ways to avoid a deadlock situation where one process had part of what it needed but was stuck because the process holding the other needed resource was in turn waiting for the first process to finish. His algorithms for allowing multiple processes (or processors) to take turns gaining access to memory or other resources would become fundamental for the design of new computing architectures.

During the 1970s, Dijkstra immigrated to the United States, where he became a research fellow at Burroughs, one of the major manufacturers of mainframe computers. During this time he helped launch the "structured programming" movement. His paper "GO TO Considered Harmful" criticized the use of that unconditional "jump" instruction because it made programs hard to read and verify. The newer structured languages such as Pascal and C affirmed Dijkstra's belief in avoiding or discouraging such haphazard program flow (see STRUCTURED PROGRAMMING).

Dijkstra spent the last decades of his career as a professor of mathematics at the University of Texas at Austin, where he held the Schlumberger Centennial Chair in Computer Science. Dijkstra had some unusual quirks for a computer scientist. His papers were handwritten with a fountain pen, and he did not even own a personal computer until late in life.

In 1972 Dijkstra won the Association for Computing Machinery's Turing Award. After his death on August 6, 2002, in Nuenen, The Netherlands, the ACM renamed its award for papers in distributed computing as the Dijkstra Prize.

Perhaps Dijkstra's greatest testament, however, is found in the millions of lines of computer code that are better organized and easier to maintain because of the widespread adoption of structured programming.

Further Reading
Dijkstra, Edsger Wybe. *A Discipline of Programming.* Upper Saddle River, N.J.: Prentice Hall, 1976.
————. *Selected Writings on Computing: A Personal Perspective.* New York: Springer, 1982.
————, and W. H. J. Feijen. *A Method of Programming.* Reading, Mass.: Addison-Wesley, 1988.
Shasha, Dennis, and Cathy Lazere. *Out of Their Minds: The Lives and Discoveries of 15 Great Computer Scientists.* New York: Springer-Verlag, 1995.

disabled persons and computing

The impact of the personal computer upon persons having disabilities involving sight, hearing, or movement has been significant but mixed. Computers can help disabled people communicate and interact with their environment, better enabling them to work and live in the mainstream of society. At the same time, changes in computer technology can, if not ameliorated, exclude some disabled persons from fuller participation in a society where computer access and skills are increasingly taken for granted.

COMPUTERS AS ENABLERS

Computers can be very helpful to disabled persons. With the use of text-to-speech software, blind people can have online documents read to them. (With the aid of a scanner, printed materials can also be input and read aloud.) Persons with low vision can benefit from software that can present text in large fonts or magnify the contents of the screen. Text can also be printed (embossed) in Braille. Deaf or hearing-impaired persons can now use e-mail or instant messaging software for much of their communication needs, largely replacing the older and more cumbersome teletype (TTY and TTD) systems. As people who have seen presentations by physicist Stephen Hawking know, even quadriplegics who have only the use of head or finger movements can input text and have it spoken by a voice synthesizer. Further, advances in coupling eye movements (and even brain wave patterns) to computer systems and robotic extensions offer hope that even profoundly disabled persons will be able to be more self-sufficient.

CHALLENGES

Unfortunately, changes in computer technology can also cause problems for disabled persons. The most pervasive problem arose when text-based operating systems such as MS-DOS were replaced by systems such as Microsoft Windows and the Macintosh that are based on graphic icons and the manipulation of objects on the screen. While text commands and output on the older system could be easily turned into speech for the visually impaired, everything, even text, is actually graphics on a Windows system. While it is possible to have software "hook into" the operating system to read text within Windows out loud, it is much more difficult to provide an alternative way for a blind person to find, click on, drag, or otherwise manipulate screen objects. Thus far, while Microsoft and other operating system developers have built some "accessibility" features such as screen magnification into recent versions of their products, there is no systematic, integrated facility that would allow a blind person to have the same facility as a sighted person.

The growth of the World Wide Web also poses problems for the visually impaired, since many Web pages rely on graphical buttons for navigation. Software plug-ins can provide audio cues to help with screen navigation. While Web browsers usually have some flexibility in setting the size of displayed fonts, some newer features (such as CASCADING STYLE SHEETS) can remove control over font size from the user.

Because most computer systems today use graphical user interfaces, the failure to provide effective access may be depriving blind and visually impaired persons of employment opportunities. Meanwhile, the computer industry, educational institutions, and workplaces face potential challenges under the Americans with Disabilities Act (ADA), which requires that public and workplace facilities be made accessible to the disabled. Some funding through the Technology-Related Assistance Act has been provided to states for promoting the use of adaptive technology to improve accessibility.

Further Reading
Adaptive Computer Products. Available online. URL: http://www. makoa.org/computers.htm. Accessed July 9, 2007.
Better Living through Technology. Available online. URL: http://www.bltt.org/. Accessed July 9, 2007.
Gnome [Linux]. Accessibility Project. Available online. URL: http://developer.gnome.org/projects/gap/. Accessed July 9, 2007.
Microsoft Accessibility. Available online. URL: http://www.microsoft.com/enable/default.aspx. Accessed July 9, 2007.
Slatin, John M., and Sharron Rush. *Maximum Accessibility: Making Your Web Site More Usable for Everyone.* Boston: Addison-Wesley Professional, 2003.
Thatcher, Jim, et al. *Web Accessibility: Web Standards and Regulatory Compliance.* New York: Springer, 2006.
Vanderheiden, Gregg C. *Making Software More Accessible for People with Disabilities.* Collingdale, Penn.: Diane Pub. Co., 2004.

disaster planning and recovery

Most businesses, government offices, or other organizations are heavily dependent on having continuous access to their data and the hardware, network, and software necessary to work with it. Activities such as procurement (see SUPPLY CHAIN MANAGEMENT), inventory, order fulfillment, and customer lists are vital to day-to-day operations. Any disaster that might disrupt these activities, whether natural (such as an earthquake or severe weather) or human-made (see COMPUTER VIRUS and CYBERTERRORISM), must be planned for. Such planning is often called "business continuity planning."

The most basic way to protect against data loss is to maintain regular backups (see BACKUP AND ARCHIVE SYSTEMS). On-site backups can protect against hardware failure, and can consist of separate storage devices (see NETWORKED STORAGE) or the use of redundant storage within the main system itself (see RAID). However, for protection against fire or other larger-scale disaster, it is also necessary to have regular off-site backups, whether using a dedicated facility or an online backup service.

To protect against power failure or interruption, one or more uninterruptible power supplies (UPS) can be used, and possibly a backup generator to deal with longer-term outages. All equipment should also have surge protection to avoid damage from power fluctuations.

Of course anything that can minimize the chance of disaster happening or the extent of its effects should also be part of disaster planning. This can include structural reinforcement, physical security, firewalls and antivirus software, and fire alarms and suppression systems.

DISASTER PLANNING
Despite the best precautions, disasters will continue to happen. Organizations whose continued existence depends on their data and systems need to plan systematically how they are going to respond to foreseeable risks, and how they are going to recover and resume operations. Planning for disasters involves the following general steps:

- specify the potential costs and other impacts of loss of data or access

- use that data to prioritize business functions or units

- assess how well facilities are currently being protected

- determine what additional hardware or services (such as additional file servers, attached storage, or remote backup) should be installed

- develop a comprehensive recovery plan that specifies procedures for dealing with various types of disasters or extent of damage, and including immediate response, recovery or restoration of data, and resumption of normal services

- develop plans for communicating with customers, authorities, and the general public in the event of a disaster

- specify the responsibilities of key personnel and provide training in all procedures

- arrange ahead of time for sources of supplies, additional support staff, and so on

- establish regular tests or drills to verify the effectiveness of the plan and to maintain the necessary skills

Recent natural disasters as well as the 9/11 terrorist attacks have spurred many organizations to begin or enhance their disaster planning and recovery procedures.

Further Reading
Benton, Dick. "Disaster Recovery: A Pragmatist's Viewpoint." *Disaster Recovery Journal,* Winter 2007, pp. 79–81. Available online. URL: http://www.drj.com/articles/win07/2001-16.pdf. Accessed September 13, 2007.
Disaster Recovery Guide. Available online. URL: http://www.disaster-recovery-guide.com/. Accessed September 13, 2007.
Disaster Recovery World: The Business Continuity Planning & Disaster Recovery Planning Directory. Available online. URL: http://www.disasterrecoveryworld.com/. Accessed September 13, 2007.
Snedaker, Susan. *Business Continuity and Disaster Recovery Planning for IT Professionals.* Rockland, Md.: Syngress, 2007.
Wallace, Michael. *The Disaster Recovery Handbook: A Step-by-Step Plan to Ensure Business Continuity and Protect Vital Operations, Facilities, and Assets.* New York: AMACOM, 2004.

disk array   *See* RAID.

distance education

Distance education (also called distance learning or virtual learning) is the use of electronic information and communication technology to link teachers and students without their being together in a physical classroom.

Distance education in the form of correspondence schools or classes actually began as early as the mid-19th century with teaching of the Pitman Shorthand writing method. Later, correspondence classes became part of Chautauqua, a movement to educate the rural and urban working classes, taking advantage of the growing reach of mail service through Rural Free Delivery. In correspondence schools, each lesson is typically mailed to the student, who completes the required work and returns it for grading. A certificate is awarded upon completion of course requirements. A few universities (such as the University of Wisconsin) also began to offer correspondence programs.

By the middle of the 20th century, radio and then television was being used to bring lectures to students. This increased the immediacy and spontaneity of teaching. The invention of videotape in the 1970s allowed leading teachers to create customized courses geared for different audiences. However, the ability of students to interact with teachers remained limited.

In the 1960s computers also began to be used for education. One of the earliest and most innovative programs was PLATO (Programmed Logic for Automatic Teaching Operations), which began at the University of Illinois but was later expanded to hundreds of networked terminals. PLATO in many ways pioneered the combining of text, graphics, and sound—what would later be called multimedia. PLATO also provided for early forms of both e-mail and computer bulletin boards.

Meanwhile, with the development of ARPANET and eventually the Internet, a new platform became available for delivering instruction. By the mid-1990s, courses were being delivered via the Internet (see WORLD WIDE WEB).

MODERN DISTANCE EDUCATION

As broadband Internet access becomes the norm, more Internet-based learning environments are taking advantage of video conferencing technology, allowing teachers and students to interact face to face. This helps answer a common objection by critics that distance education cannot replicate the personal and social dimensions of face-to-face education.

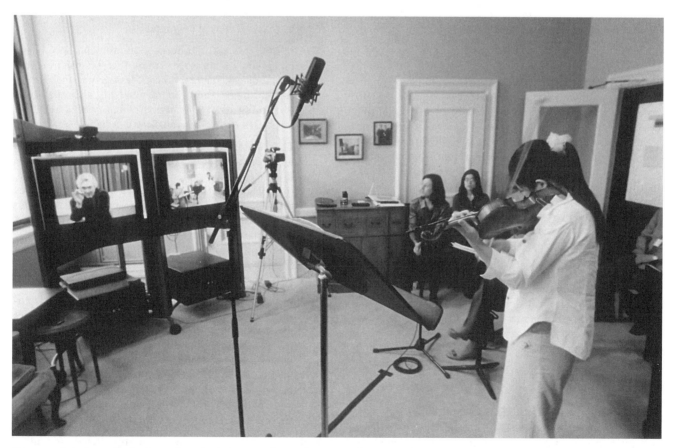

Distance education technologies such as this Polycom video conferencing software enable teachers and students to see, talk, and interact with each other. Here, Manhattan School of Music student Wu Jie of the Zukerman Performance Program demonstrates her violin technique to Maestro Zukerman. (PHOTO BY ANDREW LEPLEY FOR BUSINESS WIRE VIA GETTY IMAGES)

Another way this objection is sometimes addressed by universities is by having a period of physical residency (perhaps a few weeks) as part of the semester.

New platforms for distance education continue to emerge. Class content including lectures has been formatted for delivery to mobile devices such as iPods (see PDA and SMARTPHONE). Another intriguing idea is to establish the classroom within an existing virtual world, such as the popular game *Second Life* (see ONLINE GAMES.) Here students and teachers can meet "face to face" through their virtual embodiments (avatars). It seems only a matter of time before entire universities will exist in such burgeoning alternative worlds.

Further Reading

Bates, A. W. (Tony). *Technology, E-Learning and Distance Education.* 2nd ed. New York: Routledge, 2005.

Distance Education at a Glance. University of Idaho. Available online. URL: http://www.uidaho.edu/eo/distglan. Accessed September 13, 2007.

Distance Learning. About.com. Available online. URL: http://distancelearn.about.com/. Accessed September 13, 2007.

Moore, Michael Grahame, ed. *Handbook of Distance Education.* 2nd ed. Lawrence Mahwah, N.J.: Erlbaum, 2007.

Peterson's Online Degrees and Distance Learning Programs. Available online. URL: http://www.petersons.com/distancelearning/. Accessed September 13, 2007.

Simonson, Michael, et al. *Teaching and Learning at a Distance: Foundations of Distance Education.* 3rd ed. Upper Saddle River, N.J.: Prentice Hall, 2005.

distributed computing

This concept involves the creation of a software system that runs programs and stores data across a number of different computers, an idea pervasive today. A simple form is the central computer (such as in a bank or credit card company) with which thousands of terminals communicate to submit transactions. While this system is in some sense distributed, it is not really decentralized. Most of the work is done by the central computer, which is not dependent on the terminals for its own functioning. However, responsibilities can be more evenly apportioned between computers (see CLIENT-SERVER COMPUTING).

Today the World Wide Web is in a sense the world's largest distributed computing system. Millions of documents stored on hundreds of thousands of servers can be accessed by millions of users' Web browsers running on a variety of personal computers. While there are rules for specifying addresses and creating and routing data packets (see INTERNET and TCP/IP), no one agency or computer complex controls access to information or communication (such as e-mail).

ELEMENTS OF A DISTRIBUTED COMPUTING SYSTEM

The term *distributed computer system* today generally refers to a more specific and coherent system such as a database where data objects (such as records or views) can reside on any computer within the system. Distributed computer systems generally have the following characteristics:

- The system consists of a number of computers (sometimes called *nodes*). The computers need not necessarily use the same type of hardware, though they generally use the same (or similar) operating systems.

- Data consists of logical objects (such as database records) that can be stored on disks connected to any computer in the system. The ability to move data around allows the system to reduce bottlenecks in data flow or optimize speed by storing the most frequently used data in places from which it can be retrieved the most quickly.

- A system of unique *names* specifies the location of each object. A familiar example is the DNS (Domain Naming System) that directs requests to Web pages.

- Typically, there are many processes running concurrently (at the same time). Like data objects, processes can be allocated to particular processors to balance the load. Processes can be further broken down into *threads* (see CONCURRENT PROGRAMMING). Thus, the system can adjust to changing conditions (for example, processing larger numbers of incoming transactions during the day versus performing batches of "housekeeping" tasks at night).

- A remote procedure call facility enables processes on one computer to communicate with processes running on a different computer.

- In inter-process communication protocols specify the processing of "messages" that processes use to report status or ask for resources. Message-passing can be asynchronous (not time-dependent, and analogous to mailing letters) or synchronous (with interactive responses, as in a conversation).

- The capabilities of each object (and thus the messages it can respond to or send) are defined in terms of an interface and an implementation. The interface is like the declaration in a conventional program: It defines the types of data that can be received and the types of data that will be returned to the calling process. The implementation is the code that specifies how the actual processing will be done. The hiding of implementation details within the object is characteristic of object-oriented programming (see CLASS).

- A distributed computing environment includes facilities for managing objects dynamically. This includes lower-level functions such as copying, deleting, or moving objects and systemwide capabilities to distribute objects in such as way as to distribute the load on the system's processors more evenly, to make backup copies of objects (replication), and to reclaim and reorganize resources (such as memory or disk space) that are no longer allocated to objects.

Three widely used systems for distributed computing are Microsoft's DCOM (Distributed Component Object Model), OMG's Common Object Request Broker Architecture (see MICROSOFT .NET and CORBA), and Sun's Java/

Remote Method Invocation (Java/RMI). While these implementations are quite different in details, they provide most of the elements and facilities summarized above.

APPLICATIONS

Distributed computing is particularly suited to applications that require extensive computing resources and that may need to be scaled (smoothly enlarged) to accommodate increasing needs (see GRID COMPUTING). Examples might include large databases, intensive scientific computing, and cryptography. A particularly interesting example is SETI@home, which invites computer users to install a special screen saver that runs a distributed process during the computer's idle time. The process analyzes radio telescope data for correlations that might indicate receipt of signals from an extraterrestrial intelligence (see COOPERATIVE PROCESSING).

Besides being able to marshal very large amounts of computing power, distributed systems offer improved fault tolerance. Because the system is decentralized, if a particular computer fails, its processes can be replaced by ones running on other machines. Replication (copying) of data across a widely dispersed network can also provide improved data recovery in the event of a disaster.

Further Reading

Farley, Jim, and Mike Loukides. *Java Distributed Computing.* Sebastopol, Calif.: O'Reilly, 1998.
Garg, Vijay K. *Concurrent and Distributed Computing in Java.* New York: Wiley, 2004.
———. *Elements of Distributed Computing.* New York: Wiley, 2002.
Goff, Max K. *Network Distributed Computing: Fitscapes and Fallacies.* Upper Saddle River, N.J.: Prentice Hall, 2003.
MacDonald, Matthew. *Microsoft .NET Distributed Applications: Integrating XML Web Services and .NET Remoting.* Redmond, Wash.: Microsoft Press, 2003.
Obasanjo, Dare, and Sanjay Bhatia. "An Introduction to Distributed Object Technologies." Available online. URL: http://www.25hoursaday.com/_IntroductionToDistributedComputing.html. Accessed February 1, 2008.
Shan, Yen-Ping, Ralph H. Earle, and Marie A. Lenzi. *Enterprise Computing with Objects: from Client-Server Environments to the Internet.* Reading, Mass.: Addison-Wesley, 1997.
SETI@Home. Available online. URL: http://setiathome.ssl.berkeley.edu/. Accessed August 14, 2007.

DNS (domain name system)

The operation of the Internet requires that each participating computer have a unique address to which data packets can be routed (see INTERNET and TCP/IP). The Domain Name System (DNS) provides alphabetical equivalents to the numeric IP addresses, giving the now familiar-looking Web addresses (URLs), e-mail addresses, and so on.

The system uses a set of "top-level" domains to categorize these names. One set of domains is based on the nature of the sites involved, including: .com (commercial, corporate), .edu (educational institutions), .gov (government), .mil (military), .org (nonprofit organizations), .int (international organizations), .net (network service providers, and so on).

The other set of top-level domains is based on the geographical location of the site. For example, .au (Australia), .fr (France), and .ca (Canada). (While the United States has the .us domain, it is generally omitted in practice, because the Internet was developed in the United States).

INTERNET COUNTRY CODES (PARTIAL LIST)

AD	Andorra
AE	United Arab Emirates
AF	Afghanistan
AG	Antigua and Barbuda
AI	Anguilla
AL	Albania
AM	Armenia
AN	Netherlands Antilles
AO	Angola
AQ	Antarctica
AR	Argentina
AS	American Samoa
AT	Austria
AU	Australia
AW	Aruba
AZ	Azerbaijan
BA	Bosnia and Herzegovina
BB	Barbados
BD	Bangladesh
BE	Belgium
BF	Burkina Faso
BG	Bulgaria
BH	Bahrain
BI	Burundi
BJ	Benin
BM	Bermuda
BN	Brunei Darussalam
BO	Bolivia
BR	Brazil
BS	Bahamas
BT	Bhutan
BV	Bouvet Island
BW	Botswana
BY	Belarus
BZ	Belize
CA	Canada
CC	Cocos (Keeling) Islands
CF	Central African Republic
CG	Congo
CH	Switzerland
CI	Côte d'Ivoire (Ivory Coast)
CK	Cook Islands
CL	Chile
CM	Cameroon
CN	China
CO	Colombia
CR	Costa Rica

CS	Czechoslovakia (former)		IR	Iran
CU	Cuba		IS	Iceland
CV	Cape Verde		IT	Italy
CX	Christmas Island		JM	Jamaica
CY	Cyprus		JO	Jordan
CZ	Czech Republic		JP	Japan
DE	Germany		KE	Kenya
DJ	Djibouti		KG	Kyrgyzstan
DK	Denmark		KH	Cambodia
DM	Dominica		KI	Kiribati
DO	Dominican Republic		KM	Comoros
DZ	Algeria		KN	Saint Kitts and Nevis
EC	Ecuador		KP	Korea (North)
EE	Estonia		KR	Korea (South)
EG	Egypt		KW	Kuwait
EH	Western Sahara		KY	Cayman Islands
ER	Eritrea		KZ	Kazakhstan
ES	Spain		LA	Laos
ET	Ethiopia		LB	Lebanon
FI	Finland		LC	Saint Lucia
FJ	Fiji		LI	Liechtenstein
FK	Falkland Islands (Malvinas)		LK	Sri Lanka
FM	Micronesia		LR	Liberia
FO	Faroe Islands		LS	Lesotho
FR	France		LT	Lithuania
FX	France, Metropolitan		LU	Luxembourg
GA	Gabon		LV	Latvia
GB	Great Britain (UK)		LY	Libya
GD	Grenada		MA	Morocco
GE	Georgia		MC	Monaco
GF	French Guiana		MD	Moldova
GH	Ghana		MG	Madagascar
GI	Gibraltar		MH	Marshall Islands
GL	Greenland		MK	Macedonia
GM	Gambia		ML	Mali
GN	Guinea		MM	Myanmar
GP	Guadeloupe		MN	Mongolia
GQ	Equatorial Guinea		MO	Macau
GR	Greece		MP	Northern Mariana Islands
GS	S. Georgia and S. Sandwich Isls.		MQ	Martinique
GT	Guatemala		MR	Mauritania
GU	Guam		MS	Montserrat
GW	Guinea-Bissau		MT	Malta
GY	Guyana		MU	Mauritius
HK	Hong Kong		MV	Maldives
HM	Heard and McDonald Islands		MW	Malawi
HN	Honduras		MX	Mexico
HR	Croatia (Hrvatska)		MY	Malaysia
HT	Haiti		MZ	Mozambique
HU	Hungary		NA	Namibia
ID	Indonesia		NC	New Caledonia
IE	Ireland		NE	Niger
IL	Israel		NF	Norfolk Island
IN	India		NG	Nigeria
IO	British Indian Ocean Territory		NI	Nicaragua
IQ	Iraq		NL	Netherlands

NO	Norway		TR	Turkey
NP	Nepal		TT	Trinidad and Tobago
NR	Nauru		TV	Tuvalu
NT	Neutral Zone		TW	Taiwan
NU	Niue		TZ	Tanzania
NZ	New Zealand (Aotearoa)		UA	Ukraine
OM	Oman		UG	Uganda
PA	Panama		UK	United Kingdom
PE	Peru		UM	US Minor Outlying Islands
PF	French Polynesia		US	United States
PG	Papua New Guinea		UY	Uruguay
PH	Philippines		UZ	Uzbekistan
PK	Pakistan		VA	Vatican City State (Holy See)
PL	Poland		VC	Saint Vincent and the Grenadines
PM	St. Pierre and Miquelon		VE	Venezuela
PN	Pitcairn		VG	Virgin Islands (British)
PR	Puerto Rico		VI	Virgin Islands (U.S.)
PT	Portugal		VN	Viet Nam
PW	Palau		VU	Vanuatu
PY	Paraguay		WF	Wallis and Futuna Islands
QA	Qatar		WS	Samoa
RE	Reunion		YE	Yemen
RO	Romania		YT	Mayotte
RU	Russian Federation		YU	Yugoslavia
RW	Rwanda		ZA	South Africa
SA	Saudi Arabia		ZM	Zambia
SB	Solomon Islands		ZR	Zaire
SC	Seychelles		ZW	Zimbabwe
SD	Sudan			
SE	Sweden			
SG	Singapore			
SH	St. Helena			
SI	Slovenia			
SJ	Svalbard and Jan Mayen Islands			
SK	Slovak Republic			
SL	Sierra Leone			
SM	San Marino			
SN	Senegal			
SO	Somalia			
SR	Suriname			
ST	Sao Tome and Principe			
SU	USSR (former)			
SV	El Salvador			
SY	Syria			
SZ	Swaziland			
TC	Turks and Caicos Islands			
TD	Chad			
TF	French Southern Territories			
TG	Togo			
TH	Thailand			
TJ	Tajikistan			
TK	Tokelau			
TM	Turkmenistan			
TN	Tunisia			
TO	Tonga			
TP	East Timor			

DOMAINS AND ADDRESSES

A complete Internet address generally consists of a word representing the name of the organization or company, possibly followed by the name of a department or division. This is followed by the top-level domain. Here are some examples:

well.com The Well conferencing system, a business in the U.S.

acm.org The Association for Computing Machinery, a non-profit professional organization

state.gov United States Department of State

berkeley.edu University of California, Berkeley

www2.physics.ox.ac.uk Department of Physics, Oxford University, Oxfordshire, United Kingdom.

To access a service at a given site, the host address is prefixed to indicate the server or service. Most commonly, this is www for World Wide Web. Thus www.well.com indicates the Web server at the well.com host, while ftp.well.com would indicate the ftp (file transfer protocol) server. (In some cases, if there is no prefix, www will be assumed.)

A complete Web address or URL (Uniform Resource Locator) also includes a prefix for the protocol to be used (see WORLD WIDE WEB). Most commonly this is http:// (for hypertext transfer protocol), though most Web browsers will treat this as the default and not require that it be typed. ftp:// can be used to access ftp servers via the Web. Finally, a URL must include the path to the directory that actually contains the HTML document or other resource, as well as its filename. Thus a complete address for a hypothetical user's home page might be:

http://www.BigUniversity.edu/users/tomr/index.html

INTERNAL ADDRESSING

When a Web user types such an address, the Web browser connects to a nearby *name server*. This program translates the name into an IP (Internet Protocol) address. The address consists of four 8-bit numbers called *tuples*, separated by periods. For example, the domain name www.well. com currently translates to 208.178.101.2. The first number represents one of five classes of networks, with the first three classes (A-C) organized according to the number and size of networks and D and E being reserved for one-to-many "broadcast" transmissions and experimentation respectively.

To obtain a domain name, a person or organization contacts one of several registration services accredited by ICANN (the nonprofit Internet Corporation for assigned Names and Numbers). Each name must be unique. Considerable legal disputation has occurred when someone not connected with a company has registered a domain containing that company's name. The tremendous growth of e-commerce has made distinctive or easy-to-remember domain names a scarce and valuable commodity. Foreseeing this, some speculators bought up attractive domains in the hope (sometimes realized) of selling them to corporations at a huge profit. Anti–"domain squatting" laws were passed in reaction. In other cases, disgruntled employees or consumers have registered domains for Web sites critical of major corporations such as airlines and telephone companies. In the courts, this pits the right of free speech against the right of a company to control the use of its name.

EXPANDING THE SYSTEM

The expansion of the Internet has strained the capacity of the existing DNS. The shortage of "name space" is being addressed by the release of IP Version 6, which replaces the 32-bit addresses with 128-bit ones. In addition, in November 2000 ICANN announced the creation of seven new top-level domains: .aero (air transport), .biz (business), .coop (cooperatives), .info (general-purpose), .museum (museums), .name (personal sites), and .pro (professionals such as lawyers, accountants, and physicians). However, the situation is muddled by the existence of competing proposals and the use of unofficial DNS systems that provide their own domains (but require special software for access, since they are not recognized by regular DNS servers).

Perhaps a more fundamental issue is the adopting of a system designed by English speakers to a world that increasingly seeks international access and standards (see INTERNATIONALIZATION). The problem is how to meet local needs without creating new barriers through incompatible addressing schemes. The proposal being implemented as of

the mid-2000s is called Internationalizing Domain Names in Applications (IDNA). This standard includes an algorithm by which address labels written using the many character sets and diacritical marks in the world's languages (as rendered in Unicode) can be translated to the standard ASCII characters used by the existing DNS (see CHARACTERS AND STRINGS).

Further Reading
Albitz, Paul, and Cricket Liu. *DNS and BIND*, 4th ed. Sebastopol, Calif.: O'Reilly, 2001.
ICANN (Internet Corporation for Assigned Names and Numbers). Available online. URL: http://www.icann.org/. Accessed August 14, 2007.
"Internationalization of Domain Names: A History of Technology." Internet Society. www.isoc.org/pubpolpillar/docs/i18n-dns-chronology.pdf. Accessed July 10, 2007.
InterNIC. Available online. URL: http://www.internic.net/. Accessed August 14, 2007.
InterNIC Registry WhoIs [domain look-up]. Available online. URL: http://www.internic.net/whois.html. Accessed August 14, 2007.

documentation of program code

Computer system documentation can be divided into two main categories based upon the intended audience. Manuals and training materials for users focus on explaining how to use the program's features to meet the user's needs (see DOCUMENTATION, USER). This entry, however, focuses on the creation of documentation for programmers and others involved in software development and maintenance (see also TECHNICAL WRITING).

Software documentation can consist of *comments* describing the operation of a line or section of code. Early programming with its reliance on punched cards had only minimal facilities for incorporating comments. (Some of the proponents of COBOL thought that the language's English-like syntax would make additional documentation unnecessary. Like the similar claim that trained programmers would no longer be needed, the reality proved otherwise.)

After the switch from punchcard input to the use of keyboards, adding comments became easier. For example, a comment in C looks like this:

```
printf("Hello, world\n");
    /* Display the traditional message */
```

while C++ uses comments in this form:

```
cout << "Hello, World";
    // This is also a comment
```

Each language provides a particular symbol or set of symbols for separating comments from executable code. The compiler ignores comments when compiling the program.

While proper commenting can help people understand a program's functions, the coding style should also be one that promotes clarity. This includes the use of descriptive and consistent names for variables and functions. This can also be influenced by the conventions of the operating system: For example, Windows has many special data structures that should be used consistently.

In addition to the commented source code, external documentation is usually provided. Design documents can range from simple flowcharts or outlines to detailed specifications of the program's purpose, structure, and operations. Rather than being considered an afterthought, documentation has been increasingly integrated into the practice of software engineering and the software development process. This practice became more prevalent during the 1960s and 1970s when it became clear that programs were not only becoming larger and more complex, but also that significant programs such as business accounting and inventory applications were likely to have to be maintained or revised for perhaps decades to come. (The lack of adequate documentation of date-related code in programs of this vintage became an acute problem in the late 1990s. See Y2K PROBLEM.)

DOCUMENTATION TOOLS

As programmers began to look toward developing their craft into a more comprehensive discipline, advocates of structured programming placed an increased emphasis not only on proper commenting of code but on the development of tools that could automatically create certain kinds of documentation from the source code. For example, there are utilities for C, C++, and Java (javadoc) that will extract information about class declarations or interfaces and format them into tables. Most software development environments now include features that cross-reference "symbols" (named variables and other objects). The combination of comments and automatically generated documentation can help with maintaining the program as well as being helpful for creating developer and user manuals.

While programmers retain considerable responsibility for coding standards and documentation, larger programming staffs typically have specialists who devote their full time to maintaining documentation. This includes the logging of all program change requests and the resulting new distributions or "patches," the record of testing and retesting of program functions, the maintenance of a "version history," and coordinating with technical writers in the production of revised manuals.

Further Reading
Barker, Thomas T. *Writing Software Documentation: A Task-Oriented Approach.* 2nd ed. New York: Longman, 2002.
Goodliffe, Pete. *Code Craft: The Practice of Writing Excellent Code.* San Francisco: No Starch Press, 2007.
Knuth, Donald E. *Literate Programming.* Stanford, Calif.: Center for the Study of Language and Information, 1992.
Rüping, Andreas. *Agile Documentation: A Pattern Guide to Producing Lightweight Documents for Software Projects.* Hoboken, N.J.: Wiley, 2003.
Society for Technical Communication. Available online. URL: http://www.stc.org/. Accessed July 9, 2007.

documentation, user

As computing moved into the mainstream of offices and schools beginning in the 1980s and accelerating through the 1990s, the need to train millions of new computer users spawned the technical publishing industry. In addition to the manual that accompanied the software, third-party publishers produced full-length books for beginners and advanced users as well as "dictionaries" and reference manuals (see also TECHNICAL WRITING). A popular program such as WordPerfect or (today) Adobe Photoshop can easily fill several shelves in the computer section of a large bookstore.

A number of publishers targeted particular audiences and adopted distinctive styles. Perhaps the best known is the IDG "Dummies" series, which eventually diversified its offerings from computer-related titles to everything from home remodeling to investing. Berkeley, California, publisher Peachpit Press created particularly accessible introductions for Windows and Macintosh users. At the other end of the spectrum, publishers Sams, Osborne, Waite Group, and Coriolis targeted the developer and "power user" community and the eclectic, erudite volumes from O'Reilly grace the bookshelves of many UNIX users.

ONLINE DOCUMENTATION

During the 1980s, the lack of a multitasking, window-based operating system limited the ability of programs to offer built-in (or "online") documentation. Traditionally, users could press the F1 key to see a screen listing key commands and other rudimentary help. However, both the Macintosh and Windows-based systems of the 1990s included the ability to incorporate a standardized, hypertext-based help system in any program. Users could now search for help on various topics and scroll through it while keeping their main document in view. Another facility, the "wizard," offered the ability to guide users step by step through a procedure.

The growth of the use of the Web has provided a new avenue for online help. Today many programs link users to their Web site for additional help. Even help files stored on the user's own hard drive are increasingly formatted in HTML for display through a Web browser. Additional sources of help for some programs include training videos and animated presentations using programs such as PowerPoint.

By the late 1990s, printed user manuals were becoming a less common component in software packages. (Instead, the manual was often provided as a file in the Adobe Acrobat format, which reproduces the exact appearance of printed material on the screen.) The computer trade book industry has also declined somewhat, but the bookstore still offers plenty of alternatives for users who are more comfortable with printed documentation.

Further Reading
Casabona, Helen. "From Good to Great—The Finer Points of Writing User Documentation." www.stc.org/confproceed/1995/PDFs/PG437440.PDF. Accessed July 9, 2007.
"How to Publish a Great User Manual." Available online. URL: http://www.asktog.com/columns/017ManualWriting.html. Accessed July 9, 2007.
Kukulska-Hulme, Agnes. *Language and Communication: Essential Concepts for User Interface and Documentation Design.* New York: Oxford University Press, 1999.
Online Technical Writing [textbook]. Available online. URL: http://www.io.com/~hcexres/textbook/acctoc.html. Accessed July 9, 2007.

Society for Technical Communication. Available online. URL: http://www.stc.org/. Accessed July 9, 2007.

document model

Most early developers and users of desktop computing systems thought in terms of application programs rather than focusing on the documents or other products being created with them. From the application point of view, files are opened or created, content (text or graphics) is created, and the file is then saved. There is no connection between the files except in the mind of the user. The dominant word processors of the 1980s (such as WordStar and WordPerfect) were designed as replacements for the typewriter and emphasized the efficient creation of text (see WORD PROCESSING). Users who wanted to work with other types of information had to run completely separate applications, such as dBase for databases or Lotus 1-2-3 for spreadsheets. Working with graphics images (to the extent it was possible with early PCs) required still other programs.

This "application-centric" way of thinking suited program developers at a time when most computer systems (such as those running MS-DOS) could run only one program at a time. But increasing processor power, memory, and graphics display capabilities during the late 1980s made it possible to create an operating system such as Microsoft Windows that could display text fonts and formatting, graphics and other content in the same window, and run several different program windows at the same time (see MULTITASKING). In turn, this made it possible to present a model that was more in keeping with the way people had worked in the precomputer era.

In the new "document model," instead of thinking in terms of individual application programs working with files, users could think in terms of creating documents. A document (such as a brochure or report) could contain formatted text, graphics, and data brought in from database or spreadsheet programs. This meant that in the course of working with a document users would actually be invoking the services of several programs: a word processor, graphics editor, database, spreadsheet, and perhaps others. To the user, however, the focus would be on a screen "desktop" on which would be arranged documents (or projects), not on the process of running individual programs and loading files.

IMPLEMENTING THE DOCUMENT MODEL

There are two basic approaches to maintaining documents. One is to create large programs that provide all of the features needed, including word processing, graphics, and data management (see APPLICATION SUITE). While such tight integration can (ideally at least) create a seamless working environment with a consistent user interface, it lacks flexibility. If a user needs capabilities not included in the suite (such as, perhaps the ability to create an HTML version of the document for the Web), one of two cumbersome procedures would have to be followed. Either the operating system's "cut and paste" facilities might be used to copy data from another application into the document (possibly with formatting or other information lost in the process), or possibly the document could be saved in a file format that could be read by the program that was to provide the additional functionality (again with the possibility of losing something in the translation).

LINKING AND EMBEDDING

A more sophisticated approach is to create a protocol that applications could use to call upon one another's services. The Windows COM (Component Object Model) uses a technology formerly called OLE (Object Linking and Embedding). Using this facility, someone working on a document in Microsoft Word could "embed" another object such as an Excel spreadsheet or an Access database into the current document (which becomes the *container*). When the user double-clicks on the embedded object, the appropriate application is launched automatically, and the user sees the screen menus and controls from that application instead of those in Word. (One can also think of Word in this example being the client and Excel or Access as the server—see CLIENT-SERVER COMPUTING). All work done with the embedded object is automatically updated by the server application and everything is stored in the same document file. Alternatively, an application may be *linked* rather than embedded. In that case, the container document simply contains a pointer to the file in the other application. Whenever that file is changed, all documents that are linked to it are updated. Object embedding thus preserves a document-centric approach but works with any applications that support that facility, regardless of vendor. The Macintosh operating system offers a similar facility. Apple and IBM attempted unsuccessfully to create a competing standard called Open-Doc. This should not be confused with the more recent Open Document standard from the popular open-source application Open Office. Meanwhile Microsoft's COM, gradually introduced during the later 1990s, has been largely superseded by .NET (see MICROSOFT .NET). This reflects a shift in emphasis from a document model (within a sin-

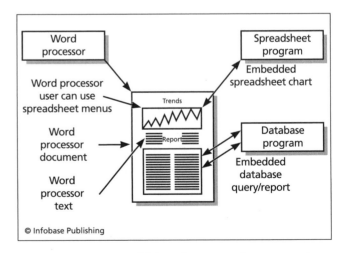

© Infobase Publishing

The Document Object Model (DOM) treats a Web page as an object that can be manipulated using a variety of scripting languages.

gle computer) to a more comprehensive "network object model."

Document and object models are also increasingly important for working on the Web. This can be seen in the increasing use of XML documents and the Document Object Model (see XML and DOM). This involves the use of a consistent programming interface (see API) by which many applications can create or process XML documents for data communication or display.

Further Reading
Bornestein, Niel M. *.NET and XML*. Sebastapol, Calif.: O'Reilly Media, 2003.
"Document Object Model (DOM)." World Wide Web Consortium. Available online. URL: http://www.w3.org/DOM/. Accessed August 12, 2007.
Lowry, Juval. *Programming .NET Components*. 2nd ed. Sebastapol, Calif.: O'Reilly Media, 2005.
"Open Document Format for Office Applications: OASIS Standard." Available online. URL: http://docs.oasis-open.org/office/v1.0. Accessed July 10, 2007.

DOM (Document Object Model)

The Document Object Model (DOM) is a way to represent a Web document (see HTML and XML) as an object that can be manipulated using code in a scripting language (see JAVASCRIPT). The DOM was created by the World Wide Web Consortium (W3C) as a way to standardize methods of manipulating Web pages at a time when different browsers used different access models. The full specification is divided into four levels (0 through 3). By 2005, most DOM specifications were supported by the major Web browsers.

Using DOM, a programmer can navigate through the hierarchical structure of a document, following links or "descending" into forms and user-interface objects. With DOM one can also add HTML or XML elements, as well as load, save, or format documents.

Code can also be written to respond to a number of "events," including user keyboard or mouse activity and interactions with specific user-interface elements and HTML forms. For example, the "mouseover" event will be triggered when the user moves the mouse cursor over a defined region. The code can then perform an action such as popping up a box with explanatory text. The "submit" event will be triggered when the user has finished filling in a form and clicked the button to send it to the Web site. When an event occurs, the event object is used to pass detailed information about it to the program, such as which key or button was pressed, the location of the mouse pointer, and so on.

Although learning the DOM methods and how to use them takes some time, and familiarity with JavaScript is helpful, the syntax for accessing DOM methods should be familiar to anyone who has used an object-oriented programming language. Here are some simple sample statements.

```
Get the document with the specified ID:
  document.getElementById(ID)
Get the element with the specified tag:
  document.getElementByTagName(tagname)
```

```
Get the specified attribute (property) of
the specified element:
  myElement.getAttribute(attributeName)
Create an element with the specified tag and
reference it through a variable:
  var myElementNode = document.
  createElement(tagname)
```

EVALUATION
Although dynamic HTML (DHTML) also has an object model that can be used to access and manipulate individual elements, DOM is more comprehensive because it provides access to the document as a whole and the ability to navigate through its structure.

By providing a uniform way to manipulate documents, DOM makes it easier to write tools to process them in a series of steps. For example, database programs and XML parsers can produce DOM document "trees" as output, and an XSLT (XML style sheet processor) can then be used to format the final output.

For working with XML, another popular alternative is the Simple API for XML (SAX). The SAX model is quite different from DOM in that the former "sees" a document as a stream of events (such as element nodes) and the parser is programmed to call methods as events are encountered. DOM, on the other hand, is not a stream but a tree that can be entered arbitrarily and traversed in any direction. On the other hand, SAX streams do not require that the entire document be held in memory, and processing can sometimes be faster.

Further Reading
Document Object Model FAQ. World Wide Web Consortium. Available online. URL: http://www.w3.org/DOM/faq.html. Accessed September 16, 2007.
Heilmann, Christian. *Beginning JavaScript with DOM Scripting and Ajax: From Novice to Professional*. Berkeley, Calif.: APress, 2006.
Keith, Jeremy. *DOM Scripting: Web Design with JavaScript and the Document Object Model*. Berkeley, Calif.: APress, 2005.
Robie, Jonathan. "What Is the Document Object Model?" World Wide Web Consortium. Available online. URL: http://www.w3.org/TR/WD-DOM/introduction.html. Accessed September 14, 2007.
Sambells, Jeffrey, and Aaron Gustafson. *Advanced DOM Scripting: Dynamic Web Design Techniques*. Berkeley, Calif.: APress, 2007.

DOS *See* MS-DOS.

Dreyfus, Hubert
(1929–)
American
Philosopher, Cognitive Psychologist

As the possibilities for computers going beyond "number crunching" to sophisticated information processing became clear starting in the 1950s, the quest to achieve artificial intelligence (AI) was eagerly embraced by a number of innovative researchers. For example, Allen Newell, Herbert Simon, and Cliff Shaw at the RAND Corporation, attempted to write programs that could "understand" and intelligently manipulate symbols rather than just literal numbers or characters. Similarly, MIT's Marvin Minsky (see MINSKY, MARVIN) was attempting to build a robot that could not only perceive its environment, but in some sense understand and manipulate it. (See ARTIFICIAL INTELLIGENCE and ROBOTICS.)

Into this milieu came Hubert Dreyfus, who had earned his Ph.D. in philosophy at Harvard. Dreyfus had specialized in the philosophy of perception (how meaning can be derived from a person's environment) and phenomenology (the understanding of processes). When Dreyfus began to teach a survey course on these areas of philosophy, some of his students asked him what he thought of the artificial intelligence researchers who were taking an experimental and engineering approach to the same topics the philosophers had discussed abstractly.

Philosophy had attempted to explain the process of perception and understanding (see also COGNITIVE SCIENCE). One tradition, the rationalism represented by such thinkers as Descartes, Kant, and Husserl took the approach of formalism and attempted to elucidate rules governing the process. They argued that in effect the human mind was a machine (albeit a wonderfully complex and versatile one). The opposing tradition, represented by the phenomenologists Wittgenstein, Heidegger, and Merleau-Ponty, took a holistic approach in which physical states, emotions, and experience were inextricably intertwined in creating the world that people perceive and relate to.

If computers, which at that time had only the most rudimentary "senses" and no emotions could perceive and understand in the way humans did, then the rules-based approach of the rationalist philosophers would be vindicated. But when Dreyfus had examined the AI efforts, he wrote a paper titled "Alchemy and Artificial Intelligence." His comparison of AI to alchemy was provocative in that it suggested that like the alchemists, the modern AI researchers had met with only limited success in manipulating their materials (such as by teaching computers to perform such intellectual tasks as playing checkers and even proving mathematical theorems). However, Dreyfus concluded that the kind of flexible, intuitive, and ultimately robust intelligence that characterizes the human mind couldn't be matched by any programmed system. Each time AI researchers demonstrated the performance of some complex task, Dreyfus examined the performance and concluded that it lacked the essential characteristics of human intelligence. Dreyfus expanded his paper into the book *What Computers Can't Do.* Meanwhile, critics complained that Dreyfus was moving the goal posts after each play, on the assumption that "if a computer did it, it must not be true intelligence."

Two decades later, Dreyfus reaffirmed his conclusions in *What Computers Still Can't Do,* while acknowledging that the AI field had become considerably more sophisticated in creating systems of emergent behavior (such as neural networks).

Currently a professor in the Graduate School of Philosophy at the University of California, Berkeley, Dreyfus continues his work in pure philosophy (including a commentary on phenomenologist philosopher Martin Heidegger's *Being and Time*) while still keeping an eye on the computer world in his latest publication, *On the Internet.*

Further Reading
Dreyfus, Hubert. *What Computers Can't Do: A Critique of Artificial Reason.* New York: Harper and Row, 1972.
———. *What Computers Still Can't Do.* Cambridge, Mass.: MIT Press, 1992.
Dreyfus, Hubert, and Stuart Dreyfus. *Mind over Machine: the Power of Human Intuitive Expertise in the Era of the Computer.* Rev. ed. New York: Free Press, 1988.
Henderson, Harry. *Artificial Intelligence: Mirrors for the Mind.* New York: Chelsea House, 2007.

DRM   *See* DIGITAL RIGHTS MANAGEMENT.

DSL (digital subscriber line)

DSL (digital subscriber line) is one of the two most prevalent forms of high-speed wired access to the Internet (see BROADBAND and CABLE MODEM). DSL can operate over regular phone lines (sometimes called POTS or "plain old telephone service"). DSL takes advantage of the fact that existing phone lines can carry frequencies far beyond the narrow band used for voice telephony. When installing DSL, the phone company must evaluate the quality of existing lines to determine how many frequency bands are usable, and thus how much data can be transmitted. Further, because the higher the frequency the shorter the distance the signal can travel, the available bandwidth drops as one gets farther from the central office or a local DSL access Multiplexer (DSLAM).

Typical DSL services can range in speed from 128 kbps to 3 Mbps. Many providers offer higher speeds at additional cost. Speeds quoted are generally maximums; actual speed may be less due to poor line quality or greater distance from the central office.

The most common form of DSL is ADSL (asymmetric DSL), which has much higher download speeds than upload speeds. This is generally not a problem, since most users consume much more content than they generate. The lower frequencies are generally reserved for regular voice and fax service. A single DSL modem can serve multiple users in a local network by being connected to a router.

As more people move from land-line phone service to cellular, there has been greater demand for offering so-called naked DSL—DSL without traditional phone service. DSL can also be provided over optical fiber (see FIBER OPTICS).

Note that an older and lower-bandwidth version of the technology called ISDN (Integrated Services Digital Network) is still in use, but has largely been superseded by DSL/ADSL.

ALTERNATIVES TO DSL
Cable is still more popular than DSL, though the latter has closed the gap somewhat. The fact that the two services can both provide fast Internet access (mostly) through existing

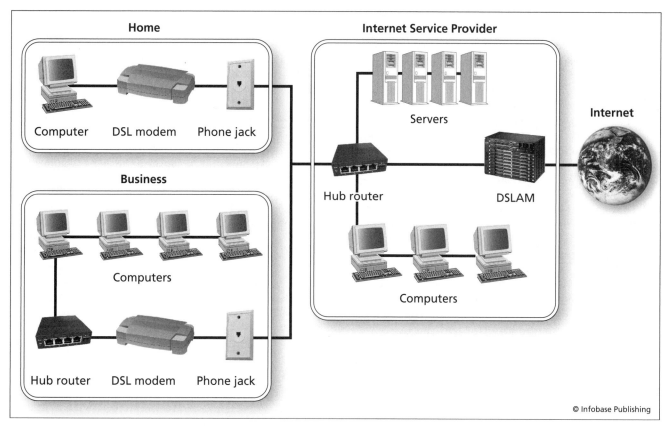

Home

Computer   DSL modem   Phone jack

Business

Computers

Hub router   DSL modem   Phone jack

Internet Service Provider

Servers

Hub router   DSLAM

Internet

Computers

© Infobase Publishing

DSL uses special modems to convert between computer data and signals that can travel over ordinary phone lines. This technology is widely used to provide broadband Internet access.

infrastructure has created considerable competition. Thus a cable provider can now offer telephone service via the Internet (see VOIP) at the same time a phone provider using DSL can offer movies and television programming streamed over the network. The fact that in many locations DSL and cable providers are in competition can result in lower rates or more attractive "bundles" of services for consumers.

On average, cable modem speeds are somewhat faster than DSL; however, cable speeds can degrade as more users are added to a circuit. Although both services have had their share of glitches, they now both tend to be quite reliable.

Further Reading

Golden, Philip, Herve Dedieu, and Krista S. Jacobsen, eds. *Implementation and Applications of DSL Technology*. Boca Raton, Fla.: CRC Press, 2007.

Mitchell, Bradley. "DSL vs. Cable: Modem Comparison." Available online. URL:http://compnetworking.about.com/od/dslvscablemodem/a/dslcablecompare.htm. Accessed September 16, 2007.

Reynolds, Janice. *A Practical Guide to DSL: High-Speed Connections for Local Loop and Network*. New York: CMP Books, 2001.

Smith, Roderick W. *Broadband Internet Connections: A User's Guide to DSL and Cable*. Upper Saddle River, N.J.: Addison-Wesley Professional, 2007.

DTP   *See* DESKTOP PUBLISHING.

DVR (digital video recording)

A digital video recorder (DVR) records digital television broadcasts and stores them on a disk (see HARD DISK and CD-ROM AND DVD-ROM). DVRs first appeared as commercial products in 1999 in Replay TV and TiVo, the latter becoming the most successful player in the field.

A DVR works with digital signals and discs rather than tape used by the video cassette recorders (VCRs) that had become popular starting in the 1980s. The digital recorder has several advantages over tape:

- much larger capacity, limited only by hard drive size

- instant (random) access to any recorded programming without having to go forward or backward through a tape

- the ability to "time shift" within a live broadcast, including pausing and instant replay

- the ability to skip over commercials

- digital special effects

DVR AND INTEGRATED ENTERTAINMENT

Besides what it can do with the program itself, the other big advantage of DVR technology stems from the fact that it produces digital data in a standard format (usually an

MPEG file) that is fully compatible with PCs and other computing devices. Indeed, by installing one or more TV tuners or "cable cards" (for access to digital cable signals) to a PC, one need only add suitable software to turn a Windows, Macintosh, or Linux PC into a versatile DVR. Alternatively, many cable and satellite TV services are offering set-top boxes with built-in DVRs.

Services such as TiVo also provide access to an online program schedule (for a monthly charge). This works with features that allow the user to scan for and review program listings and to arrange, for example, to record all new episodes of a weekly series as they arrive. DVRs with dual tuners allow for recording two live programs simultaneously, or recording one while watching another.

DVR technology is also now being used for closed-circuit television (CCTV) surveillance systems, due to superior storage and playback capabilities. Similar technology is also found in digital video cameras (camcorders).

DVRs are part of a landscape where entertainment that used to be confined to television broadcast, cable, or satellite systems can now be received digitally over the Internet. Since DVRs produce digital output, recorded programs can be easily shared over the Internet, such as by posting on the popular YouTube site, possibly leading to loss of revenue for the original providers (see INTELLECTUAL PROPERTY AND COMPUTING). In response, HBO and other providers have argued for requiring that DVRs recognize content that is flagged as "copy never" and refuse to copy such programs.

Another problem for providers is the growing number of DVR users who have the ability to easily skip over commercials. Attempts are being made to make commercials shorter and more entertaining, or to rely more on product placement within the programming itself.

Further Reading

DVR Buying Guide. Available online. URL: http://products. howstuffworks.com/dvr-buying-guide.htm. Accessed September 16, 2007.

ReplayTV. Available online. URL: http://www.replaytv.com/. Accessed September 16, 2007.

TiVo. Available online. URL: http://www.tivo.com/. Accessed September 16, 2007.

E

eBay

eBay Inc. (NASDAQ symbol: EBAY) is the world's largest online auction and shopping site. The first appearance of the auction service was in 1995 as AuctionWeb, part of the personal Web site of Pierre Omidyar (see AUCTIONS, ONLINE and OMIDYAR, PIERRE). Omidyar was surprised at how rapidly the auction service (which was initially free) grew. After he imposed a modest listing fee, Omidyar found himself receiving thousands of dollars in small checks, and decided that online auctions could become a full-time business.

In September 1997, with Jeff Skoll now on board as president, AuctionWeb officially became eBay. When the company went public in 1998 (at the height of the first "Internet boom"), Omidyar and Skoll became instant millionaires. Meanwhile, eBay took on Margaret (Meg) Whitman as its new CEO, and under her leadership the company has expanded rapidly through its first decade.

eBay also seeks new markets and revenue through strategic acquisitions. These include payment services such as the very popular PayPal, other e-commerce sites such as Half.com, shopping.com, and rent.com, and even Skype, the Internet phone service. eBay's net revenue for 2007 was $7.67 billion.

ONLINE AUCTIONS

Auctions remain at the core of eBay's business, with millions of items in dozens of categories being listed and sold each day. Offerings can range from factory equipment (in the Business & Industrial category) to books, toys, sports memorabilia—even cars and, in a limited fashion, real estate. There are now hundreds of small- to medium-size businesses who derive their revenue from eBay, whether selling their own merchandise, acting as agents for others, or selling software or templates for managing auctions.

eBay does not charge any buyer's fees, but makes its money by charging the seller for each listing and then a percentage of the selling price. As of 2007 eBay has regional operations in more than 20 countries, including China and India. (Yahoo, a distant second to eBay in online auctions, discontinued its U.S. auction site in mid-2007.)

BEYOND AUCTIONS

In recent years eBay has increasingly tried to build a more "traditional" online shopping experience in parallel with its auctions. The Buy It Now feature allows a seller to list an item at a fixed price, either instead of auctioning the item or as an option that can be exercised if there have been no auction bids. Sellers can organize their offerings into regular "stores" to make it easier for customers to browse their merchandise. (Many traditional stores, such as antiques or collectibles dealers, now offer some of their items via their eBay store.) eBay Express, introduced in 2006, adds convenience by allowing users to buy selected items from multiple sellers using a standard online shopping cart.

Like Amazon.com, eBay has focused considerable attention on developing more ways for users to comment on their purchases and otherwise contribute content (see USER-CREATED CONTENT). The most important mechanism is feedback, which lets buyers summarize their opinions of a transaction after its completion. The feedback system has been recently expanded and structured to allow users to give specific ratings on aspects of the transaction, such as accuracy of description and shipping cost and speed. Although not perfect (feedback can be "pumped up" by setting up phony transactions between two accounts), the system does allow buyers to exercise a certain amount of caution before bidding on an expensive item from an unknown seller. eBay also offers various forms of "consumer protection" if items are not received or are substantially not as described.

Not surprisingly in a marketplace of this size, there is opportunity for various forms of fraud, including sale of counterfeit, defective, or lower-grade merchandise and, on the part of buyers, credit card fraud. eBay has been criticized for not policing fraud adequately. Generally, the service has maintained the position that it is only a facilitator of transactions. If it had to guarantee the authenticity of merchandise, it would have to operate like a conventional auction house, with the attendant fees. However, eBay has solicited the help of experts in fields such as coins and stamps to help them identity counterfeit or misdescribed items.

eBay provides a number of forums for user comments, including discussion boards and chat rooms. Users can also write reviews and guides to help, for example, novice collectors who might find themselves overwhelmed by the coin or stamp listings. In mid-2006 eBay expanded its "community content" to include an eBay Community Wiki (see WIKIS AND WIKIPEDIA) and eBay blogs (see BLOGS AND BLOGGING).

eBay is always trying to make it easier to match users' specific needs with the thousands of potentially relevant offerings. Providing recommendation information (including user-generated recommendations) is another way to make shopping easier and more satisfying, as has been shown by Amazon.com. Another possible way to get a bigger share of users' day-to-day purchases is to make eBay available on mobile devices as well as linking it to sites such as Facebook, where young people in particular spend much of their time (see SOCIAL NETWORKING.)

Long-time eBay CEO Meg Whitman stepped down in March 2008, while calling for innovation to reinvigorate a company that many observers now consider to be staid and "old school" in the age of Web 2.0. Whitman's successor, John Donahoe, has announced a new fee structure and new ways of searching for and displaying listings—developments that have provoked some controversy in the seller community.

Further Reading

Bergstein, Brian. "Middle-aged eBay in for Changes." *Associated Press/San Francisco Chronicle,* June 18, 2007. p. C3.

Cohen, Adam. *The Perfect Store: Inside eBay.* Boston: Little, Brown, 2002.

Collier, Marsha. *eBay for Dummies.* 5th ed. Hoboken, N.J.: Wiley, 2006.

———. *Starting an eBay Business for Dummies.* Hoboken, N.J.: Wiley, 2007.

eBay. Available online. URL: http://www.ebay.com. Accessed September 16, 2007.

e-books and digital libraries

An e-book is a book whose text is stored in digital form and can be read on a PC or a handheld reading device. Since most books today are created on word processors and typesetting systems, it is easy for a publisher to create an electronic version. Older books that exist only in printed form can be scanned and converted to text (see OPTICAL CHARACTER RECOGNITION).

An e-book has a number of advantages over its printed counterpart. The text can be searched and can include links to sections or even to documents on the World Wide Web. Reading software or devices can easily enlarge text for the visually handicapped, or read it in a synthesized voice. Since only bits need to be moved around, e-books save trees as well as the cost of manufacturing, transporting, warehousing, and displaying conventional books.

There are some disadvantages. Many people are not comfortable reading large amounts of text at a computer. Portable reading devices that may be more convenient are relatively expensive and not standardized. There is no universal format for e-books, so some software or readers may not be able to read all e-books.

As of 2008 the e-book landscape may be in the process of being reshaped. Amazon's Kindle book reader is the latest attempt to marry e-books to handheld devices. Weighing less than a paperback book, the Kindle can download books and other content directly over a cellular broadband connection and display text using an "electronic ink" technology that simulates print. Amazon is offering a large selection of e-books including electronic versions of current best sellers at prices several dollars below that of the hardback version.

Authors and publishers, like other content creators, may have to deal with the illicit copying and distribution of text in digital form, as happened with the last *Harry Potter* book even before its publication in 2007. Some e-books contain a form of copy protection (see DIGITAL RIGHTS MANAGEMENT). This, as with video and music, can lead to compatibility problems.

A number of e-publishers as well as conventional publishers now offer books online, most commonly as PDF (portable document format) files. A hybrid service, "publish on demand," keeps the book on file and prints and ships bound copies as they are ordered, eliminating the problem of remainders. In the future, so-called digital paper (a thin membrane that can display text), may be used to create a more booklike reading experience.

DIGITAL LIBRARIES

A digital library is to e-books what a conventional library is to printed books. Sometimes called an electronic library or virtual library, digital libraries can be created in a variety of ways. Printed books can now be scanned and digitized rapidly. Google has said that it can scan 3,000 volumes a day using a proprietary system. (This is not necessary, of course, for books that were originally created in digital form.)

Advantages of digital libraries include the following:

- There is never a shortage of copies or the need for a reader to wait for access.

- Many digital libraries allow full searching of the text of all volumes. Libraries can also use a common data format (such as "Open Archives.") to make their material searchable throughout the Internet.

- Many older, hard-to-find books can be made more "discoverable" and accessible.

Project Gutenberg is one of the oldest and best-known digital library projects, dating back to 1971. Most of the collection consists of scanned or transcribed texts of public domain (no longer subject to copyright) books. As of late 2007, Project Gutenberg had more than 17,000 different titles in its collection.

Of course more recent books are covered by copyright. In order to include copyrighted books in a digital library, some sort of compensation to the copyright holder generally needs to be made, and it is unclear how that might be implemented in a way that preserves free access.

There are also what might be called "digital pseudo-libraries" such as Google Book Search. Google has been scanning part or all of the collections of universities such as Stanford, Harvard, and Oxford as well as the New York Public Library. Google provides full access to public domain books (or those for which permission has been obtained from the publisher). For copyrighted books there is a limited ability to search by keyword and view a limited number of pages. Amazon.com's "Search inside the Book" works rather similarly, but only with books for which the publisher has granted permission.

Google's initiative has aroused some controversy because, according to traditional practice, someone wanting access to a copyrighted work beyond "fair use" is supposed to obtain permission. Google has reversed this presumption, allowing publishers who do *not* want their material to be available to opt out. The Authors Guild of America and the Association of American Publishers have separately sued Google for copyright infringement. Google argues that the limited amount of text provided for copyrighted books falls within the fair use provisions of copyright law. The authors and publishers, however, point to the fact that Google is copying the whole text of the book in order to allow for searching.

If the legal issues can be settled in such a way as to allow robust digital libraries, the benefits for researchers will be considerable. Google already offers a "my library" feature that users can use to search for books they already know and organize and search them digitally.

Further Reading

Google Book Search. Available online. URL: http://books.google.com/. Accessed September 16, 2007.

Hirschhorn, Michael. "The Hapless Seed: Publishers and Authors Should Stop Cowering. Google Is Less Likely to Destroy the Book Business Than to Slingshot It into the 21st Century." *Atlantic Monthly,* June 2007, p. 134 ff.

Kelly, Kevin. "Scan This Book!" *New York Times Magazine,* May 14, 2006. pp. 42–49, 64, 71.

Kresh, Diane, ed. *The Whole Digital Library Handbook.* Chicago: American Library Association, 2007.

Lesk, Michael. *Understanding Digital Libraries.* 2nd ed. San Francisco: Morgan Kaufmann, 2004.

Project Gutenberg. Available online. URL: http://www.gutenberg.org/wiki/Main_Page. Accessed September 16, 2007.

Thompson, Bob. "Google Wants to Digitize Every Book. Publishers Say Read the Fine Print First." *Washington Post,* August 12, 2006, p. D1.

Eckert, J. Presper
(1919–1995)
American
Computer Engineer

J. Presper Eckert played a key role in the design of what is often considered to be the first general-purpose electronic digital computer, then went on to pioneer the commercial computer industry. An only child, Eckert grew up in a prosperous Philadelphia family that traveled widely and had many connections with Hollywood celebrities such as Douglas Fairbanks and Charlie Chaplin. He was a star student in his private high school and also did well at the University of Pennsylvania, where he graduated in 1941 with a degree in electrical engineering and a strong mathematics background.

Continuing at the university as a graduate student and researcher, Eckert met an older researcher, John Mauchly. They found they shared a deep interest in the possibilities of electronic computing, a technology that was being spurred by the needs of war research. After earning his master's degree in electrical engineering, in 1942 Eckert joined Mauchly in submitting a proposal to the Ballistic Research Laboratory of the Army Ordnance Department for a computer that could be used to calculate urgently needed firing tables for guns, bombs, and missiles. The Army granted the contract, and they organized a team that grew to 50 people. Begun in April 1943, their ENIAC (Electronic Numerical Integrator and Computer) was finished in 1946. While it was too late to aid the war effort, the room-size machine filled with 18,000 vacuum tubes demonstrated the practicability of electronic computing. Its computation rate of 5,000 additions per second far exceeded other calculators of the time.

With some input from mathematician John von Neumann, Eckert and Mauchly began to develop a new machine, EDVAC, for the University of Pennsylvania (see VON NEUMANN, JOHN). While this effort was still under way, they formed their own business, the Eckert-Mauchly Computer Corporation and began to develop the BINAC (BINary Automatic Computer), which was intended to be a (relatively) compact and lower-cost version of ENIAC. This machine demonstrated a key principle of modern computers—the storage of program instructions along with data. The ability to store, manipulate, and edit instructions vastly increased the flexibility and ease of use of computing machines (see HISTORY OF COMPUTING).

By the late 1940s, Eckert and Mauchly began to develop Univac I, the first commercial implementation of the new computing technology. When financial difficulties threatened to sink their company in 1950, it was acquired by

Remington Rand. Working as a division within that company, the Eckert-Mauchly team completed Univac I in time for the computer to make a remarkably accurate forecast of the 1952 presidential election results.

Eckert continued with the Sperry-Rand Corporation (later called Univac and then Unisys Corporation) and became a vice president and senior technical adviser. He retired in 1989. He received an honorary doctorate from the University of Pennsylvania in 1964. In 1969, he was awarded the National Medal of Science, the nation's highest award for achievement in science and engineering.

Further Reading

Eckstein, P. "Presper Eckert." *IEEE Annals of the History of Computing* 18, vol. 1, Spring 1996, 25–44.
McCartney, Scott. *Eniac: the Triumphs and Tragedies of the World's First Computer.* New York: Berkley Books, 1999.
Smithsonian Institution. National Museum of American History. "Presper Eckert Interview." Available online. URL: http://americanhistory.si.edu/collections/comphist/eckert.htm. Accessed August 14, 2007.

e-commerce

Since the introduction of credit cards and the beginning of banking automation in the 1960s, computers and communications networks have played an increasing role in the infrastructure of commerce (see BANKING AND COMPUTERS). Some businesses also established proprietary networks (for example, to allow pharmacies to order drugs directly from suppliers).

Electronic sales directly to consumers were pioneered by "teletex," such as the French Minitel, as well as such services as CompuServe and America Online. However, these services were proprietary, meaning that businesses could only market to subscribers. The widespread adoption of the Internet in the mid-1990s (see WORLD WIDE WEB) created an open and potentially much larger marketplace.

The first e-commerce boom came in the late 1990s, when enthusiasm about the seeming potential for unlimited profits drove numerous online startups, often with poorly conceived business plans that assumed that rapid expansion and low prices would result in gaining control of a particular sector and achieving a dominant (and profitable) position. Among the numerous casualties of the "dot-bust" of 2000–2001 was WebVan, a company that sold and delivered groceries directly to consumer's homes.

While the bursting of the "dot-com bubble" was painful to investors, entrepreneurs, and workers, recovery was soon underway. The recovery was aided by the steady growth of Internet users (particularly those with broadband connections), innovative software for interacting with consumers and analyzing transaction information, and the coming of age of a generation that had virtually grown up online.

Today e-commerce is a steadily growing sector, and it is increasingly international, fed by nearly 1.5 billion Internet users worldwide. (China, with more than 250 million Web users, has become the world's largest online market.)

Meanwhile in the United States in 2007 total consumer retail sales on the Internet reached $136 billion, up nearly 20 percent over the previous year. According to a report from Forrester Research, online retail revenues (excluding travel-related services) will pass $250 billion by 2011. Surveys show that about 80 percent of American Internet users have bought something online, while many users who buy products off-line originally searched for information about them online.

The most popular e-commerce sectors today include the selling of books, music and movies, travel-related services, electronics, clothing, luxury goods, and medications. (In 2006, online buyers actually spent more money on clothing than on computers and related products.) A number of other online activities can be considered part of e-commerce, although they are usually not included in retailing statistics (see AUCTIONS, ONLINE; ONLINE GAMBLING; ONLINE GAMES; and SOCIAL NETWORKING).

INFRASTRUCTURE

Successful e-commerce depends on a complex array of services, facilities, and software. For marketing and consumer communications, see ONLINE ADVERTISING and CUSTOMER RELATIONSHIP MANAGEMENT. Behind the scenes, transaction data is constantly being collected and analyzed to determine the success of the marketing program and to "personalize" the customer experience and allow for targeted marketing (see COOKIES and DATA MINING).

The actual transaction processing requires shopping cart software and a connection to the credit card processing infrastructure (see DIGITAL CASH). Specialized forms of selling require additional software and support systems (see, for example, AUCTIONS, ONLINE). An ongoing e-business must also deal with functions shared by "brick and mortar" (traditional) stores: inventory control, ordering from suppliers (see SUPPLY CHAIN MANAGEMENT), taxes, payroll, and so on. The broader e-commerce sector also includes businesses that do not target consumers but, rather, the needs of business itself—so-called business to business or B2B.

SECURITY AND PRIVACY

One continuing obstacle to the growth of e-commerce has been consumers' concerns about the theft or misuse of personal information gathered as part of the shopping process. This can involve either fake Web sites (see PHISHING AND SPOOFING) or legitimate businesses that sell information about customers without their knowledge or consent (see PRIVACY IN THE DIGITAL AGE). According to a report from Gartner Research, more than $900 million in e-commerce sales during 2006 was lost because of consumers' security concerns, and about a billion dollars more in sales was lost because customers decided not to buy online at all.

TRENDS

E-commerce is maturing even as it continues to evolve. Some trends in the second half of the 2000 decade reflect changes in what is presented to the consumer, how it is delivered, and how users can participate in ways other than simply viewing content and selecting products:

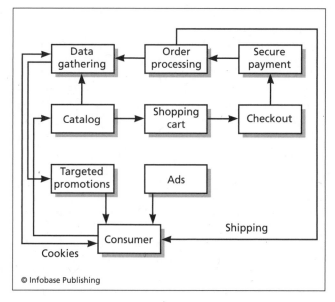

© Infobase Publishing

E-commerce involves far more than just advertising and selling goods and services. In a typical e-commerce system a "shopping cart" records consumers' selections. The items ordered must be processed against inventory and prepared for shipping. Meanwhile, information about the user's selections and viewing is fed into a database from which patterns of consumer behavior can be extracted. Some techniques of information gathering raise privacy concerns, however.

- delivery of richer and more interactive multimedia experience, catering to the widespread availability of broadband connections

- integration of marketing using programming interfaces (see MASHUPS) with popular online services such as Google Maps, online game worlds, and social networking sites (see ONLINE GAMES and SOCIAL NETWORKING)

- increasing participation of consumers in developing the quality of the shopping experience, such as through user product reviews and blogs (see USER-CREATED CONTENT)

- increased emphasis on serving rapidly growing foreign markets, such as India and China

- the spread of e-commerce to new mobile platforms (see PDA and SMARTPHONE)

Further Reading

Combe, Colin. *Introduction to e-Business: Management and Strategy.* Burlington, Mass.: Elsevier, 2006.
eCommerce Info Center. Available online. URL: http://www.ecominfocenter.com/. Accessed July 15, 2007.
eCommerce Times. Available online. URL: http://www.ecommercetimes.com. Accessed July 16, 2007.
Forrester Research. Available online. URL: http://www.forrester.com. Accessed July 10, 2007.
Grant, Graeme. "Trends in 2007 Online Retailing." Available online. URL: http://www.destinationcrm.com/articles/default.asp?ArticleID=6938. Accessed July 10, 2007.
Jupiter Research. Available online. URL: http://www.jupiterresearch.com. Accessed July 10, 2007.
Laudon, Kenneth, and Carol Traver. *E-Commerce: Business, Technology, Society.* 3rd ed. Upper Saddle River, N.J.: Prentice Hall, 2006.

education and computers

Computers are widely used in educational institutions from elementary school to college. While computers have had as yet little impact on the structure or organization of schools, educational software and the use of the Internet has had a growing impact on how education is delivered.

HISTORY

During the 1950s and early 1960s, computer resources were generally too scarce, expensive, and cumbersome to be used for teaching, although universities aspired to have computers to aid their graduate and faculty researchers. However, during the 1960s computer engineers and educators at the Computer-based Education Research Laboratory at the University of Illinois, Urbana, formed a unique collaboration and designed a computer system called PLATO. The PLATO system used mainframe computers to deliver instructional content to up to 1,000 simultaneous users at terminals throughout the University of Illinois and other educational institutions in the state. PLATO pioneered the interactive approach to instruction and the use of graphics in addition to text. The PLATO system was later marketed by Control Data Corporation (CDC) for use elsewhere. During this time Stanford University also set up a system for delivering computer-assisted instruction (CAI) to users connected to terminals throughout the nation. (See COMPUTER-AIDED INSTRUCTION.)

By the early 1980s, microcomputers had become relatively affordable and capable of running significant educational software including graphics. Apple Computer's Apple II became an early leader in the school market, and the introduction of the Macintosh in 1984 with the Hypercard scripting language inspired many teachers and other enthusiasts to create their own educational software. By the early 1990s, IBM compatible PCs with Windows were catching up. Commercially available computer games (such as *Civilization* or *Railroad Tycoon*) also offered ways to enrich social studies and other classes (see COMPUTER GAMES).

The advent of the World Wide Web and graphical Web browsing in the mid-1990s spurred schools to connect to the Internet. The Web offered the opportunity for educators to create resources that could be accessed by colleagues and students anywhere in the world. The use of Web portals such as Yahoo!, library catalogs, and online encyclopedias gave teachers and students potential access to a far greater variety of information than could possibly be found in textbooks. The Web also offered the opportunity for students at different schools to participate in collaborative projects, such as community surveys or environmental studies.

APPLICATIONS

Educational applications of computing can be divided into several categories, as summarized in the following table.

While small compared to the business market, the educational software industry is a significant market, targeting both schools and parents seeking to improve their children's academic performance. However, the educational use of computers extends far beyond specialized software. Schools are in effect a major industry in themselves, requiring much of the same support software as large businesses.

TRENDS

The growth of the World Wide Web has led to some shift of emphasis away from stand-alone, CD-ROM based applications running on local PCs or networks. Educators are excited about the possibilities for online collaboration. Public concern about children achieving an adequate level of technical skill (see COMPUTER LITERACY) has fueled an increasing commitment of funds for computer hardware, software, and networking for schools.

Some visionaries speak of a 21st-century "virtual school" that has no classroom in the conventional sense, but uses the Internet and conferencing software to bring teachers and students together. While there has been only limited experimentation in creating virtual secondary schools, thousands of university courses are now offered online, and many degree programs are now available. Some institutions such as the University of Phoenix have made such "distance learning" a core part of their growth strategy.

Several factors have caused other observers to have misgivings about the rush to get schools onto the "information superhighway." Many schools lack adequate physical facilities and teacher training. Under those circumstances other priorities might deserve precedence over the installation of technology that may not be effectively utilized. At the same time, the lagging in access to technology by minorities and the poor may suggest that schools must play a significant role in providing such access and enabling the coming generation to catch up (see DIGITAL DIVIDE).

The debate over how best to use technology in the schools also reflects fundamental theories about teaching and learning. Critics of information technology such as Clifford Stoll (see STOLL, CLIFFORD) have reacted against the mechanical, rote nature of much educational software. They also decry the hype of some advocates who have suggested technology as a panacea for the problems of low performance, poor motivation, and lack of accountability in many schools.

Some advocates of computer use agree with the criticism of uncreative and poorly planned "e-learning" programs, but argue that the answer is to use technology that helps good teachers unlock creativity. For example, Seymour Papert and his LOGO language are based on "constructivist" principles where students learn through doing (see PAPERT, SEYMOUR and LOGO). From this point of view, "computer literacy" should not be a focus in itself, but one outcome of a program that creates literate and capable learners (see COMPUTER LITERACY.)

APPLICATION AREA	USERS	EXAMPLES
Computer-aided instruction (CAI)	Generally high school and up	Course modules for science, social studies, etc. Students evaluated and materials presented on the basis of student performance (see COMPUTER-AIDED INSTRUCTION).
Drill-and-practice	Elementary school students	Sets of math problems, geography quizzes, etc. Sounds or graphics used for reward for correct answers.
Online collaborative learning	Elementary and high school students	Students from different schools use e-mail or chat to coordinate a project, such as creating a Web site about local environmental issues.
Online classes	Mainly college and adults	Students participate remotely through videoconferencing, chat, e-mail, etc. (see DISTANCE EDUCATION).
Educational simulations	Junior high and older students	Gamelike programs that simulate real-world problems, such as managing a city to investing in the stock market. Often commercially available games can be used.
Reference and resources	Elementary and older students	Online encyclopedias and knowledge bases (see also WIKIS AND WIKIPEDIA); specialized references on CD-ROM or DVD; online reference and bibliographical databases; library catalogs; and Web site of universities, museums, and government agencies.
General software applications	Students and teachers	Use of general-purpose software such as word processors, publishing, or presentation programs for creating class projects and reports. Also use of e-mail, chat, and blogs for collaboration and after-hours communication between students and teachers.
Administrative applications	Teachers and administrators	Use of specialized or general-purpose software to maintain attendance, grades, and other class and school administration functions.

Further Reading

Cuban, Larry. *Oversold & Underused: Computers in the Classroom.* Cambridge, Mass.: Harvard University Press, 2001.

Global Schoolhouse. Available online. URL: http://www.globalschoolnet.org. Accessed July 19, 2007.

November, Alan. *Empowering Students with Technology.* Arlington, Ill.: SkyLight Professional Development, 2001.

Paley, Amit R. "Software's Benefits on Tests in Doubt: Study Says Tools Don't Raise Scores." *Washington Post,* April 5, 2007, p. 1.

Pflaum, William D. *The Technology Fix: The Promise and Reality of Computers in Our Schools.* Alexander, Va.: Association for Supervision and Curriculum Development, 2004.

Richardson, Will. *Blogs, Wikis, Podcasts, and Other Powerful Web Tools for Classrooms.* Thousand Oaks, Calif.: Corwin Press, 2006.

Stoll, Clifford. *High Tech Heretic: Why Computers Don't Belong in the Classroom and Other Reflections by a Computer Contrarian.* New York: Doubleday, 1999.

"Technology Integration." Education World. Available online. URL: http://www.educationworld.com/a_tech/. Accessed July 19, 2007.

Warlick, David. *Classroom Blogging: A Teacher's Guide to the Blogosphere.* Morrisville, N.C.: Lulu.com, 2005.

education in the computer field

Education and training in computer-related fields runs the gamut from courses in basic computer concepts in adult education or junior college programs to postgraduate programs in computer science and engineering. Curricula can be roughly divided into the following areas

- computer literacy and applications
- computer science
- information systems

COMPUTER LITERACY AND APPLICATIONS

There is a general consensus that basic knowledge of computer terminology and mastery of widely used types of software will be essential for a growing number of occupations (see COMPUTER LITERACY). The elementary and junior high school curriculum now generally includes computer classes or "labs" where students learn the basics of word processing, spreadsheets, databases, graphics software, and use of the World Wide Web. There may also be introductory courses in programming, usually featuring easy-to-use programming languages such as Logo or BASIC.

Some high schools offer a track geared toward preparation for college studies in computer science. This track may include courses in more advanced languages such as C++ or Java. Because of public interest and marketability, courses in graphics (such as use of Adobe Photoshop), multimedia, and Web design are also increasingly popular. Adult education and community college programs feature a similar range of courses. Many of today's adult workers went to school at a time when personal computers were not readily available and computer literacy was not generally emphasized. The career prospects of many older workers are thus increasingly limited if they don't receive training in basic computer skills.

Technical or vocational schools offer tightly focused programs that are geared toward providing a set of marketable skills, often in conjunction with gaining industry certifications (see CERTIFICATION OF COMPUTER PROFESSIONALS).

COMPUTER SCIENCE

In the early 1950s, knowledge of computing tended to have an ad hoc nature. On the practical level, computing staffs tended to train newcomers in the specific hardware and machine-level programming languages in use at a particular site. On the theoretical level, programmers in scientific fields were likely to come from a background in electronics, electrical engineering, or similar disciplines.

As it became clear that computers were going to play an increasingly important role, courses specific to computing were added to curricula in mathematics and engineering. By the late 1950s, however, leading people in the computing field had become convinced that a formal curriculum in computer science was necessary for further advance in an increasingly sophisticated computing arena (see COMPUTER SCIENCE). By the early 1960s, efforts at the University of Michigan, University of Houston, Stanford, and other institutions had resulted in the creation of separate graduate departments of computer science. By the mid-1960s, the National Academy of Sciences and the President's Science Advisory Committee had both called for a major expansion of efforts in computer science education to be aided by federal funding. During the 1970s and 1980s, mathematical and engineering societies (in particular the Association for Computing Machinery (ACM) and Institute for Electrical and Electronic Engineering (IEEE)) worked to established detailed computer science curricula that extended to undergraduate study. By 2000, there were 155 accredited programs in computer science in the United States.

INFORMATION SYSTEMS

The traditional computer science curriculum emphasizes theoretical matters such as algorithm and program design and computer architecture. Hiring managers in corporate information systems departments have observed that computer science graduates often have little experience in such practical considerations as systems analysis, or the designing of computer systems to meet business requirements. There has also been an increasing need for systems administrators, database administrators, and networking professionals who are well versed in the management and maintenance of particular systems.

In response to demand from industry, many universities have instituted degree programs in information systems (sometimes called MIS or Management Information Systems) as an alternative to computer science. While these programs include some study of theory, they focus on practical considerations and often include internships or other practical work experience. Some programs offer more ambitious students a dual track leading to an MBA.

CHALLENGES

There has always been a gap between the emphases in computer and information science programs and the needs of a rapidly changing marketplace. However, additional challenges face education in the computer field today. The number of undergraduate computer science degrees awarded in Ph.D.-granting universities in the United States has steadily declined since 2000. In part this may be a delayed reaction to the decline in employment of programmers early in the decade (due to the bursting of the "dot-com bubble") that has since leveled off but has not significantly grown (see EMPLOYMENT IN THE COMPUTER FIELD). This, together with the outsourcing of many jobs (see GLOBALISM AND THE COMPUTER INDUSTRY) may have in turn discouraged young people from entering the field.

At the same time, many observers insist that prospects are good for educators and students who can target emerging high-demand skills. These include areas such as computer security, data mining, bioinformatics, Web content management, and even aspects of business management. Educators will be challenged to strike a balance between a comprehensive treatment of concepts that have many potential applications and the need to provide specific skills that are in demand in the market.

Further Reading

ACM-IEEE Joint Task Force on Computing Curricula. "Computer Science. 2001." Available online. URL: http://acm.org/education/curric_vols/cc2001.pdf. Accessed July 22, 2007.

———. "Information Systems." 2001 Available online. URL: http://www.acm.org/education/is2002.pdf. Accessed July 22, 2007.

Anthes, Gary. "Computer Science Looks for a Remake." *Computerworld*. May 1, 2006. Available online. URL: http://www.computerworld.com/careers/story/0,10801,110959,00.html. Accessed July 22, 2007.

Computer Science Directory. Available online. URL: http://csdir.org/. Accessed July 22, 2007.

Denning, Peter J. "Great Principles in Computing Curricula." *ACM SIGCSE Bulletin*, vol. 36, March 2004. Available online. URL: http://portal.acm.org/citation.cfm?id=1028174.971303. Accessed July 22, 2007.

Greening, Tony, ed. *Computer Science Education in the 21st Century*. New York: Springer-Verlag, 2000.

Open Directory Project. "Computer Science." Available online. URL: http://dmoz.org/Computers/Computer_Science/. Accessed July 22, 2007.

e-government

Just as the way business is organized and conducted has been profoundly changed by information and communications technology, the operation of government at all levels has been similarly affected. The term *e-government* (or electronic government) is a way of looking at these changes as a whole and of considering how government uses (or might use) various computer applications.

The use of information technology in government can involve changes in the organization and internal communications of government departments, changes in how services are delivered to the public, and providing new ways for the public to interact with the agency.

Internally, government agencies have many of the same information management and sharing needs as private enterprises (see DATA MINING, DATABASE ADMINISTRATION,

E-MAIL, GROUPWARE, PERSONAL INFORMATION MANAGER, and PROJECT MANAGEMENT SOFTWARE). However, government agencies are likely to have to adapt their information systems to account for complex, specialized regulations (both those the agency administers and others it is subject to). The standards of openness and accountability are generally different from and stricter than those that apply to private organizations.

A major focus of e-government is in expanding agencies' presence on the Web and making government sites more useful. This can include providing summaries of regulations or other complicated information, offering online assistance, allowing filing of tax or other forms electronically, and helping with applications such as for Social Security or Medicare. Where applicants must physically visit the office, a computerized system can make it easy to make appointments to reduce time waiting in line (a welcome option now offered by many state departments of motor vehicles).

IMPLEMENTATION

Obtaining employees with the necessary skills for maintaining sophisticated information systems and modern dynamic Web sites is not easy. The government hiring process tends to be cumbersome and slow to respond to changing needs. Government must often compete with a private sector that is willing to pay high prices for top talent.

In many cases, adopting comprehensive e-government would require a rethinking of an agency's purpose and priorities. There is also a tension between the Web culture, which focuses on linking information across conventional boundaries, and the tendency of bureaucracies to compartmentalize and centrally control information. Nevertheless, even without fundamentally restructuring how agencies operate, there has been considerable success with bringing information to the public through a central portal (USA.gov, formerly FirstGov).

Once a service is offered, it has to be promoted. While some services (such as "e-filing" of tax returns) can be readily promoted for their convenience, other services are more obscure or may be of interest only to a narrow constituency.

SOCIAL AND POLITICAL IMPACT

A survey by the Hart-Teeter poll found that respondents considered the most important potential benefit of e-government to be greater government accountability; the second was greater access to information; and, perhaps surprisingly, convenience came third.

One criticism of e-government initiatives is that they often lack central coordination and may be implemented without keeping in mind the need of an agency to provide uniform, consistent, and impartial treatment to all citizens. For example, if an agency focuses its resources on developing its Web site, people who lack online access may come to feel that they are receiving "second class" service (see DIGITAL DIVIDE). This is particularly unfortunate because the unconnected people are likely to be in poor and isolated communities that are most likely to be in need of government services.

As with private enterprise, there can also be important online privacy issues: Information that has been collected digitally is easy to transfer to other agencies or even (as in the case of DMV information in some states) sold to private companies. Having a clearly spelled-out privacy policy is crucial.

Besides keeping private what people expect to be private, government agencies must also provide information that helps ensure public accountability. Information collected by government agencies is often subject to the Freedom of Information Act (FOIA). This may require that data be provided in a format that is readily accessible.

Further Reading

Briefing Book Outline: E-Government. Advisory Committee to the Congressional Internet Caucus. Available online. URL: http://www.netcaucus.org/books/egov2001/. Accessed September 18, 2007.

Center for Technology in Government. Available online. URL: http://www.ctg.albany.edu. Accessed September 18, 2007.

Cordella, Antonio. "E-Government: Towards the E-bureaucratic Form?" *Journal of Information Technology* 22 (2007): pp. 265–274.

Government Computerization (Open Directory). Available online. URL: http://www.dmoz.org/Society/Issues/Science_and_Technology/Computers/Government_Computerization/. Accessed September 18, 2007.

LaVigne, Mark. "E-Government: Creating Tools of the Trade." Available online. URL: http://www.netcaucus.org/books/egov2001/pdf/e-govt.pdf. Accessed September 18, 2007.

Rosencrance, Linda. "User Satisfaction with Federal Government Web Sites down Slightly." *Computerworld,* September 18, 2007. Available online. URL: http://www.computerworld.com/action/article.do?command=viewArticleBasic&articleId=9037079. Accessed September 18, 2007.

USA.gov: "Government Made Easy" [Official U.S. Government Portal]. Available online. URL: http://www.usa.gov/. Accessed September 18, 2007.

Eiffel

Eiffel is an interesting programming language developed by Bertrand Meyer and his company Eiffel Software in the 1980s. The language was named for Gustav Eiffel, the architect who designed the famous tower in Paris. The language and accompanying methodology attracted considerable interest at software engineering conferences.

Eiffel fully supports (and in some ways pioneered) programming concepts found in more widely used languages today (see CLASS and OBJECT-ORIENTED PROGRAMMING). Syntactically, Eiffel emphasizes simple, reusable declarations that make the program easier to understand, and tries to avoid obscure or lower-level code such as compiler optimizations.

PROGRAM STRUCTURE

An Eiffel program is called a "system," emphasizing its structure as a set of classes that represent the types of real-world data that need to be processed. A simple class might look like this:

```
class
    COUNTER
```

```
feature—access counter value
    total: INTEGER
feature—manipulate counter value
    increment is—increase counter by one
        do
            total :- total + 1
        end
    decrement is—decrease counter by one
        do
            total := total - 1
        end
    reset is—reset counter to zero
        do
            total := 0
        end
end
```

(In this listing language, keywords are in bold and user-defined objects are in italics. This formatting will be done automatically as the user enters the text.) Once the class is defined, making an instance of it is very simple:

```
my_counter COUNTER

create my_counter
```

The Eiffel compiler itself compiles to an intermediate "bytecode" that, in the final stage, is compiled into C, taking advantage of the ready availability of optimized C compilers.

A unique feature of Eiffel is the ability to set up "contracts" that specify in detail how classes will interact with one another. (This goes well beyond the usual declarations of parameters and enforcement of data types.) For example, with the COUNTER class an "invariant" can be declared such that total >= 0. This means that this condition must always remain true no matter what. A method can also require that the caller meet certain conditions. After processing and before returning to the caller, the method can ensure that a particular condition is true. The point of these specifications is that they make explicit what a given unit of code expects and what it promises to do in return. This can also improve program documentation.

IMPLEMENTATION AND USES

Eiffel's proponents note that it is more than a language: It is designed to provide consistent ways to revise and reuse program components throughout the software development cycle. The current implementation of Eiffel is available for virtually all platforms and has interfaces to C, C++, and other languages. This allows Eiffel to be used to create a design framework for reusing existing software components in other languages. Eiffel's consistent object-oriented design also makes it useful for documenting or modeling software projects (see MODELING LANGUAGES).

Eiffel was developed around the same time as C++. Eiffel is arguably cleaner and superior in design to the latter language. However, two factors led to the dominance of C++: the ready availability of inexpensive or free compilers and the existence of thousands of programmers who already knew C. Eiffel ended up being a niche language used for teaching software design and for a limited number of commercial applications using the EiffelStudio programming environment.

Eiffel has been recognized for its contributions to the development of object-oriented software design, most recently by the Association for Computing Machinery's 2006 Software System Award for Impact on Software Quality.

Further Reading

"Eiffel in a Nutshell." Available online. URL: http://archive.eiffel. com/eiffel/nutshell.html. Accessed September 19, 2007.

Eiffel Zone. Available online. URL: http://eiffelzone.com/. Accessed September 19, 2007.

"Introduction to Eiffel" [Flash presentation]. Available online. URL: http://www.eiffel.com/developers/presentations/eiffel_ introduction/player.html?slide=. Accessed September 19, 2007.

Meyer, Bertrand. *Eiffel: The Language.* Englewood Cliffs, N.J.: Prentice Hall, 1991.

———. *Object-Oriented Software Construction.* 2nd ed. Upper Saddle River, N.J.: Prentice Hall, 2000.

Wiener, Richard. *An Object-Oriented Introduction to Computer Science Using Eiffel.* Upper Saddle River, N.J.: Prentice Hall, 1996.

Electronic Arts

Electronic Arts (NASDAQ symbol: ERTS) is a pioneering and still prominent maker of games for personal computers (see COMPUTER GAMES). Its fortunes largely mirror those of the game industry itself.

In 1982 Trip Hawkins and several colleagues left Apple Computer and founded a company called Amazin' Software. The company was founded with the goal of making "software that makes a personal computer worth owning." Hawkins also had an ambitious goal of turning it into a billion-dollar company, but this goal would not be achieved until the mid-1990s. Meanwhile, after considerable internal debate, the company changed its name to Electronic Arts in late 1982. This name reflects Hawkins's belief that computer games were an emerging art form and that their developers should be respected as artists. This would be reflected in game box covers that looked like record jackets and prominently featured the names of the developers.

In 1983 EA published three games for the Atari 800 computer that typified playability and diversity. *Archon* combined chesslike strategy with arcade-style battles; *Pinball Construction Set* let users create and play their own layouts; and the unique *M.U.L.E.* was a deceptively simple game of strategic resources—and one of the first multiplayer video games. EA titles published in the later 1980s include an exploration game *Seven Cities of Gold*, the graphically innovative space conquest game *Starflight*, and the role-playing series *The Bard's Tale.*

In its early years the company published games developed by independent programmers, but in the late 1980s it began to develop some games in house. EA sought out innovative games and promoted them directly to retailers. While it was difficult at first to market often-obscure games to stores, as the games became successful and regular retail channels were established, EA's revenue began to outpace that of competitors. (Hawkins left in 1991 to found the game company 3DO.)

CHALLENGES AND CRITICISM

By the 2000s EA, now under Larry Probst, had suffered loss of its once-dominant position in what had become an increasingly diverse industry. EA was criticized by some investment analysts for declining to follow the trend toward ultraviolent, M-rated games such as *Grand Theft Auto,* though the company later softened that stand. In recent years the company's big sellers have been its graphically intense and realistic sports simulations, notably *John Madden Football.* (Besides the NFL, EA has contracts with NASCAR, FIFA [soccer], and the PGA and Tiger Woods.)

In 2007 EA announced that it would come out with Macintosh versions of many of its top titles. However, critics have noted that the company seems to be publishing fewer original titles in favor of yearly updates (particularly in their sports franchises).

Along with much of the game industry, EA has increasingly focused on console games (see GAME CONSOLE). EA currently develops games for the leading consoles; in fact, about 43 percent of EA's 2005 revenue came from sales for the Sony PlayStation2 alone. (Total revenue in 2008 was $4.02 billion.) EA has also been expanding into online games, starting in 2002 with an online version of *The Sims,* a "daily life simulator." (See ONLINE GAMES.)

Some critics have objected to EA's practice of buying smaller companies in order to get control over their popular games, and then releasing versions that had not been properly tested. Perhaps the most-cited example is EA's acquiring of Origin Systems and its famous *Ultima* series of role-playing games. Once acquired, EA produced two new titles in the series that many gamers consider to not be up to the *Ultima* standard.

The company has also been criticized for requiring very long work hours from developers; it eventually settled suits from game artists and programmers demanding compensation for unpaid overtime.

EA has shown continuing interest in promoting the profession of game development. In 2004 the company made a significant donation toward the development of a game design and production program at the University of Southern California.

Meanwhile, founder Hawkins has founded a company called Digital Chocolate, focusing on games for mobile devices.

Further Reading

EA.com Web site. Available online. URL: http://www.ea.com. Accessed September 19, 2007.

EA Sports Home Page. Available online. URL: http://www. easports.com/. Accessed September 19, 2007.

"Mobile Games: Way beyond Phone Tag" [Interview with Trip Hawkins]. *Business Week Online,* April 3, 2006. Available online. URL: http://www.businessweek.com/technology/ content/apr2006/tc20060403_840834.htm. Accessed September 19, 2007.

Waugh, Eric-Jon Rossel. "A Short History of Electronic Arts." *Business Week Online*, August 25, 2006. Available online. URL: http://www.businessweek.com/innovate/content/aug2006/id20060828_268977.htm. Accessed September 19, 2007.

electronic voting systems

There are a variety of ways to electronically register, store, and process votes. In recent years older manual systems (paper ballots or mechanical voting machines) have been replaced in many areas with systems ranging from purely digital (touch screens) to hybrid systems where marked paper ballots are scanned and tabulated by machine. However, voting systems have been subject to considerable controversy, particularly following the Florida debacle in the 2000 U.S. presidential election.

The criteria by which voting systems are evaluated include:

- how easy it is for the voter to understand and use the system

- accessibility for disabled persons

- whether the voter's intentions are accurately recorded

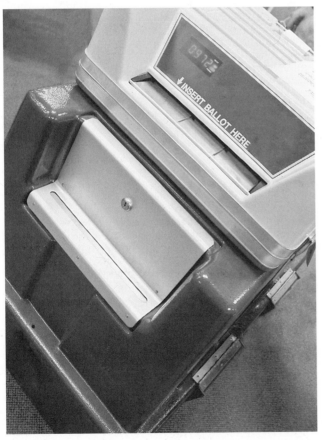

There are several types of electronic voting systems, such as this box that automatically tallies specially marked ballots. Common concerns include the potential for tampering and the need to provide for independent verification of results. (LISA MCDONALD/ISTOCKPHOTO)

- the ability to make a permanent record of the vote

- prevention of tampering (physical or electronic)

- provisions for independent auditing of the votes in case of dispute

The degree to which a given system meets these criteria can vary considerably because of both design and implementation issues.

EARLY SYSTEMS

The earliest form of voting system consisted of paper ballots marked and tabulated entirely by hand. The first generation of "automatic" voting systems involved mechanical voting machines (where votes were registered by pulling levers). Next came two types of hybrid systems where votes were cast mechanically but tabulated automatically. These systems used punch cards (see PUNCHED CARDS AND PAPER TAPE) or "marksense" or similar systems where the voter filled in little squares and the ballots were then scanned and tabulated automatically.

The ultraclose and highly disputed 2000 U.S. presidential election "stress-tested" voting systems that most people had previously believed were reasonably accurate. The principal problems were the interpretation of punch cards that were not properly punched through (so-called dimpled or hanging chads) and the fact that some ballot layouts proved to be confusing or ambiguous. Two types of voting systems have been proposed as replacements for the problematic earlier technology.

TOUCHSCREEN

This type of system uses a screen display that can be directly manipulated by the voter (see TOUCHSCREEN). In the most common type, called DRE (direct-recording electronic), a computer program interprets and tabulates the vote as it is cast, storing an image in a removable memory unit and (usually) printing out a copy for backup. After voting is complete, the memory module can be sent to the central counting office. (Alternatively, votes can be transmitted over a computer network in batches throughout the day.) In a few cases, voting has also been implemented through secure Internet sites.

OPTICAL SCAN

Concern about potential tampering with computers has led many jurisdictions to begin to replace touchscreen systems with optical-scan systems, where the voter marks a sturdy paper ballot. (About half of U.S. counties now use optical-scan systems.) The advantage of optical systems is that the voter physically marks the ballot and can see how he or she has voted, and after tabulation the physical ballots are available for review in case of problems. However, optical-scan ballots must be properly marked using the correct type of pencil, or they may not be read correctly. Unlike the touchscreen, it is not possible to give the voter immediate feedback so that any errors can be corrected. Optical-ballot systems may cost more because of paper and printing costs for the ballots, which may have to be prepared in several languages. However this cost may be offset by not having to develop or validate the more complicated software needed for all-electronic systems.

Whatever system is used, federal law requires that visually or otherwise disabled persons be given the opportunity, wherever possible, to cast their own vote in privacy. With optical-scan ballots, this is accommodated with a special device that plays an audio file listing the candidates for each race, with the voter pressing a button to mark the choice. However, disability rights advocates have complained that existing systems still require that another person physically insert the marked ballot into the scanner. Touchscreen systems, however, with the aid of audio cues, can be used by visually disabled persons without the need for another person to be present. They are thus preferred by some advocates for the disabled.

REFORMS AND ISSUES

In response to the problems with the 2000 election, Congress passed the Help America Vote Act in 2002. Since then, the federal government has spent more than $3 billion to help states replace older voting systems—in many cases with touchscreen systems.

The biggest concern raised about electronic voting systems is that they, like other computer systems, may be susceptible to hacking or manipulation by dishonest officials. In 2007 teams of researchers at the University of California–Davis were invited by the state to try to hack into its voting systems. For the test, the researchers were provided with full access to the source code and documentation for the systems, as well as physical access. The hacking teams were able to break into and compromise every type of voting system tested. In their report, the researchers outlined what they claimed to be surprisingly weak electronic and physical security, including flaws that could allow hackers to introduce computer viruses and take over control of the systems.

Manufacturers and other defenders of the technology have argued that the testing was unrealistic and that real-world hackers would not have had nearly as much information about or access to the systems. (This may underestimate the resourcefulness of hackers, as shown with other systems, such as the phone system and computer networks.)

Another issue is who will be responsible for independently reviewing the programming (source) code for each system to verify that it does not contain flaws. Manufacturers generally resist such review, considering the source code to be proprietary. (A possible alternative might be an open-source voting system. Advocates of open-source software argue that it is safer precisely because it is open to scrutiny and testing—see OPEN-SOURCE MOVEMENT.)

One common response to these security concerns is to require that all systems generate paper records that can be verified and audited. Some defenders of existing technology say that adding a parallel paper system is unnecessarily expensive and introduces other problems such as printer failures. They argue that all-electronic systems can be made safer and more secure, such as through the use of encryption. (A proposed compromise would be for the machine to print out a simple receipt with a code that the voter could use to verify online that the vote was tabulated.)

As of 2007, 28 states had passed laws requiring that voting systems produce some sort of paper receipt or record that shows the voter what has been voted and that can be used later for an independent audit or recount,

Although control of elections is primarily a state or local responsibility, the federal government does have jurisdiction over elections for federal office. As a practical matter, any changes in voting technology or procedures mandated by Congress for federal elections will end up being used in local elections as well.

In 2007, congressional leaders decided not to require a major overhaul of the nation's election systems until at least 2012. However, the inclusion of some sort of paper record is being mandated for the 2008 election. For users of touchscreen systems, the simplest way to accommodate this is to add small paper-spool printers, but some states have complained that their systems would require more-expensive accommodations.

Meanwhile, a lively debate continues in many states and other jurisdictions about how to meet the need for accessible but secure voting systems without breaking the budget.

Further Reading

Drew, Christopher. "Accessibility Isn't Only Hurdle in Voting System Overhaul." *New York Times,* July 21, 2007. Available online. URL: http://www.nytimes.com/2007/07/21/washington/21vote.html. Accessed September 20, 2007.

Open Voting Consortium. Available online. URL: http://www.openvotingconsortium.org/. Accessed September 20, 2007.

Rubin, Aviel. *Brave New Ballot: The Battle to Safeguard Democracy in the Age of Electronic Voting.* New York: Morgan Road Books, 2006.

Saltman, Roy G. *The History and Politics of Voting Technology: In Quest of Integrity and Public Confidence.* New York: Palgrave Macmillan, 2006.

United States Election Assistance Commission. "2005 Voluntary Voting System Guidelines." Available online. URL: http://www.eac.gov/voting%20systems/voting-system-certification/2005-vvsg. Accessed September 20, 2007.

"Verified Voting: Mandatory Manual Audits of Voter-Verified Paper Records." Available online. URL: http://www.verifiedvoting.org/. Accessed September 20, 2007.

Wildemuth, John. "State Vote Machines Lose Test to Hackers." *San Francisco Chronicle,* July 28, 2007, p. A-1. Available online. URL: http://www.sfgate.com/cgi-bin/article.cgi?f=/c/a/2007/07/28/MNGP6R8TJO1.DTL. Accessed September 20, 2007.

e-mail

Electronic mail is perhaps the most ubiquitous computer application in use today. E-mail can be defined as the sending of a message to one or more individuals via a computer connection.

DEVELOPMENT AND ARCHITECTURE

The simplest form of e-mail began in the 1960s as a way that users on a time-sharing computer system could post and read messages. The messages consisted of text in a file that was accessible to all users. A user could simply log into the

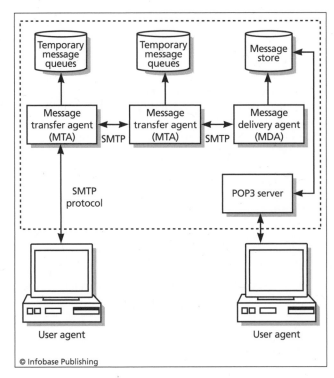

Transmission of an e-mail message depends on widely used protocols such as SMTP, which controls message format and processing, and POP3, which handles interaction between mail servers and client programs. As long as the formats are properly followed, users can employ a wide variety of mail programs (agents), and service providers can use a variety of mail server programs.

system, open the file, and look for messages. In 1971, however, the ARPANET (ancestor of the Internet—see INTERNET) was used by researchers at Bolt Beranek and Newman (BBN) to send messages from a user at one computer to a user at another. The availability of e-mail helped fuel the growth of the ARPANET through the 1970s and beyond.

As e-mail use increased and new features were developed, the question of a standardized protocol for messages became more important. By the mid-1980s, the world of e-mail was rather fragmented, much like the situation in the early history of the telephone, where users often had to choose between two or more incompatible systems. Apranet (or Internet) users used SMTP (Simple Mail Transport Protocol) while a competing standard (OSI MHS, or Message Handling System) also had its supporters. Meanwhile, the development of consumer-oriented online services such as CompuServe and America Online threatened a further balkanization of e-mail access, though systems called *gateways* were developed to transport messages from one system to another.

By the mid-1990s, however, the nearly universal adoption of the Internet and its TCP/IP protocol had established SMTP and the ubiquitous Sendmail mail transport program as a uniform infrastructure for e-mail. The extension of the Internet protocol to the creation of intranets has largely eliminated the use of proprietary corporate e-mail systems.

Instead, companies such as Microsoft and Google compete to offer full-featured e-mail programs that include group-oriented features such as task lists and scheduling (see also PERSONAL INFORMATION MANAGER).

E-MAIL TRENDS

The integration of e-mail with HTML for Web-style formatting and MIME (for attaching graphics and multimedia files) has greatly increased the richness and utility of the e-mail experience. E-mail is now routinely used within organizations to distribute documents and other resources. However, the addition of capabilities has also opened security vulnerabilities. For example, Microsoft Windows and the popular Microsoft Outlook e-mail client together provide the ability to run programs (scripts) directly from attachments (files associated with e-mail messages). This means that it is easy to create a virus program that will run when an enticing-looking attachment is opened. The virus can then find the user's mailbox and mail copies of itself to the people found there. E-mail has thus replaced the floppy disk as the preferred medium for such mischief. (See COMPUTER VIRUS.)

Beyond security issues, e-mail is having considerable social and economic impact. E-mail has largely replaced postal mail (and even long-distance phone calls) as a way for friends and relatives to keep in touch. As more companies begin to use e-mail for providing routine bills and statements, government-run postal systems are seeing their first-class mail revenue drop considerably. Despite the risk of viruses or deception and the annoyance of electronic junk mail (see SPAM), e-mail has become as much a part of our way of life as the automobile and the telephone.

Further Reading
Costales, Bryan, and Eric Allman. *Sendmail.* 3rd ed. Sebastapol, Calif.: O'Reilly, 2002.
Sendmail Consortium. Available online. URL: http://www.send mail.org/. Accessed July 22, 2007.
Shipley, David, and Will Schwalbe. *Send: The Essential Guide to Email for Office and Home.* New York: Knopf, 2007.
Song, Mike, Vicki Halsey, and Tim Burress. *The Hamster Revolution: How to Manage Your Email Before It Manages You.* San Francisco: Bennett-Koehler, 2007.

embedded system

When people think of a computer, they generally think of a general-purpose computing system housed in a separate box, for use on the desk or as a laptop or hand-held device. However, the personal computer and its cousins are only the surface of a hidden web of computing capability that reaches deep into numerous devices used in our daily lives. Modern cars, for example, often contain several specialized computer systems that monitor fuel injection or enhance the car's grip on the road under changing conditions. Many kitchen appliances such as microwaves, dishwashers, and even toasters contain their own computer chips. Communications systems ranging from cell phones to TV satellite dishes include embedded computers. Most important, embedded systems are now essential to the operation of critical infrastructure

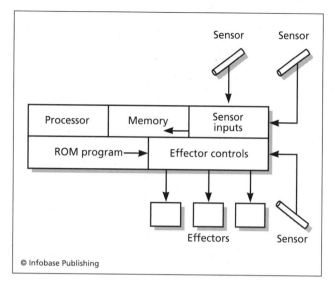

© Infobase Publishing

An embedded system is a computer processor that is part of a "real-world" device that must interact with its environment. Sensor inputs (such as torque or pressure sensors) provide real-time data about conditions faced by the device (such as a vehicle). This data is processed by the onboard processor under the control of a permanent (ROM) program, and commands are issued to the effector controls, which might, for example, apply braking pressure.

such as medical monitoring systems and power transmission networks. (The potential vulnerability of embedded systems to the Y2K date-related problems was a major concern in the months leading up to 2000, especially because many embedded systems might have to be replaced rather than just reprogrammed. In the event, it turned out that there were relatively few date-dependent systems and only minor disruptions were experienced. See Y2K PROBLEM.)

CHARACTERISTICS OF EMBEDDED SYSTEMS

What most distinguishes an embedded system from a desktop computer is not that it is hidden inside some other device, but that it runs a single, permanent program whose job it is to monitor and respond to the environment in some way. For example, an oven controller would accept a user input (the desired temperature), monitor a sensor or thermostat, and control the heat to ensure that the correct temperature is being maintained. Embedded systems are thus similar to robots in that they sense and manipulate their environment.

Architecturally, an embedded system typically consists of a microprocessor, some *nonvolatile memory* (memory that can maintain its contents indefinitely), sensors (to receive readings from the environment), signal processors (to convert inputs into usable information), and "effectuators" (switches or other controls that the embedded system can use to change its environment). In practice, an embedded system may not have its own sensors or effectors, but instead interface with other systems (such as avionics or steering).

Programmers of embedded systems often use special compilers or languages that are particularly suited for creating embedded software (see ADA and FORTH).

Because available memory is limited, embedded program code tends to be compact. Since embedded systems are often responsible for critical infrastructure, their operating programs must be carefully debugged. Designers try to make programs "robust" so they can respond sensibly to unexpected conditions or at least "fail gracefully" in a way least likely to cause damage. Other strategies to improve the reliability of embedded systems include the use of overdesigned, fault-tolerant components (as in the military "milspec") and the use of separate, redundant systems so that a failing system can be "locked out" and processing can continue elsewhere.

Further Reading

Catsoulis, John. *Designing Embedded Hardware.* 2nd ed. Sebastapol, Calif.: O'Reilly, 2005.
Embedded.com: "Thinking inside the Box." Available online. URL: http://www.embedded.com/. Accessed July 22, 2007.
Embedded Systems. Dr. Dobb's Portal. Available online. URL: http://www.ddj.com/dept/embedded/. Accessed July 22, 2007.
Norergaard, Tammy. *Embedded Systems Architecture: A Comprehensive Guide for Engineers and Programmers.* Burlington, Mass.: Newnes/Elsevier, 2005.
Simon, David E. *An Embedded Software Primer.* Upper Saddle River, N.J.: Addison-Wesley/Pearson Education, 1999.

employment in the computer field

The number of computer-related positions has grown rapidly over the past few decades. According to the U.S. Bureau of Labor Statistics, by the mid-1990s the fastest-growing professions in the United States included systems analysts, computer scientists, and computer engineers. By the mid-2000s, computer-related occupations were still near the top of the list, which by then also included network and communications analysts (second only to "home health aids").

Computer-related employment can be broken down into the following general categories:

- hardware design and manufacturing, including computer systems, peripherals, communications and network hardware, and other devices

- the software industry, ranging from business applications to consumer software, games, and entertainment

- the administrative sector (systems administration, network administration, database administration, computer security, and so on)

- the Web sector, including ISPs, Web hosts and page developers, and e-commerce applications

- the support sector, including training and education, computer book publishing, technical support, and systems repair and maintenance

In addition to these "pure" computer-related jobs, there are many other positions that involve working with PCs. These include word processing/desktop publishing, statistics, scientific research, accounting and billing, shipping, retail sales and inventory, and manufacturing. (See also PROGRAMMING AS A PROFESSION.)

JOB MARKET CONSIDERATIONS

In the late 1990s, a number of sources forecast a growing gap between the number of positions opening in computer-related fields and the number of new people entering the job market (estimates of the gap's size ranged into the hundreds of thousands nationally). Particularly in the Internet sector, demand for programmers and system administrators meant that new college graduates with basic skills could earn unprecedented salaries, while experienced professionals could often become highly paid consultants. Despite the growing emphasis on computing in secondary and higher education, computer science and engineering candidates were in particularly short supply. As a result, many companies received permission to hire larger numbers of immigrants from countries such as India.

The "dot crash" of 2001–2002 saw a sharp if temporary decline in demand for computer professionals, particularly in the Web and e-commerce sectors, but it impacted hardware sales as well. The industry then saw a resurgence, but with an emphasis on somewhat different skill sets. Skills in strong demand toward the end of the 2000 decade include:

- detection, prevention, and investigation of computer attacks (see COMPUTER CRIME AND SECURITY and COMPUTER FORENSICS)

- improvements in operating system and software security

- use of open-source software and operating systems (see OPEN-SOURCE MOVEMENT and LINUX)

- surveillance and physical security (see BIOMETRICS)

- transaction analysis for both security and marketing applications (see DATA MINING)

- e-commerce applications and management (see CUSTOMER RELATIONSHIP MANAGEMENT)

- rapid development of efficient, highly interactive Web services (see AJAX, WEB 2.0 AND BEYOND, and SCRIPTING LANGUAGES)

- hardware and software for mobile and wireless devices (particularly delivery and integration of media)

- content management for Web sites and media services

- scientific computing, particularly genetic and biological applications (see BIOINFORMATICS)

On the other hand, with the successful passing of the Y2K crisis, the outlook for mainframe programmers (particularly using COBOL) is increasingly dim. Prospects are also poor for certain operating, network, and database systems with declining market share (such as OS/2, Novell networking, and some older database systems). It is true that as baby boomer programmers retire, there will be some demand for maintenance or conversion of obsolescent systems. Finally, as global trends toward outsourcing and relocating of lower-level support and even programming continue, it may become harder for domestic workers to begin to climb the IT ladder.

Socially, the key challenges that must be met to ensure a healthy computer-related job market are the improvement of education at all levels (see EDUCATION AND COMPUTERS) and the increasing of ethnic and gender diversity in the field (which is related to the fostering of more equal educational opportunity), and adapting to changes in the global economy (see GLOBALISM AND THE COMPUTER INDUSTRY).

Further Reading
Brandel, Mary. "12 IT Skills that Employers Can't Say No To." *Computerworld,* July 11, 2007. Available online. URL: http://www.computerworld.com/action/article.do?command=viewArticleBasic&articleId=9026623. Accessed July 22, 2007.
Farr, Michael. *Top 100 Computer and Technical Careers.* 3rd ed. St. Paul, Minn.: JIST Works, 2006.
Henderson, Harry. *Career Opportunities in Computers and Cyberspace.* 2nd ed. New York: Facts On File, 2004.
Information Technology Jobs in America. New York: Info Tech Employment Publications, 2007.
U.S. Department of Labor. Occupational Outlook Handbook, 2006–07 edition. Available online. URL: http://www.bls.gov/oco/. Accessed July 22, 2007.
Vocational Information Center. Computer Science Career Guide. Available online. URL: http://www.khake.com/page17.html. Accessed July 22, 2007.

emulation

One consequence of the universal computer concept (see VON NEUMANN, JOHN) is that in principle any computer can be programmed to imitate the operation of any other. An emulator is a program that runs on one computer but accurately processes instructions written for another (see also MICROPROCESSOR and ASSEMBLER). For example, fans of older computer games can now download emulation programs that allow modern PCs to run games originally intended for an Apple II microcomputer or an Atari game machine. Emulators allowing Macintosh and Linux users to run Windows programs have also achieved some success.

In order to work properly, the emulator must set up a sort of virtual model of the target microprocessor, including appropriate registers to hold data and instructions and a suitably organized segment of memory. While carrying out instructions in software rather than in hardware imposes a considerable speed penalty, if the processor of the emulating PC is much faster than the one being emulated, the emulator can actually run faster than the original machine.

An entire hardware and software environment can also be emulated; this is called a virtual machine. For example, programs such as VMware can be used to run Windows, Linux, and BSD UNIX, each in a separate "compartment" that appears to be a complete machine, with all the necessary hardware drivers and emulated facilities.

The term *virtual machine* can also refer to language such as Java, where programs are first compiled into a platform-independent intermediate "byte code," which is then run by a Java virtual machine that produces the instructions needed for a given platform.

In the past, emulation was sometimes used to allow programmers to develop software for large, expensive mainframes while using smaller machines. Emulators can also consist of a combination of specially-designed chips and software, as in the case of the "IBM 360 on a chip" that became available for the IBM PCs.

The term *emulation* is also sometimes used to refer to a program that accurately simulates the operation of a hardware device. For example, when printers that included hardware for processing the PostScript typographical language were expensive, programs were developed that could process the PostScript instructions in the PC itself and then send the output as graphics to a less expensive printer.

Further Reading

Comparison of Virtual Machines. Wikipedia. Available online. URL: http://en.wikipedia.org/wiki/Comparison_of_virtual_machines. Accessed July 23, 2007.

Smith, Jim, and Ravi Nair. *Virtual Machines: Versatile Platforms for Systems and Processes.* San Francisco: Morgan Kaufmann, 2005.

VMware. Available online. URL: http://www.vmware.com. Accessed July 23, 2007.

encapsulation

In the earliest programming languages, any part of a program could access any other part simply by executing an instruction such as "jump" or "goto." Later, the concept of the subroutine helped impose some order by creating relatively self-contained routines that could be "called" from the main program. At the time the subroutine is called, it is provided with necessary data in the form of global variables or (preferably) parameters, which are variable references or values passed explicitly when the subroutine is called. When the subroutine finishes processing, it may return values by changing global variables or changing the values of variables that were passed as parameters (see PROCEDURES AND FUNCTIONS).

While an improvement over the totally unstructured program, the subroutine mechanism has several drawbacks. If it is maintained as part of the main program code, one programmer may change the subroutine while another programmer is still expecting it to behave as previously defined. If not properly restricted, variables within the subroutine might be accessed directly from outside, leading to unpredictable results. To minimize these risks, languages such as C and Pascal allow variables to be defined so that they are "local"—that is, accessible only from code within the function or procedure. This is a basic form of encapsulation.

The class mechanism in C++ and other object-oriented languages provides a more complete form of encapsulation (see OBJECT-ORIENTED PROGRAMMING, CLASS, and C++). A class generally includes both private data and procedures or methods (accessible only from within the class) and *public* methods that make up the interface. Code in the main program uses the class interface to create and manipulate new objects of that class.

Encapsulation thus both protects code from uncontrolled modification or access and hides information (details) that programmers who simply want to use functionality don't need to know about. Thus, high-quality classes can be designed by experts and marketed to other developers who can take advantage of their functionality without having to "reinvent the wheel."

Further Reading

Berard, Edward V. "Abstraction, Encapsulation, and Information Hiding." Available online. URL: http://www.itmweb.com/essay550.htm. Accessed July 23, 2007.

Booch, G. *Object-Oriented Analysis and Design with Applications.* 3rd ed. Upper Saddle River, N.J.: Addison-Wesley, 2007.

Müller, Peter. "Introduction to Object-Oriented Programming Using C++." Available online. URL: http://www.gnacademy.org/uu-gna/text/cc/Tutorial/tutorial.html. Accessed August 14, 2007.

Poo, Danny, and Derek Kiong. *Object-Oriented Programming and Java.* 2nd ed. New York: Springer-Verlag, 2007.

encryption

The use of encryption to disguise the meanings of messages goes back thousands of years (the Romans, for example, used *substitution ciphers,* where each letter in a message was replaced with a different letter). Mechanical cipher machines first came into general use in the 1930s. During World War II the German Enigma cipher machine used multiple rotors and a configurable plugboard to create a continuously varying cipher that was thought to be unbreakable. However, Allied codebreakers built electromechanical and electronic devices that succeeded in exploiting flaws in the German machine (while incidentally advancing computing technology). During the cold war Western and Soviet cryptographers vied to create increasingly complex cryptosystems while deploying more powerful computers to decrypt their opponent's messages.

In the business world, the growing amount of valuable and sensitive data being stored and transmitted on computers by the 1960s led to a need for high-quality commercial encryption systems. In 1976, the U.S. National Bureau of Standards approved the Data Encryption Standard (DES), which originally used a 56-bit key to turn each 64-bit chunk of message into a 64-bit encrypted ciphertext. DES relies upon the use of a complicated mathematical function to create complex permutations within blocks and characters of text. DES has been implemented on special-purpose chips that can encrypt millions of bytes of message per second.

PUBLIC-KEY CRYPTOGRAPHY

Traditional cryptosystems such as DES use the same key to encrypt and decrypt the message. This means that the key must be somehow transmitted to the recipient before the latter can decode the message. As a result, security may be compromised. However, the same year DES was officially adopted, Whitfield Diffie and Martin Hellman proposed a very different approach, which became known as public-key cryptography. In this scheme each user has two keys, a private key and a public key. The user publishes his or her public key, which enables any interested person to send the user an encrypted message that can be decrypted only by using the user's private key, which is kept secret. The system is more secure because the private key is never trans-

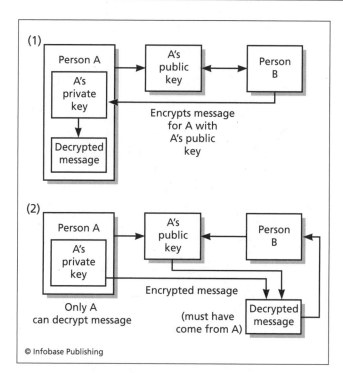

(1)

Person A

A's private key

A's public key

Person B

Decrypted message

Encrypts message for A with A's public key

(2)

Person A

A's private key

A's public key

Person B

Only A can decrypt message

Encrypted message

(must have come from A)

Decrypted message

© Infobase Publishing

Public key encryption allows users to communicate securely without having to exchange their private keys. In part 1, person A publishes a public key, which can be used by anyone else (such as person B) to encrypt a message that only person A can read. In part 2, person A encrypts a message with his or her private key. Since this message can only be encrypted using person A's public key, person B can use the published public key to verify that the message is indeed from person A.

mitted. Further, a user can distribute a message encrypted with his or her private key that can be decrypted only with the corresponding public key. This provides a sort of signature for authenticating that a message was in fact created by its putative author.

In 1978, Ron Rivest, Adi Shamir, and Leonard Adelman announced the first practical implementation of public-key cryptography. This algorithm, called RSA, became the prevailing standard in the 1980s. While keys may need to be lengthened as computer power increases, RSA is likely to remain secure for the foreseeable future.

LEGAL CHALLENGES

Until the 1990s, the computer power required for routine use of encryption was generally beyond the reach of most small business and consumer users, and there was little interest in a version of the RSA algorithm for microcomputers. Meanwhile, the U.S. federal government tried to maintain tight controls over encryption technology, including prohibitions on the export of encryption software to many foreign countries.

However, the growing use of electronic mail and the hosting of commerce on the Internet greatly increased concern about security and the need to implement an easy-to-use form of encryption. In 1990, Philip Zimmermann wrote

an RSA-based email encryption program that he called Pretty Good Privacy (PGP). However, RSA, Inc. refused to grant him the necessary license for its distribution. Further, FBI officials and sympathetic members of Congress seemed poised to outlaw the use of any form of encryption that did not include a provision for government agencies to decode messages.

Believing that people's liberty and privacy were at stake, Zimmermann gave copies of PGP to some friends. The program soon found its way onto computer bulletin boards, and then spread worldwide via Internet newsgroups and ftp sites. Zimmermann then developed PGP 2.0, which offered stronger encryption and a modular design that made it easy to create versions in other languages. The U.S. Customs Department investigated the distribution of PGP but dropped the investigation in 1996 without bringing charges. (At about the same time a federal judge ruled that mathematician Daniel Bernstein had the right to publish the source code for an encryption algorithm without government censorship.)

Government agencies eventually realized that they could not halt the spread of PGP and similar programs. In the early 1990s, the National Security Agency (NSA), the nation's most secret cryptographic agency, proposed that standard encryption be provided to all PC users in the form of hardware that became known as the Clipper Chip. However, the hardware was to include a "back door" that would allow government agencies and law enforcement (presumably upon fulfilling legal requirements) to decrypt any message. Civil libertarians believed that there was far too much potential for abuse in giving the government such power, and a vigorous campaign by privacy groups resulted in the mandatory Clipper Chip proposal being dropped by the mid-1990s in favor of a system called "key escrow." This system would require that a copy of each encryption key be deposited with one or more trusted third-party agencies. The agencies would be required to divulge the key if presented with a court order. However, this proposal has been met with much the same objections that had been made against the Clipper Chip.

In the early 21st century, the balance is likely to continue to favor the code-makers over the code-breakers. While it is rumored that the NSA can use arrays of supercomputers to crack any encrypted message given enough time, and a massive eavesdropping system called Echelon for analyzing message traffic has been partially revealed, as a practical matter most of the world now has access to high-quality cryptography. Only radically new technology (see QUANTUM COMPUTING) is likely to reverse this trend.

Further Reading

Cobb. Chey/ *Cryptography for Dummies.* Hoboken, N.J.: John Wiley & Sons, 2004.

Henderson, Harry. *Privacy in the Information Age* (Library in a Book) 2nd ed. New York: Facts On File, 2006.

"The International PGP Home Page." Available online. URL: http://www.pgpi.org/. Accessed February 2, 2008.

Levy, Stephen. *Crypto.* New York: Viking, 2001.

Singh, Simon. *The Code Book: the Science of Secrecy from Ancient Egypt to Quantum Cryptography.* New York: Anchor Books, 2000.

Engelbart, Douglas
(1925–   )
American
Computer Engineer

Douglas Engelbart invented key elements of today's graphical user interface, including the use of windows, hypertext links, and the ubiquitous mouse. Engelbart grew up on a small farm near Portland, Oregon, and acquired a keen interest in electronics. His electrical engineering studies at Oregon State University were interrupted by wartime service in the Philippines as a radar technician. During that time he read a seminal article by Vannevar Bush entitled "As We May Think." Bush presented a wide-ranging vision of an automated, interlinked text system not unlike the development that would become hypertext and the World Wide Web (see BUSH, VANNEVAR).

After returning to college for his Ph.D. (awarded in 1955), Engelbart worked for NACA (the predecessor of NASA) at the Ames Laboratory. Continuing to be inspired by Bush's vision, Engelbart conceived of a computer display that would allow the user to visually navigate through information displays. Engelbart received his doctorate in electrical engineering in 1955 at the University of California, Berkeley, taught there a few years, and then went to the Stanford Research Institute (SRI), a hotbed of futuristic ideas. In 1962, Engelbart wrote a seminal paper of his own, titled "Augmenting Human Intellect: A Conceptual Framework." In this paper Engelbart emphasized the computer not as a mere aid to calculation, but as a tool that would enable people to better visualize and organize complex information to meet the increasing challenges of the modern world. The hallmark of Engelbart's approach to computing would continue to be his focus on the central role played by the user.

In 1963, Engelbart left SRI and formed his own research lab, the Augmentation Research Center. During the 1960s and 1970s, he worked on implementing linked text systems (see HYPERTEXT AND HYPERMEDIA). In order to help users interact with the computer display, he came up with the idea of a device that could be moved to control a pointer on the screen. Soon called the "mouse," the device would become ubiquitous in the 1980s.

Engelbart also took a key interest in the development of the ARPANET (ancestor of the Internet) and adapted his NLS hypertext system to help coordinate network development. (However, the dominant form of hypertext on the Internet would be Tim Berners-Lee's World Wide Web—(see BERNERS-LEE, TIM.) In 1989, Engelbart founded the Bootstrap Institute, an organization dedicated to improving the collaboration within organizations, and thus their performance. During the 1990s, this nurturing of new businesses and other organizations would become his primary focus.

Engelbart received the MIT-Lemelson Award and the A.M. Turing Award in 1997 and the National Medal of Technology in 2000. Public recognition of Engelbart's work and ideas about human-computer interaction was also reflected in a Stanford University symposium called "Engelbart's Unfinished Revolution."

Further Reading
Bardini, Thierry. *Bootstrapping: Douglas Engelbart, Coevolution, and the Origins of Personal Computing.* Stanford, Calif.: Stanford University Press, 2000.
Bootstrap Institute. Available online. URL: http://www.bootstrap.org. Accessed July 23, 2007.
"Engelbart's Unfinished Revolution." Stanford University Symposium. Available online. URL: http://www.itmweb.com/essay550.htm. Accessed July 23, 2007.
"Internet Pioneers: Doug Engelbart." Available online. URL: http://www.ibiblio.org/pioneers/Engelbart.html. Accessed April 18, 2008.

Engelberger, Joseph
(1925–   )
American
Entrepreneur, Roboticist

Joseph Engelberger and George Devol created the first industrial robot, revolutionizing the assembly line. Engelberger went on to develop other robots that can work in hospitals and other settings while tirelessly promoting industrial robotics.

Engelberger was born on July 26, 1925, in New York City. During World War II he was selected for a special program where promising students were paid to study physics at Columbia University. Just after the war he worked as an engineer on early nuclear tests in the Pacific. He also worked on aerospace and nuclear power projects. After completing his military duties, Engelberger attended Columbia University's School of Engineering and earned B.S. (1946) and M.S. degrees in physics and electrical engineering. This solid background in science and engineering would shape Engelberger's practical approach to robot design.

A number of technologies of the 1940s and 1950s contributed to the later development of robotics. The war had greatly increased the development of automatic controls and servomechanisms that allow for precise positioning and manipulation of machine parts. The rise of nuclear power and the need to safely handle radioactive materials also spurred the development of automatic controls. Engelberger began to develop business ventures in the automation field, starting a company called Consolidated Controls.

In the mid-1950s Engelberger met George Devol, an inventor who had patented a programmable transfer machine. This was a device that could automatically move components from one specified position to another, such as in a die-casting machine that formed parts for automobiles. Engelberger realized that Devol's machine could, with some additional extensions and capabilities, become a robot that could be programmed to work on an assembly line.

In 1956 Engelberger and Devol founded Unimate, Inc.—the world's first industrial robot company. Their robot, also called Unimate, is essentially a large "shoulder" and arm. The shoulder can move along a track to position the arm near the materials to be manipulated. The arm can be equipped with a variety of specialized grasping "hands" to suit the task. The robot is programmed to perform a set of repetitive motions. It is also equipped with various devices for aligning the workpiece (the object to be manipulated) and to make small adjustments for variations.

In spring 1961 the first Unimate robot began operations on the assembly line at the General Motors Plant in Turnstedt, a suburb of Trenton, New Jersey. Most of the factory's 3,000 human workers welcomed the newcomer because Unimate would be doing a job involving the casting of car doors and other parts from molten metal—hot, dangerous work. That first Unimate worked for nearly 10 years, tirelessly keeping up with three shifts of human workers each day.

In 1980 Engelberger published *Robotics in Practice*. This book, together with *Robotics in Service* (1988), became a standard textbook that defined the growing robotics industry. The two titles also marked a shifting of Engelberger's focus from industrial robots to service robots—robots that would do their jobs not in factories, but in workplaces such as warehouses or hospitals.

In the 1980s Engelberger founded HelpMate Robotics, Inc. The company's most successful product has been the HelpMate robot. The robot is designed to dispatch records, laboratory samples, and supplies throughout a busy hospital. HelpMate does not follow a fixed track. Rather, it is programmed to visit a succession of areas or stations and makes its own way, using cameras to detect and go around obstacles. HelpMate can even summon an elevator to go to a different floor!

Along with other robotics entrepreneurs, Engelberger is also looking toward a time when robots will be able to perform a number of useful tasks in the home. In particular, Engelberger sees great potential for robots in helping to care for the growing population of elderly people who need assistance in the tasks of daily life. He points out that no government or insurance company can afford to hire a full-time human assistant to enable older people to continue to live at home. However, a suitable robot could fetch things, remind a person when it is time to take medication, and even perform medical monitoring and summon help if necessary.

Joseph Engelberger's achievements in industrial and service robotics have won him numerous plaudits and awards from the industry. He has also received honorary doctorates from five institutions, including Carnegie Mellon University in Pittsburgh—one of the great centers of robotics research in the United States.

Since 1977, the Robotics Industries Association has presented the annual Joseph F. Engelberger Award to honor the most significant innovators in the science and technology of robotics. Engelberger was elected to the National Academy of Engineering in 1984. He also received the Progress Award of the Society of Manufacturing Engineers and the Leonardo da Vinci Award of the Society of Mechanical Engineers, as well as the 1982 American Machinist Award. In 1992 Engelberger was included in the London *Sunday Times* series on "The 1000 Makers of the 20th Century." Japan has awarded him the Japan Prize for his key role in the establishment of that nation's thriving robotics industry. In 2000 Engelberger delivered the keynote address to the World Automation Congress, which was also dedicated to him. In 2004 he received the IEEE Robotics and Automation Award.

Further Reading
Brain, Marshall. "Robotic Nation." Available online. URL: http://marshallbrain.com/robotic-nation.htm. Accessed May 3, 2007.
Engelberger, Joseph F. *Robotics in Practice: Management and Applications of Industrial Robots*. New York: AMACOM, 1980.
———. *Robotics in Service*. Cambridge, Mass.: MIT Press, 1989.
———. "Whatever Became of Robotics Research." Robotics Online. http://www.roboticsonline.com/public/articles/articlesdetails.cfm?id=769. Accessed May 3, 2007.
Henderson, Harry. *Modern Robotics: Building Versatile Machines*. New York: Chelsea House, 2006.
Nof, Shimon Y. *Handbook of Industrial Robotics*. 2nd ed. New York: Wiley, 1999.
Robotics Online. Available online. URL: http://www.roboticsonline.com/. Accessed May 3, 2007.

enterprise computing

This concept refers to the organization of data processing and communications across an entire corporation or other organization. Historically, computing technology and infrastructure often developed at different rates in the various departments of a corporation. For example, by the 1970s, departments such as payroll and accounting were making heavy use of electronic data processing (EDP) using mainframe computers. The introduction of the desktop computer in the 1980s often resulted in operations such as marketing, corporate communications, and planning being conducted using a disparate assortment of software, databases, and document repositories. Even the growing use of networking often meant that an enterprise had several different networks with at best rudimentary intercommunication.

The movement toward enterprise computing, while often functioning as a buzzword for the selling of new networking and knowledge management technology, conveys a real need both to manage and leverage the growing information resources used by a large-scale enterprise. The infrastructure for enterprise computing is the network, which today is increasingly built using Internet protocol (see TCP/IP), although legacy networks must often still be supported. Enterprise-oriented software uses the client-server model, with an important decision being which operating systems to support (see CLIENT-SERVER COMPUTING).

The need for flexibility in making data available across the organization is leading to a gradual shift from the older relational database (RDBMs) to object-oriented databases (OODBMs). One advantage of object-oriented databases is that it is more scalable (able to be expanded without running into bottlenecks) and data can be distributed dynamically to take advantage of available computing resources. (An alternative is the central depository. See DATA WAREHOUSE.) The dynamic use of storage resources is also important (see DISK ARRAY).

The payoffs for a well-integrated enterprise information system go beyond efficiency in resource utilization and information delivery. If, for example, the marketing department has full access to data about sales, the data can be analyzed to identify key features of consumer behavior (see DATA MINING).

Further Reading
Bernard, Scott A. *An Introduction to Enterprise Architecture Planning.* 2nd ed. Bloomington, Ind.: AuthorHouse, 2005.
Blanding, Steve, ed. *Enterprise Operations Management Handbook.* 2nd ed. Boca Raton, Fla.: CRC Press, 1999.
Carbone, Jane. *IT Architecture Toolkit.* Upper Saddle River, N.J.: Prentice Hall PTR, 2004.
Zachman Institute for Framework Advancement. Available online. URL: http://www.zifa.com/. Accessed July 29, 2007.

entrepreneurs in computing

Much publicity has been given to figures such as Microsoft founder and multibillionaire Bill Gates, who turned a vest-pocket company selling BASIC language tapes into the dominant seller of operating systems and office software for PCs. Historically, however, the role of key entrepreneurs in the establishment of information technology sectors repeats the achievements of such 19th- and early 20th-century technology pioneers as Thomas Edison and Henry Ford. There appear to be certain times when scientific insight and technological capability can be translated into businesses that have the potential to transform society while making the pioneers wealthy.

Like their counterparts in earlier industrial revolutions, the entrepreneurs who created the modern computer industry tend to share certain common features. In positive terms one can highlight imagination and vision such as that which enabled J. Presber Eckert and John Mauchly to conceive that the general-purpose electronic computer could find an essential place in the business and scientific world (see ECKERT, J. PRESPER and MAUCHLY, JOHN). In the software world, observers point to Bill Gates's intense focus and ability to create and market not just an operating system but also an approach to computing that would transform the office (see GATES, WILLIAM, III). The Internet revolution, too, was sparked by both an "intellectual entrepreneur" such as Tim Berners-Lee, inventor of the World Wide Web (see BERNERS-LEE, TIM) and by Netscape founders Mark Andreessen and Jim Clark, who turned the Web browser into an essential tool for interacting with information both within and outside of organizations.

While technological innovation is important, the ability to create a "social invention"—such as a new vehicle or plan for doing business, can be equally telling. At the beginning of the 21st century, the World Wide Web, effectively less than a decade old, is seeing the struggle of entrepreneurs such as Amazon.com's Jeff Bezos, eBay's Pierre Omidyar, and Yahoo!'s Jerry Yang to expand significant toeholds in the marketing of products and information into sustainable businesses.

Historically, as industries mature, the pure entrepreneur tends to give way to the merely effective CEO. In the computer field, however, it is very hard to sort out the waves of innovation that seem to follow close upon one another. Some sectors, such as the selling of computer systems (a sector dominated by entrepreneurs such as Michael Dell [Dell Computers] and Compaq's Rod Canion) seem to have little remaining scope for innovation. In other sectors, such as operating systems (an area generally dominated by Microsoft), an innovator such as Linus Torvalds (developer of Linux) can suddenly emerge as a viable challenger. And as for the Internet and e-commerce, it is too early to tell whether the pace of innovation has slowed and the shakeout now under way will lead to a relatively stable landscape. (Note: a number of other biographies of computer entrepreneurs are featured in this book. For example, see ANDREESSEN, MARC; BEZOS, JEFFREY P.; ENGELBERGER, JOSEPH; MOORE, GORDON E.; and OMIDYAR, PIERRE.)

Further Reading
Cringely, R. X. *Accidental Empires.* New York: Harper, 1997.
Henderson, Harry. *A to Z of Computer Scientists.* New York: Facts On File, 2003.
———. *Communications and Broadcasting.* (Milestones in Science and Invention). New York: Chelsea House, 2006.
Jager, Rama Dev and Rafael Ortiz. *In the Company of Giants.* New York: McGraw-Hill, 1997.
Malone, Michael S. *Betting It All: The Technology Entrepreneurs.* New York: Wiley, 2001.
———. *The Valley of Heart's Delight: A Silicon Valley Notebook, 1963–2001.* New York: Wiley, 2001.
Reid, R. H. *Architects of the Web.* New York: John Wiley, 1997.
Spector, Robert. *amazon.com: Get Big Fast.* New York: Harper-Business, 2000.

enumerations and sets

It is sometimes useful to have a data structure that holds specific, related data values. For example, if a program is to perform a particular action for data pertaining to each day of the week, the following Pascal code might be used:

```
type Day is (Monday, Tuesday, Wednesday,
Thursday, Friday, Saturday, Sunday)
```

Such a data type (which is also available in Ada, C, and C++) is called an *enumeration* because it enumerates, or "spells out" each and every value that the type can hold.

Once the enumeration is defined, a looping structure can be used to process all of its values, as in:

```
var Today: Day;
for Today: = Monday to Sunday do (some state-
ments)
```

Pascal, C, and C++ do not allow the same item to be used in more than one enumeration in the same name space (area of reference). Ada, however, allows for "overloading" with multiple uses of the same name. In that case, however, the name must be qualified by specifying the enumeration to which it belongs, as in:

```
If Day = Days ('Monday') . . .
```

As far as the compiler is concerned, an enumeration value is actually a sequential integer. That is, Monday = 0, Tuesday = 1, and so on. Indeed, built-in data types such as Boolean are equivalent to enumerations (false = 0, true = 1) and in a sense the integer type itself is an enumeration consisting of 0, 1, 2, 3, . . . and their negative counterparts. Pascal also includes built-in functions to retrieve the preceding value in the enumeration (pred), the following element (succ), or the numeric position of the current element (ord).

The main advantage of using explicit enumerations is that a constant such as "Monday" is more understandable to the program's reader than the value 0. Enumerations are frequently used in C and C++ to specify a limited group of items such as flags indicating the state of device or file operation.

Unlike most other languages Pascal and Ada also allow for the definition of a *subrange,* which is a sequential portion of a previously defined enumeration. For example, once the Day type has been defined, an Ada program can define subranges such as:

```
subtype Weekdays is Days range Monday . .
Friday;
subtype Weekend is Days range Saturday . .
Sunday;
```

SETS

The set type (found only in Pascal and Ada) is similar to an enumeration except the order of the items is not significant. It is useful for checking to see whether the item being considered belongs to a defined group. For example, instead of a program checking whether a character is a vowel as follows:

```
if (char = 'a') or (char = 'e') or (char = 'i')
or (char = 'o') or (char = 'u') . . .
```

the program can define:

```
type Vowels = (a, e, i, o, u);
if char in Vowels . . .
```

Further Reading
Sebesta, Robert W. *Concepts of Programming Languages.* 8th ed. Boston: Pearson, 2007.

ergonomics of computing

Ergonomics is the study of the "fit" between people and their working environment. Because computers are such a significant part of the working life of so many people, finding ways for people to maximize efficiency and reduce health risks associated with computer use is increasingly important.

Since the user will be looking at the computer monitor for hours on end, it is important that the display be large enough to be comfortably readable and that there be enough contrast. Glare on the monitor surface should be avoided. It is recommended that the monitor be placed so that the top line of text is slightly below eye level. A distance of about 18 inches to two feet (roughly arm's length) is recommended. There has been concern about the health effects of electromagnetic radiation generated by monitors. Most new monitors are designed to have lower emissions.

While the "standard" keyboard has changed little in 20 years of desktop computing, there have been attempts at innovation. One, the Dvorak keyboard, uses an alternative arrangement of letters to the standard "QWERTY." Although it is a more logical arrangement from the point of view of character frequency, studies have generally failed to show sufficient advantage that would compensate for the effort of retraining millions of typists. There have also been specially shaped "ergonomic" keyboards that attempt to bring the keys into a more natural relationship with the hand (see KEYBOARD).

The use of a padded wrist rest remains controversial. While some experts believe it may reduce strain on the arm and neck, others believe it can contribute to Carpal Tunnel Syndrome. This injury, one of the most serious repetitive stress injuries (RSIs), is caused by compression of a nerve within the wrist and hand.

Because of reliance on the mouse in many applications, experts suggest selecting a mouse that comfortably fits the hand, with the buttons falling "naturally" under the fingers. When moving the mouse, the forearm, wrist, and fingers should be kept straight (that is, in line with the mouse). Some people may prefer the use of an alternative pointing device (such as trackball or "stub" within the keyboard itself, often found in laptop computers).

A variety of so-called ergonomic chairs of varying quality are available. Such a chair can be a good investment in worker safety and productivity, but for best results the chair must be selected and adjusted after a careful analysis of the individual's body proportions, the configuration of the workstation, and the type of applications being used. In general, a good ergonomic chair should have an adjustable seat and backrest and feel stable rather than rickety.

The operating system and software in use are also important. Providing clear, legible text, icons or other controls and a consistent interface will contribute to the user's overall sense of comfort, as well as reducing eyestrain. It is also important to try to eliminate unnecessary repetitive motion. For example, it is helpful to provide shortcut key combinations that can be used instead of a series of mouse movements. Beyond specific devices, the development of an integrated design that reduces stress and improves usability is part of what is sometimes called human factors research.

In March 2001, President Bush cancelled new OSHA standards that would have further emphasized reporting and mitigating repetitive stress and musculo-skeletal disorders (MSDs). However, the legal and regulatory climate is likely to continue to place pressure on employers to take ergonomic considerations into account.

Further Reading
Coe, Marlana. *Human Factors for Technical Communicators.* New York: Wiley, 1996.
Dul, Jan, and Bernard Weerdmeester. *Ergonomics for Beginners: A Quick Reference Guide.* 2nd ed. Grand Rapids, Mich.: CRC Press, 2001.
"Ergonomic Design for Computer Workstations." Available online. URL: http://www.ergoindemand.com/computer-workstation-ergonomics.htm. Accessed July 30, 2007.

Salvendy, Gavriel. *Handbook of Human Factors and Ergonomics.* 3rd ed. New York: Wiley, 2006.

U.S. Department of Labor. Occupational Safety & Health Administration. "Computer Workstations." Available online. URL: http://www.osha.gov/SLTC/etools/computerworkstations/index.html. Accessed July 30, 2007.

Vredenberg, Karel, Scott Isensee, and Carol Righi. *User-Centered Design: an Integrated Approach.* Upper Saddle River, N.J.: Prentice Hall, 2001.

error correction

Transmitting data involves the sending of bits (ones and zeros) as signaled by some alternation in physical characteristics (such as voltage or frequency). There are a number of ways in which errors can be introduced into the data stream. For example, electrical "noise" in the line might be interpreted as spurious bits, or a bit might be "flipped" from one to zero or vice versa. Generally speaking, the faster the rates at which bits are being sent, the more sensitive the transmission is to effects that can cause errors.

While a few wrong characters might be tolerated in some text messages or graphics files, binary files representing executable programs must generally be received perfectly, since random changes can make programs fail or produce incorrect results. Data communications engineers have devised a number of methods for checking the accuracy of data transmissions.

The simplest scheme is called *parity.* A single bit is added to each eight-bit byte of data. In even parity, the extra (parity) bit is set to one when the number of ones in the byte is odd. In odd parity, a one is added if the data byte has an even number of ones. This means that the receiver of the data can expect it to be even or odd respectively. When the byte arrives at its destination, the receiving program checks the parity bit and then counts the number of ones in the rest of the byte. If, for example, the parity is even but the data as received has an odd number of ones, then at least one of the bits must have been changed in error. Parity is a fast, easy way to check for errors, but it has some unreliability. For example, if there were *two* errors in trans-

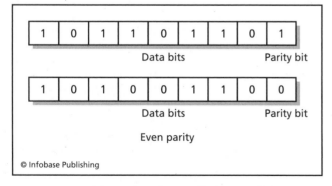

Even parity

© Infobase Publishing

For even parity, if the number of ones in the byte is odd, the parity bit is set to one to make the total number of ones even. Odd parity would work the same way, except the parity bit would be set when necessary to ensure an odd number of ones.

mission such that a one became a zero and a different zero became a one, the parity would be unchanged and the error would not be detected.

The checksum method offers greater reliability. The binary value of each block of data is added and the sum is sent along with the block. At the destination, the bits in the block are again added to see if they still match the sum. A variation, the *cyclical redundancy check* or CRC, breaks the data into blocks and divides them by a fixed number. The remainder for the division for each block is appended to the block and the calculation is repeated and checked at the destination. Today most modem control software implements parity or CRC checking.

A more sophisticated method called the *Hamming Code* offers not only high reliability but also the ability to automatically correct errors. In this scheme the data and check bits are encrypted together to create a code word. If the word received is not a valid code word, the receiver can use a series of parity checks to find the original error. Increasing the ratio of redundant check bits to message bits improves the reliability of the code, but at the expense of having to do more processing to encrypt that data and requiring more time to transmit it.

Further Reading

"Error-Correcting Code." Wolfram MathWorld. Available online. URL: http://mathworld.wolfram.com/Error-CorrectingCode.html. Accessed July 31, 2007.

Fung, Francis Yein Chei. "A Survey of the Theory of Error-Correcting Codes." Available online. URL: http://cadigweb.ew.usna.edu/~wdj/teach/ecc/codes.html. Accessed July 31, 2007.

Wicker, S. B. *Error Control Systems for Digital Communication and Storage.* Upper Saddle River, N.J.: Prentice Hall, 1995.

error handling

An important characteristic of quality software is its ability to handle errors that arise in processing (also called *run-time errors* or "exceptions"). Before it is released for general use, a program should be thoroughly tested with a variety of input (see QUALITY ASSURANCE, SOFTWARE). When errors are found, the soundness of the algorithm and its implementation must be checked, as well as the program logic (see ALGORITHM). Interaction between the program and other programs (including the operating system) as well as with hardware must also be considered. (See BUGS AND DEBUGGING.)

However, even well-tested software is likely to encounter errors. Therefore a program intended for widespread use must include instructions for dealing with errors, anticipated or otherwise. The process of error handling can be divided into four stages: validation, detection, communication, and amelioration.

Data validation is the first line of defense. At the "front end" of the program, data being entered by a user (or read from a disk file or communications link) is checked to see whether it falls within the prescribed parameters. (In the case of a program such as a data management system, the user interface plays an important role. Data input fields can be designed so that they accept only valid characters.

On-line help and error messages can explain to users why a particular input is invalid.)

However, data validation can ensure only that data falls within the generally acceptable parameters. Some particular combination or context of data might still be erroneous, and calculations performed within the program can also produce errors. Some examples include a divisor becoming zero (not allowable mathematically) or a number overflowing or underflowing (becoming too large or too small for register or memory space allotted for it).

Error communication is generally handled by a set of error codes (special numeric values) returned to the main program by the function used to perform the calculation. In addition, errors that arise in file processing (such as "file not found") also return error codes. For example, suppose there is a division function in C++

```
double Quotient(double dividend, double
divisor) throw(ZERODIV)
{
  if (0.0 == divisor)
    throw ZERODIV();
  return dividend / divisor;
}
```

In C++ "throw" means to post an error that can be "caught" by the appropriate error-handling routine. Thus, the corresponding "catch" code might have:

```
catch( ZERODIV )
{
  cout << "Division by zero error!" << endl;
}
```

Once an error has been detected and communicated, decision statements (branches or loops) can check for the presence of error codes and execute appropriate instructions based on what is encountered. (In object-oriented languages such as C++ special classes and objects are often used to handle errors.)

Many simple utility programs respond to errors by issuing an error message and then quitting. However, many real-world applications must be able to respond to errors and continue processing (for example, a program reading data from a scientific instrument may have to deal with the occasional "outlier" or a strange value caused by a burst of interference). Depending on circumstances, the error amelioration code might simply reject the erroneous data or result, ask for the data to be re-sent, or keep a log or statistics of the number and kind of errors encountered. More sophisticated approaches based on mathematical error analysis are also possible.

Further Reading

Dony, Christophe. *Advanced Topics in Exception Handling Techniques*. New York: Springer, 2006.

"Exceptions" [Java tutorials]. Available online. URL: http://java.sun.com/docs/books/tutorial/essential/exceptions/. Accessed July 31, 2007.

Romanovsky, Alexander, et al., eds. *Advances in Exception Handling Techniques*. New York: Springer, 2001.

Soulle, Juan. "Exceptions" [C++]. Available online. URL: http://www.cplusplus.com/doc/tutorial/exceptions.html. Accessed July 31, 2007.

expert systems

An expert system is a computer program that uses encoded knowledge and rules of reasoning to draw conclusions or solve problems. Since reasoning (as opposed to mechanical calculation) is a form of intelligent behavior, the field of expert systems (also called knowledge representation or knowledge engineering) is part of the broader field of AI (see ARTIFICIAL INTELLIGENCE).

HISTORY AND APPLICATIONS

By the end of the 1950s, early research in artificial intelligence was producing encouraging results. A number of tasks associated with human reasoning seemed to be well within the capabilities of computers. Early checkers and chess programs, while far from expert level, were steadily improving. Computer programs were proving geometry theorems. One of the most important AI pioneers, John McCarthy, declared that in principle all human knowledge could be encoded in such a way that programs could "understand" and reason from that knowledge to new conclusions.

Two disparate approaches to achieving AI gradually emerged. In the early 1960s, many researchers tried to generalize the automated reasoning process so that a program could analyze and solve a wide variety of problems, much in the way a human being can. The resulting programs were indeed flexible, but it was difficult to work with anything other than simplified problems. (The SHRDLU program, for example, worked in an abstract world of blocks on a table.)

The other approach was to try to provide exhaustively specified rules for dealing with a more narrowly defined realm of knowledge. The DENDRAL program, developed in the mid-1960s by Edward A. Feigenbaum and associates, was designed to analyze the mass spectra of organic molecules according to theories employed by chemists (see FEIGENBAUM, EDWARD). It eventually became clear that the key to the success for such program lay more in the "capturing" and encoding of expert knowledge than in the development of more flexible methods of reasoning. The methods for encoding and working with the knowledge were refined and further developed into a variety of expert systems during the 1970s.

In the 1980s, expert system technology became mature enough to leave the laboratories and play a role in industry. Two early applications were Digital Equipment Corporation's XCON, which automatically configured minicomputers from component parts at a rate and accuracy far surpassing that of human engineers. Another, Dipmeter Advisor, used real-time data to predict the dip (tilt) of rock layers in a drill bore. (This information was crucial for determining the feasibility of an oil or gas well.)

Today expert systems are a mature technology (and indeed, the most tangible success of AI research in practical applications). Expert systems are used in applications as diverse as engine troubleshooting, diagnosis of rare diseases, and investment analysis.

ANATOMY OF AN EXPERT SYSTEM

An expert system has two main components, a *knowledge base* and an *inference engine*. The knowledge base consists of a set of assertions (facts) or of rules expressed as if . . . then statements that specify conditions that, if true, allow a particular inference to be drawn (see PROLOG). The inference engine accepts new assertions or queries and tests them against the stored rules. Because satisfying one rule can create a condition that is to be tested by a subsequent rule, chains of reasoning can be built up. If the reasoning is from initial facts to an ultimate conclusion, it is called forward chaining. If a conclusion is given and the goal is to prove that conclusion, there can be backward chaining from the conclusion to the assertions (similar to axioms in mathematical proofs).

While some rules are ironclad (for example, if a closed straight figure has three sides, it's a triangle) in many real-world applications it is necessary to take a probabilistic approach. For example, experience might suggest that if a customer buys reference books there is a 40 percent chance the customer will also buy a related CD-ROM product. Thus, rules can be given weights or *confidence factors* and as the rules are chained, a cumulative probability for the conclusion can be generated and some threshold probability for asserting a conclusion can be specified. (See also FUZZY LOGIC and UNCERTAINTY).

While rules-based inference systems are relatively easy to traverse automatically, they may lack the flexibility to codify the knowledge needed for complex activities (such as automatic analysis of news stories). An alternative approach involves the construction of a knowledge base consisting of *frames*. A frame (also called a *schema*) is an encoded description of the characteristics and relationships of entities. For example, an expert system designed to analyze court cases might have frames that describe the roles and interests of the defendant, defense counsel, prosecutor, and so on, and other frames describing the trial and sentencing process. Using this knowledge, the system might be able to predict what sort of plea agreement a particular defendant might reach with the state. While potentially more robust than a rules-based system, a frames-based system faces the twin challenges of building and maintaining a complex and open-ended knowledge base and of developing methods of reasoning more akin to generalized artificial intelligence (see ARTIFICIAL INTELLIGENCE).

TRENDS

Expert systems (particularly of the rules-based variety) now have an established place in business, industry, and science. The field of genomics and genetic engineering, widely seen as the "technology of the 21st century" may be a par-

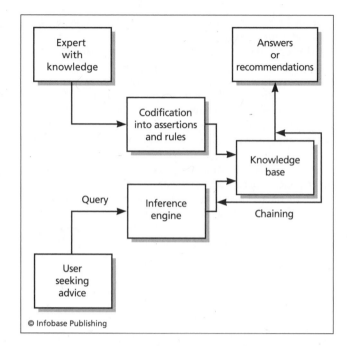

Building an expert system requires that the knowledge of experts be "captured" in the form of a series of assertions and rules called a knowledge base. Once the knowledge base is established, users seeking advice can use an inference engine to examine the knowledge base for valid conclusions that can be expressed as recommendations, often with varying degrees of confidence.

ticularly fruitful applications area for analytical expert systems. Another promising area is the use of expert systems for e-commerce marketing analysis (see DATA MINING). An emerging emphasis in expert system development is the use of object-oriented concepts (see OBJECT-ORIENTED PROGRAMMING) and distributed database and knowledge sharing technology to build and maintain large knowledge bases more efficiently.

Further Reading

"Expert Systems." American Association for Artificial Intelligence. Available online. URL: http://www.aaai.org/AITopics/html/expert.html. Accessed July 31, 2007.

Giarrartano, Joseph C., and Gary D. Riley. *Expert Systems: Principles and Programming.* 4th ed. Boston: Thomson Course Technology, 2004.

"Introduction to Expert Systems." Available online. URL: http://www.expertise2go.com/webesie/tutorials/ESIntro/. Accessed July 31, 2007.

Jackson, Peter. *Introduction to Expert Systems.* 3rd ed. Reading, Mass.: Addison-Wesley, 1998.

F

fault tolerance

Fault tolerance is a design concept that recognizes that all computer-based systems will fail eventually. The question is whether a system as a whole can be designed to "fail gracefully." This means that even if one or more components fail, the system will continue to operate according to its design specifications, even if its speed or throughput must decrease.

METHODS AND IMPLEMENTATIONS

There are a number of ways to make a system more fault tolerant. Individual components such as hard drives can be composed of multiple units so that the remaining units can take over if one fails (see also RAID). If each key component has at least one backup, then there should be time to replace the primary before the backup also fails.

Another way to achieve fault tolerance is to provide multiple paths to successful completion of the task. In fact, this is how packet-switched networks like the Internet work (see TCP/IP). If one communications link is down or too congested, packets are given an alternative routing.

Fault diagnosis software can also play an important role both in determining how to respond to a problem (beyond any automatic response) and for providing data that will be useful later to system administrators or technicians. Some fault diagnosis systems can use elaborate rules (see EXPERT SYSTEMS) to pinpoint the cause of a fault and recommend a solution.

The amount of fault tolerance to be provided for a system depends on a number of factors:

- How important is it that the system *not* fail?

- How critical is a given component to the operation of the system?

- How likely is it that a given component will fail? (Mean time between failures, or MBTF)

- How expensive is it to make the component or system fault tolerant?

A related concept is fail-safe. While fault tolerance emphasizes continued operation despite one or more failures, fail-safe emphasizes the ability to shut down safely in case of an unrecoverable failure. With computer-based systems, fail-safe design can use redundant systems (as in avionics) to perform calculations, with a failing system "outvoted" if necessary by the good ones. In most cases there should also be a provision to alert the pilot or operator in time to take over operations from the automatic system.

Another common example of fail-safe is modern operating systems that create a "journal" of pending operations to files that can be used to restore the integrity of the system after a power failure or other abrupt shutdown (see FILE SYSTEM.)

Further Reading

Isermann, Rolf. *Fault-Diagnosis Systems: An Introduction from Fault Detection to Fault Tolerance.* New York: Springer, 2006.
Koren, Israel, and C. Mani Krishna. *Fault-Tolerant Systems.* San Francisco: Morgan Kauffman, 2007.
National Institute of Standards and Technology. "A Conceptual Framework for System Fault Tolerance." Available online. URL:

http://hissa.nist.gov/chissa/SEI_Framework/framework_
1.html. Accessed September 20, 2007.
Pullum, Laura L. *Software Fault Tolerance: Techniques and Implementation.* Norwood, Mass.: Artech House, 2001.

Feigenbaum, Edward

(1936–)
American
Computer Scientist

Edward Feigenbaum is a pioneer artificial intelligence
researcher, best known for his development of expert systems (see ARTIFICIAL INTELLIGENCE). Feigenbaum was born
in Weehawken, New Jersey. His father, a Polish immigrant,
died before Feigenbaum's first birthday. His stepfather, an
accountant and bakery manager, was fascinated by science
and regularly brought young Edward to the Hayden Plane-
tarium's shows and to every department of the vast Museum
of Natural History. The electromechanical calculator his
father used to keep accounts at the bakery particularly fascinated Edward. His interest in science gradually turned to
a perhaps more practical interest in electrical engineering.

While at the Carnegie Institute of Technology (now
Carnegie Mellon University), Feigenbaum was encouraged to venture beyond the more mundane curriculum to
the emerging field of computation. He became interested
in John von Neumann's work in game theory and decision making and also met Herbert Simon, who was conducting pioneering research into how organizations made
decisions (see VON NEUMANN, JOHN). This in turn brought
Feigenbaum into the early ferment of artificial intelligence
research in the mid-1950s. Simon and Alan Newell had just
developed Logic Theorist, a program that simulated the
process by which mathematicians proved theorems through
the application of heuristics, or strategies for breaking problems down into simpler components from which a chain of
assertions could be assembled leading to a proof.

Feigenbaum quickly learned to program IBM mainframes and then began writing AI programs. For his doctoral thesis, he explored the relation of artificial problem
solving to the operation of the human mind. He wrote a
computer program that could simulate the human process of perceiving, memorizing, and organizing data for
retrieval. Feigenbaum's program, the Elementary Perceiver
and Memorizer (EPAM), was a seminal contribution to AI.
Its "discrimination net," which attempted to distinguish
between different stimuli by retaining key bits of information, would eventually evolve into the *neural network* (see
NEURAL NETWORK). Together with Julian Feldman, Feigenbaum edited the 1962 book *Computers and Thought,* which
summarized both the remarkable progress and perplexing
difficulties encountered during the field's first decade.

During the 1960s, Feigenbaum worked to develop systems that could perform induction (that is, derive general
principles based on the accumulation of data about specific
cases). Working on a project to develop a mass spectrom-
eter for a Mars probe, Feigenbaum and his fellow researchers became frustrated at the computer's lack of knowledge

about basic rules of chemistry. Feigenbaum then decided
that such rules (or knowledge) might be encoded in such
a way that the program could apply it to the data being
gathered from chemical samples. The result in 1965 was
Dendral, the first of what would become a host of successful and productive expert systems (see EXPERT SYSTEM). A
further advance came in 1970 with Meta-Dendral, a program that could not only apply existing rules to determine
the structure of a compound, it could also compare known
structures with the existing database of rules and infer new
rules, thus improving its own performance.

During the 1980s, Feigenbaum coedited the four-volume
Handbook of Artificial Intelligence. He also introduced expert
systems to a lay audience in two books, *The Fifth Generation*
(co-authored with Pamela McCorduck) and *The Rise of the
Expert Company.*

Feigenbaum combined scientific creativity with entrepreneurship in founding a company called IntelliGenetics
and serving as a director of Teknowledge and IntelliCorp.
These companies pioneered the commercialization of
expert systems. In doing so, Feigenbaum and his colleagues
publicized the discipline of "knowledge engineering"—the
capturing and encoding of professional knowledge in medicine, chemistry, engineering, and other fields so that it can
be used by an expert system. In what he calls the "knowledge principle" he asserts that the quality of knowledge in
a system is more important than the algorithms used for
reasoning. Thus, Feigenbaum has tried to develop knowledge bases that might be maintained and shared as easily as
conventional databases.

Remaining active in the 1990s, Feigenbaum was second
president of the American Association for Artificial Intel-
ligence and (from 1994 to 1997) chief scientist of the U.S.
Air Force. In 1995, Feigenbaum received the Association for
Computing Machinery's prestigious A. M. Turing Award.
Founder of the Knowledge Systems Laboratory at Stanford
University, Feigenbaum remains a professor emeritus of
computer science at that institution.

Further Reading

Feigenbaum, Edward, Julian Feldman, and Paul Armer, eds. *Com-
puters and Thought.* Cambridge, Mass.: MIT Press, 1995.
Feigenbaum, Edward, Pamela McCorduck, and H. Penny Nii. *The
Rise of the Expert Company: How Visionary Companies are
Using Artificial Intelligence to Achieve Higher Productivity and
Profits.* New York: Vintage Books, 1989.
Henderson, Harry. *Artificial Intelligence: Mirrors for the Mind.* New
York: Chelsea House, 2007.
Shasha, Dennis, and Cathy Lazere. *Out of Their Minds: The Lives
and Discoveries of 15 Great Computer Scientists.* New York:
Copernicus/Springer-Verlag, 1995.

fiber optics

A fiber optic (or optical fiber) cable transmits photons
(light) instead of electrons. Depending on the diameter of
the cable, the light is guided either by total internal reflec-
tion or as a waveguide (manipulating refraction). These
principles were known as early as the mid-19th century and
began to be used in the 20th century for such applications

as dental and medical illumination and in experiments in transmitting images for television.

DEVELOPMENT

Optical fiber in its modern form was developed in the 1950s. The glass fiber through which the light passes is surrounded by a transparent cladding designed to provide the needed refractive index to keep the light confined. The cladding in turn is surrounded by a resin buffer layer and often an outer jacket and plastic cover. Fiber used for communication is flexible, allowing it to bend if necessary.

Early optical fiber could not be used for practical communication because of progressive attenuation (weakening) of the light as it traveled. However, by the 1970s the attenuation was being reduced to acceptable levels by removing impurities from the fibers. Today the light signals can travel hundreds of miles without the need for repeaters or amplifiers. In the 1990s a new type of optical fiber (photonic crystal) using diffraction became available. This kind of fiber is particularly useful in applications that require higher power signals.

COMMUNICATIONS AND NETWORK APPLICATIONS

Optical fiber has several advantages over ordinary electric cable for communications and networking. The signals can travel much farther without the need for a repeater to boost the signal. Also, the ability to modulate wavelengths allows optical fiber to carry many separate channels, greatly increasing the total data throughput. Optical fiber does not emit RF (radio frequency) energy, a source of "cross talk" (interference) in electrical cable. Fiber is also more secure than electrical cable because it is hard for an eavesdropper to tap.

Today fiber is used for most long-distance phone lines and Internet connections. Many cable television systems are upgrading from video cable to fiber because of its greater reliability and ability to carry more bandwidth and enhanced data services.

The last area where electrical (copper) cable predominates is in the "last mile" between main lines and houses or buildings, and within local networks. However, new buildings and higher-end homes often include built-in fiber. Increasingly, phone companies are upgrading service by bridging the last mile through fiber-to-the-home (FTTH) networks. While requiring a considerable investment, FITH allows phone companies to replace relatively slow DSL with faster (higher bandwidth) service better suited to deliver video, data, and phone service simultaneously (see BANDWIDTH, CABLE MODEM, and DSL). As of 2008, 3.3 million American homes had fiber connections, mainly through Verizon's FIOS service. It is expected that FTTH will be built into many new housing developments.

In 2007 Corning announced the development of "nanostructured" optical fibers that can be bent more sharply (such as around corners) without loss of signal. Corning is working with Verizon to develop easier and cheaper ways to provide FTTH.

(Related optical principles can also be applied to computer design. See OPTICAL COMPUTING.)

Further Reading

The Fiber Optic Association. "User's Guide to Fiber Optic System Design and Installation." Available online. URL: http://www.thefoa.org/user/. Accessed September 20, 2007.

"Fiber Optics: The Basics of Fiber Optic Cable: A Tutorial." Available online. URL: http://www.arcelect.com/fibercable.htm. Accessed September 20, 2007.

Fiber to the Home Council. Available online. URL: http://www.ftthcouncil.org/. Accessed September 20, 2007.

Hecht, Jeff. *Understanding Fiber Optics.* 5th ed. Upper Saddle River, N.J.: Prentice Hall, 2005.

Palais, Joseph C. *Fiber Optic Communications.* 5th ed. Upper Saddle River, N.J.: Prentice Hall, 2004.

file

At bottom, information in a computer is stored as a series of bits, which can be grouped into larger units such as bytes or "words" that represent particular numbers or characters. In order to be stored and retrieved, a collection of such binary data must be given a name and certain attributes that describe how the information can be accessed. This named entity is the file.

FILES AND THE OPERATING SYSTEM

Files can be discussed at three levels, the physical layout, the operating system, and the application program. At the physical level, a file is stored on a particular medium. (See FLOPPY DISK, HARD DISK, CD-ROM, and TAPE DRIVES.) On disk devices a file takes up a certain number of sectors, which are portions of concentric tracks. (On tape, files are usually stored as contiguous segments or "blocks" of data.)

The file system is the facility of the operating system that organizes files (see OPERATING SYSTEM). For example, on DOS and older Windows PCs, there is a file allocation table (FAT) that consists of a linked list of clusters (each cluster consists of a fixed number of sectors, varying with the overall size of the disk). When the operating system is asked to access a file, it can go through the table and find the clusters belonging to that file, read the data and send it to the requesting application. Modern file systems further organize files into groups called folders or directories, which can be nested several layers deep. Such a hierarchical file system makes it easier for users to organize the dozens of applications and thousands of files found on today's PCs. For example, a folder called Book might have a subfolder for each chapter, which in turn contains folders for text and illustrations relating to that chapter.

Besides storing and retrieving files, the modern file system sets characteristics or attributes for each file. Typical attributes include write (the file can be changed), read (the file can be accessed but not changed), and archive (which determines whether the file needs to be included in the next backup). In multi-user operating systems such as UNIX there are also attributes that indicate ownership (that is, who has certain rights with regard to the file). Thus a file may be executable (run as a program) by anyone, but writeable (changeable) only by someone who has "superuser" status (see also DATA SECURITY).

The current generation of file systems for PCs includes additional features that promote efficiency and particularly data integrity. Versions of Windows starting with NT, 2000, and XP come standardly with NTFS, the "New Technology File System," which includes journaling, or the keeping of a record of all transactions affecting the system (such as deleting or adding a file). In the event of a mishap such as a power failure, the transactions can be "replayed" from the journal, ensuring that the file system reflects the actual current status of all files. NTFS also uses "metadata" that describes each file or directory. Database principles can thus be applied to organizing and retrieving files at a higher level.

Linux (based on UNIX) uses a single file system hierarchy that incorporates all devices in the system. (The network file system, NFS, effectively extends the hierarchy to all machines on the local network.) The popular Linux ext3 file system also includes journaling.

FILES AND APPLICATIONS

The ultimate organization of data in a file depends on the application. A typical approach is to define a data record with various fields. The program might have a loop that repeatedly requests a record from the file, processes it in some way, and repeats until the operating system tells it that it has reached the end of the file. This would be a *sequential* access; a program can also be set up for *random* access, which means that an arbitrary record can be requested and that request will be translated into the correct physical location in the file. The two approaches can be combined in ISAM (Indexed Sequential Access Method), where the records are stored sequentially but fields are indexed so a particular record can be retrieved.

Since files such as graphics (images), sound, and formatted word processing documents can only be read and used by particular applications, files are often given names with extensions that describe their format. When a Windows user sees, for example, a Microsoft Word document, the filename will have a .DOC extension (as in chapter.doc) and will be shown with an icon registered by the application for such files. Further, a file association will be registered so that when a user opens such a file the Word program will run and load it.

From a user interface point of view, the use of the file as the main unit of data has been criticized as not corresponding to the actual flow of most kinds of work. While from the computer's point of view, the user is opening, modifying, and saving a succession of separate files, the user often thinks in terms of working with documents (which may have components stored in a number of separate files.) Thus, many office software applications offer a document-oriented or project-oriented view of data that hides or minimizes the details of individual files (see DOCUMENT MODEL).

Further Reading

Callaghan, Brent. *XFS Illustrated*. Reading, Mass.: Addison-Wesley Professional, 1999.
Matloff, Norman. "File Systems in UNIX." Available online. URL: http://heather.cs.ucdavis.edu/~matloff/UnixAndC/Unix/FileSyst.html. Accessed August 1, 2007.
Nagar, Rajeev. *Windows NT File System Internals*. Reprint. Amherst, N.H.: OSR Press, 2006.
Pate, Steven. *UNIX File Systems: Evolution, Design, and Implementation*. New York: Wiley, 2003.

file server

The growth in desktop computing since the 1980s has resulted in much data being moved from mainframe computers to desktop PCs, which are now usually linked by networks. While a network enables users to exchange files, there remains the problem of storing large files or collections of files (such as databases) that are too large for a typical PC hard drive or that need to be accessed and updated by many users.

The common solution is to obtain a computer with large, fast disk drives (see also RAID). This computer, the file server, is equipped with software (often included with the networking package) that serves (provides) files as requested by users or applications on the other PCs on the network. (See also CLIENT-SERVER COMPUTING.) The specifics of configuring the server for optimum efficiency, providing adequate security, and arranging for backup or archiving varies with the particular network operating system in use (the most popular environments are Windows NT, Vista, and the various versions of UNIX and Linux).

The file server has many advantages over storing the files needed by each user on his or her own PC. By storing the files on a central server, ordinary users' PCs do not need to have larger, more expensive disk drives. Central storage also makes it easier to ensure that backups are run regularly (see BACKUP AND ARCHIVE SYSTEMS).

There are some potential problems with this approach. With central storage, a failure of the file server could bring work throughout the network to a halt. (The use of RAID with its redundant "mirror" disks is designed to prevent the failure of a single drive from making data inaccessible). As the network and/or size of the data store gets larger, multiple servers are usually used. The performance of a file server is also greatly affected by the efficiency of the caching mechanism used (see CACHE).

As the amount of data that must be stored increases, organizations will consider storage area network (SAN) and network attached storage (NAS, see NETWORKED STORAGE) technologies. SAN makes it easier for numerous users to share a resource such as an automated tape library or disk RAID, while NAS is an efficient way to allow files to be centrally stored but readily shared.

All but the simplest servers require special software or extensions to the operating system. For example, Microsoft Windows Server is essentially a version of Windows with built-in facilities for managing a file or application server, including the ability to organize "clusters" of servers and balance the load of requests. Linux often comes in server versions as well, though this is basically simply a distribution preconfigured with the programs needed to manage servers (such as Samba).

Meanwhile, the reason for having a file server is changing. Cost of storage is much less of an issue for smaller offices with the recent availability of high-capacity drives (500 GB or more) starting at approximately $100. However, a central server still may offer better security and can serve as a central repository from which documents or source code can be "checked out" and updated in an orderly way (version control).

Further Reading

"Designing and Deploying File Servers." Microsoft TechNet. Available online. URL: http://technet2.microsoft.com/ windowsserver/en/library/42befce4-7c15-4306-8edc-a80b 8c57c67d1033.mspx. Accessed August 1, 2007.

Eckstein, Robert, David Collier-Brown, and Peter Kelly. *Using Samba*. 2nd ed. Sebastapol, Calif.: O'Reilly Media, 2003.

Matthews, Martin S. *Windows Server 2003: A Beginner's Guide*. 2nd ed. Berkeley, Calif.: McGraw-Hill Osborne, 2003.

"Samba: Opening Windows to a Wider World." Available online. URL: http://us3.samba.org. Accessed August 1, 2007.

Tulloch, Mitch. *Introducing Windows Server 2008*. Redmond, Wash.: Microsoft Press, 2007.

file-sharing and P2P networks

File-sharing services allow participants to provide access to files on their personal computers, such as music or video. In turn, the user can browse the service to find and download material of interest. The structure is generally that of a peer-to-peer (P2P) network with no central server.

The first major file-sharing service was Napster. This was a P2P network but had a central server that provided the searchable list of files and locations—but not the files themselves, which were downloaded from users' PCs. Napster was forced to close in 2001 by legal action from copyright holders (see INTELLECTUAL PROPERTY AND COMPUTING). A new but unrelated for-pay service opened later under the same name.

Because of the legal vulnerability of centralized-list P2P services, a new model was developed, typified by Gnutella. This is a fully P2P model with both indexing and data decentralized in nodes throughout the network. As of mid-2006, Gnutella and similar services such as Kazaa had an estimated 10 million users.

BITTORRENT

Many services today use the popular BitTorrent file-sharing protocol. A BitTorrent client (either the program of that name or another compatible one) can transmit or receive any type of data. To share a file, the client creates a "torrent"—a small file that contains metadata describing the file and an assignment to a "tracker." The tracker is another computer (node) that coordinates the distribution of the file. Although this sounds complicated and a request takes longer to set up than an ordinary HTTP connection, the advantage is that once set up, downloading is efficiently managed even for files for which there is high demand. The downloading client connects to multiple clients that provide pieces of the desired file. Because of its efficiency, BitTorrent allows for distribution of substantial amounts of data at low cost, particularly since the system "scales up" automatically without having to provide extra resources. BitTorrent is currently being used for a variety of legally distributed material, including video, sound, and textual content (see BLOGS AND BLOGGING, PODCASTING, and RSS).

LEGAL ISSUES

Because of their frequent use to share copyrighted music, video, or other material, a variety of organizations of copyright owners have sued file-sharing services and/or their users. The biggest problem for the courts is to determine whether there is "substantial non-infringing use"—that is, the service is being used to exchange legal data.

Some file-sharing services have been accused of distributing malware (viruses or spyware) or of being used to distribute material that is illegal per se (such as child pornography).

In response to litigation threats, file-sharing services have tended to become more decentralized, and some have features that increase anonymity of users (see ANONYMITY AND THE INTERNET) or use encryption.

Further Reading

BitTorrent. Available online. URL: http://www.bittorrent.com/. Accessed September 20, 2007.

Gardner, Susannah, and Kris Krug. *BitTorrent for Dummies*. Hoboken, N.J.: Wiley, 2006.

Gnutella. Available online. URL: http://www.gnutella.com/. Accessed September 20, 2007.

Roush, Wade. "P2P: From Internet Scourge to Savior." *Technology Review*. December 15, 2006. Available online. URL: http://www.technologyreview.com/Biztech/17904. Accessed September 21, 2007.

Schmidt, Aernout, Wilfred Dolfsma, and Wim Keuvelaar. *Fighting the War on File Sharing*. Cambridge: Cambridge University Press, 2007.

Silverthorne, Sean. "Music Downloads: Pirates—or Customers?" [Q&A with Felix Oberholzer-Gee]. Harvard Business School Working Knowledge. Available online. URL: http://hbswk. hbs.edu/item/4206.html. Accessed September 20, 2007.

Wang, Wallace. *Steal This File Sharing Book: What They Won't Tell You About File Sharing*. San Francisco: No Starch Press, 2004.

file transfer protocols

With today's networked PCs and the use of e-mail attachments it is easy to send a copy of a file or files from one computer to another, because networks already include all the facilities for doing so. Earlier, many PCs were not networked but could be connected via a dial-up modem. To established the connection, a terminal program running on one PC had to negotiate with its counterpart on the other machine, agreeing on whether data would be sent in 7- or 8-bit chunks, and the number of parity bits that would be included for error-checking (see ERROR CORRECTION). The sending program would inform the receiving program as to the name and basic type of the file. For binary files (files intended to be interpreted as literal binary codes, as with executable programs, images, and so on) the contents would be sent unchanged. For text files, there might be the issue of which character set (7- bit or 8-bit ASCII) was being used, and whether the ends of lines were to be marked with a CR (carriage return) character, an LF (linefeed), or both (see CHARACTERS AND STRINGS).

IMPLEMENTATIONS

Once the programs agree on the basic parameters for a file transfer, the transfer has to be managed to ensure that it completes correctly. Typically, files are divided into blocks of data (such as 1K, or 1024 bytes each). During the 1970s, Ward Christensen developed Xmodem, the first widely used file transfer program for PCs running CP/M (and later, MS-DOS and other operating systems). Xmodem was quite reliable because it incorporated a checksum (and later, a more advanced CRC) to check the integrity of each data block. If an error is detected, the receiving program requests a retransmission.

The Ymodem program adds the capability of specifying and sending a batch of files. Zmodem, the latest in this line of evolution, automatically adjusts for the amount of errors caused by line conditions by changing the size of the data blocks used and also includes the ability to resume after an interrupted file transfer. Another widely used file transfer protocol is Kermit, which has been implemented for virtually every platform and operating system. Besides file transfer, Kermit software offers terminal emulation and scripting capabilities. However, despite their robustness and capability, Zmodem and Kermit have been largely supplanted by the ubiquitous Web download link.

In the UNIX world, the ftp (file transfer protocol) program has been a reliable workhorse for almost 30 years. With ftp, the user at the PC or terminal connects to an ftp server on the machine that has the desired files. A variety of commands are available for specifying the directory, listing the files in the directory, specifying binary or text mode, and so on. While the traditional implementation uses typed text commands, there are now many ftp clients available for PCs that use a graphical interface with menus and buttons and allow files to be selected and dragged between the local and remote machines.

Even though many files can now be downloaded through HTML links on Web pages, ftp is still the most efficient way to transfer batches of files, such as for uploading content to a Web server.

Further Reading

"FTP New User Guide." Available online. URL: http://www.ftp-planet.com/ftpresources/basics.htm. Accessed August 4, 2007.

Loshin, Peter. *Big Book of Internet File Transfer RFCs.* San Diego, Calif.: Morgan Kaufmann, 2000.

Pike, Mary Ann, and Noel Estabrook. *Using FTP.* Indianapolis: Que, 1995.

"What Is Kermit?" Columbia University. Available online. URL: http://www.columbia.edu/kermit/kermit.html. Accessed August 4, 2007.

film industry and computing

Anyone who compares a science fiction film of the 1960s or 1970s with a recent offering will be struck by the realism with which today's movie robots, monsters, or aliens move against vistas of giant starships and planetary surfaces. The computer has both enhanced the management of cinematic production processes and made possible new and startling effects.

The role of the computer in film begins well before the first camera rolls. Writers can use computers to write scripts, while specialized programs can be used to lay out storyboards. Using 3D programs somewhat like CAD (drafting) programs, set designers can experiment with the positioning of objects before deciding on a final design and obtaining or creating the physical props. For mattes (backgrounds against which the characters will be shot in a scene), a computer-generated scene can now be inserted directly into the film without the need for an expensive, hand-painted backdrop.

Similarly, animation and special effects can now be rendered in computer animation form and integrated into the storyboard so that the issues of timing and combining of effects can be dealt with in the design stage. The actual effects can then be created (such as by using extremely realistic computer-controlled puppets and models together with computer generated imagery, or CGI) with the assurance that they will properly fit into the overall sequence. The ability to combine physical modeling, precise control, and added textures and effects can now create a remarkably seamless visual result in which the confrontation between a beleaguered scientist and a vicious velociraptor seems quite believable.

Just as the physical and virtual worlds are frequently blended in modern moviemaking, the traditional categories of visual media have also merged. Disney's fully animated films such as *The Lion King* benefit from the same computer-generated lighting and textures as the filming of live actors. Using 3D graphics engines, computer game scenes are now rendered with almost cinematic quality (see COMPUTER GAMES). Even characters from old movies can be digitally combined (composited) with new footage. (Of course, the artistic value of such efforts may be controversial.)

Computer technology, now relatively inexpensive, can also give the generally lower budget world of television access to higher-quality effects. As computers continue to become more powerful yet cheaper, amateur or independent filmmakers are gaining abilities previously reserved to big Hollywood studios.

The delivery of film and video has also been greatly affected by digitization. Classic movies can be digitized to rescue them from deteriorating film stock, while videos can be delivered digitally over cable TV systems or over the Internet. The ability to easily copy digital content does raise issues of piracy or theft of intellectual property (see INTELLECTUAL PROPERTY AND COMPUTING).

More recently, digital camcorders (and video modes in digital cameras and even cell phones) are making access to basic "film" technology a part of everyday life. A few minutes browsing a video-sharing site (see YOUTUBE) reveals a wide variety of documentary and creative productions ranging from the equivalent of the old "home movie" to professional quality.

Further Reading

Harris, Tom. "How Digital Cinema Works." Available online. URL: http://entertainment.howstuffworks.com/digital-cinema.htm. Accessed August 4, 2007.

Masson, Terrence, ed. *CG 101: A Computer Graphics Industry Reference.* 2nd ed. Williamstown, Mass.: Digital Fauxtography, 2007.

McKernan, Brian. *Digital Cinema: The Revolution in Cinematography, Post-Production, and Distribution.* New York: McGraw-Hill, 2005.

"Milestones in Film History: Greatest Visual and Special Effects." Available online. URL: http://www.filmsite.org/visualeffects. html. Accessed August 4, 2007.

Sawicki, Mark. *Filming the Fantastic: A Guide to Visual Effects Cinematography.* Burlington, Mass.: Focal Press, 2007.

Slone, Michael. *Special Effects: How to Create a Hollywood Film on a Home Studio Budget.* Studio City, Calif.: Michael Wiese Productions, 2007.

Swartz, Charles S, ed. *Understanding Digital Cinema: A Professional Handbook.* Burlington, Mass.: Focal Press, 2005.

Willis, Holly. *New Digital Cinema: Reinventing the Moving Image.* London: Wallflower Press, 2005.

financial software

Large businesses use complex database systems, spreadsheets, and other applications for activities such as accounting, planning/forecasting, and market research (see BUSINESS APPLICATIONS OF COMPUTERS). Here we will consider the variety of consumer and small business software applications that are available to help with the planning and management of financial activities, such as

- home budgeting and money management

- investment and retirement planning

- college financing

- tax planning and filing

- home buying or selling

- basic accounting, inventory, and other activities for small business

Basic home money management programs (such as the popular Quicken) handle the budgeting and recording of daily and monthly expenses. The program can usually also interface with on-line banking services (see BANKING AND COMPUTERS) as well as exporting data to tax filing software.

For small or home-based businesses, programs such as QuickBooks can provide basic management of inventory, sales, taxes, expenses, and other functions. There are also niche programs for applications such as managing on-line auctions or Web-based sales.

For financial planning, there are a variety of programs (ranging from small free or shareware utilities available on-line to full commercial packages) that offer special calculators, graphs, and other aids for planning for the future. For example, the future value of a savings account at various points can be calculated given the interest rate, or the full cost of a loan or mortgage similarly calculated. Full-featured programs usually include helpful explanations of the various types of financial instruments. Some programs conduct an "interview" where the program asks the user about his or her objectives, priorities, or tolerance for risk, and then recommends a course of action. Such programs can be

helpful even though they lack the experience and breadth of knowledge available to a human financial planner.

Tax preparation software is perhaps the fastest-growing consumer financial application. Programs normally must be purchased each year to incorporate the latest changes in tax law. An important incentive has been created by the Internal Revenue Service encouraging electronic filing of tax returns by promising speedier refunds to "e-filers."

TRENDS

Publishers of respected guidebooks (such as for college admissions and financial aid) are creating electronic versions that can be easier to use and more up to date than the printed counterpart. Meanwhile, many Web sites are offering utilities such as financial calculators, implemented in Java and run on-line without any software having to be downloaded by the user. The services can be offered to attract users for paid services or simply to acquire e-mail addresses for solicitation. Users should be cautious about revealing sensitive identification or financial data to unknown on-line sites.

The growth in small and home-based businesses is likely to continue in an economy that continues to offer new opportunities while reducing job security. While starting a small business is always an uncertain enterprise, easy-to-use accounting software offers the budding entrepreneur a better chance of being able to stay on top of expenses during the crucial first months of business.

The growing complexity of financial choices available to average consumers and the need for more people to take responsibility for their retirement planning is likely to increase the range and capability of financial planning applications in the future.

Further Reading

Financial Software Reviews—Personal Finance Software. Available online. URL: http://financialsoft.about.com/od/reviews-financesoftware/Software_Reviews_Personal_Finance_Software.htm. Accessed August 4, 2007.

Heady, Robert K., Christy Heady, and Hugo Ottolenghi. *The Complete Idiot's Guide to Managing Your Money.* 4th ed. Indianapolis: Alpha Books, 2005.

Ivens, Kathy. *Quickbooks 2007: The Official Guide.* Berkeley, Calif.: Osborne/McGraw-Hill, 2007.

Nelson, Stephen L. *Quicken 2007 for Dummies.* Hoboken, N.J.: Wiley, 2007.

finite-state machine

There are many calculations or other processes that can be described using a specific series of states or conditions. For example, the state of a combination lock depends not only on what numeral is being dialed or punched at the moment, but on the numbers that have been previously entered. An even simpler example is a counter (such as a car odometer), whose next output is equal to one increment plus its current setting. In other words, a state-based device has an inherent "memory" of previous steps.

In computing, a program can be set up so that each possible input, when combined with the current state, will

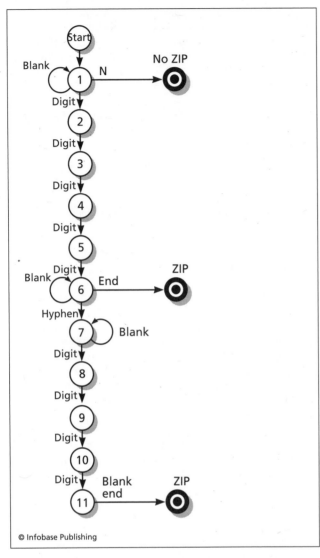

© Infobase Publishing

This diagram shows a finite-state representation of a ZIP code. The arrows link each state (within a circle) to its possible successor. In this simple example each digit must be followed by another digit until the fifth digit, which can either be followed by a blank (indicating a five-digit ZIP code) or four more digits for a 9-digit ZIP.

result in a specified output. That output becomes the new state of the machine. (Alternatively, the machine can be set so that only the current state determines the output, without regard to the previous state.) This is supported by the underlying structure of the logic switching within computer circuits as well as the "statefulness" of all calculations. (Given n, $n+1$ is defined, and so on.) Alan Turing showed that combining the state mechanism with an infinite memory (conceptualized as an endless roll of tape) amounted to a universal computer—that is, a mechanism that could perform any valid calculation, given enough time (see TURING, ALAN).

The idea of the sequential (or state) machine is closely related to *automata*, which are entities whose behavior is controlled by a state table. The interaction of such automata can produce astonishingly complex patterns (see CELLULAR AUTOMATA).

APPLICATIONS
Many programs and operating systems are structured as an endless loop where an input (or command) is processed, the results returned, the next input is processed, and so on, until an exit command is received. A *mode* or state can be used to determine the system's activity. For example, a program might be in different modes such as waiting for input, processing input, displaying results, and so on. The program logic will refer to the current state to determine what to do next and at some point the logic will *transition* the system to the next state in the sequence. The validity of some kinds of programs, protocols, or circuits can therefore be proven by showing that there is an equivalent finite-state machine—and thus that all possible combinations of inputs have been accounted for.

Finite-state machines have many other interesting applications. Simple organisms can be modeled as a set of states that interact with the environment (see ARTIFICIAL LIFE). The lower-level functions of robots can also be represented as a set of interacting finite-state machines. Even video game characters often use FSMs to give them a repertoire of plausible behavior.

Further Reading
Finite State Machine Editor [software]. Available online. URL: http://fsme.sourceforge.net/. Accessed August 4, 2007.
Meyer, Nathaniel. "Finite State Machine Tutorial." Available online. URL: http://www.generation5.org/content/2003/FSM_Tutorial.asp. Accessed August 4, 2007.
Wagner, Ferdinand, et al. *Modeling Software with Finite State Machines: A Practical Approach.* Boca Raton, Fla.: CRC Press, 2006.

firewall
The vulnerability of computer systems to malicious or criminal attack has been greatly increased by the growing number of connections between computers (and local networks) and the worldwide Internet (see COMPUTER CRIME AND SECURITY, INTERNET, and TCP/IP). The widespread use of permanent broadband connections by consumers (such as DSL and cable modem links) has increased the risk to home users. Intruders can use "port scanning" programs to determine what connections a given system or network has open, and can use other programs to snoop and steal or destroy sensitive data.

A firewall is a program (or combination of software and hardware) that sits between a computer (or local network) and the Internet. Typical firewall functions include:

- Examining incoming data packets and blocking those that include commands to examine or use unauthorized ports or IP addresses

- Blocking data packets that are associated with common hacking techniques such as "trojans" or "backdoor" exploitations

- Hiding all the internal network addresses on a local network, presenting only a single address to the outside world (this is also called NAT, or Network Address Translation)

- Monitoring particular applications such as ftp (file transfer protocol) and telnet (remote login), restricting them to certain addresses. Often a special address called a proxy is established rather than allowing direct connections between the outside and the local network.

Firewalls are usually configured by providing a rule that specifies what is to be done based on the origin address or other characteristics of an incoming packet. Because connections made by local programs to the outside can also compromise the system, rules are also created for such applications. The firewall package may come with a set of default rules for common applications and situations. When something not covered by the rules happens, the user will be prompted and guided to establish a new rule.

Modern firewalls are "stateful," meaning that they keep track not only of the source and destination of individual packets but their context (including originating application). Microsoft Windows Vista has improved the operating system's built-in firewall, at the expense of added complexity. Zone Labs's ZoneAlarm is another popular PC firewall. Linux provides a default firewall called iptables, which can be configured by a variety of applications. For added protection, users of broadband Internet connections should not connect their PC directly to the Internet. Rather, an inexpensive wired or wireless router that includes a built-in firewall can be connected on one side to the cable or DSL modem and on the other side to one or more computers in the local network.

Internet security packages for home users often combine a firewall with other services such as virus protection, parental control, and blocking of objectionable content or advertising.

Further Reading

Home PC Firewall Guide. Available online. URL: http://www.firewallguide.com/. Accessed August 4, 2007.
Komar, Brian, Ronald Beekelaar, and Joern Wettern. *Firewalls for Dummies.* 2nd ed. New York: Wiley, 2003.
Noonan, Wes, and Ido Dubrawsky. *Firewall Fundamentals.* Indianapolis: Cisco Press, 2006.
ZoneAlarm. Available online. URL: http://www.zonealarm.com. Accessed August 4, 2007.
Zwicky, Elizabeth D., Simon Cooper, and D. Brent Chapman. *Building Internet Firewalls.* 2nd ed. Sebastopol, Calif.: O'Reilly Media, 2000.

FireWire

FireWire is a high-speed serial interface used by personal computers and digital audio and video equipment. (The name FireWire is an Apple brand name, but it is used generically. Technically it is the IEEE 1394 Serial Bus.)

FireWire was developed in the 1990s by the IEEE P1394 Working Group with substantial funding from Apple and help from engineers from major corporations including IBM, Digital Equipment Corporation (DEC), Sony, and Texas Instruments. In 1993 it was hailed as the "most significant new technology" by *Byte* magazine.

FireWire was intended to replace Apple's parallel SCSI (Small Computer System Interface). (Sony's implementation, called I.Link, omits the two power pins in favor of a separate power connector.) However, because Apple asked for $1.00 per port in patent royalties, Intel instead developed a faster version of the universal serial bus (see USB) and that, rather than FireWire, is the standard port on most Windows machines.

Common uses for FireWire include connecting digital video (such as camcorder) devices, audio devices, and some data storage devices. FireWire is favored over USB 2.0 for many professional applications because of its higher speed and power distribution capabilities. However, it is more expensive than USB 2.0, which provides sufficient speed for many consumer peripherals such as digital cameras and printers.

Further Reading

Anderson, Don, and MindShare, Inc. *FireWire System Architecture.* 2nd ed. Boston: Addison-Wesley Pearson Education, 1998.
FireWire (Apple Developer Connection). Available online. URL: http://developer.apple.com/hardwaredrivers/firewire/index.html. Accessed September 20, 2007.

flag

A flag is a variable that is used to specify a particular condition or status (see VARIABLE). Usually a flag is either true or false. For example, a flag Valid_Form could be set to true before the input form is processed. If the validation check for any data field fails, the flag would be set to false. After the input procedure has ended, the main program would check the Valid_Form flag. If it's true, the data on the form is processed (for example, continuing on to the payment process). If the flag is false, the input form might be redisplayed with errors or omissions highlighted.

Flags can be combined to check multiple conditions. For example, suppose the input form routine also looked up the customer's account and checked to make sure the customer was approved for purchasing. The test for this might read:

```
If Valid_Form and Valid_Customer then
// continue processing else
// display error messages
```

In such cases, the flags are combined using the appropriate *and* or *or* operators (see BOOLEAN OPERATORS).

While flags are often used inside a routine to keep track of processing, modern programming practice discourages the use of "global" flags at the top level of the program. As with other global variables, such flags are vulnerable to being unpredictably changed or to having two parts of the program check the same flag without being able to rely on its state. (Thus a routine relies on a global flag being true but calls another routine that sets the flag to false without the original routine checking it again.) If several routines (or even programs) are being run at the same time, the situation gets even more complicated and a semaphore that can be controlled by one process at a time is more appropriate (see CONCURRENT PROGRAMMING). However, a main program that sets a flag to indicate the program mode and does not allow the flag to be changed by routines within the program is relatively safe.

Flags can also have more than two valid conditions, such as for specifying a number of possible states for a file or device. This usage is found mostly in operating systems.

Further Reading
"C++ I/O Flags." Available online: URL: http://www-control.eng. cam.ac.uk/~pcr20/www.cppreference.com/cppio_flags.html. Accessed August 4, 2007.

"Class Flags" [Java]. Available online. URL: http://java.sun.com/ j2ee/sdk_1.3/techdocs/api/javax/mail/Flags.html. Accessed August 4, 2007.

Myers, Gene. "Becoming Bit Wise." C-Scene Issue 09. Available online. URL: http://www.gmonline.demon.co.uk/cscene/cs9/ cs9-02.html. Accessed April 28, 2008.

Vincent, Alan. "Flag Variables, Validation and Function Control." Available online. URL: http://wsabstract.com/javatutors/ valid1.shtml. Accessed February 4, 2008.

flash and smart mobs

A flash mob is a spontaneously organized public gathering facilitated by ubiquitous mobile communications (see especially TEXTING AND INSTANT MESSAGING). The earliest flash mobs were a mixture of whimsy and social experiment. The first reported example, coordinated by Bill Wasik, senior editor of *Harper's Magazine,* occurred in June 2003 when a hundred people suddenly showed up on the ninth floor of Macy's in New York City, claiming to be shopping for a "love rug."

SMART MOBS

Smart mobs are similar in organization to flash mobs but tend to be more purposeful and enduring forms of social organization. The phenomenon was first described by Howard Rheingold in his book *Smart Mobs: The Next Social Revolution* (see RHEINGOLD, HOWARD). Rheingold describes several examples of smart mobs, including teenage "thumb tribes" in Tokyo and Helsinki, Finland (named for their use of tiny thumb-operated keyboards on cell phones). Their typical activities included organizing impromptu raves or converging on rock stars or other celebrities.

Smart mobs took on a more political bent in 1999 with spontaneously organized, fast-moving antiglobalization protests in Seattle. Police had considerable difficulty containing the protests, their communications and coordination capabilities not being equal to the task.

Another political smart mob occurred in 2001 when protesters in the Philippines used text messaging to organize demonstrations against the government of President Joseph Estrada. The protests grew rapidly, and Estrada was soon forced from office (see POLITICAL ACTIVISM AND THE INTERNET). Smart mob techniques were also used starting in 2003 to coordinate protests against the Iraq War. As

wireless communication continues to become ubiquitous, aspects of smart mob organization can be expected to turn up in future mass movements.

Even as the term has faded from public use, flash mobs have continued to flourish, appealing to a desire to have fun while striking out against an overly regimented consumer society. The term *urban playground movement* has also been used for the promotion of such gatherings.

Further Reading
Berton Justin. "Flash Mob 2.0: Urban Playground Movement Invites Participation." *San Francisco Chronicle,* November 10, 2007. Available online. URL: http://www.sfgate.com/cgi-bin/ article.cgi?f=/c/a/2007/11/10/MNMVT8UM9.DTL. Accessed April 23, 2008.

"Howard Rheingold: Smart Mobs." *Edge,* July 16, 2002. Available online. URL: http://www.edge.org/3rd_culture/rheingold/ rheingold_print.html. Accessed September 21, 2007.

Packer, George. "Smart-Mobbing the War." *New York Times,* March 9, 2003. Available online. URL: http://query.nytimes.com/ gst/fullpage.html?res=9D02E5D61F3CF93AA35750C0A9 659C8B63. Accessed September 21, 2007.

Rheingold, Howard. *Smart Mobs: The Next Social Revolution.* Cambridge, Mass.: Basic Books, 2002.

———. Smart Mobs Web site. Available online. URL: http://www. smartmobs.com/. Accessed September 21, 2007.

Schwartz, John. "New Economy: In the Tech Meccas, Masses of People, or 'Smart Mobs,' are Keeping in Touch through Wireless Devices." *New York Times,* July 22, 2002. Available online. URL: http://query.nytimes.com/gst/fullpage.html?sec =technology&res=98 02E5D61638F931A15754C0A9649C8B 63. Accessed March 28, 2008.

flash drive

A flash or "thumb" drive is a small data storage device that uses semiconductor flash memory rather than a disk drive. It is connected to a digital device using the universal serial bus (see USB). Because most computers, digital cameras, and other digital devices have USB ports, a flash drive is a convenient way to provide up to 16 GB (as of 2007) of low power, rewritable memory. Flash drives first appeared in late 2000.

Flash drives can use a separate USB cable (useful when several devices need to be connected to closely spaced USB ports) or simply have a connector that plugs directly into the port. Many people who regularly work with several computers carry their backup data or even a complete operating system (such as Linux) on a flash drive, perhaps connected to their keyring.

In Windows Vista some recent flash drives can be used to provide an additional system memory cache through a feature called ReadyBoost.

Flash drives can also be built into portable devices, including video and audio players. A competing technology (particularly found in digital cameras and PDAs) is the Secure Digital (SD) memory card developed by Matsushita, SanDisk, and Toshiba, which offers comparable capacity but is proprietary and requires a special interface.

For high-security applications, flash drives can include built-in encryption or fingerprint readers (see BIOMETRICS). However, as with other readily portable media, unsecured flash drives containing sensitive data pose a real risk to many organizations.

Under development by SanDisk and Intel are larger flash drives (32 or 80 GB) suitable for replacing hard drives in laptops. The benefits are lower weight and power consumption.

Further Reading
Axelson, Jan. *USB Mass Storage: Designing and Programming Devices and Embedded Hosts.* Madison, Wis.: Lakeview Research, 2006.

Oreskovic, Alexei. "Intel Prepares Flash Attack." The Street. com, September 9, 2007. Available online. URL: http://www.thestreet.com/s/intel-prepares-flash-attack/newsanalysis/techsemis/10380471.html. Accessed September 20, 2007.

Tyson, Jeff. "How Flash Memory Works." Available online. URL: http://developer.apple.com/hardwaredrivers/firewire/index.html. Accessed September 20, 2007.

flat-panel display

The traditional computer display uses a cathode ray tube (CRT) like that in a television set (see MONITOR). The flat-panel display is an alternative used in most laptop computers and many desktop systems. The most common type uses a liquid crystal display (LCD). The display consists of a grid of cells with one cell for each of the three colors (red, green, and blue) for each pixel.

The LCD cells are sandwiched between two polarizing filter layers that consist of many fine parallel grooves. The two filters are set so that the grooves on the second are rotated 90 degrees with respect to the first. By default, the light is polarized by the first filter, twisted by the liquid crystals so it is parallel to the grooves of the second filter, and thus passes through to be seen by the viewer. (For color displays, the light is first passed through one of three color filters to make it red, green, or blue as set for that pixel.) However, if current is applied to a crystal cell, the crystals realign so that the light passes through them without twisting. This means that the second polarizing filter now blocks the light and the cell appears opaque (or dark) to the viewer.

Color LCD displays can use two different mechanisms for sending the current through the crystals. In passive matrix displays, the current is timed so that it briefly charges the correct crystal cells. The charges fade quickly, making the image look dim, and the display cannot be refreshed quickly because of the persistence of ghost images. This means that such displays do not work well with games or other programs with rapidly changing displays.

In an active matrix display, each display cell is controlled by its own thin film transistor (TFT). These displays are sharper, brighter, and can be refreshed more frequently, allowing better displays for animations and games. However, fabrication costs for TFT displays are higher, and the displays are also vulnerable to having a few transistors fail, leading to permanent dark spots on the display. Active matrix displays also use more power, reducing battery life on laptop PCs. A general disadvantage of flat panel displays is that their pixel dimensions are fixed, so setting the display to a resolution smaller than its full dimensions usually results in an unsatisfactory image.

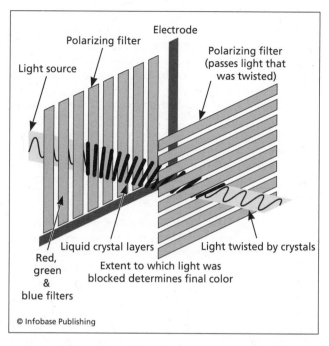

When the current is off, the liquid crystals remain twisted so the light passes through both polarizing panels and illuminates the display. However, when current is applied, the crystals straighten out, causing the light to be blocked by the second polarizing panel.

As newer technologies bring down the cost of flat-panel LCD displays they are increasingly being seen on desktop PCs, where they have the advantage of taking up much less space than conventional monitors while drawing less power.

Further Reading
"The PC Technology Guide. Flat panel displays." Available online. URL: http://www.pctechguide.com/43FlatPanels.htm. Accessed August 14, 2007.

White, Ron. *How Computers Work.* 8th ed. Indianapolis: Que, 2005.

floppy disk

Until the mid-1990s, the floppy disk or diskette was the primary method for distributing software and providing removable data storage for personal computers. Diskettes first appeared in the late 1960s on IBM minicomputers, and became more widespread on a variety of minicomputers and early microcomputers during the 1970s.

The now obsolete 8-inch and 5-¼ inch disks were made from Mylar with a metal oxide coating, the assembly being housed in a flexible cardboard jacket (hence the term "floppy disk"). The more compact 3.5-inch diskettes first widely introduced with the Apple Macintosh in 1984 became the standard type for all PCs by the 1990s. These diskettes are no longer truly "floppy" and come in a rigid plastic case.

A typical floppy disk drive has a controller with two magnetic heads so that both sides of the diskette can be used to hold data. The surface is divided into concentric tracks that are in turn divided into sectors. (For more on disk organization, see HARD DRIVE.) The heads are precisely positioned to the required track/sector location using stepper motors under control of the disk driver. The data capacity of a disk depends on how densely tracks can be written on it. Today's 3.5-inch diskettes typically hold 1.44 MB of data.

In recent years, drive technology has advanced so that many more tracks can be precisely written in the same amount of surface. The result is found in products such as the popular Zip disks, which can hold 100 MB or even 250 MB, making them comparable in capacity and speed with older, smaller hard drives.

Since the late 1990s, the traditional floppy disk has become less relevant for most users. With more computers connected to networks, the use of network copying commands or e-mail attachments has made it less necessary to exchange files via floppy, a practice dubbed "sneaker-net." When data needs to be backed up or archived, the high-capacity USB drive, tape, or writable CD is a more practical alternative to low-capacity floppies. (See BACKUP AND ARCHIVE SYSTEMS.) With its iMac line, Apple actually discontinued including a floppy drive as standard equipment. In PC-compatible laptops, a floppy drive is often available as a plug-in module that can be alternated with other devices. Desktop systems still sometimes come with a single 3.5-inch drive.

Further Reading
White, Ron. *How Computers Work.* 8th ed. Indianapolis: Que, 2005.

flowchart

A flowchart is a diagram showing the "flow" or progress of operations in a computer program. Flowcharting was one of the earliest aids to program design and documentation, and a plastic template with standard flowcharting symbols was a common programming accessory. Today CASE (computer-aided software engineering) systems often include utilities that can automatically generate flowcharts based on the control structures and procedure calls found in the program code (see CASE).

The standard flowchart symbols include blocks of various shapes that represent input/output, data processing, sorting and collating, and so on. Lines with arrows indicate the flow of data from one stage or process to the next. A diamond-shaped symbol indicates a decision to be made by the program. If the decision is an "if" (see BRANCHING STATEMENTS) separate lines branch off to the alternatives. If the decision involves repeated testing (see LOOP), the line returns back to the decision point while another line indicates the continuation of processing after the loop exits. Devices such as printers and disk drives have their own symbols with lines indicating the flow of data to or from the device.

Complex software systems can employ several levels of flowcharts. For example, a particular routine within a program might have its own flowchart. The routine as a whole

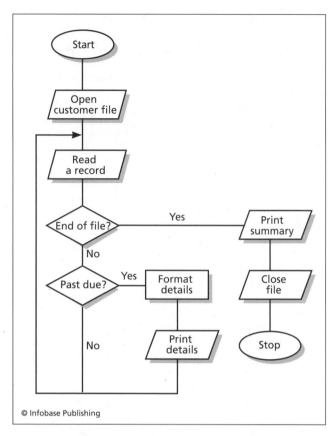

© Infobase Publishing

A flowchart uses a set of simple symbols to describe the steps involved in a data processing operation. The parallelograms indicate an input/output operation (such as reading or writing a file). The "decision diamonds" have yes and no branches depending on the result of a test or comparison.

would then appear as a symbol in a higher-level flowchart representing the program as a whole. Finally, a *system chart* might show each program that is run as part of an overall data processing system.

While still useful, flowcharting is often supplemented by other techniques for program representation (see PSEUDO-CODE). Also, modern program design tends to shift the emphasis from charting the flow of processing to elucidating the properties and relationships of objects (see OBJECT-ORIENTED PROGRAMMING).

Further Reading
Boillot, M. H., G. M. Gleason, and L. W. Horn. *Essentials of Flow-charting.* New York: WCB/McGraw-Hill, 1995.

font

In computing, a font refers to a typeface that has a distinctive appearance and style. In most word processing, desktop publishing, and other programs the user can select the point size at which the font is to be displayed and printed (in traditional typography each point size would be considered to be a separate font). Operating systems such as Windows and Macintosh usually come with an assortment of fonts, and applications can register additional fonts to make them available to the system.

Fonts are often presented as a "family" that includes the same type design with different *attributes* such as bold-face and italic. The spacing of letters could be uniform (monospace) as in the Courier font often used for printing computer program code or proportional (as with most text fonts). For proportional fonts the design can include *kerning,* or the precise fitting together of adjacent letters for a more attractive appearance. Fonts are also described as serif if they have small crossbars on the ends of letters such as at the end of the crossbar on a T in the Times Roman font. Other fonts such as Arial lack the tiny bars and are called sans serif (without serif).

There are two basic ways to store font data in the computer system. Bitmapped fonts store the actual pattern of tiny dots that make up the letters in the font. This has the advantage of allowing each letter in each point size to be precisely designed. The primary disadvantage is the amount of memory and system resources required to store a font in many point sizes. In practice, this consideration results in only a relatively few fonts and sizes being available.

The alternative, an *outline or vector* font uses a "page description language" such as Adobe PostScript or TrueType to provide graphics commands that specify the drawing of each letter in a font. When the user specifies a font, the text is rendered by processing the graphics commands in an interpreter. Since the actual bitmap doesn't need to be stored and all point sizes of a font can be generated from one description, outline fonts save memory and disk space (although they require additional processor resources for rendering). While sophisticated scaling techniques are used to maintain a pleasing appearance as the font size changes, outline fonts will not look as polished as bitmapped fonts that are hand-designed at each point size. (For use of fonts see TYPOGRAPHY, COMPUTERIZED.)

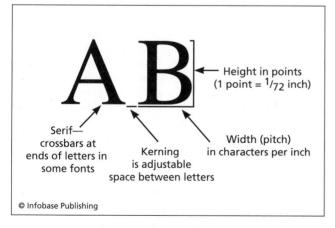

© Infobase Publishing

Strictly speaking, a particular type design is called a typeface, and a font is a rendering of a typeface with specified characteristics such as height in points and possibly width or pitch in characters per inch. Thus there are usually many fonts for each typeface.

Further Reading

Aaron, B. *TrueType Display Fonts.* San Francisco: Sybex, 1993.
Adobe Systems. *Adobe Type Library Reference Book.* 2nd ed. Mountain View, Calif.: Adobe Press, 2003.
Headley, Gwyn. *The Encyclopedia of Fonts.* London: Cassell Illustrated, 2005.
Lupton, Ellen. *Thinking with Type: A Critical Guide for Designers, Writers, Editors & Students.* New York: Princeton Architectural Press, 2007.
Microsoft Typography. Available online. URL: http://www.microsoft.com/typography/default.mspx. Accessed August 4, 2007.
TrueType Typography. Available online. URL: http://www.truetype-typography.com/. Accessed August 4, 2007.

Forth

The unusual Forth programming language was designed by Charles H. Moore in 1970. An astronomer, Moore was interested in developing a compact language for controlling motors to drive radio telescopes and other equipment.

LANGUAGE STRUCTURE

Forth has a very simple structure. The Forth system consists of a collection of *words*. Each word is a sequence of operations (which can include other existing words). For example, the DUP word makes a copy of a data value. Data is held by a stack. For example, the arithmetic expression written as 2 + 3 in most languages would be written in Forth as + 2 3. When the + operator (which in Forth is a pre-defined word) executes, it adds the next two numbers it encounters (2 and 3) together, and puts the sum on the stack (where in turn it might be fetched for further processing by the next word in the program (see STACK)). This representation is also called *postfix notation* and is familiar to many users of scientific calculators.

The words in the dictionary are "threaded" or linked so that each word contains the starting address of the next one. The Forth interpreter runs a simple loop where it fetches the next *token* (one or more characters delimited by spaces) and scans the dictionary to see if it matches a defined word (including variables). If a word is found, the code in the word is executed. If no word is found, the interpreter interprets the token as a numeric constant, loads it on the stack, and proceeds to the next word.

A key feature of Forth is its extensibility. Once you have defined a word, the new word can be used in exactly the same way as the predefined words. The various forms of *defining words* allow for great control over what happens when a new word is created and when the word is later executed. (In many ways Forth anticipated the principles of object-oriented programming, with words as objects with implicit constructors and methods. A well-organized Forth program builds up from "primitive" operations to the higher-level words, with the program itself being the highest-level word.)

Forth has always attracted an enthusiastic following of programmers who appreciate a close communion with the flow of data in the machine and the ability to precisely tailor programs. The language is completely interactive, since any word can be typed at the keyboard to execute it and display the results. Forth was also attractive in the early days of microcomputing because the lack of need for a sophisticated interpreter or compiler meant that Forth systems could run comfortably on systems that had perhaps 16K or 64K of available RAM.

Forth never caught on with the mainstream of programmers, however. Its very uniqueness and the unusual mindset it required probably limited the number of people willing to learn it. While Forth programs can be clearly organized, badly written Forth programs can be virtually impossible to read. However, Forth is sometimes found "under the hood" in surprising places (for example, the PostScript page description language is similar to Forth) and the language still has a considerable following in designing hardware control devices (see EMBEDDED SYSTEMS).

Further Reading
Brodie, L. *Starting FORTH.* 2nd ed. Upper Saddle River, N.J.: Prentice Hall, 1987.
———. *Thinking FORTH.* 2nd ed. Upper Saddle River, N.J.: Prentice Hall, 1994.
Forth Interest Group. Available online. URL: http://www.forth.org. Accessed August 14, 2007.
Rather, Elizabeth D. *Forth Application Techniques.* Hawthorne, Calif.: FORTH, Inc., 2006.

FORTRAN

As computing became established throughout the 1950s, the need for a language that could express operations in a more "human-readable" language began to be acutely felt. In a high-level language, programmers define variables and write statements and expressions to manipulate them. The programmer is no longer concerned with specifying the detailed storage and retrieval of binary data in the computer, and is freed to think about program structure and the proper implementation of algorithms.

Fortran (FORmula TRANslator) was the first widely used high-level programming language. It was developed by a project begun in 1954 by a team under the leadership of IBM researcher John Backus. The goal of the project was to create a language that would allow mathematicians, scientists, and engineers to express calculations in something close to the traditional notation. At the same time, a compiler would have to be carefully designed so that it would produce executable machine code that would be nearly as efficient as the code that would have been created through the more tedious process of using assembly languages. (See COMPILER and ASSEMBLER.)

The first version of the language, Fortran I, became available as a compiler for IBM mainframes in 1957. An improved (and further debugged version) soon followed. Fortran IV (1963) expanded the number of supported data types, added "common" data storage, and included the DATA statement, which made it easier to load literal numeric values into variables. This mature version of Fortran was widely embraced by scientists and engineers, who created immense libraries of code for dealing with calculations commonly needed for their work.

By the 1970s, the structured programming movement was well under way. This school of programming emphasized dividing programs into self-contained procedures into which data would be passed, processed, and returned. The use of unconditional branches (GOTO statements) as was common in Fortran was now discouraged. A new version of the language, Fortran 77 (or F77), incorporated many of the new structural features. The next version, Fortran 90 (F90), added support for recursion, an important technique for coding certain kinds of problems (see RECURSION). Mathematics libraries were also modernized. FORTRAN 2003 contains a number of new features, including support for modern programming structures (see OBJECT-ORIENTED PROGRAMMING) and the ability to interface smoothly with programs written in the C language. A relatively minor further revision has the tentative name FORTRAN 2008.

SAMPLE PROGRAM

The following simple example illustrates some features of a traditional FORTRAN program:

```
INTEGER INTARRAY(10)
INTEGER ITEMS, COUNTER, SUM, AVG
SUM = 0
READ *, ITEMS
DO 10 COUNTER = 1, ITEMS
   READ *, INTARRAY(COUNTER)
   SUM = SUM + INTARRAY(COUNTER)
10 CONTINUE
   AVG = SUM / ITEMS
   PRINT 'SUM OF ITEMS IS: ', SUM
   PRINT 'AVERAGE IS: ', AVG
STOP
END
```

The program creates an array holding up to ten integers (see ARRAY). The first number it reads is the number of items to be added up. It stores this in the variable ITEMS. A DO loop statement then repeats the following two statements once for each number from 1 to the total number of items. Each time the two statements are executed, COUNTER is increased by 1. The statements read the next number from the array and add it to the running total in SUM. Finally, the average is calculated and the sum and average are printed.

Like its contemporary, COBOL, Fortran is viewed by many modern programmers as a rather clumsy and anachronistic language (because of its use of line number references, for example). However, there is a tremendous legacy of tested, reliable Fortran code and powerful math libraries. (For example, a Fortran program can call library routines to quickly get the sum or cross-product of any array or matrix.) These features ensure that Fortran has continuing appeal and utility to users who are more concerned with getting fast and accurate results than with the niceties of programming style.

Further Reading
Chapman, Stephen J. *Fortran 95/2003 for Scientists & Engineers.* 3rd ed. New York: McGraw-Hill, 2007.
Chivers, Ian, and Jane Sleightholme. *Introduction to Programming with Fortran: With Coverage of Fortran 90, 95, 2003 and 77.* New York: Springer, 2005.

Page, Clive. "Clive Page's List of Fortran Resources." Available online. URL: http://www.star.le.ac.uk/~cgp/fortran.html. Accessed August 4, 2007.

Reid, John. "The Future of Fortran." Available online. URL: http://www.ieeexplore.ieee.org/iel5/5992/27213/01208645.pdf. Accessed August 4, 2007.

fractals in computing

Fractals and the related idea of chaos have profoundly changed the way scientists think about and model the world. Around 1960, Benoit Mandelbrot noticed that supposedly random economic fluctuations were not distributed evenly but tended to form "clumps." As he investigated other sources of data, he found that many other things exhibited this odd behavior. He also discovered that the patterns of distribution were "self-similar"—that is, if you magnified a portion of the pattern it looked like a miniature copy of the whole. Mandelbrot coined the term *fractal* (meaning fractured, or broken up) to describe such patterns. Eventually, a number of simple mathematical functions were found to exhibit such behavior in generating values.

Fractals offered a way to model many phenomena in nature that could not be handled by more conventional geometry. For example, a coastline that might be measured as 1,600 miles on a map might be many thousands of miles

A Mandelbrot fractal generated using Adobe PhotoShop and the KPT (Kai's Power Tools) Fraxplorer filter. (LISA YOUNT)

when measured on local maps, as the tiny inlets at every bay and beach are measured. Fractal functions could replicate this sort of endless generation of detail in nature.

Fractals showed that seemingly random or chaotic data could form a web of patterns. At the same time, Mandelbrot and others had discovered that the pattern radically depended on the precise starting conditions: A very slight difference at the start could generate completely different patterns. This "sensitive dependence on initial conditions" helped explain why many phenomena such as weather (as opposed to overall climate) resisted predictability.

COMPUTING APPLICATIONS

Many computer users are familiar with the colorful fractal patterns generated by some screen savers. There are hundreds of "families" of fractals (beginning with the famous Mandelbrot set) that can be color-coded and displayed in endless detail. But there are a number of more significant applications. Because of their ability to generate realistic textures at every level of detail, many computer games and simulations use fractals to generate terrain interactively. Fractals can also be used to compress large digital images into a much smaller equivalent by creating a mathematical transformation that preserves (and can be used to recreate) the essential characteristics of the image. Military experts can use fractal analysis either to distinguish artificial objects from surrounding terrain or camouflage, or to generate more realistic camouflage. Fractals and chaos theory are likely to produce many surprising discoveries in the future, in areas ranging from signal analysis and encryption to economic forecasting.

Further Reading

Fractal Resources & Links. Available online. URL: http://home.att.net/~Novak.S/resources.htm. Accessed August 4, 2007.

Gleick, J. *Chaos: The Making of a New Science.* New York: Viking, 1987.

Mandelbrot, Benoit. *The Fractal Geometry of Nature.* New York: W.H. Freeman, 1982.

Peitgen, Heinz-Otto, Hartmunt Jüurgens, and Dietmar Saupe. *Chaos and Fractals.* 2nd ed. New York: Springer, 2004.

functional languages

Most commonly used computer languages such as C++ and FORTRAN are *imperative* languages. This means that a statement is like a "sentence" in which the value of an expression or the result of a function is used in some way, such as assigning it to another variable or printing it. For example:

```
A = cube(3)
```

passes the parameter 3 to the cube function, which returns the value 27, which is then assigned to the variable A.

In a functional language, the values of functions are not assigned to variables (or stored in intermediate locations as functions are evaluated). Instead, the functions are manipulated directly, together with data items (atoms) arranged in lists. The earliest (and still best-known) functional language is LISP (see LISP). Programming is accomplished by defining and arranging functions until the desired processing is accomplished. (The decision making accomplished by branching statements in imperative languages is accomplished by incorporating conditionals in function definitions.)

Many functional languages (including LISP) for convenience incorporate some features of imperative languages. The ML language, for example, includes data type declarations. A similar language, Haskell, however, eschews all such imperative features.

APPLICATIONS

Functional languages have generally been used for specialized purposes, although they can in principle perform any task that an imperative language can. APL, which is basically a functional language, has devotees who appreciate its compact and powerful syntax for performing calculations (see APL). LISP and its variants have long been favored for many artificial intelligence applications, particularly natural language processing, where its representation of data as lists and the facility of its list-processing functions seems a natural fit.

Proponents of functional languages argue that they free the programmer from having to be concerned with explicitly setting up and using variables. In a functional language, problems can often be stated in a more purely mathematical way. Further, because functional programs are not organized as sequentially executed tasks, it may be easier to implement parallel processing systems using functional languages.

However, critics point out that imperative languages are much closer to how computers actually work (employing actual storage locations and sequential operation) and thus produce code likely to be much faster and more efficient than that produced by functional languages.

Further Reading

Bird, R. *Introduction to Functional Programming Using Haskell.* 2nd ed. London: Prentice Hall, 1998.

Hudak, Paul. *The Haskell School of Expression: Learning Functional Programming through Multimedia.* New York: Cambridge University Press, 2000.

Michaelson, Greg, Phil Trinder, and Hans-Wolfgang. Loidl, eds. *Trends in Functional Programming.* 2 vols. Portland, Oreg.: Intellect, 2000.

Thompson, S. Haskell. *Miranda: The Craft of Functional Programming.* 2nd ed. Reading, Mass.: Addison-Wesley, 1996.

fuzzy logic

At bottom, a data bit in a computer is "all or nothing" (1 or 0). Most decisions in computer code are also all or nothing: Either a condition is satisfied, and execution takes one specified path, or the condition is not satisfied and it goes elsewhere. In real life, of course, many situations fall between the cracks. For example, a business might want to treat a credit applicant who almost qualifies for "A" status different from one who barely made "B." While a program could be refined to include many gradations between B and A, another approach is to express the degree of "closeness" (or certainty) using fuzzy logic.

In 1965, mathematician L. A. Zadeh introduced the concept of the *fuzzy set.* In a fuzzy set, a given item is not simply either a member or not a member of a specified set. Rather, there is a degree of membership or "suitability" somewhere between 0 (definitely not a member) and 1 (definitely a member). A program using fuzzy logic must include a variety of rules for determining how much certainty to assign in a given case. One way to create rules is to ask experts in a given field (such as credit analysis) to articulate the degree of certainty or confidence they would feel in a given set of circumstances. For physical systems, data can also be correlated (such as the relationship of temperature to the likelihood of failure of a component) and used to create a rule to be followed by, for example, a chemical process control system.

Fuzzy logic is particularly applicable to the creation of programs (see EXPERT SYSTEM) that are better able to cope with uncertainty and the need to weigh competing factors in coming to a decision. It can also be used in engineering to allow designers to specify which factors they want to tightly constrain (such as for safety reasons) and which can be allowed more leeway. The system can then come up with optimized design specifications. Fuzzy logic has also been applied to areas such as pattern recognition and image analysis where a number of uncertain observations must often be accumulated and a conclusion drawn about the overall object.

Further Reading

Bojadziev, George, and Maria Bojadziev. *Fuzzy Logic for Business, Finance, and Management.* 2nd ed. Hackensack, N.J.: World Scientific Publishing Co., 2007.

"Fuzzy Logic Tutorial." *Encoder* (Seattle Robotics Society). March 1998. Available online. URL: http://www.seattlerobotics. org/encoder/mar98/fuz/flindex.html. Accessed August 4, 2007.

Mendel, Jerry M. *Uncertain Rule-Based Fuzzy Logic Systems: Introduction and New Directions.* Upper Saddle River, N.J.: Prentice Hall PTR, 2000.

Nguyen, Hung T., and Elbert A. Walker. *A First Course in Fuzzy Logic.* 3rd ed. Bocal Raton, Fla.: Chapman & Hall/CRC, 2006.

Sanchez, E. *Fuzzy Logic and the Semantic Web.* Amsterdam: Elsevier, 2006.

game consoles

Game consoles are computer devices dedicated to (or primarily used for) playing video games. The earliest such devices appeared in the 1970s from Magnavox and then Atari, and could only play simple games like *Pong* (a crude simulation of ping-pong). Slightly later systems began to feature cartridges that allowed them to play a greater variety of games.

After a shakeout in the late 1970s, Atari revived the video game industry with its hit game *Space Invaders*. However, this was followed by another industry crash as the market became glutted by often imitative and inferior games. The next leader was Nintendo, with its own hit, *Super Mario Brothers*. By the end of the 1980s another Japanese firm, Sega, had entered the American market.

During the 1990s consoles grew in power and graphic sophistication. CDs (and later DVDs) replaced cartridges and allowed for larger, more complex games. By the end of the decade the main competitors were the Sony Playstation 2 and the Nintendo 64 (indicating a 64-bit processor) and GameCube. Meanwhile Microsoft entered the game console market with its Xbox, which featured a more PC-like architecture including a built-in hard drive (soon also adopted by Sony).

NEW TECHNOLOGIES

A technology with applications far beyond games is the "cell chip" technology introduced by Sony in its PlayStation 3, introduced in late 2006. The Sony cell chip has seven cores and can reach nearly supercomputer-scale speeds (and in fact is being used to create impromptu supercomputers).

Sony also opted to include a high-definition "Blu-ray" DVD player in the PS3, strengthening its application as a media device as well as a gaming device, with Microsoft and its Xbox 360 initially opting for the ultimately unsuccessful HD DVD.

Nintendo's Wii, the third major competitor as of 2007, innovates in a different area: the user interface. The Wii comes with a controller that can track both where it is pointing and how it is being used, allowing for rather realistic sports and combat simulations.

Further Reading

Amirch, Dan. *PlayStation 2 for Dummies.* New York: Hungry Minds, 2001.

Farkas, Bart G. *The Nintendo Wii Pocket Guide.* Berkeley, Calif.: Peachpit Press, 2007.

Forster, Winnie. *The Encyclopedia of Game Machines.* Utting, Germany: Game Plan, 2005.

Johnson, Brian. *Xbox 360 for Dummies.* Hoboken, N.J.: Wiley, 2006.

Kent, Steven L. *The Ultimate History of Video Games.* New York: Three Rivers Press, 2001.

Kim, Ryan. "New Era of Game Devices Arrives: Sony and Nintendo Meet the Challenge of Microsoft's Xbox." *San Francisco Chronicle,* November 13, 2006, p. F-1. Available online. URL: http://sfgate.com/cgi-bin/article.cgi?file=/c/a/2006/11/13/BUGS1MAGAN1.DTL. Accessed September 21, 2007.

Nintendo. Available online. URL: http://www.nintendo.com. Accessed September 21, 2007.

Playstation (Sony). Available online. URL: http://www.us.playstation.com/. Accessed September 21, 2007.

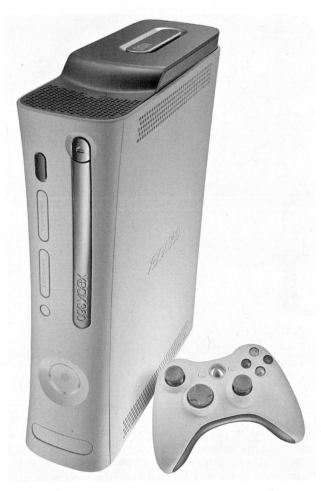

Game consoles such as this Microsoft Xbox are now more powerful than many desktop computers. (MICROSOFT CORPORATION)

Takahashi, Dean. *The Xbox 360 Uncloaked: The Real Story Behind Microsoft's Next-Generation Video Game Console.* Spiderworks, 2006.
Xbox (Microsoft). Available online. URL: http://www.microsoft.com/xbox/. Accessed September 21, 2007.

Gates, William, III (Bill)

(1955–)
American
Entrepreneur, Programmer

Bill Gates built Microsoft, the dominant company in the computer software field and in doing so, became the world's wealthiest individual, with a net worth measured in the tens of billions. Born on October 28, 1955, to a successful professional couple in Seattle, Gates's teenage years coincided with the first microprocessors becoming available to electronics hobbyists.

Gates showed both technical and business talent as early as age 15, when he developed a computerized traffic-control system. He sold his invention for $20,000, then dropped out of high school to work as a programmer for TRW for

the very respectable salary of $30,000. By age 20, Gates had returned to his schooling and become a freshman at Harvard, but then he saw a cover article in *Popular Electronics*. The story introduced the Altair, the first commercially available microcomputer kit.

Gates believed that microcomputing would soon become a significant industry. To be useful, however, the new machines would need software, and Gates and his friend Paul Allen began by creating an interpreter for the BASIC language that could run in only 4 KB of memory, making it possible for people to write useful applications without having to use assembly language. This first product was quite successful, although to Gates's annoyance it was illicitly copied and distributed for free.

In 1975, Gates and Allen formed the Microsoft Corporation. Most of the existing microcomputer companies, including Apple, Commodore, and Tandy (Radio Shack) signed agreements to include Microsoft software with their machines. However, the big breakthrough came in 1980, when IBM decided to market its own microcomputer. When negotiations for a version of CP/M (then the dominant operating system) broke down, Gates agreed to supply IBM with a new operating system. Buying one from a small Seattle company, Microsoft polished it a bit and sold it as MS-DOS 1.0. Sales of MS-DOS exploded as many other companies rushed to create "clones" of IBM's hardware, each of which needed a copy of the Microsoft product.

In the early 1980s, Microsoft was only one of many thriving competitors in the office software market. Word processing was dominated by such names as WordStar and WordPerfect, Lotus 1-2-3 ruled the spreadsheet roost, and dBase II dominated databases (see WORD PROCESSING, SPREADSHEET, and DATABASE MANAGEMENT SYSTEM). But Gates and Microsoft used the steady revenues from MS-DOS to undertake the creation of Windows, a much larger operating system that offered a graphical user interface (see USER INTERFACE). While the first versions of Windows were clumsy and sold poorly, by 1990 Windows (with versions 3.1 and later, 95 and 98) had become the new dominant OS and Microsoft's annual revenues exceeded $1 billion (see MICROSOFT WINDOWS). Gates relentlessly leveraged both the company's technical knowledge of its own OS and its near monopoly in the OS sector to gain a dominant market share for the Microsoft word processing, spreadsheet, and database programs.

By the end of the decade, however, Gates and Microsoft faced formidable challenges. The growth of the Internet and the use of the Java language with Web browsers offered a new way to develop and deliver software, potentially getting around Microsoft's operating system dominance (see JAVA). That dominance, itself, was being challenged by Linux, a version of LINUX created by Finnish programmer Linus Torvalds (see LINUX). Gates responded that Microsoft, too, would embrace the networked world and make all its software fully integrated with the Internet and distributable in new ways.

However, antitrust lawyers for the U.S. Department of Justice and a number of states began legal action in the late 1990s, accusing Microsoft of abusing its monopoly status by virtually forcing vendors to include its software with their systems. In 2000, a federal judge agreed with the government. In November 2002, an appeals court accepted a proposed settlement that would not break up Microsoft but would instead restrain a number of its unfair business practices.

Gates's personality often seemed to be in the center of the ongoing controversy about Microsoft's behavior. Positively, he has been characterized as having incredible energy, drive, and focus in revolutionizing the development and marketing of software.

On the other hand, Gates has been unapologetic about his dominance of the market. During the 1990s he often appeared defensive and abrasive in giving legal depositions or making public statements. As an executive, he has at times shown little tolerance for what he considers to be incompetence or shortsightedness on the part of subordinates.

There is another face to Bill Gates: He is one of the leading philanthropists of our time. In 2000 he and his wife founded the Bill and Melinda Gates Foundation. The foundation's endowment was about $33 billion by 2006, and Warren Buffet pledged to nearly double that through

Bill Gates is the multibillionaire cofounder of Microsoft Corporation, the leader in operating systems and software for personal computers. The company has faced antitrust actions since the late 1990s. (MICROSOFT CORPORATION)

stock donations. The foundation gives over $800 million a year to global health programs (including vaccination programs), supports a variety of global development efforts, and donates money and software to libraries and educational institutions. In June 2006 Gates announced that he would be withdrawing from involvement in the day-to-day affairs of Microsoft, in order to devote more time to philanthropy.

Since 2004, Gates has been featured on *Time* magazine's annual list of 100 most influential people. In 2005, the magazine made Gates, along with his wife and U2's lead singer Bono, "Persons of the Year." Gates has also received four honorary doctorates.

Further Reading

Bank, David. *Breaking Windows: How Bill Gates Fumbled the Future of Microsoft*. New York: Free Press, 2001.

Bill & Melinda Gates Foundation. Available online. URL: http://www.gatesfoundation.org/default.htm. Accessed August 5, 2007.

Lesinski, Jeanne M. *Bill Gates (Biography A&E)*. Rev. ed. Minneapolis: Lerner Publications, 2007.

Lowe, Jane C. *Bill Gates Speaks: Insights from the World's Greatest Entrepreneur*. New York: Wiley, 1998.

Markoff, John. "Exit, Pursued by 1,000 Bears (Microsoft Corp.'s Bill Gates)" *New York Times*, July 30, 2007, p, C1.

Wallace, James, and Jim Erickson. *Hard Drive: Bill Gates and the Making of the Microsoft Empire*. New York: HarperBusiness, 1993.

genetic algorithms

The normal method for getting a computer to perform a task is to specify the task clearly, choose the appropriate approach (see ALGORITHM), and then implement and test the code. However, this approach requires that the programmer first know the appropriate approach, and even when there are many potentially suitable algorithms, it isn't always clear which will prove optimal.

Starting in the 1960s, however, researchers began to explore the idea that an evolutionary approach might be adaptable to programming. Biologists today know that nature did not begin with a set of highly optimized algorithms. Rather, it addressed the problems of survival through a proliferation of alternatives (through mutation and recombination) that are then subjected to natural selection, with the fittest (most successful) organisms surviving to reproduce. Researchers began to develop computer programs that emulated this process.

A genetic program consists of a number of copies of a routine that contain encoded "genes" that represent elements of algorithms. The routines are given a task (such as sorting data or recognizing patterns) and the most successful routines are allowed to "reproduce" by exchanging genetic material. (Often, further "mutation" or variation is introduced at this stage, to increase the range of available solutions.) The new "generation" is then allowed to tackle the problem, and the process is repeated. As a result, the routines become increasingly efficient at solving the given problem, just as organisms in nature become more perfectly adapted to a given environment.

APPLICATIONS

Variations of genetic algorithms or "evolutionary programming" have been used for many applications. In engineering development, a virtual environment can be set up in which a simulated device such as a robot arm can be allowed to evolve until it is able to perform to acceptable specifications. (NASA has also used genetic programs competing on 80 computers to design a space antenna.) Different versions of an expert system program can be allowed to compete at performing tasks such as predicting the behavior of financial markets. Finally, a genetic program is a natural way to simulate actual biological evolution and behavior in fields such as epidemiology (see also ARTIFICIAL LIFE).

Further Reading

"Bibliography on Genetic Programming." Available online. URL: http://liinwww.ira.uka.de/bibliography/Ai/genetic.programming.html. Accessed August 5, 2007.

Brown, Chappell. "Darwin's Ideas Evolve Design." *EE Times*, February 6, 2006. Available online. URL: http://www.eetimes.com/showArticle.jhtml?articleID=178601156. Accessed August 5, 2007.

Eiben, A. E., and J. E. Smith. *Introduction to Evolutionary Computing*. New York: Springer, 2003.

Genetic-programming.org [Resources]. Available online. URL: http://www.genetic-programming.org/. Accessed August 5, 2007.

"The GP Tutorial." Available online. URL: http://www.geneticprogramming.com/Tutorial/. Accessed August 5, 2007.

Keats, Jonathon. "John Koza Has Built an Invention Machine." *Popular* Science, April 2006. Available online. URL: http://www.popsci.com/popsci/science/0e13af26862ba010vgnvcm1000004eecbccdrcrd.html. Accessed August 5, 2007.

Langdon, William B., and Riccardo Poli. *Foundations of Genetic Programming*. New York: Springer, 2002.

Riolo, Rick, and Bill Worzel, eds. *Genetic Programming Theory and Practice*. Norwell, Mass.: Kluwer Academic Publishers, 2003.

Geographical Information Systems

Cartography, or the art of mapmaking, has been transformed in many ways by the use of computers. Traditionally, mapmaking was a tedious process of recording, compiling, and projecting or plotting information about the location, contours, elevation, or other characteristics of natural geographic features or the demographic or political structure of human communities.

Instead of being transcribed from the readings of surveying instruments, geographic information can be acquired and digitized by sensors such as cameras aboard orbiting satellites. The availability of such extensive, detailed information would overwhelm any manual system of transcribing or plotting. Instead, the Geographical Information System (GIS, first developed in Canada in the 1960s) integrates sensor input with scanning and plotting devices, together with a database management system to compile the geographic information.

The format in which the information is stored is dependent on the scope and purpose of the information system. A detailed topographical view, for example, would have physical coordinates of latitude, longitude, and elevation. On the other hand, a demographic map of an urban area might

Annual rainfall in inches, Square County

© Infobase Publishing

A raster grid showing annual rainfall totals in inches for mythical Square County. Raster data is easy to work with, but the "coarseness" of the grid means that it does not capture much local variation or detail.

have regions delineated by ZIP code or voting precinct, or by individual address.

Geographic data can be stored as either a raster or a vector representation. A raster system divides the area into a grid and assigns values to each cell in the grid. For example, each cell might be coded according to its highest point of elevation, the amount of vegetation (ground cover) it has, its population density, or any other factor of interest. The simple grid system makes raster data easy to manipulate, but the data tends to be "coarse" since there is no information about variations within a cell.

Unlike the arbitrary cells of the raster grid, a vector representation is based upon the physical coordinates of actual points or boundaries around regions. Vector representation is used when the actual shapes of an entity are important, as with property lines. Vector data is harder to manipulate than raster data because geometric calculations must be made in order to yield information such as the distance between two points.

The power of geographic information systems comes from the ability to integrate data from a variety of sources, whether aerial photography, census records, or even scanned paper maps. Once in digital form, the data can be represented in a variety of ways for various purposes. A sophisticated GIS can be queried to determine, for example, how much of a proposed development would have a downhill gradient and be below sea level such that flooding might be a problem. These results can in turn be used by simulation programs to determine, for example, whether release of a chemical into the groundwater from a proposed plant site might affect a particular town two miles away. Geographic information systems are thus vital for the management of a variety of complex systems that are distributed over a geographical area, such as water and sewage systems, power transmission grids, and traffic control systems. Other applications include emergency planning (and evacuation routes) and the long-term study of the effects of global warming trends.

FROM INFORMATION TO NAVIGATION

The earliest use of maps was for facilitating navigation. The development of the Global Positioning System (GPS) made it possible for a device to triangulate readings from three of 24 satellites to pinpoint the user's position on Earth's surface within a few meters (or even closer in military applications). The mobile navigation systems that have now become a consumer product essentially use the current physical coordinates to look up information in the onboard geographical information system. Depending on the information stored and the user's needs, the resulting display can range from a simple depiction of the user's location on a highway or city street map to the generating of detailed driving directions from the present location to a desired location. As these systems are fitted with increasingly versatile natural language systems (and perhaps voice-recognition capabilities), the user will be able to ask questions such as "Where's the nearest gas station?" or even "Where's the nearest French restaurant rated at least three stars?"

Further Reading

Burrough, P.A., and R. McDonnell. *Principles of Geographical Information Systems.* 2nd ed. New York: Oxford University Press, 1998.
Demers, M. N. *Fundamentals of Geographic Information Systems.* New York: Wiley, 1997.
Longley, P. A., (and others) eds. *Geographical Information Systems: Principles, Techniques, Management and Applications.* New York: Wiley, 1998.
Ramadan, K. "The Use of GPS for GIS Applications." http://www.geogr.muni.cz/lgc/gis98/proceed/RAMADAN.html
U.S. Geological Survey. "Geographic Information Systems." http://www.usgs.gov/research/gis/title.html.

globalization and the computer industry

Globalization can be described as a group of trends that are breaking down the boundaries between national and regional economies, making countries more dependent on one another, and resulting in the freer flow of labor and resources. These trends have been praised by free trade advocates and decried by proponents of labor rights and environmentalism. However one feels about them, it is clear that global trends are reshaping the computer and information industry in many ways, and pose significant challenges.

Global trends that affect computer technology, software, and services include:

- offshoring, or the continuing movement of manufacturing of high-value components (and whole systems) from the industrialized West to regions such as Asia

- outsourcing—moving functions (such as technical support) from a company's home country to areas where suitable labor forces are cheaper (see EMPLOYMENT IN THE COMPUTER FIELD)

- removal of traditional intermediaries such as brokers and agents, with some of their functions being taken over by software (see SOFTWARE AGENT)

- decentralized networks (of which the Internet itself is the most prominent example) and the tendency of information to flow freely and quickly despite barriers such as censorship

- virtualization—creation of work groups or whole companies that are distributed across both space and time (24 hours), coordinated by the Internet and mobile communications (see VIRTUALIZATION)

- increasing use of open-source and collaborative models of software and information development (see OPEN SOURCE)

- blurring of the distinction between consumers and producers of information (see SOCIAL NETWORKING and USER-CREATED CONTENT)

These global trends can be divided roughly into three categories: movement of labor and resources, restructuring of markets, and changes in the nature and flow of production.

MOVEMENTS AND SHIFTS

Offshoring is the movement of manufacturing operations from the traditional developed industrial nations (such as the United States and Europe) to developing nations (see also DEVELOPING NATIONS AND COMPUTING). The principal motivation for this (as well as outsourcing, the movement of corporate functions and services) is lower labor and related costs. India, with its large population of well-trained, English-speaking workers, was the first beneficiary of these trends in the 1990s. Many major U.S. computer companies such as IBM, Intel, Microsoft, and HP have made major investments in software development operations in India. However, it should be noted that offshoring/outsourcing is a truly global trend, with other industrialized nations taking advantage of similar situations, particularly where there are language compatibilities. Thus Japanese companies have invested heavily in China, while Germans and other Europeans have preferred to look toward Eastern Europe.

Besides lower costs, outsourcing can speed development by taking advantage of differences in time zones, allowing for coordinated 24-hour production cycles.

FREE-TRADE CONTROVERSIES

Proponents of these trends generally include them under the umbrella of "free trade." Their arguments include:

- greater productivity through more efficient tapping of talent and resources

- improvement in the standard of living in developing nations

- lower prices for goods and services in developed nations

- spurring innovation through competition and the movement of displaced workers to higher-value jobs

Opponents point to a number of serious problems and issues, including:

- downward wage pressure and/or unemployment as workers in developed nations are displaced by offshore workers

- difficulty in retraining displaced workers

- lack of adequate protective regulations and labor rights for workers in developing nations

- potential deterioration in the quality of services (such as technical support) after outsourcing

- risks of dependence on offshore supply sources in times of crisis

RESTRUCTURING OF MARKETS

Computer-related businesses must also deal with the effects of globalization on the market for hardware, software, and services. Lower-cost offshore manufacturing has helped contribute to making many computer systems and peripherals into commodity items. This certainly benefits consumers (consider the ubiquitous $100 or less computer printer). However, it becomes more difficult to extract a premium for a brand as opposed to a generic name. Some companies have responded by relentless efforts to maximize efficiency in manufacturing (for example, see DELL, INC.), while a few others have maintained a reputation for style or innovation (see APPLE, INC.). Consumers have increasingly objected, however, to the difficulty in dealing with offshore technical support.

While the power of the Internet has opened many new ways of reaching potential customers around the world, dealing with a global marketplace brings considerable added complications, such as the need to deal with different regulatory systems (such as the European Union). In some areas (notably Asia) there is also the problem of unauthorized copying of software and media products (see SOFTWARE PIRACY AND COUNTERFEITING).

NEW WAYS OF WORKING

A global, connected economy is not only changing where work is done, but also *how* it is done. If a software developer, for example, has operations in the United States, Europe, India, and China, at any time of day there will be work going on somewhere. With the complexity and speed of operations, managers in the United States may have to keep quite long and irregular hours in order to have real-time communication with counterparts abroad. This interaction is made possible by a variety of technologies, including Internet-based phone and video conferencing and, of course, e-mail. However, this is not without added stress. Overall operations can be structured to take advantage of the time zone differences. Code or documents written in Bangalore might be reviewed and revised in Silicon Valley the same day.

Global trends are likely to continue and even accelerate as the computer and information industry continues to develop around the world. While technology can help deal with some of the challenges, there are many larger economic and political issues involved, and whether they can be satisfactorily resolved may ultimately have the greatest impact on the industry.

Further Reading

Blinder, Alan E. "Offshoring: The Next Industrial Revolution?" *Foreign Affairs* 85 (March/April 2006): 113–128.

Carmel, Erran, and Paul Tjia. *Offshoring Information Technology: Sourcing and Outsourcing to a Global Workforce.* New York: Cambridge University Press, 2005.

Currie, Wendy. *The Global Information Society.* New York: Wiley, 2000.

Friedman, Thomas L. *The World Is Flat: A Brief History of the Twenty-First Century.* 2nd rev. ed. New York: Picador, 2007.

Quinn, Michelle. "Working Around the Clock." *Los Angeles Times*, June 19, 2007. Available online. URL: http://www.latimes.com/news/la-fi-timezone19jun19,1,7843626.story?ctrack=1&cset=true. Accessed September 21, 2007.

Samii, Massood, and Gerald D. Karush, eds. *International Business & Information Technology.* New York: Routledge, 2004.

Sood, Robin. *IT, Software, and Services: Outsourcing and Offshoring: The Strategic Plan with a Practical Viewpoint.* Austin, Tex.: AiAiYo Books, 2005.

Steger, Manfred B. *Globalization: A Very Short Introduction.* New York: Oxford University Press, 2003.

Google

Google Inc. (NASDAQ symbol: GOOG) has built a business colossus by focusing on helping users find what they are looking for on the Internet while selling advertising targeted at those same users. By 2006, "to google" could be found in dictionaries as a verb meaning to look up anyone or anything online.

Google was founded by two Stanford students (see BRIN, SERGEY and PAGE, LARRY) who, for their doctoral thesis, had described a Web search algorithm that could give a better idea of the likely relevance of a given site based on the number of sites that linked to it. The two students implemented a search engine based on their ideas and hosted it on the Stanford Web site, where its popularity soon irritated the university's system administrators. In 1998 their business was incorporated as Google, Inc., and moved to the archetypal Silicon Valley entrepreneur's location—a friend's garage. However, as the company attracted investment capital and grew rapidly, it moved to Palo Alto and then its present home in Mountain View.

Google's initial public stock offering was in 2004, and the market's enthusiastic response made many senior employees instant millionaires. Google's steady growth in subsequent years has kept its stock in demand, reaching a record peak of $560 in September 2007. (In 2006 Google was added to the S&P 500 Index.)

SEARCH AND ITS LARGER CONTEXT

People tend to think of Google as a search engine. Actually, it is better to think of it as an ever-expanding network of Web-based services that include general and specialized searches but also tools for content creation and collaboration.

It is true that search and the accompanying advertising are the core of Google's revenue and thus the engine that drives its proliferation. In 2000 Google adopted keyword-based advertising. (This was not a new idea, but Google was the first to really make it work.) Basically, advertisers bid for the right to have their ad accompany the results of a search that contains a given word, on a per "click through" basis—that is, how often the user clicks on the ad to go to the advertiser's site. Advertisers are prioritized according to how much they bid, their previous click-through rate, and their ad's relevance to the search. If someone searches, for example, "widget" and Acme Widget Co. is in line for placement, the Acme ad is shown. If the user then clicks on it, Acme makes a payment to Google (and hopes to some business).

The power of keyword-based and other "contextual" advertising is that, by definition, any accompanying ad is targeted to someone who is quite probably already looking for what one is selling. And what makes this such a revenue-maker for Google is that, since the company serves over half of all Web searches, anyone wanting an ad to reach the biggest share of its potential audience will have to turn to Google.

Google's ability to offer more precisely targeted advertising has been enhanced in several ways:

- AdSense, which can be installed on a Web site where it displays ads keyed to the site's content. Revenue is shared by Google and the site owner.

- Advertisers can specify an AdWord and Google will place it on participating sites in its "content network" that it believes are relevant. The advertiser pays per thousand viewings of the ads ("impressions").

- Specialized shopping-oriented searches such as Google Product Search, which returns lists of sellers and a price comparison.

- Searches can also be local (particularly useful for mobile devices) and results can be keyed to maps.

OTHER APPLICATIONS

Google has greatly expanded beyond its core business of search and accompanying advertising. In general, the company has been emphasizing acquiring or developing tools that help users create content and collaborate. These offering include:

- Blogger, an easy-to-use blogging tool (see BLOGS AND BLOGGING)

- JotSpot, developer of wiki collaboration tools (see WIKIS AND WIKIPEDIA)

- YouTube, the largest video-sharing service, acquired by Google in 2006 (see YOUTUBE)

- Gmail, a free e-mail service

- Google Apps, which provides a Web-based office environment including a calendar and Google Docs & Spreadsheets. (The standard edition is free and represents a competitive challenge for Microsoft Office, particularly for small businesses and simpler applications.)

In addition to office and collaboration tools, Google has several other prominent applications that do not easily fit in one category:

- Google News provides a constantly updated newspaperlike format that groups stories under headlines.

- Google Book Search offers access to thousands of public-domain books and summaries or limited previews of copyrighted works (see E-BOOKS AND DIGITAL LIBRARIES)

- Google Maps and Google Earth are vast troves of map information, satellite imagery, and even street-level views of some cities.

A key to the growth of Google's new Web services is that many of them come with programming interfaces that can be used to integrate them into Web sites and applications. It is relatively easy, for example, to combine maps and data about stores or other locations (see MASHUPS).

CRITICISM

As of mid-2008 Google had more than 19,500 full-time employees. The company's workplace culture at its Mountain View "Googleplex" is famous for its gourmet food, elaborate recreation center, and other perks. (In 2007 *Fortune* magazine rated Google first in the nation as a place to work.)

Google has a market capitalization of about $180 billion, ahead of such giants as Hewlett-Packard and IBM. In 2008 Google took in $16.6 billion, with $4.2 billion in profit. Google's impact on the online world has been immense. As of mid-2007 Google was processing 54 percent of all Internet search requests, followed distantly by Yahoo! at 20 percent and Microsoft at 13 percent.

Google sets a high standard for itself. Its mission statement is "to organize the world's information and make it universally accessible and useful." A corporate motto is "don't be evil" in the pursuit of success. A number of critics have suggested, however, that Google has fallen short of its standards in a number of respects:

- Google Book Search had led to accusations of copyright violations by publishers and authors. Google has also been accused of benefiting from rampant copying of copyrighted content on its YouTube subsidiary.

- Google has been criticized for aiding China in censoring search results (see CENSORSHIP AND THE INTERNET).

- The detailed imagery available from Google Earth has been criticized by some nations on security grounds, and street-level views have raised privacy questions.

- Some Google practices, including the extensive use of cookies and analysis of users' e-mail and other content, have also aroused privacy concerns (see COOKIES and DATA MINING).

- Google has also been criticized for keeping its PageRank system secret, making it hard to determine if it is treating users fairly.

In 2007 Google acquired DoubleClick for $3.1 billion. Although the combination of the leading search company and a major online advertising service provoked concerns about a possible monopoly, the acquisition was approved by U.S. and European regulators.

While Google continues to be a subject of both admiration and debate, it is clear that it has placed powerful tools and enormous new resources in the hands of Web users around the world.

Further Reading

Battelle, John. *The Search: How Google and Its Rivals Rewrote the Rules of Business and Transformed Our Culture.* New York: Portfolio, 2005.

Davis, Harold. *Google Advertising Tools: Cashing In with AdSense, AdWords, and the Google APIs.* Sebastopol, Calif.: O'Reilly, 2006.

Kopytoff, Verne. "Who's Afraid of Google? Firms in Silicon Valley and beyond Fear Search Giant's Plans for Growth." *San Francisco Chronicle.* May 11, 2007, p. A-1. Available online. URL: http://sfgate.com/cgi-bin/article.cgi?f=/c/a/2007/05/11/MNGRIPPB2N1.DTL. Accessed September 22, 2007.

Marshall, Perry, and Bryan Todd. *Ultimate Guide to Google AdWords: How to Access 100 Million People in 10 Minutes.* Irvine, Calif.: Entrepreneur Media, 2007.

Vise, David A., and Mark Malseed. *The Google Story: Inside the Hottest Business, Media and Technology Success of Our Time.* Canada: Random House/Delta, 2006.

government funding of computer research

While the popular version of the story of the information age tends to focus on lone inventors in garages or would-be entrepreneurs working out of college dorm rooms, many of the fundamental technologies underlying computers and networks have been the results of government-funded projects.

ENIAC, the first operational full-scale electronic digital computer, was an Army Ballistic Research Laboratory project developed during and just after World War II. Early computers were also sponsored and used by the army and navy in areas such as guided missile development, and in national laboratories such as Los Alamos, where nuclear weapons were being developed. (Later the Atomic Energy Agency and its successor in the Department of Energy would play a similar role in obtaining computers, in particular developing an appetite for the more powerful machines—see SUPERCOMPUTER.)

The Office of Naval Research (ONR) played an important role in developing the underlying theory and design for computer architecture (see VON NEUMANN, JOHN), as well as sponsoring many of the early conferences on computer science, helping the discipline emerge.

As the cold war got underway, an increasing amount of funding went to military-related technology. Since computers were becoming essential for designing or operating complex technologies in aerospace, weapons systems, and other areas, it is not surprising that computer scientists have received a significant share of government research dollars.

A pattern of cooperation emerged between government agencies and companies such as Univac and particularly IBM, who were creating the computer industry. AT&T Bell Laboratories (see BELL LABS) received support for communications and semiconductor technology. Leading-edge research funded for military purposes tended to turn up five or ten years later in new generations of commercial products.

Begun in the late 1950s, one of the biggest defense computing projects was the ambitious (but only marginally successful) SAGE automated air defense system. It began with Whirlwind, the first computer designed for multitasking and continuous, real-time operation and data storage using magnetic core memory. Equally innovative were the user consoles, which pioneered such features as CRT-based output and a touch interface using a light pen.

DEFENSE ADVANCED RESEARCH PROJECTS AGENCY (DARPA)

Established in 1958 and sometimes known as the Advanced Research Projects Agency (ARPA), this agency through its Information Processing Technology Office has funded or contributed to some of the most important developments of the information age, including:

- time-sharing computer and operating systems (MIT Project Mac)

- packet-switched networks; the Internet (implemented as ARPANET)

- NLS, an early hypertext system (see HYPERTEXT)

- artificial intelligence topics including speech recognition

ARPA was unusual as a government agency in its agile management. Managers were given considerable latitude to bring together the most innovative computer scientists and turn them loose with a minimum of bureaucratic oversight.

FUNDING ACADEMIC RESEARCH AND COMPUTER SCIENCE

Although military-related research has been the largest portion of government funding for computer science, other government agencies have also played important roles. Vannevar Bush worked tirelessly to create a new national research infrastructure, and this eventually bore fruit in the National Science Foundation (NSF). Starting in the 1960s the NSF began with a focus on providing computer support for the sciences, but soon concluded that university researchers were being crippled by lack of both computers and people who could design software. The agency began to directly support the funding of university computer purchases and the development of computer science programs. By 1970 the NSF was also supporting the development of computer networks as a way for institutions to share resources. NSF funding for computer science and related activities continued to grow. In the mid-1980s NSF set up the National Center for Supercomputing Applications (NCSA), which in turn set up regional centers from which researchers could tap into supercomputer power through a high-speed network.

INDUSTRIAL COMPETITIVENESS

By the 1980s strong competitive threats to the U.S. computer industry (notably from Japan) and some government funding began to go to helping the American industry coordinate its research. An example is SEMATECH, the semiconductor manufacturing research consortium. (DARPA also played an important role in the development of VLSI [very large-scale integration] circuits.)

Another effort of this era was the Strategic Computing Initiative, which was also in part a response to Japanese developments—their Fifth Generation Computer Program. SCI aimed to develop hardware and software for advanced artificial intelligence projects, starting with a military focus, such as autonomous vehicles, voice-controlled "glass cockpit" aircraft interfaces, and expert systems for battle management.

Although there is always fluctuation and changing political priorities, there is no reason to believe that government funding will not continue to play a very important role in computer-related research and development. There will also continue to be debates over the uses to which governments put computing technology, particularly in the military, intelligence, and national security areas.

Further Reading

Defense Advanced Research Projects Agency (DARPA). Available online. URL: http://www.arpa.mil/. Accessed September 22, 2007.

National Center for Supercomputing Applications (NCSA). Available online. URL: http://www.ncsa.uiuc.edu/. Accessed September 22, 2007.

National Research Council. *Funding a Revolution: Government Support for Computing Research.* Washington, D.C.: National Academies Press, 1999. Available online. URL: http://www.nap.edu/readingroom/books/far/contents.html. Accessed September 22, 2007.

National Science Foundation. "Exploring the Frontiers of Computing." Available online. URL: http://www.nsf.gov/dir/index.jsp?org=CISE. Accessed September 22, 2007.

Redmond, Kent C., and Thomas M. Smith. *From Whirlwind to MITRE: The R&D Story of the SAGE Air Defense Computer.* Cambridge, Mass.: MIT Press, 2000.

Roland, Alex, and Philip Shiman. *Strategic Computing: DARPA and the Quest for Machine Intelligence, 1983–1993.* Cambridge, Mass.: MIT Press, 2002.

graphics card

Prior to the late 1970s, most computer applications (other than some scientific and experimental ones) did not use graphics. However, the early microcomputer systems such as the Apple II, Radio Shack TRS-80, and Commodore PET could all display graphics, either on a monitor or (with the aid of a video modulator) on an ordinary TV set. While primitive (low resolution; monochrome or just a handful of colors) this graphics capability allowed for a thriving market in games and educational software.

The earliest video displays for mainstream PCs provided basic text display capabilities (such as the MDA, or monochrome display adapter, with 25 lines of text up to 80 characters per line) plus the ability to create graphics by setting the color of individual pixels. The typical low-end graphics card of the early 1980s was the CGA (Color Graphics Adapter), which offered various modes such as 320 by 200 pixels with four colors. Computers marketed for professional use offered the EGA (Enhanced Graphics Adapter), which could show 640 by 350 pixels at 16 colors.

The ultimate video display standard during the time of IBM dominance was the VGA (Video Graphics Array), which offered a somewhat improved high resolution of 640 by 480 pixels at 16 colors, with an alternative of a lower 320 by 280 pixels but with 256 colors. Because of its use of a color palette containing index values, the 256 colors can actually be drawn from a range of 262,144 possible choices. VGA also marked a break from earlier standards because in order to accommodate such a range of colors it had to convert digital information to analog signals to drive the monitor, rather than using the digital circuitry found in earlier monitors.

Modern video cards can be loosely described as implementing SVGA (Super VGA), but there are no longer discrete standards. Typical display resolutions for desktop PCs today are 1024 by 768 or 1280 by 1024 pixels. (Laptops traditionally have had a lower-resolution 800 by 600 display, but many are now comparable to desktop displays.) The range of colors is vast, with up to 16,777,216 possible colors stored as 32 bits per pixel.

Storing 32 bits (4 bytes) for each of the pixels on a 1024 by 768 screen requires more than 3 megabytes. However, this is just for static images. Games, simulations, and other applications use moving 3D graphics. Since a computer screen actually has only two dimensions, mathematical algorithms must be used to transform the representation

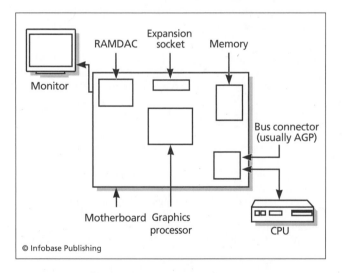

The basic parts of a graphics card. The card is connected to the CPU by the bus (often a special bus called the AGP, or Accelerated Graphics Port). Graphics data can be generated by the CPU and transferred directly to the graphics card's memory, but most cards today perform a lot of the graphics processing using the card's own on-board processor for sophisticated 3D, textures, shading, and other effects.

of objects so they look as if they have three dimensions, appearing in proper perspective, with regard to what objects are behind other objects, and with realistic lighting and shading (see COMPUTER GRAPHICS).

Traditionally, all of the work of producing the actual screen data was undertaken by the PC's main processor, executing instructions from the application program and display driver. By putting a separate processor on the video card (called a video accelerator), together with its own supply of memory (now up to 256 MB), the main system was freed from this burden. A new high-bandwidth connection between the PC motherboard and the graphics card became available with the development of the AGP (Accelerated Graphics Port). (See BUS.) Memory used on video cards is also optimized for video operations, such as by using types of memory such as Video RAM (VRAM) that do not need to be refreshed as frequently.

Increasingly, the algorithms for creating realistic images (such as lighting, shading, and texture mapping) are now supported by the software built into the video card. Of course, the applications program needs a way to tell the graphics routines what to draw and how to draw it. In systems running Microsoft Windows, a program function library called Direct3D (part of a suite called DirectX) has become the standard interface between applications and graphics hardware. Video card manufacturers in turn have optimized their cards to carry out the kinds of operations implemented in DirectX. (A nonproprietary standard called OpenGL has also achieved some acceptance, particularly on non-Windows systems.)

In evaluating video cards, the tradeoff is between the extent to which advanced graphic features are supported and the number of frames per second that can be calculated and sent to the display. If the processing becomes too complicated, the frame rate will slow down and the display will appear to be jerky instead of smooth.

Further Reading

"Graphics and Displays." Tom's Hardware. Available online. URL: http://www.tomshardware.com/graphics/index.html. Accessed August 5, 2007.

Jones, Wendy. *Beginning DirectX 10 Game Programming*. Boston: Course Technology, 2007.

Luna, Frank. *Introduction to 3D Game Programming with DirectX 9.0c: A Shader Approach*. Plano, Tex.: Wordware Publishing, 2006.

Sanchez, Julio, and Maria P. Canton. *The PC Graphics Handbook*. Boca Raton, Fla.: CRC Press, 2003.

Shreiner, Dave, et al. *OpenGL Programming Guide*. 5th ed. Upper Saddle River, N.J.: Addison-Wesley Professional, 2005.

"Video Cards." PC Guide. Available online. URL: http://www.pcguide.com/ref/video/index.htm. Accessed August 5, 2007.

graphics formats

Broadly speaking, a graphics file consists of data that specifies the color of each pixel (dot) in an image. Since there are many ways this information can be organized, there are a variety of graphics file formats. The most important and widely used ones are summarized below.

BMP (WINDOWS BITMAP)

In a bitmap format there is a group of bits (i.e. a binary value) that specifies the color of each pixel. Windows provides standard bitmap (BMP) formats for 1-bit (2 colors or monochrome), 4-bit (16 colors), 8-bit (256 colors), or 24-bit (16 million colors). The Windows bitmap format is also called a DIB (device-independent bitmap) because the stored colors are independent of the output device to be used (such as a monitor or printer). The relevant device driver is responsible for translating the color to one actually used by the device. Because it is "native" to Windows, BMP is widely used, especially for program graphics resources.

Bitmap formats have the advantage of storing the exact color of every pixel without losing any information. However, this means that the files can be very large (from hundreds of thousands of bytes to several megabytes for Windows screen graphics). BMP and other bitmap formats do support a simple method of compression called run-length encoding (RLE), where a series of identical pixels is replaced by a single pixel and a count. Bitmap files can be further compressed through the use of utilities such as the popular Zip program (see DATA COMPRESSION).

EPS

EPS (Encapsulated PostScript) is a vector-based rather than bitmap (raster) format. This means that an EPS file consists not of the actual pixel values of an image, but the instructions for drawing the image (including coordinates, colors, and so on). The instructions are specified as a text file in the versatile PostScript page description language. This format is usually used for printing, and requires a printer that supports PostScript (there are also PostScript renderers that run entirely in software, but they tend to be slow and somewhat unreliable).

GIF

GIF, or Graphics Interchange Format, is a bitmapped format promulgated by CompuServe. Instead of reserving enough space to store a large number of colors in each pixel, this format uses a color table that can hold up to 256 colors. Each pixel contains a reference (index into) the color table. This means that GIF works best with images that have relatively few colors and for applications (such as Web pages) where compactness is important. GIF also uses compression to achieve compactness, but unlike the case with JPEG it is a lossless compression called LZW. There is also a GIF format that stores simple animations.

JPEG

JPEG, which stands for Joint Photographic Experts Group, is widely used for digital cameras because of its ability to highly compress the data in a color graphics image, allowing a reasonable number of high-resolution pictures to be stored in the camera's onboard memory. The compression is "lossy," meaning that information is lost during compression (see DATA COMPRESSION). At relatively low compression ratios (such as 10:1, or 10 percent of the original image size) changes in the image due to data loss are unlikely to be perceived by the human eye. At higher ratios (approaching 100:1) the image becomes seriously degraded. JPEG's ability to store thousands of colors (unlike GIF's limit of 256) makes the format particularly suitable for the subtleties of photography.

PCX

PCX is a compressed bitmap format originally used by the popular PC Paintbrush program. In recent years it has been largely supplanted by BMP and TIFF.

TIFF

TIFF, or Tagged Image File Format, is also a compressed bitmap format. There are several variations by different vendors, which can lead to compatibility problems. Implementations can use various compression methods, generally leading to ratios of 1.5 to 1 to about 2 to 1.

Further Reading

Brown, C. Wayne, and Barry J. Shepherd. *Graphics File Formats Reference and Guide.* Greenwich, Conn.: Manning Publications, 1995.
Miano, John. *Compressed Image File Formats.* Upper Saddle River, N.J.: Addison-Wesley Professional, 1999.
"Web Style Guide: Graphics." Available online. URL: http:// webstyleguide.com/graphics/. Accessed August 5, 2007.
Murray, James D., and William vanRyper. *Encyclopedia of Graphics File Formats.* 2nd ed. (on CD-ROM). Sebastopol, Calif.: O'Reilly, 1996.
"Wotsit's Format: The Programmer's Resource." Available online. URL: http://www.wotsit.org/. Accessed August 14, 2007.

graphics tablet

While conventional pointing devices (see MOUSE) are quite satisfactory for making selections and even manipulating objects, many artists prefer the control available only through a pen or pencil, which allows the angle and pressure of the stylus tip to be varied, creating precise lines and shading. A graphics tablet (also called a *digitizing tablet*) is a device that uses a specially wired pen or pencil with a flat surface (tablet). Besides tracking the location of the pen and translating it into X/Y screen coordinates, the tablet also has pressure sensors (depending on sensitivity, the tablet

Many graphics tablets use a stylus or pen. The system can track the pen's position and, often, the amount of pressure being exerted, and draw the line accordingly.

can recognize 256, 512, or 1024 levels of pressure). In combination with buttons on the pen, the pressure level can be used to control the line thickness, transparency, or color. In addition, the driver software for some graphics tablets includes additional functions such as the ability to program the pen to control features of such applications as Adobe Photoshop.

The tablet is connected to the PC (usually through a USB port). The pen may be connected to the tablet by a tether, or it may be wireless. If the pen has an onboard battery, it can provide additional features at the expense of weight and the need to replace batteries occasionally.

A variant implementation uses a small "puck" instead of a pen. The puck, which can be moved smoothly over the tablet surface, often has a window with crosshairs in the center. This makes it particularly useful for tracing detailed drawings such as in engineering applications.

Many artists find that wielding a pen with a graphics tablet offers not only finer control, but also more natural and less fatiguing method of input than with the mouse.

Further Reading

Chastain, Sue. "Before You Buy a Graphics Tablet." Available online. URL: http://graphicssoft.about.com/od/aboutgraphics/ a/graphicstablets.htm. Accessed August 5, 2007.
Kolle, Iril C. *Graphics Tablet Solutions.* Cincinnati, Ohio: Muska & Lipman, 2001.
Threinen-Pendarvis, Cher. *The Photoshop and Painter Artist Tablet Book: Creative Techniques in Digital Painting.* Berkeley, Calif.: Peachpit Press, 2004.

green PC

This is a general term for features that reduce the growing environmental impact of the manufacture or use of computers. This impact has several aspects: energy consumption, resource consumption, e-waste, and pollution and greenhouse emissions.

ENERGY CONSUMPTION

The greatest part of a typical computer system's power consumption is from the monitor, followed by the hard drive and CPU. It follows that considerable energy can be saved if these components are powered down when not in use. On the other hand, most users do not want to go through the whole computer startup process several times a day. One solution is to design a computer system so that it turns off many components when not in use but is still able to restore full function in a few seconds.

When applied to a personal computer, the federally adopted Energy Star designation indicates a computer system that includes an energy saving mode that can power down the monitor, hard drive, or CPU after a specified period elapses without user activity, such that the inactive system consumes no more than 30 watts. In the ultimate energy-saving feature a suspend mode saves the current state of the computer's memory (and thus of program operation) to a disk file. When the user presses a key (or moves the mouse), the computer "wakes up" and reloads its memory contents from the disk, resuming operation where it left off. By 2000, virtually all new PCs were Energy Star compliant, though many users fail to actually enable the power-saving features.

In July 2007 stricter Energy Star specifications for desktop PCs were adopted. Power supplies must now be at least 80 percent efficient. Meanwhile, the International Energy Agency has been promoting an initiative to reduce power consumption of idle PCs (and other appliances) to 1 watt or less.

RESOURCE CONSUMPTION

Computers consume a variety of resources, starting with their manufacturing and packaging. Resource consumption can be reduced by building more compact units and by designing components so they can be more readily stripped and recycled or reused. Adopting reusable storage media (such as rewritable CDs), recycling printer toner cartridges, and changing office procedures to minimize the generation of paper documents are also ways to reduce resource consumption.

E-WASTE

In recent years the disposal of obsolete computers and other electronic equipment ("e-waste") has been both a growing concern and a business opportunity. There are many toxic substances in electronics components, including lead, mercury, and cadmium. Processing e-waste to recover raw materials is expensive, so greater emphasis has been placed on disassembling machines and reusing or refurbishing their individual components. Meanwhile, many communities have banned disposing of e-waste in regular trash, and some have offered opportunities to drop off e-waste at no or minimal charge. States such as California have also instituted a recycling fee that is collected upon sale of devices such as CRT monitors and televisions.

POLLUTION AND GREENHOUSE EMISSIONS

Fabrication of computer chips in more than 200 large plants around the world involves a variety of toxic chemicals and waste products. The Silicon Valley alone is home to 29 toxic sites under the EPA's Superfund Program. The shift of much of semiconductor and computer component manufacturing to countries such as China that have less strict pollution controls has also exacerbated what has become a global problem.

Whether through regulation or enlightened self-interest, companies that want to reduce future emissions can use several strategies. Manufacturing equipment and processes can be modified so they create fewer toxic substances or at least keep them from getting into the environment. Nontoxic (or less toxic) materials can be substituted where possible—for example, use of ozone-depleting chlorofluorocarbons (CFCs) as cleaning agents has been largely eliminated. Finally, waste can be properly sorted and disposed of, and recycled wherever feasible.

Like other major manufacturing sectors, the computer industry is also faced with the need to reduce the amount of the greenhouse gases (particularly CO_2) contributing to global warming. This mainly means further reducing the energy consumption of new PCs. In June 2007 a number of major players, including Google, Intel, Dell, Hewlett-Packard, Microsoft, and Sun, established the Climate Savers Computing Initiative. Going beyond Energy Star, the program is expected to reduce power consumption equivalent to 54 million tons of greenhouse gases annually—about the same as that produced by 11 million cars or 20 large coal-fired power plants.

Further Reading

Brandon, John. "Build a Green PC." *ExtremeTech,* March 2, 2007. Available online. URL: http://www.extremetech.com/article2/0,1697,2097765,00.asp. Accessed August 5, 2007.

Cascio, Jamais. *Green PC: How to Dispose of Unwanted Tech Equipment without Hassles, and Where to Find Great New Environmentally Friendly Gear.* PC World, May 22, 2006. Available online. URL: http://www.pcworld.com/article/id,125708/article.html. Accessed August 5, 2007.

"Cleaner Computers? Industry to Cut Carbon." (AP/MSNBC). June 12, 2007. Available online. URL: http://www.msnbc.msn.com/id/19203144/. Accessed August 5, 2007.

Esty, Daniel, and Andrew S. Winston. *Green to Gold: How Smart Companies Use Environmental Strategy to Innovate, Create Value, and Build Competitive Advantage.* New Haven, Conn.: Yale University Press, 2006.

Kuehr, Ruediger, and Eric Williams, eds. *Computers and the Environment: Understanding and Managing Their Impacts.* Norvell, Mass.: Kluwer Academic Publishers, 2003.

Weil, Nancy. "The Realities of Green Computing." *PC World,* August 3, 2007. Available online. URL: http://www.pcworld.com/article/id,135509/article.html. Accessed August 5, 2007.

grid computing

Grid or cluster computing involves the creation of a single computer architecture that consists of many separate computers that function much like a single machine. The computers are usually connected using fast networks (see LOCAL AREA NETWORK). The purpose of the arrangement can be to provide redundant processing in case of system failures, to dynamically balance a fluctuating work load, or to split large computations into many parts that can be performed simultaneously. This latter approach to "high-performance computing" creates the virtual equivalent of a very large and powerful machine (see SUPERCOMPUTER).

ARCHITECTURE

Grid and cluster architectures often overlap, but the term *grid* tends to be applied to a more loosely coordinated structure where the computers are dispersed over a wider area (not a local network). In a grid, the work is usually divided into many separate packets that can be processed independently without the computers having to share data. Each task can be completed and submitted without waiting for the completion of any other task. Clusters, on the other hand, more closely couple computers to act more like a single large machine.

The first commercially successful product based on this architecture was the VAXcluster released in the 1980s for DEC VAX minicomputers. These systems implemented parallel processing while sharing file systems and peripherals.

In 1989 an open-source cluster solution called Parallel Virtual Machine (PVM) was developed. These clusters could mix and match any computers that could connect over a TCP/IP network (i.e., the Internet).

CURRENT IMPLEMENTATIONS AND APPLICATIONS

Clusters made from hundreds of desktop-class computer processors can achieve supercomputer levels of performance at comparatively low prices. An example is the System X supercomputer cluster at Virginia Tech, which generates 12.25 TFlops (trillion floating point operations per second) from 1100 Apple XServe G5 dual-processor desktops running Mac OS X.

Additional savings and flexibility can be found in Beowulf clusters, which use standard commodity PCs running open-source operating systems (such as Linux) and software such as the Globus Toolkit.

Another type of implementation is the "ad hoc" computer grid. These are projects where users sign up to receive and process work packets using their PC's otherwise idle time. Examples include SETI@Home (search for extraterrestrial intelligence) and Folding@Home (protein-folding calculations). For more on this type of arrangement, see COOPERATIVE PROCESSING.

Although there has been some recent interest in enterprise grids, most grid computing applications are in science. The world's most powerful computer grid, TeraGrid, is funded by the National Science Foundation and ties together major supercomputing and advanced computing installations at universities and government laboratories. Current applications for TeraGrid include weather and climate forecasting, earthquake simulation, epidemiology, and medical visualization.

Further Reading

Globus Toolkit Homepage. Available online. URL: http://www.globus.org/toolkit/. Accessed September 23, 2007.

Haynos, Matt. "Perspectives on Grid: Grid Computing—Next-Generation Distributed Computing." IBM, January 27, 2004. Available online. URL: http://www.ibm.com/developerworks/grid/library/gr-heritage/. Accessed September 23, 2007.

Kacsuk, Peter, Thomas Fahringer, and Zsolt Nemeth. *Distributed and Parallel Systems: From Cluster to Grid Computing.* New York: Springer Science/Business Media, 2007.

Kopper, Karl. *Linux Enterprise Cluster: Build a Highly Available Cluster with Commodity Hardware and Free Software.* San Francisco: No Starch Press, 2005.

Plaszczak, Pawel, and Richard Wellner, Jr. *Grid Computing: The Savvy Manager's Guide.* San Francisco: Morgan Kaufman, 2006.

Robbins, Stuart. *Lessons in Grid Computing: The System Is a Mirror.* New York: Wiley, 2006.

TeraGrid. Available online. URL: http://www.teragrid.org/. Accessed September 23, 2007.

groupware

When PCs were first introduced into the business world, they tended to be used in isolation. Individual workers would prepare documents such as spreadsheets and database reports and then print them out and distribute them as memos, much in the way of traditional paper documents.

However, as computers began to be tied together into local area networks (see LOCAL AREA NETWORK) in the 1980s, focus began to shift toward the use of software to facilitate communication, coordination, and collaboration among workers. This loosely defined genre of software was dubbed groupware.

Popular groupware software suites such as Lotus Notes and Microsoft Exchange generally offer at least some of the following features:

- e-mail coordination, including the creation of group or task-oriented mail lists

- shared calendar, giving each participant information about all upcoming events

- meeting management, including scheduling (ensuring compatibility with everyone's existing schedule) and facilities booking

- scheduling tasks with listing of persons responsible for each task, progress (milestones met), and checking off completed tasks

- real-time "chat" or instant message capabilities

- documentation systems that allow a number of people to make comments on the same document and see and respond to each other's comments

- "whiteboard" systems that allow multiple users to draw a diagram or chart in real time, with everyone able to see and possibly modify it

Groupware is increasingly integrated with the Internet, with documents and shared resources (calendars, schedules, and so on) implemented in HTML as Web pages or Web-linked databases. (See also PERSONAL INFORMATION MANAGER.)

An attractive alternative to locally installed groupware is a suite of collaboration and productivity applications delivered directly via the Web and accessible using only a Web browser. Google introduced such a package called Google Apps in 2007. It has a free basic version but is expected to offer fee-based enhanced services for larger organizations.

Groupware is likely to be an increasingly important aspect of institutional information processing in a global, mobile economy. With workgroups often geographically distributed (as well as including telecommuters), traditional face-to-face meetings become increasingly impractical as well as often being considered wasteful and inefficient. New forms of collaboration are supplementing the traditional e-mail and conferencing (see BLOGS AND BLOGGING and WIKIS AND WIKIPEDIA). Wikis are particularly interesting in that they can not only track current resources, but also provide a knowledge base with lasting value.

Further Reading

Andriessen, J. H. Erik. *Working with Groupware: Understanding and Evaluating Collaboration Technology.* New York: Springer, 2003.

Boles, David. *Google Apps Administrator Guide: A Private-Label Web Workspace.* Boston: Course Technology, 2007.

Cavalancia, Nick. *Microsoft Exchange Server 2007: A Beginner's Guide.* 2nd ed. Berkeley, Calif.: McGraw-Hill Osborne Media, 2007.

Google Apps. Available online. URL: https://www.google.com/a/. Accessed August 5, 2007.

Gookin, Dan. *Google Apps for Dummies.* Hoboken, N.J.: Wiley, 2008.

Morimoto, Rand, et al. *Microsoft Exchange Server 2007 Unleashed.* Indianapolis: Sams, 2007.

Munkvold, Bjorn Erik, et al. *Implementing Collaboration Technologies in Industry.* New York: Springer, 2003.

Udell, Jon. *Practical Internet Groupware.* Sebastapol, Calif.: O'Reilly, 1999.

Grove, Andrew S.

(1936–)
Hungarian-American
Entrepreneur

Andrew Grove is a pioneer in the semiconductor industry and builder of Intel, the corporation whose processors now power the majority of personal computers. Grove was born András Gróf on September 2, 1936, in Budapest to a Jewish family. Grove's family was disrupted by the German occupation of Hungary later in World War II. Andrew's father was conscripted into a work brigade and then into a Hungarian formation of the German army. Andrew and his mother, Maria, had to hide from the Nazi roundup in which many Hungarian Jews were sent to death in concentration camps.

Although the family survived and was reunited after the war, Hungary had come under Soviet control. Andrew, now 20, believed his freedom and opportunity would be very limited, so he and a friend made a dangerous border crossing into Austria. Grove came to the United States, where he lived with his uncle in New York and studied chemical engineering. He then earned his Ph.D. at the University of California at Berkeley and became a researcher at Fairchild Semiconductor in 1963 and then assistant director of development in 1967. He soon became familiar with the early work toward what would become the integrated circuit, key to the microcomputer revolution that began in the 1970s and wrote a standard textbook (*Physics and Technology of Semiconductor Devices*).

In 1968, however, he joined colleagues Robert Noyce and Gordon Moore in leaving Fairchild and starting a new company, Intel. Grove switched from research to management, becoming Intel's director of operations. He established a management style that featured what he called "constructive confrontation"—a vigorous, objective discussion where opposing views could be aired without fear of reprisal. Critics, however, sometimes characterized the confrontations as more harsh than constructive.

Grove became a formidable competitor. In the late 1970s, it was unclear whether Intel (maker of the 8008, 8080, and subsequent processors) or Motorola (with its 68000 processor) would dominate the market for microprocessors to run the new desktop computers. Grove emphasized the training and deployment of a large sales force, and by the time the IBM PC debuted in 1982, it and its imitators would all be powered by Intel chips.

During the 1980s, Grove would be challenged to be adaptable when Japanese companies eroded Intel's share of the DRAM (memory) chip market, often "dumping" product below their cost. Grove decided to get Intel out of the memory market, even though it meant downsizing the company until the growing microprocessor market made up for the lost revenues. In 1987, Grove had weathered the storm and become Intel's CEO. He summarized his experience of the rapidly changing market with the slogan "only the paranoid survive."

During the 1990s, Intel introduced the popular Pentium line, having to overcome mathematical flaws in the first version of the chip and growing competition from Advanced Micro Devices (AMD) and other companies that made chips compatible with Intel's. Grove also had to fight prostate cancer, apparently successfully, and relinquished his CEO title in 1998, remaining chairman of the board.

Through several books and numerous articles, Grove has had considerable influence on the management of modern electronics manufacturing. He has received many industry awards, including the IEEE Engineering Leadership Recognition award (1987), and the AEA Medal of Achievement award (1993). In 1997, he was CEO of the Year (*CEO* magazine) and *Time* magazine's Man of the Year.

Further Reading

"Andy Grove." Intel. Available online. URL: http://www.intel.com/pressroom/kits/bios/grove.htm. Accessed August 5, 2007.

Grove, Andrew S. *High Output Management.* 2nd ed. New York: Vintage Books, 1995.

———. *One-on-One with Andy Grove.* New York: Putnam, 1987.

———. *Only the Paranoid Survive.* New York: Currency Doubleday, 1996.

Tedlow, Richard S. *Andrew Grove: The Life and Times of an American.* New York: Penguin/Portfolio, 2006.

H

hackers and hacking

Starting in the late 1950s, in computer facilities at MIT, Stanford, and other research universities people began to encounter persons who had both unusual programming skill and an obsession with the inner workings of the machine. While ordinary users viewed the computer simply as a tool for solving particular problems, this peculiar breed of programmers reveled in extending the capabilities of the system and creating tools such as program editors that would make it easier to create even more powerful programs. The movement from mainframes that could run only one program at a time to machines that could simultaneously serve many users created a kind of environmental niche in which these self-described *hackers* could flourish. Indeed, while administrators sometimes complained that hackers took up too much of the available computer time, they often depended on them to fix the bugs that infested the first versions of time-sharing operating systems. Hackers also tended to work in the wee hours of the night while normal users slept.

Early hackers had a number of distinctive characteristics and tended to share a common philosophy, even if it was not always well articulated:

- Computers should be freely accessible, without arbitrary limits on their use (the "hands-on imperative").

- "Information wants to be free" so that it can reach its full potential. Conversely, government or corporate authorities that want to restrict information access should be resisted or circumvented.

- The only thing that matters is the quality of the "hack"—the cleverness and utility of the code and what it lets computers do that they could not do before.

- As a corollary to the above, the reputation of a hacker depends on his (it was nearly always a male) work—not on age, experience, academic attainment, or anything else.

- Ultimately, programming was a search for truth and beauty and even a redemptive quality—coupled with the belief that technology can change the world.

Hackers were relatively tolerated by universities and sometimes prized for their skills by computer companies needing to develop sophisticated software. However, as the computer industry grew, it became more concerned with staking out, protecting, and exploiting intellectual property. To the hacker, however, intellectual property was a barrier to the unfettered exploration and exploitation of the computer. Hackers tended to freely copy and distribute not only their own work but also commercial systems software and utilities.

During the late 1970s and 1980s, the microcomputer created a mass consumer software market, and a new generation of hackers struggled to get the most out of machines that had a tiny amount of memory and only rudimentary graphics and sound capabilities. Some became successful game programmers. At the same time a new term entered the lexicon, *software piracy* (see SOFTWARE PRIVACY AND COUNTERFEITING). Pirate hackers cracked the copy protection on games

219

and other commercial software so the disks could be copied freely and exchanged at computer fairs, club meetings, and on illicit bulletin boards (where they were known as "warez"). (See COPY PROTECTION and INTELLECTUAL PROPERTY AND COMPUTING.)

The growing use of on-line services and networks in the 1980s and 1990s brought new opportunities to exploit computer skills to vandalize systems or steal valuable information such as credit card numbers. The popular media used the term *hacker* indiscriminately to refer to clever programmers, software pirates, and people who stole information or spread viruses across the Internet. The wide availability of scripts for password cracking, Web site attacks, and virus creation means that destructive crackers often have little real knowledge of computer systems and do not share the attitudes and philosophy of the true hackers who sought to exploit systems rather than destroy them.

During the 1980s, a new genre of science fiction called *cyberpunk* became popular. It portrayed a fractured, dystopian future where elite hackers could "jack into" computers, experiencing cyberspace directly in their mind, as in William Gibson's *Neuromancer* and *Count Zero*. In such tales the hacker became the high-tech analog of the cowboy or samurai, a virtual gunslinger who fought for high stakes on the newest frontier (see SCIENCE FICTION AND COMPUTING). Meanwhile, lurid stories about such notorious real-world hackers (see MITNICK, KEVIN) brought the dark side of hacking into popular consciousness.

By the turn of the new century, the popular face of hacking was again changing. Some of the most effective techniques for intruding into systems and for stealing sensitive information (see COMPUTER CRIME and IDENTITY THEFT) have always been psychological rather than technical. What started as one-on-one "social engineering" (such as posing as a computer technician to get a user's password) has been "industrialized" in the form of e-mails that frighten or entice recipients into supplying credit card or bank information (see SPAM and PHISHING AND SPOOFING). Criminal hackers have also linked up with more-traditional criminal organizations, creating rings that can efficiently turn stolen information into cash.

In response to public fears about hackers' capabilities, federal and local law enforcement agencies have stepped up their efforts to find and prosecute people who crack or vandalize systems or Web sites. Antiterrorism experts now worry that well-financed, orchestrated hacker attacks could be used by rogue nations or terrorist groups to paralyze the American economy and perhaps even disrupt vital infrastructure such as power distribution and air traffic control (see COUNTERTERRORISM and INFORMATION WARFARE). In this atmosphere the older, more positive image of the hacker seems to be fading—although the free-wheeling creativity of hacking at its best continues to be manifested in cooperative software development (see OPEN SOURCE).

Further Reading

2600 magazine. Available online. URL: http://www.2600.com. Accessed February 2, 2008.

Erickson, Jon. *Hacking: The Art of Exploitation*. 2nd ed. San Francisco: No Starch Press, 2007.

Gibson, William. *Neuromancer*. West Bloomfield, Mich.: Phantasia Press, 1986.

Hafner, Katie, John Markoff. *Cyberpunk: Outlaws and Hackers on the Computer Frontier*. New York: Simon & Schuster, 1991.

Harris, Shon, et al. *Gray Hat Hacking: The Ethical Hacker's Handbook*. Berkeley, Calif.: McGraw-Hill/Osborne Media, 2004.

Levy, Stephen. *Hackers: Heroes of the Computer Revolution*. New York: Doubleday, 1984.

Littman, J. *The Fugitive Game: On-line with Kevin Mitnick*. Boston: Little, Brown, 1996.

Mitnick, Kevin, and William L. Simon. *The Art of Deception: Controlling the Human Element of Security*. Indianapolis: Wiley, 2002.

———. *The Art of Intrusion: The Real Stories behind the Exploits of Hackers, Intruders & Deceivers*. Indianapolis: Wiley, 2005.

Raymond, Eric. *The New Hacker's Dictionary*. 3rd ed. Cambridge, Mass.: MIT Press, 1996.

handwriting recognition

While the keyboard is the traditional means for entering text into a computer system, both designers and users have long acknowledged the potential benefits of a system where people could enter text using ordinary script or printed handwriting and have it converted to standard computer character codes (see CHARACTERS AND STRINGS). With such a system people would not need to master a typewriter-style keyboard. Further, users could write commands or take notes on handheld or "palm" computers the size of a small note pad that are too small to have a keyboard (see PORTABLE COMPUTERS). Indeed, such facilities are available to a limited extent today.

A handwriting recognition system begins by building a representation of the user's writing. With a pen or stylus system, this representation is not simply a graphical image but includes the recorded "strokes" or discrete movements that make up the letters. The software must then create a representation of features of the handwriting that can be used to match it to the appropriate character templates. Handwriting recognition is actually an application of the larger problem of identifying the significance of features in a pattern.

One approach (often used on systems that work from previously written documents rather than stylus strokes) is to identify patterns of pixels that have a high statistical correlation to the presence of a particular letter in the rectangular "frame" under consideration. Another approach is to try to identify groups of strokes or segments that can be associated with particular letters. In evaluating such tentative recognitions, programs can also incorporate a network of "recognizers" that receive feedback on the basis of their accuracy (see NEURAL NETWORK). Finally, where the identity of a letter remains ambiguous, lexical analysis can be used to determine the most probable letter in a given context, using a dictionary or a table of letter group frequencies.

IMPLEMENTATION AND APPLICATIONS

A number of handheld computers beginning with Apple's Newton in the mid-1990s and the now popular Palm devices and BlackBerry have some ability to recognize handwriting. However, current systems can be frustrating to use

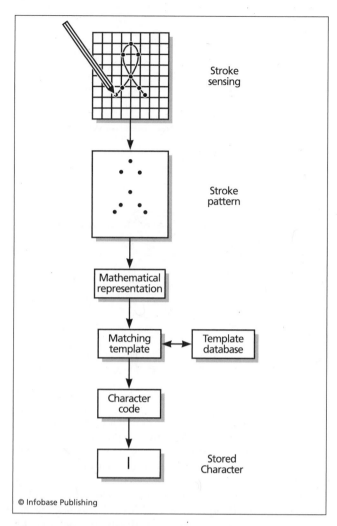

One approach to handwriting recognition involves the extraction of a stroke pattern and its comparison to a database of templates representing various letters and symbols. Ultimately the corresponding ASCII character is determined and stored.

because accuracy often requires that users write very carefully and consistently or (as in the case of the Palm) even replace their usual letter strokes with simplified alternatives that the computer can more easily recognize. If the user is allowed to use normal strokes, the system must be gradually "trained" by the user giving writing samples and confirming the system's guess about the letters. As the software becomes more adaptable and processing power increases (allowing more sophisticated algorithms or larger neural networks to be practical) users will be able to write more naturally and systems will gain more consumer acceptance. (One step in this direction is the Tablet PC, a notepad-sized computer with a digitizer tablet and a stylus and handwriting recognitions software, included in Windows XP and expanded in Windows Vista. Programs such as Microsoft OneNote use handwriting recognition to allow users to incorporate handwritten text into notes that can be organized and quickly retrieved.)

Currently, handwriting recognition is used mainly in niche applications, such as collecting signatures for delivery services or filling out "electronic forms" in applications where the user must be mobile and relatively hands-free (such as law enforcement).

Further Reading
Crooks, Clayton E. II. *Developing Tablet PC Applications.* Hingam, Mass.: Charles River Media, 2003.
"Handwriting Recognition." TechRepublic Resources. Available online. URL: http://search.techrepublic.com.com/search/handwriting+recognition.html. Accessed August 6, 2007.
Liu, Zhi-Qiang, Jin-Hai Cai, and Richard Buse. *Handwriting Recognition: Soft Computing and Probabilistic Approaches.* New York: Springer, 2003.
Matthews, Craig Forrest. *Absolute Beginner's Guide to Tablet PCs.* Indianapolis: Que, 2003.
Taylor, Paul. "Cast Off Your Keyboard." *Financial Times/FT.com* August 2, 2007. Available online. URL: http://www.ft.com/cms/s/04546598-410d-11dc-8f37-0000779fd2ac.html. Accessed August 6, 2007.
Van West, Jeff. "Using Tablet PC: Handwriting Recognition 101." Available online. URL: http://www.microsoft.com/windowsxp/using/tabletpc/getstarted/vanwest_03may28hanrec.mspx. Accessed August 6, 2007.
Zimmerman, W. Frederick. *Complete Guide to OneNote.* Berkeley, Calif.: Apress, 2003.

haptic interfaces

Most interfaces between users and computer systems involve the equivalent of switches—keyboard keys or mouse buttons. These interfaces cannot respond to degrees of pressure (for an exception, see GRAPHICS TABLET). Further, there is no feedback returned to the user through the interface device—the key or mouse does not "push back."

Haptic (from the Greek word for "touch") interfaces are different in that they do register the pressure and motion of touch, and they often provide touch feedback as well.

Force-feedback systems use movement of the control as a way to provide feedback to the operator. A common example is the control stick in an aircraft that begins to vibrate as the aircraft approaches a stall (where it would lose control). This provides immediate feedback to the pilot using the device by which he or she is already controlling the plane.

More sophisticated forms of force feedback are used in remote-controlled devices for manipulation or exploration. The first application was developed in the 1950s for handling radioactive materials. Today a combination of position and movement sensing and force feedback can be used with special gloves to enable users to grasp and heft 3D virtual objects while getting a sense of their weight, shape, and even texture.

In games, haptic joysticks and other controls such as steering wheels can provide sensations such as resistance to a car's turn or the sensation of a bat hitting a ball. The Nintendo Wii game console comes with a controller that tracks the direction and speed of its movement along with a set of simple but engrossing sports games to show its capabilities.

Some emerging or near-future uses of haptic technology include:

- remote surgery, where the surgeon can feel the resistance of tissues and the location of anatomical features

- use of haptic technology to provide robots with more humanlike gripping capabilities

- 3D sculpture in a virtual 3D world modeling the characteristics of different materials and tools

Like virtual reality itself, haptics is currently found in niche applications such as entertainment, control, and training systems. Besides the expense of the technology itself, there is the need for specialized programming. However, the time may come when haptic support, like mouse and pen support, is included in operating systems and widely available programming libraries.

Further Reading

Burdea, Grigore C. *Force and Touch Feedback for Virtual Reality.* New York: Wiley, 2006.
Haptic Interface Research Laboratory (Purdue University). Available online. URL: http://www.ecn.purdue.edu/HIRL/. Accessed September 23, 2007.
McLaughlin, Margaret L., Joao P. Hespanha, and Gaurav S. Sukhatme. *Touch in Virtual Environments: Haptics and the Design of Interactive Systems.* Upper Saddle River, N.J.: Prentice Hall, 2002.
Zyga, Lisa. "Artists 'Draw on Air' to Create 3D Illustrations." Available online. URL: http://www.physorg.com/news109425896.html. Accessed September 23, 2007.

hard disk

Even after decades of evolution in computing, the hard disk drive remains the primary means of fast data storage and retrieval in computer systems of all sizes. The disk itself consists of a rigid aluminum alloy platter coated with a magnetic oxide material. The platter can be rotated at speeds of more than 10,000 rpm. A typical drive consists of a stack of such platters mounted on a rotating spindle, with a read/write head mounted above each platter.

Early hard drive heads were controlled by a stepper motor, which positioned the head in response to a series of electrical pulses. (This system is still used for floppy drives.) Today's hard drives, however, are controlled by a voice-coil actuator, similar in structure to an audio speaker. The coil surrounds a magnet. When a current enters the coil, it generates a magnetic field that interacts with that of the permanent magnet, moving the coil and thus the disk head. Unlike the stepper motor, the voice coil is continuously variable and its greater precision allows data tracks to be packed more tightly on the platter surface, increasing disk capacity.

The storage capacity of a drive is determined by the number of platters and the spacing (and thus number) of tracks that can be laid down on each platter. Capacities have steadily increased while prices have plummeted: In 1980, for example, a hard drive for an Apple II microcomputer cost more than $1,000 and held only 5 MB of data. As of 2007 internal hard drives with a capacity of 500 GB or more cost around a $150.00.

Data is organized on the disk by dividing the tracks into segments called *sectors.* When the disk is prepared

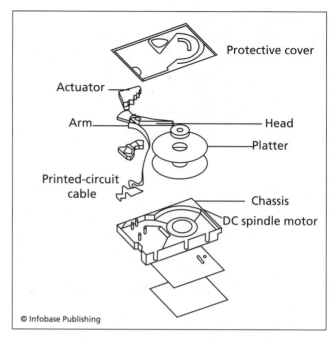

Parts of a typical hard disk drive. Many hard drives have multiple heads and platters to allow for storage of larger amounts of data.

to receive data (a process called *formatting*), each sector is tested by writing and reading sample data. If an error occurs, the operating system marks the sector as unusable (virtually any hard disk will have at least a few such bad sectors).

The set of vertical corresponding tracks on the stack of platters that make up the drive is called a *cylinder.* Since the drive heads are connected vertically, if a head is currently reading or writing for example sector 89 on one platter, it is positioned over that same sector on all the others. Therefore, the operating system normally stores files by filling the full cylinder before going to a new sector number.

Another way to improve data flow is to use *sector interleaving.* Because many disk drives can read data faster than the operating system can read it from the disk's memory buffer, data is often stored by skipping over adjacent sectors. Thus, instead of storing a file on sectors 1, 2, and 3, it might be stored on sectors 1, 3, and 5 (this is called a 2:1 interleave). Moving the head from sector 1 to sector 3 gives the system enough time to process the data. (Otherwise, by the time the system was ready to read sector 2, the disk would have rotated past it and the system would have to wait through a complete rotation of the disk.) Newer CPUs are often fast enough to keep up with contiguous sectors, avoiding the need for interleaving.

Data throughput tends to decrease as a hard drive is used. This is due to *fragmentation.* The operating system runs out of sufficient contiguous space to store new files and has to write new files to many sectors widely scattered on the disk. This means the head has to be moved more often, slowing data access. Using an operating system (or third party) defragmentation utility, users can periodically reorganize their hard drive so that files are again stored in contiguous sectors.

Files can also be reorganized to optimize space rather than access time. If an operating system has a minimum cluster size c4K, a single file with only 32 bytes of data will still consume 4,096 bytes. However, if all the files are written together as one huge file (with an index that specifies where each file begins) that waste of space would be avoided. This is the principle of *disk compression*. Disk compression does slow access somewhat (due to the need to look up and position to the actual data location for a file) and the system becomes more fragile (since garbling the giant file would prevent access to the data in perhaps thousands of originally separate files). The low cost of high capacity drives today has made compression less necessary.

INTERFACING HARD DRIVES

When the operating system wants to read or write data to the disk, it must send commands to the *driver*, a program that translates high-level commands to the instructions needed to operate the *disk controller*, which in turn operates the motors controlling the disk heads. The two most commonly used interfaces for PC internal hard drives today are both based on the ATA (Advanced Technology Attachment) standard. The older standard is PATA (parallel ATA), also called IDE (Integrated Drive Electronics) or EIDE (Enhanced IDE). Increasingly common today is SATA, or serial ATA. Another alternative, more commonly used on servers, is SCSI (Small Computer System Interface). SCSI is more expensive but has several advantages: It has the ability to organize incoming commands for greater efficiency and also features greater flexibility (an EIDE controller can connect only two hard drives, while SCSI can "daisy chain" a large number of disk drives or other peripherals). In practice, the two interfaces perform about equally well. USB (Universal Serial Bus) is frequently used to interface with external hard drive units (see USB).

The capacity continues to increase, with data able to be written more densely or perhaps in multiple layers on the same disk surface. Denser storage also offers the ability to make drives more compact. Already hard drives with a diameter of about an inch have been built by IBM and others for use in digital cameras.

The proliferation of multimedia (including video) and the growth of databases has fed a voracious appetite for hard drive space. Disks with a capacity of 1 TB (terabyte, or trillion bytes) were starting to come onto the market by 2007. For larger installations, disk arrays (see RAID) offer high capacity and data-protecting redundancy.

Perpendicular hard drive recording technology recently developed by Hitachi aligns the magnetic "grains" that hold bits of data vertically instead of horizontally, allowing for a considerably higher data density (and thus capacity, for a given size disk). Hitachi suggests that eventually 1 TB can be stored on a 3.5" disk.

Drive speeds (and thus data throughput) have also been increasing, with more users choosing 7200 rpm rather than the formerly standard 5400 rpm drives. (There are drives as fast as 15,000 rpm, but for most applications the benefits of higher speed drop off rapidly.)

Another factor in data access time and throughput is the use of a dedicated memory device (see CACHE) to "pre-fetch" data likely to be needed. Windows Vista allows memory from some USB memory sticks (see FLASH DRIVE) to work as a disk cache. "Hybrid" hard drives directly integrating RAM and drive storage are also available.

Further Reading

Jacob, Bruce, Spencer Ng, and David Wang. *Memory Systems: Cache, DRAM, Disk*. San Francisco: Morgan Kaufmann, 2007.
"Perpendicular Hard Drive Recording Technology." Available online. URL: http://www.webopedia.com/DidYouKnow/Computer_Science/2006/perpendicular_hard_drive_technology.asp. Accessed August 6, 2007.
"Storage." Tom's Hardware Guide. Available online. URL: http://www.tomshardware.com/storage/index.html. Accessed August 6, 2007.
"What's Inside a Hard Drive?" Available online. URL: http://www.webopedia.com/DidYouKnow/Hardware_Software/2002/InsideHardDrive.aspdYouKnow/Hardware_Software/2002/InsideHardDrive.asp. Accessed August 6, 2007.

hashing

A *hash* is a numeric value generated by applying a mathematical formula to the numeric values of the characters in a string of text (see CHARACTERS AND STRINGS). The formula is chosen so that the values it produces are always the same length (regardless of the length of the original text) and are very likely to be unique. (Two different strings should not produce the same hash value. Such an event is called a *collision*.)

APPLICATIONS

The two major application areas for hashing are information retrieval and cryptographic certification. In databases, an index table can be built that contains the hash values for the key fields and the corresponding record number for each field, with the entries in hash value order. To search the database, an input key is hashed and the value is compared with the index table (which can be done using a very fast binary search). If the hash value is found, the corresponding record number is used to look up the record. This tends to be much faster than searching an index file directly.

Alternatively, a "coarser" but faster hashing function can be used that will give the same hash value to small groups (called *bins*) of similar records. In this case the hash from the search key is matched to a bin and then the records within the bin are searched for an exact match.

In cryptography an encrypted message can be hashed, producing a unique fixed-length value. (The fixed length prevents attackers from using mathematical relationships that might be discoverable from the field lengths.) The hashed message can then be encrypted again to create an electronic signature (see CERTIFICATE, DIGITAL). For long messages this is more efficient than having to apply the signature function to each block of the encrypted message, yet

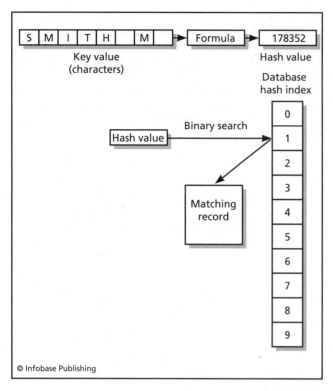

© Infobase Publishing

To search a hashed database, the hashing formula is first applied to the search key, yielding a hash value. That value can then be used in a binary search to quickly zero in on the matching record, if any.

the unique relationship between the original message and the hash maintains a high degree of security.

Finally, hashing can be used for error detection. If a message and its hash are sent together, the recipient can hash the received text. If the hash value generated matches the one received, it is highly likely the message was received intact (see also ERROR CORRECTION).

Further Reading
Partow, Arash. "General Purpose Hash Function Algorithms." Available online. URL: http://www.partow.net/programming/hashfunctions/. Accessed August 6, 2007.
Pieprzyk, Josef, and Babak Sadeghiyan. *Design of Hashing Algorithms*. New York: Springer-Verlag, 1993.

health, personal   *See* PERSONAL HEALTH INFORMATION MANAGEMENT.

heap
In operating systems and certain programming languages (such as LISP), a *heap* is a pool of memory resources available for allocation by programs. The memory segments (sometimes called *cells*) can be the same size or of variable size. If the same size, they are linked together by pointers (see LIST PROCESSING). Memory is then allocated for a variable by traversing the list and setting the required number of cells to be "owned" by that variable. (While some languages such as Pascal and C use explicit memory allocation or deallocation functions, other languages such as LISP use a separate runtime module that is not the responsibility of the programmer.)

Deallocation (the freeing up of memory no longer needed by a variable so it can be used elsewhere) is more complicated. In many languages several different pointers can be used to refer to the same memory location. It is therefore necessary not only to disconnect a given pointer from the cell, but to track the total number of pointers connected to the cell so that the cell itself is deallocated only when the last pointer to it has been disconnected. One way to accomplish this is by setting up an internal variable called a *reference counter* and incrementing or decrementing it as pointers are connected or disconnected. The disadvantages of this approach include the memory overhead needed to store the counters and the execution overhead of having to continually check and update the counters.

An alternative approach is *garbage collection*. Here the runtime system simply connects or disconnects pointers as required by the program's declarations, without making an attempt to reclaim the disconnected ("dead") cells. If and when the supply of free cells is exhausted, the runtime system takes over and begins a three-stage process. First, it provisionally sets the status indicator bit for each cell to show that it is "garbage." Each pointer in the program is then traced (that is, its links are followed) into the heap, and if a valid cell is found that cell's indicator is reset to "not garbage." Finally, the garbage cells that remain are linked back to the pool of free cells available for future allocation. The chief drawback of garbage collection is that the more cells actually being used by the program, the longer the garbage-collecting process will take (since all of these cells have to be traced and verified). Yet it is precisely when most cells are in use that garbage collection is most likely to be required.

The need for garbage collection has diminished in many programming environments because modern computers not only have large amounts of memory, most operating systems also implement virtual memory, which allows a disk or other storage device to be treated as an extension of main memory.

Note: the term *heap* is also used to describe a particular type of binary tree. (See TREE.)

Further Reading
Lafore, Robert. *Data Structures & Algorithms in Java*. 2nd ed. Indianapolis: Sams, 2002.
Preiss, Bruno R. "Heaps and Priority Queues." Available online. URL: http://www.brpreiss.com/books/opus4/html/page352.html. Accessed August 6, 2007.
Sebesta, Robert W. *Concepts of Programming Languages*. 8th ed. Boston: Addison-Wesley, 2007.

help systems
In the early days of computing, the programmers of a system tended to also be its users and were thus intimately familiar with the program's operation and command set.

If not a programmer, the user of a mainframe program was probably at least a well-trained operator who could work with the aid of a brief summary or notes provided by the programmer. However, with the beginnings of office automation in the 1970s and the growing use of desktop computers in office, home, and school in the 1980s, increasingly complex programs were being put in the hands of users who often had only minimal computer training (see COMPUTER LITERACY).

While programs often came with one or more tutorial or reference manuals, designers realized that offering help through the program itself would have some clear advantages. First, the user would not have to switch attention from the computer screen to look things up in a manual. Second, the help system could be programmed to not only provide information, but also to help the user find the information needed in a given situation. For example, related topics could be linked together and a searchable index provided.

IMPLEMENTATION

Programs running under the text-based MS-DOS of the 1980s tended to have only rudimentary help screens (often invoked by pressing the F1 key). Generally, these were limited to brief summaries of commands and associated key combinations. However, with the growing use of Microsoft Windows (and the similar Macintosh interface), a more complete and versatile help system was possible. Since these systems allowed multiple windows to be displayed on the screen, the user could consult help information while still seeing the program's main screen. This allowed for trying a recommended procedure and observing the results.

Windows and Macintosh help systems also featured highlighted links in the text that could be used to jump to related topics (see HYPERTEXT AND HYPERMEDIA). A topic word can also be typed into an index box, bringing up any matching topics. If all else fails, the entire help file could be indexed so that any word could be used to find matching topics.

More recent Windows programs also include *wizards*. A wizard is a step-by-step procedure for accomplishing a particular task. For example, if a Microsoft Word user wants to learn how to format text into multiple columns, the help system can offer a wizard that takes the user through the procedure of specifying the number of columns, column size, and so on. The steps can even be applied directly to the document with the wizard "driving" the program accordingly.

Recently, many programs have implemented their help in the form of Web pages, stored either on the user's computer or at the vendor's Web site (see HTML). HTML has the advantage that it is now a nearly universal format that can be used on a variety of platforms and (if hosted on a Web site) the help can be continually improved and updated. (Microsoft's latest version of HTML Help has supplanted its original WinHelp, which is no longer supported by Vista.)

A variety of shareware and commercial help authoring systems such as RoboHelp are available to help developers create help in Windows or HTML format. UNIX systems, which have always included an on-line manual, now typically offer HTML-based help as well.

With printed documentation being increasingly eschewed for cost-cutting reasons, users of many programs today must depend on the help system as well as on on-line documents (such as PDF files) and Web-based support.

Further Reading

Hackos, JoAnn T., and Dawn M. Steven. *Standards for Online Communication: Publishing Information for the Internet/World Wide Web/Help Systems/Corporate Intranets.* New York: Wiley, 1997.

Heng, Christopher. "Free Help Authoring, Manual and Documentation Writing Tools." Available online. URL: http://www.thefreecountry.com/programming/helpauthoring.shtml. Accessed August 6, 2007.

"Microsoft HTML Help 1.4 SDK." Available online. URL: http://msdn2.microsoft.com/en-us/library/ms670169.aspx. Accessed August 6, 2007.

Weber, Jean Hollis. *Is the Help Helpful? How to Create Online Help That Meets Your Users' Needs.* Whitefish Bay, Wisc.: Hentzenwerke Publishing, 2004.

hexadecimal system

The base 16 or hexadecimal system is a natural way to represent the binary data stored in a computer. It is more compact than binary because four binary digits can be replaced by a single "hex" digit.

The following table gives the corresponding decimal, binary, and hex values from 0 to 15:

DECIMAL	BINARY	HEX
0	0	0
1	0001	1
2	0010	2
3	0011	3
4	0100	4
5	0101	5
6	0110	6
7	0111	7
8	1000	8
9	1001	9
10	1010	A
11	1011	B
12	1100	C
13	1101	D
14	1110	E
15	1111	F

Note that decimal and hex digits are the same from 0 to 9, but hex uses the letters A–F to represent the digits corresponding to decimal 10–15. The system extends to higher numbers using increasing powers of 16, just as decimal uses powers of 10: For example, hex FF represents binary 11111111 or decimal 255. Many of the apparently arbitrary numbers encountered in programming can be better understood if one realizes that they correspond to convenient groupings of bits: FF is eight bits, sufficient to hold a single character (see CHARACTERS AND STRINGS). In low-level programming memory addresses are also usually given in hex (see ASSEMBLER).

Further Reading
Matz, Kevin. "Introduction to Binary and Hexadecimal." Available online. URL: http://www.comprenica.com/atrevida/atrtut01.html. Accessed August 7, 2007.

history of computing

With the digital computer now more than 60 years old, there has been growing interest in its history and development. Although it would take a library of books to do the subject justice, providing a summary of the main themes and trends of each decade of computing will give readers of this book some helpful context for understanding the other entries.

EARLY HISTORY

In a sense, the idea of mechanical computation emerged in prehistory when early humans discovered that they could use physical objects such as piles of stones, notches, or marks as a counting aid. The ability to perform computation beyond simple counting extends back to the ancient world: For example, the abacus developed in ancient China could still beat the best mechanical calculators as late as the 1940s (see CALCULATOR). The mechanical calculator began in the West in the 17th century, most notably with the machines created by philosopher-scientist Blaise Pascal. Other devices such as "Napier's bones" (ancestor of the slide rule) depended on proportional logarithmic relationships (see ANALOG COMPUTER).

While the distinction between a calculator and true computer is subtle, Charles Babbage's work in the 1830s delineated the key concepts. His "analytical engine," conceived but never built, would have incorporated punched cards for data input (an idea taken over from the weaving industry), a central calculating mechanism (the "mill"), a memory ("store"), and an output device (printer). The ability to input both program instructions and data would enable such a device to solve a wide variety of problems (see BABBAGE, CHARLES).

Babbage's thought represented the logical extension of the worldview of the industrial revolution to the problem of calculation. The computer was a "loom" that wove mathematical patterns. While Babbage's advanced ideas became largely dormant after his death, the importance of statistics and information management would continue to grow with the development of the modern industrial state in Europe and the United States throughout the 19th century. The punch card as data store and the creation of automatic tabulation systems would reemerge near the end of the century (see HOLLERITH, HERMAN).

During the early 20th century, mechanical calculators and card tabulation and sorting machines made up the data processing systems for business, while researchers built special-purpose analog computers for exploring problems in physics, electronics, and engineering. By the late 1930s, the idea of a programmable digital computer emerged in the work of theoreticians (see TURING, ALAN and VON NEUMANN, JOHN).

1940s

The highly industrialized warfare of World War II required the rapid production of a large volume of accurate calcula-tions for such applications as aircraft design, gunnery control, and cryptography. Fortunately, the field was now ripe for the development of programmable digital computers. Many reliable components were available to the computer designer including switches and relays from the telephone industry and card readers and punches (manufactured by Hollerith's descendant, IBM), and vacuum tubes used in radio and other electronics.

Early computing machines included the Mark I (see AIKEN, HOWARD), a huge calculator driven by electrical relays and controlled by punched paper tape. Another machine, the prewar Atanasoff-Berry Computer (see ATANASOFF, JOHN) was never completed, but demonstrated the use of electronic (vacuum tube) components, which were much faster than electromechanical relays. Meanwhile, a German inventor built a programmable binary computer that combined a mechanical number storage mechanism with telephone relays (see ZUSE, KONRAD). Zuse also proposed building an electronic (vacuum tube) computer, but the German government decided not to support the project.

During the war, British and American code breakers built a specialized electronic computer called Colossus, which read encoded transmissions from tape and broke the code of the supposedly impregnable German Enigma machines.

The most viable general-purpose computers were developed by J. Presper Eckert and John Mauchly starting in 1943 (see ECKERT, J. PRESPER and MAUCHLY, JOHN). The first, ENIAC, was completed in 1946 and had been intended to perform ballistic calculations. While its programming facilities were primitive (programs had to be set up via a plugboard), ENIAC could perform 5,000 arithmetic operations per second, about a thousand times faster than the

John Mauchly and Presper Eckert, Jr., shown with a portion of their ENIAC computer. The ENIAC is often considered to be the first general-purpose electronic digital computer. (HULTON ARCHIVE / GETTY IMAGES)

electromechanical Mark I. ENIAC had about 19,000 vacuum tubes and consumed as much power as perhaps a thousand modern desktop PCs.

1950s

The 1950s saw the establishment of a small but viable commercial computer industry in the United States and parts of Europe. Eckert and Mauchly formed a company to design and market the UNIVAC, based partly on work on the experimental EDVAC. This new generation of computers would incorporate the key concept of the *stored program:* Rather than the program being set up by wiring or simply read sequentially from tape or cards, the program instructions would be stored in memory just like any other data. Besides allowing a computer to fetch instructions at electronic rather than mechanical speeds, storing programs in memory meant that one part of a program could refer to another part during operation, allowing for such mechanisms as branching, looping, the running of subroutines, and even the ability of a program to modify its own instructions.

The UNIVAC became a hit with the public when it was used to correctly predict the outcome of the 1952 presidential election. Government offices and large corporations began to look toward the computer as a way to solve their increasingly complex data processing needs. Forty UNIVACs were eventually built and sold to such customers as the U.S. Census Bureau, the U.S. Army and Air Force, and insurance companies. Sperry (having bought the Mauchly-Eckert company), Bendix, and other companies had some success in selling computers (often for specialized applications), but it was IBM that eventually captured the broad business market for mainframe computers.

The IBM 701 (marketed to the government and defense industry) and 702 (for the business market) incorporated several emerging technologies including a fast electronic (tube) memory that could store 4,096 36-bit data words, a rotating magnetic drum that could store data that is not immediately needed, and magnetic tape for backup. The IBM 650, marketed starting in 1954, became the (relatively) inexpensive workhorse computer for businesses (see MAINFRAME). The IBM 704, introduced in 1955, incorporated magnetic core memory and also featured floating-point calculations.

1960s

The 1960s saw the advent of a "solid state" computer design featuring transistors in place of vacuum tubes and the use of ferrite magnetic core memory (introduced commercially in 1955). These innovations made computers both more compact (although they were still large by modern standards), more reliable, and less expensive to operate (due to lower power consumption.) The IBM 1401 was a typical example of this new technology: It was compact, relatively simple to operate, and came with a fast printer that made it easier to generate data.

There was a natural tendency to increase the capacity of computers by adding more transistors, but the hand-wiring of thousands of individual transistors was difficult and expensive. As the decade progressed, however, the con-cept of the integrated circuit began to be implemented in computing. The first step in that direction was to attach a number of transistors and other components to a ceramic substrate, creating modules that could be handled and wired more easily during the assembly process.

IBM applied this technology to create what would become one of the most versatile and successful lines in the history of computing, the IBM System/360 computer. This was actually a series of 14 models that offered successively greater memory capacity and processing speed while maintaining compatibility so that programs developed on a smaller, cheaper model would also run on the more expensive machines. Compatibility was ensured by devising a single 360 instruction set that was implemented at the machine level by *microcode* stored in ROM (read-only memory) and optimized for each model. By 1970 IBM had sold more than 18,000 360 systems worldwide.

By the mid-1960s, however, a new market segment had come into being: the minicomputer. Pioneered by Digital Equipment Corporation (DEC) with its PDP line, the minicomputer was made possible by rugged, compact solid-state (and increasingly integrated) circuits. Architecturally, the mini usually had a shorter data word length than the mainframe, and used indirect addressing (see ADDRESSING) for flexibility in accessing memory. Minis were practical for uses in offices and research labs that could not afford (or house) a mainframe (see MINICOMPUTER). They were also a boon to the emerging use of computers in automating manufacturing, data collection, and other activities, because a mini could fit into a rack with other equipment (see also EMBEDDED SYSTEMS). In addition to DEC, Control Data Corporation (CDC) produced both minis and large high-performance machines (the Cyber series), the first truly commercially viable supercomputers (see SUPERCOMPUTER).

In programming, the main innovation of the 1960s was the promulgation of the first widely-used, high-level programming languages, COBOL (for business) and FORTRAN (for scientific and engineering calculations), the result of research in the late 1950s. While some progress had been made earlier in the decade in using symbolic names for quantities and memory locations (see ASSEMBLER), the new higher-level languages made it easier for professionals outside the computer field to learn to program and made the programs themselves more readable, and thus easier to maintain. The invention of the COMPILER (a program that could read other programs and translate them into low-level machine instructions) was yet another fruit of the stored program concept.

1970s

The 1970s saw minis becoming more powerful and versatile. The DEC VAX ("Virtual Address Extension") series allowed larger amounts of memory to be addressed and increased flexibility. Meanwhile, at the high end, Seymour Cray left CDC to form Cray Research, a company that would produce the world's fastest supercomputer, the compact, freon-cooled Cray-1. In the mainframe mainstream, IBM's 370 series maintained that company's dominant market share in business computing.

The most striking innovation of the decade, however, was the microcomputer. The microcomputer (now often called the "computer chip") combined three basic ideas: an integrated circuit so compact that it could be laid on a single silicon chip, the design of that circuit to perform the essential addressing and arithmetic functions required for a computer, and the use of microcode to embody the fundamental instructions. Intel's 4004 introduced in late 1971 was originally designed to sell to a calculator company. When that deal fell through, Intel started distributing the microprocessors in developer's kits to encourage innovators to design computers around them. Soon Intel's upgraded 8008 and 8080 microprocessors were available, along with offerings by Rockwell, Texas Instruments, and other companies.

Word of the microprocessor spread through the electronic hobbyist community, being given a boost by the January 1975 issue of *Popular Electronics* that featured the Altair computer kit, available from an Albuquerque company called MITS for about $400. Designed around the Intel 8080, the Altair featured an expansion BUS (an idea borrowed from minis).

The Altair was hard to build and had very limited memory, but it was soon joined by companies that designed and marketed ready-to-use microcomputer systems, which soon became known as personal computers (PCs). By 1980, entries in the field included Apple (Apple II), Commodore (Pet), and Radio Shack (TRS-80). These computers shared certain common features: a microprocessor, memory in the form of plug-in chips, read-only memory chips containing a rudimentary operating system and a version of the BASIC language, and an expansion bus to which users could connect peripherals such as disk drives or printers.

The spread of microcomputing was considerably aided by the emergence of a technical culture where hobbyists

Integrated circuit (IC) chips for memory and control were making for increasingly powerful, compact, and reliable computer components. The microprocessor supplied the remaining ingredient needed for a true desktop personal computer.

and early adopters wrote and shared software, snatched up a variety of specialized magazines, talked computers in user groups, and evangelized for the cause of widespread personal computing.

Meanwhile, programming and the art of software development did not stand still. Innovations of the 1970s included the philosophy of structured programming (featuring well-defined control structures and methods for passing data to and from subroutines and procedures). New languages such as Pascal and C, building on the earlier Algol, supported structured programming design to varying degrees (see STRUCTURED PROGRAMMING). Programmers on college campuses also had access to UNIX, a powerful operating system containing a relatively simple kernel, a shell for interaction with users, and a growing variety of utility programs that could be connected together to solve data processing problems (see UNIX). It was in this environment that the government-funded ARPANET developed protocols for communicating between computers and allowing remote operation of programs. Along with this came e-mail, the sharing of information in newsgroups (Usenet), and a growing web of links between networks that would eventually become the Internet (see INTERNET).

1980s

In the 1980s, the personal computer came of age. IBM broke from its methodical corporate culture and allowed a design team to come up with a PC that featured an open, expandable architecture. Other companies such as Compaq legally created compatible systems (called "clones"), and "PC-compatible" machines became the industry standard. Under the leadership of Bill Gates, Microsoft gained control of the operating system market and also became the dominant competitor in applications software (particularly office software suites).

Although unable to gain market share comparable to the PC and its clones, Apple's innovative Macintosh, introduced in 1984, adapted research from the Xerox PARC laboratory in user interface design. At a time when PC compatibles were still using Microsoft's text-based MS-DOS, the Mac sported a graphical user interface featuring icons, menus, and buttons, controlled by a mouse (see USER INTERFACE). Microsoft responded by developing the broadly similar Windows operating environment, which started out slowly but had become competitive with Apple's by the end of the decade.

The 1980s also saw great growth in networking. University computers running UNIX were increasingly linked through what was becoming the Internet, while office computers increasingly used local area networks (LANs) such as those based on Novell's Netware system. Meanwhile, PCs were also being equipped with modems, enabling users to dial up a growing number of on-line services ranging from giants such as CompuServe to a diversity of individually run bulletin board systems (see BULLETIN BOARD SYSTEMS).

In the programming field a new paradigm, object-oriented programming (OOP) was offered by languages such as SMALLTALK and C++, a variant of the popular C language. The new style of programming focused on programs as embodying relationships between objects that are responsible for both private data and a public interface represented by methods, or capabilities offered to users of the object. Both structured and object-oriented methods attempted to keep up with the growing complexity of large software systems that might incorporate millions of lines of code. The federal government adopted the Ada language with its ability to precisely manage program structure and data operations. (See OBJECT-ORIENTED PROGRAMMING and ADA.)

1990s

By the 1990s, the PC was a mature technology dominated by Microsoft's Windows operating system. UNIX, too, had matured and become the system of choice for university computing and the worldwide Internet. Although the potential of the Internet for education and commerce was beginning to be explored, at the beginning of the decade the network was far from friendly for the average consumer user.

This changed when Tim Berners-Lee, a researcher at Geneva's CERN physics lab, adapted hypertext (a way to link documents together) with the Internet protocol to implement the World Wide Web. By 1994, Web browsing software that could display graphics and play sounds was available for Windows-based and other computers (see WORLD WIDE WEB and WEB BROWSER). The remainder of the decade became a frenzied rush to identify and exploit business plans based on e-commerce, the buying and selling of goods and services on-line (see E-COMMERCE). Meanwhile, educators demanded Internet access for schools.

In the office, the Intranet (a LAN based on the Internet TCP/IP protocol) began to supplant earlier networking schemes. Belatedly recognizing the threat and potential posed by the Internet, Bill Gates plunged Microsoft into the Web server market, included the free Internet Explorer browser with Windows, and vowed that all Microsoft programs would work seamlessly with the Internet.

Moore's Law, the dictum that computer power roughly doubles every 18 months, continued to hold true as PCs went from clock rates of a few tens of MHz to more than 1 GHz. RAM and hard disk capacity kept pace, while low-cost color printers, scanners, digital cameras, and video systems made it easier than ever to bring rich media content into the PC and the on-line world.

BEYOND 2000

The new decade began with great hopes, particularly for the Web and multimedia "dot-coms," but their stocks, inflated by unsustainable expectations, took a significant dip in 2000–2001. By the middle of the decade the computing industry had largely recovered and in many ways was stronger than ever. On the Web, new software approaches (see AJAX, APPLICATION SERVICE PROVIDER, and SERVICE-ORIENTED ARCHITECTURE) are changing the way services and even applications are delivered. The integration of search engines, mapping, local content, and user participation (see BLOGGING, USER-CREATED CONTENT, and SOCIAL NETWORKING) is changing the relationship between companies and their customers.

In hardware, Moore's law is now expressed not through faster single processors, but using processors with two, four, or more processing "cores," challenging software designers (see MULTIPROCESSING). Mobile computing is one of the strongest areas of growth (see PDA and SMARTPHONE), with devices combining voice phone, text messaging, e-mail, and Web browsing.)

The industry continues to face formidable challenges ranging from mitigating environmental impact (see GREEN PC) to the shifting of manufacturing and even software development to rapidly growing countries such as India and China (see GLOBALISM AND THE COMPUTER INDUSTRY).

Thus far, each decade has brought new technologies and methods to the fore, and few observers doubt that this will be true in the future.

Note: for a more detailed chronology of significant events in computing, see Appendix 1: "Chronology of Computing." For more on emerging technologies, see TRENDS AND EMERGING TECHNOLOGIES.

Further Reading
Allan, Roy A. *A History of the Personal Computer: The People and the Technology.* London, Ont.: Allan Publishing, 2001.
Campbell-Kelly, Martin. *From Airline Reservations to Sonic the Hedgehog: A History of the Software Industry.* Cambridge, Mass.: MIT Press, 2004.
Ceruzzi, Paul E. *A History of Modern Computing.* 2nd ed. Cambridge, Mass.: MIT Press, 2003.
Chandler, Alfred D., Jr. *Inventing the Electronic Century: The Epic Story of the Consumer Electronics and Computer Industries, with a New Preface.* Cambridge, Mass.: Harvard University Press, 2005.
Computer History Museum. Available online. URL: http://www.computerhistory.org/. Accessed June 10, 2007.
Ifrah, Georges. *The Universal History of Computing: From the Abacus to the Quantum Computer.* New York: Wiley, 2002.
NetHistory. Available online. URL: http://www.nethistory.info/index.html. Accessed August 7, 2007.

Hollerith, Herman
(1860–1929)
American
Inventor

Herman Hollerith invented the automatic tabulating machine, a device that could read the data on punched cards and display running totals. His invention would become the basis for the data tabulating and processing industry. Hollerith was born in Buffalo, New York, and graduated from the Columbia School of Mines. After graduation, he went to work for the U.S. Census as a statistician. Among other tasks he compiled vital statistics for Dr. John Shaw Billings, who suggested to Hollerith that using punched cards and some sort of tabulator would help the Census Department keep up with the growing volume of demographic statistics.

Hollerith studied the problem and decided that he could build a suitable machine. He went to MIT, where he taught mechanical engineering while working on the machine, which was partly inspired by an earlier device that had used a piano-type roll rather than punched cards as input.

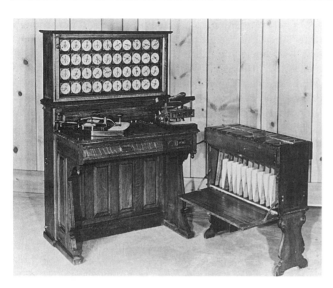

The Hollerith tabulator and sorter box, invented by Herman Hollerith and used in the 1890 U.S. census. It "read" cards by passing them through electrical contacts. (HULTON ARCHIVE / GETTY IMAGES)

The peripatetic Hollerith soon got a job with the U.S. Patent Office, partly to learn the procedures he would need to follow to patent his tabulator. He applied for several patents, including one for the punched-card tabulator. He tested the device with vital statistics in Baltimore, New York, and the state of New Jersey.

Hollerith's mature system included a punch device that a clerk could use to record variable data in many categories on the same card (a stack of cards could also be prepunched with constant data, such as the number of the census district). The cards were then fed into a device something like a small printing press. The top part of the press had an array of spring-loaded pins that connected to tiny pots of mercury (an electrical conductor) in the bottom. The pins were electrified. Where a pin encountered a punched hole in the card, it penetrated through to the mercury, allowing current to flow. The current created a magnetic field that moved the corresponding counter dial forward one position. The dials could be read after a batch of cards was finished, giving totals for each category, such as an ethnicity or occupation. The dials could also be connected to count multiple conditions (for example, the total number of foreign-born citizens who worked in the clothing trade).

Aided by Hollerith's machines, a census unit was able to process 7,000 records a day for the 1890 census, about ten times the rate in the 1880 count. Starting around 1900, Hollerith brought out improved models of his machines that included such features as an automatic (rather than hand-fed) card input mechanism, automatic sorters, and tabulators that boasted a much higher speed and capacity. Hollerith machines soon found their way into government agencies involved with vital statistics, agricultural statistics, and other data-intensive matters, as well as insurance companies and other businesses.

Facing vigorous competition and in declining health, Hollerith sold his patent rights to the company that eventually evolved into IBM, the company that would come to dominate the market for tabulators, calculators, and other office machines. The punched card, often called the Hollerith card, would become a natural choice for computer designers and would remain the principal means of data and program input for mainframe computers until the 1970s.

Further Reading

Austrian, G. D. *Herman Hollerith: Forgotten Giant of Information Processing.* New York: Columbia University Press, 1982.

Kistermann, F. W. "The Invention and Development of the Hollerith Punched Card." *Annals of the History of Computing,* 13, 245–259.

Russo, Mark. "Herman Hollerith: The World's First Statistical Engineer." Available online. URL: http://www.history.rochester.edu/steam/hollerith. Accessed August 7, 2007.

home office

The widespread use of the personal computer and associated peripherals such as printers has made it more practical for many people to do at least part of their work from their homes. In addition to traditional freelance occupations such as writing and editing, many other businesses including consulting, design, and sales can now be conducted from a home office. Computer hardware and software makers began to target a distinctive market niche that is sometimes referred to as SOHO (Small Office / Home Office), thus including both actual home offices and small commercial offices.

As a market, the SOHO has somewhat different requirements than the large offices traditionally served by major computer vendors:

- Relatively modest PCs as compared to heavy-duty file servers or workstations

- Peripherals shared by two or more PCs (although the plummeting price of printers made it common to provide each PC with its own printer)

- The need for a small "footprint"—that is, minimizing the space taken up by the equipment. Multifunction peripherals (typically incorporating printer, scanner, copier, and perhaps a fax machine) are a popular solution to this requirement.

- A simple local network (see LOCAL AREA NETWORK) with shared Internet access

- Low-end or midrange software (such as Microsoft Works or Office Small Business edition as opposed to the full-blown Office suite)

- Application for collaboration and productivity delivered via the Web (such as Google Apps) may also be an attractive alternative.

- Available installation and support (since many home users lack technical hardware or system administration skills)

Although the home or small office remains a significant market segment, specific targeting to the segment has become more difficult. With falling PC prices and increasing capabilities, there is little difference today between a mid-level "consumer" computer system and the kinds of systems previously marketed for home office use.

Further Reading

Attard, Janet. *The Home Office and Small Business Answer Book.* 2nd ed. New York: Owl Books, 2000.

Ivens, Kathy. *Home Networking for Dummies.* 4th ed. Hoboken, N.J.: Wiley, 2007.

Orloff, Erica, and Kathy Levinson. *The 60-Second Commute: A Guide to Your 24/7 Home Office Life.* Upper Saddle River, N.J.: Prentice Hall, 2003.

Slack, S. E. *CNET Do-It-Yourself Digital Home Office Projects.* Berkeley, Calif.: McGraw-Hill Osborne Media, 2007.

Small Business Computing. Available online. URL: http://www. smallbusinesscomputing.com/. Accessed August 7, 2007.

SOHO Computing. Available online. URL: http://www. sohocomputing.info/. Accessed August 7, 2007.

Hopper, Grace Murray

(1906–1992)

American

Computer Scientist

Grace Brewster Murray Hopper was an innovator in the development of high-level computer languages in the 1950s and 1960s. She is best known for her role in the development of COBOL, which became the premier language for business data processing.

Hopper was born in New York City. She graduated with honors with a B.A. in mathematics and physics from Vassar College in 1928, and went on to receive her M.A. and Ph.D. in mathematics at Yale University. She taught at Vassar from 1931 to 1943, when she joined the U.S. Naval Reserve at the height of World War II. As a lieutenant (J.G.), she was assigned to the Bureau of Ordnance, where she worked in the Computation Project at Harvard under pioneer computer designer Howard Aiken (see AIKEN, HOWARD). She became one of the first "coders" (that is, programmers) for the Mark I. After the war, Hopper worked for a few years in Harvard's newly established Computation Laboratory. In 1949, however, she became senior mathematician at the Eckert-Mauchly Corporation, the world's first commercial computer company, where she helped with program design for the famous UNIVAC. She stayed with what became the UNIVAC division under Remington Rand (later Sperry Rand) until 1971.

While working with UNIVAC, Hopper's main focus was on the development of programming languages that could allow people to use symbolic names and descriptive statements instead of binary codes or the more cryptic forms of assembly language (see ASSEMBLER). In 1952, she developed A-0, the first COMPILER (that is, a program that could translate language statements to the corresponding low-level machine instructions). She then developed A-2 (a compiler that could handle mathematical expressions), and then in 1957 she developed Flow-Matic. This was the first compiler

Grace Murray Hopper created the first computer program compiler and was instrumental in the design and adoption of COBOL. When she retired, she was the first woman admiral in U.S. Navy history. (UNISYS CORPORATION)

that worked with English-like statements and was designed for a business data processing environment.

In 1959, Hopper joined with five other computer scientists to plan a conference that would eventually result in the development of specifications for a "Common Business Language." Her earlier work with Flow-Matic and her design input played a key role in the development of what would become the COBOL language.

Hopper retained her Navy commission and even after her retirement in 1966 she was recalled to active duty to work on the Navy's data processing needs. She finally retired in 1986 with the rank of rear admiral. Hopper spoke widely about data processing issues, especially the need for standards in computer language and architecture, the lack of which she said cost the government billions of dollars in wasted resources. Admiral Hopper died on January 1, 1992, in Arlington, Virginia.

Hopper received numerous awards and honorary degrees, including the National Medal of Technology. (The navy named a suitably high-tech Aegis destroyer after her in 1996.) The Association for Computing Machines (ACM) created the Grace Murray Hopper Award to honor distinguished young computer professionals. Hopper has become a role model for many girls and young women considering careers in computing.

Further Reading

"Grace Brewster Murray Hopper" [biography]. St. Andrews University [Scotland] School of Mathematics and Statistics. Available online. URL: http://www-history.mcs.st-andrews. ac.uk/Biographies/Hopper.html. Accessed August 7, 2007.

Grace Hopper Celebration of Women in Computing. Available online. URL: http://gracehopper.org. Accessed August 7, 2007.

Marx, Christy. *Grace Hopper: The First Woman to Program the First Computer in the United States.* New York: Rosen Publishing Group, 2003.

Williams, Kathleen Broome. *Grace Hopper: Admiral of the Cyber Sea.* Annapolis, Md.: Naval Institute Press, 2004.

HTML, DHTML, and XHTML

In developing the World Wide Web, Tim Berners-Lee (see BERNERS-LEE, TIM) had to provide several basic facilities. One was a protocol, HTTP, for requesting documents over the network (see WORLD WIDE WEB). Another was a system of links between documents (see HYPERTEXT AND HYPERMEDIA). The third was a way to embed instructions in the pages so that the Web browser could properly display the text and graphics. Berners-Lee created HTML (Hypertext Markup Language) for this purpose. It is based on the more elaborate SGML (Standard Generalized Markup Language).

The basic "statement" in HTML is the *tag.* Tags are delimited by angle brackets (<>). Tags that affect a document or section of a document come in pairs, with the second member of the pair preceded by a slash. For example, the tags

```
<HTML>

</HTML>
```

indicate the beginning and end of an HTML document, while <BOLD> and </BOLD> delimit text that should be rendered in boldface.

Besides specifying such things as headings, font, font size, and typestyles, HTML includes tags for Web-related functions. One of the most useful is the A, or "anchor" tag. As with some other HTML tags, the A tag is used with *attributes* that further specify what it so be done. The A tag is usually used with the <HREF> or Hypertext Reference attribute, which specifies a document that is to be linked to the current document so that the user can click on a highlight to go there. For example:

```
<A HREF="http://www.MySite.Pages/
Glossary">Glossary of Computer Terms</A>
```

specifies a link to a particular page at a particular site. The link will appear in the browser as the highlighted text **Glossary of Computer Terms.** If clicked, the browser will load the HTML page titled Glossary.

IMPLEMENTATION AND EXTENSIONS

Inserting HTML tags by hand is a tedious and error-prone process (for example, it's easy to omit a bracket or a slash or add "illegal" spaces within tags). Fortunately, there are now many HTML editor programs that let users insert the appropriate elements much in the way word processors make it easy to specify fonts and formatting. (Indeed, programs such as Microsoft Word allow users to convert and save documents in HTML format.)

HTML has been extended in a number of ways. First, new features have been added to later versions of the lan-

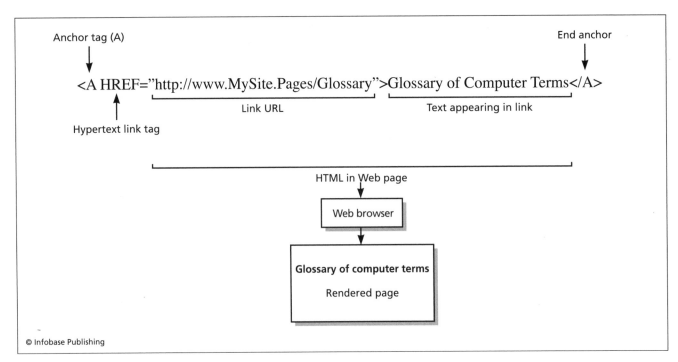

An HTML hyperlink embedded in a Web page. The anchor link gives the address (URL) of the linked page, as well as specifying the text that will appear in the link, which will be rendered by the Web browser in a special color or font.

guage, including better support for frames, columns, tables, and other formats. Browser developers have also adopted a system that allows document authors to define general styles to ensure consistent document appearance (see CASCADING STYLE SHEETS). Style sheets can inherit styles from other style sheets, allowing an organization to create general style sheets that can then be refined to create specialized styles for particular types of documents. The latest version of HTML (as of 2007) is 4.01, with 5.0 still in draft.

Dynamic HTML (DHTML) is a set of techniques that allow otherwise fixed ("static") HTML pages to be changed as users are viewing them. A scripting language (see, for example, JAVASCRIPT) is used to change the specifications (usually via the style sheet). The programming interface to the Web page is the document object model (see DOM). DHTML can be used, for example, to create drop-down menus or "rollover" buttons that change as the mouse navigates over them. Even simple games have been written in DHTML to run in Web browsers. DHTML should be distinguished from other dynamic techniques such as server-side scripting (see PERL and PHP), which changes the page *before* it is presented to the user, and asynchronous techniques that can change a part of a page without reloading it (see AJAX).

XHTML is essentially a rewriting of HTML according to the syntax of the Extensible HyperText Markup Language (see XML). Because of the stricter syntax rules for XML, XHTML cannot use many of the earlier free-form structures of HTML. However, because XML has become so prevalent a means for connecting Web pages to data sources, there are many XML tools that XHTML authors can use for parsing and syntax checking. As of 2007, XHTML 1.1 is the prevailing standard, but a draft 2.0 version represents a more thorough break from the elements of the original HTML.

Further Reading

Freeman, Eric, and Elisabeth Freeman. *Head First HTML with CSS and XHTML.* Sebastapol, Calif.: O'Reilly Media, 2005.
Goodman, Danny. *Dynamic HTML: The Definitive Reference.* 3rd ed. Sebastapol, Calif.: O'Reilly Media, 2006.
Lloyd, Ian. *Build Your Own Website the Right Way Using HTML & CSS.* Lancaster, Calif.: SitePoint, 2006.
Musicano, Chuck, and Bill Kennedy. *HTML & XHTML: The Definitive Guide.* 6th ed. Sebastapol, Calif.: O'Reilly Media, 2006.
Olsson, Tommy. "Bulletproof HTML: 37 Steps to Perfect Markup." Sitepoint. Available online. URL: http://www.sitepoint.com/article/html-37-steps-perfect-markup. Accessed August 7, 2007.
Tittel, Ed, and Mary Burmeister. *HTML 4 For Dummies.* 5th ed. Hoboken, N.J.: Wiley, 2005.

hypertext and hypermedia

Most computer users today are familiar with the concept of hypertext, even if they don't often use the term itself. Each time a Web user clicks on a link on a Web page, he or she is using hypertext. Most on-line help systems also use hypertext to take the reader from one topic to another, related topic. The term *hypermedia* acknowledges modern systems' use of many kinds of resources other than plain text, including still images, videos, and sound recordings.

In a traditional document, the reader is generally assumed to proceed sequentially from the beginning to the end. (Although there may well be footnotes or cross-references within the document, these are generally experienced as temporary divergences from the primary, sequential narrative.) Generally speaking, each reader might be expected to acquire roughly the same set of facts from the document.

In a hypertext document, however, the links between topics create multiple potential paths for readers. To the extent the author has provided links between all related topics, the reader is free to pursue his or her particular interests rather than being bound by a sequential structure imposed by the author. For example, in a document that discusses various organisms in an ecology and the effects of climate and vegetation, one reader might choose to explore one organism in depth, following links from it to other resources devoted to that organism (including outside Web pages, images, videos, and so on). Another reader might be interested specifically in the effects of rainfall on the ecology as a whole and follow a completely different set of links to sites having climatological data.

HISTORY AND DEVELOPMENT

In 1945, a time when the very first digital computers were coming on-line, Vannevar Bush, a pioneer designer of analog computers, proposed a mechanism he called the Memex (see BUSH, VANNEVAR). This system would link portions of documents to allow retrieval of related information. The proposal was impracticable in terms of the very limited capacity of computers of the time. By the 1960s, when computers had become more powerful (and the minicomputer was beginning to be a feasible purchase for libraries and schools), another visionary, Theodore Nelson, coined the terms *hypertext* and *hypermedia.* He suggested that networking (a technology then in its infancy) could allow for what would eventually amount to a worldwide database of interconnected information. Nelson developed his specifications for a system he called Xanadu, but he was unable to create a working version of the system until the late 1990s. However, in 1968 Douglas Engelbart (also known as the inventor of the computer mouse) demonstrated a more limited but workable hypertext system called NLS/Augment.

During the 1970s and 1980s, a variety of hypertext systems were created for various platforms, including Guide and Toolbook for MS-DOS and Windows PCs. Perhaps the most influential system was Hypercard, developed for Apple's Macintosh. While Hypercard did not have a complete set of facilities for creating hypertext, the flexible, programmable, linkable "cards" could be used to implement hypertext documents. Many encyclopedias and other reference products on CD-ROM began to implement some form of hypertext links.

The true explosion of hypertext came with the development and growth of the World Wide Web throughout the 1990s. Hypertext on the Web is implemented through the use of HTTP (HyperText Transport Protocol) over the Internet's TCP/IP protocol and by coding documents in HTML (Hypertext Markup Language). (See HTML, INTERNET, TCP/IP, and WORLD WIDE WEB.)

IMPLEMENTATION

A hypertext document consists of nodes. A node can be a part of a document that conveys a logical "chunk" of information, such as the text that would be under a particular heading in a traditional document. In some systems nodes can be grouped together as a *composite*—for example, the second-level headings under a first-level heading might be considered nodes making up a single composite.

The text contains *links*. A link specifies an *anchor* or specific location to which it points. The user normally doesn't see the anchor, but rather the *marker,* which is some form of highlighting (such as a different color) that indicates that an area is a link that can be clicked on. (In systems such as the Web, link markers need not be textual. Small pictures are often used as visual link markers.) Web browsers and other hypertext programs often supplement the use of links with various navigation aids. These can include buttons for traversing back or forward through a list of recently visited links, a *history* list from which previous links can be selected, and *bookmarks* that allow the user to save and descriptively label important links for easier future access.

Hypertext is becoming the dominant paradigm for presenting technical or other reference information. With less-structured text, hypertext links are usually considered to be supplemental to the traditional structure. The term *hypermedia* refers to the linking of nontextual material—images, videos, sound files, even Java applets and other programs. (Since both hypertext and hypermedia are now so ubiquitous, the terms themselves seem to be used less frequently except in an academic context.)

Hypertext perhaps achieves its fullest power when it is used for collaborative expression and research. Without being able to easily link to what is being discussed, blogs would just be static diaries (see BLOGS AND BLOGGING). Wikis, too, depend on linking not only to reference existing, related entries, but to "grow" the tree of knowledge with "stubs" being put in to encourage other contributors to flesh out related topics (see WIKIS AND WIKIPEDIA). Despite suggestions to the contrary, hypertext seems to be problematic with regard to fiction, unless a work is constructed as an explicit hypertext. If hypertext literature becomes popular, it will require that both authors and readers radically change their role and expectations with regard to the text.

Further Reading

Bromme, Rainer, and Elmar Stahl. *Writing Hypertext and Learning.* Kidlington, Oxford, U.K.: Elsevier Science, 2002.

Bush, Vannevar. "As We May Think." *Atlantic Monthly* 176, 101–108. Available online. URL: http://www.theatlantic.com/doc/194507/bush.

Landow, George P. *Hypertext 3.0: Critical Theory and New Media in an Era of Globalization.* 3rd ed. Baltimore: Johns Hopkins University Press, 2006.

McCann, Jerome. *Radiant Textuality: Literature after the World Wide Web.* New York: Palgrave Macmillan, 2001.

Nelson, Theodore. *Computer Lib/Dream Machines.* Rev. ed. Chicago: Hugo's Books, 1987.

Snyder, I. *Hypertext: the Electronic Labyrinth.* New York: New York University Press, 1997.

I

IBM

International Business Machines is familiarly known as IBM (which is its NYSE symbol) or the nickname "Big Blue." Arguably it is the world's oldest information technology company, with its roots in card tabulation and other business machines in the late 19th century (see HOLLERITH, HERMANN and PUNCHED CARDS AND PAPER TAPE). Under president Thomas J. Watson Sr., IBM developed what would become known as the "IBM card" and machinery to manage the huge amounts of data required by the U.S. Social Security system starting in the mid-1930s. However, IBM would later be criticized for providing the same technology to Nazi Germany, where it would be used to help round up Jews for the Holocaust. On the other hand, IBM calculating machines were a very necessary part of the Allied war effort, including the development of the atomic bomb.

In the 1950s, cold war–related defense work gave IBM access to new technologies, including the multiuser, real-time architecture needed for the SAGE air defense computer (see GOVERNMENT FUNDING OF COMPUTER RESEARCH.)

Despite UNIVAC's head start, IBM dominated the commercial computer industry from the mid-1950s at least until the 1970s (see MAINFRAME). The keystone product was the IBM/360 and later IBM/370 mainframe systems. IBM did not sell just hardware: It provided complete solutions in the form of hardware, operating systems, other software, and peripherals. Because of its dominance, it was hard for small innovators to gain traction, and many people in the university hacker culture felt about IBM as many of their descendants feel about Microsoft today. (IBM's dress code with its dark suits reassured business managers but added to the company's conformist image.)

RETRENCHMENT

IBM went on to set the standard for the most common type of personal computer in the 1980s (see IBM PC). However, the decade would also bring a gradual decline of IBM's dominant role. On the desktop, IBM quickly outpaced Apple (despite the latter's innovation—see MACINTOSH). However, it became legally possible and profitable to build "clone" PCs that could run the same software as the IBM PC, and often faster and at lower cost. In the 1990s the growing use of networks of increasingly powerful desktop machines would erode the mainframe market. Finally, in 2004 IBM sold its PC business (including the well-regarded Thinkpad series of laptops) to Lenovo, a Chinese company.

Today IBM remains a major seller of computer servers particularly targeted to Internet businesses. The company has also achieved success through designing chips for videogame units (see GAME CONSOLES). However, the company's overall focus is mainly on business consulting, software (including database and collaborative products), management services, and the exploitation of its vast trove of patents. IBM has also enthusiastically embraced open software and contributed a considerable amount of code to the programming community, such as the Eclipse program development system (see LINUX and OPEN SOURCE).

IBM remains the largest computer-related company (after HP). In 2007 the company earned $7 billion on revenue of $98.8 billion.

Further Reading
Bashe, Charles J., et al. *IBM's Early Computers.* Cambridge, Mass.: MIT Press, 1985.
Birth of the IBM PC (IBM Archives). Available online. URL: http://www-03.ibm.com/ibm/history/exhibits/pc25/pc25_birth.html. Accessed September 23, 2007.
Black, Edwin. *IBM and the Holocaust: The Strategic Alliance between Nazi Germany and America's Most Powerful Corporation.* New York: Three Rivers Press, 2001.
Garr, Doug. *IBM Redux: Lou Gerstner & the Business Turnaround of the Decade.* New York: HarperBusiness, 1999.
IBM Corporation. Available online. URL: http://www.ibm.com. Accessed September 23, 2007.
Pugh, Emerson W. *Memories that Shaped an Industry: Decisions Leading to IBM System/360.* Cambridge, Mass.: MIT Press, 2000.
Pugh, Emerson W., Lyle R. Johnson, and John H. Palmer. *IBM's 360 and Early 370 Systems.* Cambridge, Mass.: MIT Press, 1991.
Soltis, Frank G. *Fortress Rochester: The Inside Story of the IBM Series.* Loveland, Colo.: 29th Street Press, 2001.

IBM PC

By 1981, a small but vigorous personal computer (PC) industry was offering complete desktop computer systems. Apple's Apple II offered color graphics and expandability through an "open architecture"—slots into which cards designed by third-party vendors could be plugged. While the Apple II had its own DOS (disk operating system) as did Radio Shack's TRS-80, most microcomputers sold in the business market used CP/M, an operating system developed by Gary Kildall and his company Digital Research.

Meanwhile, IBM, the world's largest computer company (see IBM), had quietly created a special team headed by Phillip ("Don") Estridge and tasked with designing a personal computer. Unlike the case with the company's mainframe development, the team was given considerable freedom in choosing architecture and components—but they were told they would have to have a machine ready for the market in one year.

Because of the short time frame, the team chose third-party components already well established in the market, including the monitor, floppy disk drive, and a printer. Unlike Apple and most other companies, IBM created two separate video display systems, one monochrome (MDA) for sharp text for business applications and the three-color CGA system for the game and education markets (see GRAPHICS CARD).

The IBM team also adopted standards from the emerging microcomputer industry instead of trying to use existing mainframe standards. For example, they used the ASCII code to represent characters, not the EBCDIC code used on IBM mainframes. They also chose the Intel 8086 and 8088 microprocessors, which had an instruction set similar to that of the Intel 8080 used in many CP/M systems (see MICROPROCESSOR). This would make it easy for software developers to create IBM PC versions of their software quickly so that the new machine would have a repertoire of business software.

One might have expected that IBM would also adopt a version of CP/M as the PC's operating system, taking advan-

tage of the closest thing to an existing industry standard. However, CP/M was relatively expensive, and negotiations with Digital Research stumbled, leaving an opening for a much smaller company, Microsoft, to sell a DOS based on software it had licensed from Seattle Computer Products. While IBM did offer CP/M and another operating system based on the UC San Diego Pascal development system, Microsoft DOS, which became known as PC-DOS (and later MS-DOS), was cheapest and effectively became the default offering (see MS-DOS).

When IBM officially announced its PC in April 1981, Apple took out full-page ads "welcoming" the new competitor to what it considered to already be a mature industry. But by the end of 1983, a million IBM PCs had been sold, dwarfing Apple and other brands. From then on, while Apple would go on to announce its distinctive Macintosh in 1984, the IBM machine would set the industry standard. To most people, "PC" would mean "IBM PC."

OPEN STANDARDS AND EXPANSION

As more businesses bought IBM PCs, the company steadily expanded the machine's capabilities to meet the demands of the business environment. The next model, the PC-XT, introduced in 1982, included a hard disk drive and more system memory. As software became more demanding, the need for a faster and more capable processor also became apparent. In 1984, IBM responded with the PC-AT, which used the Intel 80286 processor, combining the faster processor with a wider (16-bit) and faster data bus (see BUS).

However, IBM would not have the market to itself. A consequence of the use of an open, expandable architecture and "off the shelf" processor and other components is that other companies could market PCs that were compatible with IBM's (that is, they could run the same operating system and applications software). Although competitors could not legally make a simple copy of the read-only memory (ROM) BIOS, the code that enabled the components to communicate, they could reverse-engineer a functional equivalent. The first major competitor in what became known as the "PC Clone" market was Compaq, which also offered an improved video display and a transportable model. Zenith, Tandy (Radio Shack), and HP also offered "name-brand" PC clones.

In 1987, IBM tried to establish a proprietary standard by introducing the PS/2 line, which featured a 3.5-inch floppy drive (standard PC compatibles used 5.25-inch drives), a new high-resolution graphics standard (VGA), a new system bus (MCA or Microchannel Architecture), and a new operating system (OS/2). Despite some technical advantages, the PS/2 achieved only modest success. Since the card slots were incompatible with the previous standard, existing expansion products could not be used. Microsoft soon came out with a new operating environment, Windows, which while inferior in multitasking capabilities to OS/2 was easier to use (see USER INTERFACE and MICROSOFT WINDOWS).

By the 1990s, it was clear that IBM no longer controlled the standards for PCs. (Indeed, IBM soon abandoned the PS/2 MCA architecture and returned to the earlier standard, which competitors had never left.) Instead, the industry incrementally built upon what had become known as the ISA (Industry Standard Architecture), supplementing it with a new kind of expansion card connector called PCI. Currently, IBM is in the second tier in PC sales behind industry leaders Dell and Compaq, having a market share comparable to Hewlett-Packard and Gateway. IBM also did relatively well in the laptop computer sector with its Thinkpad series, before selling it to Lenovo.

Today's industry standards are effectively determined by two companies: the chip-maker Intel and the software giant Microsoft. Indeed, "standard" PCs are now often called "Wintel" machines. The direct-order giant Dell and its competitors HP and Lenovo dominate the "commodity PC" market. However, by creating a standard that was flexible enough for two decades of PC development, IBM made a lasting contribution to computing comparable to its innovations in the mainframe arena.

Further Reading

Dell, Deborah A,. and J. Jerry Purdy. *Thinkpad: a Different Shade of Blue.* Indianapolis, Ind.: Sams, 1999.

Dell, Michael. *Direct from Dell: Strategies that Revolutionized an Industry.* New York: HarperBusiness, 2000.

Gilster, Ron. *PC Hardware: a Beginner's Guide.* New York: McGraw-Hill, 2001.

Hoskins, Jim, and Bill Wilson. *Exploring IBM Personal Computers.* 10th ed. Gulf Breeze, Fla.: Maximum Press, 1999.

Ling, Zhigun, and Martha Avery. *The Lenovo Affair: The Growth of China's Computer Giant and Its Takeover of IBM-PC.* New York: Wiley, 2006.

identity in the online world

There are two aspects of identity in cyberspace, both of which are intriguing but problematic: *Outer identity* is the name or other descriptors that are identified by other people as belonging to a particular person, and *inner identity* is a person's sense of who or what he or she "really is."

Users of online systems such as chat rooms or games have the ability to use a variety of names (pseudonyms) or to be effectively anonymous (see ANONYMITY AND THE INTERNET). In games, the identity used by a player is represented by a virtual representation called an avatar. Other players (through their own avatars) will encounter the avatar and identify it by physical appearance, behavior, and what it tells about itself (the "back story").

While opportunities to do this emerged in the 1970s with paper-and-dice role-playing games such as the very popular *Dungeons and Dragons,* there are significant differences between online identity and these earlier games. People played "D&D" in person, so it was relatively easy to maintain a distinction between a character a person was "running" and the person himself or herself. Also, these role-playing sessions were fixed in time and place: After slaying the dragon, the players went home. Indeed even the term "role-playing" made the comfortable assumption that the activity was a pretend, make-believe identity assumed by the player.

Virtual game worlds began in the 1980s with text-based MUDS (multi-user dungeons) and similar online environments. Today game worlds are graphically immersive and persistent. Although there are games focused on the traditional battles and quests, others such as *Second Life* are best described not as games at all but literal second or alternative lives that persons can participate in for hours a day. In these worlds an avatar can own property and make commitments, even a virtual form of marriage. In many cases in-game goods and money can actually be exchanged for "real world" money. And crucially, unlike the D&D encounter, in these virtual worlds the "real person" behind an avatar need never be revealed.

CONSTRUCTING IDENTITIES

The online world invites people to construct and try out identities. Because of the vital role they play in people's sense of self and their social interactions, sexual or gender identity is a particularly important issue. The online world has some clear advantages for persons who are experimenting with different identities (such as transgender). A man, for example, can create a female avatar that really looks female. Further, people can act out sexual encounters without the possible physical consequences of violence or disease. On the other hand, people can still be hurt psychologically, and online relationships can take on added risks and challenges by eventually becoming physical ones.

There are also venues where there can be "hybrid" identities. In a site such as MySpace, a person can construct the kind of "face" he or she wants to present to the world and interact with the pages of other people. Here the online identity is often tied with a physical one (potentially creating vulnerability) but need not be (creating the potential for deception).

Young people in particular will have to deal with the opportunities and challenges of multiple virtual identities. On the one hand, young people are very adaptable, especially to new technologies. On the other hand, youth and particularly adolescence has always been a time of inner conflicts and a search for lasting identity (see YOUNG PEOPLE AND COMPUTING).

The deeper philosophical and psychological implications of cyberspace are intriguing. According to some modern psychological theories (such as Marvin Minsky's "society of mind"), the mind does not consist of a single ego perhaps in conflict with unconscious forces, but rather, many separate "agents" that interact as they seek various goals. From that point of view the online world expands that model into social space and may lead to a world in which each physical person may have many virtual persons associated with it.

Online identities are becoming a fertile area of research in psychology and sociology. Pioneering work has been done by psychologist Sherry Turkle, who has explored differing male and female styles of relationship to technology, how technology affects children, and other issues.

The social and legal implications of online identity are equally challenging. Can an avatar be sued? Can one avatar commit a criminal act (perhaps even rape) against another? Might an avatar have privacy rights and the right of publicity? The legal system has hardly begun to consider such questions, and they are becoming more urgent as everything from meetings to concerts takes place in virtual worlds. It is possible that eventually online worlds will be allowed to create their own internal legal systems, perhaps subject to "metarules" about how they are to be enforced within the context of physical jurisdictions (see CYBERLAW).

Further Reading
Cooper, Robbie. *Alter Ego: Avatars and Their Creators.* London: Chris Boot, 2007.
Schroeder, Ralph, and Ann-Sofie Axelsson, eds. *Avatars at Work and Play: Collaboration and Interaction in Shared Virtual Environments.* New York: Springer, 2007.
"Sexual Identity Online." MYCyclopedia of New Media. Available online. URL: http://wiki.media-culture.org.au/index.php/Sexual_Identity_Online. Accessed September 23, 2007.
Thomas, Angela. *Youth Online: Identity and Literacy in the Digital Age.* New York: Peter Lang, 2007.
Turkle, Sherry. *Life on the Screen: Identity in the Age of the Internet.* New York: Simon & Schuster, 1995.
———. *The Second Self: Computers and the Human Spirit.* Twentieth Anniversary ed. Cambridge, Mass.: MIT Press, 2005.

identity theft

Identity theft is essentially the impersonation of someone in order to gain use of their resources or, occasionally, to escape the consequences of previous criminal behavior. The most common motive for identity theft is to gain access to a person's financial resources, such as credit cards or checking accounts, or to obtain credit or services. (Sometimes a distinction is made between identity theft, where the victim's identity is assumed and effectively becomes the perpetrator's identity, and identity fraud, where information is only used long enough to complete particular transactions.)

Identity thieves must first obtain the necessary information to pose as their victim. This can be done by physically obtaining such items as checks, receipts, credit offers, and so on from the trash or mail. Information such as name, address, account numbers, and the ultimate prize, the Social Security number, can then be used, for example, to apply for credit in the victim's name, or buy goods and have them shipped to the perpetrator's address.

People can minimize the risk of physical identity theft by securing their mail and shredding sensitive documents. However, the fastest-growing venue for identity theft is online. The online world presents additional opportunities to the criminals, the necessity of new precautions, and difficult challenges for law enforcement.

Digital information useful for identity theft can be obtained in a variety of ways. It can be physically stolen in the form of laptops or portable storage devices or obtained electronically by breaking into and compromising computer systems (see COMPUTER CRIME AND SECURITY). Programs can exploit operating-system or software flaws to travel from one networked PC to another and e-mail information back to the perpetrator (see COMPUTER VIRUS). Finally, users can be coerced, enticed, or otherwise tricked into providing the information (such as passwords) themselves, via authentic-looking institutional Web sites (see PHISHING AND SPOOFING).

INCIDENCE AND PREVENTION

According to various surveys, the incidence of identity theft increased substantially between 2001 and 2003. There are conflicting views of recent trends. Data for 2006 from Javelin Strategy and Research suggests a decrease (10.1 million U.S. adult victims in 2003 and 8.9 million in 2006). However, data from the Federal Trade Commission records 246,035 actual complaints of identity theft in 2006, making it by far the number one item on its list of consumer fraud complaints. (Of these, 25 percent reflected credit card fraud, and phone/utilities fraud and bank fraud each represented 16 percent.)

To give some further perspective, according to the Internet Crime Complaint Center (a joint program of the FBI and the National White Collar Crime Center), identity theft amounted to only 1.6 percent of reported cyber crimes. (Credit card or check fraud without confirmed identity theft added up to 9.7 percent.) Nevertheless, however measured, it is clear that identity theft remains a very serious problem.

Until recent years, response to identity theft complaints by law enforcement tended to be ineffectual and frustrating to victims. This was probably due to a combination of circumstances, including many police officers being unfamiliar with the nature of the crime or technology involved, unsure about how to proceed, and not even certain they had jurisdiction. This situation has improved considerably, however, with national organizations, greater interagency cooperation (including between federal, state, and local agencies), and strong and explicit laws against identity theft and fraud. (The Identity Theft and Assumption Deterrence Act of 2003 now makes possession of "any means of identification" to "knowingly transfer, possess, or use without lawful authority" a federal crime.)

The main goal for consumers, however, should be prevention. Steps that can greatly reduce the chance of becoming a victim of online identity theft include:

- Keep security software (antivirus, antispam, antispyware) up to date.

- Do not click on links in e-mail that purports to be from a financial institution, government agency, online merchant or auction service. Use the browser's address box to go directly to the relevant site.

- Do not post addresses, account numbers, or Social Security numbers online, including chat rooms or social networking sites. Teach children likewise, and consider installing software that can block the posting of such information.

- Make sure that the financial institutions and merchants that one uses have acceptable privacy policies and policies for dealing with "data breaches" and other loss of sensitive information.

- If you suspect you have been victimized, go to a site such as the Identity Theft Resource Center or the Privacy Rights Clearinghouse to learn how to stop further losses and reestablish credit and accounts.

Further Reading
Cullen, Terry. *The Wall Street Journal Complete Identity Theft Guidebook: How to Protect Yourself from the Most Pervasive Crime in America.* New York: Three Rivers Press, 2007.
"How Many Identity Theft Victims Are There? What Is the Impact on Victims?" Privacy Rights Clearinghouse. Available online. URL: http://www.privacyrights.org/ar/idtheftsurveys.htm. Accessed September 23, 2007.
Identity Theft Resource Center. Available online. URL: http://www.idtheftcenter.org/. Accessed September 23, 2007.
"Identity Theft Tops FTC Complaints for 2006." Consumeraffairs.com, February 8, 2007. Available online. URL: http://www.consumeraffairs.com/news04/2007/02/ftc_top_10.html. Accessed September 23, 2007.
Internet Crime Complaint Center. Available online. URL: http://www.ic3.gov/. Accessed September 23, 2007.
Privacy Rights Clearinghouse. Available online. URL: http://www.privacyrights.org/. Accessed September 23, 2007.

image processing

Image processing is a general term for the manipulation of a digitized image to produce an enhanced or more convenient version. Some of the earliest applications were in the military (aerial and, later, satellite reconnaissance) and in the space program. The military and space programs had a great need for extracting as much useful information as possible from images that were often gathered under extreme or marginal conditions. They also needed to make cameras and other hardware components simultaneously more compact and more efficient, and generally had the funds to pay for such specialized developments.

Once developed, higher-quality image processing systems found their way into other applications such as domestic surveillance and medical imaging. The development of cameras that could directly turn light into digitized images (see PHOTOGRAPHY, DIGITAL) made image processing seamless by avoiding the necessity of scanning images from traditional film.

Image processing applications can be divided into three general categories: enhancement, interpretation, and maintenance.

ENCHANCEMENT

Enhancement includes bringing out objects of interest (such as enemy vehicles or a particular rock formation on Mars) from the surrounding background by enhancing contrast or applying appropriate filters to block out the background. More sophisticated filters can also be used to compensate for defects in the original image, such as "red-eye," blur, and loss of focus. Today's image processing programs, such as the popular Adobe Photoshop, make relatively sophisticated image manipulation techniques available to interested amateurs as well as professionals. More sophisticated image enhancement techniques include the creation of 3D images based upon the differences calculated from a number of photos shot from slightly different angles.

A considerable amount of image enhancement takes place even before the photo is taken. Today's versatile cameras (see PHOTOGRAPHY, DIGITAL) include a variety of modes that are preset for different scenarios such as indoor portrait or low light. After the picture is taken, photo management programs (often bundled with the camera or even included in the operating system) not only help organize photos, but also provide simple ways to crop or enhance them.

INTERPRETATION

Interpretation refers to manipulation designed to help human observers obtain more and better information from the image. For example, "false color" can be used to heighten otherwise imperceptible color differences in the original image, or to translate nonvisual information (such as heat or radio emission levels) into visual terms.

Artificial intelligence algorithms can also be employed to automatically analyze images for features of interest (see PATTERN RECOGNITION and COMPUTER VISION). In fields such as military reconnaissance this might allow a high volume of imagery to be prescreened, with images meeting certain criteria "flagged" for the attention of human interpreters.

MAINTENANCE

Maintenance includes archiving of images, often with the aid of compression to reduce the amount of storage space required (see DATA COMPRESSION). It can also include the restoration of images that may have been degraded (as from chemical decomposition of stored film.) This can be done either by creating a reversible mathematical model of the degradation process (thus, for example, restoring colors that have changed through oxidation or other processes) or by creating a model of how the image was formed in the first place and comparing its output to the existing image.

Further Reading
GIMP, the GNU Image Manipulation Program. Available online. URL: http://www.gimp.org/. Accessed August 8, 2007.
Gonzalez, Rafael, and Richard E. Woods. *Digital Image Processing.* 3rd ed. Upper Saddle River, N.J.: Prentice Hall, 2007.
Hanbury, Allan. "A Short Introduction to Digital Image Processing." Available online. URL: http://www.prip.tuwien.ac.at/~hanbury/intro_ip/. Accessed August 8, 2007.
MathWorks Image Processing. Available online. URL: http://www.mathworks.com/applications/imageprocessing/. Accessed August 8, 2007.
Seul, Michael, Lawrence O'Gorman, and Michael J. Sammon. *Image Analysis: Description, Examples, and Code.* New York: Cambridge University Press, 2000.
Ward, Al. *PhotoShop for Right-Brainers: Photo Manipulation.* Alameda, Calif.: SYBEX, 2004.

information design

Information design is concerned with arranging and presenting information in ways that enable viewers to use it efficiently and with the greatest benefit. This discipline can be said to have begun in the 19th century with the development of diagrams and maps that present the relationship between two or more variables. These included John Snow's map of London showing the locations of cholera outbreaks in the 1850s, and a striking 1861 diagram by Joseph Minard that related the geographical progress of Napoléon's 1812 invasion of Russia with the diminishing size of the French forces. These early examples coincided with a time when industrial society was becoming increasingly complex and populous, and both government and business needed new ways to visualize statistics. Other products to which information design contributes became important in the following century: traffic and transit signs, product warning labels, and product manuals, to name a few.

Some of the basic considerations for information design include:

- effectiveness at presenting relevant information
- selection and arrangement of information
- balance of attractiveness and clarity
- proper use of the medium (size, materials, etc.)

Of course the designer has additional constraints, such as the purpose of the design (advertising, product documentation, report, etc.), policies of the client, any applicable regulations (such as for warning labels), and so on.

FROM PHYSICAL TO DIGITAL

Moving from the world of print to the Web brings new resources and challenges to the information designer. Web design has many advantages over print—powerful layout tools and perhaps templates, the availability of animation or other effects, the ability to adapt to different audiences, and, above all, interactivity. However, each of these features brings additional choices—not only font and text size, but background, use of images, whether to include animation (such as Flash), and how to design clear and easy-to-use forms and other interactive features. Further, designs may have to adapt to a variety of platforms (large desktop screens, laptops, PDAs, and mobile devices) and provide for users who have visual impairments or other disabilities (for more, see WEB PAGE DESIGN). For information displays designed to provide "at a glance" summaries and alerts about problems, see DIGITAL DASHBOARD.

Although these concerns may seem far afield from the classic principles of graphic design, they actually represent technological extensions of them. It is easy to get lost in the particulars of designing, for example, Web pages showing statistical charts, without having thought about whether the charts themselves show information clearly and accurately in the scales and proportions used.

Further Reading

Digital Web Magazine. Available online. URL: http://www.digital-web.com/. Accessed September 23, 2007.

Few, Stephen. *Information Dashboard Design: The Effective Visual Communication of Data.* Sebastapol, Calif.: O'Reilly Media, 2006.

Information Design Journal. John Herndon, Va.: Benjamins Publishing Company, 1979. Free sample available online. URL: http://www.benjamins.com/cgi-bin/t_bookview.cgi?bookid=IDJDD%2012%3A1. Accessed September 23, 2007.

Lipton, Ronnie. *The Practical Guide to Information Design.* New York: Wiley, 2007.

Löwgren, Jonas, and Erik Stolterman. *Thoughtful Interaction Design: A Design Perspective on Information Technology.* 2nd ed. Cambridge, Mass.: MIT Press, 2007.

Tufte, Edward R. *Envisioning Information.* Cheshire, Conn.: Graphics Press, 1990.

———. *The Visual Display of Quantitative Information.* 2nd ed. Cheshire, Conn.: Graphics Press, 2001.

information retrieval

While much attention is paid by system designers to the representation, storage and manipulation of information in the computer, the ultimate value of information processing software is determined by how well it provides for the effective retrieval of that information. The quality of retrieval is dependent on several factors: hardware, data organization, search algorithms, and user interface.

At the hardware level, retrieval can be affected by the inherent seek time of the device upon which the data is stored (such as a hard disk), the speed of the central processor, and the use of temporary memory to store data that is likely to be requested (see CACHE). Generally, the larger the database and the amount of data that must be retrieved to satisfy a request, the greater is the relative importance of hardware and related system considerations.

Data organization includes the size of data records and the use of indexes on one or more fields. An index is a separate file that contains field values (usually sorted alphabetically) and the numbers of the corresponding records. With indexing, a fast binary search can be used to match the user's request to a particular field value and then the appropriate record can be read (see HASHING).

There is a tradeoff between storage space and ease of retrieval. If all data records are the same length, random access can be used; that is, the location of any record can be calculated essentially by multiplying the record's sequence number by the fixed record length. However, having a fixed record size means that records with shorter data fields must be "padded," wasting disk space. Given the low cost of disk storage today, space is generally less of a consideration.

The search algorithms used by the program can also have a major impact on retrieval speed (see SORTING AND SEARCHING). As noted, if a binary search can be done against a sorted list of fields or records, the desired record can be found in only a few comparisons. At the opposite extreme, if a program has to move sequentially through a whole database to find a matching record, the average number of comparisons needed will be half the number of records in the file. (Compare looking up something in a book's index to reading through the book until you find it.)

Real-world searching is considerably more complex, since search requests can often specify conditions such as "find e-commerce but not amazon.com" (see BOOLEAN OPERATORS). Searches can also use wildcards to find a word stem that might have several different possible endings, proximity requirements (find a given word within so many words of another), and other criteria. Providing a robust set of search options enables skilled searchers to more precisely focus their searches, bringing the number of results down to a more manageable level. The drawback is that complex search languages result in more processing (often several intermediate result sets must be built and internally com-

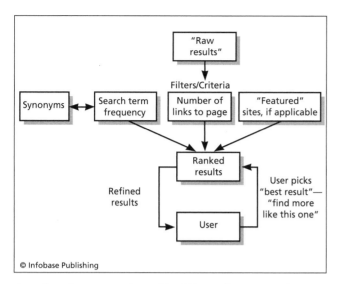

© Infobase Publishing

A number of criteria can be used by Web search engines to determine the likely relevance of search results. Perhaps the most important tool, however, is feedback from the user.

pared to one another). There is also more likelihood that searchers will either make syntax errors in their requests or create requests that do not have the intended effect.

While database systems can control the organization of data, the pathways for retrieval and the command set or interface, the World Wide Web is a different matter. It amounts to the world's largest database—or perhaps a "metabase" that includes not only text pages but file resources and links to many traditional database systems. While the flexibility of linkage is one of the Web's strengths, it makes the construction of search engines difficult. With millions of new pages being created each week, the "web-crawler" software that automatically traverses links and records and indexes site information is hard pressed to capture more than a diminishing fraction of the available content. Even so, the number of "hits" is often unwieldy (see SEARCH ENGINE).

A number of strategies can be used to provide more focused search results. The title or full text of a given page can be checked for synonyms or other ideas often associated with the keyword or phrase used in the search. The more such matches are found, the higher the degree of *relevance* assigned to the document. Results can then be presented in declining order of relevance score. The user can also be asked to indicate a result document that he or she believes to be particularly relevant. The contents of this document can then be compared to the other result documents to find the most similar ones, which are presented as likely to be of interest to the researcher.

Information retrieval from either stand-alone databases or the Web can also be improved by making it unnecessary for users to employ structured query languages (see SQL) or even carefully selected keywords. Users can simply type in their request in the form of a question, using ordinary language: For example, "What country in Europe has the largest population?" The search engine can then translate the question into the structured queries most likely to elicit documents containing the answer. Ask Jeeves (retired as of 2006) and similar search services have thus far been only modestly successful with this approach.

On a large scale, systematic information retrieval and analysis (see DATA MINING) has become increasingly sophisticated, with applications ranging from e-commerce and scientific data analysis to counterterrorism. Artificial intelligence techniques (see PATTERN RECOGNITION) play an important role in cutting-edge systems.

Finally, encoding more information about content and structure within the document itself can provide more accurate and useful retrieval. The use of XML and work toward a "semantic Web" offers hope in that direction (see BERNERS-LEE, TIM; SEMANTIC WEB; and XML).

Further Reading
Bell, Suzanne S. *Librarian's Guide to Online Searching.* Westport, Conn.: Libraries Unlimited, 2006.
Chakrabarti, Soumen. *Mining the Web: Discovering Knowledge from Hypertext Data.* San Francisco: Morgan Kaufmann, 2002.
Grossman, David A., and Ophir Frieder. *Information Retrieval: Algorithms and Heuristics.* 2nd ed. Norwell, Mass.: Springer, 2004.
"Information Retrieval Research." Search Tools Consulting. Available online. URL: http://www.searchtools.com/info/info-retrieval.html. Accessed August 8, 2007.
Meadow, Charles T., et al. *Text Information Retrieval Systems.* 3rd ed. Burlington, Mass.: Academic Press, 2007.

information theory

Information theory is the study of the fundamental characteristics of information and its transmission and reception. As a discipline, information theory took its impetus from the ideas of Claude Shannon (see SHANNON, CLAUDE).

In his seminal paper "A Mathematical Theory of Communication" published in the *Bell System Technical Journal* in 1948, Shannon analyzed the redundancy inherent in any form of communication other than a series of purely random numbers. Because of this redundancy, the amount of information (expressed in binary bits) needed to convey a message will be less than the number in the original message. It is because of redundancy that data compression algorithms can be applied to text, graphics, and other types of files to be stored on disk or transmitted over a network (see DATA COMPRESSION).

Shannon also analyzed the unpredictability or uncertainty of information as it is received—that is, the number of possibilities for the next bit or character. This is related to the number of possible symbols, but since all symbols are usually not equally likely, it is actually a sum of probabilities. Shannon used the physics term *entropy* to refer to this measure. It is important because it makes it possible to analyze the probability of error (caused by such things as "line noise") in a communications circuit. Shannon's basic formula is:

$$C = B\log_2(1 + P / N)$$

where the channel capacity C is in bits per second, B is the bandwidth, P the signal power, and N the Gaussian noise power.

Shannon found that if as long as the actual data transmission rate is less than the channel capacity C, an error-correcting code can be devised to ensure that any desired accuracy rate is achieved (see ERROR CORRECTION). A related formula can also be used to find the lowest transmission power needed given a specified amount of noise.

The influence of Shannon and his disciples on computing has been pervasive. Information theory provides the fundamental understanding needed for applications in data compression, signal analysis, data communication, and cryptography—as well as problems in other fields such as the analysis of genetic mutation or variation.

Further Reading

Cover, Thomas, and Joy A. Thomas. *Elements of Information Theory*. 2nd ed. Hoboken, N.J.: Wiley, 2006.

Gray, R. M. *Entropy and Information Theory*. New York: Springer, 1990. Available online. URL: http://ee.stanford.edu/~gray/it.pdf. Accessed August 8, 2007.

Hankerson, Darrel, Greg A. Harris, Jr., and Peter D. Johnson. *Introduction to Information Theory and Data Compression*. 2nd ed. Boca Raton, Fla.: CRC Press/Chapman & Hall, 2003.

IEEE Information Theory Society. Available online. URL: http://www.itsoc.org/. Accessed August 8, 2007.

Shannon, Claude. "A Mathematical Theory of Communication." *Bell System Technical Journal* 27 (July, October 1948): 379–423, 523–656. Available online. URL: http://plan9.bell-labs.com/cm/ms/what/shannonday/shannon1948.pdf. Accessed August 8, 2007.

information warfare

Information warfare has many aspects and can be fought on many levels. On the battlefield, it can involve collecting tactical or strategic intelligence and protecting one's own channels of communication. Conversely, it can involve disrupting the enemy's means of communication, blocking the enemy's intelligence gathering, spreading disinformation, and trying to disrupt their decision process. Beyond the battlefield, media (including the Internet) can be used for propaganda purposes.

All of these objectives today involve the use of digital information and communications systems. Examples include:

- analysis of enemy communications using both automatic tools and human analysts

- cryptography and signal analysis

- protection of computer and network infrastructure

- attacks and disruptions on enemy information infrastructure, both military and civilian (such as denial-of-service attacks on Web sites)

- use of Web sites to spread disinformation or propaganda

HISTORY AND DEVELOPMENT

Information warfare is as old as warfare itself, with such things as ruses designed to trick or confuse enemy sentries or lighting many fires to convince the enemy that one's army was much larger than in reality. Wiretapping and spoof messages began with the telegraph in the mid-19th century, and eavesdropping and other tricks with radio were used in World War I. These arts had greatly increased in scale and sophistication by World War II—an entire fake army corps was "created" to deceive the Germans prior to the D-day invasion.

Information warfare involving computers has been used in recent conflicts. The active phase of the U.S. attack in the first Gulf War in 1991 began with systematic destruction and disruption of Iraqi information and command-and-control assets through targeted attacks. As a result, the still largely intact Iraqi military was left blind as to the coming flank attack by U.S. forces.

In 2007 a series of coordinated attacks by unknown parties paralyzed much of Estonia's Web-based government and business structures following a dispute with Russia. To many observers this represents a model for "strategic" information warfare that might be used in future conflicts. (Note that the techniques used in information warfare by the national military and the kinds of cyber attacks that might be favored by terrorists overlap. For the latter, see CYBERTERRORISM.)

Further Reading

Armistead, E. Leigh. *Information Warfare: Separating Hype from Reality*. Washington, D.C.: Potomac Books, 2007.

Greenmeier, Larry. "Estonian 'Cyber Riot' Was Planned, but Mastermind Still a Mystery." *InformationWeek*, August 3, 2007. Available online. URL: http://www.informationweek.com/showArticle.jhtml?articleID=201202784. Accessed September 23, 2007.

Johnson, L. Scott. "Toward a Functional Model of Information Warfare." Available online. URL: http://bss.sfsu.edu/fischer/IR%20360/Readings/Information%20War.htm. Accessed September 23, 2007.

Libicki, Martin C. *Conquest in Cyberspace: National Security and Information Warfare*. New York: Cambridge University Press, 2007.

Rattray, Gregory J. *Strategic Warfare in Cyberspace*. Cambridge, Mass.: MIT Press, 2001.

Waltz, Edward. *Information Warfare: Principles and Operations*. Boston: Artech House, 1998.

Input/Output (I/O)

While the heart of a computer is its central processing unit or CPU (the part that actually "computes"), a computer must also have a "circulatory system" through which data moves between the CPU, the main memory, input devices (such as a keyboard or mouse), output devices (such as a printer), and mass storage devices (such as a hard or floppy disk drive). Input/Output or I/O processing is the general term for the management of this data flow (see also BUS, PARALLEL PORT, SERIAL PORT, and USB).

I/O processing can be categorized according to how a request for data is initiated, what component controls the process, and how the data flows between devices. In most early computers the CPU was responsible for all I/O activities (see CPU). Under program control, the CPU initiated a data transfer, checked the status of the device (or area of memory) that would be sending or receiving the data, and monitored the flow of data until it was complete. While this arrangement simplified computer architecture and reduced the cost of memory units or peripheral devices (at a time when computer hardware was hand-built and relatively costly), it also meant that the CPU could perform no other processing until I/O was complete.

In most modern computers, responsibility for I/O has largely been removed from the CPU, freeing it to concentrate on computation. There are several ways to implement such architecture. One method that has been used on microcomputers since their earliest day is *interrupt-driven I/O*. This means that the CPU has separate circuits on which a device requesting I/O service can "post" a request. The CPU periodically checks the circuits for an interrupt request (IRQ). If one is found, it can send a query to each device on a list until the correct one is found (the latter is called *polling*). Alternatively, the overhead involved in polling can be eliminated by having the IRQ include either a device identification number or a memory address that contains an interrupt service routine (this is called *vectored* interrupts). While interrupts alone do not free the CPU of the need to manage the I/O, they do remove the overhead of having to frequently check all devices for I/O.

The actual I/O process can also be moved out of the CPU through the use of direct memory access (DMA). Here a separate control device takes over control of the system from the CPU when I/O is requested. It then transfers data directly between a device (such as a hard disk drive) and a buffer in main memory. Although the CPU is idle during this process, the transfer is accomplished much more quickly because the full capacity of the bus can be used to move data rather than having to be shared with the flow of program instructions in the CPU.

A more sophisticated I/O control device is called a *channel*. A channel controller can operate completely independently of the CPU without requiring that the CPU become idle during a transfer. Channels can also act as a sort of specialized CPU or *coprocessor,* running program instructions to monitor the data transfer. There are also channels capable of monitoring and controlling several devices simultaneously (this is called *multiplexing*). The use of channels in mainframes such as the IBM 360 and its descendants is one reason why mainframes still perform a workhorse role in high-volume data processing.

In microcomputers the trend has also been toward offloading I/O from the CPU and the main bus to separate controllers or channels. For example, the AGP (accelerated graphics port) found on most modern PCs acts as a channel between main memory and the graphics controllers (see GRAPHICS CARD). This means that as a program generates graphics data it can be automatically transferred from memory to the graphics controllers without any load on the CPU, and over a bus that is faster than the main system bus.

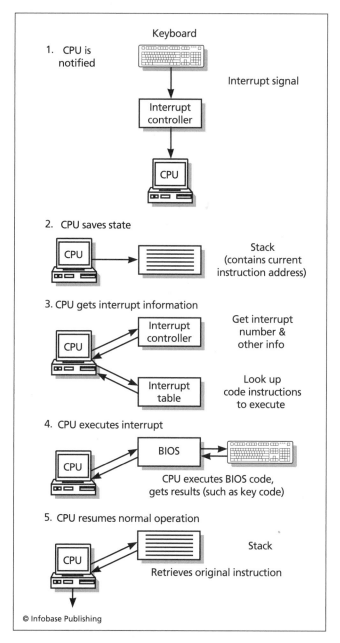

Steps in processing an interrupt request (IRQ) in a PC. (1) The device requesting attention signals the Interrupt Controller, which in turn sends a special signal to the CPU. (2) The CPU saves its state (including internal data and the address of the current instruction) to a stack. (3) The CPU gets the interrupt number and other information from the Interrupt Controller, then looks up a set of instructions for processing that particular interrupt. (4) The CPU executes the interrupt processing code, which generally links to BIOS code for handling a device such as the keyboard. (5) The CPU reloads its state information from the stack and resumes the interrupted processing.

Further Reading

Buchanan, William. *Applied PC Interfacing, Graphics and Interrupts.* Reading, Mass.: Addison-Wesley, 1999.

Karbo, Michael B. "About the PC I/O System." Available online. URL: http://www.karbosguide.com/hardware/module5a1a.htm. Accessed August 8, 2007.

White, Ron, and Timothy Downs. *How Computers Work.* 8th ed. Indianapolis: Que, 2005.

installation of software

While not often covered in computer science or software engineering courses, the process of getting a program to work on a given computer is often nontrivial. In the early days of PCs, installation generally involved simply copying the main program file and any needed settings files to a disk directory and possibly setting up the appropriate driver for the user's printer. (A cryptic user interface sometimes made the latter procedure a frequent occasion for technical support calls.) Users generally did not have to make many choices about what components to install or where to put them. On the other hand, installation programs sometimes made changes to a user's system without notification or the ability to "back out."

The ascension of Microsoft Windows to dominance as a PC operating system improved the installation process in several ways. Since the operating system and device drivers written by hardware vendors took over responsibility for installing and configuring printers and other devices, users generally didn't have to worry about configuring programs to work with specific hardware. Particularly with Windows 95 and later versions, a standard "installation kit" allows software developers to provide a familiar, step-by-step installation procedure to guide users. Generally, installation consists of an introductory screen, legal agreement, and the opportunity to choose a hard drive folder for the program. A moving "progress bar" then shows the files being copied from the installation CD to the hard drive. A "readme" file giving important considerations for using the program is usually provided. Increasingly, software registration is done by launching the user's Web browser and directing it to the vendor's Web site where a form is presented.

The installation of drivers accompanying new hardware such as a printer or scanner has been simplified even more through the "Plug and Play" feature in modern versions of Windows. This allows the system to automatically detect the presence of a new device and either install the driver automatically or prompt the user to insert a disk or CD (see PLUG AND PLAY).

Installation becomes a much more complicated matter when an enterprise has to install from tens to hundreds or thousands of copies of a program on employees' PCs. While small businesses may simply buy consumer-packaged software and install one copy on each PC, large businesses generally obtain a site license allowing a certain number of installations (or in some cases, unlimited on-site installations). Organizations must monitor the number of installations of a particular program package to ensure that licensing agreements are not violated while trying to use available software assets as efficiently as possible. (This is sometimes called *software asset management* or SAM.)

An automated installation script can be used to install a copy of the same software on each PC on the company's network—or a utility can be used to copy an exact hard disk image, including fully configured operating system and applications, to each PC. Alternatively, it is possible to buy networked versions of some programs. In this case the application actually runs on a server and is accessed from (but not copied to) each user's PC. This technique has also been adopted to provide consumers with an alternative to stand-alone installation (see APPLICATION SERVICE PROVIDER).

Installation is only the first part of the story, of course. Most significant programs will experience a steady flow of minor version updates as well as security patches. For individual users, setting the program to update automatically (if possible) or periodically checking for updates may be sufficient. For organizations, the task of making sure all the deployed copies of the software are up to date can be nontrivial, although tools such as Microsoft System Center can help. (Many Linux distributions such as Ubuntu can automatically retrieve all updates for installed packages.)

Linux and UNIX systems have also evolved more sophisticated installation systems in order to keep up with today's more complex applications and distributions. One common solution used by Red Hat and other Linux vendors is a "package" system where the user selects programs and features and the system identifies the components (packages) that must be installed to enable them.

Further Reading

Browne, Christopher. "Linux System Configuration Tools." Available online. URL: http://linuxfinances.info/info/linuxsysconfig.html. Accessed August 8, 2007.

Chappell, David. *Introducing Microsoft System Center.* Available online. URL: http://download.microsoft.com/download/7/A/1/7A1C88D7-B91A-4114-AF8D-852B481D5E F7/Introducing%20System%20Center.doc. Accessed August 8, 2007.

Habib, Irfan. "New Approaches to Linux Package Management." Linux.com November 10, 2005. Available online. URL: http://www.linux.com/feature/49405. Accessed August 8, 2007.

Honeycutt, Jerry. *Microsoft Windows Desktop Deployment Resource Kit.* Redmond, Wash.: Microsoft Press, 2003.

Mehler, Kerrie, Cameron Fuller, and John Joyner. *Microsoft System Center Operations Manager 2007 Unleashed.* Indianapolis: Sams, 2007.

Wilson, Phil. *The Definitive Guide to Windows Installer.* Berkeley, Calif.: Apress, 2004.

Intel Corporation

Intel Corporation (NASDAQ symbol: INTC) is the world's largest manufacturer of semiconductors or "computer chips." The company was founded in 1968 as Integrated Electronics Corporation by Robert Noyce and Gordon Moore (see CHIP and MOORE, GORDON).

Until the early 1980s Intel made most of its revenue from manufacturing SRAM memory chips (see MEMORY). When the Japanese had made significant inroads into the semiconductor market, Intel turned to microprocessors, which it had introduced in 1971 and which formed the basis for the development of the desktop or personal computer (see MICROPROCESSOR and PERSONAL COMPUTER). During the 1980s, Intel 8086/8088 processors and their successors (286, 386, 486) and the associated chipsets were being used in the dominant "Wintel" (Microsoft Windows plus Intel) PC architecture. By the middle of the 1990s, Intel dominated the microprocessor market with its Pentium series chips, overcoming a mathematical flaw in some of the latter.

COMPETITION

Around 2000, Intel's dominance began to be challenged. The power of modern processors allowed for the development of lower-cost commodity PCs, and when Intel continued its progression toward increased power, competitors, particularly AMD (see ADVANCED MICRO DEVICES), were able to gain greater market share with its less expensive CPUs.

In higher power chips (particularly dual- and multi-core chips with more than one processor), Intel seems to have the edge in the middle of the first decade of the new century, although AMD is coming on strong. Meanwhile Intel and Apple in 2006 made a deal to replace the PowerPC chip in the Macintosh with Intel chips.

Intel has struggled with corporate reorganization and lower sales of chipsets and motherboards (even while continuing with strong sales of its dual-core and quad-core processors). After a decline of 42 percent from 2005 to 2006, Intel's net income increased to about $7 billion in 2007. However, its workforce has continued to decline from 102,500 in mid-2006 to 86,300 in 2007. However, Intel is expecting to produce more quad-core processors, new laptop components (including flash memory instead of hard drives), and other innovations in a very competitive market.

Further Reading

Coleman, Bob, and Logan Shrine. *Losing Faith: How the Grove Survivors Led the Decline of Intel's Corporate Culture.* 2007.

Colwell, Robert P. *The Pentium Chronicles: The People, Passion, and Politics Behind Intel's Landmark Chips.* Hoboken, N.J.: Wiley, 2006.

Grove, Andrew S. *Only the Paranoid Survive: How to Exploit the Crisis Points That Challenge Every Company.* New York: Dell, 1996.

Intel Corporation. Available online. URL: http://www.intel.com. Accessed September 23, 2007.

intellectual property and computing

Intellectual property can be defined as the rights the creator of an original work (such as an invention or a book) has to control its reproduction or use. Developers of new computer hardware, software, and media content must be able to realize a return on their time and effort. This return is threatened by the ease with which programs and data on disks can be illicitly copied and redistributed. Several legal mechanisms can be used to deter such behavior.

LEGAL PROTECTION MECHANISMS

Intellectual property represented by the design of new hardware can be protected through the patent system. A patent gives the inventor the exclusive right to sell or license the invention for 20 years after the date of filing. The basic requirements for a device to be patentable are that it represents an actual physical device or process and that it be sufficiently original and useful. A mere idea for a device, a mathematical formula, or a law of nature is not patentable in itself. In computing, a patent can be given for an actual physical device that meets the originality and usefulness requirements. Software that works with that device to control a physical process can be part of the patent, but an algorithm is not patentable by itself.

In practice, however, the situation is much murkier and more problematic. Patents are viewed as a key strategic resource (and financial asset) by companies such as IBM (which holds 40,000 patents and earns $1 billion a year by licensing them), and in the decade between 1995 and 2005 the annual number of patent applications filed rose 73 percent to 409,532. This has led to a considerable backlog in the Patent Office, and critics suggest that many patents are granted without being properly examined, such as for the existence of "prior art" (previous uses of similar technology).

Large companies often complain that so-called patent trollers obtain patents that may be relevant enough to cause infringement or invalidate a later patent, and then threaten the company with litigation if they are not paid. (Small patent holders in turn complain that large companies sometimes ignore or underpay them because they assume that the patent holder cannot afford litigation.) Many companies, including eBay, Research in Motion (maker of the Blackberry PDA), and Microsoft have been embroiled in patent suits.

Major computer companies such as Google, IBM, and Apple are supporting the Patent Reform Act of 2007. The law would tighten the standards for getting a patent and make it easier to challenge the patent later.

As of mid-2008 the bill remained stalled in the Senate. Meanwhile, a federal court had overturned new patent regulations that sought to streamline the application process by reducing the amount of supporting materials submitted.

Because of these restrictions, most software is protected by copyright rather than by patent. A computer program is considered to be a written work akin to a book. (After all, a computer program can be thought of as a special type of narrative description of a process. When compiled into executable code and run on a suitable computer, a program has the ability to physically carry out the process it describes.)

Like other written works, a program has to be sufficiently original. Once copyrighted, protection lasts for the life of the author (programmer) plus 70 years. (Works made for hire are covered for 95 years from first publication or 120 years from creation.) Given the pace of change in computing, such terms are close to "forever." While not strictly necessary, registration of the work with the U.S. Copyright Office and the inclusion of a copyright statement serve as effective legal notice and prevent infringers from claiming that they did not know the work was copyrighted.

Content (that is, text or multimedia materials) presented in a computer medium can be copyrighted in the same way as its traditional printed counterpart. However, in 1996 the U.S. Supreme Court declared that a program's user interface as such could not be copyrighted (see *Lotus Development Corp. v. Borland International*, U.S. 94-2003).

Computer programs have also received protection as trade secrets. Under the Uniform Trade Secrets Act, as adopted in many states, a program can be considered a trade secret if gaining economic value from it depends upon it not being generally known to competitors, and that "reasonable effort" is undertaken to maintain its secrecy. The familiar confidentiality and non-disclosure agreements signed by many employees of technical firms are used to enforce such secrecy.

FIRST AMENDMENT ISSUES
In a few cases the government itself has sought to limit access to software, citing national security. In the 1996 case of *Bernstein v. U.S.*, however, the courts ultimately ruled that computer program code was a form of writing protected by the First Amendment, so government agencies seeking to prevent the spread of strong encryption software could not prevent its publication.

However, First Amendment arguments have been less effective in challenging private software protection mechanisms. In 2001 a U.S. District judge ruled that Princeton University computer scientist Edward Felten and his colleagues had no legal basis to challenge provisions of the Digital Millennium Copyright Act (DMCA). The scientists had claimed that a letter from the Recording Industry Association of America (RIAA) had cast a "chilling effect" on their research into DVD-protection software by threatening them with legal action if they published academic papers about copy protection software used by online music services. The RIAA had withdrawn its letter, and the courts ruled there was no longer anything to sue about. Critics of the decision claim that it still leaves the academics in a sort of legal limbo since there is no guarantee that they would not be sued if they published something.

In another widely watched case the U.S. Court of Appeals in New York affirmed a ruling that Eric Corley, editor of the hacker magazine *2600* could not publish the code for DeCSS, a program that would allow users to read encrypted DVD disks, bypassing publisher's restrictions. The Court said that the DMCA did not infringe upon First Amendment rights. This decision would appear to conflict with *Bernstein*, although the latter has to do with government censorship, not copyright. The Supreme Court is likely to hear one or more computer-related copyright cases in the years to come.

FAIR USE AND COPY PROTECTION
Although the purchase of software may look like a simple transfer of ownership, most software is accompanied by a license that actually grants only the right to use the program under certain conditions. For example, users are typically not allowed to make copies of the program and run the program on more than one computer (unless the license is specifically for multiple uses). However, as part of "fair use" users are allowed to make an archival or backup copy to guard against damage to the physical media.

Until the 1990s, it was typical for many programs (particularly games) to be physically protected against copying (see COPY PROTECTION). Talented hackers or "software pirates" are usually able to defeat such measures, and "bootleg" copies of programs outnumber legitimate copies in some Asian markets, for example (see SOFTWARE PIRACY AND COUNTERFEITING). Copy protection and/or encryption

is also typically used for some multimedia products such as DVD movies.

CHALLENGES OF NEW MEDIA
By the mid-2000 decade, the biggest intellectual property battles were not about esoteric program codes but rather revolved around how to satisfy the ordinary home consumer's appetite for music and video while preserving producers' revenues. Increasingly, music and even video is being downloaded rather than being bought in commercial packaging at the local store.

In the *Sony v. Universal* case (1984) the Supreme Court ruled that manufacturers of devices such as VCRs were not liable for their misuse if there were "substantial non-infringing uses"—such as someone making a copy of legally possessed media for their own use. However, in 2005 the Supreme Court ruled that Grokster, a decentralized file-sharing service, could be held liable for the distribution of illegally copied media if it "actively induced" such copying.

By 2006 media industry lobbyists (particularly the Recording Industry Institute of America, or RIAA) were promoting a number of bills in Congress that would further restrict consumers' rights to use media. Such measures might include requiring that devices be able to detect "flagged" media and refuse to copy it (see DIGITAL RIGHTS MANAGEMENT), as well as adding stricter provisions to the Digital Millennium Copyright Act (DMCA). These measures are opposed by cyber-libertarian groups such as the Electronic Frontier Foundation and consumer groups such as the Home Recording Rights Coalition.

Further Reading
Chabrow, Eric. "The U.S. Patent System in Crisis." *Information-Week,* February 20, 2006. Available online. URL: http://www.informationweek.com/story/showArticle.jhtml?articleID=180204145. Accessed August 12, 2007.
Electronic Frontier Foundation. "Unintended Consequences: Seven Years under the DMCA." April 2006. Available online. URL: http://www.eff.org/IP/DMCA/unintended_consequences.php. Accessed August 8, 2007.
Gilbert, Jill. *The Entrepreneur's Guide to Patents, Copyrights, Trademarks, Trade Secrets & Licensing.* New York: Berkley Books, 2004.
Home Recording Rights Coalition. Available online. URL: http://www.hrrc.org/. Accessed August 8, 2007.
"Intellectual Property Law News." FindLaw. Available online. URL: http://news.findlaw.com/legalnews/scitech/ip/. Accessed August 8, 2007.
Klemens, Ben. *Math You Can't Use: Patents, Copyright, and Software.* Washington, D.C.: Brookings Institution, 2006.
LaPlante, Alice. "Media Distribution Rights: Here Come the Judges (and Congress)." *InformationWeek,* June 29, 2006. Available online. URL: http://www.informationweek.com/story/showArticle.jhtml?articleID=189700173. Accessed August 8, 2007.
Wilson, Lee. *Fair Use, Free Use, and Use by Permission: How to Handle Copyrights in All Media.* New York: Allworth Press, 2005.

internationalization and localization
Internationalization and localization are ways to adapt computer software (often created in the United States or Europe) to other languages and cultures. The abbreviations

I18n and L10n are sometimes used for internationalization and localization, respectively (the numbers in each word refer to the number of letters in the alphabet between the letters). The two processes are complementary.

Internationalization involves designing programs so they will be as easy as possible to adapt to a variety of cultural settings. For example, the Unicode character set is preferred because it can accommodate most of the world's alphabets and many other characters. Program code can also be modularized such that date, time, and other formats for different countries can be loaded in and used as desired.

Localization involves changing a number of aspects of a software product (including user interface elements and online help) to reflect the language and culture of the intended market. Some of this is fairly straightforward: formats for numbers, currency, date, and time; text collation and sorting order; and use of the keyboard (including special keys). To the extent the program has been appropriately generalized (internationalized), it becomes easier to localize it for each setting.

Other aspects of localization can be subtler. Icons, for example, may have to be changed because their supposedly "universal" meaning would not translate well into the local culture. Documentation may have to change wording to avoid conveying ideas that may be confusing or even offensive. Even more substantial localization may be required if the target environment (such as the education system) is substantially different from that in the country where the software was written. Generally this cannot be done automatically: the program must be reviewed by someone who is knowledgeable about the target language or culture.

Further Reading

Esselink, Bert. *A Practical Guide to Localization*. Philadelphia: John Benjamins Publisher, 2000.
Internationalization (I18n) Activity. World Wide Web Consortium. Available online. URL: http://www.w3.org/International/. Accessed September 23, 2007.
Smith-Ferrier, Guy. *.NET Internationalization: The Developer's Guide to Building Global Windows and Web Applications*. Upper Saddle River, N.J.: Addison-Wesley, 2007.

Internet

The Internet is the worldwide network of all computers (or networks of computers) that communicate using a particular protocol for routing data from one computer to another (see TCP/IP). As long as the programs they run follow the rules of the protocol, the computers can be connected by a variety of physical means including ordinary and special phone lines, cable, fiber optics, and even wireless or satellite transmission.

HISTORY AND DEVELOPMENT

The Internet's origins can be traced to a project sponsored by the U.S. Defense Department. Its purpose was to find a way to connect key military computers (such as those controlling air defense radar and interceptor systems). Such a system would require a great deal of redundancy, routing communications around installations that had been destroyed by enemy nuclear weapons. The solution was to break data up into individually addressed packets that could be dispatched by routing software that could find whatever route to the destination was viable or most efficient. At the destination, packets would be reassembled into messages or data files.

By the early 1970s, a number of research institutions including the pioneer networking firm Bolt Beranek and Newman (BBN), Stanford Research Institute (SRI), Carnegie Mellon University, and the University of California at Berkeley were connected to the government-funded and administered ARPANET (named for the Defense Department's Advanced Research Projects Agency). Gradually, as use of the ARPANET's protocol spread, gateways were created to connect it to other networks such as the National Science Foundation's NSFnet. The growth of the network was also spurred by the creation of useful applications including e-mail and Usenet, a sort of bulletin-board service (see the Applications section below).

Meanwhile, a completely different world of online networking arose during the 1980s in the form of local bulletin boards, often connected using a store-and-forward system called FidoNet, and proprietary online services such as CompuServe and America Online. At first there were few connections between these networks and the ARPANET, which had evolved into a general-purpose network for the academic community under the rubric of NSFnet. (It was possible to send e-mail between some networks using special gateways, but a number of different kinds of address syntax had to be used.)

In the 1990s, the NSFnet was essentially privatized, passing from government administration to a corporation that assigned domain names (see DOMAIN NAME SYSTEM). However, the impetus that brought the Internet into the daily consciousness of more and more people was the development of the World Wide Web by Tim Berners-Lee at the European particle research laboratory CERN (see BERNERS-LEE, TIM and WORLD WIDE WEB). With a standard way to display and link text (and the addition of graphics and multimedia by the mid-1990s), the Web is the Internet as far as most users are concerned (see WEB BROWSER). What had been a network for academics and adventurous professionals became a mainstream medium by the end of the decade (see also E-COMMERCE).

APPLICATIONS

A number of applications are (or have been) important contributors to the utility and popularity of the Internet.

- E-mail was one of the earliest applications on the ancestral ARPANET and remains the single most popular Internet application. Standard e-mail using SMTP (Simple Mail Transport Protocol) has been implemented for virtually every platform and operating system. In most cases once a user has entered a person's e-mail address into the "address book," e-mail can be sent with a few clicks of the mouse. While failure of the outgoing or destination mail server can still block transmission of a message, e-mail today has a high degree of reliability (see E-MAIL).

- Netnews (also called Usenet, for UNIX User Network) is in effect the world's largest computer bulletin board. It began in 1979, when Duke University and the University of North Carolina set up a simple mechanism for "posting" text files that could be read by other users. Today there are tens of thousands of topical "newsgroups" and millions of messages (called articles). Although still impressive in its quantity of content, many Web users now rely more on discussion forums based on Web pages (see NETNEWS AND NEWSGROUPS).

- Ftp (File Transport Protocol) enables the transfer of one or more files between any two machines connected to the Internet. This method of file transfer has been largely supplanted by the use of download links on Web pages, except for high-volume applications (where an ftp server is often operated "behind the scenes" of a Web link). FTP is also used by Web developers to upload files to a Web site (see FILE TRANSFER PROTOCOLS).

- Telnet is another fundamental service that brought the Internet much of its early utility. Telnet allows a user at one computer to log into another machine and run a program there. This provided an early means for users at PCs or workstations to, for example, access the Library of Congress catalog online. However, if program and file permissions are not set properly on the "host" system, telnet can cause security vulnerabilities. The telnet user is also vulnerable to having IDs and passwords stolen, since these are transmitted as clear (unencrypted) text. As a result, some online sites that once supported telnet access now limit access to Web-based forms. (Another alternative is to use a program called "secure shell" or ssh, or to use a telnet client that supports encryption.)

- Gopher was developed at the University of Minnesota and named for its mascot. Gopher is a system of servers that organize documents or other files through a hierarchy of menus that can be browsed by the remote user. Gopher became very popular in the late 1980s, only to be almost completely supplanted by the more versatile World Wide Web.

- WAIS (Wide Area Information Service) is a gateway that allows databases to be searched over the Internet. WAIS provided a relatively easy way to bring large data resources online. It, too, has largely been replaced by Web-based database services.

- The World Wide Web as mentioned above, is now the main means for displaying and transferring information of all kinds over the Internet. Its flexibility, relative ease of use, and ubiquity (with Web browsers available for virtually all platforms) has caused it to subsume most earlier services. The utility of the Web has been further enhanced by the development of many search engines that vary in thoroughness and sophistication (see WORLD WIDE WEB and SEARCH ENGINE).

- Streaming Media protocols allow for a flow of video and/or audio content to users. Player applications for Windows and other operating systems, and growing use of high-speed consumer Internet connections (see BROADBAND) have made it possible to present "live" TV and radio shows over the Internet.

- E-commerce, having boomed in the late 1990s and crashed in the early 2000s, continued to grow and proliferate later in the decade, finding new markets and applications and spreading into the developing world (see E-COMMERCE).

- Blogs and other forms of online writing have become prevalent among people ranging from elementary school students to corporate CEOs (see BLOGS AND BLOGGING).

- Social networking sites such as MySpace and Facebook are also very popular, particularly among young people (see SOCIAL NETWORKING).

- Wikis have become an important way to share and build on knowledge bases (see WIKIS AND WIKIPEDIA).

- The integration of the Internet with traditional channels of communications is proceeding rapidly (see PODCASTING, INTERNET RADIO, and VOIP).

Even as it begins to level off in the United States, worldwide Internet usage continues to grow rapidly. Asia now has more than twice as many users as North America, although the latter still has more than five times the penetration (percentage of population).

In the United States more than half of Internet users have high-speed Internet connections (see BROADBAND), and the trend in other developed countries is similar. Broadband is both required by and contributes to the appetite of Web users for music, streaming video, and other rich media content (see STREAMING and MUSIC AND VIDEO DISTRIBUTION, ONLINE).

Now in its fourth decade, the Internet is not without daunting challenges. A major one is security—see COMPUTER CRIME AND SECURITY, COMPUTER VIRUS, CYBERTERRORISM, and INFORMATION WARFARE. Users also want protection from privacy abusers and online predators (see PRIVACY IN THE DIGITAL AGE, IDENTITY THEFT, PHISHING AND SPOOFING, and CYBERSTALKING AND HARASSMENT).

For other issues and challenges involving the Internet, see CENSORSHIP AND THE INTERNET, INTERNET ARCHITECTURE AND GOVERNANCE, INTERNET ACCESS POLICY, and DIGITAL DIVIDE.

In the longer term what we call the Internet today is likely to become so ubiquitous that people will no longer think of it as a separate system or entity. Household appliances, cars, cell phones, televisions, and virtually every other device used in daily life will communicate with other devices and with control systems using Internet protocols.

In effect, people may eventually live "inside" a World Wide Web.

Further Reading

Hafner, Katie, and Matthew Lyon. *Where Wizards Stay Up Late: The Origins of the Internet.* New York: Simon & Schuster, 1996.

Internet Society. Available online. URL: http://www.isoc.org. Accessed August 8, 2007.

Internet World Stats. Available online. URL: http://www.internetworldstats.com/stats.htm. Accessed August 8, 2007.

Okin, J. R. *The Internet Revolution: The Not-for-Dummies Guide to the History, Technology, and Use of the Internet.* Winter Harbor, Me.: Ironbound Press, 2005.

Segaller, Stephen. *Nerds 2.0.1: A Brief History of the Internet.* New York: TV Books, 1998.

Zakon, Robert H. "Hobbes' Internet Timeline v8.2." Available online. URL: http://www.zakon.org/robert/internet/timeline/. Accessed August 8, 2007.

Internet applications programming

The growth of the Internet and its centrality in business, education, and other fields has led many programmers to specialize in Internet-related applications. These can include the following:

- low-level infrastructure (networking [wired and wireless], routing, encryption support, and so on)

- Web servers and related software

- e-commerce infrastructure (see E-COMMERCE)

- interfacing with databases

- data analysis and extraction (see DATA MINING)

- support for searching (see SEARCH ENGINE)

- autonomous software to navigate the net (see SOFTWARE AGENT)

- Internet-based communications (see TEXTING AND INSTANT MESSAGING and VoIP)

- systems to deliver text and media (see STREAMING, PODCASTING, RSS)

- support for collaborative use of the Internet (see BLOGS AND BLOGGING, SOCIAL NETWORKING, and WIKIS AND WIKIPEDIA)

- security software (firewalls, intrusion analysis, etc.)

Internet applications programmers use a variety of languages and other programming tools (often in combination) to implement these applications. Some of the most common are:

- C++ is generally used for fundamental applications, particularly those that must work at the system level and for which speed and efficiency are prerequisites. Examples would include Web servers and browsers and some browser plug-ins (see C++).

- Java has largely supplanted C++ as a general-purpose language for programming small applications ("applets") that are hosted by Web sites and run on the user's browser. With a syntax that differs in only a few respects from C++, Java can also be used to write standalone applications (see JAVA).

- HTML is not really a full-fledged programming language, but it defines the layout and formatting of Web pages, as well as providing for hyperlinks and the embedding of applications. In many cases, HTML no longer has to be coded directly but can be generated from word processor-like page design programs (see DNTML, HTML, and XHTML).

- Extensible markup language (see XML) has become the preferred format for structuring a variety of data both for automatic processing (see SEMANTIC WEB) and for feeding dynamic Web pages (see AJAX).

- Scripting languages are an important tool for Internet and Web development. CGI (Common Gateway Interface) is a facility that allows scripts to control the interaction between HTML forms on a Web page and other programs such as databases (see CGI). CGI scripts are written in scripting languages (see JAVASCRIPT, PERL, PHP, PYTHON, and SCRIPTING LANGUAGES). Use of CGI is being gradually supplanted by applets written in Java as well as other scripting languages such as JavaScript and VBScript.

- Active Server Pages (ASP) is a facility that uses Windows ActiveX components to process scripts created in Visual Basic, which in turn create HTML pages "on the fly" and send them to the user's Web browser.

- Microsoft's recent .NET initiative represents an attempt to integrate Internet connectivity and distributed operation into the programming framework for all major languages.

- Similar technologies are available for other platforms such as Linux (see AJAX and DOCUMENT OBJECT MODEL)

TRENDS

Experienced programmers will continue to be needed for creating and extending the infrastructure for the Internet and Web and for providing increasingly powerful and easy-to-use tools for developing Web sites. However, the wide variety of tools now available means that people with less experience will be able to design and implement attractive and effective Web pages, plugging in functionality such as online shopping, conferencing, and site-specific search engines. If web development follows the same course as traditional programming, predictions that specialized programmers will no longer be needed will prove premature. At the same time, generalist web developers will be able to do more.

Further Reading

Andersson, Eve, Philip Greenspun, and Andrew Grumet. *Software Engineering for Internet Applications.* Cambridge, Mass.: MIT Press, 2006.

Connolly, Randy. *Core Internet Application Development with ASP. NET 2.0.* Upper Saddle River, N.J.: Prentice Hall, 2007.

Dunaev, Sergei. *Advanced Internet Programming: Technologies & Applications.* Hingham, Mass.: Charles River Media, 2001.

HTML/Web Programming Resources. Available online. URL: http://www.sandhills.cc.nc.us/html.html. Accessed August 8, 2007.

Moore, Dana, Raymond Budd, and Edward Benson. *Professional Rich Internet Applications: AJAX and Beyond.* Indianapolis: Wrox, 2007.

Web Programming Resources. Available online. URL: http://www.webreference.com/programming/. Accessed August 8, 2007.

Internet censorship   *See* CENSORSHIP AND THE INTERNET.

Internet cafés and "hot spots"

Internet cafés (also called cyber cafés) are public places where computers are connected to the Internet and available for use for a fee by the hour or minute. Many Internet cafés also sell coffee and food. Combining the social ambience of a traditional coffee shop and the attraction of the Internet, many Internet cafés acquire a regular clientele from students to adults.

These venues appeared first in the mid-1990s and spread rapidly as their popularity grew. While the most common activities at most Internet cafés are to check e-mail, send text messages, or browse the Web, some locations specialize in gaming (see ONLINE GAMES), providing more powerful machines running games over a local network. Such gaming centers have been particularly popular in Asia.

Internet cafés have grown most rapidly in countries that are becoming more urban and industrial but where many people cannot yet afford their own computers. The most striking example is China, which had 113,000 Internet cafés as of 2007. In keeping with its strict policies, however, the Chinese government closely monitors activity at Internet cafés (see CENSORSHIP AND THE INTERNET).

Internet cafes are particularly common in countries such as China, where Internet access is still relatively rare in homes. In many cases such facilities have given way to simple "hot spots," where users can wirelessly connect their own laptops or PDAs. (QIN YING/ PANORAMA/THE IMAGE WORKS)

HOT SPOTS

The number of dedicated Internet cafés in the United States and many other highly developed countries has been declining in recent years. This is largely due to the growing number of people who connect to the Internet through their own laptops and other mobile devices (see PDA and SMARTPHONE). Thus many locations, including coffee chains such as Starbucks, do not provide machines, but simply offer wireless Internet access (see WIRELESS AND MOBILE COMPUTING). Areas where one can make such a wireless connection are called "hot spots." Today virtually all major hotels and airports provide hot spots; there is normally a fee for access as with Internet cafés. (The fee is collected by routing all access through a portal.) However, a number of venues offer free Wifi access.

Users of Internet cafés or hot spots should be aware that they are sharing an ad hoc network with strangers and may be exposed to malicious software. Passwords or other sensitive data may be "sniffed" using special software. It is therefore generally a good idea not to conduct financial transactions or otherwise send sensitive information when connected to such venues, unless one has provided for encryption or can access a virtual private network. Additionally, users connecting their own machines to a hot spot should have up-to-date firewall and antivirus software.

Further Reading

Bradley, Tony, and Becky Waring. "Complete Guide to Wi-Fi Security." Available online. URL: http://www.jiwire.com/wi-fi-security-traveler-hotspot-1.htm. Accessed September 9, 2007.

Café Touch. Available online. URL: http://www.cafetouch.com/. Accessed September 9, 2007.

Cyber cafés [directory]. Available online. URL: http://www.cybercafes.com/. Accessed September 9, 2007.

Internet Café Guide. Available online. URL: http://www.internet-cafe-guide.com/. Accessed September 9, 2007.

Wi-Fi Hotspot Finder. Available online. URL: http://www.jiwire.com/search-hotspot-locations.htm. Accessed September 9, 2007.

Internet organization and governance

The Internet is remarkable as a modern institution in that, while the technology was developed with considerable government funding, the Net as we know it today is remarkably free of externally imposed authority or regulation. This is in sharp contrast with earlier communications technologies such as the telegraph and telephone, which were generally tightly regulated or even run by a government department such as the Post Office. In part this was due to the complexity of the technology and the fact that many political leaders had little familiarity with it and its implications. (Also, the speed of growth has been overwhelming in recent years, considering that the World Wide Web in its modern form was scarcely a decade and a half old as of 2008.)

INSTITUTIONS OF SELF-GOVERNANCE

While the Internet is not rigidly controlled, the need for interoperability and orderly advances in technology has led to the emergence of several organizations that provide standards and guidance. The most important of these is the World Wide Web Consortium (W3C). Other technical organizations include the Internet Engineering Task Force (IETF) and the Internet Corporation for Assigned Names and Numbers (ICANN), the latter of which administers the domain system (seen DOMAIN NAME SYSTEM). The domain registries in turn are run by many different institutions and agencies.

GROWING ROLE OF GOVERNMENTS?

Many of the key innovators of the Internet have loosely shared a somewhat anarchic or libertarian viewpoint, and reinforced it with the claim that the decentralized architecture of the Internet itself resists imposition of rules from outside. (Thus the saying, "the Internet sees censorship as a failure and routes around it.")

However, recently some writers such as LAWRENCE LESSIG of Stanford Law School have called for a reappraisal. Lessig argues that the Internet is far from ungovernable and that indeed such an important institution *must* be regulated. The question is how to regulate it wisely, shaping its architecture to support freedom, democracy, and other desirable values.

In 2003 and 2005, the United Nations brought together many government representatives who raised many issues about what they saw as inadequacies of the privately run Internet (for example, in the assigning of domain names) and a perceived bias toward American interests. The United Nations has established an Information and Communication Technologies (ICT) Task Force to carry on these meetings, which will be called the Internet Governance Forum (IGF). Other international institutions such as the International Telecommunications Union (ITU) have sometimes come into conflict with the Internet's self-governing bodies.

Within the United States there continues to be strong resistance to imposing new regulations on the Internet, in part because of fear of constricting one of the most important and fastest growing sectors of the economy.

The conflict between the Internet's self-governing culture and the needs and desires of political institutions will no doubt continue. Sometimes the conflict can be very sharp, as with China's blocking of Internet content that it finds objectionable (see CENSORSHIP AND THE INTERNET). Other issues are perhaps deeper, such as the question of how to enforce criminal laws or economic regulations that were designed for a world made of brick and steel.

Further Reading
Goldsmith, Jack, and Tim Wu. *Who Controls the Internet? Illusions of a Borderless World.* New York: Oxford University Press, 2006.
Internet Governance Project. Available online. URL: http://www. internetgovernance.org/. Accessed September 23, 2007.
Lessig, Lawrence. *Code Version 2.0.* New York: Basic Books, 2006.
MacLean, Don, ed. *Internet Governance: A Grand Collaboration.* New York: United Nations ICT Task Force, 2004.
World Wide Web Consortium. Available online. URL: http://www. w3.org/. Accessed September 23, 2007.

Internet radio

Internet radio is the provision of radio broadcast content over the Internet (see STREAMING). Basically, the digitized sound files of the broadcasts can be accessed and played using widely available software such as Windows Media Player or RealPlayer. Internet radio began in the mid-1990s, and today an increasing number of broadcast stations are offering their programming in this form, allowing them to reach audiences far beyond the reach of their signal. Some stations stream live (during the actual broadcast), while others make programs available for download. (For automatic downloading of broadcasts, see PODCASTING). There are also "radio stations" that provide their content only via the Internet. Internet radio should not be confused with satellite or cable radio, which carry conventional radio signals in real time.

For the user, Internet radio expands the selection of stations available from a few dozen over the air to hundreds or thousands. Potentially this allows for the support of specialized stations that have been struggling for audiences in traditional markets—examples might be stations broadcasting jazz or alternative music, political advocacy, or programming in less widely spoken languages.

Of course there still remains the question of how commercial Internet radio can support itself. Many on-air stations simply include their advertising in the Internet stream (although this can be sometimes ineffective if the ad refers solely to a local business). Some stations sell subscriptions or charge a fee for each program.

Regular radio stations must pay royalties to performers whose music is played on the air. Until recently, such fees have been minimal (or even ignored) for Internet radio. A major issue arose in 2007 when the U.S. Copyright Royalty Board approved a steep increase in the royalties for music on Internet radio. Many smaller Internet radio stations have protested that the increased fees would put them out of business as well as hurting many independent performers who depend on this medium to get their work heard. However, a number of stations have been able to negotiate reductions or caps on these fees on an ad hoc basis.

Further Reading
Heberlein, L. A. *The Rough Guide to Internet Radio.* London: Rough Guides, 2002.
Hoeg, Wolfgang, and Thomas Lauterbach, eds. *Digital Audio Broadcasting: Principles and Applications of Digital Radio.* 2nd ed. New York: Wiley, 2003.
Internet Radio Guide. Available online. URL: http://www. windowsmedia.com/Mediaguide/Radio. Accessed September 23, 2007.
Lee, Eric. *How Internet Radio Can Change the World: An Activist's Handbook.* Lincoln, Nebr.: iUniverse, 2004.
Web Radio [directory]. Available online. URL: http://www. radio-directory.com/. Accessed September 23, 2007.

Internet service provider (ISP)

An Internet service provider is any organization that provides access to the Internet. While nonprofit organizations such as universities and government agencies can be considered to be ISPs, the term is generally applied to a commercial, fee-based service.

Typically, a user is given an account that is accessed by logging in through the operating system's Internet connection facility by supplying a user ID and password. Once connected, the user can run Web browsers, e-mail clients, and other programs that are designed to work with an Internet connection. Most ISPs now charge flat monthly fees ranging from $20 or so for dial-up access to around $40–$60 for high-speed cable or DSL connections (see BROADBAND). Some services such as America Online and CompuServe include ISP service as part of a package that also includes such features as software libraries, discussion forums, and instant messaging. Online services tend to be more expensive than "no frills" ISP services.

Most personal ISP accounts include a small allotment of server space that users can use to host their personal Web pages. There are generally extra charges for larger allotments of space, for sites that generate high traffic, and for commercial sites. Business-oriented ISPs typically provide a more generous starting allotment along with more extensive technical support and more reliable and higher-capacity servers that are managed 24 hours a day.

The rapid growth in Internet use in the mid-1990s encouraged many would-be entrepreneurs to start ISPs. However, with so many providers entering the field and with the price for basic Internet connections falling, it soon became apparent that the survival prospects for "generic" ISPs would be poor. People entering the business today strive to provide added-value services such as superior Web page hosting facilities, hosting blogs or wikis, or to focus on specialized services for particularly industries (such as real estate).

Today's ISPs also face a variety of legal challenges, including customer privacy vs. the war on terrorism (see PRIVACY IN THE DIGITAL AGE), responsibility for copyright infringement (see INTELLECTUAL PROPERTY AND COMPUTING), and possible liability for online defamation, harassment, or worse (see CYBERSTALKING AND HARASSMENT).

Further Reading

Berkowitz, Howard C. *Building Service Provider Networks*. New York: Wiley, 2002.
"Everything You Wanted to Know About Internet Service Providers." Available online. URL: http://www.ispconsumerguide.com/. Accessed February 6, 2008.
"ISP Liability." BitLaw. Available online. URL: http://www.bitlaw.com/internet/isp.html. Accessed August 8, 2007.
Nguyen, John V. *Designing ISP Architectures*. Upper Saddle River, N.J.: Prentice Hall, 2002.

interpreter

An interpreter is a program that analyzes (parses) programming commands or statements in a high-level language (see PROGRAMMING LANGUAGES), creates equivalent executable instructions in machine code (see ASSEMBLER) and executes them. An interpreter differs from a compiler in that the latter converts the entire program to an executable file rather than processing and executing it a statement at a time (see COMPILER).

Many earlier versions of the BASIC programming language were implemented as interpreters. Since an interpreter only has to hold one program statement at a time in memory, it could run on early microcomputers that had only a few tens of thousands of bytes of system memory. However, interpreters run programs considerably more slowly than a compiled program would run. One reason is that an interpreter "throws away" each source code statement after it interprets it. This means that if a statement runs repeatedly (see LOOP), it must be re-interpreted each time it runs. A compiler, on the other hand, would create only one set of machine code instructions for the loop and then move on. Also, because a compiler keeps the entire program in memory, it can analyze the relationship between multiple statements and recognize ways to rearrange or substitute them for greater efficiency.

Interpretation can also be used to bridge differences in hardware platforms. For example, in the UCSD Pascal system developed in the 1970s, an interpreter first translates the Pascal source code into a standardized "P-code" (pseudocode) for a generic processor called a P-machine. To

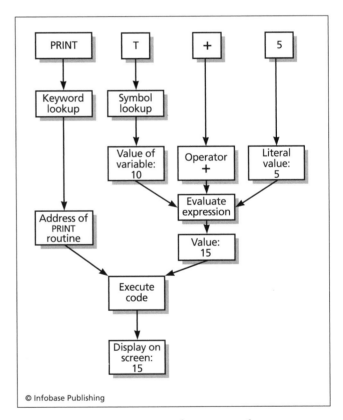

An interpreter scans a program code or command statement to determine what each token (word or symbol) represents. Keywords such as PRINT are looked up in a dispatch table that contains instructions for dealing with that function. Variables are looked up in a symbol table that gives their current value. Values and operators make up expressions that are interpreted to yield their final value. In this case the final value of 15 is given as data to the PRINT routine, which is executed to put the number 15 on the screen.

run the program on a particular actual machine, a second interpreter translates the P-code into specific executable machine instructions for that machine. Today Java uses a similar idea. A Java programming system translates source code into an intermediate "bytecode," which is interpreted by a Java Virtual Machine, usually running with a Web browser.

In practice, with today's high-speed computers and graphical operating environments, interpretative and compilation functions are often seamlessly integrated into a programming environment where code is checked for syntax as it is entered, incrementally compiled (such that only changed code is recompiled), and the programmer receives the same kind of rapid feedback that was the hallmark of the early BASIC interpreters (see PROGRAMMING ENVIRONMENT). Purely interpretive systems survive mainly in the form of text command processors for operating systems (see SHELL).

Further Reading

Craig, Ian. *The Interpretation of Object-Oriented Programming Languages*. 2nd ed. New York: Springer, 2002.

Mack, Ronald. *Writing Compilers and Interpreters: An Applied Approach Using C++*. 2nd ed. New York: Wiley, 2006.

Watt, David, and Deryck Brown. *Programming Language Processors in Java: Compilers and Interpreters*. Upper Saddle River, N.J.: Prentice Hall, 2000.

iRobot Corporation

iRobot is an innovative company based in Burlington, Massachusetts, that makes robots for home use (the Roomba robotic vacuum cleaner and its floor-washing cousin Scooba) to military robots such as various PackBot models designed for reconnaissance, bomb disposal, and other dangerous tasks.

iRobot was founded by robotics pioneer Rodney Brooks of MIT's Artificial Intelligence Lab (see BROOKS, RODNEY) and two former MIT students, Helen Greiner and Colin Angle. The company was founded in 1991 and incorporated in 2000. Its first product was My Real Baby, a realistic (and complicated) animated doll that proved to be too expensive for the toy market. Roomba, on the other hand, was released in 2002 and has met with considerable success—2 million units had been sold by May 2006. Besides Scooba, Roomba has been joined by Dirt Dog (a workshop cleaner and picker-upper) and Verro, a pool cleaner. iRobot has also produced an educational/hobby robot called iRobot Create.

iRobot has done considerable work for the military, based on work in the 1990s with robots that crawled or rolled on tanklike tracks and were equipped with grasping devices and other attachments. The PackBot series comes in models adaptable to a variety of military tasks, and has been used in Iraq and Afghanistan.

In 2007 iRobot released a redesigned, more durable version of Roomba. Meanwhile cofounder Colin Angle has said that the company is looking at many exciting future applications, including industrial cleaning, mining, and oil exploration. In the home, Roomba may be joined by outdoor robots that can mow the lawn and trim the hedges.

iRobot is a midsized company whose revenue has grown from $54.3 million in 2003 to $227 million in 2007, with a gross profit of $82.6 million and 423 employees.

Further Reading

Henderson, Harry. *Modern Robotics: Building Versatile Machines*. New York: Chelsea House, 2006.

iRobot. Available online. URL: http://www.irobot.com. Accessed September 23, 2007.

Lombardi, Candace. "iRobot's Angle on the Future: More Profit." CNET News, August 23, 2007. Available online. URL: http://www.news.com/iRobots-Angle-on-the-future-More-profit/2008-11394_3-6204031.html. Accessed September 24, 2007.

Pereira, Joseph. "Natural Intelligence: Helen Greiner Thinks Robots Are Ready to Become Part of the Household." *Wall Street Journal* Classroom Edition, October 2002. Available online. URL: http://www.wsjclassroomedition.com/archive/02oct/COVR_ROBOT.htm. Accessed September 24, 2007.

Roush, Wade. "Will Home Robots Ever Clean Up?" *Technology Review*, March 3, 2006. Available online. URL: http://www.technologyreview.com/article/16542/. Accessed September 24, 2007.

Java

Java is a computer language similar in structure to C++. Although Java is a general-purpose programming language, it is most often used for creating applications to run on the Internet, such as Web servers. A special type of Java program called an applet can be linked into Web pages and run on the user's Web browser (see APPLET).

As an object-oriented language, Java uses classes that provide commonly needed functions including the creation of user interface objects such as windows and buttons (see CLASS and OBJECT-ORIENTED PROGRAMMING). A variety of sets of classes ("class frameworks") are available, such as the AWT (Abstract Windowing Toolkit).

PROGRAM STRUCTURE

A Java program begins by importing or defining classes and using them to create the objects needed for the program's functions. Code statements then create the desired output or interaction from the objects, such as drawing a picture or putting text in a window. Here is a simple Java applet program:

```
import java.applet.Applet;
import java.awt.Graphics;
public class HelloWorld extends Applet {
  public void paint(Graphics g) {
    g.drawString("Hello world!", 50, 25);
  }
}
```

The first two lines import (bring in) standard classes. The applet class is the foundation on which applet programs are built. The AWT (Abstract Windowing Toolkit) is a set of classes that provide a graphical user interface.

The program then declares a new class called Hello-World and specifies that it is built on (extends) the applet class.

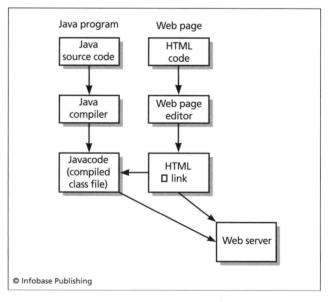

© Infobase Publishing

After an embedded Java program (called an applet) is compiled, its executable file (Javacode) is stored on the Web server, together with the HTML file for the Web page to which the program is linked.

254

Next is a declaration for a method (procedure for doing something) called paint. This method uses a graphics object g that includes various capabilities for drawing things on the screen. Finally, the program uses the graphic object's predefined drawstring method to draw a string of text.

To develop this program, the programmer compiles it with the Java compiler. He or she then creates an HTML page that includes a tag that specifies that this code is to be run when the link is activated (see HTML).

DEVELOPMENT OF JAVA

Java was created by James Gosling (1955–   ). It began as an in-house project at Sun Microsystems to design a language that could be used to program "smart" consumer devices such as an interactive television. When this project was abandoned, Gosling, Bill Joy, and other developers realized that the language could be adapted to the rapidly growing Internet. Developers of Web pages needed an easier way to create programs that could run when the page was accessed by a user. By implementing user controls on Web pages, the designers could give Web users the ability to interact online in much the same way they interact with objects on the screen on a Macintosh or Windows PC.

ADVANTAGES

Java has largely fulfilled this promise for Web developers. C++ programmers have an easy learning curve to Java, since the two languages have very similar syntax and a similar use of classes and other object-oriented features. On the other hand, programmers who don't know C++ benefit from Java being more streamlined than C++. For example, Java avoids the necessity to use pointers (see POINTERS AND INDIRECTION) and uses classes as the consistent building block of program structure. Software powerhouses such as Microsoft (until recently) and IBM have joined Sun in promoting Java.

Another much-touted feature of Java is its platform independence. The language itself is separate from the various operating system platforms. For each platform, a Java Virtual Machine (JVM) is created, which interprets or compiles the code generated by the Java compiler so it can run on that platform.

For security, Java applets run within a "sandbox" or restricted environment so the user is protected from malicious Java programs. (For example, programs are not allowed to access the user's disk or to connect the user's machine to another Web site.) Web browsers can also be set to disable the running of Java applets.

A MATURE TECHNOLOGY

Sun Java comes in two basic "flavors": the Java 2 Standard Edition (J2SE) for Microsoft Windows, Sun (Solaris), and Linux, and the Enterprise Editions (J2EE), which includes features needed in large, complex environments. Microsoft developed its own dialect of Java for Windows, but effectively abandoned it as a result of legal action by Sun. (Companies are allowed to develop Java implementations for various platforms, so long as they pass Sun's strict validation process.)

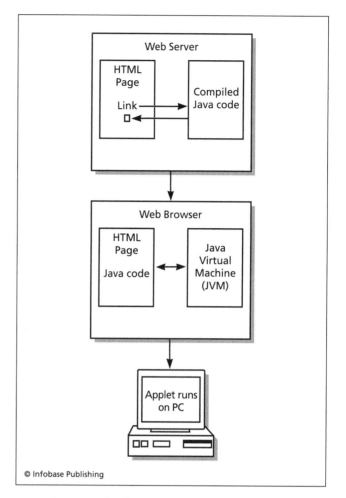

© Infobase Publishing

To run the Java applet, the user loads the linked page in the Web browser. The applet may then run automatically, or it may be connected to a particular link or a control such as a button. Once activated, the applet is downloaded by the Web browser, which then runs its Javacode using a module called a Java Virtual Machine (JVM). There is a separate JVM for each type of computer system.

Java has paid particular attention to building reusable software components. "JavaBeans" package a number of related objects (classes) into a unit that can be accessed through a standard set of methods and automatically queried for information about their contents.

Today powerful and well-documented Java programming interfaces are available for working with Web services. While client-side Java applets run in the Web browser, Java Server Pages (JSPs) embed code in an HTML page. The code is compiled into a server-side application or "servlet." XML processing and database access is provided through the Java API for XML (JAX).

Java has largely fulfilled its promise of bringing mainstream object-oriented programming to a wide variety of platforms. The language is now often taught as a first language instead of C or C++. However, the idea of a single dominant language seems to be no longer applicable in the rapidly evolving world of software development.

In 2006 Sun Microsystems announced that it would make Java's source code freely available (see OPEN SOURCE). In part this may be an attempt to maintain Java's position among programmers, some of whom have shifted their attention from Java to Microsoft's own offshoot of C++ (see C#). However, Java's greatest challenge seems to be in the Web programming area, where it faces increasing competition from more agile languages (see, for example, RUBY) as well as a variety of scripting languages that may be easier to learn and quicker to use for many applications.

Further Reading

Arnold, Ken, James Golsing, and David Holmes. *The Java Programming Language.* 4th ed. Upper Saddle River, N.J.: Prentice Hall, 2005.

Burd, Barry. *Beginning Programming with Java for Dummies.* 2nd ed. Hoboken, NJ: Wiley, 2005.

Gosling, James. "Is Java Getting Better with Age?" [interview]. Cnet News. Available online. URL: http://news.com.com/2008-7345_3-6022062.html. Accessed April 10, 2007.

Krill, Paul. "Java Facing Pressure from Dynamic Languages." *InfoWorld,* March 25, 2006. Available online. URL: http://www.infoworld.com/article/06/03/25/76803_HNjavapressure_1.html. Accessed April 10, 2007.

McGovern, James, et al. *Java Web Services Architecture.* San Francisco: Morgan Kaufmann, 2003.

Schildt, Herbert. *Swing: A Beginner's Guide.* New York: McGraw-Hill, 2007.

JavaScript

JavaScript is one of several popular languages that can enable Web pages to interact with users more quickly and efficiently (see VBSCRIPT, PHP, and SCRIPTING LANGUAGES). The language first appeared in the mid-1990s' Netscape 2 browser under the name LiveScript. Technically, JavaScript is the Sun Microsystems trademark for its implementation of a standard called ECMAScript. Despite the name, JavaScript is not directly related to the Java programming language.

In its early years JavaScript was perhaps a victim of its own success. Having a relatively easy-to-use scripting language provided an easier way to add features such as 3D buttons and pop-up windows to formerly humdrum Web forms. However, as with an earlier generation's fondness for multiple fonts, early JavaScript programmers were often prone to add unnecessary and confusing clutter to Web pages. Besides sometimes annoying users, early JavaScript also suffered from significant differences in how it was implemented by the major browsers. As a result, Netscape users were sometimes stymied by JavaScript written for Microsoft Internet Explorer, and vice versa. Finally, browser flaws have sometimes allowed JavaScript to be used to compromise security such as by installing malware-infested "browser helpers." As a result, many security experts began to recommend that users disable JavaScript execution in their browsers.

USING JAVASCRIPT

JavaScript syntax and language constructs are similar to those of C, with the addition of basic object-oriented features (see OBJECT-ORIENTED PROGRAMMING). The language itself has no capabilities for manipulating the environment (such as input/output). Instead, JavaScript calls upon an "engine" written for each host environment (normally a Web browser). The engine implements features designed to control how a Web page interacts with the user, such as the display of windows and controls such as menus, buttons, or toolbars. JavaScript can also be used to validate a Web form in the browser before it is submitted to the server. In general, "browser side" JavaScript processing reduces the load on Web servers while allowing pages to respond quickly, such as by changing graphics as the user's mouse pointer passes over parts of the page.

The principal interface between JavaScript and HTML pages is the Document Object Model (see HTML and DOCUMENT OBJECT MODEL). A World Wide Web Consortium (W3C) standard defines the DOM functions, and most browsers now consistently support Levels 1 and 2 of these standards. However, there are many Web users who cannot run standard JavaScript, such as users with visual disabilities (see DISABLED PERSONS AND COMPUTING), users of some mobile browsers (such as for PDAs or smart phones), or users who have simply disabled JavaScript for security reasons. Therefore, when JavaScript is used for essential page functions (such as form processing), the developer should provide an alternative way for the user to perform the relevant task. (In the case of disabled users, this may be a legal requirement.)

Traditionally, JavaScript code has been embedded directly in the containing HTML page, using tags like the following:

```
<!DOCTYPE HTML PUBLIC "-//W3C//DTD HTML
4.01//EN"
"http://www.w3.org/TR/html4/strict.dtd">
<html dir="ltr" lang="en">
<head>
<title>JavaScript Example</title>
<body>
<script type="text/JavaScript">
     var Name = prompt ("Enter your
     name","");
   alert(Name);
</script>
</body>
</html>
```

When a JavaScript-enabled browser encounters this code, a text box will prompt the user for a name, which is stored in the variable Name and then displayed in an alert box.

In modern Web design to XHTML standards, however, just as formatting information is kept in a separate document (see CASCADING STYLE SHEETS) JavaScript code is also maintained in a separate file and simply linked to within the HTML page:

```
<script type="text/javascript"
src="mainscript.js"></script>
```

JavaScript can do much more than just display information or process forms. JavaScript can access a variety of Web services (such as databases and search engines) and create custom pages in response to user actions (see also AJAX). JavaScript can also be embedded in applications other than Web browsers: for example, the Adobe Acrobat and Reader and even operating-system scripting (such as Microsoft's JScript and JScript.NET). Although attention in recent years seems to have shifted more to languages such as PHP, JavaScript remains a widely used and powerful Web design tool.

Further Reading

Flanagan, David. *JavaScript: The Definitive Guide.* 5th ed. Sebastapol, Calif.: O'Reilly, 2006.

Heilmann, Christian. *Beginning JavaScript with DOM Scripting and Ajax.* Berkeley, Calif.: APress, 2005.

JavaScript at Webdeveloper.com. Available online. URL: http://www.webdeveloper.com/javascript/. Accessed September 24, 2007.

JavaScript.com: The Definitive JavaScript Resource. Available online. URL: http://www.javascript.com/. Accessed September 24, 2007.

Wilton, Paul, and Jeremy McPeak. *Beginning JavaScript.* 3rd ed. Indianapolis: Wiley, 2007.

job control language

In the early days of computing, data processing generally had to be done in batches. By modern standards the memory capacity of the computer was very limited (see MAINFRAME). Typically, programs had to be loaded one at a time from punch cards or tape. The data to be processed by each program also had to be made available by being mounted on a tape drive or inserted as a stack of cards into the card reader (see PUNCHED CARDS AND PAPER TAPE). After the program ran, its output would consist of more data cards or tape, which might in turn be used as input for the next program.

For example, a series of programs might be used to read employee time cards and calculate the payments due after various items of withholding. That data might in turn be input into a program to print the payroll checks and another program to print a summary report.

In order for all this to work, the computer's operating system must be told which files (on which devices) are to be used by the program, the memory partition in which the program is to be run, the device to which output will be written or saved, and so on. This is done by giving the computer instructions in job control language (JCL). (In the punch card days, the JCL cards were put at the top of the deck before the cards with the instructions for the program itself.)

For a simple example, we will use some elements of IBM MVS JCL. In this version of job control language the general form for all statements is

```
//name operation operands comment
```

where *name* is a label that can be used to reference the statement from elsewhere, *operation* indicates one of a set of defined JCL language commands, *operand* is a series of values to be passed to the system, and *comment* is optional explanatory text.

The three basic types of statement found in most job control languages are JOB, EXEC, and DD. The JOB statement identifies the job and the user running it and sets up some parameters to specify the handling of the job.

```
//JOB,CLASSPROJ1,GROUP=J999996,USER=P999995,
//PASSWORD=?
```

This statement passes information to the system that identifies the job name, group as assigned by the facility, and user ID. The PASSWORD parameter is given a question mark to indicate that it will be prompted for at the terminal. Other parameters can be used to specify such matters as the amount of computer time to be allocated to the job and the way in which any error messages will be displayed.

The EXEC statement identifies the program to be run. Some systems can also have a library of stored JCL procedures that can also be specified in the EXEC statement. This means that frequently run jobs can be run without having to specify all the details each time. An example EXEC statement is:

```
//Datasort EXEC BINSORT,BUFFER=256K
```

Here the statement is labeled Datasort so it can be referenced from another part of the program. The procedure to be executed is named BINSORT, and it is passed a parameter called BUFFER with a value of 256K (presumably this is the amount of memory to be used to hold data to be sorted).

One or more DD (Data Definition) statements are used to specify sets (sources) of data to be used by the program. This includes a specification of the type (such as disk or tape) and format of the data. It also includes instructions specifying what is to happen to the data set. For example, the data set might be old (existing) or newly created by the program. It may also be temporary (possibly to be passed on to the next program) or permanent ("cataloged").

Since interactive, multitasking operating systems such as Windows and UNIX are now the norm in most computing, JCL is used less frequently today. However, it is still needed in large computer installations running operating systems such as IBM MVS (see MAINFRAME) and for some batch processing of scientific or statistical programs (such as in FORTRAN or SAS).

Further Reading

Brown, Gary Deward. *System 390 Job Control Language.* 4th ed. New York: John Wiley, 1998.

Malaga, Ernie, and Ted Holt. *Complete CL: The Definitive Control Language Programming Guide.* 3rd ed. Carlsbad, Calif.: Midrange Computing, 1999.

Jobs, Steven Paul

(1955–)
American
Entrepreneur

Steve Jobs was cofounder of Apple Computer and shaped the development and marketing of its distinctive Macintosh personal computer (see APPLE CORPORATION). Jobs showed an enthusiastic interest in electronics starting in his high school years and gained experience through summer work at Hewlett-Packard, one of the dominant companies of the early Silicon Valley. In 1974, he began to work for pioneer video game designer Nolan Bushnell at Atari. He also became a key member of the Homebrew Computer Club, a group of hobbyists who designed their own microcomputer systems using early microprocessors.

Meanwhile, Jobs's friend Steve Wozniak had developed plans for a complete microcomputer system that could be built using a single-board design and relatively simple circuits (see WOZNIAK, STEVEN). In it Jobs saw the potential for a standardized, commercially viable microcomputer system. They formed a company called Apple Computer (named apparently for the vanished orchards of Silicon Valley) and built a prototype they called the Apple I. Although they could only afford to build a few dozen of the machines, they made a favorable impression on the computer enthusiast community. By 1977, they were marketing a more complete and refined version, the Apple II.

Unlike kits that could be assembled only by experienced hobbyists, the Apple II was ready to use "out of the box." It included a cassette tape recorder for storing programs. When connected to a monitor or an ordinary TV, the machine could create color graphics that were dazzling compared to the monochrome text displays of most computers. Users could buy additional memory (the first model came with only 4K of RAM) as well as cards that could drive devices such as printers or add other capabilities.

The ability to run a program called VisiCalc (see SPREAD-SHEET) propelled the Apple II into the business world, and about 2 million of the machines were eventually sold. In 1982, when *Time* magazine featured the personal computer as its "man of the year," Jobs's picture appeared on the cover. As he relentlessly pushed Apple forward, supporters pointed to Jobs's charismatic leadership, while detractors said that he could be ruthless when anyone disagreed with his vision of the company's future.

However, 1982 also brought industry giant IBM into the market. Its 16-bit computer was more powerful than the Apple II, and IBM's existing access to corporate purchasing departments resulted in the IBM PC and its "clones" quickly dominating the business market (see IBM PC).

Jobs responded to this competition by designing a PC with a radically different user interface, based largely on work during the 1970s and the Xerox PARC laboratory. The first version, called the Lisa, featured a mouse-driven graphical user interface that was much easier to use than the typed-in commands required by the Microsoft/IBM DOS. While the Lisa's price tag of $10,000 kept it out of the mainstream market, its successor, the Macintosh, attracted millions of users, particularly in schools, although the IBM PC and its progeny continued to dominate the business market (see MACINTOSH). Meanwhile, Jobs had recruited John Sculley, former CEO of PepsiCo, to serve as Apple's CEO.

After a growing divergence with Sculley over management style and Apple's future priorities, Jobs left the company in 1985. Using the money from selling his Apple stock, Jobs bought a controlling interests in Pixar, a graphics studio that had been spun off from LucasFilm. He also founded a company called NextStep. The company focused on high-end graphics workstations that used a sophisticated object-oriented operating system. However, while its software (particularly its development tools) was innovative, the company was unable to sell enough of its hardware and closed that part of the business in 1993.

In 1997, Jobs returned as CEO of Apple. By then the company was struggling to maintain market share for its Macintosh line in a world that was firmly in the "Wintel" (Windows on Intel-based processors) camp. He had some success in revitalizing Apple's consumer product line with the iMac, a colorful, slim version of the Macintosh. He also focused on development of the new Mac OS X, a blending of the power of UNIX with the ease-of-use of the traditional Macintosh interface.

BEYOND THE MAC

At the beginning of the new century, Jobs and Apple made bold moves beyond the company's traditional strengths. The Power PC chip in the Mac was phased out in favor of Intel chips, the same hardware that runs Microsoft Windows machines. (Indeed, the Mac was also given a utility that allowed it to run Windows.) This potentially opened the Mac to a much wider range of software.

The biggest move, however, was into media, first with powerful video-authoring software for home users as well as professionals, then with the tiny iPod that redefined the portable media player (see MUSIC AND VIDEO PLAYERS, DIGITAL). At the same time, Apple entered the digital music business in a big way with the iTunes store (see MUSIC AND VIDEO DISTRIBUTION, ONLINE). In 2007 Apple charged into the mobile communications market (see SMARTPHONE) with the innovative if expensive iPhone. So far the market has responded positively to Jobs's initiatives, with Apple stock increasing in value more than 10 times between 2003 and 2006.

While Jobs is brash and unconventional (reflecting his countercultural roots), critics have accused him of egotism and of having an overly aggressive (and abrasive) managerial style. Jobs has also been the subject of lingering investigations into his receiving discounted Apple stock options, failing to report the resulting taxable income, and correspondingly overstating Apple's earnings. In December 2006 Apple's internal investigation cleared Jobs of responsibility for these issues, and the options were never exercised. Whatever the future brings, Steve Jobs has an assured place in the history of entrepreneurship and innovation in computing.

Further Reading

"Bill Gates and Steve Jobs" [on-stage interview]. *All Things Digital,* May 30, 2007. Available online. URL: http://d5.allthingsd.com/20070530/d5-gates-jobs-interview/. Accessed August 11, 2007.

Jobs, Steve. "Steve Jobs: Oral History" [interview]. April 20, 1995. Available online. URL: http://www.cwheroes.org/archives/histories/jobs.pdf. Accessed August 11, 2007.

Markoff, John. *What the Dormouse Said: How the Sixties Counterculture Shaped the Personal Computer Industry.* New York: Penguin, 2005.

Young, Jeffrey S., and William L. Simon. *iCon: Steve Jobs, the Greatest Second Act in the History of Business.* Hoboken, N.J.: Wiley, 2005.

journalism and computers

The pervasive use of computers and the Internet has changed the practice of journalism in many ways. This entry will focus on the general impact of technology on the creation and dissemination of news content. For discussion of software used in the production of publications, see DESKTOP PUBLISHING, and WORD PROCESSING. For the role that journalism plays in the computer industry, see JOURNALISM AND THE COMPUTER INDUSTRY.

RESEARCH AND NEWSGATHERING

The gathering of on-scene information at newsworthy events began to change in the 1980s, when notebook-sized portable computers became available. Instead of having to "file" stories with the newspaper by telegraph or phone, the reporter could write the piece and send it to the newspaper's computer using a phone connection (see MODEM) or later, Internet-based e-mail.

The ability of reporters (particularly investigative reporters) to do in-depth research has been greatly enhanced by the Internet. Traditionally, reporters looking for background material for an assignment could consult printed reference works, their publications' archives of printed articles (the "morgue"), and various public records, usually in paper form. This process was necessarily slow, and it was difficult to widen research to include a greater variety of sources while still remaining timely.

Today most publications produce and store their material electronically and make it available online. Reporters thus have virtually instant access to articles written by their colleagues around the world. Instead of having to rely on a few press releases, position papers, or wire stories, reporters can search the Internet to delve more deeply into the underlying source material, such as original documents or statistics. An increasing number of public records are also available online.

CHANGING STANDARDS AND NEW CHALLENGES

After being submitted electronically, reporters' stories can be edited, revised as necessary, and submitted to the computer-controlled typesetting systems that have now become standard in most publications. Besides saving production costs, computer-based newspaper production also makes it easier to make last-minute changes as well as to create special editions that include regional news.

However, at the same time the greater use of information technology has made print journalists more productive, it has also contributed to trends that continue to challenge the viability of print journalism itself. The nature of the Internet poses new challenges to reporter-researchers. The accuracy of traditionally published books or articles is backed implicitly by the reputation of the publisher as well as that of the author. By offering a wide variety of materials produced outside the mainstream publishing process, often by unknown authors, the Internet can provide a much greater diversity of viewpoints (see also WIKIS AND WIKIPEDIG). The downside is that the reporter-researcher has little assurance of the veracity or accuracy of facts given on unknown Web sites. This creates a greater burden of fact checking in responsible journalism or, alternatively, a relaxation of the traditional standards. (The most famous example of the latter is Matt Drudge, a self-made Internet-based journalist who sometimes dramatically "scooped" his more plodding colleagues but did not adhere to the old journalistic standard of finding two independent sources for each key fact.)

The use of the Internet as both a research tool and a medium of publication is also bound up with the ever-accelerating pace of the "news cycle," or the time it takes for a story to be disseminated and responded to. Broadcast journalism with the advent of 24-hour news networks such as CNN has steadily increased the pace of the broadcast news cycle. Many newspapers and magazines have found having Web sites to be a competitive necessity. The Internet potentially combines the immediacy of broadcast journalism with the ability to use text to convey information in depth. The organization of Web pages (see HYPERTEXT AND HYPERMEDIA) avoids the physical limitations of the printed medium.

In addition to Web sites that mirror and expand the contents of printed newspapers, a number of distinctive Internet-only sites emerged in the mid to late 1990s. Examples include salon.com, an "online newsmagazine" that also includes regular featured columnists and discussion forums. However, the downturn in the Internet-based economy in 2002 made the original idea of having free access supported by advertising less viable. Such sites are now trying to convert to a subscription-based model similar to that of print-based publications, but it is unclear whether they will be able to attract enough paying subscribers.

NEW ALTERNATIVES AND NEW QUESTIONS

The Internet is rapidly changing not only how journalism is produced, but how it is delivered—and indeed, the role and future of the profession itself. Broadcast journalism, already greatly changed by the advent of cable TV networks in the 1990s, has now found itself needing to deliver programs through new channels (see PODCASTING, INTERNET RADIO, and MUSIC AND VIDEO DISTRIBUTION, ONLINE). With "broadcasts" available any time at user request, the news cycle has essentially vanished into a 24/7 reality where wave upon wave of stories is constantly flowing and changing.

The more profound change, though, is in who gets to practice and define journalism. Everyone it seems has something to say online (see BLOGS AND BLOGGING). Bloggers who cover current events (especially politics) at their best represent the latest incarnation of "citizen journalism" (see POLITICAL ACTIVISM AND THE INTERNET). However, issues of objectivity (and the line between activist and journalist) have been raised, as has the question of what legal protections for journalists should apply to bloggers and online news reporters.

In addition to blogs, photo and video sharing sites (see, for example, YOUTUBE) now widely distribute material, often quite controversial, that might once have been ignored by mainstream media. For their part, many mainstream journalists now also maintain blogs through which readers can respond to stories of the day.

At the same time, in an era when a stream of both images and the printed word is on tap 24 hours a day, print journalism faces a shrinking market and the need to justify itself to consumers. The industry has responded since the 1970s by an increasing number of mergers of metropolitan daily newspapers as well as the merging of newspapers into broader-based media companies. Many people have grown up with the daily routine of a newspaper at the breakfast table, and there is still a cachet for prestigious publications such as the *New York Times* and the *Wall Street Journal*. Futurists have predicted that newspapers might eventually be delivered to "electronic book" devices, perhaps through a wireless connection (see E-BOOKS AND DIGITAL LIBRARIES). This might combine the immediacy of the Internet with the physical convenience and portability of a newspaper.

Further Reading

Cornfield, Michael. "Buzz, Blogs, and Beyond: The Internet and the National Discourse in the Fall of 2004." Pew Center for Internet & American Life. Available online. URL: http://www.pewinternet.org/ppt/BUZZ_BLOGS__BEYOND_Final05-16-05.pdf. Accessed August 11, 2007.

Daily KOS. Available online. URL: http://www.dailykos.com. Accessed August 11, 2007.

Gillmor, Dan. *We the Media: Grassroots Journalism by the People, for the People.* Sebastapol, Calif.: O'Reilly, 2004.

Meyer, Philip. *The Vanishing Newspaper: Saving Journalism in the Information Age.* Columbia, Mo.: University of Missouri Press, 2004.

Pew Center for Civic Journalism. Available online. URL: http://www.pewcenter.org. Accessed August 11, 2007.

Quinn, Stephen, and Vincent Filak, eds. *Convergent Journalism: An Introduction—Writing and Producing across Media.* Burlington, Mass.: Focal Press, 2005.

Salon.com. Available online. URL: http://www.salon.com. Accessed August 1, 2007.

Slate.com. Available online. URL: http://www.slate.com. Accessed August 1, 2007.

Wulfemeyer, K. Tim. *Online Newswriting.* Ames, Iowa: Blackwell Publishing, 2006.

journalism and the computer industry

Developments in the computer industry and user community have been chronicled by a great variety of printed and on-line publications. As computer science began to emerge as a discipline in the late 1950s and 1960s, academically oriented groups such as the Association for Computing Machinery (ACM) and Institute for Electrical and Electronics Engineers (IEEE) began to issue both general and special-interest journals. Meanwhile, the computer industry developed both computer science-oriented publications (such as the *IBM Systems Journal*) and independent industry periodicals such as *Datamation*.

The development of microcomputer systems in the mid- to late-1970s was accompanied by a proliferation of varied and often feisty publications. *Byte* magazine, which coined the term *PC* in 1976, became a respected trade publication that introduced new technologies while showcasing what programmers could do with the early systems. The weekly newspaper *InfoWorld* provided more immediate and detailed coverage of industry developments, and was joined by similar publications such as *Information Week* and *Computerworld*. Meanwhile, technically savvy programmers and do-it-yourself engineers turned to such publications as the exotically named *Dr. Dobbs' Journal of Computer Calisthenics and Orthodontia* (eventually shortened to *Dr. Dobbs' Journal*). Many groups of people who owned particular systems (see USER GROUPS) also published their own newsletters with technical tips.

The success of the IBM PC family of computers established a broad-based consumer computing market. It was accompanied by the success of *PC Magazine,* which addresses a wide spectrum of both general consumers and "power users." As the revenue for the PC industry grew in the 1990s, the trade publications grew fatter with advertising. The popularity of the Internet and particularly the World Wide Web in the latter part of the decade provided niches for a spate of new publications including *Internet World* and *Yahoo! Internet Life.* At the same time, many traditional publications began to offer expanded content via Web sites. For example, Ziff Davis, publisher of *PC Magazine* and other computer magazines created ZDNet, which offered a large amount of content from the magazines plus expanded news and extensive shareware and utility libraries.

Like earlier technological developments, the PC and the Internet have also spawned cultural expressions. The culture growing around the Internet and a generation of young programmers, artists, and writers saw expression in another genre of publications, ranging from small, eclectic printed or Web "zines" to the slick *Wired* magazine.

FROM PRINT TO ONLINE

Many of the pressures on mainstream journalism also apply to computer industry journalism. As computer hardware became a commodity with lower profit margins, and with the shift to e-commerce and online activity, many print magazines have folded or at least shrunk. In 1998 the venerable *Byte* became an online-only publication, a path finally followed by *InfoWorld* in 2007.

Online sites such as ZDNET and CNET now carry in-depth news and product reviews. Slashdot ("New for Nerds") is particularly popular among programmers. As with mainstream journalism, blogs also play an important part in professional and industry journalism in the computing field.

Further Reading

CNET. Available online. ULR: http://www.cnet.com. Accessed August 11, 2007.

"Computer Industry: Trade Magazines." Yahoo.com. Available online. URL: http://dir.yahoo.com/Business_and_Economy/

Business_to_Business/Computers/Industry_Information/ Trade_Magazines/. Accessed August 11, 2007.

"Tag: Computer Industry" [blogs]. Available online. URL: http:// it.wordpress.com/tag/computer-industry. Accessed August 11, 2007.

"Top 100 Computer and Software Magazines." Available online. URL: http://netvalley.com/top100mag.html. Accessed August 11, 2007.

ZDNET. Available online. URL: http://www.zdnet.com. Accessed August 11, 2007.

Joy, Bill
(1955–)
American
Software Engineer, Entrepreneur

Bill Joy developed many of the key utilities used by users and programmers on UNIX systems (see UNIX). He then became one of the industry's leading entrepreneurs and later, a critic of some aspects of computer technology.

As a graduate student in computer science and electrical engineering at the University of California at Berkeley in the 1970s, Joy worked with UNIX designer Ken Thompson (1943–) to add features such as virtual memory (paging) and TCP/IP networking support to the operating system (the latter work was sponsored by DARPA, the Defense Advanced Research Projects Agency). These developments eventually led to the distribution of a distinctive version of UNIX called Berkeley Software Distribution (BSD), which rivaled the original version developed at AT&T's Bell Laboratories. The BSD system also popularized features such as the C shell (a command processor) and the text editors "ex" and "vi." (See SHELL.)

As opposed to the tightly controlled AT&T version, BSD UNIX development relied upon what would become known as the open-source model of software development (see OPEN-SOURCE MOVEMENT). This encouraged programmers at many installations to create new utilities for the operating system, which would then be reviewed and integrated by Joy and his colleagues. BSD UNIX gained industry acceptance and was adopted by the Digital Equipment Corporation (DEC), makers of the popular VAX series of minicomputers.

In 1982, Joy left UC Berkeley and co-founded Sun Microsystems, a company that became a leader in the manufacture of high-performance UNIX-based workstations for scientists, engineers, and other demanding users. Even while becoming a corporate leader, he continued to refine UNIX operating system facilities, developing the Network File System (NFS), which was then licensed for use not only on UNIX systems but on VMS, PC-DOS, and Macintosh systems. Joy's versatility also extended to hardware design, where he helped create the Sun SPARC reduced instruction set (RISC) microprocessor that gave Sun workstations much of their power.

In the early 1990s, Joy turned to the growing world of Internet applications and embraced Java, a programming language created by James Gosling (see JAVA). He devel-

oped specifications, processor instruction sets, and marketing plans. Java became a very successful platform for building applications to run on Web servers and browsers and to support the needs of e-commerce. As Sun's chief scientist since 1998, Joy has led the development of Jini, a facility that would allow not just PCs but many other "Java-enabled" devices such as appliances and cell phones to communicate with one another.

Recently, however, Joy has expressed serious misgivings about the future impact of artificial intelligence and related developments on the future of humanity. Joy remains proud of the achievements of a field to which he has contributed much. However, while rejecting the violent approach of extremists such as Unabomber Theodore Kaczynski, Joy points to the potentially devastating unforeseen consequences of the rapidly developing capabilities of computers. Unlike his colleague Ray Kurzweil's optimistic views about the coexistence of humans and sentient machines, Joy points to the history of biological evolution and suggests that superior artificial life forms will displace humans

Bill Joy made key contributions to the Berkeley Software Distribution (BSD) version of UNIX, including developing its Network File System (NFS). As a cofounder of Sun Microsystems, Joy then helped develop innovative workstations and promoted Java as a major language for developing Web applications. (BILL JOY, KLEINER PERKINS CAULFIELD & BYERS)

who will be unable to compete with them. He believes that given the ability to reproduce themselves, intelligent robots or even "nanobots" (see NANOTECHNOLOGY) might soon be uncontrollable.

Joy also expresses misgivings about biotechnology and genetic engineering, seen by many as the dominant scientific and technical advance of the early 21st century. He has proposed that governments develop institutions and mechanisms to control the development of such dangerous technologies, drawing on the model of the agencies that have more or less successfully controlled the development of nuclear energy and the proliferation of nuclear weapons for the past 50 years. (For contrasting views see KURZWEIL, RAY and SINGULARITY, TECHNOLOGICAL.)

In 2003 Joy left Sun and became a venture capitalist, specializing in technologies and projects to combat what he sees as serious global dangers, such as pandemic disease and the possibility of bioterrorism.

Joy received the ACM Grace Murray Hopper Award for his contributions to BSD UNIX before the age of 30. In 1993, he was given the Lifetime Achievement Award of the USENIX Association, "For profound intellectual achievement and unparalleled services to the UNIX community."

Further Reading

Brown, John Seely and Paul Duguid. "A Response to Bill Joy and the Doom-and-Gloom Technofuturists." Available online. URL: http://www.aaas.org/spp/rd/ch4.pdf. Accessed August 12, 2007.

Joy, Bill. "Why the Future Doesn't Need Us." *Wired,* 8.04, April 2000. Available online. URL: http://www.wired.com/wired/archive/8.04/joy.html. Accessed August 12, 2007.

Joy, Bill, et al. *The Java Language Specification.* 3rd ed. Upper Saddle River, N.J.: Prentice Hall, 2005. Available online. URL: http://java.sun.com/docs/books/jls/download/langspec-3.0.pdf. Accessed August 12, 2007.

O'Reilly, Tim. "A Conversation with Bill Joy." O'Reilly Network, February 12, 2001. Available online. URL: http://www.openp2p.com/pub/a/p2p/2001/02/13/joy.html. Accessed August 12, 2007.

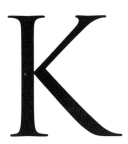

Kay, Alan
(1940–)
American
Computer Scientist

Alan Kay developed a variety of innovative concepts that changed the way people use computers. Because he devised ways to have computers accommodate users' perceptions and needs, Kay is thought by many to be the person most responsible for putting the "personal" in personal computers. Kay also made important contributions to object-oriented programming, changing the way programmers organized data and procedures in their work.

Kay's father developed prostheses (artificial limbs) and his mother was an artist and musician. These varied perspectives contributed to Kay's interest in interaction with and perception of the environment. In the late 1960s, while completing work for his Ph.D. at the University of Utah, Kay developed his first innovations in both areas. He helped Ivan Sutherland with the development of a program called Sketchpad that enabled users to define and control onscreen objects, while also working on the development of Simula, a language that helped introduce new programming concepts (see SIMULA and OBJECT-ORIENTED PROGRAMMING). Indeed, Kay coined the term *object-oriented* in the late 1960s. He viewed programs as consisting of objects that contained appropriate data that could be manipulated in response to "messages" sent from other objects. Rather than being rigid, top-down procedural structures, such programs were more like teams of cooperating workers. Kay also worked on parallel programming, where programs carried out several tasks simultaneously (see CONCURRENT PROGRAMMING). He likened this structure to musical polyphony, where several melodies are sounded simultaneously.

Kay participated in the Defense Advanced Research Projects Agency (DARPA)—funded research that was leading to the development of the Internet. One of these DARPA projects was FLEX, an attempt to build a computer that could be used by nonprogrammers through interacting with onscreen controls. While the bulky technology of the late 1960s made such machines impracticable, FLEX incorporated some ideas that would be used in later PCs, including multiple onscreen windows.

During the 1970s, Kay worked at the innovative Xerox Palo Alto Research Center (PARC). Kay designed a laptop computer called the Dynabook, which featured high-resolution graphics and a graphical user interface. While the Dynabook was only a prototype, similar ideas would be used in the Alto, a desktop personal computer that could be controlled with a new pointing device, the mouse (see ENGELBART, DOUGLAS). A combination of high price and Xerox's less than aggressive marketing kept the machine from being successful commercially, but Steven Jobs (see JOBS, STEVEN) would later use its interface concepts to design what would become the Macintosh.

On the programming side Kay developed Smalltalk, a language that was built from the ground up to be truly object-oriented (see SMALLTALK). Kay's work showed that there was a natural fit between object-oriented programming and an object-oriented user interface. For example, a button in a screen window could be represented by a button object in the program, and clicking on the screen button could send a message to the button program object, which would be programmed to respond in specific ways.

After leaving Xerox PARC in 1983, Kay briefly served as chief scientist at Atari and then moved to Apple, where he worked on Macintosh and other advanced projects. In 1996, Kay became a Disney Fellow and Vice President of Research and Development at Walt Disney Imagineering. In 2001 Kay founded Viewpoints Research Institute, a non-profit organization devoted to developing advanced learning environments for children. One such project is Squeak, a streamlined but powerful version of Smalltalk that Kay started developing in 1995. Another, eToys, is a multiplatform, media-rich, environment that can be used for education or "just" play. Behind it all is Kay's continuing effort to do no less than reinvent programming and peoples' relationship to computer environments.

Kay's numerous honors include the ACM Turing Award (2003) for contributions to object-oriented programming and the Kyoto Prize (2004).

Further Reading

Alter, Allan E. "Alan Kay: The PC Must Be Revamped—Now." CIO Insight. Available online. URL: http://www.cioinsight.com/article2/0,1540,2089567,00.asp. Accessed August 1, 2007.

Gasch, Scott. "Alan Kay." Available online. URL: http://ei.cs.vt.edu/~history/GASCH.KAY.HTML. Accessed August 12, 2007.

Kay, Alan. "The Early History of Smalltalk." In Thomas J. Bergin, Jr., and Richard G. Gibson, Jr., eds. *History of Programming Languages II*. New York: ACM; Reading, Mass.: Addison-Wesley, 1996.

Shasha, Dennis, and Cathy Lazere, eds. *Out of their Minds: The Lives and Discoveries of 15 Great Computer Scientists*. New York: Copernicus, 1995.

Viewpoints Research Institute. Available online. URL: http://www.vpri.org/. Accessed August 12, 2007.

kernel

The idea behind an operating system kernel is that there is a relatively small core set of "primitive" functions that are necessary for the operation of system services (see also OPERATING SYSTEM). These functions can be provided in a single component that can be adapted and updated as desirable. The fundamental services include:

- Process control—scheduling how the processes (programs or threads of execution within programs) share the CPU, switching execution between processes, creating new processes, and terminating existing ones (see MULTITASKING).

- Interprocess communication—sending "messages" between processes enabling them to share data or coordinate their data processing.

- Memory management—allocating and freeing up memory as requested by processes as well as implementing virtual memory, where physical storage is treated as an extension of main (RAM) memory. (See MEMORY MANAGEMENT.)

- File system services—creating, opening, reading from, writing to, closing, and deleting files. This includes main-

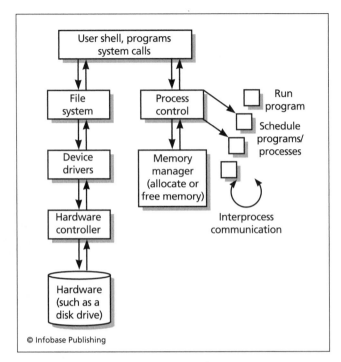

The kernel is an intermediary between users and programs and the hardware system. It provides the functions necessary for allocating and controlling processes and system resources.

taining a structure (such as a list of nodes) that specifies the relationship between directories and files. (See FILE.)

In addition to these most basic services, some operating systems may have larger kernels that include security functions (such as maintaining different classes of users with different privileges), low-level support for peripheral devices, and networking (such as TCP/IP).

The decision about what functions to include in the kernel and which to provide through device drivers or system extensions is an important part of the design of operating systems. Many early systems responded to the very limited supply of RAM by designing a "microkernel" that could fit entirely in a small amount of memory reserved permanently for it. Today, with memory a relatively cheap resource, kernels tend to be larger and include functions that are paged dynamically into and out of memory.

In the UNIX world (and particularly with LINUX) the kernel is constantly being improved through informal collaborative efforts. Many Linux enthusiasts regularly install new versions of the kernel in order to stay on the "leading edge," while more conservative users can opt for waiting until the next stable version of the kernel is released.

Further Reading

Bovet, Daniel, and Marco Cesati. *Understanding the Linux Kernel*. 3rd ed. Sebastapol, Calif.: O'Reilly, 2005.

Love, Robert. *Linux Kernel Development*. 2nd ed. Indianapolis: Novell Press, 2005.

Torvalds, Linus. "LinuxWorld: The Story of the Linux Kernel." *linuxtoday*. Available online. URL: http://www.linuxtoday.com/developer/1999032500910PS. Accessed August 12, 2007.

keyboard

Although most of today's personal computers feature a point-and-click graphical interface (see USER INTERFACE and MOUSE) the keyboard remains the main means for entering text and other data into computer applications. The modern computer keyboard traces its ancestry to the typewriter, and the layout of its alphabetic and punctuation keys remains that devised by typewriter pioneer Christopher Latham Sholes in the late 1860s.

The principal difference in operation is that while a typewriter needs only to transfer the impression of a key through a ribbon onto a piece of paper, the computer keyboard must generate an electrical signal that uniquely identifies each key. This technology dates back to the 1920s with the adoption of the teletypewriter (often known by the brand name Teletype), which allowed operators to type text at a keyboard and send it over telephone lines to be printed. The transmissions used the Baudot character code, which used five binary (off or on) positions to encode letters and characters. This gave way to the ASCII code in the 1960s (see CHARACTERS AND STRINGS) at about the time that remote time-sharing services allowed users to interact with computers through a Teletype connection.

The modern personal computer keyboard was standardized in the mid-1980s when IBM released the PC AT. This expanded keyboard now has 101 or 102 keys. It supplements the standard typewriter keys with cursor-control (arrow) keys, scroll control keys (such as Page Up and Page Down), a dozen function keys that can be assigned to commands by software, and a separate calculator-style keyboard for numeric data entry. During the 1990s, Microsoft introduced a few extra keys for Windows-specific functions.

The advent of laptop (or notebook) computers required some compromises. The keys are generally smaller, although on the better units they are still far enough apart to allow for comfortable touch-typing. Laptops often combine the function keys and cursor control keys with the regular keys, using a special "Fn" key to shift between them.

In recent years, there has been some interest in adopting an alternative key layout devised by August Dvorak in the 1950s. The theory behind this layout was that arranging the keyboard so the most commonly used keys were directly under the fingers would be more efficient than the Sholes layout, which legend claims was devised primarily to slow down typists to a speed that early typewriters could handle without jamming. However, researchers have generally been unable to find a significant improvement in either performance or ergonomics between use of the standard and Dvorak layouts, and the latter has not caught on commercially.

Concern with repetitive strain injury (RSI) has led to experiments in designing a keyboard more suited to the human wrist and hand (see ERGONOMICS OF COMPUTING). Some designs such as the Microsoft Natural Keyboard divide the layout into left and right banks of keys and angle them toward one another to reduce strain on the wrists. An extreme form of the design actually breaks the keyboard into two pieces. Such extreme designs have not found wide acceptance.

The Microsoft Natural Multimedia Keyboard features access to functions needed by today's computer users along with an ergonomic layout designed to help reduce typing stress. (MICROSOFT CORPORATION)

It is possible that the further development of voice recognition software might allow spoken dictation to supplant the keyboard for data entry. Currently, however, such technology is limited in speed and accuracy (see SPEECH RECOGNITION AND SYNTHESIS).

With the increasingly popular mobile devices (see PDA and SMARTPHONE), keyboards are sometimes dispensed with entirely. For light data entry (such as for e-mail and text messaging), a small version of the standard keyboard can be used. (In such cases users can type with their thumbs.) With touch-sensitive screens on mobile devices, a "virtual keyboard" can be displayed on the screen; however, the lack of tactile feedback means this data-entry method takes some getting used to. One can also obtain a keyboard that can wirelessly connect to such a device to allow for more extensive data entry (see BLUETOOTH).

Further Reading

"Computer Keyboard Design." Cornell University Ergonomics Web. Available online. URL: http://ergo.human.cornell.edu/ahtutorials/ckd.htm. Accessed August 12, 2007.

Lundmark, Torbjorn. *Qwirky Qwerty: The Story of the Keyboard @ Your Fingertips.* Sydney, Australia: New South Wales University Press, 2002.

Kleinrock, Leonard

(1934–   )
American
Engineer, Computer Scientist

Every day billions of e-mails, text messages, and media files are sent over the worldwide Internet. The infrastructure that allows the efficient transmission of this vast data traffic is largely based on the system of packet-switching and routing invented by Leonard Kleinrock.

Kleinrock was born in 1934 and grew up in New York City. When he was only six years old Kleinrock built a crystal radio, the first of many electronics projects, built from cannibalized old radios and other equipment. Kleinrock attended the Bronx High School of Science, home of many of the nation's top future engineers. However, when it came time for college, the family had no money to pay for his higher education, so he attended night courses at the City College of New York while working as an electronics technician and later as an engineer. Kleinrock graduated first in his class in 1957 and earned a fully paid fellowship to the Massachusetts Institute of Technology.

At MIT Kleinrock became interested in finding ways for computers and their users to communicate with each other. The idea of computer networking was in its infancy, but he submitted a proposal in 1959 for Ph.D. research in network design.

In 1961 Kleinrock published his first paper, "Information Flow in Large Communication Nets." Existing telephone systems did what was called "circuit switching": To establish a conversation, the caller's line is connected to the receiver's, forming a circuit that existed for the duration of the call. This also meant that the circuit would not be available to anyone else, and that if something was wrong with the connection there was no way to route around the problem.

Kleinrock's basic idea was to set up data connections that would be shared among many users as needed. Instead of the whole call (or data transmission) being assigned to a particular circuit, it would be broken up into packets that could be sent along whatever circuit was the most direct. If there was a problem, the packet could be resent on an alternative route. This form of "packet switching" provided great flexibility as well as more efficient use of the available circuits. Kleinrock further elaborated his ideas in his dissertation, for which he was awarded his Ph.D. in 1963. The following year MIT published his book *Communications Nets,* the first full treatment of the subject.

Kleinrock joined the faculty at the University of California, Los Angeles. In 1968 the Defense Department's Advanced Research Projects Agency (ARPA) asked him to design a packet-switched network that would be known as ARPANET. The computers on the network would be connected using special devices called Interface Message Processors (IMPs). The overall project was under the guidance and supervision of one of Kleinrock's MIT office mates, Lawrence Roberts.

On October 29, 1969, Kleinrock and his assistants sent the first data packets between UCLA and Stanford over phone lines. Their message, the word "login," was hardly as dramatic as Alexander Graham Bell's "Watson, come here, I need you!" Nevertheless, a form of communication had been created that in a few decades would change the world as much as the telephone had done a century earlier.

The idea of computer networking did not catch on immediately, however. Besides requiring a new way of thinking about the use of computers, many computer administrators were concerned that their computers might be swamped with users from other institutions, or that they might ultimately lose control over the use of their machines. Kleinrock worked tirelessly to convince institutions to join the nascent network. By the end of 1969 there were just four ARPANET "nodes": UCLA, the Stanford Research Institute,

UC Santa Barbara, and the University of Utah. By the following summer, there were ten.

During the 1970s Kleinrock trained many of the researchers who would advance the technology of networking. While Kleinrock's first network was not the Internet we know today, it was an essential step in its development. In successfully establishing communication using the packet-switched ARPANET, Kleinrock showed that such a network was practicable.

By the early 1990s Kleinrock was looking toward a future where most network connections were wireless and accessible through a variety of computerlike devices such as handheld "palmtop" computers, cell phones, and others not yet imagined. In such a network the intelligence or capability is distributed throughout, with devices communicating seamlessly so the user no longer need be concerned about what particular gadget he or she is using. By the middle of the following decade, much of this vision had become reality.

Although his name is not well known to the general public, Kleinrock has won considerable recognition within the technical community. This includes Sweden's L. M. Ericsson Prize (1982), the Marconi Award (1986), and the National Academy of Engineering Charles Stark Draper Prize (2001).

Further Reading

Hafner, Katie, and Matthew Lyon. *Where Wizards Stay Up Late: The Origins of the Internet.* New York: Simon and Schuster, 1996.
Kleinrock, Leonard. "Information Flow in Large Communication Nets." Available online. URL: http://www.lk.cs.ucla.edu/LK/Bib/REPORT/PhD/proposal.html. Accessed May 3, 2007.
"Leonard Kleinrock's Home Page." Available online. URL: http://www.lk.cs.ucla.edu/. Accessed May 3, 2007.

knowledge representation

The earliest concern of computer science was the representation of "raw" data such as numbers in programs (see DATA TYPES). Such data can be used in calculations, and actions taken based on tests of data values, using branching (IF) or looping structures.

However, facts are more than data. A fact is an *assertion,* for example about a relationship, as in "Joe is a son of Mike," often expressed in a form such as son (Joe, Mike). Implications can also be defined as proceeding from facts, such as

```
son (Joe, Mike) implies father (Mike, Joe) or
son (Joe, Mike) and son (Mike, Phil) implies
grandson (Joe, Phil)
```

While it can be expressed in a variety of different forms of notation, this *predicate calculus* forms the basis for many automated reasoning systems that can operate on a "knowledge base" of assertions, prove the validity of a given assertion, and even generate new conclusions based upon existing knowledge (see also EXPERT SYSTEMS).

An alternative form of knowledge representation used in artificial intelligence programs is based on the idea of frames. A frame is a structure that lists various characteristics or relationships that apply to a given individual or class. For example, the individual "cat" might have a frame that includes characteristics such as "warm-blooded" and "bears live young." In turn, these characteristics are also assigned to the class "mammal" such that any individual having those characteristics belongs to that class. A program can then follow the linkages and conclude that a cat is a mammal. Linkages can also be diagrammed as a "semantic network" in a structure called a directed graph, with the lines between nodes labeled to show relationships.

Knowledge representation systems have different considerations depending on their intended purpose. A KR system in an academic research setting might be intended to demonstrate *completeness*: that is, it can generate all possible conclusions from the facts given. However, expert systems designed for practical use usually do not attempt to generate all possible conclusions (which might be computationally impracticable) but to generate *useful* conclusions that are likely to serve the needs of the knowledge consumer.

It is also important to note that epistemology (the theory of knowledge) plays an important role in understanding and evaluating KR systems. As an example, the assertion "Mary believes she is 600 years old" might be a fact (Mary is observed to hold such a belief), but the contents of the belief are presumably not factual. The context of this belief might also be different if Mary is an adult as opposed to being a five-year-old child. Similarly, ontological (state of being) considerations can also complicate the evaluation of assertions. For example, should a fire be treated as an object in itself, a process, or an attribute of a burning object? Knowledge representation thus intertwines philosophy and computer science.

The booming interest in extracting new patterns from data (see DATA MINING) and the effort to encode more knowledge into Web documents (see ONTOLOGIES AND DATA MODELS, SEMANTIC WEB, and XML) all involve applications of knowledge representation.

Perhaps the most ambitious knowledge representation project (and the longest-lasting one) has been Cyc (short for Encyclopedia). Headed by AI researcher Douglas Lenat, the object of Cyc is to create a massive network representing the relationships and characteristics of millions of objects and concepts found in peoples' daily lives and work. Ideally a wide variety of programs (both specialized and general purpose) will be able to use this knowledge base (see EXPERT SYSTEM). Projects such as Cyc and the Web Ontology Language (OWL) also offer the possibility of a much more intelligent Web search (see SEARCH ENGINE) as well as systems that can automatically summarize news stories and other material.

Further Reading

Brachman, Ronald, and Hector Levesque. *Knowledge Representation and Reasoning.* San Francisco: Morgan Kaufmann, 2004.
Cycorp. Available online. URL: http://www.cyc.com/. Accessed August 12, 2007.
Davis, Randall, Howard Shrobe, and Peter Szolovits. "What Is a Knowledge Representation?" *AI Magazine* 14 (1993): 17–33. Available online. URL: http://groups.csail.mit.edu/medg/ftp/psz/k-rep.html. Accessed August 12, 2007.
Lacy, Lee W. *Owl: Representing Information Using the Web Ontology Language.* Victoria, B.C., Canada: Trafford, 2005.
Makahfi, Pejman. "Introduction to Knowledge Modeling." Available online. URL: http://www.makhfi.com/KCM_intro.htm. Accessed August 12, 2007.

Knuth, Donald

(1938–   )
American
Computer Scientist

Donald Knuth has contributed to many aspects of computer science, but his most lasting contribution is his monumental work, *The Art of Computer Programming,* which is still in progress.

Born in Milwaukee on January 10, 1938, Knuth's initial background was in mathematics. He received his master's degree at the Case Institute of Technology in 1960 and his Ph. D. from the California Institute of Technology (Caltech) in 1963. As a member of the Caltech mathematics faculty Knuth became involved with programming and software engineering, serving both as a consultant to the Burroughs Corporation and as editor of the Association for Computing Machinery (ACM) publication *Programming Languages.* In 1968, Knuth confirmed his change of career direction by becoming professor of computer science at Stanford University.

In 1971, Knuth published the first volume of *The Art of Computer Programming* and received the ACM Grace Murray Hopper Award. His broad contributions to the field as well as specific work in the analysis of algorithms and computer languages garnered him the ACM Turing Award, the most prestigious honor in the field. Knuth also did important work in areas such as LR (left-to-right, rightmost) parsing, a context-free parsing approach used in many program language interpreters and compilers (see PARSING).

However, Knuth then turned away from writing for an extended period. His primary interest became the development of a sophisticated software system for computer-generated typography. He developed both the TeX document preparation system and METAFONT, a system for typeface design that was completed during the 1980s. TeX found a solid niche in the preparation of scientific papers, particularly in the fields of mathematics, physics, and computer science where it can accommodate specialized symbols and notation.

Knuth did return to *The Art of Computer Programming* and by the late 1990s he had completed two more of a projected seven volumes. With his broad interests and contributions and "big picture" approach to the evaluation of programming languages, algorithms, and software engineering methodologies, Knuth can fairly be described as one of the "Renaissance persons" of the computer science field. His numerous awards include the ACM Turing Award (1974), IEEE Computer Pioneer Award (1982), American Mathematical Society's Steele Prize (1986), and the IEEE's John von Neumann Medal (1995).

Further Reading

Frenkel, Karen A. "Donald E. Knuth: Scholar with a Passion for the Particular." *Profiles in Computing, Communications of the ACM,* vol. 30, no. 10, October 1987.

———. *The Art of Computer Programming.* 3rd ed. vols. 1–3. Reading, Mass.: Addison-Wesley, 1998.

Knuth, Donald E. *Literate Programming.* Stanford, Calif.: Center for the Study of Language and Information, 1992.

———. *Things a Computer Scientist Rarely Talks About.* Stanford, Calif.: CSLI Publications, 2001.

Slater, Robert. *Portraits in Silicon.* Cambridge, Mass.: MIT Press, 1987.

Kurzweil, Ray

(1948–)
American
Inventor, Futurist

Ray Kurzweil began his career as an inspired inventor who brought words to the blind and new kinds of sounds to musicians. Drawing upon his experience with the rapid

Prolific inventor and futurist Ray Kurzweil believes that technology will soon take an exponential leap called "the singularity." (MELANIE STETSON FREEMAN / *THE CHRISTIAN SCIENCE MONITOR* / GETTY IMAGES)

progress of technology, Kurzweil then wrote a series of books that predicted a coming breakthrough into a world shared by advanced intelligent machines and enhanced human beings.

Kurzweil was born on February 12, 1948, in Queens, New York, to an extremely talented family. Kurzweil's father, Fredric, was a concert pianist and conductor. Kurzweil's mother, Hanna, was an artist, and one of his uncles was an inventor. By the time he was 12, Kurzweil was building and programming his own computer. He wrote a statistical program that was so good that IBM distributed it as well as a music-composing program. The latter earned him first prize in the 1964 International Science Fair and a meeting with President Lyndon B. Johnson in the White House. Kurzweil even appeared on the television show *I've Got a Secret.*

In 1967 Kurzweil enrolled in the Massachusetts Institute of Technology, majoring in computer science and literature. By the time he received his B.S. in 1970, Kurzweil had met some of the most influential thinkers in artificial intelligence research, including Marvin Minsky, whom he looked to as a mentor (see MINSKY, MARVIN). Kurzweil had become fascinated with the use of AI to aid and expand human potential. In particular, he focused on pattern recognition, or the ability to classify or recognize patterns such as the letters of the alphabet on a page of text.

Early character-recognition technology had been limited because it could only match very precise shapes, making it impractical for reading most printed material. Kurzweil, however, used his knowledge of expert systems and other AI principles to develop a program that could use general rules and relationships to "learn" to recognize just about any kind of text (see OCR). This program, called Omnifont, would be combined with the flatbed scanner (which Kurzweil invented in 1975) to create a system that could scan text and convert the images into the corresponding character codes, suitable for use with programs such as word processors.

A chance conversation with a blind fellow passenger on a plane convinced Kurzweil that he could build a machine that could scan text and read it out loud. Kurzweil would combine his scanning technology with a speech synthesizer (see SPEECH RECOGNITION AND SYNTHESIS). Kurzweil had to create an expert system with hundreds of rules for properly voicing the words in the text.

In 1976 Kurzweil was able to announce the Kurzweil Reading Machine (KRM). Soon after the machine's debut, Kurzweil struck up a friendship with the legendary blind pop musician Stevie Wonder. They shared an interest in musical instruments and music synthesis. Existing analog synthesizers were very versatile, but their output sounded "thin" and artificial compared to the rich overtones in the sound of a piano or guitar. Kurzweil was able to create a much more realistic synthesizer sound using digital rather than analog technology.

The first Kurzweil synthesizer, the K250, was released in 1983. His machine was the result of considerable research in digitally capturing and representing the qualities of notes from particular instruments, including the "attack," or initial building of sound, the "decay," or decline in the sound, the sustain, and the release (when the note is ended.) The resulting sound was so accurate that professional orchestra conductors and musicians could not distinguish the synthesized sound from that of the real instruments.

Throughout the 1980s and 1990s Kurzweil applied his boundless inventiveness to a number of other challenges, including speech recognition. The reverse of voice synthesis, speech recognition involves the identification of phonemes (and thus words) in speech that has been converted into computer sound files. Kurzweil sees a number of powerful technologies being built from voice recognition and synthesis, including telephones that automatically translate speech and devices that can translate spoken words into text in real time for deaf people. He also believes that the ability to control computers by voice command, which is currently rather rudimentary, should also be greatly improved.

TECHNOLOGICAL APOCALYPSE?

During the 1990s, though, much of Kurzweil's interest turned from inventing the future to considering its likely course. His 1990 book *The Age of Intelligent Machines* offered a popular account of how AI research would change many human activities. In 1999 Kurzweil published *The Age of Spiritual Machines*. It made the provocative claim that, by the middle of the 21st century, machine intelligence would surpass that of humans. Kurzweil revisited the topic in his latest book, *The Singularity Is Near* (2005). The title seems to consciously echo the apocalyptic language of a prophet predicting the last judgment or the coming of a messiah. The word "singularity" is intended to describe the effects of relentless, ever-increasing technological progress that eventually reaches a sort of "critical mass" and changes the world beyond all recognition (see SINGULARITY, TECHNOLOGICAL).

As he depicts life in 2009, 2019, 2029, and finally 2099, Kurzweil portrays a world in which sophisticated AI personalities become virtually indistinguishable from humans and can serve people as assistants, advisers, and even lovers. Meanwhile, neural implants will remove the obstacles of handicaps such as blindness, deafness, or lack of mobility (see NEURAL INTERFACE). Other implants will greatly enhance human memory, allow for the instant download of knowledge, and function as "natural" extensions to the brain. (For critics of such "strong AI" claims see DREYFUS, HUBERT and WEIZENABUM, JOSEPH.)

Kurzweil continues to engage in provocative projects. Under the slogan "live long enough to live forever," he is researching and marketing various supplements intended to promote longevity, and he reportedly monitors his own diet and bodily functions carefully.

Whatever the future brings, Ray Kurzweil has become one of America's most honored inventors. Among other awards, he has been elected to the Computer Industry Hall of Fame (1982) and the National Inventors Hall of Fame (2002). He has received the ACM Grace Murray Hopper Award (1978), Inventor of the Year Award (1988), the Louis Braille Award (1991), the National Medal of Technology (1999), and the MIT Lemelson Prize (2001).

Further Reading

Henderson, Harry. *Artificial Intelligence: Mirrors for the Mind.* New York: Chelsea House, 2007.

Joy, Bill. "Why the Future Doesn't Need Us." *Wired Magazine* 8 (April 2000). Available online. URL: http://www.wirednews.com/wired/archive/8.04/joy.html. Accessed May 5, 2007.

Kurzweil, Raymond. The *Age of Spiritual Machines: When Computers Exceed Human Intelligence.* New York: Putnam, 1999.

———. *The Singularity Is Near.* New York: Viking, 2005.

KurzweilAI.net. Available online. URL: http://www.kurzweilai.net/. Accessed May 5, 2007.

Kurzweil Technologies. Available online. URL: http://www.kurzweiltech.com/ktiflash.html. Accessed May 5, 2007.

Richards, Jay, ed. *Are We Spiritual Machines?: Ray Kurzweil vs. the Critics of Strong AI.* Seattle, Wash.: Discovery Institute, 2001.

L

LAN **LAN** *See* LOCAL AREA NETWORK.

language translation software

Anyone who has learned a new language has also gained an appreciation for how difficult it is to translate from one language to another while preserving the intent, meaning, and context of the original. Not surprisingly, developing software to perform this task, often called "machine translation" (MT), has also proven to be difficult. (For a more general discussion of how languages can be represented or studied using a computer, see LINGUISTICS AND COMPUTING.)

RULES-BASED APPROACHES

There are several approaches that can be taken to automatic language translation. A rules-based system parses the original text to construct an intermediate representation. The program then "transfers" the represented structure to an equivalent structure in the target language, drawing upon extensive lexicons (dictionaries) containing such things as phrase structures, word structures (morphology), and semantics (meanings). Developing this extensive knowledge base and the rules for manipulating it is the most challenging part of developing rules-based language translation systems. (For more on the general process of computer "understanding" of language, see NATURAL LANGUAGE PROCESSING.)

Generally a translation produced by a rules-based system will be intelligible to a speaker of the target language, who will be able to understand the broad meaning of the original text. However, it is likely to sound "awkward" and miss certain nuances.

A simplified approach is based on a dictionary of words or phrases and their meanings. Each source word or phrase is simply looked up and converted to its equivalent in the target language. Because it does not deal with grammatical structure or context, this method is not very satisfactory except perhaps for translating simple lists or catalogues.

STATISTICAL APPROACHES

The other main approach to automatic translation relies on statistical analysis of a large body of text (corpus) that is already translated into two languages. For example, the Bayes theorem (see BAYESIAN ANALYSIS) can be used to estimate the probability that string A in French (for example, "c'est un chien") will occur in the English version as string A' "it's a dog."). Depending on the application, the same approach can be applied word for word, phrase for phrase, or sentence for sentence. Statistical approaches have had good success (particularly if the corpus is both representative and sufficiently extensive). However, since it is based on probability, there is always a chance that a segment of text will be given the most likely translation rather than the meaning intended by the writer.

EVALUATION AND APPLICATIONS

There are a number of features in real human languages and usage that are challenging for translation software to deal with. Words can be ambiguous due to multiple meanings, and phrases can be syntactically ambiguous. (A famous example is, "Time flies like an arrow; fruit flies like a banana.") Rules-based translation software can attempt to include rules for determining which word or sentence meaning is intended, while statistically based programs can try to determine the probability that a given word or phrase in a given context has a certain meaning. Idioms or words that are not in the program's dictionary can also cause problems. (For example, Babel Fish translates "He already had two strikes against him" literally, losing the nuance based on the baseball reference.)

There are a variety of translation software packages in use today. The oldest is SYSTRAN, which was developed during the cold war of the 1960s to translate Russian scientific and technical documents, and later has been used by the European Union to work with documents in the union's various languages. Today SYSTRAN is the engine behind such popular Web sites from AltaVista (Babel Fish) and Google Language Tools. These services can translate text or whole Web pages (with varying degrees of success).

Simple handheld translation devices with phrases commonly needed by travelers are also available. More sophisticated devices (see SPEECH RECOGNITION AND SYNTHESIS) that can facilitate two-way conversations are also being developed for applications such as military interrogation and civil affairs.

Further Reading

Babel Fish (AltaVista). Available online. URL: http://babelfish.altavista.com/. Accessed September 25, 2007.

Hutchins, W. John. "Compendium of Translation Software." June 2007. Available online. URL: http://www.hutchinsweb.me.uk/Compendium.htm. Accessed September 25, 2007.

Hutchins, W. John and Harold L. Somers. *Introduction to Machine Translation*. Burlington, Mass.: Academic Press, 1992.

Trujillo, Arturo. *Translation Engines: Techniques for Machine Translation*. New York: Springer, 1999.

Lanier, Jaron
(1960–)
American
Computer Scientist, Inventor

Jaron Lanier pioneered the technology of virtual reality that is gradually having an impact on areas as diverse as entertainment, education, and even medicine.

Lanier was born on May 3, 1960, in New York City, although the family would soon move to Las Cruces, New Mexico. Lanier's father was a cubist painter and science writer and his mother a concert pianist (she died when the boy was nine years old). Living in a remote area, the precocious Lanier learned to play a large variety of exotic musical instruments and created his own science projects.

Lanier dropped out of high school, but fortunately sympathetic officials at New Mexico State University let him take classes there when he was only 14 years old. Lanier even received a grant from the National Science Foundation to let him pursue his research projects. Although fascinated by computers (and their possibilities as an aid to music

and other expressive arts), Lanier had a sporadic academic career, taking him to Bard College, where he dropped out of their computer music program.

However, by the mid-1980s Lanier had gotten back into computing by creating sound effects and music for Atari video games and writing a commercially successful game of his own called *Moondust*. He developed a reputation as a rising star in the new world of game design.

Lanier then began to experiment with ways to immerse the player more fully in the game experience. Using money from game royalties, he joined with a number of experimenters and built a workshop in his house. One of these colleagues was Tom Zimmermann, who had designed a "data glove" that could send commands to a computer based on hand and finger positions.

As the 1980s progressed, investors became increasingly interested in the new technology, and Lanier was able to expand his operation considerably, working on projects for NASA, Apple Computers, Pacific Bell, Matsushita, and other companies.

Lanier then coined the term "virtual reality" to describe the experience created by this emerging technology. A user wearing a special helmet has a computer-generated scene projected such that the user appears to be "within" the world created by the software. The world is an interactive one: Using gloves and body sensors, when the user walks in a particular direction the world shifts just as it would when walking in the "real" world. The gloves appear as the user's "hands" in the virtual world, and objects in that world can be grasped and manipulated much like real objects. In effect, the user has been transported to a different world created by the VR software (see VIRTUAL REALITY).

Virtual reality technology had existed in some form long before Lanier; it perhaps traces its roots back to the first mechanical flight simulators built during World War II. However, existing systems such as those used by NASA and the Air Force were extremely expensive, requiring powerful mainframe computers. They also lacked flexibility—each system was built for one particular purpose, and the technology was not readily transferable to new applications. Lanier's essential achievement was to use the new, inexpensive computer technology of the 1980s to build versatile software and hardware that could be used to create an infinite variety of virtual worlds.

Unfortunately the hippylike Lanier (self-described as a "Rastafarian hobbit" because of his dreadlocks) did not mesh well with the big business world into which his initial success had catapulted him. Lanier had to juggle numerous simultaneous projects as well as becoming embroiled in disputes over his patents for VR technology. In 1992 Lanier lost control of his patents to a group of French investors whose loans to VPL Research had not been paid, and he was forced out of the company he had founded.

During the 1990s Lanier founded several new companies to develop various types of VR applications. These include the Sausalito, California, software company Domain Simulations and the San Carlos, California, company New Leaf Systems, which specialized in medical applications for VR technology. Another company, New York-based Original Ventures, focuses on VR-based entertainment systems.

From 1997 to 2001, Lanier was chief scientist of Advanced Network and Services, a developer of the Internet2 (advanced high-speed networking) project, as well as serving as lead scientist of the National Tele-Immersion Initiative, a coalition of universities developing applications for Internet2. From 2001 to 2004 Lanier was also a visiting scientist at Silicon Graphics, Inc., doing fundamental research on tele-immersion and telepresence. Since 2004 Lanier has been a fellow at the International Computer Science Institute at UC Berkeley and since 2006, an interdisciplinary scholar-in-residence at Berkeley.

FUTURIST AND TECHNOLOGY PUNDIT

Lanier's humanistic and artistic background is reflected in the stance he has taken in recent years toward the technology he helped create. He has a column called "Jaron's World" in *Discover* magazine and regularly contributes to other publications such as *Edge*. In his writings Lanier has criticized the tendency to see the Internet as some sort of collective intelligence, warning that the individual might be in danger of being overwhelmed (see FLASH MOB). Lanier has also coined the term "cybernetic totalism" to refer to the tendency to put the constructs of the computer world ahead of the full dimensionality of human experience.

Besides writing and lecturing on virtual reality, Lanier is active as both a musician and an artist. In 1994 he released his CD *Instruments of Change*. Lanier's paintings and drawings have also been exhibited in a number of galleries.

Further Reading

Cave, Damien. "Artificial Stupidity: Virtual Reality Pioneer Jaron Lanier Says Computers Are Too Dumb to Take Over the World." Salon.com, Oct. 4, 2000. Available online. URL: http://archive.salon.com/tech/feature/2000/10/04/lanier/index.html. Accessed February 6, 2008.
"Homepage of Jaron Lanier." Available online. URL: http://www.jaronlanier.com/. Accessed September 25, 2007.
Lanier, Jaron. "The Hazards of the New Online Collectivism." *Edge.* Available online. URL: http://www.edge.org/3rd_culture/lanier06/lanier06_index.html. Accessed September 25, 2007.
———. "One Half of a Manifesto." Available online. URL: www.edge.org/3rd_culture/lanier/lanier_index.html. Accessed September 25, 2007.
Steffen, Alan. "What Keeps Jaron Lanier Awake at Night: Artificial Intelligence, Cybernetic Totalism, and the Lack of Common Sense." *Whole Earth* (Spring 2003): 24–29. Available online. URL: http://www.wholeearthmag.com/ArticleBin/111-5.pdf. Accessed September 25, 2007.

laptop computer

A laptop is a portable computer that contains all components (keyboard, display, motherboard, drives, etc.) in a single (usually hinged) case. In general, a laptop can perform the same tasks as a desktop computer, though not necessarily as quickly. (Laptops that have the full power and

A traditional laptop computer with a clamshell case. Laptops are now differentiated into lighter, more compact "notebooks" and somewhat larger and heavier "desktop replacement" units. Meanwhile, smaller hand held devices can replace laptops for some functions.

capacity of a desktop are sometimes called "desktop replacements," while smaller, lighter, but less powerful machines are called "notebooks." For even smaller or lighter computers, see PDA and TABLET PC.)

TYPICAL COMPONENTS

A typical laptop computer in 2007 has the following components:

- a processor such as Intel Core 2 duo or a version (such as Pentium M) optimized for wireless and lower-power consumption

- one or two gigabytes (GB) of system memory

- hard drive (80 to 160 GB capacity)

- combo CD/DVD optical drive with read-and-write capabilities

- LCD flat panel display (widescreen format) from 14 inches to 17 inches

- graphics card or integrated graphics

- wireless networking

- keyboard with touch pad and/or pointing stick (to simulate the mouse)

- six- or nine-cell lithium ion or lithium polymer battery

Modern laptops are well supplied with USB and network (Ethernet) ports. Many include readers for memory cards (such as SD cards). Additional capabilities can be provided by means of PC cards or Express cards. Most laptops run the same operating systems (such as Windows Vista or Mac OS X) as their desktop counterparts.

While portable and convenient, laptops do have some disadvantages compared to desktops: They cost more for a given level of performance; they are more difficult to repair; and they are more attractive to thieves.

DEVELOPMENT AND TRENDS

The idea of small, portable personal computers goes back to the Dynabook concept developed at Xerox PARC in the 1970s (see also KAY, ALAN). The first "portable" computers were often more aptly described as "luggable," having more the form factor of a suitcase than that of today's laptops. Nevertheless, the first commercially successful portable computers, the Osborne 1 (1981) and the Compaq Portable (1983), began to show the feasibility of portable computing. (At the other end of the size spectrum, the successful Radio Shack TRS-80 Model 100 established the utility of the notebook-sized computer.) In the 1980s true laptops from companies such as Zenith and Toshiba with the familiar clamshell design emerged, running PC-compatible MS-DOS and, later, Windows applications. (Apple entered the market with the Macintosh Portable in 1989, followed by the PowerBook series, introduced in 1991.)

Most improvements in laptops in the 1990s and beyond have been incremental (more storage, sharper displays, more efficient batteries, and so on). Wireless (see BLUETOOTH and WIRELESS AND MOBILE COMPUTING) connectivity is now standard. Laptop development has bifurcated somewhat, with higher-end machines rivaling desktops for media, gaming, and other applications, while notebooks often become lighter, sometimes forgoing optical and even hard drives in favor of network connectivity and flash memory storage. Specially "ruggedized" laptops are used by the military on battlefields and in other harsh environments. Meanwhile, PDAs and smart phones capable of e-mail, Web browsing, and light data entry offer an alternative to laptops for people who are on the road frequently.

Further Reading
Gookin, Dan. *Laptops for Dummies*. 2nd ed. Hoboken, N.J.: Wiley, 2006.
Laptop Magazine. Available online. URL: http://www.laptopmag.com/index.htm. Accessed September 26, 2007.
Laptops [resources and reviews]. Available online. URL: http://reviews.cnet.com/laptops.html. Accessed September 26, 2007.
Miller, Michael. *Your First Notebook PC*. Indianapolis: Que, 2008.
Sandler, Corey. *Upgrading & Fixing Laptops for Dummies*. Indianapolis: Wiley, 2006.
Wilson, James E. *Vintage Laptop Computers: First Decade, 1980–89*. Denver, Colo.: Outskirts Press, 2006.

law enforcement and computers

Besides his superb reasoning skills, perhaps Sherlock Holmes's most important asset was his extensive collection of notes that provided a cross-referenced index to London's criminal underworld. Today computer applications have given law enforcers investigative, forensic, communication, tactical, and management tools that Holmes and his rivals in the old Scotland Yard could not have imagined.

For the officer on the street, the ability to obtain auto license, stolen vehicle, or outstanding warrant information in near real-time provides a much better picture of the potential risk in making stops or arrests. Other "tactical" technology includes new devices for homing in on gunshots and the growing use of remote-controlled robots for bomb disposal and hostage negotiations (see ROBOTICS). A more controversial area is the use of CCTV (closed-circuit TV) surveillance cameras in public places, advocated as a crime deterrent but raising concerns about privacy and intrusive social control.

If a criminal case is opened, a variety of software applications come into play. These include case management programs for keeping track of evidence and witness interviews. Evidence must be properly logged at all times to maintain a legally defensible chain of custody against accusations of tampering.

The investigation of a crime involves many computerized forensic aids. Besides automated matching of fingerprints and, increasingly other physical data (see BIOMETRICS), records can also be searched to detect patterns such as crimes with related modus operandi (MOs). The ability to access information from other jurisdictions and to interface federal, state, and local agencies is also very important, particularly for cases involving organized crime, interstate fugitives, and terrorism.

Since data stored on computers is an increasingly prevalent form of evidence, law enforcement specialists must also employ tools to recover data that may have been partially erased or encrypted by suspects (see COMPUTER FORENSICS). Computers can be more active instruments of crime (see COMPUTER CRIME AND SECURITY). Such traditional tools as wiretapping must be adapted to new forms of communication such as e-mail while addressing concerns about civil liberties and privacy (see PRIVACY IN THE DIGITAL AGE).

High-level planning for law enforcement budgets and priorities requires access to detailed crime statistics. At the national level, the Justice Department's Bureau of Justice Statistics is a definitive information source. Law enforcers, like other professionals, increasingly use Web sites, chat areas, and e-mail lists to discuss computer-related law enforcement issues with colleagues.

Law enforcement agencies also use the same "bread and butter" software needed by any substantial organization, including word processing, spreadsheet, payroll, and other accounting programs.

Further Reading
Boba, Rachel. *Crime Analysis and Crime Mapping*. Thousand Oaks, Calif.: Sage Publications, 2005.
Chu, James. *Law Enforcement Information Technology: A Managerial, Operational, and Practitioner Guide*. Grand Rapids, Mich.: CRC Press, 2001.
Foster, Raymond E. *Police Technology*. Upper Saddle River, N.J.: Prentice Hall, 2004.
Goold, Benjamin J. *CCTV and Policing: Public Area Surveillance and Police Practices in Britain*. New York: Oxford University Press, 2004.
Gottschalk, Petter. *Knowledge Management Systems in Law Enforcement: Technologies and Techniques*. Hershey, Pa.: Idea Group Publishing, 2006.
Pattvina, April, ed. *Information Technology and the Criminal Justice System*. Thousand Oaks, Calif.: Sage Publications, 2004.

legal software

Modern law offices rely heavily on software to manage cases and records, to perform legal research, and to prepare pleadings and other documents. Many of these functions can be included in a legal software suite such as Amicus Attorney. Some typical law office management modules include the following:

- client file, which provides links to all events, tasks, time and billing, and so on involving each client

- general contact file with contact information for other people the office deals with regularly (such as court clerks)

- calendar for managing appointments, meetings, and deadlines

- time tracking and billing

- document management (often interfaces with suites such as Microsoft Office)

RESEARCH TOOLS

Legal research is easier (but in some ways more complex) than in the days of going through dusty files in the local courthouse or poring through a law library. The most used online database for legal research is LexisNexis, which contains two parts: Lexis (focusing on legal documents) and Nexis (for business research). Some of the most important Lexis content is:

- text of all U.S. statutes and laws

- U.S. published case opinions

- public records including property records, liens, and licenses

- laws and opinions for many non-U.S. jurisdictions

- articles from law journals

A free service called LexisOne provides a subset of U.S. legal decisions. Lexis also has a File & Serve service that allows for documents to be filed with courts or served upon participating firms. Nexis complements Lexis for many investigations because it offers news articles, particularly those relating to business activities. Other commercial legal information services include Westlaw and Loislaw. A free compilation of legal information that can provide an alternative for researching laws and cases is provided by the Legal Information Institute at Cornell University.

Finding citations or news is only part of the task of the legal researcher. The organization and management of all the data needed for any legal specialty is challenging. Printed reference books are cumbersome and can be quickly outdated. Recently a number of legal writers have been using wikis (see WIKIS AND WIKIPEDIA) as a tool for collaboration in creating online legal references. For example, the Internet Law Treatise sponsored by the Electronic Frontier Foundation covers a variety of legal issues relating to the use of the Internet. The Legal Information Institute at Cornell Law School is collaborating with experts to create a complete legal dictionary and encyclopedia in wiki form.

Further Reading

Bernstein, Paul. "Winning Software for Winning Cases." *Trial,* 37 (November 1, 2001): 82 ff.

Delaney, Stephanie. *Electronic Legal Research: An Integrated Approach.* Albany, N.Y.: Delmar (Thomson), 2002.

Internet Law Treatise (Electronic Frontier Foundation.) Available online. URL: http://ilt.eff.org/index.php/Table_of_Contents. Accessed September 26, 2007.

Law Office Computing Magazine. Available online. URL: http://www.lawofficecomputing.com. Accessed September 26, 2007.

LexisNexis. Available online. URL: http://www.lexisnexis.com. Accessed September 26, 2007.

Payne Consulting. *Microsoft Word 2002 for Law Firms.* Roseville, Calif.: Prima Publishing, 2001.

Wex (Legal Information Institute at Cornell Law School). Available online. URL: http://www.law.cornell.edu/wex/index.php/Main_Page. Accessed September 26, 2007.

Lessig, Lawrence
(1961–)
American
Law Professor and Writer

Law professor Lawrence Lessig is a pioneer in developing legal theories that deal with some of the most difficult issues emerging in the online world (see, for example, INTELLECTUAL PROPERTY AND COMPUTING).

Lessig was born on June 3, 1961, in Rapid City, South Dakota, but grew up in Williamsport, Pennsylvania. As a student at Yale Law School in the 1970s, Lessig, though previously president of Pennsylvania's Teenage Republicans, became interested in liberal values following the Watergate scandal and his exposure to authoritarian communist regimes during a summer trip to Eastern Europe.

After graduating from Yale, Lessig clerked for U.S. Supreme Court Justice Antonin Scalia, an articulate conservative with whom he could debate a variety of issues. When he began to teach law himself at the University of Chicago in 1991, he also began to incorporate issues arising in cyberspace in his lectures. In one article Lessig criticized the Communications Decency Act for forcing sites to block access to adult pornography in order to protect children. Eventually the Supreme Court agreed and overturned the law.

In the late 1990s Lessig served as a special master to the Supreme Court in the Microsoft antitrust case. This time Lessig sided with the government, agreeing that Microsoft had used its near monopoly in operating systems to bundle its own Web browser to the detriment of rival Netscape.

THE CREATIVE COMMONS

In recent years Lessig has undertaken to promote a more comprehensive set of legal principles aimed at protecting privacy, expression, and other fundamental rights in cyberspace. Many of the early and more radical Internet advocates saw the new medium as a libertarian or anarchist "free zone" that needed to be protected from any government interference. Lessig, however, has argued in his books *Code and Other Laws of Cyberspace* (2000) and *Code Version 2.0* (2006) that the online world needs a combination of technical architecture, private initiative, and reasonable regulation.

In his book *Free Culture* (2004) Lessig celebrates the creativity and freedom of expression of the Internet but warns that the Net may soon be "locked down" by the power of the corporate media. Lessig is therefore a strong supporter of "net neutrality," the proposed policy that would prohibit Internet service providers from charging different rates for different content providers or types of content (see NET NEUTRALITY). Without this policy, advocates believe that large corporations will gradually squeeze smaller, independent voices off the Web by making it harder for them to access users. In effect, what had been a shared "commons" would become property bounded by fences, much as common pastures were once turned into private farms.

According to Lessig, another obstacle to a vigorous online creative culture is the present copyright system. Under this system there is a presumption that permission is required for most usage of a work. This makes it difficult for creators to confidently use all the tools for working with existing content to create new expressions (see MASHUPS).

To protect free expression, Lessig founded an organization called Creative Commons in 2001, which has developed a new kind of license. Under this license the creator can specify what users can do with the work—copy it, create derivative works, and so on. Thus far Creative Commons licenses have mainly been applied to online works such as images shared on photo-sharing sites. Because applying for a Creative Commons license includes providing descriptors (metadata) about the work, people looking for material to be used in their own work can easily determine what they are allowed to do with a given work.

In 1997 Lessig left his professorship at the University of Chicago Law School to become a professor at Harvard Law School (1997–2000), and then Stanford (2000–). Lessig has been a guest lecturer at many universities and other institutions around the world. He also serves as a board member for many important cyberspace institutions, including the Electronic Frontier Foundation and Creative Commons.

In early 2008 Lessig announced that he would take on a challenge perhaps even more daunting than preserving Internet freedom—the battle against what he sees as pervasive corruption in the political system. He has proposed the creation of a grassroots movement that would encourage all incumbents and candidates to pledge to stop accepting contributions from lobbyists, to stop putting special interest "pork" in legislation, and to conduct publicly financed campaigns.

Lessig has received a number of academic awards as well at the Editor's Choice award from *Linux Journal* (2002), was named one of Fifty Top Innovators by *Scientific American* (2002), and received the Free Software Foundation Award (2003).

Further Reading

Creative Commons. Available online. URL: http://creativecommons.org/. Accessed September 26, 2007.

Lessig, Lawrence. *Code: Version 2.0* New York: Basic Books, 2006.

———. *Free Culture: The Nature and Future of Creativity.* New York: Penguin Books, 2004.

Lessig blog. Available online. URL: http://www.lessig.org/. Accessed September 26, 2007.

O'Brien, Chris. "Stanford Law Prof Wants Society to Clean Up Its Act." *San Jose Mercury News,* September 9, 2007. Available online. URL: http://www.mercurynews.com/business/ci_6843973. Accessed September 26, 2007.

libraries and computing

The library is the institution traditionally charged with the collection and distribution of humanity's collective heritage of written information. It is thus not surprising that the development of modern information technology has meant that libraries have had to undergo pervasive changes in their practices and responsibilities.

One of the earliest applications for automation in libraries was cataloging. By the 1960s, the ever-increasing volume of books and serials (periodicals) published each year was placing a growing burden on the manual cataloging system. Under this system, catalogers at large libraries (and particularly the Library of Congress) prepared a catalog record for each new publication. These records were distributed by the Library of Congress in the form of catalog card proof slips. These, as well as compiled card images from other libraries, could be used by each library to prepare catalog records for its own holdings.

As mid-size computers became more affordable, it became practicable for at least large library systems to put their catalog records on-line. In 1968, the MARC (Machine Readable Cataloging) standard was first promulgated. A MARC record uses specific, numbered fields to describe the elements of a book, such as its catalog card number, main entry, title, imprint, collation (pagination), and subject headings.

At first, MARC records were distributed mainly on magnetic tape in place of card slips. However, by the late 1970s large on-line cataloging systems such as OCLC (On-line College Library Center) and RLIN (Research Library Information Network) were enabling libraries to search for and download cataloging information in real time, and in turn upload their own original catalog records to the shared database. This greatly reduced redundant cataloging effort. If a library receives a new book, a library assistant can search for a preexisting catalog record. The record can then be easily modified for local use, such as by adding a call number and holdings information. The problem of authorities (standardized entries for names) is also made more manageable by being able to check entries on-line.

By the 1980s, the next logical step was under way: The card catalog began to be replaced by a wholly electronic catalog, enabling library patrons to search the catalog at a terminal. Besides saving money, the on-line catalog also offers researchers many more ways to search for materials: for example, they can use keywords and not rely only on titles and subject headings.

Along with cataloging, libraries began to automate their circulation and acquisitions systems as well. As these systems become integrated, libraries can both monitor the demand (finding materials that are in heavy use and need additional copies) and speed up the supply, by integrating the acquisitions system with ordering systems maintained by book distributors.

However, while most librarians consider the computer to be a boon to their profession, there are criticisms and further challenges. Nicholson Baker, for example, has decried the abandonment of information in card catalogs that was not carried over into electronic form. Baker has also criticized the replacement of bound archives of periodicals with microfilm, which is often of poor quality and prone to deterioration. The storage of publications on computer media has also met with concerns that the physical durability of the media has not been sufficiently investigated, and that in a rapidly changing technological world data formats can become obsolete, no longer supported, and potentially unreadable (see also BACKUP AND ARCHIVE SYSTEMS).

The growth of the World Wide Web has also presented libraries with both opportunities and challenges. Catalogers and reference librarians are struggling to find new ways to categorize and retrieve the always-changing and ephemeral content of Web pages. Meanwhile, librarians have faced not only funding and training issues in providing expanded public Web access in libraries, but have also had to deal with demands that Web content be filtered to protect children from objectionable content. (The American Library Association opposes such filtering as a form of censorship.)

Besides being a source of Internet connectivity for students and people who cannot afford their own computer, today's libraries provide a wide variety of media products, including audio CDs, audio and video tapes, and DVDs. Increasingly, though, modern librarians are moving away from the idea of a library as a repository of resources and are placing greater emphasis on providing guidance and starting points for users who are seeking to navigate the often-overwhelming Web. Although they face many challenges, libraries seem to be succeeding in the task of reinventing themselves.

Further Reading

American Library Association. Office for Information Technology Policy. Available online. URL: http://www.ala.org/oitp. Accessed August 13, 2007.

Baker, Nicholson. *Double Fold: Libraries and the Assault on Paper.* New York: Vintage Books, 2002.

Burke, John J. *Neal-Schuman Library Technology Companion: A Basic Guide for Library Staff.* 2nd ed. New York: Neal Schuman Publications, 2006.

Hanson, Kathlene, and H. Frank Cervone. *Using Interactive Technologies in Libraries (LITA Guide).* New York: Neal Schuman Publishers, 2007.

Library and Information Technology Association. Available online. URL: http://www.lita.org/ala/lita/litahome.cfm. Accessed August 13, 2007.

library, program

Programming is a labor-intensive activity, especially when the time required to test, debug, and verify the operation of the program code is included. It is not surprising, then, that even the earliest programmers sought ways to reuse the code for commonly needed operations such as data

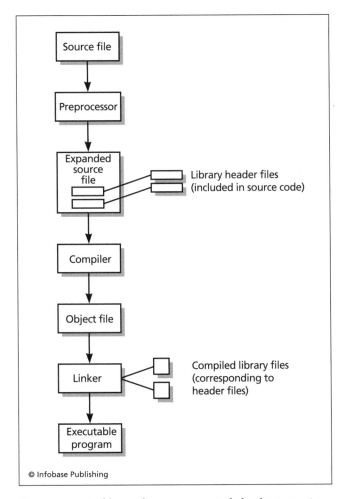

To use a program library, the programmer includes the appropriate header file in the source code. After the source code is compiled, the linker links it to the compiled object code file corresponding to the header file, creating a single executable file.

input, sorting, calculation, and formatting rather than writing it from scratch. If a well-organized collection or library of program routines is available, developers of new applications can concentrate on the aspects particular to the current problem and use the library code for routine operations.

In the mainframe world, the use of program libraries was also mandated by the limited amount of main memory available. A data processing task was often accomplished by retrieving a series of card decks or tapes from the library and mounting them in turn. Intermediate results could be passed between programs under the control of a special script (see JOB CONTROL LANGUAGE).

Some programming languages, notably C and its descendants C++ and Java, are designed to provide a small core of essential features (such as control structures, data types, and operators). Other functions, such as math routines, data I/O (input/output), and formatting are provided in library files that are invoked by programs that need particular features. There are several advantages to this approach. The

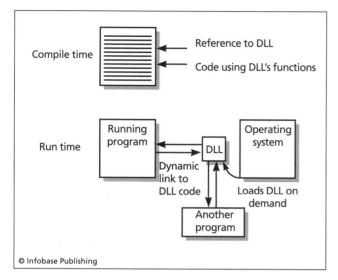

Dynamic linking is an alternative approach to library use. The program is compiled with a reference to the library, but it is not linked to the library code until the program is actually running. Since several different running programs can link to the same dynamic link library (DLL), memory is saved.

core language is kept simple because it doesn't have to deal with issues such as the actual storage of data in memory that are dependent on the particular architecture of each type of machine. To "port" the language to a new machine, specialists in its architecture can implement the standard library functions. In addition to the standard libraries included with the compiler, programmers are also free to create additional libraries to support particular applications such as graphics.

With a traditional library the library routines invoked in the source code are included in the final executable file. With most modern operating systems, however, many programs are active in memory at the same time (see MULTITASKING). Storing the same commonly used routines (such as standard I/O) with each program wastes memory. Therefore, operating systems such as Microsoft Windows use *dynamic linking*. This means that instead of compiling the library code into the program to create the executable file, the program links to the library at execution time. If another program is using the library, the new program links to the same copy in memory rather than having to store another copy. (Dynamically linked libraries [DLLs] include special code to keep track of the invocation of the library functions by each separate program.)

Further Reading

Josuttis, Nicolai M. *The C++ Standard Library: A Tutorial and Reference.* Upper Saddle River, N.J.: Addison-Wesley, 1999.

Loosemore, Sandra, et al. *GNU C Library Application Fundamentals.* Boston: GNU Press, 2004.

———. et al. *GNU C Library System & Network Applications.* Boston: GNU Press, 2004.

Lundh, Fredrik. *Python Standard Library.* Sebastapol, Calif.: O'Reilly, 2001.

Licklider, Joseph Carl Robnett
(1915–1990)
American
Computer Scientist, Psychologist

Most of the early computer pioneers came from backgrounds in mathematics or engineering. This naturally led them to focus on the computer as a tool for computation and information processing. Joseph Licklider, however, brought an extensive background in psychology to the problem of designing interactive computer systems that could provide better communication and access to information for users.

Licklider was born on March 11, 1915, in St. Louis, Missouri. During the 1930s, he attended Washington University in St. Louis, earning B.A. degrees in psychology, mathematics, and physics. He then concentrated on psychology for his graduate studies, earning an M.A. at Washington University and then receiving his Ph.D. from the University of Rochester in 1942.

While at Rochester, Licklider participated in a study group led by Norbert Wiener, pioneer in the new field of cybernetics, in the late 1940s. This brought him into contact with emerging computer technology and its exciting prospects for the future. In turn, Licklider's psychology background allowed him a perspective quite different from the mathematical and engineering background shared by most early computer pioneers.

Cybernetics emphasized the computer as a system that could interact in complex ways with the environment. Licklider added an interest in human-computer interaction and communication. He began to see the computer as a sort of "amplifier" for the human mind. He believed that humans and computers could work together to solve problems that neither could successfully tackle alone. The human could supply imagination and intuition, while the computer provided computational "muscle." Ultimately, according to the title of his influential paper, it might be possible to achieve a true "Man-Computer Symbiosis."

During the 1950s, Licklider taught psychology at the Massachusetts Institute of Technology, hoping eventually to establish a full-fledged psychology department that would elevate the concern for what engineers call "human factors." From 1957 to 1962 he also served in the private sector as a vice president for engineering psychology at Bolt Beranek and Newman, the company that would become famous for pioneering networking technology.

In 1962, the federal Advanced Research Projects Agency (ARPA) appointed Licklider to head a new office focusing on leading-edge development in computer science. Licklider soon brought together research groups that included in their leadership three of the leading pioneers in artificial intelligence: John McCarthy, Marvin Minsky, and Allen Newell (see ARTIFICIAL INTELLIGENCE; MCCARTHY, JOHN; and MINSKY, MARVIN). By promoting university access to government funding, Licklider also fueled the growth of computer science graduate programs at major universities such as Carnegie Mellon University, University of California at Berkeley, Stanford University, and the Massachusetts Institute of Technology.

In his research activities, Licklider focused his efforts not so much on AI as on the development of interactive computer systems that could promote his vision of human-computer symbiosis. This included time-sharing systems, where many users could share a large computer system, and networks that would allow users on different computers to communicate with one another. He believed that the cooperative efforts of researchers and programmers could develop complex programs more quickly than teams limited to a single agency or corporation (see also OPEN-SOURCE MOVEMENT).

Licklider's efforts to focus ARPA's resources on networking and human-computer interaction would provide the resources and training that would, in the late 1960s, begin the development of what would become the Internet. Licklider spent the last two decades of his career teaching at MIT. Before his death in 1990, he presciently predicted that by 2000 people around the world would be linked in a global computer network.

Further Reading

"Internet Pioneers: J. C. R. Licklider." Available online. URL: http://www.ibiblio.org/pioneers/licklider.html. Accessed August 13, 2007.

Licklider, J. C. R. "Man-Computer Symbiosis." *IRE Transactions on Human Factors in Electronics,* vol. HFE-1, March 4–11, 1960. Available online. URL: http://www.memex.org/licklider.pdf. Accessed August 13, 2007.

Licklider, J. C. R., and Robert W. Taylor. "The Computer as a Communication Device." *Science and Technology,* April, 1968. Available online. URL: http://gatekeeper.dec.com/pub/DEC/SRC/publications/taylor/licklider-taylor.pdf. Accessed August 13, 2007.

Waldrop, M. Mitchell. *The Dream Machine: J. C. Licklider and the Revolution That Made Computing Personal.* New York: Viking, 2001.

linguistics and computing

The study of human language and advances in computer science have been closely intertwined. The field of computational linguistics uses computer systems to investigate the structure of natural language. In turn, the area of natural language processing involves the creation of software that can apply linguistic principles to process written or spoken human language (see NATURAL LANGUAGE PROCESSING, LANGUAGE TRANSLATION SOFTWARE, and SPEECH RECOGNITION AND SYNTHESIS).

As simple low-level instruction codes began to evolve into complex high-level programming language, language designers had to struggle to give precise, complete, and unambiguous definitions for the language's structure. This is essential for language users to be confident that their programs will yield the desired results. It is also important that developers trying to implement a language on different hardware platforms and operating systems have rigorous language specifications so the compiler on the new system will produce programs equivalent to those on the system where the language was first developed.

When computer scientists turned to linguistics for help in defining programming languages, they found the work of Noam Chomsky, perhaps the 20th century's preeminent linguist, to be particularly helpful. Chomsky developed a concept of formal language in which grammar could be specified as a series of rules built up a level at a time. For example, at the lowest level, there is an alphabet from which recognized words are generated. Next there are rules for generating phrases (such as a noun phrase consisting of a noun with optional adjectives and a verb phrase consisting of a verb with optional adverbs). In turn, phrases can be combined to form sentences.

Because grammatical structures are created by applying rules to strings of symbols (words), the result is called a *generative grammar.* Chomsky sought to apply this concept of a "transformational generative grammar" as a universal structure applicable to all human languages. Meanwhile, computer scientists could use formal grammar rules to define the valid statements in programming languages (see also BACKUS-NAUR FORM). This in turn allows a compiler *parser* to break down high-level language statements and convert them into low-level instruction codes that can actually be executed by the CPU (see ASSEMBLER and PARSING).

As new languages and more powerful hardware gave computers increased power to deal with complex systems, computer scientists (and artificial intelligence researchers in particular) applied themselves to the problem of computer processing of human languages. Success in this field might lead not only to computer systems that humans could communicate with far more naturally, but also to automatic machine translation that could, for example, allow an English speaker and a Chinese speaker to communicate via e-mail.

However, developers of natural language systems face formidable challenges. Most fundamentally, while computers process symbols using a restrictive, deterministic procedure that Chomsky classifies as *finite state* (see FINITE STATE MACHINE), human languages must be understood using the more complex transformational grammar. The language processing system must therefore have rules that can cope with the often ambiguous structure of actual human speech. (For example, does the word *fly* in a given sentence mean an insect, a baseball batted high in the air, or perhaps a zippered opening in one's trousers?)

One way to limit the problem is to deal with a restricted realm of discourse. For example, a natural language "front end" to a database might assume that all input nouns refer to entities that exist in the database, such as employees, positions, salaries, and so on. It then becomes a matter of translating a query such as "How many employees in the human resources department make more than $50,000 a year" into something like:

```
find quantity (employee.department = "human
resources") and (employee.salary > 50,000)
```

Understanding unrestricted text such as that found in newspaper stories is much more complex, since fewer assumptions can be made about the subject of the discourse. Here the AI concept of frames can prove useful. A frame is a sort of script that describes the elements of life's common events or transactions. For example, suppose a news story begins "Joe X was arrested yesterday for the murder of Sarah Y. He was arraigned today and bail was denied." A system reading the story might see "arrested" and see that it links to an internal frame called "crime." The crime frame might have slots for "accused person," "charge," "victim," and "custodial status." The system could then interpret the story as indicating that Joe is the accused person, murder is the charge, Sarah is the victim. For the custodial status the system might look to another frame called "arraignment" that includes the rule that if bail is allowed and paid, the person's status is "released until trial" while if the bail is either not allowed or not paid, the status is "in custody."

Computational linguistics and natural language processing are likely to be of increasing interest in years to come. With the World Wide Web bringing the world's languages into more pervasive contact, the ability to translate or automatically summarize Web pages and e-mail will be very marketable. It is also likely that advanced, secret research in the field is also being carried out by organizations such as the National Security Agency (NSA), which monitor worldwide communications.

Further Reading
Hausser, Roland. *Foundations of Computational Linguistics: Human-Computer Communication in Natural Language.* 2nd ed. New York: Springer, 2001.

Lawler, John. *Using Computers in Linguistics: A Practical Guide.* New York: Routledge, 1998.

"Linguistics, Natural Language, and Computational Linguistics Meta-Index." Stanford University Natural Language Processing. Available online. URL: http://www-nlp.stanford.edu/links/linguistics.html. Accessed August 13, 2007.

Mitkov, Ruslan. *The Oxford Handbook of Computational Linguistics.* New York: Oxford University Press, 2005.

Linux

Linux is an increasingly popular alternative to proprietary operating systems. Its development sprang from two sources. First was the creation of open-source versions of UNIX utilities (see UNIX) by maverick programmer Richard Stallman as part of the GNU ("Gnu's not UNIX") project during the 1980s. Although these tools were useful, the kernel, or basic set of operating system functions, was still missing (see KERNEL). Starting in 1991, another creative programmer, Linus Torvalds, began to release open-source versions of the UNIX kernel (see TORVALDS, LINUS). The combination of the kernel and utilities became known as Linux (a combination of Linus and UNIX), though Stallman and his supporters believe that GNU/Linux is a more accurate name.

DEVELOPMENT AND DISTRIBUTIONS

As an open-source product, Linux is continually being developed by a community of thousands of loosely organized programmers. (The further development of the kernel itself is more closely supervised by Torvalds and a system of review that he set up.) New versions of the Linux kernel are released frequently, including support (drivers) for new devices and refinements in other features.

A distribution or "distro" is a package consisting of a Linux kernel, standard utilities, and a variety of other software such as office and graphics programs, Web-related

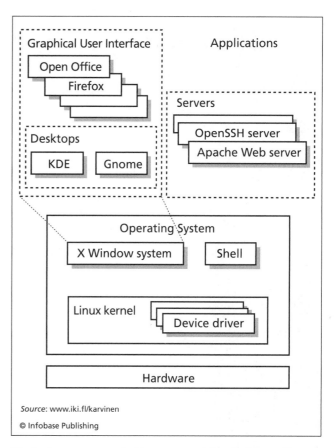

The basic components of a Linux system. A distribution, or "distro," combines the latest version of the common kernel with a window manager, selected software, and, perhaps, custom features.

programs, and so on. Some distributions such as Novell and Red Hat are geared toward business use and provide fee-based support and consulting (Red Hat spun off Fedora as a free user-supported distribution). One of the most popular distributions as of the mid-2000s is Ubuntu. Named for an African word meaning "humanity toward others" and funded by millionaire Mark Shuttleworth, Ubuntu combines a business-oriented component (through Canonical Ltd.) and a large and enthusiastic community of desktop users from all walks of life.

USING LINUX

Linux is very versatile and probably runs on more kinds of devices than any other operating system. These include supercomputer clusters, Web and file servers, desktops (including PCs designed for Windows and Macs), laptops, PDAs, and even a few smart phones. The Linux programmer has many programming languages and environments to choose from, including C++, Java, Perl, PHP, and Ruby. Thousands of open-source programs have been written for or ported to Linux, including OpenOffice.org (a suite comparable to Microsoft Office), databases (such as MySQL), and Apache, the most popular Web server.

Although Linux rapidly gained a significant share in server applications, early versions of Linux for ordinary desktop

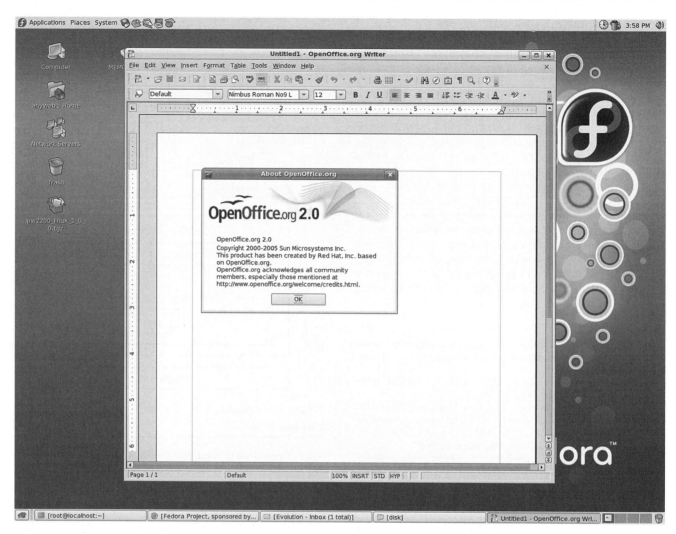

A Linux system running Open Office, a full-featured (and free) office software suite (SUN MICROSYSTEMS)

users were criticized as being hard to install and to configure for various types of hardware. However, current versions of Linux have an installation experience that is comparable to that of Windows, and such details as disk partitioning and setting up networks can often be handled automatically. (There can still be problems with some devices such as wireless cards for laptops, but even there things have improved considerably.) A Linux distribution such as Ubuntu is now a viable alternative to Windows unless one has to use certain programs (such as PhotoShop or many games) that do not have Linux versions. However, such options as dual-booting, emulation, or virtual machines offer the ability to use both Linux and Windows on the same machine.

Further Reading

Hill, Benjamin Mako, et al. *The Official Ubuntu Book.* 3rd ed. Upper Saddle River, N.J.: Prentice Hall, 2008.

Linux.com. Available online. URL: http://www.linux.com/. Accessed September 26, 2007.

Linux Journal. Available online. URL: http://www.linuxjournal.com/. Accessed September 26, 2007.

Matthew, Neil, and Richard Stones. *Beginning Linux Programming.* 4th ed. Indianapolis: Wrox, 2007.

Red Hat. Available online. URL: http://www.redhat.com/. Accessed September 26, 2007.

Sery, Pal G. *Ubuntu Linux for Dummies.* Hoboken, N.J.: Wiley, 2007.

Siever, Ellen, et al. *Linux in a Nutshell.* 5th ed. Sebastapol, Calif.: O'Reilly, 2005.

Sobell, Mark G. *A Practical Guide to Red Hat Linux: Fedora Core and Red Hat Enterprise Linux.* 3rd ed. Upper Saddle River, N.J.: Prentice Hall, 2006.

Ubuntu. Available online. URL: http://www.ubuntulinux.org/. Accessed September 26, 2007.

LISP

As interest in AI (see ARTIFICIAL INTELLIGENCE) developed in the early 1950s, researchers soon became frustrated by the low-level computer languages of the day, which emphasized computation and other manipulation of numbers rather than the processing of symbolic data.

At the 1956 Dartmouth Summer Research Project on Artificial Intelligence, a gathering that brought together the key early pioneers in the field, John McCarthy presented his concepts for a different kind of computer language. Such a language, he believed, should be able to deal with mathematical functions in their own terms—by manipulating symbols, not just calculating numbers.

Together with Marvin Minsky, McCarthy began to implement a language called LISP (for "list processor"). (See MCCARTHY, JOHN and MINSKY, MARVIN.) As the name suggests, the language uses lists to store data (see LIST PROCESSING) and features many functions for manipulating list elements. List can consist of single elements (called "atoms") as in

(A B C D)

but lists can also include other lists, as in

(A (B C) D)

Each list item is stored as a "node" containing both a pointer to its data value and a pointer to the next item in the linked list. The LISP system typically includes housekeeping functions such as "garbage collection," where the memory from discarded list items is returned to the free memory pool for later allocation.

LISP programs look forbidding at first sight because they tend to have many nested parentheses. However, expressions and functions are actually constructed in a much simpler way than in most other languages. Without the need for complicated parsing, the LISP interpreter (called "eval" because it evaluates its input) looks at the stream of data and first asks whether the next item is a constant (such as a number, quoted symbol, string, quoted list, or keyword). If so, its value is returned. Otherwise, the interpreter checks to see if the item is a defined variable and, if so, returns its value. Finally, the interpreter checks to see if there is a list. If so, the list is considered to be a function call followed by its arguments. The function is called, given the data, and the result is returned.

The following table shows some items in a list program and how the interpreter will evaluate them:

EXAMPLES OF LISP ITEMS

TYPE OF ITEM	EXAMPLE	EVALUATION
integer	24	24
float	5.5	5.5
ratio	3/4	0.75
keyword	defun	defines function
quoted integer	'24	24
quoted list	'(3 1 4 1 5)	(3 1 4 1 5)
boolean	nil	false
function call	+2 4	6
variable	a	its value
quoted variable	'a	a

Languages such as Algol, Pascal or C emphasize statements and procedures. LISP, on the other hand, was the first FUNCTIONAL language (see FUNCTIONAL LANGUAGES). The heart of a LISP program is functions that are evaluated together with their arguments. LISP includes many built-in, or *primitive* functions. Besides the usual mathematical operations, there are primitives for basic list-processing functions. For example, the list function creates a list from its arguments: (list 1 2 3) returns the list (1 2 3), while the cons function inserts an atom into the beginning of a list, and the append function tacks it onto the end. Programmers define their own functions using the defun keyword.

LISP has two other features that make it a powerful and flexible language for manipulating symbols and data. LISP allows for recursive functions (see RECURSION). For example, the following function raises a variable x to the power y:

```
(defun power (x y)
  (if (= y 0) 1
    (* x (power x (1- y)))))
```

Here the if expression checks to see whether y is 0. If not, the second expression invokes the function (power) itself, which performs the same test. The result is that the function keeps calling itself, storing temporary values, until y gets down to 0. It then "winds itself back up," multiplying x by itself y times.

But perhaps the most interesting feature of LISP is that it makes no distinction between programs (functions) and data. Since a function call and its arguments themselves constitute a list, a function can be fed as data to other functions. This makes it easy to write programs that modify their own operation.

LANGUAGE DEVELOPMENT

LISP quickly caught on with artificial intelligence researchers, and the version called LISP 1.5 was considered robust enough for writing large-scale applications. While "mainstream" computer scientists often used Algol and its descendants as a universal language for expressing algorithms, LISP became the lingua franca for AI people.

However, a number of dialects such as Mac-LISP diverged as versions were written to support new hardware or were promoted by companies such as LMI and Symbolics. While researchers liked the interactive nature of interpreted LISP (where functions could be defined and immediately tried out at the keyboard), practical applications required compilation into machine language to achieve adequate speed.

A widely used LISP variant is Scheme, developed by MIT researchers in the mid-1970s. Scheme simplifies LISP syntax (while still preserving the spirit), and at the same time generalizes further by allowing functions to have all the capabilities of data entities. That is, functions can be passed as parameters to other functions, returned as values, and assigned to variables and lists.

By the 1980s, personal computers became powerful enough to run LISP, and the proliferation of LISP variants running on different platforms led to a standardization movement that resulted in Common LISP in 1984. Common LISP combines the features of many existing dialects, includes a rich variety of data types, and also makes greater allowance for the imperative, sequential programming approach of languages such as C. It thus accommodates varying styles of programming. It is widely available today in both commercial and shareware versions.

Seymour Papert created a LISP-like language called Logo, which has been used to teach sophisticated computer science ideas to young students (see LOGO).

Further Reading
Dybvig, R. Kent. *The SCHEME Programming Language.* 3rd ed. Cambridge, Mass.: MIT Press, 2003.
"Lisp Information and Resources." Available online. URL: http://www.lispmachine.net/. Accessed August 13, 2007.
Queinnec, Christian. *Lisp in Small Pieces.* New York: Cambridge University Press, 1996.
Seibel, Peter. *Practical Common Lisp.* Berkeley, Calif.: Apress, 2005.

list processing

A list is a series of data items that can be accessed sequentially by following links from one item to the next. Lists can be very useful for ordering or sorting data items and for storing them on a stack or queue.

There are two general approaches to constructing lists. In a data list used with procedural programming languages such as C, each list item consists of a structure consisting of a data member and a pointer. The pointer, called "next," contains the address of the next item. A program can easily "step through" a list by starting with the first item, processing its data, then using the pointer to move to the next item, continuing until some condition is met or the end of the list is reached.

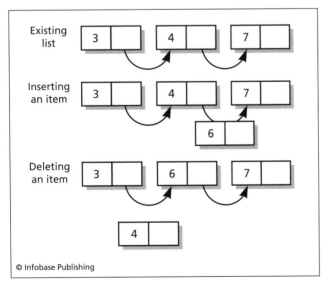

A singly linked list. Each node (item) includes a value and a pointer to the next node. Inserting a new node is simply a matter of adjusting the pointer of an existing node to point to the new node, with the new node's pointer in turn pointing to the next item (or the end of the list). A node is removed by disconnecting its pointer.

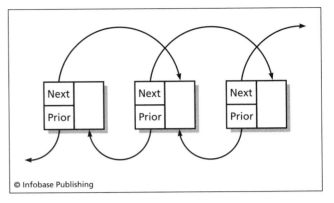

A doubly linked list. Each node has two pointers, one to the next item and one to the prior (preceding) item. While doubly linked lists use more memory, they can be processed more quickly because they can be traversed in either direction.

In LISP-type languages, however, a more general structure is used, since essentially all data is part of a list. Here each item is a *node* that can contain a pointer to any valid object and a pointer to the next node. One advantage of this scheme is that since fixed-length data fields are not used, the list can be "hooked up" to objects of varying sizes and types. This can also use memory more efficiently, though at the cost of additional processing being needed to periodically reclaim memory ("garbage collection").

Besides traversing (stepping through) a list by following its "next" pointer, the basic list-processing operations are insertion and deletion. It is easy to insert a new element into a list: You first move to the item after which the new item is to be inserted. Next, you connect that item's "next" pointer (link) to the new item. You then connect the new item's next link to the item that originally followed the insertion point. Deleting an item is even simpler: You "snip out" the item by connecting the item that originally linked to it to the item that was originally after it.

Sometimes lists are set up so that each item has two pointers: one to the next item and one to the previous one. Such *doubly linked lists* can be traversed in either direction, making retrieval faster in some situations, though at the cost of storing the extra pointers. Lists are also used to implement some specialized data structures (see STACK and QUEUE).

APPLICATIONS

Lists are generally used to provide convenient access to relatively small amounts of data where flexibility is required. Unlike an array, a list need use only as much memory as it needs to accommodate the current number of items (including their associated pointers). A LISP-style node list can be even more flexible in that items with varying sizes and types of data can be included in the same list. Lists are thus a more flexible way to implement such things as look-up tables. (See also ARRAY.)

Further Reading
Covington, Michael A. "Some Recursive List Processing Algorithms in Lisp." Available online. URL: http://www.ai.uga.

edu/mc/LispNotes/RecursiveListProcessingAlgorithmsIn-Lisp.pdf. Accessed August 13, 2007.

"Linked List Basics [In C/C++]" Available online. URL: http://cslibrary.stanford.edu/103/LinkedListBasics.pdf. Accessed August 13, 2007.

Seibel, Peter. "They Called It LISP for a Reason: List Processing." Available online. URL: http://www.gigamonkeys.com/book/they-called-it-lisp-for-a-reason-list-processing.html. Accessed August 13, 2007.

local area network (LAN)

Starting in the 1980s, many organizations sought to connect their employees' desktop computers so they could share central databases, share or back up files, communicate via e-mail, and collaborate on projects. A system that links computers within a single office or home, or a larger area such as a building or campus, is called a local area network (LAN). (Larger networks linking branches of an organization throughout the country or world are called wide area networks, or WANs. See NETWORK.)

HARDWARE ARCHITECTURE

There are two basic ways to connect computers in a LAN. The first, called Ethernet, was developed by a project at the Xerox Palo Alto Research Center (PARC) led by Robert Metcalfe. Ethernet uses a single cable line called a *bus* to which all participating computers are connected. Each data packet is received by all computers, but processed only by the one it is addressed to. Before sending a packet, a computer first checks to make sure the line is free. Sometimes, due to the time delay before a packet is received by all computers, another computer may think the line is free and start trans-

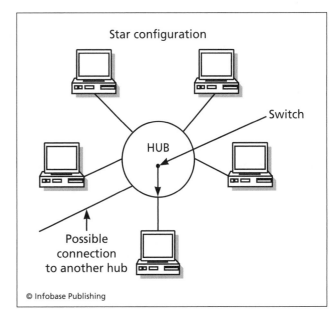

The Star network configuration uses a central hub to which each PC is attached. To extend the network (such as into other offices), the hubs can be connected to one another so they function as switches. When a token arrives that is addressed to one of its PCs, the hub will route it to the appropriate machine.

mitting. The resulting *collision* is resolved by having both computers stop and wait varying times before resending.

Because connecting all computers to a single bus line is impractical in larger installations, Ethernet networks are frequently extended to multiple offices by connecting a bus in each office to a switch, creating a subnetwork or segment (this is sometimes called a *star topology*). The switches are then connected to a main bus. Packets are first routed to the switch for the segment containing the destination computer. The switch then dispatches the packet to the destination computer. Another advantage of this *switched Ethernet* system is that more-expensive, high-bandwidth cable can be used to connect the switches to move the packets more quickly over greater distances, while less-expensive cabling can be used to connect each computer to its local switch.

An alternative way to arrange a LAN is called *token ring*. Instead of the computers being connected to a bus that ends in a terminator, they are connected in a circle where the last computer is connected to the first. Interference is prevented by using a special packet called the token. Like the use of a "talking stick" in a tribal council, only the computer holding the token can transmit at a given time. After transmitting, the computer puts the token back into circulation so it can be grabbed by the next computer that wants to send data.

LAN SOFTWARE

Naturally there must be software to manage the transmission and reception of data packets. The structure of a packet (sometimes called a *frame*) has been standardized with a preamble, source and destination addresses, the data itself,

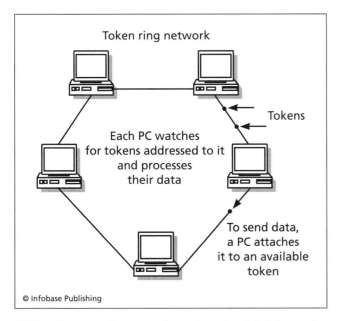

A Token Ring network connects the machines in a "chain" around which messages called tokens travel. Any PC can "grab" a passing token and attach data and the address of another PC to it. Each PC in turn watches for tokens that are addressed to it.

a checksum, and two special layers that interface with the differing ways that Ethernet and token ring networks physically handle the packets.

The low-level processing of data packets must also be interfaced with the overall operating system so that, for example, a user on a desktop PC can "see" folders and files on the file server and whole files can be transferred between server and desktop PC. From the 1980s to the mid-1990s the most common LAN operating system for DOS and later Windows-based PCs was Novell Netware, while Macintosh users used AppleTalk. Later versions of Windows (notably Windows NT) then incorporated their own networking support, and Netware use declined somewhat.

The tremendous popularity of the Internet (particularly the Web) starting in the mid-1990s propelled the Internet protocol (see TCP/IP) into the forefront of networking. Today's business and home computers use essentially the same tools to connect to the global Internet and to one another. (The term *Intranet,* once used to distinguish local TCP/IP networks from the Internet, is now pretty much obsolete.)

Meanwhile, the technologies used to implement this universal networking have proliferated. While the Internet is most commonly delivered to homes and businesses via wires (see CABLE MODEM and DSL), wireless networking has replaced cable for many local networks, including most home networks (see WIRELESS AND MOBILE COMPUTING), with the hub of the network being an inexpensive router and wireless access point.

Further Reading

Briere, Danny, Pat Hurley, and Edward Ferris. *Wireless Home Networking for Dummies.* Hoboken, N.J.: Wiley Publishing, 2006.
Komar, Brian. *Sams Teach Yourself TCP/IP Networking in 21 Days.* 2nd ed. Indianapolis: Sams, 2002.
Lowe, Doug. *Networking for Dummies.* Hoboken, N.J.: Wiley Publishing, 2007.
Spurgeon, Charles. *Ethernet: The Definite Guide.* Sebastopol, Calif.: O'Reilly Media, 2000.
Tittel, Ed, Earl Follis, and James E. Gaskin. *Networking with NetWare for Dummies.* 4th ed. Foster City, Calif.: IDG Books, 1998.

Logo

Logo is a derivative of LISP (see LISP) that preserves much of that language's list processing and symbolic manipulation power while offering simpler syntax, easier interactivity, and graphics capabilities likely to appeal to young people. Logo has often been used as a first computer language for students in elementary and junior high school grades. As Harold Abelson noted in his *Apple Logo* primer in 1982, "Logo is the name for a philosophy of education and a continually evolving family of programming languages that aid in its realization."

Logo was developed starting in 1967 by educator Seymour Papert and his colleagues at Bolt, Beranek and Newman, Inc. Papert, a mathematician and AI pioneer, had became interested in devising an education-oriented computer language after working with developmental psychol-

ogist Jean Piaget. Papert focused particularly on Piaget's emphasis on "constructivism"—the idea that people learn mainly by fitting new concepts into an existing framework built from the experience of daily life. Papert came to believe that abstract computer languages such as FORTRAN or even BASIC were hard for children to assimilate because their algebraic formulas and syntax had little in common with daily activities such as walking, playing, drawing, or making things.

For example, most computer languages implement graphics using statements that specify screen points using Cartesian coordinates (X, Y). A square, for example, might be drawn by statements such as:

```
PLOT 100, 100
LINETO 150, 100
LINETO 150, 150
LINETO 100, 150
LINETO 100, 100
```

While familiarity with the coordinate system eventually allows one to visualize this operation, it is far from intuitive.

Papert, however, includes a "turtle" in his Logo language. The turtle was originally an actual robot that could be programmed to move around; in most systems today it is represented by a cursor on the screen. As the turtle moves, it uses a "pen" to leave a "trail" that draws the graphic.

With turtle commands, a square can be drawn by:

```
FD 50 (that is, forward 50)
RT 90 (turn right 90 degrees)
FD 50
RT 90
FD 50
RT 90
FD 50
RT 90
```

Here, the student programmer can easily visualize walking and turning until he or she arrives back at the starting point. In keeping with Piaget's theories, the learning is congruent with the physical world and daily activities.

Logo includes control structures similar to those in other languages, so the above program can be rewritten as simply:

```
REPEAT 4 [FD 50 RT 90]
```

Logo is much more than a set of simple drawing commands, however. Students can also be encouraged to use the list-processing commands to create everything from computer-generated poetry to adventure games. Unlike LISP's obscurely named commands such as car and cdr, Logo's list commands are readily understandable. For example, first returns the first item in a list, while butfirst returns all of the list except the first item.

Logo procedures are introduced by the to keyword, implying that the programmer is "teaching the computer" how to do something. For example, a procedure to draw a square with a variable size and starting position might look like this:

```
to square :X :Y :Size
setxy :X :Y
repeat 4 [fd :Size rt 90]
end
```

Logo has been steadily enhanced over the years, and includes not only a full set of math functions, but also many versions include special sound, graphics, and multimedia functions for Windows or Macintosh systems. By the mid-1980s, Logo had been combined with the popular LEGO building toy to create LEGO Logo. This popular kit enables students to build and control a variety of robots and other gadgets.

By the 1990s, Logo had to some extent become a casualty to the pressure on educators to provide "real world" programming skills using languages such as C++ or Java. However, Logo using educators have continued to flourish in parts of Europe, Japan, and Latin America. Logo has also been energized by the development of two recent versions. MicroWorlds Logo took advantage of the Macintosh interface to provide a full-featured multimedia environment, and it was later adapted for Windows systems. Another version, StarLogo, emphasizes parallel processing concepts, and is able to control thousands of separate turtles that can be programmed to simulate behaviors such as bird flocks or traffic flows. As Brian Harvey's books show, Logo's accessible, interactive nature continues to make it a good choice for teaching computer science to adults as well.

Further Reading

Harvey, Brian. *Computer Science Logo Style.* Vols. 1–3, 2nd ed. Cambridge, Mass.: MIT Press, 1997.
LCSI Microworlds. Available online. URL: http://www.microworlds.com/. Accessed August 13, 2007.
Logo Foundation (MIT). Available online. URL: http://el.media.mit.edu/Logo-foundation/. Accessed August 13, 2007.
Papert, Seymour. *Mindstorms: Children, Computers, and Powerful Ideas.* New York: Basic Books, 1993.
"StarLogo on the Web." MIT Dept. of Education. Available online. URL: http://education.mit.edu/starlogo/. Accessed August 13, 2007.
"Welcome to MSWLogo." Available online. URL: http://www.softronix.com/logo.html. Accessed August 13, 2007.

loop

If computers were merely fast sequential calculators, they would still be of some use. However, much of the power of the computer comes from its ability to carry out repetitive tasks without supervision. The *loop* is the programming language structure that controls such activities. Virtually every language has some form of loop construct, with variations in syntax ranging from the relatively English-like COBOL and Pascal to the more cryptic C. We will use BASIC for our examples, since its syntax is easy to read.

The standard while loop performs the specified actions *as long as* the specified condition is true. For example:

```
While NOT EOF (Input_File)
  Read_Record
  Process_Record
Wend
Print "Done!"
```

This loop first checks to see whether the end of the input file (opened earlier) has been reached. If not, it reads and processes a record (using procedures defined elsewhere). The "Wend" marks the end of the statements controlled by the loop. When the end of the file is reached, the test fails (returns false) and control skips to the statement following Wend. See the accompanying flowchart for a visual depiction of the operation of this loop.

A variant form of while loop performs the test *after* executing the enclosed instructions. For example:

```
Do
  Print "Enter a number: "
  Input Number
  Print "You entered: ";Number
While (Number <> 0)
Print "I'm Done!"
```

This loop will display each number the user enters, then test it for zero. After a zero is encountered, control will skip to the final print statement.

Note that because this second form of while loop does not perform the test until it has performed the specified actions at least once, it would not be appropriate for the first example. In that case, the loop would attempt to get a record before discovering it had reached the end of the file, and an error would result.

The for loop is useful when an action is to be repeated for each of a limited series of cases. For example, this loop would print out the ASCII characters corresponding to the codes from 32 through 65:

```
For CharVal = 32 to 65 Step 1
  Print Char$(CharVal)
```

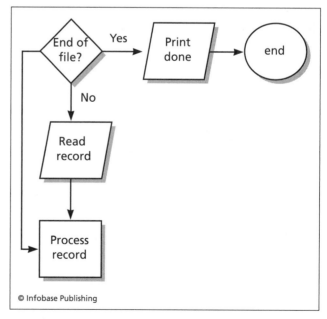

Flowchart for a loop that reads and processes records until it reaches the end of the file. Programmers must make sure that the end condition of a loop is properly defined, or the loop may run endlessly, "hanging" the program.

```
Next CharVal
```

Here Char$ is a function that output the character corresponding to the supplied ASCII character code. The step clause specifies the interval over which the variable within the loop is to be incremented. Here it's not strictly necessary, since it defaults to 1.

Loops of all sorts can be "nested" so that an inner loop executes completely for each step of the outer loop. For example:

```
For Vertical = 0 to 767
   For Horizontal = 0 to 1023
      Print_Pixel (Vertical, Horizontal)
   Next Horizontal
Next Vertical
```

Here the program will move across each line of the screen, printing the contents of each pixel. Each time the inner loop finishes, the outer loop increments, moving the scanning down to the next line. Indention is used to make the relation between the outer and inner loops clear.

In programming loops it's important to frame the test conditions correctly so that they terminate appropriately. An "endless loop" can cause a program to "hang" indefinitely. However, some programs do code an endless outer loop to indicate the program is to run indefinitely unless closed by the operating system. For example, the loop

```
While (1)
   ' Instructions go here
Wend
```

will execute indefinitely, since the value one is equivalent to "true."

Since many programs spend most of their time repeatedly executing loops, programmers seeking to improve the performance of their code pay especial attention to the code within the body of a loop. Any code such as a variable assignment, conversion, or calculation that needs to be done only once should be moved outside the loop.

Further Reading

"Control Flow." Wikipedia. Available online. URL: http://en.wikipedia.org/wiki/Control_flow. Accessed August 13, 2007.

Sebesta, Robert W. *Concepts of Programming Languages*. 8th ed. Boston: Addison-Wesley, 2007.

Lua

Lua is a scripting language created by three programmers at the Pontifical University of Rio de Janeiro, Brazil. (The word *Lua* is Portuguese for "Moon"). The language has begun to attract some attention, particularly among game and Web programmers.

Lua has simple syntax and can support both traditional (imperative) programming and functional programming (see FUNCTIONAL LANGUAGES). Many of the features such as inheritance and name spaces that are built into most object-oriented languages are not part of Lua, but can be created through the language's extension methods (see OBJECT-ORIENTED PROGRAMMING).

A simple function in Lua looks like this:

```
function factorial(n)
if n = 0 then
   return 1
end
return n * factorial(n - 1)
end
```

Note the lack of required semicolons.

Besides a simple assortment of built-in types, Lua uses tables to create complex, user-defined types. Tables include pairs of values and keys. The key can either be a number (creating the equivalent of an array in other languages), or a string. For example, the following table consists of three values indexed by strings:

```
coin = { quarter = 25, dime = 10, nickel = 5,
   penny = 1 }
printer (coin["dime"])       -- prints 10
```

By including functions and the data they use into a table, classes similar to those used in object-oriented languages can also be created.

IMPLEMENTATION AND USE

Lua programs are compiled to an intermediate form (bytecode) that runs on a Lua virtual machine for each platform. Lua is intended to work closely with C programs, transferring data via a stack.

Because of its compact runtime packages, Lua is often included in applications to provide an interface for editing or extending the application. This is particularly true of games such as *World of Warcraft*.

Further Reading

Ierusalimschy, Roberto. *Programming in Lua*. 2nd ed. Rio de Janeiro, Brazil: Lua.org, 2006.

Jung, Kurt, and Aaron Brown. *Beginning Lua Programming*. Indianapolis: Wiley, 2007.

Lua.org. Available online. URL: http://www.lua.org. Accessed September 26, 2007.

M

Macintosh

Since its inception in 1984, Apple's Macintosh line of personal computers has offered a distinctive, innovative alternative to the more mainstream IBM-compatible PCs. When the Macintosh came out, it was billed as the computer "for the rest of us." Unlike the text-based, command-driven DOS-based IBM PC and its "clones," the "Mac" offered an interface that consisted of menus, folders, and icons that could be manipulated by clicking and dragging (see USER INTERFACE and MOUSE). The system came out of the box with a paint/draw program and a word processor that could show documents using the actual font sizes and styles that would appear in printed text. This "WYSIWYG" (What You See Is What You Get) feature quickly made the Mac the machine of choice for desktop publishers and graphic artists. The Mac also met with some success in the educational market, where the way had been paved by the earlier Apple II.

However, there were factors would limit the Mac to a minority market share. The first models ran slowly. Although its Motorola 68000 processor was comparable to the Intel 80286 used by the IBM XT and AT series, the need to draw extensive graphics placed a heavier burden on the Mac's CPU.

Marketing decisions also proved to be problematic. The IBM PC had an "open architecture." Clone makers were able to legally produce machines that were functionally equivalent, and Microsoft was able to license to clone manufacturers essentially the same DOS operating system that IBM used. This created a robust market as manufacturers competed with added features or lower prices.

Apple, on the other hand, jealously guarded the Apple's hardware and the ROM (read only memory) that held the key operating system code. Apple made only a brief and half-hearted attempt to license the Mac OS to third parties in 1995, and by then it was probably too late. Apple CEO Steve Jobs (see JOBS, STEVE) kept prices relatively high, betting that the Mac's unique operating system and interface would entice people to buy the more expensive machine.

But something of a vicious circle set in. Since the Mac used a unique operating system, developing new applications (or porting existing ones) to the Mac was expensive. And since the Mac market represented only a small fraction of the PC-compatible market, developers were reluctant to create such software. Some flagship products such as Aldus PageMaker and Adobe Photoshop did cater to the Mac's graphic strengths. In general, however, the PC-compatible owner had a far wider range of software to choose from, and businesses were traditionally more comfortable with IBM equipment, even if IBM didn't make it.

Microsoft helped develop some successful Mac software, including versions of its Word and Office programs. But Microsoft CEO Bill Gates responded to the Mac's interface advantages over MS-DOS by developing a new operating environment, Windows. Apple sued, claiming that Microsoft had gone beyond the license it had negotiated with Apple for use of elements of the Mac interface. By the early 1990s, however, Apple had lost the lawsuit. While the early versions of Windows were clumsy and met with little success, version 3.0 and, later, Windows 95 succeeded in providing a user experience that was increasingly close to that achieved by the Mac.

A Power Mac G5 with a 30-inch Apple Cinema HD display. (APPLE COMPUTER)

Apple kept trying to innovate and carve out a larger market share, designing both Power Macs that used the PowerPC RISC (reduced instruction set) microprocessor and PowerBook laptops. Toward the end of the 1990s, Apple tried to address the low end of the market with the iMac, a colorful, sleek machine packed with features such as home video editing, and achieved modest success in attracting new customers.

NEW INSIDES, NEW DIRECTIONS?

In 2000 Apple began to revamp what was becoming a somewhat aging architecture. First came the replacement of the operating system with a UNIX variant (see UNIX) while retaining the user interface (see OS X). In 2006 Apple began to transition the Mac's processor from the Motorola/IBM Power PC to the Intel chips that power Windows-compatible PCs. The use of a widely available chip (and provision for Windows via the "Boot Camp" utility) may make the Mac cheaper and more attractive to mainstream PC users. Meanwhile the Macintosh OS X operating system continued to evolve from the "Tiger" edition to "Leopard," released in Fall 2007.

While the Mac continues to be a niche market, sales have been strong, increasing steadily each year. However, in recent years the Mac has been overshadowed somewhat as an Apple icon, supplanted by the iPod and in 2007 by the iPhone.

Further Reading

Levy, Steven. *Insanely Great: The Life and Times of Macintosh, the Computer that Changed Everything.* New York: Penguin, 2000.
Litt, Samual A., et al. *Mac OS X Bible, Tiger Edition.* Indianapolis: Wiley, 2005.
MacDailyNews. Available online. URL: http://www.macdailynews.com/. Accessed August 14, 2007.
Mac Observer. Available online. URL: http://www.macobserver.com/. Accessed August 14, 2007.
Macworld: The Mac Experts. Available online. URL: http://www.macworld.com/. Accessed August 14, 2007.
Pogue, David, and Adam Goldstein. *Switching to the Mac: The Missing Manual, Tiger Edition.* Sebastapol, Calif.: O'Reilly Media, 2005.

macro

For both programmers and ordinary users, the ability to "package" a group of instructions so that it can be invoked with a single command can save a lot of effort. The term *macro* is used for such instruction packages in a variety of contexts.

In the early days, programmers had to work with low-level machine instructions (see ASSEMBLER). Developers soon realized that a program could be used to write other programs. This program, called a macro assembler, lets the programmer write a group of instructions such as:

```
COMPARE macro

LOAD %1 ' load first data item
STOREX  ' store in register X
LOAD %2 ' load second data item
STOREY  ' store in register Y
CMPXY   ' compare X and Y registers

endm
```

Now, if the programmer wants to compare the contents of two memory locations (say COUNTER and LIMIT), he or she can write simply:

```
COMPARE COUNT LIMIT
```

The assembler replaces COMPARE with the sequence of instructions above, substituting COUNT and LIMIT for %1 and %2.

Macros are also used in some higher-level languages, notably C. A module called the macro processor performs the required substitutions into the source code before the code is parsed and compiled. For example, a C programmer might include this macro in a program file:

```
#define IS_LOWERCASE(x) (( (x)>='a') && (
(x) <='z') )
```

Somewhere in the program there might appear a statement such as:

```
if IS_LOWERCASE (Letter)
```

The macro processor will replace this with:

```
if (( (Letter)>='a') && ((Letter) <='z') )
```

This saves typing as well as reducing the chance of a typo creating a hard-to-find bug.

Macros are similar to procedures and functions (see PROCEDURES AND FUNCTIONS) in that they let the programmer treat a group of instructions as a single unit, simplifying coding. However, a given procedure appears in the code only once, although it may be called upon from many different parts of the program. A macro, on the other hand, is not "called." Each time it is mentioned, the macro is *replaced* by the corresponding instructions. Thus macros increase the size of the source code.

Many programmers today prefer using functions with the appropriate code rather than macros. Using functions saves space, since each function's code need only appear once. Although there is some processing overhead at runtime in calling the function, the function approach also ensures that the data sent to the function will be checked to make sure it is of the proper type. The macro, on the other hand, usually leaves it up to the programmer to make sure the data type being used is appropriate.

APPLICATION MACROS

The term *macro* is also used with applications software. Here it can mean a series of commands (such as cursor positioning or text formatting) that are recorded and assigned to a certain key combination. For example, a word processor user might define a macro called Letter and record the keystrokes and/or mouse movements needed to open a new document, insert a letterhead from a file, update the date, insert a salutation, and position the cursor to continue writing the letter. The recorded keystrokes might be assigned to the key combination Control + L.

More elaborate macros can be written to automate complex tasks in spreadsheets and word processors. Microsoft provides an entire language, Visual Basic for Applications (VBA), for writing macros for its Office products.

Further Reading

Gonzalez, Juan Pablo, et al. *Office VBA Macros You Can Use Today: Over 100 Amazing Ways to Automate Word, Excel, PowerPoint, Outlook and Access.* Uniontown, Ohio: Holy Macro! Books, 2005.

Jelen, Bill. *VBA and Macros for Microsoft Office Excel 2007.* Indianapolis: Que, 2007.

Kochan, Stephen. *Programming in C.* 3rd ed. Indianapolis: Sams, 2004.

"Preprocessor Directives." Available online. URL: http://www.cplusplus.com/doc/tutorial/preprocessor.html. Accessed August 14, 2007.

Maes, Pattie
(1961–   )
Belgian/American
Computer Scientist

Pattie Maes is a pioneer in the creation of software agents, intelligent programs that work with people to help them find what they need online, whether it is relevant news stories, a vacation itinerary, or a good place for a romantic dinner for two in San Francisco.

Born June 1, 1961, in Brussels, Belgium, Maes was interested in science (particularly biology) from an early age. She received bachelor's (1983) and doctoral (1987) degrees in computer science and artificial intelligence from the University of Brussels.

In 1989 Maes moved from Belgium to the Massachusetts Institute of Technology, where she joined the Artificial Intelligence Lab. There she worked with an innovative researcher who had created swarms of simple but intriguing insectlike robots (see BROOKS, RODNEY). Two years later Maes became an associate professor at the MIT Media Lab, famed for innovations in how people interact with computer technology (see MIT MEDIA LAB). There she founded the Software Agents Group to promote the development of a new kind of computer program.

These programs (see SOFTWARE AGENT) have considerable autonomy and intelligence. Like a human travel or real estate agent, an agent program must have detailed knowledge of the appropriate area of expertise; the ability to ask the client questions about preferences, priorities, and constraints; and the ability to find the best deals and negotiate with service providers.

Maes's goal has been to create software agents who think and act much like their human counterparts. To carry out a task using an agent, the user does not have to specify exactly how it is to be done. Rather, the user describes the task, and the software engages in a dialog with the user to obtain the necessary guidance.

A software travel agent would know—or ask about—such things as how much the user wants to spend and whether he or she prefers sites involving nature, history, or adventure. It would also ask about and take into consideration constraints of budget, travel time, comfort, and so on. The software agent would then use its database and procedures to put together an itinerary based on the user's needs and desires. It would not only know where to find the best fares and rates, it would also know how to negotiate with hotels and other services. Indeed, it might negotiate with *their* software agents.

In 1995 Maes cofounded Firefly Networks, a company that attempted to create commercial applications for software agent technology. Although the company was bought by Microsoft in 1998, one of its ideas—"collaborative filtering"—can be experienced by visitors to sites such as Amazon.com. Users in effect are given an agent whose job it is to provide recommendations for books and other media. The recommendations are based upon observing not only what items the user has already purchased, but also what else has been bought by people who bought those same items. More advanced agents can also tap into feedback resources such as user book reviews on Amazon or auction feedback on eBay (see SOCIAL NETWORKING).

A listing of Maes's current research projects at MIT conveys many aspects of and possible applications for software agents. These include the combining of agents with interactive virtual reality, using agent technology to create characters for interactive storytelling, the use of agents to match people with the news and other information they are most likely to be interested in, an agent that could be sent into an online market to buy or sell goods, and even a "Yenta" agent that would introduce people who are most likely to make a good match.

Maes has participated in many high-profile conferences such as AAAI (American Association for Artificial Intelligence) and ACM Siggraph, and her work has been featured in numerous magazine articles. She was one of 16 modern "visionaries" chosen to speak at the 50th anniversary of the ACM. She has also been repeatedly named by *Upside* magazine as one of the 100 most influential people for development of the Internet and e-commerce. *Time Digital* featured her in a cover story and selected her as a member of its "cyber elite." *Newsweek* put her on its list of 100 Americans to be watched for in the year 2000. That same year the Massachusetts Interactive Media Council gave her its Lifetime Achievement Award.

Further Reading
D'inverno, Mark, and Michael Luck, eds. *Understanding Agent Systems*. 2nd. ed. New York: Springer Verlag, 2003.
Maes, Pattie. "Intelligence Augmentation: A Talk with Pattie Maes." Available online. URL: http://www.edge.org/3rd_culture/maes./ Accessed May 5, 2007.
Software Agents group at MIT Media Lab. http://agents.mit.edu. Accessed May 5, 2007.

mainframe

In the era of vacuum tube technology, all computers were large, room-filling machines. By the 1960s, the use of transistors (and later, integrated circuits), enabled the production of smaller (roughly, refrigerator-sized) systems (see MINICOMPUTER). By the late 1970s, desktop computers were being designed around newly available computer chips (see MICROPROCESSOR). Although they, too, now use integrated circuits and microprocessors, the largest scale machines are still called mainframes.

The first commercial computer, the UNIVAC I (see ECKERT, J. PRESPER and MAUCHLY, JOHN) entered service in 1951. These machines consisted of a number of large cabinets. The cabinet that held the main processor and main memory was originally referred to as the "mainframe" before the name was given to the whole class of machines.

Although the UNIVAC (eventually taken over by Sperry Corp.) was quite successful, by the 1960s the quintessential mainframes were those built by IBM, which controlled about two-thirds of the market. The IBM 360 (and in the 1970s, the 370) offered a range of upwardly compatible systems and peripherals, providing an integrated solution for large businesses.

Traditionally, mainframes were affordable mainly by large businesses and government agencies. Their main application was large-scale data processing, such as the census, Social Security, large company payrolls, and other applications that required the processing of large amounts of data, which were stored on punched cards or transferred to magnetic tape. Programmers typically punched their COBOL or other commands onto decks of punched cards that were submitted together with processing instructions (see JOB CONTROL LANGUAGE) to operators who mounted the required data tapes or cards and then submitted the program cards to the computer.

By the late 1960s, however, *time-sharing* systems allowed large computers to be partitioned into separate areas so that they can be used by several persons at the same time. The punched cards began to be replaced by Teletypes or video terminals at which programs or other commands could be entered and their results displayed or printed. At about the same time, smaller computers were being developed by Digital Equipment Corporation (DEC) with its PDP series (see MINICOMPUTER).

With increasingly powerful minicomputers and later, desktop computers, the distinction between mainframe, minicomputer, and microcomputer became much less pronounced. To the extent it remains, the distinction today is more about the bandwidth or amount of data that can be processed in a given time than about raw processor performance. Powerful desktop computers combined into networks have taken over many of the tasks formerly assigned to the largest mainframe computers. With a network, even a large database can be stored on dedicated computers (see FILE SERVER) and integrated with software running on the individual desktops.

Nevertheless, mainframes such as the IBM System/390 are still used for applications that involve processing large numbers of transactions in near real-time. Indeed, many of the largest e-commerce organizations have a mainframe at the heart of their site. The reason is that while the raw processing power of high-end desktop systems today rivals that of many mainframes, the latter also have high-capacity *channels* for moving large amounts of data into and out of the processor.

Early desktop PCs relied upon their single processor to handle most of the burden of input/output (I/O). Although PCs now have I/O channels with separate processors (see BUS), mainframes still have a much higher data throughput. The mainframe can also be easier to maintain than a network, since software upgrades and data backups can be handled from a central location. On the other hand, a

The IBM System/360 was the most successful mainframe in computer history. It was actually a "family" of upwardly compatible machines. (IBM CORPORATE ARCHIVES)

system depending on a single mainframe also has a single point of vulnerability, while a network with multiple mirrored file servers can work around the failure of an individual server.

Further Reading

Butler, Janet G. *Mainframe to Client-Server Migration: Strategic Planning Issues and Techniques.* Charleston, S.C.: Computer Technology Research Corporation, 1996.

Ebbers, Mike, Wayne O'Brien, and Bill Ogden. *Introduction to the New Mainframe: z/OS Basics.* Raleigh, N.C.: IBM Publications, 2007. Available online. URL: ftp://www.redbooks.ibm.com/redbooks/SG246366/zosbasics_textbook.pdf. Accessed August 14, 2007.

"Mainframe Programming: Some Useful Resources for Practitioners of the Craft." Available online. URL: http://www.oberoi-net.com/mainfrme.html. Accessed August 14, 2007.

Prasad, N. S. *IBM Mainframes: Architecture and Design.* 2nd ed. New York: McGraw-Hill, 1994.

Pugh, Emerson W., Lyle R. Johnson, and John H. Palmer. *IBM's 360 and Early 370 Systems.* Cambridge, Mass.: MIT Press, 1991.

management information system

The first large-scale use of computers in business in the late 1950s and 1960s focused on fundamental data processing. Companies saw computers primarily as a way to automate such functions as payroll, inventory, orders, and accounts payable, hoping to keep up with the growing volume of data in the expanding economy while saving labor costs associated with manual methods. The separate data files and programs used for basic business functions were generally not well integrated and could not be easily used to obtain crucial information about the performance of the business.

By the 1970s, the growing capabilities of computers encouraged executives to look for ways that their information systems could be used to competitive advantage. Clearly, one possibility was that reporting and analysis software could be used to help them make faster and better decisions, such as about what products or markets to emphasize. To achieve this, however, the "data processing department" had to be transformed into a "management information system" (MIS) that could allow analysis of business operations at a variety of levels.

THE MIS PYRAMID

If one thinks of the information infrastructure of an enterprise as being shaped like a pyramid, the bottom of the pyramid consists of the transactions themselves, where products and services are delivered, and the supporting point of sale, inventory, and distribution systems that keep track of the flow of product.

The next layer up begins the process of integration and operational control. For example, previously separate sales and inventory system (perhaps updated through a daily batch process) now become part of an integrated system where a sale is immediately reflected in reduced inventory, and the inventory system is in turn interfaced with the order system so more of a product is ordered when it goes out of stock.

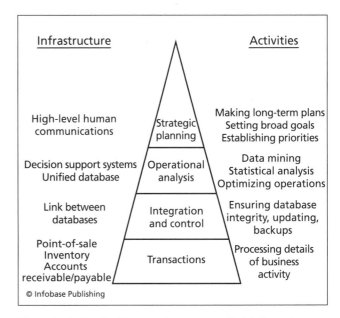

The activities involved in managing an enterprise's information infrastructure can be drawn as a pyramid. The raw material of transactions at the bottom are stored in databases. Moving up the pyramid, these data sources are integrated and refined to provide better information about business operations as well as material for operational analysis and strategic planning.

The next layer can be called the operational analysis layer. Here such functions as sales, inventory, and ordering aren't simply connected; they are part of the same system of databases. This means that both simple and complex queries and analysis can be run against a database containing every type of transaction that the business engages in. In addition to routine reports such as sales by region or product line, market researchers or strategic planners can receive the data they need to answer questions such as:

- What products are staying on the shelf the longest?

- What is the ratio between profitability and shelf space for particular items?

- What is the relationship between price reductions, sales, and profits for a certain category of items?

The goal of this layer is to help managers identify the variables that affect the performance of their store or other business division and to determine how to optimize that performance (see also DIGITAL DASHBOARD).

The very top level can be called the strategic planning layer. Here top-level executives are interested in the overall direction of the business: determining which divisions of a company should receive the greatest long-term investment, and which perhaps should be phased out. For example:

- Which kind of sales are growing the fastest: in-store, mail-order catalog, or Internet on-line store?

- How is our market share trending compared to various classes of competitors?

- How are sales trending with regard to various types (demographics) of customers?

SOFTWARE SUPPORT

There are many considerations to choosing appropriate software to support the users who are trying to answer questions at the various levels of management. At minimum, to create a true management information system, the information from daily transactions must be made accessible to a variety of query or analysis programs.

In the past three decades many established businesses have had to go through a painful process of converting a variety of separate databases and "legacy software" (often written in COBOL in the 1960s or 1970s) into a modern relational database such as Oracle or Microsoft Access. Sometimes a company has decided that the cost of rewriting software and converting data is simply too high, and instead, opts for a patchwork of utility programs to convert data from one program to another.

The growth of networking in the 1980s and Web-based intranets in the 1990s required that the old model of a large, centralized data repository accessed directly by only a few users be replaced by a less centralized model, sometimes going as far as using a distributed database system where data "objects" can reside throughout the network yet be accessed quickly by any user (see DATABASE MANAGEMENT SYSTEM). An alternative is the data repository that includes queries and other tools (see DATA WAREHOUSE).

FUTURE OF MIS

With the prominence of the Internet and e-commerce today, MIS has had to cope with an even more complex and fast-moving world. On the one hand, widespread e-commerce enables the capturing of more detailed data about transactions and consumer behavior in general. New tools for analyzing large repositories of data (see DATA MINING) make it possible to continually derive new insights from the recent past. It is thus clear that information is not just a tool but also a corporate asset in itself. On the other hand, fierce competition and often shrinking profit margins in e-commerce have placed increasing pressure on MIS departments to find the greatest competitive advantage in the shortest possible time.

The importance of MIS has also been reflected in its place in the corporate hierarchy. The top-level executive post of Chief Information Officer (CIO) has perhaps not yet achieved parity with the Chief Financial Officer (CFO), but healthy budgets for MIS even in constrained economic times testify to its continuing importance.

Further Reading

Kroenke, David. *Using MIS.* Upper Saddle River, N.J.: Prentice Hall, 2005.

Laudon, Kenneth C., and Jane P. Laudon. *Management Information Systems: Managing the Digital Firm.* 10th ed. Upper Saddle River, N.J.: Prentice Hall, 2006.

"MIS Resources." Available online. URL: http://www3.uakron.edu/management/index.html. Accessed August 14, 2007.

Turban, Efraim, et al. *Information Technology for Management.* 6th ed. New York: Wiley, 2007

map information and navigation systems

A variety of online services use the integration of maps and databases to provide detailed information ranging from weather forecasts to traffic conditions to local shopping and restaurants. Increasingly, these services can be customized to the user's needs. Further, when combined with global positioning system (GPS) devices, the map display can be focused on the user's current location, providing navigation and/or "points of interest" information. (For mapping systems primarily designed for scientific or other analytical use, see GEOGRAPHICAL INFORMATION SYSTEMS.)

MAPQUEST

MapQuest has its roots in the Cartographic Services Division of R. R. Donnelley, a leading maker of printed maps. The company first went online in 1996, was renamed MapQuest in 1999, and was acquired by America Online (AOL) in 2000.

The basic services offered by MapQuest are street maps of a user-specified location, and driving routes between an origin and a destination. In recent years the service has been elaborated to allow users to customize routes, to obtain location-related "Yellow Pages" service from AOL, and to receive maps and driving directions on PDAs and mobile phones.

GOOGLE MAPS AND GOOGLE EARTH

Arriving on the Web in 2005 was Google Maps, a more sophisticated and versatile mapping service. There are four types of map view: street map, actual satellite or aerial photo, street map overlaid on photo, and street-level photo views (in selected cities.) Besides specifying a particular location for the map, users can enter queries such as "pizza in Berkeley" to highlight locations where the pies are available.

A related application is Google Earth, which was based on a product acquired by Google in 2004. Google Earth is available for PCs running Windows, Mac OS, and Linux, and shows detailed imagery of most terrain at 15-meters resolution or smaller, with considerably more detailed imagery of some cities. Views have also been enhanced to provide a better 3D visualization of features such as the Grand Canyon or Mount Everest, as well as a significant number of major buildings. In 2007 Google added sky views as well as surface views of the Moon and Mars.

Like other Google services, Google Maps and Google Earth offer extensive interfaces that can be used to link maps and imagery with data from other programs. For example, Wikipedia articles that include coordinate tags will now be automatically linked to the corresponding content from Google Earth. (For more on the creation of new applications through combining existing services, see MASHUPS.)

Because mapping services (particularly Google) have featured relatively high-resolution aerial and even street-level photographic views, some government agencies around the world have complained that the service is providing too much detail of military or other sensitive installations. (This is also a potential terrorism concern.) Also, privacy advocates are concerned that actual images of identifiable persons show up in the street-level imagery.

Google has responded to security concerns by blurring the imagery of some U.S. locations, presumably at government request. They have also argued that pictures of people who are in public places months or years earlier are not a real privacy concern.

MOBILE NAVIGATION SYSTEMS

Mobile navigation systems can provide maps, driving directions, and sometimes additional information such as traffic conditions and advisories. The system can be either built into the dashboard (as with many higher-end vehicles) or available as a mounted unit such as those from Garmin, Tom Tom, and Magellan (see CARS AND COMPUTING).

Mobile navigation systems link the user's current location (obtained through the GPS system) to the unit's stored database of maps and other information, such as local points of interest. (Some units have backup dead-reckoning systems based on the car's motion, for use when GPS signals are lost or distorted because of buildings or other obstacles.)

An alternative to in-car systems is the smartphone or PDA equipped with GPS and navigation software. These have the advantage of also being useful for pedestrians or hikers.

Users should look for navigation systems that have features such as:

- large, clear, readable display

- overhead display and display from driver's point of view

- uncluttered user interface to avoid distracting the driver

- voice announcements of driving directions and other information

- comprehensive maps and database including the ability to load supplemental coverage for other areas

An important and sometimes overlooked issue with mobile navigation systems is the need to design the display and user interface so as to minimize distraction. A combination of large displays without unnecessary complexity and the use of spoken driving directions can help. A more controversial approach is to disable many functions of the system (such as entering new destinations) while the car is in motion.

Further Reading

Car GPS (Navigation) Reviews. CNET. Available online. URL: http://reviews.cnet.com/4566-3430_7-0.html. Accessed September 30, 2007.

Crowder, David A. *Google Earth for Dummies*. Hoboken, N.J.: Wiley, 2007.

Google Maps. Available online. URL: http://maps.google.com. Accessed September 29, 2007.

Hofmann-Wellenhof, Bernhard, Klaus Legat, and Manfred Wieser. *Navigation: Principles of Positioning and Guidance*. New York: Springer, 2003.

"Introduction to In-Car Navigation." Crutchfield Advisor. Available online. URL: http://www.crutchfieldadvisor.com/ISEO-rgbtcspd/learningcenter/car/navigation.html. Accessed September 30, 2007.

Mapquest. Available online. URL: http://www.mapquest.com/. Accessed September 29, 2007.

Purvis, Michael, Jeffrey Sambells, and Cameron Turner. *Beginning Google Maps Applications with PHP and Ajax: From Novice to Professional*. Berkeley, Calif.: APress, 2006.

marketing of software

The way software has been produced and marketed has changed considerably in the past five decades. In the nascent computer industry of the 1950s, commercial software was developed and marketed by the manufacturers of computer systems—firms such as Univac (later Sperry-Univac), Burroughs, and of course, IBM (see MAINFRAME). However, a separate (third-party) software industry emerged as early as 1955 with the founding of Computer Usage Corporation (CUC) by two former IBM employees. Nevertheless, the primary competition was between hardware manufacturers, with software seen as part of the overall package.

By the early 1960s, larger software companies emerged such as Computer Science Corporation (CSC) and Electronic Data Systems (EDS), which became an empire under the energetic, albeit often controversial leadership of H. Ross Perot, as well as the European giant SAP. These companies specialized in providing customized software solutions for users who could not meet their needs with the software library offered by the maker of their computer system (see also BUSINESS APPLICATIONS OF COMPUTERS).

By the 1970s, however, vendor-supplied and contracted custom software alternatives were being increasingly accompanied by "off the shelf" software packages. By 1976, 100 software products from 64 software companies had reached the $1 million mark in sales.

The 1980s saw the emergence of a completely new sector: desktop computer (PC) users. Traditionally, software had been marketed to programmers or managers, but now individual users or office managers could buy and install word processing programs, spreadsheets, database, and other programs. At the same time, a market for software for use in the home and schools, particularly education, personal creativity, and game programs required new methods of marketing. For the first time ads for software began to appear on TV and in general-interest magazines.

While large businesses still required custom-made software, most small to medium businesses looked for powerful and integrated office software solutions (see APPLICATION SUITE, OFFICE AUTOMATION, and GATES, WILLIAM). By the mid-1990s, Microsoft's Office suite had dominated this market, although that dominance began to be threatened to some extent by software written in Java or hosted on LINUX systems (see also OPEN SOURCE MOVEMENT).

The growth in the Internet (see E-COMMERCE) has also offered new venues for the marketing and distribution of software. Sites such as ZDNet and CNet have to some extent displaced computer magazines as sources for product reviews. These sites also offer extensive libraries of "try before you buy" software (see SHAREWARE), some of which is trial versions of full-blown commercial products. The local "mom and pop" PC software store has largely vanished, with software now marketed mainly by chain stores such as Electronics Boutique or CompUSA, and increasingly, through Web-based stores, often established by the chains, as well as the giant on-line bookstore Amazon.com.

Another trend impacting the traditional package model of software delivery is the hosting and serving of applications online (see APPLICATION SERVICE PROVIDER) using a subscription model. Google has offered a free suite of office and communications applications online, including components that can be used off-line. This and other emerging offerings may portend further splitting of the software market into high-end or specialized applications on the one hand, and added-value or premium versions of free software on the other.

Further Reading

Campbell-Kelly, Martin. *From Airline Reservations to Sonic the Hedgehog: A History of the Software Industry.* Cambridge, Mass.: MIT Press, 2004.

Computer History Museum. Software Industry SIG. "Preserving the History of the Software Industry." Available online. URL: http://www.softwarehistory.org/. Accessed August 14, 2007.

Cusumano, Michael. *The Business of Software: What Every Manager, Programmer, and Entrepreneur Must Know to Thrive and Survive in Good Times and Bad.* New York: Free Press, 2004.

Hasted, Edward. *Software that Sells: A Practical Guide to Developing and Marketing Your Software Project.* Indianapolis: Wiley, 2005.

Plunkett, Jack W. *Plunkett's InfoTech Industry Almanac.* Houston, Tex.: Plunkett Research, 2007.

mashups

Today the creative world has blurred the boundaries that once separated works of art. Songs are sampled and remixed from earlier songs. *Star Wars* fans create new chapters in the saga by remixing existing footage and adding their own new footage and effects. The fluidity and ease of manipulation of all digital data, regardless of its original source, has made it easier than ever before to reuse, repurpose, or reinvent content. It is not surprising, then, that software itself can be mixed and matched to create new applications called mashups.

Mashups are new applications created by putting together data or features from existing applications, such as maps and databases. Many Web applications are designed to make their services available to other programs (see SERVICE-ORIENTED ARCHITECTURE and WEB 2.0). This can be done at the programming level by providing an application programming interface (see API), or even through simpler facilities that can be used easily by nonprogrammers.

A number of major Web sites and applications provide resources for mashups, including Google (particularly Google Maps), Amazon, eBay, Flickr, YouTube (image and video sharing), the "social bookmarking" site del.iciou.us, and many others. Some of these services (and third parties) have provided mashup editors to simplify the process of creating mashups.

As a simple example, suppose one wants to create a map display showing rentals in San Francisco by neighborhood, color coded by rent range. Google Maps can generate maps for any area and plot points on them, given coordinates in a standard tag. Craigslist has rental ads including addresses. To create a mashup, a "screenscraper" can be used to extract addresses and rents from the ads, and the colored points can then be plotted on a map of the city via Google.

While the most common types of mashups are created by and for ordinary users, mashup techniques can also be used in business applications. For example, data from several sources (such as Web feeds [see RSS]) or various databases can be brought together and provided with an easy-to-use interface (see DIGITAL DASHBOARD).

Mashups can also be considered to be an aspect of the emerging new information economy. Developers may be finding it in their interest to provide APIs and services suitable for mashups because, in turn, the mashups increase the use of the original program. By providing these services, developers are also contributing to a "digital commons" that benefits all.

Further Reading

Feiler, Jesse. *How to Do Everything with Web 2.0 Mashups.* Emeryville, Calif.: McGraw Hill-Osborne, 2007.

Google Mashup Editor. Available online. URL: http://code.google.com/gme/. Accessed October 1, 2007.

"How to Create a Mashup." Mashup Awards. Available online. URL: http://mashupawards.com/create/. Accessed October 1, 2007.

Lewis, Andre, et al. *Beginning Google Maps Applications with Rails and Ajax: From Novice to Professional.* Berkeley, Calif.: Apress, 2007.

Microsoft Popfly. Available online. URL: http://www.popfly.com/. Accessed October 1, 2007.

QEDWiki (IBM). Available online. URL: http://services.alphaworks.ibm.com/qedwiki/. Accessed October 1, 2007.

Yahoo! Pipes. Available online. URL: http://pipes.yahoo.com/pipes/. Accessed October 1, 2007.

Yee, Raymond. *Pro Web 2.0 Mashups: Remixing Data and Web Services.* Berkeley, Calif.: Apress, 2008.

mathematics of computing

The roots of modern computer science lie in an interest in rapid computation. Simple mechanical calculators (see CALCULATOR) may date back to ancient times; however, it is the work of mathematicians Blaise Pascal (1623–1662) and Gottfried Leibniz (1646–1716) that gave rise to the first practical mechanical calculators. By the mid-19th century, Charles Babbage (1791–1871) had conceptualized and designed mechanical computers that included the essential features (programs, processor, memory, input/output) of the modern digital computer (see BABBAGE, CHARLES). His motivation was the need for rapid, accurate calculation of statistical tables made necessary by the manufacturing economy of the Industrial Revolution. By the end of the century, the volume of such data had increased to the point where mechanical calculators and tabulators (see HOLLERITH, HERMAN) had become the only practical way to keep up.

Mathematically, a computer can be seen as a way to rapidly and automatically execute procedures that have been proven to lead to reliable solutions to a problem (see ALGORITHM). Once computers came on the scene, mathematical principles for verifying or proving algorithms would acquire new practical importance.

By the early 20th century, however, mathematicians were beginning to examine the problem of determining what propositions were provable, and in 1931 Kurt Godel published a proof that any mathematical system necessarily allowed for the formation of propositions that could not be proven using the axioms of that system. An analogous question was determining what problems were computable. Working independently, two researchers (see CHURCH, ALONZO and TURING, ALAN) formulated models that could be used to test for computability. Turing's model, in particular, provided a theoretical construct (the Turing Machine) that could, using combinations of a few simple operations, calculate anything that was computable.

By the 1940s, electromechanical (relays) or electronic (tube) switching elements made it possible to build practical high-speed computers. Computer circuit designers could draw upon the advances in symbolic logic in the 19th century (see BOOLEAN OPERATORS). Boolean logic, with its true/false values, would prove ideal for operating computers constructed from on/off switched elements.

The mathematical tools of the previous 150 years could now be used to design systems that could not only calculate but also manipulate symbols and achieve results in higher mathematics (see the next entry, MATHEMATICS SOFTWARE).

MATHEMATICS AND MODERN COMPUTERS

A variety of mathematical disciplines bear upon the design and use of modern computers. Simple or complex algebra using variables in formulas is at the heart of many programs ranging from financial software to flight simulators. Indeed, one of the most enduring scientific and engineering languages takes its name from the process of translating formulas into computer instructions (see FORTRAN).

Geometry, particularly the analytical geometry based upon the coordinate system devised by Rene Descartes (1596–1650) is fundamental to computer graphics displays, where the screen is divided into X (vertical) and Y (horizontal) axes. Modern graphics systems have added 3D depiction and sophisticated algorithms to allow the rapid display of complex objects. Beyond graphics, the Cartesian insight that converted geometry into algebra makes a variety of geometrical problems accessible to computation, including the finding of optimum paths for circuit design. Design of computer and network architectures also involves the related field of topology. The fascinating field of fractal geometry has found use in computer graphics and data storage techniques (see FRACTALS IN COMPUTING).

Aspects of number theory, often considered the most abstract branch of mathematics, have found surprising relevance in computer applications. These include randomization (random number generation) and the factoring of large numbers, which is crucial for cryptography.

Mathematics also bears on computer networking with regard to communications theory (see BANDWIDTH and SHANNON, CLAUDE) and techniques for error correction.

THE COMPUTER'S CONTRIBUTION TO MATHEMATICS

Mathematics as a discipline is thus essential to its younger sibling, computer science. In turn, however, computer science and technology have enriched the pursuit of mathematical truth in surprising ways. As early as 1956, a program called *Logic Theorist,* written by Herbert Simon (1916–2001) and Allen Newell (1927–1992) demonstrated how a program (that is, a collection of algorithms) could prove mathematical propositions given axioms and rules. While these early programs worked on a somewhat hit-or-miss basis, later theorem-solving programs produced solutions different from the standard ones known to mathematicians, and sometimes more elegant. Thus the computer, which began as an aid to calculation, became an aid to symbol manipulation and to some extent an independent creative source.

Further Reading
"Computers & Math News." ScienceDaily. Available online. URL: http://www.sciencedaily.com/news/computers_math/. Accessed August 14, 2007.
Henderson, Harry. *Modern Mathematics: Powerful Patterns in Nature and Society.* New York: Chelsea House, 2007.
Maxfield, Clive, and Alan Brown. *The Definitive Guide to How Computers Do Math.* New York: Wiley-Interscience, 2005.
McCullough, Robert. *Mathematics for Computer Technology.* 3rd ed. Engelwood, Colo.: Morton Publishing, 2006.
Took, D. James, and Norma Henderson. *Using Information Technology in Mathematics Education.* New York: Haworth Press, 2001.
Vince, John. *Mathematics for Computer Graphics.* 2nd ed. New York: Springer, 2005.

mathematics software

As explained in the preceding article, computer science looked to mathematics to create and verify its algorithms. In turn, computer software has greatly aided many levels of mathematical work, ranging from simple calculations to manipulation of symbols and abstract forms.

At the simplest level, computers overlap the functions of simple electronic calculators. Indeed, operating systems such as Microsoft Windows and UNIX systems include calculator utilities that can be used to solve problems requiring a basic four function or more elaborate scientific calculator.

The true power of the computer became more evident to ordinary users when spreadsheet software was introduced commercially in 1979 with VisiCalc (see SPREADSHEET). Spreadsheets make it easy to maintain and update summaries and other reports generated by formulas. Later versions of spreadsheet programs such as Lotus 1-2-3 and Microsoft Excel have the ability to create a wide variety of plots and charts to show relationships between variables in visual terms.

Moving from simple formulas to the manipulation of symbolic quantities (as in algebra), the Association for Computing Machinery (ACM) classification system describes several broad areas of computer-aided mathematics. These include *numerical analysis* (techniques for solving, linear, non-linear, and differential equations), *discrete mathematics* (combinatorial and graph theory), and probability and statistics.

There are two general approaches to mathematical software. One is the creation of libraries of routines or procedures that address particular kinds of problems. A programmer who is creating software that must deal with particular mathematical problems can link these routines to the program, call the procedures with appropriate variables or data, and return the results to the main program for further processing (see PROCEDURES AND FUNCTIONS). The language FORTRAN is still widely used for developing mathematics libraries, and there is a legacy of tens of thousands of routines available. Modern systems have the ability to link these procedures to programs written in more recent languages such as C.

The advantage of using program libraries is that they don't require learning new programming techniques. Each routine can be treated as a "black box." However, it is often desirable to work with traditional mathematical notation (what one might see on a blackboard in a calculus class, rather than typed into computer code). A stand-alone software package such as Mathcad, Matlab, or Mathematica can automatically simplify or solve algebraic expressions or perform hundreds of traditional mathematical procedures. For statistical analysis, programs such as SPSS can apply all of the standard statistical tests to data and provide a large variety of graphics.

Further Reading

Field, Andy. *Discovering Statistics Using SPSS.* 2nd ed. Thousand Oaks, Calif.: SAGE Publications, 2005.

Griffith, Arthur. *SPSS for Dummies.* Hoboken, N.J.: Wiley, 2007.

Netlib Repository of Mathematical Software, Papers, and Databases. Available online. URL: http://www.netlib.org/. Accessed August 14, 2007.

Press, William H., et al. *Numerical Recipes.* 3rd ed. New York: Cambridge University Press, 2007.

Ruskeepaa, Heikki. *Mathematica Navigator: Mathematics, Statistics, and Graphics.* Burlington, Mass.: Elsevier Academic Press, 2004.

Wellin, Paul, Richard Gaylord, and Samuel Kamin. *An Introduction to Programming with Mathematica.* New York: Cambridge University Press, 2005.

Wolfram, Stephen. *The Mathematica Book.* 5th ed. Champaign, Ill.: Wolfram Media, 2003.

Wolfram Mathematica Home Page. Available online. URL: http://www.wolfram.com/. Accessed August 14, 2007.

Mauchly, John William

(1907–1980)
American
Inventor, Computer Scientist

John Mauchly was codesigner of the earliest full-scale digital computer, ENIAC, and its first commercial successor,

Univac (see also ECKERT, J. PRESPER). His and Eckert's work went a long way toward establishing the viability of the computer industry in the early 1950s.

Mauchly was born on August 30, 1907, in Cincinnati, Ohio. He attended the McKinley Technical High School in Washington, D.C., and then began his college studies at Johns Hopkins University, eventually changing his major from engineering to physics. The spectral analysis problems he tackled for his Ph.D. (awarded in 1932) and in postgraduate work required a large amount of painstaking calculation. So, too, did his later interest in weather prediction, which led him to design a mechanical computer for harmonic analysis of weather data (see ANALOG COMPUTER). He also learned about binary switching circuits ("flip-flops") and experimented with building electronic counters, which used vacuum tubes and were much faster than counters using electromagnetic relays.

Mauchly taught physics at Ursinus College in Philadelphia from 1933 to 1941. On the eve of World War II, however, he went to the University of Pennsylvania's Moore School of Engineering and took a course in military applications of electronics. He then joined the staff and began working on contracts to prepare artillery firing tables for the military. Realizing how intensive the calculations would be, in 1942 he wrote a memo proposing that an electronic calculator be built to tackle the problem. The proposal was rejected at first, but by 1943 table calculation by mechanical methods was falling even further behind. Herman Goldstine, who had been assigned by the Aberdeen Proving Ground to break the bottleneck, approved the calculator project.

With Mauchly providing theoretical design work and J. Presper Eckert heading the engineering effort, the Electronic Numerical Integrator and Computer, better known as ENIAC, was completed too late to influence the outcome of the war. However, when the machine was demonstrated in February, 1946, it showed that a programmable electronic computer was not only about a thousand times faster than an electromechanical calculator, it could be used as a general-purpose problem-solver that could do much more than existing calculators.

Mauchly and Eckert left the Moore School after a dispute about who owned the patent for the computer work. They jointly founded what became known as the Eckert-Mauchly Computer Corporation, betting on Mauchly's confidence that there was sufficient demand for computers not only for scientific or military use, but for business applications as well. By 1950, however, they were struggling to sell and build their improved computer, Univac, while fulfilling existing government contracts for a scaled-down version called BINAC. In 1950, they sold their company to Remington Rand, while continuing to work on Univac. In 1952, Univac stunned the world by correctly predicting the presidential election results on election night long before most of the votes had come in.

Early on, Mauchly saw the need for a better way to write computer programs. Univac and other early computers had been programmed through a mixture of rewiring, setting of switches, and entering numbers into registers. This made programming difficult, tedious, and error-prone. Mauchly wanted a way that variables could be represented symbolically: for example, Total rather than a register number such as 101. Under Mauchly's supervision William Schmitt wrote what became known as Brief Code. It allowed two-letter combinations to stand for both variables and operations such as multiplication or exponentiation. A special program read these instructions and converted them to the necessary register and machine operation commands (see INTERPRETER). While primitive compared to later languages (see ASSEMBLER and PROGRAMMING LANGUAGES), Brief Code represented an important leap forward in making computers more usable.

Mauchly stayed with Remington Rand and its successor Sperry Rand until 1959, but then left over a dispute about the marketing of the Univac. He continued his career as a consultant and lecturer. Mauchly and Eckert also became embroiled in a patent dispute arising from their original work with ENIAC. Accused of infringing Sperry Rand's ENIAC patents, Honeywell claimed that the ENIAC patent was invalid, with another computer pioneer, John Atanasoff, claiming that Mauchly and Eckert had obtained crucial ideas after visiting his laboratory in 1940 (see ATANASOFF, JOHN VINCENT).

In 1973, Judge Earl Richard Larson ruled in favor of Atanasoff and Honeywell. However, many historians of the field give Mauchly and Eckert the lion's share of the credit because it was they who had built full-scale, practical machines.

Mauchly played a key role in founding the Association for Computing Machinery (ACM), one of the field's premier professional organizations. He served as its first vice president and second president. He received many tokens of recognition from his peers, including the Howard Potts Medal of the Franklin Institute. In turn, the ACM established an Eckert-Mauchly award for excellence in computer design. John Mauchly died on January 8, 1980.

Further Reading

"John W. Mauchly and the Development of the ENIAC Computer: An Exhibition in the Department of Special Collections, Van Pelt Library, University of Pennsylvania." Available online. URL: http://www.library.upenn.edu/exhibits/rbm/mauchly/jwmintro.html. Accessed August 14, 2007.

McCartney, Scott. *ENIAC: The Triumphs and Tragedies of the World's First Computer.* New York: Berkeley Books, 1999.

Stern, N. "John William Mauchly: 1907–1980," *Annals of the History of Computing* (April 1980): 100–103.

McCarthy, John
(1927–   )
American
Computer Scientist, AI Pioneer

Starting in the 1950s, John McCarthy played a key role in the development of artificial intelligence as a discipline, as well as developing LISP, the most popular language in AI research.

John McCarthy was born on September 4, 1927, in Boston, Massachusetts. He completed his B.S. in mathematics at the California Institute of Technology, then earned his Ph.D. at Princeton University in 1951. During the 1950s, he held teaching posts at Stanford University, Dartmouth College, and the Massachusetts Institute of Technology.

Although he seemed destined for a prominent career in pure mathematics, he encountered computers while working during the summer of 1955 at an IBM laboratory. He was intrigued with the potential of the machines for higher-level reasoning and intelligent behavior (see ARTIFICIAL INTELLIGENCE). The following year he put together a conference that brought together people who would become key AI researchers, including Marvin Minsky (see MINSKY, MARVIN). He proposed that "the study is to proceed on the basis of the conjecture that every aspect of learning or any other feature of intelligence can in principle be so precisely described that a machine can be made to simulate it. An attempt will be made to find how to make machines use language, form abstractions and concepts, solve kinds of problems now reserved for humans, and improve themselves."

Mathematics had well-developed symbolic systems for expressing its ideas. McCarthy decided that if AI researchers were to meet their ambitious goals, they would need a programming language that was equally capable of expressing and manipulating symbols. Starting in 1958, he developed LISP, a language based on lists that could flexibly represent data of many kinds and even allowed programs to be fed as data to other programs (see LISP). LISP would be used in the coming decades to code most AI research projects, and McCarthy continued to play an important role in refining the language, while moving to Stanford in 1962, where he would spend the rest of his career.

McCarthy also contributed to the development of ALGOL, a language that would in turn greatly influence modern procedural languages such as C. He also helped develop new ways for people to use computers. Consulting with Bolt, Beranek and Newman (the company that would later build the beginnings of the Internet), he helped design time-sharing, a system that allowed many users to share the same computer, bringing down the cost of computing and making it accessible to more people. He also sought to make computers more interactive, designing a system called THOR, which used video display terminals. Indeed, he pointed the way to the personal computer in a 1972 paper on "The Home Information Terminal."

In 1971, McCarthy received the prestigious A. M. Turing award from the Association for Computing Machinery. In the 1970s and 1980s, he taught at Stanford and remained a prominent spokesperson for AI, arguing against critics such as philosopher Hubert Dreyfus (see DREYFUS, HUBERT), who claimed that machines could never achieve true intelligence. In 2000 McCarthy retired from Stanford, where he remains a Professor Emeritus. In 2003 McCarthy received the Benjamin Franklin Medal in Computer and Cognitive Science.

Further Reading
"John McCarthy's Home Page." Available online. URL: http://www-formal.stanford.edu/jmc/. Accessed August 14, 2007.
McCarthy, John. "The Home Information Terminal." *Man and Computer: Proceedings of the International Conference, Bordeaux, France, 1970.* Basel: S. Karger, 1972, 48–57.
———. "Philosophical and Scientific Presuppositions of Logical AI," in McCarthy, H. J. and Vladimir Lifschitz, eds., *Formalizing Common Sense: Papers by John McCarthy.* Norwood, N.J.: Ablex, 1990.
McCorduck, Pamela. *Machines Who Think.* 2nd ed. Natick, Mass.: A. K. Peters, 2004.

measurement units used in computing

Newcomers to the computing world often have difficulty mastering the variety of ways in which computer capacity and performance are measured. A good first step is to look at the most common metric prefixes that indicate the magnitude of various units (see table).

COMMON METRIC PREFIXES USED IN COMPUTING

PREFIX	MAGNITUDE
kilo	10^3 (1 thousand)
mega	10^6 (1 million)
giga	10^9 (1 billion)
tera	10^{12} (1 trillion)
milli	10^{-3} (1 thousandth)
micro	10^{-6} (1 millionth)
nano	10^{-9} (1 billionth)
pico	10^{-12} (1 trillionth)

Strictly speaking, most computer measurements are based on the binary system, using powers of two. Thus *kilo* actually means 2^{10}, which is actually 1,024, and *mega* is actually 2^{20}, or 1,048,576. However, this distinction is generally not important for gaining a sense of the magnitudes involved. In 1998, the International Electrotechnical Commission promulgated a new set of prefixes for these base two computer-related magnitudes, such that for example, mebi- is supposed to be used instead of mega-. There is little evidence thus far that this scheme is being widely adopted.

We will now consider some of the main areas in which computer capacity or performance is measured.

STORAGE CAPACITY

The smallest unit of information, and thus of data storage, is a bit (*binary digit*). A bit can be either 1 or 0 and is physically represented in different ways according to the memory or storage device being used. On most computers the most-used storage unit is the *byte,* which contains eight bits. Since this represents eight binary digits, or 2^8, a byte can hold values from 0 to 255 (decimal). The following table gives some typical units of storage.

DATA STORAGE UNITS

UNIT	TYPICAL USE
bit	Processor data handling capacity. Most processors today can handle 32 or 64 bits at a time.
byte (8 bits)	Holds an ASCII character value or a small number, 0–255.
kilobyte	Used to measure RAM (random access memory) and floppy disk capacity for early PCs.
megabyte	RAM capacity in older PCs; hard drive capacity in older PCs.
gigabyte	Memory and drive capacity in modern PCs.
terabyte	Large modern hard drives; drive arrays (see RAID).

GRAPHICS

Printed output is generally measured in dots per inch (dpi). Screen images and images used in digital photography are measured in pixels or megapixels. However, the amount of data needed to specify (and thus store) a pixel in an image depends on the number of colors and other information to be stored. (See GRAPHICS FORMATS.)

PROCESSOR SPEED

Processor speed is measured in millions of cycles per second (megahertz or MHz). The earliest microprocessors had speeds measured in 1–2 MHz or so. PCs of the 1980s ranged from about 8 to 50 MHz. In the 1990s, speeds ramped up to the hundreds of MHz, and in today's systems PC speeds are often measured in gigahertz (GHz).

CALCULATION SPEED

The speed at which a computer can perform calculations depends on more than raw processor speed. For example, a processor that can store or fetch 32-bit numbers can perform many calculations faster than one with only a 16-bit capacity even if the two processors have the same clock speed in cycles per second.

Calculation speed is often measured in "flops" or floating-point operations per second (see NUMERIC DATA), or for modern processors, megaflops. While this measurement is often touted in product literature, savvy users look to more reliable benchmarks that re-create actual conditions of use, including calculation-intensive, data transfer intensive, or graphics-intensive operations.

DATA COMMUNICATIONS AND NETWORKING

The speed at which data can be transferred over a modem or network connection is measured in bits per second (BPS). A related term, *baud,* was used (somewhat inaccurately) with the earlier modems (see BANDWIDTH and MODEM).

TYPICAL DATA TRANSFER SPEEDS IN BITS PER SECOND (BPS)

DEVICE	APPROXIMATE SPEED
Ethernet	(10 base) 10 Mbps
Fast Ethernet	100 Mbps
Gigabit Ethernet	1 Gbps
V.90 dial-up modem	56 Kbps
ISDN phone line	64 Kbps
DSL (ADSL)	1–24 Mbps
Cable modem (DOCSIS)	10–160 Mbps
Bluetooth 2.0 (see BLUETOOTH)	3 Mbps
802.11b wireless (see WIRELESS COMPUTING)	11 Mbps
802.11g wireless (see WIRELESS COMPUTING)	54 Mbps
802.11n wireless	540 Mbps

Further Reading
How Many? A Dictionary of Units of Measurement. Available online. URL: http://www.unc.edu/~rowlett/units/. Accessed August 14, 2007.
Online [unit] Conversion. Available online. URL: http://www.onlineconversion.com/. Accessed August 14, 2007.
SG Bits/Bytes Conversion Calculator. Available online. URL: http://www.speedguide.net/conversion.php. Accessed August 14, 2007.

media center, home

In recent years many families have acquired a plethora of media and devices to play it—CD and DVD players, radios, and of course TVs. Meanwhile, the family has likely also acquired one or more PCs, which are also capable of playing digital audio and video from various sources. The media center is a way of integrating all of these media into one centrally located device, the PC, and ideally being able to serve it on demand anywhere in the home.

Modern PCs already have optical drives (see CD-ROM and DVD-ROM). TV signals can be received using a TV tuner card or, for digital cable signals, a "cable card." (High-definition TV or HDTV is becoming increasingly popular.) There are also tuners for AM/FM radio. Of course audio and video files can also be received directly over the Internet (see INTERNET RADIO; MUSIC AND VIDEO DISTRIBUTION, ONLINE; and STREAMING).

MEDIA STORAGE AND DISTRIBUTION

Once media is received, it can be stored on one or more hard drives (see also DVR). The media can also be distributed to remote speaker units via the wired (or more often wireless) network. Remote controls are usually provided to allow the system to be controlled from anywhere in the building.

Of course software is needed to integrate these devices and functions. Microsoft provides the Windows Media Center for Vista and the Windows XP Media Center Edition. (There is also third-party software for Windows.) Linux systems can run programs such as MythTV (free) or the commercial SageTV. For the Macintosh, a project called CenterStage was under development as of 2008.

Prebuilt media centers with the PC and all the necessary inputs and outputs are also available. They are often sold as part of home theater systems.

Further Reading
Ballew, Joli. *How to Do Everything with Windows Vista Media Center.* Emeryville, Calif.: McGraw-Hill/Osborne, 2007.
Knoppmyth (Knoppix and MythTV Linux media center). Available online. URL: http://www.knoppmythwiki.org/?id=KnoppmythWiki. Accessed October 1, 2007.
Layton, Julia. "How Media-Center PCs Work." Available online. URL: http://computer.howstuffworks.com/media-center-pc.htm. Accessed October 1, 2007.
Smith, Stewart, and Michael Still. *Practical MythTV: Building a PVR and Media Center PC.* Berkeley, Calif.: Apress, 2007.
Windows Media Center (Vista). Available online. URL: http://www.microsoft.com/windows/products/windowsvista/features/details/mediacenter.mspx. Accessed October 1, 2007.

medical applications of computers

Since health care delivery is a business (indeed, one of the largest sectors of the economy), any hospital, health plan, or independent medical practice involves much the same software as any other large business. This includes accounts receivable and payable, payroll, and supplies inventory. Both general and customized industry software can be used for these functions; however, this article focuses on applications specific to medicine.

MEDICAL INFORMATION SYSTEMS

The management of information specifically related to medical care is sometimes called *medical informatics*. The type of information gathered depends on many factors including the type of institution, ranging from a small doctor's office to a large clinic to a full hospital and the nature and scope of the treatment provided. However, one can make some generalizations.

For outpatients, the required information includes an extensive medical record for each patient, including records of medical tests and their results, prescriptions and their status, and so on. For hospital patients, there are also admissions records, an extensive list of itemized charges, and records that must be maintained for public health or other governmental purposes. Hospitals increasingly use customized, integrated hospital information systems (HIS) that integrate billing, medical records, and pharmacy.

Additional record keeping needs arise from the mechanisms used to pay for health care. Each health payment system, whether government-run (such as Medicare or the Veterans' Administration) or a private health maintenance organization (HMO), has extensive rules and procedures about how each surgery, treatment, test, or medication can be submitted for payment. The software must be able to use recognized classifications systems such as the DSM-IV (Diagnostic and Statistical Manual of Mental Disorders).

CLINICAL INFORMATION MANAGEMENT

The modern hospital generates extensive real-time data about the condition of patients, particularly those in critical or intensive care or undergoing surgery. Many hospitals have bedside or operating room terminals where physicians or nurses can review summaries of data such as vital statistics (blood pressure, heart function, and so on). Data can also be entered or reviewed using handheld computers (see PORTABLE COMPUTERS). The ultimate goal of such systems is to provide as much useful information as possible without overwhelming medical personnel with data entry and related tasks that might detract from patient care.

In 2001, a new group, the Patient Safety Institute, was formed in an attempt to create a nationwide standardized format for electronic patient records. This would make it possible for emergency personnel to download a patient's record into a handheld computer and access potentially lifesaving information such as medications and allergies.

There has also been some progress in medical decision support systems. Going beyond data summarization, such systems can analyze changes or trends in medical data and highlight those of clinical significance. Such systems can also aid in the compilation of medical charts or possibly compile portions of the chart automatically for later review.

DIAGNOSTIC AND TREATMENT SYSTEMS

The diagnosis and treatment of many conditions has been profoundly enhanced by the use of computer-assisted medical instruments. At the beginning of the last century the use of X-rays revolutionized the imaging of the anatomy of living things. X-rays, however, were limited in detail and depth of imaging. Techniques of tomography, involving synchronized movement of the X-ray tube and film, were then developed to create a sharp focus deeper within the target structure. The development of computerized tomography (CT or CAT) scanning in the 1970s used a different and more effective approach: A beam of X rays is swept through the target area while computerized radiation detectors precisely calculate the absorption of radiation, and thus the density of the tissue or other structure at each point. This results in a highly detailed image that can be viewed as a series of layers or combined into a three-dimensional holographic display.

Another widely used imaging technique is positron emission tomography (PET) scanning, which tracks the radiation emission from a short-lived radioisotope injected into the patient. It is particularly helpful for studying the flow of blood or gas and other physiological or metabolic changes. Magnetic resonance imaging (MRI) uses the absorption and re-emission of radio waves in a strong magnetic field to identify the characteristic signature of the hydrogen nucleus (i.e. a proton) in water within the body, and thus delineate the surrounding structures.

Besides controlling the scanning process (especially in CAT scanning), the computer is essential for the creation and manipulation of the resulting images. A typical image processing (IP) system is actually an array of many individual processors that perform calculations and comparisons on parts of the image to enhance contrast and extract information that can lead to a more precise depiction of the area of interest. The resulting images (consisting of an array of pixels and associated information) can be further enhanced in a variety of ways using video processing software. Other software using pattern recognition techniques can be programmed to look for tumors or other anomalous structures (see IMAGE PROCESSING).

TRENDS

Medical informatics is likely to be a strong growth area in coming decades. As the population ages, demand for medical care will increase. At the same time, there will be growing pressure to control costs. Although technology is expensive, there is a general belief that information can be leveraged to provide more cost-effective treatment and management of health care delivery.

Medical systems are likely to become more integrated. There have been proposals to create permanent, extensive electronic medical records that patients might even "wear" in the form of a small implanted chip. However, concern about the consequences of violation of privacy and misuse of medical information (such as by employers or insurers) raises significant challenges (see also PRIVACY IN THE DIGITAL AGE).

There are many exciting possibilities for computer-assisted medical treatment. It may eventually be possible to provide all the detail of a CAT scan or MRI while a medical procedure is being performed. At any rate, surgeons will be able to see ever more clearly what they are doing, and robot-controlled surgical instruments (such as lasers) are already operating with a precision that cannot be matched by human hands. Such instrumentation also allows for the possibility that skilled surgeons might be able to operate through telepresence, bringing lifesaving surgery to remote areas (see TELEPRESENCE).

Information technology (and the World Wide Web in particular) is also giving patients more data and choices

This NASA project is developing a "smart" probe that could provide instant analysis of breast tumors to guide surgeons in their work. Such instruments could make surgery more accurate and effective, as well as reducing unnecessary operations. (NASA PHOTO)

about prospective drugs and treatments. (See PERSONAL HEALTH INFORMATION MANAGEMENT.)

Further Reading

Chen, Hsinchun, et al., eds. *Medical Informatics: Knowledge Management and Data Mining in Biomedicine.* New York: Springer, 2005.

LinuxMedNews [open-source medical software]. Available online. URL: http://www.linuxmednews.com/. Accessed August 14, 2007.

Medical Software (Yahoo! Directory). Available online. URL: http://dir.yahoo.com/Business_and_Economy/Business_to_Business/Health_Care/Software/Me dical/. Accessed August 14, 2007.

Shortliffe, Edward H., and James J. Cimino, eds. *Biomedical Informatics: Computer Applications in Health Care and Biomedicine.* New York: Springer, 2006.

Sullivan, Frank, and Jeremy Wyatt. *ABC of Health Informatics.* Malden, Mass.: Blackwell, 2006.

Wooton, Richard. *Introduction to Telemedicine.* 2nd ed. London: Royal Society of Medicine Press, 2006.

memory

Generally speaking, memory is a facility for temporarily storing program instructions or data during the course of processing. In modern computers the main memory is random access memory (RAM) consisting of silicon chips. Today's personal computers typically have from between 64MB (megabytes) and 512MB of main memory.

DEVELOPMENT OF THE TECHNOLOGY

In early calculators "memory" was stored as the positions of various dials. Charles Babbage conceived of a "store" of such dials that could hold constants or other values needed during processing by his Analytical Engine (see BABBAGE, CHARLES).

A number of forms of memory were used in early electronic digital computers. For example, a circuit with an inherent delay could be used to store a series of pulses that could be "refreshed" every fraction of a second to maintain the data values. The Univac I, for example, used a mercury delay line memory. Researchers also experimented with cathode ray tubes (CRTs) to store data patterns.

The most practical early form of memory was the ferrite core, which consisted of an array of tiny donut-shaped magnets, crisscrossed by electrical lines so that any element can be addressed by row and column number. By converting data into appropriate voltage levels, the magnetic state of the individual elements can be switched on and off to represent 1 or 0. In turn, a current can be passed through any element to read its current state—although the element must then be remagnetized. Ferrite cores were relatively fast but expensive, and "core" became programmers' shorthand for the amount of precious memory available.

By the 1960s, the use of transistors and integrated circuits made electronic solid-state memory systems possible. Since then, the MOSFET (Metal Oxide Semiconductor Field Effect Transistor) using CMOS (Complementary Metal Oxide) fabrication has been the dominant way to implement DRAM (dynamic random access memory). Here "dynamic" means that the memory must be "refreshed" by applying current after data is read in each cycle, and "random access" means that any desired memory location can be accessed directly rather than requiring locations to be read sequentially.

Static RAM is used in some computer components where maximum memory speed is desirable. Static memory is faster because it does not need to be refreshed after each reading cycle. However, it is also considerably more expensive.

Memory performance is also dependent on how quickly locations in the memory can be addressed. The earliest forms of DRAM required that the row and column of the desired memory location be sent in separate cycles. EDO (Extended Data Out) and more recent technologies allow the row to be requested one time, and then just the column given for adjacent or nearby locations. Timing and pipelining techniques can also be used to start a new request while the previous one is still being processed.

For SDRAM (synchronous DRAM), memory speed is limited by the inherent response time of the memory chip, but also by the number of clock cycles per second initiated by the data bus (see BUS). Double data rate (DDR) SDRAM is able to use both the "rising" and "falling" part of the clock cycle to transfer data, doubling throughput. It is being superseded by DDR2, which achieves another doubling because it can run the data-transfer bus at twice the system clock speed. However, this does increase latency (the time needed to begin an access). On the horizon is DDR3, which can run the bus at four times clock speed—yet another doubling. Possible future memory technologies include "spintronics," or the use of the quantum state or "spin" of electrons to hold data. The speed, compactness, and reliability of this technology could exceed current devices by a factor of hundreds to thousands.

As memory gets faster, it continues to get cheaper (at least for all but the latest technology). At the same time, memory demands continue to increase. Today's PCs generally come with 1 GB (billion bytes) of RAM, and 2 GB or more is often recommended, particularly for memory-hungry operating systems such as Microsoft Windows Vista and applications such as Adobe PhotoShop and video editing.

Another popular kind of memory is "flash" (nonvolatile) memory that does not require power to maintain its contents. This kind of memory is used in a wide variety of devices, including digital cameras, PDAs, and media players (see also FLASH DRIVE).

In actual systems, a small amount of faster memory (see CACHE) is used to hold the data that is most likely to be immediately needed. A proper balance between primary and secondary cache and main memory in the system chipset makes it less necessary to use the fastest, most expensive form of main memory.

Many computers also have ROM (Read-Only Memory) or PROM (Programmable Read-Only Memory). This memory holds permanent system settings and data (see BIOS) that are needed during the startup process (see BOOT SEQUENCE).

Further Reading

Jacob, Bruce, Spencer Ng, and David Wang. *Memory Systems: Cache, DRAM, Disk.* San Francisco: Morgan Kaufmann, 2007.

Tom's Hardware: Motherboards and RAM. Available online. URL: http://www.tomshardware.com/motherboard/index.html. Accessed August 15, 2007.

"Ultimate Memory Guide." Available online. URL: http://www.kingston.com/tools/umg/default.asp. Accessed August 15, 2007.

memory management

Whatever memory chips or other devices are installed in a computer, the operating system and application programs must have a way to allocate, use, and eventually release portions of memory. The goal of memory management is to use available memory most efficiently. This can be difficult in modern operating environments where dozens of programs may be competing for memory resources.

Early computers were generally able to run only one program at a time. These machines didn't have a true operating system, just a small loader program that loaded the application program, which essentially took over control of the machine and accessed and manipulated the memory. Later systems offered the ability to break main memory into several fixed partitions. While this allowed more than one program to run at the same time, it wasn't very flexible.

VIRTUAL MEMORY

From the very start, computer designers knew that main memory (RAM) is fast but relatively expensive, while secondary forms of storage (such as hard disks) are slower but relatively cheap. Virtual memory is a way to treat such auxiliary devices (usually hard drives) as though they were part of main memory. The operating system allocates some storage space (often called a swapfile) on the disk. When programs allocate more memory than is available in RAM, some of the space on the disk is used instead. Because RAM and disk are treated as part of the same address space (see ADDRESSING), the application requesting memory doesn't "know" that it is not getting "real" memory. Accessing the disk is much slower than accessing main memory, so programs using this secondary memory will run more slowly.

Virtual memory has been a practical solution since the 1960s, and it has been used extensively on PCs running operating systems such as Microsoft Windows. However, with prices of RAM falling drastically in the new century, there is likely to be enough main memory on the latest systems available to run most popular applications.

MEMORY ALLOCATION

Most programs request memory as needed rather than a fixed amount being allocated as part of program compilation. (After all, it would be inefficient for a program to try to guess how much memory it would need, and possibly tie up memory that could be used more efficiently by other programs.) The operating system is therefore faced with the task of matching the available memory with the amounts being requested as programs run.

One simple algorithm for memory allocation is called *first fit*. When a program requests memory, the operating system looks down its list of available memory blocks and allocates memory from the first one that's large enough to fulfill the request. (If there is memory left over in the block after allocation, it becomes a new block that is added to the list of free memory blocks.)

As a result of repeated allocations using this method, the memory space tends to become fragmented into many leftover small blocks of memory. As with fragmentation of files on a disk, memory fragmentation slows down access, since the hardware (see MEMORY) must issue repeated instructions to "jump" to different parts of the memory space.

Using alternative memory allocation algorithms can reduce fragmentation. For example, the operating system can look through the entire list (see HEAP) and find the *smallest* block that is still large enough to fulfill the allocation request. This *best fit* algorithm can be efficient. While it still creates fragments from the small leftover pieces, the fragments usually don't amount to a significant portion of the overall memory.

The operating system can also enforce standard block sizes, keeping a "stockpile" of free blocks of each permitted size. When a request comes in, it is rounded to the nearest amount that can be made from a combination of the standard sizes (much like making change). This approach, sometimes called the *buddy system*, means that programs may receive somewhat more or less memory than they want, but this is usually not a problem.

RECYCLING MEMORY

In a multitasking operating system, programs should release memory when it is no longer needed. In some programming environments memory is released automatically when a data object is no longer valid (see VARIABLE), while in other cases memory may need to be explicitly freed by calling the appropriate function.

Recycling is the process of recovering these freed-up memory blocks so they are available for reallocation. To reduce fragmentation, some operating systems analyze the free memory list and combine adjacent blocks into a single, larger block (this is called *coalescence*). Operating systems that use fixed memory block sizes can do this more quickly because they can use constants to calculate where blocks begin and end.

Many more sophisticated algorithms can be used to improve the speed or efficiency of memory management. For example, the operating system may be able to receive information (explicit or implicit) that helps it determine whether the requested memory needs to be accessed extremely quickly. In turn, the memory management system may be designed to take advantage of particular processor architecture. Combining these sources of knowledge, the memory manager might decide that a particular requested memory block be allocated from special high speed memory (see CACHE).

While RAM is now cheap and available in relatively large quantities even on desktop PCs, the never-ending race between hardware resources and the demands of ever larger database and other applications guarantees that memory management will remain a concern of operating system designers. In particular, distributed database systems where data objects can reside on many different machines in the network require sophisticated algorithms that take not only memory speed but also network load and speed into consideration.

Further Reading

Blunden, Bill. *Memory Management: Algorithms and Implementations in C/C++*. Plano, Tex.: Wordware Publishing, 2002.

Jones, Richard, and Rafael D. Lins. *Garbage Collection: Algorithms for Automatic Dynamic Memory Management*. New York: Wiley, 1996.

"The Memory Management Reference Beginner's Guide: Overview." Available online. URL: http://www.memorymanagement. org/articles/begin.html. Accessed August 15, 2007.

message passing

In the early days of computing, a single program usually executed sequentially, with interruptions for calls to various procedures or functions that would perform data processing tasks and then return control to the main program (see PROCEDURES AND FUNCTIONS). However, by the 1970s UNIX and other operating systems had introduced the capability of running several programs at the same time (see MULTITASKING). Additionally, it became common to create a large program that would manage data and smaller programs that could link users to that service (see CLIENT-SERVER COMPUTING). Further, programs themselves began to be organized in a new way (see OBJECT-ORIENTED PROGRAMMING). A program now consisted of a number of entities (objects) representing data and methods (things that can be done with the data).

Thus, both at the operating system and application level it became necessary to have various objects communicate with one another. For example, a client program requests a service from the server. The server performs the required service and reports its completion. The mechanism by which information can be sent from one program to another (or between objects in a program) is called *message passing*.

In one message-passing scheme, two objects (such as client and server) agree on a standard memory location called a *port*. Each program checks the port regularly to see if a message (containing instructions, data, or an address where data can be found) is pending. In turn, outgoing messages can be left at the port.

The client-server idea can be found within operating systems as well. For example, there can be a component devoted to providing file-related services, such as opening or reading a file (see FILE). An application that wants to open a file leaves an appropriate message to the operating system. The operating system has a message dispatcher that examines incoming messages and routes them to the correct component (the file system manager in this case).

Within an object-oriented program, an object is sent a message by invoking one of its *methods* (Smalltalk and other languages) or *member functions* (C++ or Java). For example, suppose there's an object call Speaker that represents the system's internal speaker. As part of a user alert procedure, there might be a call to

```
Speaker.Beep (500)
```

which might be defined to mean "sound a beep for 500 milliseconds."

There are a number of issues involved in setting up message-passing systems. For example, it is convenient for many programs or objects to use the same port or other facility for leaving and retrieving messages, but that means the operating system must spend additional time routing or dispatching the messages. On the other hand, if two objects create a *bound* port, no others can use it, so each can assume that any message left there is from the other object.

During the 1992–1994 period, a standard called MPI (Message Passing Interface) was established by a group of more than 40 industry organizations. It has since been superseded by MPI-2.

Further Reading

Gropp, William, Ewing Lusk, and Rajeev Thakur. *Using MPI: Portable Parallel Programming With the Message-Passing Interface*. 2nd ed. Cambridge, Mass.: MIT Press, 1999.

"The Message Passing Interface (MPI) Standard." Available online. URL: http://www-unix.mcs.anl.gov/mpi/. Accessed February 6, 2000.

Petzold, Charles. *Programming Windows: the Definitive Guide to the Win32 API*. 5th ed. Redmond, Wash.: Microsoft Press, 1998.

Quinn, Michael J. *Parallel Programming in C with MPI and OpenMP*. New York: McGraw-Hill Education, 2003.

microprocessor

A microprocessor is an integrated circuit chip that contains all of the essential components for the central processing unit (CPU) of a microcomputer system such as a personal computer.

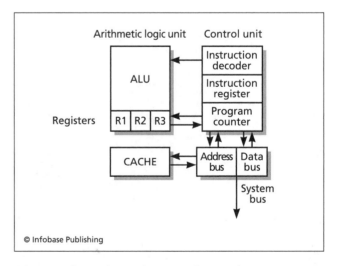

Schematic of a simple microprocessor. The control unit is responsible for fetching and decoding instructions, as well as fetching or writing data to memory. The Arithmetic Logic Unit (ALU) does the actual computing (including arithmetic and logical comparisons). The registers hold data being currently used by the ALU, while the cache contains instructions that have been pre-fetched because they are likely to be needed soon.

Microprocessor development began in the 1960s when a new company called Intel was given a contract to develop chips for programmable calculators for a Japanese firm. Marcian E. "Ted" Hoff headed the project. He decided that rather than hard-wiring most of the calculator logic into the chips, he would create a general-purpose chip that could read instructions and data, perform basic arithmetic and logical functions, and transfer data between memory and internal locations called registers.

The resulting *microprocessor,* when combined with some RAM (random access memory), some preprogrammed ROM (Read Only Memory), and an input/output (I/O) chip constituted a tiny but complete CPU, soon dubbed "a computer on a chip." This first microprocessor, the Intel 4004, had only a few thousand transistors, could handle data only 4 bits at a time, and ran at 740 KHz (about one three-thousandth the speed of the latest Pentium IV chips).

Intel gradually refined the chip, giving it the logic circuits to enable it to perform additional instructions, more internal stack and register space, and 8 KB of space to store programs. The 8008 could handle 8 bits of data at a time, while the 8080 became the first microprocessor that was capable of serving as the CPU for a practical microcomputer system. Its descendants, the 8088 and 8086 (16-bit) powered industry-standard IBM-compatible PCs. Meanwhile, other companies such as Motorola (68000), Zilog (Z-80), and MOS Technology (6502) powered competing PCs from Apple, Atari, Commodore, and others.

With the dominance of the IBM PC and its clones (see IBM PC), the Intel 80 × 86 series in turn dominated the microprocessor market. (The *x* refers to successive digits, as in 80286, 80386, and 80486.) At the next level this nomenclature was replaced by the Pentium series, which is up to the Pentium 4 as of 2002.

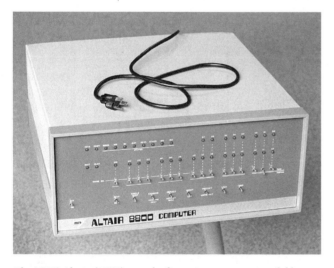

The MITS Altair (1975) was the first microcomputer available commercially. It was generally purchased in kit form. While the Altair did not have much processing capacity, it aroused great interest and inspired other computer builders such as Apple's Steve Wozniak and Steve Jobs. (CHRISTOPHER FITZGERALD / THE IMAGE WORKS)

According to a famous dictum called Moore's Law, the density (number of transistors per cubic area) and speed (in terms of clock rate) of microprocessors has roughly doubled every 18 months to two years. Intel expects to be making microprocessors with 1 billion transistors by 2007.

MICROPROCESSOR AND MICROCOMPUTER

A microcomputer is a system consisting of a microprocessor and a number of auxiliary chips. The microprocessor chip serves as the central processing unit (CPU). It contains a clock that regulates the flow of data and instructions (each instruction takes a certain number of clock cycles to execute). There is also an index register that keeps track of the instruction being executed. A small number of locations called *registers* within the CPU allow for storing or retrieving the data being used by instructions much more quickly than retrieval from main memory (RAM).

Typically, the instruction register advances to the next instruction. The instruction is fetched, decoded, and sent to the CPU's ALU (arithmetic logic unit) for processing. Data needed to be processed by the instruction are either fetched from a register or, through an *address register,* fetched from RAM. (Some processors store one operand for an arithmetic operation in a special register called the *accumulator.*)

Floating-point operations (those involving numbers that can include decimal points) require special registers that can keep track of the decimal position. Until the mid-1990s, many systems used a separate microprocessor called a *coprocessor* to handle floating point operations. However, later chips such as the Pentium series integrate floating point operations into the main chip.

In order to function as the heart of a microcomputer, the CPU must communicate with a variety of other devices by interacting with special controller chips. For example, there is a bus interface chip (see BUS) that decodes memory addresses and routes requests to the appropriate devices on the motherboard. When data is requested from memory, a memory controller must physically fetch the data from RAM (see MEMORY). There is also a cache controller that interfaces with one or two levels of high-speed cache memory (see CACHE). The algorithms implemented in the cache controller aim to have the next instructions and the most-likely needed data already in the cache when the CPU requests them.

Other devices such as disk drives, modems, printers, and video cards are all connected to the CPU through input/output (I/O) interfaces that connect to the system bus. Most of the devices connected to the bus have their own microprocessors. Software (see DEVICE DRIVER) translates high-level programming instructions (such as to open a file) to the appropriate device commands.

The CPU and many other devices also contain ROM (read only memory) chips that have permanent basic instructions stored on them (see BIOS). This enables the CPU and other devices to perform the necessary actions to enter into communication when the system starts up (see BOOT SEQUENCE).

NEW FEATURES EMERGE

Improvements in microprocessors during the 1980s included wider data paths and the ability to address a larger amount of memory. For example, the Intel 80386 was the first 32-bit processor for PCs and could address 4 GB of memory. (Earlier processors such as the 80286 had to divide memory into segments or use paging to swap memory in and out of a smaller space to make it look like a larger one.) Over the years microprocessors tended to add more built-in cache memory, enabling them to have more instructions or data ready for immediate use.

Another way to get more performance out of a microprocessor is to increase the speed with which instructions can be executed. One technique, called *pipelining,* breaks the processor into a series of segments, each of which can execute a particular operation. Instead of waiting until an instruction has been completely executed and then turning to the next one, a pipelined microprocessor moves the instruction from segment to segment as its operations are executed, with following instructions moving into the vacated segments. As a result, two or more instructions can be undergoing execution at the same time.

In addition to pipelining, the Pentium series and other recent chips can have instructions executing simultaneously using different arithmetic logic units (ALUs) or floating-point units (FPUs).

Another way to improve instruction processing is to use a simpler set of instructions. First introduced during the 1980s for minicomputers and high-end workstations (such as the Sun SPARC series), reduced instruction set computer (RISC) chips have smaller, more uniform instructions that can be more easily pipelined, as well as many registers for holding the results of the intermediate processing. During the 1990s, RISC concepts were also adopted in PC processor designs such as the 80486 and Pentium (see REDUCED INSTRUCTION SET COMPUTER).

The latest major development has been the multicore microprocessor, which has two, four, or more separate processing units. The Intel Core Duo and Core 2 Duo chips and similar processors from AMD are now included in most new PCs.

The equivalent of a supercomputer on a chip is on the way. Cisco's 192-core Metro chip powers its most capable network routers, while Nvidia's GeForce 8800 graphics processor sports 128 cores. In addition to these specialized processors, in early 2007 Intel demonstrated a prototype 80-core processor that could form the basis of a new generation of general-purpose processors.

Another significant multicore processor architecture is the Cell chip, developed by Sony, IBM, and Toshiba. This chip includes a multithreaded (able to run multiple streams of code) controller processor plus numerous architectural features that maximize efficiency and throughput. The first appearance of this 2 "teraflop" (trillion calculations per second) chip was not on a scientific computer, but rather the Sony Play Station 3—see GAME CONSOLE.

Multicore processors create new challenges for programmers who have to create code that will apportion program tasks efficiently among the cores (see MULTIPROCESSING).

In the new century, it is unclear when physical limitations will eventually slow down the tremendous rate of increase in microprocessor power. As the chips get denser and smaller, more heat is generated with less surface through which it can be removed. At still greater densities, quantum effects may also begin to be a problem. On the other hand, new technologies might take the elements of the processor down to a still smaller level (see MOLECULAR COMPUTING and NANOTECHNOLOGY).

While the stand-alone desktop, laptop, or handheld computer is the most visible manifestation of the microprocessing revolution, there are hundreds of "invisible" microprocessors in use for every visible computer. Today microprocessors help monitor and control everything from home appliances to cars to medical devices (see EMBEDDED SYSTEMS).

Further Reading

Kim, Ryan. "New Era of Game Devices Arrives: Sony and Nintendo Meet the Challenge of Microsoft Xbox." *San Francisco Chronicle,* November 13, 2006, p. F1, F6.

Markoff, John. "Intel Prototype May Herald a New Age of Processing." *New York Times,* February 12, 2007.

Sperling, Ed. "Special Report: Inside the New Multicore Processors." *Electronic News,* April 13, 2007. Available online. URL: http://www.edn.com/article/CA6434384.html. Accessed August 15, 2007.

Stokes, John. *Inside the Machine: An Illustrated Introduction to Microprocessors and Computer Architecture.* San Francisco: No Starch Press, 2006.

Tom's Hardware. "CPU." Available online. URL: http://www.tom-shardware.com/cpu/. Accessed August 15, 2007.

Microsoft Corporation

Microsoft Corporation (NASDAQ symbol: MSFT) is the world's largest computer software company, with almost 80,000 employees worldwide and annual revenue exceeding $51 billion.

Microsoft was founded in the mid-1970s by Bill Gates (see GATES, WILLIAM) in order to sell his version of the BASIC programming language for early microcomputers such as the Altair 8800. The BASIC software was moderately successful, but it would be an operating system called MS-DOS (or PC-DOS) that would catapult Microsoft to industry leadership, thanks to an agreement with IBM, which introduced what would become the industry standard personal computer in 1982 (see IBM PC).

In the mid-1980s Microsoft partnered with IBM to develop OS/2, which was intended to be a more powerful multitasking operating system to replace DOS. However, Microsoft's real interest was in the development of Windows (see MICROSOFT WINDOWS), which first became successful with version 3.0 in 1990. Meanwhile, Microsoft leveraged its experience with Windows to release Microsoft Office, which soon displaced WordPerfect, the previous market leader.

SHIFTING STRATEGIES AND LEGAL ISSUES

Many observers have noted that Microsoft was slow in appreciating the importance of networking and particularly the World Wide Web in the mid-1990s. Novell was the market leader in networking at the time, and Netscape's graphical browser had brought millions of users to the Web. Bill Gates himself announced that the company would embark on a "net-centered" strategy, and this was reflected in the development of Windows NT, software for enterprise network and Web servers, and the Internet Explorer browser, which dominated the desktop by the end of the decade.

The continuing dominance of Microsoft operating systems and office applications on the desktop provided the cash flow that gave the company the resources to catch up and then dominate almost any market it chose. However, this same dominance raised legal issues that would be litigated through the late 1990s and beyond. Microsoft was accused of using its knowledge of unpublished Windows internal code to give products such as Microsoft Office an advantage over competitors. A more prominent accusation was that Microsoft's "bundling" of products such as Internet Explorer with Windows amounted to an unfair advantage over competitors such as Netscape, since Explorer would appear to be "free" to consumers. A series of civil actions under the name *United States v. Microsoft* resulted in a 2001 settlement with the U.S. Department of Justice that required Microsoft to share all information about its Windows API (see APPLICATIONS PROGRAMMING INTERFACE) with competitors for at least five years. This result was controversial, with defenders of Microsoft arguing that the company had done no more than compete effectively by using the results of its own previous work, while opponents argued that Microsoft's coercive monopoly power had scarcely been dented by the settlement. In 2008 the software giant continued to struggle with legal pressures. A European Union court has upheld previous rulings that the company had engaged in monopolistic and anticompetitive practices.

Legal controversies aside, by the mid-2000s Microsoft was facing some serious challenges, particularly from the popularity of free and open-source software (see OPEN-SOURCE MOVEMENT). This is particularly true of the Web server market, where the combination of the Apache Web server and Linux has gained a major market share. Meanwhile, on the desktop, Windows Vista (released in January 2007) did not sell as well as predicted during its first six months. The Apple Macintosh is maintaining its small but significant market share, and even Linux distributions such as Ubuntu are beginning to appear as an option on new PCs. Microsoft's flagship Office suite is facing competition from products such as Open Office and particularly the Web-based Google Apps. (In 2007 Microsoft began to roll out Office Live Workspace, offering extensions of Office applications rather than a complete suite.) In other areas, the Microsoft Network (MSN) online service has struggled, while the company has done better with its Xbox gaming console as well as the best-selling game *Halo*.

Despite some stumbling and many controversies, Microsoft's vast resources and many ongoing research projects (with a $6 billion annual budget) make it likely the company will continue to adapt and sometimes innovate, remaining a strong competitor for many years to come. For example, the company is now putting more resources into Web search technology, an area that has been dominated by Google, as well as eyeing applications as diverse as home media servers and social networking. Microsoft has also sought to expand its Internet presence by acquiring Yahoo!, the extensive but aging Web portal. (However, the first acquisition attempt was rebuffed, and future plans remain uncertain as of mid-2008).

Further Reading
Bank, David. *Breaking Windows: How Bill Gates Fumbled the Future of Microsoft*. New York: Free Press, 2001.
Blakely, Rhys. "Microsoft's Chief Executive Has Seen the Future—and the Future is Advertising: Steve Ballmer's Plans for the Computer Software Giant Include Taking on Yahoo! and Google in Their Own Internet Territory." Times Online (U.K.). Available online. URL: http://business.timesonline.co.uk/tol/business/industry_sectors/technology/article2570485.ece. Accessed October 2, 2007.
Microsoft Corporation. Available online. URL: http://www.microsoft.com. Accessed October 2, 2007.
Microsoft Timeline. Available online. URL: http://www.thocp.net/companies/microsoft/microsoft_company.htm. Accessed October 2, 2007.
Microsoft Watch (eWeek). Available online. URL: http://www.microsoft-watch.com/. Accessed October 2, 2007.
Page, William H., and John E. Lopatka. *The Microsoft Case: Antitrust, High Technology, and Consumer Welfare*. Chicago: University of Chicago Press, 2007.
Slater, Robert. *Microsoft Rebooted: How Bill Gates and Steve Ballmer Reinvented Their Company*. New York: Portfolio, 2004.
Wallace, James, and Jim Erickson. *Hard Drive: Bill Gates and the Making of the Microsoft Empire*. New York: Wiley, 1992.

Microsoft .NET

Microsoft .NET is a programming platform (see CLASS and OBJECT-ORIENTED PROGRAMMING) that is intended to provide a clear and consistent way for applications written in a variety of languages such as C++, C#, and Visual Basic to access Windows functions and to interact with other programs and services on the same machine or over the Internet.

.NET consists of the following main parts:

- Base Class Library of data types and common functions (such as file manipulation and graphics) that is available to all .NET languages

- Common Language Runtime, which provides the code that applications need to run within the operating system, manage memory, and so forth ("Common language" means it can be used for any .NET programming language.)

- ASP .NET, a class framework for building dynamic Web applications and services (the latest version of ASP—see ACTIVE SERVER PAGES)

- ADO .NET, a class framework that allows programs to access databases and data services

The latest version (as of 2008) is .NET Framework 3.5 and is built into Windows Vista and Windows Server 2008. New components include:

- Windows Presentation Foundation, providing a user interface based on 3D graphics, with objects described using Microsoft's XAML markup language (see XML)

- Windows Communication Foundation, providing ways for .NET programs to communicate locally or over the network

- Windows Workflow Foundation, for structuring and automating tasks and transactions

- Windows CardSpace, for managing digital identities in transactions

PLATFORMS

In the relationship between language and runtime libraries, Microsoft .NET, particularly when used with the C# language (see C#), is similar to the use of Java and its libraries as in the Java Enterprise Edition (EE). For Windows, .NET has the advantage of being built specifically for that operating system; however, Java has the advantage of running on all major platforms, including not only Windows, but also Mac OS X and Linux, as well as being an open-source platform. (However, the open-source Mono project has developed a partial implementation of .NET for non-Windows as well as Windows platforms.)

Further Reading

Boehm, Anne. *Murach's ADO .NET 2.0 Database Programming with VB 2005.* Fresno, Calif.: Murrach, 2007.

Chappell, David. *Understanding .NET.* 2nd ed. Upper Saddle River, N.J.: Addison-Wesley Professional, 2006.

MacDonald, Matthew. *Beginning ASP.NET 3.5 in VB 2008: From Novice to Professional.* Berkeley, Calif.: APress, 2007.

.NET Framework Developer Center. Available online. URL: http://msdn2.microsoft.com/en-us/netframework/default.aspx. Accessed October 2, 2007.

Troelsen, Andrew. *Pro C# with .NET 3.0, Special Edition.* Berkeley, Calif.: Apress, 2007.

Walther, Stephen. *ASP .NET Unleashed.* Indianapolis: Sams, 2006.

Microsoft Windows

Often simply called Windows, Microsoft Windows refers to a family of operating systems now used on the majority of personal computers. Windows PCs run Intel or Intel-compatible microprocessors and use IBM-compatible hardware architecture.

HISTORY AND DEVELOPMENT

By 1984, the IBM PC and its first "clones" from other manufacturers dominated the market for personal computers, quickly overtaking the previously successful Apple II and various machines running the CP/M operating system. Through a combination of initiative and luck, Microsoft CEO Bill Gates had licensed what became its MS-DOS operating system to IBM, while retaining the rights to license it also to the clone manufacturers (see also GATES, WILLIAM).

However, 1984 also brought Apple back into contention with the Macintosh. Using a graphical user interface (GUI) largely based on research done at Xerox's Palo Alto Research Center (PARC) in the 1970s, the Macintosh was strikingly more attractive and user-friendly than the all-text, command-line driven MS-DOS. As third parties began to offer GUI alternative to DOS, Microsoft rushed to complete its own GUI, called Windows. Although it was actually announced well before the coming of the Macintosh, Windows 1.0 was not released until 1985. Its poor fonts, graphics, and window operation made it compare unfavorably to the Macintosh. Through the rest of the 1980s, Microsoft struggled to improve Windows. The acceptance of Windows was aided by several large software manufacturers such as Aldus (PageMaker) writing software for the new operating system as well as Microsoft's designing or porting its own software such as the Excel spreadsheet.

Windows 3.0, released in 1990, was considerably improved and began to attract significant numbers of users away from MS-DOS—based programs. Microsoft was also greatly aided by its ability to leverage its operating system dominance to make it economically imperative for PC manufacturers to "bundle" Windows with new PCs.

About the same time, Microsoft had been working with IBM on a system called OS/2. Unlike Windows, which was actually a program running "on top of" MS-DOS, OS/2 was a true operating system that had sophisticated capabilities such as multitasking, multithreading, and memory protection. Microsoft eventually broke off its relationship with IBM, abandoned OS/2, but incorporated some of the same features into a new version of Windows called NT (New Technology), first released in 1993. NT, which progressed through several versions, was targeted at the high-end server market, while the consumer version of Windows continued to evolve incrementally as Windows 95 and Windows 98 (released in those respective years). These versions included improved support for networking (including TCP/IP, the Internet standard) and a feature called "PLUG AND PLAY" that allowed automatic installation of drivers for new hardware.

Toward the end of the century, Microsoft began to merge the consumer and server versions of Windows. Windows 2000 incorporated some NT features and provided somewhat greater security and stability for consumers. With Windows XP, released in 2001, the separate consumer and NT versions of Windows disappeared entirely, to be replaced by home and "professional" versions of XP.

Introduced in early 2007, Microsoft Windows Vista includes a number of new features, including a 3D user interface ("Aero"), easier and more robust networking, built-in multimedia capabilities (such as photo management and DVD authoring), improved file navigation, and desktop search. Perhaps the most important, if problematic, feature is enhanced security, including User Account Control, which halts suspect programs and requests permission for them to continue. Although this makes it harder for malware to get a foothold, many users find the constant "nags" to be annoying. (As of 2008 adoption of Vista has been slower than expected, with many users opting to remain with Windows XP.)

The next version of Windows, with the working name Windows 7, should be released around 2010. Its focus

Introduced in 2007, Microsoft Windows Vista features bet-
ter security, a 3D look, new search facilities, and multimedia
features. (MICROSOFT CORPORATION)

appears to be a combination of "back to basics" (a response
to the sluggish performance of Vista) and more seamless
user access to data and media from a variety of sources.

USER'S PERSPECTIVE

From the user's point of view, Windows is a way to control
and view what is going on with the computer. The user
interface consists of a standard set of objects (windows,
menus, buttons, sliders, and so on) that behave in generally consistent ways. This consistency, while not absolute,
reduces the learning curve for mastering a new application.
Programs can be run by double-clicking on their icon on
the underlying screen (called the desktop), or by means of
a set of menus.

Windows users generally manage their files through
a component called Windows Explorer or My Computer.
Explorer presents a treelike view of folders on the disk.
Each folder can contain either files or more folders, which
in turn can contain files, perhaps nested several layers deep.
Folders and files can be moved from place to place simply
by clicking on them with the mouse, moving the mouse
pointer to the destination window or folder, and releasing the button (this operation is called dragging). Another
useful feature is called a context menu. Accessed by clicking with the right-hand mouse button, the menu brings
up a list of operations that can be done with the currently
selected object. For example, a file can be renamed, deleted,
or sent to a particular destination.

Windows includes a number of features designed to
make it easier for users to control their PC. Most settings
can be specified through windows called *dialog boxes*,
which include buttons, check boxes, or other controls. Most
programs also use Windows's Help facility to present help
pages using a standard format where related topics can be
clicked. Most programs are installed or uninstalled using
a standard "wizard" (step-by-step procedure), and wizards
are also used by many programs to help beginners carry out
more complex tasks (see HELP SYSTEMS).

MULTITASKING

From the programmer's point of view, Windows is a mul-
titasked, event-driven environment (see MULTITASKING).
Programmers must take multitasking into account in recognizing that certain activities (such as I/O) and resources
(such as memory) may vary with the overall load on the
system. Responsible programs allocate no more memory
than they need, and release memory as soon as it is no
longer needed. If the pool of free memory becomes too low,
Windows starts swapping the least recently used segments
of memory to the hard drive. This scheme, called virtual
memory, allows a PC to run more and larger programs than
would otherwise be possible, but since accessing the hard
drive takes considerably longer than accessing RAM, the
system as a whole starts slowing down.

Windows also has a rather small amount of memory
reserved for its GDI (Graphics Device Interface), a system
used for displaying graphical interface objects such as icons.
If this resource pool (which has been made somewhat more
flexible in later versions of Windows) runs out, the system
can grind to a halt.

PROGRAMMING PERSPECTIVE

Programmers moving to Windows from more traditional
systems (such as MS-DOS) must also deal with a new paradigm called event-driven programming. Most traditional
programs are driven by an explicit line of execution through
the code—do this, make this decision, and depending on it,
do that—and so on. Windows programs, however, typically
display a variety of menus, buttons, check boxes, and other
user controls. They then wait for the user to do something.
The user thus has considerable freedom to move about in
the program, performing tasks in different orders.

A Windows program, therefore, is driven by *events*. An
event is generally some form of user interaction such as
clicking on a menu or button, moving a slider, or typing
into a text box. The event is conveyed by a message (see
MESSAGE PASSING) that Windows dispatches to the affected
object. For example, if the user presses down (clicks) the left
mouse button while the mouse pointer is over a window, a
WM_BUTTONDOWN message is sent to that window.

Each of these interface objects (collectively called *controls*) has a message-handling procedure that identifies
the message. The object must then have appropriate program code that responds to each possible type of event. For
example, if the user clicks on the File menu and then clicks
on Open, the code will display a standard dialog box that
allows for selecting the file to be opened.

Fortunately for the programmer, Windows provides
developers with a large collection of types of windows, dialog boxes, and controls that can be displayed using a function call. For example, this code (after some preliminary
declarations), displays a type of window called a list box:

```
HWND MyWindow;
hMyWindow = CreateWindow("LISTBOX","Availabl
e Services",
   WS_CHILD|WS_VISIBLE,
   0,0,100,200
hwndParent,NULL,hINst,NULL);
```

Here the various parameters passed to the CreateWindow function specify the type of window, window title, characteristics, and location. The function returns a "window handle," which is a pointer that holds the window's address and allows it to be accessed later.

Most Windows programming environments, including C++ and particularly, Visual Basic, now let program designers avoid having to specify code such as the above to create windows and other objects. Instead, the programmer can click and drag various objects onto a design screen to establish the interface that will be seen by the program's user. The programmer can then use Properties settings to specify many characteristics of the screen objects without having to explicitly program them.

Microsoft and third-party developers also provide readymade programming code in *dynamic link libraries* (DLLs). These resources (see LIBRARY, PROGRAM) can be called by any application, which can then use any object or function defined in the library. Windows also provides a facility called OLE (Object Linking and Embedding). This lets an application such as a word processor "host" another application such as a spreadsheet. Thus, the Microsoft Word, for example, can embed a Microsoft Excel spreadsheet into a document, and the spreadsheet can be worked with using all the usual Excel commands. In other words, OLE lets applications make their features, controls, and functionality accessible to other applications. Indeed, collections of controls are often packaged as OCX (OLE controls) and sold to developers.

Despite all this available help, Windows presents a steep learning curve for many programmers. There are hundreds of functions for handling interface objects, drawing graphics, managing files, controlling devices, and other tasks. With the growing use of object-oriented programming languages (see OBJECT-ORIENTED PROGRAMMING and C++) in the late 1980s and 1990s, Microsoft devised the Microsoft Foundation Classes (MFC). This framework defines all of the interface objects and other entities (such as data structures) as C++ classes.

Using MFC, a programmer, instead of calling a function to create a window, creates an object of a particular Window class. To customize a window, the programmer can use inheritance to derive a new window class. The various functions for controlling windows are then defined as member functions of the window class. This use of object-oriented, class-based design organizes much of the great hodgepodge of Windows functions into a logical hierarchy of objects and makes it easier to master and to use.

For example, using the traditional Windows API (see APPLICATIONS PROGRAMMING INTERFACE) one puts a text string into a list box using this code:

```
LRESULT LRes;
LRes = SendMessage(hMyListBox,LB_
ADDSTRING,0,"Network Services");
```

(LRes is a number that will hold a code that says whether the item was successfully added)

Using MFC, this code can be rewritten as:

```
CListBox * pListBox;
int nRes;
nRes = pListBox->AddString ("Network Services");
```

Here a pointer is declared to an object of the ListBox class, and a member function of that class, AddString, is then called. While this code may not look simpler, it uses a consistent object-oriented approach.

The new common framework for Windows programming is called .NET. Closely integrated with the latest versions of Windows (XP SP2 and Vista), the class framework has been revamped and expanded. .NET provides a common language runtime (CLR) for access from different languages such as C++, C#, and Visual Basic .NET. (See MICROSOFT .NET.)

TRENDS

By just about any standard Microsoft Windows has achieved remarkable success, capturing and largely holding the lion's share of the PC operating system market. However, Windows has been persistently criticized on grounds of reliability and security. Perhaps feeling the pressure from users and potential regulators, Microsoft has placed greater emphasis on security in recent years; Windows Vista integrates security much more tightly into the structure of the system. However, as long as Windows is the most widely used operating system, it will continue to be the biggest target for creators of viruses and other malware.

Microsoft has included powerful facilities that allow Windows applications to be controlled by other applications or remotely (see SCRIPTING LANGUAGES). Unfortunately, these facilities have proven to be quite vulnerable to computer viruses that can use them to damage systems connected to the Internet. There seems to be a never-ending race between developers of program "patches" designed to plug security holes and inventive, albeit malicious virus writers.

Windows continues to face a variety of challenges. The ability to deliver applications directly through Web browsers on any platform may make it less compelling for a user with simple computing needs to pay the premium for a Windows-based PC. (For example, Google now delivers basic word processing, spreadsheet, email, and other applications—see APPLICATION SERVICE PROVIDER.) Linux, too, may be gradually gaining a greater share on the desktop. Versions such as the popular and frequently updated Ubuntu now install about as easily as Windows, provide a similar user interface, and include a variety of software, including Open Office (see LINUX and OPEN-SOURCE MOVEMENT).

While Windows still remains the dominant PC operating system with tens of thousands of applications and at least several hundred million users around the world, it is likely that the PC operating systems of 2020 will be as different from today's Windows as the latter is from the MS-DOS of the early 1980s.

Further Reading
Bellis, Mary. "The Unusual History of Microsoft Windows." About. com. Available online. URL: http://inventors.about.com/od/mstartinventions/a/Windows.htm. Accessed August 15, 2007.

Bott, Ed, Carl Siechert, and Craig Stinson. *Microsoft Windows Vista Inside Out.* Redmond, Wash.: Microsoft Press, 2007.

Minasi, Mark, and Byron Hynes. *Administering Windows Vista Security: The Big Surprises.* Indianapolis: Wiley/SYBEX, 2007.

Simpson, Alan. *Alan Simpson's Windows XP Bible.* 2nd ed. Indianapolis: Wiley, 2005.

———. *Alan Simpson's Windows Vista Bible.* Indianapolis: Wiley, 2007.

Smith, Ben, and Brian Komar. *Microsoft Windows Security Resource Kit.* 2nd ed. Redmond, Wash.: Microsoft Press, 2005.

"Windows Products and Technologies History." Microsoft. Available online. URL: http://www.microsoft.com/windows/ WinHistoryIntro.mspx. Accessed August 15, 2007.

middleware

Often two applications that were originally created for different purposes must later be linked together in order to accomplish a new purpose. For example, a company selling scientific instruments may have a large database of product specifications, perhaps written in COBOL some years ago. The company has now started selling its products on the Internet, using its Web server and e-commerce applications (see E-COMMERCE). Prospective customers of the Web site need to be able to access detailed information about the products. Unfortunately, the Web software (perhaps written in Java) has no easy way to get information from the company's old product database. Rather than trying to convert the old database to a more modern format (which might take too long or be prohibitively expensive), the company may choose to create a *middleware* application that can mediate between the old and new applications.

There are a variety of types of middleware applications. The simplest and most general type of facility is the RPC (Remote Procedure Call), which allows a program running on a client computer to execute a program running on the server. DCE (Distributed Computing Environment) is a more robust and secure implementation of the RPC concept that provides file-related other operating system services as well as executing remote programs.

More elaborate architectures are used to link complex applications such as databases where a program running on one computer on the network must get data from a server. For example, an Object Request Broker (ORB) is used in a CORBA (Common Object Request Broker Architecture) system to take a data request generated by a user and find servers on the network that are capable of fulfilling the request (see CORBA).

Middleware is often inserted into a program to allow for better monitoring or control of distributed processing. For example a TP (transaction processing) monitor is a middleware program that keeps track of a transaction that may have to go through several stages (such as point of sale entry, credit card processing, and inventory update). The TP monitor can report whether any stage of the transaction processing failed (see TRANSACTION PROCESSING).

Middleware can also be put in charge of *load balancing*. This means distributing transactions so that they are evenly apportioned among the servers on the network, in order to avoid creating delays or bottlenecks.

While use of middleware may not be as "clean" a solution as designing an integrated system from the bottom up, the economic realities of a fast-changing information environment (particularly with regard to deployment on the Web) often makes middleware an adequate second-best choice.

Further Reading

Britton, Chris, and Peter Bye. *IT Architectures and Middleware: Strategies for Building Large, Integrated Systems.* 2nd ed. Upper Saddle River, N.J.: Addison-Wesley Professional, 2004.

Myerson, Judith M. *The Complete Book of Middleware.* Boca Raton, Fla.: CRC Press, 2002.

Puder, Arno, Kay Römer, and Frank Pilhofer. *Distributed Systems Architecture: A Middleware Approach.* San Francisco: Morgan Kaufman, 2006.

"What Is Middleware?" ObjectWeb. Available online. URL: http://middleware.objectweb.org/. Accessed August 15, 2007.

military applications of computers

War has always been one of the most complex of human enterprises. Even leaving actual combat aside, the U.S. military and defense establishment constitute a huge employer, research and training agency, and transportation network. Managing all these activities require sophisticated database, inventory, tracking, and communications systems. When thousands of private defense contractors of varying sizes are considered as part of the system, the complexity and scope of the enterprise become even larger.

Specifically, military information technology applications can be divided into the following broad areas: logistics, training, operations, and battle management.

LOGISTICS

It is often said that colonels worry about tactics while generals preoccupy themselves largely with logistics. Logistics is the management of the warehousing, distribution, and transportation systems that supply military establishments and forces in the field with the equipment and fuel they need to train and to fight. Logistics within the United States is analogous to similar problems for very large corporations. The same bar codes, point of use terminals, and other tracking, inventory, and distribution systems that Amazon.com uses to get books quickly to customers while avoiding excessive inventory are, in principle, applicable to modernizing military logistic systems.

An added dimension emerges when logistical support must be supplied to forces operating in remote countries, possibly in the face of efforts by an enemy to disrupt supply. Such considerations as efficient loading procedures to accommodate limited air transport capacity, prioritization of shipping to provide the most urgently needed items, and transportation security can all come into play. (The military has pioneered the use of retinal scanners and other systems for controlling access to sensitive areas. See BIOMETRICS.)

The need for mobility and compactness makes laptops and even palmtops the form factors of choice. Military or "milspec" versions of computer hardware are generally built with more rugged components and greater resistance to heat, moisture, or dust.

TRAINING

The use of automated systems to provide training goes back at least as far as the World War II era Link trainer, which used automatic controls and hydraulics to place trainee pilots inside a moveable cockpit that could respond to their control inputs. Today computer simulations with sophisticated graphics and control systems can provide highly realistic depictions of flying a helicopter or jet fighter or driving a battle tank. The military has even adapted commercial flight simulators for training purposes. Simulations can also cover Special Forces operations and tactical decision making. Indeed, many real-time simulations (RTS) sold as popular commercial games and avidly played by young people already contain enough realistic detail to be adopted by the military as is. For example, the game Rainbow Six, based on operations in Tom Clancy novels, simulates tactical counterterrorism operations. In turn, the U.S. Army has used a simulation game called Full Spectrum Warrior to give young gamers a taste of the military life.

OPERATIONS

Aircraft, ships, and land vehicles used by the military have been fitted with a variety of computerized systems. The "glass cockpit" in aircraft is replacing the increasingly unmanageable maze of dials and switches with information displays that can keep the pilot focused on the most crucial information while making other information readily available. Traditional keyboards and joystick-type controllers can be replaced by touch screens and even by systems that can understand a variety of voice commands (see SPEECH RECOGNITION AND SYNTHESIS). Similar control interfaces can be used in tanks or ships.

Robotics offers a variety of intriguing possibilities for extending the reach of military forces while minimizing casualties. Remote-control robots can be used to clear minefields, disarm roadside bombs, or perform reconnaissance. (The Predator armed reconnaissance drone was first used successfully in anti-terrorist operations in Afghanistan in 2002.) Armed robots could assault enemy strong points without risking soldiers. The development of autonomous robots that can plan their own missions, select targets, and make other decisions is a longer-term prospect that depends on the application of artificial intelligence in the extremely challenging and chaotic battlefield environment.

BATTLE MANAGEMENT

Battle management is the ability to gather, synthesize, and present crucial information about the environment around the military unit and enable military personnel to make rapid, accurate decisions about threats and the best way to neutralize them.

The earliest example, the SAGE (Strategic Air Ground Environment) computer system, resulted from a massive development effort in the 1950s that strained the capacity of early vacuum tube-based computers to its limit. The purpose of SAGE was to provide an integrated tracking and display system that could give the Strategic Air Command (SAC) complete real-time information about any Soviet nuclear bomber strikes in progress against the continental United States. Descendents of this system were able to track ballistic missiles.

The Aegis system first deployed aboard selected navy ships in the 1970s is a good example of a tactical battle management system on a somewhat smaller scale. Aegis is a computerized system that can integrate information from sophisticated shipboard radar and sonar arrays as well as receiving and merging data from other ships and reconnaissance assets (such as helicopters). The captain of an Aegis cruiser or destroyer therefore has a real-time picture showing the locations, headings, and speeds of friendly and enemy ships, aircraft, and missiles. The system can also automatically distribute the available munitions to most effectively engage the most threatening targets.

Ultimately, the military hopes to give each unit in the field and even individual soldiers a battle management display that would pinpoint enemy vehicles and other activity. Unpiloted drone aircraft such as the Predator can loiter over the battlefield and feed video and other data into the battle management system.

While the ability to transmit and process large amounts of information can lead to strategic or tactical advantage, it also demands increased attention to security. If an enemy can jam the information processing system, its advantages could be lost at a crucial moment. Worse, if an enemy can "spoof" the system or introduce deceptive data, the military's information system could become a weapon in the enemy's hands (see COMPUTER CRIME, ENCRYPTION, INFORMATION WARFARE AND SECURITY).

BEYOND THE BATTLEFIELD

Today's military faces the challenge of diverse types of conflict (including counterinsurgency and peacekeeping), the need to interact with cultures that may be unfamiliar to most soldiers, and the need to deal with the psychological as well as physical casualties of war. A number of innovative applications of simulation and information technology are being developed.

In 2006 the U.S. military began to use a game called "Tactical Iraqi" in which soldiers must learn not only conversational phrases, but the difference between appropriate and culturally insensitive gestures and actions.

Another simulation, created at the University of Southern California, uses VR technology (see VIRTUAL REALITY) to place soldiers suffering from posttraumatic stress disorder (PTSD) back into the combat environment under controlled conditions. The goal is to gradually desensitize the person to the traumatic sights, sounds, and events.

On the information and intelligence front, the need to translate and interpret massive amounts of material in many languages in near real time has led the Defense Advanced Research Projects Agency (DARPA) to begin to develop a system that would use separate "engines" for translation, interpretation, and summarization.

Further Reading
Evans, Nicholas D. *Military Gadgets: How Advanced Technology Is Transforming Today's Battlefield.* Upper Saddle River, N.J.: Prentice Hall, 2004.

"How Military Robots Work." HowStuffWorks. Available online. URL: http://science.howstuffworks.com/military-robot.htm. Accessed August 15, 2007.

Jardin, Xeni. "VR Goggles Heal Scars of War." *Wired*, August 22, 2005. Available online. URL: http://www.wired.com/science/discoveries/news/2005/08/68575. Accessed August 15, 2007.

Roland, Alex, and Philip Shiman. *Strategic Computing: DARPA and the Quest for Machine Intelligence, 1983–1993*. Cambridge, Mass.: MIT Press, 2002.

Strachan, Ian W. *Jane's Simulation and Training Systems, 2005–2006*. Alexandria, Va.: Jane's Information Group, 2005.

Vargas, Jose Antonio. "Virtual Reality Prepares Soldiers for Real War." *Washington Post*, February 14, 2006, p. A01.

Vizard, Frank, and Phil Scott. *21st Century Soldier: The Weaponry, Gear, and Technology of the Military in the New Century*. New York: Popular Science/Time Inc., 2002.

minicomputer

The earliest general-purpose electronic digital computers were necessarily large, room-size devices. In the 1960s, however, the replacement of tubes with transistors (and gradually, integrated circuits) gave designers the choice of either keeping computers large and packing more processing and memory capacity into them, or making smaller computers that still had considerable power. The latter option led to the *minicomputer* as contrasted with the larger mainframe (see MAINFRAME).

Compared to mainframes, minicomputers often handled data in smaller "chunks" (such as 16 bits as compared to 32 or 64) and had a smaller memory capacity. Minicomputers also tended to have more limited input/output (I/O) capacity. However, while large businesses still needed mainframes to handle their large databases and volume of transactions, the minicomputer offered a relatively low cost (tens of thousands of dollars rather than hundreds of thousands), computing facility for scientific laboratories, university computing centers, industrial control, and various specialized needs.

The pioneering and most successful minicomputer company was the Digital Equipment Corporation (DEC). In 1960, DEC introduced its PDP-1, which was followed in 1965 by the quite successful PDP-8, which sold for only $18,000. By the early 1970s, DEC had been joined by competitors such as Data General and the availability of integrated memory circuits (RAM) and microprocessors packed more speed and capacity into each succeeding model.

The minicomputer had several important effects on the development of computer science and the "computer culture" as a whole (see HACKERS AND HACKING). Minicomputers gave university students direct, interactive access to computers through time-sharing, Teletype terminals, or CRT display terminals. Because minicomputers usually lacked the extensive (and expensive) software packages that came with mainframes, university users developed and eagerly swapped software such as program editors and debuggers. This cooperative effort achieved its most striking result in the development of the UNIX operating system.

The reader has probably noticed that this article refers to minicomputers in the past tense. The minicomputer didn't really disappear, but rather was transmogrified. By the late 1980s and certainly the 1990s, the personal desk-

Minicomputers such as this DEC PDP-8 brought computing power to many academic and scientific institutions for the first time. They also encouraged a culture of cooperative software development that led to such innovations as the UNIX operating system. (PAUL PIERCE COMPUTER COLLECTION)

top computer had taken advantage of more powerful microprocessors and ever more densely packed memory chips to create workstations that rivaled or exceeded the power of established minicomputers. Eventually, the minicomputer as a category virtually disappeared, its functions taken over by machines such as the powerful graphics workstations developed by companies such as Sun Microsystems and Silicon Graphics and today's forests of Web and file servers.

Further Reading

Fottral, Jerry. *Mastering the AS/400: A Practical Hands-On Guide*. Loveland, Colo.: 29th Street Press, 2000.

Hoskins, Jim, and Roger Dimmick. *Exploring IBM eServer Series*. 11th ed. Gulf Breeze, Fla.: Maximum Press, 2003.

PDP-1 Computer Exhibit. Computer History Museum. Available online. URL: http://www.computerhistory.org/pdp-1/. Accessed August 15, 2007.

Rifkin, Glenn, and George Harrar. *The Ultimate Entrepreneur: The Story of Ken Olsen and Digital Equipment Corporation*. Chicago: Contemporary Books, 1988.

Schein, Edgar H. *DEC Is Dead, Long Live DEC: The Lasting Legacy of Digital Equipment Corporation*. San Francisco: Berrett-Koehler, 2004.

Minsky, Marvin Lee

(1927–)
American
Computer Scientist

Starting in the 1950s, Marvin Minsky played a key role in the establishment of ARTIFICIAL INTELLIGENCE (AI) as a discipline. Combining cognitive psychology and computer science, Minsky developed ways to make computers function in "brain-like" ways (see NEURAL NETWORK) and then developed provocative insights about how the human brain might be organized.

Marvin Minsky was born in New York City on August 9, 1927. His father was a medical doctor, and Marvin proved to be a brilliant science student at the Bronx High School of Science and the Phillips Academy. Although he majored in mathematics at Harvard, he also showed a strong interest in biology and psychology. In 1954, he received his Ph.D. in mathematics at Princeton. In 1956, he was a key participant in the seminal Dartmouth conference that established the goals of the new discipline of artificial intelligence.

One of the most important of those goals was to explore the relationship between thinking in the human brain and the operation of computers. Earlier in the century, research into the electrical activities of neurons (the brain's information-processing cells) had led to speculation that the brain functioned something like an intricate telephone switchboard, carrying information through millions of tiny connections. During the 1940s, researchers had begun to experiment with creating electronic circuits that mimicked the activity of neurons.

In 1957, Fran Rosenblatt built a device called a perceptron. It consisted of a network of electronic nodes that can transmit and respond to signals that function much like nerve stimuli in the brain (see NEURAL NETWORK). For example, a perceptron could "recognize" shapes by selectively reinforcing the stimuli from light hitting an array of photocells. In 1969, Minsky and Seymour Papert co-authored a very influential book on the significance and limitations of perceptron research. Their work not only spurred research into neural networks and their possible practical applications, but also proved a strong impetus for the new field of *cognitive psychology,* bridging the study of human mental processes and the insights of computer science (see COGNITIVE SCIENCE).

Meanwhile, Minsky had joined with John McCarthy (see MCCARTHY, JOHN) to found the Artificial Intelligence Laboratory at the Massachusetts Institute of Technology (MIT). In moving from basic perception to the higher order ways in which humans learn, Minsky developed the concept of frames. Frames are a way to categorize knowledge about the world, such as how to plan a trip. Frames can be broken into subframes. For example, the trip-planning frame might have subframes about air transportation, hotel reservations, and packing. Minsky's frames concept became a key to the construction of expert systems that today allow computers to advise on such topics as drilling for oil or medical diagnosis (see EXPERT SYSTEMS and KNOWLEDGE REPRESENTATION). In the 1970s, Minsky and his colleagues at MIT designed robotic systems to test the ability to use frames to accomplish simpler tasks, such as navigating around the furniture in a room.

Minsky believed that the results of research into simulating cognitive behavior had fruitful implications for human psychology. In 1986, Minsky published *The Society of Mind*. This book suggests that the human mind is not a single entity (as classical psychology suggests) or a system with a small number of often-warring subentities (as psychoanalysis asserted). It is more useful, Minsky suggests, to think of the mind as consisting of a multitude of independent agents that deal with different parts of the task of living and interact with one another in complex ways. What we call mind or consciousness, or a sense of self is, therefore, what emerges from this ongoing interaction.

Minsky continues his exploration of human psychology and cognition with his latest book, *The Emotion Machine*. He has suggested that emotions are actually just alternative ways of thinking and accessing mental resources. In effect, the mind solves problems by looking among its "scripts" for those that seem applicable to the current situation, and then reflecting on them and revising as necessary.

Minsky continues his research at MIT. He has received numerous awards, including the ACM Turing Award (1969) and the International Joint Conference on Artificial Intelligence Research Excellence Award (1991).

Further Reading
Henderson, Harry. *Artificial Intelligence: Mirrors for the Mind*. New York: Chelsea House, 2007.
Marvin Minsky's Home Page. Available online. URL: http://web.media.mit.edu/~minsky/. Accessed April 10, 2007.
Minsky, Marvin. *The Emotion Machine: Commonsense Thinking, Artificial Intelligence, and the Future of the Human Mind*. New York: Simon & Schuster, 2006.
———. *The Society of Mind*. New York: Simon & Schuster, 1988.

MIT Media Lab

While often associated with innovations in computer interfaces and use of new technology, the Media Lab at the Massachusetts Institute of Technology (MIT) is actually a part of the School of Architecture and Planning. This origin is perhaps reflected in the organization's multidisciplinary research, including not only computer science and technology but cognitive science, learning, art, and design.

The lab was founded in 1985 by Nicholas Negroponte and former MIT President Jerome Wiesner (see NEGROPONTE, NICHOLAS). As of 2006 the lab's directorship was assumed by Frank Moss. The lab is funded mainly by corporate donations, though some projects receive government funding or are done in partnership with other schools or other parts of MIT. There is some ongoing tension between the specific needs and desires of corporate sponsors and the lab's research interests, and over the disposition of intellectual property created by projects.

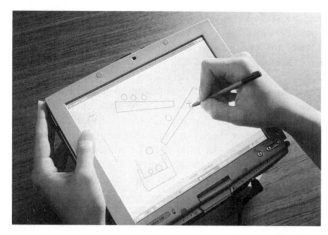

The MIT Media Lab has devised a variety of new ways for people to use computers. This is an innovative laptop sketchbook that creates animations directly from drawings. (SAM OGDEN / PHOTO RESEARCHERS, INC.)

EMPHASES AND PROJECTS

The focus of most of the lab's diverse projects is on finding innovative and productive new ways for people to use computers and related technology. Recently there has been an emphasis on more practical applications such as aiding "disabled, disadvantaged, [and] disenfranchised" people in becoming pioneers in using technology that everyone may use someday. The "One Laptop per Child" project to develop inexpensive computers for developing countries is also a part of this effort.

As of 2007 there were 27 separate research groups at the lab, including the following:

- Object-Based Media—objects that can "understand" and describe their environment

- Personal Robots—robots that interact with people socially (see BREAZEAL, CYNTHIA)

- Computing Culture—relationships among art, technology, and culture

- Molecular Machines—logical and mechanical devices using molecular-scale parts

- Software Agents—programs that can serve as assistants for human activities

- Ambient Intelligence—interfaces that are "pervasive, intuitive, and intelligent" (see MAES, PATTIE)

- Society of Mind—applying models of human cognitive processing to machines (see MINSKY, MARVIN)

- Affective Computing—developing computers that can recognize and respond intelligently to human emotion

Further Reading

Bourzac, Katherine. "Media Lab Courts Corporate Funding." *Technology Review,* February 21, 2006. Available online. URL: http://www.technologyreview.com/Biztech/16383/?a=f. Accessed October 2, 2007.

————. "The Media Lab's New Pilot." *Technology Review.* Available online. URL: http://www.technologyreview.com/article/16851/. Accessed October 2, 2007.

Maeda, John. *Maeda @ Media.* New York: Universe Publications, 2001.

The Media Lab: Inventing a Better Future. Available online. URL: http://www.media.mit.edu/. Accessed October 2, 2007.

Negroponte, Nicholas. *Being Digital.* New York: Vintage Books, 1996.

Mitnick, Kevin D.

(1963–)
American
Computer Cracker/Hacker, Consultant

Once notorious for breaking into computers and stealing information, Kevin Mitnick later became a consultant and author on computer security.

Mitnick was born October 6, 1963, in Van Nuys, California. With little parental supervision and few other friends, Mitnick became involved with "phone phreaks," people who had learned to manipulate the long-distance phone system. However, Mitnick soon turned his attention to breaking into computer systems. Mitnick first got in trouble in high school for breaking into the school district's computer system. He also allegedly broke into the North American Air Defense Command computer, though fortunately without starting a nuclear war as in the movie *War Games.* Despite being caught stealing Bell System technical manuals and put on probation, Mitnick continued breaking into computers. In 1989 he received a one-year prison sentence for breaking into computers at MCI and Digital Equipment Corporation. After getting out he violated his probation by stealing more Bell documents, and a warrant was issued for his arrest.

Mitnick then went underground, eluding authorities for two years and using a variety of fake identities. However, when Mitnick broke into the computer of physicist and computer security expert Tsutomu Shimomura and stole a large number of documents and programs, and later taunted him on the phone, Shimomura resolved to track down the intruder. Shimomura and several other experts set up a tracking program at The Well (a computer conferencing system where Mitnick had stashed the stolen material). Mitnick attempted to disguise his location by routing calls through a phone company switching office in Raleigh, North Carolina, but when Shimomura figured out that Mitnick was calling from Raleigh, he and a Sprint phone technician drove around Raleigh scanning for the calls from Mitnick's cellular modem, tracking him to his apartment building. They then called federal agents, who arrested Mitnick.

Mitnick became a cause célèbre in the hacker community. The controversy was heightened by two books written about the case, one by Shimomura and *New York Times* journalist John Markoff and the other by Jonathan Littman, who argued that the charges against Mitnick were overinflated and government prosecutors overzealous.

After serving a total of five years in prison (four and a half before he was actually tried), Mitnick was released in January 2000 on condition that he not use any form of computer network. (Mitnick appealed this restriction and it was later lifted.) Meanwhile, Mitnick then wrote two books describing both technical and psychological or "social engineering" methods used by hackers, and giving advice on how computer owners can protect themselves. Mitnick currently owns his own computer security company.

Further Reading

Goodell, Jeff. *The Cyberthief and the Samurai: The True Story of Kevin Mitnick—and the Man Who Hunted Him Down.* New York: Dell, 1996.
Littman, Jonathan. *The Fugitive Game: Online with Kevin Mitnick.* Boston: Little, Brown, 1996.
Mitnick, Kevin D. *The Art of Deception: Controlling the Human Element of Security.* Indianapolis: Wiley, 2002.
———. *The Art of Intrusion: The Real Stories behind the Exploits of Hackers, Intruders & Deceivers.* Indianapolis: Wiley, 2005.
Shimomura, Tsutomo, and John Markoff. *Takedown: The Pursuit and Capture of Kevin Mitnick, America's Most Wanted Computer Outlaw—by the Man Who Did It.* New York: Hyperion, 1996.

modeling languages

Most significant modern software projects are not simply programs, however large, but complex systems of programs or modules. Such systems have to be designed and fully described before they can be coded. Traditional methods may be adequate for simple programs (see FLOWCHART and PSEUDOCODE), but they do not capture many aspects of design and behavior. When used for software projects and information systems, modeling languages allow for components and their relationships to be described and diagrammed systematically.

UML

Unified Modeling Language, or UML, is the most widely used modeling language for software projects. UML describes software in three ways: the functions of the system as seen by the user; the system's objects, attributes, and relationships (see CLASS and OBJECT-ORIENTED PROGRAMMING); and how the system behaves, as seen by how objects interact and how their state changes. A variety of diagrams can be used to summarize this information:

- activity—describes processes and data flow, as in business transactions
- class—shows classes and data types and their relationships
- communication—the messages (data) exchanged between classes
- components—the major parts of the system
- composite structure—the internal structure of a class or component

- deployment—where the system is executed, including hardware and software servers
- interaction overview—a way to show the overall flow of control
- object—objects and relationships at a particular point in time
- package—organization of elements of the model into packages, showing dependencies
- sequence—how messages are organized chronologically
- state machine—the possible states an object or interaction can have, and how each type of input changes the state (see FINITE-STATE MACHINE)
- timing—how the state of an object changes over time as it responds to events
- use case—actors and actions (such as a customer making a purchase)

Some critics believe that UML can be overused, leading to large, complex descriptions and numerous diagrams that can be almost as hard to work with as the code itself. Further, the UML itself has to be maintained, being revised and expanded as the design and code change. Integrating modeling functions into programming environments and providing a seamless path from model to specification to code is a possible alternative, though hard to realize in practice.

Further Reading

Chonoles, Michael Jesse, and James A. Schardt. *UML 2 for Dummies.* New York: Wiley, 2003.
"Introduction to the Diagrams of UML 2.0." Agile Modeling. Available online. URL: http://www.agilemodeling.com/essays/umlDiagrams.htm. Accessed October 3, 2007.
Miles, Russ, and Kim Hamilton. *Learning UML 2.0.* Sebastapol, Calif.: O'Reilly, 2006.
UML Forum. Available online. URL: http://www.uml-forum.com/. Accessed October 3, 2007.
UML Resource Page. (Object Management Group). Available online. URL: http://www.uml.org/. Accessed October 3, 2007.

modem

As computers proliferated and users experienced an increasing need to exchange data and communicate, it became logical to tap into the telephone system, a communications technology that already linked millions of places around the world.

The problem is that the conventional telephone is an analog rather than digital device. It converts sound (such as speech) into continuously varying electrical signals. Computers, on the other hand, use discrete pulses of on/off (binary) data. However, it proved relatively easy to build a device that could "modulate" the data pulses, imposing them on a sort of carrier wave and thus converting them into electrical signals that could travel along telephone lines. At the other end of the line a corresponding device could "demodulate" that telephone signal, converting it back into data pulses. This "modulator-demodulator" device is known as a *modem* for short.

A modem contains both the modulator and demodulator circuit, with a connection to a cable and a phone jack on one side and a connection to the computer on the other. The computer connection can be provided by connecting to a standard port on the outside of the PC (see SERIAL PORT) or by mounting the modem on a card that slides into the PC's internal bus (see BUS) and connects to the outside phone line through a jack. The modem must also have a component that generates the dialing pulses needed to establish a phone connection.

The first modems for PCs appeared in the early 1980s and were very slow by modern standards, transmitting data at 300 bps (bits per second). However, speed steadily improved, reaching 1,200, 2,400, 9,600 and so on up to 56,000, which is about the maximum practical speed for this technology over ordinary phone lines.

Phone lines are far from hermetically sealed, and random fluctuations called "line noise" can sometimes be misinterpreted by the modem as part of the data signal, leading to errors. However, modern modems include sophisticated error-correcting protocols (see ERROR CORRECTION) and can automatically negotiate with each other to reduce data transmission speed over noisy lines. Data compression techniques also make it possible to have an effectively greater transmission speed by packing more information into less data. In the 1990s, there were some problems caused by competing standards, but today most modems meet the International Telecommunications Union (ITU) v.90 standard for 56 kbps transmission. The modem is now a reliable, stable commodity included as standard equipment in most new PCs.

Modems have met with increasing competition as a means to connect homes to the Internet. Data can be transmitted over video cable or special phone lines (such as DSL or ADSL) at 20–30 times faster than for a modem on an ordinary phone line (see BROADBAND). However, besides being two to three more times expensive than typical dial-up services, broadband technologies tend to be concentrated in urban areas. Nevertheless, the versatile modem is becoming a secondary means of data communication for most users.

Further Reading

Banks, Michael A. *The Modem Reference: The Complete Guide to PC Communications.* 4th ed. Medford, N.J.: CyberAge Books, 2000.

Brain, Marshall. "How Modems Work." Available online. URL: http://www.howstuffworks.com/modem.htm. Accessed August 15, 2007.

Glossbrenner, Alfred, and Emily Glossbrenner. *The Complete Modem Handbook.* New York: MIS Press, 1995.

molecular computing

While the electronic digital computer is by far the most prevalent type of calculating device in use today, it is also possible to build computational devices that exploit natural laws and processes to solve problems (see ANALOG COMPUTER). One of the most intriguing approaches is based upon chemistry and biology rather than electronics.

Consider that all living things possess a detailed "database system" of coded information, namely, the DNA sequences that define their genetic code. DNA consists of strands composed of four bases: adenine (A), cytosine (C), guanine (G), and thymine (T). There are a variety of ways in which biologists can "sequence" a strand of DNA, that is, determine the order of bases in it. It is also relatively easy to make many copies of a given chain by using the polymerase chain reaction (PCR) technique.

This stockpile of coded DNA strands can be used to solve combinatorial problems. This type of problem becomes exponentially harder to solve through "brute force" computation as the number of elements increases. An example is the famous "Traveling Salesman Problem." Here the goal is to determine a route that visits all of a list of cities while visiting each city only once.

As Leonard Edelman pointed out in his 1994 article in *Science,* a DNA-based approach to the traveling salesman problem begins by assigning two sets of four bases to each city. Next, a similar DNA combination is assigned to each available direct route between two cities, using half (four bases) of the sequence assigned to the respective cities. That is, if one city is coded TCGTAGCT and another city is coded GCATTAAG, then a route from the first city to the second would be coded TCGTTAAG.

When binding one DNA strand to another, T always binds with A, and C always binds with G. Therefore a "complement" can be defined that will bind with a given DNA string. For example, the complement of TCGTAGCT would be AGCATCGA.

Next, the strands representing the complements for the cities are mixed with the ones representing routes. If a city complement runs into a route containing that city, they bind together. The other end (representing the other end of the route) might then encounter another route strand, thus extending the route to a third city and so on, until there are strands representing potentially complete routes to all the

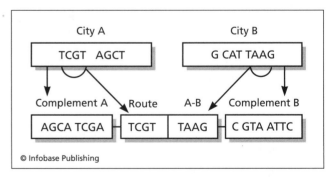

© Infobase Publishing

Molecular computing takes advantage of the properties of molecules such as DNA to create what is in effect a massive array of parallel processors. In this example, DNA strands can be coded to represent cities and possible routes between them so that they will chemically solve the Traveling Salesman Problem.

cities. After the mixing and combining is completed, separation and sequencing techniques can be used to find the shortest strand that includes all the cities. This represents the solution to the problem.

The attractiveness of molecular computing lies in its being "massively parallel" (see MULTIPROCESSING). Although molecular operations are individually much slower than electronics, DNA strands can be replicated and assembled in great numbers, potentially allowing them to go through quintillions (10^{18}) of combinations at the same time. In 1996, Dan Boneh designed an approach using DNA combinations that could be used to break the Data Encryption Standard (DES) encryption scheme by testing huge numbers of keys simultaneously.

In 2002 researchers at the Weizmann Institute of Science in Rehovot, Israel, announced that they had constructed a DNA computer that could perform 330 teraflops (trillions of operations per second). Two years later Weizmann researchers described their new DNA computer, which could be used to diagnose and treat cancer on the cellular level.

Although this application suggests the potential power in molecular computing, the approach has significant drawbacks. There are many ways that damage can occur to DNA strands during combination and processing, leading to errors. Even for the combinatorial problems that are molecular computing's strong suit, conventional electronic computers using large arrays of parallel processors are able to offer comparable power and a much easier interface. However, molecular computing illustrates the rich way in which information and information processing are embedded in nature and the potential for harnessing it for practical applications.

Further Reading

Amos, Martyn. *Genesis Machines: The New Science of Biocomputing.* New York: Overlook, 2008.
———. *Theoretical and Experimental DNA Computation.* New York: Springer, 2005.
———, ed. *Cellular Computing.* New York: Oxford University Press, 2004.
Calude, Christian, and Gheorghe Păun. *Computing with Cells and Atoms: An Introduction to Quantum, DNA and Membrane Computing.* New York: Routledge, 2001.
Păun, Gheorghe, Grzegorz Rozenberg, and Arto Salomaa. *DNA Computing: New Computing Paradigms.* New York: Springer, 1998.
Ryu, Will. "DNA Computing: A Primer." Ars Technica. Available online. URL: http://arstechnica.com/reviews/2q00/dna/dna-1.html. Accessed August 15, 2007.

monitor

As designers strove to make computers more interactive and user-friendly, the advantages of the cathode ray tube (CRT) already used in television became clear. Not only could text be displayed without wasting time and resources on printing but the individually addressable dots (pixels) could be used to create graphics. While such displays were used occasionally in defense and research systems in the 1950s, the first widespread use of CRT video monitors came with the new generation of smaller computers developed in the 1960s (see MINICOMPUTER). Since such computers were often used for scientific, engineering, industrial control, and other real-time applications, the combination of video display and keyboard (i.e., a Video Display Terminal, or VDT) was a much more practical way for users to oversee the activities of such systems. (This oversight function also led to the term *monitor*.)

A monitor can be thought of as a television set that receives a converted digital signal rather than regular TV programming. To send an image to the screen, the PC first assembles it in a memory area called a video buffer (modern video cards can store up to 64 MB of complex graphics data. See COMPUTER GRAPHICS). Ultimately, the graphics are stored as an array of memory locations that represent the colors of the individual screen dots, or pixels. The video card then sends this data through a digital to analog converter (DAC), which converts the data to a series of voltage levels that are fed to the monitor.

The monitor has electron "guns" that are aimed according to these voltages. (A monochrome monitor has

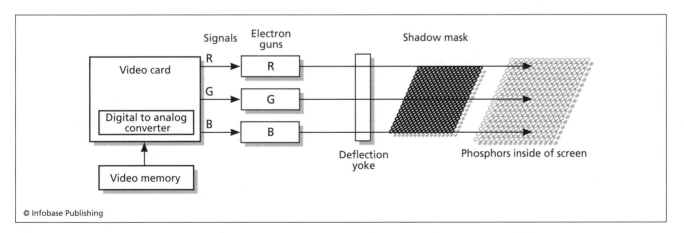

A standard computer monitor works much like an ordinary color TV set. The difference is that the signal is derived not from a broadcast program, but from the contents of video memory as processed and converted by the computer's graphics card.

only one gun, while a color monitor, like a color TV, has separate guns for red, blue, and green). The electrons from the guns pass through a lattice called a shadow mask, which keeps the beams properly separated and aligned. Each pixel location on the inner surface of the CRT is coated with phosphors, one that responds to each of the three colors.

The intensity of the beam hitting each color determines the brightness of the color, and the mixture of the red, blue, and green color levels determines the final color of the pixel. (Today's graphics systems can generate more than 16.7 million different colors, although the human eye cannot make such fine distinctions.)

The beam sweeps along a row of pixels and then turns off momentarily as it is refocused and set to the next row. The process of scanning the whole screen in this way is repeated 60 times a second, too fast to be noticed by the human eye. Less expensive monitors were sometimes designed to skip over alternate lines on each pass so that each line is refreshed only 30 times a second. This *interlaced* display can have noticeable flicker, and falling prices have resulted in virtually all current monitors being noninterlaced.

Another factor influencing the quality of a CRT monitor is the size of the screen area devoted to each pixel. The spacing in the shadow mask that defines the pixel areas is called the *dot pitch*. A smaller dot pitch allows for a sharper image.

During the 1980s, emerging video standards offered increasing screen resolution and number of colors, starting with the first IBM PC color displays at 320 × 200 pixels, 4 colors up to video graphics array (VGA) displays at 1024 × 768 pixels and at least 256 colors. The latter is considered the minimum standard today, with some displays going as high as 1600 × 1200 with millions of colors.

Meanwhile, the CRT monitor became a commodity item with steadily falling prices. A 19-inch color monitor now costs only a few hundred dollars. Ergonomically, it is important for the combination of display size and resolution to be set to avoid eyestrain. There has been some concern about users receiving potentially damaging nonionizing radiation from CRT displays, but studies have generally been unable to confirm such effects. Modern monitors are generally designed to minimize this radiation.

CRT displays are too bulky and power-hungry for laptop or handheld devices, which generally use liquid crystal displays (LCDs). In recent years large LCD displays suitable for desktop systems have also declined in price, and are rapidly becoming the display of choice even for regular PCs (see FLAT-PANEL DISPLAY).

Further Reading

Carmack, Carmen, and Jeff Tyson. "How Computer Monitors Work." Available online. URL: http://www.howstuffworks.com/monitor.htm. Accessed August 15, 2007.

Goldwasser, Samuel M. "Notes on the Troubleshooting and Repair of Computer and Video Monitors." Available online. URL: http://www.repairfaq.org/sam/monfaq.htm. Accessed August 15, 2007.

Moore, Gordon E.
(1929–)
American
Entrepreneur

The microprocessor chip is the heart of the modern computer, and Gordon Moore deserves much of the credit for putting it there. His insight into the computer chip's potential and his business acumen and leadership would lead to the early success and market dominance of Intel Corporation.

Moore was born on January 3, 1929, in the small coastal town of Pescadero, California, south of San Francisco. His father was the local sheriff and his mother ran the general store. Young Moore was a good science student, and he attended the University of California, Berkeley, receiving a B.S. in chemistry in 1950. He then went to the California Institute of Technology (Caltech), earning a dual Ph.D. in chemistry and physics in 1954. Moore thus had a sound background in materials science that would help prepare him to evaluate the emerging research in transistors and semiconductor devices that would begin to transform electronics in the later 1950s.

After spending two years doing military research at Johns Hopkins University, Moore returned to the West Coast to work for Shockley Semiconductor Labs in Palo Alto. However, Shockley, who would later share in a Nobel Prize for the invention of the transistor, alienated many of his top staff, including Moore, and they decided to start their own company, Fairchild Semiconductor, in 1958.

Moore became manager of Fairchild's engineering department and, the following year, director of research. He worked closely with Robert Noyce, who was developing a revolutionary process for placing the equivalent of many transistors and other components onto a small chip.

Moore and Noyce saw the potential of this integrated-circuit technology for making electronic devices including clocks, calculators, and especially computers vastly smaller yet more powerful. In 1965 he formulated what became widely known in the industry as Moore's law. This prediction suggested that the number of transistors that could be put in a single chip would double about every year (later it would be changed to 18 months or two years). Remarkably, Moore's law would still hold true into the 21st century, although as transistors get ever closer together, the laws of physics begin to impose limits on current technology.

Moore, Noyce, and Andrew Grove found that they could not get along well with the upper management in Fairchild's parent company, and decided to start their own company, Intel Corporation, in 1968, using $245,000 plus $2.5 million from venture capitalist Arthur Rock (see GROVE, ANDREW and INTEL CORPORATION). They made the development and application of microchip technology the centerpiece of their business plan. Their first products were RAM (random access memory) chips (see CHIP).

Seeking business, Intel received a proposal from Busicom, a Japanese firm, for 12 custom chips for a new calculator. Moore and Grove were not sure they were ready to undertake such a large project, but then Ted Hoff, one of their first employees, suggested that they could build a chip that had a general-purpose central processing unit (CPU) that could be programmed with whatever instructions were needed for each application. With the support of Moore and other Intel leaders, the project got the go-ahead. The result was the microprocessor, and it would revolutionize not only computers, but just about every sort of electronic device (see MICROPROCESSOR).

Under the leadership of Moore, Grove, and Noyce, the 1980s would see Intel established as the leader in microprocessors, starting when IBM chose Intel microprocessors for its hugely successful IBM PC. IBM's competitors, such as Compaq, Hewlett-Packard, and later Dell, would also use Intel microprocessors for most of their PCs.

In his retirement, Moore enjoyed fishing at his summer home in Hawaii while being active as a philanthropist. Moore gave a record-setting $600 million donation to Caltech in 2001, and in 2003 Moore and his wife, Betty, set up a $5 billion foundation focusing on environmental and social initiatives. Moore has also had a long-time interest in SETI, or the search for extraterrestrial intelligence.

Moore has been awarded the prestigious National Medal of Technology (1990), the IEEE Founders Medal, the W. W. McDonnell Award, as well as the Presidential Medal of Freedom (2002). In 2003 Moore was elected a fellow of the American Association for the Advancement of Science.

Further Reading
Burgelman, Robert, and Andrew S. Grove. *Strategy Is Destiny*. New York: Simon & Schuster, 2001.
"Calibrating Gordon Moore." *Caltech News* 36 (2002). Available online. URL: http://pr.caltech.edu/periodicals/CaltechNews/articles/v36/moore.html. Accessed May 5, 2007.
"Laying Down the Law." *Technology Review* 104 (May 2001): 65. Available online. URL: http://www.technologyreview.com/Infotech/12403/. Accessed October 3, 2007.
Mann, Charles. "The End of Moore's Law?" *Technology Review* 103 (May 2000): 42. Available online. URL: http://www.technologyreview.com/Infotech/12090/. Accessed October 3, 2007.

motherboard

Large computers generally had separate large cabinets to hold the central processing unit (CPU) and memory (see MAINFRAME). Personal computers, built in an era of integrated electronics, use a single large circuit board to serve as the base into which chips and expansion boards are plugged. This base is called the motherboard.

The motherboard has a special slot for the CPU (see MICROPROCESSOR). Data lines (see BUS) connect the CPU to RAM (see MEMORY) and various device controllers. Besides compactness, use of a motherboard minimizes the use of possibly fragile cable connections. It also provides expansion capability. Assuming its pins are compatible with the slot and it is operationally compatible, a PC user can plug a more powerful processor into the slot on the motherboard, upgrading performance. Memory expansion is also provided using a row of memory sockets. Memory, originally inserted as rows of separate chips plugged into individual sockets, is now provided in single modules called DIMMs that can be easily slid into place.

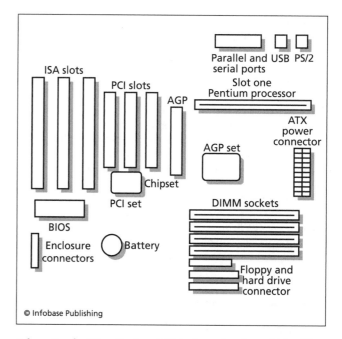

Schematic of a PC motherboard. Note the sockets into which additional RAM memory chips (DIMM) modules can be inserted, as well as the slots for ISA and PCI standard expansion cards.

The motherboard also generally includes about six general-purpose expansion slots. These follow two different standards, ISA (industry standard architecture) and PCI (peripheral component interconnect) with PCI now predominating (see BUS). These slots allow users to mix and match such accessories as graphics (video) cards, disk controllers, and network cards. Additionally, the motherboard includes a chip that stores permanent configuration settings and startup code (see BIOS), a battery, a system clock, and a power supply.

The most important factors in choosing a motherboard are the type and speeds of processor it can accommodate, the bus speed, the BIOS, system chipset, memory and device expansion capacity, and whether certain features (such as video) are integrated into the motherboard or provided through plug-in cards. Generally, users must work within the parameters of their system's motherboard, although knowledgeable people who like to tinker can buy a motherboard and build a system "from scratch" or keep their current peripheral components and upgrade the motherboard.

Further Reading
Palmer, Charlie. *How to Build Your Own PC: Save a Buck and Learn a Lot*. West St. Paul, Minn.: HCM Publishing, 2005.
Rosenthal, Morris. *Build Your Own PC*. 4th ed. Emeryville, Calif.: McGraw-Hill/Osborne, 2004.
Soderstrom, Thomas. "Beginner's Guide to Motherboard Selection." *Tom's Hardware*. Available online. URL: http://www.tomshardware.com/2006/07/26/beginners_guide_to_motherboard_selection/. Accessed August 15, 2007.
Wilson, Tracy V. "How Motherboards Work." Available online. URL: http://www.howstuffworks.com/motherboard.htm. Accessed August 15, 2007.

Motorola Corporation

Motorola Corporation (NYSE symbol: MOT) is a venerable American manufacturer of communications and other electronic equipment, including computers and cell phones. The company was founded in 1928 by Paul and Joseph Galvin as Galvin Manufacturing Corporation. The name *Motorola* arose in the early 1930s when the company began manufacturing car radios ("motor" as in car plus "ola" as in Victrola), and the company's name was officially changed to Motorola Corporation in 1947. Many of the company's subsequent products would relate to radio, such as police car radios, walkie-talkies, and cordless phones. Motorola introduced the first "brick" cell phone in 1983. Today Motorola is best known for stylish cell phones with names such as RAZR and KRZR.

Motorola also played an important role in building the global satellite communications network through the Iridium Company in the late 1990s. However, the company filed for bankruptcy when it could not attract enough telecommunications companies to use its services.

MICROPROCESSORS

Though the market came to be dominated by Intel (see INTEL CORPORATION), Motorola was an important manufacturer of microprocessors in the 1980s and 1990s. Motorola's 68000 series micrprocessors and later PowerPC series (developed jointly with IBM) were used in several computer systems of the early 1980s, including the Commodore Amiga and the Atari ST, as well as workstation terminals (Sun) and UNIX systems. The greatest consumer impact, however, would be its use in the Apple Macintosh, starting in 1984.

The later Power PC (PPC) series, launched in 1993, is a RISC processor (reduced instruction set, see RISC). This line of processors would be used in the Power Mac and other Macintosh systems until Apple adopted Intel chips in 2006.

Motorola's fortunes declined in the early to mid 2000s. In 2001 Motorola spun off its defense-related business to General Dynamics. It spun off its computer chip manufacturing division in 2004 as Freescale Semiconductors, and in 2007 Motorola sold its embedded communications chip unit to Emerson Electric. Motorola has said it would focus on its core communications business.

Despite strong demand for its cell phones and other mobile devices, in 2006 Motorola earned 42.9 billion in revenue, but its profits were down 48 percent from the previous year. This was attributed to strong price competition. In 2007 the company said it would cut 3,500 of its 66,000 employees.

Further Reading
Motorola Web site. Available online. URL: http://www.motorola.com/. Accessed October 3, 2007.
Pande, Meter S., Robert P. Neumann, and Roland R. Cavanagh. *The Six Sigma Way: How GE, Motorola, and Other Top Companies Are Honing Their Performance.* New York: McGraw-Hill, 2000.
Petrakis, Harry Mark. *The Founder's Touch: The Life of Paul Galvin of Motorola.* 3rd ed. Chicago: Motorola University Press, 1991.
Schoenborn, Guenter. *Entering Emerging Markets: Motorola's Blueprint for Going Global.* Rev. ed. New York: Springer, 2006.
Wray, William C., Joseph D. Greenfield, and Ross T. Bannatyne. *Using Microprocessors and Microcomputers: The Motorola Family.* 4th ed. Upper Saddle River, N.J.: Prentice Hall, 1998.

mouse

Traditionally, computers were controlled by typing in commands at the keyboard. However, as far back as the mid-1960s researchers had begun to experiment with providing users with more natural ways to interact with the machine. In 1965, Douglas Engelbart at the Stanford Research Institute (SRI) devised a small box that moved over the desk on wheels and was connected to the computer by a cable. As the user moved the box around, it sent signals representing its motion. These signals in turn were used to draw a pointer on the screen. Engelbart found that this system was less taxing on users than alternative such as light pens or joysticks (see ENGELBART, DOUGLAS).

This device, dubbed a "mouse," remained largely a laboratory novelty. In the 1970s, however, Xerox designed a mouse-driven graphical user interface for its Alto system, which saw only limited use. In 1984, however, Apple intro-

This Microsoft wireless optical mouse eliminates moving parts and wires for smooth, accurate, reliable performance. (MICROSOFT CORPORATION)

duced the mouse to millions of users of its Macintosh. By the early 1990s, millions more users were switching their IBM-compatible PCs from text commands (see MS-DOS) to the mouse-driven Windows interface. (See MICROSOFT WINDOWS.) Today a desktop PC without a mouse would be as unthinkable as one without a keyboard.

Meanwhile, the mouse became smaller and sleeker. Instead of wheels, the contemporary mouse uses a rolling ball that turns two adjacent rollers inside the mouse. A mouse pad with a special surface is generally used to provide uniform traction. A newer type of mouse uses optical sensors instead of rollers to sense its changing position, and does not require a mouse pad. Some mice are also cordless, using infrared or wireless data connections.

Since mice are generally impracticable for laptop use (see PORTABLE COMPUTERS), designers have offered a variety of alternatives. These include a trackball (a rolling ball built into the keyboard), a touch-sensitive finger pad, or a small stub that can be moved like a joystick by the fingertip.

Most mice now have at least two buttons. Generally, the left button is used for selecting objects, opening menus, or launching programs. The right button is used to bring up a menu of actions that can be done with the selected object. Activating a button is called *clicking*. It is the operating system that assigns significance to clicking or double-clicking (clicking twice in rapid succession) or *dragging* (holding a button down while moving the pointer). Some mice have a third button and/or a small wheel that can be used to scroll the display, but only certain software recognizes these functions.

Further Reading
Brain, Marshall, and Carmen Carmack. "How Computer Mice Work." Available online. URL: http://computer.howstuffworks.com/mouse.htm. Accessed August 15, 2007.
Pang, Alex Soojung-Kim. "The Making of the Mouse." American Heritage. Available online. URL: http://www.americanheritage.com/articles/magazine/it/2002/3/2002_3_48.shtml. Accessed August 15, 2007.

MS-DOS
The MS-DOS operating system became standard for personal computers built by IBM and its imitators (see IBM PC) during the 1980s. Today it has been virtually displaced by various versions of Microsoft Windows (see MICROSOFT WINDOWS). However, MS-DOS is important as an expression of both the limitations of the first generation of personal computers and the remarkable patience and ingenuity of its developers and users.

DEVELOPMENT
By the end of the 1970s, there were a number of rudimentary operating systems for personal computers that used a variety of microprocessors. Generally, their capabilities were limited to loading and running programs and providing basic file organization and access.

The most sophisticated early PC operating system was CP/M, developed by Gary Kildall's Digital Research for machines based on the Intel 8008 microprocessor. CP/M offered more advanced capabilities such as the ability

to use not only floppy but also hard disks, and included improved commands for listing file directories. CP/M even offered rudimentary programming tools, such as an editor and assembler, as well as an expandable architecture that allowed programmers to write utilities that could be in effect added to the operating system (see ASSEMBLER).

In one of computer history's greatest missed opportunities, Kildall and IBM failed to come to an agreement in 1980 for creating a version of CP/M for the IBM PC, which was being developed using the new 16-bit 8086 processor. IBM turned instead to Bill Gates and Microsoft, who had achieved something of a reputation for their widely used BASIC language package for personal computers (see GATES, WILLIAM). Gates agreed to provide IBM with an operating system, and did so by buying a program called QDOS ("quick and dirty operating system"), which had been developed by Tim Paterson of Seattle Computer Products. This program was released for the IBM PC as PC-DOS in 1981. However, Microsoft did not sell IBM an exclusive license, so when "clone" makers proved able to legally build IBM-compatible machines, Microsoft could sell them a generic version called MS-DOS. As the PC market boomed, this provided Microsoft with a large revenue stream, and the company never looked back.

FEATURES
MS-DOS offered a rather "clean" design that separates the operating system into three parts. There is a hardware-independent I/O system (stored as the file MSDOS.SYS), which processes requests from programs for access to disk files or to other devices such as the screen. The routines needed to actually communicate with the devices are stored in a separate file, IO.SYS, which is written by each computer manufacturer. (As users from the early 1980s remember, "PC-compatible" machines often had proprietary variations in areas such as video.) Finally, the command processor (COMMAND.COM) displays the once familiar C:\> prompt and waits for the user to type commands. For example, the DIR command followed by a path specification such as C:\TEMP lists the contents of that directory. Programs, too, can be run by typing their names at the prompt.

The MS-DOS file system, which remained largely unchanged until the most recent versions of Windows, uses a FAT (file allocation table) to indicate the disk allocation units or "clusters" assigned to each file. Starting with MS-DOS 2.0 in 1983, a hierarchical scheme of directories and subdirectories was introduced, allowing for better organization of the larger amount of space on hard disks.

One interesting feature of MS-DOS is the ability to load a program into memory and keep it available even while other programs are in use. This "terminate and stay resident" (TSR) function was soon used by enterprising developers to provide utilities such as notepads, calendars or shortcuts (see MACRO) that users could activate through special key combinations.

Users, however, had to struggle to keep enough memory free for their applications, resident programs, and device drivers. A combination of CPU addressing limitations and the high price of memory meant that early IBM PCs had a maximum of 640 kB of memory to hold the operating system and application programs. A trick called "expanded memory" was developed to allow data to be swapped back and forth between the 640 kB of usable memory and the 1–2 MB of additional memory that became available in the later 1980s.

By the early 1990s, MS-DOS (then up to version 6.0) was offering an alternative command processor (called DOS-SHELL) that included some mouse operations, better support for larger amounts of memory, and the ability to switch between different application programs. However, by that time Windows 3.0 was proving increasingly successful, and by 1995 most new PCs were being shipped with Windows. Many new users scarcely used MS-DOS at all. Finally, with the advent of Windows NT, 2000, and XP, the MS-DOS program code that still lurked within the process of running Windows disappeared entirely.

Further Reading
"Information and Help with Microsoft DOS." Available online. URL: http://www.computerhope.com/msdos.htm. Accessed August 15, 2007.
Paterson, Tim. "An Inside Look at MS-DOS." *Byte,* vol. 8, no. 6, June 1983, 250–252. Available online. URL: http://www.patersontech.com/Dos/Byte/InsideDos.htm. Accessed February 6, 2008.

multimedia
The earliest computers produced only numeric output or text (which itself actually consists of numbers—see CHARACTERS AND STRINGS). During the 1960s, CRT graphics (see MONITOR) came into limited use, mainly on computers used for scientific and engineering applications (see MINICOMPUTER). However, most business computer users continued to receive only textual output. A notable exception in the 1970s was PLATO, a system of networked educational computer terminals that combined text, graphics, and sound. It is this combination that became known as *multimedia.*

While much less powerful than mainframes or minicomputers, the hobbyist and early commercial PCs (see GRAPHICS CARD) of the late 1970s generally did have the capability of producing simple monochrome or color graphics on a monitor or TV screen. The Apple Macintosh, first released in 1984, was a considerable leap forward: Its user interface was inherently graphical, with even text being rendered as graphic bitmaps (see MACINTOSH).

The arrival of the PC greatly encouraged the development of entertainment software (see COMPUTER GAMES) as well as educational programs. As PCs became more powerful and gained hard drives and, by the late 1980s, CD-ROM drives (see CD-ROM), it became practical to put extensive multimedia content on systems in the home and school. One popular application has been encyclopedias, where the text from the printed version can be enhanced with graphics such as photographs, maps, and charts. Besides being more compelling and easier to use than the printed version, multimedia encyclopedias can be updated easily through annual upgrades, as well as allowing for linking to Web sites that can further amplify or update the content.

Encyclopedias and other educational programs also benefited from the use of links that the user can click with the mouse, bringing up additional or related information or illustrations (see HYPERTEXT AND HYPERMEDIA). Bill Atkinson's Hypercard, released for the Macintosh in 1987, provided a multimedia "construction set" that could be used by nonprogrammers to create simple hyperlinked presentations, educational programs, and even games. Hypertext and linking are the "glue" that binds multimedia into an integrated experience.

Multimedia business presentations are now routinely created using software such as Microsoft PowerPoint, then projected at meetings. While simple presentations can emulate the traditional "slide show," one-upmanship inevitably leads to more elaborate animations.

MULTIMEDIA AND DAILY LIFE
DVD-ROM drives, with about six times the storage capacity of CDs, now make it practical to include video or even feature-length movies as part of a PC multimedia package. Meanwhile, the video capabilities of PCs continue to grow, with many PCs as of 2008 having 256 MB or more of video memory. Combined with processors running at up to 2.5 GHz, this allows computer-generated graphics to rival the quality of live video.

However, the most important trend is probably the delivery of online multimedia content (see INTERNET, ONLINE SERVICES, and WORLD WIDE WEB). The widespread marketing of the Mosaic and Netscape browsers (see WEB BROWSER) in the mid-1990s changed the Internet from an arcane, text-driven experience to a multimedia platform. The ability to deliver a continuous "feed" of video and audio (see MUSIC AND VIDEO DISTRIBUTION, ONLINE and STREAMING) allows content such as TV news reports to be carried with full video and radio broadcasts carried "live" with good fidelity. Newspapers and broadcast outlets are increasingly investing in online versions of their content, viewing a Web presence as a business necessity. As more Internet users gain access to high-speed cable and DSL services (see BROADBAND), multimedia is becoming as pervasive a part of the computing experience as television is in daily life.

Many facets of that daily life are likely to be affected by multimedia technology in coming years. The ability to deliver real-time, high-quality multimedia content, as well as the use of cameras (see VIDEOCONFERENCING and WEB CAM) has made "virtual" meetings not only possible but also routine in some corporate settings. When applied to lectures, this technology can facilitate "distance learning" where teachers work with students without them occupying the same room (see DISTANCE EDUCATION and EDUCATION AND COMPUTERS). Video "chat" services and immersive, pervasive online games have become important social outlets for many people, with the experience becoming ever more realistic (see ONLINE GAMES and VIRTUAL REALITY).

Already, the concept of multimedia is becoming less distinctive precisely because it is so pervasive. Today's Web users expect to see images, video, and sound, whether as part of a news story or an educational presentation, and multimedia is appearing on all sorts of new platforms (see

MUSIC AND VIDEO PLAYERS, DIGITAL; SMARTPHONES; and DIGITAL CONVERGENCE). Further, equipped with digital cameras and camcorders (even cell-phone cameras), together with easy-to-use editing software, more and more people are becoming not just consumers of multimedia, but creators as well (see USER-CREATED CONTENT and YOUTUBE).

Further Reading

Coorough, Calleen, and James E. Shuman. *Multimedia for the Web: Creating Digital Excitement.* Boston: Thomson Course Technology, 2005.

Lauer, David, and Stephen Pentak. *Design Basics: Multimedia Edition.* 6th ed. Belmont, Calif.: Wadsworth Publishing, 2006.

MIT Media Lab. Available online. URL: http://www.media.mit.edu. Accessed August 15, 2007.

"Multimedia and Authoring Resources on the Internet." Northwestern University Library. Available online. URL: http://www.library.northwestern.edu/dms/multimedia.html. Accessed August 15, 2007.

Packer, Randall, and Ken Jordan. *Multimedia: From Wagner to Virtual Reality.* Expanded ed. New York: Norton, 2002.

Vaughan, Tay. *Multimedia: Making It Work.* 7th ed. New York: McGraw-Hill/Osborne Media, 2006.

multiprocessing

One way to increase the power of a computer is to use more than one processing unit. In early computers (see MAINFRAME) a single processor handled both program execution and input/output (I/O) operations. In the late 1950s, however, machines such as the IBM 709 introduced the concept of *channels,* or separate processing units for I/O operations. In such systems the central processor sends a set of I/O commands (such as to read a file into memory) to the channel, which has its own processor for carrying out the operation.

True multiprocessing, however, involves the use of more than one central processing unit (CPU). One successful design, Control Data Corporation's CDC 6600 (1964), contained both multiple arithmetic/logic units (the part of the CPU that does calculations) and multiple controllers for I/O and memory access control. IBM soon added multiprocessing capability to its 360 line of mainframes.

Multiprocessing can be either asymmetric or symmetric. Asymmetric multiprocessing essentially maintains a single main flow of execution with certain tasks being "handed over" by the CPU to auxiliary processors. (For example, the Intel 80386 processor could be purchased with an additional floating-point processor, allowing such calculations to be performed using more efficient hardware. When the Pentium line was developed, floating-point was integrated into the main CPU).

Symmetric multiprocessing (SMP) has multiple, full-fledged CPUs, each capable of the full range of operations. The processors share the same memory space, which requires that each processor that accesses a given memory location be able to retrieve the same value. This *coherence* of memory is threatened if one processor is in the midst of a memory access while another is trying to write data to that same memory location. This is usually handled by a "locking" mechanism (see CONCURRENT PROGRAMMING) that prevents two processors from simultaneously accessing the same location.

A subtler problem occurs with the use by processors of separate internal memory for storing data that is likely to be needed (see CACHE). Suppose CPU "A" reads some data from memory and stores it in its cache. A moment later, CPU "B" writes to that memory location, changing the data. At this point the data in "A's" cache no longer matches that in the actual memory. One way to deal with this problem is called *bus snooping.* Each CPU includes a controller that monitors the data line (see BUS) for memory locations being used by other CPUs. When it sees an address that refers to an area of memory currently being stored in the cache, the controller updates the memory from the cache. This write operation sends a signal that lets other CPUs know that any cached data they have for that location is no longer valid. This means the other CPUs will go back to memory and reread the current data.

Alternatively, all CPUs can be given a single shared cache. While less complicated, this approach limits the number of CPUs to the maximum data-handling capacity of the bus.

Larger-scale multiprocessing systems consist of latticelike arrays of hundreds or even thousands of CPUs, which are referred to as *nodes.* Indeed, small *clusters* of CPUs using the architecture given above can be connected together to form larger arrays. Each cluster can have its own shared memory cache. Because accessing memory at a remote node takes considerably longer than accessing

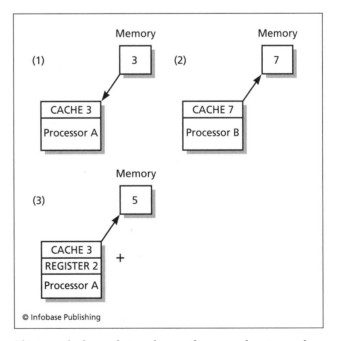

This example shows what can happen if processes do not properly manage a shared memory resource. At (1) processor A retrieves 3 from the memory location. At (2) processor B copies 7 from its cache to that same memory location. Finally, at (3) processor A adds the 3 it had retrieved to a 2 in its register, storing 5 back in a location where processor B probably expects there to still be a 7.

"local" memory within the cluster, maintaining coherence through bus monitoring is impracticable. Instead, memory is usually organized into data objects that are *distributed* optimally to reduce the necessity for remote access, and the objects are shared by CPUs requesting them through a directory system.

MULTIPROGRAMMING

In order for a program to take advantage of the ability to run on multiple CPUs, the operating system must have facilities to support multiprocessing, and the program must be structured so its various tasks are most efficiently distributed among the CPUs. These separate tasks are generally called *threads*. A single program can have many threads, each executing separately, perhaps on a different CPU, although that is not required.

The operating system can use a number of approaches to scheduling the execution of processes or threads. It can simply assign the next idle (available) CPU to the thread. It can also give some threads higher priority for access to CPUs, or let a thread continue to "own" its CPU until it has been idle for some specified time.

The use of threads is particularly natural for applications where a number of activities must be carried on simultaneously. For example, a scientific or process control application may have a separate thread reading the data being returned from each instrument, another thread monitoring for alarm conditions, and other threads generating graphical output.

Threads also allow the user to continue interacting with a program while the program is busy carrying out earlier requests. For example, the user of a Web browser can continue to use menus or navigation buttons while the browser is still loading graphics needed for the currently displayed Web page. A search program can also launch separate threads to send requests to multiple search engines or to load multiple pages.

Support for multiprogramming and threads can now be found in versions of most popular programming languages, and some languages such as Java are explicitly designed to accommodate it.

Multiprogramming often uses groups or clusters of separate machines linked by a network. Running software on such systems involves the use of communications protocols such as the MPI (message-passing interface). This programming interface has been widely deployed on many platforms for use with languages such as C/C++ and FORTRAN. Another popular programming interface is OpenMP, which features the allocation of execution threads and the distribution of work among them.

A MULTIPROCESSED WORLD

The demand for software that can efficiently use multiple processors is likely for some time to outstrip the supply of programmers who can "think in parallel." One reason is that today most new PCs have two processing "cores," with four-core systems available and more to come (see MICROPROCESSOR). This means that many mainstream applications will eventually need to be rewritten for the new hardware environment. Another factor is that as the price per processor continues to decline and high-end multiprocessing machines are reaching 1 "petaflop" (1 quadrillion operations per second), many supercomputer applications will also need to be rewritten.

Meeting this demand not only takes training, it also takes appropriate languages and other tools. In recent years, therefore, the Defense Advanced Research Projects Agency (DARPA) has funded research in "High Productivity Computer Systems" by such companies as Cray, Sun, and IBM. New languages built "from the ground up" for multiprocessing include Sun's Fortress, Cray's Chapel, and IBM's new project, code-named X10. Ultimately, systems for developing multiprocessing software should take most of the architectural details off the hands of the programmer, allowing performance to smoothly "scale up" with the increasing number of processors.

Further Reading

Anthes, Gary. "Languages for Supercomputing Get 'Suped' Up." *Computerworld,* March 12, 2007. Available online. URL: http://www.computerworld.com/action/article.do?command=viewArticleBasic&articleId=283477. Accessed April 5, 2007.
Culler, David E., and Jaswinder Pal Singh. *Parallel Computer Architecture: A Hardware/Software Approach*. San Francisco: Morgan Kaufmann, 1999.
Dongara, Jack, et al., eds. *Sourcebook of Parallel Computing*. San Francisco: Morgan Kaufmann, 2003.
Feldman, Michael. "Our Manycore Future." HPC Wire. Available online. URL: http://www.hpcwire.com/hpc/1295541.html. Accessed April 5, 2007.
Merritt, Rick. "Where Are the Programmers? Enrollment Wanes Just as Computer Scientists Grapple with Problem of Parallelism." *EE Times,* March 12, 2007. Available online. URL: http://www.eetimes.com/showArticle.jhtml?articleID=197801653. Accessed April 5, 2007.
Quinn, Michael J. *Parallel Programming in C with MPI and OpenMP*. New York: McGraw-Hill, 2003.

multitasking

Users of modern operating systems such as Microsoft Windows are familiar with multitasking, or running several programs at the same time. For example, a user might be writing a document in a word processor, pause to check the e-mail program for incoming messages, type a page address into a Web browser, then return to writing. Meanwhile, the operating system may be running a number of other programs tucked unobtrusively into the background, such as a virus checker, task scheduler, or system resource monitor.

Each running program "takes turns" using the PC's central processor. In early versions of Windows, multitasking was *cooperative,* with each program expected to periodically yield the processor to Windows so it could be assigned to the next program in the queue. One weakness of this approach is that if a program crashes, the CPU might be "locked up" and the system would have to be rebooted. However, Windows NT, 2000, and XP (as well as operating systems such as UNIX) use preemptive multitasking. The operating system assigns a "slice" of processing (CPU) time to a program and then switches it to the next program regardless of what

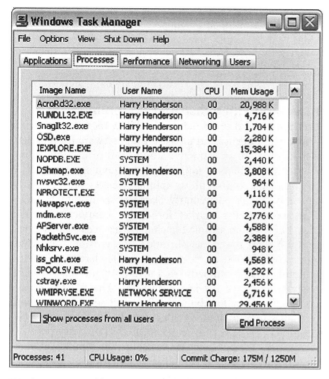

Windows users can bring up a window listing all processes or tasks running on the system, and shut down any task that has stopped responding to input.

might be happening to the previous program. Thus, if a program "crashes," the CPU will still be switched to the next program, and the user can maintain control of the system and shut down the offending program.

Systems with preemptive multitasking often give programs or tasks different levels of priority that determine how big a slice of CPU time they will get. For example, the "active" program (in Windows, the one whose window has been selected for interaction by the user) will be given preference over a background program such as a print spooler. Also, the operating system can more intelligently assign CPU time according to what a given program is doing. Thus, if a program is waiting for user input, it may be given only an occasional slice of CPU time so it can check to see whether input has been received. (The user, after all, is millions of times slower than the CPU.) When some input (such as a menu selection) is ready for processing, the program can be given higher priority.

Priority can be expressed in two different ways. One way is to move a program up in the list of running tasks (see QUEUE). This ensures it gets a turn before any lower-priority task. The other way is to have turns of varying length, with the higher-priority program getting a longer turn.

Even operating systems with preemptive multitasking can provide facilities that programs can use to communicate their own sense of their priority. In UNIX systems, this is referred to as "niceness." A "nice" program gives the operating system permission to interrupt lengthy calculations so other programs can have a turn, even if the

program's priority would ordinarily entitle it to a greater share of the CPU.

Multitasking should be distinguished from two several similar-sounding terms. Multitasking refers to entirely separate programs taking turns executing on a single CPU. *Multithreading,* on the other hand, refers to separate pieces of code within a program executing simultaneously but sharing the program's common memory space. Finally, *multiprocessing* or parallel processing refers to the use of more than one CPU in a system, with each program or thread having its own CPU (see MULTIPROCESSING).

Further Reading

Silberschatz, Abraham, Peter Baer Galvin, and Greg Gagne. *Operating System Concepts.* 7th ed. New York: Wiley, 2004.
Tanenbaum, Andrew S., and Albert S. Woodhull. *Operating Systems Design and Implementation.* 3rd ed. Upper Saddle River, N.J.: Prentice Hall, 2006.

music, computer

Computers have had a variety of effects on the performance, rendering, and composition of music. At the same time, the sound capabilities of standard personal computers have improved greatly, and music and other sounds have become an integral part of games and educational software (see MULTIMEDIA).

After the invention of the vacuum tube, a number of electronic instruments were devised. The best known is the theremin, invented by Lev Termin, a Russian physicist, in 1919. The instrument consists of a vacuum tube connected to two antennas. The player varies the pitch and volume of its eerie sound by moving his or her hands near the antennas.

Some composers became fascinated by electronic music, both for its sense of modernity and its promise of breaking the bonds of traditional form and instrumentation. In 1953, German composer Karlheinz Stockhausen (1928–2007) founded an Electronic Music Studio in Cologne and created electronic works.

Meanwhile, inventors experimented with electronic synthesizers such as the RCA MKI and MKII, which used vacuum tubes and could be programmed with punched paper tape. The advent of solid-state circuitry in the 1960s made synthesizers far more reliable and compact. The Moog synthesizer in particular became a staple of leading-edge rock and avant-garde music. It was now time for the computer to catch up to the potential of electronic sound.

In the 1970s, digital music synthesizers with keyboards and microprocessor-controlled sound generation became available to adventurous (and fairly well-to-do) musicians. RAY KURZWEIL's digital music synthesis system, introduced in 1984, achieved a new level of sonic realism by using programming stored in read-only memory (ROM) to emulate subtle characteristics such as attack and timbre, realistically re-creating the sounds of many types of orchestral instruments.

Computer music synthesis enabled composers to experiment with algorithmic composition. That is, they could use programs to create new works by combining randomization with the permutation of patterns (serialism). Compositions have also been based on applying mathematical structures (such as fractals) and the concepts being discovered by computer scientists, including adaptive structures such as neural nets and genetic algorithms.

Like most avant-garde music, computer music composition remained largely unknown to most people. However, the technology of music synthesis was to become democratically available to everyday musicians as well. As the personal computer began to bring increasingly powerful microprocessors to consumers, it became practicable to in effect add a music synthesizer to the PC. The musical instrument digital interface, MIDI, provides a protocol for connecting traditional musical instruments such as pianos and guitars to a personal computer. MIDI specifies the pitch, volume, attack (how a note increases to maximum volume), and decay (how it dies away). The musician then uses the instrument as an input device, with the notes played being recorded as MIDI data. Different tracks can then be edited (such as to transpose to a different key), and combined in various ways to create complete compositions. Because MIDI stores instructions, not actual digitized sound, it is a quite compact way to store music. MIDI brought the synthesizer within reach of just about any serious musician—and many amateurs.

PC sound cards can play sound in two ways. *Wave Table Synthesis* uses a table of stored digital samples of notes played by various instruments, and algorithmically manipulates them to reproduce the MIDI-encoded music. *FM Synthesis* attempts to create waves that replicate the intended sounds, based on a model of what happens in a given instrument. It is less faithful to the original sound, since it does not capture the detailed "texture" of a digital sample.

Today's PCs have sound cards that can handle both playback of audio CDs and rendering of digitized and synthesized sounds. The cards have the capacity to support many simultaneous voices (polyphony) as well as rendering speech faithfully. While early PCs tended to have only tiny internal speakers, most PCs today come with speakers (often including subwoofers and even multiple speakers for "surround sound") comparable to midrange home stereo systems.

Of course great hardware would not be very useful without software that can help even beginning composers turn their ideas into sound. One example is GarageBand for the Macintosh, which makes it easy to make compositions from sampled and sequenced loops together with music played using sampled and synthesized instrument sounds and a MIDI keyboard. (Sony ACID Pro offers similar features for Microsoft Windows users.)

Further Reading
Burns, Kristine H. "History of Electronic and Computer Music Including Automatic Instruments and Composition Machines." Available online. URL: http://eamusic.dartmouth.edu/~wowem/electronmedia/music/eamhistory.html. Accessed August 15, 2007.
Freebyte Music Zone. Available online. URL: http://www.freebyte.com/music/. Accessed August 15, 2007.
Manning, Peter. *Electronic and Computer Music.* New York: Oxford University Press, 2004.
Nelson, Mark. *Getting Started in Computer Music.* Boston: Thomson Course Technology, 2005.
White, Paul. *Basic MIDI.* London: Sanctuary, 2004.
Williams, Ryan. *Windows XP Digital Music for Dummies.* Hoboken, N.J.: Wiley, 2004.

music and video distribution, online

Since most audio and much video is now recorded in digital format, the Internet and media-player software for a variety of platforms are an attractive way to sell or otherwise distribute the products of musicians and moviemakers.

For a time, online file swapping (see FILE-SHARING AND P2P NETWORKS), particularly Napster, seemed to be massively eroding the market for online commercial music sales. Legal action against file-sharing services and (starting in 2003) their users has curtailed this erosion somewhat, with a considerable number of former file-sharers switching to buying paid downloads. As a result, several major online music stores have become successful. The most common model sells songs for about a dollar each—sometimes more for higher quality audio or files that do not have copy protection (see DIGITAL RIGHTS MANAGEMENT).

Apple's iTunes Music store debuted in 2003 and soon became the market leader. The combination of the iTunes store, the iTunes media player software (available for both Macintosh and PC), and the very popular iPod (see MUSIC AND VIDEO PLAYERS, DIGITAL) has been very successful. As of 2007 iTunes still had the largest selection of music available (about 6 million songs), and had sold more than 3 billion songs. The service also sells videos (television episodes, music videos, short films, and feature-length movies) at varying prices.

Rhapsody, a service that predates iTunes, offers a subscription model: The user has unlimited streaming access to the music as long as the monthly fee is paid as well as the pay-per-track option.

ALTERNATIVE MODELS
There are also alternatives to the big, label-controlled music services. A number of services now bring together independent musicians and their audience. There are also some innovative pricing models. Arnie Street, for example, starts out with uploaded music being available free, but then gradually raises the price (up to 98 cents) as more people download it. The service also offers user participation (keyword tagging by users) and social-networking features. Another service, eListeningPost, lets musicians post music that people can download as a preview (playable a limited number of times) or buy using PayPal. The service also helps musicians build their fan base by collecting e-mail addresses.

VIDEO
Video-sharing sites are very popular (see YOUTUBE). TV networks are now providing selected episodes of popular shows online for free, hoping to entice more regular viewers. However, only 7 percent of users surveyed by the Pew Internet & American Life Project in 2007 said they

had paid for any online video content. This may gradually change, particularly as more users move from basic broadband connections to enhanced, higher-speed ones (see BROADBAND). Already video sales through iTunes and Amazon's new Unbox (and smaller services) are expected by Forrester Research to generate $279 million in revenue for 2007. However, some analysts believe that the business model for selling video will soon shift to something more like that of premium cable channels, with streaming video available by subscription.

Further Reading

Arnie Street. Available online. URL: http://amiestreet.com/. Accessed October 3, 2007.

Bove, Tony, and Cheryl Rhodes. *iPod & iTunes for Dummies.* 5th ed. Hoboken, N.J.: Wiley, 2007.

eListeningPost. Available online. URL: http://elisteningpost.com/. Accessed October 3, 2007.

iTunes. Available online. URL: http://www.apple.com/itunes/store. Accessed October 3, 2007.

Madden, Mary. "Artists, Musicians, and the Internet." Pew Internet & American Life Project, December 5, 2004. Available online. URL: http://www.pewinternet.org/pdfs/PIP_Artists.Musicians_Report.pdf. Accessed October 3, 2007.

———. "Online Video." Pew Internet & American Life Project, July 25, 2007. Available online. URL: http://www.pewinternet.org/pdfs/PIP_Online_Video_2007.pdf. Accessed October 3, 2007

Madden, Mary, and Lee Rainie. "Music and Video Downloading Moves beyond P2P." Pew Internet & American Life Project, March 2005. Available online. URL: http://www.pewinternet.org/pdfs/PIP_Filesharing_March05.pdf. Accessed October 3, 2007.

Muchmore, Michael. "New Ways to Get Music." *ExtremeTech.,* December 10, 2006. Available online. URL: http://www.extremetech.com/article2/0,1697,2070636,00.asp. Accessed October 3, 2007.

Rhapsody. Available online. URL: http://www.rhapsody.com/home.html. Accessed October 3, 2007.

music and video players, digital

One characteristic of the rapidly evolving digital world is the ability to play a variety of media (music, photos, video) on many devices, ranging from desktop PCs to smart phones and, of course, iPods and MP3 players (see DIGITAL CONVERGENCE). The ability to organize and play media requires suitable software and hardware.

On desktop and laptop PCs, media-playing software is available for all operating systems. Examples include Windows Media Player, Apple iTunes, and RealPlayer (which also has a Linux version). This software typically includes these features:

- plays most types of media files (see GRAPHICS FORMATS and SOUND FILE FORMATS)

- plays content on CD/DVD, files on the hard drive, or content being received directly from the Internet (see CD-ROM AND DVD-ROM; MUSIC AND VIDEO DISTRIBUTION, ONLINE; and STREAMING)

- controls are modeled on those found on DVD players, with customizable appearance

- creates a music library that can be searched or reorganized

- can create playlists and queues and can play back songs in order, shuffled (randomized), or according to other preferences

- can obtain additional information about media (tracks, albums, and so on) online

PORTABLE PLAYERS

First unveiled in 2001, the Apple iPod is the best-selling example of a portable media player (often called a "digital audio player"). Its compact, stylish design and simple user interface quickly caught on, even though the first model was only compatible with the Macintosh and it used Apple's AAC format rather than MP3. Later iPods added capacity, larger screens, and features (such as being able to play video), while Apple also offered the inexpensive iPod Shuffle. A competitor is the Microsoft Zune player and music store, which, however, as of mid-2007 had made little headway against market leader Apple. A third well-reviewed choice is the Creative Labs Zen series. A variety of other portable players are available. Higher capacity units use tiny hard drives (up to 160 GB capacity or so), while smaller capacity models use flash memory (2–32 GB) instead. A variety of other handheld devices can play music and video (see PDA and SMARTPHONE).

With the ubiquitous use of iPods and other players, particularly by young people, some health and safety concerns have been raised. If played too loudly for long periods of time through headphones or earbuds, the devices may cause hearing damage. Drivers and pedestrians may also be

The Apple iPod is the most popular portable digital media player, featuring a simple, effective interface and the ability to play music and show video. (APPLE CORPORATION)

at greater risk if the music they are listening to cuts off the sound of approaching vehicles.

Further Reading

Johnson, Brian. *Zune for Dummies.* Hoboken, N.J.: Wiley, 2007.

Kelby, Scott. *The iPod Book: Doing Cool Stuff with the iPod and the iTunes Store.* 3rd. ed. Berkeley, Calif.: Peachpit Press, 2006.

MP3 Players (CNet Reviews). Available online. URL: http://reviews.cnet.com/Music/2001-6450_7-0.html. Accessed October 3, 2007.

"MP3 Players: The Basics and History." Available online. URL: http://www.mp3playerlimelight.com/. Accessed October 3, 2007.

Rathbone, Andy. *MP3 for Dummies.* New York: Hungry Minds, 2001.

nanotechnology

Ordinary refining and manufacturing involve the use of grinding, cutting, heating, application of chemicals, and other processes that affect large numbers of atoms or molecules at once. These processes are necessarily imprecise: Some atoms or molecules will end up unprocessed or somehow out of alignment. The resulting material will thus fall short of its maximum theoretical strength or other characteristics.

In a talk given in 1959, physicist Richard Feynman suggested that it might be possible to manipulate atoms individually, spacing them precisely. As Feynman also pointed out, the implications for computer technology are potentially very impressive. A current commercial DIMM memory module about the size of a person's little finger holds about 250 megabytes (MB) worth of data. Feynman calculated that if 100 precisely arranged atoms were used for each bit of information, the contents of all the books that have ever been written (about 10^{15} bits) could be stored in a cube about 1/200 of an inch wide, just about the smallest object the unaided human eye can see. Further, although the density of computer logic circuits in microprocessors is millions of times greater than it was with the computers of 1959, computers built at the atomic scale would be billions of times smaller still. Indeed, they would be the smallest (or densest) computers possible short of one that used quantum states within the atoms themselves to store information (see QUANTUM COMPUTING). "Nanocomputers" could also efficiently dissipate heat energy, overcoming a key problem with today's increasingly dense microprocessors.

Feynman offered some possible methods of manufacture, and discussed some of the obstacles that would have to be overcome to do engineering at a molecular or atomic scale. These include lubrication, the effects of heat, and electrical resistance. He invited adventurous high school students to develop science projects to explore this new technology.

The idea of atomic-level engineering lay largely dormant for about two decades. Starting with a 1981 paper, however, K. Eric Drexler began to flesh out proposed structures and methods for a branch of engineering he termed *nanotechnology*. (The "nano" refers to a nanometer, or one billionth of a meter.) Research in nanotechnology today focuses on two broad areas: assembly and replication. Assembly is the problem of building tools (called assemblers) that can deposit and position individual atoms. Since such tools would almost certainly be prohibitively expensive to manufacture individually, research has focused on the idea of making tools that can reproduce themselves. This area of research began with John von Neumann's 1940s concept of self-replicating computers (see VON NEUMANN, JOHN). If an assembler can assemble other assemblers from the available "feedstock" of atoms, then obtaining the number of assemblers necessary to manufacture the intended product would be no problem. (As science fiction writers have pointed out, the ultimate problem would be making sure the self-reproducing assemblers do not get out of control and start turning everything around them, potentially the whole Earth, into more of themselves.)

COMPUTING APPLICATIONS

Science fiction aside, there are several potential applications of nanotechnology in the manufacture of computer components. One is the possible use of carbon nanotubes in place of copper wires as conductors in computer chips. As chips continue to shrink, the connectors have also had to get smaller, but this in turn increases electrical resistance and reduces efficiency. Nanotubes, however, are not only superb electrical conductors, they are also far thinner than their copper counterparts. Intel Corporation has conducted promising tests of nanotube conductors, but it will likely be a number of years before they can be manufactured on an industrial scale.

An obstacle to manufacturing carbon nanotubes is that each newly made batch is a mixture of "metallic" (conducting) and semiconducting tubes of different diameters. Manufacturing, however, requires tubes that meet strict requirements. Fortunately researchers at Northwestern University in 2006 developed a way to sort the tubes by adding substances that changed their density according to both their diameter and their electrical conductivity.

Another alternative is "nanowires." One design consists of a germanium core surrounded by a thin layer of crystalline silicon. Nanowires are easier to manufacture than nanotubes, but their performance and other characteristics may make them less useful for general-purpose computing devices.

The ultimate goal is to make the actual transistors in computer chips out of nanotubes instead of silicon. An important step in this direction was achieved in 2006 by IBM researchers who created a complete electronic circuit using a single carbon nanotube molecule.

Further Reading

Booker, Richard D. and Earl Boysen. *Nanotechnology for Dummies.* Hoboken, N.J.: Wiley, 2005.

Bullis, Kevin. "Nanotube Computing Breakthrough: A Method for Sorting Nanotubes by Electronic Properties Could Help Make Widespread Nanotube-Based Electronics a Reality." *Technology Review,* October 30, 2006. Available online. URL: http://www.technologyreview.com/Nanotech/17672/. Accessed August 16, 2007.

———. "Nanowire Transistors Faster than Silicon." *Technology Review,* June 20, 2006. Available online. URL: http://www.technologyreview.com/Nanotech/17008/. Accessed August 16, 2007.

Edwards, Steven A. *The Nanotech Pioneers: Where Are They Taking Us?* New York: Wiley, 2006.

Kanellos, Michael. "Intel Eyes Nanotubes for Future Chip Designs." CNET News, November 10, 2006. Available online. URL: http://news.com.com/2100-1008_3-6134437.html. Accessed August 16, 2007.

Korkin, Anatoli, et al., eds. *Nanotechnology for Electronic Materials and Devices.* New York: Springer, 2007.

Nanotech Web. Available online. URL: http://nanotechweb.org/. Accessed August 16, 2007.

natural language processing

Since at least the days of Hal 9000 and early *Star Trek,* the computer of the future was supposed to be able to understand what people wanted, when expressed in ordinary language and not programming code. Computer scientists have been working on this capability, called natural language processing (NLP), for decades.

NLP is a multidisciplinary field that draws from linguistics and computer science, particularly artificial intelligence (see also LINGUISTICS AND COMPUTING and SPEECH RECOGNITION AND SYNTHESIS). In terms of linguistics, a program must be able to deal with words that have multiple meanings ("wind up the clock" and "the wind is cold today") as well as grammatical ambiguities (in the phrase "little girl's school" is it the school that is little, the girls, or both?). Of course each language has its own forms of ambiguity.

Programs can use several strategies for dealing with these problems, including using statistical models to predict the likely meaning of a given phrase based on a "corpus" of existing text in that language (see LANGUAGE TRANSLATION SOFTWARE).

As formidable as the task of extracting the correct (literal) meaning from text can be, it is really only the first level of natural language processing. If a program is to successfully summarize or draw conclusions about a news report from North Korea, for example, it would also have to have a knowledge base of facts about that country and/or a set of "frames" (see MINSKY, MARVIN) about how to interpret various situations such as threat, bluff, or compromise.

APPLICATIONS

There are a variety of emerging applications for NLP, including the following:

- voice-controlled computer interfaces (such as in aircraft cockpits)

- programs that can assist with planning or other tasks (see SOFTWARE AGENTS)

- more-realistic interactions with computer-controlled game characters

- robots that interact with humans in various settings such as hospitals

- automatic analysis or summarization of news stories and other text

- intelligence and surveillance applications (analysis of communication, etc.)

- data mining, creating consumer profiles, and other e-commerce applications

- search-engine improvements, such as in determining relevancy

Further Reading

Jackson, Peter, and Isabelle Moulinier. *Natural Language Processing for Online Applications: Text Retrieval, Extraction and Categorization.* 2nd ed. Philadelphia: John Benjamins, 2007.

Kao, Anne, and Steve R. Poteet, eds. *Natural Language Processing and Text Mining.* New York: Springer, 2007.

Manning, Christopher D., and Hinrich Schütze. *Foundations of Statistical Natural Language Processing.* Cambridge, Mass.: MIT Press, 1999. Available online. URL: http://www-nlp.stanford.edu/fsnlp/. Accessed October 4, 2007.

Natural Language Toolkit. Available online. URL: http://nltk.sourceforge.net/index.php/Main_Page. Accessed October 4, 2007.

Resources for Text, Speech and Language Processing. Available online. URL: http://www.cs.technion.ac.il/~gabr/resources/resources.html. Accessed October 4, 2007.

Negroponte, Nicholas
(1944–)
American
Computer Scientist

As founder and longtime director of the MIT Media Lab, Nicholas Negroponte has overseen and contributed to some of the most creative developments in human-computer interaction and interface design.

Born in 1943, the son of a Greek shipping magnate, Negroponte grew up in New York City. He attended the Massachusetts Institute of Technology (MIT), earning his master's degree in architecture in 1966 and joining the faculty. The following year Negroponte founded the MIT Architecture Machine Group, which focused on developing new ways for people to interact with computers. In 1985, Negroponte and Jerome Wiesner founded the MIT Media Lab, which has become world famous as a center of research into new media and innovative computer interfaces (see MIT MEDIA LAB).

Negroponte made a different contribution to the new computer culture in 1992 when he became a key investor in *Wired Magazine,* where he also contributed a column until 1998. Many of the ideas in these columns were reworked into Negroponte's 1995 book *Being Digital.* This book was widely influential in its predictions of a coming world where information and entertainment would become a pervasive web and people would interact actively with the new media (see DIGITAL CONVERGENCE and UBIQUITOUS COMPUTING). Negroponte's slogan is "move bits, not atoms," meaning that the new economy will be focused more on information and media than physical production. Some critics, however, have argued that Negroponte's work was filled with a naive utopianism that did not consider the potential difficulties and social consequences of the new technology.

As Negroponte observed how venture capitalists were pursuing the digital revolution of the 1990s, he began to seek similar funding for the Media Lab. This was controversial, since the lab had a strong academic culture, with its reluctance to become too involved with corporate agendas. In 2000 Negroponte stepped down as director of the Media Lab, gradually becoming less involved in the ongoing reorganization of the institution. In 2006 he also relinquished his post as chairman, though he has retained his post as professor at MIT.

ONE LAPTOP PER CHILD
In recent years Negroponte has focused his efforts on designing and distributing low-cost laptop PCs to millions of children in developing nations. (The project is called "One Laptop per Child.") In 2005 at the World Summit on the Information Society held in Tunis, Negroponte unveiled a $100 laptop called the Children's Machine. However, in the next few years commitments from participating nations have been slower than anticipated. Undaunted, Negroponte in 2007 announced a new way to distribute the machines—make them such an attractive buy that consumers in developed countries would be willing to pay a few hundred dollars for *two* of them—one for the consumer and one to go to a student in a developing country.

Negroponte also continues to be active as an investor or board member in technology startups as well as being a board member of Motorola and a member of the editorial board of the *Wall Street Journal.*

Further Reading
Hamm, Steve. "Give a Laptop and Get One." *BusinessWeek,* September 24, 2007. Available online. URL: http://www.businessweek.com/technology/content/sep2007/tc20070923_960941.htm. Accessed October 4, 2007.
Negroponte, Nicholas. *Being Digital.* New York: Vintage Books, 1996.
———. "Creating a Culture of Ideas." *Technology Review,* February 2003. Available online. URL: http://www.technologyreview.com/Biztech/13074/. Accessed October 4, 2007.
Nicholas Negroponte (home page). Available online. URL: http://web.media.mit.edu/~nicholas/. Accessed October 4, 2007.
Pogue, David. "Laptop with a Mission Widens Its Audience." *New York Times,* October 4, 2007. Available online. URL: http://www.nytimes.com/2007/10/04/technology/circuits/04pogue.html. Accessed October 5, 2007.

netiquette
As each new means of communication and social interaction is introduced, social customs and etiquette evolve in response. For example, it took time before the practice of saying "hello" and identifying oneself became the universal way to initiate a phone conversation.

By the 1980s, a system of topical news postings (see NETNEWS AND NEWSGROUPS) carried on the Internet was becoming widely used in universities, the computer industry, and scientific institutions. Many new users did not understand the system, and posted messages that were off topic. Others used their postings as to insult or attack ("flame") other users, particularly in newsgroups discussing perennially controversial topics such as abortion. When a significant number of postings in a newsgroup are devoted to flaming and counter-flaming, many users who had sought civilized, intelligent discussion leave in protest.

In 1984, Chuq von Rospach wrote a document entitled "A Primer on How to Work with the Usenet Community." It and later guides to net etiquette or "netiquette" offered useful guidelines to new users and to more experienced users who wanted to facilitate civil discourse. These suggestions include:

- Learn about the purpose of a newsgroup before you post to it. If a group is moderated, understand the moderator's guidelines so your postings won't be rejected.

- Before posting, follow some discussions to see what sort of language, tone, and attitude seems to be appropriate for this group.

- Do not post bulky graphics or other attachments unless the group is designed for them.

- Avoid "ad hominem" (to the person) attacks when discussing disagreements.

- Do not post in ALL CAPS, which is interpreted as "shouting."

- Check your postings for proper spelling and grammar. On the other hand, avoid "flaming" other users for their spelling or grammar errors.

- When replying to an existing message, include enough of the original message to provide context for your reply, but no more.

- If you know the answer to a question or problem raised by another user, send it to that user by e-mail. That way the newsgroup doesn't get cluttered up with dozens of versions of the same information.

In 1994, a firm of immigration attorneys enraged much of the online community by posting messages offering their services in each of the thousands of different newsgroups. "Spam" was born. Technically savvy users responded by creating "cancelbots" or programs that attempt to detect and automatically delete postings containing spam. Today, spam is mainly conveyed by e-mail, with mail servers and client programs offering various options for blocking it (see SPAM).

NETIQUETTE IN THE 21ST CENTURY
In the new century, newsgroups and traditional conferencing systems have diminished in importance, but e-mail is more pervasive than ever, and a variety of new online media have emerged (see, for example, BLOGS AND BLOGGING). Many of the tried-and-true rules for newsgroup postings apply as well to other media, but there are also new considerations.

As many politicians and business executives have learned to their dismay, e-mail must be assumed to be essentially as permanent as a handwritten letter. Similarly, blogs, postings to sites such as MySpace (see SOCIAL NETWORKING), and other online content can be copied, linked to, archived, or otherwise persist for many years. Today's intemperate remarks may emerge years later when a prospective employer "googles" a job candidate.

Blogs are meant to link and be linked to, so issues of properly crediting material and respecting copyright can be important. This can also apply to contributions to content-sharing sites and to articles for wikis (see WIKIS AND WIKIPEDIA); Wikipedia has evolved a rather comprehensive set of standards whereby readers can "flag" content that is problematic.

Further Reading
Housley, Sharon. "Blog and RSS Feed Etiquette." Available online. URL: http://www.small-business-software.net/blog-etiquette.htm. Accessed August 16, 2007.
Kallos, Judith. *Because Netiquette Matters! Your Comprehensive Guide to E-mail Etiquette and Proper Technology Use.* Philadelphia: Xlibris 2004.
McKay, Dawn Rosenberg. "Email Etiquette." Available online. URL: http://careerplanning.about.com/od/communication/a/email_etiquette.htm. Accessed August 16, 2007.
Netiquette Home Page. Available online. URL: http://www.albion.com/netiquette/. Accessed August 16, 2007.
Strawbridge, Matthew. *Netiquette: Internet Etiquette in the Age of the Blog.* Cambridge, U.K.: Software Reference Ltd., 2006.
Von Rospach, Chuq. "A Primer on How to Work with the Usenet Community." Available online. URL: http://faqs.cs.uu.nl/na-dir/usenet/primer/part1.html. Accessed August 16, 2007.

Net Neutrality
In recent years there has been growing concern that Internet users may eventually be treated differently by service providers depending on the kind of data they download or the kind of application programs they use online. Advocates of network (or net) neutrality (see for example CERF, VINCENT) want legislation that would bar cable, DSL, or other providers (see BROADBAND and INTERNET SERVICE PROVIDER) from making such distinctions, such as by charging content providers higher fees for high volumes of data or even blocking certain applications. Advocates of net neutrality believe that, since there are rather limited choices for broadband Internet service, discrimination on the basis of Web content could lead to a loss of freedom for consumers and providers alike.

Critics of the net neutrality proposal tend to discount such concerns. One analogy they use is traditional mail. Users can choose different types of shipping service, but having overnight service available does not mean that packages cannot be delivered using cheaper means. Likewise, they believe that the market can provide "tiers" of Internet service without disenfranchising any providers or users.

Increasing concern about the issue began in 2005 when the Federal Communications Commission announced that broadband (cable and DSL) Internet would be treated under the less stringent Title I information service under the Communications Act of 1934, rather than being treated under Title II as a "common carrier" like traditional phone service. At the same time, the agency issued policy guidelines that promoted free access, consumer choice, and competition. However these guidelines have no legal force.

In June 2007 the Federal Trade Commission (FTC) more or less sided with the critics of net neutrality by urging regulators to be careful about imposing rules that would prevent providers from innovating in offering premium services. Meanwhile two proposed net neutrality bills failed to pass Congress in 2006. However, in July 2008 the FCC in a 3-2 decision ordered Comcast, the largest U.S. cable service provider, to stop degrading service to users who used file-sharing protocols.

It should be noted that a number of rules restricting certain kinds of Internet access already exist. Major service providers have agreements called "peering arrangements" that specify how certain kinds of transmissions will be handled. Many service providers also block certain data ports to reduce the spread of spam by insecure systems or try to restrict the use of peer-to-peer (P2P) systems (see FILE-SHARING AND P2P NETWORKS).

In the long run a balance will likely be struck between providers' need to control traffic to maintain efficiency and quality of service (QoS) and the rights of users to exchange information and resources freely.

Further Reading

"Crackdown: Comcast Blocks Peer-to-Peer Web Traffic." *Portfolio.com*, October 19, 2007. Available online. URL: http://www.portfolio.com/views/blogs/daily-brief/2007/10/19/crackdown-comcast-blocks-peer-to-peer-web-traffic. Accessed October 21, 2007.

Gilroy, Angela A. *Net Neutrality: Background and Issues.* Congressional Research Service, May 16, 2006. Available online. URL: http://fas.org/sgp/crs/misc/RS22444.pdf. Accessed October 25, 2007.

Leonard, Thomas M., and Randolph J. May, eds. *Net Neutrality or Net Neutering: Should Broadband Internet Services Be Regulated?* New York: Springer, 2006.

"Network Neutrality in the United States." Wikipedia. Available online. URL: http://en.wikipedia.org/wiki/Network_neutrality_in_the_US. Accessed October 21, 2007.

Nuechterlein, Jonathan E., and Philip J. Weiser. *Digital Crossroads: American Telecommunications Policy in the Internet Age.* Cambridge, Mass.: MIT Press, 2007.

netnews and newsgroups

Originally called Usenet and originating in the UNIX user community in the late 1970s, netnews is distributed today over the Internet in the form of thousands of newsgroups devoted to just about every imaginable topic.

DEVELOPMENT

By the late 1970s, researchers at many major universities were using the UNIX operating system (see UNIX). In 1979, a suite of utilities called UUCP was distributed with the widely used UNIX Version 7. These utilities could be used to transfer files between UNIX computers that were linked by some form of telephone or network connection.

Two Duke University graduate students, Tom Truscott and Jim Ellis, decided to set up a way in which users on different computers could share a collection of files containing text messages on various topics. They wrote a simple set of shell scripts that could be used for distributing and viewing these message files. The first version of the news network linked computers at Duke and at the University of North Carolina. Soon these programs were revised and rewritten in the C language and distributed to other UNIX users as the "A" release of the News software.

During the 1980s, the news system was expanded and features such as moderated newsgroups were added. As the Internet and its TCP/IP protocol (see TCP/IP) became a more widespread standard for connecting computers, a version of News using the NNTP (Network News Transmission Protocol) over the Internet was released in 1986. Netnews is a mature system today, with news reading software available for virtually every type of computer.

STRUCTURE AND FEATURES

Netnews postings are simply text files that begin with a set of standard headers, similar to those used in e-mail. (Like e-mail, news postings can have binary graphics or program files attached, using a standard called MIME, for Multipurpose Internet Mail Extensions.)

The files are stored on news servers—machines that have the spare capacity to handle the hundreds of gigabytes of messages now posted each week. The files are stored in a typical hierarchical UNIX fashion, grouped into approximately 75,000 different newsgroups.

As shown in the following table, the newsgroups are broken down into 10 major categories. The names of individual groups begin with the major category and then specify subdivisions. For example, the newsgroup comp.sys.ibm.pc deals with IBM PC-compatible personal computers, while comp.os.linux deals with the Linux operating system.

MAIN DIVISIONS OF NETNEWS NEWSGROUPS

CATEGORY	COVERAGE
alt	An alternative system with its own complete selection of topics.
biz	Business-related discussion, products, etc.
comp	Computer hardware, software and operating systems.
humanities	Arts and literature, philosophy, etc.
misc.	Various topics that don't fit in another category.
news	Announcements and information relating to the news system itself.
rec	Sports, games, and hobbies.
sci	The sciences.
soc	Social and cultural issues.
talk	Current controversies and debates.

DISTRIBUTION AND READING

The servers are linked into a branching distribution system. Messages being posted by users are forwarded to the nearest major regional "node" site, which in turn distributes them to other major nodes. In turn, when messages arrive at a major node from another region, they are distributed to all the smaller sites that share the *newsfeed*. Due to the volume of groups and messages, many sites now choose to receive only a subset of the total newsfeed. Sites also determine when messages will expire (and thus be removed from the site).

There are dozens of different news reading programs that can be used to view the available newsgroups and postings. On UNIX systems, programs such as elm and tin are popular, while other newsreaders cater to Windows, Macintosh, and other systems. Major Web browsers such as Netscape and Internet Explorer offer simplified news reading features. To use these news readers, the user accesses a newsfeed at an address provided by the Internet Service Provider (ISP). There are also services that let users simply navigate through the news system by following the links on a Web page. The former service called DejaNews, now

Google Groups, is the best-known and most complete such site.

Further Reading
Google Groups. Available online. URL: http://groups.google.com. Accessed August 16, 2007.
Hauben, Michael, and Ronda Hauben. *Netizens: On the History and Impact of Usenet and the Internet.* Los Alamitos, Calif.: IEEE Computer Society Press, 1997.
Lueg, Christopher, and Danyel Fisher, eds. *From Usenet to CoWebs: Interacting with Social Information Spaces.* London: Springer, 2003.
Pfaffenberger, Bryan. *The USENET Book: Finding, Using, and Surviving Newsgroups on the Internet.* Reading, Mass.: Addison-Wesley, 1995.
Spencer, Henry, and David Lawrence. *Managing Usenet.* Sebastopol, Calif.: O'Reilly, 1998.

network

In the 1940s, the main objective in developing the first digital computers was to speed up the process of calculation. In the 1950s, the machines began to be used for more general data-processing tasks by governments and business. By the 1960s, computers were in use in most major academic, government, and business organizations. The desire for users to share data and to communicate both within and outside their organization led to efforts to link computers together into networks.

Computer manufacturers began to develop proprietary networking software to link their computers, but they were limited to a particular kind of computer, such as a DEC PDP minicomputer, or an IBM mainframe. However, the U.S. Defense Department, seeing the need for a robust, decentralized network that could maintain links between their computers under wartime conditions, funded the development of a protocol that, given appropriate hardware to bridge the gap, could link these disparate networks (see INTERNET, LOCAL AREA NETWORK).

NETWORK ARCHITECTURE

Today's networks are usually defined by open (that is, nonproprietary) specifications. According to the OSI (open systems interconnection) model, a network can be considered to be a series of seven layers laid one atop another (see DATA COMMUNICATION).

The physical layer is at the bottom. It specifies the physical connections between the computers, which can be anything from ordinary phone lines to cable, fiber optic, or wireless. This layer specifies the required electrical characteristics (such as voltage changes and durations) that constitute the physical signal that is recognized as either a 1 or 0 in the "bit stream."

The next layer, called the data link layer, specifies how data will be grouped into chunks of bits (frames or packets) and how transmission errors will be dealt with (see ERROR CORRECTION).

The network layer groups the data frames as parts of a properly formed data packet and routes that packet from the sending node to the specified destination node. A variety of routing algorithms can be used to determine the most efficient route given current traffic or line conditions.

The transport layer views the packets as part of a complete transmission of an object (such as a Web page) and ensures that all the packets belonging to that object are sorted into their original sequence at the destination. This is necessary because packets belonging to the same message may be sent via different routes in keeping with traffic or line conditions.

The session layer provides application programs communicating over the network with the ability to initiate, terminate, or restart an interrupted data transfer.

The presentation layer ensures that data formats are consistent so that all applications know what to expect. This layer can also provide special services (see ENCRYPTION and DATA COMPRESSION).

Finally, the application layer gives applications high-level commands for performing tasks over the network, such as FILE TRANSFER PROTOCOL (ftp).

Most modern operating systems support this model. The Internet protocol (see TCP/IP) has become the lingua franca for most networking, so modern versions of Microsoft Windows and the Macintosh Operating System as well as all versions of UNIX provide the services that applications need to make and manage TCP/IP connections.

Networks that link computers remotely (such as over phone lines) are sometimes called wide area networks, or WANs. Networks that link computers within an office, home, or campus, usually using cables, are called local area networks (LANs). See LOCAL AREA NETWORK for more details about LAN architecture and software.

TRENDS

It has become the norm for desktop and portable computers to have access to the Internet. A computer from which one cannot send or receive e-mail or view Web pages almost gives the perception of being crippled, because so many applications now assume that they can access the network. For example, the latest antivirus programs regularly check their manufacturer's Web site and download the latest virus definitions and software patches. Recent versions of Windows, too, include a built-in update facility that can obtain security patches and newer versions of device drivers.

The flip side of the power of networking to keep every PC (and its user) up to date is the vulnerability to both intrusion attempts and viruses (see COMPUTER CRIME AND SECURITY). Virtually all networks include a layer of software whose job it is to attempt to block intrusions and protect sensitive information (see FIREWALL).

Besides attending to security, network administrators and engineers must continually monitor the traffic on the network, looking for bottlenecks, such as an often-requested database being stored on a file server with a relatively slow hard drive. Besides upgrading key hardware, another approach to relieve congestion is to adopt a distributed database (see DATABASE MANAGEMENT SYSTEM) that stores "data objects" throughout the network and can dynamically relocate them to improve access.

The growing appetite for data-rich applications such as high-fidelity audio and video (see STREAMING and MULTIMEDIA) tends to put a strain on the capacity of most networks. In response, institutional users look to optical fiber and other high capacity connections (see BANDWIDTH), while home users are rapidly switch in from dial-up service on regular phone lines (see MODEM) to DSL phone lines and cable.

While existing network architectures have worked remarkably well, they were designed for only a small fraction of today's traffic. There have been a number of initiatives and proposals for higher capacity networks and for integrating new features (such as security and e-mail sender verification). For a review of these developments, see INTERNET ARCHITECTURE AND GOVERNANCE.

Further Reading

Derfler, Frank J., Jr., and Les Freed. *How Networks Work.* 7th ed. Indianapolis: Que, 2004.
Donahue, Gary. *Network Warrior.* Sebastapol, Calif.: O'Reilly Media, 2007.
Komar, Brian. *Sams Teach Yourself TCP/IP Networking in 21 Days.* 2nd ed. Indianapolis: Sams, 2002.
Kozierok, Charles. "The TCP/IP Guide." Available online. URL: http://www.tcpipguide.com/free/index.htm. Accessed August 16, 2007.
Tanenbaum, Andrew S. *Computer Networks.* 4th ed. Upper Saddle River, N.J.: Prentice Hall, 2002.

networked storage

With huge databases, e-commerce and other Web servers, and even home media centers, more data needs to be served over networks than ever before. There are two common ways to provide storage for databases and other resources on a network.

A network attached storage (NAS) unit can be thought of as a dedicated data storage unit that is available to all users of a network. Unlike a traditional dedicated file storage unit (see FILE SERVER), a NAS unit typically has an operating system and software designed specifically (and only) for providing data storage services. The actual storage is usually provided by an array of hard drives (see RAID). Files on the NAS are accessed through protocols such as SMB (server message block), common on Windows networks, and NFS (network file system), used on many UNIX and some Linux networks. In recent years smaller, lower-cost NAS devices have become available for smaller networks, including home networks, where they can store music, video, and other files (see also MEDIA CENTER PC).

STORAGE AREA NETWORK (SAN)

Although it sounds similar, a storage area network (SAN) does not function as its own file server. Rather, it attaches storage modules such as hard drives or tape libraries to an existing server so that it appears to the server's operating system as though it were locally attached. Typically the protocol used to attach the storage is SCSI (see SCSI), but the physical connection is fiber or high-speed Ethernet. The emphasis for SAN applications is the need for fast access to data, such as in large online databases, e-mail servers, and high-volume file servers. SANs offer great flexibility, since storage can be expanded without changing the network structure, and a replacement server can quickly be attached to the storage in case of hardware failure.

Further Reading

Bird, David. "Storage Basics: Storage Area Networks." Available online. URL: http://www.enterprisestorageforum.com/sans/features/article.php/981191. Accessed October 5, 2007.
NAS. Network World. Available online. URL: http://www.networkworld.com/topics/nas.html. Accessed October 5, 2007.
Network Attached Storage Reviews and Price Comparisons. PC Magazine. Available online. URL: http://www.pcmag.com/category2/0,1738,677853,00.asp. Accessed October 5, 2007.
Poelker, Christopher, and Alex Nikitin. *Storage Area Networks for Dummies.* New York: Wiley, 2003.
Preston, W. Curtis. *Using SANs and NAS.* Sebastopol, Calif.: O'Reilly, 2002.
Tate, Jon, Fabiano Lucchese, and Richard Moore. *Introduction to Storage Area Networks.* 4th ed. IBM Redbooks. Available online. URL: http://www.redbooks.ibm.com/redbooks/pdfs/sg245470.pdf. Accessed October 5, 2007.

neural interfaces

In the kind of science fiction sometimes called "cyberpunk," people are able to "jack in" or connect their brains directly to computer networks. Because of this direct input into the brain (or perhaps the optic and other sensory nerves), a person who is jacked in experiences the virtual world as fully real, and can (depending on the world's rules) manipulate it with his or her mind. This kind of all-immersive virtual reality is still science fiction, but today people are beginning to control computers and artificial limbs directly with their minds.

NEUROPROSTHETICS

Neuroprosthetics is the creation of artificial limbs or sensory organs that are directly connected to the nervous system. The first (and most widely used) example is the cochlear implant, which can restore hearing by taking sound signals from a microphone and converting them to electrical impulses that directly stimulate auditory nerves within the cochlea, a part of the inner ear. Similarly, experimental retinal implants that stimulate optic nerves are beginning to offer crude but useful vision to certain blind patients.

Research in connecting the brain to artificial arms or legs is still in its early stages, but scientists using microelectrode arrays have been able to record signals from the brain's neurons and correlate them to different types of motor movements. In a series of experiments at Duke University, researchers first trained a monkey to operate a joystick to move a shape in a video game. They then recorded and analyzed the signals produced by the monkey's brain while playing the game, and correlated them with the motor movements in the joystick. Next, they replicated these movements with a robotic arm as the monkey moved the joystick. Finally, they were able to train the monkey to move the robotic arm without using the joystick at all, simply by "thinking" about the movements.

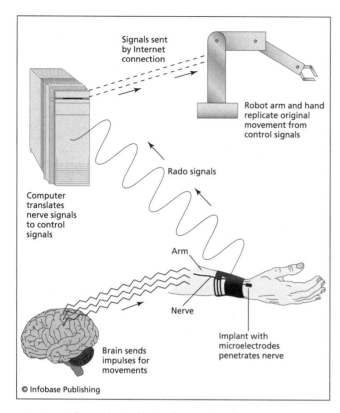

Signals sent by Internet connection

Robot arm and hand replicate original movement from control signals

Rado signals

Computer translates nerve signals to control signals

Arm

Nerve

Brain sends impulses for movements

Implant with microelectrodes penetrates nerve

© Infobase Publishing

Experimental neural interfaces link nerve impulses to a computer, allowing users to control computers (and even remote robots) literally by thinking.

Human subject are now performing similar feats. The next step is to build robotic limbs that can be controlled by the person thinking in a certain way. Ideally, a person should be able to think about clenching a hand or tapping an index finger and have the prosthetic hand replicate those movements. One obvious application for this technology is to enable quadriplegics who have little or no motion capability to control wheelchairs or other devices mentally.

FUTURE BRAIN IMPLANTS

As more is learned about the detailed functioning of neuronal networks inside the brain, "cognitive prosthetics" may become feasible. One example might be computer memory modules that might act as a surrogate or extension of human memory, perhaps helping compensate for loss of memory due to age or disease. (Early experiments on interfacing to the hippocampus, a part of the brain important for forming memories, have been underway since 2003.)

Other possibilities might include processors that could give a person the ability to think about a mathematical problem and "see" the answer, or to search databases or the Web simply by visualizing or thinking about the information desired.

Further Reading

"Brain Implants Move at the Speed of Thought." *WebMD Medical News,* April 15, 2004. Available online. URL: http://www.webmd.com/stroke/news/20040415/Brain-Implants.Accessed October 5, 2007.

Cooper, Huw, and Louise Craddock, eds. *Cochlear Implants: A Practical Guide.* 2nd ed. Hoboken, N.J.: Wiley, 2006.

Eisenberg, Anne. "What's Next: Don't Point, Just Think: The Brain Wave as Joystick." *New York Times,* March 28, 2002. Available online. URL: http://query.nytimes.com/gst/fullpage.html?res=9C01E7D8103BF93BA15750C0A96 49C8B63. Accessed October 5, 2007.

Graham-Rowe, Duncan. "World's First Brain Prosthesis Revealed." *New Scientist,* March 12, 2003. Available online. URL: http://www.newscientist.com/article/dn3488.html. Accessed October 5, 2007.

He, Bin, ed. *Neural Engineering.* New York: Kluwer Academic, 2005.

"New Prosthetic Devices Will Convert Brain Signals into Action." *Science Daily,* October 4, 2007. Available online. URL: http://www.sciencedaily.com/releases/2007/10/071003130747.htm. Accessed October 5, 2007.

neural network

When digital computers first appeared in the late 1940s, the popular press often referred to them as "electronic brains." However, computers and living brains operate very differently. The human brain contains about 100 billion neurons, and each neuron can form connections to as many as a thousand neighboring ones. Neurons respond to electronic signals that jump across a gap (called a synapse) and into electrodelike dendrons. The incoming signals form combinations that in turn determine whether the neuron becomes "excited" and in turn emits a signal through its axon. Clumps of neurons, therefore, act as networks that in effect sum up incoming signals and develop a response to them. That is, they "learn."

In a conventionally operated computer, the "neurons" (memory locations) are not inherently connected, and the central processing unit (CPU) uses arbitrary, interchangeable memory locations for storing data. Algorithms written by a programmer and implemented in instructions executed by the CPU impose cognition, to the extent one can speak of it in computers. In the brain, however, cognition seems to be something that emerges from the cooperating activities and connections of the neurons in response to sense stimuli, and possibly the creation of agentlike entities, as described in Marvin Minsky's book *The Society of Mind.*

Alan Turing and John von Neumann (see TURING, ALAN and VON NEUMANN, JOHN) had established the universality of the computer. That is, any calculation or logical operation that can be performed at all can be performed by an appropriate computer program. This means that the "brain" model of a network of interconnected neurons can also be implemented in a computer. During the 1940s, Warren S. McCulloch and Walter Pitts developed an electronic "neuron" in the form of a binary (on/off) switch that could be linked into networks and used to perform logical functions.

During 1950s, Marvin Minsky, working at the MIT Artificial Intelligence Laboratory (see MINSKY, MARVIN) further developed these concepts, and Frank Rosenblatt developed a classic form of neural network called a Perceptron. This consists of a network of processing elements (that is, functions), each of which are presented with weighted inputs (called *vectors*) from which it calculates an output value of either true (1) or false (0). The designer of the system knows what the correct output should be. If a given element (or node) produces the correct output, no changes are made. If it produces the wrong output, however, the weights given for each input are changed by some increment, plus a further adjustment or "bias" factor. This adjustment is repeated for all units as necessary until the output is correct. In other words, each neuron is constantly adapting the way it evaluates its inputs and thus its output, and that output is in turn being fed into the evaluation process of the neighboring neurons. (In practice, a neural network can have several layers of processing units, with one layer providing inputs to the next.)

For example, suppose a neural network is being trained to recognize objects based on the light being received from an array of sensors. The sensor readings are interpreted by a number of "neurons," which should output 1 if part of the desired object exists at the location scanned by its sensor. At first there will be many false readings—points at which part of the object is not recognized, or is falsely recognized. However, after many cycles of adjustment this "supervised learning" process results in a neural network that has a high probability of being able to identify all objects of a given general form. What is significant here is that a *generalized* ability has been achieved, and it has emerged without any specific programming being required!

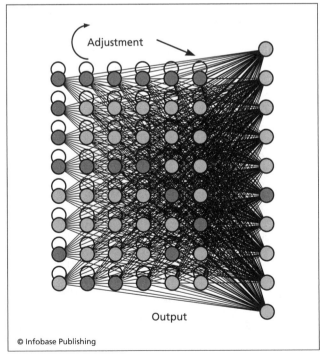

In a computer neural network the "neurons" or nodes are "trained" to detect a pattern by being reinforced when they successfully register it.

Neural networks have been making their way into commercial applications. They can be used to help robots recognize the key components of their environment (see ROBOTICS and COMPUTER VISION), for interpreting spoken language (see SPEECH RECOGNITION AND SYNTHESIS), and for problems in classification and statistical analysis (see DATA MINING). In general, the neural network approach is most useful for applications where there is no clear algorithmic approach possible—in other words, applications that deal with the often "fuzzy" realities of daily life (see FUZZY LOGIC).

Further Reading

Bishop, Christopher M. *Neural Networks for Pattern Recognition.* New York: Oxford University Press, 1995.

Haykin, Simon. *Neural Networks: A Comprehensive Foundation.* 2nd ed. Upper Saddle River, N.J.: Prentice Hall, 1998.

McNellis, Paul D. *Neural Networks in Finance: Gaining Predictive Edge in the Market.* Burlington, Mass.: Elsevier Academic Press, 2005.

"Neural Networks & Connectionist Systems." Association for the Advancement of Artificial Intelligence. Available online. URL: http://www.aaai.org/AITopics/html/neural.html. Accessed August 16, 2007.

nonprocedural languages

Most computer languages are designed to facilitate the programmer declaring suitable variables and other data structures, then encoding one or more procedures for manipulating the data to achieve the desired result (see DATA TYPES and PROCEDURES AND FUNCTIONS). A further refinement is to join data and data manipulation procedures into *objects* (see OBJECT-ORIENTED PROGRAMMING).

However, since the earliest days of computing, programmers and language designers have tried to create higher-level, more abstract ways to specify what a program should do. Such higher-level specifications are, after all, easier for people to understand. And if the computer can do the job of translating a high-level specification such as "Find all the customers who haven't bought anything in 30 days and send them this e-mail message" into the appropriate procedural steps, people will be able to spend less time coding and debugging the program.

It is actually best to think of a continuum that has at one end highly detailed procedures (see ASSEMBLER) and at the other end an English-like syntax like that given above. Already in an early language like FORTRAN the emphasis is moving away from the details of how you multiply numbers and store the result to simply specifying the operation much like the way a mathematician would write it on a blackboard. such as T = I + M. COBOL can render such specifications even more readable, albeit verbose: ADD I TO M GIVING T, for example. However, these languages are still essentially procedural.

Some languages are less procedural in that they hide most of the details (or subprocedures) involved in carrying out the desired operation. For example, in modern database languages such as SQL what would be a procedure (or a set of procedures) in some languages is treated as a *query* at a high level (see SQL). For example:

```
select customer where (today - customer.
lastpurchasedate) > 30
```

Programming packages such as Mathematica are also nonprocedural in that they allow for problems to be stated using the same symbolic notation that mathematicians employ, and many standard procedures for solving or transforming equations are then carried out automatically.

Other examples of relatively nonprocedural languages include logic-programming languages (see PROLOG and EXPERT SYSTEMS) and languages where the desired results are built up from defining functions rather than through a series of procedural steps (see LISP and FUNCTIONAL LANGUAGES).

Further Reading

Abraham, Paul W., et al. *Functional, Concurrent and Logic Programming Languages.* Vol. 4 of *Handbook of Programming Languages,* edited by Peter H. Salus. Indianapolis: Macmillan Technical Publishing, 1998.

Gilmore, Stephen, ed. *Trends in Functional Programming.* Portland, Ore.: Intellect, 2005.

Truitt, Thomas D., Stuart B. Mindlin, and Tarralyn A. Truitt. *An Introduction to Nonprocedural Languages: Using NPL.* New York: McGraw-Hill, 1983.

numeric data

Text characters and strings can be stored rather simply in computer memory, such as by devoting 8 bits (one byte) or 16 bits to each character. The storage of numbers is more complex because there are both different formats and different sizes of numbers recognized by most programming languages.

Integers (whole numbers) have the simplest representation, but there are two important considerations: the total number of bits available and whether one bit is used to hold the sign.

Since all numbers are stored as binary digits, an *unsigned* integer has a range from 0 to 2^{bits} where "bits" is the total number of bits available. Thus if there are 16 bits available, the maximum value for an integer is 65535. If negative numbers are to be handled, a signed integer must be used (in most languages such as C, C++, and Java, an integer is signed unless unsigned is specified). Since one bit is used to hold the sign and each bit doubles the maximum size, it follows that a signed integer can have only half the range above or below zero. Thus, a 16-bit signed integer can range from -32,768 to 32,767.

One complication is that the available sizes of integers depend on whether the computer system's native data size is 16, 32, or 64 bits. In most cases the native size is 32 bits, so the declaration "int" in a C program on such a machine implies a signed 32-bit integer that can range from -2^{31} or -2,147,483,647 to $2^{31}-1$, or 2,147,483,647. However, if one is using large numbers in a program, it is important to check that the chosen type is large enough. The *long* specifier is often used to indicate an integer twice the normal size, or 64 bits in this case.

FLOATING POINT NUMBERS

Numbers with a fractional (decimal) part are usually stored in a format called floating point. The "floating" means that the location of the decimal point can be moved as necessary to fit the number within the specified digit range. A floating point number is actually stored in four separate parts. First comes the *sign*, indicating whether the number is negative or positive. Next comes the *mantissa,* which contains the actual digits of the number, both before and after the decimal point. The *radix* is the "base" for the number system used. Finally, the *exponent* determines where the decimal point will be placed.

For example, the base 10 number 247.35 could be represented as 24735×10^{-2}. The -2 moves the decimal point at the end two places to the left. However, floating-point numbers are *normalized* to a form in which there is just one digit to the left of the decimal point. Thus, 247.35 would actually be written 2.4735×10^2. This system is also known as scientific notation.

As noted earlier, actual data storage in modern computers is always in binary, but the same principle applies. According to IEEE Standard 754, 32-bit floating-point numbers use 1 bit for the sign, 8 bits for the exponent, and 23 bits for the mantissa (also called the *significand*, since it expressed the digits that are significant—that is, guaranteed not to be "lost" through overflow or underflow in processing). The *double precision* float, declared as a "double" in C programs, uses 1, 11, and 52 bits respectively.

Programmers who use relatively small numbers (such as currency amounts) generally don't need to worry about loss of precision. However, if two numbers being multiplied are large enough, even though both numbers fit within the 32-bit size, their product may well generate more digits than can be held within the 23 bits available for the mantissa. This means that some precision will be lost. This can be avoided to some extent by using the "double" size.

Since floating-point calculations use more processor cycles (see MICROPROCESSOR) than integer calculations, processor designers have paid particular attention to improving floating-point performance. Indeed, processors are often rated in terms of "megaflops" (millions of floating-point operations per second) or even "gigaflops" (billions of flops).

Further Reading

"IEEE Standard for Floating Point Arithmetic." Available online. URL: http://www.psc.edu/general/software/packages/ieee/ieee.html. Accessed August 16, 2007.

"Numeric Data Types and Expression Evaluation [in C]." Available online. URL: http://www.psc.edu/general/software/packages/ieee/ieee.html. Accessed August 16, 2007.

Sebesta, Robert W. *Concepts of Programming Languages.* 8th ed. Boston: Addison-Wesley, 2007.

"Type Conversion and Conversion Operators in C#." Available online. URL: http://www.psc.edu/general/software/packages/ieee/ieee.html. Accessed August 16, 2007.

"XML Schema Numeric Data Types." Available online. URL: http://www.psc.edu/general/software/packages/ieee/ieee.html. Accessed August 16, 2007.

object-oriented programming (OOP)

During the last two decades the way in which programmers view the data structures and functions that make up programs has significantly changed. In simplified form the earliest approach to programming was roughly the following:

- Determine what results (or output) the user needs.

- Choose or devise an algorithm (procedure) for getting that result.

- Declare the variables needed to hold the input data.

- Get the data from the file or user input.

- Assign the data to the variables.

- Execute the algorithm using those variables.

- Output the result.

While this type of approach often works well for small "quick and dirty" programs, it becomes problematic as the complexity of the program increases. In real-world applications data structures (such as for a customer record or inventory file) are accessed and updated by many different routines, such as billing, inventory, auditing, summary report generation, and so on. It is easy for a programmer working on one part of the program to make a change in a data field specification (such as changing its size or underlying data type) without other programmers finding out. Suddenly, other parts of the program that relied on the original definitions start to "break," giving errors, or worse, silently produce incorrect results.

During the 1970s, computer scientists advocated a variety of reforms in programming practices (see STRUCTURED PROGRAMMING) in an attempt to make code both more readable and safer from unwanted side effects. For example, the "goto" or arbitrary jump from one part of the program to another was discouraged in favor of strictly controlled iterative structures (see LOOP). Also encouraged was the declaration of local variables that could not be changed from outside the procedure in which they were defined.

DEVELOPMENT OF OBJECT-ORIENTED LANGUAGES

However, a more radical programming paradigm was also in the making. In existing languages, there is no inherent connection between data and the procedures that operate upon that data. For example, the employee record may be declared somewhere near the beginning of the program, while procedures to update fields in the record, copy the record, print the record, and so on may well be found many pages deeper into the program.

A new approach, object-oriented programming is based on the fact that in daily life we interact with thousands of objects. An object, such as a ball, has properties (such as size and color) and capabilities (such as bouncing). In interacting with an object, we use its capabilities. It is much more natural to think of an object as a whole than to have its properties and capabilities jumbled together with those of other objects.

Simula 67, developed in the late 1960s, was the first object-oriented language (see SIMULA). It was followed in the 1970s by Smalltalk, a language developed at the Xerox

PARC laboratory, home of innovative research in graphical user interfaces. Smalltalk, like Windows today, treats each window, menu, and other control on the screen as an object (see SMALLTALK). Finally, during the 1980s C++ came into prominence, adding the essential features of object-oriented programming to the already very popular C language. Today most popular mainstream languages, including C++, Java, and Visual Basic, are object-oriented (see C++ AND JAVA). Many specialized database languages are also object-oriented.

ELEMENTS OF OBJECT-ORIENTED PROGRAMMING
The various object-oriented languages differ somewhat in capabilities, and of course in syntax. However, being object-oriented generally implies that the language has the following features.

CLASSES AND OBJECTS
An object is defined using a template called a class. A class contains both the data needed to characterize the object and the procedures (sometimes called methods or member functions) needed to work with the object (see CLASS). Thus, there could be a class for circles to be drawn on a graphics display. The class might include as its data the x and y coordinates for the center of the circle, the size of the radius, whether the circle is filled, the color to be used for filling, and so on. (See C++ for more examples.)

When the program needs to use an object of the class, it declares it in the same way it would an ordinary built-in data type such as an integer. Languages such as C++ provide for a special function called a constructor that can be used to define the processing needed when a new object is created—for example, memory allocation and setting initial values for variables.

To access data or functions within a class, the name of an object of that class is used, followed by a variable or function. Thus, if there's a class called *circle*, a program might specify the following:

```
MyCircle Circle; // Declare an object of the
    Circle class
MyCircle.X = 100; // X coordinate on screen
MyCircle.Y = 50; // Y coordinate on screen
MyCircle.Radius = 25; // Radius in pixels
MyCircle.Filled = True; // A Boolean con-
    stant equal to 1
MyCircle.FillColor = Blue; // a previously
    defined color constant
```

Once these specifications have been made, the circle can be drawn by calling upon its "draw" method or member function:

```
MyCircle.Draw;
```

The designer of a class can choose to restrict access to certain data items or functions, using a keyword such as private or protected. For example, instead of having the part of the program that uses the class directly set the x and y coordinates, it could keep those variables private and

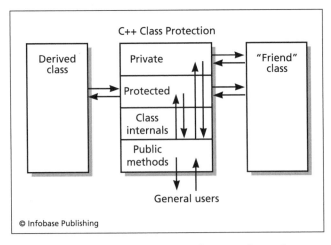

In the object-oriented C++ programming language data within a class can be restricted in several ways. Private data can be accessed only from within the class itself, or from another class declared to be a "friend" of the containing class. Protected data has these forms of access, plus it can also be accessed from any class derived from the containing class. Finally, Public data or functions (methods) can be accessed from anywhere in the program, and provides the interface by which the class is used.

instead provide a method called SetPos. The class might then take the coordinates specified by the user and adjust them to fit the screen dimensions. The Draw method would then use the adjusted internal coordinates rather than those supplied originally by the user.

INHERITANCE
Many objects are more elaborate or specialized variations of more basic objects. For example, in Microsoft Windows the various kinds of dialog boxes are specialized versions of the general Window class. Therefore, the specialized version is created by declaring it to be derived from a "base class." Put another way, the specialized class *inherits* the basic data and functions available in the base (parent) class. The programmer can then add new data or functions or modify the inherited ones to create the necessary behavior for the specialized class.

Languages such as C++ allow for a class to be derived from more than one base class. This is called multiple inheritance. For example, a Message Window class might inherit its overall structure from the Window class and its text-display capabilities from the Message class. However, it can sometimes be difficult to keep the relationships between multiple classes clear. The Java language takes the alternative approach of being limited to only single inheritance of classes, but allowing interfaces (specifications of how a class interacts with the program) to be multiply inherited.

POLYMORPHISM AND OVERLOADING
Different kinds of objects often have analogous methods. For example, suppose there is a series of classes that represent various polygons: square, triangle, hexagon, and

so forth. Each class has a method called "perimeter" that returns the total distance around the edges of the object. If each of these classes is derived from a base polygon class, each class inherits the base class's perimeter method and adapts it for its own use. Thus, a square might calculate its perimeter simply by multiplying the length of a side by four, while the rectangle would have to add up different-sized pairs of sides, and so on.

Similarly, the same operator in a language can have different meanings depending on what data types it is being applied to. The plus (+) operator, for example, is defined in most languages so that various types of integers or floating-point values can be added (see NUMERIC DATA).

Object-oriented languages such as C++ allow operators to be given additional definitions so they can handle additional data types, including classes defined by the user. For example, what might adding the string "object" and the string "oriented" yield? The most sensible answer is a new string that contains both of the original strings: "object oriented." If one defines a String class, then one can also define the + operator as a member function of that class, such that when something like String1 + String2 is encountered, the expression will be evaluated as the combination (concatenation) of the two strings. The + operator is said to have been *overloaded* for use with the String class.

ENCAPSULATION
The ability to keep the detailed workings of a class private promotes program reliability (see ENCAPSULATION). Software developers can create well-organized libraries of classes that other programmers can use simply by referring to the interface specifications (see LIBRARY, PROGRAM). Encapsulation also makes programs more readable. Once one understands the capabilities of the objects, it is relatively easy to understand the overall operation of the program without getting bogged down in details. Object-oriented programming takes the encapsulation achieved through the earlier structured programming movement and makes it more integral to the language structure.

TRENDS
Object-oriented programming was initially decried as a fad by some critics. The initial learning curve for traditionally trained programmers and the overhead that made early implementations of languages such as Smalltalk run slowly inhibited acceptance of the new paradigm at first. However, the introduction of C++ by Bjarne Stroustrup provided a fairly easy path for C programmers into the object-oriented world. For example, the class was syntactically similar to the familiar struct.

The movement toward object-oriented programming and design was also spurred by the more or less coincidental popularity of graphical user interfaces such as Microsoft Windows. Since these systems are built upon event-driven programming using a variety of coexisting objects, the object-oriented class approach fit such operating systems much more naturally. Thus, during the late 1980s and 1990s,

many Windows programmers began to use the Microsoft Foundation Classes (MFC) as their way to structure their access to the operating system. Similarly, popular languages for Web development (see for example JAVA and C#) are thoroughly object-oriented, and even most scripting languages also contain object-oriented features.

An object-based approach also fits more naturally into environments where programs and data may be running on many interconnected computers (see NETWORK and MULTIPROCESSING). Treating the client and server programs as interacting objects thus makes sense, as does treating databases as collections of data objects (see DATABASE MANAGEMENT SYSTEM). The object-oriented approach can also be applied at a higher level of abstraction in designing systems (see DESIGN PATTERNS and MODELING LANGUAGES).

Further Reading
Hamilton, J. P. *Object-Oriented Programming with Visual Basic .NET.* Sebastapol, Calif.: O'Reilly Media, 2002.
Josuttis, Nicolai. *Object-Oriented Programming in C++.* New York: Wiley, 2002.
"Lesson: Object-Oriented Programming Concepts." The Java Tutorials. Available online. URL: http://www.psc.edu/general/software/packages/ieee/ieee.html. Accessed August 16, 2007.
Lutes, Kyle, Alka Harriger, and Jack Purdum. *An Information Systems Approach to Object-Oriented Programming Using Microsoft Visual C# .NET.* Boston: Course Technology, 2005.
Prata, Stephen. *C++ Primer Plus.* 5th ed. Indianapolis: Sams, 2004.
Weisfeld, Matt. "The Evolution of Object-Oriented Languages." Available online. URL: http://www.developer.com/design/article.php/3493761. Accessed August 16, 2007.

office automation
The transition from manual to mechanical to electronic processing of information in the office spanned most of the 20th century. In the previous century, the typewriter allowed for the mechanical production of letters and other documents by skilled workers, accommodating (and perhaps encouraging) a growing amount of paperwork. At the turn of the century the card tabulator (see HOLLERITH, HERMAN and PUNCHED CARDS AND PAPER TAPE) began the mechanization of information processing.

During the first half of the 20th century, mechanical or electro-mechanical calculators made by such companies as Burroughs came into more widespread use by bookkeepers and clerks (see CALCULATOR). Meanwhile, one company, International Business Machines (IBM) came to dominate the area of card sorting and tabulating equipment.

When digital computers first came into commercial use in the 1950s, they were too large and expensive to be used in ordinary offices. Bookkeepers and other workers did not deal with computers directly, but were supported by data processing departments or outside service bureaus for what became known as electronic data processing, or EDP.

By the 1970s, the advent of the microprocessor made desk-size information processing systems possible (see MICROPROCESSOR). The first widespread application was the dedicated word processing system, of which the most successful version was developed by An Wang. These systems provided for typing and printing documents and storing them in a file system (see WORD PROCESSING).

During the 1980s, the general-purpose desktop computer (see PERSONAL COMPUTER) became powerful enough to supplant the dedicated word-processing system. Besides providing word-processing functions through ever more versatile versions of programs such as WordPerfect, WordStar, and Microsoft Word, the PC could also run programs to support bookkeeping, accounting, mailing list, and other functions (see DATABASE MANAGEMENT SYSTEM and SPREADSHEET). Gradually, many of these separate programs were merged into office suites such as Microsoft Office (see APPLICATION SUITE). Using a suite meant that information could be easily transferred between word-processing documents, spreadsheets, and database files, facilitating the generation of many kinds of reports and presentations.

Later in the 1980s, two new aspects of office automation began to emerge: communication and collaboration. The use of special hardware and software to connect PCs within an office or throughout the organization (see NETWORK and LOCAL AREA NETWORK) made new applications possible. E-mail began to replace printed memos or phone calls as the preferred way for workers and management to communicate. Programs such as Lotus Notes and Microsoft Outlook added features such as the ability of workers to share a common calendar of tasks, while scheduling software offered more elaborate ways to keep track of large, detailed team projects (see PROJECT MANAGEMENT SOFTWARE).

Today a variety of tools are available for facilitating collaboration. Most word-processing software now offers a feature called revision marking, which lets various editors and reviewers comment on or make revisions to a document. The author can then merge the revisions into a new draft. "Whiteboard" programs let several users on the network work simultaneously on the same virtual screen, drawing diagrams or making outlines.

TRENDS

Even as desk space was being cleared for the first office PCs, pundits began to claim that the "paperless office" was at hand. Actually, the first stages of automation contributed to an increase in the use of paper. On the one hand, word processors and other programs made it easier to generate documents and keep them up to date. On the other hand, the documents were all printed on paper—in part because the ability to share them electronically was nonexistent or rudimentary, and in part because many workers, particularly senior executives, still preferred to work with paper.

The growth of networking made it possible for more people to distribute documents electronically, while higher-resolution video displays made it easier to view pages on the screen. During the 1990s, the inexpensive document scanner (see SCANNER) made it practicable to scan incoming paper documents into text files (see OPTICAL CHARACTER RECOGNITION). While the office is not yet paperless, the tide of paper may now be receding at last.

The ubiquity of the Internet and the use of the HTML format for documents (see HTML and LAN) characterize the latest phase in the evolution of office automation. Many corporate procedure manuals and other resources are now being stored on company Web sites where they can be updated easily and consulted with the aid of search engines. Databases to which workers need shared access are also being hosted through Web sites. HTML and XML are emerging as common formats for exchanging documents between systems, along with Adobe's Portable Document Format (PDF), which offers a faithful reproduction of the printed page.

Changes in how the Internet is being used for communication and collaboration are also having an impact on the office. In particular, blogs are being used as a way for key people to keep coworkers updated (see BLOGS AND BLOGGING), and wikis can be an effective way for building a common knowledge base for both employees and customers (see WIKIS AND WIKIPEDIA).

Many workers can now access the full resources of the office through laptop computers and Internet connections. Workers on the go can also use handheld or palm computers such as the PalmPilot (see PDA) to access e-mail, calendar, and other information. The growing use of video-conferencing over the Internet using inexpensive cameras and broadband connections is also promoting the "virtual meeting" (see VIDEO CONFERENCING).

Further Reading
Brown, M. Katherine, Brenda Huettner, and Char James-Tanny. *Managing Virtual Teams: Getting the Most from Wikis, Blogs, and Other Collaborative Tools*. Plano, Tex.: Wordware Publishing, 2007.

Greenbaum, Joan. *Windows on the Workplace: Technology, Jobs, and the Organization of Office Work*. New York: Monthly Review Press, 2004.

Mobile Office Technology. Available online. URL: http://mobileoffice.about.com/. Accessed August 16, 2007.

Obringer, Lee Ann. "How Virtual Offices Work." Available online. URL: http://communication.howstuffworks.com/virtual-office.htm. Accessed August 16, 2007.

Sellen, Abigail J., and Richard H. R. Harper. *The Myth of the Paperless Office*. Cambridge, Mass.: MIT Press, 2002.

Scoble, Robert, and Shel Israel. *Naked Conversations: How Blogs Are Changing the Way Businesses Talk with Customers*. Hoboken, N.J.: Wiley, 2006.

Wibbels, Andy. *Blogwild!: A Guide for Small Business Blogging*. New York: Penguin Group, 2006.

Omidyar, Pierre

(1967–   )
French-Iranian/American
Entrepreneur, Inventor

One of the most remarkable stories of the development of e-commerce has been the online auction pioneered by Pierre Omidyar and the hugely successful eBay auction site he founded (see ONLINE AUCTIONS.)

Omidyar was born on June 27, 1967, in Paris. His family is of Iranian descent. While working in his high school library Omidyar encountered his first computer and soon wrote a program to catalog books. Omidyar enrolled at Tufts University to study computer science. However, after three years he became bored with classes and went to work

Pierre Omidyar founded eBay, the world leader in online auctions. (ACEY HARPER / TIME LIFE PICTURES / GETTY IMAGES)

as a programmer. Omidyar helped develop a drawing program for the new Apple Macintosh, but after a year returned to finish his degree, which he received in 1988. He then went to work for Claris, a subsidiary of Apple. There he developed MacDraw, a very popular application for the Macintosh.

By 1991 Omidyar had become interested in an emerging application, "pen computing," which uses a special pen and tablet to allow computer users to enter text in ordinary handwriting, which would be recognized and converted to text by special software. Omidyar and three partners formed a company called Ink Development to work on pen computing technology. However, the market for such software was slow to develop. The partners changed their company name to eShop and their focus to e-commerce, the selling of goods and services online. But e-commerce would not become big until the mid-1990s when the graphical browser made the Web attractive and easy to use. Meanwhile Omidyar also did some graphics programming for the movie effects company General Magic.

Omidyar retained his interest in e-commerce, with a particular focus on finding new markets in which buyers and sellers could meet. Online auctions offered one such mechanism, and Omidyar created a site called AuctionWeb. AuctionWeb was based on a simple idea: Let a user put up something for bid, and have the software keep track of the bids from other users until the ending time is reached, with the highest bid being the winner.

At first Omidyar made AuctionWeb free for both buyers and sellers, but as the site exploded in popularity he began to charge sellers a small fee to cover his Internet service costs. As the months passed, thousands of dollars in small checks began to pour in. Using $1 million he received from Microsoft for the sale of his former company eShop, Omidyar decided to expand his auction site into a full-time business.

Thanks to the Web, it was now possible to run an auction without cataloger, auctioneer, or hotel room. The job of describing the item could be given to the seller, and of course digital photos or scanned images could be used to show the item to potential bidders. The buyer would pay the seller directly, and the seller would be responsible for shipping the item.

Because overhead costs are essentially limited to maintaining the Web site and developing the software, the company could charge sellers about 2 percent instead of the 10–15 percent demanded by traditional auction houses. Buyers would pay no fees at all. And because the cost for selling is so low, sellers could sell items costing as little as a few dollars, while regular auction houses generally avoid lots worth less than $50–$100.

With the aid of business partner and experienced Web programmer Jeff Skoll, Omidyar revamped and expanded the site, renaming it eBay (combining the "e" in electronic with the San Francisco Bay near which they lived). Unlike the typical Web business that promised investors profit sometime in the indefinite future, eBay made money from the first quarter and just kept making more.

Through their relationship with a venture capital firm, Benchmark, Omidyar and Skoll gained not only $5 million for expansion but the services of Meg Whitman, an experienced executive who had compiled an impressive track record with firms such as FTD (the flower delivery service), the toy company Hasbro, Procter & Gamble, and Disney. eBay's growth continued: by the end of 1997 about 150,000 auctions were being held each day.

In 1998 they decided to take the company public. By the time the first trading day ended, Omidyar's stock was worth $750 million, and Whitman and the other key players had also done very well.

One possible weakness in the eBay model was that it relied heavily on trust by the seller and especially the buyer. What if a buyer won an item only to receive something that was not as described or, worse, never received anything at all? But while this happened in a small number of cases, Omidyar through his attention to building communities for commerce had devised an interesting mechanism called "feedback." Both sellers and buyers were encouraged to post brief evaluations of each transaction, categorized as positive, neutral, or negative. A significant number of negative feedbacks served as a warning signal, so both sellers and buyers had an incentive to fulfill their part of the bargain. The system was not perfect, but the continued patronage of several million users suggested that it worked. (An escrow system was also made available for use with more expensive items.)

As the new century dawned, Omidyar became less personally involved with eBay. In 1998 he had stepped down as CEO, the post going to Whitman. In 2004 Omidyar and his wife, Pam, turned their attention to the Omidyar Network, a new structure that replaces the traditional foundation with a decentralized approach combining nonprofit and for-profit initiatives focusing on empowering individuals and communities. Omidyar has also been investing in microfinancing (the making of small loans directly to poor entrepreneurs in developing countries).

Further Reading

Cohen, Adam. "Coffee with Pierre: Creating a Web Community Made Him Singularly Rich." *Time*, December 27, 1999, p. 78 ff.

———. *The Perfect Store: Inside eBay.* New York: Little, Brown & Company, 2002.

Ericksen, Gregory K. *Net Entrepreneurs Only: 10 Entrepreneurs Tell the Stories of Their Success.* New York: Wiley, 2000.

Pierre's Web (blog). http://pierre.typepad.com/. Accessed May 6, 2007.

Sachs, Adam. "The Millionaire No One Knows." *Gentlemens' Quarterly,* May 2000, p. 235.

online advertising

In the late 1990s "banner ads" started to appear on Web sites, and other forms of advertising soon followed. Companies rushed into the online world, either with the belief that it had unlimited potential for finding new customers, or out of fear that the competition would get there first. Unfortunately it was hard to measure the actual effectiveness of ads, and Web sites (such as for publications) that looked to third-party advertising as a source of income found the outlook bleak in the wake of the bursting of the "dot-com bubble" of the early 2000 decade.

Only a few years later, however, advertisers using new business models and targeting techniques have made online advertising not only a viable business, but a rapidly growing one. (According to the Interactive Advertsing Bureau, Internet advertising revenue in the United States in 2007 was $21.2 billion, up 26 percent from 2005.)

The effects of the online advertising revolution are rippling outward, impacting traditional advertising media such as newspapers (in particular see CRAIGSLIST), magazines, and even television.

PLATFORMS AND TYPES OF ADS

There are many different applications that can be accompanied by different types of advertising. These include e-mails (free e-mail services usually include an ad in every message), newspapers and other publications (often with ads related to the subject of an accompanying article), and even blogs (see BLOGS AND BLOGGING). Indeed, the most popular blogs can actually make a reasonable income from advertising.

Types of ads include the following:

- Banner ads are contained in rectangles, often at the top of the Web page. (Sometimes they can mimic dialog boxes from the operating system.) They still account for about half of all online advertising, and can appear on sites of all types.

- Pop-up or pop-under ads appear above or beneath the current window, respectively.

- Floating ads appear over the main page content, often moving across the screen.

- Interstitial ads are displayed before the requested content (such as an article or video) is shown. They run for a specified period of time, although they can sometimes be closed by the viewer.

Many ads are animated; some even contain video clips. There are also ads formatted for mobile devices, including text messages sent to cell phones.

ECONOMICS OF ONLINE ADVERTISING

A company or organization can of course advertise its own products or services on its Web site. Alternatively, a site can arrange with an online advertiser to carry ads for other peoples' goods or services, in exchange for a fee. The advertiser in turn gets paid by the company whose ads are being run. The payment can be calculated in a variety of ways: CPM (cost per thousand people who see the ad), the number of sales leads, or the number of people who actually buy something.

As the first luster of the Web began to wear off, corporate advertising departments increasingly wanted better measurements of the exposure their ads were receiving, and wanted ads that were better targeted to people more likely to "click through" to the advertiser's site. Since it involves people who are already looking for specific things, Web search is an effective and profitable activity to be linked to contextually related ads. Google in particular has been very successful in auctioning or selling the opportunity to have one's ad appear in the results of a search request containing a specific keyword (see GOOGLE).

Another way that Google and other large search engines or portals can make money from advertisers is through "affiliate marketing"; Google's version is called Ad Sense. Participating Web sites are indexed, and the resulting keywords are matched with ads awaiting placement. The site carrying the ad generally gets a per-click payment. However, the problem of "click fraud" has also arisen: Scammers can set up an affiliate site and then use special software to generate the clicks, while making them come from a variety of sources. Despite these problems, in 2006 about 40 percent of revenue from online advertising was attributed to search-related ads.

While search engine usage perhaps provides the most direct indication of consumer interests, considerable attention has also been focused on developing systems that can track where a given individual goes on a large e-commerce site (see COOKIES), and look for clues about likely future purchases (see DATA MINING).

MAINTAINING USER INTEREST

The dark side of online advertising is found in programs that are surreptitiously installed on users' PCs and then download and display advertising from shady Web operations (see SPYWARE AND ADWARE). While many users now regularly run programs to block such malware, even legitimate online advertising can irritate users, particularly when ads are too prominent, float over (and block) text, or lurk behind the browser window. Modern Web browsers have ad-blocking features that work with varying degrees of effectiveness. As with TV, online advertisers increasingly have to cope with impatient users who do not have to look at ads unless they actually want to.

Advertisers can employ several strategies to keep users willing to look at ads. One is to make the ad unobtrusive and brief, and on the way to something the user really wants to see. In 2007 YouTube began such advertising. Another is to provide free versions of software or services that, in exchange for being free, require the user to put up with some screen real estate being devoted to ads. Finally, as with TV, advertising can be woven into the content itself, such as in online computer games.

A sensitive area is the attempt to balance advertisers' desire to know as much as possible about consumers' interests and buying habits with the same consumers' concern about protecting their privacy (see PRIVACY IN THE DIGITAL AGE).

Further Reading

"Click Fraud: The Dark Side of Online Advertising." *Business Week,* October 2, 2006. Available online. URL: http://www.businessweek.com/magazine/content/06_40/b4003001.htm. Accessed October 5, 2007.

Davis, Harold. *Google Advertising Tools: Cashing In with AdSense, AdWords, and the Google APIs.* Sebastopol, Calif.: O'Reilly, 2006.

Interactive Advertising Bureau. Available online. URL: http://www.iab.net/. Accessed October 5, 2007.

Plummer, Joe, et al. *The Online Advertising Playbook: Proven Strategies and Tested Tactics from the Advertising Research Foundation.* Hoboken, N.J.: Wiley, 2007.

Scott, David Meerman. *The New Rules of Marketing & PR: How to Use News Releases, Blogs, Podcasting, Viral Marketing & Online Media to Reach Buyers Directly.* Hoboken, N.J.: Wiley, 2007.

Search Engine Marketing Professional Organization (SEMPO). Available online. URL: http://www.sempo.org. Accessed October 5, 2007.

Sloan, Paul. "The Quest for the Perfect Online Ad: Web Advertisers Are Moving beyond Search, Using Powerful Science to Figure Out What You Want." *Business 2.0* [Magazine]. Available online. URL: http://money.cnn.com/magazines/business2/business2_archive/2007/03/01/8401043/index.htm. Accessed October 5, 2007.

online frauds and scams

In the old days con men and scammers went to where there were a lot of people with loose cash and where anonymity was the order of the day—perhaps a carnival or fair. Today in all too many cases the Internet fills this bill. With millions of inexperienced new users coming online in recent years, the opportunities for frauds and scams are significant, as is the problem of fighting such crime. In 2007 the Internet Crime Complaint Center (a partnership between the FBI and the National White Collar Crime Center) logged its one-millionth complaint. Of the 461,096 cases referred to law enforcement agencies, the estimated dollar loss is $647.1 million, with a median loss of $270 per complaint.

Many online frauds represent adaptations of traditional criminal practices to the online world. E-mail (see SPAM) carries offers for dubious cures for mostly imagined sexual ills, or for prescription drugs at too-good-to-be-true prices, or for "genuine replica Rolex watches." Internet auction sites also offer a venue for selling fakes and counterfeits of various sorts. The primary protections for the consumer are knowledge about the goods in question and taking advantage of community resources such as feedback provided by other buyers (see also AUCTIONS, ONLINE and EBAY).

Entire fake businesses can appear online, complete with professional-quality Web sites. If a prospective purchaser has never heard of the company, checking with the Better Business Bureau, or looking for a certification such as Trust-E, is a good idea. (Scammers can also impersonate legitimate businesses in order to get personal information from customers—see PHISHING AND SPOOFING.)

Investments are another fertile area for online scammers. These include "pump and dump" schemes where chatroom or blog postings are used to "talk up" some obscure stock and then cash in when investors start buying it and raising the price. Pyramid schemes and multilevel-marketing (MLM) programs where money from new participants is used to pay back earlier investors also appear from time to time.

A common theme of victimization seems to be that many Web users seem to suspend their usual skepticism and caution when they go online. This is perhaps due to the relative unfamiliarity of the online world and the lack of experience in evaluating products, investments, or services.

A variety of other frauds and scams appear online or via e-mail with some frequency:

- the "419" or "Nigerian money letter" that promises a rich cut for helping facilitate a money transfer for a distressed official

- fraudulent charitable solicitations, particularly after such disasters as the Asian tsunami or Hurricane Katrina

- adoption and marriage scams

- educational fraud, such as worthless degrees offered by unaccredited institutions

- dubious employment schemes or "home businesses" involving preparing mailings or medical billing

- services that offer to "repair" bad credit ratings

- tax-avoidance schemes, often based on nonexistent legal claims or loopholes

FIGHTING ONLINE FRAUD

Because perpetrators are hard to track down (see ANONYMITY AND THE INTERNET), and because of the ability to endlessly create new Web sites and e-mails, it is hard to control this form of crime (see COMPUTER CRIME AND SECURITY). However, considerable resources are now being brought to bear, with significant success. Depending on the type of fraud, federal agencies such as the Securities and Exchange Commission (SEC), Federal Trade Commission (FTC), and the Food and Drug Administration (FDA) will investigate, and agencies such as the FBI will pursue perpetrators. Every state also has an office of consumer protection or consumer affairs, and local district attorneys may become involved when perpetrators are operating in their area or victimizing residents.

Private agencies also play an important role. Besides the Better Business Bureau, most industries or professions have some form of certification of products or practices. There are also professional services that will authenticate collectibles such as stamps, coins, and sports cards.

Government and private agencies also offer a variety of consumer education materials that explain common frauds and suggest ways to shop prudently for goods or services.

Further Reading

Federal Bureau of Investigation. "Internet Fraud." Available online. URL: http://www.fbi.gov/majcases/fraud/internetschemes.htm. Accessed October 6, 2007.

Henderson, Harry. *Internet Predators (Library in a Book)* New York: Facts On File, 2005.

Internet Crime Complaint Center. Available online. URL: http://www.ic3.gov/. Accessed October 6, 2007.

Securities and Exchange Commission. "Internet Fraud: How to Avoid Internet Investment Scams." Available online. URL: http://www.sec.gov./investor/pubs/cyberfraud.htm. Accessed February 7, 2008.

Silver Lake Editors. *Phishing, Spoofing, ID Theft, Nigerian Advance Schemes, Investment Frauds, False Sweethearts: How to Recognize and Avoid Internet Era Rip-offs.* Aberdeen, Wash.: Silver Lake Publishing, 2006.

online gambling

Despite its illegality in the United States, Internet-based gambling has been very popular—by 2004 more than 20 million Americans had tried some form of online gambling, and in 2005 they bet about $5.9 billion.

Online casinos appeared in 1995, but at first they could only be played "for fun," with no actual money changing hands. That soon changed: In 1996, InterCasino appeared—it would be the first of hundreds of online casinos, sports bookmakers, and other types of gambling. Generally these operations are based outside of the United States—Caribbean islands such as Antigua and Curaçao are popular locations.

Online casinos offer traditional table games such as blackjack, roulette, and craps. Generally odds and payoffs are comparable to those at traditional casinos. Assuming the game is honest and properly programmed, the house's revenue comes from a percentage of the amount bet—blackjack having the lowest house percentage and roulette the greatest. Slot machines (which give an even higher percentage to the house) can also be simulated online.

Although occasional cases of software programmed to cheat have been documented, a more common problem is failure to pay winnings promptly, or at all. Recourse is difficult, since the casino is offshore and the activity is illegal for U.S. players. Players can, however, consult lists of so-called rogue casinos to be avoided. Some players cheat as well, typically by opening multiple accounts in order to get the "signing bonus."

ONLINE POKER

Online poker has become very popular, particularly games such as Texas Hold'Em. Estimated revenues from online poker in the United States were $2.4 billion in 2005. Unlike the case with casino games, online poker players play against each other, not the house. The house's revenue comes from a "rake," or percentage, of the pot. Many sites offer organized tournaments, and some online players have gone on to win traditional tournaments. (The aptly named Chris Moneymaker won an online tournament, qualifying him to enter the 2004 World Series of Poker, which he went on to win.)

Like online casinos, online poker is illegal in the United States. Proponents argue that while any given hand is random, poker in the long run is a game of skill, not chance. A group called the Poker Players Alliance has been lobbying to exempt poker from Internet gambling laws.

A third type of online gambling is sports betting, which is legal in many countries but only in Nevada in the United States. The Web has also given sports bettors a forum for discussing (or arguing about) teams and their prospects.

LEGAL AND OTHER ISSUES

In 1998 the federal government charged more than 20 Americans with operating gambling services in violation of the Federal Wire Act, which prohibits wagering over the phone lines used for most Internet transmissions. Most of the charges were subsequently dropped or plea-bargained, with only one casino operator serving 17 months in federal prison. In 2002 a federal appeals court ruled that while the Wire Act applied to sports betting, it did not apply to online betting on games of chance. However, subsequent legal ambiguity has led major Internet services such as Google and Yahoo! to remove online gambling advertisements from their sites. Meanwhile, a suit by the Casino City gambling portal on First Amendment grounds was dismissed, although other legal challenges were underway in 2007.

In recent years antigambling activists have adopted an indirect strategy of going after the infrastructure used for gambling transactions. In 2006 Congress passed the Unlawful Internet Gambling Enforcement Act, which prohibits U.S. credit card companies and banks from transferring funds to or from Internet gambling sites. (One of the arguments used by proponents was that terrorists might be using online gambling sites to launder money.)

Another issue raised by online gambling opponents is that the high-speed, highly interactive (click-and-response) nature of online games of chance made it easier for people prone to gambling addiction to get and stay "hooked." Particular concern has been raised about teens who decide to gamble using parents' credit cards. However, studies such as the British Gambling Prevalence Survey 2007 have suggested that the growing popularity of online gambling has not led to an increase in the rate of gambling addiction.

On the other hand, congressional liberals such as Rep. Barney Frank (Dem.-Massachusetts) have sponsored legislation that would legalize (and tax) Internet gambling, and provide for programs to deal with underage and compulsive gambling. Opponents have charged that the legalization measure is being backed by major "brick and mortar" casinos who want a piece of the online action, as well as the credit card companies, which would also get a piece of each transaction. (As of 2007 neither this nor other attempts to legalize online gambling in the United States have been passed.)

Further Reading

Dunnington, Angus. *Gambling Online.* Hassocks, West Sussex, U.K.: D&B Publishing, 2004.

Norton, Kate. "Online Gambling Hedges Its U.S. Bets." *Business Week,* August 21, 2006. Available online. URL: http://www.businessweek.com/globalbiz/content/aug2006/gb20060821_544446.htm. Accessed October 9, 2007.

Somach, Tom. "Gambling Gold Rush? A Congressional Push Last Year Stopped Many Americans from Playing the Games Online, but the Law May Be Changed." *San Francisco Chronicle,* July 2, 2007, p. C1–2.

Vogel, J. Philip. *Internet Gambling: How to Win Big Online Playing Bingo, Poker, Slots, Lotto, Sports Betting & Much More.* New York: Black Dog & Leventhal Publishers, 2006.

Woellert, Lorraine. "A Web Gambling Fight Could Harm Free Trade." *Business Week,* August 12, 2007, p. 43. Available online. URL: http://www.businessweek.com/magazine/content/07_33/b4046041.htm. Accessed October 9, 2007.

online games

Online games today range from elaborate war games to open-ended fantasy worlds to virtual universes that mirror "real-world" activities, including economics, politics, and even education.

The first online games appeared in the late 1970s on PLATO, an educational network, as well as on the early Internet of the 1980s. These MUDs (multiuser dungeons) were generally based on pen-and-paper role-playing games of the time, notably *Dungeons & Dragons.* These games were

Second Life is not a "game," but a virtual world that now includes just about every known human activity—its money is even exchangeable for real-world cash. (COPYRIGHT 2006, LINDEN RESEARCH INC., ALL RIGHTS RESERVED)

text based, with players typing their characters' actions and dialog while the changing world as seen by the players was similarly described. By the early 1990s, however, MUDs had spun off many variants. Many were still "hack n' slash" dungeon games (which were also offered on America Online and other commercial services). Many of these MUD-like games such as AOL's *Neverwinter Nights* offered simple graphics. Meanwhile other games began to offer more sophisticated social interactions as well as the ability of players to make their own additions to the game world, including buildings.

MASSIVELY MULTIPLAYER ONLINE ROLE-PLAYING GAMES (MMORPGS)

Today's online games feature a "persistent world" hosted on one or more servers that grows and develops from day to day and in which the "avatars" or representatives of thousands of players interact with game-generated creatures or one another, using client software. Players can spend hundreds of hours helping their characters develop skills, increasing their levels through experience points gained from successful combat or other activities. Players (and their characters) frequently form organizations such as guilds or clans, because the tougher challenges generally require the cooperation of different types of classes of characters (fighters, healers, and magic-users).

Modern MMORPGs began in the late 1990s with such titles as *Ultima Online* and *EverQuest.* The most popular MMORPG in the mid-2000s was *World of Warcraft.*

FROM GAMES TO ALTERNATIVE WORLDS

Humans are social primates, and they tend to bring their full repertoire of behavior to any new situation. Even games such as *World of Warcraft* or *Everquest* are not entirely about combat and character skills: they are also about alliance, trust, betrayal, and bonding.

Back in the 1980s psychologists began to write about the social interactions that were emerging in MUDs and how players perceived their virtual world (see TURKLE, SHERRY). However *Second Life,* launched by Linden Lab in 2003, is not a game at all, but a complete virtual world in which participants, called "residents" (through their avatars) can do just about anything—play and be entertained, have relationships (including virtual sex), but also conduct more mundane businesses and meetings and even attend university courses.

The ability to do nearly anything also means the ability to do things that may be offensive and even illegal. Indeed, an emerging issue is how "real world" laws apply to these virtual worlds. In *Second Life,* residents buy and sell in-world real estate and goods, using a currency called Linden Dollars (L$). These L$ and U.S. dollars can be traded at the rate (as of early 2007) of 270 L$ to one dollar U.S. This means that residents in the virtual world can actually run profitable businesses (or make investments) that can be cashed out for "real" money. Further, the avatars, property, and other in-world creations developed by users remain their intellectual property, not that of Linden Labs.

The close and growing ties between virtual worlds such as *Second Life* and "real" world society raises many legal and even social issues:

- Should income made in the virtual world be taxable?

- If residents of a virtual world make contracts with one another, are they enforceable? If so, who has jurisdiction? (See CYBERLAW.)

- Is the virtual world itself subject to national laws, or might it eventually acquire a form of sovereignty? (Already a few nations have "virtual embassies" within *Second Life*.)

Meanwhile, representatives of major companies ranging from Microsoft and Google to *Second Life's* Linden Labs have proposed making online identities and avatars "portable" so that a person could use them in his or her online games and virtual communities (see VIRTUAL COMMUNITY).

Further Reading
Castronova, Edward. *Synthetic Worlds: The Business and Culture of Online Games.* Chicago: University of Chicago Press, 2006.
Jennings, Scott. *Massively Multiplayer Games for Dummies.* Hoboken, N.J.: Wiley, 2005.
Rice, Robert A., Jr. *MMO Evolution.* Morrisville, N.C.: Lulu.com, 2006.
Terdiman, Daniel. "Tech Titans Seek Virtual World Interoperability." *CNet News.* October 12, 2007. Available online. URL: http://www.news.com/Tech-titans-seek-virtual-world-interoperability/2100-1043_3-6213148. html. Accessed October 13, 2007.
v3image. *A Beginner's Guide to Second Life.* Las Vegas, Nev.: Arche-Books, 2007.

online investing
As with shoppers, investors have increasingly been attracted to the interactivity and ease of online transactions. In addition to allowing stocks to be bought or sold with just a few clicks, online brokers (also called discount brokers) charge much lower transaction fees than their traditional counterparts, typically less than $10 per trade.

Some online brokers, such as E*Trade, Scottrade, and TD Ameritrade, were established as Internet brokers. However, traditional brokerages such as Charles Schwab and Waterhouse have also opened online discount brokerages.

In addition to fast, inexpensive trading, many online brokers also offer a variety of resources and tools, including stock quotes and charts, research reports, and screening programs to help investors pick the mutual funds or individual investments that meet their objectives. For more sophisticated investors, some brokers offer simulations for testing investment strategies and programmed trading, which will execute buy or sell orders automatically depending on specified conditions.

Online brokers can specialize, seeking customers who want to make frequent trades but do not need other support, or investors who are interested in obtaining IPOs (initial public offerings) of up-and-coming companies. Some brokers may emphasize mutual funds and cater to retire-ment accounts, while others might offer government or corporate bonds, foreign stocks, "penny stocks," or more exotic investments.

The interactivity and low transaction costs in online investing may encourage people to become involved in highly speculative penny stocks, options, day trading, foreign exchange markets, and other areas that are not suitable for most individual investors. While there is a great deal of useful information available online, it is a good idea to begin by discussing investment goals and potential risks with a trusted financial adviser.

TRENDS
Since trading fees have gone down about as far as they can go and still allow for profitability, online brokerages are increasingly competing by offering distinctive features and enhanced customer service. In the course of rapid expansion, service has become somewhat uneven: A 2006 J.D. Powers survey found that 41 percent of investors had encountered at least one problem with accessing their accounts or executing a trade.

Besides trying to improve reliability, online brokers are also branching out by offering financial planning and other personal services for their larger investors, and some are opening retail outlets where people can actually see a broker.

Further Reading
Choosing a Broker. Yahoo! Finance. Available online. URL: http://biz.yahoo.com/edu/ed_broker.html. Accessed October 17, 2007.
Davidson, Alexander. *The Complete Guide to Online Stock Market Investing.* 2nd ed. Philadelphia: Kogan Page, 2007.
Krantz, Matt. *Investing Online for Dummies.* 6th ed. Hoboken, N.J.: Wiley, 2008.
Parmar, Neil. "Finding the Best Broker." *SmartMoney.* July 10, 2007. Available online. URL: http://www.smartmoney.com/brokers/index.cfm. Accessed October 17, 2007.

online job searching and recruiting
In the old days, people found jobs by word of mouth or by reading newspaper classified ads. While word of mouth (or at least e-mail) can still be very useful for finding job leads, increasingly both employers and job seekers are turning first to a variety of online sites. (Indeed, as of mid-2007 one large site, Monster.com, claimed to have more than 73 million resumes in its database and 42 million job seekers per month.)

There are a number of large sites that list thousands of jobs at any given time. Examples include Monster.com, JobCentral, and CareerJournal (from *The Wall Street Journal*). Meanwhile, many of the "career classifieds" from newspapers have been replaced by postings on Craigslist, which has a number of regional sites and covers buy/sell, apartment rentals, and other types of ads as well (see CRAIGSLIST).

In evaluating a job site it is important to get a feel for the kinds of jobs offered and the target audience, such as professionals, recent graduates, white-collar or service-sector jobs, and so on. Other important features to look for include:

- powerful search or filtering capability, such as by type of job or employer, keywords in job description, or locality
- the ability to put one's resume online and edit or update it as needed.
- the ability to have several versions of one's resume tailored to different types of jobs
- automatic e-mail alerts about newly added jobs that meet the user's criteria
- privacy protections so that contact information from resumes is not used for marketing or other nonemployment purposes
- lack of fees to job seekers (normally employers are the service's source of revenue)

Job seekers can use job search engines such as Career Builder that will search the major job-finding sites and/or employers' own sites according to the user's criteria.

In addition to dedicated job-hunting sites and recruiting agencies, a less formal but rapidly growing trend is the meeting of employers and would-be employees through sites such as Facebook (see SOCIAL NETWORKING), where people often freely describe their interests. Employers in turn are increasingly searching online for information about applicants, which can cause a problem if the results include "indiscreet" writings or perhaps photos, perhaps dating back to high school. (On the other hand, there are also social networks such as LinkedIn that specialize in business contacts.)

Finally, online job seekers should beware of fake "job offers" that ask for information such as social security numbers (see ONLINE FRAUDS AND SCAMS).

Further Reading
CareerBuilder. Available online. URL: http://www.careerbuilder. com. Accessed October 21, 2007.
Craigslist. Available online. URL: http://www.craigslist.com. Accessed October 21, 2007.
Dikel, Margaret Riley, and Frances E. Roehm. *Guide to Internet Job Searching 2006–2007 Edition.* New York: McGraw-Hill, 2006.
Job-Hunt: The Guide to Finding Employment Online. Available online. URL: http://www.job-hunt.org/job-search.html. Accessed October 21, 2007.
Kerber, Ross. "Online Job Hunters Grapple with Misuse of Personal Data." *Boston Globe,* October 1, 2007. Available online. URL: http://www.boston.com/business/globe/articles/2007/10/01/online_job_hunters_grapple_with_misuse_of_personal_data/. Accessed October 21, 2007.
Monster.com. Available online. URL: http://www.monster.com. Accessed October 21, 2007.
Napoli, Lisa. "New Job-Seeking Tool? It's the Network." *Marketplace* (American Public Media). October 19, 2007. Available online. URL: http://marketplace.publicradio.org/display/web/2007/10/19/online_job_networking. Accessed October 21, 2007.
USAJOBS [federal job information]. Available online. URL: http://www.usajobs.gov/. Accessed October 21, 2007.

online research

The proliferation of online databases, information services (see ONLINE SERVICES) and Web sites has made more information accessible to more people than ever before. At the same time, the complexity of the online world challenges researchers to develop a new set of skills to cope with it.

It is useful to divide online offerings into three broad categories: specialized databases, online information services, and the Web as a whole (see WORLD WIDE WEB). Each of these areas requires a somewhat different approach by the online researcher.

A common research task is to find and evaluate books or articles on a given subject. Most local libraries have their catalogs online, and the world's largest library catalog, that of the Library of Congress (LC), is also available in several forms on the Web.

Newspaper and magazine articles can be found in a number of general-purpose databases such as InfoTrac. These databases can be searched in public libraries: Remote access is generally restricted to the library's cardholders. These records can consist of a bibliographic description only (that is, author, title, periodical, issue date, and so on) or can include an abstract or in many cases the full text of the article. In addition, most major newspapers now offer free access to recent articles on their Web site, with older articles available for a nominal fee. Magazines, too, frequently offer selected articles or their complete contents online.

Using the search facility for an online catalog or periodical database is generally simple, particularly if an author or title is known. For subject searching, some familiarity with LC subject headings is helpful. However, the ability of most systems to search for matching words in titles or subjects means that the researcher can be quickly led to the correct subject in most cases.

Another way to get tables of contents, jacket copy, and reviews of books is to browse the online catalogs of major booksellers, particularly Amazon.com and BarnesandNoble.com. Publishers' Web sites are another good way to get information about books, particularly new or forthcoming titles.

Journalists need a broad familiarity with online research tools and use computers and online services in many facets of their work (see JOURNALISM AND COMPUTERS). Researchers looking for specialized articles in fields such as law or medicine need more rigorous skills.

Most legal research is done using databases such as LexisNexis. These databases are expensive but indispensable to practitioners. However, students and others who can't afford this access can still find U.S. Supreme Court, Court of Appeals, and many state court decisions online, thanks to the efforts of organizations such as the Legal Information Institute at Cornell Law School. Because of the complexity of multiple jurisdictions and the need to trace chains of precedent ("shepardizing"), online legal research has become an increasingly important paraprofessional task.

Medical research is similarly complex, due to the thousands of precise terms for conditions, procedures, and drugs. The sheer volume of articles (MEDLINE has more than 11 million citations dating back to the 1960s) can make it hard to find and evaluate the most relevant material.

By far the most extensive information resource today is the World Wide Web with its millions of sites and pages of information. There are two basic approaches to finding material on the Web. The first is to use a search engine by typing in keywords or phrases (see SEARCH ENGINE). Even though search engines such as Google index only a modest fraction of the available pages on the Web, a search on a topic such as "database design" can yield from thousands to millions of possible "hits." Most search engines do attempt to rank results in decreasing order of matching or relevance.

An alternative approach is to browse the categorized list of topics presented by a site such as Yahoo! (www.yahoo.com) or About.com (www.about.com). The advantage of this approach is that the site's researchers have selected the links for each topic that they believe to be the most valuable, and the number of possibilities is likely to be more manageable (see PORTAL).

The tremendous increase in personal expression and collaboration on the Web is opening new channels of information (see BLOGS AND BLOGGING, USER-CREATED CONTENT, and WIKIS AND WIKIPEDIA). Wikipedia, for example, has some articles that are as reliable and fully documented as those found in a traditional encyclopedia, while others might be best described as "works in progress." The researcher must decide whether a given article or posting is definitive or perhaps just usefully suggestive of further resources.

Online research remains more an art than a science. The researcher must choose the appropriate tools—bibliographical resources, specialized databases, information services, search engines, and portals—and evaluate and integrate the results so they are useful for a given question or project. Students and researchers now have unprecedented access to information, but sophisticated critical thinking skills must be employed. In particular, it can be difficult to evaluate the background or credentials of the people behind Web sites that are not associated with recognized media outlets or other organizations.

Further Reading

Dornfest, Rael, Paul Bausch, and Tara Calishain. *Google Hacks: Tips and Tools for Finding and Using the World's Information.* 3rd ed. Sebastopol, Calif.: O'Reilly, 2006.

Hock, Randolph. *The Extreme Searcher's Internet Handbook: A Guide for the Serious Searcher.* 2nd ed. Medford, N.J.: Information Today, 2007.

Internet Public Library. Available online. URL: http://www.ipl.org/. Accessed August 16, 2007.

Research and Documentation Online. Available online. URL: http://www.dianahacker.com/resdoc/. Accessed August 16, 2007.

Schlein, Alan M. *Find It Online.* 4th ed. Tempe, Ariz.: Facts on Demand Press, 2004.

Tomaiuolo, Nicholas, Steve Coffman, and Barbara Quint. *The Web Library: Building a World Class Personal Library with Free Web Resources.* Medford, N.J.: Information Today, 2004.

online services

The ability of PC owners to connect to remote computers (see MODEM) led to the proliferation of both free and commercial online information services during the 1980s. At one end of the spectrum were bulletin board systems (BBS), many run by hobbyists on PCs connected to a few phone lines (see BULLETIN BOARD SYSTEMS). They offered users the ability to read and post messages on various topics as well as to download or contribute software (see also SHAREWARE).

The growing number of connected PC owners soon offered entrepreneurs a potential market for a commercial online information service. One of the oldest, CompuServe, had actually been started in 1969 as a business time-sharing computer system. In 1979, it launched a service for home computer users, offering e-mail and technical support forums. By the mid-1980s, the service had added an online chat service called CB Simulator (see CHAT, ONLINE) as well as news content. The service's greatest strength, however, remained its forums, which offered technical support for just about every sort of computer hardware or software, together with download libraries containing system patches, drivers, utilities, templates, macros, and other add-ons.

By then, however, the online service market had become quite competitive. While CompuServe focused on computer-savvy users, America Online (AOL), founded in 1985 by Steve Case, targeted the growing legion of new PC users who needed an easy-to-navigate interface. AOL grew steadily, reaching a million customers in 1994 (see AMERICA ONLINE). AOL chat groups became very popular, spawning a vigorous online culture while raising controversies about sexual content in some chat "rooms." A third service, Prodigy, also catered to the new user.

Meanwhile, the World Wide Web and the advent of graphical Web browsers such as Netscape and Microsoft Internet Explorer in the mid-1990s led millions of users to connect to the Internet (see INTERNET, WEB BROWSER, and WORLD WIDE WEB). Internet service providers (ISPs) offered direct, no-frills access to the Web. CompuServe and AOL soon offered their users access to the Internet as well. However, accessing the Web through an online information service was usually more expensive, and often slower, than using an ISP and a Web browser directly. Additionally, free Web portal services such as Yahoo! began to offer extensive information resources of their own.

The Internet thus threatened to shrink the market for the commercial online services. AOL fought back in the late 1990s by cutting its monthly rates to make them competitive with ISPs, flooding the mails with free disks and trial offers, bundling introductory packages with new computer systems, and promoting added-value information services such as stock quotes. In 1998, the market consolidated when AOL bought CompuServe, continuing to run the latter as a subsidiary targeted at more sophisticated users. The same year AOL bought Netscape to gain access to its browser technology. Finally, AOL merged with Time-Warner, hoping to leverage the latter's huge media resources, such as by offering classic TV fare. However, the flagship online service continued to struggle in the 2000s, essentially abandoning the ISP part of its business. Meanwhile CompuServe, after peaking in the 1990s, gradually shrank to a shadow of its former self. Even mighty Microsoft has had trouble growing its Microsoft Network (MSN), reinventing it in 1999 as a Web portal and then trying to integrate it more closely with its operating system and software products as "Windows Live" as well as providing services such as instant messaging, blogging, and picture sharing.

The long-term prospects for AOL and other commercial online services are uncertain. Many of the advantages these services had until the late 1990s have diminished. For example, the once mutually incompatible e-mail systems of online services have been replaced by standard Internet e-mail protocols, so there is little advantage to using a particular service for e-mail. Users can obtain e-mail accounts from a variety of ISPs or through free Web-based services such as hotmail.com. Content such as news, video, and music (see STREAMING) is available from many Web sites, and most companies now offer extensive online technical support for their products. At the same time, attempts to support content-rich sites through either advertising or a subscription model have largely foundered. For services such as AOL, the ultimate question is whether the parts of the service still form a sufficiently compelling whole.

Further Reading

America Online. Available online. URL: http://www.aol.com. Accessed August 16, 2007.
Bourne, Charles P. *A History of Online Information Services, 1963–1976.* Cambridge, Mass.: MIT Press, 2003.
Kaufeld, John. *AOL for Dummies.* Hoboken, N.J.: Wiley, 2004.
Microsoft Network. Available online. URL: http://www.msn.com. Accessed August 14, 2007.
Swisher, Kara. *There Must Be a Pony in Here Somewhere: The AOL Time Warner Debacle.* New York: Three Rivers Press, 2004.

ontologies and data models

A persistent problem in artificial intelligence (see ARTIFICIAL INTELLIGENCE) is how to provide a software system with a model that it can use to reason about a particular subject or domain. A data model or ontology basically consists of classes to which the relevant objects might belong, relationships between classes, and attributes that objects in that class can possess. (For implementation of these ideas within programming languages, see CLASSES and OBJECT-ORIENTED PROGRAMMING.)

For example, a business ontology might include classes such as:

- Entity—a business or person

- Supplier—an Entity that provides wholesale goods or services

- Customer—an Entity that buys the company's goods or services

- Contractor—an Entity that performs work for the company on contract

In the above list it can be seen that the last three classes all include as their parent or "superclass" the class Entity. Another way to put this is to say that the Entity class "subsumes" the last three classes. These relationships can be easily shown in tree diagrams, with the most general or "universal" class at the top and the more specialized classes extending downward and outward. The process of defining related classes and specifying criteria for the inclusion of an object in a class is called "partitioning." (Readers familiar with set theory will also note that the language of sets, subsets, and inclusion also works well with this scheme.)

Classes can have other types of relationships. For example, a class can be defined as being "part of" a structure built from several classes. For example, a Customer might be part of a Transaction class.

Attributes are assigned to classes as appropriate. Note in the example above that when attributes such as contact information are defined for the Entity class, they will also apply to the descendant classes Supplier, Customer, and Contractor.

IMPLEMENTATION

Ontologies can be used to provide guidance to a variety of types of programs (for example, see EXPERT SYSTEM, NATURAL LANGUAGE PROCESSING, and SOFTWARE AGENT). Thus if an automatic news summarizer program encounters a story that includes references to opposing lawyers and legal issues, it could apply an ontology that defines the likely relationship of the participants in the case.

Creating useful ontologies is quite labor intensive in terms of the human thinking and coding involved. However, there have been substantial efforts in recent years to create anthologies for many fields, particularly in biology and genetics. The Web Ontology Language (OWL) is a popular tool for creating ontologies that can be used to make Web content more understandable to programs (see SEMANTIC WEB).

Meanwhile, an ambitious and long-running project called Cyc (for Encyclopedia) under the direction of Douglas Lenat has been engaged in creating what amounts to vast ontologies for many of the domains included in everyday human life as well as specialized fields of knowledge. A large portion of this work has been made available as open source.

Further Reading

CYCorp. Available online. URL: http://www.cyc.com/. Accessed October 21, 2007.
Gasevic, Dragan, Dragan Djuric, and Vladan Devedzic. *Model Driven Architecture and Ontology Development.* New York: Springer, 2006.
Macy, Lee W. *OWL: Representing Information Using the Web Ontology Language.* Victoria, B.C., Canada: Trafford Publishing, 2005.
Nigro, Hector Oscar, Sandra Gonzalez Cisaro, and Daniel Xodo, eds. *Data Mining with Ontologies: Implementations, Findings, and Frameworks.* Hershey, Penn.: Idea Group, 2007.
Web Ontology Language (OWL), World Wide Web Consortium. Available online. URL: http://www.w3.org/2004/OWL/. Accessed October 21, 2007.

open-source movement

For a long time programmers have released programs as freeware meaning that users did not have to buy or license the software. There is also "try before you buy" software (see SHAREWARE). However, while freeware sometimes includes not only the executable program but the source code (the actual program instructions), most shareware and virtually all other commercially distributed software does not. As a result, users wishing to fix, modify, or extend the software are generally at the mercy of the company that owns and distributes it.

In university and research computing environments, however, it has been common for programmers to freely share and extend utilities such as program editors. Indeed, much of the necessary software for the earliest minicomputers of the 1960s was created by clever, energetic hackers (see HACKERS AND HACKING). Because the source code (usually on paper tape) was freely distributed, people could easily create and distribute new (and presumably, improved) versions. Having source code also made it possible to "port" software to a newly released machine without having to wait for the relatively ponderous efforts of the official developers.

In particular, although the licensing of the two major versions of the UNIX operating system were controlled by AT&T's Bell Laboratories and the University of California's Berkeley Software Distribution (BSD) respectively, much UNIX software including programming languages (see PERL and PYTHON) and the Web's most popular server, Apache, have been distributed using an *open source* model.

The best-known open-source effort is the GNU Project created by Richard Stallman (1953–). GNU, a recursive acronym meaning "GNU's Not UNIX," is a collection of software that provides much of the functionality of AT&T's UNIX without being subject to the latter's licensing fees and restrictions. When creating his own open source version of UNIX (see LINUX), Linus Torvalds (see TORVALDS, LINUS) and his colleagues drew upon the considerable base of software already created by GNU.

According to Stallman and many other advocates, "open source" software is not necessarily free. What is required is that users receive the full source code (or have it readily available for free or at nominal charge). Users are free to modify or expand the source code to create and distribute new versions of the software. Following a legal mechanism that Stallman calls "copyleft," the distributor of open-source software must allow subsequent recipients the same freedom to revise and redistribute. However, not all software that is billed as open source follows all of Stallman's requirements, including being copylefted. Formally, open-source software is generally licensed according to various versions of the General Public License (GPL). The latest version, GPL3, released in 2007, has been controversial. Among other things, it more aggressively attempts to prevent open-source software from being restricted or otherwise hampered by being combined with patented software or proprietary hardware.

Open-source software has the potential for providing diversity and alternatives in a world where some categories such as PC operating systems and office software are dominated by one or a few large companies. Indeed, sometimes companies have converted an existing product to open source, as is the case with Sun Microsystems and Star Office, a suite that runs under Linux. Netscape also resorted to open source as part of an unsuccessful attempt to fight off Microsoft for dominance of the browser market in the mid to late 1990s. By making a product open source, a company may hope to tap into the volunteer effort of many talented programmers to improve or expand the program. The company is still free to create proprietary software upon the "base" of a successful open source product. Moderately successful companies such as Linux distributor Red Hat have a business plan based upon providing superior packaging, technical support, and customized solutions around its Linux distribution.

While some critics have questioned whether viable business models can be built directly upon open-source software, there is little doubt that open-source development has made a substantial contribution to the infrastructure of the computer industry. Linux runs about a third of all Web servers, and products such as the Apache Web server and MySQL database are also in widespread use, as is the Eclipse integrated development environment.

Many advocates see open source as part of a larger philosophy and even a social movement (see USER-CREATED CONTENT). They believe that by creating value through collaboration and sharing, open source may challenge classical economics based on scarcity and competition.

Further Reading
Babcock, Charles. "Open Source Software: Who Gives and Who Takes?" *InformationWeek*. May 15, 2006. Available online. URL: http://www.informationweek.com/story/showArticle.jhtml?articleID=187202790. Accessed August 16, 2007.

DiBona, Chris, Danese Cooper, and Mark Stone, eds. *Open Sources 2.0: The Continuing Evolution*. Sebastapol, Calif.: O'Reilly, 2005.

DiBona, Chris, Sam Ockman, and Mark Stone, eds. *Open Sources: Voices from the Open Source Revolution*. Sebastapol, Calif.: O'Reilly, 1999.

Enterprise Open Source (EOS) Directory. Available online. URL: http://www.eosdirectory.com/. Accessed August 16, 2007.

LaMonica, Martin. "'Free' is the New 'Cheap' for Software Tools." *CNET News*. Available online. URL: http://news.com.com/2100-7344_3-6032986.html. Accessed August 16, 2007.

Ohloh Open Source Directory. Available online. URL: http://www.ohloh.net/. Accessed August 16, 2007.

Rosen, Lawrence. *Open Source Licensing: Software Freedom and Intellectual Property Law*. Upper Saddle River, N.J.: Prentice Hall, 2004.

Stallman, Richard. "Richard Stallman Sets the Free Software Record Straight" [interview with Jennifer LeClaire]. *Linux Insider*. Available online. URL: http://www.linuxinsider.com/story/50122.html. Accessed August 14, 2007.

Weber, Steven. *The Success of Open Source*. Cambridge, Mass.: Harvard University Press, 2005.

operating system

An operating system is an overarching program that manages the resources of the computer. It runs programs and provides them with access to memory (RAM), input/output devices, a file system, and other services. It provides application programmers with a way to invoke system services, and gives users a way to control programs and organize files.

DEVELOPMENT

The earliest computers were started with a rudimentary "loader" program that could be used to configure the system to run the main application program. Gradually, a more sophisticated way to schedule and load programs, link programs together, and assign system resources to them was developed (see JOB CONTROL LANGUAGE and MAINFRAME).

As systems were developed that could run more than one program at a time (see MULTITASKING), the duties of the operating systems became more complex. Programs had to be assigned individual portions of memory and prevented from accidentally overwriting another program's memory area. A technique called *virtual memory* was developed to enable a disk drive to be treated as an extension of the main memory, with data "swapped" to and from the disk as necessary. This enabled the computer to run more and/ or larger applications. The operating system, too, became larger, amounting to millions of bytes worth of code.

During the 1960s, time sharing became popular particularly on new smaller machines such as the DEC PDP series (see MINICOMPUTER), allowing multiple users to run programs and otherwise interact with the same computer. Operating systems such as Multics and its highly successful offshoot UNIX developed ways to assign security levels to files and access levels to users. The UNIX architecture featured a relatively small *kernel* that provides essential process control, memory management, and file system services, while drivers performed the necessary low-level control of devices and a shell provided user control. (See UNIX, KERNEL, DEVICE DRIVER, and SHELL.)

Starting in the late 1970s, the development of personal computers recapitulated in many ways the earlier evolution of operating systems in the mainframe world. Early microcomputers had a program loader in read-only memory (ROM) and often rudimentary facilities for entering, running, and debugging assembly language programs.

During the 1980s, more complete operating systems appeared in the form of Apple DOS, CP/M, and MS-DOS for IBM PCs. These operating systems provided such facilities as a file system for floppy or hard disk and a command-line interface for running programs or system utilities. These systems could run only one program at a time (although exploiting a little-known feature of MS-DOS allowed additional small programs to be tucked away in memory).

As PC memory increased from 640 kB to multiple megabytes, operating systems became more powerful. Apple's Macintosh operating system and Microsoft Windows could manage multiple tasks. Today personal computer operating systems are comparable in sophistication and capability to those used on mainframes. Indeed, PCs can run UNIX variants such as the popular Linux.

COMPONENTS

While the architecture and features of operating systems differ considerably, there are general functions common to almost every system. The "core" functions include "booting"

the system and initializing devices, process management (loading programs intro memory assigning them a share of processing time), and allowing processes to communicate with the operating system or one another (see KERNEL). Multiprogramming systems often implement not only processes (running programs) but also *threads,* or sections of code within programs that can be controlled separately.

A memory management scheme is used to organize and address memory, handle requests to allocate memory, free up memory no longer being used, and rearrange memory to maximize the useful amount (see MEMORY MANAGEMENT).

There is also a scheme for organizing data created or used by programs into files of various types (see FILE). Most operating systems today have a hierarchical file system that allows for files to be organized into directories or folders that can be further subdivided if necessary. In operating systems such as UNIX, other devices such as the keyboard and screen (console) and printer are also treated like files, providing consistency in programming. The ability to redirect input and output is usually provided. Thus, the output of a program could be directed to the printer, the console, or both.

In connecting devices such as disk drives to application programs, there are often three levels of control. At the top level, the programmer uses a library function to open a file, write data to the file, and close the file. The library itself uses the operating system's lower-level input/output (I/O) calls to transfer blocks of data. These in turn are translated by a *driver* for the particular device into the low-level instructions needed by the processor that controls the device. Thus, the command to write data to a file is ultimately translated into commands for positioning the disk head and writing the data bytes to disk.

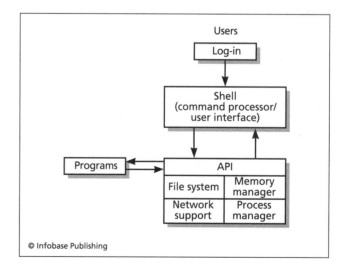

A typical operating system processes user commands or actions using an interface (such as a shell). Both user commands and requests from application programs communicate with the operating system through the application Programming Interface (API), which provides services such as file, memory, process, and network management.

Operating systems, particularly those designed for multiple users, must also manage and secure user accounts. The administrator (or sometimes, ultimately, the "super user" or "root") can assign users varying levels of access to programs and files. The owners of files can in turn specify whether and how the files can be read or changed by other users (see DATA SECURITY).

In today's highly networked world most operating systems provide basic support for networking protocols such as TCP/IP. Applications can use this facility to establish network connections and transfer data over the local or remote network (see NETWORK).

The operating system's functions are made available to programmers in the form of program libraries or an application programming interface (API). (See LIBRARY, PROGRAM and APPLICATION PROGRAMMING INTERFACE.)

The user can also interact directly with the operating system. This is done through a program called a *shell* that accepts and responds to user commands. Operating systems such as MS-DOS and early versions of UNIX accepted only typed-in text commands. Systems such as Microsoft Windows and UNIX (through facilities such as XWindows) allow the user to interact with the operating system through icons, menus, and mouse movements. Application programmers can also provide these interface facilities through the API. This means that programs from different developers can have a similar "look and feel," easing the learning curve for users.

ISSUES AND TRENDS

As the tasks demanded of an operating system have become more complex, designers have debated the best overall form of architecture to use. One popular approach, typified by UNIX, is to use a relatively small kernel for the core functions. A community of programmers can then write the utilities needed to manage the system, performing tasks such as listing file directories, editing text, or sending e-mail. New releases of the operating system then incorporate the most useful of these utilities. The user also has a variety of shells (and thus interfaces) available.

The kernel approach makes it relatively easy to port the operating system to a different computer platform and then develop versions of the utilities. (Kernels were also a necessity when system memory was limited and precious, but this consideration is much less important today.)

Designers of modern operating systems face a number of continuing challenges:

- security, in a world where nearly all computers are networked, often continuously (see COMPUTER CRIME AND SECURITY and FIREWALL)

- the tradeoff between powerful, attractive functions such as scripting and the security vulnerabilities they tend to present

- the need to provide support for new applications such as streaming audio and video (see STREAMING)

- ease of use in installing new devices (see DEVICE DRIVER and PLUG AND PLAY)

- The continuing development of new user-interface concepts, including alternative interfaces for the disabled and for special applications (see USER INTERFACE and DISABLED PERSONS AND COMPUTING)

- the growing use of multiprocessing and multiprogramming, requiring coordination of processors sharing memory and communicating with one another (see MULTIPROCESSING AND CONCURRENT PROGRAMMING)

- distributed systems where server programs, client programs, and data objects can be allocated among many networked computers, and allocations continually adjusted or balanced to reflect demand on the system (see DISTRIBUTED COMPUTING)

- the spread of portable, mobile, and handheld computers and computers embedded in devices such as engine control systems (see LAPTOP COMPUTER, PDA, and EMBEDDED SYSTEM). (Sometimes the choice is between devising a scaled-down version of an existing operating system and designing a new OS that is optimized for devices that may have limited memory and storage capacity.)

Further Reading

Bach, Maurice J. *The Design of the UNIX Operating System*. Englewood Cliffs, N.J.: Prentice Hall, 1986.

Ritchie, Dennis M. "The Evolution of the UNIX Time-Sharing System." *Lecture Notes in Computer Science #79: Language Design and Programming Methodology,* New York: Springer-Verlag, 1980. Available online. URL: http://cm.bell-labs.com/cm/cs/who/dmr/hist.html. Accessed August 14, 2007.

Silberschatz, Abraham, Peter Baer Galvin, and Greg Gagne. *Operating System Concepts*. 7th ed. New York: Wiley, 2004.

operators and expressions

All programming languages provide operators to specify arithmetic functions. Some of them, such as addition +, subtraction -, multiplication ×, and division ÷, are familiar from elementary school arithmetic (although the asterisk rather than the traditional x is used for multiplication in program code, to avoid confusion with the letter x). Additional operators found in languages such as C, C++, and Java include % (modulus, or remainder after division), ++ (adds one and stores the result back into the operand), and -- (decrement; subtracts one and stores the result back into the operand).

Operands are data items such as variables, constants, or literals (actual numbers) that are operated on by the operator. An operator is called unary if it takes just one operand (the increment operator ++ is an example). An operator that takes two operands is considered to be binary, and this is true of most arithmetic operations such as addition, multiplication, subtraction, and division.

A combination of operands and operators constitutes an arithmetic expression that evaluates to a particular value when the program runs. Thus in the C statement:

```
Total = SubTotal + SubTotal Tax × Tax_Rate;
```

the value of the SubTotal Tax is multiplied by the value of the variable Tax_Rate, the result is added to the value of SubTotal, and the result of the entire expression is stored in the variable Total. Compilers generally parse arithmetic expressions by converting them from an "infix" form (as in A + B) to a "postfix" form (as in + A B), resolving them into a simple form that is ready for conversion to machine code.

OPERATOR PRECEDENCE

The preceding example raises an important question. How does one know that the subtotal is to be multiplied by the tax rate and then the result added to the subtotal, as opposed to adding the subtotal and tax and multiplying the result by the tax rate? The former procedure is intuitively correct to human observers, but since computers lack intuition, specific rules of precedence are defined for operators. These rules, which are similar for all computer languages, tell the compiler that when code is generated for arithmetic operations, multiplications and divisions are carried out first (moving from left to right), and then additions and subtractions are resolved in the same way. The rules of precedence do become more complex when the relational, logical, and assignment operators are included. Finally, expressions can be enclosed in parentheses to overrule precedence and force them to be evaluated. Thus in the expression (A + B) * C the addition will be carried out before the multiplication.

Generally speaking, the levels of precedence for most languages are as follows:

1. scope resolution operators (specify local v. global versions of a variable)
2. invoking a method from a class, array subscript, function call, increment or decrement
3. size of (gets number of bytes in an object), address and pointer dereference, other unary operators (such as "not" and complement); creation and deallocation functions; type casts
4. class member selection through a pointer
5. multiplication, division, and modulus
6. addition and subtraction
7. left and right shift operators
8. less than and greater than
9. equal and not equal operators
10. bitwise operators (AND, then exclusive OR, inclusive OR)
11. logical operators (AND, then OR)
12. assignment statements

The basic arithmetic operators are built into each programming language, but many of the newer object-oriented languages such as C++ allow for programmer-defined operators and a process called overloading in which the same operator can be defined to work with several different kinds of data. Thus the + operator can be extended so that if it is given character strings instead of numbers, it will "add" the strings by combining (concatenating) them.

Further Reading

"Operators in C and C++." Wikipedia. Available online. URL: http://en.wikipedia.org/wiki/Operators_in_C_and_C++. Accessed August 16, 2007.

Sebesta, Robert W. *Concepts of Programming Languages.* 8th ed. Upper Saddle River, N.J.: Addison-Wesley, 2007.

Stroustrup, Bjarne C. *The C++ Programming Language.* 3rd ed. Reading, Mass.: Addison-Wesley, 1997. See chap. 6 "Expressions and Statements" and chap. 11 "Operator Overloading."

"Summary of Operators." Java Tutorials. Available online. URL: http://java.sun.com/docs/books/tutorial/java/nutsandbolts/opsummary.html. Accessed August 16, 2007.

"Summary of Operators in Java." Available online. URL: http://sunsite.ccu.edu.tw/java/tutorial/java/nutsandbolts/opsummary.html.

optical character recognition (OCR)

Today it is easy to optically scan text or graphics printed on pages and convert it into a graphical representation for storage in the computer (see SCANNER). However, a shape such as a letter *c* doesn't mean anything in particular as a graphic. Optical character recognition (OCR) is the process of identifying the letter or other document element that corresponds to a given part of the scanned image and converting it to the appropriate character (see CHARACTERS AND STRINGS). If the process is successful, the result is a text document that can be manipulated in a word processor, database, or other program that handles text. Raymond Kurzweil (1948–) marketed the first commercially practicable general-purpose optical character recognition system in 1978.

Once the document page has been scanned into an image format, there are various ways to identify the characters. One method is to use stored *templates* that indicate the pattern of pixels that should correspond to each character. Generally, a threshold of similarity is defined so that an exact match is not necessary to classify a character: The template most similar to the character is chosen. Some systems store a set of templates for each of the fonts most commonly found in printed text. (Recognizing cursive writing is a much more complex process: See HANDWRITING RECOGNITION.)

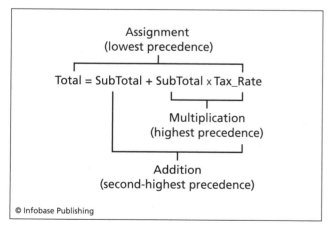

A compiler or interpreter processes a program statement by applying its operators in order of precedence. Here the multiplication is done first, and then its result is used in the addition. The assignment (=) operator has the lowest precedence and is applied last, assigning the value of the entire expression to the variable Total.

A more generalized method uses structural features (such as "all t's have a single vertical line and a shorter crossbar line") to classify characters. To analyze a character, the different types of individual features are identified and then compared to a set of rules to determine the character corresponding to that particular combination of features. Sometimes thresholds or "fuzzy logic" are used to decide the *probable* identity of a character.

OCR systems have improved considerably, the process also being speeded up by today's faster processors. Most scanners are sold with OCR software that is perhaps 95 percent accurate, with higher end systems being more accurate still. This is certainly good enough for many purposes, although material that is to be published or used in legal documents should still be proofread by human beings.

Further Reading

More, Shunji, Hirobumi Nishida, and Hiromitsu Yamada, eds. *Optical Character Recognition.* New York: Wiley, 1999.
Rice, S. V., G. Nagy and T. A. Nartker. *Optical Character Recognition: An Illustrated Guide to the Frontier.* Boston: Kluwer, 1999.

optical computing

Light is the fastest thing in the universe, and the science and technology of optics have developed greatly since the invention of the laser in the 1960s. It is not surprising, therefore, that computer designers have explored the possibility of using optics rather than electronics for computation and data storage.

An early idea was to use a grid of laser beams to create logical circuits, exploiting the ability of one laser to be used to "quench" or switch off another one. However, creating a large number of tiny laser beams proved impracticable, as did managing the heat created by the process. However, by the 1980s, experimenters were interacting "microlasers" with semiconductors, exploiting quantum effects. This brought the energy (and heat) problem under control while vastly increasing the potential density of the optical circuitry.

The incredible rate at which conventional silicon-based electronic circuitry continued to increase in density and capacity has limited the incentive to invest in the large-scale research and development that would be needed to develop a complete optical computer with processor and a corresponding optical memory technology.

Instead, current research is exploring the possibility of combining the best features of the optical and electronic system. Silicon chips have a limited surface for connecting data inputs, while light can carry many more channels of data through micro-optics. It may be possible to couple a micro-optic array to the surface of the silicon chip in such a way that the chip could have the equivalent of thousands of connecting pins to transmit data. In March 2007 IBM unveiled a prototype hybrid chip that combines optical and semiconductor technology to achieve eight times the data transfer rate of conventional technologies.

The value of optics is more conclusively demonstrated in data transmission and storage technology. Fiber optic cables are being used in many cases to carry large quanti-ties of data with very high capacity (see FIBER OPTICS) and may gradually supplant conventional network cable in more applications. The use of lasers to store and read information is seen in CD-ROM and DVD-ROM technology, which has replaced the floppy disk as the ubiquitous carrier of software and handy backup medium (see CD-ROM AND DVD-ROM).

Further Reading

Goswami, Debabrata. "Optical Computing." *Resonance,* June 2003. Available online. URL: http://www.ias.ac.in/resonance/June2003/pdf/June2003p56-71.pdf. Accessed August 16, 2007.
Knight, Will. "Laser Chips Could Power Petaflop Computers." *New Scientist.com.* March 21, 2006. Available online. URL: http://www.newscientist.com/article.ns?id=dn8876. Accessed August 16, 2007.
"Now, Just a Blinkin' Picosecond!: NASA Scientists Are Working to Solve the Need for Computer Speed Using Light Itself to Accelerate Calculations and Increase Data Bandwidth." Science & NASA. Available online. URL: http://science.nasa.gov/headlines/y2000/ast28apr_1m.htm. Accessed August 16, 2007.
Saleh, Bahaa E. A., and Malvin Carl Teich. *Fundamentals of Photonics.* 2nd ed. New York: Wiley Interscience, 2007.

Oracle Corporation

Founded in 1977, Oracle Corporation (NASDAQ symbol: ORCL) is a leading developer of business database software (see DATABASE MANAGEMENT SYSTEM) as well as systems for other enterprise operations (see CUSTOMER RELATIONSHIP MANAGEMENT and SUPPLY CHAIN MANAGEMENT). These functions are integrated through a structure called Oracle Information Architecture that can coordinate the operations of servers and storage systems (see GRID COMPUTING). Besides selling software, a major part of Oracle's business is providing consulting and support for fitting the software to the needs of corporate customers, as well as training (through Oracle University) and distributed application services (Oracle on Demand). In 2007 Oracle had $18 billion in sales, netting $4.74 billion in profit. The company had over 73,000 employees.

Since its founding, Oracle's CEO has been the dynamic though often controversial Larry Ellison, who recognized the importance of relational databases (with their ability to connect information from many sources) as a way to meet the growing information needs of modern business. In the 1970s IBM was the dominant leader in relational databases for mainframe computers, but when personal computers running Windows became prevalent around 1990, IBM was slow to enter the new market. Ellison and competitors such as Sybase and Informix were able to carve out strong niches, with Oracle coming out on top by the end of the decade. (However, by the 2000s IBM's DB2 for UNIX/Linux and Windows and Microsoft SQL Server [for Windows only] were strong competitors, with open-source products MySQL and PostgreSQL also gaining attention—see SQL.)

In recent years Oracle has also expanded through acquisition, picking up other software companies, including PeopleSoft; Retek, Inc.; and Siebel Systems, for a combined total of over $16 billion. In 2007 Oracle filed a lawsuit against its major competitor in business management applications (see SAP), charging them with unfair practices.

Further Reading

Allen, Christopher, Simon Chatwin, and Catherine A. Creary. *Introduction to Relational Databases and SQL Programming.* Burr Bridge, Ill.: McGraw-Hill Technology Education, 2004.

Gerald, Bastin, Nigel King, and Dan Natcher. *Oracle E-Business Suite Manufacturing & Supply Chain Management.* Berkeley, Calif.: McGraw-Hill/Osborne, 2002.

Oracle Corporation. Available online. URL: http://www.oracle.com/index.html. Accessed October 25, 2007.

Rittman, Mark. "Oracle Information Architecture Explained." Available online. URL: http://www.dba-oracle.com/oracle_news/2004_8_11_rittman.htm. Accessed October 25, 2007.

Stackowiak, Robert, Joseph Rayman, and Rick Greenwald. *Oracle Data Warehousing and Business Intelligence Solutions.* Indianapolis: Wiley, 2007.

Symonds, Matthew. *Softwar: An Intimate Portrait of Larry Ellison and Oracle.* New York: Simon & Schuster, 2003.

OS X

Jaguar, panther, tiger, and leopard—these and other names of sleek big cats represent versions of Apple's Macintosh operating system, OS X (pronounced "OS 10"—see APPLE CORPORATION and MACINTOSH). Unlike the previous Mac OS, OS X, while broadly maintaining Apple's user interface style (see USER INTERFACE), is based on a version of UNIX called OpenStep, developed by NeXT starting in the 1980s (see UNIX). OS X development began when Steve Jobs returned as Apple CEO in 1997 (see JOBS, STEVEN PAUL) and the company bought NeXT, acquiring the software. The first version, OS X 10.0, or Cheetah, was released in 2001, but the system was not widely used until 10.1 (Puma) was released later the same year.

At the core of OS X is a free and open-source version of UNIX called "Darwin," with a kernel called XNU. On top of this Apple built a distinctive and subtly colorful user interface called Aqua and a new version of the Macintosh Finder file and program management system.

APPLICATIONS AND DEVELOPMENT

Today OS X includes a variety of useful software packages—some free and some optional. These include iLife (digital media management), iWork (productivity), and Front Row (home media center). OS X10.5 also includes Time Machine, an automatic backup system that can restore files (including deleted files) as well as earlier system settings.

For software developers, OS X provides an integrated development environment called "Xcode," which works with modified open-source compilers for major programming languages, including C, C++, and Java. Further, because OS X is UNIX-based, many UNIX and Linux programs can be recompiled to run on it. Since mid-2005 Apple (and OS X) have been transitioning from the earlier IBM/Motorola processors to Intel processors. This transition was largely complete by 2007, though OS X 10.5 (Leopard) still provides support for applications written for the PowerPC.

OS X has been well received by critics, and together with its bundled software has made the Macintosh a popular platform for users who want a seamless computing experience, particularly with regard to graphics and media.

Further Reading

LeVitus, Bob. *Mac OS X Leopard for Dummies.* Hoboken, N.J.: Wiley, 2007.

Mingis, Ken, and Michael DeAgonia. "In Depth: Apple's Leopard Leaps to New Heights." *Computerworld,* October 25, 2007. Available online. URL: http://www.computerworld.com/action/article.do?command=viewArticleBasic&articleId=9043838. Accessed October 26, 2007.

Pogue, David. *Mac OS X Leopard: The Missing Manual.* Sebastapol, Calif.: Pogue Press/O'Reilly, 2007.

Singh, Amit. "A History of Apple's Operating Systems." Available online. URL: http://www.kernelthread.com/mac/oshistory/. Accessed October 26, 2007.

———. *Mac OS X Internals: A Systems Approach.* Upper Saddle River, N.J.: Addison-Wesley Professional, 2006.

P

Page, Larry
(1973–)
American
Entrepreneur

Together with business partner SERGEY BRIN, Larry (Lawrence Edward) Page revolutionized the role of Web search in the modern Internet economy by developing the Google search engine and building an industry-leading company around it.

Page was born in East Lansing, Michigan, into a very computer-oriented family (both his parents were computer scientists and his brother became a computer engineer). It was perhaps not surprising that Page received a BSE in engineering at the University of Michigan in 1995, then entered the doctoral program in computer science at Stanford University. There he met Sergey Brin (see BRIN, SERGEY). The two students soon developed an interest in the burgeoning Web, particularly in finding a better way to search for information. The result was their collaboration on a page-ranking system that prioritized search results based on the popularity of sites as shown by the number of links to them (see also SEARCH ENGINE). The other half of Page and Brin's achievement was in developing advertising models that would turn users' Web interests into revenue. Their key insight was that sellers would be eager to advertise to people who had already shown by searching that they were interested in particular products or services.

By the early 2000s Google dominated Web search, which would became a springboard to many other services, including local searching, maps, and even free online office applications (see GOOGLE). In 2004 Page and Brin became rich when Google offered its stock to the public (by 2006 Page's net worth was estimated at $16.6 billion, just behind Brin).

Google has profoundly changed the way people use the Web, so much that "to google" has become a verb for searching online. However, the company's size and dominant position in Web search advertising has raised concerns among some critics that Google might be gaining too much control over the market (see ONLINE ADVERTISING). Meanwhile in 2001 Page and Brin hired Eric Schmidt to become Google's CEO, giving Page more time to pursue other interests, one of which is his investment in Tesla Motors, developer of an advanced electric vehicle that can go up to 250 miles on one battery charge. Another interest of Page and Brin is spurring the private development of space travel, such as through Google's Lunar X Prize, announced in September 2007. It would award $20 million to the first team to land and successfully operate a lunar surface rover.

Page's impact on the Internet economy has been widely recognized. In 2004 he was inducted into the National Academy of Engineering for his work in developing the Google search engine. In 2005 *Time* included Page in its list of the world's 100 most influential people, and in 2007 *PC World* placed him at number one on a list of the most important people on the Web.

Further Reading
Battelle, John. *The Search: How Google and Its Rivals Rewrote the Rules of Business and Transformed Our Culture.* New York: Portfolio, 2006.

Carr, David F. "Brin, Page Show No Signs of Slowing Down." eWeek.com. March 15, 2007. Available online. URL: http://www.eweek.com/article2/0,1895,2104091,00.asp. Accessed October 27, 2007.

Google. Available online. URL: http://www.google.com. Accessed October 27, 2007.

Papert, Seymour
(1928–)
South African/American
Computer Scientist

Seymour Papert is an artificial intelligence pioneer and innovative educator who has brought computer science to a wider audience, especially young people.

Papert was born in Pretoria, South Africa, on March 1, 1928. He attended the University of Witwatersrand, earning his bachelor's degree in mathematics in 1949 and Ph.D. in 1952. As a student he became active in the movement against the racial apartheid system, which was becoming entrenched in South African society. His unwillingness to accept the established order and his willingness to be an outspoken activist would serve him well later when he took on the challenge of educational reform.

Papert went to Cambridge University in England and earned another Ph.D. in 1952, then did mathematics research from 1954 to 1958. During this period artificial intelligence, or AI, was taking shape as researchers began to explore the possibilities for using increasingly powerful computers to create or at least simulate intelligent behavior. In particular, Papert worked closely with another AI pioneer in studying neural networks and perceptrons (see MINSKY, MARVIN). These devices made electronic connections much like those between the neurons in the human

Seymour Papert has made it his life work to create computer facilities (such as the Logo language) that reflect the psychology of learning and enable even young students to experiment with "powerful ideas." (BILL PIERCE / TIME LIFE PICTURES / GETTY IMAGES)

brain. By starting with random connections and reinforcing appropriate ones, a computer could actually learn a task (such as solving a maze) without being programmed with instructions.

Papert's and Minsky's research acknowledged the value of this achievement, but in their 1969 book *Perceptrons* they also suggested that this approach had limitations, and that researchers needed to focus not just on the workings of brain connections but upon how information is actually perceived and organized.

This focus on cognitive psychology came together with Papert's growing interest in the process by which human beings assimilated mathematical and other concepts. From 1958 to 1963 he worked with Jean Piaget, a Swiss psychologist and educator. Piaget had developed a theory of learning that was quite different from that held by most educators. Traditional educational theory tended to view children as being incomplete adults who needed to be "filled up" with information.

Piaget, however, observed that children did not think like defective adults. Rather, children at each stage of development had characteristic forms of reasoning that made perfect sense in terms of the tasks at hand. Piaget believed that children best developed their reasoning skills by being allowed to exercise them freely and learn from their mistakes, thus progressing naturally.

In the early 1960s Papert went to the Massachusetts Institute of Technology, where he cofounded the MIT Artificial Intelligence Laboratory with Minsky in 1965. He also began working with children and developing computer systems better suited for allowing them to explore mathematical ideas.

The tool that he created to enable this exploration was the LOGO computer language (see LOGO). Logo provided a visual, graphical environment at a time when most programming resulted in long, hard-to-read printouts. At the center of the Logo environment is the "turtle," which can be either a screen cursor or an actual little robot that can move around on the floor, tracing patterns on paper. Young students can give the turtle simple instructions such as FORWARD 50 or RIGHT 100 and draw everything from squares to complicated spirals. As students continued to work with the system, they could build more complicated programs by writing and combining simple procedures. As they work, the students are exploring and grasping key ideas such as repetition and recursion (the ability of a program to call itself repeatedly).

In creating Logo, Papert believed that he had demonstrated that "ordinary" students could indeed understand the principles of computer science and explore the wider vistas of mathematics. But when he saw how schools were mainly using computers for rote learning, he began to speak out more about problems with the education system. Building on Piaget's work, Papert called for a different approach. Papert often makes a distinction between "instructivism," or the imparting of information to students, and "constructivism," or a student learning by doing.

Papert "retired" from MIT in 1998, but remains very active, as can be seen from the many Web sites that describe his work. Papert lives in Blue-Hill, Maine, where he teaches at the University of Maine. He has also established the Learning Barn, a laboratory for exploring innovative ideas in education. Papert has worked on ballot initiatives to have states provide computers for all their students, as well as working with teenagers in juvenile detention facilities. Today, educational centers using Logo and other ideas from Papert can be found around the world.

In December 2006, while attending a conference in Hanoi, Papert was struck by a motorcycle and suffered serious brain injuries. As of 2008 his rehabilitation is progressing well.

Papert has received numerous awards including a Guggenheim fellowship (1980), Marconi International fellowship (1981), the Software Publishers Association Lifetime Achievement Award (1994), and the Computerworld Smithsonian Award (1997).

Further Reading

Abelson, Harold, and Andrea DiSessa. *Turtle Geometry: The Computer as a Medium for Exploring Mathematics.* Cambridge, Mass.: MIT Press, 1981.

Harvey, Brian. *Computer Science Logo Style.* (3 vols.) 2nd ed. Cambridge, Mass.: MIT Press, 1997.

Papert, Seymour. *The Children's Machine: Rethinking School in the Age of the Computer.* New York: Basic Books, 1993.

———. *The Connected Family: Bridging the Digital Generation Gap.* Marietta, Ga.: Longstreet Press, 1996.

———. *Mindstorms: Children, Computers and Powerful Ideas.* 2nd ed. New York: Basic Books, 1993.

"Seymour Papert." Available online. URL: http://www.papert.org/. Accessed May 6, 2007.

parallel port

There are two basic ways to send data from a computer to a peripheral device such as a printer. A single wire can be used to carry the data one bit at a time (see SERIAL PORT), or multiple parallel wires can be used to send the bits of a data word or byte simultaneously.

Serial ports have the advantage of needing only one line (wire), but sending a byte (eight-bit word) requires waiting for each of the eight bits to arrive in succession at the destination. With a parallel connection, however, the eight bits of the byte are sent simultaneously, each along its own wire, so parallel ports are generally faster than serial ports. Also, since the data is transmitted simultaneously, the protocol for marking the beginning and end of each data byte is simpler. On the other hand, parallel cables are more expensive (since they contain more wires) and are generally limited to a length of 10 feet or so because of electrical interference between the parallel wires.

The original parallel interface for personal computers was designed by Centronics, and a later version of this 36-pin connector remains popular today. Later, IBM designed a 25-pin version. In addition to the wires carrying data, additional wires are used to carry control signals.

Most modern parallel ports use two more advanced interfaces, EPP (Enhanced Parallel Port) or ECP (Extended Capabilities Port). Besides allowing for data transmission up to 10 times faster than the original parallel port, these enhanced ports allow for bi-directional (two-way) communications. This means that a printer can send signals back to the PC indicating that it is low on toner, for example. Printer control software running on the PC can therefore display more information about the status of the printer and the progress of the printing job. Besides printers, the parallel interface has also been used to connect external CD-ROM and other storage devices.

Although early PCs often provided their parallel port connectors on plug-in expansion cards, most PCs today have two parallel connectors built into the motherboard. In recent years the faster and more flexible Universal Serial Bus (see USB) interface has increasingly replaced the parallel port for printers, scanners, digital cameras, external storage drives, and many other devices.

Further Reading

Parallel Port Central. Available online. URL: http://www.lvr.com/parport.htm. Accessed August 17, 2007.

"Parallel Port Configuration." Available online. URL: http://www.geocities.com/nozomsite/parallel.htm. Accessed August 17, 2007.

Tyson, Jeff. "How Parallel Ports Work." Available online. URL: http://computer.howstuffworks.com/parallel-port.htm. Accessed August 17, 2007.

parsing

Just as a speaker or reader of English must be able to recognize the significance of words, phrases, and other components of sentences, a computer program must be able to "understand" the statements, commands, or other input that it is called upon to process.

For example, an interpreter for the BASIC language must be able to recognize that

```
PRINT "End of Run"
```

contains a previously defined command or keyword (PRINT) and that the quote marks enclose a string of characters that are to be interpreted literally rather than standing for something else. Once the type of element or data item is recognized, then the appropriate procedure can be called upon for processing it. (See COMPILER and INTERPRETER.)

Similarly, a command processor (see SHELL) for an operating system such as UNIX will look at a line of input such as

```
ls -l /bin/MyProgs
```

and recognize that ls is an executable utility program. It will pass the rest of the command line to the ls program, which is then executed. In turn, ls must parse its command line and recognize that -l is a particular option that controls how the directory listing is displayed, and /bin/MyProgs is a pathname that specifies a particular directory location in the file system.

To parse its input, the language or command interpreter begins by looking in the program language or command statement for tokens. (This process is called lexical analysis.) A token is normally defined as a series of one or more characters separated by "whitespace" (blanks, carriage returns, and so on). A token is thus analogous to a word in English.

The series of tokens is then sent to the parser. The parser's job is to identify the significance of each token and to group the tokens into properly formed statements. Generally, the parser first checks the tokens for keywords—words such as "if" or "loop" that have a special meaning in a particular programming language. (In the BASIC example, PRINT is a keyword: In many other languages such functions are external rather than being part of the language itself.) As keywords (and punctuation symbols such as the semicolon used at the end of statements in C and Pascal) are identified, the parser uses a set of rules to determine the overall structure of the statement. For example, a language might define an if statement as follows:

```
If <Boolean-expression> then <statement>
else <statement>
```

This means that when the parser encounters an "if" it will expect to find between that word and "then" an expression that can be tested for being true or false (see BOOLEAN OPERATORS). Following "then," it will expect to find a complete statement. If it finds the optional keyword "else," that word will be followed by an alternative statement. Thus in the statement

```
If Total > Limit Print "Overflow" else Print
Total
```

The elements would be broken down as follows:

```
If               keyword

Total > Limit    Boolean expression

Print            keyword

"Overflow"       String literal (characters
                 to be printed)

else             keyword

Print            keyword

Total            variable
```

When writing a parser, the programmer depends on a precise and exhaustive description of the possible legal constructs in the language (see also BACKUS-NAUR FORM). In turn, these rules are turned into procedures by which the parser can construct a representation of the relationships between the tokens. This representation is often represented as an upside-down tree, rather like the sentence diagrams used in English class.

In general form, an expression, for example, can be diagrammed as consisting of one or more terms (variables, constants, or literal values) or other expressions separated by operators.

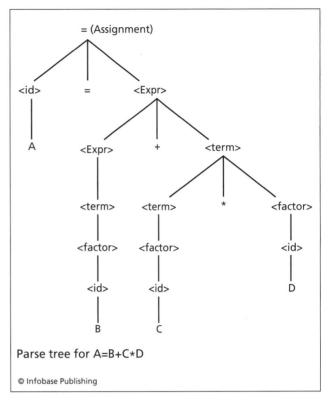

Parse tree for A=B+C*D

© Infobase Publishing

A parse tree for the statement A = B + C × D. Notice how the expression on the right-hand side of the equals (assignment) sign is eventually parsed into the component identifiers and operators.

Notice that these diagrams are often recursive. That is, the definition of an expression can include expressions. The number of levels that can be "nested" is usually limited by the compiler if not by the definition of the language.

The underlying rules must be constructed in such a way that they are not ambiguous. That is, any given string of tokens must result in one, and only one parse tree.

Once the elements have been extracted and classified, a compiler must also analyze the nonkeyword tokens to make sure they represent valid data types, any variables have been previously defined, and the language's naming conventions have been followed (see COMPILER).

Fortunately, people who are designing command processors, scripting languages, and other applications requiring parsers need not work from scratch. Tools such as YACC (a grammar definition compiler) and BISON and ANTLR (parser generators) are available for UNIX and other platforms.

Further Reading

Aho, Alfred V., et al. *Compilers: Principles, Techniques, & Tools.* 2nd ed. Boston: Pearson/Addison-Wesley, 2007.

Bowen, Jonathan P., and Peter T. Breuer. "Razor: The Cutting Edge of Parser Technology." Oxford University Computing Laboratory. Available online. URL: http://www.jpbowen.com/pub/toulouse92.pdf. Accessed August 17, 2007.

Donnelly, Charles, and Richard M. Stallman. *Bison Manual: Using the YACC-Compatible Parser Generator.* Boston, Mass.: GNU Press, 2003.

Levine, John R., Tony Mason, and Doug Brown. *lex & yacc*. Sebastapol, Calif.: O'Reilly, 1995.

Louden, Kenneth C. *Compiler Construction: Principles and Practice*. Boston: PWS Publishing, 1997.

Metsker, Steven John. *Building Parsers with Java*. Boston: Pearson/Addison-Wesley, 2001.

Pascal

By the early 1960s, computer scientists had become increasingly concerned with finding ways to better organize or structure programs. Indeed, one language (see ALGOL) had already been developed in part to demonstrate and encourage sound programming practices, including the proper use of control structures (see LOOP and STRUCTURED PROGRAMMING). However, Algol lacked a full range of data types and other features needed for practical programming, while arguably being too complex and inconsistent to serve as a good teaching language.

Niklaus Wirth at ETH (the Swiss Federal Institute of Technology) worked during the mid-1960s with a committee that was trying to overcome the problems with Algol and make the language more practical and attractive to computer manufacturers and users. However, Wirth gradually became disillusioned with the committee's unwieldy results, and proceeded to develop a new language, Pascal, announcing its specifications in 1970.

Pascal both streamlined Algol and extended it. Besides providing support for character, Boolean, and set data types, Pascal allows users to define new data types by combining the built-in types. This feature is particularly useful for defining a "record" type that, for example, might combine an employee's name and job title (characters), ID number (a long integer), and salary (a floating-point number). The rigorous use of data types also extends to the way procedures are called and defined (see PROCEDURES AND FUNCTIONS).

Pascal attracted much interest among computer scientists and educators by providing a well-defined language in which algorithms could be expressed succinctly. The acceptance of Pascal was also aided by its innovative compiler design. Unlike the machine-specific compilers of the time, the Pascal compiler did not directly create machine code. Rather, its output was "P-code," a sort of abstract machine language (see also PSEUDOCODE). A run-time system written for each computer interprets the P-Code and executes the appropriate machine instructions. This meant that Pascal compilers could be "ported" to a particular model of computer simply by writing a P-Code Interpreter for that machine. This strategy would be used more than two decades later by the creators of a popular language for Web applications (see JAVA).

STRUCTURE OF A PASCAL PROGRAM

The following simple program illustrates the basic structure of a program in Pascal. (The words in bold type are keywords used to structure the program.) The program begins with a Type section that declares user-defined data types. These can include arrays, sets, and records (composite types that can include several different basic types of data). Here an array of up to 10 integers (whole numbers) is defined as a type called IntList.

The Var (variable) section then declares specific variables to be used by the program. Variables can be defined using either the language's built-in types (such as integer) or types previously defined in the Type section. An important characteristic of Pascal is that user-defined types must be defined before they can be used in variable declarations, and variables in turn must be declared before they can be used in the program. Some programmers found this strictness to be confining, but it guards against, for example, a typographical error introducing an undefined variable in place of the one intended. Today most languages enforce the declaration of variables before use.

The word *begin* introduces the executable part of the program. The variables needed for the loop are first initialized by assigning them a value of zero. Note that in Pascal := (colon and equals sign) is used to assign values. The outer *if* statement is used to ensure that the user does not input an invalid number of items. The *for* loop then reads each input value, assigns it to its place in the array, and keeps a running total. That total is then used to compute the average, which is output by the writeln (write line) statement.

```
program FindAvg (input, output);
    type IntList = array [1 . . 10] of integer;
    var
        Ints: IntList;
        Items, Count, Total, Average: integer;
begin
    Average := 0;
    Total := 0;
    Readln (Items);
    If ((Items > 0) and (Items <= 10)) then
        begin
        for Count := 1 to Items do
            begin
            readln (Ints [Count]);
            Total := Total + Ints [Count]
            end;
        Average := Total / Items;
        Writeln ('The average of the items is:',
        Average)
        end
else
        Writeln ('Error: Number of items must be
        between 1 and 10')
end.
```

IMPACT OF PASCAL

Pascal achieved modest commercial success. The P-Code idea was embraced by the UCSD P-System developed by the University of California at San Diego. In the late 1970s and early 1980s, the P-System brought the benefits of Pascal's structured programming to users of computers such as the Apple II, for which the only alternatives had been machine-language or a poorly structured version of BASIC. Later in the 1980s, Borland International came out with

Turbo Pascal. This compiler used direct compilation rather than P-Code, sacrificing portability for speed and efficiency. It included an integrated programming environment that made development much cheaper and easier than with existing "bulky" and expensive compilers such as those from Microsoft. Turbo became very popular and eventually included language extensions that supported object-oriented programming. But Pascal became best known in its role as a first language for teaching programming and for expressing algorithms.

However, by 1990 the tide had clearly turned in favor of C and C++. These languages used a more cryptic syntax than Pascal and lacked the latter's rigorous data typing mechanism. Systems programmers in particular preferred C's ability to get "close to the machine" and manipulate memory directly without being confined by type definitions. C had also received a big boost because its developers were also among the key developers of UNIX, a very popular operating system in campus computing environments.

During the 1990s, C, C++, and Java even began to supplant Pascal for computer science instruction. Nevertheless, by encouraging structured programming concepts and helping educate a generation of computer scientists, Pascal made a lasting impact on the computer field. Wirth continued his work with the development of Modula-2 and Oberon, which were confined mainly to the academic world. However, Pascal also was a major influence on the development of Ada, a language endorsed by the U.S. federal government that combines structured programming with object-oriented features and the ability to manage extensive packages of routines (see ADA).

Further Reading

Free Online Pascal and Delphi Tutorials and Documentation. Available online. URL: http://www.thefreecountry.com/documentation/onlinepascal.shtml. Accessed August 17, 2007.

Free Pascal Compiler. http://www.freepascal.org/

Jensen, Kathleen, Niklaus Wirth, and A. Mickel. *Pascal User Manual and Report: ISO Pascal Standard.* 4th ed. New York: Springer-Verlag, 1991.

Koffman, Elliot B. *Turbo Pascal.* 5th update ed. Reading, Mass.: Addison-Wesley, 1997.

Rachele, Warren. *Learn Object Pascal with Delphi.* Plano, Tex.: Wordware Publishing, 2000.

Wirth, Niklaus. *Programming in Modula-2.* 3rd, corr. ed. New York: Springer-Verlag, 1985.

pattern recognition

After many years of effort researchers have been able to create systems that can recognize particular human faces (see COMPUTER VISION). On the other hand, any normal six-month-old child can effortlessly recognize familiar faces (such as parents). The fundamental task of turning raw data (whether from senses, instruments, or computer files) into recognizable objects or drawing inferences is called pattern recognition. Pattern recognition is at the heart of many areas of research and application in computing (see ARTIFICIAL INTELLIGENCE and DATA MINING). Despite the challenge in getting machines to do what comes naturally for biological organisms, the potential payoffs are immense.

A pattern-recognition system begins with data, whether stored or real-time (such as from a robot's camera). The first task in turning potentially billions of bytes of data into meaningful objects is to extract features from what is likely a high proportion of redundant or irrelevant data. (With visual images, this often involves finding edges that define shapes.) The extracted features are then classified to determine what objects they might represent. This can be done by comparing structures to templates or previously classified data or by applying statistical analysis to determine the likely correlation of the new data to existing patterns (see BAYESIAN ANALYSIS).

Pattern recognition often includes learning algorithms as well; indeed, the field is often considered to be a subtopic of machine learning. For example, classification systems can be refined by "training" them and reinforcing successful determinations (see NEURAL NETWORK).

APPLICATIONS

There are numerous applications of pattern recognition, often as part of intelligent systems used in such areas as linguistics (see LANGUAGE TRANSLATION SOFTWARE), communications, intelligence and surveillance, identity verification (see BIOMETRICS), and the analysis of credit card transaction patterns for signs of fraud. Some examples are shown in the following table:

DATA	PROCEDURES	RESULTS
speech	phonemes, transition rules	text
handwritten address	character classification	identified postal address
handwritten check (ATM)	character classification	identify amount of deposit
general text	grammar and syntax	structure and meaning
e-mail	identify characteristics Bayesian filter	spam detection
facial image	feature templates, statistics	identified person
biometric (retina, fingerprint, etc.)	feature extraction and template comparison	verified identity

Further Reading

Bishop. Christopher M. *Pattern Recognition and Machine Learning.* New York: Springer, 2006.

Duda, Richard O., Peter E. Hart, and David G. Stork. *Pattern Classification.* New York: Wiley, 2001.

International Association of Pattern Recognition. Available online. URL: http://www.iapr.org/. Accessed November 3, 2007.

Pattern Recognition. American Association for Artificial Intelligence. Available online. URL: http://www.aaai.org/AITopics/html/pattern.html. Accessed November 4, 2007.

Recognition Technology and Pattern Analysis. Available online. URL: http://alumnus.caltech.edu/~dave/pattern.html. Accessed November 4, 2007.

Theodoridis, Sergios, and Konstantinos Koutroumbas. *Pattern Recognition*. 3rd ed. San Diego, Calif.: Academic Press, 2006.

PDA (personal digital assistant)

The first stage in making computing available away from the office desk was the development of "portable" and then laptop computers in the 1980s (see LAPTOP COMPUTER). Laptops, however, are relatively heavy and bulky, and thus not suitable for activities such as making notes at meetings or keeping track of appointments while on the go. The logical solution to that need was to develop a computer small enough to carry in a pocket or purse. The first handheld computer to achieve widespread recognition was Apple's Newton, which the company referred to as a "personal digital assistant." This term, usually abbreviated to PDA, became a generic category with the introduction of the Palm Pilot, which first appeared in 1996, followed by the seemingly ubiquitous RIM Blackberry in 1999.

FEATURES AND USES

Modern PDAs have sharp, readable displays, even given the limited screen size. The role of the mouse is taken by navigation buttons, and the ability to select items on the screen by touch or using a stylus (see TOUCHSCREEN). (Some PDAs include small keyboards that can be typed on using two fingers or thumbs.) The operating system (such as Palm OS, Windows Mobile, or even Linux) is in read-only memory, and working memory is provided, expandable through the use of SD (Secure Digital) or Compact Flash memory cards. Wireless connectivity provides access to the Internet and for transferring data between the PDA and a regular PC (see BLUETOOTH and WIRELESS AND MOBILE COMPUTING). A synchronization program installed on the PC can be used to ensure that the latest version of each file will be stored on both devices. (This also allows larger programs on the PC to work on and update data from the PDA—see PERSONAL INFORMATION MANAGER.)

Typical PDA applications include an appointment calendar, address book for contacts, a simple note-taking program (see HANDWRITING RECOGNITION), and increasingly, e-mail and a special Web browser designed for small displays. Many PDAs can also use their Bluetooth connection to place calls through suitably equipped cellphones.

PDAs can also be used for specialized applications that involve the need to receive or update data while driving or walking. Examples include navigation (with the use of a GPS device), delivery services, warehouse inventory management, reading utility meters, taking orders electronically in restaurants, and maintaining patient records in hospitals. Besides allowing for the recording of data, PDAs can also include task-specific references such as prescription drug databases or a medical dictionary.

As with many other things in computing, the boundaries of the PDA category are becoming more fluid. While software-enhanced mobile phones evolved separately (see SMARTPHONE), increasingly PDA functions and telephony are being seamlessly integrated into a single device, as with Palm's Treo and especially Apple's 2007 introduction of the iPhone, which also introduced an innovative "multitouch" interface that can respond to natural finger gestures such as flicking, sliding, or pinching.

Further Reading

Al-Ubaydli, Mohammad. *The Doctor's PDA and Smartphone Handbook: A Guide to Handheld Healthcare*. London: Royal Society of Medicine Press, 2006.

Carlson, Jeff, and Agen G. N. Schmitz. *Palm Organizers*. 4th ed. Berkeley, Calif.: Peachpit Press, 2004.

Hormby, Tom. "Early History of Palm." Silicon User. Available online. URL: http://siliconuser.com/?q=node/17. Accessed November 4, 2007.

Kao, Robert, and Dante Sarigumba. *BlackBerry for Dummies*. New York: Wiley, 2007.

Palm. Available online. URL: http://www.palm.com/us/. Accessed November 4, 2007.

PDA Reviews. Brighthand. Available online. URL: http://www.brighthand.com/. Accessed November 4, 2007.

RIM (Research in Motion). Available online. URL: http://www.rim.com/. Accessed November 4, 2007.

PDF (portable document format)

The PDF (portable document format) created by Adobe Systems has become a very common way to make documents available in a way that preserves the appearance of the original.

When PDF first came out in the early 1990s it was not very suitable for use on the Web. PDF documents could only be viewed using expensive proprietary software, they could not include embedded links (and thus could not be hypertext), and they were large enough to be slow for downloading on the dial-up connections of the time.

All this had changed by the end of the decade: Adobe distributes the free Adobe Reader and plug-ins for all major platforms and browsers.

OPERATION

The PDF specifications are open source, so anyone can write software to create or read documents in the format. PDF includes three elements: a subset of the PostScript page description language (see POSTSCRIPT), a system for specifying and embedding common fonts (or referring to other fonts), and a system for "packaging" the text and graphics descriptions into a file in compressed form. Later versions of the PDF specification also allow users to interact with the document, such as by filling in fields in a form or adding annotations to the text. PDF also includes support for tags (see XML) and descriptors that can be used with programs such as screen readers for the blind.

PDF also includes support for encrypting documents so they can only be read with a password, and for controlling whether the document can be copied or printed, though this depends on the user's software understanding and obeying the restrictions.

Although creating and editing PDF documents originally required the relatively expensive Adobe Acrobat software, there are now a number of free or low-cost editors and other PDF utilities for Windows, Mac OS X, and Linux/UNIX platforms.

Further Reading

Adobe Creative Team. *Adobe Acrobat 8 Classroom in a Book.* San Jose, Calif.: Adobe Press, 2007.

Adobe Reader (free download). Available online. URL: http://www.adobe.com/products/acrobat/readstep2_allversions.html. Accessed November 4, 2007.

Lowagie, Bruno. *iText in Action: Creating and Manipulating PDF.* New York: Manning Publications, 2007.

Padova, Ted. *Adobe Acrobat 8 PDF Bible.* Indianapolis: Wiley, 2007.

Planet PDF: The Home of the PDF Community. Available online. URL: http://www.planetpdf.com/. Accessed November 4, 2007.

Perl

The explosive growth of the World Wide Web has confronted programmers with the need to find ways to link databases and other existing resources to Web sites. The specifications for such linkages are found in the Common Gateway Interface (see CGI). However, the early facilities for writing CGI scripts were awkward and often frustrating to use.

Back in 1986, UNIX developer Larry Wall had created a language called Perl (Practical Extraction and Report Language). There were already ways to write scripts for simple data processing (see SCRIPTING LANGUAGES) as well as a handy pattern-manipulation language (see AWK). However, Wall wanted to provide a greater variety of functions and techniques for finding, extracting, and formatting data. Perl attracted a following within the UNIX community. Since much Web development was being done on UNIX-based systems by the mid- and late-1990s, it was natural that many webmasters and applications programmers would turn to Perl to write their CGI scripts.

As with many UNIX scripting languages, Perl's syntax is broadly similar to C. However, the philosophy behind C is to provide a sparse core language with most functionality being handled by standard or add-in program libraries. Perl, on the other hand, starts with most of the functionality of UNIX utilities such as sed (stream editor), C shell, and awk, including the powerful regular expressions familiar to UNIX users. The language also includes a "hash" data type (a collection of paired keys and values) that makes it easy for a program to maintain and check lists such as of Internet hosts and their IP addresses (see HASHING).

Wall made it a point to solicit and respond to feedback from Perl users, often by adding features or functions. Wall's approach has been to provide as much practical help for programmers as possible, rather than worrying about the language being well-defined, consistent, and thus easy to learn. For example, in most languages, to make something happen only if a certain condition is not true, one writes something like this:

```
If ! (test for valid data)
   Print Error-Msg;
Else Process_Data;
```

In Perl, however, one can use the "unless" clause. It looks like this:

```
Unless (Test for invalid data) {
   Process_Data;
}
```

Syntactically, the unless clause does not provide anything more than using an If and Else would, and it involves learning a different structure. However, it has the practical benefit of making the program a little easier to read by keeping the emphasis on what the program expects to be doing, not on the possible error. Similarly, Perl offers an "until" loop:

```
Until (Condition is met) {
   Do something;
}
```

In C, one would have to say

```
While (Condition is not met) {
   Do something;
}
```

This "Swiss army knife" approach to providing language features has been criticized by some computer scientists as encouraging undisciplined and hard-to-verify programming. However, Perl's many aficionados see the language as the versatile, essential toolbox for the ever-challenging world of Web programming. As the language evolved through the late 1990s, it also added a full set of object-oriented features (see OBJECT-ORIENTED PROGRAMMING).

SAMPLE PERL PROGRAM

The following very simple code illustrates a Perl program that reads some lines of data from a file and prints them out. The first line tells UNIX to execute the Perl interpreter. The file name data.txt is assigned to the string variable $file. The file is then opened and assigned to the variable INFO. A single statement (not a loop) suffices to assign all the lines in the file to the array @lines. The "foreach" statement is a compact form of For loop that assigns each line in the array to the string variable $line and then prints it to the screen as HTML.

```
#!/usr/local/bin/perl
$file = 'data.txt';
open(INFO, "<$file" ) ;
@lines = <INFO> ;
foreach $line (@lines)
{
   print "\n <P> $line </P>" ;
}
# DONE
```

Although now somewhat overshadowed by newer scripting languages for Web development (see, for example, PYTHON and PHP), Perl is a mature technology in widespread use, particularly for data extraction, conversion, and manipulation. The Comprehensive Perl Archive Network (CPAN) has over 12,000 modules that are freely available to programmers.

Further Reading

Comprehensive Perl Archive Network (CPAN). Available online. URL: http://www.cpan.org/. Accessed August 17, 2007.

Conway, Damian. *Perl Best Practices*. Sebastapol, Calif.: O'Reilly, 2005.

Lee, James. *Beginning Perl*. 2nd ed. Berkeley, Calif.: Apress, 2004.

Schwartz, Randall L., Tom Phoenix, and Brian D. Foy. *Learning Perl*. 4th ed. Sebastapol, Calif.: O'Reilly, 2005.

Wall, Larry, Tom Christiansen, and Jon Orwant. *Programming Perl*. 3rd ed. Sebastapol, Calif.: O'Reilly, 2000.

personal computer (PC)

The development of the "computer chip" (see MICROPROCESSOR) and the increasing use of integrated circuit technology made it possible by the mid-1970s to begin to think about designing small computers as office machines or consumer devices that could be individually owned or used. In about a decade the personal computer, or PC, would become well established in many businesses and a growing number of homes. After another decade, it became almost as ubiquitous as TV sets and microwaves. Parallel developments in hardware, software, operating systems, and accessory devices made this revolution possible.

The first commercial "personal computer" was the MITS Altair, a microcomputer kit built around an Intel 8080 microprocessor. Building the kit required considerable skill with electronics assembly, but enthusiasts (including a young Bill Gates) were soon writing software and designing add-on modules for the kit (see GATES, WILLIAM). A variety of publications, notably *Byte* magazine, as well as the Homebrew Computer Club gave hobbyists a forum for sharing ideas.

By the late 1970s, personal computing was starting to become accessible to the general public. The Altair enthusiasts had moved on to more powerful systems that offered such amenities as floppy disk drives and an operating system (CP/M, developed by Gary Kildall). Meanwhile, less technically experienced people could also begin to experiment with personal computing, thanks to the complete, ready-to-run PCs being offered by Radio Shack (TRS-80), Commodore (Pet), and in particular, the Apple II.

In order to make serious inroads into the business world, however, the PC needed useful, reliable software. WordStar and later WordPerfect made it possible to replace expensive special-purpose word processing machines (such as those made by Wang) with the more versatile PC. One of the biggest spurs to business use of PCs, however, was an entirely new category of software—the spreadsheet. Dan Bricklin's VisiCalc (see SPREADSHEET) would make the PC attractive to accountants and corporate planners.

The watershed year in personal computing was 1981 because it brought the computer giant IBM into the PC arena (see IBM PC). The IBM PC had a somewhat more powerful processor and could hold more memory than the Apple II, but its main advantage was that it was backed by IBM's decades-long reputation in office machines. Businesses were used to buying IBM products, and conversely, many corporate buyers believed that if IBM was offering desktop computers, then PCs must be useful business machines.

IBM (like Apple) had adopted the idea of open architecture—the ability for third companies to make plug-in cards to add functions to the machine. Thus, the IBM PC became the platform for a burgeoning hardware industry. Further, it turned out that other companies could reverse-engineer the internal code that ran the system hardware (see BIOS) without infringing IBM's legal rights. This meant that companies could make "clones" or IBM-compatible machines that could run the same software as the genuine IBM PC. The first clone manufacturers (such as Compaq) sometimes improved upon IBM such as by offering better graphics or faster processors. However, by the late 1980s the trend was toward companies competing through lower prices for roughly equivalent performance. Facing a declining market share, IBM tried to introduce a new architecture, called *microchannel,* that provided a mainframelike bus architecture for more efficient input/output control. However, whatever technical advantages the new system (called PS/2) might have, the market voted against it by continuing to buy the ever more powerful clones built on the original IBM architecture.

Lower prices and more attractive options led to a growing number of users, which in turn encouraged greater investment in software development. By the mid-1980s, Lotus (headed by Mitch Kapor) dominated the spreadsheet market with its Lotus 1-2-3, while WordPerfect dominated in word processing.

However, Microsoft, whose MS-DOS (or PC-DOS) had become the standard operating system for IBM-compatible PCs, introduced a new operating environment with a graphical user interface (see MICROSOFT WINDOWS). By the mid-1990s, Windows had largely supplanted DOS. Microsoft also committed resources and exploited its intimate knowledge of the operating system to achieve dominance in office software through MS Word, MS Excel (spreadsheet), and MS Access (database).

At the margins Apple's Macintosh (introduced in 1984 and steadily refined) has retained a significant following, particularly in education, publishing, and graphic arts applications (see MACINTOSH). Although Windows now provides a similar user interface, Mac enthusiasts believe their machine is still easier to use (and more stylish), and often see it as a badge for those who "think different."

PC TRENDS

When graphical Web browsing made the Internet widely accessible in the mid-1990s, the demand for PCs increased accordingly. The desire for e-mail, Web browsing, and help with children's homework led many families to purchase their first PCs. By 2000, about two-thirds of American children had access to computers at home, and virtually all schools had at least some PCs in the classroom. Using sophisticated manufacturing and order processing systems, companies such as Dell and Gateway sell PCs directly to consumers and businesses, largely displacing the neighborhood computer store. These efficiencies (and lower prices for memory, processors, and other hardware) have brought the cost for a basic home PC down to less than $500, while the capabilities available for those willing to spend $1,500 or so continued to increase. PC users now expect to be able to play CD- and DVD-based multimedia while hearing good quality sound.

A number of challenges to the growth of the PC industry have also emerged. As more and more of the activity of PC users began to focus on the Internet, some companies began to host office applications on servers (see APPLICATION SERVICE PROVIDER). Some pundits began to say that with applications being moved to remote servers or offered over the corporate LAN, the PC on the desk could be stripped down considerably. The "network PC" could make do with a slower processor, less memory, and no hard drive, since all data could be stored on the server.

Generally, however, the attempts to supplant the full-featured, general-purpose PC have made little progress. One reason is that the cost of complete PC has declined so much that the supposed cost savings of a network PC or Internet appliance have become less significant. Further, privacy issues and the desire of people to have control over their own data are often cited as arguments in favor of the PC.

Ironically, the PC industry's greatest challenge may come from its very success. As more and more households in the United States and other developed countries have PCs, it becomes harder to maintain the sales rate. By the early 2000s, the power of recent PCs had become so great that the desire to upgrade every few years may have become less compelling and the recent economic downturn has hit the computer industry particularly hard. So far it looks like the fastest-growing areas in computer hardware no longer involve the traditional desktop PC, but handheld (palm) computers (see PDA) and the embedding of more powerful computer capabilities into other machines such as automobiles (see EMBEDDED SYSTEMS).

Further Reading

Freiberger, Paul, and Michael Swaine. *Fire in the Valley: The Making of the Personal Computer.* New York: McGraw-Hill, 1999.
Long, Larry. *Personal Computing Demystified.* Emeryville, Calif.: McGraw-Hill/Osborne, 2004.
Polsson, Ken. "Chronology of Personal Computers." Available online. URL: http://www.islandnet.com/~kpolsson/comphist/. Accessed August 17, 2007.
Thompson, Robert, and Barbara Fritchman Thompson. *Building the Perfect PC.* Sebastapol, Calif.: O'Reilly, 2006.
———. *Repairing and Upgrading Your PC.* Sebastapol, Calif.: O'Reilly, 2006.
White, Ron, and Timothy Edward Downs. *How Computers Work.* 8th ed. Indianapolis: Que, 2005.

personal health information management

Health care is at once a complex endeavor with many players, a vast industry, and a major expense of individuals, businesses, and governments. At the center of it all stands the prospective patient (or consumer) seeking to maintain or restore health.

While the health care industry has long been a major user of computer technology (see MEDICAL APPLICATIONS OF COMPUTERS), the modern Web has brought a variety of services (many free) that can help health care consumers learn more about conditions and treatments and compare hospitals, doctors, and other providers.

MEDICAL INFORMATION SITES

In today's health care environment patients often have only a few minutes to ask their doctor important questions about their condition and possible treatments. Patients often feel they have been left on their own when it comes to obtaining detailed information. According to surveys by the Pew Internet & American Life Project, by the end of 2005 about 20 percent of Web users were reporting that the Internet "has greatly improved the way they get information about health care." Further, 7 million users had reported that Web sites had "played a crucial or important role in coping with a major illness."

A variety of Web sites ranging from comprehensive and excellent to dubious (at best) offer health-related information. In evaluating them, it is important to determine who sponsors the site and what is the source of the information provided. The very extensive WebMD site, for example, is reviewed for accuracy by an independent panel of experts. One of the foremost medical institutions, the Mayo Clinic, also has an authoritative site. The site OrganizedWisdom. com offers a search engine that emphasizes information that has been reviewed by doctors for accuracy, while eliminating low-quality or duplicative results.

Even if information is accurate, however, users may often lack the necessary background or context for interpreting it correctly. Understanding the results of medical studies, for example, requires some knowledge of how studies are designed, the population used, and the statistical significance and applicability of the results. As a practical matter, therefore, patients should not make any major decisions about diet, medication, or treatment options without consulting a medical professional. Attempts at self-diagnosis can be particularly problematic.

SUPPORT GROUPS AND PROVIDER RATINGS

On the other hand, carefully chosen online information can be very useful and can even improve outcomes. Patients can learn what questions to ask their physicians, and may even be able to suggest relevant information of which the physician is unaware.

During treatment, patients can find emotional and practical support online. In keeping with the trend toward online social cooperation (see SOCIAL NETWORKING and USER-CREATED CONTENT) a number of sites are helping consumers find or create support groups. Such groups have long been important, particularly for patients with conditions such as cancer or serious chronic disease. For example, DailyStrength.org offers 500 online support groups for a great variety of conditions. Users can create online journals to describe their daily struggles and can send supportive messages and "hugs." According to a 2007 report by the Pew Internet & American Life Project, about half of adults with chronic conditions use the Internet regularly and extensively to help them manage their treatment and life issues.

Selecting a compatible medical professional is another area where online sites can help prospective patients. User ratings have proven helpful on Amazon.com and other sites for a variety of products and services (for example, Yelp. com and the popular Angie's List). A site called RateMDs.

com has applied the same mechanism to allow patients to anonymously rate their doctors. (As with other user-provided reviews, however, one needs to be aware of the possibility that the reviews do not constitute a representative sample of consumer experience.) Patients can also personally share their experiences via a YouTube-like site called ICYou.com.

Although social networking and content-sharing sites have been most popular among the younger generation, the increasing adoption of these venues by older adults and seniors is likely to fuel growth in online health-related services in years to come, as is the continuing need to find cost-effective ways of serving growing patient populations.

Further Reading

Colliver, Victoria. "For These Startups, Patients Are a Virtue." *San Francisco Chronicle*, October 1, 2007, p. C1. Available online. URL: http://www.sfgate.com/cgi-bin/article.cgi?f=/c/a/2007/10/01/BUDKSGAF4.DTL. Accessed November 5, 2007.

Cullen, Rowena. *Health Information on the Internet: A Study of Providers, Quality, and Users.* Westport, Conn.: Praeger, 2006.

DailyStrength. Available online. URL: http://dailystrength.org/. Accessed November 5, 2007.

Fox, Susannah. "E-Patients with a Disability or Chronic Disease." Pew Internet & American Life Project, October 8, 2007. Available online. URL: http://www.pewinternet.org/pdfs/EPatients_Chronic_Conditions_2007.pdf. Accessed November 6, 2007.

Lewis, Deborah, et al., eds. *Consumer Health Informatics: Informing Consumers and Improving Health Care.* New York: Springer, 2005.

Madden, Mary, and Susannah Fox. "Finding Answers Online in Sickness and in Health." Pew Internet & American Life Project. Available online. URL: http://www.pewinternet.org/pdfs/PIP_Health_Decisions_2006.pdf. Accessed November 6, 2007.

Mayo Clinic. Available online. URL: http://www.mayoclinic.com/. Accessed November 5, 2007.

Organized Wisdom. Available online. URL: http://organizedwisdom.com/. Accessed November 5, 2007.

RateMDs.com. Available online. URL: http://www.ratemds.com. Accessed November 5, 2007.

WebMD. Available online. URL: http://www.webmd.com/. Accessed November 5, 2007.

personal information manager (PIM)

A considerable amount of the working time of most businesspeople is taken up not by primary business tasks but in keeping track of contacts, phone conversations, notes, meetings, deadlines, and other information needed to plan or coordinate activities. Software designers have responded to this reality by creating software to help manage personal information.

Early PC users improvised ways of using available software applications for tracking their activities. For example, a spreadsheet with text fields might be used to record and sort contacts and their associated information such as phone numbers or data could be organized in tables in word processor documents. However, such improvisations can be awkward to use. Loading a full-sized word processor or spreadsheet application takes time (and until Windows and other multitasking solutions came along, only one pro-

gram could be run at a time). Further, it is hard to integrate information or keep track of the "big picture" with several different kinds of information stored in different formats with different programs.

What was needed was a single application that could integrate the personal information and make it accessible without the user having to shut down the main application program. The first successful PIM was Borland Sidekick, first released in 1984. Although MS-DOS was designed to run only a single program at a time, it had an obscure feature that allowed additional small programs to be loaded into memory where they could be triggered using a key combination. Taking advantage of this feature, Sidekick allowed someone while using, for example, a word processor, to pop up a note-taking window, an address book, calendar, telephone dialer, calculator, or other features. When Microsoft Windows replaced DOS, it became possible to run more than one full-fledged application at a time. PIMs could then become full-fledged applications in their own right, and offer additional features.

As e-mail became more common on local networks in the later 1980s and via the Internet in the 1990s, PIM features began to be integrated with e-mail programs such as Microsoft Outlook and Netscape Navigator's communications facilities. New features included the automatic creation of journal entries from various activities and the creation of "rules" for recognizing and routing e-mail messages with particular senders or subjects. A variety of freeware and shareware PIMs are available for users who want an alternative to the commercial products, and a number of PIMs are available for Macintosh and Linux-based systems. Web-based personal information management tools can make it particularly easy to coordinate a widely scattered workforce, since each user merely has to access the serving Web site. Recently, low-cost (or even free) Web-based applications that include PIM as well as productivity features have been introduced—for example, Google Apps.

The growth of handheld (or palm) computers (see PDA) and more sophisticated cell phones has created a need to provide PIM features for these devices (see SMARTPHONE). Since the capacity of handheld devices is limited compared to desktop PCs, there is also a need for software to allow easy transfer of information between portable devices and desktop PCs. This can be done with a serial, USB, or even wireless connection.

In the future, the PIM is likely to become an integrated system that operates on a variety of handheld and desktop devices and seamlessly maintains all information regardless of how it is received. There will also be greater ability to give voice commands (such as to dial a person or to ask for information about a contact), and to have messages read aloud (see SPEECH RECOGNITION AND SYNTHESIS). The software is also likely to include sophisticated "agents" that can be instructed to carry out such tasks as prioritizing messages or returning routine calls (see SOFTWARE AGENT).

Further Reading

Boyce, Jim, Beth Sheresh, and Doug Sheresh. *Microsoft Outlook 2007 Inside Out.* Redmond, Wash.: Microsoft Press, 2007.

Jones, William. *Personal Information Management.* Seattle: University of Washington Press, 2007.

Lineberger, Michael. *Total Workday Control: Using Microsoft Outlook.* San Ramon, Calif.: New Academy Publishers, 2006.

"Over 160 Free Personal Information Managers." Available online. URL: http://www.lifehack.org/articles/lifehack/over-160-free-personal-information-managers.html. Accessed August 17, 2007.

philosophical and spiritual aspects of computing

When modern digital computing emerged in the 1940s, it evolved from two roots: engineering (particularly electrical engineering) and mathematics. The goals of the earliest computer designers were focused naturally enough on computing, although several early thinkers (see BUSH, VANNEVAR; SHANNON, CLAUDE; and TURING, ALAN) had already begun to think of computers as symbol-processing and knowledge-retrieving machines, not just number crunchers.

As computer scientists began to become more concerned about the structure of data and the modeling of real-world objects in computer languages (see OBJECT-ORIENTED PROGRAMMING), they began to wrestle with some areas long familiar to philosophers. As data structure involved into knowledge representations, epistemology (the philosophical investigation of the meaning and accessibility of knowledge) became more relevant, particularly in developing systems for artificial intelligence and machine learning. Also relevant is ontology (the nature and relationship of entities—see ONTOLOGIES AND DATA MODELS), particularly with regard to the modern effort to encode relationships between items of knowledge into Web pages (see SEMANTIC WEB).

THE COMPUTER AS PHILOSOPHICAL LABORATORY

Beyond investigating the potential for applying philosophical ideas to knowledge engineering, many philosophers have also taken increasing notice of the possibilities that artificial intelligence, highly complex dynamic structures (particularly the Internet), and human-computer interaction offer for investigating long-standing and often seemingly intractable philosophical problems.

One of the knottiest problems is the nature of something that people experience during every waking moment—consciousness, that awareness of being an "I" or "self" that is experiencing both an inner world of memories and thoughts and the outer world conveyed by the senses. One reason why the problem of consciousness is so difficult to resolve is that cognitive scientists and philosophers lack the ability to compare human consciousness with other possible consciousness. (Some "higher" animals may be conscious in some sense, but they cannot tell us about it.) However, as AI programs attempt to model aspects of human cognition, they can help us find similarities and possible differences between the way computers and people "think." Of course philosophers take a wide variety of positions on the question of whether there is anything ultimately distinctive about what we call consciousness, and whether comput-

ers or robots might someday become truly conscious. (For examples of differing views see DREYFUS, HUBERT; KURZWEIL, RAYMOND; and MCCARTHY, JOHN.)

Finally, a number of writers have related developments in modern computing to ultimate philosophical or spiritual concerns. For example, the World Wide Web can be compared to the world-girdling "noosphere" of evolving knowledge described by theologian-paleontologist Pierre Teilhard de Chardin in the mid-20th century. Thus there has been considerable speculation (and perhaps hype) about a new form of collective consciousness emerging through the interaction of people as well as increasingly intelligent programs on the Net. On the other hand, the experience of immersive online environments (see ONLINE GAMES and VIRTUAL REALITY) revisits a question that goes back to Descartes in the 17th century—whether what we perceive as reality might actually be an illusion—and this question resonates with the works of Western Gnostics and Eastern Buddhists, not to mention Hollywood's *The Matrix.*

The dialog among philosophy, spiritual practice, and the rapidly changing computer world is likely to remain fascinating.

Further Reading

Davis, Erik. *Techgnosis: Myth, Magic, and Mysticism in the Age of Information.* New York: Harmony Books, 1998.

Floridi, Luciano. *Philosophy and Computing: An Introduction.* New York: Routledge, 1999.

———, ed. *Philosophy of Computing and Information.* Malden, Mass.: Blackwell, 2004.

Foerst, Anne. *God in the Machine: What Robots Teach Us about Humanity and God.* New York: Dutton, 2004.

Hayles, N. Katherine. *How We Became Posthuman: Virtual Bodies in Cybernetics, Literature, and Informatics.* Chicago: University of Chicago Press, 1999.

International Association for Computing and Philosophy. Available online. URL: http://www.ia-cap.org/. Accessed November 6, 2007.

Irwin, William. *The Matrix and Philosophy.* Chicago: Open Court, 2002.

Tetlow, Philip. *The Web's Awake: An Introduction to the Field of Web Science and the Concept of Web Life.* Hoboken, N.J.: Wiley, 2007.

phishing and spoofing

Just about anyone with an e-mail account has received messages purporting to be from a bank, a popular e-commerce site such as Amazon or eBay, or even a government agency. Typically the message warns of a problem (such as a suspended account) and urges the recipient to click on a link in the message. If the user does so, what appears to look like the actual site of the relevant institution is actually a "spoof," or fake site. If the user goes on to enter information such as account numbers or passwords in order to fix the "problem," the information actually goes to the operator of the fake site, where it can be used for fraudulent purchases or even impersonation (see IDENTITY THEFT). The bogus site can also attempt to download viruses, spyware, keyloggers, or other forms of "malware" to the unwitting user's computer.

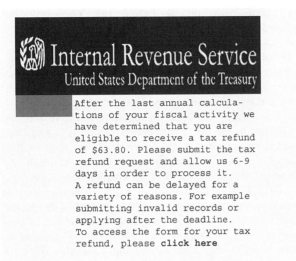

After the last annual calculations of your fiscal activity we have determined that you are eligible to receive a tax refund of $63.80. Please submit the tax refund request and allow us 6-9 days in order to process it.
A refund can be delayed for a variety of reasons. For example submitting invalid records or applying after the deadline.
To access the form for your tax refund, please **click here**

Regards,
Internal Revenue Service

"Phishing" messages such as this fake IRS e-mail try to trick users into clicking on links to equally bogus Web sites that can steal personal information or infect computers with viruses.

This all-too-common scenario is called "phishing," alluding to "fishing" for unwary users with various sorts of bait, with the f changed to ph in keeping with traditional hacker practice. Phishing is similar to other techniques for manipulating people through deception, fear, or greed that hackers often refer to as "social engineering." Unlike one-on-one approaches, however, phishing relies on the ability to send large quantities of e-mail at virtually no cost (see SPAM), the availability of simple techniques for disguising both e-mail addresses and Web addresses (URLs), and the ease with which the appearance of a Web site can be convincingly replicated.

Although e-mail is the most common "hook" for phishing, any form of communication, including text or instant messages, can be used. Recently sites such as MySpace have become targets for automated phishing expeditions that changed links on pages to point to fraudulent sites (see SOCIAL NETWORKING).

DEFENSES AND COUNTERMEASURES

Wary users have a number of ways to reduce their chance of being "phished." Some signs of bogus messages include:

- The message is addressed generically ("dear PayPal user") or to the user's e-mail address rather than the account name.

- The text of the message contains spelling errors or poor grammar.

- The URL shown for a link in the message (perhaps via a "tool tip") does not match the institution's real Web address.

There are even interactive games such as "Anti-Phishing Phil" that users can play to test their ability to detect phishing attempts.

Unfortunately, modern phishers are becoming increasingly sophisticated. Some phishing messages can be personalized, using the target's actual name. URLs can be disguised so that discrepancies do not appear. When in doubt, the safest thing to do is always to access the institution by typing (not copying) its name directly in the Web browser rather than clicking on a link in e-mail. (In a practice called "pharming," a legitimate Web site can in effect be hijacked so that normal user accesses will be diverted to the fraudulent site. Users have no real defense against pharming; this is a matter for security professionals at the relevant Web sites.)

Fortunately there are ways in which software can help detect and block most phishing attempts. A good spam filter is the first line of defense and can block many phishing messages from getting to the user in the first place. Anti-phishing features are also increasingly included in Web browsers, or available as plug-ins. Thus "blacklists" of known phishing sites can be checked in real time and warnings given, or the site's address can be blocked from access by the system. Web sites can also introduce an added layer of security: Bank of America, for example, asks users to select and label one of several images offered by the bank. The image and label are subsequent displayed as part of the log-in process. If the user does not see the image and the user's label, then the site is presumably not the real bank site.

LEGISLATIVE RESPONSE

Phishing has been one of the fastest-growing types of online crime in recent years (see COMPUTER CRIME AND SECURITY and ONLINE FRAUDS AND SCAMS). By mid-2007 the Anti-Phishing Working Group (an association of financial institutions and businesses) was reporting the appearance of more than 30,000 new phishing sites per month (the largest number operating from China), though a site typically stays online for only a few days. Phishing contributed significantly to the $49 billion cost of identity theft in 2006 as estimated by Javelin Research. Further, industry surveys have suggested that phishing has aroused considerable consumer concern, slowing down the adoption or continued use of some financial services (see BANKING AND COMPUTERS).

In response to this growing concern, the U.S. Federal Trade Commission filed its first civil suit against a suspected phisher in 2004. The United States and other countries have also arrested phishing suspects, generally under some form of wire fraud statute. Starting in 2004, anti-phishing bills have been introduced in Congress, though none had passed as of 2007. However, the CAN-SPAM Act of 2003 was used in 2007 to convict a defendant accused of sending thousands of phishing e-mails purporting to be from America Online (AOL). Many states have also introduced anti-phishing legislation.

Further Reading

Anti-Phishing Working Group. Available online. URL: http://www.antiphishing.org/. Accessed November 6, 2007.

Jacobson, Markus, and Steven Myers, eds. *Phishing and Counter-measures: Understanding the Increasing Problem of Electronic Identity Theft.* Hoboken, N.J.: Wiley-Interscience, 2006.

James, Lance. *Phishing Exposed: Uncover Secrets from the Dark Side.* Rockland, Mass.: Syngress Publishing, 2005.

Lininger, Rachael, and Russell Dean Vines. *Phishing: Cutting the Identity Theft Line.* Indianapolis: Wiley, 2005.

PhishTank. Available online. URL: http://www.phishtank.com/. Accessed November 6, 2007.

Sheng, Steve, et al. "Anti-Phishing Phil: The Design and Evaluation of a Game That Teaches People Not to Fall for Phish." Carnegie Mellon University. Available online. URL: http://cups.cs.cmu.edu/soups/2007/proceedings/p88_sheng.pdf. Accessed November 6, 2007. (The game itself is available at http://cups.cs.cmu.edu/antiphishing_phil/.)

U.S. Federal Trade Commission. "FTC Consumer Alert: How Not to Get Hooked by a 'Phishing' Scam." October 2006. Available online. URL: http://www.ftc.gov/bcp/edu/pubs/consumer/alerts/alt127.pdf. Accessed February 8, 2008.

photography, digital

For more than 150 years photography has depended on the use of film made from light-sensitive chemicals. However, digital photography, first developed in the 1970s, emerged in the late 1990s as a practical, and in some ways superior, alternative to traditional photography.

The basic idea behind digital photography is that light (photons) can create an electrical charge in certain materials. In 1969, engineers at Bell Labs invented a light-sensitive semiconductor that became known as a charge-coupled device (CCD). The original intention of the developers was to use an array of CCDs to make a compact black-and-white video camera for the videophone, a device that did not prove commercially viable. However, astronomers were soon using CCD arrays to capture images too faint for the human eye or even for conventional film.

Digital photography remained confined to such specialized applications until the mid-1990s. By then, the growing use of multimedia and the World Wide Web made digital photography an attractive alternative for getting images online quickly, avoiding the need to scan traditional prints or negatives. At the same time, cheaper, more powerful processors and larger capacity memory storage made good quality digital cameras more viable as a consumer product.

A digital camera uses the same type of lenses and optical systems as a conventional camera. Instead of falling upon film, however, the incoming light strikes an array of CCD "photosites." Each photosite represents one picture element, or pixel, which will appear as a tiny dot in the resulting picture. (Camera resolution is typically measured in millions of pixels, or "megapixels.")

The surface of the array contains an abundance of free electrons. As light strikes a photosite, it creates a charge that draws and concentrates nearby electrons. The voltage at a photosite is thus proportional to the intensity of the light striking it. The charge of each row of photosites is transferred to a corresponding read-out register, where it is amplified to facilitate measurement.

The camera uses an analog-to-digital converter (see ANALOG AND DIGITAL) to convert the amplified voltages to digital numeric values. Early consumer digital cameras typically used 8-bit values, limiting the camera to a range of gradated intensity from 0 to 255. However, many cameras today use up to 12 bits, giving a range of 0 to 4096.

The CCD mechanism itself measures only light intensities, not colors. To obtain color, many cameras use a red, green, or blue (RGB) filter at each photosite. (Some manufacturers use cyan, yellow, green, and magenta filters instead.) Since each photosite registers only a single color, interpolation algorithms must be used to estimate the actual color of each pixel by using laws of color optics and comparing the colors and intensity of the adjacent pixels.

New high-end cameras are starting to eschew interpolation in favor of using a complete, separate CCD array for

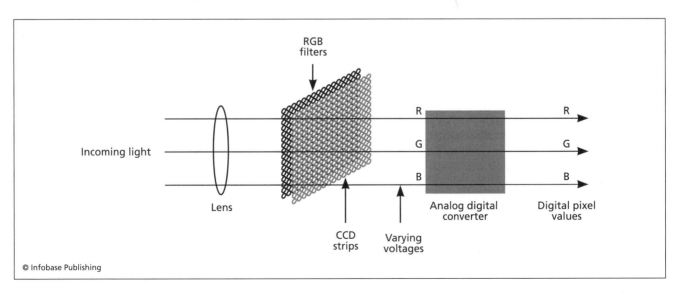

Digital cameras use a charge-coupled device (CCD) to convert incoming light to varying voltages that are digitized to create pixel values. They have largely replaced traditional film cameras for most applications.

each of the three RGB colors and thus making and combining three complete exposures that directly capture the colors. This produces the best possible color accuracy but is more expensive.

The final image data is stored using a standard file format, usually JPEG (see GRAPHICS FORMATS). The most commonly used storage medium is an insertable "flash" memory module. The major competing memory card standards are CompactFlash, SD, and Sony Memory Stick. Storage capacities run to 4 GB. As with regular RAM, the cost of flash memory has declined considerably in recent years.

In determining the adequacy of the camera's storage capacity, the user must also consider the camera's resolution (number of pixels) and whether images will be compressed before storage (see DATA COMPRESSION). While a certain amount of compression can be achieved without discernable degradation of the image, more drastic "lossy" compression sacrifices image quality for compactness. It should also be noted that as image resolution (and thus file size) increases, the time needed to process and store each image will also increase, limiting how rapidly successive exposures can be made.

Most digital cameras have a USB connector (see USB), making it easy to upload the stored images from the camera to a PC. Once in the PC, images can be edited or otherwise manipulated using the basic photo editing software usually included with the camera or a full-featured professional product such as Adobe Photoshop.

The same trends that have brought more capability per dollar spent on digital cameras have been even more evident in printers (see PRINTER). Using resolutions of 2880 dots per inch or more and special papers, digital camera users can make prints with a quality similar to that produced by traditional photo developers. Computer prints are more subject to color fading over time than are conventional prints, although some printer manufacturers now offer toner that will resist fading for 25 years or more.

FUTURE TRENDS

High-end consumer digital cameras reached the 8–10 megapixel range by 2008, allowing for images that can be "blown up" to 10 by 12 inches or larger while retaining image quality comparable to conventional photos. Professional-grade digital cameras ("digital SLRs") are rated at 10 megapixels or more. The need for such cameras for professional work arises not only from the higher resolution requirements but also because these cameras have the very high-quality optics used in fine 35 mm cameras, as well as having a greater variety of available specialty lenses. (Consumer digital "superzoom" cameras, however, do offer zoom lenses roughly comparable to those for low-cost 35-mm cameras.) The quality and convenience of digital photography ensure that digital cameras will supplant conventional cameras for most consumer and many professional applications. Many digital cameras also have the ability to shoot short video sequences. The ubiquity of digital cameras and digital video (even in many cell phones) has had important social consequences by facilitating transmission of pictures of disasters,

political gaffes, and other events often outside the mainstream media (see USER-CREATED CONTENT and YOUTUBE).

Digital camcorders will also become more widely used. Their resolution is generally from about a quarter million pixels to a million pixels—considerably lower than for digital still cameras, but adequate and likely to improve. Digital video cameras are also rated according to lux value, indicating the minimum light level for satisfactory recording. Most digital videos store the captured image to tape (either MiniDV or Hi-8), but some newer cameras use built-in recordable DVD disks instead (see CD-ROM AND DVD-ROM). The ability to digitally edit video direct from the camera is also an important advantage.

Further Reading

Busch, David D. *Digital SLR Cameras & Photography for Dummies.* 2nd ed. Hoboken, N.J.: Wiley, 2007.
Etchells, Dave. "Finding the Right Digital Camera." Imaging Resource Newsletter. Available online. URL: http://www.imaging-resource.com/TIPS/BUYGD/BUYGUID.HTM. Accessed August 17, 2007.
King, Julie Adair. *Digital Photography for Dummies.* 5th ed. Hoboken, N.J.: Wiley, 2005.
Silva, Robert. "Digital Camcorder Formats." Available online. URL: http://hometheater.about.com/od/camcorders/a/camformats_2.htm. Accessed August 17, 2007.
Wilson, Tracy V., K. Nice, and G. Gurevich. "How Digital Cameras Work." Available online. URL: http://www.howstuffworks.com/digital-camera.htm. Accessed August 17, 2007.

PHP

PHP is a very popular scripting language primarily used for creating dynamic Web pages (see AJAX and SCRIPTING LANGUAGES). PHP originated in 1994 as a way for Danish programmer Rasmus Lerdorf to replace a set of Perl scripts used to manage his own Web page—hence the original name "personal home page." Lerdorf released the first version together with a "form interpreter" in 1995. In 1997 the language parser was rewritten by Israeli developers Zeev Suraski and Andi Gutmans, who launched PHP3 in 1998; since then the initials PHP have (recursively) stood for PHP: Hypertext Processor. In 2004 the current version, PHP5, was released. As the language has evolved, it has improved in its support for objects (see OBJECT-ORIENTED PROGRAMMING) as well as in its connectivity to MySQL and other database and Web-application coordination technologies.

PHP normally runs on a Web server and processes PHP code, which is often embedded within Web pages (see HTML). The classic Hello World program would look like this:

```
<? php
echo "Hello, World";
?>
```

The PHP processor parses only the code within the delimiters <? and ?>. (An alternative set of delimiters is <script language ='php'> </script>.

Besides being embedded in HTML pages, PHP can be used interactively at the command line, where it has replaced older languages such as awk, Perl, or shell scripting for many users. PHP can also be linked to user-interface libraries (such as GTK+ for Linux/UNIX) to create applications that run on the client machine rather than the server.

PHP has a basic set of data types plus one called "resource" that represents data processed by special functions that return images, text files, database records, and so on. Additionally, PHP5 provides full support for objects, including private and protected member variables, constructors and destructors, and other features similar to those found in C++ and other languages.

There are numerous libraries of open-source objects and functions that enable PHP scripts to perform common Internet tasks, including accessing database servers (such as MySQL) as well as extensions to the language to handle popular Web formats such as Adobe Flash animation. Programmers have access to a wide range of PHP resources through PEAR (the PHP Extension and Application Repository).

The combination of sophisticated features and easy interactive scripting has made PHP the language of choice for many Web developers, who use it as part of the group of technologies called LAMP, for Linux, Apache (Web server), MySQL (database), and PHP.

Further Reading

Achour, Mehdi, et al. *PHP Manual*. Available online. URL: http://www.php.net/manual/en/. Accessed November 7, 2007.

Lerdorf, Rasmus, Kevin Tatroe, and Peter MacIntyre. *Programming PHP*. 2nd ed. Sebastapol, Calif.: O'Reilly Media, 2006.

PEAR-PHP Extension and Applications Repository. Available online. URL: http://pear.php.net/. Accessed November 7, 2007.

PHP [official Web site]. Available online. URL: http://php.net/. Accessed November 7, 2007.

Zandstra, Matt. *PHP Objects, Patterns, and Practice*. Berkeley, Calif.: Apress, 2004.

PL/I

By the early 1960s, two programming languages were in widespread use: FORTRAN for scientific and engineering applications and COBOL for business computing. However, applications were becoming larger and more complex, calling for a wider variety of capabilities. For example, scientific programmers needed to provide data-processing and reporting capabilities as well as computation. Business programmers, in turn, increasingly needed to work with formulas and statistics and needed floating-point and other number formats.

Language developers thus began to look toward a general-purpose language that could be equally at home with words, numbers, and data files. Meanwhile, IBM was preparing to replace its previously separate scientific and business computer systems with the versatile System/360. They and one of their user groups, SHARE, formed a joint committee to develop a new language for this new machine.

At first the designers thought in terms of extending FORTRAN to provide better text and data-processing capabilities, so they designated the new language FORTRAN VI. However, their focus soon changed to designing a completely new language, which was known until 1965 as NPL (New Programming Language). Because this acronym already stood for Britain's National Physical Laboratory, the name of the language was changed to PL/I (Programming Language I).

LANGUAGE FEATURES

PL/I has been described as the "Swiss army knife of languages" because it provides so many features drawn from disparate sources. The basic block structure and control structures (see LOOP and BRANCHING STATEMENT) were adapted from Algol, a relatively small language that had been devised by computer scientists as a model for structured programming (see ALGOL) and is also similar to Pascal (see PASCAL). Blocks can be nested, and variables declared within a block can be accessed only within that block and its nested blocks, unless declared explicitly otherwise.

PL/I includes a particularly rich variety of data types and can specify even the number of digits for numeric data. A PICTURE clause similar to that in COBOL can be used to specify exact layout. However, the language takes a more pragmatic approach than Algol or Pascal; data need not be declared and will be given default characteristics based on context. Input/Output (I/O) is built into the language rather than provided in an external library, and the flexible options include character, streams of characters, blocks, and records with either sequential or random access.

In general, PL/I provides more control over the low-level operation of the machine than Algol or even successors such as C. For example, there is an unusual amount of control over how variables are stored, ranging from STATIC (present throughout the life of the program) to AUTOMATIC (allocated and deallocated as the containing block is entered and exited) to CONTROLLED, where memory must be explicitly allocated and freed. Pointers allow memory locations to be manipulated directly. PL/I also provided more elaborate facilities for handling exceptions (errors) arising from hardware condition, arithmetic, file-handling, or other conditions.

EXAMPLE PROGRAM

The following program executes a DO loop and counts from one to the number of items specified. It then outputs the total of the numbers and their average.

```
COUNTEM: PROCEDURE OPTIONS (MAIN);
DECLARE (ITEMS, COUNTER, SUM, AVG) FIXED;
ITEMS = 10;
SUM = 0;
DO COUNTER = 1 TO ITEMS;
  SUM = SUM + COUNTER;
END;
AVG = SUM / ITEMS;
PUT SKIP LIST ("TOTAL OF ");
PUT ITEMS;
PUT ("ITEMS IS ");
PUT TOTAL;
PUT SKIP LIST ("THE AVERAGE IS: ");
PUT AVG;
END COUNTEM;
```

IMPACT OF THE LANGUAGE

Because of its many practical features and its availability for the popular IBM 360 mainframes, PL/I enjoyed considerable success in the late 1960s and 1970s. The language was later ported to most major platforms and operating systems. When personal computers came along, PL/I became available for IBM's OS/2 operating system as well as for Microsoft's DOS and Windows, although the language never really caught on in those environments.

Computer scientists such as structured programming guru Edsger Dijkstra decried PL/I's lack of a clear, well-defined structure. In his Turing Award Lecture in 1972, Dijkstra opined that "I absolutely fail to see how we can keep our growing programs firmly within our intellectual grip when by its sheer baroqueness the programming language—our basic tool, mind you!—already escapes our intellectual control." (See DIJKSTRA, EDSGER.)

On a practical level the sheer number of features in the language meant that truly mastering it was a lengthy process. A language like C, on the other hand, had a much simpler "core" to master even though it was less versatile. PL/I also tended to retain the mainframe associations from its birth at IBM, while C grew up in the world of minicomputers and the UNIX community and proved more suitable for PCs. Nevertheless, PL/I provided many examples that language designers could use in attempting to design better implementations.

Further Reading

The Essentials of PL/I Programming Language. Piscataway, N.J.: Research and Education Association, 1993.
Hughes, Joan Kirby. *PL/I Structured Programming.* 3rd ed. New York: Wiley, 1986.
"PL/I Frequently Asked Questions (FAQ)." Available online. URL: http://www.faqs.org/faqs/computer-lang/pli-faq/. Accessed August 17, 2007.
"The PL/I Language." Available online. URL: http://home.nycap.rr.com/pflass/pli.htm. Accessed February 9, 2008.
Sebesta, Robert W. *Concepts of Programming Languages.* 8th ed. Boston: Pearson Addison-Wesley, 2007.

Plug and Play

In early MS-DOS systems installation of new hardware such as a printer often had to be performed manually by copying files (see DEVICE DRIVER) to the hard drive from floppies and then making specified settings to the system configuration files AUTOEXEC.BAT and CONFIG.SYS. These settings often involved unfamiliar concepts such as interrupts (IRQs) and DMA (direct memory access) channels.

When Windows came along, device manufacturers generally provided an installation program that takes care of copying the files and making the necessary changes to the system registry. However, there was still the problem of ensuring that one had a driver compatible with the version of the operating system in use, and users were sometimes asked to make choices for which they were not prepared (such as choosing which port to use).

By the mid-1990s, Intel was promoting a standard for the automated detection and configuration of devices. Known as Plug and Play (PnP), this standard was incorporated in versions of Microsoft Windows starting with Windows 95 (see MICROSOFT WINDOWS). The required hardware support soon appeared on PC motherboards and expansion cards.

With Plug and Play the user simply connects a printer, scanner, or other device to the PC. Windows detects that a device has been connected and queries it for its official name and other information. If necessary, Windows can then prompt the user for a disk containing the appropriate driver or even search for a driver on a Web site.

The concept of Plug and Play extends beyond the Windows world, however. In recent years there has been interest in developing a Universal Plug and Play (UPnP) protocol by which a variety of devices could automatically configure themselves with any of a variety of different networks. This would be particularly helpful for home users who are increasingly setting up small networks so they can share broadband Internet connections, as well as the growing number of users who want their desktop PC to work with handheld (palm) computers and other devices. Microsoft supports UPnP in versions of Windows starting with ME and XP.

Further Reading

Bigelow, Stephen J. *The Plug & Play Book.* New York: McGraw Hill, 1999.
Shanley, Tom. *Plug and Play System Architecture.* Reading, Mass.: Addison-Wesley, 1995.
Universal Plug and Play Forum. Available online. URL: http://www.upnp.org/. Accessed August 17, 2007.

plug-in

A number of applications programs include the ability for third-party developers to write small programs that extend the main program's functionality. For example, thousands of "filters" (algorithms for transforming images) have been written for Adobe Photoshop. These small programs are called *plug-ins* because they are designed to connect to the main program and provide their service whenever it is desired or required.

Perhaps the most commonly encountered plug-ins are those available for Web browsers such as Firefox, Netscape, or Internet Explorer. Plug-ins can enable the browser to display new types of files (such as multimedia). Many standard programs for particular kinds of files are now provided both as stand-alone applications and as browser plug-ins. Examples include Adobe (PDF document format), Apple QuickTime (graphics, video, and animation), RealPlayer (streaming video and audio), and Macromedia Flash (interactive animation and presentation). These and many other plug-ins are offered free for the downloading, in order to increase the number of potential users for the formats and thus the market for the development packages.

One of the most useful plug-ins found in most browsers is one that allows the browser to run Java applets (see JAVA). In turn, Java is often used to write other plug-ins.

Beyond such traditional workhorses, a number of innovative browser plug-ins have appeared, particularly for the increasingly popular Firefox browser. For example, there are plug-ins that enable the user to view and work with the HTML and other elements of the page being viewed. Another popular area is plug-ins that make it easier to capture and organize material from Web pages, going well beyond the standard favorites or bookmark facility.

Including plug-in support for an application enables volunteer or commercial third-party developers to in effect increase the feature set of the main application, which in turn benefits the original developer. In the broader perspective, plug-ins are a way to harness the collaborative spirit found in open-source development, creating a community that is continually improving applications tools and making them more versatile. (The open-source Eclipse programming environment is a good example.)

Further Reading

Add-Ons for Internet Explorer. Available online. URL: http://www.windowsmarketplace.com/category.aspx?bcatid=834&tabid=1/. Accessed August 17, 2007.

Benjes-Small, Candice M., and Melissa L. Just. *The Library and Information Professional's Guide to Plug-ins and Other Web Browser Tools.* New York: Neal Schuman, 2002.

Clayberg, Eric, and Dan Rubel. *Eclipse: Building Commercial-Quality Plug-ins.* 2nd ed. Upper Saddle River, N.J.: Addison-Wesley Professional, 2006.

Drafahl, Jack, and Sue Drafahl. *Plug-ins for Adobe Photoshop: A Guide for Photographers.* Buffalo, N.Y.: Amherst Media, 2004.

Firefox Add-ons: Common Plugins for Firefox. Available online. URL: https://addons.mozilla.org/en-US/firefox. Accessed August 17, 2007.

Google Desktop Gadgets [plug-ins] Available online. URL: http://desktop.google.com/plugins/?hl=en. Accessed August 17, 2007.

Pilgrim, Mark. *Greasemonkey Hacks: Tips & Tools for Remixing the Web with Firefox.* Sebastapol, Calif.: O'Reilly, 2005.

podcasting

Podcasting (from iPod plus broadcasting) lets users subscribe to and automatically download regularly distributed content (such as radio broadcasts) over the Internet. The media files can be stored on an Apple iPod or other media player (see MUSIC AND VIDEO PLAYERS, DIGITAL), personal computer, or other device (see SMARTPHONE). Podcasting became popular starting around 2004–05 and has become widely used by individuals and organizations.

Typically, files to be podcast are put on a Web server. The URLs for the files and other information (such as episode titles) is provided in files called feeds, using a format such as RSS or Atom (see RSS). The user installs client software (such as iPodder), browses the feeds (such as through an online directory), and decides what to subscribe to. The software then periodically checks the feeds, obtains the URLs of the latest files, and downloads them automatically. The software can, if desired, then transfer the downloaded files to a portable media player, such as over a USB connection.

APPLICATIONS

There are many sources of podcasts. News organizations can provide regular audio or video podcasts as a supplement to regular text material. Podcasting also offers a way for a small news organization or independent journalist to build an audience using equipment as simple as a microphone and perhaps a video camera. Podcasts also provide a way for political organizations to keep in touch with supporters (and perhaps supply them with talking points). Any source of periodically distributed audio or video can be a candidate for podcasting. These include class lectures, corporate communications, and even religious services.

Further Reading

Geoghegan, Michael W., and Dan Hlass. *Podcast Solutions: The Complete How-To Guide to Getting Heard around the World.* Berkeley, Calif.: Apress, 2005.

Juice: The Cross-Platform Podcast Receiver. Available online. URL: http://juicereceiver.sourceforge.net/. Accessed November 7, 2007.

King, Kathleen P., and Mark Gura. *Podcasting for Teachers: Using a New Technology to Revolutionize Teaching and Learning.* Charlotte, N.C.: Information Age Publishing, 2007.

Mack, Steve, and Mitch Ratcliffe. *Podcasting Bible.* Indianapolis: Wiley, 2007.

Morris, Tee, and Evo Terra. *Podcasting for Dummies.* Hoboken, N.J.: Wiley, 2006.

Podcast Alley. Available online. URL: http://www.podcastalley.com/. Accessed November 7, 2007.

Podcasting News. Available online. URL: http://www.podcasting-news.com/. Accessed November 7, 2007.

pointers and indirection

The memory in a computer is accessed by numbering the successive storage locations (see ADDRESSING). When a programmer declares a variable, the compiler associates its name with a location in available memory (see VARIABLE). If the variable is used in an expression, when the expression is evaluated, the variable's name is replaced by its current value—that is, with the contents of the memory location associated with the variable. Thus, the expression Total + 10 is evaluated as "the contents in the address associated with Total" plus 10.

Sometimes, however, it is useful to have the general capability to access memory locations without assigning explicit variables. This is done through a special type of variable called a *pointer*. The only difference between pointers and regular variables is that the value stored in a pointer is not the data to be ultimately used by the program. Rather, it is the address of that data. Here are some examples from C, a language that famously provides support for pointers:

```
Int MyVar;        // Declare a regular variable
Int *MyPtr;       // Declare a pointer to an
   integer
                  // (int) variable
MyVar = 10;       // Set the value of MyVar
   to 10
MyPtr = &MyVar;   // Store the address of
   MyVar in
                  // the pointer MyPtr
```

In C, an asterisk in front of a variable name indicates that the variable is a pointer to the type declared. In the

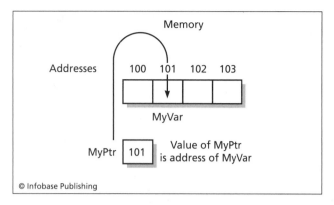

© Infobase Publishing

A pointer is a variable whose value is an address location. Here MyPtr holds the address 101.

second line above, therefore, MyPtr is a pointer to an integer variable. This means that the address of any integer variable can be stored in MyPtr. The last line uses the & (ampersand) to represent the address of the variable MyVar. Therefore, it stores that address in MyPtr.

Examining the lines above, one sees that the variable MyVar has the value 10. The pointer variable MyPtr has the value of whatever machine address contains the contents of the variable MyVar. In an expression, putting an asterisk in front of a pointer name "dereferences" the pointer. This means that it returns not the address stored in the pointer, but the value stored at the address stored in the pointer (see the diagram). Therefore if one writes:

```
AnotherVar = * MyPtr;
```

What is the value of AnotherVar? The answer is the current value of MyVar (whose address had been stored in MyPtr)—that value, as assigned earlier, is 10.

The general concept of storing the address of another variable in a variable is called indirection, or indirect addressing. It was first used in assembly language to work with index registers—special memory locations in a processor that store memory addresses.

USES FOR POINTERS

Although the concept may seem esoteric, pointers have a number of uses. For example, suppose one has a buffer (perhaps storing video graphics data) and one wants to copy it from one area to another. One could declare the buffer to be an array (see ARRAY) and then reference each element, or memory location and copy it. However, this would be rather awkward. Instead, one can declare a pointer, set it to the starting address of the buffer, and then simply use a loop to increment the pointer, pointing in turn to each location in the buffer.

A similar approach applies to strings in C and related languages. A string of characters in C is declared as an array of char. In an array, the name of the array is actually a pointer to the first data location. It is therefore easy to manipulate strings by getting their starting address by referencing the name and then using one or more pointers to step through the data locations. For example, the following function copies the contents of one string into another:

```
strcpy(char *s1,char *s2)
{
  while (*s2)
    *s1++ = *s2++;
}
```

The function takes two strings, s1 and s2, declared as pointers to char. It then steps (increments) them (using the ++ operator) so that the value in each location in s2 is copied into the corresponding location in s1. The loop exits when the value at s2 is 0 (null), indicating that the end of string marker has been reached.

Another common use for pointers is in memory allocation. Typically, a program requests memory by giving the memory allocation function a pointer and the amount of memory requested. The function allocates the memory and then returns the starting address of the new memory in the pointer, so the program knows how to access that memory.

Pointers are also useful for passing a "bulky" variable such as a data record to a procedure or function. Suppose, for example, a program needs to pass a 65,000 byte record to a procedure for printing a report. If it passes the actual record, the system has to make a copy of the whole record, tying up memory. If, instead, a pointer to the record is used, only the address is passed. The procedure can then access the record at that address without having to make a copy.

In C and some other languages it is even possible to have a pointer that points to another pointer. A common case is an array of strings, such as

```
Char Form [80] [20];
```

representing a form that has 20 lines of 80 characters. Each line is an array of characters and the form as a whole is thus "an array of arrays of characters." Therefore, to dereference (get the value of) a character one would first dereference the line, and then the column.

PROBLEMS WITH POINTERS

Pointers may be useful, but they are also prone to causing programming problems. The simplest one is failing to distinguish between a pointer and its value. For example, suppose one writes:

```
Total = Total + MyPtr;
```

intending to add the value of the variable pointed to by MyPtr to Total. Unfortunately, the asterisk (dereferencing operator) has been inadvertently omitted, so what gets added to Total is the machine address stored in MyPtr!

Another problem comes when a pointer is used to allocate memory, the memory is later deallocated, but the pointer is left pointing to it.

Because pointers can potentially access any location in memory (or at least attempt to), some computer scientists view them as more dangerous than useful. It's true that most things one might want to do with a pointer can be accomplished by alternative means. One attempt to tame pointers is found in C++, which offers the "reference" data type. A reference is essentially a constant pointer that once assigned to a variable always dereferences that variable and cannot be pointed anywhere else. Java has gone even further by not including traditional pointers at all.

Further Reading
Jensen, Ted. "A Tutorial on Pointers and Arrays in C." Available online. URL: http://home.netcom.com/~tjensen/ptr/pointers.htm. Accessed August 17, 2007.
Parlante, Nick. "Pointers and Memory." Available online. URL: http://cslibrary.stanford.edu/102/PointersAndMemory.pdf. Accessed August 17, 2007.
Sebesta, Robert W. *Concepts of Programming Languages.* 8th ed. Boston: Pearson Addison-Wesley, 2007.
Soulle, Juan. "Pointers [in C++]." Available online. URL: http://www.cplusplus.com/doc/tutorial/pointers.html. Accessed August 17, 2007.

political activism and the Internet

Although newspapers and particularly television remain the most popular sources used by voters to obtain information about candidates and issues, reports by the Pew Internet & American Life Project found that online media was used by about a third of American voters in the 2006 midterm elections, and about 15 percent used it as their primary information source. (The latter rate was about 35 percent among young people who had access to broadband Internet connections at home.) The researchers also found that about half of the online users had sought information not available elsewhere, while 41 percent believed that newspapers and television did not provide them with all the information they wanted.

It is true that much of the political information users find online is news that originated with mainstream print or broadcast news outlets. However, a growing role is also being played by blogs, issue-oriented Web sites, or sites created by candidates themselves, including profiles on the MySpace social networking site.

A surprising number of people who look to the Internet for political information participate actively, with about a quarter engaging in blogs or other online postings, whether expressing their own opinions or forwarding e-mail or reposting material. As users become more active (see USER-CREATED CONTENT), they are even becoming part of "official" debates, as in 2007 when primary candidates were asked questions submitted as 30-second YouTube videos.

ADVANTAGES AND PITFALLS FOR CANDIDATES

For political candidates and campaigns, the Internet is a mixed blessing. Advantages include:

- can reach a large number of people at relatively low cost

- can bypass a possibly indifferent mainstream media and reach people directly

- provides ways to organize and motivate supporters (see BLOGS AND BLOGGING, PODCASTING, and SOCIAL NETWORKING)

- allows for easier fund-raising, including potentially millions of small donations

The first major candidate to put together a campaign based on these principles was Howard Dean, who for a time was frontrunner for the 2004 Democratic presidential nomination. In the run-up to the 2008 race, libertarian Republican Ron Paul, while barely registering in the polls, startled the mainstream media by raising more than $4 million in one day from thousands of supporters organized on the Web.

However, there are pitfalls for politicians in the digital age as well. It is hard to control or coordinate self-organized activists, who may adopt positions that contradict the candidate's stated platform or engage in intemperate attacks. (In 2007 a video "mashup" by a Barack Obama supporter portraying Hillary Clinton as "big brother" in the famous 1984 Apple Macintosh commercial led to denials that the Obama campaign had anything to do with it.)

Further, the legions of independent bloggers virtually guarantee that "stumbles" that might have been missed or ignored by traditional media will be featured in blogs or displayed on YouTube for millions to ponder. (An example was Virginia senator George Allen, whose use of an obscure racial epithet *macaca* may have cost him reelection in 2006 when it was captured by a video blogger.) It is unclear whether the intense 24-hour scrutiny will force candidates to become ever more tightly scripted in their public activities so as to avoid "macaca moments."

Some critics also suggest that the Internet may actually weaken democracy in some ways. Because of the increasing ability to personalize or customize what news one sees and whom one converses with, people could end up being simply confirmed in their beliefs and isolated from larger dialog. Extremist groups already use Web sites not only to recruit people, but to keep followers motivated and focused on their issues, while in effect filtering out opposing views. The creation of such isolated constituencies, able to choose to see only the kinds of things that make them comfortable, could be bad for democracy. (This could be called a form of self-censorship, as opposed to outwardly imposed censorship, as in China—see CENSORSHIP AND THE INTERNET.) On the other hand, the sheer amount and variety of information available may make it hard for people to cut themselves off in this way.

Despite these misgivings, the importance of the Web for political activism and campaigns is clear. No campaign, whether political or issue advocacy, can afford not to have a quality Web site and staff who are adept at the new media and forms of communication, expression, and social networking.

Further Reading
Chadwick, Andrew. *Internet Politics: States, Citizens, and New Communication Technologies.* New York: Oxford University Press, 2006.
Garofoli, Joe. "Blogger Fest a Magnet for Liberal Politicos." *San Francisco Chronicle,* July 29, 2007, p. A1. Available online. URL: http://sfgate.com/cgi-bin/article.cgi?f=/c/a/2007/07/29/MNGRVR91RU1.DTL. Accessed November 8, 2007.
Guynn, Jessica. "Growing Internet Role in Election." *San Francisco Chronicle,* June 4, 2007, p. C-1. Available online. URL: http://sfgate.com/cgi-bin/article.cgi?f=/chronicle/archive/2007/06/04/BUGI6Q5L181.DTL. Accessed November 8, 2007.

Hogarth, Paul. "Hillary, Obama and the YouTube Election." *Beyond-Chron,* March 21, 2007. Available online. URL: http://www. beyondchron.org/news/index.php?itemid=4322. Accessed November 8, 2007.

Jakoda, Karen A. B., ed. *Crossing the River: The Coming of Age of the Internet in Politics and Advocacy.* Philadelphia: Xlibris, 2005.

Nagourney, Adam. "Internet Injects Sweeping Change into U.S. Politics." *New York Times,* April 2, 2006, p. 1 ff. Available online. URL: http://www.nytimes.com/2006/04/02/washington/02campaign.html?_r=1&or ef=slogin. Accessed November 8, 2007.

Rainie, Lee, and John Horrigan. "Election 2006 Online." Pew Internet & American Life Project, 2007. Available online. URL: http://www.pewinternet.org/pdfs/PIP_Politics_2006.pdf. Accessed November 8, 2007.

TechPresident: Personal Democracy Forum. Available online. URL: http://www.techpresident.com/. Accessed November 8, 2007.

popular culture and computing

Computer technology first came to public consciousness with the wartime ENIAC and the first commercial machines such as Univac in the early 1950s. The war had shown the destructive side of new technologies (particularly atomic power), but corporate and government leaders were soon promoting their beneficial prospects. Just as atomic energy advocates promised to provide power that was abundant, cheap, and clean, the computer, or "giant brain" was touted for its ability to solve problems that had been beyond human capabilities.

OMINOUS MACHINES

However, the computer, too, had its shadow in the popular consciousness. With their mysterious flashing lights and white-coated programmer/priests, mainframe computers were often seen as modern embodiments of the "mad scientist" trope, as in the movie *Colossus: The Forbin Project* (1970), where American and Soviet supercomputers joined forces to take over the world. Artificial intelligence also usurped humanity in the more mystical *2001: A Space Odyssey* (1968).

On the domestic front, the mainframe computer also became a symbol of misgivings about the bureaucratic state and corporate conformity. The romantic comedy film *Desk Set,* featuring Katharine Hepburn as a beleaguered corporate librarian, at first seems to confirm these fears, only to reveal that the computer had been misunderstood and would bring about a happier future for all. (IBM, incidentally, provided much of the technical support for the film.)

The counterculture of the 1960s seemed much less sanguine about the digital future. To many of the generation of activists starting with the Free Speech Movement in 1964, computers were the tools of the military-industrial complex, and computing facilities were sometimes picketed or even physically attacked.

However, a computer-savvy wing of the counterculture was also rising (see HACKERS AND HACKING). Activists began to see the machines as a tool for community organization and communication, as in 1973 with Community Memory, the first computer bulletin board system, accessed by teletype terminals.

GETTING PERSONAL

By the late 1970s the personal computer had arrived. On the one hand, PCs would seem not to fit the mainframe stereotype. After all, the desktop machines are small and designed to be accessible helpers in everyday life and work. Still, they could be connected to networks and perhaps used to take over the Pentagon's doomsday weapons—as in the movie *War Games* (1983). As fear of what malicious or criminal hackers could do took a more practical turn in the 1990s, such movies as *The Net* and *Sneakers* created a higher-tech incarnation of the spy thriller. Finally, the series of movies beginning with *The Matrix* extrapolated from the ultrarealistic movie effects and games of the coming century to raise the question of whether consensus reality could actually be a huge computer simulation.

Meanwhile, the figure of the computer "geek" or "nerd" has become a staple character in movies and TV shows—clever, socially inept, but indispensable for keeping the modern world running. In some eyes, the entrepreneurial success of Silicon Valley and the dot-coms placed Bill Gates and his colleagues in the same mold as Thomas Edison and Henry Ford a century earlier.

DIGITIZATION OF CULTURE

By the turn of the new century the network that had been portrayed as the domain of hackers and spies had become the all-pervasive World Wide Web. Today computers and the Internet are not only reflected in American popular culture—they are profoundly reshaping it. Computer games (particularly see ONLINE GAMES) have become vast, persistent social worlds, as are sites like MySpace and Facebook (see SOCIAL NETWORKING).

With the blending of formerly distinct media (see DIGITAL CONVERGENCE) and the fluid sharing and re-creation of images (see USER-CREATED CONTENT and MASHUPS), the digital world now permeates mainstream culture—or, one might say, the culture itself has become digitized. Meanwhile the line between fact and fiction, creator and viewer, expert and amateur has become increasingly blurred.

Further Reading
Fishwick, Marshall William. *Probing Popular Culture: On and Off the Internet.* Binghamton, N.Y.: Haworth Press, 2004.

Friedman, Ted. *Electric Dreams: Computers in American Culture.* New York: NYU Press, 2005.

King, Brad, and John Borland. *Dungeons and Dreamers: The Rise of Computer Game Culture from Geek to Chic.* Emeryville, Calif.: McGraw-Hill/Osborne, 2003.

"Machines (and more) in Movies, Books and Music." Berkshire Publishing Group. Available online. URL: http://www.berkshirepublishing.com/HumanComputerInteractionAndPopCulture/list.asp. Accessed August 17, 2007.

Nelson, Theodore H. *Computer Lib/Dream Machines: You Can and Must Understand Computers Now.* Chicago: Nelson, 1974. (Expanded, reprinted by Microsoft Press, 1987).

Polsson, Ken. "Personal Computer References in Pop Culture." Available online. URL: http://www.islandnet.com/~kpolsson/comppop/. Accessed August 17, 2007.

portal

The legion of new World Wide Web users who went online in the mid-1990s could easily navigate and "surf" the Web, using browsers such as Netscape and Internet Explorer (see WEB BROWSER). However, the lack of a reliable starting point and a systematic way to find information often led to frustration. Search engines such as AltaVista and Lycos (see SEARCH ENGINE) provided some help, but there was no single guide that could present the most useful information at a glance.

Meanwhile, in 1994, two graduate students, Jerry Yang and David Filo, had begun to circulate an organized listing of their favorite Web sites by e-mail. When the list proved very popular, they decided they could make a business out of providing a Web site that could serve as a topical guide to the Web. The result was Yahoo!, the most successful of what would come to be called *Web portals* (see YAHOO).

Yahoo! and other portals such as MSN (Microsoft Network), Excite, American Online (AOL), and Lycos generally provide a listing organized by topic and subtopic. For example, the general topic "Computers and Internet" in Yahoo! is divided into many subtopics such as communications and networking, hardware, software, and so on. Many topics are further subdivided until, at the bottom, there is a list of actual Web links that can be clicked upon to take the user directly to the relevant site.

The advantage of using a portal over using a search engine is that the links on a portal have generally been selected for quality, relevance, and usefulness. The disadvantage is the flip side of that selectivity: The links may reflect the tastes, agenda, or commercial interests of the portal developers and thus exclude important points of view. When seeking to learn more about a subject, many researchers therefore both work "inward" from a portal and "outward" via a search engine (see ONLINE RESEARCH).

To gain a competitive edge and raise revenue, portals typically include a considerable amount of advertising. Some portals also charge companies for being included or featured in listings or displays. General-purpose portals usually also contain such information as current news, stock prices, weather, and other timely information in an attempt to become their user's default page. Portals (particularly Yahoo!) have also sought to become more attractive (and profitable) by including such services as travel, financial services, games, and auctions.

Some portals emphasize particular approaches to information. For example, About.com goes beyond simply listing links to providing extensive guides to hundreds of subjects in a sort of newsletter format. There are also portals designed to serve particular constituencies, such as professional groups, industries, or hobby or interest groups. Companies can also create "enterprise portals" that can help employees keep in touch with developments and share information. Such portals often serve as the Web-based interface to the corporate local area network (LAN).

As with other information content providers, commercial portal developers have struggled to obtain enough revenue to keep up with the need to expand and compete in new areas. It is unclear whether the market will support more than a handful of large consumer portals in the long run, but both commercial and specialized portals have become an important part of the way most people access the Web.

Further Reading

About.com. Available online. URL: http://www.about.com. Accessed August 17, 2007.

Angel, Karen. *Inside Yahoo!: Reinvention and the Road Ahead.* New York: Wiley, 2002.

"Frequently Asked Questions about Portals (FAQs)." Traffick. Available online. URL: http://www.traffick.com/article.asp?aID=9. Accessed August 17, 2007.

Hock, Randolph. *Yahoo! to the Max: An Extreme Searcher Guide.* Medford, N.J.: Information Today, 2005.

Kastel, Berthold. *Enterprise Portals for the Business and IT Professional.* Sarasota, Fla.: Competitive Edge International, 2003.

Linwood, Jeff, and Dave Minter. *Building Portals with the Java Portlet API.* Berkeley, Calif.: Apress, 2004.

Sullivan, Dan. *Proven Portals: Best Practices for Planning, Designing, and Developing Enterprise Portals.* Upper Saddle River, N.J.: Addison-Wesley Professional, 2003.

Utvich, Michael, Ken Milhous, and Yana Beylinson. *1 Hour Web Site: 120 Professional Web Templates and Skins to Let You Create Your Own Web Sites—Fast.* Hoboken, N.J.: Wiley, 2007.

Yahoo! Available online. URL: http://www.yahoo.com. Accessed August 17, 2007.

PostScript

Early computer printers were limited to one or a few built-in fonts, either stamped on typewriter style keys on daisy wheels, or stored as patterns in the printer's software (with dot matrix printers). In the mid-1970s, when Xerox researchers were developing the laser printer, they realized they needed an actual programming language that could describe fonts, graphics, and other elements that could be printed on the more versatile new printers. PARC researchers developed InterPress; meanwhile two of them, John Warnock and Chuck Geschke, founded their own company in 1982 (see ADOBE SYSTEMS). They then created a more streamlined version of InterPress that they called PostScript. The first printer to include built-in PostScript capability was Apple's LaserWriter, in 1985. PostScript soon became the standard for a burgeoning industry (see DESKTOP PUBLISHING).

Because PostScript is an actual programming language (for a somewhat similar language, see FORTH), software such as word processors can include functions that turn a text document into a PostScript document, ready for printing. A PostScript interpreter in the printer (or even in another application) interprets the PostScript commands to re-create the document. The commands specify rasters (combinations of straight lines and curves), which can be scaled and transformed to provide the specified output, including fonts, which can be enhanced by including "hints" to help the system identify key features. This processor is thus sometimes called a Raster Image Processor (RIP).

DECLINE

By the late 1990s, however, PostScript was declining in use. In part this was because of the advent of cheaper ink-jet printers, which used simpler (and cheaper) software. Further, PostScript's role as a standard format for distributing documents has been largely replaced by one of Adobe's other standards, the Portable Document Format (see PDF). However, PostScript-equipped laser printers are still favored for heavy-duty printing jobs, because the document processing can be done in the printer instead of adding to the burden of the main CPU.

Further Reading

Adobe Systems Incorporated. *PostScript Language Reference.* 3rd ed. Reading, Mass.: Addison-Wesley, 1999. Available online. URL: http://partners.adobe.com/public/developer/en/ps/PLRM.pdf. Accessed November 8, 2007.

———. *PostScript Language Tutorial and Cookbook.* Upper Saddle River, N.J.: Addison-Wesley Professional, 2007. Available online. URL: http://www-cdf.fnal.gov/offline/PostScript/BLUEBOOK.PDF. Accessed November 8, 2007.

Weingartner, Peter. "A First Guide to PostScript." Available online. URL: http://www.tailrecursive.org/postscript/postscript.html. Accessed November 8, 2007.

presentation software

Whether at a business meeting or a scientific conference, the use of slides or transparencies has been largely replaced by software that can create a graphic presentation. Generally, the user creates a series of virtual "slides," which can consist of text (such as bullet points) and charts or other graphics. Often there are templates already structured for various types of presentations, so the user only needs to supply the appropriate text or graphics. There are a variety of options for the general visual style, as well as for transitions (such as dissolves) between slides. Another useful feature is the ability to time the presentation and provide cues for the speaker. Finished presentations can be shown on a standard monitor screen (if the audience is small) or output to a screen projector.

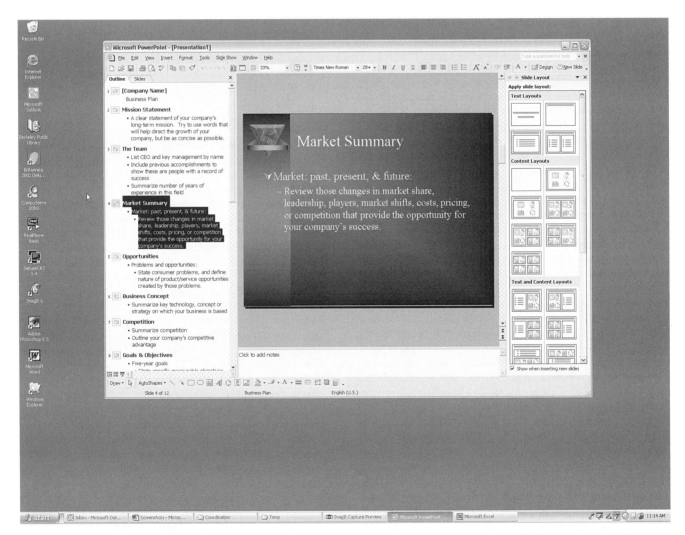

Microsoft PowerPoint is an example of presentation software. Such software uses a "slideshow" metaphor in which screens corresponding to slides can be created and arranged on a timeline for playing. Many types of special effects are also available.

Microsoft PowerPoint is the most widely used presentation program. It includes the ability to import Excel spreadsheets, Word documents, or other items created by Microsoft Office suite applications. The user can switch between outline view (which shows the overall structure of the presentation) to viewing individual slides or working with the slides as a collection.

There are a number of alternatives available including Apple's Keynote and Open Office, which includes a presentation program comparable to PowerPoint. Another alternative is to use HTML Web-authoring programs to create the presentation in the form of a set of linked Web pages. (PowerPoint and other presentation packages can also convert their presentations to HTML.) Although creating presentations in HTML may be more difficult than using a proprietary package and the results may be somewhat less polished, the universality of HTML and the ability to run presentations from a Web site are strong advantages of that approach.

A number of observers have criticized the general sameness of most business presentations. Some presentation developers opt to use full-fledged animation, created with products such as Macromedia Director.

Further Reading

Impress: More Power to Your Presentations. Available online. URL: http://www.openoffice.org/product/impress.html. Accessed August 17, 2007.

Keynote (Apple iWork). Available online. URL: http://www.apple.com/iwork/keynote/. Accessed August 17, 2007.

Lowe, Doug. *PowerPoint 2007 for Dummies.* Hoboken, N.J.: Wiley, 2007.

Rutledge, Patrice-Anne, Geetesh Bajaj, and Tom Muccolo. *Special Edition Using Microsoft Office PowerPoint 2007.* Indianapolis: Que, 2006.

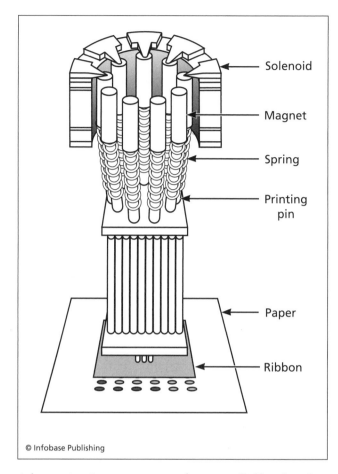

A dot-matrix printer uses an array of pins controlled by solenoids. Each character has a pattern of pins that are pushed against a typewriter-like ribbon to form the character on the paper.

printers

From the earliest days of computing, computer users needed some way to make a permanent record of the machine's output. Although results of a program could be punched onto cards or saved to magnetic tape or some other medium, at some point data has to be readable by human beings. This fact was recognized by the earliest computer and calculator designers: Charles Babbage (see BABBAGE, CHARLES) designed a printing mechanism for his never-finished computing "engine," and Williams Burroughs patented a printing calculator in 1888.

TYPEWRITER-LIKE PRINTERS

The large computers that first became available in the 1950s (see MAINFRAME) used "line printers." These devices have one hammer for each column of the output. A rapidly moving band of type moves under the hammers. Each hammer strikes the band when the correct character passes by. Printing is therefore done line by line, hence the name. Line printers were fast (600 lines per minute or more) but like the mainframes they served, they were bulky and expensive.

The typewriter offered another point of departure for designing printers. A few early computers such as the BINAC (an offshoot of ENIAC) used typewriters rigged with magnetically controlled switches (solenoids). However, a more natural fit was with the Teletype, invented early in the 20th century to print telegraph messages. Since the Teletype is already designed to print from electrically transmitted character codes, it was easy to rig up a circuit to translate the contents of computer data into appropriate codes for printing. (Since the Teletype could send as well as receive messages, it was often used as a control terminal for computer operators or for time-sharing computer users into the 1970s.)

The daisy-wheel printer was another typewriter-like device. It used a movable wheel with the letters embedded in slim "petals" (hence the name). It was slow (about 10 characters a second), noisy, and expensive, but it was the only affordable alternative for early personal computer users who required "letter-quality" output.

DOT-MATRIX PRINTERS

The dot-matrix printer, which came into common use in the 1980s, uses a different principle of operation than typewriter-style printers. Unlike the latter, the dot-matrix printer does not form solid characters. Instead, it uses an array of magnetically controlled pins (9 pins at first, but 24 on later models). Each character is formed by pressing the appropriate pins into a ribbon that pushes into the paper, leaving a pattern of tiny dots.

Besides being relatively inexpensive, dot-matrix printers are versatile in that a great variety of character styles or fonts can be printed (see FONT), either by loading different sets of bitmaps. Likewise, graphic images can also be printed. However, because the characters are made of tiny dots, they don't have the crisp, solid look of printed type.

LASER AND INK-JET PRINTERS

The majority of printers used today use laser or ink-jet technology. Both combine the versatility of dot-matrix with the letter quality of typewriter-style printers. Xerox introduced the first laser printer in the 1970s, although the technology was too expensive for most users at first.

The laser printer converts data from the computer into signals that direct the laser beam to hit precise, tiny areas of a revolving drum. The drum is covered with a charged (usually negative) film. The areas hit by the laser, however,

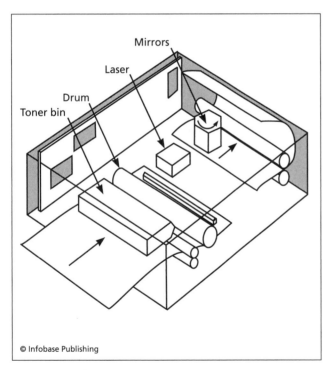

A laser printer uses a mirror-controlled laser beam to strike small spots on a rotating drum (called an OPC or Organic Photoconducting Cartridge) that had been given an electrical charge (usually positive) by a corona wire. The spots where the light beam hit are given an opposite charge (usually negative). The drum is then coated with a powdery toner that is charged opposite to the places where the light hit, so the toner clings to the drum to form the patterns of the characters or graphics. A piece of paper is then given a strong negative charge so it can pull the toner off the drum as it passes under it. Finally, heated rollers called fusers bind the toner to the paper to form the final image.

gain the opposite charge. As the drum continues to revolve, toner (a black powder) is dispensed. Because the toner is given a charge opposite to the places where the laser hit, the toner sticks to those places. Meanwhile, the paper is drawn into the drum. Because the paper is given the same charge as that produced by the laser beam (but stronger), the toner is pulled from the dots on the drum to the corresponding parts of the paper, forming the characters or graphics. A heating system then fuses the toner to the paper to make the image permanent. Meanwhile, the drum is discharged and the printer is ready for the next sheet of paper.

Color laser printers are also available, although they are still relatively expensive. They work by using four revolutions of the drum for each sheet of paper, depositing appropriate amounts of black, magenta, cyan, and yellow toner.

Laser printers fell in price throughout the 1990s (to $500 or so), but were soon rivaled by a different technology, the ink-jet printer.

The ink-jet printer has a print head that contains an ink cartridge for each primary printing color. Each cartridge has 50 nozzles, each thinner than a human hair. To print, the appropriate nozzles of the appropriate colors are subjected to electric current, which goes through a tiny resistor in the nozzle. An intense heat results for a few microseconds, long enough to create a tiny bubble that in turn forces a droplet of ink onto the page.

Ink-jet printers are generally slower than lasers, although fast enough for most purposes. Although the ink-jet is like the dot-matrix in producing tiny dots, the dots are much finer. With output at up to 2,880 dots per inch, the resulting characters are virtually indistinguishable from type-printing. Using high resolution and special papers, ink-jet printers can now also produce photo prints comparable to those created by traditional processes.

An interesting offshoot of ink-jet printing technology can be found in the development by HP of skin patches that can deliver controlled doses of drugs using tiny, virtually painless needles. The tiny droplets of drugs are transported in much the same way as ink goes from cartridge reservoir to page.

TRENDS

By the end of the 1990s, the ink-jet printer was declining steeply in price, and today quite capable units can be purchased for as little as $30 or so. Because of their greater speed, however, lasers are still used for higher-volume printing operations. "Multifunction" units combining printer, scanner, copier, and fax functions are also popular and cost less than a printer alone did only a few years ago.

Advocates of office automation have long predicted the "paperless office," but so far computers and their printers have churned out more paper, not less. However, there are some trends that might eventually reverse this course. Development of practical "electronic books" (page-size displays that can hold thousands of pages of text) may reduce the need for printed output (see E-BOOKS AND DIGITAL LIBRARIES). Another possible replacement for printing is "electronic ink," a sheet of paper with charged ink held in suspension. The text or graphics on the page can be changed electronically, so it can be reused indefinitely. Finally, the ability to access data anywhere on handheld or laptop computers may also reduce the need to make printouts.

Further Reading

Harris, Tom. "How Laser Printers Work." Available online. URL: http://home.howstuffworks.com/laser-printer.htm. Accessed August 17, 2007.

"Printer Buying Guide" (Cnet Reviews). Available online. URL: http://reviews.cnet.com/4520-7604_7-1016838-1.html. Accessed August 17, 2007.

"Printers: The Essential Buying Guide." *PC Magazine.* Available online. URL: http://www.pcmag.com/article2/0,1895,1766,00.asp. Accessed August 17, 2007.

Tyson, Jeff. "How Inkjet Printers Work." Available online. URL: http://www.howstuffworks.com/inkjet-printer.htm. Accessed August 17, 2007.

privacy in the digital age

Quoted in Fred H. Cate's *Privacy in the Information Age,* legal scholar Alan F. Westin has defined privacy as "the claim of individuals, groups, or institutions to determine for themselves when, how, and to what extent information about themselves is communicated to others."

Since the mid-19th century, advances in communications technology have raised new problems for people seeking to protect privacy rights. During the Civil War telegraph lines were tapped by both sides. In 1928, the U.S. Supreme Court in *Olmstead v. U.S.* refused to extend Fourth Amendment privacy protections to prevent federal agents from tapping phone lines without a warrant. Almost 50 years later, the court would revisit the issue in *Katz v. U.S.* and rule that telephone users did have an "expectation of privacy." The decision also acknowledged the need to adapt legal principles to the realities of new technology.

In the second half of the 20th century the growing use of computers would raise two basic kinds of privacy problems: surveillance and misuse of data.

SURVEILLANCE AND ENCRYPTION

Since much sensitive personal and business information is now transmitted between or stored on computers, such information is subject to new forms of surveillance or interception. Keystrokes can be captured using surreptitiously installed software and e-mails can be intercepted from servers or a user's hard drive. Many employers now routinely monitor employees' computer activity at work, including their use of the World Wide Web. When this practice is challenged, courts have generally sided with the employer, accepting the argument that the computers at work exist for business purposes, not private communications, and thus do not carry much of an expectation of privacy. Employers, however, have been encouraged to spell out their employee monitoring or surveillance policies explicitly. Outside the workplace, some protection is offered by the Electronic Communications Privacy Act (ECPA), passed in 1986.

Shadowy accounts about a secret system called Echelon have suggested that the National Security Agency has in place a massive system that can intercept worldwide communications ranging from e-mail to cell phone conversations. Apparently, rooms full of supercomputers can sift through this torrent of communication, looking for key words that might indicate a threat to the United States or its allies. (Much communication is in "clear" text; the ability of the government to crack strong encryption is unclear.)

Technology can be used to penetrate privacy, but it can also be used to safeguard it (see ENCRYPTION). Public key encryption programs such as Pretty Good Privacy (PGP) can encode text so that it cannot be read without a very-hard-to-crack key. The U.S. government, whose agencies enjoyed powerful surveillance capabilities, initially fought to suppress the use of encryption, but a combination of unfavorable court decisions and the ability to spread software across the Internet has pretty much decided the battle in favor of encryption users.

In the aftermath of the terrorist attacks of September 11, 2001, the federal government pressed for expanded surveillance powers, some of which were granted in the USA PATRIOT Act of 2001. (The Foreign Intelligence Surveillance Act [FISA] regulates wiretapping of U.S. persons to obtain foreign intelligence information, requiring that a warrant be obtained from a secret court. In 2008 after revelations that the administration was engaging in warrantless domestic surveillance outside of FISA, Congress passed an amendment that required FISA permission to wiretap Americans living abroad.) Computerized surveillance and identification systems (see BIOMETRICS) are also likely to be expanded in airports in other public places as part of the "War on Terrorism."

INFORMATION PRIVACY

Many privacy concerns arise not from the activities of spy or police agencies, but from the potential for the misuse of the many types of personal information now collected by businesses or government agencies. As far back as 1972, the Advisory Committee on Automated Personal Data Systems recommended the following standards to the secretary of the Department of Health, Education, and Welfare:

1. There must be no personal data record-keeping systems whose very existence is secret.
2. There must be a way for an individual to find out what information about him/her is on record and how it is used.
3. There must be a way for an individual to correct or amend a record of identifiable information about him/her.
4. There must be a way for an individual to prevent information about him/her that was obtained for one purpose from being used or made available for other purposes without his/her consent.
5. Any organization creating, maintaining, using, or disseminating records of identifiable personal data must guarantee the reliability of the data for their intended use and must take precautions to prevent misuse of the data.

The Federal Privacy Act of 1974 generally implemented these principles with regard to data maintained by federal agencies. Later, federal laws have attempted to address particular types of information, including school records, medical records, and video rentals.

However, much of the information collected from people results from commercial transactions or other interactions with businesses, particularly via the Internet. Although encrypted processing systems have reduced the chance that a credit card number submitted to a store will be stolen, so-called identity thieves may be able to obtain credit reports under false pretenses or collect enough information about a person from various databases (including Social Security numbers). With that information, the thief can take out credit cards in the person's name and run up huge bills (see IDENTITY THEFT and PHISHING AND SPOOFING). While the direct financial liability from identity theft is capped, the psychological impact and the effort required for victims to rehabilitate their credit standing can be considerable. In a few cases the same techniques have been used by stalkers, sometimes with tragic consequences.

The ability of Web sites to track where a visitor clicks by means of small files called "cookies" has also disturbed many people (see COOKIES). As with the recording of purchase information by supermarkets and other stores, businesses justify the practice as allowing for targeted marketing that can provide consumers with information likely to be of interest to them. (Many e-mail addresses are also gathered to be sold for use for unsolicited e-mail—see SPAM.) An even more intrusive technique involves the surreptitious installation of software on the user's computer for purposes of displaying advertising content or gathering information. In turn, programmers have distributed free utilities for identifying and removing such "adware" or "spyware."

While such consumer tracking is not as dangerous as identity theft, it feels like an invasion of privacy to many people as well as a source of insecurity, particularly because there are as yet few regulations governing such practices. However, in response to such concerns many businesses have put "privacy statements" on their Web sites, explaining what information about visitors will be collected and how it may be used. Businesses that meet standards for disclosure of their privacy practices can also display the seal of approval of organizations such as TRUSTe.

Many privacy advocates, however, believe that self-regulation is not sufficient to truly protect consumer privacy. They support strong new regulations, including "opt-in" provisions that would require businesses to receive explicit permission from the consumer before collecting information.

PRIVACY AND PERVASIVE COMPUTING
Beyond the Web and e-commerce, new challenges to privacy are emerging (see UBIQUITOUS COMPUTING). In the movie *Minority Report*, stores instantly mine data about approaching consumers and project personalized holographic ads in front of their eyes. While that technology is happily not here yet, many of the component pieces are (see DATA MINING and RFID). Add global positioning (GPS) tracking to the mix, and another important part

of privacy is threatened: "locational privacy." Certainly one can envisage situations where knowing not only who someone is but *where* they are can increase vulnerability to abuse.

In response to these pervasive threats to privacy, many advocates continue to push for regulation of data gathering and ways to hold people legally responsible for misuse of personal information. However, some writers such as science fiction writer and futurist David Brin argue that the battle for privacy is already lost, but the battle for transparency and mutual accountability may still be won—if the watched can watch the watchers.

Further Reading
Brin, David. *The Transparent Society*. Reading, Mass.: Addison-Wesley, 1998.
Cate, Fred H. *Privacy in the Information Age*. Washington, D.C.: Brookings Institution, 1997.
Electronic Frontier Foundation. Available online. URL: http://www.eff.org. Accessed August 17, 2007.
Electronic Privacy Information Center. Available online. URL: http://www.epic.org. Accessed August 17, 2007.
Henderson, Harry. *Privacy in the Information Age (Library in a Book)*. 2nd ed. New York: Facts On File, 2006.
Hunter, Richard. *World without Secrets: Business, Crime and Privacy in the Age of Ubiquitous Computing*. New York: Wiley 2002.
Monmonier, Mark. *Spying with Maps: Surveillance Technologies and the Future of Privacy*. Chicago: University of Chicago Press, 2002.
Solove, Daniel. *The Digital Person: Technology and Privacy in the Information Age*. Rev. ed. New York: NYU Press, 2006.

procedures and functions
From the earliest days of programming, programmers and language designers realized that it would be very useful to organize programs so that each task to be performed by the program had its own discrete section of code. After all, a program will often have to perform the same task, such as sorting or printing data, at several different points in its processing. Instead of writing out the necessary code instructions each time they are needed, why not write the instructions just once and have a mechanism by which they can be called upon as needed? Such callable program sections have been known as procedures, subroutines, or subprograms.

The simplest sort of subroutine is found in assembly languages and early versions of BASIC or FORTRAN. In BASIC, for example, a GOSUB statement contains a line number. When the statement is encountered, execution "jumps" to the statement with that line number, and continues from there until a statement such as RETURN is encountered. For example:

```
10 TOTAL = 10

20 GOSUB 40

30 END

40 PRINT "The total is: ";

50 PRINT TOTAL

60 RETURN
```

Here execution jumps from line 20 to line 40. After lines 40–60 are executed, the program returns to line 30, where it ends.

PROCEDURES WITH PARAMETERS

The simple subroutine mechanism has some disadvantages, however. The subroutine gets the information it needs from the main part of the program implicitly through the global variables that have been defined (see VARIABLE). If it needs to return information, it does it by changing the value of one or more of these global variables. The problem is that many different subroutines may be relying upon the same variables and at the same time changing them, leading to unpredictable results. Modern programming practice therefore generally avoids using global variables as much as possible.

Most high-level languages today (including Pascal, C/C++, Java, and modern versions of BASIC) define subprograms as procedures that pass information through specified parameters. For example, a procedure in Pascal might be defined as:

```
Procedure PrintChar (CharNum : integer);
```

This procedure has one parameter, an integer that specifies the number of the character to be printed (see CHARACTERS AND STRINGS).

The main program can call the procedure by giving its name and an appropriate character number. For example:

```
PrintChar (32);
```

The code within the procedure does not work with the parameter CharNum directly. Rather, it receives a copy that it can use. Thus, the procedure might include the statements:

```
Writeln ('Character number: ', CharNum );
Writeln (chr (CharNum));
```

The program will print the character number and then print the character itself on the next line (for character number 32 this will actually be a blank).

This typical way of using parameters is called *passing by value*. However, it is possible to pass a parameter to a procedure and have the parameter itself used rather than working with a copy. This is called "passing by reference." Pascal uses the var keyword for this purpose, while C passes a pointer to the variable (see POINTERS AND INDIRECTION), and C++ and Java prefix the variable name with an ampersand (&). For example, suppose one has a C function defined as follows:

```
int ByTwo (int * Val)
{
  Val = Val * 2;
}
```

In the following statements in the calling program:

```
Int Value, NewValue;
Value = 10;
NewValue = By Two (Value);
```

NewValue would be set to 20 because the actual variable Value has been multiplied by two inside the ByTwo function.

FUNCTIONS

A function is a procedure that returns its results as a value in place of the function name in the calling statement. For example, a function in C to raise a specified number to a specified power might be defined like this:

```
int Power (int base, int exp)
```

(C and related languages don't use a keyword like Pascal's procedure or function because in C all procedures are functions.)

This definition says that the Power function takes two integer parameters, base and exp, and returns an integer value.

Suppose somewhere in the program there are the following statements:

```
Int Base = 8;

Int Dimensions = 3;
Size = Power (Base, Dimensions);
```

The variable Size will receive the value of Power (8, 3) or 512.

Although the syntax for using procedures or functions varies by language, there are some principles that are generally applicable. The type of data expected by a procedure should be carefully defined (see DATA TYPES). Modern compilers generally catch mismatches between the type of data in the calling statement and what is defined in the procedure declaration. Procedures should also be checked for unwanted "side effects," which they can minimize by not using global variables.

Procedures and functions relating to a particular task are often grouped into separate files (sometimes called units or modules) where they can be compiled and linked into a program that needs to use them (see LIBRARY, PROGRAM).

Object-oriented languages such as C++ think of procedures in a somewhat different way from the examples shown here. While a traditional program sees procedures as blocks of code to be invoked for various purposes, an object-oriented program sees procedures as "methods" or capabilities of the program's various objects (see OBJECT-ORIENTED PROGRAMMING).

Further Reading
"Introduction to Python: Functions." Available online. URL: http://www.penzilla.net/tutorials/python/functions/. Accessed August 17, 2007.
Kernighan, Brian W., and Dennis M. Ritchie. *The C Programming Language*. 2nd ed. Englewood Cliffs, N.J.: Prentice Hall, 1988.
"Procedures and Functions" [in VBScript]. Available online. URL: http://www.functionx.com/vbscript/Lesson06.htm. Accessed August 17, 2007.
Sebesta, Robert W. *Concepts of Programming Languages*. 8th ed. Boston: Pearson Addison Wesley, 2007.
"Subroutines" [in Perl]. Available online. URL: http://www.comp.leeds.ac.uk/Perl/subroutines.html. Accessed August 17, 2007.

programming as a profession

All computer applications depend upon the ability to direct the machine to perform instructions such as fetching or storing data, making logical comparisons, or performing calculations. Although practical electronic computers first began to be built in the 1940s, it took considerable time for programming to emerge as a distinct profession. The first programmers were the computer designers themselves, followed by people (often women) recruited from clerical persons who were good at mathematics. With machines like ENIAC, programming was more like setting up a complicated piece of factory machinery than like writing. Switches or plugboards had to be set, and numeric instruction codes punched on cards to instruct the machine to move each piece of data from one location to another or to perform an arithmetic or logical operation.

Two factors led to greater recognition for the art or craft of programming. First, as more computers were built and put to work for various purposes, more programmers were needed, as well as more attention to their training and management. Second, as programs became larger and more complex, a number of high-level languages such as COBOL and FORTRAN came into use (see PROGRAMMING LANGUAGES). Besides making it easier to write programs, having just a few languages in widespread use made skills more readily transferable from one computer installation to another. And as with any profession, programming developed bodies of knowledge and practice.

At the same time, advances in language development would raise a recurrent question: Are professional programmers really necessary? Since FORTRAN looked a lot like ordinary mathematical notation, couldn't scientists and engineers just write the programs they need without hiring specialists for the job? Similarly, some enthusiasts led managers to think that with COBOL accountants (or even managers) could write their own business programs.

Sometimes part-time or "amateur" programming did prove to be practicable, particularly for scientists who found that writing a quick FORTRAN routine to solve a problem was easier than trying to explain the problem to a professional programmer. However, the professional programmer's job was never really in danger. Businesspeople were less inclined to try to learn COBOL and entrust something like the company's payroll processing to ad hoc efforts. In addition, the programs that controlled the operation of the computer itself, which became known as operating systems, required both arcane knowledge and the ability to design, verify, test, and debug increasingly complex systems (see SOFTWARE ENGINEERING).

DEVELOPMENT OF PRACTICE

In response to this growing complexity, computer scientists approached the improvement of programming practice on several levels. New languages developed in the 1960s and 1970s featured well-defined control structures, data types, and procedure calls (see ALGOL, PASCAL, C, DATA TYPES, LOOP, and STRUCTURED PROGRAMMING). The management of programming teams and the factors affecting productivity were examined by pioneers such as Frederick Brooks, author of *The Mythical Man-Month*, and IBM sponsored workshops and study groups.

While many mainframe business programmers continued to write and maintain programs written in the older languages (such as COBOL), starting in the 1970s a new generation of systems and applications programmers used C and worked in a different environment—campus minicomputers running UNIX. Unlike the hierarchical, systematic approach of the "mainframe culture," the minicomputer programmers tended toward a decentralized but cooperative approach (see OPEN-SOURCE MOVEMENT and HACKERS and HACKING).

When the personal computer revolution began to arrive at the end of the 1970s, much of the evolution of programming culture would be recapitulated. Since early microcomputer systems had very limited memory, programmers who wanted to get useful work out of machines such as the Apple II had to work mainly in assembly language or write quick and dirty programs in a limited dialect of BASIC. The hobbyists and early adopters often knew little about the academic world of computer science and software engineering, but they were good at wringing the most out of each clock cycle and byte of memory.

As personal computers gained in power and capability through the 1980s, programmers were able to use higher-level languages such as C. Applications such as word processors, spreadsheets, and graphics programs became more complex, and programmers had to work in larger teams like their mainframe counterparts.

At the same time, the sharp demarcation between programmer and user became less distinct with the personal computer. Many users who were not professional programmers used applications software that included programmable features, such as spreadsheets and simple data bases (see MACRO and SCRIPTING LANGUAGE). New languages such as Visual Basic let even relatively inexperienced programmers plug in user interfaces and other components and create useful programs (see PROGRAMMING ENVIRONMENT).

Each sector of programming seems to go through a cycle of improvisation and innovation followed by standardization and professionalization. Just as the early ENIAC programmers evolved into the organized hierarchy of corporate programming departments, the individuals and small groups who wrote the first personal computer software evolved into large teams using sophisticated software to track to the modules, versions, and development steps of major programming projects. Similarly, when the explosion of the World Wide Web starting in the mid-1990s brought a new demand for people who could code HTML, CGI, and Java, much of the most interesting work was done by individuals and small companies. But if history repeats itself, the Internet applications field will undergo the same process of professionalization, with increasingly elaborate standards and expectations (see CERTIFICATION OF COMPUTER PROFESSIONALS).

Throughout the history of programming, visionaries have announced that the time was coming when most if not all programming could be automated. All a person will have to do is give a reasonably coherent description of the desired results and the required program will be coded by some form of artificial intelligence (see EXPERT SYSTEMS, GENETIC ALGORITHMS, and NEURAL NETWORK). But while users have now been given the ability to do many things that formerly required programming, it seems there is still a demand for programmers who can move the bar another step higher. The profession continues to evolve without any signs of impending extinction.

Further Reading

Brooks, Frederick. *The Mythical Man-Month, Anniversary Edition: Essays on Software Engineering.* Reading, Mass.: Addison-Wesley, 1995.

Ceruzzi, Paul. *A History of Modern Computing.* Cambridge, Mass.: MIT Press, 1998.

Henderson, Harry. *Career Opportunities in Computers and Cyberspace.* 2nd ed. New York: Facts On File, 2004.

Kohanski, Daniel. *Moths in the Machine: The Power and Perils of Programming.* New York: St. Martin's Press, 2000.

Ullmann, Ellen. *Close to the Machine: Technophilia and its Discontents.* San Francisco: City Lights Books, 1997.

programming environment

The first programmers used pencil and paper to sketch out a series of commands, or punched them directly on cards for input into the machine. But as more computer resources became available, it was a natural thought that programs could be used to help programmers create other programs. The availability of Teletype or early CRT terminals on time-sharing systems by the 1960s encouraged programmers to write simple text editing programs that could be used to create the computer language source code file, which in turn would be fed to the compiler to be turned into an executable program (see TERMINAL and TEXT EDITOR). The assemblers and BASIC language implementations on the first personal computers also included simple editing facilities.

More powerful programming editors soon evolved, particularly in academic settings. One of the best known is EMACS, an editor that contains its own LISP-like language that can be used to write macros to automatically generate program elements (see LISP and MACRO). With the many other utilities available in the UNIX operating system, programmers could now be said to have a programming environment—a set of tools that can be used to write, compile, run, debug, and analyze programs.

More tightly integrated programming environments also appeared. The UCSD "p-system" brought together a program editor, compiler, and other tools for developing Pascal programs. While this system was somewhat cumbersome, in the mid-1980s Borland International released (and steadily improved) Turbo Pascal. This product offered what became known as an "integrated development interface" or IDE. Using a single system of menus and windows, the

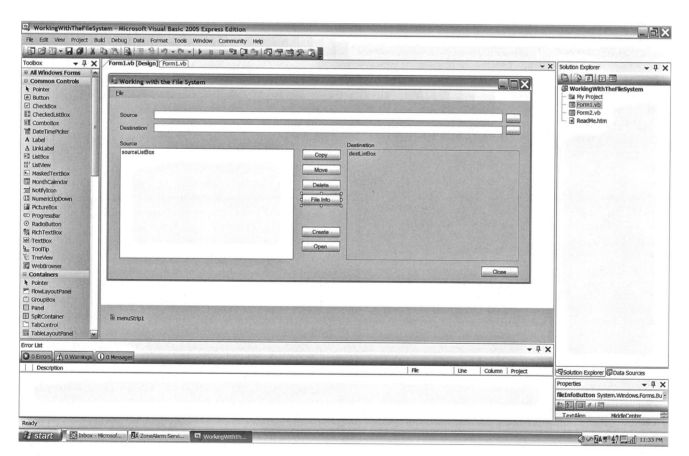

As the name suggests, Microsoft Visual Basic provides a visual programming environment in which the controls that make up a program's user interface can be placed on a form. Various properties (characteristics) of the controls can then be set, and program code is then written and attached to govern how the objects will behave.

programmer could edit, compile, run, and debug programs without leaving the main window.

The release of Visual Basic by Microsoft a few years later brought a full graphical user interface (GUI). Visual Basic not only ran in Windows, it also gave programmers the ability to design programs by arranging user interface elements (such as menus and dialog boxes) on the screen and then attaching code and setting properties to control the behavior of each interface object. This approach was soon extended by Microsoft to development environments for C and C++ (and later, Java) while Borland released Delphi, a visual Pascal development system. Today visual programming environments are available for most languages. Indeed, many programming environments can host many different languages and target environments. Examples include Microsoft's Visual Studio .NET and the open-source Eclipse, which can be extended to new languages via plug-ins.

Modern programming environments help the programmer in a number of ways. While the program is being written, the editor can highlight syntax errors as soon as they're made. Whether arising during editing or after compilation, an error message can be clicked to bring up an explanation, and an extensive online help system can provide information about language keywords, built-in functions, data types, or other matters. The debugger lets the programmer trace the flow of execution or examine the value of variables at various points in the program.

Most large programs today actually consists of dozens or even hundreds of separate files, including header files, source code files for different modules, and resources such as icons or graphics. The process of tracking the connections (or dependencies) between all these files, which used to require a list called a *makefile* can now be handled automatically, and relationships between classes in object-oriented programs can be shown visually.

Researchers are working on a variety of imaginative approaches for future programming environments. For example, an interactive graphical display (see VIRTUAL REALITY) might be used to allow the programmer to in effect walk through and interact with various representations of the program.

Further Reading

Burd, Barry. *Eclipse for Dummies.* Hoboken, N.J.: Wiley, 2004.

Hladni, Ivan. *Inside Delphi 2006.* Plano, Tex.: Wordware Publishing, 2006.

Kernighan, Brian W., and Rob Pike. *The UNIX Programming Environment.* Englewood Cliffs, N.J.: Prentice Hall, 1984.

Parsons, Andrew, and Nick Randolph. *Professional Visual Studio 2005.* Indianapolis: Wiley, 2006.

programming languages

There are many ways to represent instructions to be carried out by a computer. With early machines like ENIAC, programs consisted of a series of detailed machine instructions. The exact movement of data between the processor's internal storage (registers) and internal memory had to be specified, along with the appropriate arithmetic operations.

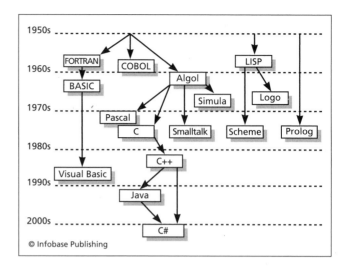

The evolution of a few major programming languages through five decades. There are actually hundreds of different programming languages that have seen at least some use in the past 50 years.

This lowest level, least abstract form of programming languages is hardest for humans to understand and use.

The first step toward a more symbolic form of programming is to use easy-to-remember names for instructions (such as ADD or CMP for "compare") as well as to provide labels for storage locations (variables) and subroutines (see PROCEDURES AND FUNCTIONS). The file of symbolic instructions (called source code) is read by a program called an assembler (see ASSEMBLER), which generates the low-level instructions and actual memory addresses to be used by the program. Because of its ability to closely specify machine operations, assembly language is still used for low-level hardware control or when efficiency is at a premium.

Most languages in use today are higher-level. The mainstream of programming languages consists of languages that are procedural in nature. That is, they specify a main set of instructions that are executed in sequence, although the program can branch off (see BRANCHING STATEMENTS) or repeat a series of statements until a condition is satisfied (see LOOP). A program can also call a set of instructions defined elsewhere in the program. Constant or variable data is declared to be of a certain type such as integer or character (see DATA TYPES) before it is used. There are also rules that determine what parts of a program can access what data (see VARIABLE). For examples of procedural languages, see ALGOL, BASIC, C, COBOL, FORTRAN, and PASCAL.

A variant of procedural languages is the object-oriented language (see OBJECT-ORIENTED PROGRAMMING). Such languages (see C++, JAVA, and SMALLTALK) still use sequential execution and procedures, but the procedures are "packaged" together with relevant data into objects. In order to display a picture, for example, the program will call upon a particular object (created from a class of such objects) to execute its display function with certain parameters such as location and dimensions.

NONPROCEDURAL LANGUAGES

Although the bulk of today's software is written using procedural languages, there are some important languages constructed using quite different paradigms. LISP, for example, is a powerful language used in artificial intelligence applications. LISP is written by putting together layers of functions that carry out the desired processing (see NONPROCEDURAL LANGUAGES, LISP, and FUNCTIONAL LANGUAGES). There are also "logic programming" languages, of which Prolog is best known (see PROLOG). Here a chain of logical steps is constructed such that the program can traverse it to find the solution of a problem.

CONTEXT AND CHANGE IN PROGRAMMING LANGUAGES

Because of the amount of effort it takes to truly master a major programming language, most programmers are fluent in only a few languages and developers tend to standardize on one or two languages. The store of tried-and-true code and lore built up by the programming community tends to make it disadvantageous to radically change languages. Thus, FORTRAN and COBOL, although more than 40 years old, are still in considerable use today. C, which is about 30 years old, has been gradually supplanted by C++ and Java, but the latter languages represent an object-oriented evolution of C, intentionally designed to make it easy for programmers to make the transition. (Smalltalk, which was designed as a "pure" object-oriented language, never achieved widespread use in commercial development.)

Similarly, when programmers had to cope with parallel processing (programs that can have several threads of execution going at the same time), they have tended to favor "parallelized" versions of familiar languages rather than wholly new ones (see CONCURRENT PROGRAMMING and PARALLEL PROCESSING).

While the basic elements of computer languages tend to persist in the same recognizable forms, the way programmers experience their use of languages has changed considerably through the use of modern visual integrated development environments (see PROGRAMMING ENVIRONMENT). A variety of languages have also been designed for tasks such as data management, interfacing Web pages, and system administration (see SCRIPTING LANGUAGES, AWK, PERL, PHP, and PYTHON).

Further Reading

Bergin, Thomas J., and Richard G. Gibson, eds. *History of Programming Languages-II.* Reading, Mass.: Addison-Wesley/ACM Press, 1996.
Sebesta, Robert W. *Concepts of Programming Languages.* 8th ed. Boston: Pearson Addison-Wesley, 2007.

project management software

Whether a project involves only a few people in the same department or thousands of people and several years, there is a variety of software to help managers plan and monitor the status of their projects.

At the simplest level, PIM software (see PERSONAL INFORMATION MANAGER) can be used by an individual to monitor simple personal projects. Such software generally includes the ability to record the description, priority, due date, and reminder date for a task.

Project management software is generally used to plan larger projects involving many persons or teams. A complex project must first be broken down into tasks. (Large projects often have subprojects as an intermediate entity.) Next, dependencies must be taken into account. For example, the user testing program for a software product can't begin until a usable preliminary ("alpha" or "beta") version of the program is available. The various "resources" assigned to a subproject or task must also be tracked, including personnel and number of hours assigned and budget allocations. In tracking personnel assigned to a project, their availability (who is on vacation and who is assigned to what location) must also be considered.

Once the scheduling and priorities are arranged, the inevitable divergences between what was planned and what is actually happening must be monitored. Good project management software provides many tools for the purpose. Available charts and reports often include:

- Gantt charts that use bars to show the duration and percentage of completion of the various overlapping subprojects or tasks.

- PERT (Program Evaluation and Review Technique) charts that show each subproject or task as a rectangular "node" with information about the task. The connections between nodes show the relationships (dependencies) between the items. PERT charts are usually used at the beginning stages of planning.

- Analysis tools that show critical paths and bottlenecks (places where one or more tasks falling behind might threaten large portions of the project). Generally, the more preceding items a task is dependent on, the more likely that task is to fall behind.

- Tools for estimating the probability for completion of a given task based on the probabilities of tasks it is dependent on, as well as other factors such as the likelihood of certain resources becoming available.

- A system of alerts or "stoplights" that show slowdowns, potential problems, or areas where work has stopped completely. These can be set to be triggered when various specified conditions occur.

- Integration between project management and budget reporting so tasks and the project as a whole can be monitored in relation to budget constraints.

- Integration between the project management software and individual schedules kept in PIM software such as Microsoft Outlook or in handheld computers (PDAs) such as the PalmPilot.

- Integration between project management and software for scheduling meetings.

Given the scope and pace of today's business, scientific, and other projects, project management software is often a vital tool. However, using too elaborate a project tracking system for a relatively small and well-defined project may divert time and energy away from the work itself. Fortunately, a wide variety of project management programs are available, ranging from full-fledged products such as Microsoft Project or Primavera Project Planner to simpler shareware or free products.

Further Reading

Marmel, Elaine. *Microsoft Office Project 2007 Bible*. Indianapolis: Wiley, 2007.

Muir, Nancy C. *Microsoft Office Project 2007 for Dummies*. Hoboken, N.J.: Wiley, 2007.

Open Workbench: Open-Source Project Scheduling for Windows. Available online. URL: http://www.openworkbench.org/. Accessed August 17, 2007.

"Project Management Software." Wikipedia. Available online. URL: http://en.wikipedia.org/wiki/Project_management_software. Accessed August 17, 2007.

Prolog

Since the 1950s, researchers have been intrigued by the possibility of automating reasoning behavior, such as logical inference (see ARTIFICIAL INTELLIGENCE). A number of demonstration programs have been written to prove theorems starting from axioms or assumptions. In 1972, French researcher Alain Colmerauer and Robert Kowalski at Edinburgh University created a logic programming language called Prolog (for Programmation en Logique) as a way of making automated reasoning and knowledge representation more generally available.

A conventional procedural program begins by defining various data items, followed by a set of procedures for manipulating the data to achieve the desired result. A Prolog program, on the other hand, begins with a set of facts (axioms) that are assumed to be true. (This is sometimes called declarative programming.)

For example, the fact that Joe is the father of Bill would be written:

```
Father (Joe, Bill).
```

The programmer then defines logical rules that apply to the facts. For example:

```
father (X, Y) :- parent (X, Y), is male (X)
grandfather (X, Y) :- father (X, Z), parent (Z, Y)
```

Here the first assertion says that a person X is the father of Y if he is the parent of Y and is male. The second assertion says that X is Y's grandfather if he is the father of a person Z who in turn is a parent of Y.

When a program runs, it processes queries, or assertions whose truth is to be proven. Using a process called unification, the Prolog system looks for facts or rules that apply to the query and attempts to create a logical chain leading to proving the query is true. If the chain breaks (because no matching fact or rule can be found), the system "backtracks" by looking for another matching fact or rule from which to attempt another chain.

Prolog aroused considerable interest among artificial intelligence researchers who were hoping to create a powerful alternative to conventional programming languages for automating reasoning. This interest was further spurred by the Japanese Fifth Generation Computer Program of the 1980s, which sought to create logical supercomputers and made Prolog its language of choice. Although some such machines were built, the idea never really caught on. However, Borland International (makers of the highly successful Turbo Pascal) released a Turbo Prolog that made the language more accessible to students using PCs, although it used some nonstandard language extensions.

Despite its commercial success being limited, Prolog has been used in a number of areas of artificial intelligence research. Its rules-based structure is naturally suited for expert systems, knowledge bases, and natural language processing (see EXPERT SYSTEMS and KNOWLEDGE REPRESENTATION). It can also be used as a prototyping language for designing systems that would then be recoded in conventional languages for speed and efficiency.

Further Reading

Bratck, Ivan. *Prolog Programming for Artificial Intelligence*. 3rd ed. Reading, Mass.: Addison-Wesley, 2000.

Clocksin, W. F., and C. S. Mellish. *Programming in Prolog: Using the ISO Standard*. 5th ed. New York: Springer, 2003.

Fisher, J. R. "Prolog: Tutorial." Available online. URL: http://www.csupomona.edu/~jrfisher/www/prolog_tutorial/contents.html. Accessed August 17, 2007.

Sterling, Leon, and Ehud Shapiro. *The Art of Prolog: Advanced Programming Techniques*. 2nd ed. Cambridge, Mass.: MIT Press, 1994.

SWI-Prolog [free Prolog for Windows, Linux, and Mac OS]. Available online. URL: http://www.swi-prolog.org/. Accessed August 17, 2007.

pseudocode

Because humans generally think on a higher (or more abstract) level than that provided by even relatively high-level programming languages such as BASIC or Pascal, it is sometimes suggested that programmers use some form of pseudocode to express how the program is intended to work. Pseudocode can be described as a language that is more natural and readable than regular programming languages, but sufficiently structured to be unambiguous. For example, the following pseudocode describes how to calculate the cost of wall-to-wall carpet for a room:

```
Get room length (in feet)
Get room width
Multiply length by width to get area (in
square feet)
Get price of carpet per square foot
Multiply price/sq. foot by area to get total
cost.
```

Pseudocode generally includes the basic control structures used in programming languages (see BRANCHING

STATEMENTS and LOOP) but is not concerned with small details of syntax. For example, this pseudocode might determine whether to charge sales tax for an online purchase:

```
Get customer's state of residence
If state is "CA" then
   Tax = Price * .085
   Total = Price + Tax
End If
```

Once the pseudocode has been written and reviewed, the statements can be recoded in the programming language of choice. For example, the preceding example might look like this in C:

```
If (state == "CA") {
   Tax = Price * .085;
   Total = Price + Tax;
}
```

The term *pseudocode* can also be applied to "intermediate languages" that provide a generic, machine-independent representation of a program. For example, in the UCSD Pascal system the language processor generates a "p-code" that is turned into actual machine language by an interpreter written for each of the different types of computer supported. Today Java takes a similar approach.

Further Reading
Bailey, T. E., and Kris Lundgaard. *Program Design with Pseudocode.* 3rd ed. Pacific Grove, Calif.: Brooks/Cole, 1989.
Daviduck, Brent. "Introduction to Programming in C++: Algorithms, Flowcharts and Pseudocode." Available online. URL: http://www.allclearonline.com/applications/DocumentLibraryManager/upload/program_intro .pdf. Accessed August 17, 2007.
Gilberg, Richard F,. and Behrouz A. Forouzan. *Data Structures: a Pseudocode Approach with C++.* 2nd ed. Boston: Course Technology, 2004.
Neapolitan, Richard E. *Foundations of Algorithms Using C++ Pseudocode.* 2nd ed. Sudbury, Mass.: Jones and Bartlett, 1998.

psychology of computing

Computing is a complex, pervasive, and increasingly vital human activity. It is not surprising that human psychology can play an important role in many aspects of computer use.

Since the 1960s psychology (in particular see COGNITIVE SCIENCE) has contributed to the structuring of interaction between computer systems and users (see USER INTERFACE). It is important to note the significant differences between how computers and humans perceive and process information: computers are extremely fast in processing in a highly structured setting (e.g., a program). The human brain, on the other hand, while thousands of times slower, is thus far greatly superior in coping with loosely structured data through pattern recognition, the making of analogies, and generalization. A number of researchers (see, for example, LICKLIDER, J. C. R.) have promoted the idea of creating a human-computer synergy where the structure of the system takes advantage of both the machine's computational and data-retrieval abilities and the human user's ability to work with the larger picture. Such research is continuing as autonomous software (see SOFTWARE AGENT) and is beginning to interact with Web users.

PSYCHOLOGY OF CYBERSPACE
The Internet and its perception as a shared cyberspace adds new dimensions to the psychology of computing. In fact, the emphasis here is not on computation per se but on the representation of ideas and images, communication, social interaction, and identity. In particular, pioneering work (see TURKLE, SHERRY) has illuminated ways in which online interactions affect identity and sense of self—even encouraging the assumption of multiple identities (see IDENTITY IN THE ONLINE WORLD and ONLINE GAMES). Indeed, virtual worlds such as *Second Life* offer new ways to study the formation of communities and social interactions.

On the positive side, it has been argued that cyberspace has encouraged people (particularly adolescents) to experiment with new identities in a relatively safe environment, but lack of inhibition and experience can lead to risky behavior such as involvement with sexual predators. The very fact that many people (particularly the young) may spend several hours a day or more immersed in the online world has also led to concerns; some psychologists have even suggested that "Internet addiction disorder" (IAD) be included as an official mental disorder similar to compulsive gambling. However, as of 2007, the American Medical Association has not recommended that IAD be classified as a mental disorder, and the American Society of Addiction Medicine has resisted such a status. Generally, excessive or inappropriate use of the Internet has been seen as a symptom of more traditional diagnoses such as obsession or compulsion.

Further Reading
Card, Stuart K., Thomas P. Moran, and Allen Newell, eds. *The Psychology of Human-Computer Interaction.* Grand Rapids, Mich.: CRC, 1986.
Joinson, Adam N. *Understanding the Psychology of Internet Behaviour: Virtual Worlds, Real Lives.* New York: Palgrave Macmillan, 2003.
Suler, John. "Computer and Cyberspace Addiction." *International Journal of Applied Psychoanalytic Studies* 1 (2004): 359–382. Available online. URL: http://www-usr.rider.edu/~suler/psycyber/cybaddict.html. Accessed November 8, 2007.
———. "The Psychology of Cyberspace." Available online. URL: http://www-usr.rider.edu/~suler/psycyber/psycyber.html. Accessed November 8, 2007.
Turkle, Sherry. *Life on the Screen: Identity in the Age of the Internet.* New York: Touchstone, 1995.
———. *The Second Self: Computers and the Human Spirit.* Twentieth anniversary ed. Cambridge, Mass.: MIT Press, 2005.
Wallace, Patricia. *The Psychology of the Internet.* New York: Cambridge University Press, 1999.
Weinberg, Gerald. *The Psychology of Computer Programming.* Silver anniversary ed. New York: Dorset House, 1998.
Whitty, Monica T., and Adrian N. Carr. *Cyberspace Romance: The Psychology of Online Relationships.* New York: Palgrave Macmillan, 2006.

punched cards and paper tape

In 1804, the French inventor Joseph-Marie Jacquard invented an automatic weaving loom that used a chain of punched cards to control the pattern in the fabric. A generation later, a British inventor (see BABBAGE, CHARLES) decided that punched cards would be a suitable medium for inputting data into his proposed mechanical computer, the Analytical Engine.

Although Babbage's machine was never built, by 1890 an American inventor was using an electromechanical tabulating machine to process census data punched into cards (see HOLLERITH, HERMAN). Card tabulating machines were improved and marketed by International Business Machines (IBM) throughout the first part of the 20th century. IBM would also create the 80-column standard punched card that would become familiar to a generation of programmers.

Later machines included features such as mechanical sorting, enhanced arithmetic functions, and the ability to group cards by a particular criterion and print subtotals, counts, or other information about each group. Although these machines were not computers, they did introduce the idea of automated data processing.

During the 1930s, a number of companies introduced punch card tabulators that could work with alphanumeric data (that is, letters as well as numbers). With these expanded capabilities, punch card systems could be used to keep track of military recruits, taxpayers, or customers (such as insurance policy holders). IBM emphasized the new machines' features by calling them "accounting machines" instead of tabulators.

While tabulators and calculators using punched cards gave a taste of the power of automated data processing, they had a very limited programming ability. For example, they could not make more than very simple comparisons or decisions, and could not repeat steps under program control (looping). The desire to create a general-purpose data processing system led in the 1940s to the development of the electronic computer.

When the first computers were developed, it was natural to turn to the existing punched cards and their machinery for a medium for inputting data and program instructions into the new machines. Because computers contained working memory, the program could be stored in its entirety during processing, enabling looping, subroutines, and other ways to control processing. Because the amount of available memory or "core" was severely limited, not much data could be stored inside the computer. However, complicated processing could be broken into a series of steps where a program was loaded and run, the input data cards read and processed, and the intermediate results punched onto a set of output cards. The card could then be input to another program to carry out the next phase.

By the 1970s, however, faster and easier to use media such as magnetic tape and disk drives were being employed for program and data storage. Instead of having to use a keypunch machine to create each program statement, programmers could type their commands at a terminal, using a text editor (see PROGRAMMING ENVIRONMENT). Even the government began to phase out punched cards. Today some "legacy" punch card systems are maintained, and there is sometimes a need to read and convert archival data in punch card form.

Ironically, this workhorse of early data processing would surface again in the U.S. presidential election of 2000, when problems with the interpretation of partly punched "chads" on ballot punch cards would lead to great controversy.

Further Reading

Cardamation Company. Available online. URL: http://www.cardamation.com/. Accessed August 17, 2007.

Dyson, George. "The Undead: The Little Secret That Haunts Corporate America: A Technology That Won't Go Away." *Wired* 7.03 (March 1999): 141–145, 170–172. Available online. URL: http://www.wired.com/wired/archive/7.03/punchcards.html. Accessed August 17, 2007.

Philips, N. V. "Everything About Punch Cards." Available online. URL: http://www.museumwaalsdorp.nl/computer/en/punchcards.html. Accessed August 17, 2007.

Province, Charles M. "IBM Punch Cards in the Army." Available online. URL: http://www.geocities.com/pattonhq/ibm.html. Accessed February 9, 2008.

Python

Created by Guido van Rossum and first released in 1990, Python is a relatively simple but powerful scripting language (see SCRIPTING LANGUAGES and PERL). The name comes from the well-known British comedy group Monty Python.

Python is particularly useful for system administrators, webmasters, and other people who have to link various files, data source, or programs to perform their daily tasks. The language currently has a small but growing (and quite enthusiastic) following.

Python dispenses with much of the traditional syntax used in the C family of languages. For example, the following little program converts a Fahrenheit temperature to its Celsius equivalent:

```
temp = input("Farenheit temperature:")
print (temp-32.0) *5.0/9.0
```

Without the semicolons and braces found in C and related languages, Python looks rather like BASIC. Also note that the type of input data doesn't have to be declared. The runtime mechanism will assume it's numeric from the expression found in the print statement. Python programs thus tend to be shorter and simpler than C, Java, or even Perl programs. The simple syntax and lack of data typing does not mean that Python is not a "serious" language, however. Python contains full facilities for object-oriented programming, for example.

Python programs can be written quickly and easily by trying commands out interactively and then converting the script to bytecode, a machine-independent representation that can be run on an interpreter designed for each machine environment. Alternatively, there are translation programs that can convert a Python script to a C source file that can then be compiled for top speed.

Perl is still a popular scripting language for UNIX and Web-related applications. Perl contains a powerful built-in regular expression and pattern-matching mechanism, as well as many other built-in functions likely to be useful for practical scripting. Python, on the other hand, is a more generalized and more cleanly structured language that is likely to be suited for a wider variety of applications, and it is more readily extensible to larger and more complex applications.

Further Reading

Lutz, Mark. *Learning Python*. 3rd. ed. Sebastapol, Calif.: O'Reilly, 2007.

———. *Programming Python*. 3rd ed. Sebastapol, Calif.: O'Reilly, 2006.

Python Programming Language—Official Web site. Available online. URL: http://www.python.org/. Accessed August 17, 2007.

Zelle, John M. *Python Programming: An Introduction to Computer Science*. Wilsonville, Ore.: Franklin, Beedle & Associates, 2004.

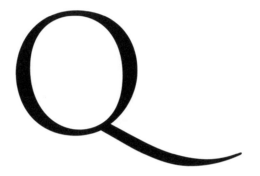

quality assurance, software

Modern software programs are large and complex, and contain many interrelated modules. If a program is not thoroughly tested before it goes into service, it may contain errors that can result in serious consequences (see RISKS OF COMPUTING).

In the early days of computing, programmers generally tested their code informally and nonsystematically. The assumption was that after the program was given to the users any problems that arose could be fixed through "patches" or replacement versions containing bug fixes. Today, however, it is increasingly recognized that assuring the quality and reliability of software requires a systematic, comprehensive process that begins when software requirements are first specified and continues after the program has been released.

Any program is designed to meet the needs of a specific type of users for specific applications. Therefore, the first step must be to make sure that users are able to communicate their requirements and that the software engineers understand the users' needs and concerns. Detailed written specifications, flowcharts, and other depictions of the program can be reviewed by user representatives (see FLOWCHART and CASE). The specifications can be further explored by creating a prototype or demonstration of the program's features (see PRESENTATION SOFTWARE). Since a prototype can be dynamic and let users have simulated interactions with the program, it may reveal usability problems that would be hard to spot from charts or documentation. The result of this initial verification process should be that the users agree that the program will do what they need and that they will be comfortable using it.

In moving from design to implementation (writing the actual code), the developers must first choose an appropriate approach (see ALGORITHM) and data representation. Choosing an algorithm that is known to be sound is preferable, but if an algorithm must be modified (or a new one developed), developers may be able to take advantage of mathematical techniques that will suggest, if not totally prove, the algorithm's accuracy and reliability.

As the programmers write the code, they should try to use best practices (see SOFTWARE ENGINEERING). Doing so ensures that the code will be readable and organized in such a way that the source of a problem area can be identified easily, and any "fix" that must be made will be less likely to have unforeseen side effects.

Developers can also include special code that will facilitate testing. This code can include assertions—statements that test specified conditions (such as variable values) at key points in the program, displaying appropriate messages if the values are not within the proper range. Large, complex programs can also include diagnostic modules that give the developers a sort of virtual console that they can use to monitor conditions while the program is running, or "drill into" particular areas for closer inspection (see BUGS AND DEBUGGING).

Although a certain amount of testing and debugging can and should be done while the code is being written, more extensive and systematic testing is usually performed after a preliminary version of the program has been completed. (This is sometimes called an *alpha version*.) There are two basic approaches to designing the tests. "White box" tests use the developer's knowledge of the code to design test data that will test all of the program's structural features (see PROCE-

DURES AND FUNCTIONS, BRANCHING STATEMENTS, and LOOP). The testers may be aided by mathematical analysis that identifies "partitions" or ranges within the data that should result in a particular execution path being taken through the program. "Fault coverage" tests can also be designed to test for various specific types of errors (such as input/output, numeric overflow, loss of precision, and so on).

A shortcoming of white box tests is that because the tester knows how the program works, he or she may unconsciously select mainly "reasonable" data or situations. (It has been observed that users are under no such compulsion!) One way to compensate for this bias is to also perform "black box" tests. These tests assume no knowledge of the inner workings of the program. They approach the program from the outside, submitting data (or otherwise interacting with the program) either through the user interface or using an automated process that simulates user input. The tester tries to generate as wide a variety of input data as possible, often by using randomization techniques. The result is that the ability of the program to deal with "unreasonable" data will also be tested, and unforeseen situations may arise and have to be dealt with.

Once this cycle of testing and fixing problems is finished, the program will probably be given to a selected group of users who will operate it under field conditions—that is, in the same sort of environment the program will be used once it is sold or deployed. This process is sometimes called *beta testing*. (Game companies have traditionally relied upon the willingness of gamers to test a new game in exchange for getting to play it sooner.)

The priority (and thus the resources) devoted to testing will vary according to many factors, including

- the complexity of the program (and thus the likelihood of problems)

- the presence of strong competitors who could take advantage of significant problems with the program

- the potential financial impact or legal exposure from bugs or problems

- the ability to "amortize" the costs of developing testing tools and procedures over a number of years as new versions of the program are developed

The Holy Grail for quality assurance would be to develop powerful artificially intelligent automatic testing programs that could analyze a program and develop and execute a variety of thorough tests. However, such a program would itself be very complex, difficult and expensive to develop, and subject to its own bugs. Nevertheless, a number of organizations (notably, IBM) have devoted considerable attention to the problem.

Further Reading

Ginac, Frank P. *Customer-Oriented Software Quality Assurance.* Upper Saddle River, N.J.: Prentice Hall PTR, 1997.
Godbole, Nina S. *Software Quality Assurance: Principles and Practice.* Pangbourne, U.K.: Alpha Science International, 2004.
Schulmeyer, G. Gordon, ed. *Handbook of Software Quality Assurance.* 4th ed. Boston: Artech House, 2007.
Software QA and Testing Resource Center. Available online. URL: http://www.softwareqatest.com/. Accessed August 17, 2007.

quantum computing

The fundamental basis of electronic digital computing is the ability to store a binary value (1 or 0) using an electromagnetic property such as electrical charge or magnetic field.

However, during the first half of the 20th century, physicists discovered the laws of quantum mechanics that apply to the behavior of subatomic particles. An electron or photon, for example, can be said to be in any one of several "quantum states" depending on such characteristics as spin. In 1981, physicist Richard Feynman came up with the provocative idea that if quantum properties could be "read" and set, a computer could use an electron, photon, or other particle to store not just a single 1 or 0, but a number of values simultaneously. The simplest case, storing two values at once, is called a "qubit" (short for "quantum bit"). In 1985, David Deutsch at Oxford University fleshed out Feynman's ideas by creating an actual design for a "quantum computer," including an algorithm to be run on it.

At the time of Feynman's proposal, the techniques for manipulating individual atoms or even particles had not yet been developed (see NANOTECHNOLOGY), so a practical quantum computer could not be built. However, during the 1990s considerable progress was made, spurred in part by the suggestion of Bell Labs researcher Peter Shor, who outlined a quantum algorithm that might be used for rapid factoring of extremely large integers. Since the security of modern public key cryptography (see ENCRYPTION) depends on the difficulty of such factoring, a working quantum computer would be of great interest to spy agencies.

The reason for the tremendous potential power of quantum computing is that if each qubit can store two values simultaneously, a register with three qubits could store eight values, and in general, for n qubits one can operate on 2^n values simultaneously. This means that a single quantum processor might be the equivalent of a huge number of separate processors (see MULTIPROCESSING). Clearly many problems that have been considered not practical to solve (see COMPUTABILITY AND COMPLEXITY) might be tackled with quantum computers.

However, the practical problems involved in designing and assembling a quantum computer are expected to be very formidable. Although scientists during the 1990s achieved the ability to arrange individual atoms, the precise placement of atoms and even individual particles such as photons would be difficult. Furthermore, as more of these components are assembled in very close proximity, it becomes more likely that they will interfere with one another, causing "decoherence," where the superimposed values "break down" to a single 1 or 0, thus causing loss of information. However, some researchers are hopeful that standard mathematical techniques (see ERROR CORRECTION) could be used to keep this problem in check. For example, redundant components could be used so that even if one decoheres, the others could be used to regenerate the information.

Another approach is to use a large number of quantum components to represent each qubit. In 1998, Neil Gershenfeld and Isaac L. Chuang reported successful experiments using a liquid with nuclear magnetic resonance (NMR) technology. Here each atom in a molecule (for

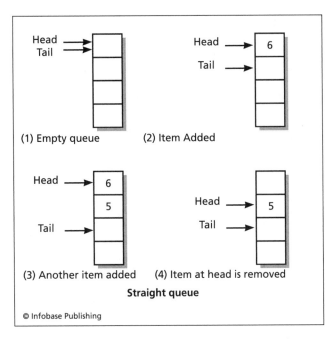

Straight queue

© Infobase Publishing

In an empty queue, the head and tail pointers point to the first cell in memory. To add a value, it is placed at the cell pointed to by the head pointer, and the tail pointer is moved up one cell. If an item is removed, the head and tail pointers are moved down one place. (Note that items must be added at the tail and removed at the head.)

example, chloroform), would represent one qubit, and a large number of molecules would be used, for redundancy. Since each "observation" (that is, setting or reading data) affects only a few of the many molecules for each qubit, the stability of the information in the system is not compromised. However, this approach is limited by the number of atoms in the chosen molecule—perhaps to 30 or 40 qubits.

There are many potential applications for quantum computing. While the technology could be used to crack conventional cryptographic keys, researchers have suggested that it could also be used to generate unbreakable keys that depend on the "entanglement" of observers and what they observe. The sheer computational power of a quantum computer might make it possible to develop much better computer models of complex phenomena such as weather, climate, the economy—or of quantum behavior itself.

Further Reading

Burda, Ioan. *Introduction to Quantum Computation*. Boca Raton, Fla.: Universal Publishers, 2005.

Kaye, Phillip, Raymond LaFlamme, and Michele Mosca. *An Introduction to Quantum Computing*. New York: Oxford University Press, 2007.

Quantum Computer News. Science Daily. Available online. URL: http://www.sciencedaily.com/news/computers_math/quantum_computers/. Accessed August 17, 2007.

West, Jacob. *The Quantum Computer*. Available online. URL: http://www.cs.caltech.edu/~westside/quantum-intro.html. Accessed August 17, 2007.

queue

A queue is basically a "line" of items arranged according to priority, much like the customers waiting to check out in a supermarket. Many computer applications involve receiving, tracking, and processing requests. For example, an operating system running on a computer with a single processor must keep track of which application should next receive the processor's attention. A print spooler holds documents waiting to be printed. A web or file server must keep track of requests for Web pages, files, or other services. Queues provide an orderly way to process such requests. Queues can also be used to efficiently store data in memory until it can be processed by a relatively slow device such as a printer (see BUFFERING).

As a data structure, a queue is a type of list (see LIST PROCESSING). New items are inserted at one end and removed (deleted) from the other end. This contrasts with a stack, where all insertions and deletions are made at the same end (see STACK). Just as the next person served at the supermarket is the one at the head of the line, the end of a queue from which items are removed is called the head or front. And just as new people arriving at the supermarket line join the end of the line, the part of the queue where new items are added is called the tail or rear. Since the first item in line is the first to be removed, a queue is called a FIFO (first in, first out) structure.

To create a queue, a program first allocates a block of memory. It then sets up to pointers (see POINTERS AND INDIRECTION). One pointer stores the address of the item at the head of the queue; the other has the address of the item at the tail. When the queue starts out, it is empty. This means that both the head and tail pointer start out pointing to the same location.

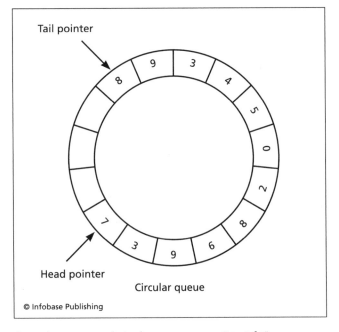

Circular queue

© Infobase Publishing

A circular queue works in the same way as a "straight" queue, except that when the last cell in the allotted memory block is reached, the pointer or data "wraps around" to the first cell.

To add an item, the tail pointer is moved back one location and the item is stored there. To remove an item, the head pointer is simply moved back one location. (The data that had been pointed to by the head pointer can be either retrieved or discarded, depending on the application.)

In actuality it's not quite so simple. As items are added to the queue, the tail pointer keeps moving back in memory with the head pointer trailing behind as items are deleted. If the queue is sufficiently active (many items are being added and removed), the queue will end up "crawling" through memory somewhat like a worm until all the memory is consumed.

In a real line at the supermarket, as a customer leaves the checkout stand, each of the persons in line moves up one space. In a computer queue this could be accomplished by moving each item up one location whenever an item is removed at the head. However, having to move all the data items each time one is changed would be very inefficient. Instead, one could allow the head of the queue to move only up to some specified location. At that point, the head is moved back to the beginning of the memory block, and thus the space that had been vacated by the tail as it moved up is reutilized. In effect this wraps the memory around into a circle, so this is called a *circular queue*.

Further Reading

Brookshear, J. Glenn. *Computer Science: An Overview.* 6th ed. Reading, Mass.: Addison-Wesley, 2000.

Skiena, Steven S. "Priority Queues." Available online. URL: http://www2.toki.or.id/book/AlgDesignManual/BOOK/BOOK3/NODE130.HTM. Accessed August 17, 2007.

Suh, Eric. "The Queue Data Structure." Available online. URL: http://www.cprogramming.com/tutorial/computerscience theory/queue.html. Accessed August 17, 2007.

R

RAID (redundant array of inexpensive disks)

Computer storage is relatively cheap today (see HARD DISK), but having continued access to data in the event of hardware failure is essential to any enterprise. RAID, or redundant array of inexpensive disks, is a way to turn plentiful storage into higher reliability and/or speed of access. RAID works by turning a group of drives into a single logical unit; the operating system need not deal with this internal organization, but simply reads or writes data as usual.

To improve reliability, data can be *mirrored,* or copied to two or more disks. While the data obviously takes up more space, the advantage is that the data remains intact and recoverable if any one drive fails. Further reliability can be achieved by storing redundant data (such as parity bits or Hamming codes), to diagnose and fix some disk problems (see ERROR CORRECTION and FAULT TOLERANCE).

To achieve greater speed of data access, data can be "striped," where a file is broken into pieces, with each piece stored on a sector on a different drive. Thus instead of the head of a single drive having to jump around to multiple sectors to read the data, the heads on all the drives can simultaneously read many parts of the file, which are then assembled into the proper order.

LEVELS AND COMPROMISES

By combining mirroring, error correction, and/or striping, different "levels" of RAID can be implemented to suit different needs. There are various trade-offs: Striping can increase access speed, but uses more storage space and, by increasing the number of disks, also increases the chance that one will fail. Implementing error correction can make failure recoverable, but slows data access down because data has to be read from more than one location and compared.

The most commonly used RAID levels are:

- RAID 0—striping data across disks, higher speed but no error correction; failure of any disk can make data unrecoverable

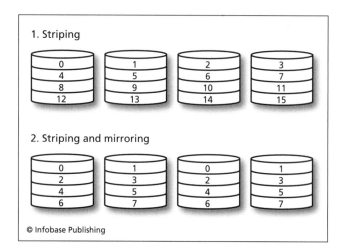

© Infobase Publishing

Striping spreads data across several disk drives so that a single head movement on each drive can fetch a large amount of data. Mirroring duplicates each sector of data on a second disk drive, ensuring that if one drive fails the data can still be retrieved. Combinations of both techniques are often used, trading space for reliability (or vice versa).

- RAID 1—mirroring (data stored on at least two disks), data intact as long as one disk is still operating

- RAID 3 and 4—striping plus a dedicated disk for parity (error checking)

- RAID 5—striping with distributed parity; data can be restored automatically after a failed disk is replaced

- RAID 6—like RAID 5 but with parity distributed so that data remains intact unless more than two drives fail

In actuality, RAID configurations can be very complex, where different levels can be "layered" above one another, with each treating the next as a virtual drive, until one gets down to the actual hardware. Although RAID is often implemented using a physical (hardware) controller, operating systems can also create a virtual RAID structure in software, interposed between the logical drive as seen by the read/write routines and the physical drives.

Although RAID is most commonly used with large shared storage units (see FILE SERVER and NETWORKED STORAGE), with the drastic decline in hard drive prices, simple RAID configurations (such as two mirrored drives) are also appearing in higher-end desktop PCs.

Further Reading

Leider, Joel. "How to Select a RAID Disk Array." Enterprise Storage Forum. Available online. URL: http://www.enterprisestorage-forum.com/hardware/features/article.php/726491. Accessed November 8, 2007.

"Redundant Array of Inexpensive Disks (RAID)." PC Guide. Available online. URL: http://pcguide.com./ref/hdd/perf/raid/index.htm. Accessed November 8, 2007.

random number generation

Computer applications such as simulations, games, and graphics applications often need the ability to generate one or more random numbers (see SIMULATION and COMPUTER GAMES). Random numbers can be defined as numbers that show no consistent pattern, with each number in the series neither affected in any way by the preceding number, nor predictable from it.

One way to get random digits is to simply start with an arbitrary number with a specified number of digits, perhaps 10. This first number is called the seed. Multiply the seed by a constant number of the same length, and take that number of digits off the right end of the product. The result becomes the new seed. Multiply it by the original constant to generate a new product, and repeat as often as desired. The result is a series of digits that appear randomly distributed as though generated by throwing a die or spinning a wheel. This type of algorithm is called a congruential generator.

The quality of a random number generator is proportional to its period, or the number of numbers it can produce before a repeating pattern sets in. The period for a congruential generator is approximately 2^{32}, quite adequate for many applications. However, for applications such as very large-scale simulations, different algorithms (called shift-register and lagged-Fibonacci) can be used, although

these also have some drawbacks. Combining two different types of generators produces the best results. The widely used McGill Random Number Generator Super-Duper combines a congruential and a shift-register algorithm.

Generating a random number series from a single seed will work fine with most simulations that rely upon generating random events under the control of probabilities (Monte Carlo simulations). However, although the sequence of numbers generated from a given seed is randomly distributed, it is always the same series of numbers for the same seed. Thus, a computer poker game that simply used a given seed would always generate the same hands for each player. What is needed is a large collection of potential seeds from which one can be more or less randomly chosen. If there are enough possible seeds, the odds of ever getting the same series of numbers become vanishingly small.

One way to do this is to read the time (and perhaps date) from the computer's system clock and generate a seed based on that value. Since the clock value is in milliseconds, there are millions of possible values to choose from. Another common technique is to use the interval between the user's keystrokes (in milliseconds). Although they are not perfect, these techniques are quite adequate for games.

So-called true random number generators extract random numbers from physical phenomena such as a radioactive source (the HotBits service at Fourmilab in Switzerland) or even atmospheric noise as detected by a radio receiver. For the ultimate in random numbers, researchers have looked to quantum processes that are inherently random. In 2007 researchers at an institute in Zagreb, Croatia, began to offer the Quantum Random Bit Generator Service, which is keyed to unpredictable emissions of photons in a semiconductor. The output of most random number services can be interfaced with MATLAB and other popular mathematical software packages.

Further Reading

Gentle, James E. *Random Number Generation and Monte Carlo Methods.* 2nd ed. New York: Springer, 2004.

HotBits: Genuine Random Numbers, Generated by Radioactive Decay. Available online. URL: http://www.fourmilab.ch/hotbits/. Accessed August 18, 2007.

"Introduction to Randomness and Random Numbers." Available online. URL: http://www.random.org. Accessed August 18, 2007.

real-time processing

There are many computer applications (such as air traffic control or industrial process control) that require that the system respond almost immediately to its inputs.

In designing a real-time system there are always two questions to answer: Will it respond quickly enough most of the time? How much variation in response time can we tolerate? A system that responds to real-time environmental conditions (such as the amount of traction or torque acting on a car's wheels) needs to have a sampling rate and a rate of processing the sampled data that's fast enough so that the system can correct a dangerous condition in time. The responsiveness required of course varies with the situation and with the potential consequences of failure. An air traffic control system may be able to take a few seconds between processing radar samples, but it better get it right in time. Systems like this where real-time response is absolutely crucial are sometimes called "hard real-time systems."

Other systems are less critical. A streaming audio system has to keep its buffer full so it can play in real time, but if it stutters once in a while, no one's life is in danger. (And since download rates over the Internet can vary for many reasons it's not realistic to expect too perfect a level of performance.) A slower "soft real-time system" like a bank's ATM system should be able to respond in tens of seconds, but if it doesn't, the consequences are mainly potential loss of customers and revenue. A fairly wide variation in response time may be acceptable as long waits don't occur often enough to drive away too many customers.

To put together the system, the engineer must look at the inherent speed of the sampling device (such as radar, camera, or simply the keyboard buffer). The speed of the processor(s) and the time it takes to move data to and from memory are also important. The structure, strengths, and weaknesses of the host operating system can also be a factor. Some operating systems (including some versions of UNIX) feature a guaranteed maximum response time for various operating system services. This can be used to help calculate the "worst case scenario"—that combination of inputs and the existing state of the system that should result in the slowest response.

Another approach available in most operating systems is to assign priority to parts of the processing so that the most critical situations are guaranteed to receive the attention of the system. However, things must be carefully tuned so that even lower priority tasks are accomplished in an acceptable length of time.

The design of the data structure or database used to hold information about the process being monitored is also important. In most databases the age of the data is not that important. For a payroll system, for example, it might be sufficient to run a program once in a while to weed out people who are no longer employees. For a nuclear power plant, if data is getting too old such that it's not keeping up with current condition, some sort of alarm or even automatic shutdown might be in order. With a system that has softer constraints (such as an automatic stock trading system), it may be enough to be able to get most trades done within a specified time and to gather data about the performance of the system so the operators can decide whether it needs improvement.

Real-time systems are increasingly important because of the importance of the activities (such as air traffic control and power grids) entrusted to them, and because of their pervasive application in everything from cars to cell phones to medical monitors (see EMBEDDED SYSTEM and MEDICAL APPLICATIONS OF COMPUTERS). The systems also tend to be increasingly complex because of the increasing interconnection of systems. For example, many real-time systems have to interact with the Internet, with communications services, and with ever more sophisticated multimedia display systems. Further, many real-time systems must use multiple processors (see MULTIPROCESSING), which can increase the robustness and reliability of the system but also the complexity of its architecture, and thus the difficulty in determining and ensuring reliability.

Further Reading

Buttazzo, Giorgio C. *Hard Real-Time Computing Systems: Predictable Scheduling Algorithms and Applications*. 2nd ed. New York: Springer, 2005.
Cheng, Albert M. K. *Real-Time Systems: Scheduling, Analysis, and Verification*. Hoboken, N.J.: Wiley, 2002.
Resources for Real-Time Computing. TechRepublic. Available online. URL: http://search.techrepublic.com.com/search/real-time+computing.html. Accessed August 19, 2007.

recursion

Even beginning programmers are familiar with the idea that a series of program statements can be executed repeatedly as long as (or until) some condition is met (see LOOP). For example, consider this simple function in Pascal. It calculates the factorial of an integer, which is equal to the product of all the integers from 1 to the number. Thus factorial 5, or 5! = 1 * 2 * 3 * 4 * 5 = 120.

```
Function Factorial (n: integer) : integer
Begin
    i: integer;
    For i = 1 to n do
        Factorial := Factorial * i;
End.
```

If the main program has the line:

```
Writeln (Factorial (5));
```

then the 5 is sent to the function, where the loop simply multiplies the numbers from 1 to 5 and returns 120.

However, it is also possible to have a function call *itself* repeatedly until a specified condition is met. This is called

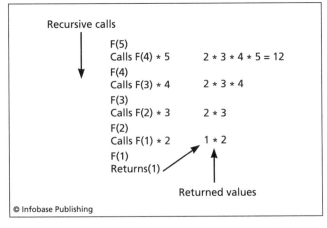

*In recursion, a procedure calls itself until some defined condition is met. In this example of a Factorial procedure, F(1) is defined to return 1. Once it does, the returned value is plugged into its caller, which then returns the value of 1 * 2 to its caller, and so on.*

recursion, and it allows for some compact but powerful coding. A recursive version of the Factorial function in Pascal might look like this:

```
function Factorial (n:integer) :integer
begin
  if (n = 1) then
    Factorial := 1
  else
    Factorial := Factorial (n - 1) * n;
end;
```

Why does this work? An alternative way to define a factorial is to say that the factorial of a number is that number times the factorial of one less than the number. Thus, the factorial of 5 is equal to 5 * 4! or 5 * 4 * 3 * 2 * 1. But in turn the factorial of 4 would be equal to 4 * (3 * 2 * 1), and so on down to the factorial of 1, which is simply 1. Thus, in general terms the factorial of n is equal to n * factorial (n - 1).

What happens if this function is called by the program statement:

```
Writeln ("Factorial of 5 is "); Factorial (5)
```

First, the Factorial function is called with the value 5 assigned to n. The If statement checks and sees that n is not 1, so it calls factorial (i.e., itself) with the value of n - 1, or 4. This new instance of the factorial function gets the 4, sees that it is not 1, and calls factorial again with n - 1, which is now 4 - 1 or 3. This continues until n is 2, at which point factorial 1 is called. But this time n is 1, so it returns the value of 1 rather than calling itself yet again.

Now the returned value of 1 replaces the call to Factorial (n - 1) in the preceding instance of Factorial (where n had been 2). That 1 is therefore multiplied by 2, and 1 * 2 = 2 is returned to the preceding instance, where n had been 3. Now that 2 gets multiplied by 3 and returned to the instance where n had been 4. This continues until we're back at the first call to factorial 5, where the value of 4 * 3 * 2 * 1 now gets multiplied by that 5, giving 120, or factorial 5. (See the accompanying diagram for help in visualizing this process.)

A RECURSIVE SORTING ALGORITHM

In the preceding example recursion does no more than a simple loop could, but many problems lend themselves more naturally to a recursive formulation. For example, suppose you have an algorithm to merge (combine) two lists of integers that have been sorted into ascending values. The procedure simply takes the smaller of the two numbers at the front of the two lists until one list runs out of numbers (any numbers in the remaining list can then simply be included).

Using an English-like syntax, one can write a recursive procedure to sort a list of numbers by calling itself repeatedly, then using the Merge procedure:

```
Procedure Sort
Begin
  If the list has only one item, return
  Else
    Sort the first half of the list
```

```
    Sort the second half of the list
    Merge the two sorted lists
  End If
End (Sort)
```

Sort will call itself until one of the lists has only one item (which by definition is "sorted"), and the Merge procedure will build the sorted list.

To implement recursion, the run-time system for the language must use an area of memory (see STACK) to temporarily store the values associated with each instance of a function as it calls itself. Depending on the implementation, there may be a limit on how many levels of recursion are allowed, or on the size of the stack. (However, the plentiful supply of available memory on most systems today makes this less of an issue.)

The first generation of high-level computer languages (such as FORTRAN and COBOL) did not allow recursion. However, the second generation of procedural languages starting around 1960 with Algol, as well as successors such as Pascal and C do allow recursion. The LISP language (see LISP and FUNCTIONAL LANGUAGES) uses recursive definitions extensively, and recursion turns out to be very useful for processing the grammars for artificial and natural languages (see PARSING). Recursion can also be used to generate interesting forms of graphics (see FRACTALS IN COMPUTING).

Further Reading
Hillis, W. Daniel. *The Pattern in the Stone: The Simple Ideas that Make Computers Work.* New York: Basic Books, 1998.
McHugh, John. "The Animation of Recursion." Available online. URL: http://www.animatedrecursion.com/home. Accessed August 19, 2007.
Roberts, E. S. *Thinking Recursively.* New York: Wiley, 1986.

reduced instruction set computer (RISC)

All things being equal, the trend in computer design is to continually add new features. There are several reasons why this is the case with computer processors:

- to create a "family" of upwardly compatible computers (see COMPATABILITY AND PORTABILITY)

- to make a new machine more competitive with existing systems, or to give it a competitive advantage

- to make it easier to write compilers for popular languages

- to allow for more operations to be done with one (or a few) instructions rather than requiring many instructions

There are certainly exceptions to the trend toward complexity. The minicomputer, for example, represented in some ways a simplification of the exiting mainframe design. It didn't have as many ways of working with memory (see ADDRESSING) and lacked the multiple input/output "channels" and their separate processors. But once minicomputers were introduced and achieved success, the same competitive and other pressures led their designers to start adding complexity.

One way processor designers coped with the demand for more complicated instructions was to give the main processor a microprocessor with its own set of simple instructions. When the main processor received one of the complex instructions, it would be executed by being broken down into simpler instructions or "microcode" to be executed by the sub-processor.

This approach gave processor designers greater flexibility. It also made things easier for compiler designers, because the compiler could translate higher-level language statements into fewer, more complex instructions, leaving it to the hardware with its micro engine to break them down into the ultimate machine operations. However, it also meant that the processor had to decode and execute more instructions in every processor cycle, making it less efficient and slower and losing some of the benefits of the faster processors that were becoming available.

In 1975, John Cocke and his colleagues at IBM decided to build a new minicomputer architecture from the ground up. Instead of using complex instructions and decoding them with a micro engine, they would use only simple instructions that could be executed one per cycle. The clock (and thus the cycle time) would be much faster than for existing machines, and the processor would use pipelining so it could decode the next instruction while still executing the previous one. Similarly, in many cases the next item of data needed could be fetched at the same time the data from the previous step was being written (stored). This approach became known as reduced instruction set computing (RISC), because the number of instructions had been reduced compared to exiting systems, which then became known as complex instruction set computing (CISC).

Since the RISC system had only simple instructions, compilers could no longer use many complicated but handy instructions. The compiler would have to take over the job of the micro engine and break all statements down into the basic instructions. It became important that the compiler be able to generate the optimal set of instructions by analyzing how data would have to be moved around in the machine's registers and memory. In other words, RISC hardware gained higher performance through simplification at the hardware level but at the cost of making compilers more complicated. Fortunately, both hardware and software designers were able to meet the challenge and in the process learn how to get the most out of new technology.

RISC would also play a part in the design of the microprocessors that began to power personal computers. For example, the DEC Alpha, a "pure" RISC chip introduced in 1992, provided a level of power that made it suitable for high-performance workstations. Another successful RISC-based development has been the SPARC (Scalable Processor ARCHitecture) developed by Sun Microsystems for servers, computer clusters, and workstations.

Perhaps the most interesting development, however, has been the gradual application of RISC principles to mainstream processors such as the Intel 80×86 series used in most personal computers today. Increasingly, the recent Pentium series chips, while supporting their legacy of CISC instructions, are processing them using an inner architecture that uses RISC principles and takes advantage of pipelining, as well as using more registers and a larger data cache. However, the sheer increase in clock cycle speed and performance in the newer chips has made the old tradeoff between complicated and simple instructions less relevant.

Further Reading

Dandamudi, Sivarama P. *Guide to RISC Processors for Programmers and Engineers: Introduction to Assembly Language Programming for Pentium and RISC Processors.* New York: Springer, 2005.
Knuth, Donald E. *MMIX—A RISC Computer for the New Millennium.* Vol. 1, fascicle 1 of *Art of Computer Programming.* Upper Saddle River, N.J.: Addison-Wesley Professional, 2005.

regular expression

Many users of UNIX and the old MS-DOS are familiar with the ability to use "wildcards" to find filenames that match specified patterns. For example, suppose a user wants to list all of the TIF graphics files in a particular directory. Since these files have the extension .tif, a UNIX ls command or a DOS dir command, when given the pattern *.tif, will match and list all the TIF files. (One does have to be aware of whether the operating system in question is case-sensitive. UNIX is, while MS-DOS is not.)

The specification *.tif tells the command "match all files whose names consist of one or more characters and that end with a period followed by the letters tif." It is one of many possible regular expressions. (See the accompanying table for more examples.) The asterisk here is a "metacharacter." This means that it is not treated as a literal character, but as a pattern that will be matched in a specified way.

Most operating systems that have command processors (see SHELL) allow for some form of regular expressions, but don't necessarily implement all of the metacharacters. UNIX provides the most extensive use for regular expressions (see UNIX). UNIX has an operating system facility called *glob* that expands regular expressions (that is, substitutes for them whatever matches) and passes them on to the many UNIX tools or utilities designed to work with regular expressions. These tools include editors such as ex and vi, the character translation utility (tr), the "stream editor" (sed), and the string-searching tool grep. For example, sed can be used to remove all blank lines from a file by specifying

```
sed 's/^$/d' list.txt
```

This command finds all lines with no characters (^$) in the file list.txt and deletes them from the output. Even more extensive use of pattern-matching with regular expressions is found in many scripting languages (see SCRIPTING LANGUAGES, AWK, and PERL).

It is true that most of today's computer users don't enter operating system commands in text form but instead use menus and manipulate icons (see USER INTERFACE and MICROSOFT WINDOWS). If such a user wants to change one word to another throughout a word processing document, he or she is likely to open the Edit menu, select Find,

METACHARACTERS IN REGULAR EXPRESSIONS

METACHARACTER	MEANING
. (period)	Matches any single character in that position
?	Matches zero or one of any character
*	Matches zero or more of the preceding character (thus * matches any number of characters)
+	Matches one or more of the preceding character (thus 9+ matches 9, 99, 999, etc.)
[ ]	Matches any of the characters enclosed by the brackets
–	Specifies a range of characters. Placing the range in brackets will match any character within the range. For example, [0–9] matches any digit, [A–Z] matches any uppercase character, and [A–Za–z] matches any alphabetic character.
\	"Quotes" the following character. If it is a metacharacter, the following character will be treated as an ordinary character. Thus \? matches an actual question mark.
^	Matches the beginning of a line
$	Matches the end of a line

and type the "before" and "after" words into a dialog box. However, even in such cases if the user has some familiarity with regular expressions, more sophisticated substitutions can be accomplished. In Microsoft Word, for example, a variety of wildcards (i.e., metacharacters) can be used for operations that would be hard to accomplish through mouse selections.

Further Reading

Friedl, Jeffrey. *Mastering Regular Expressions.* 3rd ed. Sebastapol, Calif.: O'Reilly, 2006.

Regular Expressions Tutorial, Tools & Languages, Examples, Books & Reference. Available online. URL: http://www.regular-expressions.info/. Accessed August 19, 2007.

Stubblebine, Tony. *Regular Expression Pocket Reference: Regular Expressions for Perl, Ruby, PHP, Python, C, Java and .NET.* Sebastapol, Calif.: O'Reilly, 2007.

Watt, Andrew. *Beginning Regular Expressions.* Indianapolis: Wiley, 2005.

research laboratories in computing

The value of creating and maintaining environments for long-term research in computer science and engineering has long been recognized by academic institutions, industry organizations, and corporations.

ACADEMIC RESEARCH INSTITUTIONS

Artificial intelligence and robotics have been the focus of many academic computer science research facilities (see ARTIFICIAL INTELLIGENCE and ROBOTICS). They are examples of areas that show great potential but that demand a substantial investment in long-term research. There are many research organizations in the AI field, but a few stand out as particularly important examples.

The Massachusetts Institute of Technology (MIT) Artificial Intelligence Lab has a wide-ranging program but has emphasized robotics and related fields such as computer vision and language processing.

The MIT Media Lab has become well known for work with new media technologies and the digital and graphical representation of data. However, in recent years it has expanded its focus to the broader area of human-machine interaction and the pervasive presence of intelligent devices in the home and larger environment.

The Stanford Artificial Intelligence Laboratory (SAIL) played an important role in the development of the LISP language (see LISP) and other AI research. Today Stanford's important role in AI is continued by its Robotics Laboratory and the Knowledge Systems Laboratory. Carnegie Mellon University also has a number of influential AI labs and research projects.

On the international scene Japan has had strong research programs in academic and industrial AI, such as the Neural Computing Center at Keio University and the Knowledge-Based Systems Laboratory at Shizuoka University. There are a number of important AI research groups in the United Kingdom, such as at Cambridge, Oxford, King's College, and the University of Edinburgh (where the logic language Prolog was developed).

Some of the most interesting research sometimes emerges from outside the main concerns of an institution. The World Wide Web, for example, was developed by Tim Berners-Lee (see BERNERS-LEE, TIM) while he was working with the coordination of scientific computing at CERN, the giant European particle physics laboratory.

CORPORATE RESEARCH INSTITUTIONS

The challenging nature of computer applications and the competitiveness of the industry have also led a number of major companies to underwrite permanent research institutions. Much corporate-funded research has gone into developing the basic infrastructure of computing rather than to the more esoteric topics pursued by academic departments. However, corporations have also funded "pure" research that may have little short-term application but can ultimately lead to new technologies.

The concept of the industrial laboratory is often attributed to Thomas Edison, whose famous Menlo Park, New Jersey, facility (founded in 1876) put experimentation and development of new inventions on a systematic, continuous basis. Instead of an invention forming the basis for a company, Edison saw invention itself as the core business.

A similar approach motivated the founding of BELL LABORATORIES. Bell Labs would play a direct role in making modern digital electronics possible when three of its researchers, John Bardeen, Walter H. Brattain, and William B. Shockley invented the transistor in 1947.

On the software side, Bell supported the work of Claude Shannon, whose fundamental theorems of information transmission would become a key to the design of the computer networks (see SHANNON, CLAUDE). The development of the UNIX operating system at Bell in the early 1970s (see RITCHIE, DENNIS and UNIX) would provide much of the infrastructure that would be used for computing at universities and other research institutions and ultimately in the development of the Internet. Similarly, Ritchie and Thompson also developed C, the language that together with its offshoots C++ and Java would become the most widely used general-purpose programming languages for the rest of the century and beyond.

IBM built its first research lab in 1945, beginning a network that would eventually include facilities in Switzerland, Israel, Japan, China, and India. IBM research has generally focused on core hardware and software technologies, including the development of the first hard drive in 1956 and the development of the FORTRAN language by John Backus in 1957 (see FORTRAN). Other IBM innovations have included online commerce (the SABRE airline reservation system), the relational database, and the first prototype RISC (reduced instruction set computer).

Xerox is best known for its photocopiers and printers, but in the late 1960s the company decided to try to diversify its products by recasting itself as developer of a comprehensive "architecture of information" in the office. During the 1970s, its Palo Alto Research Center (PARC) invented much of the technology (such as the mouse, graphical user interface, and notebook computer) that would become familiar to consumers a decade later in the Macintosh and Microsoft Windows.

In 1991, Microsoft, then a medium-sized company, established its Microsoft Research division, which has since grown to include four laboratories in Redmond, Washington, the San Francisco Bay Area, Cambridge, England, and Beijing. The labs maintain close ties with universities, and their research areas have included data mining and analysis, geographic information systems (Terraserver), natural language processing, and computer conferencing and collaboration.

The role of government agencies in funding computer-related research should not be overlooked (see GOVERNMENT FUNDING OF COMPUTER RESEARCH). The Internet evolved from a project funded by the Department of Defense's ARPA (Advanced Research Projects Agency) in 1968 (see INTERNET). The network architecture and hardware in turn were developed by a contractor, Bolt, Beranek and Newman (BBN). In the late 1970s, the Defense Department would issue contracts for development of the Ada computer language. Other projects funded by Defense and other government agencies can be found in areas such as robotics, autonomous vehicles, and mapping systems.

COORDINATING RESEARCH

Two large professional organizations for computer scientists and engineers, the Association for Computing Machinery (ACM) and the Computer Society of the Institute of Electrical and Electronics Engineers (IEEE), serve as clearinghouses and disseminators of research. The Computing Research Association (CRA) brings together more than 200 North American university computer science departments, government-funded research institutions, and corporate research laboratories. Its goal is to improve the opportunities for and quality of research and education in the computer field. (For contact information for these and other selected computer-related organizations, see Appendix IV.)

OTHER TYPES OF RESEARCH

The social impact of computing technology is also the subject of considerable ongoing research. Topics include consumer behavior, the use of media, and sociological analysis of online communities. A particularly useful effort is the extensive surveys and overviews produced by the Pew Center for the Internet and American Life project. The computer hardware, software, and e-commerce sectors are of course also the subject of research by economists, experts in organizational behavior, investment analysts, and so on.

Further Reading
Computing Research Association. Available online. URL: http://www.cra.org/. Accessed August 19, 2007.
Open Directory Project. Computer Science Research Institutes. Available online. URL: http://www.dmoz.org/Computers/Computer_Science/Research_Institutes/. Accessed August 19, 2007.
Pew Internet & American Life Project. Available online. URL: http://www.dmoz.org/Computers/Computer_Science/Research_Institutes/. Accessed August 19, 2007.
TRN's Research Directory: A Worldwide Listing of Technology Research Laboratories. Available online. URL: http://www.trnmag.com/Directory/directory.html. Accessed August 19, 2007.

reverse engineering

Back in the days of mechanical clocks, curious kids would sometimes take a clock apart to try to figure out how it worked. A few were even able to reassemble the clock correctly—these youngsters were likely to become engineers! With software, reverse engineering is the process of "taking apart" software and analyzing its operation without having access to the program code itself. Among other possibilities, reverse engineering may allow one to:

- provide equivalent functions without violating copyright laws

- emulate one operating system within another (see EMULATION)

- determine a file format so other programs can use it as well (interoperability)

- document the operation of a program whose documentation is lost or no longer available

- determine whether a competing product violates one's patents or copyrights

TECHNIQUES

Reverse engineering can be thought of as running the development process backwards (see SOFTWARE DEVELOPMENT). Instead of starting with the specification of the system and writing code, one starts with the operating program and constructs a detailed description of its organization. Several general techniques can be used:

- disassembly (turning the machine-level code into somewhat higher-level code with symbolic labels, etc.) (see ASSEMBLER)

- decompilation (which attempts to turn machine code into a higher-level language such as C) (see COMPILER)

- systematically supplying data of various types and analyzing the program's response (this is especially used when analyzing communications protocols)

Perhaps the most significant example of reverse engineering occurred in the early 1980s when competitors reverse engineered the built-in code (see BIOS) that controlled the low-level functions of the original IBM PC, thus enabling the manufacture of legal "clones" by such companies as Compaq. This was done by creating a "clean room" staffed with engineers who had no involvement with IBM and were not privy to any of the internal secrets of the BIOS.

Reverse engineering has been widely used to provide open-source implementations of formerly proprietary technologies. Examples include Samba (Windows SMB file sharing), Open Office (similar to Microsoft Office), Mono (Windows .NET API), and especially Windows emulators for Linux such as Wine.

Generally, under the Digital Millennium Copyright Act of 1998, courts have been sympathetic to reverse engineering that enables users to exercise what would be considered "fair use" under copyright laws or to provide more widespread compatibility with other products. However, reverse engineering may be illegal when the intent is to bypass software "locks" (see COPY PROTECTION) in order to make illegal copies, or when the machine code is copied or manipulated (such as by decompiling).

There are a number of ways in which reverse engineering (or similar practices) can be applied to technology other than software. Perhaps the most unusual example was the successful reconstruction of an ancient Greek astronomical calculator called the Antikythera mechanism. In general, the process of reverse engineering, by spreading knowledge of how to access and interface systems and provide functionality, ultimately contributes to the development of new technology and software.

Further Reading

Eilam, Eldad. *Reversing: Secrets of Reverse Engineering.* Indianapolis: Wiley, 2005.
James, Dick. "Reverse Engineering Delivers Product Knowledge, Aids Technology Spread." Available online. URL: http://electronicdesign.com/Articles/Index.cfm?AD=1&ArticleID=11966. Accessed November 11, 2007.
Musker, David C. "Reverse Engineering." Available online. URL: http://www.jenkins-ip.com/serv/serv_6.htm. Accessed November 11, 2007.
Perry, Mike, and Nasko Oskov. "Introduction to Reverse Engineering Software." Available online. URL: http://www.acm.uiuc.edu/sigmil/RevEng/. Accessed November 11, 2007.
Raja, Vinesh, and Kiran J. Fernandes, eds. *Reverse Engineering: An Industrial Perspective.* New York: Springer, 2008.

RFID (radio frequency identification)

For some years now people have become used to swiping credit or debit cards to buy things in stores, or have used magnetic cards to access transit systems. Increasingly, however, the information needed for identification, whether of goods in a warehouse or customers in a store, is being scanned wirelessly using radio frequency identification (RFID) systems.

An RFID system uses a tag or card (see SMART CARD) that is able to store and modify information in memory, together with a tiny antenna and transmitter for communicating the information.

PASSIVE VS. ACTIVE

Passive RFID tags have no power supply; the power induced by the reading signal is used to transmit the response. Because this power is very small, passive tags can only be read at distances from about 4 inches (10 cm) to a few yards (meters), depending on the antenna size and type. The main advantage of passive tags is that the lack of a battery makes them small, lightweight, and inexpensive, making them ideal for attaching to merchandise (they have also been embedded under the skin of pets and, in a few cases, even people). Smart cards for use in transit systems and similar applications are also passive; the system is activated by "tagging" or bringing the card near the reader.

Active RFID tags have their own battery. Their advantage is that they are able to initiate communication with the reader, and the signal they send is much stronger, more reliable, and with greater range (up to about 1,500 feet [500 m]). The stronger signal allows for communication in rougher environments (such as outdoors for tracking cattle or shipping containers).

There is also a sort of hybrid called a semipassive tag. This also has a battery, but only uses it for internal processing, not sending signals. The tag can gather information (such as logging temperature) and send it when queried by a reader.

Current uses for RFID tags and cards of various types include:

- automatic fare payments systems for transit systems

- automatic toll payments for bridges and turnpikes

- automatic book checkout systems for libraries, where it reduces repetitive strain injury (RSI) in staff and simplifies checking shelves

- student ID cards

- passports (RFID has been included in new U.S. passports since 2006)

- tracking cattle, including determining the origin of unhealthy animals

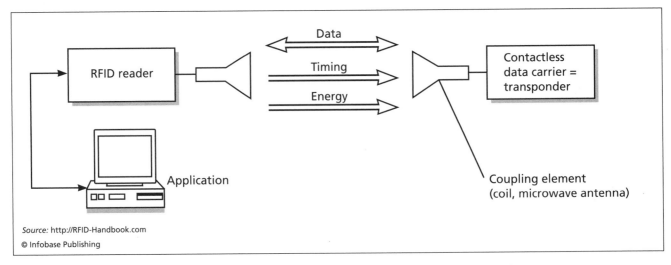

Parts of an RFID system. Depending on whether the chip is active or passive, the reader can be inches or yards away.

- identification chips placed beneath the skin of pets
- experimental human RFID implants (pioneered by British computer scientist Kevin Warwick) and now used by VIP customers in a few nightclubs
- tracking goods from original shipment to inventory (Wal-Mart now requires its major suppliers to include RFID labels with shipments)
- scientific sensors, such as seismographic instruments

PRIVACY AND SECURITY ISSUES

The benefits of RFID technology are numerous: better inventory control (see SUPPLY CHAIN MANAGEMENT); more secure passports and other forms of ID; faster, easier access to transportation systems; and potentially, the avoidance of mishaps in hospitals, such as the wrong patient receiving a drug or procedure.

However, there are privacy and security concerns that remain to be fully resolved. The primary threat is that unauthorized persons could illicitly obtain information or track people or goods, for purposes ranging from simple larceny to identity theft. Privacy rights organizations have also raised concerns that information about consumer purchases could be used for unwanted marketing (or sold to third parties), while information about a library patron's reading habits could trigger unwarranted government investigations in the name of fighting terrorism.

There is an incentive to produce RFID cards and tags that are resistant to unauthorized reading or tampering. A cryptographic protocol can be used such that no information will be sent or received unless the reader and tag "know" the correct keys. Another possibility is to create a device that can "jam" reading attempts in the device's vicinity, perhaps protecting a customer's grocery cart from being scanned. Finally, RFID cards can be put inside in a sleeve of material that blocks the signals. However, cryptographic and other security technologies raise the cost of RFID devices and may make them impracticable for some applications.

In September 2006 the National Science Foundation awarded a $1.1 million grant to the RFID Consortium for Security and Privacy to study potential risks and safeguards for the technology. That same year a group of major corporations together with the National Consumers League released a draft set of standards and guidelines for best practices in using RFID, with broader scope than the existing EPC (electronic product code) standards.

A Radio Frequency ID (RFID) "chip" from 3M. RFID is finding many applications, but has also raised privacy concerns. (3M CORPORATION)

Further Reading

EPC Global. Available online. URL: http://www.epcglobalinc.org/. Accessed November 12, 2007.

Feder, Barbara. "Guidelines for Radio Tags Aim to Protect Buyer Privacy." *New York Times,* May 1, 2006. Available online. URL: http://query.nytimes.com/gst/fullpage.html?res=9805E FDF113FF932A35756C0A96 09C8B63. Accessed November 12, 2007.

Glover, Bill, and Himanashu Bhatt. *RFID Essentials*. Sebastapol, Calif.: O'Reilly, 2006.

Newitz, Annalee. "The RFID Hacking Underground." *Wired,* May 2006. Available online. URL: http://www.wired.com/wired/archive/14.05/rfid.html. Accessed November 12, 2007.

"Privacy Best Practices for Deployment of RFID Technology, Interim Draft." CDT Working Group on RFID, May 1, 2006. Available online. URL: http://www.nclnet.org/advocacy/technology/rfid_guidelines_05012006.htm. Accessed November 12, 2007.

Sweeney, Partick J., II. *RFID for Dummies*. Hoboken, N.J.: Wiley, 2005.

Rheingold, Howard

(1947–)
American
Writer

On his Web site, Howard Rheingold says that he "fell into the computer realm from the typewriter dimension, then plugged his computer into his telephone and got sucked into the net." A prolific writer, explorer of the interaction of human consciousness and technology, and chronicler of virtual communities, Rheingold has helped people from students to businesspersons to legislators understand the social significance of the Internet and communications revolution.

Born on July 17, 1947, in Phoenix, Arizona, he was later educated at Reed College in Portland, Oregon, but lived and worked for most of his life in the San Francisco Bay Area. A child of the counterculture, his interests included the exploration of consciousness and cognitive psychology. His books in this area would include *Higher Creativity* (written with Willis Harman, 1984), *The Cognitive Connections* (written with Howard Levine, 1986), and *Exploring the World of Lucid Dreaming* (written with Stephen LaBerge, 1990). In 1994 he updated the *Whole Earth Catalog,* a remarkable resource book by Stewart Brand that had become a bible for the movement toward a more self-sufficient and human-scale life in the 1970s.

Rheingold bought his first personal computer, mainly because he thought word processing would make his work as a writer easier. In 1983 he bought a modem and was soon intrigued by the thousands of PC bulletin board systems that were an important way to share files and ideas in the days before the World Wide Web (see BULLETIN BOARD SYSTEMS). Interacting with these often tiny cyberspace villages helped Rheingold explore his developing ideas about the nature and significance of virtual communities.

In 1985 Rheingold joined The WELL (Whole Earth 'Lectronic Link), a unique and remarkably persistent community that began as an unlikely meeting place of Deadheads (*Grateful Dead* fans) and computer hackers. Compared to most bulletin boards, the WELL was more like the virtual equivalent of the cosmopolitan San Francisco Bay Area.

The sum and evaluation of these experiences can be found in what is perhaps Rheingold's most seminal book, *The Virtual Community* (1993; revised, 2000), which represents both a participant's and an observer's tour through the online meeting places that had begun to function as communities (see VIRTUAL COMMUNITY). Rheingold chronicled the romances, feuds ("flame wars"), and growing pains that made The WELL seem much like a small town or perhaps an artist's colony that just happened to be in cyberspace.

In addition to The WELL, Rheingold also explores MUDs (Multi-User Dungeons) and other elaborate online fantasy role-playing games, NetNews (also called Usenet) groups, chat rooms, and other forms of online interaction (see CONFERENCING SYSTEMS and NETNEWS AND NEWSGROUPS). Rheingold continues to manage the Brainstorms Community, a private Web-conferencing community that allows for thoughtful discussions about a variety of topics.

Rheingold saw the computer (and computer networks in particular) as a powerful tool for creating new forms of community. The original edition of his book *Tools for Thought* (1985 and revised 2000), with its description of the potential of computer-mediated communications, seems prescient today after a decade of the Web. Rheingold's *Virtual Reality* (1991) introduced that immersive technology.

Around 1999 Rheingold started noticing the emergence of a different kind of virtual community—a mobile, highly flexible, and adaptive one. In his book *Smart Mobs*, Rheingold gives examples of groups of teenagers coordinating their activities by sending each other text messages on their cell phones (see FLASH MOBS). Rheingold believes that the combination of mobile and network technology may be creating a social revolution as important as that triggered by the PC in the 1980s and the Internet in the 1990s.

In 1996 Rheingold launched Electric Minds, an innovative company that tried to offer virtual-community-building services while attracting enough revenue from contract work and advertising to become self-sustaining and profitable in about three years. He received financing from the venture capital firm Softbank. However, the company failed, and Rheingold came to believe that there was a fundamental mismatch between the profit objectives of most venture capitalists and the patience needed to cultivate and grow a new social enterprise. Rheingold then started a more modest effort, Rheingold Associates.

According to Rheingold and coauthor Lisa Kimball, some of the benefits of creating such communities include the ability to get essential knowledge to the community in times of emergency, to connect people who might ordinarily be divided by geography or interests, to "amplify innovation," and to "create a community memory" that prevents important ideas from getting lost. Rheingold continues to both create and write about new virtual communities, working through such efforts as the Cooperation Commons (a collaboration with the Institute for the Future). He is also a nonresident Fellow of the Annenberg School for Communication.

Rheingold's writings have garnered a variety of awards. In 2003 *Utne Reader* magazine gave an Independent Press Award for a blog based on *Smart Mobs*. That year Rheingold also gave the keynote speech for the annual Webby Awards for Web-site design.

Further Reading
Cooperation Commons. Available online. URL: http://www.cooperationcommons.com/. Accessed November 12, 2007.

Hafner, Katie. *The Well: A Story of Love, Death & Real Life in the Seminal Online Community.* New York: Carol & Graf, 2001.

Howard Rheingold [home page]. Available online. URL: http://www.rheingold.com. Accessed November 12, 2007.

Kimball, Lisa, and Howard Rheingold. "How Online Social Networks Benefit Organizations." Available online. URL: http://www.rheingold.com/Associates/onlinenetworks.html. Accessed November 12, 2007.

Rheingold, Howard. *Smart Mobs: The Next Social Revolution.* New York: Basic Books, 2003.

———. *Tools for Thought: The History and Future of Mind-Expanding Technology.* 2nd rev. ed. Cambridge, Mass.: MIT Press, 2000.

———. *The Virtual Community: Homesteading on the Electronic Frontier.* Revised ed. Cambridge, Mass.: MIT Press, 2000. [The first edition is also available online. URL: http://www.rheingold.com/vc/book/. Accessed November 12, 2007.]

Smart Mobs Blog. Available online. URL: http://www.smartmobs.com/. Accessed November 12, 2007.

The WELL Available online. URL: http://www.well.com. Accessed November 12, 2007.

risks of computing

Programmers and managers of software development are generally aware of the need for software to properly deal with erroneous data (see ERROR HANDLING). They know that any significant program will have bugs that must be rooted out (see BUGS AND DEBUGGING). Good software engineering practices and a systematic approach to assuring the reliability and quality of software can minimize problems in the finished product (see SOFTWARE ENGINEERING and QUALITY ASSURANCE, SOFTWARE). However, serious bugs are not always caught, and sometimes the consequences can be catastrophic. For example, in the Therac 25 computerized X-ray cancer treatment machine, poorly thought-out command entry routines plus a counter overflow resulted in three patients being killed by massive X-ray overdoses. The overdoses ultimately occurred because the designers had removed a physical interlock mechanism they believed was no longer necessary.

Any computer application is part of a much larger environment of humans and machines, where unforeseen interactions can cause problems ranging from inconvenience to loss of privacy to potential injury or death. Seeing these potential pitfalls requires thinking beyond the specifications and needs of a particular project. For many years the Usenet newsgroup comp.risks (and its collected form, Risks Digest) have chronicled what amounts to an ongoing symposium where knowledgeable programmers, engineers, and others have pointed out potential risks in new technology and suggested ways to minimize them.

UNEXPECTED SITUATIONS

A common source of risks arises from designers of control systems failing to anticipate extreme or unusual environmental conditions (or interactions between conditions). This is a particular problem for mobile robots, which unlike their tethered industrial counterparts must share elevators, corridors, and other places with human beings. For example, a hospital robot was not designed to recognize when it was blocking an elevator door—a situation that could have blocked a patient being rushed into surgery. A basic principle of coping with unexpected situations is to try to design a fail-safe mode that does not make the situation worse. For example, an automatic door should be designed so that if it fails it can be opened manually rather than trapping people in a fire or other disaster.

UNANTICIPATED INTERACTIONS

The more systems there are that can respond to external inputs, the greater the risk that a spurious input might trigger an unexpected and dangerous response. For example, the growing number of radio-controlled (wireless) devices have great potential for unexpected interactions between different devices. In one case reported to the Risks Forum, a Swedish policeman's handheld radio inadvertently activated his car's airbag, which slammed the radio into him. Several military helicopters have crashed because of radio interference. Banning the use of electronic devices at certain times and places (for example, aboard an aircraft that is taking off or landing) can help minimize interference with the most safety-critical systems.

At the same time, regulations themselves introduce the risk that people will engage in other forms of risky behavior in an attempt to either follow or circumvent the rule. For example, the Japanese bullet train system imposed a stiff penalty for operators who failed to wear a hat. In one case an operator left the train cabin to retrieve his hat while the train kept running unsupervised. This minor incident actually conceals two additional sorts of risks—that of automating a system so much that humans no longer pay attention, and the inability of the system to sense the lack of human supervision.

UNANTICIPATED USE OF DATA

The growing number of different databases that track even the intimate details of individual lives has raised many privacy issues (see PRIVACY IN THE DIGITAL AGE). Designers and maintainers of such databases had some awareness of the threat of unauthorized persons breaking into systems and stealing such data (see COMPUTER CRIME AND SECURITY). However, most people were surprised and alarmed by the new crime of identity theft, which began to surface in significant numbers in the mid- to late-1990s (see IDENTITY THEFT).

It turned out that while a given database (such as customer records, bank information, illicitly obtained DMV records, and so on) usually did not have enough information to allow someone to successfully impersonate another's identity, it was not difficult to use several of these sources together to obtain, for example, the information needed to apply for credit in another's name. In particular, while most people guarded their credit card numbers, they tended not to worry as much about Social Security numbers (SSN). However, since many institutions use the SSN to index their records, the number has become a key for unlocking personal data.

Further, as more organizations put their records online and make them Web-accessible, the ability of hackers, private investigators (legitimate or not), and "data brokerage" services to quickly assemble a dossier of sensitive information on any individual was greatly increased. Here we have a case where a powerful tool for productivity (the Internet) also becomes a facilitator for using the vulnerabilities in any one system to compromise others.

In an increasingly networked and technologically-dependent world, the anticipation and prevention of computer risks has become very important. To the extent companies may be legally liable for the more direct forms of risk, there is more incentive for them to devote resources to risk amelioration. However, many computer-related risks are at least as much social as technological in nature, and are beyond the scope of concern of any one company or organization. Social risks ultimately demand a broader social response.

Technology itself can be used to help ameliorate technological risks. Artificial intelligence techniques (see EXPERT SYSTEMS and NEURAL NETWORK) might be used improve the ability of a system to adapt to unusual conditions. However, any such programming then becomes prone to bugs and risks itself.

So far, the most successful way to deal with the broad range of computer risks has been through human collaboration as facilitated by the Internet. Through venues such as the Risks Forum computer-mediated communications and collaboration allows for the pooling of human intelligence in the face of the growing complexity of human inventiveness.

Further Reading

Comp.risks [access via Google Groups]. Available online. URL: http://groups.google.com/group/comp.risks/topics. Accessed August 19, 2007.

Glass, Robert L. *Software Runaways: Monumental Software Disasters.* Upper Saddle River, N.J.: Prentice Hall, 1997.

Neumann, Peter G. *Computer-Related Risks.* Reading, Mass.: Addison-Wesley, 1994.

Peterson, Ivars. *Fatal Defect: Chasing Killer Computer Bugs.* New York: Vintage Books, 1996.

Ritchie, Dennis

(1941–   )
American
Computer Scientist

Together with Ken Thompson, Dennis Ritchie developed the UNIX operating system and the C programming language—two tools that have had a tremendous impact on the world of computing for three decades.

Ritchie was born on September 9, 1941, in Bronxville, New York. He was exposed to communications technology and electronics from an early age because his father was director of the Switching Systems Engineering Laboratory at BELL LABORATORIES. (Switching theory is closely akin to computer logic design.) Ritchie attended Harvard University and graduated with a B.S. in physics. However, by then his interests had shifted to applied mathematics and in particular, the mathematics of computation, which he later described as "the theory of what machines can possibly do" (see COMPUTABILITY AND COMPLEXITY). For his doctoral thesis he wrote about recursive functions (see RECURSION).

Together with Ken Thompson, Dennis Ritchie developed the UNIX operating system and the C programming language, two of the most important developments in the history of computing. (PHOTO COURTESY OF LUCENT TECHNOLOGIES' BELL LABS)

This topic was proving to be important for the definition of new computer languages in the 1960s (see ALGOL).

In 1967, however, Ritchie decided that he had had enough of the academic world. Without finishing the requirements for his doctorate, he started work at Bell Labs, his father's employer. Bell Labs has made a number of key contributions to communications and information theory (see RESEARCH LABORATORIES IN COMPUTING).

By the late 1960s, computer operating systems had become increasingly complex and unwieldy. As typified by the commercially successful IBM System/360, the operating system was proprietary, had many hardware-specific functions and tradeoffs in order to support a family of upwardly compatible computer models, and was designed with a top-down approach.

During his graduate studies, however, Ritchie had encountered a different approach to designing an operating system. A new system called Multics was being designed jointly by Bell Labs, MIT, and General Electric. Multics was quite different from the batch-processing world of mainframes: It was intended to allow many users to share a computer. He had also done some work with MIT's Project

Mac. The MIT computer students, the original "hackers" (in the positive meaning of the term), emphasized a cooperative approach to designing tools for writing programs. This, too, was quite different from IBM's highly structured and centralized approach.

Unfortunately, the Multics project itself grew increasingly unwieldy. Bell Labs withdrew from the Multics project in 1969. Ritchie and his colleague Ken Thompson then decided to apply many of the same principles to creating their own operating system. Bell Labs wasn't in a mood to support another operating system project, but they eventually let Ritchie and Thompson use a DEC PDP-7 minicomputer. Although small and already obsolete, the machine did have a graphics display and a Teletype terminal that made it suitable for the kind of interactive programming they preferred. They decided to call their system UNIX, punning on Multics by suggesting something that was simpler and better integrated.

Instead of designing from the top down, Ritchie and Thompson worked from the bottom up. They designed a way to store data on the machine's disk drive (see FILE), and gradually wrote the necessary utility programs for listing, copying, and otherwise working with the files. Thompson did the bulk of the work on writing the operating system, but Ritchie did make key contributions such as the idea that devices (such as the keyboard and printer) would be treated the same way as other files. Later, he reconceived data connections as "streams" that could connect not only files and devices but applications and data being sent using different protocols. The ability to flexibly assign input and output, as well as to direct data from one program to another, would become hallmarks of UNIX.

When Ritchie and Thompson successfully demonstrated UNIX, Bell Labs adopted the system for its internal use. UNIX turned out to be ideal for exploiting the capabilities of the new PDP-11 minicomputer. As Bell licensed UNIX to outside users, a unique community of user-programmers began to contribute their own UNIX utilities (see OPEN-SOURCE MOVEMENT).

In the early 1970s, Ritchie also collaborated with Thompson in creating C, a streamlined version of the earlier BCPL and CPL languages. C would be a "small" language that was independent of any one machine but could be linked to many kinds of hardware thanks to its ability to directly manipulate the contents of memory. C became tremendously successful in the 1980s. Since then, C and its offshoots C++ and Java became the dominant languages used for most programming today.

Ritchie still works at Bell Labs's Computing Sciences Research Center. (When AT&T spun off many of its divisions, Bell Labs became part of Lucent Technologies.) Ritchie developed an experimental operating system called Plan 9 (named for a cult sci-fi movie). Plan 9 attempts to take the UNIX philosophy of decentralization and flexibility even further, and is designed especially for networks where computing resources are distributed.

Ritchie has received numerous awards, often given jointly to Thompson. These include the ACM Turing Award (1985), the IEEE Hamming Medal (1990), the Tsutomu Kanai Award (1999), and the National Medal of Technology (also 1999).

Further Reading
Dennis Ritchie Home Page. Available online. URL: http://www. cs.bell-labs.com/who/dmr/. Accessed August 27, 2007.
Kernighan, B. W., and Dennis M. Ritchie. *The C Programming Language.* Upper Saddle River, N.J., Prentice Hall, 1978. (A second edition was published in 1989.)
Lohr, Steve. *Go To.* New York: Basic Books, 2001.
Plan 9 from Bell Labs. 4th ed. Available online. URL: http://plan9. bell-labs.com/plan9/. Accessed August 19, 2007.
Ritchie, Dennis M., and Ken Thompson. "The UNIX Time-Sharing System." *Communications of the ACM* 17, 7 (1974): 365–375.
Slater, Robert. *Portraits in Silicon.* Cambridge, Mass.: MIT Press, 1987.

robotics

The idea of the automaton—the lifelike machine that performs intricate tasks by itself—is very old. Simple automatons were known to the ancient world. By the 18th century, royal courts were being entertained by intricate humanlike automatons that could play music, draw pictures, or dance. A little later came the "Turk," a chess-playing automaton that could beat most human players.

However, things are not always what they seem. The true automatons, controlled by gears and cams, could play only whatever actions had been designed into them. They could not be reprogrammed and did not respond to changes in their environment. The chess-playing automaton held a concealed human player.

True robotics began in the mid-20th century and has continued to move between two poles: the pedestrian but useful industrial robots and the intriguing but tentative creations of the artificial intelligence laboratories.

INDUSTRIAL ROBOTS

In 1921, the Czech playwright Karel Capek wrote a play called *R.U.R.* or *Rossum's Universal Robots*. Robot is a Czech word that has been translated as work(er), serf, or slave. In the play the robots, which are built by factories to work in other factories, eventually revolt against their human masters.

During the 1960s, real robots began to appear in factory settings (see also ENGELBERGER, JOSEPH). They were an outgrowth of earlier machine tools that had been programmed by cams and other mechanisms. An industrial robot is basically a movable arm that ends in a "hand" called an end effector. The arm and hand can be moved by some combination of hydraulic, pneumatic, electrical, or mechanical means. Typical applications include assembling parts, welding, and painting. The robot is programmed for a task either by giving it a detailed set of commands to move to, grasp, and manipulate objects, or by "training" the robot by moving its arm, hand, and effectors through the required motions, which are then stored in the robot's memory. By the early 1970s, Unimation, Inc. had created a profitable business from selling its Unimate robots to factories.

The early industrial robots had very little ability to respond to variations in the environment, such as the "work

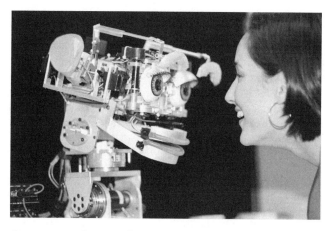

An experimental NASA robot arm. (NASA PHOTO)

piece" that the robot was supposed to grasp being slightly out of position. However, later models have more sophisticated sensors to enable them to adjust to variations and still accomplish the task. The more sophisticated computer programs that control newer robots have internal representations or "frames of reference" to keep track of both the robot's internal parameters (angles, pressures, and so on) and external locations in the work area.

MOBILE ROBOTS AND SERVICE ROBOTS

Industrial robots work in an extremely restricted environment, so their world representation can be quite simple. However, robots that can move about in the environment have also been developed. Military programs have developed automatic guided vehicles (AGVs) with wheels or tracks, capable of navigating a battlefield and scouting or attacking the enemy (see MILITARY APPLICATIONS OF COMPUTERS). Space-going robots including the Sojourner Mars rover also have considerable onboard "intelligence," although their overall tasks are programmed by remote commands.

Indeed, the extent to which mobile robots are truly autonomous varies considerably. At one end is the "robot" that is steered and otherwise controlled by its human operator, such as law enforcement robots that can be sent into dangerous hostage situations. (Another example is the robots that fight in arena combat in the popular *Robot Wars* shows.)

Moving toward greater autonomy, we have the "service robots" that have begun to show up in some institutions such as hospitals and laboratories. These mobile robots are often used to deliver supplies. For example, the Help-Mate robot can travel around a hospital by itself, navigating using an internal map. It can even take an elevator to go to another floor.

Service robots have had only modest market penetration, however. They are relatively expensive and limited in function, and if relatively low-wage more versatile human labor is available, it is generally preferred. For now mobile robots and service robots are most likely to turn up in specialized applications in environments too dangerous for

human workers, such as in the military, law enforcement, handling of hazardous materials, and so on.

SMART ROBOTS

Robotics has always had great fascination for artificial intelligence researchers (see ARTIFICIAL INTELLIGENCE). After all, the ability to function convincingly in a real-world environment would go a long way toward demonstrating the viability of true artificial intelligence.

Building a smart, more humanlike robot involves several interrelated challenges, all quite difficult. These include developing a system for seeing and interpreting the environment (see COMPUTER VISION) as well a way to represent the environment internally so as to be able to navigate around obstacles and perform tasks.

One of the earliest AI robots was "Shakey," built at the Stanford Research Institute (SRI) in 1969. Shakey could navigate only in a rather simplified environment. However, the "Stanford Cart," built by Hans Moravec in the late 1970s could navigate around the nearby campus without getting into too much trouble.

An innovative line of research began in the 1990s at MIT (see BROOKS, RODNEY and BREAZEAL, CYNTHIA). Instead of a "top down" approach of programming robots with explicit logical rules, so-called behavior-based robotics works from the bottom up, coupling systems of sensors and actuators that each have their own simple rules, from which can emerge surprisingly complex behavior. The MIT "sociable robots" Cog and Kismet were able to explore the world and learn to interact with people in somewhat the way a human toddler might.

Today it is possible to buy an AI robot for one's home, in the form of toys such as Sony's AIBO robot dog, which can emulate various doggy behaviors such as chasing things and communicating by body language. Some robot toys not only have an extensive repertoire of behavior and vocalizations, but also can learn to some extent (see NEURAL NETWORK.)

It is also possible to experiment with robotics at home or school, thanks to kits such as the LEGO Logo, which combines a popular building set with a versatile educational programming language (see LOGO).

FUTURE APPLICATIONS

A true humanoid robot with the kind of capabilities written about by Isaac Asimov and other science fiction writers is not in sight yet. However, there are many interesting applications of robots that are being explored today. These include the use of remote robots for such tasks as performing surgery (see TELEPRESENCE) and the application of robotics principles to the design of better prosthetic arms and legs for humans (bionics). Farther afield is the possibility of creating artificial robotic "life" that can self-reproduce (see ARTIFICIAL LIFE).

Further Reading
Breazeal, Cynthia. *Designing Sociable Robots.* Cambridge, Mass.: MIT Press, 2004.
Brooks, Rodney A. *Flesh and Machines: How Robots Will Change Us.* New York: Pantheon Books, 2002.

Henderson, Harry. *Modern Robotics: Building Versatile Machines.* New York: Chelsea House, 2006.

Humanoid Robotics Group [MIT Artificial Intelligence Laboratory]. Available online. URL: http://www.ai.mit.edu/projects/humanoid-robotics-group/. Accessed August 19, 2007.

Menzel, Peter, and Faith D'Alusio. *Robo Sapiens: Evolution of a New Species.* Cambridge, Mass.: MIT Press, 2001.

Nof, Shimon Y. *Handbook of Industrial Robotics.* 2nd ed. New York: Wiley, 1999.

Pires J. Norberto. *Industrial Robots Programming: Building Applications for the Factories of the Future.* New York: Springer, 2007.

Schraft, Rolf Dieter, and Gernot Schmierer. *Service Robots: Products, Scenarios, Visions.* Natick, Mass.: A. K. Peters, 2000.

Severin, E. Oliver. *Robotic Companions: Mentorbots and Beyond.* New York: McGraw-Hill, 2004.

Tesler, Pearl. *Universal Robots: The History and Workings of Robotics.* TheTech Museum. Available online. URL: http://www.thetech.org/exhibits/online/robotics/universal/index.html. Accessed August 19, 2007.

RPG (Report Program Generator)

Many business computer programs written for mainframe computers involved reading data from files, performing relatively simple procedures, and outputting printed reports. During the 1960s, some people believed that COBOL, a general-purpose (but business-oriented) computer language, would be easy enough for nonprogrammers to use (see COBOL). Although this turned out not to be the case, IBM did succeed in creating RPG (Report Program Generator), a language designed to make it easier for programmers (including beginners) to generate business reports.

Most COBOL programs read data, perform tests and calculations, and print the results. RPG, first released in 1964 for use with the new System/360 mainframe and the smaller System/3, simplifies this process and eliminates most writing of program code statements.

A "classic" RPG program is built around the "RPG cycle," consisting of three stages. During the input stage, the input device(s), file type, access specifications, and data record structure are specified. (These specifications can be quite elaborate.) The heart of the program specifies calculations to be performed with the various data fields, while the output section specifies how the results will be laid out in report form, including such things as headers, footers, and sections.

Subsequent versions of RPG added more features. RPG-IV, released in 1994, includes the ability to define subroutines, for example. IBM has also released VisualAge RPG, which allows for the creation and running of RPG programs in the Microsoft Windows environment. There are also tools for interfacing RPG programs with various database systems and to use RPG for writing Web-based (CGI) programs.

Further Reading

Cozzi, Robert. *The Modern RPG IV Language.* 4th ed. Lewisville, Tex.: MC Press Online, 2006.

Martin, Jim. *Free-Format RPG IV: How to Bring Your RPG Programs into the 21st Century.* Lewisville, Tex.: MC Press Online, 2005.

Meyers, Bryan, and Jef Sutherland. *VisualAge for RPG by Example.* Loveland, Colo.: Duke Press, 1998.

RSS (Really Simple Syndication)

Web sites such as news providers and blogs (see BLOGS AND BLOGGING) are constantly posting new material. While readers can periodically visit a site to look for new material, an increasingly popular option is to subscribe to a "Web feed" and receive the latest information automatically. The most commonly used tool for Web feeds is RSS, which can stand for Really Simple Syndication, Rich Site Summary, or RDF Site Summary, depending on the format used.

The data in an RSS feed can include article titles, summaries, excerpts (such as the first paragraph), or the complete article or posting. Feeds can also include multimedia such as graphics, video, or sound. The data (and any linked material) is formatted using standard markup elements (see HTML and XML). The following is an excerpt of a simple RSS feed provided by the RSS Advisory Board:

```
<?xml version="1.0"?>
<rss version="2.0">
  <channel>
    <title>Liftoff News</title>
    <link>http://liftoff.msfc.nasa.gov/</link>
    <description>Liftoff to Space Exploration.</description>
    <language>en-us</language>
    <pubDate>Tue, 10 Jun 2003 04:00:00 GMT</pubDate>
    <lastBuildDate>Tue, 10 Jun 2003 09:41:01 GMT</lastBuildDate>
    <docs>http://blogs.law.harvard.edu/tech/rss</docs>
    <generator>Weblog Editor 2.0</generator>
    <managingEditor>editor@example.com</managingEditor>
    <webMaster>webmaster@example.com</webMaster>
    <item>
      <title>Star City</title>
      <link>http://liftoff.msfc.nasa.gov/news/2003/news-starcity.asp</link>
      <description>How do Americans get ready to work with Russians aboard the International Space Station? They take a crash course in culture, language and protocol at Russia's &lt;a href="http://howe.iki.rssi.ru/GCTC/gctc_e.htm"&gt;Star City&lt;/a&gt;.</description>
      <pubDate>Tue, 03 Jun 2003 09:39:21 GMT</pubDate>
      <guid>http://liftoff.msfc.nasa.gov/2003/06/03.html#item573</guid>
    </item>
  </channel>
</rss>
```

As part of the process of setting up a feed on the Web server, the feed is "published" so that it can be found and read using a client program called a reader or aggregator (the latter can combine feeds or organize them in a newspaper-like format for convenience). RSS readers can be stand-alone applications or be included with many modern Web browsers and e-mail clients. Alternatively, Web-based readers or aggregators such as NewsGator Online can allow feeds to be read using any Web browser. Readers of Web pages can find RSS feeds by looking for a "subscribe" icon or the words RSS or XML. Specialized search engines such as Bloglines can also help users find interesting feeds. Additionally, information on the server can also be used by software to automatically deliver the latest content (see PODCASTING).

HISTORY AND DEVELOPMENT

Forerunners of RSS go back to the mid-1990s, with RDF Site Summary first appearing in 1999 for use on Netscape's portal. The adoption of RSS by the *New York Times* in 2002 greatly aided the popularization of the format, as did the growing number of blogs that needed a way for contributors and readers to keep in touch. Today Web browsers such as Internet Explorer, Mozilla Firefox, and Safari support RSS. File-sharing services such as BitTorrent can be combined with RSS to deliver content automatically to users' hard drives. An offshoot of RSS called Atom has been less widely adopted, but offers better compatibility with XML standards and better management of multimedia content.

Further Reading
Bloglines. Available online. URL: http://www.bloglines.com/. Accessed November 19, 2007.
Calishain, Tara. *Information Trapping: Real-Time Research on the Web.* Berkeley, Calif.: New Riders, 2007.
Finkelstein, Ellen. *Syndicating Web Sites with RSS Feeds for Dummies.* Hoboken, N.J.: Wiley, 2005.
RSS Advisory Board. Available online. URL: http://www.rssboard.org. Accessed November 19, 2007.
Sherman, Chris. "What Is RSS, and Why Should You Care?" *Search Engine Watch,* August 30, 2005. Available online. URL: http://searchenginewatch.com/showPage.html?page=3530926. Accessed November 19, 2007.

RTF (Rich Text Format)

Rich Text Format was developed in the later 1980s by programmers at Microsoft. Its purpose is to allow for interchange of documents between Microsoft Word and other software, while preserving the original formatting.

An RTF file is itself a plain text file containing the document text enclosed in control codes that determine the formatting. For example:

```
{\rtf1\ansi{\fonttbl{\f0\froman\fprq2\
fcharset0 Times New Roman;}}\f0\pard
This is some {\b bold} text.\par
}
```

The backslash starts a control code, such as for specifying a font or a style, such as Times New Roman and bold in the example. Curly brackets { } enclose the text to be affected by the control code. Thus the example above would be rendered in a word processor as:

This is some **bold** text.

Although RTF is an 8-bit format, special escape sequences can be used to specify 16-bit Unicode characters, such as for non-Roman alphabets.

Libraries and utilities are available for reading and writing RTF from most popular programming languages, including Perl, PHP, and Ruby.

In practice, RTF created by word processors tends to contain many control codes needed to ensure compatibility with older programs, making the files bulky and not practicable to edit directly. However, saving a file in RTF is a good way to ensure that a document can be used by recipients who may have, for example, older versions of Word. (It is quite typical for the latest default Word format to not be compatible with earlier versions.)

Further Reading
Burke, Sean M. *RTF Pocket Guide.* Sebastapol, Calif.: O'Reilly, 2003.
Microsoft Corporation. Word 2007: Rich Text Format (RTF), Specification, Version 1.9. Available online. URL: http://www.microsoft.com/downloads/details.aspx?FamilyID=DD422B8D-FF06-420 7-B476-6B5396A18A2B. Accessed November 13, 2007.

Ruby

Ruby is a versatile yet consistent programming language that has become popular in recent years, particularly for Web development. Designed by Yukihiro Matsumoto and first released in 1995, Ruby has a compact syntax familiar to many users of Perl and other scripting languages (see PERL and SCRIPTING LANGUAGE), avoiding, for example, the need to declare variable types. However, Ruby is also a thoroughgoing object-oriented language somewhat like Smalltalk (see SMALLTALK). Matsumoto has stressed that the design of the language is intended to stress being natural and enjoyable for the programmer, rather than focusing on the needs of the machine.

STRUCTURE

In Ruby, every data type is an object (see CLASS, DATA TYPE, and OBJECT), even those defined as primitive types in other languages, such as integers and Booleans. Although one can use the traditional procedural method of defining variables and then working with them, they are still implicitly treated as part of the root object called "Self."

Every function is a method that belongs to some class. Thus -5.abs invokes the absolute value method on the integer -5, returning 5. Similarly, "wireless wombat".length would return the length of the string, 15. Ruby includes many built-in methods for working with data structures such as arrays and hashes, and there are many additional libraries and applications available.

There are Ruby interpreters for all major operating systems. In addition to reading and executing a program from a file, as with many scripting languages, Ruby can also be used interactively to test statements:

```
% ruby eval.rb
ruby> puts "Hello, world."
Hello, world.
   nil
ruby> exit
```

Here the Ruby interpreter is told to run eval.rb, a special program that interactively evaluates statements and expressions. The *puts* command puts (outputs) the string *Hello, world*. The evaluator then reports that the *puts* method returned no value (nil).

Although Ruby is traditionally an interpreted language, a version that will produce byte code for a virtual machine (similar to Java) is in development, and a more direct compiler is certainly possible.

RUBY ON RAILS

The most popular programming environment for Ruby is Ruby on Rails, an open-source application framework aimed particularly at writing programs that connect Web sites to databases. The framework is based on the model-view controller approach (separating data access and logic from the user interface) and includes "scaffolding" that can be quickly filled in to provide data-driven Web sites with basic functionality. Developers can also create plug-ins to extend the built-in packages.

Further Reading

Baird, Kevin. *Ruby by Example: Concepts and Code*. San Francisco: No Starch Press, 2007.

Burd, Barry. *Ruby on Rails for Dummies*. Hoboken, N.J.: Wiley, 2007.

Cooper, Peter. *Beginning Ruby: From Novice to Professional*. Berkeley, Calif.: Apress, 2007.

"Ruby: A Programmer's Best Friend." Available online. URL: http://www.ruby-lang.org/en/. Accessed November 13, 2007.

Slagell, Mark. "Ruby User's Guide." Available online. URL: http://www.mentalpointer.com/ruby/index.html. Accessed November 13, 2007.

Stewart, Bruce. "An Interview with the Creator of Ruby." O'Reilly Linux devcenter, November 29, 2001. Available online. URL: http://www.linuxdevcenter.com/pub/a/linux/2001/11/29/ruby.html. Accessed November 13, 2007.

Thomas, Dave. *Programming Ruby: The Pragmatic Programmer's Guide*. 2nd ed. Raleigh, N.C.: Pragmatic Bookshelf, 2004.

S

SAP

SAP (NYSE symbol: SAP) is a German acronym for *Systeme, Anwendungen, und Produkete in der Datenverarbeitung* ("Systems, Applications, and Products in Data Processing"). Five former IBM engineers in Germany founded the company in 1972.

Although unfamiliar to the American public, unlike IBM and Microsoft, SAP is the world's largest business software company, and fourth-largest software provider in general (behind Microsoft, IBM, and Oracle). The company operates worldwide through three geographical divisions.

APPLICATIONS AND PRODUCTS

SAP specializes in Enterprise Resource Planning (ERP), enhancing a corporation's ability to manage its key assets and needs and to plan for the future. This software consists of three tiers: the database, an application server, and the client. Early versions of this software were designed to run on mainframes. Other major products include:

- SAP NetWeaver, which integrates all other SAP modules using modern open-standard Web technologies (see SERVICE-ORIENTED ARCHITECTURE)
- Customer Relationship Manager (see CRM)
- Supply Chain Management (see SUPPLY CHAIN MANAGEMENT)
- Supplier Relationship Management
- Human Resource Management System
- Product Lifestyle Management

- Exchange Infrastructure
- Enterprise Portal
- SAP Knowledge Warehouse

CHALLENGES

SAP has recognized for some time that while its base of large Fortune 500 companies has given it steady income, changing trends in business have been limiting the software giant's growth. In particular, the trend has been toward smaller, simpler, more scalable applications that can be integrated with modern Web services. In September 2007 SAP announced SAP Business ByDesign, a flexible set of enterprise management services that are delivered over the Web. However, it remains to be seen how well SAP will be able to compete with more agile companies such as NetSuite and Salesforce.com, and whether the company will be able to upgrade its existing large company user base without disaffecting it.

SAP's major competitor in the United States is Oracle (see ORACLE), which has sued SAP in 2007 for unfairly downloading and using patches and support materials from Oracle and using them to support former Oracle customers. SAP and Oracle have generally had quite different growth strategies: SAP grows by expanding and extending its own products, while Oracle has grown mainly through acquiring other companies. However, in October 2007 SAP acquired Business Objects, a leader in "business intelligence" systems, for $6.8 billion. This may signal SAP's willingness to engage in further strategic acquisitions.

Further Reading
McDonald, Kevin, et al. *Mastering the SAP Business Information Warehouse: Leveraging the Business Intelligence Capabilities of SAP NetWeaver.* 2nd ed. Indianapolis: Wiley, 2006.
Ricadela, Aaron. "SAP's Down-Market Gamble." *Business Week,* September 19, 2007. Available online. URL: http://www.businessweek.com/technology/content/sep2007/tc20070919_181869.htm. Accessed October 5, 2007.
SAP.com. Available online. URL: http://www.sap.com/usa/index.epx. Accessed October 8, 2007.
"SAP History: From Start-Up Software Vendor to Global Market Leader." Available online. URL: http://www.sap.com/company/history.epx. Accessed October 8, 2007.
Vogel, Andreas, and Jan Kimbell. *mySAP for Dummies.* Hoboken, N.J.: Wiley, 2004.
Woods, Dan, and Jeffrey Word. *SAP NetWeaver for Dummies.* Hoboken, N.J.: Wiley, 2004.

satellite Internet service

As with television, satellite Internet service can provide access to areas (such as remote locations, ships, or land vehicles) where wired service is not available (see BROADBAND). Besides the satellite, the system includes a terrestrial facility that has two connections: routers and proxy servers that manage the flow of traffic to and from the Internet, and an "uplink" transmitter that communicates with the satellite. In addition, there may be a connection to the public telephone network.

Each user has a satellite dish and associated equipment similar to those used for receiving satellite TV, though the dish is larger and existing TV dishes cannot be used. In the Northern Hemisphere, the user must have an unobstructed view of the southern sky (most satellites orbit over the equator). The equipment is also adapted for use on ships and recreational vehicles.

The user also has a modem (either external or on a card in the PC) to convert the satellite signals to data, and software supplied by the satellite service.

There are two types of systems for sending data from the user back to the Internet. In a dial-return system, the user has a conventional telephone dial-up modem that connects by phone to a hub at the terrestrial facility. Downloading is at broadband speeds (comparable to low-end DSL or cable), but uploading is at dial-up speeds. (This is not usually a problem unless the user is uploading large files.) In a two-way system, the user has a transmitter that sends data directly back to the satellite. This is usually several times faster than dial-up, but is more expensive.

Because of the time it takes signals to travel between a satellite and the ground, all satellite Internet systems have a built-in delay, or latency. Satellites in geosynchronous orbit (about 22,000 miles [35,405.6 km] high) have wider coverage but higher latency, while using lower orbits reduces latency but requires more satellites to provide continuous coverage. Latency can be problematic for applications such as Internet telephony (see VOIP).

Although small compared with that for cable or DSL, the satellite user base is growing, particularly in areas and countries that lack wired infrastructure. Users who have cable or DSL available in their neighborhood would have little reason to obtain satellite Internet, since the initial and monthly costs are considerably higher and download speeds are somewhat slower. Also, the need to use encrypted virtual private networks (VPN) to secure business data can lower effective speeds substantially. Finally, though the satellites themselves are very reliable, satellite service is subject to interruption during heavy rain or snow.

Further Reading
Brodkin, Jon. "Satellite Services and Telecommuting Not Always a Pretty Mix." *NetworkWorld,* July 24, 2007. Available online. URL: http://www.networkworld.com/news/2007/072407-satellites-for-telecommuting.html. Accessed November 14, 2007.
"How Does Satellite Internet Operate?" HowStuffWorks. Available online. URL: http://computer.howstuffworks.com/question606.htm. Accessed November 14, 2007.
Kota, Sastri L., Kaveh Pahlavan, and Pentti Leppanen. *Broadband Satellite Communication for Internet Access.* Norwell, Mass.: Kluwer Academic Publishers, 2004.
Nutter, Ron. "Getting More Performance from a Satellite Internet System." *Network World,* April 16, 2007. Available online. URL: http://www.networkworld.com/columnists/2007/041607nutter.html. Accessed November 14, 2007.

scanner

In order for a computer to work with information, the information must be digitized—converted to data that application programs can recognize and manipulate (see CHARACTERS AND STRINGS). Computer users have thus been confronted with the task of converting millions of pages of printed words or graphics into machine-readable form. Since it is expensive to re-key text (and impractical to redraw images), some way is needed to automatically convert the varying shades or colors of the text or images into a digitized graphics image that can be stored in a file.

This is what a scanner does. The scanner head contains a charge-coupled device (CCD) like that used in digital cameras (see PHOTOGRAPHY, DIGITAL). The CCD contains thousands or millions of tiny regions that can convert incoming light into a voltage level. Each of these voltage levels, when amplified, will correspond to one pixel of the scanned image. (A color scanner uses three different diodes for each pixel, each receiving light through a red, green, or blue filter.)

The operation of the head depends on the type of scanner. In the most common type, the flatbed scanner, a motor moves the head back and forth across the paper, which lies facedown on a glass window. In a sheet-fed scanner, the head remains stationery and the paper is fed past it by a set of rollers. Finally, there are handheld scanners, where the job of moving the scanner head is performed by the user moving the scanner back and forth over the page.

The resolution of a scanner depends on the number of pixels into which it can break the image. The color depth depends on how many bits of information that it can store per pixel (more information means more gradations of color or gray). Resolutions of 2,400 dots per inch (dpi) or more are now common, with up to 36 bit color depth, allowing for about 68.7 billion colors or gradations (see COLOR IN COMPUTING).

Besides considerations of resolution and color depth, the quality of a scanned image depends on the quality of the scanner's optics as well as on how the page or other object reflects light. As anyone who has browsed eBay listings knows, the quality of scans can vary considerably. Most scanners come with software that allows for the scanner to be controlled and adjusted from the PC, and image-editing software can be used to further adjust the scanned image.

Even if the input is a sheet of text, the scanner's output is simply a graphical image. Special software must be used to interpret scanned images of text and identify which characters and other features are present (see OPTICAL CHARACTER RECOGNITION). Since such software is not 100 percent accurate, human proofreaders may have to inspect the resulting documents.

Like printers, scanners have become quite inexpensive in recent years. Quite serviceable units are available for around $100 or so. (Popular multifunction devices often include scanner, copier, fax, and printer capabilities. A scanner can be used as a copier or fax by sending its output to the appropriate mechanism.)

Many home users now use scanners to digitize images for use in personal Web pages, online auctions, and other venues. Since sheet-fed scanners can only process individual sheets (not books, magazines, or objects) they are now less popular. Handheld scanners are somewhat tedious to use and require a steady hand, so they are generally used only in special circumstances where a flatbed scanner is not available. For capturing images of three-dimensional objects it is often easier to use a digital camera than a scanner.

Specialized scanners are also available. For example, although many flatbed scanners have a holder for scanning film negatives, a dedicated film scanner (costing perhaps $500) is a better choice if one wants to scan and possibly retouch or restore photographs. There are also high-end drum-type scanners that can scan at resolutions of 10,000 dpi or more.

Further Reading
Busch, David D. *Mastering Digital Scanning with Slides, Film, and Transparencies.* Boston: Muska & Lipman, 2004.
Chambers, Mark L. *Scanners for Dummies.* Hoboken, N.J.: Wiley, 2004.
PC Tech. Guide: Scanners. Available online. URL: http://www.pctechguide.com/55Scanners.htm. Accessed August 20, 2007.

scheduling and prioritization

Often in computing, a fixed resource must be parceled out among a number of competing users. The most obvious example is the operating system's scheduling the running of programs. Most computers have a single central processor (CPU) to execute programs. However, today virtually all operating systems (except for certain dedicated applications—see EMBEDDED SYSTEM) are expected to have many programs available simultaneously. For example, a Microsoft Windows user might have a word processor, spreadsheet, e-mail program, and Web browser all open at the same time. Not only might all of these programs be carrying out tasks or waiting for the user's input, but dozens of "hidden" system programs are also running in the background, providing services such as network support, virus protection, and printing services (see MULTITASKING).

In this environment each executing program (or "process") will be in one of three possible states. It may be actively executing (that is, its code is being run by the CPU). It may be ready to execute—that is, "wanting" to perform some activity but needing access to the CPU. Or, the program may be "blocked"—that is, not executing and unable to execute until some external condition is met. Blockage is usually caused by an input/output (I/O) operation. An example would be a program that's waiting for data to finish loading from a file.

In this sort of single-processor multiple-program system, the simplest arrangement is to have the operating system dole out fixed amounts of execution time to each program. Each program that indicates that it's ready to run gets placed in a list (see QUEUE) and given its turn. When the amount of time fixed for a turn has passed, the operating system saves the program's "state" in the processor—the contents of the registers, address pointed to by the pointer to the next instruction to be executed, and so on. This stored information can be considered to be a "virtual processor." When the program's turn comes around again, the processor is reloaded with the contents of the virtual processor and execution continues where it had left off.

USE OF PRIORITY

The above scheme assumes that all programs should have equal priority. In other words, that the timely completion of one program is not more important than that of another, or that no program should be "bumped up" in the queue for some reason. In reality, however, most operating systems do give some programs preference over others.

For example, suppose the word processor has just received a user's mouse click on a menu. The next program in the queue for execution, however, is an antivirus program that's checking all the files on the hard drive for possible viruses. The latter program is important, but since the user is not waiting for it to finish, a delay in its execution won't cause a significant problem. The user, on the other hand, is expecting the menu just clicked on to open almost instantly, and will become irritated with even a short delay. Therefore, it makes sense for the operating system to give a program that's responding to immediate user activity a higher priority than a program that's carrying out tasks that don't require user intervention.

There are other times when a program must (or should) be given a higher priority. A program may be required to complete a task within a guaranteed time frame (for example, to dispatch emergency services personnel). The operating system must therefore provide a way that the program can request priority execution.

In general, an operating system that supports real-time applications or that requires great attention to efficiency in using valuable devices may need a much more sophisticated scheduling algorithm that factors in the availability of key devices or services and adjusts program priorities in order to minimize bottlenecks and guarantee that the system's response will be within required parameters. Indeed, the method used for assigning priorities may actually be changed in response to changes in the various "loads" on the system. Sophisticated systems may also include programs that can predict the likely future load on the system in order to adjust for it as quickly as possible.

SCHEDULING MULTIPROCESSOR SYSTEMS

These general principles also apply to systems where more than one processor is available (see MULTIPROCESSING), but there is the added complication of deciding where the scheduling program will be run. In a multiprocessing system that has one "master" and many "slave" processors, the scheduling program runs on the master processor. This arrangement is simple, but it means that when a slave processor wants to schedule a program it must wait until the scheduling program gets its next time-slice on the master processor.

One alternative is to allow any processor that has free time to run the scheduling algorithm. This is harder to set up because it requires a mechanism to make sure two processors do not try to run the scheduling program at the same time, but it smooths out the bottleneck that would arise from relying on a single processor.

A variant of this approach is "distributed scheduling." Here each processor runs its own scheduling program. All the schedulers share the same set of information about the status and queuing of processes on the system, and a locking mechanism is used to prevent two processors from changing the same information at the same time. This approach is easiest to "scale up" since added processors can come with their own scheduling programs.

Two trends in recent years have changed the emphasis in scheduling algorithms. One is the continuing drop in price per unit of processing power and memory. This means that maximum efficiency in using the hardware can often give way in favor of catering to the user's convenience and perceptions by giving more priority to interaction with the user. The other development is the growing use of systems where much of the burden of graphics and interactivity is placed on the user's desktop, thus simplifying the complexity of scheduling for the server (see CLIENT-SERVER COMPUTING).

Principles of scheduling and priority can be applied in areas other than computer operating systems. Scheduling human activities (such as factory work) adds further complications such as the dependence of one task upon the prior performance of one or more other tasks (see PROJECT MANAGEMENT SOFTWARE) and the "just-in-time" scheduling for minimizing the investment in materials or inventory.

Further Reading

Brucker, Peter. *Scheduling Algorithms*. 5th ed. New York: Springer-Verlag, 2007.

Leung, Joseph Y-T., ed. *Handbook of Scheduling Algorithms, Models, and Performance Analysis*. Boca Raton, Fla.: CRC Press, 2004.

Pinedo, M. *Scheduling: Theory, Algorithms, and Systems*. 2nd ed. Upper Saddle River, N.J.: Prentice Hall, 2001.

science fiction and computing

The image of the mechanical brain or "knowledge engine" has a surprisingly long history in Western literature. As far back as Jonathan Swift's *Gulliver's Travels* (1726), we find a gigantic engine that can create books on every conceivable subject. While this was a satirical jab at thinkers who were ushering in a rational, mechanistic cosmos, the idea that the cunning mechanical automatons being created for the amusement of princes might someday think did not seem so far-fetched. This belief would be strengthened in the coming two centuries by the triumph of the Industrial Revolution. In Jules Verne's *Paris in the Twentieth Century* (written in 1863), giant calculating machines and facsimile transmissions were used to coordinate business activities.

As early as the beginning of the 20th century, writers had been exploring what might happen if some combination of artificial brains and robots offered the possibility of catering to all human needs. In E. M. Forster's "The Machine Stops," published in 1909, people no longer even have to leave their insectlike cells because even their social needs are provided through machine-mediated communication not unlike today's Internet. In the 1930s and 1940s, other writers such as John W. Campbell and Jack Williamson wrote stories in which a worldwide artificial intelligence became the end point of evolution, with humans either becoming extinct or living static, pointless lives.

Science fiction writers had also been considering the ramifications of a related technology, robotics. The term *robot* came from Karel Čapek's *R.U.R.* (Rossum's Universal Robots). Although the robot had a human face, it could have inhuman motives and threaten to become Earth's new master, displacing humans. Isaac Asimov offered a more benign vision, thanks to the "laws of robotics" embedded in his machines' very circuitry. The first law states, "A robot shall not harm a human being or, through inaction, cause a human being to come to harm." In the real world, of course, artificial intelligence had no such built-in restrictions (see ARTIFICIAL INTELLIGENCE).

Science fiction of the "Golden Age" of the pulp magazines had only limited impact on popular culture as a whole. Once actual computers arrived on the scene, however, they became the subject for movies as well as novels. D. F. Jones's novel *Colossus: The Forbin Project* (1966), which became a film in 1970, combined cold war anxiety with fear of artificial intelligence. Joining forces with its Soviet counterpart, Colossus fulfills its orders to prevent war by taking over and instituting a world government. Similarly, Hal in the film *2001: A Space Odyssey* (based on the work of Arthur C. Clarke) puts its own instinct for self-preservation ahead of the frantic commands of the spaceship's crew. However, the artificial can also strive to be human, as in the 2001 movie *A.I.*

During the 1940s and 1950s science fictional computers tended to be larger, more powerful versions of existing mainframes, sometimes aspiring to godlike status. However, in Murray Leinster's book *A Logic Named Joe* (1946), a "Logic" is found in every home, complete with keyboard and television screen. All the Logics are connected to a huge relay circuit called the Tank, and the user can obtain everything from TV broadcasts to weather forecasts or even

Perhaps the most famous science fiction movie of all is 2001: A Space Odyssey (1968). However, the millennial year passed without either an AI like Hal or passenger space lines. (©ARENAPAL / TOPHAM / THE IMAGE WORKS)

the answers to history trivia questions. Although the Logic is essentially an electronic-mechanical system, its functionality is startlingly similar to that achieved by the Internet almost half a century later.

Writers such as William Gibson (*Neuromancer*) and Vernor Vinge (*True Names*) later began to explore the world mutually experienced by computer users as a setting where humans could directly link their minds to computer-generated worlds (see VIRTUAL REALITY). A new elite of cyberspace masters were portrayed in a futuristic adaptation of such archetypes as the cowboy gunslinger, samurai, or ninja. Unlike the morally unambiguous world of the old western movies, however, the novels and movies with the new "cyberpunk" sensibility are generally set in a jumbled, fragmented, chaotic world. That world is often dominated by giant corporations (reflecting concerns about economic globalism) and is generally dystopian.

Meanwhile as cyberspace continues to become reality, cyberpunk has lost its distinctiveness as a genre. Gibson's latest work (and that of other writers such as Bruce Sterling and Vernor Vinge) is more apt to explore ways of communicating and networking that belong to just the day after tomorrow, if not already appearing (particularly among young people) today.

CYBERPUNK AND BEYOND

As personal computers and networking began to burgeon in the 1980s, the focus began to shift from computers as "characters" to the ways in which people interact with, and are changed by, new technology.

Although the term *cyberspace* was introduced by writer William Gibson in his 1982 short story "Burning Chrome," the word did not come into greater prominence until his 1984 novel *Neuromancer,* where it was described as "a consensual hallucination experienced daily by billions. . . ." (See CYBERSPACE.) It became the arena for a new style of science

fiction called cyberpunk, where outlaws and murderous corporations duel on the virtual frontier. Beyond the high-tech chases, questions of the ultimate meaning of cyberspace and of reality itself emerge, as in the *Matrix* trilogy of movies, or in the ultimate transformation of consciousness in human and machine (see SINGULARITY, TECHNOLOGICAL).

Further Reading

Asimov, Isaac, Patricia S. Warrick, and Martin H. Greenberg, eds. *Machines That Think: The Best Science Fiction Stories about Robots and Computers.* New York: Holt, Rinehart, and Winston, 1984.
Computer in Science Fiction. Available online. URL: http://www.technovelgy.com/ct/Science_List_Detail.asp?BT=Computer. Accessed November 14, 2007.
Conklin, Groff, ed. *Science-Fiction Thinking Machines: Robots, Androids, Computers.* New York: Vanguard Press, 1954.
Franklin, H. Bruce. "Computers in Fiction," 2000. Available online. URL: http://andromeda.rutgers.edu/~hbf/compulit.htm. Accessed November 14, 2007.
Frenkel, James, ed. *True Names by Vernor Vinge and the Opening of the Cyberspace Frontier.* New York: Tor, 2001.
Gibson, William. *Neuromancer.* New York: Ace Books, 1984.
———. *Pattern Recognition.* New York: G. P. Putnam's Sons, 2003.
Vinge, Vernor. *Rainbow's End.* New York: Tor, 2006.
Warrick, Patricia S. *The Cybernetic Imagination in Science Fiction.* Cambridge, Mass.: MIT Press, 1980.

scientific computing applications

From microbiology to plasma physics, modern science would be impossible without the computer. This is not because the computer has replaced the scientific method of observation, hypothesis, and experiment. Modern scientists essentially follow the same intellectual procedures as did Galileo, Newton, Darwin, and Einstein. Rather, understanding of the layered systems that make up the universe has now reached so complex and detailed a level that there is too much data for an individual human mind to grasp. Further, the calculations necessary to process the data usually can't be performed by unaided humans in any reasonable length of time. This can be caused either by the inherent complexity of the calculation (see COMPUTABILITY AND COMPLEXITY) or the sheer amount of data (as in DNA sequencing; see BIOINFORMATICS and DATA MINING).

INSTRUMENTATION

Some apparatus such as particle accelerators are complicated enough to make it expedient to control the operation by computer. It is simply more convenient to have instruments such as spectrocopes process samples automatically under computer control and produce printed results.

Most instruments for gathering data use electronics to turn physical measurements into numeric representations (see ANALOG AND DIGITAL and DATA ACQUISITION). The modern instrument's built-in processor and software performs preliminary processing that used to have to be done later in the lab. This can include scaling the data to an appropriate range of values, eliminating "noise" data, and providing an appropriate time framework for interpreting the data. Use of electronics also enables the data to be transmitted from a remote location (telemetry). See SPACE EXPLORATION AND COMPUTERS.

DATA ANALYSIS

The analysis of data to obtain theoretical understanding of the processes of nature also greatly benefits from the power of computers ranging from ordinary PCs to high-performance scientific workstations to large supercomputers. The possible significance of variables can be determined by statistical techniques (see also STATISTICS AND COMPUTING).

The fundamental task in understanding any system is to isolate the significant variables and determine how they affect one another. In many cases this can be done by solving differential equations, where a dependent variable changes as a result of changes in one or more independent variables. For example, the classical Maxwell theory of wave behavior is a system of differential equations that could be used to understand, for example, how radar waves will bounce off an object with a given shape and reflectivity. However, real-world objects have complicating factors: A given problem may include aspects of wave behavior, electromagnetic interaction, deformation of material, and so on. While the great scientists of the late 19th to mid-20th century could develop elegant formulas showing key relationships in nature, the interaction of many different phenomena often requires much more formidable computation that must be applied to many individual components.

It might be considered fortunate that the computer came along at about the time that it was required for further scientific progress. However, another way to look at it is to note that much of the pressure that led to investment in the development of computers came from that very need for computational resources, albeit primarily for wartime projects.

SIMULATION AND VISUALIZATION

Even if scientists have a basic understanding of a system, it may be hard to determine what the overall results of the interaction of the many particles (or other elements) in the system will be. This is true, for example, in the analysis of events taking place in nuclear reactors. Fortunately computers can apply the laws of the system to each of many particles and determine the resulting actions from their aggregate behavior (see SIMULATION). Simulation is particularly important in fields where actual experiments are not possible because of distance or time. Thus, a hypothesis about the formation of the universe can be tested by applying it to a set of initial conditions believed to reflect those at or near the time of the big bang.

However, even the most skilled scientists have trouble relating numbers to the shape and interaction of real-world objects. Computers have greatly aided in making it possible to visualize structures and phenomena using high-resolution 3D color graphics (see COMPUTER GRAPHICS). Features

Computer processing of photographic or scanned data can provide detailed information about our environment. In this NASA test project, aerial and satellite imagery is analyzed to yield information about the ripeness of grapes in a vineyard, as well as moisture, soil conditions, and plant disease. (NASA PHOTO)

of interest can be enhanced, and arbitrary ("false") colors can be used to visually show such things as temperature or blood flow. These techniques can also be used to create interactive models where scientists can, for example, combine molecules in new ways and have the computer calculate the likely properties of the result. Finally, computer visualization and modeling can be used both to teach science and to give the general public some visceral grasp of the meaning of scientific theories and discoveries.

Further Reading

Heath, Michael T. *Scientific Computing: An Introductory Survey.* 2nd ed. New York: McGraw-Hill, 2002.

Jahne, Bernd. *Practical Handbook of Image Processing for Scientific Applications.* 2nd ed. Boca Raton, Fla.: CRC Press, 2004.

Langtangen, H. P., A. M. Bruaset, and E. Quak, eds. *Advances in Software Tools for Scientific Computing.* New York: Springer, 2000.

Linux Software: Scientific Applications. Available online. URL: http://linux.about.com/od/softscience/Linux_Software_ Scientific_Applications.htm. Accessed August 20, 2007.

Oliveira, Suely, and David E. Stewart. *Writing Scientific Software: A Guide to Good Style.* New York: Cambridge University Press, 2006.

Scientific Computing and Numerical Analysis FAQ. Available online. URL: http://www.mathcom.com/corpdir/techinfo.mdir/index.html. Accessed August 20, 2007.

scripting languages

There are several different levels at which someone can give commands to a computer. At one end, an applications programmer writes program code that ultimately results in instructions to the machine to carry out specified processing (see PROGRAMMING LANGUAGES and COMPILER). The result is an application that users can control in order to get their work done.

At the other end, the ordinary user of the application uses menus, icons, keystrokes, or other means to select program features in order to format a document, calculate a spreadsheet, create a drawing, or perform some other task. Today most users also control the operating system by using a graphical user interface to, for example, copy files.

However, there is an intermediate realm where text commands can be used to work with features of the operating system, to process data through various utility programs, and to create simple reports. For example, a system administrator may want to log the number of users on the system at various times, the amount of disk capacity being used, the number of hits on various pages on a Web server, and so on. (See SYSTEM ADMINISTRATOR.) It would be expensive and time-consuming to write and compile full-fledged application programs for such tasks, particularly if changing needs will dictate frequent changes in the processing.

The use of the operating system shell and shell scripting (see SHELL) has traditionally been the way to deal with automating routine tasks, especially with systems running UNIX. However, the complexity of modern networks and in particular, the Internet, has driven administrators and programmers to seek languages that would combine the quick, interactive nature of shells, the structural features of

full-fledged programming languages, and the convenience of built-in facilities for pattern-matching, text processing, data extraction, and other tasks. The result has been the development of a number of popular scripting languages (see AWK, PERL, and PYTHON).

WORKING WITH SCRIPTING LANGUAGES

Although the various scripting languages differ in syntax and features, they are all intended to be used in a similar way. Unlike languages such as C++, scripting languages are interpreted, not compiled (see INTERPRETER). Typically, a script consists of a number of lines of text in a file. When the file is invoked (such as by someone typing the name of the language followed by the name of the script file at the command prompt), the script language processor parses each statement (see PARSER). If the statement includes a reference to one of the language's internal features (such as an arithmetic operator or a print command), the appropriate function is carried out. Most languages include the basic types of control structures (see BRANCHING STATEMENTS and LOOP) to test various variables and direct execution accordingly.

The trend in higher-level languages has been to require that all variables be declared to be used for particular kinds of data such as integer, floating-point number, or character string (see DATA TYPES). Scripting languages, however, are designed to be easy to use and scripts are relatively simple and easy to debug. Since the consequences of errors involving data types are less likely to be severe, scripting languages don't require that variables be declared before they are used. The language processor will make "common sense" assumptions about data. Thus if an integer such as 23 and a floating-point number like 17.5 must be added together, the integer will be converted to floating point and the result will be expressed as the floating-point value 40.5.

Similarly, scripting languages take a relaxed view about scope, or the parts of a program from which a variable's value can be accessed. Scripting languages do provide for some form of subroutine or procedure to be declared (see PROCEDURES AND FUNCTIONS). Generally, variables used within a subroutine will be considered to be "local" to that subroutine, and variables declared outside of any subroutine will be treated as global.

With compilers for regular programming languages, a great deal of attention must be paid to creating fast, efficient code. A scientific program may need to optimize calculations so that it can tackle cutting-edge problems in physics or engineering. A commercial application such as a word processor must implement many features to be competitive, and yet be able to respond immediately to the user and complete tasks quickly.

Scripting languages, on the other hand, are typically used to perform housekeeping tasks that don't place much demand on the processor, and that often don't need to be finished quickly. Because of this, the relative inefficiency of on-the-fly interpretation instead of optimized compilation is not a problem. Indeed, by making it easy for users to write and test programs quickly, the interpreter makes it much easier for administrators and others to create simple but useful tools for monitoring the system and extracting necessary information. Scripting languages can also be used to quickly create a prototype version of a program that will be later recoded in a language such as C++ for efficiency.

Scripting languages were originally written for operating systems that process text commands. However, with the popularity of Microsoft Windows, Macintosh, and various UNIX-based graphical user interfaces, many users and even system administrators now prefer a visual scripting environment. For example, Microsoft Visual Basic for Windows (and the related Visual Basic for Applications and VBScript) allow users to write simple programs that can harness the features of the Windows operating system and user interface and take advantage of prepackaged functionality available in ActiveX controls (see BASIC). In these visual environments the tasks that had been performed by script files can be automated by setting up and linking appropriate objects and adding code as necessary.

WEB SCRIPTING

Aside from shell programming, the most common use of scripting languages today is to provide interactive features for Web pages and to tie forms and displays to data sources. On the Web server, such technologies as ASP (see ACTIVE SERVER PAGES) use scripts embedded in (or called from) the HTML code of the page. On the client side (i.e., the user's Web browser) languages such as JavaScript and VBScript can be used to add features.

(For specific scripting languages, see AWK, JAVASCRIPT, LUA, VBSCRIPT, PERL, PHP, and TCL. For general-purpose languages that have some features in common with scripting languages, see PYTHON and RUBY.)

Further Reading
Barron, David. *The World of Scripting Languages*. New York: Wiley, 2000.
Brown, Christopher. "Scripting Languages." Available online. URL: http://cbbrowne.com/info/scripting.html. Accessed August 20, 2007.
Foster-Johnson, Eric, John C. Welch, and Micah Anderson. *Beginning Shell Scripting*. Indianapolis: Wiley, 2005.
Ousterhout, John K. "Scripting: Higher Level Programming for the 21st Century." *IEEE Computer*, March 1998. Available online. URL: http://www.tcl.tk/doc/scripting.html. Accessed August 20, 2007.
"Scriptorama: A Slightly Skeptical View on Scripting Languages." Available online. URL: http://www.softpanorama.org/Scripting/index.shtm. Accessed August 20, 2007.
Sebesta, Robert W. *Concepts of Programming Languages*. 8th ed. Boston: Pearson/Addison-Wesley, 2007.

search engine

By the mid-1990s, many thousands of pages were being added to the World Wide Web each day (see WORLD WIDE WEB). The availability of graphical browsing programs such as Mosaic, Netscape, and Microsoft Internet Explorer (see WEB BROWSER) made it easy for ordinary PC users to view Web pages and to navigate from one page to another. However, people who wanted to use the Web for any sort of sys-

tematic research found they needed better tools for finding the desired information.

There are basically three approaches to exploring the Web: casual "surfing," portals, and search engines. A user might find (or hear about) an interesting Web page devoted to a business or other organization or perhaps a particular topic. The page includes a number of featured links to other pages. The user can follow any of those links to reach other pages that might be relevant. Those pages are likely to have other interesting links that can be followed, and so on. Most Web users have surfed in this way: It can be fun and it can certainly lead to "finds" that can be bookmarked for later reference. However, this approach is not systematic, comprehensive, or efficient.

Alternatively, the user can visit a site such as the famous Yahoo! started by Jerry Yang and David Filo (see PORTAL and YAHOO!). These sites specialize in selecting what their editors believe to be the best and most useful sites for each topic, and organizing them into a multilevel topical index. The portal approach has several advantages: The work of sifting through the Web has already been done, the index is easy to use, and the sites featured are likely to be of good quality. However, even Yahoo!'s busy staff can examine only a tiny portion of the estimated 1 trillion or so Web pages being presented on about 175 million different Web sites (as of 2008). Also, the sites selected and featured by portals are subject both to editorial discretion (or bias) and in some cases to commercial interest.

ANATOMY OF A SEARCH ENGINE

Search engines such as Lycos and AltaVista were introduced at about the same time as portals. Although there is some variation, all search engines follow the same basic approach. On the host computer the search engine runs automatic Web searching programs (sometimes called "spiders" or "Web crawlers"). These programs systematically visit Web sites and follow the links to other sites and so on through many layers. Usually, several such programs are run simultaneously, from different starting points or using different approaches in an attempt to cover as much of the Web as possible. When a Web crawler reaches a site, it records the address (URL) and compiles a list of significant words. The Web crawlers give the results of their searches to the search engine's indexing program, which adds the URLs to the associated keywords, compiling a very large word index to the Web.

Search engines can also receive information directly from Web sites. It is possible for page designers to add a special HTML "metatag" that includes keywords for use by search engines. However, this facility can be misused by some commercial sites to add popular words that are not actually relevant to the site, in the hope of attracting more hits.

To use a search engine, the user simply navigates to the search engine's home page with his or her Web browser. (Many browsers can also add selected search engines to a special "search pane" or menu item for easier access.) The user then types in a word or phrase. Most search engines accept logical specifiers (see BOOLEAN OPERATORS) such as

AND, OR, or NOT. Thus, a search for "internet and statistics" will find only pages that have both words. Some engines also allow for phrases to be put in quote marks so they will be searched for as a whole. A search for "internet statistics" will match only pages that have these two words next to each other.

Because of the huge size of the Web, even seemingly esoteric search words can yield thousands of "hits" (results). Therefore, most search engines rank the results by analyzing how relevant they are likely to be. This can be done in a simple way by comparing the frequency with which the search terms appear on the various pages. More sophisticated search engines such as Google can determine how relevant a word or phrase seems to be because of its placement or presence in a heading or how often a site is referred to from other sites (see GOOGLE). Some search engines also offer the ability to "refine" searches by adding further words and performing a new match against the set of results.

LIMITATIONS AND FUTURE OF SEARCH ENGINES

Search engines do provide many useful "hits" for both casual and professional researchers, but the current technology does have a number of limitations. Even the most comprehensive search engines now reach and index only a small fraction of the total available Web pages. One way to maximize the number of pages searched is to use a "metasearch" program such as Copernic, which submits a user's search to many different search engines. It then collates the results, removing duplicates and attempting to rank them in relevance.

Even with "relevancy" algorithms, searches for broad, general topics are likely to retrieve many less-than-useful hits. Also, current search engines have difficulty finding image and sound (music) files, which are among the most sought-after Web content. This is because the search engine cannot recognize graphics or sound as such, only file names or extensions or text descriptions. Search engines also vary considerably in their ability to read and index files in proprietary text formats such as Microsoft Word or Adobe PDF.

Once mainly an auxiliary tool for Web portals, search engines have become a major business, and a variety of new types of search engines have proliferated. By combining search with paid advertising and delivery of a variety of services, Google in particular has become one of the Web's biggest success stories. In turn, Web-site owners have attempted to use various techniques to "optimize" or raise the ranking of their pages in search results, while Google has quietly tweaked its "page rank" algorithm to keep such efforts in check.

Two major search trends that can be seen in Google are specialized searches and local search. Google offers a variety of search options to target images, video, news, even blogs. Local search (such as Google Maps) combines maps with lists of local businesses, making it easier for users to find, for example, hotels near a given airport (see MAPPING AND NAVIGATION SYSTEMS). Services such as Google Maps Street View even provide for a street-level closeup view of

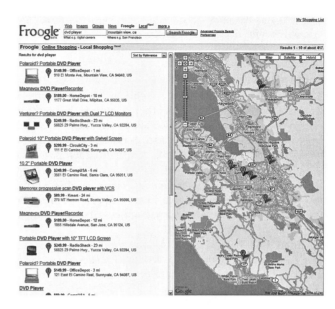

Copernic, a "metasearch engine," can pass a user's request to many different search engines and then prioritize and collate the results, weeding out duplicates. (Unfortunately some results are still much less useful than others.)

a neighborhood—almost a virtual tour, although this has raised privacy concerns. Google offers an extensive programming interface that is available to Web developers, as well as an easier-to-use facility for creating custom map displays (see MASHUPS).

In the future artificial intelligence techniques may make it possible for search engines to recognize types of images or sounds through pattern recognition. Search engines may be able to respond more appropriately to "natural language" queries such as "How many pages are there on the Web?" and find the answer, or at least Web pages that are likely to have the answer. (Current services of this type such as Ask. com tend to give hit-and-miss results.)

For now, search engines remain a useful tool, but systematic researchers should complement their results with links from portals and recommendations from authoritative sites.

Further Reading

Batelle, John. *The Search: How Google and Its Rivals Rewrote the Rules of Business and Transformed Our Culture.* New York: Penguin, 2005.

Dornfest, Rael, Paul Bausch, and Tara Calishain. *Google Hacks: Tips & Tools for Finding and Using the World's Information.* 3rd ed. Sebastapol, Calif.: O'Reilly, 2006.

Kent, Peter. *Search Engine Optimization for Dummies.* Hoboken, N.J.: Wiley, 2006.

Milstein, Sarah, J. D. Biersdorfer, and Matthew MacDonald. *Google: The Missing Manual.* 2nd ed. Sebastapol, Calif.: O'Reilly Media, 2005.

Moran, Mike, and Bill Hunt. *Search Engine Marketing, Inc.: Driving Search Traffic to Your Company's Web Site.* Upper Saddle River, N.J.: IBM Press, 2005.

Purvis, Michael, Jeffrey Sambells, and Cameron Turner. *Beginning Google Maps Applications with PHP and Ajax: From Novice to Professional.* Berkeley, Calif.: Apress, 2006.

Search Engine Watch. Available online. URL: http://searchengine-watch.com/. Accessed August 21, 2007.

semantic Web

The ever-growing World Wide Web consists of billions of linked HTML documents (and other resources), but most of the links contain no information about *why* the linkage has been made or what it might mean. Services such as Google can automatically trace the links and index each page (see SEARCH ENGINE) with the aid of "metadata" such as keywords that summarize page content. However, discovering the relationships between data items on pages, or between pages—and their meaning, or semantics—requires human scrutiny.

In his 1999 book *Weaving the Web,* World Wide Web creator Tim Berners-Lee (see BERNERS-LEE, TIM) described a new way in which Web pages might be organized in the future:

> I have a dream for the Web [in which computers] become capable of analyzing all the data on the Web—the content, links, and transactions between people and computers. A "Semantic Web," which should make this possible, has yet to emerge, but when it does, the day-to-day mechanisms of trade, bureaucracy and our daily lives will be handled by machines talking to machines. The "intelligent agents" people have touted for ages will finally materialize.

In other words, by encoding definitions of objects and their relationships into the text of Web pages, programs (see SOFTWARE AGENT) can be written to use this information to answer sophisticated questions such as "which devices from this vendor use open-source software?"

APPROACHES

The development of "machine understandable" Web resources requires that several layers of language be used. At bottom is the basic description of the structure of a document and its elements, such as titles or descriptions (see XML). Next comes RDF (Resource Description Format), which describes the relationship between data objects ("resources"). These relationships might include "a motherboard is a part of a computer" or "John owns this computer." Programs are now available to automatically create RDF statements given a database and its defined characteristics.

For these relationships to be truly useful, they must be part of a larger structure that describes their meaning. This can be provided via an RDF scheme or through the use of a language such as the Web Ontology Language (OWL)—see ONTOLOGIES AND DATA MODELS.

Programs can then query for these relationships using a language such as SPARQL.

APPLICATIONS

The semantic Web is not something that can appear overnight—after all, it will take considerable human effort to encode the information needed for machines to understand Web resources, and additional effort to code the applica-tion programs that will take advantage of that information. However, the potential payoff is huge, allowing both human and automated searchers to tackle much more sophisticated tasks.

For example, the University of Maryland is developing a prototype semantic search engine called Swoogle. It can extract information and determine relationships between documents that include RDF or OWL elements. Swoogle can also help users find appropriate ontologies for exploring a subject (see ONTOLOGIES AND DATA MODELS).

Much research needs to be done. For example, there is the problem of deriving a measure of "reliability" or "trust" based on the data sources used to answer the query, which may be scattered all over the world and represent very different kinds of sources.

Further Reading
Alesso, H. Peter, and Craig F. Smith. *Thinking on the Web: Berners-Lee, Gödel, and Turing.* New York: Wiley, 2006.
Antoniou, Grigoris, and Frank van Harmelen. *A Semantic Web Primer.* Cambridge, Mass.: MIT Press, 2004.
Davies, John, Rudi Studer, and Paul Warren. *Semantic Web Technologies: Trends and Research in Ontology-Based Systems.* Hoboken, N.J.: Wiley, 2006.
Swartz, Aaron. "The Semantic Web in Breadth." Available online. URL: http://logicerror.com/semanticWeb-long. Accessed November 13, 2007.
Swoogle [semantic Web search engine]. Available online. URL: http://swoogle.umbc.edu/index.php. Accessed November 3, 2007.

senior citizens and computing

A growing number of people 50 and older have been learning how to use computer technology and especially applications such as e-mail and Web browsing. However, a substantial number of seniors have expressed reluctance to join the digital world—as of January 2006, the Pew Internet & American Life Project found that only 34 percent of persons 65 and over were online. Some reasons why seniors have avoided the technology include the following:

- the belief that it would be too hard to learn to use it

- uncertainty about what can be done online and whether it is worth the effort

- fear of well-publicized dangers such as viruses and identity theft

- the expense of a personal computer and Internet access

Fortunately a number of these factors are gradually being ameliorated. There are numerous books and courses (such as at adult education or senior centers) that introduce the essentials of computing to seniors. Properly installed security and filtering software, together with some user education, can minimize the chances of being victimized online. Finally, Internet-capable PCs are now available for around $300 or less, though the cost of broadband access has not fallen as rapidly as that of hardware.

SENIORS AND THE INTERNET

According to research by the Pew Internet & American Life Project, for seniors who do go online, e-mail is the most popular activity (and something shared with other age groups). While teens are most prolific at adopting new technologies such as instant messaging, content sharing, and social networking, older users are less likely to adopt emerging services, but more likely to bank or make travel reservations online—perhaps reflecting their having more money for leisure travel. Older people also tend to be more avid in pursuing health information. On the other hand, buying things online seems to be equally popular with all age groups.

Computer technology can also assist seniors with the activities of daily life. At the Quality of Life Technologies Center, researchers from Pitt and Carnegie Mellon Universities are developing technologies including:

- robotic wheelchairs with arms that can manipulate objects and even assist in cooking meals

- systems to help people get out of bed, dress, bathe, and so on

- pervasive sensor networks that can monitor persons as they move around

- monitoring systems that can detect growing confusion or cognitive impairment and call for help

- systems to supervise daily activities and make sure medications are taken on time

- "coaching" software that can help maintain memory and cognitive skills, even in persons with Alzheimer's disease

(See DISABLED PERSONS AND COMPUTING.)

Further Reading

"Abby & Me." Available online. URL: http://www.abbyandme.com. Accessed November 14, 2007.
Fox, Susannah, and Mary Madden. "Are 'Wired Seniors' Sitting Ducks?" Pew Internet & American Life Project, April 2006. Available online. URL: http://www.pewinternet.org/pdfs/PIP_Wired_Senior_2006_Memo.pdf. Accessed November 14, 2007.
———. "Generations Online." Pew Internet & American Life Project, December 2005. Available online. URL: http://www.pewinternet.org/pdfs/PIP_Generations_Memo.pdf. Accessed November 13, 2007.
Rolstein, Gary. "Robotic Aids for the Disabled and Elderly." *Pittsburgh Post-Gazette*, November 14, 2007. Available online. URL: http://www.post-gazette.com/pg/07318/833537-114.stm. Accessed November 14, 2007.
Stokes, Abby. *It's Never Too Late to Love a Computer: Everything You Need to Know to Plug In, Boot Up and Get Online.* Revised ed. New York: Workman Publishing, 2005.
Stuur, Addo. *Internet and E-mail for Seniors with Windows Vista.* Visual Steps Publishing, 2006.

serial port

There are basically two ways to move data from a computer to or from a peripheral device such as a printer or modem.

A byte (8 bits) of data can be moved all at once, with each bit traveling along its own wire (see PARALLEL PORT). Alternatively, a single wire can be used to carry the data one bit at a time. Such a connection is called a *serial port*.

The serial port receives data a full byte at a time from the computer bus and uses a UART (Universal Asynchronous Receiver-Transmitter) to extract the bits one at a time and send them through the port. A corresponding circuit at the other end accumulates the incoming bits and reassembles them into data bytes.

The data bits for each byte are preceded by a start-bit to signal the beginning of the data and terminated by an stop-bit. Depending on the application, an additional bit may be used for parity (see ERROR CORRECTION). Devices connected by a serial port must "negotiate" by requesting a particular connection speed and parity setting. Failure to agree results in gibberish being received.

The official standard for serial transmission is called RS-232C. It defines various additional pins to which wires are connected, such as for synchronization (specifying when the device is ready to send or receive data) and ground. Physically, the old-style connectors are called DB-25 because they contain 25 pins (many of which are not used). Most newer PCs have DB-9 (i.e. nine pin) connectors. A "gender changer" can be used in cases where two devices both have male connectors (with pins) or female connectors (with corresponding sockets).

Because they use a single data transmission line and include error-correction, serial cables can be longer than parallel cables (25 feet or more, as opposed to 10–12 feet). Serial transmission is generally slower (at up to 115,200 bits/second) than parallel transmission. Serial connections have generally been used for such devices as modems (whose speed is already limited by phone line characteristics), keyboards, mice, and some older printers. Today the faster and more flexible USB (see UNIVERSAL SERIES BUS) is replacing serial connections for many devices including even keyboards.

Further Reading

Peacock, Craig. "Interfacing the Serial / RS232 Port." Available online. URL: http://www.beyondlogic.org/serial/serial.htm. Accessed August 21, 2007.
Tyson, Jeff. "How Serial Ports Work." Available online. URL: http://computer.howstuffworks.com/serial-port.htm. Accessed August 21, 2007.

service-oriented architecture (SOA)

The traditional model for organizing information processing, particularly in large installations, is in terms of installing and maintaining large applications that each provide many functions, and then devising ways for the applications to exchange data and otherwise coordinate with each other. As the information environment has become more complex (particularly with regard to databases and Web-related services), this approach has become more cumbersome, less flexible, and harder to maintain.

Service-oriented architecture is a new approach that focuses on services (basic functions, such as displaying and processing forms or formatting data) and provides standardized ways for them to be accessed by programs. Applications in turn are then built up by "plugging in" the required services and organizing them to meet the required logic and sequence of processing.

In order to be accessed, each service provides "metadata" (usually in XML files) that describes what data is used by a service and what it provides. The description itself can be provided using Web Services Description Language (WSDL), including network addresses (ports) for connecting to the service, the operations supported, and the abstract format of the expected data. (For more on the message protocol, see SOAP.)

There are three basic roles that must be filled in designing an SOA system: The service provider creates a service (often a Web service) and "exposes" aspects of the service and controls access to it (through security policies). The service broker provides a registry of available services and tells requesters how to connect to them. (For more on brokers, see CORBA.) Finally, the requestor in an application finds and requests services as needed.

In general SOA can be seen as part of the trend toward decentralized, loosely coupled computing (see DISTRIBUTED COMPUTING). Because all services communicate through the network, it is easy to reallocate or scale up services as needed. It is also easier to upgrade software and reuse it for new applications. (For more on combining services provided by applications, see MASHUPS.) However, SOA brings challenges of its own in terms of making services truly interoperable and conforming to standards that are still evolving.

Further Reading

Erl, Thomas. *SOA Principle of Service Design*. Boston: Pearson Education/Prentice Hall, 2007.

Hurwitz, Judith, et al. *Service Oriented Architecture for Dummies*. Hoboken, N.J.: Wiley, 2007.

Josuttis, Nicolai M. *SOA in Practice: The Art of Distributed System Design*. Sebastapol, Calif.: O'Reilly, 2007.

QAT SOA Resource Center. Available online. URL: http://www.soaresourcecenter.com/. Accessed November 14, 2007.

Service-Oriented Architecture (SOA). TechRepublic. Available online. URL: http://search.techrepublic.com.com/search/Service-Oriented+Architecture+(SOA).html. Accessed November 14, 2007.

Weerawarana, Sanjiva, et al. *Web Services Platform Architecture*. Upper Saddle River, N.J.: Prentice Hall PTR, 2005.

Shannon, Claude E.

(1916–2001)
American
Mathematician, Computer Scientist

The information age would not have been possible without a fundamental understanding of how information could be encoded and transmitted electronically. Claude Elwood Shannon developed the theoretical underpinnings for modern information and communications technology and then went on to make important contributions to the young discipline of artificial intelligence (AI).

Shannon was born in Gaylord, Michigan, on April 30, 1916. He received bachelor's degrees in both mathematics and electrical engineering at the University of Michigan in 1936. He went on to MIT, where he earned a master's degree in electrical engineering and a Ph.D. in mathematics, both in 1940. Shannon's background thus well equipped him to relate mathematical concepts to practical engineering issues.

While a graduate student at MIT, Shannon was in charge of programming an elaborate analog computer called the Differential Analyzer that had been built by Vannevar Bush (see ANALOG COMPUTER and BUSH, VANNEVAR). Actually "programming" is not quite the right word: To solve a differential equation with the Differential Analyzer, it had to be translated into a variety of physical settings and arrangements of the machine's intricate electromechanical parts.

The Differential Analyzer was driven by electrical relay and switching circuits. Shannon became interested in the underlying mathematics of these control circuits. He realized that their fundamental operations corresponded to the Boolean algebra he had studied in undergraduate mathematics classes (see BOOLEAN OPERATORS). It turned out that the seemingly abstract Boolean AND, OR and NOT operations had a practical engineering use. Shannon used the results of his research in his 1938 M.S. thesis, titled "A Symbolic Analysis of Relay and Switching Circuits." This work was honored with the Alfred Nobel prize of the combined engineering societies (this is not the same as the more famous Nobel Prize).

Claude Shannon developed the fundamental theory underlying modern data communications, as well as making contributions to the development of artificial intelligence. (LUCENT TECHNOLOGIES BELL LABS)

Along with the work of Alan Turing and John von Neumann (see TURING, ALAN and VON NEUMANN JOHN), Shannon's logical analysis of switching circuits would become essential to the inventors who would build the first digital computers in just a few years. (Demonstrating the breadth of his interests, Shannon's Ph.D. thesis would be in an entirely different application—the algebraic analysis of problems in genetics.)

In 1941, Shannon joined BELL LABORATORIES, perhaps America's foremost industrial research organization. The world's largest phone company had become increasingly concerned with how to "scale up" the burgeoning telephone system and still ensure reliability. The coming of war also highlighted the importance of cryptography—securing one's own transmissions while finding ways to break opponents' codes. Shannon's existing interests in both data transmission and cryptography neatly dovetailed with these needs.

Shannon's paper titled "A Mathematical Theory of Cryptography" would be published after the war. But Shannon's most lasting contribution would be to the fundamental theory of communication. His formulation would explain what happens when information is transmitted from a sender to a receiver—in particular, how the reliability of such transmission could be analyzed (see INFORMATION THEORY).

Shannon's 1948 paper, "A Mathematical Theory of Communication" was published in *The Bell System Technical Journal*. Shannon identified the fundamental unit of information (the binary digit, or "bit" that would become familiar to computer users). He showed how to measure the redundancy (duplication) within a stream of data in relation to the transmitting channel's capacity, or bandwidth. Finally, he showed methods that could be used to automatically find and fix errors in the transmission. In essence, Shannon founded modern information theory, which would become vital for technologies as diverse as computer networks, broadcasting, data compression, and data storage on media such as disks and CDs.

One of the unique strengths of Bell Labs is that it did not limit its researchers to topics that were directly related to telephone systems or even data transmission in general. Like Alan Turing, Shannon became interested after the war in the question of whether computers could be taught to perform tasks that are believed to require true intelligence (see ARTIFICIAL INTELLIGENCE). He developed algorithms to enable a computer to play chess and published an article on computer chess in *Scientific American* in 1950. He also became interested in other aspects of machine learning, and in 1952 he demonstrated a mechanical "mouse" that could solve mazes with the aid of a circuit of electrical relays.

The mid-1950s would prove to be a very fertile intellectual period for AI research. In 1956, Shannon and AI pioneer John McCarthy (see MCCARTHY, JOHN) put out a collection of papers titled "Automata Studies." The volume included contributions by two other seminal thinkers, John von Neumann and Marvin Minsky (see MINSKY, MARVIN).

Although he continued to do research, by the late 1950s Shannon had changed his emphasis to teaching. As Donner Professor of Science at MIT (1958–1978) his lectures inspired a new generation of AI researchers. During the same period Shannon also explored the social impact of automation and computer technology as a Fellow at the Center for the Study of Behavioral Sciences in Palo Alto, California.

Shannon received numerous prestigious awards, including the IEEE Medal of Honor and the National Medal of Technology (both in 1966). Shannon died on February 26, 2001, in Murray Hill, New Jersey.

Further Reading
Shannon, Claude Elwood. "A Chess-Playing Machine." *Scientific American,* February 1950, 48–51.
———. *Collected Papers.* Ed. N. J. A. Sloane and Aaron D. Wyner. New York: IEEE Press, 1993.
———. "A Mathematical Theory of Communication." *Bell System Technical Journal* 27; July and October, 1948, 379–423, 623–656. Available online. URL: http://plan9.bell-labs.com/cm/ms/what/shannonday/shannon1948.pdf. Accessed August 21, 2007.
Waldrop, M. Mitchell. "Claude Shannon: Reluctant Father of the Digital Age." *Technology Review,* July/Aug. 2001. Available online. URL: http://www.technologyreview.com/Infotech/12505/. Accessed August 21, 2007.

shareware and freeware

The early users of personal computers generally had considerable technical skill and a desire to write their own programs. This was partly by necessity: If one wanted to get an Apple, Atari, Commodore, or Radio Shack machine to perform some particular task, chances were one would have to write the software oneself. Commercial software was scarce and relatively expensive. However, given enough time, it was possible for hobbyists to write programs using the machine's built-in BASIC language or (with more effort) assembly language.

Programs such as utilities and games were often freely shared at gatherings of PC enthusiasts (see USER GROUPS). Many talented amateur programmers considered trying to turn their avocation into a business. However, a utility to provide better file listings or a colorful graphics program that creates kaleidoscopic images was unlikely to interest the commercial software companies who developed large programs in-house for marketing primarily to business.

In 1982, Andrew Fuegelman created a program called PC-Talk. This program provided a better way for users with modems to connect to the many bulletin board systems that were starting to spring up. Fluegelman was familiar with the common practice of public radio and TV broadcasters of soliciting pledge payments to help support their "free" service. He decided to do something similar with his program. He distributed it to many bulletin boards, where users could download it for free. However, he asked users who liked the program and wanted to continue to use it to pay him $25.

Fluegelman dubbed his method of software distribution "freeware" (because it cost nothing to try out the program). Other programmers began to use the same method with their own software. This included Jim Knopf, author of the PC-File database program, and Bob Wallace, who offered

PC-Write as a full-featured alternative to expensive commercial word processing program. Because Fluegelman had trademarked the term *freeware,* these other authors began to call their offerings *shareware.*

Today freeware means software that can be downloaded at no cost and for which there is no charge for continued use. The program may be redistributed by users as long as they don't charge for it.

Shareware, on the other hand, follows Fluegelman's original concept. The software can be downloaded for free. The user is allowed to try the program for a limited period (either a length of time such as 30 days, or a maximum number of times that the program can be run). After the trial period, the user is expected to pay the author the specified fee of continued use. (Today this is usually done through the author's Web site or a service that can accept secure credit card payments online.) Once the user pays, he or she receives either an unrestricted version of the software or frequently, an alphanumeric key that can be typed into the program to remove all restrictions. At this point the program is said to be "registered."

Users can be encouraged (or forced) to pay in various ways. Some programs keep working after the trial period, but display continual "nag" messages or remove some functionality, such as the ability to print or save one's work. ("Demos" of commercial games or other programs also have limited functionality, but cannot be registered or upgraded. They are there simply to entice consumers to buy the commercial product.)

Alternatively, some shareware authors prefer to entice their users to register by offering bonuses, such as additional features, free upgrades, or additional technical support. Sometimes (as with the RealPlayer streaming sound and video player and the Eudora e-mail program) a useful but limited "lite" version is offered as freeware, but users are encouraged to upgrade to a more full-featured "professional" version.

Shareware has been a moderately successful business for a number of program authors. For example, Phil Katz's PKZip file compression and packaging program is so useful that it has found its way onto millions of PCs, and enough users paid for the program to keep Katz in business. (PKZip and its cousin WinZip are examples of shareware programs that became so popular that they spawned commercially packaged versions.)

Shareware and freeware should be distinguished from public domain and open source software (see OPEN-SOURCE MOVEMENT). Public domain software is not only free (as with freeware), but the author has given up all rights including copyright, and users are free to alter the program's code or to use it as part of a new program. Open-source software, on the other hand, allows users free access to the software and its source code, but with certain restrictions—notably, that it not be used in some other product for which access will be restricted.

Today tens of thousands of shareware and freeware programs are available on the Internet via ftp archives, author's Web sites, and giant online libraries maintained by zdnet.com, cnet.com, tucows.com, and others.

Further Reading
Association of Shareware Professionals. Available online. URL: http://www.asp-shareware.org. Accessed August 21, 2007.
Ellis, Robert. *Handpicked Software for Mac OS X: The Best New Freeware, Shareware, and Commercial Software for Mac OS X.* Petaluma, Calif.: Futurosity, 2002.
Hasted, Edward. *Software That Sells: A Practical Guide to Developing and Marketing Your Software Project.* Indianapolis: Wiley, 2005.
Lehnert, Wendy G. *The Web Wizard's Guide to Freeware and Shareware.* Boston: Addison-Wesley, 2002.
Shareware.com. Available online. URL: http://www.shareware.com. Accessed August 21, 2007.
Tucows.com. Available online. URL: http://www.tucows.com. Accessed August 21, 2007.

shell

During the 1950s, using a computer generally meant that operators submitted batch-processing command cards (see JOB CONTROL LANGUAGE) that controlled how each program would use the computer's resources. One program ran at a time, and interaction with the user was minimal. However, when time-sharing computers began to appear in the 1960s, users gained the ability to control the computer interactively from terminals. The operating system therefore needed to have a facility that would interpret and execute the commands being typed in by the users, such as a request to list the files in a directory or to send a file to the printer. This command interpreter is called a *shell.*

To see a simple shell in action, a Windows user need only bring up a command prompt, type the word *dir,* and press Enter. A shell called command.com provides the user interface for users of IBM PC-compatible systems running MS-DOS. The command processor displays a prompt on the screen. It then interprets (see PARSING) the user's commands. If the command involves one of the shell's internal operations (such as "dir" to list a file directory), it simply executes that routine. For example the command:

```
dir temp /p
```

would be interpreted as a call to execute the dir function, passing it the name "temp" (a directory) and the /p, which dir interprets as a "switch" or instruction telling it to pause the directory listing after each screenful of text. If the command is an external MS-DOS utility such as "xcopy" (a file copying program), the shell runs that program, passing it the information (mainly file names) from the command line. Finally, the shell can run any other executable program on the system. It is then that program's responsibility to interpret and act upon any additional information that was provided.

MS-DOS also has the ability for the command.com shell to read a series of commands stored in a text file called a batch file, and having the *.bat (batch) extension. This allowed for rudimentary scripting of system housekeeping operations or other routine tasks (see SCRIPTING LANGUAGES).

UNIX SHELLS

MS-DOS largely faded away in the 1990s as more users switched to Microsoft Windows and begun to use a graphical user interface to control their machines. However, shells have achieved their greatest proliferation and elaboration with UNIX, the operating system developed by Ken Thompson and Dennis Ritchie starting in 1969 and widely used for academic, scientific, engineering, and Web applications.

UNIX shells serve the same basic purposes as the MS-DOS shell: interactive control of the operating system and the ability to run stored command scripts. However, the UNIX shells have considerably more complex syntax and capabilities.

Part of the design philosophy of the UNIX system was to place the core operating system functions in the kernel (see KERNEL and UNIX). This modular design meant that UNIX, unlike most other operating systems, did not have to commit itself to a particular form of user interface or command processor. Accordingly, a number of such processors (shells) have been developed, reflecting the programming style preferences of their originators.

The first shell to be developed was the Bourne Shell, named for its creator, Steven R. Bourne, who developed it at Bell Labs, the original home of UNIX. The Bourne shell implemented some basic ideas that are characteristic of UNIX: the ability to redirect input and output to and from files, devices or other sources (using the < and > characters), and the ability to use "pipes" (the | character) to connect the output of one command to the input of another.

The next major development was the C shell (csh). The Bourne shell used a relatively simple and clean syntax devised by its creator. As the name suggests, the C shell (developed at the University of California, Berkeley) takes its syntax from the C programming language, which was by far the most commonly used language on UNIX systems. One logical reason for this choice was that C programmers could quickly learn to write scripts with the C shell. The C shell also added support for job control (that is, moving processes between foreground and background operation) and in general was easier to use for interactively controlling programs from the command line.

UNIX users sometimes used both shells, since the simpler and more consistent syntax of the Bourne shell is generally thought to be better for writing scripts. (The two shells also reflected the split in the UNIX world between the version of the operating system provided by AT&T and the variant developed at UC Berkeley.)

David Korn at AT&T then decided to combine the best features of both shells. His Korn shell (Ksh) kept the better scripting language features from the Bourne shell but added job-control and other features from the C shell. He also added the programming language concept of functions (see PROCEDURES AND FUNCTIONS), allowing for cleaner organization of code.

Another popular shell, BASH (Bourne Again Shell) was developed by the Free Software Foundation for GNU, an open-source version of UNIX. BASH and Ksh share most features and both are compatible with POSIX, a standard specification for connecting programs to the UNIX operating system.

The surge of interest in open-source UNIX in recent years (see LINUX) has brought a new generation of shell users. Although modern Linux distributions provide a full graphical user interface (GUI)—indeed, a choice of them—such tasks as software installation and configuration often involve entering shell commands. Experienced users can also find, copy, move, or otherwise manipulate batches of files more efficiently in the shell than in using windows and mouse movements.

SHELL SCRIPTS

Regardless of the version of the shell used, shell scripts work in the same basic way. A shell script is a text file containing commands to the shell. The commands can use control statements (see BRANCHING STATEMENTS and LOOP) and invoke both the shell's internal features and the many hundreds of utility programs that are available on UNIX systems (see SCRIPTING LANGUAGES).

Once the script is written, there are two ways to execute it. One way is to type the name of the shell at the command prompt, followed by the name of the script file, as in:

```
$ sh MyScript
```

Alternatively, the chmod (change mode) command can be used to mark the script's file type as executable, and the first line of the script then contains a statement that invokes the shell, which will parse the rest of the script. The script can now be executed simply by typing its name at the command prompt (or it can be included as a command in another script).

Here is a simple example of a shell script that prints out various items of information about the user and the current session on a UNIX system:

```
#! /sbin/sh

echo My username: `whoami`
echo My current directory: `pwd`
echo
echo My disk usage:
du -k
echo
echo System status:
uptime
if test -f log.txt; then
   cat log.txt
else echo Log file not found
fi
```

The first line tells UNIX which shell to use to interpret the script (in this case the Bourne shell, sh, will be executed). The echo command simply outputs the text that follows it to the screen. "whoami" is a UNIX command that prints the user's name. The script takes advantage of an interesting UNIX feature: The whoami command is put in "backquotes" (` `). This inserts the output of the whoami command (the user name) in place of that command, and the resulting text is output by the echo command.

The du command gives the user's disk usage, while the uptime command gives some statistics about how many users are on the system and how long the system has been running. Finally, the if statement at the end of the script tests for the presence of the file log.txt. If the file exists, its contents are displayed by the "cat" command.

When "myinfo" is typed at the UNIX prompt, the output might look like the following:

```
$ myinfo
My username: hrh
My current directory: /home/h/r/hrh

My disk usage:
132 ./.nn
4 ./Mail
48 ./.elm
296 .

System status:
  7:34pm up 56 day(s), 20:39, 73 users, load
  average: 3.62, 3.45, 3.49

This is a test file.
```

Further Reading

Gite, Vivek G. "Linux Shell Scripting Tutorial." Available online. URL: http://www.freeos.com/guides/lsst/. Accessed August 21, 2007.

Kochan, Stephen, and Patrick Wood. *Unix Shell Programming*. 3rd ed. Indianapolis: Sams, 2003.

Newham, Cameron. *Learning the Bash Shell*. 3rd ed. Sebastapol, Calif.: O'Reilly Media, 2005.

Quigley, Ellie. *UNIX Shells by Example*. 4th ed. Upper Saddle River, N.J.: Prentice Hall, 2004.

Robbins, Arnold, and Bill Rosenblatt. *Learning the Korn Shell*. 2nd ed. Sebastapol, Calif.: O'Reilly Media, 2002.

Sobell, Mark G. *A Practical Guide to Linux Commands, Editors, and Shell Programming*. Upper Saddle River, N.J.: Prentice Hall, 2005.

Simonyi, Charles
(1948–)
Hungarian-American
Software Engineer, Entrepreneur

Born in Budapest, Hungary, on September 10, 1948, Charles Simonyi shaped the architecture of Microsoft's dominant software applications for many years, devised a new programming paradigm and established a company to promote it, and, along the way, became the fifth civilian "space tourist" to visit the *International Space Station*.

Simonyi's father was a professor of electrical engineering. In high school, Simonyi worked as a night watchman at a computer laboratory. When he expressed his interest, one of the engineers taught him how to program; he soon wrote a compiler and sold it to a government department. After working for a Danish company for a couple of years, Simonyi moved to the United States in 1968, attending the University of California, Berkeley, and earning a B.S. in engineering mathematics in 1972. Moving to Stanford University for graduate study, Simonyi was also hired by Xerox PARC, where he shared ideas with innovators in computer interfaces and networking. Simonyi received his Ph.D. in computer science from Stanford in 1977. In his dissertation Simonyi showed his early interest in "metaprogramming"—the development of ways to coordinate programs and provide them with a higher-level context.

In 1981 Simonyi applied directly to Bill Gates for a job (see GATES, BILL and MICROSOFT CORPORATION). At Microsoft Simonyi took charge of the development of the products that would dominate the office software market by the end of the 1980s, including Word and Excel. Simonyi also brought to Microsoft new program structure ideas that he had seen at Xerox PARC—see OBJECT-ORIENTED PROGRAMMING. At this time Simonyi also developed a standard system for naming variables that soon became known as Hungarian notation in honor of his ancestry.

The tremendous success of Simonyi as a software developer (and Microsoft's gargantuan revenue) made Simonyi independently wealthy. However, in 2002 he decided to strike out on his own, founding a company called Intentional Software with his business partner Gregor Kiczales. The company develops and promotes an approach to software design called *intentional programming*. (Simonyi had developed forerunners of this concept at Microsoft, but apparently the latter company lost interest in it, perhaps prompting Simonyi's departure.)

To develop an application, software engineers using intentional programming begin by building a "toolbox" of specific functions needed for the area in which the program is intended to operate (such as insurance or banking). Domain experts—people who have "real world" knowledge of that area—use a special editor to create a description of how the application must operate; thus the program is in a sense designed not by the programmers, but by the people who will guide its use. The program development system then connects the tools to the description to generate the final code, which can then be refined. An important feature of this process is that the specific intentions about what the program needs to do are preserved along with the code, with the result largely self-documenting. It is argued that this makes subsequent testing and modification of the software much faster and easier. The first commercial version of this development system is expected in 2008.

SPACE TOURIST AND PHILANTHROPIST

In April 2007 Simonyi, an experienced pilot, fulfilled a lifelong interest in space by riding a Russian Soyuz spacecraft to the *International Space Station*; the 10-day "vacation" cost him about $20 million. Simonyi chronicled his preparations and the trip itself via his "Nerd in Space" Web site. (Simonyi also sails in his sleek luxury yacht *Skat*.)

As a philanthropist, Simonyi established a professorship for the Public Understanding of Science at Oxford University, as well one for Innovation in Teaching at Stanford. He has given tens of millions of dollars to various programs in the arts and sciences. As of 2007 Simonyi was dating domestic arts entrepreneur and author Martha Stewart.

While it remains uncertain how successful and influential intentional programming will become, Simonyi has been hailed by Bill Gates as "one of the great programmers of all time."

Further Reading

Charles in Space. Available online. URL: http://www.charlesinspace. com/. Accessed November 14, 2007.

Greene, Stephen G. "Entrepreneur Seeks to Promote Excellence through Philanthropy." *Chronicle of Philanthropy* 16 (February 19, 2004): 20.

Intentional Software. Available online. URL: http://www.intent soft.com/. Accessed November 14, 2007.

Rosenberg, Scott. "Anything You Can Do, I Can Do Meta: Space Tourist and Billionaire Programmer Charles Simonyi Designed Microsoft Office. Now He Wants to Reprogram Software." *Technology Review,* January 2007. Available online. URL: http://www.technologyreview.com/Infotech/18047/?a=f. Accessed November 14, 2007.

Simula

One of the most interesting applications of computers is the simulation of systems in which many separate actions or events are happening simultaneously (see SIMULATION). During the 1950s, Norwegian computer scientist Kristen Nygaard began to develop a more formal way of describing and designing simulations. A typical simulation consists of a number of "objects," such as cars in a traffic flow or customers waiting in a bank line. In a bank simulation, for example, the objects (customers) would demand service from particular serving objects (teller windows). They would move in a queue and their motion would be captured at various points of time.

Nygaard used his ideas to create symbols and flow diagrams to represent the events going on in a simulation. However, existing computer languages such as Algol 60 were designed to carry out procedures sequentially and one at a time, not simultaneously. This made it difficult to write a program representing a situation in which many cars or customers were moving simultaneously.

In the early 1960s, Nygaard was joined by Ole-Johan Dahl, who had more experience with systems programming and computer language design. They worked together to create a new language that they called Simula, reflecting their emphasis on simulation programming. In designing Simula, the authors sought to create a data structure that was better suited to simultaneous actions or events. For example, in a simulation of automobile traffic, each car would be an "object" with data such as its location and speed as well as actions or capabilities such as changing speed or direction. The data for each object must be maintained separately and updated frequently.

The Algol 60 language already had a way to define code "blocks" (see PROCEDURES AND FUNCTIONS) that could contain their own local data as well as actions to be performed. Further, such blocks could be called repeatedly such that many copies could be "open" at the same time. However, these calls were still essentially sequential, not simultaneous. In their new Simula 1 language (introduced in 1965), Dahl and Nygaard created a way to simulate simultaneous processing. Even though the computer would (probably) only have a single processor such that only one copy of a block of code could be executing at a given time, Simula set up special variables for keeping track of simulated

time. Control would "jump" from one instance of a block to another such that all blocks would, for example, have their actions for the time 20:15 executed, then actions for 20:16 would be executed, and so on. A list kept track of processes in time order. Thus, Simula 1 kept all the features of Algol but made it more suitable for modeling simultaneous events (see MULTIPROCESSING).

Simula 1 was quite successful as a simulation language, but the authors soon realized that the ability to use separate invocations of a procedure to create individual "objects" had a more general application to representing data in applications other than simulations. In creating Simula 67 (the version of the language still used today), they therefore introduced the formal concept of the class as a specification that could be used to create objects of that type. They also introduced the key idea of inheritance (where one class can be derived from an earlier class), as well as a way that a derived class could redefine a procedure that it had inherited from the original (base) class (see OBJECT-ORIENTED PROGRAMMING and CLASS).

Although Simula 67 would continue to be used primarily for simulations rather than as a general-purpose programming language, its object-oriented ideas would prove to be very influential. The designers of Smalltalk and Ada would look to Simula for structural ideas, and the popular C++ language began with an effort to create a "C with classes" language along the lines of Simula. (See SMALLTALK, ADA, and C++.)

Further Reading

Holmevik, Jan-Rune. "Compiling Simula: a Study in the History of Computing and the Construction of the SIMULA Programming Languages." *STS Report* (Trondheim). Available online. URL: http://staff.um.edu.mt/jskll/simula.html. Accessed August 21, 2007.

Nygaard, K., and O.-J. Dahl. "The Development of the Simula Languages" in *The History of Programming Languages,* R. L. Wexelblat, ed. New York: Academic Press, 1981.

Pooley, R. *Introduction to Programming with Simula.* Oxfordshire, U.K.: Alfred Waller, 1987.

Sebesta, Robert W. *Concepts of Programming Languages.* 7th ed. Boston: Addison-Wesley, 2006.

Sklenar, J. "Introduction to OOP in Simula." Available online. URL: http://staff.um.edu.mt/jskll/talk.html. Accessed August 21, 2007.

simulation

A simulation is a simplified (but adequate) model that represents how a system works. The system can be an existing, real-world one, such as a stock market or a human heart, or a proposed design for a system, such as a new factory or even a space colony.

If a system is simple enough (a cannonball falling from a height, for example), it is possible to use formulas such as those provided by Newton to get an exact answer. However, many real-world systems involve many discrete entities with complex interactions that cannot be captured with a single equation. During the 1940s, scientists encountered just this problem in attempting to understand what would happen under various conditions in a nuclear reaction.

Together with physicist Enrico Fermi, two mathematicians, John von Neumann (see VON NEUMANN, JOHN) and Stanislaw Ulam, devised a new way to simulate complex systems. Instead of trying fruitlessly to come up with some huge formula to "solve" the whole system, they applied probability formulas to each of a number of particles—in effect, "rolling the dice" for each one and then observing their resulting distribution and behavior. Because of its analogy to gambling, this became known as the Monte Carlo method. It turned out to be widely useful not only for simulating nuclear reactions and particle physics but for many other activities (such as bombing raids or the spread of disease) where many separate things behave according to probabilities.

A number of other models and techniques have made important contributions to simulation. For example, the attempt to simulate the operation of neurons in the brain has led to a powerful technique for performing tasks such as pattern recognition (see NEURAL NETWORK). The application of simple rules to many individual objects can result in beautiful and dynamic patterns (see CELLULAR AUTOMATA), as well as ways to model behavior (see ARTIFICIAL LIFE). Here, instead of a system being simplified into a simulation, a simulation can be created in order to see what sort of systems might emerge.

SOFTWARE IMPLEMENTATION

Because of the number of calculations (repeated for a single object and/or applied to many objects) required for an accurate simulation, it is obviously useful for the simulation designer to have as much computer power as possible. Similarly, having many processors or a network of separate computers not only increases the available computing power, but may make it more natural to represent different objects or parts of a system by assigning each to its own processor. (This naturalness goes the other way, too: Simulation techniques can be very important in modeling or predicting the performance of computer networks including the Internet.)

However, it is also important to have programming languages and techniques that are suited for representing the simultaneous changes to objects (see also MULTIPROCESSING). Using object-oriented languages such as Simula or Smalltalk makes it easier to package and manage the data and operations for each object (see OBJECT-ORIENTED PROGRAMMING, SIMULA, and SMALLTALK).

APPLICATIONS

Simulations and simulation techniques are used for a tremendous range of applications today. Besides helping with the understanding of natural systems in physics, chemistry, biology, or engineering, simulation techniques are also applied to human behavior. For example, the behavior of consumers or traders in a stock market can be explored with a simulation based on game theory concepts. Artificial intelligence techniques (such as expert systems) can be used to give the individual "actors" in a simulation more realistic behavior.

Simulations are often used in training. A modern flight simulator, for example, not only simulates the aerodynamics of a plane and its response to the environment and to control inputs, but detailed graphics (and simulated physical motion) can make such training simulations feel very realistic, if not quite to *Star Trek* holodeck standards. Whether for flight, military exercises, or stock trading, simulations can provide a much wider range of experiences in a relatively short time than would be feasible (or safe) using the real-world activity. Simulations can also play an important part in testing software or systems or in predicating the results of business decisions or strategies.

Simulations are also frequently sold as entertainment. Many commercial strategy and role-playing games as well as vehicle simulators contain surprisingly complex simulations that make the games both absorbing and challenging (see COMPUTER GAMES and ONLINE GAMES). Such games can also have considerable educational value.

Further Reading

Gilbert, Nigel, and Klaus G. Troitzsch. *Simulation for the Social Scientist.* 2nd ed. Maidenhead, Berkshire, U.K.: Open University Press, 2005.

Laguna, Manuel, and Johan Marklund. *Business Process Modeling, Simulation, and Design.* Upper Saddle River, N.J.: Prentice Hall, 2004.

Rizzoli, Andrea Emilio. "A Collection of Modelling and Simulation Resources on the Internet." Available online. URL: http://www.idsia.ch/~andrea/simtools.html. Accessed August 21, 2007.

Ross, Sheldon M. *Simulation.* Burlington, Mass.: Elsevier Academic Press, 2006.

Shelton, Brett E., and David A. Wiley, eds. *The Design and Use of Simulation Computer Games in Education.* Rotterdam, Netherlands: Sense Publishers, 2007.

singularity, technological

The idea that an incomprehensible future is rushing down on us goes back at least as far as Alan Toffler's book *Future Shock* (1970). Toffler suggested that fundamental changes in society brought about by industrial and postindustrial developments were creating psychological stress and disorientation.

Future shock can be thought of as a steep line on a graph that represents the complexity of technological society. But what if the line were asymptotic, approaching the vertical and then disappearing? This is what science fiction writer Vernor Vinge described in the 1980s as the "technological singularity." In physics, a singularity is a place where laws break down, such as at the center of a black hole. By analogy, Vinge suggested that the development of artificial intelligence and related technologies would reach a point where intelligent machines would drive their own further development, with their design and operation far outstripping human understanding. Once intelligent machines create even more intelligent machines (and so on), more technological progress might occur in a decade or two than in the preceding thousands of years.

An obvious question is whether the singularity is in fact coming, and if so, when. Inventor and futurist RAY KURZWEIL argues that history (including the accuracy of Moore's law of doubling computational power) shows that technological progress is indeed exponential. In his book *The Singularity Is Near,* Kurzweil predicts that the threshold will be reached in the 2040s, leading to "technological change so rapid and profound it represents a rupture in the fabric of human history."

There are a number of contrary views. First, there are those who argue that there are fundamental reasons why computers will never achieve truly human-equivalent intelligence, let alone surpass it (see, for example, DREYFUS, HUBERT). Others argue that the present rate of acceleration will not necessarily continue, and that human-level AI may still be achievable, but only in centuries rather than a decades.

RESPONDING TO THE SINGULARITY

What happens if there *is* a singularity is the stuff of much speculation and science fiction. "Super AI" might lead to the development of technologies such as the ability to store or transfer the contents of a human brain, making people effectively immortal. On the other hand, superhuman intelligences might be indifferent to, or worse, hostile to, humanity. Super AI might also foster technologies such as genetic engineering or nanotechnology that have promises and dangers of their own.

There have been a number of responses to such dangers. Some critics (see, for example, JOY, BILL) urge that a limiting framework be put in place to prevent certain areas of research from getting out of hand. Others, such as Eliezer Yudkowsky of the Singularity Institute for Artificial Intelligence, want to ensure that "seed" AIs (intelligences capable of improving themselves) have safeguards and dispositions that would make them place a high regard on human interests, rather like Isaac Asimov's "three laws of robotics."

Another suggested approach is to use the growing knowledge of the detailed structure and function of the human brain to enhance or augment cognitive function. For example, a mathematician might think about a problem and seamlessly retrieve data from both personal memory and the World Wide Web, then carry out symbolic manipulations and calculations at electronic speed, all via brain implants.

A "SOFT SINGULARITY?"

While the likelihood of computer software exceeding human intelligence remains a subject for speculation and controversy, existing phenomena (and trends) in software design and computer-mediated communication (see SOCIAL NETWORKING) suggest that a new level of complexity and sophistication is rapidly emerging. As information is being increasingly coded for meaning (see SEMANTIC WEB) and programs are acting more autonomously (see ARTIFICIAL LIFE and SOFTWARE AGENT), one might say the Web is starting to understand itself, if not yet becoming conscious in the human sense. In turn, the augmentation of human cognition is already well underway. Thus many potential effects of the singularity are already significant issues.

Further Reading

Brin, David. "Singularities and Nightmares: Extremes of Optimism and Pessimism about the Human Future." Available online. URL: http://lifeboat.com/ex/singularities.and.nightmares. Accessed November 15, 2007.
Kurzweil, Raymond. *The Singularity Is Near: When Humans Transcend Biology.* New York: Viking, 2005.
KurzweilAI.net. Available online. URL: http://www.kurzweilai.net. Accessed November 15, 2007.
Lifeboat Foundation. Available online. URL: http://lifeboat.com. Accessed November 15, 2007.
Singularity Institute for Artificial Intelligence. Available online. URL: http://www.singinst.org/. Accessed November 15, 2007.
Vinge, Vernor. "The Coming Technological Singularity: How to Survive in the Post-Human Era." Available online. URL: http://www-rohan.sdsu.edu/faculty/vinge/misc/singularity.html. Accessed November 15, 2007.

Smalltalk

Working during the 1970s at the Xerox Palo Alto Research Laboratory (PARC), computer scientist Alan Kay created many ideas and devices that have found their way into today's personal computers. While designing a proposed notebook computer called the Dynabook, Kay decided to take a new approach to creating its operating system. The result would be a language (and system) called *Smalltalk.*

In developing Smalltalk, Kay built upon two important ideas. The first was that people could master the power of the computer most easily by being able to create, test, and revise programs interactively rather than having to go through the cumbersome process of traditional compilation. Seymour Papert had already created Logo, an interactive, graphics-rich language that proved especially good for teaching children surprisingly sophisticated computer science concepts. The name *Smalltalk* reflects how the first implementation of this language was also designed to be a simple, child-friendly language.

The other key idea Kay used in Smalltalk was object-oriented programming, which had first been developed in the language Simula 67 (see SIMULA and OBJECT-ORIENTED PROGRAMMING). However, instead of simply adding classes and objects to existing language features, Kay designed Smalltalk to be object-oriented from the ground up. Even the data types (such as integer and character) that are used to declare variables in traditional languages become objects in Smalltalk. Users can define new classes that are treated just like the "built-in" ones. There is no need to worry about having to declare variables to be of a certain type before they can be used; in Smalltalk variables can be associated with any object.

To get a program to perform an action, a "message" is sent to an object, which invokes one of the object's defined capabilities (methods). For a very simple example, consider the BASIC assignment statement:

```
Total = Total + 1
```

In a traditional language like BASIC, this is conceptualized as "add 1 to the value stored at the location labeled Total and store the result back in that location." In the object-oriented Smalltalk language, however, the equivalent statement would be:

```
Total <- Total + 1
```

This means "send the message + 1 to the object that is referenced by the variable called Total." This message references the + method, one of the methods that numeric objects "understand." The object therefore adds 1 to its value, and returns that value as a new object, which in turn is now referenced by the variable Total.

A "program" in Smalltalk is simply a collection of objects with the capabilities to carry out whatever processes are required. The objects and their associated variables make up the "workspace," which can be saved to disk periodically.

For the Smalltalk programmer there is no distinction between Smalltalk and the host computer's operating system. The operating system's capabilities (such as file handling) are provided within the Smalltalk system as predefined objects. Kay envisaged Smalltalk as a complete environment that could be extended by users who were not necessarily experienced programmers, and he designed its pioneering graphical user interface as a way to make it easy for users to work with the system.

Smalltalk includes a "virtual machine," whose instructions are then implemented in specific code for each major type of computer system. Because of Smalltalk's consistent structure and ability to build everything up from objects, almost all of the Smalltalk system is written in Smalltalk itself, making it easy to transplant to a new computer once the machine-specific details are provided.

Because of its elegance and consistency and its availability on personal computers, by the 1980s Smalltalk had aroused considerable interest. The language has not been widely used for mainstream applications, in part because the mechanisms needed to kept track of classes and inheritance of methods are hard to implement as efficiently as the simpler mechanisms used in traditional languages. The approach of building object-oriented features onto existing languages (as with developing C++ from C) had greater appeal to many because of efficiency and a less steep learning curve.

Nevertheless, the conceptual power of Smalltalk has made it attractive for certain AI and complex simulation projects, and it appeals to those who want a pure object-oriented approach where an application can cleanly mirror a real-world situation. Smalltalk also remains a good choice for teaching programming to children (and others). A version called *Squeak* provides a rich environment of graphics and other functions. Squeak and a number of other Smalltalk implementations are available for free download for a number of different computer systems.

Further Reading

Ducasse, Stéphane. *Squeak: Learn Programming with Robots.* Berkeley, Calif.: Apress, 2005.

Klimas, Edward J., Suzanne Skublics, and David A. Thomas. *Smalltalk with Style.* Englewood Cliffs, N.J.: Prentice Hall, 1996.

Lewis, Simon. *The Art and Science of Smalltalk.* Englewood Cliffs, N.J.: Prentice Hall, 1995.

Smalltalk. Available online. URL: http://smalltalk.org. Accessed August 21, 2007.

Squeak. Available online. URL: http://www.squeak.org/. Accessed August 21, 2007.

smart buildings and homes

A smart building, whether commercial space or a home, is one in which components ranging from HVAC (heating, ventilation, and air conditioning) to appliances, computers, communications, security, and entertainment systems are integrated into a network for easy control.

Some typical features of a smart building include the following:

- lighting that is controlled by time of day, scheduling, and occupancy sensors

- temperature and air-flow sensors to determine the amount of cooling, heating, or fresh air needed

- controls for central heating, hot water, and air conditioning systems, optimizing efficiency and minimizing energy use

- alarms for intrusion, fire, carbon monoxide/dioxide, and other hazards

- alarms indicating failure or unsafe operating conditions for various devices

- integration of alarm and status messages with communications systems, enabling users to receive them by e-mail, text message, phone, or other means

Using a secure link, the user can connect to the building via mobile phone or perhaps Internet connection and give it commands, such as to turn the heating or porch light on, close the drapes, and so on. The system can also let the remote user know who is at the door and allow for communication, or let them in.

Smart office or other buildings use many of the same technologies as smart homes, but the priorities and emphases may be different. Smart buildings are more likely to be centrally controlled and fully automated rather than allowing individuals to interact with them. (Regulatory and safety requirements are also likely to be different and more complex.)

APPLICATIONS AND QUESTIONS

The integrated controls in a smart house are potentially very useful for disabled persons or seniors who have limited mobility. Lighting could automatically be turned on as a person gets up from bed and goes to the bathroom, for example. Appliances could be controlled remotely, and even cupboards or tables could be designed to raise or lower at the touch of a button. (See DISABLED PERSONS AND COMPUTING and SENIORS AND COMPUTING.) If such systems are effective, their cost may be well worth the psychological benefits of allowing people to remain in their homes, and in comparison to the cost of assisted living or residence facilities. Smart homes could also help parents monitor toddlers or small children as well as restrict them from entering potentially hazardous parts of the house.

Critics of the smart-house concept point out that installing and integrating all the required equipment for a full

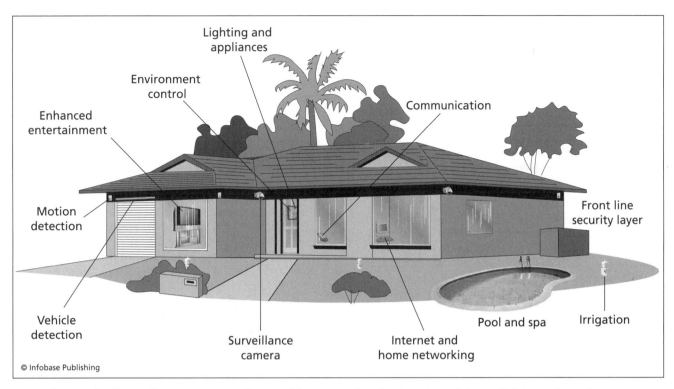

A smart house or building makes it easy to control essential functions such as heating, air conditioning, lighting, and security systems.

implementation is quite expensive. Incorporating the necessary features when building a new house would be easier, since infrastructure such as cabling can be incorporated in the building design. However, much of the technology is not fully mature. There are several standards for interconnection, including the venerable X10, Z-Wave, and Insteon—and they are incompatible with one another.

Further Reading

Briere, Danny, and Pat Hurley. *Smart Homes for Dummies.* 3rd ed. Hoboken, N.J.: Wiley, 2007.

Eisenpeter, Robert C., and Anthony Volte. *Build Your Own Smart Home.* Emeryville, Calif.: McGraw-Hill/Osborne, 2003.

"Home Automation for the Elderly and Disabled." Wikipedia. Available online. URL: http://en.wikipedia.org/wiki/Home_automation_for_the_elderly_and_disabled. Accessed November 15, 2007.

Lee, Jeanne. "Smart Homes: The Best of Today's Intelligent, Networked Home Appliances Aren't Just Cool and High-Concept. Believe It or Not, They also Make Sense." *Money,* Oct. 1, 2002, p. 120 ff.

Mitchell, Robert. "The Rise of Smart Buildings." *Computerworld,* March 14, 2005. Available online. URL: http://www.computerworld.com/networkingtopics/networking/story/0,10801,100318,00.html. Accessed November 15, 2007.

smart card

The smart card is the next generation of transaction devices. Magnetically coded credit, debit, and ATM cards have been in use for many years. These cards contain a magnetic strip encoded with a small amount of fixed data to identify the account. All the actual data (such as account balances) is kept in a central server, which is why credit cards must be validated and transactions approved through a phone (modem) link. Some magnetic strip cards such as those used in rapid transit systems are rewritable, so that, for example, the fare for the current ride can be deducted. Telephone cards work the same way. Nevertheless, these cards are essentially passive tokens containing a small amount of data. They have little flexibility.

However, since the mid-1970s it has been possible to put a microprocessor and rewritable memory into a card the size of a standard credit card. These smart cards can

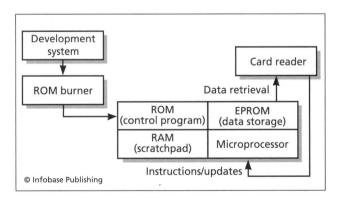

A smart card is "smart" because it does not just hold and update data, but has an embedded program and the ability to respond to a variety of requests.

store a hundred or more times the data of a magnetic strip card. Further, because they have an onboard computer (see EMBEDDED SYSTEM), they can interact with a computer at the point of service, exchanging and updating information.

Magnetic strip cards have no way to verify whether they're being used by their legitimate owner, and it is relatively easy for criminals to obtain the equipment for creating counterfeits. With a smart card, the user's PIN can be stored on the card and the terminal can require that the user type in that number to authorize a transaction. Again, the PIN can be validated without reference to a remote server.

HARDWARE AND PROGRAMMING

Besides the microprocessor and associated circuitry, the smart card contains a small amount of RAM (random access memory) to hold "scratch" data during processing, as well as up to 64 kB of ROM (read-only memory) containing the card's programming instructions. The program is created on a desktop computer and written to the ROM that is embedded in the card. Finally, the card includes up to 64 kB of EEPROM (Electrically Erasable Programmable Read Only Memory) for holding account balances and other data. This memory is nonvolatile (meaning that no power is needed to maintain it), and can be erased and rewritten by the card reader.

"Contact" cards must be swiped through the reader and are most commonly used in retail, phone, pay TV, or health care applications. "Contactless" cards need only be brought into the proximity of the reader, which communicates with it via radio signals or a low-powered laser beam. Contactless cards are more practical for applications such as collecting bridge tolls (see also RFID).

The card reader (or terminal) at the point of sale contains its own computer, which runs software that requests particular services from the card's program, including providing identifying information and balances, updating balances, and so on.

Microsoft and some other companies have introduced the PC/SC standard for programming smart cards from Windows-based systems. Another standard, Open Card, promises to be compatible with a wide range of platforms and languages, including Java. (Java, after all, descended from a project to develop a language for programming embedded systems.) However, the first commercially available Java-based smart card programming system is based on another standard called *JavaCard*.

APPLICATIONS

The same smart card might also be programmed to handle several different types of transactions, and could function as a combination phone card, ATM card, credit card, and even medical insurance card. Europe has been well ahead of the United States in adopting smart card technology, with both France and Germany beginning during the 1980s to use smart cards for their phone systems. During the 1990s, they began to develop infrastructure for universal use of smart cards for their national health care systems. In 2002, Ontario, Canada, began to replace citizenship papers with a

smart card, as well as creating a health services card. Other innovative uses for smart cards include London's city pass for tourists, which can be programmed to provide not only prepaid access but also various bonuses and promotions.

The packing of many services and the associated information onto a smart card raises greater concern that the information might be illicitly captured and abused (see PRIVACY IN THE DIGITAL AGE). Smart chips about the size of a grain of rice can be implanted beneath the skin. When scanned by hospital personnel, the patient's entire medical record can be retrieved, which can be vital for deciding which drugs to administer in an emergency when the patient is unable to communicate. However, the chips might be surreptitiously scanned by, for example, employers seeking to screen out workers with expensive medical conditions.

Smart cards (such as for digital TV access) have been counterfeited with the aid of sophisticated programs and intrusion equipment. Card makers try to design the card's circuits so that it resists intrusion and tampering and rejects programming attempts from unauthorized equipment.

Another way to prevent unauthorized use is to have the card store identifying information that can be verified through fingerprint scanners or other means (see BIOMETRICS). Smart ID and access cards are being deployed by more U.S. government agencies to control access to sensitive areas in the wake of the September 11, 2001, terrorist attacks. The newest smart cards, such as one called the Ultra Card, can hold 20 MB of information, allowing the use of much more extensive biometric data. The controversial "national ID card," if implemented, is likely to be a smart card.

A service called GSM (Global System of Mobile Communications) is gradually being adopted. Through the use of a smart card "subscriber identity module," it allows wireless phone users in any participating country to make calls and have the appropriate fees deducted. Further, the GSM can route calls to a person's number automatically to that person's handset, regardless of the country of origin and destination.

There is a very large investment in the current credit card technology, but the flexibility and potential security of smart credit and debit cards is attractive. Already some issuers have released credit cards with smart chip technology.

Further Reading

Jurgensen, Timothy M., and Scott B. Guthery. *Smart Cards: The Developer's Toolkit.* Upper Saddle River, N.J.: Prentice Hall, 2002.

Paret, Dominique. *RFID and Contactless Smart Card Applications.* New York: Wiley, 2005.

Rankl, Wolfgang, and Wolfgang Effing. *Smart Card Handbook.* 3rd ed. New York: Wiley, 2004.

Smart Card Basics. Available online. URL: http://www.smartcardbasics.com/. Accessed August 21, 2007.

smartphone

In biology, convergent evolution is when two very different types of creatures evolve similar structures or traits to cope with similar environments—for example, wings in insects, birds, and bats. Something like this has happened with hand-held mobile devices that are used to manage personal information and for communication. Personal digital assistants (see PDA) can maintain lists of phone numbers and other contact information, as well as running a variety of useful or entertaining applications. However, the first models had no provision for actually making phone calls, while later models offered the ability to make calls through a wireless connection (see BLUETOOTH) to an appropriately equipped mobile phone. This meant, though, that the user had to carry two separate devices, the PDA and the phone, somewhat defeating the objectives of portability and convenience.

Meanwhile, of course, mobile phones were a booming industry—and a very competitive one, hence the pressure to add new features. Some of these features overlap typical PDA features, such as storing contact lists, appointments, and other personal information. Further, users want to be able to not only send text messages, but read and send e-mail, and browse the Web. But providing all these applications really calls for a full-fledged (albeit compact) operating system and facilities for creating user interfaces. The result is the smartphone, which in effect is a phone that is grown into a PDA while maintaining its phone capabilities. Thus the smartphone aims to be the way to meet all communications, information management, Web, and entertainment needs in a single device. Typical features also include a camera and an audio-video media player (see MUSIC AND VIDEO PLAYERS, DIGITAL) and a small but increasingly sharp screen.

Some of the major smartphone manufacturers and their operating systems include the following:

- Symbian (Symbian OS), used by Nokia, Motorola, Samsung, and others

- Windows Mobile (enhanced Windows CE), popular in phones used in Asia

- Blackberry (RIM), the popular PDA/smartphone

- Linux, used as the base on which to build a variety of PDA/phone operating systems, including products from Motorola, Palm, and Nokia (Maemo)

- OS X (Apple), used in Apple's innovative and very popular iPhone

CONVERGENCE

As a practical matter, the PDA and smartphone categories now overlap so much that a device such as the Apple iPhone can be called either, and then some. Although Apple initially locked out full access to the iPhone operating system for developing third-party applications (and locked the phone itself to only a few providers), the overall trend in the industry is to provide more flexibility and accessibility.

Although not yet accompanied by an actual device, the 2007 announcement by Google of an open-source phone software platform called Android has been greeted by considerable interest. The product will be developed further by the Open Handset Alliance, a consortium of Google and more than 30 other companies, including T-Mobile and

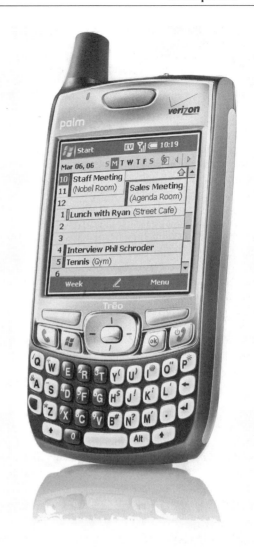

Smartphone or PDA? Sometimes it is just a matter of semantics what to call a handheld device such as this Palm Treo 700w that can make phone calls and send e-mail or text messages as well as manage information. (PALM, INC.)

Motorola. The use of open-source software should reduce costs to developers and consumers, in part by making it possible for a developer to create an application that can run on dozens of different smartphone models.

Further Reading

Ames, Patrick, and David Maloney. *Now You Know: Treo 700w Smartphone.* Berkeley, Calif.: Peachpit Press, 2007.

Best, Jo. "Analysis: What Is a Smart Phone?" Silicon.com, February 13, 2006. Available online. URL: http://networks.silicon.com/mobile/0,39024665,39156391,00.htm. Accessed November 16, 2007.

Jipping, Michael J. *Smartphone Operating System Concepts with Symbian OS.* Hoboken, N.J.: Wiley, 2007.

McPherson, Frank. *How to Do Everything with Windows Mobile.* New York: McGraw-Hill, 2006.

Open Handset Alliance. Available online. URL: http://www.open handsetalliance.com/. Accessed November 16, 2007.

Pogue, David. *iPhone: The Missing Manual*. Sebastapol, Calif.: O'Reilly, 2007.

Smartphone & Pocket PC [magazine Web site]. Available online. URL: http://www.pocketpcmag.com/defaults.asp. Accessed November 16, 2007.

SOAP

Originally standing for Simple Object Access Protocol, but now no longer an acronym, SOAP is a standard way to access Web services (see SERVICE-ORIENTED ARCHITECTURE and WEB SERVICES). In today's Web, where what appears to users to be a single site or application is usually built from many services, such a facility is essential.

Prior to SOAP, Web applications usually communicated through remote procedure calls (RPC). However there were problems with compatibility of applications running under different operating systems (and perhaps using different programming languages), as well as security problems that often led to such facilities being blocked.

SOAP, on the other hand, uses the same HTTP recognized by all Web servers and browsers (see WEB BROWSER, WEB SERVER, and WORLD WIDE WEB)—indeed, it can also use secure HTTP (https).

A SOAP request (or message) is an ordinary XML file (see XML) that includes an "envelope" element specifying it to be a SOAP message, an optional header, a body element containing the information pertaining to the function or transaction requested, and an optional fault element to specify error processing. After receiving the message, the destination server returns a message providing the requested information.

A very simple SOAP message might look like this:

```
<SOAP:Envelope
xmlns:SOAP="http://schemas.xmlsoap.org/soap/
envelope/">
  <SOAP:Body>
  <m:getPrice
xmlns:m="http://www.soapware.org/">
<itemnum>311</itemnum>
  </m:getPrice>
  </SOAP:Body>
</SOAP:Envelope>
```

The message asks for the price for item number 311.

Despite its advantages in terms of security, versatility, and readability, SOAP does have some disadvantages. The main one is that XML files can be quite lengthy, making transactions slower than with the much more compact CORBA (see CORBA).

Further Reading

Burd, Barry. *Java & XML for Dummies*. Indianapolis: Wiley, 2002.

Freeman, Adam, and Allen Jones. *Microsoft .NET XML Web Services Step by Step*. Redmond, Wash.: Microsoft Press, 2003.

"SOAP Tutorial." W3schools.com. Available online. URL: http://www.w3schools.com/soap. Accessed November 17, 2007.

"What Is SOAP" [Flash presentation]. Available online. URL: http://searchwebservices.techtarget.com/searchWebServices/downloads/what_is_soap.swf. Accessed November 17, 2007.

Zimmermann, Olaf, Mark Tomlinson, and Stefan Peuser. *Perspectives on Web Services: Applying SOAP, WSDL, and UDDI to Real-World Projects*. New York: Springer, 2003.

social impact of computing

In 2001, the Computer Professionals for Social Responsibility (CPSR) held a conference titled "Nurturing the Cybercommons, 1981–2021." Speakers looked back at the amazing explosion in computing and computer-mediated communications in the last two decades of the 20th century. They then turned to the next 20 years, discussing how computing technology offered both the potential for a more robust democracy and the threat that control of information by the few could disenfranchise the many. Their challenge was to create a "cybercommons"—a way in which the benefits of technology could be shared more equitably.

It is sobering to realize just how much happened in only two decades. The computer went from being an esoteric possession of large institutions to a ubiquitous companion of daily work and home life. At the same time, the Internet, which in 1981 had been a tool for a small number of campus computing departments and government-funded researchers, has burgeoned to a medium that is fast changing the way people buy, learn, and socialize.

The use of computing for specific applications generally brings risks along with benefits (see RISKS OF COMPUTING). Sometimes risks can go beyond a specific program into the interaction between that program and other systems. In the broadest sense, however, computer use as a human activity affects all other human activities. The ultimate infrastructure is not the computer, the software program, or even the entire Internet. Rather, it is society as a whole. There are a several dimensions along which both positive and negative possibilities can be seen.

One of the earliest hints that computers might have a broader impact on society came in 1952, when Univac's prediction of an Eisenhower election victory was relayed by anchor Walter Kronkite. (AL FENN / TIME LIFE PICTURES/GETTY IMAGES)

STRATIFICATION V. OPPORTUNITY

In the past 30 years, the computer has created millions of new jobs, ranging from webmaster to support technician to Internet café proprietor (see EMPLOYMENT IN THE COMPUTER FIELD). Millions of other jobs have been redefined: The typist has become the word processor, for example. Many other jobs have disappeared or are in the process of disappearing—such as travel agents, who have found themselves under pressure both from do-it-yourself Internet booking and the airlines deciding that they no longer needed to give agents incentives for booking.

In a rapidly changing technological and economic landscape, there are always emerging opportunities. The primacy of computer skills in the job market has, however, exacerbated a trend that was seen throughout the 20th century. New, well-paid jobs increasingly require technical training and skills—expanding the definition of "functional literacy." Throughout the second half of the century, the traditional blue-collar factory jobs that could assure a comfortable living for persons with only a high school education have become increasingly scarce. This has been the result both of increasingly competitive (and lower-priced) overseas labor and factory automation (see ROBOTICS) at home. Essentially, the well-paid tech sector and the low-paid service sector have grown rapidly, while the ground in between has eroded.

Sometimes jobs don't disappear, but are "dumbed down," becoming low-skill and low-paid. Fifty years ago, a store clerk had to be able to count up from the cash register total to the amount of money presented by the customer. Today, computerized cash registers tell the clerk exactly how much change to give (and often dispense the coins automatically). Old-style clerks had to know about prices, discounts, and special offers. Today these are handled automatically by bar codes and smart cards. Although the supermarket clerk still is moderately well paid, the ultimate end of the process is seen in the fast food clerk, who often needs only push buttons with pictures of food on them. He or she is likely to be paid little more than minimum wage. The impact of technology on jobs can even go through several stages. For example, skilled photo technicians have been replaced by the use of automated photo processing equipment. In turn, however, the growing use of digital cameras is reducing the use of film-based photography in general.

The result of these trends may well be increased social stratification. The best jobs in the information age require skills such as programming, systems analysis, or the ability to create multimedia content. However, the opportunity to acquire such skills varies and is not evenly distributed through all groups in the population (see DIGITAL DIVIDE). Although minority groups are now catching up in terms of access to computers at home and in school, disparities in the quality of education will only be magnified as technical skills increasingly correlate with good pay and benefits.

At the same time the computer offers powerful new tools for education (see EDUCATION AND COMPUTERS). Potentially, this could overcome much of the disadvantages of poverty because once the threshold of access is met, the poor person's Internet is much the same as that available to the privileged. However, mastering the necessary skills requires both provision of adequate resources and that prevailing cultural attitudes support intellectual achievement.

DEPENDENCY V. EMPOWERMENT

Computers have made people more dependent in some ways while empowering them in others. Society is increasingly dependent on computers to operate the systems that provide transportation, power, and communications infrastructure. The "y2k" scare at the end of the century proved to be unfounded, but it did give people a chance to consider what a major, prolonged failure in the information infrastructure would mean for maintaining the physical necessities of life, the viability of the economy, and the cohesion of society itself (see Y2K PROBLEM). The terrorist attacks of September 11, 2001, brought to greater public awareness the concerns about "cyberterrorism" that experts had been debating since the late 1990s. (See CYBERTERRORISM.)

At the same time, computers—and particularly the Internet—have give individuals a greater feeling of empowerment in many respects. The savvy Web user now has numerous ways to shop for everything from airline tickets to Viagra pills at prices that reflect disintermediation—the elimination of the middleman. Many people are less inclined to take the word of traditional authority figures (such as doctors) and instead are tapping into the sort of information that had been previously been accessible only to professionals. However, access to information is not the same thing as having the necessary background and skills to evaluate that information. Whether falling victim to an outright scam or simply not fully understanding the consequences of a decision, the Web user finds little in the way of a regulatory safety net. The tension between the high degree of regulation now existing in much of our society and the frontierlike qualities of cyberspace will no doubt be a major theme in the next few decades.

CENTRALIZATION V. DEMOCRACY

With new forms of media technology (such as radio and television in the 20th century), early innovators and experimenters have considerable freedom to experiment and express themselves. This freedom is largely the result of lack of pressure from powerful economic interests while the new technology is largely still "under the radar." However, as a technology matures, large corporate interests tend to consolidate the market, leaving fewer opportunities for smaller, independent operators.

By the late 1990s there was some concern that the Internet and World Wide Web were entering such a consolidation stage in the wake of such developments as the AOL/Time-Warner merger. However, while there are now large corporate presences online, the diversity of the means of expression has actually increased (see BLOGS AND BLOGGING, USER-CREATED CONTENT, WIKIS AND WIKIPEDIA, and YOUTUBE). Further, the influence of activist groups has increased to the point where any serious political campaign gives high priority to its Internet presence and the cultivation of influential bloggers (see POLITICAL ACTIVISM AND THE INTERNET).

There continue to be centralizing or antidemocratic pressures in the online world (for example, see CENSORSHIP AND THE INTERNET). There is also the conflict between the desire to protect intellectual property and the free sharing of images and other media (see DISTRIBUTION OF MUSIC AND VIDEO, ONLINE and INTELLECTUAL PROPERTY AND COMPUTING). Loss of privacy can also inhibit untrammeled political discourse (see PRIVACY IN THE DIGITAL AGE). At the same time, organizations such as the Electronic Frontier Foundation, Electronic Privacy Information Center, and Center for Democracy and Technology work to protect and advocate democratic expression.

ISOLATION V. COMMUNITY

There are many online facilities that allow individuals and groups to maintain an ongoing dialog (see CHAT, ONLINE and CONFERENCING SYSTEMS). Students at a school in Iowa can now collaborate with their counterparts in Kenya or Thailand on projects such as measuring global environmental conditions. Senior citizens who have become isolated from family members and lack access to transportation can find social outlets online.

However, critics such as Clifford Stoll believe that the growth of online communication (see also VIRTUAL COMMUNITY) may be leading to a further erosion of physical communities and a sense of neighborhood. For many years, it has been observed that people in suburbia often don't know their neighbors: The car and the phone let them form relationships and "communities" without much regard to geography. It is possible that the growth in online communities will accelerate this effect. Further, with people being able to order an increasing array of goods and services online, might the market plaza and its modern counterpart the mega mall become less of a meeting place? Even the proposal to allow people to vote online might promote democracy at the expense of the contact between citizens and the shared rituals that give people a stake in the larger community.

Thus, computer technology offers many opposing prospects and visions. The social changes that are cascading from information and communications technology are likely to be at least as pervasive in the early 21st century as the those wrought by the telephone, automobile, and television were in the 20th.

Further Reading

Association for Computing Machinery. Special Interest Group on Computers and Society. Available online. URL: http://www.sigcas.org/. Accessed August 21, 2007.

Center for Democracy & Technology. Available online. URL: http://www.cdt.org/. Accessed August 21, 2007.

Computer Professionals for Social Responsibility. Available online. URL: http://www.cpsr.org. Accessed August 21, 2007.

De Paula, Paul. *Annual Editions: Computers in Society 06/07*. Guilford, Conn.: McGraw-Hill/Duskin, 2006.

Electronic Frontier Foundation. Available online. URL: http://www.eff.org. Accessed August 21, 2007.

Kizza, Joseph Migga. *Ethical and Social Issues in the Information Age*. 2nd ed. New York: Springer, 2002.

Morley, Deborah, and Charles S. Parker. *Computers and Technology in a Changing Society*. 2nd ed. Boston: Course Technology, 2004.

Pew Internet & American Life Project. Available online. URL: http://www.pewinternet.org/. Accessed August 21, 2007.

social networking

Today, millions of people—middle, high school, and college students, but increasingly adults as well—have pages on popular Web sites such as MySpace and Facebook. These sites are significant examples of social networking: the use of Web sites and communications and collaboration technology to help people find, form, and maintain social relationships.

The origins of social networking can be traced to online venues that arose in the 1970s and 1980s, notably Usenet and, later, online chat boards (see BULLETIN BOARD SYSTEM, CONFERENCING SYSTEM, NETNEWS, and VIRTUAL COMMUNITY). In the late 1990s social networking Web sites began to appear, including Classmates.com (helping people find and communicate with former schoolmates) and SixDegrees.com, which emphasized "knows someone who knows someone who . . ." kinds of links.

By the mid-2000s the two biggest sites were Facebook and MySpace. Founded in 2006 by Mark Zuckerberg, Facebook was originally restricted to Harvard students, but eventually became open to any college student, and then high schools and even places of employment. (The name comes from a book given to incoming students in some schools to familiarize them with their peers.) As of late 2007 Facebook had more than 55 million active members and had become the seventh most visited of all Web sites.

Facebook users have profile pages that include a "wall" on which their designated circle of friends can post brief messages. (Longer or private messages similar to e-mail can also be sent.) Users can also send each other "gifts" represented by colorful icons. Finally, users in a given Facebook community can keep track of each other's status (where they are and what they are doing).

Beverly Hills, California-based, MySpace is an even larger site, near the top of the Web site popularity statistics through much of 2007. Founded in 2003, the site was created and marketed by a company called eUniverse (later Intermix), and its launch was greatly boosted by being able to tap many of eUniverse's 20 million existing subscribers. User profiles are broadly similar to those in Facebook, but are less structured and more colorful, with uploaded graphics and a blog for each user. Profiles can be elaborately customized using a variety of tools and utilities. The site has also expanded into other areas such as instant messaging (MySpaceIM), video sharing (MySpactTV), and mobile phones (MySpace Mobile).

Social network applications are also expanding behind the linking of classmates or colleagues. Companies can use social networking software to set up user groups and provide support and incentives. Medical professionals are often forming social networks to share knowledge and news—and not surprisingly, drug company representatives have moved in to make their pitch as well. Business executives and professionals can meet on LinkedIn, a site that links people only if they have an existing relationship or an "invitation" from an existing member. As further proof that the technology is maturing, about 20 percent of adult Internet users have reported visiting a social networking site in the past 30 days.

COMMERCIALIZATION

Indeed, because they are now bringing so many people together, social networking sites have become a very attractive platform for online products and businesses. Facebook, for example, is explicitly allowing selected businesses to use the site, in exchange for a portion of the revenue generated. (Even without formal relationships, many sites allow users to add code enabling third-party services.) Some utilities (often sponsored by advertising) help users make their profiles more attractive, while one called MySpacelog serves users who are anxious to see who is viewing their sites. Looming on the horizon by 2007 was Google, which is releasing OpenSocial, a set of programming interfaces that is expected to enable developers to create applications that will run on a wide variety of social networking sites.

While social networking sites generally want to encourage products that can add revenue (and value to users), some add-on applications can be problematic. In 2006 a site called Stalkerati let users automatically search for a person's profiles on popular social networking sites and consolidate them into a summary. However, the perhaps unfortunately named sited was soon blocked by MySpace and other sites, which cited privacy and security concerns. These concerns have become increasingly important as networks such as MySpace have proven attractive to spammers, identity thieves, and sexual predators. (A 2007 survey by the Pew Internet & American Life Project found that 23 percent of teens on social networks had felt "scared or uncomfortable" because of an online encounter with a stranger. However, that same report showed that many parents and teens themselves have become aware of potential risks and the need to more carefully manage where and how information is disclosed.)

Social networking is also attracting the attention of social scientists and academics: For example, the University of Michigan now has a graduate program in social computing. Meanwhile sociologist Michael Macy of Cornell University is directing a multiyear research project, funded by the National Science Foundation and Microsoft, titled "Getting Connected: Social Science in the Age of Networks."

Note: the term *social network* is also used to refer to a method of mathematical and sociological analysis of social links within organizations. Such methods can of course be applied to the online social networking sites.

Further Reading

Baloun, Karel M. *Inside FaceBook: Life, Works, and Visions of Greatness.* Victoria, B.C., Canada: Trafford, 2007.
Facebook. Available online. URL: http://www.facebook.com/. Accessed November 18, 2007.
Hupfer, Ryan, Mitch Maxson, and Ryan Williams. *MySpace for Dummies.* Hoboken, N.J.: Wiley, 2007.
Lavallee, Andrew. "At Some Schools, Facebook Evolves from Time Waster to Academic Study." *Wall Street Journal Online,* May 29, 2007. Available online. URL: http://online.wsj.com/article/SB117917799574302391.html. Accessed November 18, 2007.
Lenhart, Amanda, and Mary Madden. "Social Networking Websites and Teens: An Overview." Pew Internet & American Life Project, January 7, 2007. Available online. URL: ttp://www.pewinternet.org/pdfs/PIP_SNS_Data_Memo_Jan_2007.pdf. Accessed November 18, 2007.
———. "Teens, Privacy & Online Social Networks: How Teens Manage Their Online Identities and Personal Information in the Age of MySpace." Pew Internet & American Life Project, April 18, 2007. Available online. URL: http://www.pewinternet.org/pdfs/PIP_Teens_Privacy_SNS_Report_Final.pdf. Accessed November 18, 2007.
MySpace. Available online. URL: http://www.myspace.com/. Accessed November 18, 2007.
Shepherd, Lauren. "Social Networking Breeds Creation of Third-Party Sites." Associated Press/*San Francisco Chronicle,* June 18, 2007, p. C5.
Weber, Larry. *Marketing to the Social Web: How Digital Customer Communities Build Your Business.* Hoboken, N.J.: Wiley, 2007.

social sciences and computing

Broadly speaking, social scientists study the structure and dynamics of human societies as well as groups of all kinds. Depending on subject matter, the research can fall within one or more disciplines, for example, anthropology, psychology, economics, geography, history, political science, or sociology. As with other scientific fields, computers have greatly enhanced and expanded the ability to carry out, analyze, and communicate research findings.

APPLICATIONS

Social scientists can use a variety of software throughout the research process. For example, researchers might use the following:

- Web and bibliographical search tools to find existing research on their topic

- note-taking and concept-diagramming ("mind-mapping") software

- software to conduct polls or surveys and compile the results

- social networking analysis to better understand a group's structure and dynamics

- statistical analysis tools to analyze the findings (see STATISTICS AND COMPUTING)

- map-based systems for studying geographical aspects (see GEOGRAPHICAL INFORMATION SYSTEMS)

- modeling software to simulate the mechanism being studied, using mathematical techniques such as the Monte Carlo and Markov-Chain methods

Games and virtual worlds in particular are being used in innovative ways. Games such as the classic *SimCity* or the "social simulator" *The Sims* can be used to help students understand and experiment with economic and social dynamics. However, virtual worlds can also be studied in their own right—for example Tufts University researchers Nina Fefferman and Eric Lofgren have written a paper describing how the spread of a "virtual plague" in *Second*

Life could be studied to learn how people would be most likely to react to a real disease outbreak. And at Carnegie Mellon University, a National Science Foundation–funded project will be studying interactions in online venues as disparate as *World of Warcraft* and Wikipedia.

Further Reading

Dochartaigh, Niall O. *The Internet Research Handbook: A Practical Guide for Students and Researchers in the Social Sciences.* Thousand Oaks, Calif.: Sage Publications, 2002.

Gilbert, Nigel, and Klaus G. Troitzsch. *Simulation for the Social Scientist.* Philadelphia: Open University Press, 1999.

"The Impoverished Social Scientist's Guide to Free Statistical Software and Resources." Available online. URL: http://www.hmdc.harvard.edu/micah_altman/socsci.shtml. Accessed October 5, 2007.

Patterson, David A. "Using Spreadsheets for Data Collection, Statistical Analysis, and Graphical Representation." Available online. URL: http://web.utk.edu/~dap/Random/Order/Start.htm. Accessed October 5, 2007.

Saam, Nicole J., and Bernd Schmidt. *Cooperative Agents: Applications in the Social Science.* Norwell, Mass.: Kluwer Academic Publishers, 2001.

Summary of Survey Analysis Software (Harvard). Available online. URL: http://www.hcp.med.harvard.edu/statistics/survey-soft/. Accessed October 5, 2007.

software agent

Most software is operated by users giving it commands to perform specific, short-duration tasks. For example, a user might have a word processor change a word's typestyle to bold, or reformat a page with narrower margins. On the other hand, a person might give a human assistant higher-level instructions for an ongoing activity: for example, "Start a clippings file on the new global trade treaty and how it affects our industry."

In recent years, however, computer scientists and developers have created software that can follow instructions more like those given to the human assistant than those given to the word processor. These programs are variously called software agents, intelligent agents, or bots (short for "robots"). Some consumers have already used software agents to comb the Web for them, looking, for example, for the best online price for a certain model of digital camera. Agent programs can also assist with online auctions, travel planning and booking, and filtering e-mail to remove unwanted "spam" or to direct inquiries to appropriate sales or technical support personnel. (See also MAES, PATTIE.)

Practical agents or bots can be quite effective, but they are relatively inflexible and able to cope only with narrowly defined tasks. A travel planning agent may be able to interface with online reservations systems and book airline tickets, for example. However, the agent is unlikely to be able to recognize that a recent upsurge in civil strife suggests that travel to that particular country is not advisable.

Researchers have, however, been working on a variety of more open-ended agents that, while not demonstrably "intelligent," do appear to behave intelligently. The first program that was able to create a humanlike conversation was ELIZA. Written in the mid-1960s by Joseph Weizenbaum, ELIZA simulated a conversation with a "nondirective" psychotherapist. More recently, Internet "chatterbots" such as one called Julia have been able to carry on apparently intelligent conversations in IRC (Internet Relay Chat) rooms, complete with flirting. Other "social bots" have served as players in online games (see CHATTERBOTS).

Chatterbots are effective because they can mirror human social conventions and because much of casual human conversation contains stereotyped phrases or clichés that can be easily imitated. Ideally, however, one would want bots to be able to combine the ability to carry out practical tasks with a more general intelligence and a more "sociable" interface. This requires that the bot have an extensive knowledge base (see KNOWLEDGE REPRESENTATION) and a greater ability to understand human language (see LINGUISTICS AND COMPUTING). Small strides have been made in providing online help systems that can deal with natural language questions, as well as being able to interactively help users step through a particular tasks.

Agents or bots have also suggested a new paradigm for organizing programs. Currently, the most widely accepted paradigm treats a program as a collection of objects with defined capabilities that respond to "messages" asking for services (see OBJECT-ORIENTED PROGRAMMING). A move to "agent-oriented programming" would carry this evolution a step further. Such a program would not simply have objects that wait passively for requests. Rather, it would have multiple agents that are given ongoing tasks, priorities, or goals. One approach is to allow the agents to negotiate with one another or to put tasks "up for bid," letting agents that have the appropriate ability contract to perform the task. With each task having a certain amount of "money" (ultimately representing resources) available, the negotiation model would ideally result in the most efficient utilization of resources.

If Marvin Minsky's (see MINSKY, MARVIN) "society of mind" theory is correct and the human brain actually contains many cooperating "agents," then it is possible that systems of competing and/or cooperating agents might eventually allow for the emergence of a true artificial intelligence.

In the future, agents are likely to become more capable of understanding and carrying out high-level requests while enjoying a great deal of autonomy. Some possible application areas include data mining, marketing and survey research, intelligent Web searching, security, and intelligence gathering. However, autonomy may cause problems if agents get out of control or exhibit viruslike behavior.

Further Reading

Denison, D. C. "Guess Who's Smarter." *Boston Globe,* May 26, 2003, p. D1. Available online. URL: http://web.media.mit.edu/~lieber/Press/Globe-Common-Sense.html. Accessed August 21, 2007.

D'Inverno, Mark, and Michael Luck. *Understanding Agent Systems.* 2nd ed. New York: Springer, 2004.

Lieberman, H., et al. "Commonsense on the Go: Giving Mobile Applications an Understanding of Everyday Life." Available online. URL: http://agents.media.mit.edu/projects/mobile/BT-Commonsense_on_the_Go.pdf. Accessed August 21, 2007.

Padgham, Lin, and Michael Winikoff. *Developing Intelligent Agent Systems: A Practical Guide.* New York: Wiley, 2004.
Software Agents Group (MIT Media Lab). Available online. URL: http://agents.media.mit.edu/. Accessed August 21, 2007.

software engineering

By the late 1960s, large computer programs (such as the operating systems for mainframe computers) consisted of thousands of lines of computer code. In what became known as "the software crisis," managers of software development were facing great uncertainty about both program development schedules and the reliability of the resulting programs.

Programming had started out in the 1940s as an offshoot of mathematics, just as the building of computers was an offshoot of electrical or electronic engineering. Increasingly, however, programmers were searching for a new professional identity. What paradigm was truly appropriate? Should programmers strive to be more like mathematicians, seeking to rigorously prove the correctness of their programs? On the other hand, many programmers thought of their work as a craft, performed using individual experience and intuition, and not easily subject to standardization. Between the two poles of mathematics (or science) and craft came another possibility: engineering.

The concept of software engineering proved to be attractive. Mathematics (and science in general) are usually carried on without being immediately and directly applied to creating a particular device or process. Outside of research programs, however, computer applications were written to perform real-world tasks (such as flight control) that have real-world consequences. Thus, although the notation of a computer program resembles that of mathematics, the operation of a program more nearly resembles that of complex mechanical systems created by engineers. By attaching the label of engineering to what programmers do, advocates of software engineering hoped to develop a body of practices and standards comparable in some way to those used in engineering. Some critics, however, believe that this paradigm is inappropriate, either because they believe one should strive for the greater rigor of science or out of a preference for individual craft over standardization.

PROGRAMMING PRACTICES

One of the most pervasive contributions to software engineering has been in computer language design and coding practices. At about the same time that the concept of software engineering was being promulgated, computer scientists were advocating better facilities for defining and structuring programs (see STRUCTURED PROGRAMMING). These included well-defined control structures (see BRANCHING STATEMENTS and LOOP), use of built-in and user-defined kinds of data (see DATA TYPES), and the breaking of programs into more manageable modules (see PROCEDURES AND FUNCTIONS).

The next paradigm came in the late 1970s and had taken hold by the late 1980s (see OBJECT-ORIENTED PROGRAMMING). The ability to "hide" details of function within objects that mirrored those in the real world provided a further way to make complex programs easier to understand and maintain. The growing use of well-tested collections of procedures or objects (see LIBRARY, PROGRAM) has been essential for keeping up with the growing complexity of application programs.

Software engineers are also concerned with developing tools that will better manage the programming process and help ensure that standards are being followed (see PROGRAMMING ENVIRONMENT). The use of CASE (Computer-Aided Software Engineering) tools such as sophisticated program editors, documentation generators, class diagrammers, and version control systems has also steadily increased. Today many of these tools are available even on modest desktop computing environments (see CASE).

THE PROGRAM DEVELOPMENT PROCESS

Perhaps the most important task for software engineering has been seeking to define and improve the process by which programs are developed. In general, the overall steps in developing a program are:

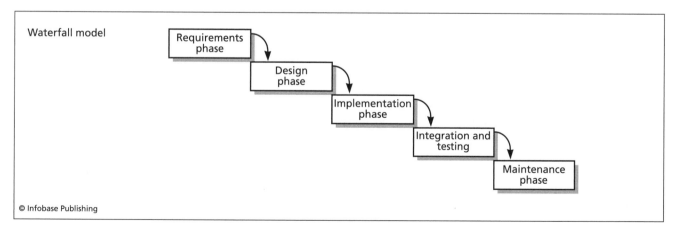

Waterfall model

Requirements phase

Design phase

Implementation phase

Integration and testing

Maintenance phase

© Infobase Publishing

The Waterfall, or Cascade, model sees software development as a more linear process going through the requirements, design, implementation, integration and testing, and maintenance phases. The results of each phase cascade down into the next.

- Detailed specification of what the program will be required to do. This can include developing a prototype and getting user's reaction to it.

- Creation of a suitable program architecture—algorithm(s) and the data types, objects, or other structures needed to implement them (see ALGORITHM).

- Coding—writing the program language statements that implement the structure.

- Verification and testing of the program using realistic data and field testing (see QUALITY ASSURANCE, SOFTWARE).

- Maintenance, or the correction of errors and adding of requested minor features (short of creating a new version of the program).

There are a number of competing ways in which to view this software development cycle. The "iterative" or "evolutionary" approach sees software development as a linear process of progress through the above steps.

The "spiral" approach, on the other hand, sees the steps of planning, risk analysis, development, and evaluation being applied repeatedly, until the risk analysis and evaluation phases result in a go/no go to finish the project.

The most commonly used approach is called *waterfall*. In it the results (output) of each stage become the input of the next stage. This approach is easiest for scheduling (see PROJECT MANAGEMENT SOFTWARE), since each stage is strictly dependent on its predecessor. However, some advocates of this approach have included the ability for a given stage to feed back to the preceding stage if necessary. For example, a problem found in implementation (coding) may require revisiting the preceding design phase.

DEVELOPING SOFTWARE ENGINEERING STANDARDS
Two organizations have become prominent in the effort to promote software engineering. The federally funded Software Engineering Institute (SEI) at Carnegie Mellon University was established in 1984. Its mission statement is to:

1. Accelerate the introduction and widespread use of high-payoff software engineering practices and technology by identifying, evaluating, and maturing promising or underused technology and practices.
2. Maintain a long-term competency in software engineering and technology transition.
3. Enable industry and government organizations to make measured improvements in their software engineering practices by working with them directly.
4. Foster the adoption and sustained use of standards of excellence for software engineering practice.

Since 1993, the IEEE Computer Society and ACM Steering Committee for the Establishment of Software Engineering as a Profession has been pursuing a set of goals that are largely complementary to those of the SEI:

1. Adopt Standard Definitions
2. Define Required Body of Knowledge and Recommended Practices (In electrical engineering, for example, electromagnetic theory is part of the body of knowledge while the National Electrical Safety Code is a recommended practice.)
3. Define Ethical Standards
4. Define Educational Curricula for (a) undergraduate, (b) graduate (MS), and (c) continuing education (for retraining and migration).

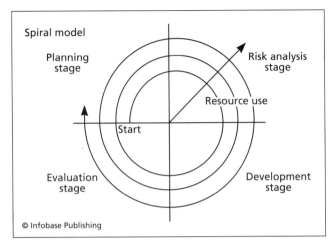

The Spiral Model visualizes software development as a process of planning, risk analysis, development, and evaluation. The cycle repeats until the project is developed to its full scope.

Further Reading
Booch, Grady. "The Promise, the Limits, the Beauty of Software." March 8, 2007. Lecture before the British Computer Society. Available online. URL: http://www.bcs.org/server.php?show=ConWebDoc.10367. Accessed August 21, 2007.

Brooks, Frederick. *The Mythical Man-Month: Essays on Software Engineering.* 20th anniversary ed. Reading, Mass.: Addison-Wesley, 1995.

Christensen, Mark J., and Richard H. Thayer. *The Project Manager's Guide to Software Engineering's Best Practices.* Los Alamitos, Calif.: IEEE Computer Society Press, 2001.

McConnell, Steve. *After the Gold Rush: Creating a True Profession of Software Engineering.* Redmond, Wash.: Microsoft Press, 1999.

Software Engineering Coordinating Committee (IEEE Computer Society and Association for Computing Machinery.) Available online. URL: http://www.acm.org/serving/se/homepage.html. Accessed August 21, 2007.

Sommerville, Ian. *Software Engineering.* 8th ed. Boston: Addison-Wesley, 2006.

software piracy and counterfeiting
According to surveys by analysis firm IDC, software piracy accounted for $7.3 billion in losses to the U.S. software industry in 2006, while reducing its expansion and thus job creation. (This is part of a larger picture in which, according to a Gallup study, 22 percent of adults in the United States reported having bought some sort of counterfeit product.) A bit of Web searching (or even reading spam in one's in-box) suggests that thousands of sites offer "cracked" software

that has been stripped of copy protection. The Business Software Alliance estimates that 35 percent of new software installed on PCs in 2006 was obtained illegally.

Although piracy can involve many forms of distribution including Web sites, file-sharing services (see FILE-SHARING AND P2P NETWORKS), and even software found on "bargain" PCs, the most visible form involves physical packages complete with box, CDs, and even holograms. These counterfeits, which range from crude to nearly indistinguishable, are often produced in full-scale factories. China has been a major source for many types of product counterfeiting, although the government has periodically cracked down on the practice. Counterfeiting has also flourished in such unlikely locales as Bangladesh and Serbia.

Industry groups also assert that the misuse of legitimately purchased software (such as running more copies than have been licensed) is also a form of piracy. The potential legal liability is enormous, so companies make rigorous policies involving software use and install monitoring systems to detect or prevent licensing violations. (For their part, industry groups have offered large cash rewards to employees who reveal their company's violations.)

COUNTERMEASURES

As perhaps the largest potential victim, Microsoft has been diligent in fighting software piracy. Recent versions of Windows, Office, and other products require that users "validate" the software, associating the license number with details of the system's hardware configuration. When the user wants to download later updates or patches, the software validation is checked. Failure of validation leads to warning messages and disabling of many features of the software.

Microsoft has also been active in suing alleged pirates, and educating consumers about the dangers of buying pirated software, which include the risk of exposure to viruses, spyware, and other harmful programs. An industry antipiracy group, the Business Software Alliance, has vigorously investigated corporate software use (often with the aid of tipsters), finding violations and making companies pay fines and buy licenses in lieu of legal action.

Meanwhile, growing pressure from the software industry has led in turn to U.S. pressure on China and other countries to go after software counterfeiting operations. In summer 2007, a joint operation by the FBI and Chinese officials led to the seizure of more than $500 million in counterfeit software.

Critics of antipiracy efforts, such as the Electronic Frontier Foundation, argue that estimates of losses from piracy assume that every pirated copy of a program represents a lost sale, ignoring the possibility that people (such as students) would not have the money to buy legitimate copies. They also point to what they consider to be heavy-handed enforcement of copyright laws and point to proposed legislation such as the Inducing Infringement of Copyrights Act, which they argue would in effect outlaw all file-sharing networks and subject people to prison sentences for minor infractions.

Further Reading

Business Software Alliance. Available online. URL: http://www.bsa.org. Accessed November 18, 2007.

Donoghue, Andrew. "Counting the Cost of Counterfeiting." CNet News, May 22, 2006. Available online. URL: http://www.news.com/Counting-the-cost-of-counterfeiting/2100-7348_3-6074831.html?tag=item. Accessed November 18, 2007.

Evers, Joris. "Fighting Microsoft's Piracy Check.:" CNet News, June 20, 2006. Available online. URL: http://www.bsa.org. Accessed November 18, 2007.

Hopkins, David, Lewis T. Kontnik, and Mark T. Turnage. *Counterfeiting Exposed: How to Protect Your Brand and Market Share.* Hoboken, N.J.: Wiley, 2003.

Plastow, Alan L. *Modern Pirates: Protect Your Company from the Software Police.* Garden City, N.Y.: Morgan James, 2006.

"Protect Yourself from Piracy." Microsoft Corporation. Available online. URL: http://www.microsoft.com/piracy/. Accessed November 18, 2007.

Sony

Sony Corporation (NYSE symbol: SNE) is the electronics business unit of Sony Group, a large Japanese multinational company that plays a leading role in worldwide electronics, games, and entertainment media (movies and music), introducing and shaping many now-familiar standards.

The company traces its origin to a radio repair shop started by Masaru Ibuka in a bombed-out building in Tokyo in 1945. He was soon joined by Akio Morita, and the men started an electronics company whose name translates in English to Tokyo Telecommunications Engineering Corporation. They started by building tape recorders, but in the early 1950s the two entrepreneurs were among the earliest to realize the potential of the transistor, marketing transistor radios starting in 1956. The devices essentially established the modern consumer electronics field, perfectly fitting with a new music fad among American teenagers—rock and roll.

With their marketing success, Ibuka and Morita realized that they needed a simple, catchy name that would appeal to Americans and other non-Japanese customers. In 1958 they came up with Sony. Although the name did not exist in any language (and thus could be made proprietary), "Sony" evokes English words such as "sound" and "sonic." (It also resembled a Japanese slang phrase "sony-sony," for something like what we would call "geeks" or "nerds" today.)

INFLUENCE ON MEDIA AND COMPUTING

One of Sony's most enduring impacts has been its establishment of standards for media and storage technologies. The company was not always successful: A famous also-ran was its Betamax videotape format, which lost out to VHS. However, the company's successful consumer products have included the following:

- Trinitron tubes for televisions and computer monitors (no longer sold in the United States)

- Walkman portable music player (1979)

- 3.5″ floppy disk (1983), which flourished until the later 1990s

- Discman CD-based music player (1984)

- Handycam camcorder and Video format (1985)

- Digital audio tape, or DAT (1987)

- Blu-ray optical disc

Sony would also become a major player in the console gaming market (see GAMING CONSOLE). In 1994 the company introduced the PlayStation, followed by later models in 2000 and 2006. Sony is also a significant seller of digital cameras, including the Mavica floppy disc (later CD), since discontinued. The company also introduced its proprietary "memory stick" for storage.

STUMBLES AND SUCCESSES
In 2005 a controversy erupted when it was revealed that Sony music CDs included as part of their copy protection (see also digital rights management) a "rootkit" that could allow PCs to be compromised. Sony eventually agreed with the Federal Trade Commission (FTC) to exchange the affected CDs and to reimburse damage to consumers' computers that might have occurred while attempting to remove the software. However, in 2007 a similar problem arose with third-party software packaged with Sony memory sticks.

Around the same time, Sony had to recall laptop batteries that had serious flaws that could cause them to overheat and catch fire. In 2006 Sony and Dell agreed to replace over 4.1 million laptop batteries—this was followed by 1.8 million Sony batteries in Apple laptops and 526,000 in IBM and Lenovo laptops.

Despite these setbacks, Sony continues to be very successful, with $70.3 billion in revenue and a net income of $1.07 billion in 2007, and about 163,000 employees worldwide.

Further Reading
Luh, Shu Shin. *Business the Sony Way: Secrets of the World's Most Innovative Electronics Giant.* New York: Wiley, 2003.
Nathan, John. *Sony.* New York: Houghton Mifflin, 1999.
Sony America. Available online. URL: http://www.sony.com/. Accessed November 18, 2007.
Sony Playstation. Available online. URL: http://www.us.playstation.com/. Accessed November 18, 2007.

sorting and searching
Because they are so fundamental to maintaining databases, the operations of sorting (putting data records in order) and searching (finding a desired record) have received extensive attention from computer scientists. A variety of different and quite interesting sorting methods have been devised (see ALGORITHM).

Any application that involves keeping track of a significant number of data records will have to keep them sorted in some way. After all, if records are simply inserted as they arrive without any attempt at order, the time it will take to find a given record will, on the average, be the time it would take to search through half the records in the database. While this might not matter for a few hundred records on a fast modern computer, it would be quite unacceptable for databases that might have millions of records.

SORTING CONSIDERATIONS
While some sorting algorithms are better than others in almost all cases, there are basic considerations for choosing an approach to sorting. The most obvious is how fast the algorithm can sort the number of records the application is likely to encounter. However, it is also necessary to consider whether the speed of the sort increases steadily (linearly) as the number of records increases, or it becomes proportionately worse. That is, if an algorithm can sort a thousand records in two seconds, will it take 20 seconds for 10,000 records, or perhaps five minutes?

In most cases one assumes that the records to be sorted are in more or less random order, but what happens if the records to be sorted are already partly sorted . . . or almost completely sorted? Some algorithms can take advantage of the partial sorting and complete the job far more quickly than otherwise. Other algorithms may slow down drastically or even produce errors under those conditions.

The range or variation in the key (the data field by which records are being sorted) may also play a role. In some cases if the keys are close together, some algorithms may be able to take advantage of that fact.

Finally, the available computer resources must be considered. Today many desktop PCs have 1 GB (gigabyte) or more of main memory (RAM), while servers or mainframes may have several GBs. If the database is small enough that it can be entirely kept in main memory, sorting is fast because any record can be accessed in the same amount of time at electronic speeds. If, however, part of the database must be kept in secondary storage (such as hard drives), the sorting program will have to be designed so that it reads a number of records from the hard drive in a single reading operation, in order to avoid the overhead of repeated disk operations. Most likely the individual batches will be read from the disk, sorted in memory, written back to disk, and then merged to sort the whole database.

SORTING ALGORITHMS
There are numerous sorting algorithms ranging from the easy-to-understand to the commonly used to the exotic and quirky. Only the highlights can be covered here; see Further Reading for sources for more detailed discussions.

SELECTION SORT
The simplest and least efficient kind of sort is called the *selection sort.* Rather like a bridge player organizing a hand, the selection sort involves finding the record with the lowest key and swapping it with the first record, then scanning back through for the next lowest key and swapping it with the second record, and so on until all the records are sorted. While this uses memory very efficiently (since the records are sorted in place), it is not only slow, but also gets worse fast. That is, the time taken to sort *n* records is proportional to n^2.

The selection approach suffers because on each pass the sort determines not only the record with the lowest key but the one with the next lowest key. However, that information is not retained. The heapsort, invented by John Williams in 1964, uses a binary tree to store a heap of sorted records (see TREE and HEAP). Once the heap is built, the tree nodes can be used to store record numbers in a corresponding array that will represent the sorted database. The heapsort is efficient because no records are physically moved, and the only memory needed is for the heap and array. The heapsort is generally considered the fastest and most reliable general-purpose sorting algorithm, with a maximum running time of log n.

BUBBLE SORT

The bubble sort is based on making comparisons and swaps. It makes the most convenient comparison possible: each record with its neighbor. The algorithm looks at the first two records. If the second has a lower key than the first, the records are swapped. The procedure continues with the second and third records, then the third and fourth, and so on through all the records, swapping pairs of adjacent records whenever they are out of order. After one pass the record with highest key will have "bubbled up to" the end of the list. The procedure is then repeated for all but the last record until the two highest records are at the end, and so on until all the records are sorted. Unfortunately, the number of comparisons and swaps that must be made makes the bubble sort as slow as the selection sort.

QUICKSORT

The quicksort improves on the basic bubble sort by first choosing a record with a key approximately midway between the lowest and highest. This key is called the *pivot*. The records are then moved to the left of the pivot if they are lower than it, and to the right if higher (that is, the records are divided into two *partitions*). The process is then repeated

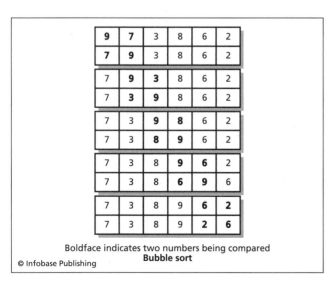

Boldface indicates two numbers being compared
Bubble sort
© Infobase Publishing

In a bubble sort, pairs of adjacent numbers are compared and switched if they are out of order. Eventually the lowest values (such as 2 in this case) will "bubble up" to the front of the list.

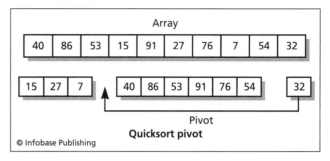

© Infobase Publishing

The Quicksort uses a value called the pivot to partition the list into two smaller lists. This process is repeated until the list has been divided and "conquered" (sorted).

to split the left side with a new pivot, and then the right side likewise. This is continued until the partition size is one, and the records are now all sorted. (Because of this repeated partitioning, quicksort is usually implemented using a procedure that calls itself repeatedly—see RECURSION.)

Devised by C. A. R. Hoare in 1962, quicksort is much faster than the bubble sort because records are moved over greater distances in a single operation rather than simply being exchanged with their neighbors. Assuming an appropriate initial pivot value is chosen, running time is proportional to the logarithm of n rather than to the square of n. The difference becomes dramatic as the size of the database increases.

INSERTION SORT

The bubble sort and quicksort are designed to work with records that are in random order. However, in many applications a database grows slowly over time. At any given time the existing database is already sorted, so it hardly makes sense to have to resort the whole database each time a new record is added.

Instead, an *insertion sort* can be used. In its simplest form, the algorithm looks sequentially through the sorted records until it finds the first record whose key is higher than that of the new record. The new record can then be inserted just before that record, much like the way a bridge player might organize the cards in a hand. (Since inserting a record and physically moving all the higher records up in memory can be time-consuming, a linked list of key values and associated record number is often used instead. (See LIST PROCESSING.) That way only the links need to be changed rather than any records being moved.

The insertion sort was improved by Donald L. Shell in 1959. His "shellsort" takes a recursive approach (like that in the quicksort), and applies the insertion sort procedure to successively smaller partitions.

Another improvement on the insertion sort is the mergesort. As the name implies, this approach begins by creating two small lists of sorted records (using a simple comparison algorithm), then merging the lists into longer lists. Merging is accomplished by looking at the two keys on the top of two lists and taking whichever is lowest until the lists are exhausted. The merge sort also lends itself to a recursive approach, and it is comparable in speed and stability to the heapsort.

HASH SORTS

All of the sorting algorithms discussed so far rely upon some form of comparison. However, it also possible to sort records by calculating their relative positions or distribution (see HASHING). In its simplest form, an array can be created whose range of indexes is equal to 1 to the maximum possible key value. Each key is then stored in the index position equal to its value (that is, a record with a key of 2314 would be stored in the array at position Array[2314]. This procedure works well, but only if the keys are all integers, the range is small enough to fit in memory, and there are no duplicate keys (since a duplicate would in effect overwrite the record already stored in that position).

A more practical approach is to use a formula (hash function) that should create a unique hash value for each key. The function must be chosen to minimize "collisions" where two keys end up with the same hash value, which creates the same problem as with duplicate keys. A hash sort is quite efficient within those constraints.

SEARCHING

Once one has a database (sorted or not), the next question is how to search for records in it. As with sorting, there are a variety of approaches to searching. The simplest and least efficient is the linear search. Like the selection sort, the linear search simply goes through the database records sequentially until it finds a matching key or reaches the end without a "hit." If there is indeed a matching record, on the average it will be found in half the time needed to process the whole database.

In most real applications the database will have been sorted using one of the methods discussed earlier. Here, the basic approach is to do a binary search. First the key in the middle record in the database is examined. The key is compared with the search key. If the search key is smaller, then any matching key must be in the first half of the database. Otherwise, it must be in the second half (unless, of course, it happens to *be* the matching key). The process is then repeated. That is, if the key is somewhere in the first half, that portion of the list is in turn split in half and its middle value is examined, and the comparison to the search key is made. Thus, the area in which the matching key must be found is progressively cut in half until either the matching key is found or there are no more records to check. Because of the power of successive division, the binary search is very quick, and doubling the size of the database means adding only one more comparison on the average.

Sometimes knowledge about the distribution of keys in the database can be used to improve even the binary search. For example, if keys are alphabetical and the search key begins with S, it is likely to be faster to pick a starting point near the end of the list rather than from the middle. A binary tree (see TREE) can be constructed from the keys in a database in order to analyze the most likely starting points for a search.

Finally, hashing (as previously discussed) can be used to quickly calculate the expected location of the desired record, provided there are no collisions.

Further Reading
Knuth, Donald E. *Art of Computer Programming, Volume 3: Searching and Sorting.* 2nd ed. Upper Saddle River, N.J.: Addison-Wesley Professional, 1998.
Ploedereder, Erhard. "The Sort Algorithm Animator V1.0." Available online. URL: http://www.iste.uni-stuttgart.de/ps/Ploedereder/sorter/sortanimation2.html. Accessed August 21, 2007.
Sedgewick, Robert. *Algorithms in C++: Parts 1–4: Fundamentals, Data Structures, Sorting, Searching.* Upper Saddle River, N.J.: Addison-Wesley, 1998.
Wilt, Nicholas. *Classical Algorithms in C++: With New Approaches to Sorting, Searching, and Selection.* New York: Wiley, 1995.

sound file formats

There are a number of ways that sound can be sampled, stored, or generated digitally (see MUSIC, COMPUTER). Here we will look at some of the most popular sound file formats.

WAV

The WAV (wave) file format is specific to Microsoft Windows. It essentially stores the raw sample data that represents the digitized audio content, including information about the sampling rate (which in turns affects the sound quality). Since WAV files are not compressed, they can consume considerable disk space.

AIFF

AIFF stands for Audio Interchange File Format, and is specific to the Apple Macintosh and to Silicon Graphics (SGI) platforms. Like WAV, it stores actual sound sample data. A variant, AIFF-C, can store compressed sound.

AU

The AU (audio) file format was developed by Sun Microsystems and is used mainly on UNIX systems, and also in Java programming.

MIDI

MIDI stands for Musical Instrument Digital Interface. Unlike most other sound formats, MIDI files don't represent sampled sound data. Rather, they represent virtual musical instruments that synthesize sound according to complex algorithms that attempt to mirror the acoustic characteristics of real pianos, guitars, or other instruments. Since MIDI is like a "score" for the virtual instruments rather than storing the sounds, it is much more compact than sampled sound formats. MIDI is generally used for music composition rather than casual listening.

MP3

MP3 is actually a component of the MPEG (Moving Picture Expert Group) multimedia standard, and stands for MPEG-1 Audio Layer 3. It is now the most popular sound format, using compression to provide a balance of sound quality and compactness that is comparable to that of standard audio CDs and suitable for most listeners. The compression algorithm relies upon psychoacoustics (the study of how people perceive the components of sound) to identify frequencies that humans can't hear, and thus may be safely discarded. The digitized sound on a CD is compressed up to 1/12 or less of its original size, so a 630 MB CD becomes about 50 MB in MP3 files.

Since most PC users now have hard drives rated in the hundreds of gigabytes (GB), it is easy to store an extensive music library in MP3 form. Most PCs now come with software that can play MP3 files (such as Windows Media Player), and there are also free and shareware programs from a variety of sources, as well as plug-ins for playing sound files directly from the Web browser.

Since MP3 is much more compact than "raw" CD format, users with inexpensive CD-RW drives can "burn" large amounts of music in MP3 form onto a single CD. This is typically done using software that "rips" the raw tracks from an audio CD and converts them to an MP3 file, which can then be stored on the PC's hard drive.

In recent years portable media players such as the iPod have become ubiquitous (see MUSIC AND VIDEO PLAYERS, DIGITAL). MP3 is the most popular format for music that is not digitally protected from copying (see DIGITAL RIGHTS MANAGEMENT). However, because MP3 involves a number of patents, it is not included by default in Linux distributions, which instead provide Ogg, a "container" that can be used for a variety of formats (see CODECS).

Further Reading
Audio File Types. Available online. URL: http://www.fileinfo.net/filetypes/audio. Accessed August 22, 2007.
Johnson, Dave, and Rick Broida. *How to Do Everything with MP3 and Digital Music.* New York: McGraw Hill Professional, 2001.
Young, Robert. *The MIDI Files.* 2nd ed. New York: Prentice Hall, 2001.

space exploration and computers

It might have been barely possible to put a satellite (or person) in orbit without the use of computers, but any more extensive exploration of space requires many types of computer applications.

HUMAN SPACE EXPLORATION

Flying to the Moon required precisely calculated and controlled "burns" to inject the Apollo spacecraft from orbit into its arcing trajectory to the Moon. The detachable Lunar Excursion Module (LEM) also had a computer on board (roughly comparable in power to something found in today's programmable calculators). Although the pilot controlled the final landing manually, the computer interpreted radar data to fix the lander's position, monitored fuel consumption, and provided other key data.

The Space Shuttle, the most complex vehicle ever built by human beings, has five onboard computer systems that control flight maneuvers (including rendezvous and docking operations), monitor and control environmental conditions, keep track of fuel, batteries, life support, and other

consumables, and provide many other functions to support the crew's tasks and experiments.

AUTOMATED SPACE EXPLORATION

Thus far, human explorers have flown no farther than the Moon. However, in the last 40 years an extensive survey of most of the solar system has been carried out by robot (that is to say, computerized) probes and landers. These probes have landed on Mars and visited every planet, as well as making close approaches to asteroids and comets.

The control computer aboard a space probe has several jobs. It must keep the probe oriented in such a way that its solar panels can receive energy from the Sun, as well as keeping an antenna pointed toward Earth so it can receive commands and return data from the probe's scientific instruments.

Starting with *Voyager 2* (a probe that is still returning data from more than 7 billion miles from Earth), space probe computers have been more autonomous, able to make attitude corrections and course corrections as needed. The onboard computer can even be reprogrammed with new instructions sent from Earth. Space probes have returned incredibly detailed pictures of the surface of the Moon and planets, preparing the way for human missions or robot landers.

Landers reach a fixed point on a planetary surface and transmit photographs, temperature, radiation, and other readings. Probes can survive only for minutes on the hostile surface of Venus, but have functioned for many months on Mars. In a remarkably ambitious mission beginning in 1976, the two Viking Mars landers were able to carry out experiments on soil samples in an unsuccessful attempt to find evidence of life while a third probe mapped the planet's surface from orbit. Besides demonstrating remarkable reliability (*Viking 2* was still operating in 1982 when it was accidentally turned off by a remote command), the mission also demonstrated the ability to coordinate surface and orbital exploration.

In July 1997, the *Mars Pathfinder* probe landed on the red planet, rolling and bouncing to a stop inside a sort of giant airbag. After deflating, the *Pathfinder* base station deployed the *Sojourner* mobile robot. This vehicle (see ROBOTICS) was controlled by operators on Earth, but because of the 10–15-minute time delay in signals arriving from Earth, the *Sojourner* had some autonomous ability to avoid collisions or other hazards. The onboard computer also had to compress and transmit images and other data. The follow-on Mars Exploration Rover (MER) program began in 2003 with the launching of two larger surface rovers dubbed *Spirit* and *Opportunity*. Landing in January 2004, the rovers have shown remarkable durability, still functioning in early 2008, far beyond their original three-month mission life.

The need to build compact computers and other electronics for space exploration helped spur the development of techniques now found in garden-variety consumer electronics. Space computers are also important for demonstrating the reliability and robustness that is necessary for applications on Earth (such as in the military). Space electronics must be shielded and "hardened" to withstand the intense solar radiation, extreme changes in temperature, and electromagnetic

How do scientists look at images that are sent back from another planet and determine what is interesting and needs further investigation? Mars rover scientists do this very task during surface mission operations. Each day, rovers send to Earth new images that the science team must examine. These images allow the scientists to think of hypotheses that relate to help the science team decide what to study and determine what experiments they will conduct. (NASA PHOTO)

fluxes or surges. Redundancy can be used where possible, but weight is always at a high premium. With the exception of certain satellites and the *Hubble Space Telescope,* space computers cannot receive on-site service visits.

Because of the high cost and risk of maintaining human life for long periods in space, it is likely that robotic probes and rovers will remain the main means for space exploration in the early 21st century.

Further Reading

Furmiss, Tim. *A History of Space Exploration and Its Future.* London: Mercury Books, 2006.

Hall, Eldon C. *Journey to the Moon: The History of the Apollo Guidance Computer.* Reston, Va.: American Institute of Aeronautics, 1996.

Mars Exploration Rover Mission (Jet Propulsion Laboratory). Available online. URL: http://marsrovers.jpl.nasa.gov/home/index.html. Accessed August 22, 2007.

Mars Pathfinder [archive]. Available online. URL: http://mpfwww.jpl.nasa.gov/MPF/index1.html. Accessed August 22, 2007.

Matloff, Gregory L. *Deep Space Probes: To the Outer Solar System and Beyond.* 2nd ed. New York: Springer, 2005.

Squyres, Steve. *Roving Mars: Spirit, Opportunity, and the Exploration of the Red Planet.* New York: Hyperion, 2005.

Spafford, Eugene H.

(1956–   )
American
Computer Scientist

Eugene (Gene) H. Spafford is a computer scientist and pioneer in network security. Spafford earned a B.A. in mathematics and computer science from the State University of New York at Brockport. He then earned M.S. (1981) and Ph.D. (1986) degrees at the Georgia Institute of Technology, with his graduate work focused on distributed operating systems.

USENET AND BEYOND

Spafford played a key role in the development of the Usenet (see NETNEWS AND NEWSGROUPS), including the backbones and connections that provided for the efficient distribution of a growing volume of news posts, as well as the system for naming newsgroups. He also created basic introductory documentation to help new users participate in the system responsibly.

On the night of November 2, 1988, sites throughout the Internet began to shut down. The culprit was a worm program (see COMPUTER VIRUS) that Spafford analyzed in a technical paper. The worm would unfortunately only be the first of a legion of worms and viruses that would infect the network, and Spafford would apply considerable effort to helping cope with them. Since then Spafford has been a computer security consultant and adviser for numerous organizations including Microsoft, Intel, the U.S. Air Force, the National Security Agency, the FBI, and the National Science Foundation.

Spafford has been on the faculty at Purdue University since 1987. In 2007, he was appointed an adjunct professor of computer science at the University of Texas at San Antonio. He is also executive director of the university's new Institute for Information Assurance.

Spafford has served on the boards of a number of professional societies, including the Computer Research Association and the U.S. Public Policy Committee of the Association for Computing Machinery (ACM). He has written several books and hundreds of papers on UNIX and Internet security and related ethical issues. Spafford became an ACM Fellow in 1997 and a Fellow of the American Association for the Advancement of Science in 1999. He was inducted as a Fellow of the Institute for Electrical and Electronics Engineers (IEEE) in 2000 and received its Technical Achievement Award in 2006. In 2007 Spafford received the ACM President's Award.

Further Reading

Garfinkel, Simson L., and Eugene H. Spafford. *Practical UNIX Security.* Sebastapol, Calif.: O'Reilly, 2003.

Rospach, Chuq von, with editing additions by Eugene H. Spafford. "A Primer on How to Work with the Usenet Community." Available online. URL: http://www.faqs.org/faqs/usenet/primer/part1/. Accessed November 18, 2007.

Spafford, Eugene H. "The Internet Worm: Crisis and Aftermath." *Communications of the ACM* 32 (June 1989): 678–687. Available online. URL: http://vx.netlux.org/lib/aes01.html. Accessed November 18, 2007.

Spaf's Home Page. Purdue University. Available online. URL: http://homes.cerias.purdue.edu/~spaf/. Accessed November 18, 2007.

spam

In a well-known 1970 sketch by the British comedy troupe Monty Python, a customer is trying to order a breakfast item that does not include Spam (the popular luncheon meat). A group of Vikings then keeps interrupting the conversation by loudly singing "Spam, lovely Spam, wonderful Spam. . . ." Segue to the mid-1990s when people (including a legal firm) began automatically posting hundreds of identical messages on Usenet (see NETNEWS AND NEWSGROUPS) groups; the sketch came to mind and the postings were quickly dubbed "spam"—although the term may actually date back to the 1980s. As news of the spam grew, some administrators and users used "cancelbots" to automatically delete the offending messages; others opposed this as censorship, and many newsgroups became effectively unreadable.

While spam can appear in any communications medium (including chat, instant messaging, and even blogs), the most prevalent type is e-mail spam, which costs U.S. businesses billions of dollars a year in processing expenditures, lost time, and damage caused by malicious software (malware) for which spam can be either a delivery vehicle or an inducement. In 2007 an estimated 90 billion spam messages were sent each day.

The fundamental driving force of spam is the fact that, given one has Internet access, sending e-mail costs essentially nothing, no matter how many messages are sent. Thus even if only a tiny number of people respond to a spam solicitation (such as for sexual-enhancement products), the result is almost pure profit for the spammer.

Besides directly making fraudulent solicitations for products that are ineffective, counterfeit, or nonexistent, spam carries two other dangers: inducements to click to visit fake Web sites (see PHISHING AND SPOOFING) and attachments containing viruses or other dangerous software (see COMPUTER VIRUS and SPYWARE AND ADWARE).

FIGHTING SPAM

Much spam is spread by first compromising thousands of systems (via viruses) and planting in them "bots," or software that can be programmed to mail spam. The controllers of "botnets" can then sell their service to spammers who want to get their message distributed widely. The spammers can also buy lists of e-mail addresses that have been "harvested" from postings, poorly secured Web sites, and so on.

Ways to stop the spread of spam include the following:

- e-mail filtering software, using a combination of text analysis by keyword or statistical correlation (see BAYESIAN ANALYSIS) and lists of Internet locations (domains) associated with spamming; filtering can be done both by service providers and individual users, or collaboratively

- tightening the technical requirements for messages to be accepted by mail servers (much spam has poorly formatted headers)

- improving techniques for blocking the viruses used by spammers to set up their bots—see COMPUTER VIRUS and FIREWALL

- attempting to shut down the infrastructure that supports spam operations, such as hosts who allow bulk

e-mail, and sellers of spamming software and illicitly gathered address lists

Spam is illegal in a number of respects. Spamming is against the "acceptable use policy" of most Internet Service Providers (ISP), though willingness to enforce these rules varies. In 2003 Congress passed the CAN-SPAM act, which bans bulk e-mail that contains misleading subject or header lines, but has been criticized for being weak and for preempting more stringent state laws. (The law also requires that messages include an opt-out provision, but spammers simply use this to verify that the e-mail address is valid.)

Although filtering software and other measures can reduce the amount of spam seen by the average user, spammers and spam-fighters continue their relentless battle with each countermeasure, leading to altering the spam to make it more likely to pass through. In the long run probably only a Net-wide authentication of all e-mail senders and/or a small per-message e-mail fee could effectively banish the scourge of spam.

Further Reading

Boutin, Paul. "Can E-mail Be Saved?" *InfoWorld,* April 16, 2004. Available online. URL: http://www.infoworld.com/article/04/04/16/16FEfuturemail_1.html. Accessed November 18, 2007.

Garretson, Cara. "12 Spam Research Projects That Might Make a Difference." *Network World,* November 2007. Available online. URL: http://www.networkworld.com/news/2007/112007-spam-research.html. Accessed April 28, 2008.

Gregory, Peter H., and Michael A. Simon. *Blocking Spam & Spyware for Dummies.* Hoboken, N.J.: Wiley, 2005.

Lee, Nicole. "How to Fight Those Surging Splogs" [spam blogs]. Wired News, October 27, 2005. Available online. URL: http://www.wired.com/culture/lifestyle/news/2005/10/69380. Accessed November 18, 2007.

Markoff, John. "Attack of the Zombie Computers Is a Growing Threat." *New York Times,* January 7, 2007. Available online. URL: http://www.nytimes.com/2007/01/07/technology/07net.html. Accessed November 18, 2007.

McWilliams, Brian S. *Spam Kings: The Real Story behind the High-Rolling Hucksters Pushing Porn, Pills, and %*@)# Enlargements.* Sebastapol, Calif.: O'Reilly, 2004.

Naughton, Philippe. "Arrest of 'Spam King' No Relief for Inboxes." *Times* (London) online, June 1, 2007. Available online. URL: http://technology.timesonline.co.uk/tol/news/tech_and_web/article1870548.ece. Accessed November 18, 2007.

Spammer-X. *Inside the Spam Cartel: Trade Secrets from the Dark Side.* Rockland, Mass.: Syngress, 2004.

Zeller, Tom. "The Fight Against Vl@gra (and Other Spam)." *New York Times,* May 21, 2006. Available online. URL: http://www.nytimes.com/2006/05/21/business/yourmoney/21spam.html?_r=1&oref=slogin. Accessed November 18, 2007.

speech recognition and synthesis

The possibility that computers could use spoken language entered popular culture with Hal 2001, the self-aware talking computer in the film *2001: A Space Odyssey.* On a practical level, the ability of users to communicate using speech rather than a keyboard would bring many advantages, such as mobile, hands-free computing and greater independence for disabled persons. Considerable progress has been made in this technology since Hal "talked" in 1968.

Speech recognition begins with digitizing the speech sounds and converting them into a standard, compact representation. The analysis can be based on matching the input sounds to one of about 200 "spectral equivalence classes" from which the representation can be created. Alternatively, algorithms can use data based on modeling how the human vocal tract produces speech sounds, and extract key features that then become the speech representation. Neural networks can also be "trained" to recognize speech features (*see* NEURAL NETWORK). The latter two approaches are potentially more flexible but also considerably more difficult, and tend to be used in research rather than in commercial voice recognition systems.

Whichever form of representation is used, it must then be matched to the characteristics of particular words or phonemes, usually with the aid of sophisticated statistical and time-fitting techniques. The simplest systems work on a word level, which may suffice if the system is restricted to a simple vocabulary and the user speaks slowly and distinctly enough. Such systems usually require that the user "train" the system by speaking selected words and phrases. The user can then control the system with a set of voice commands.

Creating a system that can handle the full range of language is much more difficult. This kind of system breaks the language down into phonemes, its basic sound constituents (English has about 40 phonemes). The system includes a stored dictionary of phoneme sequences and the corresponding words. However, "understanding" which words are being spoken is more than a matter of matching phoneme sequences to a dictionary. For one thing, the sound of the first or last phoneme in a word can change depending on the phoneme in an adjacent word.

Once the speech has been recognized, it can be converted to character data (see CHARACTERS AND STRINGS) and treated as though the text had been entered from the keyboard. This means, for example, that a user could dictate text to be placed in a word processor document as well as using voice commands to perform tasks such as formatting text. (Special words can be used to introduce and end commands.)

Voice control and dictation have been offered commercially by such companies as Dragon Systems and Kurzweil. Microsoft now includes speech recognition and synthesis facilities in the latest version of its popular office suite, Office 2007.

VOICE SYNTHESIS

The other part of the speech equation is the ability to have the computer turn character codes into spoken words. The most primitive approach is to digitally record appropriate spoken words or phrases, which can then be replayed when speech is desired. Naturally, what is spoken is limited to what is available in the recorded library, although the words and phrases can be combined in various ways. Since the combinations lack the natural transitions that speakers use, the result sounds "mechanical." Common applications include automated announcements in train stations or in prompts for voicemail systems.

To produce a synthesizer that can "speak" any natural language text, the system must have a dictionary that gives the phonemes found in each word. The 40 or so different phonemes can then be digitally recorded and the system would then identify the phonemes in each word and play them to create speech. While this solves the limited vocabulary problem, the synthesized speech is rather unnatural and hard to understand. This is because, as noted earlier, the way phonemes are sounded changes under the influence of adjacent phonemes, and these nuances are lacking in a simple phoneme playback.

More sophisticated voice synthesis systems record natural speech and identify all the possible combinations of half of a phoneme and half of an adjacent phoneme. That way the possible transition sounds are also recorded, and the resulting speech sounds considerably more natural. The drawback is that more memory and processing power are required, but these commodities are becoming increasingly cheaper.

Speech recognition and synthesis technology has made only slow inroads into the computing mainstream, such as office applications. Given the costs of hardware, software, and training, the keyboard remains more productive and cost-effective for most applications. However, voice technology does have a growing number of specialty uses, including security and access systems, speech synthesis for disabled persons who cannot see or speak, and enabling service robots to interact with people in the environment. Speech technology has also been a long-standing topic in artificial intelligence and robotics research.

Further Reading

Brown, Robert. "Exploring New Speech Recognition and Synthesis APIs in Windows Vista." MSDN Magazine. Available online. URL: http://msdn.microsoft.com/msdnmag/issues/06/01/speechinWindowsVista/. Accessed August 22, 2007.

Holmes, John, and Wendy Holmes. *Speech Synthesis and Recognition.* 2nd ed. Boca Raton, Fla.: CRC Press, 2001.

Huang, Xuedong, Alex Acero, and Hsiao-Wuen Hon. *Spoken Language Processing: A Guide to Theory, Algorithm, and System Development.* Upper Saddle River, N.J.: Prentice Hall, 2001.

Jurafsky, Daniel, and James H. Martin. *Speech and Language Processing: An Introduction to Natural Language Processing, Computational Linguistics and Speech Recognition.* Upper Saddle River, N.J.: Prentice Hall, 2000.

Speech Technology (Google Directory). Available online. URL: http://www.google.com/Top/Computers/Speech_Technology/. Accessed August 22, 2007.

Speech Technology Research, Development, and Deployment (Carnegie Mellon University). Available online. URL: http://www.speech.cs.cmu.edu/. Accessed August 22, 2007.

spreadsheet

With the possible exception of word processing, no personal computer application caught the imagination of the business world as quickly as did the spreadsheet, which first appeared as Daniel Bricklin's *VisiCalc* in 1979. VisiCalc quickly became the "killer app"—the application that could justify corporate purchases of Apple II computers. When the IBM PC began to dominate the office computing industry in the mid-1980s, it had a new spreadsheet, Lotus 1-2-3. By the end of the decade, however, Microsoft's Excel spreadsheet had come to the forefront, running on Microsoft Windows. It remains the market leader today.

HOW SPREADSHEETS WORK

A spreadsheet is basically a tabular arrangement of rows and columns that define many individual cells. Typically, the columns are lettered (A to Z, then AA, AB, and so on) while the rows are numbered. A particular cell is referenced using its column and row coordinates; thus A1 is the cell in the upper left corner of the spreadsheet.

Any cell can contain a numeric value, a formula, or a label (such as for giving a title to the spreadsheet or some section of it). Formulas reference the values in other cell locations. For example, if the formula =SUM (A1:B1) is inserted into cell C1, when the spreadsheet is calculated the sum of the contents of cells A1 and B1 will be inserted into C1. Modern spreadsheets let users select from a variety of functions (predefined formulas) for such things as interest or rates of return. Instead of having to type the individual coordinates of cells to be used in a formula, he or she can simply click on or drag across the cells to select them. Formulas can also include conditional evaluation (similar to the If statements found in programming languages—see branching statements).

Spreadsheets provide a variety of "housekeeping" commands that can be used for functions such as copying or moving a range of cells or "cloning" a cell's value into a range of cells. Large spreadsheets can be broken down into multiple linked spreadsheets to make it easier to understand and maintain.

Macros offer a powerful way to simplify and automate spreadsheet operations. A macro is essentially a set of programmed instructions to be carried out by the spreadsheet (see MACRO). One use of macros is to carry out complicated procedures by taking advantage of features similar to those found in programming languages such as Visual Basic. Macros can also be used to automate data entry into the spreadsheet and validate the data. Depending on their complexity, macros can either be typed in as a series of statements or recorded as the user takes appropriate menu and mouse actions. "Solver" utilities can also simplify the process of tweaking input variables in order to achieve a defined goal. Although spreadsheets can certainly solve many types of algebraic equations, symbolic manipulation is better handled by programs such as Mathematica (see MATHEMATICS SOFTWARE).

Besides having extensive graphics and charting capabilities, modern spreadsheets are often part of integrated office programs (see APPLICATION SUITE). Thus, a Microsoft Excel spreadsheet could obtain data from an Access database and create charts suitable for Web pages or PowerPoint presentations.

Further Reading

Balakrishnan, Nagraj, Barry Render, and Ralph M. Stair, Jr. *Managerial Decision Modeling with Spreadsheets*. 2nd ed. Upper Saddle River, N.J.: Prentice Hall, 2006.

Google Docs and Spreadsheets. Available online. URL: http://docs.google.com. Accessed August 22, 2007.

Harvey, Greg. *Microsoft Office Excel 2007 for Dummies*. Hoboken, N.J.: Wiley, 2007.

Hayden, Yvonne. *So You Need to Make a Spreadsheet: A Quick Start to Microsoft Excel 2003*. Chandler, Ariz.: Copadego Publishing, 2006.

Jelen, Bill, ed. *The Spreadsheet at 25: The 25 Year Evolution of the Invention that Changed the World*. Uniontown, Ohio: Holy Macro! Books, 2005.

Neuwirth, Ertich. "Spreadsheets, Mathematics, Science, and Statistics Education." Available online. URL: http://sunsite.univie.ac.at/Spreadsite/. Accessed August 22, 2007.

spyware and adware

Spyware and adware are two pervasive threats to computer users. Both are programs that are installed more or less surreptitiously, often accompanying an attractive-looking "free" software package or media download. Depending on how widely it is defined, as many as eight out of 10 PCs may be infected by some sort of spyware. Signs of infection can include the system slowing down or periodically freezing, Web browsers that fail to display the expected home page or search results, and the appearance of numerous unwanted pop-up windows (a sign of adware).

Ranging from least to most harmful, spyware and adware can do the following:

- Display annoying advertising that can clog up the screen or cover up information (some adware can also be spyware that uses information about the user to target advertising)

- Track Web browsing to provide information to sell to marketers (see COOKIES)

- Obtain personal information for use in identity theft

- Install keyloggers (programs that record keystrokes, such as passwords being entered) or other "back door" or "trojan" programs

STOPPING SPYWARE

Growing concern about spyware has prompted the use of antispyware programs such as Ad-Aware and Spybot-Search & Destroy, as well as a free program from Microsoft. Antispyware programs are also being included in popular security suites from companies such as Symantec and McAfee. The programs work similarly to antivirus programs, watching for suspicious behavior or "signatures" matching known spyware or adware. Depending on the program, the spyware can be blocked from executing at all or removed from the system.

The software varies considerably in effectiveness, so users may have to run several different programs to completely remove an "infestation."

Spyware has been generally given a lower priority than viruses or even spam. When challenged, spyware makers generally claim that the user authorized its installation (at least implicitly) by installing the utility or other software that contains it. Although antispyware legislation has been introduced in Congress, it has not passed as of mid-2008. However, state officials such as former New York State Attorney General Eliot Spitzer successfully sued a spyware company, winning a $7.5 million settlement.

Further Reading

"Antispyware." *PC Magazine.* Available online. URL: http://www. pcmag.com/category2/0,1738,1639157,00.asp. Accessed November 18, 2007.

Chadbrow, Eric. "Spyware and Adware Continue to Plague PCs." *InformationWeek.* March 27, 2006. Available online. URL: http://www.informationweek.com/story/showArticle.jhtml?ar ticleID=183702594. Accessed November 18, 2007.

Gregory, Peter H., and Michael A. Simon. *Blocking Spam & Spyware for Dummies.* Hoboken, N.J.: Wiley, 2005.

"Magoo's Wise Words: Guide to Eliminating Spyware." Available online. URL: http://guides\radified.com/magoo/guides/ spyware/remove_spyware_01.htm. Accessed November 18, 2007.

Shetty, Sachin. "Introduction to Spyware Keyloggers," April 14, 2005. Available online. URL: http://www.securityfocus.com/ infocus/1829. Accessed November 18, 2007.

SQL

Structured query language was originally developed in the early 1970s as a command interface for IBM mainframe databases. Today, however, SQL has become the lingua

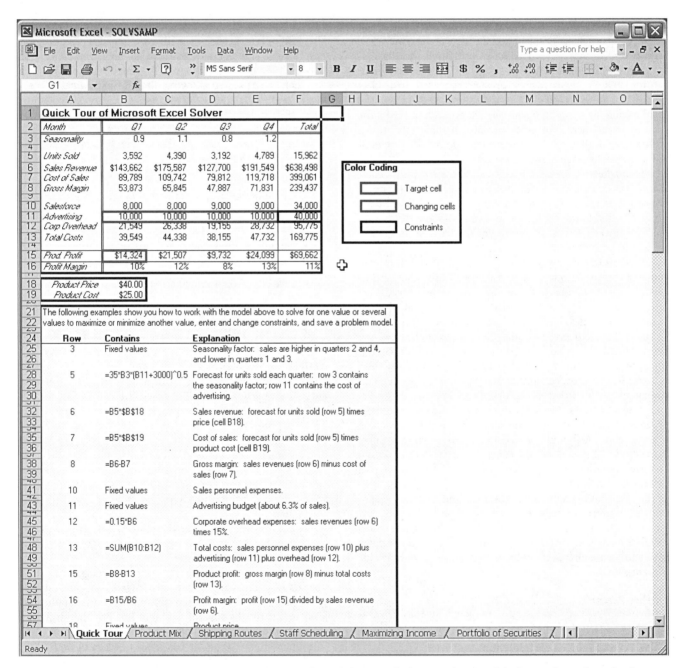

Modern spreadsheets have many sophisticated features. Microsoft Excel, for example, has a "Solver" module that can be used to solve for particular values or to maximize or minimize specified values.

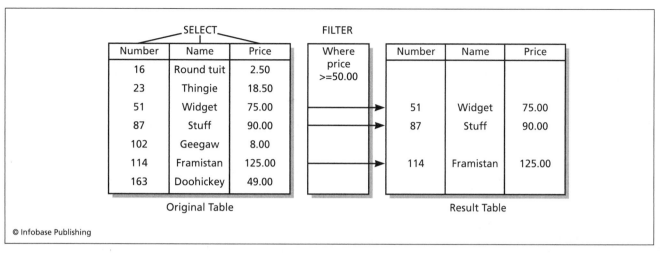

Structured Query Language (SQL) is a standardized way to query and manipulate databases. Here the statement SELECT NUMBER, NAME, PRICE WHERE PRICE >= 50.00 extracts only the records meeting that criterion.

franca for relational database systems (see DATABASE MANAGEMENT SYSTEM).

A relational database (such as Oracle, Sybase, IBM DB2, and Microsoft Access) stores data in tables called *relations*. The columns in the table describe the characteristics of an entity (corresponding to data fields). For example, in a customer database the Customer table might include attributes such as customer number, First_name, Last_Name, Street, City, Phone_number, and so on. The rows in the table (sometimes called *tuples*) represent the data records for the various customers.

Many database systems have more than one table. For example, a store's database might contain a Customers table (for information identifying a customer), an Item table (giving characteristics of an item, such as price and number in stock), and a Transaction table (whose characteristics might be customer number, date, item bought, and so on). Notice that the Transaction record contains both a customer number and an item number. It thus serves as a sort of bridge or link between the Customer and Item tables.

SQL provides commands that can be used to specify and access components of a database. For example, the INSERT and DELETE commands can be used to add or remove rows (records) from tables.

To query a database means to give criteria for selecting certain records from a table. For example, the query

```
SELECT * FROM CUSTOMERS WHERE LAST_NAME =
"Howard"
```

would return the complete records for all customers whose last name is Howard. If only selected fields are desired, they can be specified like this:

```
SELECT NUMBER, NAME, PRICE FROM ITEMS WHERE
PRICE > = 50.00
```

This query will display the Number, Name, and Price fields for all items whose price is greater than or equal to $50.00.

SQL includes many commands to further refine data processing and reporting. There are built-in mathematical functions as well as a GROUP BY command for further breaking down a report by a particular field name or value.

SQL can be used interactively by typing commands at a prompt, but database applications designed for less technical users often provide a user-friendly query form (and perhaps menus or buttons). After the user selects the appropriate fields and values, the program will then generate the necessary SQL statements and send them to the internal "database engine" for processing. The results will then be displayed for the user.

SQL procedures can be stored and managed as part of a database. SQL can also be "embedded" within a more complete programming language environment so that, for example, a Java program can perform SQL operations while using Java for processing that cannot be specified in SQL. In the mid-1990s an object-oriented version of SQL called OQL (object query language), allowing the use of that popular paradigm for database operations (see OBJECT-ORIENTED PROGRAMMING).

One of the most popular implementations of SQL is MySQL, which is privately owned and developed but available for free license on many platforms, including Windows and Linux. A number of applications are designed to work with MySQL databases: see, for example, WIKIS AND WIKIPEDIA and YOUTUBE.

Further Reading

Forta, Ben. *Sams Teach Yourself SQL in 10 Minutes*. 3rd ed. Indianapolis: Sams, 2004.

Kofler, Michael. *The Definitive Guide to MySQL 5*. 3rd ed. Berkeley, Calif.: Apress, 2005.

MySQL home page. Available online. URL: http://mysql.org/. Accessed August 22, 2007.

Rankins, Ray, et al. *Microsoft SQL Server 2005 Unleashed*. Indianapolis: Sams, 2006.

Tahaghoghi, Seyed M. M., and Hugh Williams. *Learning MySQL*. Sebastapol, Calif.: O'Reilly Media, 2006.

Taylor, Allen G. *SQL for Dummies*. 6th ed. Hoboken, N.J.: Wiley, 2006.

stack

Often a temporary storage data area is needed during processing. For example, a program that calls a procedure (see PROCEDURES AND FUNCTIONS) usually needs to pass one or more data items to the procedure. These items are specified as arguments that will be matched to the procedure's defined parameters. For example, the procedure call

```
Square (50, 50, 20)
```

could draw a square whose upper left corner is at the screen coordinates 50, 50 and whose length per side is 20 pixels.

When the compiler generates the machine code for this statement, that code will probably instruct the processor to store the numbers 50, 50, and 20 onto a stack. A stack is simply a list that represents successive locations in memory into which data can be inserted. The operation of a stack can be visualized as being rather like the spring-loaded platform onto which dishes are stacked for washing in some restaurants. As each dish (number) is added, the stack is "pushed." Because only the item "on top" (the last one added) can be removed ("popped") at any given time, a stack is described as a LIFO (last in, first out) structure. (Note that this is different from a queue, where items can be added or removed from either end [see QUEUE].)

Stacks are useful whenever nested items must be tracked. For example, a procedure might call a procedure that in turn calls another procedure. The stack can keep track of the parameters (as well as the calling address) for each pending procedure.

Stacks can also be used to evaluate nested arithmetic expressions. For example, the expression that we write in conventional (prefix) notation as

```
7 * 5 + 2
```

can be represented internally in postfix form as:

```
* + 5 7 2
```

Here one stack can be used to hold the operators (* +) and one the operands (5 7 2). The evaluation then proceeds in the following steps:

```
Pop the * from the operator stack
```

```
Since * is a binary operator (one that needs
two operands), pop the 5 and 7 from the
operand stack
```

```
Multiply 5 and 7 to get 35.
```

```
Pop the + from the operator stack.
```

```
Pop the 35 (which is now on the top of the
operand stack) and the 2
```

```
Add 35 and 2 to get 37.
```

An interesting programming language uses this stack mechanism for all processing (see FORTH). In working with stacks, it may be necessary to keep in mind any limitations on the amount of memory allocated to the stack, although a stack can also be implemented dynamically as a linked list (see LIST PROCESSING).

Further Reading
"Data Structures/Stacks and Queues." Wikibooks. Available online. URL: http://en.wikibooks.org/wiki/Data_Structures/Stacks_and_Queues. Accessed August 22, 2007.

Stallman, Richard
(1953–)
American
Computer Scientist

Richard Stallman created superb software tools—the programs that help programmers with their work. He went on to spearhead the open source movement, a new way to develop software.

Stallman was born on March 16, 1953, in New York City. He quickly showed prodigious talent for mathematics and was exploring calculus by the age of eight. Not much later, his summer camp reading included a manual for the IBM 7094 mainframe belonging to one of the counselors. Fascinated with the idea of programming languages, young Richard began writing simple programs, even though he had no access to a computer.

Fortunately, a high school honors program let him obtain some time on a mainframe, and his programming talents led to a summer job with IBM. While studying for his B.A. in physics at Harvard (which he received in 1970), Stallman found himself sneaking across town to the MIT Artificial Intelligence Lab. There he developed Emacs, a powerful text editor that could be programmed with a language modeled after LISP, the favorite language of AI researchers. While working on Emacs and other system software for the AI Lab, Stallman participated in the unique MIT "hacker culture." (During the 1970s, "hacker" still meant a creative computing virtuoso, not a cyber-criminal.)

Stallman's experience in the freewheeling, competitive yet cooperative atmosphere at MIT led him to decide in 1984 to start the Free Software Foundation, which would become his life's work. Stallman and his colleagues at the FSF worked through the 1980s to develop GNU. At the time, UNIX, the operating system of choice for most campuses and researchers, required an expensive license from Bell Laboratories. GNU (a recursive acronym for "GNU's Not UNIX") was intended to include all the functionality of UNIX but with code that owed nothing to Bell Labs. Stallman's key contributions to the project included the GNU C compiler and debugger, as well as his management of a cooperative effort in which many talented programmers would coordinate their efforts over the Internet.

By the early 1990s, most of GNU was complete except for a key component: the kernel containing the essential functions of the operating system. A Finnish programmer named Linus Torvalds decided to write the kernel and integrate it with much of the existing GNU software. The result would become known as Linux, and today it is a popular operating system that runs on many servers and workstations. While acknowledging Torvalds's efforts, Stallman insists that the operating system is more properly called GNU Linux, to reflect the large amount of GNU code it employs.

In recent years Stallman has best been known as a vigorous advocate for free software (see OPEN-SOURCE MOVEMENT) and for creating alternative structures for controlling its distribution, such as the various forms of the General Public License (GPL). Stallman has been accused of being rigid and abrasive, such as in his urging that certain terminology be used, or, in the case of the phrase "intellectual property," not used.

Stallman has received a number of important awards, including the ACM Grace Hopper Award (1990), Electronic Frontier Foundation Pioneer Award (1998), and a MacArthur Foundation fellowship (1990).

Further Reading

Free Software Foundation. Available online. URL: http://www.fsf.org/. Accessed August 22, 2007.

Stallman, Richard M. *Free Software, Free Society: Selected Essays of Richard M. Stallman.* Boston: Free Software Foundation, 2002.

Williams, Sam. *Free as in Freedom: Richard Stallman's Crusade for Free Software.* Sebastopol, Calif.: O'Reilly Media, 2002.

standards in computing

One hallmark of the maturity of a technology is the development of a variety of kinds of standards that are accepted by a majority of practitioners. There are several reasons why standards develop.

MARKETPLACE STANDARDS

In many cases, a particular product gains a prominent position in an emerging market, and would-be competitors adopt its interface and specifications. For example, the parallel port printer interface (and plug) developed by Centronics for its printers was adopted by virtually all printer manufacturers. Since it would be impracticable for computer manufacturers to provide many different parallel connectors on their machines, there was a clear market advantage in setting a standard. When a particular product (Centronics in this case) becomes that standard, it is mainly a matter of timing.

Once a marketplace standard is established, manufacturers and consumers will generally not want products that are incompatible with it. When the IBM PC and its ISA expansion card became the standard followed by many "clone" manufacturers, IBM discovered that even Big Blue flouted the standard at its peril. When IBM came out with its MCA (Microchannel Architecture) in the late 1980s, the new machines, although possessing some technical advances, did not sell as well as expected. Most people stayed with the existing IBM standard and built upwardly compatible machines upon it.

OFFICIAL STANDARDS

Some standards are developed by official bodies. For example, the International Standards Organization (ISO) has an elaborate formal process where panels of experts develop standards for a huge variety of technologies, including many relating to computing. In an increasingly global economy, international standards allow equipment (or software) from one country to be used with that from another. For example, credit cards, phone cards, and "smart cards" around the world have a common format established by ISO standards. (Standards specific to electrical and electronic engineering are developed by a similar body, the International Electrotechnical Commission, or IEC.) Standards that have become widely accepted but are not yet official ISO standards take the form of Publicly Available Specifications, or PAS. Government contracts often specify ISO standards as well as a variety of other standards developed by various government agencies. The ISO 9001 standards apply specifically to computer systems, software, and its development.

EVOLUTION OF STANDARDS

The extent of standardization within the broad information technology (IT) industry varies widely among applications. Generally, things that have been established for a long time (meaning, in computing terms, a couple decades or so) are likely to be well standardized. An example is the standards for character sets.

For areas in which new applications are emerging, practitioners tend to have less interest (or patience) with the idea of standards. For example, the World Wide Web is still relatively new, and standards for the operation of Web sites are emerging only slowly. In this case, it is mainly concern about such matters as privacy protection that has encouraged the adoption of standards for matters such as the secure transmission of credit card information on-line or privacy policies regarding the use of information obtained from Web users. The potential threat of government regulation often encourages the development of marketplace standards as an alternative.

Technical societies such as the Institute for Electrical and Electronic Engineering (IEEE) and the World Wide Web Consortium are an important forum for the discussion and development of standards.

Further Reading

Dargan, P. A. *Open Systems and Standards for Software Product Development.* Norwood, Mass.: Artech House, 2005.

Hoyle, David. *ISO 9000 Quality Systems Handbook.* 2nd. ed. Burlington, Mass.: Butterworth-Heinemann, 2006.

"ISO IEC 90003 2004 Software Standard Translated into Plain English." Paxiom Research Group. Available online. URL: http://www.praxiom.com/iso-90003.htm. Accessed August 22, 2007.

Lund, Susan K., and John W. Walz. *Practical Support for ISO 9001 Software Project Documentation: Using IEEE Software Engineering Standards.* New York: Wiley, 2006.

statistics and computing

The application of computing technology to the collection and analysis of statistics is as old as computing itself. Indeed,

Charles Babbage was an early proponent of the collection of social and economic statistics in order to understand how society was being changed by the Industrial Revolution in the early 19th century. By the end of that century, Herman Hollerith had come to the rescue of the U.S. Census Bureau by providing his card tabulation machines for the 1890 Census. (See BABBAGE, CHARLES and HOLLERITH, HERMAN.)

In the era of the mainframe, performing statistical analysis with a computer generally required writing a customized program (although the development of FORTRAN around 1960 gradually led the accumulation of an extensive library of subroutines that could be employed to perform statistical functions). Programs generally run in a batch mode, with data supplied from punched cards or tape.

When the personal computer arrived, it wasn't yet powerful enough for much statistical work, although a program such as VisiCalc (see SPREADSHEET) could be used for simple operations. Gradually, spreadsheets grew more powerful, but statisticians truly rejoiced when software packages specifically designed for statistical work began to appear.

Today there are hundreds of statistical packages available, of which the best known one for personal computers is SPSS. Most packages can be used to perform the standard forms of statistical analysis, including analysis of variance, regression analysis, discrete data analysis, time series analysis, and cluster analysis. There are also packages for specialized applications. Moving in the direction of greater generality, mathematical software such as Mathematica and MATLAB can also be used for statistical applications (see MATHEMATICS SOFTWARE). This category of software experiences steady growth because the ability to analyze data quickly and interactively is increasingly important given the growing pace of human activity, whether one is confronted with a rapidly spreading disease or a volatile economy.

Other areas related to statistical computing include the extraction of useful correlations from existing data bases (see DATA MINING) and the development of dynamic models based on probability and statistics (see SIMULATION).

Further Reading
American Statistical Association. Available online. URL: http://www.amstat.org. Accessed August 22, 2007.
Givens, Geof H., and Jennifer A. Hoeting. *Computational Statistics*. New York: Wiley-Interscience, 2005.
Griffith, Arthur. *SPSS for Dummies*. Hoboken, N.J.: Wiley, 2007.
Linnemann, Jim. "Statistical Software Resources on the Web." Available online. URL: http://www.pa.msu.edu/people/linnemann/stat_resources.html. Accessed August 22, 2007.
McKenzie, John, and Robert Goldman, Jr. *The Student Guide to MINITAB Release 14 + MINITAB Student Release 14*. Upper Saddle River, N.J.: Addison-Wesley, 2004.

Stoll, Clifford
(1950–)
American
Astrophysicist, Computer Critic

Until he became famous for tracking down a computer hacker, Clifford Stoll, born on June 4, 1950, in Buffalo, New York, was an astronomer who had received his Ph.D. from the University of Arizona in 1980. (In the 1960s and 1970s Stoll had worked as an engineer at a public radio station in Buffalo.)

In 1986, while working at the Lawrence Berkeley Laboratory as a system administrator, Stoll was asked to track down a 75-cent accounting discrepancy. As he delved into computer files, Stoll discovered that an unknown hacker had penetrated supposedly secure systems housing secret data relating to military technology. Alarmed, Stoll and his colleagues decided against immediately shutting down the intruder's accounts. Instead, they painstakingly traced him, and discovered an even more alarming possibility: that he was using the lab's computers to reach other computers operated by the military and defense contractors. Despite being virtually ignored when reporting his findings to the FBI, Stoll and his impromptu team soldiered on, even planting false data to keep the intruder's interest while continuing to trace his movements. Finally Stoll was able to get the attention of federal authorities. The intrusion was traced to a West German hacker spy ring that was selling secrets to the Soviet KGB.

Stoll's book *Cuckoo's Egg* recounted this adventure in vivid, accessible terms, and made the *New York Times* bestseller list for 16 weeks in 1990. For many readers, this was their first introduction to the vulnerabilities of computer systems.

CYBER-CRITIC
In writing and lectures, Stoll is engaging if sometimes a bit frenetic. He soon turned his iconoclastic attitude toward computers themselves, warning about the dangers of overreliance on them. Stoll's books *Silicon Snake Oil* and *High Tech Heretic* particularly target the use of computers in education. Stoll believes that the technology has been embraced as a panacea for the endemic problem of underperforming schools. However, Stoll notes that the technology is often used for superficial purposes, with little attention to reading and writing skills, while the needs of teachers and students and their vital relationship remain neglected. In turn, advocates of computers in education have criticized Stoll as being superficial and lacking understanding of what good software can really do (see COMPUTERS AND EDUCATION).

In more recent years Stoll has devoted more time to his first love, astronomy. He also has an unusual hobby: making one-sided Klein bottles.

Further Reading
Stoll, Clifford. *The Cuckoo's Egg: Tracking a Spy through the Maze of Computer Espionage*. New York: Doubleday, 1989.
———. *High-Tech Heretic: Why Computers Don't Belong in the Classroom, and Other Reflections by a Computer Contrarian*. New York: Doubleday, 1999.
———. *Silicon Snake Oil: Second Thoughts on the Information Highway*. New York: Doubleday, 1995.
"When Slide Rules Ruled." *Scientific American*, May 2006, pp. 80–87.

streaming
Web users increasingly have access to such content as news broadcasts, songs, and even full-length videos. The problem

is that the user must receive the content in real time at a steady pace, not in sputters or jerks. However, factors such as load on the Web server and network congestion between the server and user can cause delays in transmission. One way to reduce the problem would be to compress the data (see DATA COMPRESSION). However, excessive compression would compromise audio or picture quality to an unacceptable extent. Fortunately, a technology called streaming offers a way to smooth out the transmission of large amounts audio or video content (see also MULTIMEDIA).

When a user clicks on an audio or video link, the player software (or Web browser plug-in) is loaded and the transmission begins. Typically, the player stores a few seconds of the transmission (see BUFFERING), so any momentary delays in the transmission of data packets will not appear as the data starts to play. Assuming the rate of transmission remains sufficient, enough data remains in the buffer so that data can be "fed" to the playing software at a steady pace. If, however, there is too much delay due to network congestion, the playback will pause while the player refills its buffer.

The most popular media players for PCs (such as WinAmp, RealPlayer, and Windows Media Player) provide for streaming data. Despite streaming, connections of fewer than about 56 kbps are likely to result in occasional interruption of content. Together with the use of streaming, the move to faster cable or DSL connections (see BROADBAND) is improving the multimedia experience for Web users. In turn, the ability to easily access video online has fueled video-sharing services (see user-created content and YOU-TUBE). Meanwhile, the growing use of fiber and other high-speed connections into homes is beginning to make "on demand" streaming video services and IPTV (television programming delivered via the Internet) competitive with existing cable and satellite systems.

Further Reading

Follansbee, Joe. *Get Streaming!: Quick Steps to Delivering Audio and Video Online*. Burlington, Mass.: Focal Press, 2004.
"Introduction: How to Create Streaming Video." Media College. Available online. URL: http://www.mediacollege.com/video/streaming/overview.html. Accessed August 22, 2007.
IPTV news. Available online. URL: http://www.iptvnews.net/. Accessed August 22, 2007.
Mack, Steve. *Streaming Media Bible*. New York: Wiley, 2002.
Stolarz, Damien. *Mastering Internet Video: A Guide to Streaming and On-Demand Video*. Upper Saddle River, N.J.: Addison-Wesley Professional, 2004.

Stroustrup, Bjarne
(1950–   )
Danish
Computer Scientist

Bjarne Stroustrup created C++, an object-oriented successor to the popular C language that has now largely supplanted the original language.

Stroustrup was born on December 30, 1950, in Aarhus, Denmark. As a student at the University of Aarhus his inter-

ests were far from limited to computing (indeed, he found programming classes to be rather dull). However, unlike literature and philosophy, programming did offer a practical job skill, and Stroustrup began to do contract programming for Burroughs, an American mainframe computer company. To do this work, Stroustrup had to pay attention to both the needs of application users and the limitations of the machine, on which programs had to be written in assembly language to take optimal advantage of the memory available.

By the time Stroustrup received his master's degree in computer science from the University of Aarhus, he was an experienced programmer, but he soon turned toward the frontiers of computer science. He became interested in distributed computing (writing programs that run on multiple computers at the same time) and developed such programs at the Computing Laboratory at Cambridge University in England, where he earned his Ph.D. in 1979.

The 1970s was an important decade in computing. It saw the rise of a more methodical approach to programming and programming languages (see STRUCTURED PROGRAMMING). It also saw the development of a powerful and versatile new computing environment: the UNIX operating system and C programming language developed by Dennis Ritchie (see RITCHIE, DENNIS) and Ken Thompson and Bell Laboratories. Soon after getting his doctorate, Stroustrup moved to Bell Labs, where he became part of that effort.

As Stroustrup continued to work on distributed computing, he decided that he needed a language that was better than C at working with the various modules running on the different computers. He studied an early object-oriented language (see OBJECT-ORIENTED PROGRAMMING and SIMULA). Simula had a number of key concepts including the organization of a program into classes, entities that combined data structures and associated capabilities (methods). Classes and the objects created from them offered a better way to organize large programs, and was particularly

In the 1980s Bjarne Stroustrup created the object-oriented C++ language that became the most popular language for general applications programming. (BJARNE STROUSTRUP)

suited for distributed computing and parallel programming where there were many separate entities running at the same time.

However, Simula was fairly obscure, and it was unlikely that the large community of systems programmers who were using C would switch to a totally different language. Instead, starting in the early 1980s, Stroustrup decided to add object-oriented features (such as classes with member functions, user-defined operators, and inheritance) to C. At first he gave the language the rather unwieldy name of "C with Classes." However, in 1985 he changed the name to C++. (The ++ is a reference to an operator in C that adds one to its operand, thus C++ is "C with added features.")

At first some critics criticized C++ for retaining most of the non-object oriented features of C (unlike pure object languages such as Smalltalk), while others complained that the overhead required in processing classes made C++ slower than C. During the 1990s, however, C++ became increasingly popular, aided by its relatively smooth learning curve for C programmers and the development or more efficient compilers. C++ is now the most widely used general purpose computer language.

Stroustrup has been honored for his contributions to computer science. In 1993 he received the ACM Grace Hopper Award for his work on C++, and became an AT&T Fellow. After leaving AT&T Stroustrup became a professor holding the College of Engineering Chair in Computer Science at Texas A&M University. In 2004 Stroustrup received the IEEE Computer Society Computer Entrepreneur Award, and in 2005 the William Procter Prize for Scientific Achievement.

Further Reading

Bjarne Stroustrup [home page]. Available online. URL: http://parasol.tamu.edu/people/bs/. Accessed August 22, 2007.

Dolya, Aleksey. "Interview with Bjarne Stroustrup." *Linux Journal,* August 28, 2003. Available online. URL: http://www.linuxjournal.com/article/7099. Accessed August 22, 2007.

Pontin, Jason. "The Problem with Programming: Bjarne Stroustrup, the Inventor of the C++ Programming Language, Defends His Legacy and Examines What's Wrong with Most Software Code." *Technology Review,* November 28, 2006. Available online. URL: http://www.techreview.com/Infotech/17831/page1/. Accessed August 22, 2007.

Stroustrup, Bjarne. *The C++ Programming Language.* Special 3rd ed. Upper Saddle River, N.J.: Addison-Wesley, 1997.

———. *The Design and Evolution of C++.* Reading, Mass.: Addison-Wesley, 1995.

structured programming

As programs grew longer and more complex during the 1960s, computer scientists began to pay more attention to the ways in which programs were organized. Most programming languages had a statement called "GOTO" or its equivalent. This statement transfers control to some arbitrary other point in the program, as identified by a label or line number.

In 1968, computer scientist Edsger Dijkstra (see DIJKSTRA, EDSGER) sent a letter to the editor of the *Proceedings of the ACM* with the title "GO TO Statement Considered Harm-ful." In it he pointed out that the more such jumps programs made from place to place, the harder it was for someone to understand the logic of the program's operation.

The following year, Dijkstra introduced the term *structured programming* to refer to a set of principles for writing well-organized programs that could be more easily shown to be correct. One of these principles is statements such as If . . . Then . . . Else be used to organize a choice between two or more alternatives (see BRANCHING STATEMENTS) and that statements such as While be used to control repetition or iteration of a statement (see LOOP).

Other computer scientists added further principles, such as modularization (breaking down a program into separate procedures, such as for data input, different stages of processing, and output or printing). Modularization makes it easier to figure out which part of a program may be causing a problem, and to fix part of a problem without affecting other parts. A related principle, information hiding, keeps the data used by a procedure "hidden" in that procedure so that it can't be changed from some other part of the program.

Structured programming also encourages stepwise refinement, a program design process described by Niklaus Wirth, creator of Pascal. This is a top-down approach in which the stages of processing are first described in high-level terms (see also PSEUDOCODE), and then gradually fleshed out in their details, much like the writing of an outline for a book.

The principles of structured programming were soon embodied in a new generation of programming languages (see ALGOL, PASCAL, and C). Although use of well-structured language didn't guarantee good structured programming practice, it at least made the tools available.

The ideas of structured programming form a solid basis for programming style today. They have been supplemented rather than replaced by a new paradigm developed in the 1970s and 1980s (see OBJECT-ORIENTED PROGRAMMING).

Further Reading

Dhal, Ole-Johan, Edsger W. Dijkstra, and C. A. R. Hoare, eds. *Structured Programming.* New York: Academic Press, 1972.

Dijkstra, Edsger. *A Discipline of Programming.* Englewood Cliffs, N.J.: Prentice Hall, 1976.

———. "Go To Statement Considered Harmful." *Communications of the ACM* 11, no. 3 (1968): 147–148.

———. "Notes on Structured Programming." Available online. URL: http://www.cs.utexas.edu/users/EWD/ewd02xx/EWD249.PDF. Accessed August 22, 2007.

Orr, Kenneth T. *Structured Systems Development.* New York: Yourdon Press, 1977.

Sun Microsystems

Founded in 1982, Sun Microsystems (NASDAQ symbol: JAVA) has played an important role in the development of computer workstations and servers, UNIX-based operating systems, and the Java programming language (see JAVA, UNIX, and WORKSTATION).

During the 1980s, Sun was known mainly for its workstations for programmers and graphics professionals, running on its own SPARC series microprocessors. However, by the 1990s the growing power of regular desktop PCs was reducing the need for special-purpose workstations. As the Web grew starting in the 1990s, Sun's line of multiprocessing Web servers became quite successful, though the "dot-bust" of the early 2000s cut revenues.

One of Sun's founders was a key developer of UNIX software (see JOY, BILL). Sun developed its own version of UNIX (SunOS) for its workstations in the 1980s, and then joined with AT&T to develop the widely used UNIX System V Release 4, which in turn became the basis for Sun's new operating system, Solaris. (Sun has also supported the use of Linux on its hardware.)

Sun's biggest impact on software development, however, has been its development of the Java language and platform since the early 1990s. Although newer languages such as Python, PHP, and Ruby have come along to challenge it, Java, with its ability to run via "virtual machines" on all major platforms, is widely used and has a rich set of library routines and programming frameworks.

Scott McNealy, one of the company's founders, remains its chairman. Sun had $13.87 billion revenue in 2007 ($473 million net income), and employs about 36,400 people.

Further Reading

Boyous, Jon. "Java Technology: The Early Years." Sun Developer Network. Available online. URL: http://java.sun.com/features/1998/05/birthday.html. Accessed November 18, 2007.

Southwick, Karen. *High Noon: The Inside Story of Scott McNealy and the Rise of Sun Microsystems.* New York: Wiley, 1999.

Sun Microsystems. Available online. URL: http://www.sun.com/. Accessed November 18, 2007.

Sun Multimedia Center [videos]. Available online. URL: http://sunfeedroom.sun.com. Accessed November 18, 2007.

Sun Wikis. Available online. URL: http://wikis.sun.com. Accessed November 18, 2007.

supercomputer

The term *supercomputer* is not really an absolute term describing a unique type of computer. Rather, it has been used through successive generations of computer design to describe the fastest, most powerful computers available at a given time. However, what makes these machines the fastest is usually their adoption of a new technology or computer architecture that later finds its way into standard computers.

The first supercomputer is generally considered to be the Control Data CDC 6600, designed by Seymour Cray in 1964. The speed of this machine came from its use of the new, faster silicon (rather than germanium) transistors and its ability to run at a clock speed of 10 MHz (a speed that would be achieved by personal computers by the mid-1980s). Even with transistors, these machines generated so much heat that they had to be cooled by a Freon-based refrigeration system.

Cray then left CDC to form Cray Research. He designed the Cray 1 in 1976, the first of a highly successful series of supercomputers. The Cray 1 took advantage of a new technology, integrated circuits, and new architecture: vector processing, in which a single instruction can be applied

A Cray 190 A supercomputer. Seymour Cray's leading-edge machines defined supercomputing for many years. (NASA PHOTO)

to an entire series (or array) of numbers simultaneously. This innovation marked the use of parallel processing as one of the distinguishing features of supercomputers. The machine's monolithic appearance gave it a definite air of science fiction, and the first one built was installed at the secretive Los Alamos National Laboratory.

The next generation, the Cray X-MP, carried parallelism further by incorporating multiple processors (the successor, Cray Y-MP, had 8 processors, which together could perform a billion floating-point operations per second [1 gigaflop]).

Soon Cray no longer had the supercomputer field to itself, and other companies (particularly the Japanese manufacturers NEC and Fujitsu) entered the market. The number of processors in supercomputers increased to as many as 1,024 (in the 1998 Cray SV1), which can exceed 1 *trillion* floating-point operations per second (1 teraflop).

Meanwhile, processors for desktop computers (such as the Intel Pentium) also continued to increase in power, and it became possible to build supercomputers by combining large numbers of these readily available (and relatively low-cost) processors.

The ultimate in multiprocessing is the series of Connection Machines built by Thinking Machines Inc. (TMI) and designed by Daniel Hillis. These machines have up to 65,000 very simple processors that run simultaneously, and can form connections dynamically, somewhat like the process in the human brain. These "massively parallel" machines are thus attractive for artificial intelligence research. It is also possible to achieve supercomputerlike power by having many computers on a network divide the work of, for example, cracking a code or analyzing radio telescope data for signs of intelligent signals.

Programs for supercomputers must be written using special languages (or libraries for standard languages) that are designed to provide for many processes to run at the same time and that allow for communication and coordination between processing (see MULTIPROCESSING).

APPLICATIONS

Supercomputers are always more expensive and somewhat less reliable than standard computers, so they are used only when necessary. As the power of standard computers continues to grow, applications that formerly required a multimillion-dollar supercomputer can now run on a desktop workstation (a good example is the creation of detailed 3D graphics).

On the other hand, there are always applications that will soak up whatever computing power can be brought to bear on them. These include analysis of new aircraft designs, weather and climate models, the study of nuclear reactions, and the creation of models for the synthesis of proteins. The never-ending battle of organizations such as the National Security Agency (NSA) to monitor worldwide communications and crack ever-tougher encryption also demands the fastest available supercomputers (see QUANTUM COMPUTING).

ARCHITECTURE

The fastest "conventional" supercomputers as of 2007 were IBM's Blue Gene series, expected to reach a speed of 3 pflop (peta, or quadrillion floating point operations per second). Machines of this magnitude are usually destined for institutions such as the Los Alamos National Laboratory (see GOVERNMENT FUNDING OF COMPUTER RESEARCH).

However, for many applications it may be more cost-effective to build systems with numerous coordinated processors (a sort of successor to the 1980s Connection Machine). For example, the Beowulf architecture involves "clusters" of ordinary PCs coordinated by software running on UNIX or Linux. The use of free software and commodity PCs can make this approach attractive, though application software still has to be rewritten to run on the distributed processors.

Recently a new resource for parallel supercomputing came from an unlikely place: the new generation of cell processors found in game consoles such as the Sony Playstation 3. This architecture features tight integration of a central "power processor element" with multiple "synergistic processing elements." IBM is currently developing a new supercomputer called Roadrunner that will include 16,000 conventional (Opteron) and 16,000 cell processors, and is expected to reach a speed of 1 pflop.

Finally, an ad hoc "supercomputer" can be created almost for free, using software that parcels out calculation tasks to thousands of computers participating via the Internet, as with SETI@Home (searching for extraterrestrial radio signals) and Folding@Home (for protein-folding analysis). (See COOPERATIVE PROCESSING.)

Further Reading
"Blue Gene." IBM. Available online. URL: http://www-03.ibm.com/servers/deepcomputing/bluegene.html. Accessed August 22, 2007.

Gropp, William, Ewing Lusk, and Thomas Sterling. *Beowulf Cluster Computing with Linux*. 2nd ed. Cambridge, Mass.: MIT Press, 2003.

"IBM to Build World's First Cell Broadband Engine Based Supercomputer." September 6, 2006. Available online. URL: http://www-03.ibm.com/press/us/en/pressrelease/20210.wss. Accessed August 22, 2007.

Murray, C. J. *The Supermen: The Story of Seymour Cray and the Technical Wizards behind the Supercomputer*. New York: Wiley, 1995.

National Center for Supercomputing Applications (NCSA). Available online. URL: http://www.ncsa.uiuc.edu/. Accessed August 22, 2007.

National Research Council. *Getting Up to Speed: The Future of Supercomputing*. Washington, D.C.: National Academies Press, 2005.

Scientific American. *Understanding Supercomputing*. New York: Warner Books, 2002.

Top 500 Supercomputer Sites. Available online. URL: http://www.top500.org/. Accessed August 22, 2007.

supply chain management

Few consumers are aware of the complexity of the network of organizations, transportation and storage facilities, and information processing facilities that are needed to turn raw materials into finished products. The term *supply chain management* was developed in the 1980s to refer to the systematic efforts to improve the efficiency and reliability of this vital business activity. Although the details will vary with the industry, a supply chain can include the following activities:

- obtaining the raw materials or components needed for the product

- manufacturing finished products

- marketing the product

- distributing the product to retailers or other outlets

- servicing the product and supporting customers

- (increasingly) providing for the ultimate recycling or disposal of the product

At all stages of the chain, planners must take into consideration what location for operations is most advantageous and how materials will be transported, warehoused, and tracked. Potential suppliers must be evaluated for cost and reliability. Schedules must be monitored. Finally, everything should be part of a comprehensive plan that spells out the objectives and how they will be measured.

SOFTWARE

Of course such a complex process involving a great deal of information, monitoring, and decision making is ripe for software assistance. Some companies offer comprehensive solutions (see, for example, SAP), but they must still be adapted to the needs of a particular industry and manufacturer. Software must be interfaced and integrated with existing databases, management information systems, and other software. Nevertheless, in a very competitive world market, enterprises have little choice but to develop an effective way to manage and optimize their supply chains.

Further Reading
Blanchard, David. *Supply Chain Management Best Practices*. Hoboken, N.J.: Wiley, 2007.

Chopra, Sunil, and Peter Meindl. *Supply Chain Management*. 3rd ed. Upper Saddle River, N.J.: Prentice Hall, 2006.

Simchi, David, Philip Kaminsky, and Edith Simchi-Levi. *Designing and Managing the Supply Chain.* 2nd ed. New York: McGraw-Hill, 2002.

Worthen, Ben. "ABC: An Introduction to Supply Chain Management." *CIO.* Available online. URL: http://www.cio.com/article/40940. Accessed November 18, 2007.

Sutherland, Ivan Edward

(1938–   )
American
Computer Scientist

Today it is hard to think about computers without interactive graphics displays. Whether one is flying a simulated 747 jet, retouching a photo, or just moving files from one folder to another, everything is shown on the screen in graphical form. For the first two decades of the computer's history, however, computers lived in a text-only world, except for a few experimental military systems. During the 1960s and 1970s Ivan Sutherland would almost single-handedly create the framework for modern computer graphics while designing Sketchpad, the first computer drawing program.

Sutherland was born on May 16, 1938, in Hastings, Nebraska, but the family later moved to Scarsdale, New York. His father was a civil engineer, and as a young boy Sutherland was fascinated by the drawing and surveying instruments his father used. When he was about 12, Sutherland and his brother Bert got a job working for a pioneer computer scientist named Edmund Berkeley. Berkeley gave Sutherland the opportunity to play with "Simon," a suitcase-sized electromechanical computer that could add numbers as long as the total did not exceed 30. Simon eventually rewired the machine so it could divide numbers as well.

Sutherland first attended Carnegie Mellon University, where he received a B.S. in electrical engineering in 1959. The following year he earned an M.A. from the California Institute of Technology (Cal Tech). He then went to MIT to do his doctoral work under Claude Shannon at the Lincoln Laboratory (see SHANNON, CLAUDE).

At MIT Sutherland was able to work with the TX-2, an advanced (and very large) transistorized computer that was a harbinger of the minicomputers that would become prevalent later in the decade. Unlike the older mainframes, the TX-2 had a graphics display and could accept input from a light pen as well as switches that could serve something like the functions that mouse buttons do today. The machine also had 70,000 36-bit words of memory, an amount that would not be achieved by personal computers until the 1980s. Having this much memory made it possible to store the pixel information for detailed graphics objects.

Having access to this interactive machine gave Sutherland the idea for his doctoral dissertation (submitted in 1963). He developed a program called Sketchpad, which required that he develop algorithms for drawing realistic objects by plotting pixels and polygons as well as scaling objects in relation to the viewer's position. Sutherland's Sketchpad could even automatically "snap" lines into place as the user drew on the screen with the light pen. Besides drawing, Sketchpad demonstrated the beginnings of the "graphical user interface" that would be further developed by researchers at Xerox PARC in the 1970s and would reach the consumer in the 1980s.

After demonstrating Sktechpad in 1963 and receiving his Ph.D. from MIT, Sutherland took on a quite different task. He became the director of the Information Processing Techniques Office (IPTO) of the Defense Department's Advanced Research Projects Agency (ARPA)—see LICKLIDER, J. C. R. While continuing his research on graphics Sutherland thus also oversaw the work on computer time-sharing and the networking research that would eventually lead to the ARPANet and the Internet.

In 1968 Sutherland and David Evans went to the University of Utah, where they established an Information Processing Technology Office (IPTO)–funded computer graphics research program. There, Sutherland's group brought computer graphics to a new level of realism. For example, they developed the ability to place objects in front of other objects, which required intensive calculations to determine what was obscured. They also developed an idea suggested by Evans called *incremental computing.* Instead of drawing each pixel in isolation, they used information from previously drawn pixels to calculate new ones, considerably speeding up the rendering of graphics. The results began to approach the realism of a photograph. (The two researchers also founded a commercial enterprise, Evans and Sutherland, to exploit their graphics ideas. It became one of the leaders in the field.)

In 1976 Sutherland left the University of Utah to serve as the chairman of the computer science department at Cal Tech. Working with a colleague, Carver Mead, Sutherland developed a systematic concept and curriculum for integrated circuit design, which became the main specialty of the department. He would later point out that it was the important role that geometry played in laying out components and wires that had intrigued him the most.

Sutherland left Caltech in 1980 and started a consulting and venture capital firm with Bob Sproull, whom he had met years earlier at Harvard. In 1990 Sun Microsystems bought the company for its technical expertise, making it the core of Sun Labs, where Sutherland continues to work as a Sun Microsystems Fellow and vice president. Sutherland received the prestigious ACM Turing Award in 1988.

Further Reading

"An Evening with Ivan Sutherland: Research and Fun" [partial transcript and online video]. Computer History Museum, October 19, 2005. Available online. URL: http://www.mprove.de/script/05/sutherland/index.html. Accessed November 18, 2007.

Frenkel, Karen A. "Ivan E. Sutherland, 1988 A. M. Turing Award Recipient" [Interview]. *Communications of the ACM* 32 (June 1989): 711.

Sutherland, Ivan E. "Sketchpad—A Man-Machine Graphical Communication System." University of Cambridge Computer Laboratory. Technical Report No. 574 [with new preface], September 2003. Available online. URL: http://www.cl.cam.ac.uk/TechReports/UCAM-CL-TR-574.pdf. Accessed November 18, 2007.

system administrator

A system administrator is the person responsible for managing the operations of a computer facility to ensure that it runs properly, meets user needs, and protects the integrity of users' data. Such facilities range from offices with just a few users to large campus or corporate facilities that may be served by a large staff of administrators.

The system administrator's responsibilities often include:

- setting up accounts for new users

- allocating computing resources (such as server space) among users

- configuring the file, database, or local area network (LAN) servers

- installing new or upgraded software on users' workstations

- keeping up with new versions of the operating system and networking software

- using various tools to monitor the performance of the system and to identify potential problems such as device "bottlenecks" or a shortage of disk space

- ensuring that regular backups are made

- configuring network services such as e-mail, Internet access, and the intranet (local TCP/IP network)

- using tools such as firewalls and virus scanners to protect the system from viruses, hacker attacks, and other security threats (see also COMPUTER CRIME AND SECURITY)

- providing user orientation and training

- creating and documenting policies and procedures

System administrators often write scripts to automate many of the above tasks (see SCRIPTING LANGUAGES). Because of the complexity of modern computing environments, an administrator usually specializes in a particular operating system such as UNIX or Windows.

A good system administrator needs not only technical understanding of the many components of the system, but also the ability to communicate well with users—good "people skills." Larger organizations are more likely to have separate network and database administrators, while the administrator of a small facility must be a jack (or jill) of all trades.

Further Reading

Culp, Brian. *Windows Vista Administration: The Definitive Guide.* Sebastapol, Calif.: O'Reilly Media, 2007.
Frisch, Æleen. *Essential System Administration.* 3rd ed. Sebastapol, Calif.: O'Reilly Media, 2002.
———. *Essential Windows NT System Administration.* Sebastapol, Calif.: O'Reilly Media, 1998.
Information for Linux System Administration (Librenix). Available online. URL: http://librenix.com/. Accessed August 22, 2007.
Limoncelli, Thomas A., Christina J. Horgan, and Strata R. Chalup. *The Practice of System and Network Administration.* 2nd ed. Upper Saddle River, N.J.: Addison-Wesley Professional, 2007.

Nemeth, Evi, Garth Snyder, and Trent R. Hein. *Linux System Administration Handbook.* 2nd ed. Upper Saddle River, N.J.: Prentice Hall PTR, 2006.

systems analyst

The systems analyst serves as the bridge between the needs of the user and the capabilities of the computer system. The systems analyst goes into action when users request that some new application or function be provided (usually in a corporate computing environment).

The first step is to define the user's requirements and to prepare precise specifications for the program. In doing so, the systems analyst is aided by methodologies developed by computer scientists over the last several decades (see STRUCTURED PROGRAMMING and OBJECT-ORIENTED PROGRAMMING). Often flowcharts or other aids are used to help visualize the operation of the program (see also CASE).

After communicating with the user, the systems analyst must then communicate with the programmers, helping them understand what is needed and reviewing their work as they begin to design the program. Although the systems analyst may do little actual programming, he or she must be familiar with programming tools and practices. This may make it possible to suggest existing software or components that could be adapted instead of undertaking the cost and time involved with creating a new program. As a program is developed, systems analysts are often responsible for designing tests to ensure that the software works properly (see QUALITY ASSURANCE, SOFTWARE).

Depending on the organizational structure, all or part of the analysis function may be included in the job description "programmer-analyst" or included as part of the duties of a senior software engineer or manager of program development. Experienced systems analysts are likely to be called upon to participate in the evaluation of possible investments in new software or hardware, and other aspects of long-term planning for computing facilities.

Further Reading

Satzinger, John W., Robert B. Jackson, and Stephen D. Burd. *Systems Analysis & Design in a Changing World.* 4th ed. Boston: Course Technology, 2006.
Shelly, Gary B., Thomas J. Cashman, and Harry J. Rosenblatt. *Systems Analysis & Design.* 7th ed. Boston: Course Technology, 2007.
Systems Analysis Web Sites. Available online. URL: http://www.umsl.edu/~sauterv/analysis/analysis_links.html. Accessed August 22, 2007.
Whitten, Jeffrey L., and Lonnie D. Bentley. *Introduction to Systems Analysis & Design.* New York: McGraw-Hill/Irwin, 2007.
———. *Systems Analysis & Design Methods.* 7th ed. New York: McGraw-Hill/Irwin, 2005.

systems programming

Applications programmers write programs to help users work better, while systems programmers write programs to help the computer itself work better (see OPERATING SYSTEM). Systems programmers generally work for companies in the computer industry that develop operating systems, network facilities, program language compilers and other software development tools, utilities, and device drivers. However, systems programmers can also work for applications developers to help them interface their programs to the operating system or to devices (see DEVICE DRIVER and APPLICATIONS PROGRAMMING INTERFACE).

Modern operating systems are highly complex, so systems programmers tend to specialize in particular areas. These might include device drivers, software development tools, program language libraries, applications programming interfaces (APIs), and utilities for monitoring system conditions and resources. Systems programmers develop the infrastructure needed for networking, as well as multiple-processor computers and distributed computing systems. Systems programmers also play a key role when an application program must be "ported" to a different platform or simply modified to run under a new version of the operating system.

Generally, an application programmer works at a fairly high level, using language functions and APIs to have the program ask the operating system for services such as loading or saving files, printing, and so on. The systems programmer, on the other hand, must be concerned with the internal architecture of the system (such as the buffers allocated to hold various kinds of temporary data) and with how commands are constructed for disks and other devices. Generally, the systems programmer must also have a more thorough knowledge of data structures and how they are physically represented in the machine as well as the comparative efficiency of various algorithms. Because it determines how efficiently the system's resources can be used, systems programming must often be "tight" and optimized for peak performance. Thus, although lower-level assembly language is no longer used for much applications programming, it can still be found in systems programming.

Further Reading

Beck, Leland L. *System Software: An Introduction to Systems Programming.* 3rd ed. Reading, Mass.: Addison-Wesley, 1996.

Hart, Johnson M. *Windows System Programming.* 3rd ed. Upper Saddle River, N.J.: Addison-Wesley Professional, 2004.

Love, Robert. *Linux System Programming.* Sebastapol, Calif.: O'Reilly Media, 2007.

Robbins, Kay A., and Steven Robbins. *UNIX Systems Programming.* Upper Saddle River, N.J.: Prentice Hall PTR, 2003.

T

tablet PC

As the name suggests, a tablet PC is a small computer about the size of a notebook (not to be confused with a "notebook PC," which is a small, light laptop). The user can write on the screen with a stylus to take notes (for similar functionality, see GRAPHICS TABLET), draw, and make selections with stylus or fingertip.

If the user writes on the screen, software converts the writing to the appropriate characters and stores them in a file (see HANDWRITING RECOGNITION). As with some PDAs, there may also be a system of shorthand "gestures" that can be used to write more quickly. Alternatively, the user can type with stylus or fingertips on a "virtual keyboard" displayed on the screen (see TOUCHSCREEN).

A more versatile and natural interface is becoming available: "multitouch," pioneered by the Apple iPhone and Microsoft Surface, can recognize multiple motions and pressure points simultaneously. This allows the user to, for example, flick the finger to "turn a page" or use a pinching motion to "pick up" an object.

Applications for tablet PCs include many PDA-type applications (see PERSONAL INFORMATION MANAGEMENT and PDA), field note taking, inventory, and other tasks that require a device that is not encumbering. Because of its compactness, a tablet PC can also be a good reader for e-books (see E-BOOKS AND DIGITAL LIBRARIES).

Tablet PCs generally follow common specifications developed by Microsoft, and often use Windows XP Tablet PC Edition or, later, Windows Vista, which has built-in support for tablet PCs. These operating systems include support for sophisticated handwriting recognition that can be "trained" by the user and that can store handwritten input in special data formats. Voice recognition is also supported.

A "convertible" tablet PC is a hybrid in which the tablet is attached to a base containing a keyboard. The display can be used vertically (laptop style) or rotated and folded down over the keyboard for tablet use.

INTERNET TABLETS

An interesting variant is the Internet tablet, best known in Nokia's N-series. These are smaller and lighter than a tablet PC. The Nokia N810, for example, has a slide-out keyboard as well as a virtual screen keyboard. The most notable feature is the Internet browser and related applications, such as e-mail and instant messaging, and built-in wireless connections (see BLUETOOTH and WIRELESS COMPUTING). Although there is no phone, Internet-based services such as Skype can be used to place calls, or a Bluetooth-equipped mobile phone. The Nokia series uses a variant of Linux and can run a large variety of open-source applications.

Further Reading

Linenberger, Michael. *Seize the Work Day: Using the Tablet PC to Take Control of Your Work and Meeting Day.* San Ramon, Calif.: New Academy Publishers, 2004.

Stevenson, Nancy. *Tablet PCs for Dummies.* New York: Wiley, 2003.

Tablet PC Review. Available online. URL: http://www.tabletpcreview.com/. Accessed November 19, 2007.

"What Is a Tablet PC?" Microsoft, February 9, 2005. Available online. URL: http://www.microsoft.com/windowsxp/tabletpc/evaluation/about.mspx. Accessed November 19, 2007.

tape drives

Anyone who has seen computers in old movies is familiar with the row of large, freestanding tape cabinets with their spinning reels of tape. The visual cue that the computer was running consisted of the reels thrashing back and forth vigorously while rows of lights flashed on the computer console. Magnetic tape was indeed the mainstay for data storage in most large computers (see MAINFRAME) in the 1950s through the 1970s.

In early mainframes the main memory (corresponding to today's RAM chips) consisted of "core"—thousands of tiny magnetized rings crisscrossed with wires by which they could be set or read. Because core memory was limited to a few thousand bytes (kB), it was used only to hold the program instructions (see PUNCHED CARDS AND PAPER TAPE) and to store temporary working data while the program was running.

The source data to be processed by the program was read from a reel of tape on the drive. If the program updated the data (rather than just reporting on it), it would generally write a new tape with the revised data. In large facilities a person called a tape librarian was in charge of keeping the reels of tape organized and providing them to the computer operators as needed.

OPERATION

A mainframe tape drive had two reels, the supply reel and the take-up reel. Because each reel had its own motor, they could be spun at different speeds. This allowed a specified length of tape to be suspended between the two reels, serving as sort of a buffer and allowing the take-up reel to accelerate at the start of a read or write operation without danger of breaking the tape. The "buffer" tape was actually suspended in a partial vacuum, which both kept the tape taut enough to prevent snarling and allowed for air pressure sensors to activate the appropriate motor when the amount of tape in the buffer went above or below preset points.

Data was read or written by the read and write heads respectively, in units called frames. In addition to the 1 or 0 data bits, each frame included parity bits (see ERROR CORRECTION). The frames were combined into blocks, with each block having a header in front of the data frames and one or more frames of check (parity) bits following the data.

The two predominant tape formats were the IBM format, which used variable-length data blocks (and thus could not be rewritten) and the DEC format, which used fixed-length blocks, allowing data to be rewritten in place, albeit at some cost in speed and efficiency.

During the 1960s, magnetic disks (see HARD DISK) increasingly came into use, and more of the temporary data being used by programs began to be stored on disk rather than on tape. Eventually, tapes were relegated to storing very large data sets or archiving old data.

However, when the first desktop microcomputers (such as the Apple II and Radio Shack TRS-80) came along in the late 1970s and early 1980s, they, like the first mainframes, had very limited main memory and disk drives were unavailable or expensive. As a result, programs (such

A NASA automated tape library. These facilities can store trillions of bytes of data. (NASA PHOTO)

as Bill Gates's Microsoft Basic) often came on tape cassettes, and the computer included an interface allowing it to be connected to an ordinary audio cassette recorder. However, this use of tapes was quite short-lived, and was soon replaced by the floppy disk drive and later, hard drives and CD-ROM drives.

TAPES AS BACKUP DEVICES

By the 1990s, PC users generally used tapes only for making backups. A typical backup tape drive uses DAT (digital audio tape) cartridges that hold from hundreds of megabytes to several gigabytes of data. Most drives use a rotating assembly of four heads (two read and two write) that verify data as it's being written. As a backup medium, tape has a lower cost per gigabyte than disk devices. It is easy to use and can be set up to run unattended (except for periodically changing cartridges).

However, since tapes are written and read sequentially, they are not convenient for restoring selected files (see BACKUP AND ARCHIVE SYSTEMS). Many smaller installations now prefer using a second ("mirror") hard drive as backup, using disk arrays (see RAID) or using recordable CDs or optical drives for smaller amounts of data (see CD-ROM AND DVD-ROM).

Many large companies and government agencies have thousands of reels of tape stored away in their vaults since the 1960s, including data returned from early NASA space missions. As time passes, it becomes increasingly difficult to guarantee that this archived data can be successfully read. This is due both to gradual deterioration of the medium and the older data formats becoming obsolete (see BACKUP AND ARCHIVE SYSTEMS).

Further Reading

Brain, Marshall. "How Tape Drives Work." Available online. URL: http://electronics.howstuffworks.com/cassette.htm. Accessed August 22, 2007.

Haylor, Phil. *Computer Storage: A Manager's Guide.* Victoria, B.C., Canada: Trafford, 2005.

Scapicchio, Mark. "How Tape Drives Work: Tape Backup Still a Good Option." *Smart Computing,* October 2002, pp. 69–72. Available online. URL: http://www.smartcomputing.com/editorial/article.asp?article=articles/archive/r0608/13 r08/13r08.asp&guid=. Accessed August 22, 2007.

White, Ron, and Timothy Edward Downs. *How Computers Work.* 8th ed. Indianapolis: Que, 2005.

Tcl

Developed by John Ousterhout in 1988, Tcl (Tool command language) is used for scripting, prototyping, testing interfaces, and embedding in applications (see SCRIPTING LANGUAGES).

Tcl has an unusually simple and consistent syntax. A script is simply a series of commands (either built in or user defined) and their arguments (parameters). A command itself can be an argument to another command, creating the equivalent of a function call in other languages.

For example, setting the value of a variable uses the *set* command:

```
set total 0
```

The value of the variable can now be referenced as $total.

Control structures are simply commands that run other commands. A *while* loop, for example, consists of a command or expression that performs a comparison, followed by a series of commands to be executed each time it returns "true":

```
while { MoreInFile } {
   GetData
   DisplayData
}
```

In practice, many of the commands used are utilities from the operating system, usually UNIX or Linux. Tcl also includes a number of useful data structures such as associative arrays, which consist of pairs of data items such as:

```
set abbr (California)    CA
```

EXTENSIONS AND APPLICATIONS

Tcl includes a number of extensions that, for example, provide access to popular database formats such as MySQL and can interface with other programming languages such as C++ and Java. The most widely used extension is Tk, which provides a library for creating user interfaces for a variety of operating systems and languages such as Perl, Python, and Ruby.

Tcl has been described as a "glue" to connect existing applications. It is relatively easy to write and test a script interactively (often at the command line), and then insert it into the code of an application. When the application runs, the Tcl interpreter runs the script, whose output can then be used by the main application (see INTERPRETER).

Further Reading

Foster-Johnson, Eric. *Graphical Applications with Tcl & Tk.* 2nd ed. New York: Hungry Minds, 1997.

Wall, Kurt. *Tcl and Tk Programming for the Absolute Beginner.* Boston: Course Technology, 2007.

Welch, Brent B., and Ken Jones. *Practical Programming in Tcl and Tk.* 4th ed. Upper Saddle River, N.J.: Prentice Hall, 2003.

TCP/IP

Contrary to popular perception, the Internet is not e-mail, chat rooms, or even the World Wide Web. It is a system by which computers connected to various kinds of networks and with different kinds of hardware can exchange data according to agreed rules, or protocols. All the applications mentioned (and many others) then use this infrastructure to communicate.

TCP/IP (Transmission Control Protocol/Internet Protocol) provides the rules for transmitting data on the Internet. It consists of two parts. The IP (Internet Protocol) routes packets of data. The header information also includes:

- The total length of the packet. In theory packets can be as large as 65 kbytes; in practice they are limited to a smaller maximum.

- An identification number that can be used if a packet is broken into smaller pieces for efficiency in transmission. This allows the packet to be reassembled at the destination.

- A "time to live" value that specifies how many hops (movements from one intermediate host to another) the packet will be allowed to take. This is reduced by 1 for each hop. If it reaches 0, the packet is assumed to have gotten "lost" or stale, and is discarded.

- A protocol number (the protocol is usually TCP, see below).

- A checksum for checking the integrity of the header itself (not the data in the packet).

- The source and destination addresses.

The source and destination are given as IP addresses, which are 32 bits long and typically written as four sets of up to three numbers each—for example, 208.162.106.17

A NETWORK OF NETWORKS

As the name implies, the Internet is a network that connects many local networks. The IP address includes an ID for each network (called a subnet) and each host computer on the network. The arrangement and meaning of these fields differs somewhat among five classes of IP addresses. The first three classes are designed for different sizes of networks, and the latter two are used for special purposes such as "multicasting" where the same data packet is sent to multiple hosts.

Many Internet users (at home as well as in offices) are part of a local network (see LOCAL AREA NETWORK). Typically, all users on the local network share a single Internet connection, such as a DSL or cable line. This sharing

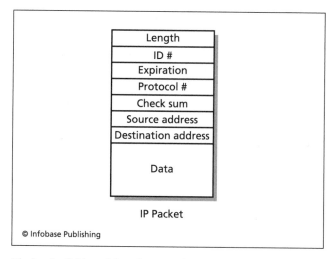

Length	
ID #	
Expiration	
Protocol #	
Check sum	
Source address	
Destination address	
Data	

IP Packet

© Infobase Publishing

The header fields and data for an IP (Internet Protocol) packet. The packets can travel over different routes to the destination address and then be reassembled in the correct order.

is enabled by having one computer (or a hardware device called a router) connected to the Internet, serving as the link between the local network and the rest of the world. A facility called Network Address Translation (NAT) assigns a private IP address to each computer on the network. When a computer wants to make an Internet connection, its outgoing packet is assigned a public IP address from a pool. When packets replying to that public address are received, they are converted back to the private address and thus routed to the appropriate user.

NAT has the benefit of providing some security against intrusion, since from the outside only the single public IP address is visible, not the private addresses of the various

machines on the network. However, using NAT (and a similar scheme called PAT that allows different hosts to use the same IP address by being assigned different port numbers) causes some slowdown because of the translation process.

Another facility, Dynamic Host Configuration Protocol (DHCP), is used to assign an arbitrary available public IP address to each host when it connects to the network. This system is now used by most DSL and cable systems, and it reduces the danger of running out of IP numbers (each network is assigned a range of numbers, and is thus limited to that many IP addresses).

A more lasting and flexible solution to address depletion is the new Internet Protocol version 6 (IPv6). The new addressing scheme allows for about 3×10^{38} addresses, more than enough to accommodate every star in the known universe if it were a networked computer! The scheme is being rolled out gradually but should be well established by the end of the 2000 decade.

DOMAIN NAME SYSTEM

Internet users typically don't have to worry about IP numbers, except perhaps when configuring their software. Instead they use alphabetic addresses, such as http://www.factsonfile.com. The Domain Name System (DNS) sets up a correspondence between the names (which include domains such as .com for commercial or .edu for educational institutions) and the IP numbers (see DNS).

TRANSMISSION CONTROL PROTOCOL

The Transmission Control Protocol (the TCP part of TCP/IP) controls the flow of packets that have been structured as described above. To use TCP, the sending computer opens a special file called a socket, which is identified by the computer's IP number plus a port number. Standard port numbers are used for the various protocols such as www (Web) and ftp (File Transfer Protocol). The receiving computer connects using a corresponding socket. TCP includes basic flow control and error-checking features similar to those used for most data transmissions. For some applications (such as connecting to the domain name server) error control is not needed, so a simpler protocol called the User Datagram Protocol is used.

THE BIG PICTURE

How does TCP/IP fit into the use of the Internet? When an application such as an e-mail program, Web browser, or ftp client makes a connection, IP packets using TCP flow control carry the requests from the client to the server and the server's response back to the client. Each application has its own protocol to specify these requests (such as for a Web page). For e-mail the protocol is SMTP (Simple Mail Transfer Protocol); Web servers and browsers use HTTP (Hypertext Transfer Protocol); and for file transfers it is FTP (File Transfer Protocol). (See also E-MAIL, HTML, HYPERTEXT AND HYPERMEDIA, and FILE TRANSFER PROTOCOLS.)

Further Reading
Hagen, Silvia. *IPv6 Essentials.* 2nd ed. Sebastapol, Calif.: O'Reilly, 2006.

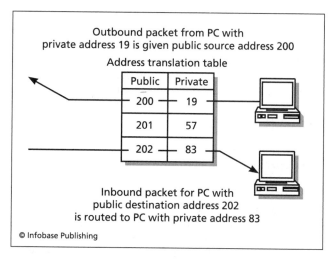

Outbound packet from PC with private address 19 is given public source address 200

Address translation table

Public	Private
200	19
201	57
202	83

Inbound packet for PC with public destination address 202 is routed to PC with private address 83

© Infobase Publishing

Network Address Translation (NAT) can protect computers on a local network by giving outgoing packets an arbitrary public source address in place of a computer's actual (private) address. Incoming packets addressed to that public address are then routed to the correct private address using a table.

IPv6 Information Page. Available online. URL: http://www.ipv6.org/. Accessed August 22, 2007.

Kozierok, Charles M. *The TCP/IP Guide: A Comprehensive, Illustrated Internet Protocols Reference.* San Francisco, Calif.: No Starch Press, 2005.

Leiden, Candace, and Marshall Wilensky. *TCP/IP for Dummies.* 5th ed. New York: Wiley, 2003.

Raz, Uri. "Uri's TCP/IP Resources List." Available online. URL: http://www.private.org.il/tcpip_rl.html. Accessed August 22, 2007.

technical support

Competition and user demand have led to modern software becoming increasingly complex and often stuffed with esoteric features. Despite improvement in programs' own built-in help systems (see HELP SYSTEMS), users will often have questions about how to perform particular tasks. There will also be times when a program doesn't perform as the user expects because the user misunderstands some feature of the program, the program has an internal flaw (see *bugs and debugging), or there is a problem in interaction between the application program, the user's operating system, or the user's hardware (see DEVICE DRIVER).*

To get help when problems arise, users often turn to the technical support facility, often called a help desk. This facility can either be internal to an organization (helping the organization's computer users with a wide range of problems), or belong to the maker of the software (and available in varying degrees to all licensed users of that software).

Large help desks often have two or more levels or tiers of assistants. The first tier assistant can respond to the simplest (and usually most common) situations. For example, a first-tier support person for a cable or DSL Internet Service Provider could tell a caller whether service has been interrupted in their area and if not, take the caller through a set of steps to reset a "hung" modem. If the situation is more complex (or the basic steps do not resolve it), the call will be "escalated" to the next tier, where a more experienced technician can address detailed software configuration issues.

Advanced technical support representatives can use tools such as remote operation software that lets them take over control of the user's PC in order to see exactly what is going on. They can also submit detailed problem reports to engineers in cases were a modification (patch) to the software might be needed.

SUPPORT ALTERNATIVES

Users who are dissatisfied with the wait for phone support or dealing with poorly trained support personnel may be able to take advantage of alternative sources of information and support. Most software companies now have Web sites that include a support section that offers services such as

- Frequently Asked Questions (FAQ) files with answers to common problems.

- A searchable "knowledge base" of articles relating to various aspects of the software, including compat-

ibility with other products, operating system issues, and so on.

- Forms or e-mail links that can be used to submit questions to the company. Typically questions are answered in one or two working days.

- A bulletin board where users can share solutions and tips relating to the software.

Web sites for publications such as *PC Magazine* and ZDNet also offer articles and other resources for working with the various versions of Microsoft Windows and popular applications.

TECHNICAL SUPPORT ISSUES

As with many other aspects of the computer industry, the changing economic climate has had an impact on technical support practices. Many companies are hoping that providing more extensive Web-based technical support will reduce the need for help desk representatives. Companies that don't want to create their own support Web sites can turn to consultants such as Expertcity.com or PCSupport.com to create and manage such services for a fee.

Another way companies have sought to reduce help desk costs is to outsource their technical support operations. Most software companies are in areas with a relatively high cost of labor. With modern communications and network services, there is no need for the help desk personnel to be at the company headquarters. Workers in less expensive parts of the United States or even in countries such as India that have a large pool of technically trained, English-speaking persons can often offer help services at a lower cost than running an in-house help desk, even when the cost of training and phone line charges are taken into account. On the other hand, there have been complaints by customers that some overseas support staff have language problems or are poorly trained, using only rote "scripts" to try to diagnose problems.

Poor technical support can lead customers to switch to competing products. While this may not be much of a concern in a rapidly expanding industry (where new customers seem to be available in abundance), the situation is different in stagnant or contracting economic conditions. Trying to reduce technical support costs may bring some short-term help to the bottom line, but in the longer run the result might be fewer customers and less revenue. An alternative approach is to consider technical support to be part of a broad effort to maintain customer loyalty; this is often called Customer Relationship Management (see CUSTOMER RELATIONSHIP MANAGEMENT). With regard to technical support, CRM is implemented by using software to better track the resolution of customer's problems as well as to use information obtained in the support process to offer the customer additional products or services custom-tailored to individual situations. With such an approach the effort to provide better technical support is seen not simply as a necessary business expense but as an investment with an expected (though hard to measure) return.

Further Reading
Best Free Technical Support Sites. Available online. URL: http://www.techsupportalert.com/best_free_tech_support_sites.htm. Accessed August 22, 2007.
Czegel, Barbara. *Help Desk Practitioner's Handbook*. New York: Wiley, 1999.
Fleischer, Joe, and Brendan Read. *The Complete Guide to Customer Support: How to Turn Technical Assistance into a Profitable Relationship*. New York: CMP Books, 2002.
Tourniaire, Françoise, and Richard Farrell. *The Art of Software Support: Design & Operation of Support Centers and Help Desks*. Upper Saddle River, N.J.: Prentice Hall, 1997.
Tyman, Dan. "Tech Nightmares: Customer Service." CNET Reviews, January 24, 2005. Available online. URL: http://reviews.cnet.com/4520-10168_7-5621441-1.html. Accessed August 22, 2007.

technical writing

Users of complex systems require a variety of instructional and reference materials, which are produced by technical writers and editors. (It should be noted that technical writing covers many areas other than computer software and systems. However, it is the latter that fall within the scope of this book.)

The traditional products produced by technical writers in the computer industry can be divided into three broad categories: software manuals, trade books, and in-house documentation for developers.

SOFTWARE MANUALS

Until the mid-1990s, just about every significant software product came with a manual (or a set of manuals). A typical manual might include an overview of the program, an introductory tutorial, and a complete, detailed reference guide to all commands or functions.

In theory, staff technical writers (or sometimes contractors) develop the manuals during the time the program is being written. They have access to the programmers for asking questions about the program's operation, and they receive updates from the developers that describe changes or added features. In practice, however, writers may not be assigned to a project until the program is almost done. The programmers, who are under deadline pressure, may not be very communicative, and the writers may have to make their best guess about some matters. The result can be a manual that is no longer "in synch" with the program's actual feature set.

Technical writers often work in a publications department with other professionals including editors, desktop publishers, and graphics specialists. While manuals can be written using an ordinary word processing program, many departments use programs such as FrameMaker that are designed for the production and management of complex documents.

In recent years, many software manufacturers have stopped including printed user manuals with their packages, or include only slim "Getting Started" manuals. As a money-saving measure the traditional documentation is often replaced by a PDF (Adobe Portable Document Format) document on the CD. There is also a greater reliance on extensive on-line help, using either a Windows or Macintosh-specific format or the HTML format that is the lingua franca of the World Wide Web. (See DOCUMENTATION, USER.)

Technical writers have thus had to learn how to construct Help files in these various formats (see AUTHORING SYSTEMS, HELP SYSTEMS, and HTML). Creation of interactive tutorials also requires knowledge of multimedia formats and even animation (such as Flash).

TRADE BOOKS

As millions of people became new computer users during the 1980s, a thriving computer book publishing industry offered users a more user-friendly approach than that usually provided in the manuals issued by the software companies. The "Dummies" books, offering bite-sized servings of information written in a breezy style and accompanied by cartoons, eventually spread beyond computers into hundreds of other fields and the format was then copied by other publishers. Publishers such as Sams, Coriolis, and particularly O'Reilly have aimed their offerings at more experienced users, programmers, and multimedia developers.

Computer trade books are often written by experienced developers and systems programmers who can offer advanced knowledge and "tips and tricks" to their less experienced colleagues. Since many technical "gurus" are not experienced writers, the best results often come from collaboration between the expert and an experienced technical writer and/or editor who can review the material for completeness, organization, and clarity.

In recent years there has been some contraction in the computer book industry. This has arisen from several sources: improved on-line help included in products; the dominance of many applications areas by a handful of products; and fewer people needing beginner-level instruction.

IN-HOUSE DOCUMENTATION

Many technical writers work within software companies or in the information systems departments of other corporations, universities, or government agencies. Their work is generally more highly structured than that of the manual or book writer. As part of a development team, a technical writer may be in charge of creating documentation describing the data structures, classes, and functions within the program. This task is aided by a variety of tools including facilities for extracting such information automatically from C++ or Java programs. The writer may also be responsible for maintaining logs that show each change or addition made to the program during each compiled version or "build."

This type of technical writing requires detailed knowledge of operating systems, programming languages, software development tools, and software engineering methodology. It also requires the ability to work well as part of a team, often under conditions of high pressure.

TECHNICAL WRITING AS A PROFESSION

Until the 1980s, few institutions offered degrees in technical writing. Programmers with an interest in writing or writers with a technical bent entered the field informally.

During the 1980s, the number of degree offerings increased, and people began to specifically prepare for the field, often by earning a computer science degree with a specialization in technical writing. Organizations such as the Society for Technical Communication have offered technical writers and editors a forum for discussing their profession, including issues relating to certification.

Further Reading
Alred, Gerald J., Charles T. Brusaw, and Walter E. Oliu. *The Handbook of Technical Writing.* 8th ed. Boston: St. Martin's Press, 2006.

Lindsell-Roberts, Sheryl. *Technical Writing for Dummies.* New York: Hungry Minds, 2001.

Pringle, Alan S., and Sarah S. O'Keefe. *Technical Writing 101: A Real-World Guide to Planning and Writing Technical Documentation.* 2nd ed. Research Triangle Park, N.C.: Scriptorium Publishing Services, 2003.

Society for Technical Communication. Available online. URL: http://www.stc.org/. Accessed August 22, 2007.

technology policy

Policy makers have found themselves increasingly confronted with difficult issues relating to the Internet, the information industry, and an economy increasingly dependent on computing and communications technology. As with other complex issues such as health care, there has been difficulty reaching a consensus or formulating comprehensive or consistent policies.

Historically many of the early innovators in modern computing (even Microsoft) have tended to keep aloof from politics and lobbying. This may have been due in part to libertarian or laissez-faire beliefs that the best thing the government could do for the information highway was to stay out of its way.

Today, however, with vital economic interests and thousands of jobs at stake, major computer companies have joined the political game with a vengeance. In turn, candidates in the run up to the 2008 presidential election have not neglected to court technology leaders. (For example, in 2008 leading Democratic candidates Hillary Rodham Clinton and Barack Obama both outlined extensive technology agendas.)

Major policy issues involving information technology industries include:

- foreign trade and the protection of intellectual property (see INTELLECTUAL PROPERTY AND COMPUTING and SOFTWARE PIRACY AND COUNTERFEITING)

- attempts to reform the patent system to prevent what is seen as dubious and expensive litigation

- the need for an increasing number of trained workers and providing a sufficient number of visas for foreign workers (see GLOBALIZATION AND THE COMPUTER INDUSTRY)

- preserving equal access to the Internet (see NET NEUTRALITY), which pits content providers against telecommunications companies

- promoting the development of a next-generation Internet infrastructure ("Internet 2")

- government support for computer research (such as through the National Science Foundation)—see GOVERNMENT FUNDING OF COMPUTER RESEARCH

- favorable treatment of online businesses with regard to taxation (often objected to by traditional brick-and-mortar businesses)—see e-commerce

- laws against computer-related fraud and other crime (see COMPUTER CRIME AND SECURITY and ONLINE FRAUDS AND SCAMS)

- Privacy regulations (see IDENTITY THEFT and PRIVACY IN THE DIGITAL AGE)

The computer industry is also involved in issues that will affect its future over the longer term, such as the need to improve math and science education in elementary and high schools, energy and environmental policy, and issues such as health care and pensions that affect all sectors of the economy.

INTERNATIONAL ASPECTS

In a global industry, American information technology policy cannot be considered without looking at the policies of other nations (both industrialized and developing) and their potential impacts. For example, China's success (or lack thereof) in protecting intellectual property has a direct impact on the revenue of major software companies. Similarly, issues of censorship (see CENSORSHIP AND THE INTERNET) create dilemmas for companies that must balance concern for human rights with the opportunity to enter huge new markets. Other important aspects of comparative technology policy include research funding, subsidies, patent and copyright law, and labor standards.

Further Reading
Aspray, William, ed. *Chasing Moore's Law: Information Technology Policy in the United States.* Raleigh, N.C.: SciTech Publishing, 2004.

Coopey, Richard, ed. *Information Technology Policy: An International History.* New York: Oxford University Press, 2004.

Marcus, Alan I., and Amy Sue Bix. *The Future Is Now: Science and Technology Policy in America since 1958.* Amherst, N.Y.: Humanity Books, 2006.

MIT Center for Technology, Policy, and Industrial Development. Available online. URL: http://web.mit.edu/ctpid. Accessed November 27, 2007.

Nuechterlein, Jonathan E., and Philp J. Weiser. *Digital Crossroads: American Telecommunications Policy in the Internet Age.* Cambridge, Mass.: MIT Press, 2007.

Ricadela, Aaron. "Technology Companies Have Much at Stake in 2008." *Business Week,* September 19, 2007. Available online. URL: http://www.businessweek.com/technology/content/sep2007/tc20070917_079427.htm. Accessed November 26, 2007.

U.S. Technology Administration. Available online. URL: http://www.technology.gov/. Accessed November 27, 2007.

telecommunications

Since its birth in the mid-20th century, the digital computer and the telephone have had a close mutual relationship. Many of the first programmable calculators and computers built in the early 1940s used relays and other components that were being manufactured for the increasingly automated phone system (see AIKEN, HOWARD). The phone industry contributed ideas as well as hardware. Scientists at the Bell Laboratories carried out fundamental research into information theory that would soon be applied to data communications (see SHANNON, CLAUDE).

As computers became more capable in the 1950s and 1960s, they began to return the favor, making possible increasing automation for the phone system. Meanwhile, computers were starting to be hooked up to telephone lines (see MODEM) so they could exchange data and allow their users to communicate (see NETWORK).

The development of a global network (see INTERNET) and its growth through the 1980s provided a universal platform for data communications. At first, the Internet was used mostly by academics and engineers, but the advent of the World Wide Web and in particular, graphical Web browsers made Internet access ubiquitous among small businesses and home users by the late 1990s.

Institutional Internet users often had fast access through dedicated phone lines (designated T-1, and so on), while homes, small businesses, and schools were limited to much slower dial-up access. This began to change in the late 1990s as alternatives to POTS ("plain old telephone service") emerged in the form of DSL (a much faster service running over regular phone lines) and cable modems that used the infrastructure that already brought TV to millions of homes.

IMPACT OF DEREGULATION

Prior to the court-ordered breakup of AT&T in 1984, the phone industry functioned in a monolithic way and was not very responsive to the needs of the growing computer networking industry.

The breakup of AT&T led to growing competition, providing a wider variety of telecommunications equipment and lower phone rates just as PC users were starting to buy modems and sign up with online services and bulletin boards. The growing deregulation movement in the 1990s (culminating in the Telecommunications Act of 1996) furthered this process by opening cable and broadcast television, radio, and other wireless communication to competition.

With more than half of American Internet users on high-speed connections (see BROADBAND), the delivery of communications and media over the Net can only grow. Wireless and mobile services (satellite, cell network, and 802.11—see WIRELESS COMPUTING) have also been growing vigorously. The result is that the "information highway" now has many lanes, with some being express lanes.

CONVERGENCE AND THE FUTURE

The ability of the Internet to transmit any sort of data virtually anywhere at relatively low cost has created new alternatives to traditional communications technologies. For example, sending digitized voice telephone calls as packets over the Internet can provide a lower-cost alternative to conventional long distance calling (see VOIP). At the same time, previously separate functions are converging

into "smarter" devices. Thus, the handheld computer and the cell phone seem to be converging into a single device that can provide data management (see SMARTPHONE). Web browsing, and communications in a single package.

Computers and communications technology will continue to grow more intertwined. Today it is increasingly hard to distinguish information technology, media content, and communications technology as being distinctive sectors. After all, a consumer can watch a movie in the theater or later on broadcast, cable, or satellite TV, rent it on commercial videotape or DVD disk (playable on PCs as well as portable players), or even view it as a streaming file direct from the Internet. Although these technologies have differing technical constraints, their end products are the same for the consumer.

This multiplicity of function means that the competitive environment is increasingly hard to predict, since there are so many possible players. The companies offering content through this variety of technologies are also increasingly intertwined.

For analysts, studying any technology requires awareness of the many possible alternatives, while studying any application means considering the many possible technological implementations. For policy makers and regulators, the challenge is to provide for such public goods as equal access, privacy, and protection of intellectual property in a communications infrastructure that is truly global in scope and evolving at a pace that frequently outdistances the political process.

Further Reading
Benjamin, Stuart Minor, et al. *Telecommunications Law and Policy.* 2nd ed. Durham, N.C.: Carolina Academic Press, 2006.
———. *Telecommunications Law and Policy, 2007 Supplement.* Durham, N.C.: Carolina Academic Press, 2007.
Goleniewski, Lillian. *Telecommunications Essentials: The Complete Global Source.* 2nd ed. Upper Saddle River, N.J.: Addison-Wesley, 2007.
Hill Associates. *Telecommunications: A Beginner's Guide.* Berkeley, Calif.: McGraw-Hill/Osborne, 2002.
"Media and Telecommunications Policy and Legislation." Moffitt Library, UC Berkeley. Available online, URL: http://www.lib.berkeley.edu/MRC/MediaPolicy.html. Accessed August 22, 2007.
Olejniczak, Stephen P. *Telecom for Dummies.* Hoboken, N.J.: Wiley, 2006.

telecommuting

Telecommuting (also called telework) is the ability to work from home or from some location other than the main office. According to a report by the nonprofit organization WorldatWork, 28.7 million people worked from home at least one day a month in 2006. (Self-employed persons, of course, have a much higher rate of working from home.)

Telecommuting was made possible by the growing capabilities of home computers and the availability of network connections that allow the worker at home to have access to most of the people and facilities that would be available if the worker were on site. Workers and companies that promote telecommuting often cite the following advantages:

- elimination of stressful, time-wasting commutes

- workers may be more productive because they have fewer office distractions, unnecessary meetings, etc.

- reduction of traffic, air pollution, and fuel costs

- greater flexibility in working hours

- the ability of working parents with small children to combine child care and work to some extent

- reduction of costs associated with office facilities

However, telecommuting has its critics in management. Some of the problems or disadvantages cited include:

- Worker productivity may decrease due to lack of sufficient discipline and workers becoming distracted at home.

- Managers may have trouble keeping track of or evaluating the activities of workers who are not physically present.

- Telecommuters may miss critical information and go "out of the loop."

- Security can be compromised, particularly through theft of laptops containing sensitive personal data.

- Possible legal liabilities and application of OSHA rules to home working situations.

It is true that telecommuting is suitable mainly for jobs that involve information processing rather than person-to-person contact, such as service jobs. However, the use of videoconferencing or Web conferencing technology increasingly makes it possible for suitably equipped telecommuters to participate in meetings almost as directly as if they were physically present (see CONFERENCING SYSTEMS and TELEPRESENCE).

In some cases involving videoconferencing or other activities that require high-powered computer systems and high bandwidth connections, telecommuters physically commute to a "satellite work center" near their home that has the appropriate equipment. This can provide some of the advantages of telecommuting such as flexibility and lower commute and office costs.

A number of issues must be worked out between workers and management for any telecommuting program, including:

- Who will pay for the equipment used by the telecommuter

- Procedures for monitoring the work

- How telecommuters will participate in meetings (either remotely or in person)

- The portion of the worker's hours involving telecommuting, and the portion requiring attendance in-house

TRENDS

Telecommuting was touted in the mid-1990s as the wave of the future. In reality, the statistics given earlier suggest while it is a viable option for a significant minority of workers, telecommuting is not growing as rapidly as had been predicted. The growing power of desktop PCs and the availability of broadband (DSL or cable) network connections should help facilitate telecommuting. In the longer term new technologies may make the distinction between telecommuters and physically present workers much less important (see TELEPRESENCE and VIRTUAL REALITY).

Further Reading

American Telecommuting Association. Available online. URL: http://www.yourata.com/index.html. Accessed August 22, 2007.

Dziak, Michael. *Telecommuting Success: A Practical Guide for Staying in the Loop While Working Away from the Office.* Indianapolis: Park Avenue, 2001.

Messmer, Ellen. "Telecommuting Security Concerns Grow." *InfoWorld,* April 18, 2006. Available online. URL: http://www.infoworld.com/article/06/04/18/77520_HNtelecommunting-security_1.html. Accessed August 22, 2007.

Zetlin, Minda. *Telecommuting for Dummies.* New York: Hungry Minds, 2001.

telepresence

An old phone company slogan asserted that "long distance is the next best thing to being there." Today technology has made the ability to "be there" a much more complete experience. It is now quite common for businesspersons to "attend" a meeting in a distant city using video cameras to see and be seen, with images and voice traveling over special leased lines or high speed Internet connections (see VIDEOCONFERENCING).

While videoconferencing and suitable software allows remote interaction and collaboration (such as being able to build a spreadsheet or diagram together), the remote participant has little ability to physically interact with the environment. He or she can't walk freely around, perhaps joining other meeting participants in an adjacent room while they have pastries and coffee. The remote participant also cannot handle physical objects such as models.

There are two basic approaches to letting persons have an unconstrained experience in a remote environment. The first is to use technology to create a virtual presence where a person can experience a simulated environment from many different angles and move freely through it while grasping and manipulating objects (see VIRTUAL REALITY). In a virtual environment each participant can be represented by an "avatar" body that can be programmed to move in response to head trackers, gloves, and other devices.

However, a virtual reality is an artificial representation of the world. A group of people having a meeting in a physical space can't interact with someone who is in virtual space except in the most rudimentary ways. To be on an equal footing, all participants would have to be in either physical or virtual space.

TELEROBOTICS

The alternative is to connect the remote participant to a mobile robot (this is sometimes called *telerobotics*). Such robots already exist, although their capabilities are limited and they are not yet widely used for meetings. RODNEY BROOKS, director of the MIT Artificial Intelligence Laboratory, foresees a not very distant future in which such robots will be commonplace.

The robot will have considerable built-in capabilities, so the person who has "roboted in" to it won't need to worry about the mechanics of walking, avoiding obstacles, or focusing vision on particular objects. Seeing and acting through the robot, the person will be able to move around an environment as freely as persons who are physically present. The operator can give general commands amounting to "walk over there" or "pick up this object" or perform more delicate manipulations by using his or her hands to manipulate gloves connected to a force-feedback mechanism.

Brooks sees numerous applications for robotic telepresence. For example, someone at work could "robot in" to his or her household robot and do things such as checking to make sure appliances are on or off, respond to a burglar alarm, or even refill the cat's food dish. Robotic telepresence could also be used to bring expertise (such as that of a surgeon) to any site around the world without the time and expense of physical travel. Indeed, robots may be the only way (for the foreseeable future) that humans are likely to explore environments far beyond Earth (see SPACE EXPLORATION AND COMPUTERS).

Further Reading

Ballantyne, Garth H., Jacques Marescaux, and Pier Cristoforo Giulianotti. *The Primer of Robotic and Telerobotic Surgery: A Basic Guide to Heart Disease.* 4th ed. Baltimore: Lippincott Williams & Wilkins, 2004.

Frere, Manuel, et al., eds. *Advances in Telerobotics.* New York: Springer, 2007.

Goldberg, Ken, ed. *The Robot in the Garden: Telerobotics and Telepistemology in the Age of the Internet.* Cambridge, Mass.: MIT Press, 2000.

NASA Space Telerobotics Program. Available online. URL: http://ranier.hq.nasa.gov/telerobotics_page/telerobotics.shtm. Accessed August 22, 2007.

Sheppard, P. J., and G. R. Walker, eds. *Telepresence.* New York: Springer, 1998.

Tele-Robotics Links [Online robots, etc.]. Available online. URL: http://queue.ieor.berkeley.edu/~goldberg/art/telerobotics-links.html. Accessed August 22, 2007.

template

The term *template* is used in a several contexts in computing, but they all refer to a general pattern that can be customized to create particular products such as documents.

In a word processing program such as Microsoft Word, a template (sometimes called a style sheet) is a document that comes with a particular set of styles for various elements such as titles, headings, first and subsequent paragraphs, lists, and so on. Each style in turn consists of various characteristics such as type font, type style (such as bold), and spacing. The template also includes properties of the document as a whole, such as margins, header, and footer.

To create a new document, the user can select one of several built-in templates for different types of documents such as letters, faxes, and reports, or design a custom template by defining appropriate styles and properties. Special sequences of programmed actions can also be attached to a template (see MACRO).

Templates can be created and used for applications other than word processing. A spreadsheet template consists of appropriate macros and formulas in an otherwise blank spreadsheet. When it is run, the template prompts the user to enter the appropriate values and then the calculations are performed. A database program can have input forms that serve in effect as templates for creating new records by inputting the necessary data.

CLASS TEMPLATES

Some programming languages use the term *template* to refer to an abstract definition that can be used to create a variety of similar classes for handling different types of data, which in turn are used to create actual objects. For example, once the programmer defines the following template:

```
template<class ANY_TYPE>
ANY_TYPE maximum(ANY_TYPE a, ANY_TYPE b)
{
    return (a > b) ? a : b;
```

This template provides any class with the maximum function, which can compare any two objects of that class and return the larger one. (See C++, CLASS, and OBJECT-ORIENTED PROGRAMMING.)

Further Reading

Abrams, David, and Aleksey Gurtovoy. *C++ Template Metaprogramming: Concepts, Tools, and Techniques from Boost and Beyond.* Upper Saddle River, N.J.: Addison-Wesley Professional, 2004.

Free Templates for Work, Home, and Play. Microsoft Office Online. Available online. URL: http://office.microsoft.com/en-us/templates/FX100595491033.aspx. Accessed August 22, 2007.

"Introduction to the Standard Template Library" [for C++]. SGI. Available online. URL: http://www.sgi.com/tech/stl/stl_introduction.html. Accessed August 22, 2007.

Krieger, Stephanie. *Advanced Microsoft Office Documents Inside Out.* 2007 ed. Redmond, Wash.: Microsoft Press, 2007.

Vandevoorde, David, and Nicolai M. Josuttis. *C++ Templates: The Complete Guide.* Upper Saddle River, N.J.: Addison-Wesley, 2002.

Utvich, Michael, Ken Milhous, and Yana Beylinson. *1 Hour Web Site: 120 Professional Templates and Skins.* Hoboken, N.J.: Wiley, 2006.

Walkenbach, John. *Microsoft Office Excel 2007 Bible.* Indianapolis: Wiley, 2007.

terminal

Throughout the 1950s, operators interacted with computers primarily by punching instructions onto cards that were then fed into the machine (see PUNCHED CARDS AND PAPER TAPE). Although this noninteractive batch processing procedure would continue to be used with mainframe computers during the next two decades, another way to use computers began to be seen in the 1960s.

With the beginning of time-sharing computer systems, several users could run programs on the computer at the same time. The users communicated with the machine by typing commands into a Teletype or similar device. Such a device is called a *terminal*.

The simple early terminals did little more than accept lines of text commands from the user and print responses or lines of output coming from the computer. However, a newer type of terminal began to replace the Teletype. It consisted of a keyboard attached to a televisionlike cathode ray tube (CRT) display. Users still typed commands, but the computer's output could now be displayed on the screen.

Gradually, CRT terminals gained additional capabilities. The text being entered was now stored in a memory buffer that corresponded to the screen and the user could use special control commands or keys to move the input cursor anywhere on the screen when creating a text file. This made it much easier for users to revise their input (see TEXT EDITING). These "smart terminals" had their own small processor and ran software that provided these functions.

During the 1970s, the UNIX operating system developed a sophisticated way to support the growing variety of terminals. It provided a library of cursor-control routines (called *curses*) and a database of terminal characteristics (called *termcap*).

When the personal computer came along, it had a keyboard, a processor, the ability to run software, and a connection for a TV or monitor. The PC thus had all the ingredients to become a smart terminal. Indeed, a modern PC is a terminal, but users don't usually have to think in those terms. The exception is when the user runs a communications program to connect to a remote computer (perhaps a bulletin board) with a modem. These programs, such as the Hyperterminal program that comes with Windows, allow the PC to emulate (work like) one of the standard terminal types such as VT-100. This ability to emulate a standard terminal means that any software that supports that physical terminal should also work remotely with a PC.

Today most interaction with remote programs is through a Web browser, although protocols such as telnet are still used to provide terminallike access to remote programs. Many commands previously entered as text lines in a terminal are now given using the mouse with menus and icons (see USER INTERFACE).

The relationship between a terminal and remote computer is analogous to that between a workstation (or desktop PC) and the network server in that the burden of processing is divided between the two devices in various ways (see CLIENT-SERVER COMPUTING). A "thin client" PC performs relatively little processing with the server doing most of the work.

Specialized terminals are still used for many applications. An ATM, for example, is a special-purpose banking terminal driven by a keypad and touchscreen.

Further Reading

Archive of Video Terminal Information. Available online. URL: http://www.cs.utk.edu/~shuford/terminal/index.html. Accessed August 22, 2007.
Free Software Foundation. The Termcap Library. Available online. URL: http://www.gnu.org/manual/termcap-1.3/termcap.html. Accessed August 22, 2007.
Linux Terminal Server Project. Available online. URL: http://www.ltsp.org/. Accessed August 22, 2007.

text editor

As noted in the previous article (see TERMINAL), an alternative to batch-processing punch card driven computer operations emerged in the 1960s in the form of text commands typed at an interactive console or terminal. At first text could be typed only a line at a time and there was no way to correct a mistake in a previous line.

Soon, however, programmers began to create text editing programs. The first editors were still line-oriented, but they stored the lines for the current file in memory. To display a previous line, the user might simply type its number. To correct a word in the line the user might type something like

```
c/fot/for
```

to change the typo "fot" to the word "for" in the current line.

Starting in the early 1970s, the UNIX system provided both a line editor (ed or ex) and a "visual editor" (vi). The latter editor works with terminals that can display full screens of text and allow the cursor to be moved anywhere on the screen. This type of editor is also called a screen editor.

Most ordinary PC users use word processors rather than text editors to create documents. Unlike a text editor, a word processor's features are designed to create output that looks as much like a printed document as possible. This includes the ability to specify text fonts and styles. However, most systems also include a simpler text editor that can be useful for making quick notes (in Windows this program is indeed called Notepad).

The primary use of text editors today is to create programs and scripts. These must generally be created using only standard ASCII characters (see CHARACTERS AND STRINGS), without all the embedded formatting commands and graphics found in word processing documents. Programmer's text editors can be very sophisticated in their own right, providing features such as built in syntax checking and formatting or (as with the Emacs editor) the ability to program the editor itself. Ultimately, however, program editors must create a source code file that can be processed by the compiler.

Text editors are also useful for writing quick, short scripts (see SCRIPTING LANGUAGES) and can be handy for writing HTML code for the Web. However, many Web pages are now designed using word processor–like programs that convert the WYSIWYG (what you see is what you get) formatting into appropriate HTML codes automatically.

Further Reading

Cameron, Debra, et al. *Learning GNU Emacs.* 3rd ed. Sebastapol, Calif.: O'Reilly, 2005.
Chassell, Robert J. *An Introduction to Programming in Emacs Lisp.* 2nd ed. Boston, Mass.: GNU Press, 2004.
Robbins, Arnold, and Linda Lamb. *Learning the vi Editor.* 6th ed. Sebastapol, Calif.: O'Reilly, 1998.
Shareware Text Editors, Word Processors, etc. Available online. URL: http://www.passtheshareware.com/c-txtwp.htm. Accessed August 22, 2007.
Smith, Larry L. *How to Use the UNIX-LINUX vi Text Editor: Tips, Tricks, and Techniques (And Tutorials Too!)* Charleston, S.C.: BookSurge, 2007.

texting and instant messaging

Although they use different devices and formats, text messaging on cell phones and PDAs and instant messaging through online services have much in common. Both involve sending short messages to other users who can receive them and reply as soon as they are online. (This is an ad hoc connection that differs from a chat room [see CHAT, ONLINE] in that the latter is an established location where people go to converse with other members. It also differs from an online discussion group [see CONFERENCING SYSTEMS and NETNEWS AND NEWSGROUPS] where messages are posted and may be replied to later, but there is no real-time communication.)

Text messaging or *texting* uses a protocol called Short Message Service (SMS), which is available with most cell phones and service plans as well as PDAs that have wireless connections. When a user sends a message to a designated recipient, it goes to a service center where it is routed to the destination phone; if that phone is not connected, the message is stored and retried later. Typically messages are limited to 160 characters, though up to six or so messages can be concatenated and treated as a longer message.

While texting did not become popular until the late 1990s, instant messaging began in the 1970s as a way for multiple users on a shared computer or network (such as a UNIX system) to communicate in real time using commands such as *send* and *talk* (the latter being more conversational—see CHAT, ONLINE). In the late 1980s and early 1990s, various dial-up services (see AMERICA ONLINE and ONLINE SERVICES) provided for sending text messages (AOL was the first to use the term *instant messages* for its facility). By the mid-1990s instant messaging was well established on the Internet, often employing a graphical user interface, as with ICQ and AOL Instant Messenger (AOL later acquired ICQ as well).

There have been efforts to allow users of different instant messaging systems to communicate with one another, but resistance on the part of the proprietary networks (often citing security concerns) has hobbled this process thus far. Instant messaging has also been implemented as an application for phones and other mobile devices (an effort headed by the Open Mobile Alliance). Generally this involves reworking the IM software to use the SMS text service to carry messages.

CULTURAL IMPACT

Between 2000 and 2004 the numbers of text messages sent worldwide soared from 17 billion to 500 billion. At about a dime a message, texting became a major source of revenue for phone companies. Since then, texting has continued to grow, particularly in parts of Europe, the Asia-Pacific region (particularly China), and Japan (where it has largely become an Internet-based service).

In the United States texting is most popular among teenagers (see YOUNG PEOPLE AND COMPUTING). It is not uncommon to see a bench full of teens talking excitedly to one another while carrying on simultaneous texting with unseen friends in what, to many adult onlookers, appears to be an incomprehensible code, their conversation perhaps ending with *ttyl* (talk to you later).

Loosely affiliated groups communicating by text (see FLASH MOBS) have organized everything from "happenings" to serious protest campaigns (as in the anti-WTO [World Trade Organization] demonstrations in Seattle in 1999 and in the Philippines uprising in 2001.)

The popularity of texting has increasingly attracted the attention of fraudsters (see PHISHING AND SPOOFING and SPAM) as well as more legitimate marketers.

Further Reading

All about Instant Messengers. Available online. URL: http://aboutmessengers.com/. Accessed November 29, 2007.

Hord, Jennifer. "How SMS Works." Available online. URL: http://communication.howstuffworks.com/sms.htm. Accessed November 29, 2007.

Le Bodic, Gwenael. *Mobile Messaging Technologies and Services: SMS, EMS and MMS.* 2nd ed. New York: John Wiley and Sons, 2005.

Rheingold, Howard. *Smart Mobs: The Next Social Revolution.* Cambridge, Mass.: Perseus Books, 2003.

SMS Speak Translator. Available online. URL: http://smspup.com/smsSpeak.php. Accessed November 29, 2007.

"Web Messengers Handbook." Available online. URL: http://web2.ajaxprojects.com/web2/newsdetails.php?itemid=35. Accessed November 29, 2007.

Torvalds, Linus

(1969–   )
Finnish
Software Developer

Linus Torvalds developed Linux, a free version of the UNIX operating system that has become the most popular alternative to proprietary operating systems.

Torvalds was born on December 28, 1969, in Helsinki, Finland. His childhood coincided with the microprocessor revolution and the beginnings of personal computing. At the age of 10, he received a Commodore PC from his grandfather, a mathematician. He learned to write his own software to make the most out of the relatively primitive machine.

In 1988, Torvalds enrolled in the University of Helsinki to study computer science. There he encountered UNIX, a powerful and flexible operating system that was a delight for programmers who liked to tinker with their computing environment. Having experienced UNIX, Torvalds could no longer be satisfied with the operating systems that ran on most PCs, such as MS-DOS, which lacked the powerful command shell and hundreds of utilities that UNIX users took for granted.

Torvalds's problem was that the UNIX copyright was owned by AT&T, which charged $5,000 for a license to run UNIX. To make matters worse, most PCs weren't powerful enough to run UNIX anyway.

At the time there was already a project called GNU underway (see OPEN SOURCE and STALLMAN, RICHARD). The Free Software Foundation was attempting to replicate all the functions of UNIX without using any of AT&T's proprietary code. This would mean that the AT&T copyright would not apply, and the functional equivalent of UNIX could be given away for free. Stallman and the FSF had already provided key tools such as the C compiler and the Emacs program editor. However, they had not yet created the heart of the operating system (see KERNEL). The kernel contains the essential functions needed for the operating system to control the computer's hardware, such as creating and managing files on the hard drive.

In 1991, Torvalds wrote his own kernel and put it together with the various GNU utilities to create what soon became known as Linux. Torvalds adopted the open source license (GPL) pioneered by Stallman and the FSF, allowing Linux to be distributed freely. The software soon spread through ftp sites on the Internet, where hundreds of enthusiastic users (mainly at universities) helped to improve Linux, adding features and writing drivers to enable it to work with more kinds of hardware.

By the mid-1990s, the free and reliable Linux had become the operating system of choice for many Web site developers. Torvalds, who still worked at the University of Helsinki as a researcher, faced an ever-increasing burden of coordinating Linux development and deciding when to release successive versions. As companies sprang up to market software for Linux, they offered Torvalds very attractive salaries, but he did not want to be locked into one particular Linux package (distribution).

Instead, in 1997 Torvalds moved to California's Silicon Valley, where he became a key software engineer at Transmeta, a company that makes Crusoe, a processor designed for mobile computing.

In 2003 Torvalds left Transmeta. In 2004 he moved to Portland, Oregon, where the Linux Foundation, a non-profit consortium dedicated to promoting the growth of Linux, supports his work. There he concentrates on guiding the continuing development of the Linux core, or kernel. Although he strongly supports open-source software, Torvalds has been criticized by some advocates for his pragmatic approach of using proprietary software when it seems to be more suitable to a given task.

Further Reading
Dibona, Chris. *Open Sources: Voices from the Open Source Revolution.* Sebastapol, Calif.: O'Reilly, 2001.
"FM Interview with Linus Torvalds: What Motivates Free Software Developers?" *First Monday,* vol. 3 (1998). Available online. URL: http://www.firstmonday.org/issues/issue3_3/torvalds/. Accessed August 23, 2007.
Linux Foundation. Available online. URL: http://www.linux-foundation.org/en/Main_Page. Accessed August 23, 2007.
Richardson, Marjorie. "Interview: Linus Torvalds." *Linux Journal,* November 1, 1999. Available online. URL: http://www.linux-journal.com/article/3655. Accessed August 23, 2007.
Torvalds, Linus, and David Diamond. *Just for Fun: The Story of an Accidental Revolutionary.* New York: HarperCollins, 2001.

touchscreen

As the name implies, a touchscreen is a screen display that can respond to various areas being touched or pressed. Invented in 1971, the first form of touchscreens to become part of daily life were found on automatic teller machines (ATMs) and point-of-sale credit card processors.

Touchscreens can detect the pressure of a finger or stylus in several ways: A "resistive" touchscreen uses two layers of electrically conductive metallic material separated by a space. When an area is touched, the two layers are electrically connected, and the change in electrical current is registered and converted to a code that identifies the location touched. Surface acoustic wave (SAW) touchscreens use an ultrasonic wave that is interrupted by a touch; capacitive touchscreens respond to the change in electron storage (capacitance) caused by contact with a human body. Various other acoustic, mechanical (strain-based), or optical systems can also be used, with the latter being particularly popular.

Touchscreens can have drawbacks ranging from problems with long fingernails to screen "keys" placed too close together for normal fingers. Responsiveness is also considerably slower than with a keyboard. Depending on the technology used, dirt or grease can also become a problem.

OTHER APPLICATIONS

In addition to dedicated uses such as banking and retailing, touchscreens are a common form of input for mobile phones, PDAs, and similar devices (see PDA and SMARTPHONE) where a "virtual" on-screen keyboard is often used for entering text. Particularly versatile systems such as Apple's iPhone combine proximity sensors with touchscreen technology in order to be able to recognize gestures such as pinching and flicking. Another example is Microsoft's "Surface" interface. (For related technologies, see GRAPHICS TABLET and TABLET PC.)

Further Reading
Baig, Edward C. *iPhone for Dummies.* New York: Wiley, 2007.
Buxton, Bill. "Multi-Touch Systems That I Have Known and Loved." Available online. URL: http://www.billbuxton.com/multitouchOverview.html. Accessed November 29, 2007.
Microsoft Surface. Available online. URL: http://www.microsoft.com/surface/. Accessed November 29, 2007.
Plaisant, Catherine. "High-Precision Touchscreens: Museum Kiosks, Home Automation and Touchscreen Keyboards." University of Maryland Human-Computer Interaction Lab, January 31, 1999. Available online. URL: http://www.cs.umd.edu/hcil/touchscreens/. Accessed November 29, 2007.
Woyke, Elizabeth. "Reach Out and Touch—A Phone Screen." Forbes.com, November 28, 2007. Available online. URL: http://www.forbes.com/technology/2007/11/28/touch-screens-phones-tech-holidaytech07-cx_ew_1128touch.html. Accessed November 29, 2007.

transaction processing

Many computer applications involve the arrival of a set of data that must be processed in a specified way. For example, a bank's ATM system receives a customer's request to deposit money together with identification of the account and the amount to be deposited. The system must accept the deposit, update the account balance, and return a receipt to the customer. This is an example of real-time transaction processing.

Some applications process transactions in batches. For example, a company may run a program once a month that generates paychecks and withholding stubs from employee records that include hours worked, number of dependents claimed, and so on. Indeed, in the ATM example, the account balance is typically not updated during the on-line transaction, but instead a batch transaction is stored. Overnight that transaction will be processed together with other transactions affecting that account (such as checks), and the balance will then be officially updated. (The program module that keeps track of the progress of transactions is called a transaction monitor.)

There are several considerations that are important in designing transaction systems. While some transactions may simply involve a request for information and do not update any files, many transactions may require that several files or database records be updated. For example, a transfer of funds from a saving account to a checking account will require that both accounts be updated: the first with a debit and the second with a credit. What happens if the computer performs the savings debit but then goes down before the checking account can be credited? The result would be an upset customer whose money seems to have disappeared.

The solution to this problem is to design a process where the various updates are done not to the actual databases but to associated temporary databases. Once these potential transactions are posted, the system issues a "commit" command to the databases. Each database must send a reply acknowledging that it's ready to perform the actual update. Only if all databases reply affirmatively is the commit command given, which updates all the databases simultaneously.

Further Reading

Lewis, Philip M., Arthur Bernstein, and Michael Kifer. *Databases and Transaction Processing: An Application-Oriented Approach.* Upper Saddle River, N.J.: Addison-Wesley, 2002.

Little, Mark, Jon Maron, and Greg Pavlik. *Java Transaction Processing: Design and Implementation.* Upper Saddle River, N.J.: Prentice Hall PTR, 2004.

tree

The tree is a data structure that consists of individual intersections called *nodes*. The tree normally starts with a single root node. (Unlike real trees, data trees have their root at the top and branch downward.) The root connects to one or more nodes, which in turn branch into additional nodes, often through a number of levels. (A node that branches downward from another node is called that node's *child* node.) A node at the bottom that does not branch any further is called a terminal node or sometimes a leaf.

Trees are useful for expressing many sorts of hierarchical structures such as file systems where the root of a disk holds folders that in turn can hold files or additional folders, and so on down many levels. (A corporate organization chart is a noncomputer example of a hierarchical tree.)

The most common type of tree used as a data structure is the binary tree. A binary tree is a tree in which no node has more than two child nodes. To move through data stored in a binary tree, a program can use two pointers, one to the current node's left child and one to its right child (see POINTERS AND INDIRECTION). The pointers can then be used to trace the paths through the nodes. If the tree represents a file that has been sorted (see SORTING AND SEARCHING), comparing nodes to the desired value and branching accordingly quickly leads to the desired record.

Alternatively, the data can be stored directly in contiguous memory locations corresponding to the successive numbers of the nodes. This method is faster than having to trace through successive pointers, and a binary search algorithm can be applied directly to the stored data. On the other hand it is easier to insert new items into a linked list (see LIST PROCESSING).

A common solution is to combine the two structures, storing the linked list in a contiguous range of memory by storing its root in the middle of the range, its left child at the beginning of the range, its right child at the end, and then repeatedly splitting each portion of the range to store each level of children. Intuitively, one can see that algorithms for processing such stored trees will take a recursive approach (see RECURSION).

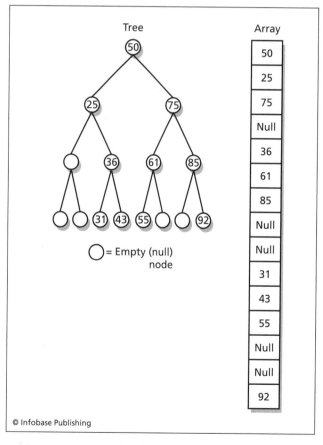

In a binary tree each node either has two branch (child) nodes or is a terminal node. Here a binary tree is shown with the equivalent representation in an array of memory locations. Notice that the numbers are stored level by level (50 in the top level, 25 and 75 in the second level, and so on.) Null would be a special value (such as -1) representing an empty node.

For efficiency it's important to keep all branches of a tree approximately the same length. A B-tree (balanced tree) is designed to automatically optimize itself in this way.

Trees lend themselves to game programs where a series of moves and their possible replies must be explored to varying levels. A chess program will typically create a tree from the current position, but use various criteria to determine which moves should be explored beyond just a few levels, thus "pruning" the game tree.

Further Reading
Brookshear, J. Glenn. *Computer Science: An Overview.* 9th ed. Boston: Addison-Wesley, 2006.
Lafore, Robert. *Data Structures & Algorithms in Java.* 2nd ed. Indianapolis: Sams, 2002.
Parlante, Nick. "Binary Trees." Available online. URL: http://cslibrary.stanford.edu/110/BinaryTrees.html. Accessed August 23, 2007.

trends and emerging technologies

Because of the complexity of computer systems, software, and business models, it is easy to "fail to see the forest for the trees." Stepping back once in a while to see what is changing (or likely to change) is recommended. Such a perspective is particularly useful for professionals who need to periodically evaluate their skills against the market, investors, venture capitalists, journalists, and educators.

It is nearly a commonplace to say that the future is coming at an accelerating rate (for the ultimate speculation, see SINGULARITY, TECHNOLOGICAL). What adds to the challenge is the way each new technology (and social adaptation) has multiple consequences, whether it is social networking, "viral marketing," or individually targeted, location-aware advertising. At the same time, attempting to distinguish short-term hype from genuine trends is always difficult—anyone in the computing field can compile a glossary of now-obsolete buzzwords. Thus reading a variety of perspectives from advocates to pundits to critics is essential.

OVERALL TRENDS

That said, the following table suggests some activities that are in transition from a traditional model to one that reflects emerging technological and social trends.

FROM	TO
desktop PC	mobile and pervasive computing
wired networks	wireless and mobile computing
separate handling of media	integrated media servers
broadcasting	user-driven, customizable channels of media
user as consumer	user as contributor and sharer
individual consumers	social networking
software as product	software as service
proprietary code	open source
interface-driven tasks	search-driven interfaces
arbitrarily organized data	semantically retrievable information.
e-commerce	integration of online and traditional channels

For more information about emerging trends, see the following entries: DIGITAL CONVERGENCE, OPEN-SOURCE MOVEMENT, SERVICE-ORIENTED ARCHITECTURE, SMART BUILDINGS AND HOMES, SOCIAL NETWORKING, UBIQUITOUS COMPUTING, USER-CREATED CONTENT, WEB 2.0 AND BEYOND, WEB SERVICES.

Further Reading
Haskin, David. "Don't Believe the Hype: The 21 Biggest Technology Flops." *Computerworld,* April 4, 2007. Available online. URL: http://www.computerworld.com/action/article.do?articleId=9012345&comm and=viewArticleBasic. Accessed November 29, 2007.
Piquepaille, Roland. "Emerging Technology Trends" [blog]. ZDNet. Available online. URL: http://blogs.zdnet.com/emergingtech/. Accessed November 29, 2007.
Plunkett Research. "Plunkett's E-Commerce & Internet Industry." Available online. URL: http://www.plunkettresearch.com/Industries/ECommerceInternet/tabid/151/Default.aspx. Accessed November 29, 2007.
———. "Plunkett's InfoTech, Computers & Software Industry." Available online. URL: http://www.plunkettresearch.com/Industries/InfoTechComputersSoftware/tabid/152/Default.aspx. Accessed November 29, 2007.
Seidensticker, Bob. *Future Hype: The Myths of Technology Change.* San Francisco: Berrett-Koehler, 2006.

trust and reputation systems

Trust and reputation are inherently connected. Participants in any transaction (such as respondents to classified ads or participants in an online auction) want assurance that they will receive the promised value in exchange for what they are giving up. The default anonymity of online transactions (see ANONYMITY AND THE INTERNET) presents a challenge: How does one obtain information about someone's reputation (how he or she has behaved previously) and thus be assured of his or her reliability?

One solution is to collect and characterize the experiences of participants in previous transactions with that person or entity. For an auction (see AUCTIONS, ONLINE and EBAY), the system can solicit and tabulate ratings ("feedback") by participants in each transaction. However, this kind of simple system must guard against being subverted by people who create false identities (fronts) and transactions and use them to inflate the feedback score.

A more sophisticated reputation system can be used for ranking Web pages, contributions (such as blogs or product reviews), or other works. Instead of giving each participant the same single "vote," the feedback is weighted according to the responder's own reputation. Thus if a number of people whose own product reviews have been highly recommended also recommend another review, that review will be much more highly rated. Examples of this sort of system include the PageRank algorithm used by Google, the consumer review site Epinions.com, and the "techie" favorite Slashdot.

Developing a trust and reputation system that is effective but unobtrusive is particularly important for the collaborative creation of content such as for search engines and wikis (see USER-CREATED CONTENT and WIKIS AND WIKIPEDIA).

Online attacks on reputation (defamation) are also a growing problem for both businesses and individuals. Even if the perpetrator is legally prosecuted or otherwise forced to stop, removing the defamatory material (and links to it) can be difficult, since the material may have been extensively cached, archived, or otherwise copied. Thus consultants in "reputation management" have begun to offer their services to the victims of undeserved negative reputations. (This includes optimizing for search engines so that negative material will be pushed to the bottom of the results.)

Further Reading

Dellarocas, Chrysanthos. "Reputation Mechanisms" [overview and resources]. University of Maryland, R. H. Smith School of Business. Available online. URL: http://www.washingtonpost. com/wp-dyn/content/article/2007/07/01/AR2007070101355. html? hpid=artslot. Accessed November 29, 2007.

Kinzie, Susan, and Ellen Nakashima. "Calling In Pros to Refine Your Google Image." Washington Post.com, July 2, 2007. Available online. URL: http://www.washingtonpost.com/ wp-dyn/content/article/2007/07/01/AR2007070101355.html? hpid=artslot. Accessed November 29, 2007.

Masum, Hassan, and Yi-Cheng Zhang. "Manifesto for the Reputation Society." *First Monday*, vol. 9 (July 2004). Available online. URL: http://www.firstmonday.org/issues/issue9_7/ masum/index.html. Accessed November 29, 2007.

Solove, Daniel J. *The Future of Reputation: Gossip, Rumor, and Privacy on the Internet.* New Haven, Conn.: Yale University Press, 2007.

Trust Metrics Evaluation Project. Available online. URL: http:// www.trustlet.org/wiki/Trust_Metrics_Evaluation_project. Accessed November 29, 2007.

Turing, Alan Mathison

(1912–1954)
British
Mathematician, Computer Scientist

Alan Turing's broad range of thought pioneered many branches of computer science, ranging from the fundamental theory of computability to the question of what might constitute true artificial intelligence.

Turing was born in London on June 23, 1912. His father worked in the Indian (colonial) Civil Service, while his mother came from a family that had produced a number of distinguished scientists. As a youth Turing showed great interest and aptitude in both physical science and mathematics. When he entered King's College, Cambridge, in 1931, his first great interest was in probability, where he wrote a well-regarded thesis on the Central Limit Theorem.

Turing's interest then turned to the question of what problems could be solved through computation (see COMPUTABILITY AND COMPLEXITY). Instead of pursuing conventional mathematical strategies, he re-imagined the problem by creating the Turing Machine, an abstract "computer" that performs only two kinds of operations: writing or not writing a symbol on its imaginary tape, and possibly moving one space on the tape to the left or right. Turing showed that from this simple set of states (see FINITE STATE MACHINE) any possible type of calculation could be constructed. His 1936 paper "On Computable Numbers"

together with another researcher's different approach (see CHURCH, ALONZO) defined the theory of computability. Turing then came to America, studied at Princeton University, and received his Ph.D. in 1938.

Turing did not remain in the abstract realm, however, but began to think about how actual machines could perform sequences of logical operations. When World War II erupted, Turing returned to Britain and went into service with the government's Bletchley Park code-breaking facility. He was able to combine his previous work on probability and his new insights into computing devices to help analyze cryptosystems such as the German Enigma cipher machine and to design specialized code-breaking machines.

As the war drew to an end, Turing's imagination brought together what he had seen of the possibilities of automatic computation, and particularly the faster machines that would be made possible by harnessing electronics rather than electromechanical relays. In 1946, he received a British government grant to build the ACE (Automatic Computing Engine). This machine's design incorporated advanced programming concepts such as the storing of all instructions in the form of programs in memory without the mechanical setup steps required for machines such as the ENIAC. Another important idea of Turing was that programs could modify themselves by treating their own instructions just like other data in memory. However, the engineering of the advanced memory system led to delays, and Turing left the project in 1948 (it would be completed in 1950). Turing also continued his interest in pure mathematics and developed a new interest in a completely different field, biochemistry.

Turing's last and perhaps greatest impact would come in the new field of artificial intelligence. Working at the University of Manchester, Turing devised a concept that became known as the Turing Test. In its best-known variation, the test involves a human being communicating via a Teletype with an unknown party that might be either another person or a computer. If a computer at the other end is sufficiently able to respond in a humanlike way, it may fool the human into thinking it is another person. This achievement could in turn be considered strong evidence that the computer is truly intelligent. Since Turing's 1950 article computer programs such as ELIZA and Web "chatterbots" have been able to temporarily fool people they encounter, but no computer program has yet been able to pass the Turing Test when subjected to extensive probing questions by a knowledgeable person.

Alan Turing had a secret that was very dangerous in that time and place: He was gay. In 1952, the socially awkward Turing stumbled into a set of circumstances that led to his being arrested for homosexual activity, which was illegal and heavily punished at the time. The effect of his trial and forced medical "treatment" suggested that his death from cyanide poisoning on June 7, 1954, was probably a suicide.

Alan Turing's many contributions to computer science were honored by his being elected a Fellow of the British Royal Society in 1951 and by the creation of the prestigious Turing Award by the Association for Computing Machinery, given every year since 1966 for outstanding contributions to computer science.

Further Reading

Henderson, Harry. *Modern Mathematicians*. New York: Facts On File, 1996.

Herken, R. *The Universal Turing Machine*. 2nd ed. London: Oxford University Press, 1988.

Hodges, A. *Alan Turing: The Enigma*. New York: Simon & Schuster, 1983. Reprinted New York: Walker, 2000.

———. "Alan Turing" [Web site]. Available online. URL: http://www.turing.org.uk/. Accessed August 23, 2007.

Turing, Alan M. "Computing Machinery and Intelligence." *Mind*, vol. 49, 1950, 433–460.

———. "On Computable Numbers, with an Application to the Entscheidungsproblem." *Proceedings of the London Mathematical Society*, vol. 2, no. 42, 1936–1937, 230–265.

———. "Proposed Electronic Calculator." In Carpenter, B. E. and R. W. Doran, eds. *A. M. Turing's ACE Report of 1946 and other Papers*. Charles Babbage Institute Reprint Series in the History of Computing, vol. 10. Cambridge, Mass.: MIT Press, 1986.

Turkle, Sherry

(1948–   )
American
Scientist, Writer

From the time cyberspace began to become a reality in the 1980s, Sherry Turkle has been a pioneer in studying the psychological, social, and existential effects of computer use and online interaction.

Turkle was born Sherry Zimmerman on June 18, 1948, in Brooklyn, New York. After graduating from Abraham Lincoln High School as valedictorian in 1965, she enrolled in Radcliffe College (which later became part of Harvard University) in Cambridge, Massachusetts. However, when she was a junior her mother died, she quarreled with her stepfather, and dropped out of Radcliffe because she was no longer able to keep up with her studies.

Turkle then went to France, which by the late 1960s had become the scene of social and intellectual unrest. A new movement called poststructuralism was offering a radical critique of modern institutions. Turkle became fascinated by its ideas and attended seminars by such key figures as Michael Foucault and Roland Barthes. In particular, personal experience and intellectual interest joined in spurring her to investigate personal identity in the modern world.

Poststructuralism and postmodernism see identity as something constructed by society or by the individual, not something inherent. The new philosophers spoke of people as having multiple identities between which they could move "fluidly." In being able to try on a new identity for herself, Turkle could see the applicability of these ideas to her own life and began to explore how they may also explain the changes that were sweeping through society.

Turkle decided to return to the United States to resume her studies. In 1970 she received an A.B. degree in social studies, summa cum laude, from Radcliffe. After working for a year with the University of Chicago's Committee on Social Thought, she enrolled in Harvard, receiving an M.A. in sociology in 1973. She went on to receive her doctorate in sociology and personality psychology in 1976, writing about the relationship between Freudian thought and the modern French revolutionary movements.

PSYCHOLOGY OF CYBERSPACE

After getting her Harvard Ph.D. Turkle accepted a position as an assistant professor of sociology at nearby MIT. Here she found a culture as exotic as that of the French intellectuals, but seemingly very different. In encountering the MIT hackers that would later be described in Steven Levy's book *Hackers*, Turkle became intrigued by the way the students were viewing all of reality (including their own emotions) through the language of computers.

For many of the MIT computer students, the mind was just another computer, albeit a complicated one. An emotional overload required "clearing the buffer," and troubling relationships should be "debugged." Fascinated, Turkle began to function as an anthropologist, taking notes on the language and behavior of the computer students. In her second book, *The Second Self: Computers and the Human Spirit* (1984), Turkle says that the computer for its users is not an inanimate lump of metal and plastic, but an "evocative object" that offers images and experiences and draws out emotions. She also explained that the computer could satisfy deep psychological needs, particularly by offering a detailed but structured "world" (as in a video game) that could be mastered, leading to a sense of power and security.

Although the computer culture of the time was largely masculine, Turkle also observed that this evocative nature of technology could also allow for a "soft approach" based on relationship rather than rigorous logic. This "feminine" approach usually met with rejection. Turkle believed that this message had to be changed if girls were not to be left behind in the emerging computer culture.

Turkle had already observed how computer activities (especially games) often led users to assume new identities, but she had mainly studied stand-alone computer use. In the 1990s, however, online services and the Internet in particular increasingly meant that computer users were interacting with other users over networks. In her 1995 book *Life on the Screen*, Turkle takes readers inside the fascinating world of the MUD, or multi-user dungeon (see online games). In this fantasy world created by descriptive text, users could assume any identity they wished. An insecure teenage boy could become a mighty warrior—or perhaps a seductive woman. A woman, by assuming a male identity, might find it easier to be assertive and would avoid the sexism and harassment often directed at females.

The immersive world of cyberspace offers promises, perils, and potential, according to Turkle. As with other media, the computer world can become a source of unhealthy escapism, but it can also give people practice in using social skills in a relatively safe environment, although there can be difficulty in transferring skills learned online to a face-to-face environment.

Although the computer has provided a new medium for the play of human identity, Turkle has pointed out that the question of identity (and the reality of multiple identities) is inherent in the postmodern world. Looking to the future, Turkle suggests that the boundary between cyberspace and so-called real life is vanishing, forcing people to confront the question of what "reality" means. Meanwhile, she remains concerned about the paradox in which people may be becoming at once more connected and more alienated than ever before.

Turkle married artificial intelligence pioneer and educator Seymour Papert; the marriage ended in divorce. She continues today at MIT as a professor of the sociology of science and as director of the MIT Initiative on Technology and Self. She has received a number of fellowships, including a Rockefeller Foundation fellowship (1980), a Guggenheim fellowship (1981), Fellow of the American Association for the Advancement of Science (1992), and World Economic Forum Fellow (2002).

In 1984 Turkle was selected Woman of the Year by *Ms.* magazine, and she has made a number of other lists of influential persons such as the "Top 50 Cyber Elite" of *Time Digital* (1997) and *Time Magazine*'s Innovators of the Internet (2000).

Further Reading
"Sherry Turkle" [Home page]. Available online. URL: http://web. mit.edu/sturkle/www/. Accessed November 29, 2007.
Turkle, Sherry. "Can You Hear Me Now?" *Forbes,* May 7, 2007, p. 176.
———. *Evocative Objects: Things We Think With.* Cambridge, Mass.: MIT Press, 2007.
———. *Life on the Screen: Identity in the Age of the Internet.* New York: Simon and Schuster, 1995.
———. *The Second Self: Computers and the Human Spirit.* New York: Simon and Schuster, 1984.

typography, computerized

The more than five-century-old art of typography (the design, arrangement, and setting of printing type) was transformed in the latter part of the 20th century by digital technology. With the exception of some traditional presses devoted to the fine book market, nearly all type used today is designed and set by computer.

Most users are familiar with the typefaces distributed with their operating system and software, such as the popular Adobe and TrueType (see ADOBE SYSTEMS and FONT). Many such font designs are based on (and sometimes named after) traditional typefaces, modified for readability using typical displays and printers.

For control of composition, there are three overlapping levels of software, ranging from easiest to use (but most limited) to most complex, versatile, and precise. Modern word processors such as Microsoft Word and Open Office provide enough control for many types of shorter documents (see WORD PROCESSING). Desktop publishing software adds facilities suitable for layout of fliers, brochures, newsletters, and similar publications that often mix text and graphics (see DESKTOP PUBLISHING).

More elaborate documents such as books, magazines, and newspapers require more sophisticated facilities to control the layout and flow of text. Some traditional choices include LaTex (for the Tex typesetting program), used particularly by scientists and other academics, and the older troff and its offshoots on UNIX systems. More recent programs include Quark, FrameMaker, PageMaker, and InDesign. Related utilities often used in digital typography include font editors (for design and modification) and utilities to convert fonts from one format to another.

Further Reading
Bringhurst, Robert. *The Elements of Typographic Style.* Version 3.0. Point Roberts, Wash.: Hartley & Marks, 2004.
Ellison, Andy. *The Complete Guide to Digital Type: Creative Use of Typography in the Digital Arts.* New York: Collins Design, 2004.
FontLab. Available online. URL: http://www.fontlab.com/i. Accessed November 30, 2007.
FontSite: Digital Typography & Design. Available online. URL: http://www.fontsite.com/. Accessed November 30, 2007.
Knuth, Donald E. *Computers and Typesetting.* 5 vols. Upper Saddle River, N.J.: Addison-Wesley Professional, 2001.
Page Layout Programs. Aeonix Publishing Group. Available online. URL: http://www.aeonix.com/pagelay.htm. Accessed November 30, 2007.
Troff: The Text Processor for Typesetters. Available online. URL: http://troff.org/. Accessed November 30, 2007.

U

ubiquitous computing

Traditionally people have thought of computers as discrete devices (such as a desktop or handheld device), used for specific purposes such as to send e-mail or browse the Web. However, many researchers and futurists are looking toward a new paradigm that many believe is rapidly emerging. Ubiquitous (or pervasive) computing focuses not on individual computers and tasks but on a world where most objects (including furniture and appliances) have the ability to communicate information. (This has also been called "the Internet of things.") This can be viewed as the third phase in a process where the emphasis has gradually shifted from individual desktops (1980s) to the network and Internet (1990s) to mobile presence and the ambient environment.

Some examples of ubiquitous computing might include:

- picture frames that display pictures attuned to the user's activities

- "dashboard" devices that can be set to display changing information such as weather and stock quotes

- parking meters that can provide verbal directions to nearby attractions

- kiosks or other facilities to provide verbal cues to guide travelers, such as through airports

- home monitoring systems that can sense and deal with accidents or health emergencies

Ubiquitous computing greatly increases the ability of people to seamlessly access information for their daily activities, but the fact that the user is in effect "embed-ded" in the network can also raise issues of privacy and the receiving of unwanted advertising or other information (see PRIVACY IN THE DIGITAL AGE).

An early center of research in ubiquitous computing was Xerox PARC, famous for its development of graphical user interfaces (particularly the work of Mark Weiser). Today a major force is MIT (see MIT MEDIA LAB), especially its Project Oxygen, which explores networks of embedded computers. This challenging research area brings together aspects of many other fields (see artificial intelligence, distributed computing, psychology of computing, smart buildings and homes, touchscreen, user interface, and WEARABLE COMPUTERS). Note that while the user's experience of ubiquitous computing might be similar in some ways to that of virtual reality, the latter puts the user into a computer-generated world, while the former uses computing power to enhance the user's connections to the outside world (see VIRTUAL REALITY).

Further Reading

Greenfield, Adam. *Everywhere: The Dawning Age of Ubiquitous Computing.* Berkeley, Calif.: New Riders, 2006.
Igoe, Tom. *Making Things Talk: Practical Methods Connecting Physical Objects.* Sebastapol, Calif.: O' Reilly, 2007.
MIT Project Oxygen. Available online. URL: http://www.oxygen.lcs.mit.edu/. Accessed November 30, 2007.
Morville, Peter. *Ambient Findability: What We Find Changes Who We Become.* Sebastapol, Calif.: O'Reilly, 2005.
Terdiman, Daniel. "Meet the Metaverse, Your New Digital Home." CNET News. Available online. URL: http://news.com.com/Meet+the+metaverse%2C+your+new+digital+home/2100-1025_3-6175973.html. Accessed November 30, 2007.

Vasilakos, Athanasios, and Witold Pedrycz, eds. *Ambient Intelligence, Wireless Networking, and Ubiquitous Computing.* Boston: Artech House, 2006.

UNIX

By the 1970s, time-sharing computer systems were in use at many universities and engineering and research organizations. Such systems, often running on computers such as the PDP series (see MINICOMPUTER), required a new kind of operating system that could manage the resources for each user as well as the running of multiple programs (see MULTITASKING).

An elaborate project called Multics had been begun in the 1960s in an attempt to create such an operating system. However, as the project began to bog down, two of its participants, Ken Thompson and Dennis Ritchie, (see RITCHIE, DENNIS) decided to create a simple, more practical operating system for their PDP-7. The result would become UNIX, an operating system that today is a widely used alternative to proprietary operating systems such as those from IBM and Microsoft.

ARCHITECTURE

The essential core of the UNIX system is the kernel, which provides facilities to organize and access files (see KERNEL and FILE), move data to and from devices, and control the running of programs (processes). In designing UNIX, Thompson deliberately kept the kernel small, noting that he wanted maximum flexibility for users. Since the kernel was the only part of the system that could not be reconfigured or replaced by the user, he limited it to those functions that reliability and efficiency dictated be handled at the system level.

Another way in which the UNIX kernel was kept simple was through device independence. This meant that instead of including specific instructions for operating particular models of terminal, printers, or plotters within the kernel, generic facilities were provided. These could then be interfaced with device drivers and configuration files to control the particular devices.

A UNIX system typically has many users, each of whom may be running a number of programs. The interface that processes user commands is called the *shell*. It is important to note that in UNIX a shell is just another program, so there can be (and are) many different shells reflecting varying tastes and purposes (see SHELL). Traditional UNIX shells include the Bourne shell (sh), C shell (csh), and Korn shell (ksh). Modern UNIX systems can also have graphical user interfaces similar to those found on Windows and Macintosh personal computers (see USER INTERFACE).

WORKING WITH COMMANDS

UNIX systems come with hundreds of utility programs that have been developed over the years by researchers working at Bell Labs and campuses such as the University of California at Berkeley (UCB). These range from simple commands for working with files and directories (such as cd to set a current directory and ls to list the files in a directory) to language compilers, editors, and text-processing utilities.

Whatever shell is used, UNIX provides several key features for constructing commands. A powerful system of patterns (see REGULAR EXPRESSION) can be used to find files that match various criteria. For a very simple example, the command

```
% ls *.doc
```

will list all files in the current directory that end in .doc. (The % represents the command prompt given by the shell.)

Most earlier operating systems used special syntax to refer to devices such as the user's terminal and the printer. UNIX, however, treats devices just like other files. This means that a program can receive its input by opening a terminal file and send its output to another file. For example:

```
% cat > note
This is a note.
^D
```

The cat (short for concatenate) command adds the user's input to a file called note. The ^D stands for Control-D, the special character that marks end-of-file. Once the command finishes, there is a file called note on the disk, which can be listed by the ls command:

```
% ls –l note
-rw——- 1 hrh well 16 Mar 25 20:16 note
```

The contents of the file can be checked by issuing another cat command:

```
% cat note
This is a note.
```

Many commands default to taking keyboard input if no input file is specified. For example, one can type sort followed by a list of words to sort:

```
% sort
apple
pear
orange
tangerine
lemon
^D
```

Once the input is finished, the sort command outputs the sorted list:

```
apple
lemon
orange
pear
tangerine
```

One of the things that makes UNIX attractive to its users is the ability to combine a set of commands in order to perform a task. For example, suppose a user on a timesharing system wants to know which other users are logged on. The who command provides this information, but it includes a lot of details that may not be of interest. Suppose one just wants the names of the current users. One way to do this is to connect the output of the who command to awk, a scripting language (see AWK and SCRIPTING LANGUAGES).

```
% who | awk ' { print $1 }'
```

Here the vertical bar (called a *pipe*) connects the output of the first command to the input of the second. Thus, the awk command receives the output of the who command. The statement print $1 tells awk to output the first column from who's output, which is just the names of the users. The first part of the list looks like this:

```
mnemonic
bernie
kryan
nanlev
goddessj
brady
demaris
techgirl
```

This is fine, but the output might be better if it were sorted. All that's needed is to add one more pipe to connect the output of the awk command to the sort command:

```
% who | awk ' {print $1}' | sort
aarong
aimee
almanac
amicus
autumn
biscuit
bradburn
brian
```

The ability to redirect input and output and to use pipes to connect commands makes it easy for UNIX users to create mini-programs called scripts to perform tasks that would require full-fledged compiled programs on other systems. For example, the preceding command could be put into a file called users, and the file could be set to be executable. Once this is done, all the user has to do to get the user list is to type users at a shell prompt. Today UNIX users have a wide choice of powerful scripting languages (see PERL and PYTHON).

UNIX THEN AND NOW

The versatility of UNIX quickly made it the operating system of choice for most campuses and laboratories, as well as for many software developers. When PCs came along in the late 1970s and 1980s, they generally lacked the resources to run UNIX, but developers of PC operating systems such as CP/M and MS-DOS were influenced by UNIX ideas including the hierarchical file system with its levels of directories, the use of a command-processing shell, and wildcards for matching filenames.

Besides hardware requirements, another barrier to the use of UNIX by home and business users was that the operating system was copyrighted by Bell Labs and a UNIX license often cost more than the PC to run it on. However, a combination of the efforts of the Free Software Foundation (see STALLMAN, RICHARD) and a single inspired programmer (see TORVALDS, LINUS) resulted in the release of Linux, an operating system that is fully functionally compatible with UNIX but uses no AT&T code and is thus free of licensing fees (see LINUX).

Although UNIX has been somewhat overshadowed by its Linux progeny, a variety of open-source versions of traditional UNIX systems have become available. In 2005 Sun Microsystems released OpenSolaris (based on UNIX System V). There is also OpenBSD, derived from the UC Berkeley Software Distribution (but with stronger security features), and available for most major platforms. Finally, the continuing influence of UNIX can also be seen in the current generation of operating systems for the Apple Macintosh (see OS X).

Further Reading
Bach, Maurice J. *The Design of the UNIX Operating System.* Upper Saddle River, N.J.: Prentice Hall, 1986.

Lucas, Michael W. *Absolute OPENBSD: UNIX for the Practical Paranoid.* San Francisco: No Starch Press, 2003.

OpenBSD. Available online. URL: http://www.openbsd.org/. Accessed August 23, 2007.

OpenSolaris Project. Available online. URL: http://www.opensolaris.org/os/. Accessed August 23, 2007.

Peek, Jerry, Grace Todino-Gonquet, and John Strang. *Learning the UNIX Operating System.* 5th ed. Sebastapol, Calif.: O'Reilly Media, 2002.

Robbins, Arnold. *UNIX in a Nutshell.* 4th ed. Sebastapol, Calif.: O'Reilly Media, 2005.

Salus, Peter H. *A Quarter Century of UNIX.* Reading, Mass.: Addison-Wesley, 1994.

"Unix Introduction and Quick Reference." Available online. URL: http://www.decf.berkeley.edu/help/handouts/unix-intro.pdf. Accessed August 23, 2007.

USB

The traditional ways to connect a computer to peripheral devices such as printers are via parallel and serial connections (see PARALLEL PORT and SERIAL PORT). Both methods are standardized and reliable, but by the mid-1990s people wanted to connect many more data-hungry devices to their PCs, including scanners, digital cameras, and external storage drives. Besides wanting faster data transfers, system designers looked for a way to connect more devices without having to add more ports to the motherboard. It would also be convenient to be able to plug or unplug devices without having to reboot the PC. The Universal Serial Bus (USB) has all these features: It is relatively fast and quite flexible.

Introduced in 1996, USB uses a four-wire cable with small rectangular connectors. Devices can be connected directly to the host USB hub built into the computer. Alternatively, a second hub can be connected to the host hub, allowing for several devices to share the same connection. (Often for convenience, monitors and other devices now include built-in USB hubs for ease in connecting other devices on the desktop.)

Two of the four wires carry power from the PCs power supply (or from a secondary powered hub) to the connected devices. The other two wires carry data. The 1s and 0s in the data are signaled by the difference in voltage between the two wires. (This tends to reduce the effects of outside electromagnetic interference, since if both wires are affected similarly, the difference between them won't change.)

When a USB device is connected, it creates a voltage change that causes the USB system in the PC to query it for identifying information. If the information indicates that the device has not been installed, the operating system begins an installation procedure that can be carried out either automatically or with a little help from the user (see PLUG AND PLAY).

Once a device is installed, its identifying information tells the USB system what data rate it can handle. (Some devices, such as keyboards, don't need to be very fast, while others, such as CD drives, place a premium on speed.) The USB system assigns each device an address. The system functions like a miniature token-ring network, sending queries or commands with tokens identifying the appropriate device. The devices respond to requests that have their token and in turn send requests when they have data to transmit.

The USB system can assign priorities to devices according to their need for an uninterrupted flow of data. A recordable CD (CD-RW) drive, for example, is sensitive to interruptions in the flow of data, so it is given a high priority. A keyboard sends only a tiny bit of data at a time, and it can get by with a low priority, requesting service as needed. Other devices such as scanners and printers may handle a large flow of data, but are not very sensitive to interruptions and can have a medium priority.

The original USB specification allowed for up to 12 MB/sec data transfers. However, the current USB 2.0 specification allows speeds up to 480 MB/sec, enough to easily handle a digital camera, scanner, and CD-RW drive simultaneously.

Today USB connections are also often used to add external hard drive storage as well as for handy "thumb drives" or "memory sticks" that make it easy to carry one's documents and even software from one machine to the next (see FLASH DRIVE). Finally, a USB-connected wireless adapter (see BLUETOOTH and WIRELESS COMPUTING) can be the most convenient way to put an older desktop or laptop PC into communication with the rest of the world.

Further Reading

Axelson, Jan. "USB Central" [Web site]. Available online. URL: http://www.lvr.com/usb.htm. Accessed August 23, 2007.
———. *USB Complete: Everything You Need to Develop Custom USB Peripherals*. 3rd ed. Madison, Wisc.: Lakeview Research, 2005.
———. *USB Mass Storage: Designing and Programming Devices and Embedded Hosts*. Madison, Wisc.: Lakeview Research, 2006.
ZDNET USB Resources. Available online. URL: http://updates.zdnet.com/tags/USB.html?t=17&s=0&o=0. Accessed August 23, 2007.

user-created content

Traditional print and broadcast media divide the world into two groups: content producers and content consumers. However, as noted by its creator Tim Berners-Lee at its very beginning, the World Wide Web had at least the potential for users to take an active role in linking existing content and contributing their own (see BERNERS-LEE, TIM and WORLD WIDE WEB). Indeed, Berners-Lee wanted Web client software to include not only browsing functions but easy ways for users to create their own Web pages.

In reality, early users faced something of a learning curve, usually having to cope with HTML to some extent, for example. But by the mid-2000s a variety of new media of communication had become readily accessible using an ordinary Web browser at sites that host the required software. The most prominent applications are blogs (see BLOGS AND BLOGGING) and wikis, particularly Wikipedia (see WIKIS AND WIKIPEDIA).

Meanwhile, inexpensive digital still and video cameras and easy-to-use editing software encouraged people to make their own media creations. Sites to enable users to upload, share, and comment on their creations have flourished (see YOUTUBE).

The growth of sites such as Facebook and MySpace (see SOCIAL NETWORKING) has also provided new ways for users from junior high school age on up to create and share content.

APPLICATIONS AND ISSUES

The effects of this ability to create as well as consume media are proving to be far-reaching. YouTube, for example, has featured elaborate documentaries and controversial political pieces. Candidates in the 2008 presidential primary campaign had to deal with a new debate format where voter's questions, rather than being read by a moderator, are presented in videos created by the voters themselves.

User-created content is also becoming significant in media and journalism. Some newspapers are even beginning to experiment with assigning stories to volunteer collaborators—possibly as a way of coping with diminishing revenues and budgets for professional journalists.

Forms of user-created content are also increasingly prevalent in the traditional broadcast media. While previously existing only in such forms as talk radio and game shows, more or less unscripted participation is now found in reality TV and the endless variants on *American Idol*.

Economists and social scientists are beginning to explore how a combination of open-source, user-created content, and mass collaboration are changing how information is assembled and used and even how products are designed (see OPEN-SOURCE MOVEMENT).

However, user-created content also raises challenges: What kind of journalistic standards might be applied to non-professional reporting and documentaries? (See JOURNALISM AND COMPUTERS.) What should be the responsibility of the owner of a venue such as YouTube for content that might violate someone's copyright or be defamatory? The creation of content that contains information about personal identity can also be problematic from a privacy and security standpoint. Finally, in venues such as online games where users may have spent hundreds of hours creating content, the question of who owns it and what they can do with it is significant.

Further Reading

Espejo, Roman. *User-Generated Content*. (At Issue Series). Farmington, Mich.: Greenhaven Press, 2007.

Garofoli, Joe. "User News Sites Offer Diverse Stories, Some Questionable Sources." *San Francisco Chronicle.* Available online. URL: http://www.sfgate.com/cgi-bin/article.cgi?f=/c/a/2007/09/12/MNPDS3RE6.DTL&hw=user+content&sn=001&sc=100. Accessed November 30, 2007.

Gillmor, Dan. *We the Media: Grassroots Journalism by the People, for the People.* Cambridge, Mass.: O'Reilly, 2004.

Hietannen, Herkko, Ville Oksanen, and Vikko Valimaki. *Community Created Content: Law, Business, and Policy.* Helsinki, Finland: Turre Publishing, 2007.

King, Brad. "User-Created Content Comes to TV." Technology Review blogs, December 5, 2006. Available online. URL: http://www.technologyreview.com/blog/editors/17485/. Accessed November 30, 2007.

Tapscott, Don, and Anthony D. Williams. *Wikinomics: How Mass Collaboration Changes Everything.* New York: Portfolio, 2006.

Vickery, Graham, and Sacha Wunsch-Vincent. *Participative Web and User-Created Content: Web 2.0 Wikis and Social Networking.* Paris: Organization for Economic Cooperation and Development, 2007.

user groups

Computer users have always had an interest in finding and sharing information about the systems they are trying to use. As early as 1955, users of the IBM 701 mainframe banded together, in this case to try to influence IBM's decisions about new software. Later, users of minicomputers made by Digital Equipment Corporation formed DECUS.

By the mid-1970s, microcomputer experimenters had organized several groups, of which the most influential was probably the Homebrew Computer Club, meeting first in a garage in Menlo Park, California, 1975. The group soon was filling an auditorium at Stanford University. Members demonstrated and explained their hand-built computer systems, argued the merits of kits such as the Altair, and later, witnessed Steve Wozniak's prototype Apple I computer.

At the other end of the scale, users of UNIX on university computer systems had formed USENIX, the UNIX user's group. A growing system of newsgroups called USENET (see NETNEWS AND NEWSGROUPS) would soon extend beyond UNIX concerns to hundreds of other topics.

Early PC users had great need for user groups. Technical support was primitive and the variety of computer books limited, so the best way to get quirky hardware or balky software to work was often to ask fellow users, read user group newsletters, or skim through the great variety of small publications that catered to users of particular systems. Users could also meet to swap public domain software disks. User groups could be formed around software as well as hardware. Thus, users could swap spreadsheet templates or discuss Photoshop techniques.

User groups have gradually become less important, or perhaps it is better to say that they have changed their mode of existence. Starting in the mid-1980s, the modem and bulletin board, on-line services such as CompuServe and later, Web sites offered more convenient access to information and software without the need to attend meetings. At the same time, the quality and reliability of hardware and software has steadily improved, even though there is always a new crop of problems.

User groups played a key role in the adoption of new technology, much as they had in earlier movements such as amateur radio. Today it might be said that every user has the opportunity to join numerous virtual user groups, although the sense of fellowship and mutual exploration may be somewhat lacking.

Further Reading

Association of Personal Computer User Groups. Available online. URL: http://www.apcug.org/. Accessed August 23, 2007.

Moen, Rick. "Linux User Group HOWTO." Available online. URL: http://www.linux.org/docs/ldp/howto/User-Group-HOWTO.html. Accessed August 23, 2007.

MUG Center: The Mac User Group Resource Site. Available online. URL: http://www.mugcenter.com/. Accessed August 23, 2007.

USENIX. Available online. URL: http://www.usenix.org/. Accessed August 23, 2007.

User Group Network. Available online. URL: http://www.usergroups.net/. Accessed August 23, 2007.

WUGNET (Windows Users Group Network). Available online. URL: http://www.wugnet.com/. Accessed August 23, 2007.

user interface

All computer designers are faced with the question of how users are going to communicate with the machine in order to get it to do what they want it to do. User interfaces have evolved considerably in 60 years of computing.

The user interface for ENIAC and other early computers consisted of switches or plugs for configuring the machine for a particular problem, followed by loading instructions from punch cards. The mainframes of the 1950s and 1960s had control consoles from which text commands could be entered (see JOB CONTROL LANGUAGE).

The time-sharing computers that became popular starting in the 1960s still used only text commands, but they were more interactive. Users could type commands to examine directories and files, and run utilities and other programs. Starting in the 1970s, UNIX provided a powerful and flexible way to combine commands to carry out a variety of tasks interactively or through batch processing (see UNIX and SHELL).

The first graphical user interfaces (GUIs) resulted from experimental work at the Xerox Palo Alto Research Center (PARC) during the 1970s. Instead of typing commands at a prompt, GUI users can use a mouse to open menus and select commands, and click on icons to open programs and files. For operations that require detailed specifications, a standard dialog box can be presented, using controls such as check boxes, buttons, text boxes, and sliders.

GUIs entered the mainstream thanks to Apple's Macintosh and Microsoft Windows for IBM-compatible PCs. By the mid-1990s, the GUI had supplanted text-based operating systems such as MS-DOS for most PC users. The strength of the GUI is that it can visually model the way users work with objects in the real world. For example, a file can be deleted by dragging it to a trash can icon and dropping it in. Dragging a slider control to adjust the volume for a sound card is directly analogous to moving a slider on a home stereo system.

Because a system like Windows or the Macintosh provides developers with standardized interface objects and conventions, users are able to learn the basics of operating a new application more quickly. Whereas in the old days different programs might use slightly different keystrokes or commands for saving a file, Windows users know that in virtually any application they can open the File menu and select Save, or press Ctrl-S.

With the growth of the World Wide Web, interface design has extended to Web pages. Generally, Web pages use similar elements to desktop GUIs, but there are some special considerations such as browser compatibility, response at differing connection speeds, and the integration of text and interactive elements.

GUIs do have some general drawbacks. An experienced user of a text-based operating system might be able to type a precise command that could find all files of a given type on the system and copy them to a backup directory. The GUI counterpart might involve opening the Search menu, typing a file specification, and making further selections and menu choices to perform the copy. Command-driven systems also provide for powerful scripting capabilities. GUI systems often allow for the recording of keystrokes or menu selections, but this is less powerful and versatile.

Another important consideration is the difficulty that people with certain disabilities may have in using GUI systems. There are a variety of possible solutions, many of which are incorporated in Microsoft Windows, Web browsers, and other software. These include screen magnifier or reader utilities for the visually impaired and alternatives to the mouse such as head tracker/pointers (see DISABLED PERSONS AND COMPUTING).

Designers of user interfaces have to consider whether the elements of the system are intuitively understandable and consistent and whether they can be manipulated in efficient yet natural ways (see also ERGONOMICS OF COMPUTING).

ALTERNATIVE AND FUTURE INTERFACES

The marketplace has spoken, and the desktop GUI is now the mainstream interface for most ordinary PC users. However, there are a variety of other interfaces that are used for particular circumstances or applications, such as:

- touchscreens (as with ATMs) (see TOUCHSCREEN)

- handwriting or written "gesture" recognition, such as on handheld computers (see HANDWRITING RECOGNITION) or for drawing tablets

- voice-controlled systems (see SPEECH RECOGNITION AND SYNTHESIS)

- trackballs, joysticks, and touchpads (used as mouse alternatives)

- virtual reality interfaces using head-mounted systems, sensor gloves, and so on (see VIRTUAL REALITY)

Because much interaction with computers is now away from the desktop and taking place on laptops, handheld, or palm computers, and even in cars, there is likely to be continuing experimentation with user interface design.

Further Reading
Arlov, Laura. *GUI Design for Dummies*. New York: Wiley, 1997.
Galitz, Wilbert O. *The Essential Guide to User Interface Design: An Introduction to GUI Design Principles and Techniques*. 3rd ed. Indianapolis: Wiley, 2007.
Hobart, John. "Principles of Good GUI Design." Available online. URL: http://www.iie.org.mx/Monitor/v01n03/ar_ihc2.htm. Accessed August 23, 2007.
Johnson, Jeff. *GUI Bloopers 2.0: Common User Interface Design Don'ts and Dos*. San Francisco: Morgan Kaufmann, 2007.
Krug, Steve. *Don't Make Me Think: A Common Sense Approach to Web Usability*. 2nd ed. Berkeley, Calif.: New Riders, 2005.
Stephenson, Neil. *In the Beginning Was the Command Line*. New York: Avon Books, 1999.

V

variable

Virtually all computer programs must keep track of a variety of items of information that can change as a result of processing. Such values might include totals or subtotals, screen coordinates, the current record in a database, or any number of other things. A *variable* is a name given to such a changeable quantity, and it actually represents the area of computer memory that holds the relevant data.

Consider the following statement in the C language:

```
int Total = 0;
```

Variables have several attributes. First, every variable has a name—Total in this case. Although this name actually refers to an address in memory, in most cases the programmer can use the much more readable name instead of the actual address.

It is possible to have more than one name for the same variable by having another variable point to the first variable's contents, or by declaring a "reference" variable (see POINTERS AND INDIRECTION).

Each variable has a data type, which might be number, character, string, a collection (such as an array), a data record, or some special type defined by the programmer (see DATA TYPES). With some exceptions (see SCRIPTING LANGUAGES) most modern programming languages require that the programmer declare each variable before it is used. The declaration specifies the variable's type—in the current example, the type is int (integer, or whole number).

A variable is usually given an initial value by using an assignment statement; in the example above the variable Total is given an initial value of 0, and the assignment is combined with the declaration. (Some languages automatically assign a default value such as 0 for a number or a null character for a string, but with other languages failure to assign a value results in the variable having as its value whatever happens to be currently stored in the memory address associated with the variable. An explicit assignment is thus always safer and more readable.)

When exactly do variables get set up, and when do they get their values? This varies with the programming language (see BINDING). With C and similar languages, a variable receives its data type when the program is compiled (compile time). The type in turn determines the range of values that the variable can hold (physically based on the number of bytes of memory allocated to it). The variable's value is actually stored in that location when the program is executed (run time).

A few languages such as APL and LISP use dynamic binding, meaning that a data type is not associated with a variable until run time. This makes for flexibility in programming, but at some cost in efficiency of storage and execution speed.

During processing, a variable's value can change through the use of operators in expressions (see OPERATORS AND EXPRESSIONS). Thus, the example value Total might be changed by a statement such as:

```
Total = Total + Subtotal;
```

When this statement is executed, the following happens:

The value of the memory location labeled "Subtotal" is obtained.

The value of the memory location labeled "Total" is obtained.

The two values are added together.

The result is stored in the location labeled "Total," replacing its former value.

SCOPE OF VARIABLES

In early programming languages variables were generally global, meaning that they could be accessed and changed from any part of the program. While this practice is convenient, it became riskier as programs became larger and more complex. One part of a program might be using a variable called Total or Subtotal to keep track of some quantity. Later, another part is written to deal with some other calculation, and uses the same names. The programmer may think of the second Total and Subtotal as being quite separate from the first, but in reality they refer to the same memory locations and any change affects both of them. Thus, it's easy to create unwanted "side effects" when using global variables.

Starting in the 1960s and more systematically during the 1970s, there was great interest in designing computer languages that could better manage the structure and complexity of large programs (see STRUCTURED PROGRAMMING). One way to do this is to break programs up into more manageable modules that each deal with some specific task (see PROCEDURES AND FUNCTIONS). Unless explicitly declared to be global, variables within a procedure or function are local to that unit of code. This means that if two procedures both have a variable called Total, changes to one Total do not affect the other.

Generally, in block-structured languages such as Pascal a variable is by default local to the block of code in which it is defined. This means it can be accessed only within that block. (Its visibility is said to be limited to that block.) The variable will also be accessible to any block that is nested within the defining block, unless another variable with the same name is declared in the inner block. In that case the inner variable supersedes the outer one, which will not be visible in the inner block.

Some languages such as APL and early versions of LISP define scope differently. Since these languages are not block structured, scope is determined not by the relationship of blocks of code but by the sequence in which functions are called. At run time each variable's definition is searched for first in the code where it is first invoked, then in whatever function called that code, then in the function that called *that* function, and so on. As with dynamic binding, dynamic scooping offers flexibility but at a considerable price. In this case, the price is that the program's effects on variables will be hard to understand, and the search mechanism slows down program execution. Dynamic scoping is thus not often used today, even in LISP.

Global variables were convenient because they allowed information generated by one part of a program to be accessed by any other. However, such accessibility can be provided in a safer, more controlled form by explicitly passing variables or their values to a procedure or function when it is called (see PROCEDURES AND FUNCTIONS).

Object-oriented languages provide another way to control or encapsulate information. Variables describing data used within a class are generally declared to be private (accessible only within the functions used by the class). Public (i.e., global) variables are used sparingly. The idea is that if another part of the program wants data belonging to a class, it will call a member function of the class, which will provide the data without giving unnecessary access to the class's internal variables.

A final concept that is important for understanding variables is that of lifetime, that is, how long the definition of a variable remains valid. For efficiency, the runtime environment must deallocate memory for variables when they can no longer be used by the program (that is go "out of scope"). Generally, a variable exists (and can be accessed) only while the block of code in which it was defined is being executed (including any procedures or functions called from that block). In the case of a variable declared in the main program, this will be until the program as a whole reaches its end statement. For variables within procedures or functions, however, the lifetime lasts only until the procedure or function ends and control is returned to the calling statement. However, languages such as C allow the special keyword *static* to be used for a variable that is to remain in existence as long as the program is running. This can be useful when a procedure needs to "remember" some information between one call and the next, such as an accumulating total.

Further Reading

Kernighan, Brian W., and Dennis Ritchie. *The C Programming Language.* 2nd ed. Englewood Cliffs, N.J.: Prentice Hall, 1988.
Sebesta, Robert W. *Concepts of Programming Languages.* 8th ed. Boston: Addison-Wesley, 2007.
Stroustrup, Bjarne. *The C++ Programming Language.* special ed. Reading, Mass.: Addison-Wesley, 2000.

VBScript

Dating back to the mid-1990s, VBScript is a scripting language developed by Microsoft and based on its popular Visual Basic programming language (see BASIC and SCRIPTING LANGUAGE). It is also part of the evolution of what Microsoft called "active scripting," based on components that allow outside access to the capabilities of applications. The host environment in which scripts run is provided through Windows (as with Windows Script Host) or within Microsoft's Internet Explorer browser.

For client-side processing, VBScript can be used to write scripts embedded in HTML pages, which interact with the standard Document Object Model (see DOM) in a way similar to other Web scripting languages (in particular, see JAVASCRIPT). However, VBScript is not supported by popular non-Microsoft browsers such as Firefox and Opera, so developers generally must use the widely compatible JavaScript instead. VBScript can also be used for processing on the Web server, particularly in connection with Microsoft's Web servers (see ACTIVE SERVER PAGES).

Because versions of Windows starting with Windows 98 include Windows Script Host, VBScripts can also be written to run directly under Windows. One unfortunate consequence was scripts containing worms (such as the I LOVE YOU worm) or other malware and mailed as attachments to unwary users.

EXAMPLES

VBScript code will be very familiar to users of Visual Basic and generally follows syntax similar to that of other object-oriented languages. The canonical "Hello World" program can be simply written as:

```
WScript.Echo "Hello World!"
```

Where WScript is the object representing the script host.

To get user input through a text box, the programmer can write code like this:

```
option explicit
dim userInput
userInput = InputBox("What is your name?:",
"Greetings")
if userInput = "" then
  Msgbox "You did not write anything or you
  pressed cancel!"
else
  MsgBox "Hello, " & userInput & ".",
  vbInformation
end if
```

Of course VBScript has libraries and interfaces to enable it to perform much more complicated tasks, such as querying databases and configuring other aspects of Windows systems through Windows Management Instrumentation (WMI) and Active Directory Services Interface (ADSI).

Although the language (and code using it) will be in use for years to come, Microsoft is no longer actively developing VBScript, having moved on to a new programming framework (see MICROSOFT .NET) and focusing on languages such as Visual Basic .NET.

Further Reading

Jones, Don. *VBScript, WMI, and ADSI Unleashed.* Indianapolis: Sams, 2007.
VBScript Sample Scripts. Available online. URL: http://cwashington.netreach.net/depo/default.asp?topic=repository&scripttype=vbscript. Accessed December 2, 2007.
VBScript Tutorial. W3Schools. Available online. URL: http://www.w3schools.com/vbscript/default.asp. Accessed December 2, 2007.
VBScript User's Guide. Microsoft Developer Network. Available online. URL: http://msdn2.microsoft.com/en-us/library/sx7b3k7y.aspx. Accessed December 2, 2007.
Wilson, Ed. *Microsoft VBScript Step by Step.* Redmond, Wash.: Microsoft Press, 2007.

videoconferencing

The growth of the global economy has meant that many companies have operations in many locations around the world. The time and expense involved in travel have encouraged the search for alternatives to face-to-face meetings (see TELEPRESENCE). The added discomfort and uncertainty related to current airline travel is likely to further spur this movement.

Basic videoconferencing is carried out by using video cameras and microphones to carry the image and voice of each person so that it can be seen by all participants. The video and sound data is digitized and transmitted between the participants' locations, using some existing communications link. Although direct satellite technology can be used, it is very expensive. A more practicable alternative is the use of a proprietary system over special phone lines (such as ISDN or DSL). Increasingly, however, broadband connections to the general Internet are used (see also VOIP). This is relatively inexpensive and flexible, but sometimes less reliable because of the effects of network congestion.

The quality of imagery depends on the system. High-end systems, which can cost tens of thousands of dollars, use large, high-definition screens or even special projection equipment that can give a 3D look to peoples' faces. Although high-end videoconferencing software and hardware can be expensive, there are now a variety of alternatives for small businesses and individual users. (As of 2002 the printing store chain Kinko's is offering videoconferencing through some of its stores for $450/hr.)

For smaller, less formal meetings there are more affordable alternatives. Products such as Microsoft NetMeeting, CuSeeMe, and Yahoo Messenger set up user accounts and a directory that makes it easy for users to connect. Other than the Internet connection, the only hardware needed is a microphone and an inexpensive camera (see WEB CAM).

Business videoconferencing systems often include the ability for participants to view and interact with software applications. This makes it possible not only to view slide shows or other presentations (see PRESENTATION SOFTWARE) but to collaborate on creating documents. An "electronic whiteboard" can be used to display not only computer text and graphics but also handwritten notes created by participants using electronic drawing pads. The system can also create a hardcopy record of documents developed during the meeting.

Besides business meetings and conferences and product roll-outs, videoconferencing can also be used for a variety of other applications including sales presentations and for conducting focus groups for market research.

Videoconferencing is also being used increasingly in education. For K-12 classes, a videoconferencing field trip can take children to a museum or science laboratory that would otherwise be too far to visit. Both docent and students can see and hear one another, as well as being able to see exhibits or experiments close up. For college students and adults, it is possible to attend classes given by eminent lecturers and participate fully just as though they were enrolled on campus (see also DISTANCE EDUCATION AND COMPUTERS).

Further Reading

Barlow, Janelle, Peta Peter, and Lewis Barlow. *Smart Videoconferencing: New Habits for Virtual Meetings.* San Francisco: Berrett-Koehler Publishers, 2002.
Pachnowski, Lynn M. "Virtual Field Trips Through Teleconferencing." *Learning & Leading with Technology* 29, no. 6 (March 2002): 10.

Prencipe, Loretta W. "Management Briefing: Do You Know the Rules and Manners of an Effective Virtual Meeting?" *Info-World* 23, no. 18 (April 30, 2001): 46.

Spielman, Sue, and Liz Winfield. *The Web Conferencing Book.* New York: AMACOM, 2003.

Videoconferencing Product Reviews from PC Magazine. Available online. URL: http://www.pcmag.com/category2/0,1874,4836,00.asp. Accessed August 23, 2007.

Winters, Floyd Jay, and Julie Manchester. *Web Collaboration Using Office XP and NetMeeting.* Upper Saddle River, N.J.: Prentice Hall, 2002.

video editing, digital

When videotape first became available in the 1950s, recorders cost thousands of dollars and could only be afforded by TV studios. Today the VCR is inexpensive and ubiquitous. However, it is hard to edit videotape. Tape is a linear medium, meaning that to find a given piece of video the tape has to be moved to that spot. Removing or adding something involves either physically splicing the tape (as is done with film) or more commonly, feeding in tape from two or more recorders onto a destination tape. Besides being tedious and limited in capabilities, "linear editing" by copying loses a bit of quality with each copying operation.

Today, however, it is easy to shoot video in digital form (see PHOTOGRAPHY, DIGITAL) or to convert analog video into digital form. Digital video is a stream of data that represents sampling of the source signal, such as from the charge-coupled device (CCD) that turns light photons into electron flow in a digital camera or digital camcorder. This process involves either software or hardware compression for storage and decompression for viewing and editing (such a scheme is called a CODEC for "compression/decompression"). The most widely used formats include DV (Digital Video) and MPEG (Motion Picture Expert Group), which has versions that vary in the amount of compression and thus fidelity.

In a turnkey system, the input source is automatically digitized and stored. In desktop video using a PC, a video capture card must be installed. The card turns the analog video signal into a digital stream. The most commonly used interface to bring video into a PC is IEE1394, better known as FireWire, which has the high bandwidth needed to transfer video data.

Once the video is captured, it can be stored in frame buffers in memory and edited in various ways using a variety of software. Expensive turnkey systems come with advanced software, while desktop video users can choose from products such as Media Studio Pro or Adobe Premiere. The editing interface usually has a timeline and thumbnails showing the location of key frames in the sequence. Individual clips can be extracted and tweaked with motion and transition effects; a variety of filters (see PLUG-IN) can be applied to the video. The accompanying sound track(s) can also be edited. Once things look right, the software is told to render (create) the finished video and save it to disk.

The ever-increasing processing power and disk capacity of today's PC is likely to make real-time video editing more feasible. This means that video can be played back directly from the edited timeline without transitions or effects having to be rendered first. Digital video cameras are also likely to increase in picture quality. Already desktop video is proving to be an affordable, viable alternative to expensive turnkey systems for many applications.

Meanwhile, like digital photography, digital video is rapidly becoming a creative medium for the masses, aided by easy-to-use basic software for Macintosh (such as iMovie) and numerous products for Windows. Another driver for the proliferation of this medium is the ease with which videos can be uploaded and shared (see USER-CREATED CONTENT and YOUTUBE).

Further Reading
Brandon, Bob. *The Complete Digital Video Guide: A Step-by-Step Handbook for Making Great Home Movies Using Your Digital Camcorder.* Pleasantville, N.Y.: Readers Digest, 2005.

Digital Video Editing. Available online. URL: http://videoediting.digitalmedianet.com/. Accessed August 23, 2007.

Digital Video Resources. Available online. URL: http://www.manifest-tech.com/links/mmtech.htm. Accessed August 23, 2007.

Goodman, Robert M., and Patrick McCrath. *Editing Digital Video: The Complete Creative and Technical Guide.* New York: McGraw-Hill, 2003.

Pogue, David. *iMovie 6 & iDVD: The Missing Manual.* Sebastapol, Calif.: O'Reilly Media, 2006.

Underahl, Keith. *Digital Video for Dummies.* Hoboken, N.J.: Wiley, 2006.

virtual community

Back in the mid-19th century, a number of technical professionals began to "chat" online without meeting physically—they were telegraph operators who relayed messages across the growing web that one author has called "The Victorian Internet." When computer networking began to grow in the 1970s, its own pioneers used facilities such as newsgroups (see NETNEWS AND NEWSGROUPS) to discuss a variety of topics. By the early 1980s, users were interacting on-line in complex fantasy games called MUDs (Multi-User Dungeons, or Dimensions) or MOOs (Muds, Object-Oriented). A little later, bulletin boards and especially systems such as the WELL (Whole Earth 'Lectronic Link) based in the San Francisco Bay Area (see BULLETIN BOARD and CONFERENCING SYSTEMS) provided long-term outlets for people to share information and interact on-line.

Looking at the WELL, a writer named Howard Rheingold introduced the term *virtual community* in a 1993 book. He explored the ways in which a sufficiently compelling and versatile technology encouraged people to form long-term contacts, form personal relationships, and carry out feuds. When on-line, participants experience such a venue as the WELL as a place that becomes almost as tangible (and often as "real") as a physical place such as a small town or corner bar.

Virtual community members who live in the same geographical area sometimes do get together physically (the WELL has had picniclike "WELL Office Parties" for many years). Members can band together to support a colleague who faces a crisis such as the life-threatening illness of a son (on the WELL, blank postings called *beams* are often used as an expression of sympathy). The virtual community can also serve as a rallying point following a physical disaster such as the 1989 earthquake in the San Francisco Bay Area. On a daily basis, virtual communities can often provide help or advice from a remarkable variety of highly qualified experts.

Virtual communities have their share of human foibles and worse. A virtual world that is compelling enough to immerse participants for hours on end is also powerful enough to engage emotions and expose vulnerabilities. For example, in a MUD called LambdaMOO one participant used descriptive language to have his game character "rape" a female character created by another participant, inflicting genuine distress. Like physical communities, virtual communities must evolve rules of governance, and actions in a virtual community can have real-world legal consequences.

Critics such as Clifford Stoll have argued that virtual communities are not only not a substitute for "true" physical community, but also may be further fragmenting neighborhoods and isolating people. (On the other hand, people who are already physically isolated, such as rural folk and the elderly or disabled, may find an outlet for their social needs in a virtual community.) Certainly the "bandwidth" in terms of human experience is less in a virtual community than in a physical community. Ideally, individuals should cultivate a mixture of virtual and physical community relationships.

Like a number of "virtual" concepts, virtual community is gradually blending into everyday life and thus becoming less distinct as an idea. Millions of people now participate in a form of virtual community through games such as *Second Life* (see ONLINE GAMES). Young people keep constantly in touch through a web of text messages (see FLASH MOBS and TEXTING AND INSTANT MESSAGING). Finally, the popularity of sites such as MySpace and Facebook may be partly due to the seamless way they bring together conventional social ties and their virtual extensions (see SOCIAL NETWORKING).

Further Reading

Barnes, S. B. *On-line Connections: Internet Personal Relationships.* Cresskill, N.J.: Hampton Press, 2000.
Dibbel, J. "Rape in Cyberspace: How an Evil Clown, a Haitian Trickster Spirit, Two Wizards and a Cast of Dozens Turned a Database into a Society." *The Village Voice*, Dec. 21, 1993, 39.
———. *My Tiny Life: Crime and Passion in a Virtual World.* New York: Henry Holt, 1998.
Powazek, Derek. *Design for Community: The Art of Connecting Real People in Virtual Places.* Berkeley, Calif.: New Riders, 2001.
Renninger, K. Ann, and Wesley Shumar, eds. *Building Virtual Communities: Learning and Change in Cyberspace.* New York: Cambridge University Press, 2002.
The Well. Available online. URL: http://www.well.com. Accessed August 23, 2007.
Rheingold, Howard. *The Virtual Community: Homesteading on the Electronic Frontier.* 2nd ed. Reading, Mass.: Addison-Wesley, 1993.
———. "Howard Rheingold Home Page." http://www.rheingold.com/
Turkle, Sherry. *Life on the Screen: Identity in the Age of the Internet.* New York: Simon & Schuster, 1995.

virtualization

One of the most powerful tools for understanding and manipulating a complex system is creating models or representations that simplify (while retaining the essentials) or that provide other useful ways of looking at the system. This ability to translate systems into representations is used in many fields, and probably dates back to the first cave paintings of our prehistoric ancestors.

In the computing field, virtualization involves the creation of a working model or representation of one system within a different system. This idea has been widely used in the field since the 1960s. Some applications of virtualization include:

- An appropriate model of a system (such as a programming framework—see APPLICATION PROGRAMMING INTERFACE) that hides unneeded details can make it easier for programmers to understand and access its functions (see DESIGN PATTERNS and MODELING LANGUAGES).

- A compiler for a language that compiles all programs to an intermediate representation (such as "byte-code"). A virtual machine running on each kind of platform can then run the code, taking care of the details required by the host hardware (see COMPILER and JAVA).

- A virtual machine created in software can be designed to perform all the functions available on a particular hardware platform or operating system, allowing software to be run on a system different from the one for which it was originally written (see EMULATION). For example, there are a number of virtualization programs (such as VMWare for PCs) that can create separate areas in memory, each running a different operating system, such as a version of Windows or Linux.

- Multiple processors or entire computers can be treated as a single entity for processing a program, with software designed to assign threads of execution to physical processors and to coordinate the use of shared data (see GRID COMPUTING).

- A physical device such as a disk drive can be made to appear as several separate devices to the operating system (for better organization of data). Similarly, many servers can run on the same physical machine. Conversely, multiple drives can appear to be a single logical device while providing redundancy and error recovery (see RAID).

- A secure "virtual private network" can be created within the larger public Internet. The virtual system takes care of encrypting and transmitting data through the physical network.

SOCIAL VIRTUALIZATION

The concept of virtualization can also be applied to how work involving computers is being conceptualized and organized in the modern world (see GLOBALIZATION AND THE COMPUTER INDUSTRY and UBIQUITOUS COMPUTING). A "virtual office" or even "virtual corporation" is a business entity that is not tied to a physical location, but uses networks, communications technology, and facilities such as video conferencing to keep workers in touch. Alternatively, several organizations can share the same physical space (such as for mail or shipping) while maintaining their separate identities.

Similarly, people can form long-lasting social networks while meeting physically seldom (if at all)—see SOCIAL NETWORKING and VIRTUAL COMMUNITY.

Further Reading
Brown, M. Katherine. *Managing Virtual Teams: Getting the Most from Wikis, Blogs, and Other Collaborative Tools.* Plano, Tex.: Wordware, 2007.
Golden, Bernard. *Virtualization for Dummies.* Indianapolis: Wiley, 2007.
Goldworm, Barb, and Anne Skamarock. *Blade Servers and Virtualization: Transforming Enterprise Computing while Cutting Costs.* Indianapolis: Wiley, 2007.
Smith, James E., and Ravi Nair. *Virtual Machines: Versatile Platforms for Systems and Processes.* San Francisco: Morgann Kaufmann, 2005.
Virtualization [news]. NetworkWorld. Available online. URL: http://www.networkworld.com/topics/virtualization.html. Accessed December 2, 2007.
Virtualization News Digest. Available online. URL: http://www.virtualization.info/. Accessed December 2, 2007.
Wolf, Chris, and Erick M. Halter. *Virtualization: From the Desktop to the Enterprise.* Berkeley, Calif.: Apress, 2005.

virtual reality

As the graphics and processing capabilities of computers grew increasingly powerful starting in the 1980s, it became possible to think in terms of creating a 3D environment that would not only appear to be highly realistic to the user, but also would respond to the user's natural motions in realistic ways.

This idea is not that new in itself. Starting as early as the 1930s, the military built mechanical flight trainers or simulators that could create a somewhat realistic experience of what a pilot would see and feel during flight. More sophisticated versions of these mechanical simulators helped the United States train the tens of thousands of pilots it needed during World War II while reducing the resources needed for actual flight hours. Today the military continues to pioneer the use of realistic computerized simulators to train

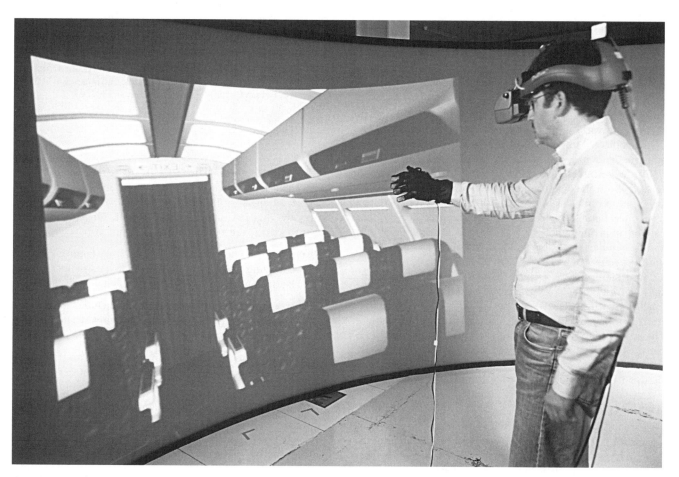

A NASA researcher wearing an early virtual reality (VR) outfit, including head-mounted display and gloves whose position can be tracked. (NASA PHOTO)

tank crews and even individual soldiers in the field (see MILITARY APPLICATIONS OF COMPUTERS).

Early simulators used "canned" graphics and could not respond very smoothly to control inputs (such as a pilot moving stick or rudder). Modern virtual reality, however, depends on the ability to smoothly and quickly generate realistic 3D graphics. At first such graphics could only be generated on powerful workstations such as those made by Sun or Silicon Graphics. However, as anyone who has recently played a computer game or simulation knows, there has been great improvement in the graphics available on ordinary desktop PCs since the mid-1990s.

A variety of software and programming tools can be used to generate 3D worlds on a PC (see COMPUTER GRAPHICS). First released in 1995, a facility called VRML (Virtual Reality Modeling Language) is now supported by many Web browsers. There are also programming extensions for Java (Java 3D).

Modern computer games thus embody aspects of virtual reality in terms of graphics and responsiveness. But true VR is generally considered to involve a near total immersion. Instead of a screen, a head-mounted display (HMD) is generally used to display the virtual world to the user while shutting out environmental distractions. Typically, slightly different images are presented to the left and right eyes to create a 3D stereo effect.

The other half of the VR equation is the way in which the user interacts with the virtual objects. Head-tracking sensors are used to tell the system where the user is looking so the graphics can be adjusted accordingly. Other sensors can be placed in gloves worn by the user. The system can thus tell where the user's hand is within the virtual world, and if the user "grasps" with the glove, the user's hand in the virtual world will grasp or otherwise interact with the virtual object. More elaborate systems involve a full-body suit studded with sensors.

To make interaction realistic, VR researchers have had to study both the operation of human senses and that of the skeleton and muscles. For a truly realistic experience, the user must be able to feel the resistance of objects (which can be implemented by a force-feedback system). Sound can be handled easily, but as of yet not much has been done with the senses of smell and taste.

In designing a VR system, there are a number of important considerations. Will the user be physically immersed (such as with an HMD), or, as in some military applications, will the user be seeing both a virtual and the actual physical world? How important is graphic realism vs. real-time responsiveness? (Opting too much for computationally intensive realism might cause unacceptable latency, or delay between a user action and the environment's response.)

APPLICATIONS

Besides military training, currently the most viable application for VR seems to be entertainment. VR techniques have been used to create immersive experiences in elaborate facilities at venues such as Disneyland and Universal Studios, and to some extent even in local arcades. VR that is accompanied by convincing physical sensations has allowed

for the creation of a new generation of roller coasters that if built physically would be too expensive, too dangerous, or even physically impossible.

However, there are other significant emerging applications for VR. When combined with telerobotic technology (see TELEPRESENCE), VR techniques are already being used to allow surgeons to perform operations in new ways. VR technology can also be used to make remote conferencing more realistic and satisfactory for participants. Clearly the potential uses for VR for education and training in many different fields are endless. VR technology combined with robotics could also be used to give disabled persons much greater ability to carry out the tasks of daily life.

In the ultimate VR system, users will be networked and able to simultaneously experience the environment, interacting both with it and one another. The technical resources and programming challenges are also much greater for such applications. The result, however, might well be the sort of environment depicted by science fiction writers such as William Gibson (see CYBERSPACE AND CYBER CULTURE).

Further Reading
Kim, Gehard Jounghyun. *Designing Virtual Reality Systems: The Structured Approach.* New York: Springer, 2005.
McMenemy, Karen, and Stuart Ferguson. *A Hitchhiker's Guide to Virtual Reality.* Wellesley, Mass.: A. K. Peters, 2007.
Sherman, William R., and Alan B. Craig. *Understanding Virtual Reality: Interface, Application, and Design.* San Francisco: Morgan Kaufman, 2003.
Sturrock, Carrie. "Virtual Becomes Reality at Stanford." *San Francisco Chronicle,* April 29, 2007, p. A1. Available online. URL: http://sfgate.com/cgi-bin/article.cgi?f=/c/a/2007/04/29/MNGFPPGVPF1.DTL. Accessed August 23, 2007.
Virtual Reality Resources. Available online. URL: http://vresources.org/. Accessed August 23, 2007.

VoIP (voiceover Internet protocol)

The basic idea of VoIP is simple: the Internet can carry packets of any sort of data (see TCP/IP), which means it can carry the digitized human voice as well, carrying ordinary phone calls. There are several ways to do this:

- a regular phone plus an adapter that connects to the computer and compresses and converts between regular analog phone signals and the digital equivalent

- a complete "IP phone" unit that includes all needed hardware and software—no computer needed, just a network connection, such as to a router

- use of the computer's own sound card and speakers with a microphone, plus software (often free)

Using that last option, VoIP service can be essentially free, regardless of distance. However, one can only call someone who is currently connected to the Internet and also has VoIP software.

Alternatively, one can subscribe to a VoIP provider such as Skype who also provides connectivity to the "plain old telephone service" (POTS). This allows calling anyone who has an ordinary phone: The VoIP provider sends the call

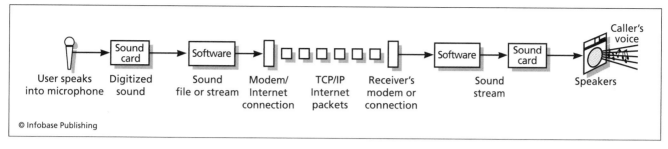

A regular telephone carries the voice as an analog signal over the phone line. For Internet (IP) telephony, however, the user's voice from the microphone is converted to a digital signal that is carried by standard Internet packets. At the destination, the packets are reassembled into a stream of digital data that is then sent to the sound card to be turned back into voice sounds to be played through the system speaker.

over the Internet to the nearest connection point, where it is placed as a regular phone call. The charges are much lower than typical long-distance plans. (For example, as of 2007 Skype charged a flat rate of $3.00 a month for unlimited calls to the United States and Canada and only a few cents per minute to most developed countries.)

Since video can also be sent over the Internet, video over IP for calling and conferencing is also becoming more common. Video does require a higher bandwidth connection than does voice.

ADVANTAGES AND DRAWBACKS

VoIP has several advantages over regular phone service. Because it uses the Internet's flexible packet-switching system, it uses bandwidth more efficiently (indeed, much conventional phone service is now carried as digital packets as well). If done through a direct computer-to-computer connection with free software, VoIP can be essentially free to the user, since the Internet connection is presumably already paid for. (It also follows that VoIP is most advantageous for long-distance calling.) Finally, VoIP can be used with wireless mobile devices, sometimes with lower cost than cell service.

At least as currently implemented, VoIP does have some disadvantages:

- Like cordless phones (but unlike traditional phones), VoIP requires that the user be connected to power. This may make the system unavailable in an emergency.

- Also, in an emergency, a 911 operator has no way to know where the caller is located geographically. This could be a problem if the caller is unable to provide this information.

- While a regular phone is a pretty simple device, VoIP requires special hardware or a PC, which might fail.

- VoIP requires a working Internet connection—in practice, a high-speed connection (see BROADBAND). Load or instability in the network could cause interruptions in calls or a lowering of voice quality.

- As with other data sent over the Internet, there are potential security concerns. Encryption can be used to secure VoIP calls, but this in turn leads to concerns

by law enforcement agencies seeking to implement eavesdropping warrants.

Despite these disadvantages, VoIP is likely to continue to become more prevalent and reliable due to the advantages of integrating with the global Internet and a wide variety of devices.

Further Reading
"History of VoIP." Available online. URL: http://www.utdallas.edu/~bjackson/history.html. Accessed December 3, 2007.

Kelly, Timothy V. *VoIP for Dummies.* Hoboken, N.J.: Wiley, 2005.

Skype. Available online. URL: http://www.skype.com/. Accessed December 3, 2007.

Valdes, Robert. "How VoIP works." Available online. URL: http://communication.howstuffworks.com/ip-telephony.htm. Accessed December 3, 2007.

Van Meggelen, Jim, Jared Smith, and Leif Madsen. *Asterisk: The Future of Telephony.* Sebastopol, Calif.: O'Reilly, 2005.

Venezia, Paul. "Open Source VoIP Makes the Business Connection." *Infoworld,* March 19, 2007. Available online. URL: http://www.infoworld.com/article/07/03/19/12FEopenvoip_1.html. Accessed December 3, 2007.

Wallingford, Ted. *Switching to VoIP.* Sebastopol, Calif.: O'Reilly, 2005.

von Neumann, John
(1903–1957)
Hungarian–American
Mathematician, Computer Scientist

John von Neumann made wide-ranging contributions in fields as diverse as pure logic, simulation, game theory, and quantum physics. He also developed many of the key concepts for the architecture of the modern digital computer and helped design some of the first successful machines.

Von Neumann was born on December 28, 1903, in Budapest, Hungary, to a family with banking interests. As a youth he showed a prodigious talent for calculation and interest in mathematics, but his father opposed his pursuing a career in pure mathematics. Therefore, when von Neumann entered the University of Berlin in 1921 and the Technische Hochschule in 1923, he earned his Ph.D. in chemical engineering. However, in 1926 he went back to Budapest and earned a Ph.D. in mathematics with a dissertation on set theory. He

John von Neumann developed automata theory as well as fundamental concepts of computer architecture such as storing programs in memory along with the data. He also did seminal work in logic, quantum physics, simulation, and game theory. (SPL / PHOTO RESEARCHERS, INC.)

would then serve as privatdozent, or lecturer, at Berlin and the University of Hamburg.

During the mid-1920s, two competing mathematical descriptions of the behavior of atomic particles were being offered by Erwin Schrödinger's wave equations and Werner Heisneberg's matrix approach. Von Neumann showed that the two theories were mathematically equivalent. His 1932 book, *The Mathematical Foundations of Quantum Mechanics,* remains a standard textbook to this day. Von Neumann also developed a new form of algebra where "rings of operators" could be used to describe the kind of dimensional space encountered in quantum mechanics.

Meanwhile, von Neumann had become interested in the mathematics of games, and developed the discipline that would later be called game theory. His "minimax theorem" described a class of two-person games in which both players could minimize their maximum risk by following a specific strategy.

COMPUTATION AND COMPUTER ARCHITECTURE

In 1930, von Neumann immigrated to the United States, where he would become a naturalized citizen and spend the rest of his career. He was made a Fellow at the new Institute

for Advanced Study at Princeton at its founding in 1933, and would serve in various capacities there and as a consultant for the U.S. government.

In the late 1930s, interest had begun to turn to the construction of programmable calculators or computers (see CHURCH, ALONZO and TURING, ALAN). Just before and during World War II, von Neumann worked on a variety of problems in ballistics, aerodynamics, and later, the design of nuclear weapons. All of these problems cried out for machine assistance, and von Neumann became acquainted both with British research in calculators and the massive Harvard Mark I programmable calculator (see AIKEN, HOWARD).

A little later, von Neumann learned that two engineers were working on a new kind of machine: an electronic digital computer called ENIAC that used vacuum tubes for its switching and memory, making it about a thousand times faster than the Mark I. Although the first version of ENIAC had already been built by the time von Neumann came on board, he served as a consultant to the project at the University of Pennsylvania's Moore School.

The earliest computers (such as the Mark I) read instructions from cards or tape, discarding each instruction as it was performed. This meant, for example, that to program a loop, an actual loop of tape would have to be mounted and controlled so that instructions could be repeated. The electronic ENIAC was too fast for tape readers to keep up, so it had to be programmed by setting thousands of switches to store instructions and constant values. This tedious procedure meant that it wasn't practicable to use the machine for anything other than massive problems that would run for many days.

In his 1945 "First Draft of a Report on the EDVAC" and his more comprehensive 1946 "Preliminary Discussion of the Logical Design of an Electronic Computing Instrument," von Neumann established the basic architecture and design principles of the modern electronic digital computer.

Von Neumann declared that in future computers the machine's internal memory would be used to store constant data and all instructions. With programs in memory, looping or other decision making can be accomplished simply by "jumping" from one memory location to another. Computers would have two forms of memory: relatively fast memory for holding instructions, and a slower form of storage that could hold large amounts of data and the results of processing. (In today's PCs these functions are provided by the random access memory [RAM] and hard drive respectively.) The storage of programs in memory also meant that a program could treat its own instructions like data and change them in response to changing conditions.

In general, von Neumann took the hybrid design of ENIAC and conceived of a design that would be all-electronic in its internal operations and store data in the most natural form possible for an electronic machine—binary, with 1 and 0 representing the on and off switching states and, in memory, two possible "marks" indicated by magnetism, voltage levels, or some other phenomenon. The logical design would be consistent and largely independent of the vagaries of hardware.

Eckert and Mauchly (see ECKERT, J. PRESPER and MAUCHLY, JOHN WILLIAM) and some of their supporters would later claim that they had already conceived of the idea of storing programs in memory, and in fact they had already designed a form of internal memory called a mercury delay line. Whatever the truth in this assertion, it remains that von Neumann provided the comprehensive theoretical architecture for the modern computer, which would become known as the von Neumann architecture. Von Neumann's reports would be distributed widely and would guide the beginnings of computer science research in many parts of the world.

Looking beyond EDVAC, von Neumann, together with Herman Goldstine and Arthur Burks, designed a new computer for the Institute for Advanced Study that would embody the von Neumann principles. The IAS machine's design would in turn lead to the development of research computers for RAND Corporation, the Los Alamos National Laboratory, and in several countries including Australia, Israel, and even the Soviet Union. The design would eventually be commercialized by IBM in the form of the IBM 701.

In his later years, von Neumann continued to explore the theory of computing. He studied ways to make computers that could automatically maintain reliability despite the loss of certain components, and he conceived of an abstract self-reproducing automaton (see CELLULAR AUTOMATA).

Von Neumann's career was crowned with many awards reflecting his diverse contributions to American science technology. These include the Distinguished Civilian Service Award (1947), Presidential Medal of Freedom (1956), and the Enrico Fermi Award (1956). Von Neumann died on February 8, 1957, in Washington, D.C.

Further Reading

Aspray, William. *John von Neumann and the Origins of Modern Computing.* Cambridge, Mass.: MIT Press, 1990.

Heims, S. J. *John von Neumann and Norbert Wiener: From Mathematics to the Technologies of Life and Death.* Cambridge, Mass.: MIT Press, 1980.

"John Louis von Neumann" [biography]. Available online. URL: http://ei.cs.vt.edu/~history/VonNeumann.html. Accessed August 23, 2007.

MacRae, Norman. *John Von Neumann: The Scientific Genius Who Pioneered the Modern Computer, Game Theory, Nuclear Deterrence, and Much More.* 2nd ed. Providence, R.I.: American Mathematical Society, 2000.

von Neumann, John. *The Computer and the Brain.* New Haven, Conn.: Yale University Press, 1958.

———. *Theory of Self-Reproducing Automata.* Edited and compiled by Arthur W. Burks. Urbana: University of Illinois Press, 1966.

Wales, Jimmy

(1966–)
American
Internet Entrepreneur

Jimmy Wales is a key force behind Wikipedia, the community-edited online encyclopedia that has become a popular stop for Web users seeking information about any of millions of topics.

Wales was born on August 7, 1966, in Huntsville, Alabama, and received his early education in a tiny private school run by his mother and grandmother. However, Wales then went to an advanced college preparatory school in Huntsville, where he was extensively exposed to computer technology. Wales went on to earn a bachelor's degree in finance from Auburn University and a master's in finance at the University of Alabama. He entered but did not complete the doctoral program, later attributing his dropping out to boredom. During the 1990s Wales became research director at Chicago Options Associates, trading so successfully in currency and interest rate options that he achieved lifetime financial security.

By that time Wales had become involved with the growing e-commerce boom. However, his first project, an "erotic search engine" called Bomis, would be controversial. Using the Bomis site, Wales and Larry Sanger then launched their first online encyclopedia, Nupedia. In 2001, however, Sanger suggested the use of wiki software (see WIKIS AND WIKIPEDIA). Wales and Sanger set up the parameters for how users would contribute, collaborate, and review articles. Wikipedia soon far outstripped Nupedia. Sanger

and Wales found themselves in frequent disagreement, and Sanger left Wikipedia in 2002.

WIKIPEDIA AND BEYOND

In 2003 Wales established the Wikimedia Foundation, a nonprofit organization to support Wikipedia and a variety of new projects based on online communities. In 2004 Wales and Angela Beesley founded a for-profit company, Wikia, Inc. Besides making it easy for individuals and communities to organize and manage their own wikis and blogs, Wikia also intends to apply the wiki collaborative principle to creating a search engine that would draw upon users' own expertise and interests and operate "transparently." Wales believes this model will prove to be superior to proprietary operations such as Google.

Wales was criticized in 2005 for editing his own biography in Wikipedia, downplaying the pornographic nature of Bomis and minimizing Sanger's role as a cofounder of Wikipedia. Wales later expressed regrets about his editing, while continuing to insist that Sanger's role was that of an employee rather than a cofounder. (Sanger later created Citizendium, an online encyclopedia that requires stricter credentials for editors.)

Politically, Wales describes himself as a passionate objectivist (follower of Ayn Rand's philosophy) and a libertarian who admires philosopher-economist F. A. Hayek (though distancing himself from the Libertarian Party). Wales's interest in decentralized, emergent organizations (such as Wikipedia and Wikia) can be seen as flowing out of his political philosophy. At the same time, the scope and powers Wales continues to exercise over Wikipedia can be unclear and subject to controversy.

In 2005 Wales became a member of the board of directors of Socialtext, a developer of wiki technology. In 2006 he also joined the board of Creative Commons, developer of new ways to share intellectual property. That same year *Time* listed Wales among 100 of the year's most influential people, and Wales received a Pioneer Award from the Electronic Frontier Foundation. Wales lives near St. Petersburg, Florida.

Further Reading

"Jimmy Wales: Free Knowledge for Free Minds" [blog]. http://blog. jimmywales.com/. Accessed May 10, 2007.

Lee, Ellen. "As Wikipedia Moves to S.F., Founder Discusses Planned Changes." *San Francisco Chronicle,* November 30, 2007. Available online. URL: http://www.sfgate.com/cgi-bin/ article.cgi?f=/c/a/2007/11/30/BUOMTKNJA.DTL. Accessed December 3, 2007.

Mangu-Ward, Katherine. "Wikipedia and Beyond: Jimmy Wales' Sprawling Vision." *Reason,* June 2007, pp. 18–29.

Tapscott, Don, and Anthony D. Williams. *Wikinomics: How Mass Collaboration Changes Everything.* New York: Penguin, 2006.

Wikipedia. Available online. URL: http://en.wikipedia.org/wiki/ Main_Page. Accessed May 10, 2007.

wearable computers

For some time, technology pundits have talked about computers being literally woven into daily life, embedded in clothing and personal accessories. However, implementations have thus far seen only limited use. For example, watches with limited computer functions (see PDA) have not proven popular—a watch large enough for input and display of information would likely be too bulky for comfort. (People have also walked about with attached webcams, although the novelty seems to have quickly worn off.)

EMERGING POSSIBILITIES

There are, however, a number of more limited wearable computers that are likely to be practical. Small cards (see RFID and SMART CARD) could provide tracking for children or others needing monitoring. Embedded sensors could be designed to detect whether an elderly person has fallen or perhaps has suffered a heart attack.

Head-mounted displays that fit into eyeglasses or goggles are already in use and can offer applications ranging from gaming (see VIRTUAL REALITY) to providing informational overlays to aid in military reconnaissance, police patrol, or firefighting. (This could also be combined with tracking and communications.) Other embedded computers might provide hands-free voice recognition or language translation.

More whimsical wearable computers could control the colors and patterns displayed by garments, perhaps varying them with the mood of the wearer.

Whimsy aside, some serious effort is now going into developing a wide range of wearable computer applications. The most prominent effort is wearIT@work, funded by the

*A fashion model wears a "Skooltool" outfit that allows information to be played through earphones or projected onto the lenses. The outfit was a collaboration between MIT researchers and fashion designers. (*SAM OGDEN / PHOTO RESEARCHERS, INC.*)*

European Union. It is developing an Open Wearable Computing Framework and standard hardware.

Further Reading

Cristol, Hope. "The Future of Wearable Computers." *The Futurist* 36 (September 1, 2002): 68.

MIT Media Lab Wearable Computing. Available online. URL: http://www.media.mit.edu/wearables/index.html. Accessed December 3, 2007.

"Wearing Technology on Your Sleeve." PhysOrg, November 26, 2007. Available online. URL: http://www.physorg.com/news 115310793.html. Accessed December 3, 2007.

WearIt@Work. Available online. URL: http://www.wearitatwork. com/. Accessed December 3, 2007.

Xu, Yangsheng, Wen Jung Li, and Ka Keung Lee. *Intelligent Wearable Interfaces.* Hoboken, N.J.: Wiley Interscience, 2007.

Web 2.0 and beyond

Somewhere between a buzzword and a genuine new paradigm, Web 2.0 refers to a number of developments that are

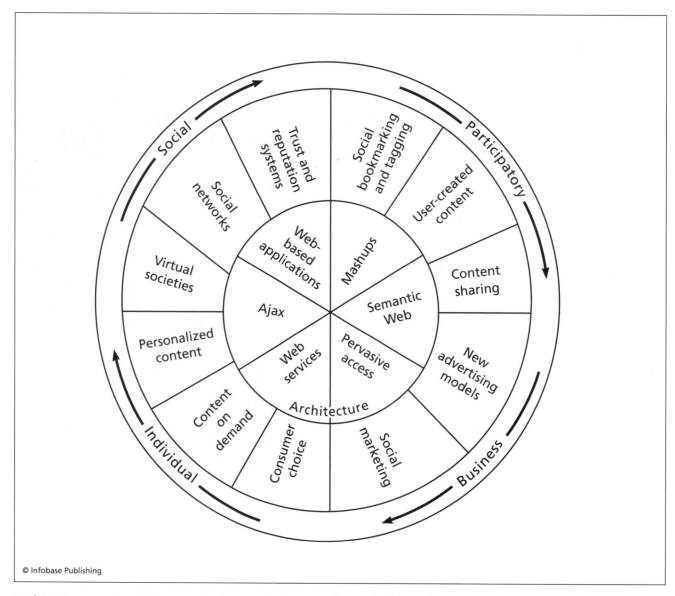

"Web 2.0" is a somewhat nebulous term, but its core technologies are changing both how information is presented on the Web and how users can create and share their own content.

changing the way content is created and presented on the Web, as well as ways in which Web users are using technology to create new communities and institutions. (The term is somewhat misleading because it seems to imply a new version of the fundamental Web software itself. It is more a change in the way the Web is perceived and used.)

The term emerged into prominence following a 2004 conference that emphasized the Web as being not just a place to offer services, but a platform upon which to build them, offering applications that are not dependent on any particular operating system. As services were built and users participated in new ways, the emerging communities would then extend the power of the Web platform even further (see SOCIAL NETWORKING and USER-CREATED CONTENT). For some often-cited examples, see CRAIGSLIST, EBAY, WIKIPEDIA, and YOUTUBE.

WEB 2.0 TOOLS

Although the most important part of Web 2.0 is its business and social models, a number of Web technologies are needed to provide the flexibility and rich interaction needed to offer a new Web experience. These include:

- dynamic, efficient generation of content (see AJAX)

- programming interfaces (see API) using structured text files (see XML)

- platforms for running applications in the browser, such as Google apps

- merging and customizing content from different sources (see MASHUPS)

- user subscription to content (see PODCASTING and RSS)

- platforms for user-created content and collaboration (see BLOGS AND BLOGGING, SOCIAL NETWORKING, and WIKIS AND WIKIPEDIA)

In some quarters the term Web 2.0 is already obsolete or relegated to a marketing buzzword, while the search is on for new ways to describe the latest developments such as, inevitably, "Web 3.0." One possible emphasis moving beyond Web 2.0 is the leveraging of the actual knowledge contained in Web pages, properly encoded and interpreted by applications (see SEMANTIC WEB and SOFTWARE AGENT).

Whatever terminology might be used, the important thing is that people are using the Web in the late 2000 decade in substantially new ways, and that the consequences are likely to spread beyond the online world to society as a whole.

Further Reading

Fost, Dan. "Digital Utopia: A New Breed of Technologists Envisions a Democratic World Improved by the Internet." *San Francisco Chronicle,* November 5, 2006, p. F1. Available online. URL: http://sfgate.com/cgi-bin/article.cgi?file=/c/a/2006/11/05/BUGIGM5A2D1.DTL. Accessed December 4, 2007.
———. "The People Who Populate Web 2.0" *San Francisco Chronicle,* November 5, 2006, p. F5. Available online. URL: http://sfgate.com/cgi-bin/article.cgi?file=/c/a/2006/11/05/BUG78M5OUA1.DTL. Accessed December 4, 2007.
Madden, Mary, and Susannah Fox. "Riding the Waves of 'Web 2.0.'" Pew Internet & American Life Project, October 5, 2006. Available online. URL: http://www.pewinternet.org/pdfs/PIP_Web_2.0.pdf. Accessed December 4, 2007.
Metz, Cade. "Web 3.0." *PC Magazine,* March 14, 2007. Available online. URL: http://www.pcmag.com/article2/0,1759,2102852,00.asp. Accessed December 4, 2007.
Solomon, Gwen, and Lynne Schrum. *Web 2.0: New Tools, New Schools.* Eugene, Ore.: ISTE, 2007.
Vossen, Gottfried, and Stephan Hagemann. *Unleashing Web 2.0: From Concepts to Creativity.* Burlington, Mass.: Morgan Kaufmann, 2007.

Web browser

The World Wide Web consists of millions of sites (see WORLD WIDE WEB and WEB SERVER) that provide hypertext documents (see HTML and WEB PAGE DESIGN) that can include not only text but still images, video, and sound. To access these pages, the user runs a Web-browsing program.

The basic function of a Web browser is to request a page by specifying its address (URL, uniform [or universal] resource locator). This request resolves to a request (HTTP, HyperText Transport Protocol) that is processed by the relevant Web server. The server sends the HTML document to the browser, which then displays it for the user. Typically, the browser stores recently requested documents and files in a local cache on the user's PCs. Use of the cache reduces the amount of data that must be resent over the Internet. However, sufficiently skilled snoopers can examine the cache to find details of a user's recent Web surfing. (Caching is also used by Internet Service Providers so they can provide frequently requested pages from their own server rather than having to fetch them from the hosting sites.)

When the Web was first created in the early 1990s (see BERNERS-LEE, TIM) it consisted only of text pages, although there were a few experimental graphical Web extensions developed by various researchers. The first graphical Web browser to achieve widespread use was Mosaic created by Marc Andreessen, developed at the National Center for Supercomputing Applications (NCSA). (See ANDREESSEN, MARC.) By 1993, Mosaic was available for free download and had become the browser of choice for PC users.

Andreessen left NCSA in 1994 to found Netscape Corporation. The Netscape Navigator browser improved Mosaic in several ways, making the graphics faster and more attractive. Netscape included a facility called Secure Sockets Layer (SSL) for carrying out encrypted commercial transactions on-line (see E-COMMERCE).

Microsoft, which had been a latecomer to the Internet boom, entered the fray with its Microsoft Internet Explorer. At first the program was inferior to Netscape, but it was steadily improved. Aided by Microsoft's controversial tactic of bundling the free browser starting with Windows 95, Internet Explorer has taken over the leading browser position with

A Web browser such as Microsoft Internet Explorer or Firefox makes it easy to find and move between linked Web pages. Browser users can record or "bookmark" favorite pages. Browser plug-ins provide support for services such as streaming video and audio. Here, part of the photo library of the National Oceanic and Atmospheric Administration is shown. (NOAA IMAGE)

about a 75 percent market share by 2001. However, a rather strong competitor later emerged in Firefox, and other browsers such as Opera and Safari also have their supporters, who feel those products are more agile, versatile, and perhaps more secure than Internet Explorer.

Some typical features of a modern Web browser include

- navigation buttons to move forward and back through recently visited pages

- tabs to switch between Web pages

- a "history" panel allowing return to pages visited in recent days

- a search button that brings up the default search engine (which can be chosen by the user)

- the ability to save page as "favorites" or "bookmarks" for easy retrieval

THE BROWSER AS PLATFORM

Today a Web user can view a live news broadcast, listen to music from a radio station, or view a document formatted to near-print quality. All these activities are made possible by "helper" software (see PLUG-IN) that gives the Web browser the capability to load and display or play files in special formats. Examples include the Adobe PDF (Portable Document Format) reader, the Windows Media Player, and RealPlayer for playing video and audio content (see STREAMING).

What makes the browser even more versatile is the ability to load and run programs from Web sites (see JAVA). Java was highly touted starting in the mid-1990s, and some observers believed that by making Web browsers into platforms capable of running any sort of software, there would be less need for proprietary operating systems such as Microsoft Windows. Microsoft has responded by trying to shift developers' emphasis from Java to its proprietary technology called .NET. Meanwhile, the tools for making Web pages more versatile and interactive continue to proliferate, including later versions of HTML and XML (see WEB PAGE DESIGN). This proliferation, as well as use of proprietary extensions can cause problems in accessing Web sites from older or less-known browsers.

The growing numbers of handheld or palm computers (see PORTABLE COMPUTERS) are accompanied by scaled-down Web browsers. These are generally controlled by touch and have a limited display size, but can provide information useful to travelers such as driving directions, weather forecasts, and capsule news or stock summaries.

Further Reading

Barker, Donald I., and Katherine T. Pinard. *Microsoft Internet Explorer 7, Illustrated Essentials*. Boston: Course Technology, 2007.

Browser Review. Available online. URL: http://www.yourhtml-source.com/starthere/browserreview.html. Accessed August 23, 2007.

Firefox 2. Available online. URL: http://www.mozilla.com/en-US/firefox/. Accessed August 23, 2007.

Opera Browser Home Page. Available online. URL: http://www.opera.com/. Accessed August 23, 2007.

Ross, Blake. *Firefox for Dummies*. Hoboken, N.J.: Wiley, 2006.

Windows Internet Explorer 7. Available online. URL: http://www.microsoft.com/windows/products/winfamily/ie/default.mspx. Accessed August 23, 2007.

webcam

Thousands of real-time views of the world are available on the Web. These include everything from the prosaic (a coffee machine at MIT) to the international (a view of downtown Paris or Tokyo) to the sublime (a Rocky Mountain sunset). All of these views are made possible thanks to the availability of inexpensive digital cameras (see PHOTOGRAPHY, DIGITAL).

To create a basic webcam, the user connects a digital camera to a PC, usually via a USB cable. A program controls the camera, taking a picture at frequent intervals (perhaps every 30 seconds or minute). The picture is received from the camera as a JPG (JPEG) file. The program then uploads the picture to the user's Web page (usually using file transfer protocol, or ftp), replacing the previous picture. Users connected to the Web site can click to see the latest picture. Alternatively, a script running on the server can update the picture automatically.

HISTORY AND APPLICATIONS

One of the earliest and most famous webcams was created by Quentin Stafford-Fraser in 1991. He later recalled that he and his fellow "coffee club" members were tired of making the long trek to the coffee room at the Cambridge University computer laboratory. It seemed that more often than not the life-giving brew so necessary to computer science had already been consumed. So they rigged a video camera, connected it to a video capture card, and fed the image into the building's local network. Now researchers working anywhere in the building could get an updated image of the coffee machine three times a minute. This wasn't technically a webcam. At the time the Web was just being developed by Tim Berners-Lee (see BERNERS-LEE, TIM). However, the camera was put on the Web in 1993, where it resided until 2001 when the laboratory housing the now-famous coffee machine was moved.

The webcam became a social phenomenon in 1996 when a college student named Jennifer Ringley started Jennicam, a webcam set up to make a continuing record of her daily life available on the Web. There were soon many imitators. Apparently this use of Webcams taps into humans' intense curiosity about the details of each other's lives—a curiosity that to some critics tips over into voyeurism and obsession. The popularity of such social webcams may have contributed to the "reality TV" phenomenon at the turn of the new century.

Webcams have many practical applications, however. People on the road can log into the Web and check to make sure everything is okay at home. A webcam also makes an inexpensive monitor for checking on infants or toddlers in another room, or checking on the behavior of a babysitter ("Nannycam").

Webcams can also serve an educational purpose. They can take viewers to remote volcanoes or the interior of an

Amazon rain forest. In a sense, viewers who saw the pictures of the Martian surface and the explorations of the Sojourner rover were using the farthest-reaching webcam of all.

Further Reading

Breeden, John, and Jason Byrne. *Guide to Webcams.* Indianapolis: Prompt Publications, 2001.

Layton, Julia. "How Webcams Work." Available online. URL: http://www.howstuffworks.com/webcam.htm. Accessed August 23, 2007.

Mobberly, Martin. *Lunar and Planetary Webcam User's Guide.* New York: Springer, 2006.

Webcam Resources. Available online. URL: http://www.resourcehelp. com/qserwebcam.htm. Accessed August 23, 2007.

Web filter

Listings of the most frequent requests typed into Web search engines usually begin with the word *sex.* Although sensational journalism of the mid-1990s sometimes unfairly portrayed the World Wide Web as nothing more than an electronic red light district, it is indisputable that there are many Web sites that feature material that most people would agree is not suitable for young people. Many parents as well as some schools, libraries, and workplaces have installed Web filter programs, marketed under names such as SurfWatch or NetNanny. Popular Internet security programs (such as those from Norton/Symantec) also include Web filter modules.

The Web filter examines requests made by a Web user (see WORLD WIDE WEB and WEB BROWSER) and blocks those associated with sites deemed by the filter user to be objectionable. There are two basic mechanisms for determining whether a site is unsuitable. The first is to check the site's address (URL) against a list and reject a request for any site on the list. (Most filter programs come with default lists; the filter user can add other sites as desired. Generally, the filter is installed with a password so only the authorized user [such as a parent] can change the filter's behavior.)

The other filtering method relies on a list of keywords associated with objectionable activities (such as pornography). When the user requests a site, the filter checks the page for words on the keyword list. If a matching word or phrase is found, the site is blocked and not shown to the user.

Each method has its drawbacks: Using a site list will miss new sites that appear between list updates, while using keywords can result in appropriate sites also being blocked. For example, a keyword filter that blocks sites with the word *breast* will probably also block a site devoted to breast cancer research, a fact often pointed out by opponents of laws requiring the use of Web filters. The list and keyword methods can be combined.

Filtering and parental control often involves more than simply blocking Web sites. Many filtering products attempt to scan and block problematic chat and e-mail messages. Another type of filtering tries to stop users (particularly children) from providing sensitive information such as their name and address online. Another common parental control feature is the ability to limit the times of day and total amount of time a child can go online.

Besides protecting children from inappropriate material at home or in a school or library, Web filters are also used in workplaces. Besides wanting to keep workers from becoming distracted, employers are concerned that allowing Internet pornography in the workplace may make them liable for creating a "hostile work environment" under sexual harassment laws.

However, civil liberties groups such as the ACLU object to the use of Web filters in public libraries on First Amendment grounds and have vigorously fought such legislation in the courts. The 1996 Communications Decency Act was declared unconstitutional by the U.S. Supreme Court, and a later law, the 1998 Child On-line Protection Act (which requires that users of adult Web sites provide proof of age) was overturned by the U.S. Supreme Court in 2004.

The next attempt at protecting children online was the 2002 Children's Internet Protection Act. This law was eventually upheld by the courts subject to the requirement that adult library users be given prompt unfiltered access to the Internet upon request.

Critics of Web filters suggest that rather using technical tools to block access to the Internet, parents and teachers should talk to children about their use of the Internet and supervise it if necessary. Another approach is to focus on Web sites that are designed especially for kids.

Further Reading

Gilbert, Alorie, and Stefanie Olsen. "Do Web Filters Protect Your Child?" CNET News. Available online. URL: http://news. com.com/Do+Web+filters+protect+your+child/2100-1032_ 3-6030200.html. Accessed August 23, 2007.

Internet Filter Reviews. Available online. URL: http://internet-filter-review.toptenreviews.com/. Accessed August 23, 2007.

Olsen, Stefanie. "Kids Outsmart Web Filters." CNET News. Available online. URL: http://news.com.com/Kids+outsmart+Web+filters/ 2009-1041_3-60 62548.html. Accessed August 23, 2007.

Parental Control Software. Available online. URL: http://www. softforyou.com/. Accessed August 23, 2007.

webmaster

There are many online services (including some free ones) that will provide users with personal Web pages. There are also programs such as Microsoft FrontPage that allow users to design Web pages by arranging objects visually on the screen and setting their properties. However, creating and maintaining a complete Web site with its many linked pages, interactive forms and interfaces to databases and other services is a complicated affair. For most moderate to large-size organizations, it requires the services of a new category of IT professional: the webmaster.

Although the mixture of tasks and responsibilities will vary with the extent and purpose of the Web site, the skill set for a webmaster can include the following:

DEVELOPING AND EXTENDING THE WEB SITE

- understanding how the Web site responds to and manages requests (see WEB SERVER)

- fluency in the basic formatting of text and other page content and the use of frames and other tools for organizing and presenting text (see HTML)

- extended formatting and content organization facilities such as Cascading Style Sheets (CSS), Dynamic HTML (DHTML), and Extensible Markup Language (see XML)

- use of graphics formats and graphics and animation programs (such as Photoshop, Flash, and Dream-Weaver)

- extending the interactivity of Web pages through writing scripts using tools such as JavaScript and PHP (see CGI, JAVASCRIPT, and PYTHON)

- dealing with platform and compatibility issues, including browser compatibility

It is hard to draw a bright line between advanced tasks for webmasters and full-blown applications designed to run on servers or Web browsers. Some additional tools for extending Web capabilities include:

- languages for Web application development (see C#, JAVA, RUBY, and VISUAL BASIC)

- Web server and browser plug-ins

- Active X controls and the Microsoft .NET framework (for Windows-based systems)

- techniques for the efficient updating of dynamic Web pages (see AJAX)

ADMINISTRATIVE TASKS

- Obtaining, organizing, and updating the content for Web pages (this may be delegated to writers, editors, or graphics specialists)

- monitoring the performance of the Web server

- ensuring site availability and response time

- recommending acquisition of new hardware or software as necessary

- using tools to gather information about how the site is being used, what parts are being visited, the effectiveness of advertising, and so on (This is particularly relevant to commercial sites, and can raise privacy issues.)

- setting up and managing facilities for online shopping (see E-COMMERCE)

- installing and using security tools (particularly important for commercial and sensitive government sites)

- developing policies and deploying tools to help protect users' privacy and to control the use of information they submit online

- working with major search engine providers to ensure that the site is presented to relevant searches

- fielding queries from users about the operation of the site

- relating the Web site operation to other concerns such as marketing, technical support, or the legal department

- developing policies for Web site use

- integrating the Web site operations into the overall corporate planning and budgeting process

The mixture of technical professional and administrator that is the webmaster makes for an always interesting and challenging career. In larger organizations there may be further differentiation of roles, with the webmaster mainly charged with operation and maintenance of the site, with the development and extension of the site handled by content providers and programmers. However, even in such cases the webmaster will need to have a general understanding of how the various features of the Web site interact and of the tools used to create and maintain them. People with webmaster skills can also work as independent consultants to set up and run Web sites for smaller businesses, schools, and nonprofit organizations.

Webmaster skills are now taught in high school, community college, vocational school, and as part of university information technology programs. However the situation with regard to certification remains somewhat chaotic, with a variety of proprietary and multivendor certifications competing for attention.

The long-term outlook for qualified webmasters remains good. Many organizations have made a fundamental commitment to use of the Web for business functions, and webmasters are needed to manage this effort.

Further Reading
American Association of Webmasters. Available online. URL: http://www.aawebmasters.com/. Accessed August 23, 2007.
Big Webmaster—Webmaster Resources. Available online. URL: http://www.bigwebmaster.com/. Accessed August 23, 2007.
Spainhour, Sebastian. *Webmaster in a Nutshell.* 3rd ed. Sebastapol, Calif.: O'Reilly, 2002.
Still, Julie. *The Accidental Webmaster.* Medford, N.J.: Information Today, 2003.

Web page design

The World Wide Web has existed for fewer than two decades, so it is not surprising that the principles and practices for the design of attractive and effective Web pages are still emerging. As seen in the preceding entry (see WEBMASTER), creating Web pages involves many skills. In addition to the basic art of writing, many skills that had belonged to separate professions in the print world now often must be exercised by the same individual. These include typography (the selection and use of type and type styles), composition (the arrangement of text on the page), and graphics. To this mix must be added nontraditional skills such as designing interactive features and forms, interfacing with other facilities (such as databases), and perhaps the incorporation of features such as animation or streaming audio or video.

However new the technology, the design process still begins with the traditional questions any writer must ask:

What is the purpose of this work? Who am I writing for? What are the needs of this audience? A Web site that is designed to provide background information and contact for a university department is likely to have a printlike format and a restrained style. Nevertheless, the designer of such a site may be able to imaginatively extend it beyond the traditional bounds—for example, by including streaming video interviews that introduce faculty members.

A site for an online store is likely to have more graphics and other attention-getting features than an academic or government site. However, despite the pressure to "grab eyeballs," the designer must resist making the site so cluttered with animations, pop-up windows, and other features that it becomes hard for readers to search for and read about the products they want.

A site intended for an organization's own use should not be visually unattractive, but the emphasis is not on grabbing users' attention, since the users are already committed to using the system. Rather, the emphasis will be on providing speedy access to the information people need to do their job, and in keeping information accurate and up to date.

Once the general approach is settled on, the design must be implemented. The most basic tool is HTML, which has undergone periodic revisions and expansions (see HTML). Even on today's large, high-resolution monitors a screen of text is not the same as a page in a printed book or magazine. There are many ways text can be organized (see HYPERTEXT AND HYPERMEDIA). A page that is presenting a manual or other lengthy document can mimic a printed book by having a table of contents. Clicking on a chapter takes the reader there. Shorter presentations (such as product descriptions) might be shown in a frame with buttons for the reader to select different aspects such as features and pricing. Frames (independently scrollable regions on a page) can turn a page into a "window" into many kinds of information without the user having to navigate from page to page, but there can be browser compatibility issues. Tables are another important tool for page designers. Setting up a table and inserting text into it allows pages to be formatted automatically.

Many sites include several different navigation systems including buttons, links, and perhaps menus. This can be good if it provides different types of access to serve different needs, but the most common failing in Web design is probably the tendency to clutter pages with features to the point that they are confusing and actually harder to use.

Although the Web is a new medium, much of the traditional typographic wisdom still applies. Just as many people who first encountered the variety of Windows or Macintosh fonts in the 1980s filled their documents with a variety of often bizarre typefaces, beginning Web page designers sometimes choose fonts that they think are "edgy" or cool, but may be hard to read—especially when shown against a purple background!

Today it is quite possible to create attractive Web pages without extensive knowledge of HTML. Programs such as FrontPage and DreamWeaver mimic the operation of a word processor and take a WYSIWYG (what you see is what you get) approach. Users can build pages by selecting and arranging structural elements, while choosing styles for headers and other text as in a word processor. These programs also provide "themes" that help keep the visual and textual elements of the page consistent. Of course, designing pages in this way can be criticized as leading to a "canned" product. People who want more distinctive pages may choose instead to learn the necessary skills or hire a professional Web page designer. A feature called Cascading Style Sheets (CSS) allows designers to precisely control the appearance of Web pages while defining consistent styles for elements such as headings and different types of text (see CASCADING STYLE SHEETS).

Most Web pages include graphics, and this raises an additional set of issues. Most users now have fast Internet connections (see BROADBAND), but others are still limited to slower dial-up speeds. One way to deal with this situation is to display relatively small, lower-resolution graphics (usually 72 pixels per inch), but to allow the user to click on or near the picture to view a higher-resolution version. Another consideration in today's wireless world is ensuring that Web pages likely to be useful to users on the go, such as a restaurant guide, display well in the small browsers found in mobile devices (see PDA and SMARTPHONE). Page designers must also make sure that the graphics they are using are created in-house, are public domain, or are used by permission.

Animated graphics (animated GIFs or more elaborate presentations created with software) can raise performance and compatibility issues. Generally, if a site offers, for example, Flash animations, it also offers users an alternative presentation to accommodate those with slower connections or without the necessary browser plug-ins.

The line between Web page design and other Web services continues to blur as more forms of media are carried online (see digital convergence). Web designers need to learn about such media technologies (see for example PODCASTING, RSS, and STREAMING) and find appropriate ways to integrate them into their pages. Web pages may also need to provide or link to new types of forums (see BLOGS AND BLOGGING and WIKIS AND WIKIPEDIA).

Further Reading
Beaird, Jason. *The Principles of Beautiful Web Design*. Lancaster, Calif.: Sitepoint, 2007.
Lopuck, Lisa. *Web Design for Dummies*. 2nd ed. Hoboken, N.J.: Wiley, 2006.
Robbins, Jennifer Niederst. *Learning Web Design: A Beginner's Guide to HTML, Graphics, and Beyond*. Sebastapol, Calif.: O'Reilly Media, 2003.
Sitepoint. Available online. URL: http://www.sitepoint.com. Accessed August 23, 2007.

Web server

Most Web users are not aware of exactly how the information they click for is delivered, but the providers of information on the Web must be able to understand and use the Web server. In simple terms, a Web server is a program running on a networked computer (see INTERNET). The server's job is to deliver the information and services that are requested by Web users.

When a user types in (or clicks on) a link in the browser window, the browser sends a HTTP request (see HTTP and WEB BROWSER). To construct the request, the browser first looks at the address (URL) in the user request. An address such as http://www.well.com/conferencing.html consists of three parts:

- The protocol, specifying the type of request. For Web pages this is normally http. In many cases this part can be omitted and the browser will assume that it is meant.

- The name of the server—in this case, www.well. com. The www indicates that it is a World Wide Web server. The rest of the server name gives the organization and the domain (.com, or commercial).

- The specific page being requested. A Web page is simply a file stored on the server, and has the extension htm or html to indicate that it is an HTML-formatted page. If no page is specified, the server will normally provide a default page such as index.html.

In order to direct the browser's request to the appropriate host and server, the browser sends the URL to a name server (see DOMAIN NAME SYSTEM). The name server provides the appropriate numeric IP address (see TCP/IP). The browser then sends an HTTP "get" request to the server's IP address.

Assuming the page requested is valid, the server sends the HTML file to the browser. The browser in turn interprets the formatting and display instructions in the HTML file and "renders" the text and graphics appropriately. It is remarkable that this whole process from user click to displayed page usually takes only a few seconds, even if the Web site is thousands of miles away and requests must be relayed through many intervening computers.

WEB SERVER FEATURES

Web servers would be simple if Web pages consisted only of static text and graphics. However, Web pages today are dynamic: They can display animations, sound, and video. They also interact with the user, responding to menus and other controls, presenting and processing forms, and retrieving data from linked databases. To do these things, the server cannot simply serve up a preformatted page, it must dynamically generate a unique page that responds to the user's actions.

This interactivity requires that the server be able to run programs (scripts) embedded in Web pages. The Common Gateway Interface (CGI) is the basic mechanism for this, though many Web page developers can now work at a higher level to create their page's interaction through scripts in languages such JavaScript. (See CGI and SCRIPTING LANGUAGES.) The task of interfacing Web pages with database facilities is often accomplished using powerful data-management languages (see PERL and PYTHON).

Windows-based servers use ASP (Active Server Pages), a facility that links the Web server to Windows ActiveX controls to access databases. The interaction is usually scripted in VB Script or JScript.

Modern Web server software also contains modules for monitoring and security—an increasingly important consideration as Web sites become essential to business and the delivery of goods and services.

One of the most popular and reliable Web servers in use today is Apache, developed in 1995 and freely distributed with Linux and other UNIX systems (there is also a Windows version). The name is a pun on "a patchy server," meaning that it was developed by adding a series of "software patches" to existing NCSA server code. Microsoft also provides its own line of Web server software that is specific to Windows.

The future should see an increasingly seamless integration between Web servers, browsers, and other applications. Microsoft has been promoting .NET, an initiative that is designed to build Internet access and interoperability into all applications, providing operating system extensions and programming frameworks (see AJAX and MICROSOFT .NET).

Beyond Microsoft's mainly proprietary efforts, another source of integration is the growing use of the Extensible Markup Language (see XML) and its offshoot SOAP (Simple Object Access Protocol) (see SOAP). The goal is to give Web documents and other objects the ability to "communicate" their content and structure to other programs, and to allow programs to freely request and provide services to one another regardless of vendor, platform, or location. As this trend progresses, the Web server starts to "disappear" as a separate entity and the provision of Web services becomes a distributed, cooperative effort (see also WEB SERVICES).

Further Reading

Apache Software Foundation. Available online. URL: http://www. apache.org/. Accessed August 23, 2007.

Aulds, Charles. *Linux Apache Web Server Administration.* 2nd ed. Alameda, Calif.: Sybex, 2002.

Braginski, Leonid. *Running Microsoft Internet Information Server.* New York: McGraw-Hill, 2000.

Jones, Brian W. *How to Host Your Own Web Server.* Morrisville, N.C.: Lulu.com, 2006.

Rosenbrock, Eric, and Eric Filson. *Setting Up LAMP: Getting Linux, Apache, MySQL, and PHP Working Together.* Alameda, Calif.: Sybex, 2004.

Silva, Steve. *Web Server Administration.* Boston: Course Technology, 2003.

Web services

A characteristic of the modern Web and its development is that much of the software is designed to offer services or capabilities that can be called upon by applications. This creation of powerful, versatile building blocks has greatly sped the evolution of Web applications (see WEB 2.0 AND BEYOND).

In order to be useful, a service must be able to understand "messages" (requests) and provide appropriate responses. The medium of exchange is a structured text file (see XML) and a standard format. Three commonly used specifications (defined by the World Wide Web Consortium, or W3C) are what was originally called Simple Object

Access Protocol (see SOAP), the Web Services Description Language (WSDL), and the Universal Description Discovery and Integration (UDDI), which can coordinate and "broker" the services. To keep requester and responder on the same page (so to speak), the W3C also provides a set of "profiles" that specify which versions of which specifications are being used. Additionally, a number of specialized specifications are under development, such as for handling considerations for security and transactions.

There are several ways in which Web services can be accessed:

- Remote Procedure Call (RPC), which generally uses WSDL and follows a format similar to the traditional way programs call upon library functions

- An organization based on the available messages rather than calls or operations (see SERVICE-ORIENTED ARCHITECTURE)

- Representational State Transfer (REST), which views applications or services as collections of "resources" with specific addresses (URLs) and specific requests using HTTP

A variety of other specifications and approaches can be used; this area is a very fluid one. Fortunately, programmers and even users (see MASHUPS) can build new Web applications without having to know the details of how the underlying services work.

Further Reading

Cerami, Ethan. *Web Services Essentials: Distributed Applications with XML-RPC, SOAP, UDDI & WSDL.* Sebastapol, Calif.: O'Reilly, 2002.
Papazoglu, Michael. *Web Services: Principles and Technology.* Upper Saddle River, N.J.: Prentice-Hall, 2007.
Richardson, Leonard, and Sam Ruby. *RESTful Web Services.* Sebastapol, Calif.: O'Reilly, 2007.
World Wide Web Consortium. Web Services Activity. Available online. URL: http://www.w3.org/2002/ws/. Accessed December 4, 2007.

Weizenbaum, Joseph
(1923–)
German-American
Computer Scientist

Joseph Weizenbaum, after writing one of the most famous programs in the history of artificial intelligence research, eventually became one of the most persistent and cogent critics of the AI project itself.

Weizenbaum was born on January 8, 1923, in Berlin to Jewish parents. Having fled Nazi Germany with his parents, in 1941 Weizenbaum enrolled in Wayne University in Detroit, Michigan. However, the following year he enlisted in the United States Army Air Corps. After the war he resumed his study of mathematics. While working as a research assistant, Weizenbaum had the opportunity to help design and build an early digital computer, and although he received his master's degree in mathematics in 1950, he would spend his career in the computer field.

From 1955 to 1963 Weizenbaum worked for General Electric's Computer Development Laboratory as a systems engineer. During this time he would oversee the design and implementation of the first integrated computerized banking system, for Bank of America.

In 1963 Weizenbaum returned to academia, joining the faculty at MIT, which had one of the nation's foremost programs in artificial intelligence research. He contributed to the development of the time-sharing computer system at MIT and early computer networks, but the work for which he would be most remembered started with his interest in getting machines to "understand" human language (see natural language processing).

In 1966 Weizenbaum and a collaborator, psychiatrist Kenneth Colby, created a remarkable program called Eliza (named for the character in Bernard Shaw's play who is taught "proper English" by Professor Henry Higgins). Eliza was remarkable not for its complexity or for innovative use of AI techniques, but for the way it used a few simple procedures to convey the impression that it was a true artificial intelligence—or indeed, a real person.

Eliza worked basically by identifying key words and basic sentence structure and then "mirroring" the user's statements back. A sample snippet of human-Eliza dialog might look like this (Eliza's words are in caps).

Men are all alike.

IN WHAT WAY?

They're always bugging us about something or other.

CAN YOU THINK OF A SPECIFIC EXAMPLE?

Well, my boyfriend made me come here.

YOUR BOYFRIEND MADE YOU COME HERE?

He says I'm depressed much of the time.

I AM SORRY TO HEAR YOU ARE DEPRESSED.

It's true. I am unhappy.

DO YOU THINK COMING HERE WILL HELP YOU NOT TO BE UNHAPPY?

I need some help, that much seems certain.

WHAT WOULD IT MEAN TO YOU IF YOU GOT SOME HELP?

Perhaps I could learn to get along with my mother.

TELL ME MORE ABOUT YOUR FAMILY.

Although the program seemed to have at best a superficial understanding of human language, Weizenbaum soon became dismayed at how readily people treated it as though it were a human being. (Indeed, Colby wanted to use a program like Eliza to automate psychotherapy.)

The result of these concerns was Weizenbaum's book *Computer Power and Human Reason,* a collection of essays that both explain the achievements of AI pioneers and points out their limitations. If, as Weizenbaum observes, "the computer programmer is creator of universes for which he alone is responsible . . . universes of almost unlimited complexity . . .," then indeed the computer scientist must *take* responsibility for his or her creations. This is the challenge that Weizenbaum believes has not been taken seriously enough.

As the 1960s progressed, the United States plunged into the Vietnam War, and racial tension crackled in the streets of major cities. Weizenbaum became increasingly concerned that technology was being used for warlike and oppressive purposes. As an activist, Weizenbaum campaigned against what he saw as the misuse of technology for military purposes such as missiles and missile defense systems. He was founder of a group called Computer Professionals against the ABM (anti-ballistic missile).

Weizenbaum does not consider himself to be a Luddite, however, and he is not without recognition of the potential good that can come from computer technology, though he believes that this potential can only be realized if humans change their attitudes toward nature and their fellow humanity.

During the 1970s and 1980s Weizenbaum not only taught at MIT, but also lectured or served as a visiting professor at a number of institutions, including the Center for Advanced Studies in the Behavioral Sciences at Stanford University (1972–73), Harvard University (1973–74), and coming full circle, the Technical University of Berlin and the University of Hamburg.

In 1988 Weizenbaum retired from MIT. That same year he received the Norbert Wiener Award for Professional and Social Responsibility from Computer Professionals for Social Responsibility (CPSR). In 1991 he was given the Namur Award of the International Federation for Information Processing. He also received European honors such as the Humboldt Prize from the Alexander von Humboldt Foundation in Germany.

Further Reading

Ben-Aaron, Diana. "Weizenbaum Examines Computers [and] Society." *The Tech* (Massachusetts Institute of Technology) vol. 105, April 9, 1985. Available online. URL: http://www-tech.mit.edu/V105/N16/weisen.16n.html. Accessed December 4, 2007.
ELIZA [running as a Java Applet]. Available online. URL: http://www.manifestation.com/neurotoys/eliza.php3. Accessed December 4, 2007.
Henderson, Harry. *Artificial Intelligence: Mirrors for the Mind*. New York: Chelsea House, 2007.
Weizenbaum, Joseph. *Computer Power and Human Reason*. San Francisco: W. H. Freeman, 1976.
"Weizenbaum: Rebel at Work" [information about and excerpts from a film by Peter Haas]. Available online. URL: http://www.ilmarefilm.org/W_E_1.htm. Accessed December 4, 2007.

Wiener, Norbert

(1894–1964)
American
Mathematician, Philosopher

Norbert Wiener developed the theory of cybernetics, or the process of communication and control in both machines and living things. His work has had an important impact both on philosophy and on design principles.

Wiener was born on November 26, 1894, in Columbia, Missouri. His father was a linguist at Harvard University, and spurred an interest in communication which the boy combined with an avid pursuit of mathematics and science (particularly biology). A child prodigy, Wiener started reading at age three, entered Tufts University at age 11, and earned his B.A. in 1909 at the age of 14, after concluding that his lack of manual dexterity made biological work too frustrating. He earned his M.A. in mathematics from Harvard only three years later, and his Harvard Ph.D. in mathematical logic just a year later in 1913. He then traveled to Europe, where he met leading mathematicians such as Bertrand Russell, G. H. Hardy, Alfred North Whitehead, and David Hilbert. When the United States entered World War I, Wiener served at Aberdeen Proving Ground, where he designed artillery firing tables.

After the war, Wiener was appointed as an instructor at MIT, where he would serve until his retirement in 1960. However, he continued to travel widely, serving as a Guggenheim Fellow at Copenhagen and Göttingen in 1926, and a visiting lecturer at Cambridge (1931–32) and Tsing-Hua University in Beijing (1935–36). Wiener's scientific interests proved to be as wide as his travels, including research into stochastic and random processes (such as the Brownian motion of microscopic particles) where he sought more general mathematical tools for the analysis of irregularity.

During the 1930s, Wiener began to work more closely with MIT electrical engineers who were building mechanical computers (see BUSH, VANNEVAR and ANALOG COMPUTER). He learned about feedback controls and servomechanisms that enabled machines to respond to forces in the environment.

During World War II, he did secret military research with an engineer, Julian Bigelow, on antiaircraft gun control mechanisms, including methods for predicting the future position of an aircraft based upon limited and possibly erroneous information.

Wiener became particularly interested in the feedback loop—the process by which an adjustment is made on the basis of information (such as from radar) to a predicted new position, a new reading is taken and a new adjustment made, and so on. (He had first encountered these concepts at MIT with his friend and colleague Harold Hazen.) The use of "negative feedback" made it possible to design systems that would progressively adjust themselves such as by intercepting a target. More generally, it suggested mechanisms by which a machine (perhaps a robot) could progressively work toward a goal.

Wiener's continuing interest in biology led him always to relate what he was learning about control and feedback mechanisms to the behavior of living organisms. He had followed the work of Arturo Rosenbleuth, a Mexican physiologist who was studying neurological conditions that appeared to result from excessive or inaccurate feedback. (Unlike the helpful negative feedback, positive feedback in effect amplifies errors and sends a system swinging out of control.)

By the end of World War II, Wiener, Rosenbleuth, the neuropsychiatrist Warren McCulloch, and the logician

Walter Pitts were working together toward a mathematical description of neurological processes such as the firing of neurons in the brain. This research, which started out with the relatively simple analogy of electromechanical relays (as in the telephone system) would eventually result in the development of neural network theory (see NEURAL NETWORK and MINSKY, MARVIN). More generally, these scientists and others (see VON NEUMANN, JOHN) had begun to develop a new discipline for which Wiener in 1947 gave the name *cybernetics*. This word is from a Greek word referring to the steersman of a ship, suggesting the control of a system in response to its environment.

The field of cybernetics attempted to draw from many sources, including biology, neurology, logic, and what would later become robotics and computer science. Wiener's 1948 book, *Cybernetics or Control and Communication in the Animal and the Machine,* was as much philosophical as scientific, suggesting that cybernetic principles could be applied not only to scientific research and engineering but also to the better governance of society. (On a more practical level Wiener also worked with Jerome Wiesner on designing prosthetics to replace missing limbs.)

Although Wiener did not work much directly with computers, the ideas of cybernetics would indirectly influence the new disciplines of artificial intelligence (AI) and robotics. However, in his 1950 book, *The Human Use of Human Beings,* Wiener warned against the possible misuse of computers to rigidly control or regiment people, as was the experience in Stalin's Soviet Union. Wiener became increasingly involved in writing these and other popular works to bring his ideas to a general audience.

Wiener received the National Medal of Technology from President Johnson in 1964. The accompanying citation praised his "marvelously versatile contributions, profoundly original, ranging within pure and applied mathematics, and penetrating boldly into the engineering and biological sciences." He died on March 18, 1964, in Stockholm, Sweden.

Further Reading

Conway, Flo, and Jim Siegelm. *Dark Hero of the Information Age: In Search of Norbert Wiener, The Father of Cybernetics.* New York: Basic Books, 2005.

Heims, Steve J. *John von Neumann and Norbert Wiener: From Mathematics to the Technologies of Life and Death.* Cambridge, Mass.: MIT Press, 1980.

Wiener, Norbert. *Cybernetics, or Control and Communication in the Animal and Machine.* Cambridge, Mass.: MIT Press, 1950 (2nd ed. 1961).

———. *The Human Use of Human Beings: Cybernetics and Society.* Boston: Houghton Mifflin, 1950. (2nd ed., Avon Books, 1970).

———. *Invention: The Care and Feeding of Ideas.* Cambridge, Mass.: MIT Press, 1993.

wikis and Wikipedia

A wiki (from the Hawaiian word for "quick") is a generally Web-based software application that allows users to collaboratively contribute and edit articles on various topics. Developed by Howard G. "Ward" Cunningham in the mid-1990s, the best-known example today is Wikipedia.

STRUCTURE AND SOFTWARE

Wiki software varies in details such as use of markup languages, programming interface, and platform. However, most wikis include the following features:

- Users can create new pages (articles) or edit existing ones.

- Pages contain links to related pages, sometimes using "wiki words" where WordsAreScrunchedTogetherWithIntialCaps.

- Simple markup can be used to create such effects as boldface, headings, or lists. The wiki software usually translates this to HTML for rendering.

- A record is kept of each contribution or edit, often displayed on a "Recent Changes" page.

- Many wikis use a database (such as MySQL) to store and retrieve pages. Some wikis simply store each page as a file, and a few (such as TiddlyWiki) store all pages together as a single document.

- Wikis can be public (open to anyone) or restricted, such as to members of an organization.

- The administrator of the wiki establishes guidelines or standards (such as for citing sources for facts) and procedures for dealing with disputes and controversial topics.

There is now a great variety of wiki software for just about every computing platform. At one end there is MediaWiki, the software used to implement Wikipedia, and "enterprise wikis" such as BrainKeeper and Twiki, providing complex features for large-scale knowledge bases. So-called personal Wikis such as DidiWiki and TiddlyWiki can be used by individuals for note-taking, research, or managing personal information.

WIKIPEDIA AND ITS CRITICS

Founded in 2001 (see WALES, JIMMY), Wikipedia is the world's largest and best-known wiki. As of mid-2008 Wikipedia had more than 2,500,000 articles in English and 7,500,000 in more than 250 other languages. At any given time there are about 75,000 people from all backgrounds and walks of life contributing or editing articles.

Wikipedia has a number of strengths. Its ubiquity and diversity enable it to cover tens of thousands of topics (including the obscure or the simply local) that would be deemed unsuitable or impracticable for traditional encyclopedias. Emerging topics, including recent news events, can be covered quickly and comprehensively (though perhaps blurring the lines between reference and journalism).

The principal problem raised by critics stems from issues of "quality control." Unlike the case with traditional encyclopedias, there are no requirements that contributors have academic training or otherwise demonstrate their expertise in their chosen area. Further, the ability of anyone to edit an article has led to "edit wars" as people on different sides

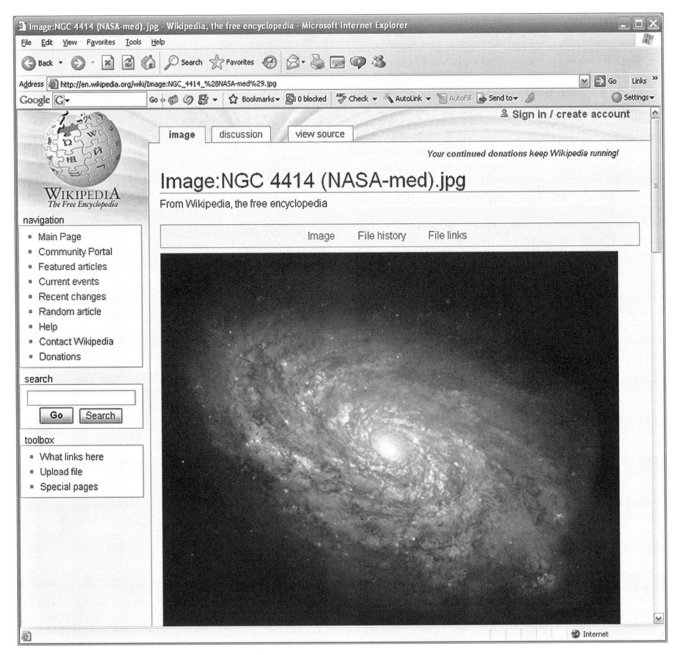

Wiki software such as the extensive and ever-growing Wikipedia, allows users to collaborate to create and update knowledge bases. Entries can include images such as the astronomical photo shown here. There is also a place for ongoing discussion of changes to the page.

of a controversial topic (or even politicians) would change articles back and forth to reflect their views.

Defenders of Wikipedia believe that the same "bottom up" writing and editing process cited by critics can also be one of the project's strengths. Each article has a record of changes, and many articles have attached discussion pages where writers can critique the page or discuss their rationales for edits. Finally, they cite a study by the journal *Nature* that found that in the science articles analyzed, Wikipedia averaged four errors while the *Encyclopaedia Britannica* was only slightly more accurate, averaging three.

Defenders of *Britannica*, however, point out that their publication has the kind of consistency that can only come through rigorous application of editorial standards. Wikipedia does have standards that writers and editors are urged to apply, such as providing a citation for every significant statement, maintaining a "neutral point of view," and refraining from including original research. Nevertheless, the quality of organization and writing does vary considerably from one article to the next.

Meanwhile innovators in Wikipedia and its community are developing new tools that may improve the reliability of the encyclopedia. One tool, WikiScanner, searches for and compiles information (such as affiliations) about wiki contributors, allowing readers to better judge their competence and motivations. Wikipedia's parent Wikimedia Foundation is also introducing a system by which previously unknown contributors will undergo a sort of probationary period while their material is scrutinized. (Eventually they would become "trusted" and their material would appear instantly, as it does now for most articles.)

Wikis have, like blogs, become a pervasive form of online communication and information sharing, and have gained considerable attention as an application for the "new" Web (see USER-CREATED CONTENT, WEB 2.0 AND BEYOND, and SOCIAL NETWORKING). Wikis are currently being used to create rapidly expanding knowledge bases (such as for technical support), to share emerging scholarship, and to promulgate documentation within an organization. Hosting services (called "wiki farms") such as Wikia offer communities wiki software and Web space, sometimes free of charge. Wiki principles are also finding their way into software such as personal information managers (see CONTENT MANAGEMENT).

Further Reading

Comparison of Wiki Software. Available online. URL: http://en.wikipedia.org/wiki/Comparison_of_wiki_software. Accessed May 11, 2007.

Ebersbach, Anja, Markus Glaser, and Richard Heigl. *Wiki: Web Collaboration*. New York: Springer, 2005.

Giles, Jim. "Wikipedia 2.0: Now with Added Trust." NewScientist. com News Service. Available online. URL: http://technology. newscientist.com/channel/tech/mg19526226.200-wikipedia-20%20-now-with -added-trust.html. Accessed December 4, 2007.

Klobas, Jane. *Wikis: Tools for Information Work and Collaboration*. Oxford: Chandos Publishing, 2006.

Lee, Ellen. "As Wikipedia Moves to S.F., Founder Discusses Planned Changes." *San Francisco Chronicle*, November 30, 2007. Available online. URL: http://www.sfgate.com/cgi-bin/article.cgi?f=/c/a/2007/11/30/BUOMTKNJA.DTL. Accessed December 3, 2007.

Leuf, Bo, and Ward Cunningham. *The Wiki Way: Quick Collaboration on the Web*. Reading, Mass.: Addison-Wesley, 2001.

MediaWiki. Available online. URL: http://www.mediawiki.org. Accessed December 4, 2007.

Wikipedia. Available online. URL: http://www.wikipedia.org. Accessed December 4, 2007.

Woods, Dan. *Wikis for Dummies*. Hoboken, N.J.: Wiley, 2007.

wireless computing

Using suitable radio frequencies to carry data among computers on a local network has several advantages. The trouble and expense of running cables (such as for Ethernet) in older buildings and homes can be avoided. With a wireless LAN (WLAN) a user could work with a laptop on the deck or patio while still having access to a high-speed Internet connection.

Typically, a wireless LAN uses a frequency band with each unit on a slightly different frequency, thus allowing all units to communicate without interference. (Although radio frequency is now most popular, wireless LANs can also use microwave links, which are sometimes used as an alternative to Ethernet cable in large facilities.)

Usually there is a network access point, a PC that contains a transceiver and serves as the network hub (it may also serve as a bridge between the wireless network and a wired LAN). The hub computer can also be connected to a high-speed Internet service via DSL or cable. It has an antenna allowing it to communicate with wireless PCs up to several hundred feet away, depending on building configuration.

Each computer on the wireless network has an adapter with a transceiver so it can communicate with the access point. The adapter can be built-in (as is the case with some handheld computers), or mounted on a PC card (for laptops) or an ISA card (for desktop PCs) or connected to a USB port.

Simple home wireless LANs can be set up as a "peer network" where any two units can communicate directly with each other without going through an access point or hub. Applications needing Internet access (such as e-mail and Web browsers) can connect to the PC that has the Internet cable or DSL connection.

A wireless LAN can make it easier for workers who have to move around within the building to do their jobs. Examples might include physicians or nurses entering patient data in a hospital or store workers checking shelf inventory.

PROTOCOLS

Several protocols or standards have been developed for wireless LANs. The most common today is IEEE 802.11b, also called WiFi with speeds up to 11 mbps (megabits per second) transmitting on 2.4 GHz (gigahertz) band. Although that would seem to be fast enough for most applications, a new alternative, 802.11n, can offer speeds up to 54 mbps. Because it uses the unlicensed 5 GHz frequency range it is not susceptible to interference from other devices.

The question of security for 802.11 wireless networks has been somewhat controversial. Obviously, wireless data can be intercepted in the same way that cell phone or other radio transmissions can. The networks come with a security feature, WEP or the newer WPA, but many users neglect to enable it, and it is vulnerable to certain types of attack. Users can obtain greater security by reducing emissions outside the building, changing default passwords and device IDs, and disabling DHCP to make it harder for snoopers to obtain a valid IP address for the network. Users can also add another layer of encryption and possibly isolate the wireless network from the wired network by using a more secure Virtual Private Network (VPN). Many of these measures do involve a tradeoff between the cost of software and administration on the one hand and greater security on the other. However, the growing popularity of wireless access should spur the development of improved built-in security.

Another wireless protocol called Bluetooth has been embedded in a variety of handheld computers, appliances, and other devices. It provides a wireless connection at speeds up to 1 MB/second (see BLUETOOTH).

MOBILE WIRELESS NETWORKING

Wireless connections can also keep computer users in touch with the Internet and their home office while they travel. Increasingly, more devices are becoming wireless capable while at the same time the functions of handheld computers, cell phones, and other devices are being merged (see also PORTABLE COMPUTERS). A new initiative called 3G (third generation) involves the establishment of ubiquitous wireless services that can connect users to the Internet (and thus to one another) from a growing number of locations.

Currently, the 3G agenda is further advanced in Europe than in the United States. One problem is that a standard protocol has not yet emerged. The leading candidates appear to be GSM (Global System for Mobile Communications), which is used by European cell phone networks, and CDMA (Code Division Multiplexing Access).

3G has different speeds ranging from 144 bps for vehicular connections to 384 kbps for personal handheld devices to 2 bps for indoor installations. All providers, ranging from cell phones to packet (IP) telephony would using a standard billing format and database so that users could operate across many sorts of services seamlessly. Another alternative is WIMAX, which can be thought of as a wider-area version of WiFi in which each base station can transmit over up to 50 kilometers. As of 2008 deployment has been slower than anticipated, with widespread coverage in U.S. cities not likely to be available for at least several years.

Ultimately, these technologies may bring a *Star Trek*–like world, with handheld devices that include not only e-mail and Web browsing capability but a "smart phone," MP3 music player, and even a digital camera.

Further Reading

Briere, Dany, Pat Hurley, and Edward Ferris. *Wireless Home Networking for Dummies.* 2nd ed. Hoboken, N.J.: Wiley, 2006.
Geier, Eric. *Wi-Fi Hotspots: Setting Up Public Wireless Internet Access.* Indianapolis: Cisco Press, 2006.
Haley, E. Phil. *Over-the-Road Wireless for Dummies.* Hoboken, N.J.: Wiley, 2006.
Kwok, Yu-Kwong Ricky, and Vincent K. N. Lau. *Wireless Internet and Mobile Computing: Interoperability and Performance.* New York: Wiley, 2007.
Wireless Networks and Mobile Computing. Available online. URL: http://www.networkworld.com/topics/wireless.html. Accessed August 23, 2007.

Wirth, Niklaus

(1934–)
Swiss
Computer Scientist

Niklaus Wirth created new programming languages such as Pascal that helped change the way computer scientists and programmers thought about their work. His work influenced later languages and ways of organizing program resources.

Wirth was born on February 15, 1934, in Winterhur, Switzerland. He received a degree in electrical engineering at the Swiss Federal Institute of Technology (ETH) in 1959, then earned his M.S. at Canada's Laval University. He went to the University of California, Berkeley, where he received his Ph.D. in 1963 and taught in the newly founded Computer Science Department at nearby Stanford University. By then he had become involved with computer science and the design of programming languages.

Wirth returned to the ETH in Zurich in 1968, where he was appointed a full professor of computer science. He had been part of an effort to improve Algol. Although Algol offered better program structures than earlier languages such as FORTRAN, the committee revising the language had become bogged down in adding many new features to the language that would become Algol-68 (see ALGOL).

Wirth believed that adding several ways to do the same thing did not improve a language but simply made it harder to understand and less reliable. Between 1968 and 1970, Wirth therefore crafted a new language, Pascal, named after the 17th-century mathematician who had built an early calculating machine.

Pascal required that data be properly defined (see DATA TYPES) and allowed users to define new types of data such as records (similar to those used in databases). It provided all the necessary control structures (see LOOP and BRANCHING STATEMENTS). Following the new thinking about structured programming (see DIJKSTRA, EDSGER) Pascal retained the "unsafe" GOTO statement but discouraged its use.

Pascal became the most popular language for teaching programming. By the 1980s, versions such as UCSD Pascal and later, Borland's Turbo Pascal were bringing the benefits of structured programming to desktop computer users. Meanwhile, Wirth was working on a new language, Modula-2. As the name suggested, the language featured the use of modules, packages of program code that could be linked to programs to extend their data types and functions. Wirth also designed a computer workstation called *Lilith*. This powerful machine not only ran Modula-2; its operating system, device drivers and all other facilities were also implemented in Modula-2 and could be seamlessly integrated, essentially removing the distinction between operating system and application programs. Wirth also helped design Modula-3, an object-oriented extension of Modula-2, as well as another language, Oberon, which was originally intended to run in built-in computers (see EMBEDDED SYSTEMS).

Looking back at the development of object-oriented programming (OOP), the next paradigm that captured the attention of computer scientists and developers after structured programming, Wirth has noted that OOP isn't all that new. Its ideas (such as encapsulation of data) are largely implicit in structured procedural programming, even if it shifted the emphasis to binding functions into objects and allowing new objects to extend (inherit from) earlier ones. But he believes the fundamentals of good programming haven't really changed in 30 years. In a 1997 interview Wirth noted that "the woes of Software Engineering are not due to lack of tools, or proper management, but largely due to lack of sufficient technical competence. A good designer must rely on experience, on precise, logical thinking; and on pedantic exactness. No magic will do."

Wirth has received numerous honors, including the ACM Turing Award (1984) and the IEEE Computer Pioneer Award (1987).

Further Reading

Pescio, Carlo. "A Few Words with Niklaus Wirth." *Software Development,* vol. 5, no. 6, June 1997. Available online. URL: http://www.eptacom.net/pubblicazioni/pub_eng/wirth.html. Accessed August 23, 2007.

Wirth, Niklaus. *Algorithms + Data Structures = Programs.* Englewood Cliffs, N.J.: Prentice Hall, 1976.

———, and Kathy Jensen. *PASCAL User Manual and Report.* 4th ed. New York: Springer-Verlag, 1991.

———. *Project Oberon: The Design of an Operating System and Compiler.* Reading, Mass.: Addison-Wesley, 1992.

———. "Recollections about the Development of Pascal." In Bergin, Thomas J., and Richard G. Gibson, eds., *History of Programming Languages-II,* 97–111. New York: ACM Press; Reading, Mass.: Addison-Wesley, 1996.

———. *Systematic Programming: An Introduction.* Reading, Mass.: Addison-Wesley, 1973.

women and minorities in computing

Although the development of computer science and technology has been an international effort, there is no doubt that the majority of contributors (particularly in the early years) were men—specifically white men. The interesting exceptions include Charles Babbage's collaborator Ada Lovelace, ENIAC's first trained programmers—all women—and of course Grace Hopper, whose COBOL revolutionized business computing. Finally, the 2007 winner of one of the field's most prestigious honors, the ACM Turing Award, is a woman, compiler developer Frances E. Allen.

Today women have gained prominent roles in all aspects of computer science as well as some high-profile posts in business (such as Carly Fiorina, former CEO of Hewlett-Packard, and Meg Whitman of eBay). However, the overall involvement of women in the higher echelons of computing remains relatively small.

Educational surveys suggest that boys and girls start out with roughly equal interest and involvement with computers, including basic courses in computer literacy and applications. However, a 2006 report from the College Board found that 59 percent of boys reported taking courses in programming, compared with 41 percent of girls. Further, the great majority of students taking advanced placement computer science exams are male.

At the college level, women made gains in the percentage of bachelor's degrees in the computer field, reaching 37 percent by the mid-1980s. However, by 2005 that percentage had declined to 22 percent. (However, women were earning 34 percent of master's degrees by 2001.) Further, the number of women working in information technology declined from 984,000 (28.9 percent) in 2000 to 908,000 (26.2 percent) in 2006.

The reasons for this decline are unclear, though some possible causes that have been suggested include the effects of the "dot-bust" and the perception that IT jobs were no longer secure, the "geek" stereotype not appealing to young women, and the availability of more attractive career paths.

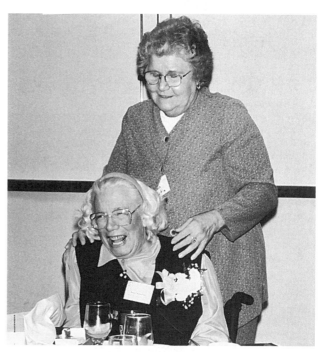

Jean Bartik (standing) and Betty Holberton answered a call for "computers" during World War II. At the time, that was the name for a clerical person who performed calculations. But these two computer pioneers, shown here at a reunion, would go on to develop important programming techniques for the ENIAC and later machines. (Courtesy of the Association for Women in Computing)

In terms of race or ethnicity, whites and Asians earn a disproportionate number of degrees in computing, although interestingly, minority women tend to earn a higher percentage than white women. Although minorities have been gradually increasing their participation in the computing field, economic disadvantage (see DIGITAL DIVIDE) and poor educational preparation continue to be obstacles for some.

EFFORTS AT CHANGE

A variety of programs have sought to interest women and minority students in computer programming and other digital careers. These can include the creation of nontraditional programming environments such as Alice, which allows students to create animated stories using scripting and 3D graphics tools. (The theory behind this is that girls are more interested in storytelling and character interaction, while traditional programming classes focused more on "shoot 'em up" games and other things of more interest to boys.) There has also been a move away from emphasis on "hard core" programming skills to a more broad-based ability to think about technology and its possible uses. (As a result of this and a certain amount of affirmative action, Carnegie Mellon raised its percentage of women computer science students from 8 percent to 40 percent.)

African Americans and other minorities have also developed a number of organizations and programs designed to promote networking among minority professionals, link applicants to job openings, and encourage professional development.

Further Reading

ACM Committee on Women in Computing. Available online. URL: http://women.acm.org/. Accessed December 4, 2007.

ADA Project. Available online. URL: http://women.cs.cmu.edu/ada/. Accessed December 4, 2007.

Association for Women in Computing. Available online. URL: http://www.awc-hq.org/. Accessed December 4, 2007.

Black Data Processing Associates. Available online. URL: http://www.bdpa.org. Accessed December 4, 2007.

Dean, Cornelia. "Computer Science Takes Steps to Bring Women to the Fold." *New York Times*, April 17, 2007. Available online. URL: http://www.nytimes.com/2007/04/17/science/17comp.html?_r=1&oref=slogin. Accessed December 4, 2007.

Margolis, Jane, and Allan Fisher. *Unlocking the Clubhouse.* Cambridge, Mass.: MIT Press, 2001.

National Center for Women and Information Technology. Available online. URL: http://www.ncwit.org/. Accessed December 4, 2007.

National Institute for Women in Trades, Technology & Science. Available online. URL: http://www.iwitts.com/. Accessed December 4, 2007.

Pinkett, Randal D. "Strategies for Motivating Minorities to Engage Computers." Carnegie Mellon University. Available online. URL: ttp://llk.media.mit.edu/papers/cmu1999.pdf. Accessed December 4, 2007.

Yount, Lisa. *A-Z of Women in Science and Math.* Rev. ed. New York: Facts On File, 2007.

word processing

Although computers are most often associated with numbers and calculation, creating text documents is probably the most ubiquitous application for desktop PCs.

The term *word processor* was actually coined by IBM in the 1960s to refer to a system consisting of a Selectric typewriter with magnetic tape storage. This allowed the typist to record keystrokes (and some data such as margin settings) on tape. Material could be corrected by being re-recorded. The tape could then be used to print as many perfect copies of the document as required. A version using magnetic cards instead of tape appeared in 1969.

The first modern-style word processor was marketed by Lexitron and Linolex. It also used magnetic tape, but it added a video display screen. Now the writer could see and correct text without having to print it first. A few years later, a new invention, the floppy disk, became the standard storage medium for dedicated word processing systems.

The word-processing systems developed by Wang, Digital Equipment Corporation, Data General, and others became a feature in large offices in the late 1970s. These systems were essentially minicomputers with screens, keyboards, and printers and running a specialized software program. Because these systems were expensive (ranging from about $8,000 to $20,000 or more), they were not affordable by smaller businesses. Typically, they were operated by specially trained personnel (who became known also as "word processors") to whom documents were funneled for processing, as with the old "typing pool."

PC WORD PROCESSING

The first microcomputer systems had very limited memory and storage capacity. However, by the late 1970s various systems using the S-100 bus and running CP/M had word-processing programs, as did the Apple II and other first-generation PCs. However, it took the entry of the IBM PC into the market in 1981 to make the PC a word-processing alternative for mainstream businesses. The machine had more memory and storage than earlier machines, and the IBM name provided reassurance to business.

A number of word-processing programs were written for the IBM PC running MS-DOS, but the market leaders were WordStar and WordPerfect. Both programs offered basic text editing and formatting, including the ability to embed commands to mark text for boldface, italic, and so on. The programs came with drivers for the more popular printers.

In 1984, the Macintosh offered a new face for word processing and other applications. Using bitmapped fonts, the Mac could show a good representation of the fonts and typestyles that would be in the printed document. This "what you see is what you get" (WYSIWYG) approach, together with the graphical user interface with mouse-driven menus meant that users did not have to learn the often obscure command key sequences used in WordStar or WordPerfect.

Microsoft then developed Windows as a graphical user interface alternative to MS-DOS for IBM-compatible PCs. By 1990, Windows was rapidly replacing DOS as the operating system of choice, and Microsoft Word was winning the battle against WordPerfect, whose Windows version was rather flawed at first.

In addition to being able to visually show fonts and formatting, Word and other modern word processors are packed with features. Some typical features today include:

- different views of the document, including an outline showing headings down to a user-specified level

- automatic table of contents and index generation

- tables and multicolumn text

- automatic formatting of bulleted and numbered lists

- built-in and user-defined styles for headings, paragraphs, and so on.

- the ability to use built-in or user-defined templates to provide starting settings for new documents (see TEMPLATE)

- the ability to record or otherwise specify a series of commands to be performed automatically (see MACRO)

- spelling and grammar checkers

- the ability to incorporate a variety of graphics image formats in the document

- automatic formatting and linking of Web hyperlinks within documents

- the ability to import and export documents in a variety of formats, including Web documents (see HTML)

- an extensive online help system including "wizards" to guide the user step-by-step through various tasks

As word processors become more extensive in their capabilities, it has become harder to distinguish them from programs designed to create precise copy for publication (see DESKTOP PUBLISHING). However, copy prepared by writers with a word processor must generally be further processed through a desktop publishing or in-house computerized typography system.

At the other end of the spectrum many users find that word processors are "overkill" for making simple notes. A variety of programs for entering simple text are available, including the Notepad program that comes with Windows. There are also applications for which plain text must be produced, without the formatting codes added by word processors. In particular, programmers often use specialized editing programs to create source code (see TEXT EDITOR).

TRENDS

Today word processing programs are generally part of an office software suite such as Microsoft Office, Corel Office, or Open Office. Documents created by other components of the suite can be embedded in word processing documents. (In Windows, object linking and embedding [OLE] is a system that allows for embedded documents to be automatically updated and to be edited using the functions of the host program. Thus, an Excel spreadsheet embedded in a Word document can be worked in place using the standard Excel interface.)

There are also features that can facilitate collaboration between workers in a networked office, such as by keeping track of revisions made by various people working on the same document.

A new alternative is the free (or low-cost) online word processor such as Google Docs & Spreadsheets and Zoho Writer. These products can be used from any Web browser and facilitate the central storage of documents for mobile users (see APPLICATION SERVICE PROVIDER).

As with other applications, word processors are increasingly being integrated with the Web, and include the ability to create HTML documents. In turn, the programs specifically designed for creating HTML documents now have many word-processor features including templates, styles, and the visual representation of the page.

Further Reading

Cox, Joyce, and Joan Preppernau. *Microsoft Office Word 2007 Step by Step.* Redmond, Wash.: Microsoft Press, 2007.

Google Docs & Spreadsheets. Available online. URL: http://www.docs.google.com. Accessed August 23, 2007.

Kunde, Brian. "A Brief History of Word Processing (Through 1986)." Available online. URL: http://www.stanford.edu/~bkunde/fb-press/articles/wdprhist.html. Accessed August 23, 2007.

Open Office.org. Available online. URL: http://www.openoffice.org/. Accessed August 23, 2007.

Petrie, Michael. "A Potted History of WordStar." Available online. URL: http://www.wordstar.org/wordstar/history/history.htm. Accessed August 23, 2007.

Zoho Writer (online word processor). Available online. URL: http://writer.zoho.com/jsp/home.jsp?serviceurl=%2Findex.do. Accessed August 23, 2007.

workstation

Like minicomputer, *workstation* is a rather slippery term whose meaning and significance has changed somewhat with the growing power of desktop PCs.

In the late 1960s and 1970s, most "personal" computing was done by individuals connected to time-sharing mainframes or minicomputers by terminals. Generally, the terminals could only display text, not graphics.

However, researchers at the Xerox Palo Alto Research Center (PARC) began to develop a more powerful computer for individual use (see ENGLEBART, DOUGLAS and KAY, ALAN). The Xerox Alto had a high-resolution bitmapped graphics display and a mouse-controlled graphical user interface. While it was expensive and not very successful commercially, the Alto set the stage for the Macintosh in 1984 and for Microsoft Windows.

Although the desktop PCs of the 1980s such as the IBM PC had some graphics capabilities, the machines lacked the capacity for graphics-intensive applications such as engineering design and the generation of movie effects. Led by Sun and Silicon Graphics (SGI), the high-performance graphics workstation emerged as a distinctive product category. These machines used relatively powerful microprocessors (such as the Sun SPARC and the MIPS) with instruction sets optimized for speed (see RISC). These systems generally ran UNIX as their operating system.

However, by the late 1990s, ordinary desktop PCs were catching up to dedicated workstations in terms of processing power and graphics features. By 2002, a desktop PC costing about $2,000 offered a 2-GB processor, 256 MB of RAM, 120 GB hard drive, and an optimized 3D graphics card that can drive displays up to 1600 by 1200 pixels or more. These systems can run Windows NT or XP, or, for users preferring UNIX, Linux offers a robust and inexpensive operating system. This sort of system rivals the capabilities of a dedicated workstation while offering all of the versatility of a general-purpose PC. As a result, the term *workstation* today refers more to a way of using a computer than to a specific class of hardware. Machines are thought of as workstations if they emphasize graphics performance and are dedicated to particular activities such as science, imaging, engineering, design (see also COMPUTER-AIDED DESIGN AND MANUFACTURING), or video editing.

Further Reading

Goldberg, Adele. *A History of Personal Workstations.* Reading, Mass.: Addison-Wesley/ACM, 1988.

Sun Microsystems. *The New User's Guide to Sun Workstation.* New York: Springer-Verlag, 1991.

World Wide Web

In little more than a decade the World Wide Web has become nearly as ubiquitous as the telephone and has become for many a preferred medium for shopping, news, entertainment, and education. Some cultural observers believe that this vast system of linked information may be having an impact on society as great as that of the invention of the printing press more than five centuries earlier.

By the beginning of the 1990s, the Internet had become well established as a means of communication between relatively advanced computer users, particularly scientists, engineers, and computer science students—primarily using UNIX-based systems (see UNIX). A number of services used the Internet protocol (see TCP/IP) to carry messages or data. These included e-mail, file transfer protocol (see FTP) and newsgroups (see NETNEWS AND NEWSGROUPS). A Wide Area Information Service (WAIS) even provided a protocol for users to retrieve information from databases on remote hosts. Another interesting service, Gopher, was developed at the University of Minnesota in 1991. It used a system of nested menus to organize documents at host sites so they could be browsed and retrieved by remote users.

Gopher was quite popular for a few years, but it would soon be overshadowed by a rather different kind of networked information service. A physicist/programmer (see BERNERS-LEE, TIM) working at CERN, the European particle physics laboratory in Switzerland, had devised in 1989 a system that he eventually called the World Wide Web (sometimes called WWW or W3). By 1990, he was running a prototype system and demonstrating it for CERN researchers and a few outside participants.

USING THE WEB

The Web consists essentially of three parts. Berners-Lee devised a markup language: that is, a system for indicating document elements (such as headers), text characteristics, and so on (see HTML). Any document could be linked to another (see HYPERTEXT AND HYPERMEDIA) by specifying that document's unique address (called a Uniform Resource Locator or URL) in a request. Berners-Lee defined the HyperText Transport Protocol, or HTTP, to handle the details needed to retrieve documents. (Although HTTP is most often used to retrieve HTML-formatted Web documents, it can also be used to specify documents using other protocols, such as ftp, news, or Gopher.)

A program (see WEB SERVER) responds to requests for documents sent over the network (usually the Internet, that is, TCP/IP). The requests are issued by a client program as a result of the user clicking on highlighted links or buttons or specifying addresses (see WEB BROWSER). The browser in turn interprets the HTML codes on the page to display it correctly on the user's screen.

At first the Web had only text documents. However, thanks to Berners-Lee's flexible design (see CLIENT-SERVER COMPUTING) new, improved Web browsers could be created and used with the Web as long as they followed the rules for HTTP. The most successful of these new browsers was Mosaic, created by Marc Andreesen at the National Center for Supercomputing Applications. NCSA Mosaic was available for free download and could run on Windows, Macintosh, and UNIX-based systems. Mosaic not only dispensed with the text commands used by most of the first browsers, it also had the ability to display graphics and play sound files. With Mosaic the text-only hypertext of the early Web rapidly became a richer hypermedia experience. And thanks to the ability of browsers to accept modules to handle new kinds of files (see PLUG-IN), the Web could also

accommodate real-time sound and video transmissions (see STREAMING).

In 1994, Andreessen left NCSA and co-founded a company called Netscape Communications, which improved and commercialized Mosaic. Microsoft soon entered with a competitor, Internet Explorer; today these two browsers dominate the market with Microsoft having taken the lead. Together with relatively low-cost Internet access (see MODEM and INTERNET SERVICE PROVIDER) these user-friendly Web browsers brought the Web (and thus the underlying Internet) to the masses. Schools and libraries began to offer Web access while workplaces began to use internal webs to organize information and organize operations. Meanwhile, companies such as the on-line bookseller Amazon.com demonstrated new ways to deliver traditional products, while the on-line auction site eBay took advantage of the unique characteristics of the on-line medium to redefine the auction.

The burgeoning Web was soon offering millions of pages, especially as entrepreneurs began to find additional business opportunities in the new medium (see E-COMMERCE). Two services emerged to help Web users make sense of the flood of information. Today users can search for words or phrases (see SEARCH ENGINE) or browse through structured topical listings (see PORTAL). Estimates from various sources suggest that as of 2007 approximately 1.2 billion people worldwide access the Web, with usage increasing most rapidly in the emerging industrial superpowers of India and China.

IMPACT AND TRENDS

The Web is rapidly emerging as an important news medium (see JOURNALISM AND THE COMPUTER INDUSTRY). The medium combines the ability of broadcasting to reach many people from one point with the ability to customize content to each person's preferences. Traditional broadcasting and publishing are constrained by limited resources and the need for profitability, and thus the range and diversity of views made available tend to be limited. With the Web, anyone with a PC and a connection to a service provider can put up a Web site and say just about anything. Millions of people now display aspects of their lives and interests on their personal Web pages (see BLOGS AND BLOGGING). The Web has also provided a fertile medium for the creation of online communities (see SOCIAL NETWORKING and VIRTUAL COMMUNITY) while contributing to significant issues (see PRIVACY IN THE DIGITAL AGE).

As the new century continues, the Web is proving itself to be truly worldwide, resilient, and adaptable to many new communications and media technologies (see DIGITAL CONVERGENCE). Nevertheless, the Web faces legal and political challenges (see CENSORSHIP AND THE INTERNET, and INTELLECTUAL PROPERTY AND COMPUTING) as well as technical challenges (see SEMANTIC WEB and WEB 2.0).

Further Reading
Berners-Lee, Tim. *Weaving the Web: The Original Design and Ultimate Destiny of the World Wide Web.* New York: Harperbusiness, 2000.

Gillies, James, and Robert Cailliau. *How the Web Was Born: The Story of the World Wide Web*. New York: Oxford University Press, 2000.

Internet Society. Available online. URL: http://www.isoc.org. Accessed August 23, 2007.

Internet World Stats. Available online. URL: http://www. internetworldstats.com/stats.htm. Accessed August 23, 2007.

Wozniak, Steven

(1950–   )
American
Computer Inventor and Engineer

Steve Wozniak, often known as "Woz," cofounded Apple computer and designed the Apple II, one of the first popular personal computers.

Born on August 11, 1950, in San Jose, California, Wozniak grew up to be a classic "electronics whiz." He built a working electronic calculator when he was 13, winning the local science fair. After graduating from Homestead High School, Wozniak tried community college but quit to work with a local computer company. Although he then enrolled in the University of California, Berkeley, to study electronic engineering and computer science, he dropped out in 1971 to go to work again, this time as an engineer at Hewlett-Packard, at that time one of the most successful companies in the young Silicon Valley.

By the mid-1970s, Wozniak was in the midst of a technical revolution in which hobbyists explored the possibilities of the newly available microprocessor or "computer on a chip." A regular attendee at meetings of the Homebrew Computer Club, Wozniak and other enthusiasts were excited when the MITS Altair, the first complete microcomputer kit, came on the market in 1975. The Altair, however, had a tiny amount of memory, had to be programmed by toggling switches to input hexadecimal codes (rather like the ENIAC), and had very primitive input/output capabilities. Wozniak decided to build a computer that would be much easier to use—and more useful.

Wozniak's prototype machine, the Apple I, had a keyboard and could be connected to a TV screen to provide a video display. He demonstrated it at the Homebrew Computer Club and among the interested spectators was his friend Steve Jobs. Jobs had a more entrepreneurial interest than Wozniak, and spurred him to set up a business to manufacture and sell the machines. Together they founded Apple Computer in June 1976. Their "factory" was Jobs's parents' garage, and the first machines were assembled by hand.

Wozniak designed most of the key parts of the Apple, including its video display and later, its floppy disk interface, which is considered a model of elegant engineering to this day. He also created the built-in operating system and BASIC interpreter, which were stored in read-only memory (ROM) chips so the computer could function as soon as it was turned on.

In 1981, just as the Apple II was reaching the peak of its success, Wozniak was almost killed in a plane crash. He took a sabbatical from Apple to recover, get married, and return to UC Berkeley (under an assumed name!) to finish his B.S. in electrical engineering and computer science.

Wozniak's life changes affected him in other ways. As Apple grew and became embroiled in the problems of large companies, "Woz" sold large amounts of his Apple stock and gave the money to Apple employees that he thought had not been properly rewarded for their work. Later in the 1980s, he produced two rock festivals that lost $25 million, which he paid out of his own money. He was quoted as saying, "I'd rather be liked than rich." He left Apple for good in 1985 and founded Cloud Nine, an unsuccessful company that designed remote control and "smart appliance" hardware.

During the 1990s, Wozniak organized a number of charitable and educational programs, including cooperative activities with people in the former Soviet Union. He particularly enjoyed classroom teaching, bringing the excitement of technology to young people. In 1985, Wozniak received the National Medal of Technology.

Further Reading

Cringely, Robert X. *Accidental Empires: How the Boys of Silicon Valley Make Their Millions, Battle Foreign Competition, and Still Can't Get a Date*. Reading, Mass.: Addison-Wesley, 1992.

Freiberger, Paul, and Michael Swaine. *Fire in the Valley: the Making of the Personal Computer*. 2nd ed. New York: McGraw-Hill, 1999.

Kendall, Martha E. *Steve Wozniak, Inventor of the Apple Computer*. 2nd rev. ed. Los Gatos, Calif.: Highland Publishing, 2001.

Woz.org [Steve Wozniak's Home Page]. Available online. URL: www.woz.org. Accessed August 27, 2007.

Wozniak, Steve, and Gina Smith. *iWoz: From Computer Geek to Cult Icon: How I Invented the Personal Computer, Co-Founded Apple, and Had Fun Doing It*. New York: W. W. Norton, 2006.

XML

Several markup languages have been devised for specifying the organization or format of documents. Today the most commonly known markup language is the Hypertext Markup Language (see HTML, DHTML, AND XHTML), which is the organizational "glue" of the Web (see WORLD WIDE WEB).

HTML is primarily concerned with rendering (displaying) documents. It describes structural features of documents (such as headers, sections, tables, and frames), but it does not really convey the structure of the information within the document. Further, HTML is not extensible—that is, one can't define one's own tags and use them as part of the language. XML, or Extensible Markup Language, is designed to meet both of these needs. In effect, while HTML is a descriptive coding scheme, XML is a scheme for creating data definitions and manipulating data within documents. (XML can be viewed as a subset of the powerful and generalized SGML, or Standard Generalized Markup Language.)

The basic building block of XML is the element, which can be used to define an entity (rather like a database record). For example, the following statement:

```
<team name="New York Yankees">
  <players>
    <player name="1">Babe Ruth</player>
    <player name="2">Lou Gehrig</player>
  </players>
</team>
```

XML text is bracketed by tags as with HTML. The "team" element has an attribute called "name" that is assigned the value "New York Yankees." (Attribute values must be enclosed in quote marks.) It also contains a nested element called *players,* which in turn defines player names, Babe Ruth and Lou Gehrig. The elements are defined at the beginning of the XML document by a DTD (Document Type Definition), or such a definition can be "included" from another file.

XML is currently supported by the leading Web browsers. In effect, it includes HTML as a subset, or more accurately XHTML (HTML conformed to XML 1.0 standards) (see HTML, DHTML, AND XHTML). Thus, XML documents can be properly rendered by browsers, while applications that are XML-enabled (or that use XML-aware ActiveX controls or similar Java facilities, for example) can parse the XML and identify the data structures and elements in the document. Together with programming languages such as Java and facilities such as SOAP (Simple Object Access Protocol), XML can be used to create applications that connect servers and documents across the Internet—it is rapidly becoming the data "glue" that holds Web sites together.

XML can be viewed as part of a trend to make data "self-describing." The ability to encode not just the structure but the logical content of documents promises a growing ability for automated agents or "bots" to take over much of the work of sifting through the Web for desired information, bringing the Web closer to the intentions of its inventor, Tim Berners-Lee (see BERNERS-LEE, TIM; SEMANTIC WEB).

Further Reading
Carey, Patrick. *New Perspectives on XML.* 2nd ed. Boston: Course Technology, 2006.

Eisenberg, J. David. "Introduction to XML." Available online. URL: http://www.digital-web.com/articles/introduction_to_xml. Accessed August 23, 2007.

Harold, Elliotte Rusty, and W. Scott Means. *XML in a Nutshell*. 3rd ed. Sebastapol, Calif.: O'Reilly Media, 2004.

Jacobs, Sas. *Beginning XML with DOM and Ajax: From Novice to Professional*. Berkeley, Calif.: Apress, 2006.

XML Resources. Available online. URL: http://www.softwareag.com/xml/Techn_Links/default.htm. Accessed August 23, 2007.

Y

Y2K problem

Sherlock Holmes once referred to a dog barking in the night. Watson, puzzled as usual, replied that no dog had barked. Holmes replied that it was the nonbarking that was significant. The same can be said about the growing concern toward the end of the 1990s that the year 2000 might bring massive, disastrous failures to many of the computer systems on which society now depended for its well-being.

Most programs written in the 1960s and 1970s (see MAINFRAME and COBOL) saved expensive memory space by storing only the second two digits of year dates. After all, dates could be understood to begin with "19" for many years to come (although some farsighted computer scientists did warn of future trouble). Eventually the century began to draw to an end.

Although much computing activity had moved onto newer systems by the 1990s, many large government and corporate computer systems were still running the original applications or their descendents. If such a program were run in the year 2000, it would have no way to distinguish a date in that year from a date in 1900. While the prospect of a centenarian being suddenly treated as a newborn was likely to be more amusing than significant, what would happen to a 30-year mortgage that was written in 1975 and intended to come due in 2005? Would people be billed based on a 70-year term? Many observers feared that some systems would actually crash because they would begin to generate nonsensical data. What, for example, might happen to an air traffic control system or automated power grid system that used dates and times to track events?

No one really knew. One problem was that there were millions of lines of code, often written by programmers who had long since retired. Nor was it simply a matter of looking for references to date fields (such as in decision statements), because of the many ways programmers could express such statements. In addition to mainframe applications, there were also the computers hardwired into devices of all kinds including cars and airplanes (see EMBEDDED SYSTEM). As with the early mainframes, these systems were often designed with limited available memory, and thus their programmers, too, may have been tempted to save bytes by lopping off the century years.

As the fateful date approached, government agencies and businesses began to invest billions of dollars and hire expensive consultants to check code for "Y2K compatibility." In the end, Y2K problems were found and fixed in the most critical systems, and the year 2000 dawned without significant mishaps. (It turned out that virtually all the embedded systems did not in fact have Y2K problems, mostly because they didn't even track year dates.)

But although the "dog didn't bark" and in retrospect some of the hype about Y2K seems excessive, it did lead to improvement in a great deal of software. Further, it increased awareness of dependence on computers for so many aspects of life—a dependence that has been cast in a harsh new light by the terrorist events of September 11, 2001 (see RISKS OF COMPUTING).

Further Reading
Crawford, Walt. "Y2K: Lessons from a Non-Event." *Online,* vol. 25, issue 2, March 2001, 73.

Finkelstein, Anthony. "Y2K: a Retrospective View." *Computing and Control Engineering Journal,* vol. 11, no. 4, August 2000, 156–159. Available online. URL: http://www.cs.ucl.ac.uk/staff/A.Finkelstein/papers/y2kpiece.pdf. Accessed August 23, 2007.

Manion, M., and W. M. Evan. "The Y2K Problem and Professional Responsibility: A Retrospective Analysis." *Technology in Society* 22 (August 2000): 351–387.

Yahoo! Inc.

Yahoo! (NASDAQ symbol: YHOO) has played an important role in the development of Web services. In 1994 Stanford students Jerry Yang and David Filo developed the first popular directory of Web sites (see PORTAL). Realizing that the millions of Web users flocking to their site provided an opportunity for advertising and services, the two partners incorporated Yahoo! in 1995. (In 1996 the company went public and raised $33.8 million, a significant amount at a time when the business potential of the Web was only beginning to be appreciated.)

Yahoo! continued to grow, and the company acquired a number of other online services, which they used to provide Web-based e-mail, Web hosting, and news. But having flown so high, Yahoo! had far to fall when the dot-com market bubble burst in 2001: A stock that had traded at around $130.00 per share fell as low as $4.06.

However, Yahoo! proved its resilience as one of the few early dot-coms to survive and has continued to thrive in the post-bubble era since 2002. The company made strategic partnerships with telecommunications companies such as BT and Verizon. Yahoo! entered a continuing struggle with another Web services powerhouse (see GOOGLE) while acquiring new media sites (such as the photo-sharing service Flickr and the social "bookmarking" service del.icio.us), and creating new services (see BLOGS AND BLOGGING and SOCIAL NETWORKING). Yahoo! also provides online storefronts, competing in that venue mainly with eBay.

Yahoo! has a strong international presence, which, however, led to a controversial case where the company provided user information to Chinese authorities that led to imprisonment of two dissidents on charges of passing state secrets. (A lawsuit by the families of the dissidents was settled by Yahoo.)

Yahoo!'s main source of revenue remains search-related advertising. The company may have received a competitive boost in 2007 with a new online advertising system called "Panama," catching up to similar technology previously deployed by Google. In fiscal 2007 Yahoo! had revenue of $6.7 billion and earned about $730 million. At the time Yahoo! had about 13,600 employees.

In 2008 Yahoo! became the target of a takeover bid by Microsoft. Although this has met with at least initial rejection, rumors continued, including the possibility that Time Warner might acquire the company and merge it with its AOL division.

Further Reading

Angel, Karen. *Inside Yahoo!: Reinvention and the Road Ahead.* New York: Wiley, 2002.

Fost, Dan. "Web Portal Works to Integrate the Companies It Has Acquired." *San Francisco Chronicle,* December 24, 2006, p. F1. Available online. URL: http://sfgate.com/cgi-bin/article.cgi?file=/c/a/2006/12/24/BUGUIN3TKS1.DTL. Accessed December 4, 2007.

Kopytoff, Verne. "'Panama System Helping Yahoo Compete." *San Francisco Chronicle,* April 8, 2007, p. D1. Available online. URL: http://sfgate.com/cgi-bin/article.cgi?f=/c/a/2007/04/08/BUGKNP3UD91.DTL. Accessed December 4, 2007.

Snell, Rob. *Starting a Yahoo! Business for Dummies.* Hoboken, N.J.: Wiley, 2006.

Wagner, Richard. *Yahoo! SiteBuilder for Dummies.* Hoboken, N.J.: Wiley, 2005.

Yahoo! Available online. URL: http://www.yahoo.com/. Accessed December 4, 2007.

young people and computing

Computers and technology play a role in the lives of most young people that many adults have difficulty comprehending. Children in industrialized countries are liable to encounter video games even before they arrive at school. Once there, they will be exposed to a considerable amount of educational software, depending on their school's affluence (see EDUCATION AND COMPUTERS). Upon returning from school, there are more sophisticated games, MySpace pages to keep updated (see SOCIAL NETWORKING), sophisticated tools for creating music and video, and, of course, the Internet in all its vast diversity. Meanwhile, a web of incessant messages (see TEXTING AND INSTANT MESSAGING) is likely to keep the youngster in touch with friends.

CHALLENGES

A major positive aspect of young peoples' involvement with computer technology is that, as with learning a second language, learning the "language" of the digital world is easiest for the young. The capabilities and opportunities for creativity offered to today's teens are astonishing—as are many of the impressive results that can be seen in young peoples' blogs, Web sites, and YouTube videos. It is also widely believed that children will need to master current and emerging technology in order to be competitive as adults.

At the same time, adults and parents in particular remain concerned about the dangers and drawbacks of teens' pervasively digital life. A 2007 survey by the Pew Internet & American Life Project found that a majority of parents whose children were online had rules about what their kids could see or play—and for how long each day. In general parents seem to be becoming somewhat less enthusiastic about their children's online activities even as the latter's positive attitude toward the technology continues to increase. Use of protective software (see WEB FILTER) is common, although tech-savvy teens have a way of staying ahead of the curve of parental restrictions.

Common parental concerns include:

- potential exposure to online sexual predators or bullying (see CYBERSTALKING AND HARASSMENT)

- viewing of inappropriate material such as pornography and highly violent games

- excessive time spent online to the detriment of study, physical activity, or sleep

While some of this concern echoes an earlier generation's misgivings about television, the online world is far more deeply embedded in daily life, and both opportunities and concerns are thus more complex (see IDENTITY IN THE ONLINE WORLD). While parents can learn more about the relevant issues, and schools can help students develop a savvy, critical attitude toward technology, communication between generations will need to be an important strategy for coping with such rapid technological change.

Further Reading

Buckingham, David, and Rebekah Willett, editors. *Digital Generations: Children, Young People, and the New Media.* Mahwah, N.J.: Lawrence Erlbaum, 2006.

Lenhart, Amanda, and Mary Madden. "Social Networking Websites and Teens: An Overview." Pew Internet & American Life Project, January 7, 2007. Available online. URL: http://www.pewinternet.org/pdfs/PIP_SNS_Data_Memo_Jan_2007.pdf. Accessed December 4, 2007.

———. "Teens and Technology: Youth Are Leading the Transition to a Fully Wired and Mobile Nation." Pew Internet & American Life Project, July 27, 2005. Available online. URL: www.pewinternet.org/pdfs/PIP_Teens_Tech_July2005web.pdf. Accessed December 4, 2007.

Macgill, Alexandra Rankin. "Parent and Teenager Internet Use." Pew Internet & American Life Project, October 24, 2007. Available online. URL: http://www.pewinternet.org/pdfs/PIP_Teen_Parents_data_memo_Oct2007.pdf. Accessed December 4, 2007.

Mazzarella, Sharon R., ed. *Girl Wide Web: Girls, the Internet, and the Negotiation of Identity.* New York: Peter Lang, 2005.

Reimer, Jeremy. "Study Shows Youth Embracing Technology Even More than Before." *Ars Technica,* August 1, 2006. Available online. URL: http://arstechnica.com/news.ars/post/20060801-7401.html. Accessed December 4, 2007.

Willard, Nancy E. *Cyber-Safe Kids, Cyber-Savvy Teens: Helping Young People Learn to Use the Internet Safely and Responsibly.* San Francisco, Calif.: Jossey-Bass, 2007.

YouTube

Since the late 1990s, Web users (particularly younger ones) have been adept at sharing media content online (see FILE-SHARING AND P2P NETWORKS). In the 2000 decade, however, the emphasis has shifted to users not merely sharing other peoples' content, but creating their own (see USER-CREATED CONTENT). The first part of the recipe was the availability of ubiquitous digital cameras and camcorders; the second part was easy-to-use video-editing software; and the third part was a Web site that could host the results.

Created in 2005 by three former PayPal employees, the video-sharing site YouTube has been the leading venue for amateur video. Although available content includes clips from movies and TV shows (some unauthorized), much of the most interesting content is original videos created and uploaded by users. Beyond just sharing or accessing content, users are encouraged to rate and comment on the videos they see, and users can also subscribe to "feeds" of new material that is likely to be of interest to them.

By 2008 more than 83 million videos were available on YouTube—and hundreds of thousands added each day.

POLITICAL INFLUENCE

Just as political pundits were beginning to notice that bloggers were creating parallel structures that rivaled the influence of the mainstream media (see JOURNALISM AND COMPUTERS), YouTube broke into the highly visual field of political advertising. Most candidates in the 2008 presidential primaries have put their statements and other videos on YouTube. However, other supporters soon got into the act, including the creator of a pro–Barack Obama ad that cast rival Hillary Rodham Clinton in the role of Big Brother in the classic "1984" Apple Macintosh commercial (see MASH-UPS). Political commentators and journalists have also been active in putting their opinions on YouTube (or commenting on those of others). Perhaps the political establishment's biggest nod to YouTube is the series of debates cosponsored by CNN and YouTube, bringing together the Republican and Democratic primary fields.

YouTube has had its share of criticism: Critics have charged the service with not sufficiently policing copyright violations and violent content (including videos of fights or bullying in schools), as well as neo-Nazi propaganda, scenes of animal abuse, and videos by anti-American insurgent groups, as well as generally tasteless exhibitionism. A few countries and some schools have responded by blocking access to the service.

If YouTube's main resource is the creativity and enthusiasm of its users, its main revenue is advertising—about $15 million per month by 2006. Don Tapscott and Anthony D. Williams, authors of the book *Wikinomics,* cite YouTube as a classic example of the new economics of mass collaboration on the Web. Google signaled its appreciation for the economic potential of YouTube by buying it for $1.65 billion in late 2006.

Further Reading

Fah, Chad. *How to Do Everything with YouTube.* Emeryville, Calif.: McGraw-Hill Osborne, 2007.

Miller, Michael. *YouTube 4 You.* Indianapolis: Que, 2007.

Sahlin, Doug. *YouTube for Dummies.* Indianapolis: Wiley, 2007.

YouTube. Available online. URL: http://www.youtube.com. Accessed December 4, 2007.

Weber, Steve. *Plug Your Business! Marketing on MySpace, YouTube, Blogs and Podcasts, and Other Web 2.0 Social Networks.* Falls Church, Va.: Weber Books, 2007.

Winograd, Morley, and Michael D. Hais. *Millennial Makeover: MySpace, YouTube, and the Future of American Politics.* Piscataway, N.J.: Rutgers University Press, 2008.

Z

Zuse, Konrad
(1910–1995)
German
Engineer, Inventor

Great inventions seldom have a single parent. Although popular history credits Alexander Graham Bell with the telephone, the almost forgotten Elisha Gray invented the device at almost the same time. And although the ENIAC is widely considered to be the first practical electronic digital computer (see ECKERT, J. PRESPER and MAUCHLY, JOHN) another American inventor built a smaller machine on somewhat different principles that also has a claim to being "first" (see ATANASOFF, JOHN). Least known of all is Konrad Zuse, perhaps because he did most of his work in a nation that was plunging the world into war.

Zuse was born on June 22, 1910, in Berlin. He studied civil engineering at the Technische Hochschule Berlin-Charlottenburg, receiving his degree in 1935. One of his tasks in engineering was performing calculations of the stress on structures such as bridges. At the time these calculations were carried out by going through a series of steps on a form over and over again, plugging in the data and calculating by hand or using an electromechanical calculator. Like other inventors before him, Zuse began to wonder whether he could build a machine that could carry out these repetitive steps automatically.

Zuse was unaware of the nearly forgotten work of Charles Babbage and that of other inventors in America and Britain who were beginning to think along the same lines (see BABBAGE, CHARLES). With financial help from his parents (and the loan of their living room), Zuse began to assemble his first machine from scrounged parts. His first machine, the Z1, was completed in 1938. The machine used slotted metal plates with holes and pins that could slide to carry out binary addition and other operations (in using the simpler binary system rather than decimal, Zuse was departing from other calculator designers).

The Z1 had trouble storing and retrieving numbers and never worked well. Undeterred, Zuse began to develop a new machine that used electromechanical telephone relays (a ubiquitous component that was also favored by Howard Aiken [see AIKEN, HOWARD]). The new machine worked much better, and Zuse successfully demonstrated it at the German Aerodynamics Research Institute in 1939.

With World War II under way, Zuse was able to obtain funding for his Z3, which was able to carry out automatic sequences from instructions (Zuse used discarded movie film instead of punched tape). The machine used 22-bit words and had 600 relays in the calculating unit and 1,800 for the memory. However, the machine could not do branching or looping the way modern computers can. It was destroyed in a bombing raid in 1944. Meanwhile, Zuse used spare time from his military duties at the Henschel aircraft company to work on the Z4, which was completed in 1949. This machine was more fully programmable and was comparable to Howard Aiken's Mark I.

By that time, however, Zuse's electromechanical technology had been surpassed by the fully electronic vacuum tube computers such as the ENIAC and its successors. (Zuse had considered vacuum tubes but had rejected them, believing that their inherent unreliability and the large numbers needed would make them impracticable for a large-scale machine.) During the 1950s and 1960s, Zuse ran a computer company, ZUSE KG, which eventually produced electronic vacuum tube computers.

Zuse's most interesting contribution to computer science would not be his hardware but a programming language called *Plankalkül* or "programming calculus." Although the language was never implemented, it was far in advance of its time in many ways. It started with the radically simple concept of grouping individual bits to form whatever data structures were desired. It also included program modules that could operate on input variables and store their results in output variables (see PROCEDURES AND FUNCTIONS). Programs were written using a notation similar to mathematical matrices.

Zuse labored in obscurity even within the computer science fraternity. However, toward the end of his life his work began to be publicized. He received numerous honorary degrees from European universities as well as awards and memberships in scientific and engineering academies. Zuse also took up abstract painting in his later years. He died on December 18, 1995.

Further Reading
Bauer, F. L., and H. Wössner. "The Plankalkül of Konrad Zuse: A Forerunner of Today's Programming Languages." *Communications of the ACM,* vol. 15, 1972, 678–685.
Lee, J. A. N. *Computer Pioneers.* Los Alamitos, Calif.: IEEE Computer Society Press, 1995.
Zuse, Konrad. *The Computer—My Life.* New York: Springer-Verlag, 1993.

APPENDIX I
BIBLIOGRAPHIES AND WEB RESOURCES

The following selections provide reference material and resources to supplement the Further Reading selections at the conclusion of the majority of entries in this book.

PRINT RESOURCES

2008 Software Industry Directory. Greenwood Village, Colo.: Webcom Communications, 2007.

Bidgoli, Hassan, ed. *The Internet Encyclopedia.* New York: Wiley, 2003.

Cortada, James W. *Historical Dictionary of Data Processing.* 3 vols. New York: Greenwood Press, 1987.

Daintith, John, and Edmund Wright. *The Facts On File Dictionary of Computer Science.* Rev. ed. New York: Facts On File, 2006.

De Palma, Paul, ed. *Annual Editions: Computers in Society 08/09.* Guilford, Conn.: McGraw-Hill/Dushkin, 2007.

Downing, Douglas, Michael Covington, and Melody Mauldin Covington. *Dictionary of Computer and Internet Terms.* 9th ed. Hauppauge, N.Y.: Barron's, 2006.

Hackett, Edward J. et al., eds. *The Handbook of Science and Technology Studies.* 3rd ed. Cambridge, Mass.: MIT Press, 2007.

Henderson, Harry. *A to Z of Computer Scientists.* New York: Facts On File, 2003.

Knee, Michael. *Computer Science and Computing: A Guide to the Literature.* Greenwich, Colo.: Libraries Unlimited, 2005.

Lee, John A. N. *International Biographical Dictionary of Computer Pioneers.* New York: Routledge, 1995.

Lubar, Stephen. *InfoCulture: The Smithsonian Book of Information Age Inventions.* Boston: Houghton Mifflin, 1993.

McGraw-Hill Encyclopedia of Science & Technology. 10th ed. New York: McGraw-Hill, 2007.

Sun Technical Publications. *Read Me First!: A Style Guide for the Computer Industry.* 2nd ed. Upper Saddle River, N.J.: Prentice Hall, 2003.

WEB RESOURCES

About.com. "Computing and Technology." Available online. URL: http://about.com/compute/. Accessed December 6, 2007.

Academic Info: Computer Science & Computer Engineering. Available online. URL: http://www.academicinfo.net/compsci.html. Accessed December 6, 2007.

ACM Computing Reviews. Available online. URL: http://www.reviews.com/home.cfm. Accessed December 6, 2007.

ACM Digital Library. Available online. URL: http://portal.acm.org/dl.cfm. Accessed December 6, 2007.

ACM Tech. News. Available online. URL: http://technews.acm.org/current.cfm. Accessed December 7, 2007.

Ars Technica. Available online. URL: http://arstechnica.com/. Accessed December 6, 2007.

ClickZ Stats. Available online. URL: http://www.clickz.com/stats/. Accessed December 6, 2007.

Cnet. Available online. URL: http://www.cnet.com. Accessed December 6, 2007.

Collection of Computer Science Bibliographies. Available online. URL: http://liinwww.ira.uka.de/bibliography/. Accessed December 7, 2007.

Computer Dictionaries, Acronyms, and Glossaries. Available online. URL: http://www.compinfo-center.com/tpdict-t.htm. Accessed December 6, 2007.

Computer History Museum. Available online. URL: http://www.computerhistory.org/. Accessed December 6, 2007.

Connected: An Internet Encyclopedia. Available online. URL: http://freesoft.org/CIE/. Accessed December 7, 2007.

FOLDOC (Free On-Line Dictionary of Computing). Available online. URL: http://foldoc.org/. Accessed December 7, 2007.

How Stuff Works. Available online. URL: http://www.howstuffworks.com/. Accessed December 6, 2007.

IDG (International Data Group). Available online. URL: http://www.idg.net/. Accessed December 7, 2007.

IEEE Annals of the History of Computing. Available online. URL: http://www.computer.org/annals/. Accessed December 6, 2007.

InfoWorld. Available online. URL: http://www.infoworld.com. Accessed December 6, 2007.

Internet News. Available online. URL: http://www.internetnews.com/. Accessed December 6, 2007.

Pew/Internet. Available online. URL: http://www.pewinternet.org. Accessed December 6, 2007.

Red Herring: The Business of Technology. Available online. URL: http://www.redherring.com/pages/pagenotfound/. Accessed December 7, 2007.

Slashdot. Available online. URL: http://slashdot.org. Accessed December 6, 2007.

Smithsonian Institution. Science and Technology Division, Information Technology. Available online. URL: http://www.si.edu/Encyclopedia_SI/science_and_technology/Information_Technology.htm. Accessed December 6, 2007.

Webopedia. Available online. URL: http://www.webopedia.com/. Accessed December 6, 2007.

Wikipedia. Available online. URL: http://en.wikipedia.org. Accessed December 6, 2007.

Yahoo! Finance. Available online. URL: http://biz.yahoo.com. Accessed December 7, 2007.

Yahoo! Science: Computer Science. Available online. URL: http://dir.yahoo.com/Science/Computer_Science/. Accessed December 6, 2007.

ZDNet. Available online. URL: http://www.zdnet.com. Accessed December 7, 2007.

APPENDIX II
A CHRONOLOGY OF COMPUTING

The following chronology lists some significant events in the history of computing. Although the first CALCULATORS (i.e., the abacus) were known in ancient times, the chronology begins with the development of modern mathematics and the first calculators in the 17th century.

1617

- John Napier published an explanation of "Napier's bones," a manual aid to calculation based on logarithms, and the ancestor to the slide rule.

1624

- William Schickard invented a mechanical CALCULATOR that can perform automatic carrying during addition and subtraction. It can also multiply and divide by repeated additions or subtractions.

1642

- Blaise Pascal invented a CALCULATOR that he calls the Pascaline. Its improved carry mechanism used a weight to allow it to carry several places. A small batch of the machines was made, but it did not see widespread use.

1673

- Gottfried Wilhelm Leibniz (co-inventor with Isaac Newton of the calculus) invented a CALCULATOR called the Leibniz Wheel. He also wrote about the binary number system that eventually became the basis for modern computation.

1786

- J. H. Muller invented a "difference engine," a machine that can solve polynomials by repeated addition or subtraction.

1822

- Charles BABBAGE designed and partially built a much more elaborate difference engine.

1832

- BABBAGE sketched out a detailed design for the Analytical Machine. This machine was to have been programmed by PUNCHED CARDS, storing data in a mechanical MEMORY, and even including a PRINTER. Although it was not built during his lifetime, Babbage's machine embodied most of the concepts used in modern computers.

1843

- Ada Lovelace provided extensive commentary on a book by BABBAGE's Italian supporter Menabrea. Besides being the first technical writer, Lovelace also wrote what might be considered the world's first computer program.

1844

- Samuel Morse demonstrated the electromagnetic telegraph by sending a message from Washington to Baltimore. The telegraph inaugurated both electric data transmission and the use of a binary character code (dots and dashes).

1850

- Amedee Mannheim created the first modern slide rule. It will become an essential accessory for engineers and scientists until the inexpensive electronic CALCULATOR arrived in the 1970s.

1854

- George Boole's book *The Laws of Thought* described what is now called Boolean algebra. BOOLEAN OPERATORS are essential for the BRANCHING STATEMENTS and LOOPS that control the operation of computer programs.

1884

- W. S. Burroughs marketed his first adding machine, beginning what will become an important CALCULATOR (and later, computer) business.

1890

- Herman HOLLERITH's PUNCHED CARD tabulator enabled the U.S. government to complete the 1890 census in record time.

1896

- HOLLERITH founded the Tabulating Machine Company, which will become the Computing, Tabulating, and Recording company (CTR) in 1911. In 1924, it will become International Business Machines (IBM).

1904

- J. A. Fleming invented the diode vacuum tube. Together with Lee de Forest's invention of the triode two years later, this development defined the beginnings of electronics, offering a switching mechanism much faster than mechanical relays.

1919

- The "flip-flop" circuit was invented by two American physicists, W. H. Eccles and R. W. Jordan. The ability of the circuit to switch smoothly between two (binary) states would form the basis for computer arithmetic logic units.

1921

- Karl Capek's play *R.U.R.* introduced the term *robot*. Robots will become a staple of science fiction "pulps" starting in the 1930s.

1930

- Vannevar BUSH's elaborate ANALOG COMPUTER, the Differential Analyzer, went into service.

1936

- Alonzo CHURCH developed the lambda calculus, which can be used to demonstrate the COMPUTABILITY of mathematical problems.
- Konrad ZUSE built his first computer, a mechanical machine based on the binary system.

1937

- Alan TURING provided an alternative (an equivalent) demonstration of COMPUTABILITY through his Turing Machine, an imaginary computer that can reduce any computable problem to a series of simple operations performed on an endless tape.
- BELL LABORATORIES mathematician George Stibitz created the first circuit that could perform addition by combining BOOLEAN OPERATORS.

1938

- In a key development in ROBOTICS, Doug T. Ross, an American engineer, created a robot that can store its experience in MEMORY and "learn" to navigate a maze.
- G. A. Philbrick developed an electronic version of the ANALOG COMPUTER.
- Working in a garage near Stanford University, William Hewlett and David Packard began to build audio oscillators. They called their business the Hewlett-Packard Company. Fifty years later, the garage would be preserved as a historical landmark.

1939

- John ATANASOFF and Clifford Berry built a small electronic binary computer called the Atanasoff-Berry Computer (ABC). A 1973, court decision would give this machine precedence over ENIAC as the first electronic digital computer.

- George Stibitz built the Complex Number Calculator, which is controlled by a keyboard and uses relays.

1940

- Claude SHANNON introduced the fundamental concepts of DATA COMMUNICATIONS theory.
- George Stibitz demonstrated remote computing by controlling his Complex Number Calculator in New York from a Teletype TERMINAL at Dartmouth College in New Hampshire.

1941

- Working in isolation in wartime Germany, Konrad ZUSE completed the Z3. Although still mechanical rather than electronic, the machine used sophisticated floating-point NUMERIC DATA.

1943

- The British-built Colossus, an electronic (vacuum tube) special-purpose computer that can rapidly analyze permutations to crack the German Enigma cipher.

1944

- Howard AIKEN completed the Harvard Mark I, a large programmable calculator (or computer) using electromechanical relays.
- John VON NEUMANN and Stanislaw Ulam developed the Monte Carlo method of probabilistic SIMULATION, a tool that would find widespread use as computer power becomes available.

1945

- ZUSE continued computer development and created a sophisticated matrix-based programming language called Plankalkül.
- Vannevar BUSH envisioned HYPERTEXT and knowledge linking and retrieval in his article "As We May Think."
- Alan TURING developed the concept of using PROCEDURES AND FUNCTIONS (subroutines) called with parameters. His team also developed the Pilot ACE (Automatic Computing Engine), which would help the development of a British computer industry.

1946

- ENIAC went into service. Developed by J. Presper ECKERT and John MAUCHLY, the machine is widely considered to be the first large-scale electronic digital computer. It used 18,000 vacuum tubes.
- In the "Princeton Reports" based upon the ENIAC work, John VON NEUMANN, together with Arthur W. Burks and Herman Goldstine described the fundamental operations of modern computers including the stored program concept—the holding of all program instructions in memory, where they can be referred to repeatedly and even manipulated like other data.

1947

- The Association for Computing Machinery (ACM) was founded.

- ECKERT and MAUCHLY formed the Eckert-Mauchly Corporation for commercial marketing of computers based on the ENIAC design.
- John VON NEUMANN began development of the EDVAC (Electronic Discrete Variable Automatic Calculator) for the U.S. government's Ballistic Research Laboratory. This machine, completed in 1952, would be the first to use programs completely stored in memory and able to be changed without physically changing the hardware.
- Richard Hamming developed ERROR CORRECTION algorithms.
- Alan TURING's paper on "Intelligent Machinery" began laying the groundwork for ARTIFICIAL INTELLIGENCE research.
- In Britain, Manchester University built the first electronic computer that can store a full program in MEMORY. It was called "baby" because it was a small test version of a planned larger machine. For its main memory it used a CRT-like tube invented by F. C. Williams.
- IBM under Thomas J. Watson, Sr. decided to enter the new computer field in a big way by beginning to develop the Selective Sequence Electronic Calculator (SSEC) as a competitor to ENIAC and the Harvard Mark I. The huge machine used thousands of both vacuum tubes and relays.
- Tom Kilburn and M. H. A. Newman invented the index register, which would be used to keep track of the current location in memory of instructions or data.
- The transistor was invented at BELL Labs by John Bardeen, Walter Brattain, and William Shockley. The solid-state device could potentially replicate all the functionality of the vacuum tube with much less size and power consumption. It would be some time before it was inexpensive enough to be used in computers, however.
- Norbert WIENER coined the term *cybernetics* to refer to control and feedback systems.
- Claude SHANNON formally introduced statistical information theory.

1949

- The Cambridge EDSAC demonstrated versatile stored-program computing. Meanwhile, ECKERT and MAUCHLY work on BINAC, a successor/spinoff of ENIAC for Northrop Aircraft Corporation.
- Frank Rosenblatt developed the perceptron, the first form of NEURAL NETWORK, for solving pattern-matching problems.
- An Wang patented "core memory," using an array of magnetized rings and wires, which would become the main memory (RAM) for many MAINFRAMES in the 1950s.

1950

- Alan TURING proposed the Turing Test as a way to demonstrate ARTIFICIAL INTELLIGENCE.
- Development began of the high-speed computers Whirlwind and SAGE for the U.S. military. The military also began to use computers to run war games or simulations.
- Claude SHANNON outlined the algorithms for a chess-playing program that could evaluate positions and perform heuristic calculations. He would build a chess-playing computer called Caissac.

- Japan began development of electronic computers under the leadership of Hideo Yamashita, who would build the Tokyo Automatic Calculator.
- Approximately 60 electronic or electromechanical computers were in operation worldwide. Each was built "by hand" as there were no production models yet.

1951

- ECKERT and MAUCHLY marketed Univac I, generally considered the first commercial computer (although the Ferranti Mark I is sometimes given co-honors).
- An Wang founded Wang Laboratories, which would become a major computer manufacturer through the 1970s.
- Grace HOPPER at Remington Rand coined the word COMPILER and began developing automatic systems for creating machine codes from higher-level instructions.

1952

- Alick Glennie developed autocode, generally considered to be the first true high-level PROGRAMMING LANGUAGE.
- Magnetic core MEMORY began to come into use.
- election night a Univac I predicted that Dwight D. Eisenhower would win the 1952 U.S. presidential election. It made its prediction an hour after the polls closed, but its findings were not released at first because news analysts insisted the race was closer.
- MANIAC was On developed to do secret nuclear research in Los Alamos.
- The IBM 701 went into production. It was one of the first computers to use magnetic TAPE DRIVES as primary means of data storage.
- IBM was accused of violating the Sherman Antitrust Act in its computer business. Litigation in one form or another would drag on until 1982.
- John VON NEUMANN described self-reproducing automata.
- The symbolic ASSEMBLER was introduced by Nathaniel Rochester.
- IBM and Remington Rand (Univac) dominated the young computer industry.

1954

- The IBM 650 was marketed. It was the first truly mass-produced computer, and relatively affordable by businesses and industries. It used a magnetic drum memory.
- In Britain, the Lyons Electronic Office (LEO) became the first integrated computer system for use for BUSINESS APPLICATIONS, primarily accounting and payroll.

1955

- Grace HOPPER created Flow-matic, the first high-level language designed for BUSINESS APPLICATIONS OF COMPUTERS.
- The Computer Usage Company (CUC) was founded by John W. Sheldon and Elmer C. Kubie. It is considered to be the first company devoted entirely to developing computer software rather than hardware.
- Bendix marketed the G-15, its competitor to the IBM 650 in the "small" business computer market.

- Users of the new IBM 704 MAINFRAME, frustrated at the lack of technical supported, formed the first computer USER GROUP, called SHARE.
- The large IBM 705 MAINFRAME is marketed by IBM. It uses magnetic core MEMORY.

1956

- The IBM 704 and Univac 1103 introduced a new generation of commercial MAINFRAMES with magnetic core storage.
- John MCCARTHY coined the term ARTIFICIAL INTELLIGENCE, or AI.
- The Dartmouth AI conference brought together leading researchers such as McCarthy, Marvin MINSKY, Herbert Simon, and Allen Newell. It would set the agenda for the field.
- Newell, Shaw, and Simon developed Logic Theorist, the first program that can prove theorems.
- A. I. Dumey described HASHING, a procedure for quickly sorting or retrieving data by assigning calculated values.
- The infant transistor industry began to grow as companies such as IBM began to build transistorized calculators.
- IBM signed a consent decree ending the 1952 antitrust complaint by restricting some of its business practices in selling mainframe computers.

1957

- John Backus and his team released FORTRAN, which would become the most widely used language for SCIENTIFIC COMPUTING APPLICATIONS.
- Digital Equipment Corporation (DEC) was founded by Ken Olsen and Harlan Anderson. The company's agenda involved the development of a new class of smaller computer, the MINICOMPUTER.
- Minicomputer development would be inspired by the MIT TX-0 computer. While not yet a "mini," the machine was the first fully transistorized computer.
- The hard drive came into service in IBM's 305 RAMAC.
- IBM developed the first dot matrix PRINTER.

1958

- The I/O interrupt used by devices to signal their needs to the CPU was developed by IBM. It would be used later in personal computers.
- China began to build computers based on Soviet designs, which in turn had been based upon American and British machines.
- Sperry Rand introduced the Univac II, a huge, powerful, and surprisingly reliable computer that used 5,200 vacuum tubes, 18,000 crystal diodes, and 184,000 magnetic cores.
- Jack Kilby of Texas Instruments built the first integrated circuit, fitting five components onto a half-inch piece of germanium.
- As the cold war continued, the U.S. Air Force brought SAGE on-line. This integrated air defense system featured REAL-TIME PROCESSING and graphics displays.

1959

- John MCCARTHY developed LISP, a language based on Alonzo CHURCH's lambda calculus and including extensive facilities for LIST PROCESSING. It would become the favorite language for ARTIFICIAL INTELLIGENCE research.
- COBOL was introduced, with much of the key work and inspiration coming from Grace HOPPER.
- IBM marketed the 7090 MAINFRAME, a large transistorized machine that could perform 229,000 additions a second. The smaller IBM 1401 would prove to be even more popular. IBM also introduced a high-speed PRINTER using type chains.
- Robert Noyce of Fairchild Semiconductor built a different type of integrated circuit, using aluminum traces and layers deposited on a silicon substrate.

1960

- Digital Equipment Corporation (DEC) marketed the PDP-1, generally considered the first commercial MINICOMPUTER.
- Control Data Corporation (CDC) impressed the industry with its CDC 1604, designed by Seymour CRAY. It offered high speed at considerably lower prices than IBM and the other major companies.
- The ALGOL language demonstrated block structure for better organization of programs. The report on the language introduced BNF (BACKUS-NAUR FORM) as a systematic description of computer language grammar.
- Donald Blitzer introduced PLATO, the first large-scale interactive COMPUTER-AIDED INSTRUCTION system. It would later be marketed extensively by Control Data Corporation (CDC).
- Paul Baran of RAND developed the idea of packet-switching to allow for decentralized information NETWORKS; the idea would soon attract the attention of the U.S. Defense Department.
- In an advance in practical ROBOTICS, the remote-operated "Handyman" robot arm and hand was put to work in a nuclear power plant.
- The U.S. Navy began to develop the Naval Tactical Data System (NTDS) to track targets and the status of ships in a combat zone.

1961

- Time-sharing computer systems came into use at MIT and other facilities. Among other things, they encouraged the efforts of the first HACKERS to find clever things to do with the computers.
- Leonard Kleinrock's paper "Information Flow in Large Communication Nets" was the first description of the packet-switching message transfer system that would underlie the INTERNET.
- Arthur Samuel's ongoing research into COMPUTER GAMES design culminated in his checkers program reaching master level. The program includes learning algorithms that can improve its play.
- The IBM STRETCH (IBM 7030) is installed at Los Alamos National Laboratory. Its advanced "pipeline" architecture allowed new instructions to begin to be processed while preceding ones were being finished. It and Univac's LARC are sometimes considered to be the first SUPERCOMPUTERS.
- IBM made a major move into scientific computing with its modular 7040 and 7044 computers, which can be used

together with the 1401 to build a "scalable" installation for tackling complex problems.

- Unimation introduced the industrial robot (the Unimate).
- Fairchild Semiconductor marketed the first commercial integrated circuit.

1962

- The discipline of COMPUTER SCIENCE began to emerge with the first departments established at Purdue and Stanford.
- MIT students created *Spacewar,* the first video COMPUTER GAME, on the PDP-1.
- On a more practical level, MIT programmers Richard Greenblatt and D. Murphy develop TECO, one of the first TEXT EDITORs.
- J. C. R. Licklider described the "Intergalactic Network," a universal information exchange system that would help inspire the development of the INTERNET.
- Douglas ENGELBART invented the computer MOUSE at SRI.
- IBM developed the SABRE online ticket reservation system for American Airlines. The system will soon be adopted by other carriers and demonstrate the use of networked computer systems to facilitate commerce. Meanwhile, IBM earned $1 billion from its computer business, which by then had overtaken its traditional office machines as the company's leading source of revenue.

1963

- Joseph WEIZENBAUM's *Eliza* program carried on natural-sounding conversations in the manner of a psychotherapist.
- Ivan Sutherland developed *Sketchpad,* the first computer drawing system.
- Reliable Metal Oxide Semiconductor (MOS) integrated circuits were perfected, and would become the basis for many electronic devices in years to come, including computers for space exploration.

1964

- The IBM System/360 was announced. It would become the most successful MAINFRAME in history, with its successors dominating business computing for the next two decades.
- IBM introduced the MT/ST (Magnetic Tape/Selectric Typewriter), considered to be the first dedicated WORD PROCESSING system. While rudimentary, it allowed text to be corrected before printing.
- Seymour CRAY's Control Data CDC 6600 is announced. When completed, it ran about three times faster than IBM's STRETCH, irritating Thomas Watson, head of the far larger IBM.
- J. Kemeny and T. Kurtz developed BASIC to allow students to program on the Dartmouth time-sharing system.
- At the other end of the scale, IBM introduced the complex, feature-filled PL/1 (Programming Language 1) for use with its System/360.
- The American National Standards Institute (ANSI) officially adopted the ASCII (American Standard Code for Information Interchange) character code.

- Paul Baran of SRI wrote a paper, "On Distributed Communication Networks," further describing the implementation of packet-switched network that could route around disruptions. The work began to attract the attention of military planners concerned with air defense and missile control systems surviving nuclear attack.
- Jean Sammet and her colleagues developed the first computer program that can do algebra.
- Gordon MOORE (a founder of Fairchild Semiconductor and later, of Intel Corporation) stated that the power of CPUs would continue to double every 18 to 24 months. "Moore's law" proved to be remarkably accurate.

1965

- IBM introduced the FLOPPY DISK (or diskette) for use with its mainframes.
- Edsgar DIJKSTRA devised the semaphore, a variable that two processes can use to synchronize their operations and aiding the development of CONCURRENT PROGRAMMING.
- The APL language developed by Kenneth Iverson provided a powerful, compact, but perhaps cryptic way to formulate calculations.
- The SIMULA language introduced what will become known as OBJECT-ORIENTED PROGRAMMING.
- The DEC PDP-8 became the first mass-produced minicomputer, with over 50,000 systems being sold. The machine brings computing power to thousands of universities, research labs, and businesses that could not afford mainframes. Designed by Edson deCastro and engineered by Gordon Bell, the PDP-8 design marked an important milestone on the road to the desktop PC.
- NASA uses an IBM onboard computer to guide Gemini astronauts in their first rendezvous in space.
- The potential of the EXPERT SYSTEM was demonstrated by *Dendral,* a specialized medical diagnostic program that began development by Edward Feigenbaum, Joshua Lederberg, and Bruce Buchanan.
- The U.S. Defense Department's ARPA (Advanced Research Projects Agency) sponsored a study of a "co-operative network of time-sharing computers." A testbed network was begun by connecting a TX-2 minicomputer at MIT via phone line to a computer at System Development Corporation in Santa Monica, California.
- Ted Nelson's influential vision of universal knowledge sharing through computers introduced the term HYPERTEXT.

1966

- In the first federal case involving COMPUTER CRIME (*U.S. v. Bennett*), a bank programmer is convicted of altering a bank program to allow him to overdraw his account.
- The first ACM Turing Award is given to Alan Perlis.
- The New York Stock Exchange automated much of its trading operations.

1967

- The memory CACHE (a small amount of fast memory used for instructions or data that are likely to be needed) was introduced in the IBM 360/85 series.

- IBM developed the first FLOPPY DISK drive.
- Seymour PAPERT introduced LOGO, a LISP-like language that would be used to teach children programming concepts intuitively.
- A chess program written by Richard Greenblatt of MIT, Mac Hack IV, achieved the playing skill of a strong amateur human player.
- Fred Brooks did early experiments in computer-mediated sense perception, laying groundwork for VIRTUAL REALITY.

1968

- Edsger DIJKSTRA's little letter entitled "GO TO Considered Harmful" argued that the GOTO or "jump" statement made programs hard to read and more prone to error. The resulting discussion gave impetus to the STRUCTURED PROGRAMMING movement. Another aspect of this movement was the introduction of the term *SOFTWARE ENGINEERING*.
- Robert Noyce, Andrew GROVE, and Gordon MOORE founded INTEL, the company that would come to dominate the MICROPROCESSOR industry by the early 1980s.
- IBM introduced the System/3, a lower-cost computer system designed for small businesses.
- Bolt, Beranek and Newman (BBN) was awarded a government contract to build "interface message processors" or IMPs to translate data between computers linked over packet-switched networks.
- Alan KAY prototyped the Dynabook, a concept that led toward both the PORTABLE COMPUTER and the graphical USER INTERFACE.
- Stanley Kubrick's movie *2001* introduced Hal 9000, the self-aware (but paranoid) computer that kills members of a deep-space exploration crew.

1969

- Ken Thompson and Dennis RITCHIE began work on the UNIX OPERATING SYSTEM. It will feature a small KERNEL that can be used with many different command SHELLS, and will eventually incorporate hundreds of utility programs that can be linked to perform tasks.
- Edgar F. Codd introduced the concept of the relational system that would form the foundation for most modern DATABASE MANAGEMENT SYSTEM.
- IBM was sued by the U.S. Department of Justice for antitrust violations. The voluminous case would finally be dropped in 1982. However, government pressure may have led the computer giant to finally allow its users to buy software from third parties, giving a major boost to the software industry.
- ARPANET is officially launched. The first four nodes of the ARPANET came online, prototyping what would eventually become the INTERNET.
- SRI researchers developed Shakey, the first mobile robot that could "see" and respond to its environment. The actual control computer was separate, however, and controlled the robot through a radio link.
- Neil Armstrong and Edwin Aldrin successfully made the first human landing on the Moon, despite problems with the onboard Apollo Guidance Computer.
- The first automatic teller machine (ATM) was put in service.

1970

- Gene AMDAHL left IBM to found Amdahl Corporation, which would compete with IBM in the mainframe "clone" market.
- An INTEL Corporation team led by Marcian E. Hoff began to develop the Intel 4004 MICROPROCESSOR.
- Digital Equipment Corporation announced the PDP-11, the beginning of a series of 16-bit minicomputers that will support time-sharing computing in many universities.
- John Conway's "Game of Life" popularized CELLULAR AUTOMATA.
- The ACM held its first all-computer chess tournament in New York City. Northeastern University's Chess 3.0 topped the field of six programs competing.
- Charles Moore began writing programs to demonstrate the versatility of his programming language FORTH.
- Xerox Corporation established the Palo Alto Research Center (PARC). This laboratory will create many innovations in interactive computing and the graphical USER INTERFACE.

1971

- Niklaus WIRTH formally announced PASCAL, a small, well-structured language that will become the most popular language for teaching COMPUTER SCIENCE for the next two decades.
- The IEEE Computer Society was founded.
- The IBM System/370 series ushered in a new generation of mainframes using densely packed integrated circuits for both CPU and MEMORY.

1972

- Dennis RITCHIE and Brian Kernighan developed C, a compact language that would become a favorite for SYSTEMS PROGRAMMING, particularly in UNIX.
- The creation of an E-MAIL program for the ARPANET included the decision to use the at (@) key as part of e-mail addresses.
- Alan KAY developed SMALLTALK, building upon SIMULA to create a powerful, seamless OBJECT-ORIENTED PROGRAMMING language and OPERATING SYSTEM. The language would eventually be influential although not widely used. Kay also prototyped the Dynabook, a notebook computer, but Xerox officials showed little interest.
- Seymour CRAY left CDC and founded Cray Research to develop new SUPERCOMPUTER.
- INTEL introduced the 8008, the first commercially available 8-bit MICROPROCESSOR.
- The 5.25-inch diskette first appeared. It would become a mainstay of personal computing until it was replaced by the more compact 3.5-inch diskette in the 1990s.
- Nolan Bushnell's Atari Corp. had the first commercial COMPUTER GAME hit, *Pong*. It and its beeping cousins would soon become an inescapable part of every parent's experience.

1973

- Alain Colmerauer and Philippe Roussel at the University of Marseilles developed PROLOG (Programming in Logic), a language that could be used to reason based upon a stored base of knowledge. The language would become popular for EXPERT SYSTEMS development.

- BELL LABORATORIES established a group to support and promulgate the UNIX OPERATING SYSTEM.
- The Ethernet protocol for LANs (LOCAL AREA NETWORKS) was developed by Robert Metcalfe.
- In a San Francisco hotel lobby Vinton CERF sketched the architecture for an Internet gateway on a napkin.
- Don Lancaster published his "TV Typewriter" design in *Radio Electronics*. It would enable hobbyists to build displays for the soon-to-be available microcomputer.
- The Boston Computer Society (BCS) was founded. It became one of the premier computer user groups.
- Gary Kildall founded Digital Research, whose CP/M OPERATING SYSTEM would be an early leader in the microcomputer field.
- A federal court declared that the Eckert-Mauchly ENIAC patents were invalid because John ATANASOFF had the same ideas earlier in his ABC computer.

1974

- The Alto graphical workstation was developed by Alan KAY and others at Xerox PARC. It did not achieve commercial success, but a decade later something very much like it would appear in the form of the APPLE MACINTOSH.
- An international computer chess tournament is won by the Russian KAISSA program, which crushed the American favorite Chess 4.0.
- Computerized product scanners were introduced in an Ohio supermarket.
- INTEL released the 8080, a MICROPROCESSOR that had 6,000 transistors, could execute 640,000 instructions per second, was able to access 64 kB of memory, and ran at a clock rate of 2 MHz.
- David Ahl's *Creative Computing* magazine began to offer an emphasis on using small computers for education and other human-centered tasks.
- Vinton CERF and Robert Kahn began to publicize their TCP/IP INTERNET protocol.
- A group at the University of California, Berkeley, began to develop their own version of the UNIX OPERATING SYSTEM.
- The 1974 Privacy Act began the process of trying to protect individual PRIVACY IN THE DIGITAL AGE.

1975

- Fred Brooks published the influential book *The Mythical Man-Month*. It explained the factors that bog down software development and focused more attention on SOFTWARE ENGINEERING and its management.
- Electronics hobbyists were intrigued by the announcement of the MITS Altair, the first complete MICROCOMPUTER system available in the form of a kit. While the basic kit cost only $395, the keyboard, display, and other peripherals were extra.
- MITS founder Ed Roberts also coined the term *PERSONAL COMPUTER*. Hundreds of hobbyists built the kits and yearned for more capable machines. Many hobbyists flocked to

meetings of the Homebrew Computer Club in Menlo Park, California.

- IBM introduced the first commercially available laser PRINTER. The very fast, heavy-duty machine was suitable only for very large businesses.
- The first ARPANET discussion mail list was created. The most popular topic for early mail lists was science fiction.
- In Los Angeles, Dick Heiser opened what is believed to be the first retail store to sell computers to "ordinary people."

1976

- Seymour CRAY's sleek, monolithlike Cray 1 set a new standard for SUPERCOMPUTERS.
- Whitfield Diffie and Martin Hellman announced a public-key ENCRYPTION system that allowed users to securely send information without previously exchanging keys.
- IBM developed the first (relatively crude) inkjet PRINTER for printing address labels.
- Shugart Associates offered a FLOPPY DISK drive to microcomputer builders. It cost $390.
- Steve WOZNIAK proposed that Hewlett-Packard fund the creation of a PERSONAL COMPUTER, while his friend Steve JOBS made a similar proposal to Atari Corp. Both proposals were rejected, so the two friends started APPLE Computer Company.
- Chuck Peddle of MOS Technology developed the 6502 MICROPROCESSOR, which would be used in the Apple, Atari, and some other early personal computers.
- Bill GATES complained about software piracy in his "Open Letter to Hobbyists." People were illicitly copying his BASIC language tapes. COPY PROTECTION would soon be used in an attempt to prevent copying of commercial programs for personal computers.
- Computer enthusiasts found an erudite forum in the magazine *Dr. Dobb's Journal of Computer Calisthenics and Orthodontia: Running Light without Overbyte*. The more mainstream *Byte* magazine also became a widely known forum for describing new projects and selling components.
- William Crowther and Don Woods at Stanford University developed the first interactive COMPUTER GAME involving an adventure with monsters and other obstacles. University administrators would soon complain that the game was wasting too much computer time.

1977

- Benoit Mandelbrot's book on FRACTALS IN COMPUTING popularized a mathematical phenomenon that would find uses in computer graphics, data compression, and other areas.
- The Data ENCRYPTION Standard (DES) was announced. Critics charged that it was too weak and probably already compromised by spy agencies.
- Vinton CERF demonstrated the versatility and extent of the Internet Protocol (IP) by sending a message around the world via radio, land line, and satellite links.
- The Charles BABBAGE Institute was founded. It would become an important resource for the study of computing history.

- Bill GATES and Paul Allen found a tiny company called MICROSOFT. Its first product was a BASIC INTERPRETER for the newly emerging PERSONAL COMPUTER systems.
- Radio Shack began selling its TRS-80 Model 1 personal computer.
- The APPLE II was released. It will become the most successful of the early (pre-IBM) PERSONAL COMPUTERs.

1978

- Diablo Systems marketed the first daisy-wheel PRINTER.
- Atari announced its Atari 400 and Atari 800 PERSONAL COMPUTERs. They offered superior graphics (for the time).
- Daniel Bricklin's VisiCalc SPREADSHEET is announced. It will become the first software "hit" for the Apple II, leading businesses to consider using personal computers.
- Ward Christiansen and Randy Suess developed the first software for BULLETIN BOARD SYSTEMS (BBS).
- The first West Coast Computer Faire was organized in San Francisco. The annual event became a showcase for innovation and a meeting forum for the first decade of personal computing.
- The BSD (Berkeley Software Distribution) version of UNIX was released by the group at the University of California, Berkeley, under the leadership of Bill JOY.
- The AWK (named for Aho, Weinberger, and Kernighan) SCRIPTING LANGUAGE appeared.

1979

- MEDICAL APPLICATIONS OF COMPUTING were highlighted when Allan M. Cormack and Godfrey N. Hounsfield received the Nobel Prize in medicine for the development of computerized tomography (CAT), creating a revolutionary way to examine the structure of the human body.
- The Ashton-Tate company began to market dBase II, a DATABASE MANAGEMENT SYSTEM that became the leader in personal computer databases during the coming decade.
- INTEL's new 16-bit processors, the 8086 and 8088, began to dominate the market.
- Hayes marketed the first MODEM, and the CompuServe ONLINE SERVICE and early bulletin boards gave a growing number of users something to connect to.
- UNIX users Tom Truscott, Jim Ellis, and Steve Bellovin developed a program to exchange news in the form of files copied between the Duke University and University of North Carolina computer systems. This gradually grew into USENET (or NETNEWS), providing thousands of topical newsgroups.
- The first networked computer fantasy game, MUD (Multi-User Dungeon), was developed.
- The first COMDEX was held in Las Vegas. It would become the PC industry's premier trade show.
- Boston's Computer Museum was founded. This perhaps signaled the computing field's consciousness of coming of age.

1980

- ADA, a modular descendent of PASCAL, was announced. The language was part of efforts by the U.S. Defense Department to modernize its software development process.

- RISC (REDUCED INSTRUCTION SET COMPUTER) microprocessor architecture was introduced.
- APPLE's initial public offering of 4.6 million shares at $22 per share sold out immediately. It was the largest IPO since that of Ford Motor Company in 1956. Apple founders, Steve JOBS and Steve WOZNIAK, became the first multimillionaires of the microcomputer generation.
- XENIX, a version of UNIX for PERSONAL COMPUTERs, was offered. It met with limited success.
- Shugart Associates announced a HARD DISK drive for personal computers. The disk stored a whopping 5 megabytes.

1981

- The IBM PC was announced. APPLE "welcomed" its competitor in ads, but the IBM machine would soon surpass its competitors as the personal computer of choice for business. Its success is aided by a version of the VisiCalc SPREADSHEET that sells more than 200,000 copies.
- Osborne introduced the PORTABLE (sort of) COMPUTER, a machine with the size and weight of a heavy suitcase.
- Apple tried to market the Apple III as a more powerful desktop computer for business, but the machine was plagued with technical problems and did not sell well.
- Digital Equipment Corporation introduced its DECmate dedicated word-processing system.
- Xerox PARC displayed the Star, a successor to the Alto with 512 kB of RAM. It was intended for use in an Ethernet NETWORK.
- A network called BITNET ("Because It's Time Network") began to link academic institutions worldwide.
- Tracy Kidder's best-selling *The Soul of a New Machine* recounted the intense Silicon Valley working culture as seen in the development of Data General's latest WORKSTATION, the Eclipse.
- Japan announced a 10-year effort to create "Fifth Generation" computing based on application of ARTIFICIAL INTELLIGENCE.

1982

- SUN MICROSYSTEMS was founded. It would specialize in high-performance WORKSTATIONS.
- AT&T began marketing UNIX (System III) as a commercial product.
- Compaq became one of the most successful makers of "clones" or IBM PC-compatible computers, introducing a portable (luggable) machine.
- The AutoCad program brought CAD (COMPUTER-AIDED DESIGN AND MANUFACTURING) to the desktop.
- The *Time* magazine "man of the year" was not a person at all—it was the PERSONAL COMPUTER!

1983

- Business use of personal computers continued to grow. WORD PROCESSING leaders WordStar and WordPerfect were joined by the first version of Microsoft Word. Lotus 1-2-3 became the new SPREADSHEET leader.
- Borland International introduced Turbo PASCAL, a speedy, easy to use programming environment for personal computers.

- An industry pundit introduced the term *vaporware* to refer to much-hyped but never-released software, such as a product called Ovation for IBM PCs.
- IBM tried to market the PC Jr., a less-expensive PC for home and school users. It failed to gain a foothold in the market.
- More successfully, IBM offered the PC XT, the first PERSONAL COMPUTER that had a built-in hard drive.
- Radio Shack introduced the Model 100, the first practical notebook computer.
- APPLE introduced the Lisa, a $10,000 computer with a graphical USER INTERFACE. Its high price and slow performance made it a flop, but its ideas would be more successfully implemented the following year in the MACINTOSH.
- John Sculley became president of Apple Computer, beginning a bitter struggle with Apple cofounder Steve JOBS.
- Richard STALLMAN began the GNU (GNU's not UNIX) project to create a version of UNIX that would not be subject to AT&T licensing.
- The movie *War Games* portrayed teenage HACKERS taking control of nuclear missile facilities.

1984

- A classic Super Bowl commercial introduced the APPLE MACINTOSH, the computer "for the rest of us." Based largely on Alan KAY's earlier work at Xerox PARC, the "Mac" used menus, icons, and a mouse instead of the cryptic text commands required by MS-DOS.
- Meanwhile, IBM introduced a more powerful personal computer, the PC/AT with the Intel 80286 chip.
- Steve JOBS leaves Apple Computer to found a company called NeXT.
- MICROSOFT CEO Bill GATES was featured on a *Time* magazine cover.
- The DOMAIN NAME SYSTEM began. It allows INTERNET users to connect to remote machines by name without having to specify an exact network path.
- British institutions develop JANET, the Joint Academic Network.
- Science fiction writer William Gibson coined the word *CYBERSPACE* in his novel *Neuromancer.* It began a new SF genre called *cyberpunk,* featuring a harsh, violent, immersive high-tech world.

1985

- Desktop publishing was fueled by several developments including John Warnock's POSTSCRIPT page description language and the Aldus PageMaker page layout program. The MACINTOSH's graphical interface gave it the early lead in this application.
- MICROSOFT WINDOWS 1.0 was released, using many of the same features as the Macintosh, although not nearly as well.
- There was increasing effort to unify the two versions of UNIX (AT&T and BSD), with guidelines including the System V Interface Standard and POSIX.
- Commodore introduces the Amiga, a machine with a sophisticated OPERATING SYSTEM and powerful color graphics. The

machine had many die-hard fans but ultimately could not survive in the marketplace.
- IBM marketed the IBM 3090, a large, powerful MAINFRAME that cost $9.3 million.
- The Cray 2 SUPERCOMPUTER broke the 1-billion-instructions-a-second barrier.
- A CONFERENCING SYSTEM called the Whole Earth 'Lectronic Link (WELL) was founded. Its earliest users are largely drawn from Grateful Dead fans and assorted techies.

1986

- The National Science Foundation funded NSFNET, which provides high-speed Internet connections to link universities and research institutions.
- Borland released a PROLOG compiler, making the ARTIFICIAL INTELLIGENCE language accessible to PC users. A PC version of SMALLTALK also appeared from another company.
- APPLE beefed up the relatively anemic MACINTOSH with the Macintosh Plus, which has more MEMORY.

1987

- Bjarne STROUSTRUP's C++ language offered OBJECT-ORIENTED PROGRAMMING in a form that was palatable to the legions of *C* programmers. The language would surpass its predecessor in the coming decade.
- Sun marked its first WORKSTATION based on RISC (REDUCED INSTRUCTION SET COMPUTING) technology.
- APPLE sold its one millionth MACINTOSH. Apple also brought out a new line of Macs (the Macintosh SE and Macintosh II) that, unlike the original Macs, were expandable by plugging in cards.
- Apple also introduced Hypercard, a simple HYPERTEXT AUTHORING SYSTEM that became popular with educators.
- IBM introduced a new line of personal computers called the PS/2. It featured a more efficient BUS called the Microchannel and some other innovations, but it sold only modestly. Most of the industry continued to further develop standards based upon the IBM PC AT.
- The Thinking Machines Corporation's Connection Machine introduced massive parallel processing. It contained 64,000 MICROPROCESSORS that could collectively perform 2 billion instructions per second.

1988

- Robert Morris Jr.'s "worm" accidentally ran out of control on the INTERNET, bringing concerns about COMPUTER CRIME AND SECURITY to public attention. The Computer Emergency Response Team (CERT) was formed in response.
- Wolfram's Mathematica program was a milestone in mathematical computing, allowing users to not merely calculate but also to solve symbolic equations automatically.
- Cray introduced the Cray Y-MP SUPERCOMPUTER. It could process 2 billion operations per second.
- IBM announced a new midrange MAINFRAME, the AS/400.
- Sandia National Laboratory began to build a massively parallel "hypercomputer" that would have 1,024 processors working in tandem.

- A consortium called the Open Software Foundation was established to promote OPEN SOURCE shared software development.

1989

- The INTERNET now had more than 100,000 host computers.
- Deep Thought defeated Danish chess grandmaster Bent Larsen, marking the first time a grandmaster had been defeated by a computer.
- INTEL announced the 80486 CPU, a chip with over a million transistors.
- Astronomer Clifford STOLL's book *The Cuckoo's Egg* recounted his pursuit of German hackers who were seeking military secrets. Stoll soon became a well-known critic of computer technology and the Internet.
- The ARPANET officially ends, having been succeeded by the NSFNET.

1990

- MICROSOFT WINDOWS became truly successful with version 3.0, diminishing the user interface advantages of the MACINTOSH.
- At SUN MICROSYSTEMS, James Gosling developed the Oak language to control EMBEDDED SYSTEMS. After the original project was canceled, Gosling redesigned the language as JAVA.
- IBM announced the System/390 MAINFRAME.
- IBM and MICROSOFT developed OS/2, an operating system intended to replace MS-DOS. Microsoft withdrew in favor of Windows, and despite considerable technical merits, OS/2 never really takes hold.
- Secret Service agents raided computer systems and bulletin boards, seeking evidence of illegal copying of a BellSouth manual, disrupting an innocent game company. In response, Mitch Kapor founded the Electronic Frontier Foundation to advocate for civil liberties of computer users. Another group, the Computer Professionals for Social Responsibility, filed a Freedom of Information Act (FOIA) request for FBI records involving alleged government surveillance of BULLETIN BOARD SYSTEMS.

1991

- The Science Museum in London exhibited a reconstruction of Charles BABBAGE's never-built difference engine.
- A Finnish student named Linus TORVALDS found that he couldn't afford a UNIX license, so he wrote his own UNIX KERNEL and combined it with GNU utilities. The result would eventually become the popular LINUX operating system.
- Developers at the University of Minnesota created Gopher, a system for providing documents over the INTERNET using linked menus. However, it was soon to be surpassed by the WORLD WIDE WEB, created by Tim BERNERS-LEE at the CERN physics laboratory in Geneva, Switzerland.
- ADVANCED MICRO DEVICES began to compete with Intel by making IBM PC-compatible CPU chips.
- APPLE and IBM signed a joint agreement to develop technology in areas that include object-oriented OPERATING SYSTEMS, multimedia, and interoperability between MACINTOSH and IBM networks.

1992

- Reports of the Michelangelo COMPUTER VIRUS frightened computer users. Although the virus did little damage, it spurred more users to practice "safe computing" and install antivirus software.
- MOTOROLA announced the Power PC, a 32-bit RISC MICROPROCESSOR that contains 28 million transistors.
- An estimated 1 million host computers were on the INTERNET. The Internet Society is founded to serve as a coordinator of future development of the network.

1993

- APPLE's Newton handheld computer created a new category of machine called the PDA, or personal digital assistant.
- MICROSOFT WINDOWS NT was announced. It is a version of the OPERATING SYSTEM designed especially for network servers.
- Steve JOBS announced that his NeXT company would abandon its hardware efforts and concentrate on marketing its innovative OPERATING SYSTEM and development software.
- Leonard Adleman demonstrated MOLECULAR COMPUTING by using DNA molecules to solve the Traveling Salesman problem.
- The Cray 3 SUPERCOMPUTER continued the evolution of that line. It could be scaled up to a 16-processor system.
- The Mosaic graphical WEB BROWSER popularized the WORLD WIDE WEB.
- The Clinton administration announced plans to develop a national "Information Superhighway" based on the INTERNET. Volunteer "Net Day" programs would begin to connect schools to the network.
- The White House established its Web site, www.whitehouse.gov.

1994

- Mosaic's developer, Marc ANDREESSEN, left NCSA and joined Jim Clark to found Netscape. Netscape soon released an improved browser called Netscape Navigator.
- APPLE announced that it would license the Mac operating system to other companies to make MACINTOSH "clones." Few companies would take them up on it, and Apple would soon withdraw the licensing offer.
- INTEL CORPORATION was forced to recall millions of dollars worth of its new Pentium CHIPs when a mathematical flaw was discovered in the floating-point routines.
- Marc Andreessen and Jim Clark founded Netscape and developed a new WEB BROWSER, Netscape Navigator. It would become the leading Web browser for several years.
- Red Hat released a commercial distribution of LINUX 1.0.
- Search engines such as Lycos and Alta Vista started helping users find Web pages. Meanwhile, a graduate student named Jerry Yang started compiling an online list of his

favorite Web sites. That list would eventually become YAHOO!

- Advertising in the form of banner ads began to appear on Web sites.

1995

- MICROSOFT WINDOWS 95 gave a new look to the operating system and provided better support for devices, including PLUG AND PLAY device configuration.
- MICROSOFT began its own on-line service, the Microsoft Network (MSN). Despite its startup icon being placed on the Windows 95 desktop, the network would trail industry leader AMERICA ONLINE, which had overtaken CompuServe and Prodigy.
- Jeff BEZOS's online bookstore, Amazon.com, opened for business. It would become the largest e-commerce retailer.
- The major online services began major promotion of access to the WORLD WIDE WEB.
- NSFNET retired from direct operation of the INTERNET, which had now been fully privatized. The agency then focused on providing new BROADBAND connections between SUPERCOMPUTER sites.
- SUN announced the JAVA language. It would become one of the most popular languages for developing applications for the World Wide Web.
- MOTOROLA announced the Power PC-602, a 64-bit CPU chip.
- Compaq ranked first in personal computer sales in the United States, followed by APPLE.
- Physicists Peter Fromherz and Alfred Stett of the Max Planck Institute of Biochemistry in Munich, Germany, demonstrated the direct stimulation of a specific nerve cell in a leech by a computer probe. This conjured visions of the "jacked-in" neural implants foreseen by science fiction writers such as William Gibson.
- The next generation of Cray SUPERCOMPUTERs, the T90 series, could be scaled up to a rate of 60 billion instructions per second.
- STREAMING (real-time video and audio) began to become popular on the WEB.
- Computer-generated imagery (CGI) was featured by Hollywood in the movie *Toy Story*.

1996

- A product called Web TV attempted to bring the WORLD WIDE WEB to home consumers without the complexity of full-fledged computers. The product achieved only modest success as the price of PERSONAL COMPUTERs continued to decline.
- The U.S. Postal Service issued a stamp honoring the 50th anniversary of ENIAC.
- The Boston Computer Society, one of the oldest COMPUTER USER GROUPS, disbanded.
- World chess champion Garry Kasparov won his first match against IBM's Deep Blue chess computer, but said the match had been unexpectedly tough.
- YAHOO! offered its stock to the public, running up the second-highest first-day gain in NASDAQ history.

- Seymour CRAY's Cray Research (a developer of SUPERCOMPUTERs) was acquired by Silicon Graphics.
- Pierre OMIDYAR turned a small hobby auction site into EBAY and was soon attracting thousands of eager sellers and buyers to the site.
- In one of its infrequent ventures into hardware, MICROSOFT announced the NetPC, a stripped-down diskless PC that would run software from a NETWORK. Such "network computers" never really caught on, being overtaken by the ever-declining price for complete PCs.

1997

- The chess world was shocked when world champion Garry Kasparov was defeated in a rematch with Deep Blue.
- A single INTERNET domain name, business.com, was sold for $150,000.
- AMAZON.COM had a successful initial public offering (IPO).
- A technology called "push" began to be hyped. It involved Web sites continually feeding "channels" of news or entertainment to user's desktops. However, the idea would fail to make much headway.
- INTERNET users banded together to demonstrate DISTRIBUTED COMPUTING by cracking a 56-bit DES cipher in 140 days.
- The Association for Computing Machinery (ACM) celebrated its 50th anniversary.

1998

- MICROSOFT WINDOWS 98 provided an incremental improvement in the OPERATING SYSTEM.
- APPLE announced the iMac, a stylish machine that rejuvenated the MACINTOSH line.
- EBAY's IPO was wildly successful, making Pierre OMIDYAR, Meg Whitman, and other eBay executives instant millionaires.
- Merger-mania hit the online service industry, with AMERICA ONLINE buying CompuServe's online service (spinning off the network facilities to WorldCom). AOL then acquired Netscape and its Web hosting technology.
- In another significant merger, Compaq acquired Digital Equipment Corporation (DEC).

1999

- Federal Judge Thomas Penfield Jackson found that MICROSOFT violated antitrust laws. The case dragged on with appeals, with the process of crafting a remedy (such as possibly the split-up of the company) still unresolved in 2002.
- Another virus, Melissa, panicked computer users.
- Some companies began to offer "free" computers to people who agreed to sign up for long-term, relatively expensive INTERNET service.
- Computer scientists and industry pundits debated the possibility of widespread computer disasters due to the Y2K PROBLEM. Companies spent millions of dollars trying to find and fix old computer code that used only two digits to store year dates.

- APPLE released OS X, a new UNIX-based OPERATING SYSTEM for the MACINTOSH.

2000

- New Year's Day found the world to be continuing much as before, with only a few scattered Y2K PROBLEMS.
- Unknown hackers, however, brought down some commercial Web sites with denial-of-service (DOS) attacks.
- AOL merged with Time-Warner, creating the world's largest media company. Critics worried about the affects of growing corporate concentration on the diversity of the INTERNET.
- MICROSOFT WINDOWS 2000 began the process of merging the consumer Windows and Windows NT lines into a single family of operating systems that would no longer use any of the underlying MS-DOS code.
- The WORLD WIDE WEB was estimated to have about 1 billion pages online.
- Tech stocks (and particularly E-COMMERCE companies) began to sharply decline as investors became increasingly skeptical about profitability.
- A growing number of Web users were beginning to switch to much faster BROADBAND connections using DSL or CABLE MODEMS.

2001

- The decline in E-COMMERCE stocks continued, with tens of thousands of jobs lost. One of the many failures was Webvan, the INTERNET grocery service. AMAZON.COM suffered losses but continued trying to expand into profitable niches. Only EBAY among the major e-commerce companies continued to be profitable.
- MICROSOFT WINDOWS XP offered consumer and "professional" versions of Windows on the same code base.
- IBM researchers created a seven "qubit" quantum computer to execute Shor's algorithm, a radical approach to factoring that could potentially revolutionize cryptography.
- Among the specters raised in the wake of the September 11 terrorist attacks was CYBERTERRORISM having the potential to disrupt vital infrastructure, services, and the economy. BIOMETRICS and more sophisticated database techniques were enlisted in the war on terrorism while civil liberties groups voiced concerns.

2002

- Wireless networking using the faster 802.11 standard became increasingly popular as an alternative to cabled or phone line networks for homes and small offices.
- Consumer digital cameras began to approach "professional" quality.
- The U.S. Supreme Court ruled that "virtual" child pornography (in which no actual children were used) was protected by the First Amendment.
- Continuing stock market declines threaten growth in the computer and INTERNET sectors.
- The music-sharing service Napster goes out of business, when it is forced to stop distributing copyrighted music.

2003

- The U.S. economy begins to recover, including the technology sector. However, there is a growing concern about jobs being "outsourced" to countries such as India and China.
- Weblogs, or BLOGS, are an increasingly popular form of online expression. Some journalists even use them to "break" major stories.
- The Recording Industry Association of America (RIAA) files hundreds of lawsuits against individual users of music file-sharing systems.
- APPLE and AMD introduce the first 64-bit microprocessors in the PERSONAL COMPUTER market.

2004

- Security remains an urgent concern as viruses and worms flood the INTERNET in vast numbers.
- SPAM also floods users' E-MAIL boxes. PHISHING messages trick users into revealing credit card numbers and other sensitive information.
- APPLE's iPod dominates the portable media player market, while its iTunes store sells over 100 million songs.
- Bloggers become a political force, winning access to major party conventions.
- Enthusiastic response to GOOGLE's initial public stock offering signals that investors may have regained confidence in the strength of the INTERNET sector.

2005

- "WEB 2.0" becomes a buzzword with WEB SERVICES being designed to be leveraged into new applications to be delivered to users' browsers.
- SONY's flawed CD copy protection leaves users vulnerable to HACKERS; consumers increasingly demand an end to restrictions on use of media they buy.
- Concerns about the security of new ELECTRONIC VOTING SYSTEMS grow.

2006

- APPLE begins selling INTEL-based Macs; meanwhile most PCs now have dual processors.
- GOOGLE buys the phenomenally successful video site YOU-TUBE for $1.65 billion.
- MICROSOFT releases its delayed Windows Vista operating system, but response is lukewarm.
- New versions of LINUX such as Ubuntu attract enthusiasts, but are slow in making inroads on the desktop.

2007

- SOCIAL NETWORKING sites such as MySpace and FaceBook are used by millions of students, but raise concerns about privacy and bullying.
- WIKIPEDIA now has more than 9 million articles in 252 languages.
- CNN and YOUTUBE join to sponsor presidential political debates, and candidates respond to questions posed in videos submitted by the public.

- GOOGLE and other free Web-based applications offer new alternatives for office software.
- APPLE introduces the iPhone and new iPods with innovative user interfaces.
- Albert Fert and Peter Grunberg receive the Nobel Prize in physics for their development of "giant magnetoresistance," a phenomenon that enables disk drives to read fainter, more densely packed magnetic signals. The result is shrinking disks and/or greater storage capacity.

2008
- A record amount of money is raised online during the presidential election campaign.
- Microsoft engages in a protracted campaign to acquire online rival Yahoo!
- Providers and advocacy groups struggle over net neutrality (equal treatment of online applications and content).

APPENDIX III
SOME SIGNIFICANT AWARDS

This appendix describes some of the major awards in computer science and technology and lists recipients as of 2001. The last names of persons with entries in this book are given in SMALL CAPITAL LETTERS.

ASSOCIATION FOR COMPUTING MACHINERY (ACM)

ACM TURING AWARD

The ACM Turing Award "is given to an individual selected for contributions of a technical nature made to the computing community. The contributions should be of lasting and major technical importance to the computer field."

ANNUAL RECIPIENTS
(A few years have joint recipients.)

1966 A. J. Perlis: "For his influence in the area of advanced programming techniques and compiler construction."

1967 Maurice V. Wilkes: "Professor Wilkes is best known as the builder and designer of the EDSAC, the first computer with an internally stored program. Built in 1949, the EDSAC used a mercury delay line memory. He is also known as the author, with Wheeler and Gill, of a volume on *'Preparation of Programs for Electronic Digital Computers'* in 1951, in which program libraries were effectively introduced."

1968 Richard Hamming: "For his work on numerical methods, automatic coding systems, and error-detecting and error-correcting codes."

1969 Marvin MINSKY [Citation not listed by ACM. However, Minsky was a key pioneer in artificial intelligence research, including neural networks, robotics, and cognitive psychology.]

1970 J. H. Wilkinson: "For his research in numerical analysis to facilitate the use of the high-speed digital computer, having received special recognition for his work in computations in linear algebra and 'backward' error analysis."

1971 John MCCARTHY: "Dr. McCarthy's lecture 'The Present State of Research on Artificial Intelligence' is a topic that covers the area in which he has achieved considerable recognition for his work."

1972 E. W. DIJKSTRA.: "Edsger Dijkstra was a principal contributor in the late 1950s to the development of the ALGOL, a high-level programming language which has become a model of clarity and mathematical rigor. He is one of the principal exponents of the science and art of programming languages in general, and has greatly contributed to our understanding of their structure, representation, and implementation. His fifteen years of publications extend from theoretical articles on graph theory to basic manuals, expository texts, and philosophical contemplations in the field of programming languages."

1973 Charles W. Bachman: "For his outstanding contributions to database technology."

1974 Donald E. KNUTH: "For his major contributions to the analysis of algorithms and the design of programming languages, and in particular for his contributions to the 'art of computer programming' through his well-known books in a continuous series by this title."

1975 Alan Newell and Herbert A. Simon: "In joint scientific efforts extending over twenty years, initially in collaboration with J. C. Shaw at the RAND Corporation, and subsequentially with numerous faculty and student colleagues at Carnegie-Mellon University, they have made basic contributions to artificial intelligence, the psychology of human cognition, and list processing."

1976 Michael O. Rabin and Dana S. Scott: "For their joint paper 'Finite Automata and Their Decision Problem,' which introduced the idea of nondeterministic machines, which has proved to be an enormously valuable concept. Their [Scott & Rabin] classic paper has been a continuous source of inspiration for subsequent work in this field."

1977 John Backus: "For profound, influential, and lasting contributions to the design of practical high-level programming systems, notably through his work on FORTRAN, and for seminal publication of formal procedures for the specification of programming languages."

1978 Robert W. Floyd: "For having a clear influence on methodologies for the creation of efficient and reliable software, and for helping to found the following important subfields of computer science: the theory of parsing, the semantics of programming languages, automatic program verification, automatic program synthesis, and analysis of algorithms."

1979 Kenneth E. Iverson: "For his pioneering effort in programming languages and mathematical notation resulting in what the computing field now knows as APL, for his contributions to the implementation of interactive systems, to educational uses of APL, and to programming language theory and practice."

1980 C. Anthony R. Hoare: "For his fundamental contributions to the definition and design of programming languages."

1981 Edgar F. Codd: "For his fundamental and continuing contributions to the theory and practice of database management systems. He originated the relational approach to database management in a series of research papers published commencing in 1970. His paper 'A Relational Model of Data for Large Shared Data Banks' was a seminal paper, in a continuing and carefully developed series of papers. Dr. Codd built upon this space and in doing so has provided the impetus for widespread research into numerous related areas, including database languages, query subsystems, database semantics, locking and recovery, and inferential subsystems."

1982 Stephen A. Cook: "For his advancement of our understanding of the complexity of computation in a significant and profound way. His seminal paper, 'The Complexity of Theorem Proving Procedures,' presented at the 1971 ACM SIGACT Symposium on the Theory of Computing, laid the foundations for the theory of NP-Completeness. The ensuing exploration of the boundaries and nature of NP-complete class of problems has been one of the most active and important research activities in computer science for the last decade."

1983 Ken Thompson and Dennis RITCHIE: "For their development of generic operating systems theory and specifically for the implementation of the UNIX operating system."

1984 Niklaus WIRTH: "For developing a sequence of innovative computer languages, EULER, ALGOL-W, MODULA and PASCAL. PASCAL has become pedagogically significant and has provided a foundation for future computer language, systems, and architectural research."

1985 Richard M. Karp: "For his continuing contributions to the theory of algorithms including the development of efficient algorithms for network flow and other combinatorial optimization problems, the identification of polynomial-time computability with the intuitive notion of algorithmic efficiency, and, most notably, contributions to the theory of NP-completeness. Karp introduced the now standard methodology for proving problems to be NP-complete which has led to the identification of many theoretical and practical problems as being computationally difficult."

1986 John Hopcroft and Robert Tarjan: "For fundamental achievements in the design and analysis of algorithms and data structures."

1987 John Cocke: "For significant contributions in the design and theory of compilers, the architecture of large systems and the development of reduced instruction set computers (RISC); for discovering and systematizing many fundamental transformations now used in optimizing compilers including reduction of operator strength, elimination of common subexpressions, register allocation, constant propagation, and dead code elimination."

1988 Ivan Sutherland: "For his pioneering and visionary contributions to computer graphics, starting with Sketchpad, and continuing after. Sketchpad, though written twenty-five years ago, introduced many techniques still important today. These include a display file for screen refresh, a recursively traversed hierarchical structure for modeling graphical objects, recursive methods for geometric transformations, and an object oriented programming style. Later innovations include a 'Lorgnette' for viewing stereo or colored images, and elegant algorithms for registering digitized views, clipping polygons, and representing surfaces with hidden lines."

1989 William (Velvel) Kahan: "For his fundamental contributions to numerical analysis. One of the foremost experts on floating-point computations. Kahan has dedicated himself to 'making the world safe for numerical computations.'"

1990 Fernando J. Corbato: "For his pioneering work organizing the concepts and leading the development of the general-purpose, large-scale, time-sharing and resource-sharing computer systems, CTSS and Multics."

1991 Robin Milner: "For three distinct and complete achievements: 1) LCF, the mechanization of Scott's Logic of Computable Functions, probably the first theoretically based yet practical tool for machine assisted proof construction; 2) ML, the first language to include polymorphic type inference together with a type-safe exception-handling mechanism; 3) CCS, a general theory of concurrency. In addition, he formulated and strongly advanced full abstraction, the study of the relationship between operational and denotational semantics."

1992 Butler W. Lampson: "For contributions to the development of distributed, personal computing environments and the technology for their implementation: workstations, networks, operating systems, programming systems, displays, security and document publishing."

1993 Juris Harmanis and Richard E. Stearns: "In recognition of their seminal paper which established the foundations for the field of computational complexity theory."

1994 Edward FEIGENBAUM and Raj Reddy: "For pioneering the design and construction of large-scale artificial intelligence systems, demonstrating the practical importance and potential commercial impact of artificial intelligence technology."

1995 Manuel Blum: "In recognition of his contributions to the foundations of computational complexity theory and its application to cryptography and program checking."

1996 Amir Pneueli: "For seminal work introducing temporal logic into computing science and for outstanding contributions to program and systems verification."

1997 Douglas ENGELBART: "For an inspiring vision of the future of interactive computing and the invention of key technologies to help realize this vision."

1998 James Gray: "For seminal contributions to database and transaction processing research and technical leadership in system implementation."

1999 Frederick P. Brooks, Jr.: "For landmark contributions to computer architecture, operating systems, and software engineering."

2000 Andrew Chi-Chih Yao: "In recognition of his fundamental contributions to the theory of computation, including the complexity-based theory of pseudorandom number generation, cryptography, and communication complexity."

2001 Ole-Johan Dahl and Kristen Nygaard: "For ideas fundamental to the emergence of object oriented programming, through their design of the programming languages Simula I and Simula 67."

2002 Ronald L. Rivest, Adi Shamir, and Leonard M. Adelman: "For their ingenious contribution for making public-key cryptography useful in practice."

2003 Alan KAY: "pioneering many of the ideas at the root of contemporary object-oriented programming languages, leading the team that developed Smalltalk, and for fundamental contributions to personal computing."

2004 Vinton G. CERF and Robert E. Kahn: "For pioneering work on internetworking, including the design and implementation of the Internet's basic communications protocols, TCP/IP, and for inspired leadership in networking."

2005 Peter Naur: "For fundamental contributions to programming language design and the definition of Algol 60, to compiler design, and to the art and practice of computer programming."

2006 Frances E. Allen: "For contributions that fundamentally improved the performance of computer programs in solving problems, and accelerated the use of high-performance computing."

2007 Edmund M. Clarke, E. Allen Emerson, and Joseph Sifakis: "For . . . developing ModelChecking into a highly effective verification technology, widely adopted in the hardware and software industries."

ECKERT-MAUCHLY AWARD

Administered jointly by ACM and IEEE Computer Society and "given for contributions to computer and digital systems architecture where the field of computer architecture is considered at present to encompass the combined hardware-software design and analysis of computing and digital systems."

ANNUAL RECIPIENTS

1979 Robert S. Barton: "For his outstanding contributions in basing the design of computing systems on the hierarchical nature of programs and their data."

1980 Maurice V. Wilkes: "For major contributions to computer architecture over three decades including notable achievements in developing a working stored-program computer, formulation of the basic principles of microprogramming, early research on cache memories, and recent studies in distributed computation."

1981 Wesley A. Clark: "For contributions to the early development of the minicomputer and the multiprocessor, and for continued contributions over 25 years that have found their way into computer networks, modular computers, and personal computers."

1982 C. Gordon Bell: "For his contributions to designing and understanding computer systems: for his contributions in the formation of the minicomputer; for the creation of the first commercial, interactive timesharing computer; for pioneering work in the field of hardware description languages; for co-authoring classic computer books and co-founding a computer museum."

1983 Tom Kilburn: "For major seminal contributions to computer architecture spanning a period of three decades. For establishing a tradition of collaboration between university and industry which demands the mutual understanding of electronics technology and abstract programming concepts."

1984 Jack B. Dennis: "For contributions to the advancement of combined hardware and software design through innovations in data flow architectures."

1985 John Cocke: "For contributions to high performance computer architecture through lookahead, parallelism and pipeline utilization, and to reduced instruction set computer architecture through the exploitation of hardware-software tradeoffs and compiler optimization."

1986 Harvey G. Cragon: "For major contributions to computer architecture and for pioneering the application of integrated circuits for computer purposes. For serving as architect of the Texas

Instruments scientific computer and for playing a leading role in many other computing developments in that company."

1987   Gene M. AMDAHL: "For outstanding innovations in computer architecture, including pipelining, instruction look-ahead, and cache memory."

1988   Daniel P. Siewiorek: "For outstanding contributions in parallel computer architecture, reliability, and computer architecture education."

1989   Seymour CRAY: "For a career of achievements that have advanced supercomputing design."

1990   Kenneth E. Batcher: "For contributions to parallel computer architecture, both for pioneering theories in interconnection networks and for the pioneering implementations of parallel computers."

1991   Burton J. Smith: "For pioneering work in the design and implementation of scalable shared memory multiprocessors."

1992   Michael J. Flynn: "For his important and seminal contributions to processor organization and classification, computer arithmetic and performance evaluation."

1993   David Kuck: "For his impact on the field of supercomputing, including his work in shared memory multiprocessing, clustered memory hierarchies, compiler technology, and application/library tuning."

1994   James E. Thornton: "For his pioneering work on high-performance processors; for inventing the 'scoreboard' for instruction issue; and for fundamental contributions to vector supercomputing."

1995   John Crawford: "In recognition of your impact on the computer industry through your development of microprocessor technology."

1996   Yale N. Patt: "For important contributions to instruction level parallelism and superscalar processor design."

1997   Robert Tomasulo: "For the ingenious Tomasulo's algorithm, which enabled out-of-order execution processors to be implemented."

1998   T. Watanabe: [Citation not available, but NEC notes that Watanabe "was a chief architect for NEC's first supercomputer, the SX-2, and is recognized for his significant contributions to the architectural design of supercomputers having multiple, parallel vector pipelines and programmable vector caches."]

1999   James E. Smith: "For fundamental contributions to high-performance microarchitecture, including saturating counters for branch prediction, reorder buffers for precise exceptions, decoupled access/execute architectures, and vector supercomputer organization, memory, and interconnects."

2000   Edward Davidson: "For his seminal contributions to the design, implementation, and performance evaluation of high-performance pipelines and multiprocessor systems."

2001   John Hennessy: "For being the founder and chief architect of the MIPS Computer Systems and contributing to the development of the landmark MIPS R2000 microprocessor."

2002   B. Ramakrishna (Bob) Rau: "For pioneering contributions to statistically scheduled instruction-level parallel processors and their compilers."

2003   Joseph A. (Josh) Fisher: "In recognition of 25 years of seminal contributions to instruction-level parallelism, pioneering work on VLIW architectures, and the formulation of the Trace Scheduling compilation technique."

2004   Frederick P. Brooks: "For the definition of computer architecture and contributions to the concept of computer families and to the principles of instruction set design; for seminal contributions in instruction sequencing, including interrupt systems and execute instructions; and for contributions to the IBM 360 instruction set architecture."

2005   Robert P. Colwell: "For outstanding achievements in the design and implementation of industry-changing microarchitectures, and for significant contributions to the RISC/CISC architecture debate."

2006   James H. Pomerene: "For pioneering innovations in computer architecture, including early concepts in cache, reliable memories, pipelining and branch prediction, for the design of the IAS computer and for the design of the Harvest supercomputer."

2007   Mateo Valero: "For extraordinary leadership in building a world class computer architecture research center, for seminal contributions in the areas of vector computing and multithreading, and for pioneering basic new approaches to instruction-level parallelism."

2008   David Patterson: "For seminal contributions to RISC microprocessor architectures, RAID storage systems design, and reliable computing, and for leadership in education and in disseminating academic research results into successful industrial products."

GRACE MURRAY HOPPER AWARD

The ACM gives this award for "the outstanding young computer professional of the year . . . selected on the basis of a single recent major technical or service contribution."

ANNUAL RECIPIENTS
Note: this award has not been given every year.

1971   Donald E. KNUTH: "For the publication in 1968 (at age 30) of Volume I of his monumental treatise 'The Art of Computer Programming.'"

1972   Paul E. Dirksen and Paul H. Kress: "For the creation of WATFOR Compiler, the first member of a powerful new family of diagnostic and educational programming tools."

1973   Lawrence M. Breed, Richard Lathwell, and Roger Moore: "For their work in the design and implementation of APL/360, setting new standards in simplicity, efficiency, reliability and response time for interactive systems."

1974   George N. Baird: "For his successful development and implementation of the Navy's COBOL Compiler Validation System."

1975   Allen L. Scherr: "For his pioneering study in quantitative computer performance analysis."

1976   Edward H. Shortliffe: "For his pioneering research which is embodied in the MYCIN program. MYCIN is a program which consults with physicians about the diagnosis and treatment of infections. In creating MYCIN, Shortliffe employed his background of medicine, together with his research in knowledge-based systems design, to produce an integrated package which is easy for expert physicians to use and extend. Shortliffe's work formed the basis for a research program supported by NIH, and has been widely studied and drawn upon by others in the field of knowledge-based systems."

1978   Raymond C. KURZWEIL: "For his development of a unique reading machine for the blind, a computer-based device that reads printed pages aloud. The Kurzweil machine is an 80-pound device that shoots a beam of light across each printed page, converts the reflected light across each printed page, converts the reflected light into digital data that is analyzed by its built-in computer, and then transformed into synthetic speech. It is expected to make reading of all printed material possible for blind people, whose reading was previously limited to material translated into Braille. The machine would not have been possible without another achievement by Kurzweil, that is, a set of rules embodied in the mini-computer program by which printed characters of a wide variety of sizes and shapes are reliably and automatically recognized."

1979   Steven WOZNIAK: "For his many contributions to the rapidly growing field of personal computing and, in particular, to the hardware and software for the Apple Computer."

1980   Robert M. Metcalfe: "For his work in the development of local networks, specifically the Ethernet."

1981   Daniel S. Bricklin: "For his contributions to personal computing and, in particular, to the design of VisCalc. Bricklin's efforts in the development of the 'Visual Calculator' provide the excellence and elegance that ACM seeks to sustain through such activities as the Awards program."

1982   Brian K. Reid: "For his contributions in the area of computerized text-production and typesetting systems, specifically Scribe which represents a major advance in this area. It embodies several innovations based on computer science research in programming language design, knowledge-based systems, computer document processing, and typography. The impact of Scribe has been substantial due to the excellent documentation and Reid's efforts to spread the system."

1984   Daniel H. H. Ingalls, Jr.: "For his work at the Xerox Palo Alto Research Center, where he was a major force, both technical and inspirational, in the development of the SMALLTALK language and its graphics facilities. He is the designer of the BITBLT primitive that is now widely used for generating images on raster-scan displays. The combination of a good idea, a good design, and very effective and careful implementation has led to BITBLT's wide acceptance in the computing community. Mr. Ingalls' research has also directly and dramatically affected the computing industry's view of what people should have in the way of accessible computing."

1985   Cordell Green: "For establishing several key aspects of the theoretical basis for logic programming and providing a resolution theorem prover to carry out a programming task by constructing the result which the computer program is to compute. For proving the constructive technique correct and for presenting an effective method for constructing the answer; these contributions providing an early theoretical basis for Prolog and logic programming."

1986   William N. JOY: "For his work on the Berkeley UNIX Operating System as a designer, integrator, and implementor of many of its advanced features including Virtual Memory, the C-shell, the vi Screen editor, and Networking."

1987   John K. Ousterhout: "For his contribution to very large scale integrated circuit computer aided design. His systems, Caesar and Magic, have demonstrated that effective CAD systems need not be expensive, hard to learn, or slow."

1988   Guy L. Steele: "For his general contributions to the development of Higher Order Symbolic Programming, principally for his advancement of lexical scoping in LISP."

1989   W. Daniel Hillis: "For his basic research on data parallel algorithms and for the conception, design, implementation and commercialization of the Connection Machine."

1990   Richard STALLMAN: "For pioneering work in the development of the extensible editor EMACS (Editing Macros)."

1991   Feng-hsuing Hsu: "For contributions in architecture and algorithms for chess machines. His work led to the creation of the Deep Thought Chess Machine, which led to the first chess playing computer to defeat Grandmasters in tournament play and the first to achieve a certified Grandmaster level rating."

1993   Bjarne STROUSTRUP: "For his early work laying the foundations for the C++ programming language. Based on the foundations and Dr. Stroustrup's continuing efforts, C++ has become one of the most influential programming languages in the history of computing."

1996   Shafrira Goldwasser: "For her early work relating computation, randomness, knowledge committee and proofs, which has shaped the foundations of probabilistic computation theory, computational number theory, and cryptography. This work is a continuing influence in design and certification of secure communications protocols, with practical applications to development of secure networks and computer systems."

1999   Wen-mei Hwu: "For the design and implementation of the IMPACT compiler infrastructure which has been used extensively both by the microprocessor industry as a baseline for product development and by academia as a basis for advanced research and development in computer architecture and compiler design."

2000   Lydia Kavraki: "For her seminal work on the probabilistic roadmap approach which has caused a paradigm shift in the area of path planning, and has many applications in robotics, manufacturing, nanotechnology and computational biology."

2001   George Necula: "For his seminal work on the concept and implementation of Proof Carrying Code, which has had a great impact on the field of programming languages and compilers and has given a new direction to applications of theorem proving to program correctness, such as safety of mobile code and component-based software."

2002   Ramakrishnan Srikant: "For his seminal work on mining association rules, which has led to association rules becoming a key data mining tool as well as part of the core syllabus in database and data mining courses."

2003   Stephen W. Keckler: "For ground-breaking analysis of technology scaling for high-performance processors that sheds new light on the methods required to maintain performance improvement trends in computer architecture, and on the design implications for future high-performance processors and systems."

2004   Jennifer Rexford: "For models, algorithms, and deployed systems that assure stable and efficient Internet routing without global coordination."

2005   Omer Reingolf: "For his work in finding a deterministic logarithmic-space algorithm for ST-connectivity in undirected graphs."

2006   Daniel Klein: "For the design of a system capable of learning a high-quality grammar for English directly from text."

2007   Vern Paxson: "For his work in measuring and characterizing the Internet."

ELECTRONIC FRONTIER FOUNDATION (EFF)

PIONEER AWARDS

The EFF gives annual "Pioneer Awards" to leaders in "expanding knowledge, freedom, efficiency, and utility."

1992   Douglas C. ENGELBART, Robert Kahn, Jim Warren, Tom Jennings, and Andrzej Smereczynski.

1993   Paul Baran, Vinton CERF, Ward Christensen, Dave Hughes, and the USENET software developers, represented by the software's originators Tom Truscott and Jim Ellis.

1994   Ivan Sutherland, Whitfield Diffie and Martin Hellman, Murray Turoff and Starr Roxanne Hiltz, Lee Felsenstein, Bill Atkinson, and the Well.

1995   Philip Zimmermann, Anita Borg, and Willis Ware.

1996   Robert Metcalfe, Peter Neumann, Shabbir Safdar, and Matthew Blaze.

1997   Marc Rotenberg, Johan "Julf" Helsingius, and (special honorees) Hedy Lamarr and George Antheil.

1998   Richard STALLMAN, Linus TORVALDS, and Barbara Simons.

1999   Jon Postel, Drazen Panic, and Simon Davies.

2000   Tim BERNERS-LEE, Phil Agre, and "Librarians Everywhere."

2001   Seth Finkelstein, Stephanie Perrin, and Bruce Ennis.

2002   Dan Gillmour, Beth Givens, Jon Johansen, and "writers of DeCSS."

2003   Amy Goodman, Eben Moglen, and David Sobel.

2004   Kim Alexander, David Dill, and Arviel Rubin.

2005   Patrick Ball, Edward Felten, and Mitch Kapor.

2006   CRAIGSLIST, Gigi Sohn, and Jimmy Wales.

2007   Yochai Benkler, Cory Doctorow, and Bruce Scheier.

2008   Mozilla Foundation, Mitchell Baker, Michael Geist, and Mark Klein.

IEEE COMPUTER SOCIETY

COMPUTER PIONEER AWARD

The IEEE Computer Society presents the Computer Pioneer Award "for significant contributions to concepts and developments in the electronic computer field which have clearly advanced the state of the art in computing." The award is given a minimum of 15 years after the achievement being awarded.

CHARTER RECIPIENTS

Howard H. AIKEN
Samuel N. Alexander
Gene M. AMDAHL

John W. Backus
Robert S. Barton
C. Gordon Bell
Frederick P. Brooks, Jr.
Wesley A. Clark
Fernando J. Corbato
Seymour R. CRAY
Edsgar W. DIJKSTRA
J. Presper ECKERT
Jay W. Forrester
Herman H. Goldstine
Richard W. Hamming
Grace M. HOPPER
Alston S. Householder
David A. Huffman
Kenneth E. Iverson
Tom Kilburn
Donald E. KNUTH
Herman Lukoff
John W. MAUCHLY
Gordon E. Moore
Allen Newell
Robert N. Noyce
Lawrence G. Roberts
George R. Stibitz
Shmuel Winograd
Maurice V. Wilkes
Konrad ZUSE

ANNUAL RECIPIENTS

(With year and achievement as cited by the Computer Society.)

1981 Jeffrey Chuan Chu: "For his early work in electronic computer logic design"

1982 Harry D. Huskey: "For the first parallel computer SWAC"

1982 Arthur Burks: "For his early work in electronic computer logic design"

1984 John Vincent ATANASOFF: "For the first electronic computer with serial memory"

1984 Jerrier A. Haddad: "For his part in the lead IBM 701 design team"

1984 Nicholas C. Metropolis: "For the first solved atomic energy problems on ENIAC"

1984 Nathaniel Rochester: "For the architecture of IBM 702 electronic data processing machines"

1984 Willem L. van der Poel: "For the serial computer ZEBRA"

1985 John G. Kemeny: "For BASIC"

1985 John MCCARTHY: "For LISP and artificial intelligence"

1985 Alan Perlis: "For computer language translation"

1985 Ivan Sutherland: "For the graphics SKETCHPAD"

1985 David J. Wheeler: "For assembly language programming"

1985 Heniz Zemanek: "For computer and computer languages—MAILUEFTERL"

1986 Cuthbert C. Hurd: "For contributions to early computing"

1986 Peter Naur: "For computer language development"

1986 James H. Pomerene: "For IAS and Harvest computers"

1986 Adriann van Wijngaarden: "For ALGOL 68"

1987 Robert E. Everett: "For Whirlwind"

1987 Reynold B. Johnson: "For RAMAC"

1987 Arthur L. Samuel: "For Adaptive non-numeric processing"

1987 Nicklaus E. WIRTH: "For PASCAL"

1988 Freidrich L. Bauer: "For computer stacks"

1988 Marcian E. Hoff, Jr.: "For microprocessor on a chip"

1989 John Cocke: "For instruction pipelining and RISC concepts"

1989 James A. Weidenhammer: "For high speed I/O mechanisms"

1989 Ralph L. Palmer: "For the IBM 604 electronic calculator"

1989 Mina S. Rees: "For the ONR Computer R&D development beginning in 1946"

1989 Marshall C. Yovits: "For the ONR Computer R&D development beginning in 1946"

1989 F. Joachim Weyl: "For the ONR Computer R&D development beginning in 1946"

1989 Gordon D. Goldstein: "For his work with the Office of Naval Research and computer R&R beginning in 1946"

1990 Werner Buchholz: "For computer architecture"

1990 C. A. R. Hoare: "For programming languages definitions"

1991 Bob O. Evans: "For compatable computers"

1991   Robert W. Floyd: "For early compilers"

1991   Thomas E. Kurtz: "For BASIC"

1992   Stephen W. Dunwell: "For project stretch"

1992   Douglas C. ENGELBART: "For human computer interaction"

1993   Erich Bloch: "For high speed computing"

1993   Jack S. Kilby: "For co-inventing the integrated circuit"

1993   Willis H. Ware: "For the design of IAS and Johnniac computers"

1994   Gerrit A. Blaauw: "In recognition of your contributions to the IBM System/360 Series of computers"

1994   Harlan B. Mills: "In recognition of contributions to Structured Programming"

1994   Dennis M. RITCHIE: "In recognition of contributions to the development of UNIX"

1994   Ken L. Thompson: "For his work with UNIX"

1995   Gerald Estrin: "For significant developments on early computers"

1995   David Evans: "For seminal work on computer graphics"

1995   Butler Lampson: "For early concepts and developments of the PC"

1995   Marvin MINSKY: "For conceptual development of artificial intelligence"

1995   Kenneth Olsen: "For concepts and development of minicomputers"

1996   Angel Angelov: "For computer science technologies in Bulgaria"

1996   Richard F. Clippinger: "For computing laboratory staff member, Aberdeen Proving Ground, who converted the ENIAC to a stored program"

1996   Edgar Frank Codd: "For the invention of the first abstract model for database management"

1996   Norber Fristacky: "For pioneering digital devices"

1996   Victor M. Glushkov: "For digital automation of computer architecture"

1996   Jozef Gruska: "For the development of computer science in former Czechoslovakia with fundamental contributions to the theory of computing and extraordinary organizational activities"

1996   Jiri Horejs: "For informatics and computer science"

1996   Lubomir Georgiev Iliev: "A founder and influential leader of computing in Bulgaria; leader of the team that developed the first Bulgarian computer; made fundamental and continuing contributions to abstract mathematics and software"

1996   Robert E. Kahn: "For the co-invention of the TCP/IP protocols and for originating the Internet program"

1996   Laszlo Kalmar: "For recognition as the developer of a 1956 logical machine and the design of the MIR computer in Hungary"

1996   Antoni Kilinski: "For pioneering work in the construction of the first commercial computers in Poland, and for the development of university curriculum in computer science"

1996   Laszlo Kozma: "For development of the 1930 relay machines, and going on to build early computers in post-war Hungary"

1996   Sergey A. Lebedev: "For the first computer in the Soviet Union"

1996   Alexej A. Lyuponov: "For Soviet cybernetics and programming"

1996   Romuald W. Marczynski: "For pioneering work in the construction of the first Polish digital computers and contributions to fundamental research in computer architecture"

1996   Grigore C. Moisil: "For polyvalent logic switching circuits"

1996   Ivan Plander: "For the introduction of computer hardware technology into Slovakia and the development of the first control computer"

1996   Arnols Reitsakas: "For contributions to Estonia's computer age"

1996   Antonin Svoboda: "For the pioneering work leading to the development of computer research in Czechoslovakia and the design and construction of the SAPO and EPOS computers"

1997   Homer (Barney) Oldfield: "For pioneering work in the development of banking applications through the implementation of ERMA, and the introduction of computer manufacturing to GE"

1997   Francis Elizabeth (Betty) Snyder-Holberton: "For the development of the first sort-merge generator for the Univac which inspired the first ideas about compilation"

1998   Irving John (Jack) Good: "For significant contributions to the field of computing as a cryptologist and statistician during World War II at Bletchley Park, as an early worker and developer of the Colossus at Bletchley Park and on the University of Manchester Mark I, the world's first stored program computer"

1999 Herbert Freeman: "For pioneering work on the first computer built by the Sperry Corporation, the SPEEDAC, and for subsequent contributions to the areas of computer graphics and image processing"

2000 Harold W. Lawson: "For inventing the pointer variable and introducing this concept into PL/I, thus providing for the first time, the capability to flexibly treat linked lists in a general-purpose high level language"

2000 Gennady Stolyarov: "For pioneering development in 'Minsk' series computers' software, of the information systems' software and applications and for data processing and data base management systems concepts dissemination and promotion"

2000 Georgy Lopato: "For pioneering development in Belarus of the 'Minsk' series computers' hardware, of the multicomputer complexes and of the 'RV' family of mobile computers for heavy field conditions"

2001 Vernon Schatz: "For the development of Electronics Funds Transfer which made possible computer to computer commercial transactions via the banking system"

2001 William H. Bridge: "For the marrying of computer and communications technology in the GE DATANET 30, putting terminals on peoples' desks to communicate with and timeshare a computer, leading directly to the development of the personal computer, computer networking and the internet"

2002 Per Brinch Hansen: "For pioneering development in operating systems and concurrent programming, exemplified by work on the RC4000 multiprogramming system, monitors, and Concurrent Pascal"

2002 Robert W. Bemer: "For meeting the world's needs for variant character sets and other symbols via ASCII, ASCII-alternate sets, and escape sequences"

2003 Martin Richards: "For pioneering system software portability through the programming language BCPL widely influential and used in academia and industry for a variety of prominent system software"

2004 Frances (Fran) E. Allen: "For pioneering work establishing the theory and practice of compiler optimization"

2005 [No award given]

2006 Arnold M. Spielberg: "For recognition of contribution to real-time data acquisition and recording that significantly contributed to the definition of modern feedback and control processes"

2006 Mamoru Hosaka: "For recognition of pioneering activities within computing in Japan"

NATIONAL MEDAL OF TECHNOLOGY AND INNOVATION

Given by the President of the United States, the National Medal of Technology and Innovation is "the highest honor bestowed by the President of the United States to America's leading innovators."

COMPUTER-RELATED RECIPIENTS
1985

AT&T BELL LABORATORIES: "For contribution over decades to modern communication systems."

Frederick P. Brooks, Jr., Erich Bloch, and Bob O. Evans, International Business Machines Corp.: "For their contributions to the development of the hardware, architecture and systems engineering associated with the IBM System/360, a computer system and technologies which revolutionized the data processing industry and which helped to make the United States dominant in computer technology for many years."

Steven P. JOBS and Steven WOZNIAK, Apple Computer, Inc.: "For their development and introduction of the personal computer which has sparked the birth of a new industry extending the power of the computer to individual users."

John T. Parsons and Frank L. Stulen, John T. Parsons Company: "For their development and successful demonstration of the numerically-controlled machine tool for the production of three-dimensional shapes, which has been essential for the production of commercial airliners and which is seminal for the growth of the robotics, CAD-CAM, and automated manufacturing industries."

1986

Bernard Gordon, Analogic Corp.: "Father of high-speed analog-to-digital conversion which has been applied to medical, analytical, computer and communications products; founder of two companies with over 2,000 employees and over $100 million in annual sales and creator of a new master's level institute located in Massachusetts to teach engineering leadership and project engineering to engineers."

Reynold B. Johnson, International Business Machines Corp.: "Introduction and development of magnetic disk storage for computers that provided access to virtually unlimited amounts of information in fractions of a second and is the basis for time sharing systems and storage of millions of records. Over $10 billion in annual sales and over 100,000 jobs arose from this development."

William C. Norris, Control Data Corp.: "Advancement of micro electronics and computer technology and creation of one of the Fortune 500—Control Data Corporation—which has over $5 billion in annual sales and over 50,000 employees."

1987

Robert N. Noyce, Intel Corp.: "For his inventions in the field of semiconductor integrated circuits, for his leading role in the establishment of the microprocessor which has led to much wider use of more powerful computers, and for his leadership of research and development in these areas, all of which have had profound consequences both in the United States and throughout the world."

1988
Robert H. Dennard, IBM T.J. Watson Research Center: "For invention of the basic one-transistor dynamic memory cell used worldwide in virtually all modern computers."

David Packard, Hewlett-Packard Company: "For extraordinary and unselfish leadership in both industry and government, particularly in widely diversified technological fields which strengthened the competitiveness and defense capabilities of the United States."

1989
Jay W. Forrester, Massachusetts Institute of Technology and Robert R. Everett, The MITRE Corp.: "For their creative work in developing the technologies and applying computers to real-time applications. Their important contributions proved vital to national and free world defense and opened a new era of world business."

1990
John V. ATANASOFF, Iowa State University (Ret.): "For his invention of the electronic digital computer and for contributions toward the development of a technically trained U.S. work force."

Jack St. Clair Kilby, Jack Kilby Co.: "For his invention and contributions to the commercialization of the integrated circuit and the silicon thermal print-head; for his contributions to the development of the first computer using integrated circuits; and for the invention of the hand-held calculator, and gate array."

John S. Mayo, AT&T Bell Laboratories: "For providing the technological foundation for information-age communications, and for overseeing the conversion of the national switched telephone network from analog to a digital-based technology for virtually all long-distance calls both nationwide and between continents."

Gordon E. MOORE, Intel Corp.: "For his seminal leadership in bringing American industry the two major postwar innovations in microelectronics—large-scale integrated memory and the microprocessor—that have fueled the information revolution."

1991
C. Gordon Bell, Stardent Computers: "For his continuing intellectual and industrial achievements in the field of computer design; and for his leading role in establishing cost-effective, powerful computers which serve as a significant tool for engineering, science and industry."

John Cocke, International Business Machines Corp.: "For his development and implementation of Reduced Instruction Set Computer (RISC) architecture that significantly increased the speed and efficiency of computers, thereby enhancing U.S. technological competitiveness."

Grace Murray HOPPER, U.S. Navy (Ret.)/Digital Equipment Corp.: "For her pioneering accomplishments in the development of computer programming languages that simplified computer technology and opened the door to a significantly larger universe of users."

1992
William H. Gates III, Microsoft Corp.: "For his early vision of universal computing at home and in the office; for his technical and business management skills in creating a world-wide technology company; and for his contribution to the development of the personal computer industry."

1993
Kenneth H. Olsen, Digital Equipment Corp.: "For his contributions to the development and use of computer technology; and for his entrepreneurial contribution to American business."

1994
[No computer-related recipients]

1995
Edward R. McCracken, Silicon Graphics, Inc.: "For his groundbreaking work in the areas of affordable 3D visual computing and super computing technologies; and for his technical and leadership skills in building Silicon Graphics, Inc., into a global advanced technology company."

IBM Team: Praveen Chaudhari, IBM TJ Watson Research Center; Jerome J. Cuomo, North Carolina State University (formerly with IBM); and Richard J. Gambino, State University of New York at Stony Brook (formerly with IBM): "For the discovery and development of a new class of materials—the amorphous magnetic materials—that are the basis of erasable, read-write, optical storage technology, now the foundation of the worldwide magnetic-optic disk industry."

1996
James C. Morgan, Applied Materials, Inc.: "For his leadership of 20 years developing the U.S. semiconductor manufacturing equipment industry, and for his vision in building Applied Materials, Inc. into the leading equipment company in the world, a major exporter and a global technology pioneer which helps enable Information Age technologies for the benefit of society."

1997
Vinton Gray CERF, MCI, and Robert E. Kahn, Corporation for National Research Initiatives: "For creating and sustaining development of Internet Protocols and continuing to provide leadership in the emerging industry of internetworking."

1998
Kenneth L. Thompson, Bell Laboratories, and Dennis M. RITCHIE, Lucent Technologies: "For their invention of UNIX® operating system and the C programming language, which together have led to enormous growth of an entire industry,

thereby enhancing American leadership in the Information Age."

1999

Raymond KURZWEIL, founder, chairman, and chief executive officer, Kurzweil Technologies, Inc.: "For pioneering and innovative achievements in computer science such as voice recognition, which have overcome many barriers and enriched the lives of disabled persons and all Americans."

Robert Taylor (Ret.): "For visionary leadership in the development of modern computing technology, including computer networks, the personal computer and the graphical user interface."

2000

Douglas C. ENGELBART, director, Bootstrap Institute: "For creating the foundations of personal computing including continuous, real-time interaction based on cathode-ray tube displays and the mouse, hypertext linking, text editing, on-line journals, shared-screen teleconferencing, and remote collaborative work. More than any other person, he created the personal computing component of the computer revolution."

The IBM CORPORATION: "For 40 years of innovations in the technology of hard disk drives and information storage products. IBM is widely recognized as the world's leader in basic data storage technologies, and holds over 2,000 U.S. patents. IBM is a top innovator of component technologies—such as flying magnetic heads (thin film heads, and magneto resistive heads), film disks, head accessing systems, digital signal processing and coding, as well as innovative hard disk drive systems. Some specific IBM inventions are used in every modern hard drive today: thin film inductive heads, MR and GMR heads, rotary actuators, sector servos and advanced disk designs. These advances outran foreign hard disk technology and enabled the U.S. industry to maintain the lead it holds today."

2001

Arun N. Netravali, Chief Scientist, Lucent Technologies and Past President of Bell Labs: "For his leadership in the field of communication systems; for pioneering contributions that transformed TV from analog to digital, enabling numerous integrated circuits, systems and services in broadcast TV, CATV, DBS, HDTV, and multimedia over the Internet; and for technical expertise and leadership, which have kept Bell Labs at the forefront in communications technology."

Jerry M. Woodall, Yale University: "For his pioneering role in the research and development of compound semiconductor materials and devices; for the invention and development of technologically and commercially important compound semiconductor heterojunction materials, processes, and related devices, such as light-emitting diodes, lasers, ultra-fast transistors, and solar cells."

2002

Calvin H. Carter, Cree, Inc.: "For his exceptional contributions to the development of silicon carbide wafers, leading to new industries in wide bandgap semiconductors and enabling other new industries in efficient blue, green, and white light, full-color displays, high-power solid-state microwave amplifiers, more efficient/compact power supplies, higher efficiency power distribution/transmission systems, and gemstones."

Carver A. Mead, California Institute of Technology: "For his pioneering contributions to microelectronics that include spearheading the development of tools and techniques for modern integrated-circuit design, laying the foundation for fabless semiconductor companies, catalyzing the electronic-design automation field, training generations of engineers that have made the United States the world leader in microelectronics technology, and founding more than twenty companies."

Team of Nick Holonyak, Jr. (University of Illinois at Urbana-Champaign), M. George Craford (Lumileds Lighting Corp.) and Russell Dean Dupuis (Georgia Institute of Technology): "For contributions to the development and commercialization of light-emitting diode (LED) technology, with applications to digital displays, consumer electronics, automotive lighting, traffic signals, and general illumination."

2003

Robert M. Metcalfe: "For leadership in the invention, standardization, and commercialization of the Ethernet."

Watts S. Humphrey: "For his vision of a discipline for software engineering, for his work toward meeting that vision, and for the resultant impact on the U.S. Government, industry, and academic communities."

2004

Ralph H. Baer: "For his groundbreaking and pioneering creation, development and commercialization of interactive video games, which spawned related uses, applications, and mega-industries in both the entertainment and education realms."

2005

Semiconductor Research Corporation: "For building the world's largest and most successful university research force to support the rapid growth and advance of the semiconductor industry; for proving the concept of collaborative research as the first high-tech research consortium; and for creating the concept and methodology that evolved into the International Technology Roadmap for Semiconductors."

Xerox Corporation: "For over 50 years of innovation in marking, materials, electronics communications, and software that created the modern reprographics, digital printing, and print-on-demand industries."

2006–2007

[No computer-related recipients]

APPENDIX IV
COMPUTER-RELATED ORGANIZATIONS

The following is a list of some important computer-related organizations, including contact information.

GENERAL COMPUTER SCIENCE ORGANIZATIONS

American Society for Information Science (http:www.asis.org/) 1320 Fenwick Lane, Suite 510, Silver Spring, MD 20910. Telephone: (301) 495-0900 e-mail: asis@asis.org

Association for Computing Machinery (http://www.acm.org/) 2 Penn Plaza, Suite 701, New York, NY 10121-0701. Telephone: (800) 342-6626 e-mail: acmhelp@acm.org

Computing Research Association (http://www.cra.org) 1100 Seventeenth Street, NW, Suite 507, Washington, DC 20036-4632. Telephone: (202) 234-2111 e-mail: webmaster@cra.org

IEEE Computer Society (http:www.computer.org) 1730 Massachusetts Ave. NW, Washington, DC 20036-1992. Telephone: (202) 371-0101 e-mail: membership@computer.org

Software Engineering Institute (http://sei.cmu.edu) 4500 Fifth Ave., Pittsburgh, PA 15213-2612. Telephone: (888) 201-4479 e-mail: customer-relations@sei.cmu.edu

APPLICATION AND INDUSTRY-SPECIFIC GROUPS

AeA (formerly American Electronics Association) (http://www.aeanet.org/) 601 Pennsylvania Avenue NW, Suite 600, North Building, Washington, DC 20004. Telephone: (202) 682-9110 e-mail: Web forms

American Association for Artificial Intelligence (http://www.aaai.org/) 445 Burgess Drive, Suite 100, Menlo Park, CA 94025-3442. Telephone: 650-328-3123 e-mail: info7contact@aaai.org

American Design Drafting Association (http://www.adda.org) 105 East Main St., Newham, TN 38059. Telephone: (731) 627-0802 e-mail: Web form

American Society for Photogrammetry and Remote Sensing (http://www.asprs.org) 5410 Grosvenor Lane, Suite 210, Bethesda, MD 20814-2160. Telephone: (301) 493-0290 e-mail: asprs@.org

American Statistical Association (http://www.amstat.org) 732 North Washington St., Alexandria, VA 22314-1943. Telephone: (703) 684-1221 e-mail: asainfo@amstat.org

Association for Library and Information Science Education (http:www.alise.org) 68 E. Wacker Place, Suite 1900, Chicago, IL 60601-7246. Telephone: (312) 795-0996 e-mail: contact@alise.org

Association for Multimedia Communication (http://www.amcomm.org) P.O. Box 10645, Chicago, IL 60610. Telephone: (773) 276-9320 e-mail: Web form

Association of American Geographers (http://www.aag.org) 1710 16th St. NW, Washington, DC 20009-3198. Telephone: (202) 234-1450 e-mail: gaia@aag.org

Association of Information Technology Professionals (http://www.aitp.org) 401 North Michigan Avenue, Suite 2400, Chicago, IL 60611-4267. Telephone: (312) 673-4793 e-mail: Web form

CAM-I (Computer-Aided Manufacturing International) (http://cami.affiniscape.com) 6836 Bee Cave, Suite 256, Austin, TX 78746. Telephone: (512) 617-6428 e-mail: Web form

Computing Technology Industry Association (CompTIA) (http://www.comptia.org) 1815 S. Meyers Road, Suite 300 Oakbrook Terrace, IL 60181-5228. Telephone: (630) 678-8300 e-mail: Web form

Digital Library Federation (http://www.digilib.org) 1755 Massachusetts Ave. NW, Suite 500, Washington, DC 20036-2124. Telephone: (202) 939-4761 e-mail: dlfinfo@clir.org

Electronics Industries Alliance (http://www.eia.org/) 2500 Wilson Blvd., Arlington, VA 22201. Telephone: (703) 907-7500 e-mail: Web form

Information Technology Association of America (http://www.itaa.org) 1401 Wilson Blvd., Suite 1100, Arlington, VA 22209. Telephone: (703) 522-5005 e-mail: Web directory

International Game Developer's Association (http://www.igda.org) 19 Mantua Road, Mt. Royal, NJ 08061. Telephone: (856) 423-2990 e-mail: contact@igda.org

International Society for Technology in Education (http://www.iste.org) 1710 Rhode Island Ave. NW, Suite 900, Washington, DC 20036. Telephone: (800) 336-5191 e-mail: iste@org

International Technology Law Association (http://www.itechlaw.org) 401 Edgewater Place, Suite 600, Wakefield, MA 01800. Telephone: (781) 876-8877 e-mail: office@itechlaw.org

International Webmasters Association (http://www.irwa.org/) 119 E. Union St., Suite F, Pasadena, CA 91103. Telephone: (626) 449-3709 e-mail: via Web links

Libraries for the Future (http://www.lff.org) 27 Union Square West, Suite 204, New York, NY 10003. Telephone: (646) 336-6236 e-mail: info@lff.org

Library and Information Technology Association (http://www.lita.org) American Library Association, 50 East Huron St., Chicago, IL 60611-2795. Telephone: (800) 545-2433 e-mail: library@ala.org

Office Automation Society International (http://www.pstcc.cc.tn.us/ost/oasi.html) 5170 Meadow Wood Blvd., Lyndhurst, OH 44124. Telephone: (216) 461-4803 e-mail: jbdyke@aol.com

Robotics Industries Association (http://www.robotics.org) 900 Victors Way, Suite 140, P.O. Box 3724, Ann Arbor, MI 48106. Telephone: (734) 994-6088 e-mail: webmaster@robotics.org

SIGGRAPH [Graphics special interest group of the Association for Computing Machinery] (http://www.siggraph.org). e-mail: Web links

Society for Information Management (http://www.simnet.org/) 401 N. Michigan Ave., Chicago, IL 60611-4267. Telephone: 312 644-6610 e-mail: info@simnet.org

Society for Modeling and Simulation International (http://www.scs.org) P.O. Box 17900 San Diego, CA 92177-7900. Telephone: (858) 277-3888 e-mail: info@scs.org

Society for Technical Communication (http://www.stc.org/) 901 N. Stuart St., Suite 904, Arlington, VA 22203-1854. Telephone: (703) 522-4114 e-mail: stc@stc.org

Software & Information Industry Association (http://www.siia.org/) 1090 Vermont Ave. NW, Sixth Floor, Washington, DC 20005-4095. Telephone: (202) 289-7442 e-mail: Web form

Telecommunications Industry Association (http://www.tiaonline.org) 2500 Wilson Blvd., Suite 300, Arlington, VA 22201-3834. Telephone: (703) 907-7700 e-mail: tia@tiaonline.org

GOVERNMENT, STANDARDS AND SECURITY ORGANIZATIONS

American National Standards Institute (ANSI) (http://ansi.org) 1819 L Street NW, 6th Floor, Washington, DC 20036. Telephone: (202) 293-8020 e-mail: info@ansi.org

Computer Emergency Response Team (CERT) (http://www.cert.org) CERT Coordination Center, Software Engineering Institute, Carnegie Mellon University, Pittsburgh, PA 15213-3890. Telephone: (412) 268-7090 e-mail: cert@cert.org

Computer Security Institute (http://www.gocsi.com/) 600 Harrison St., San Francisco, CA 94107. Telephone: (415) 947-6320 e-mail: csi@cmp.com

Defense Advanced Research Projects Agency (DARPA) (http://www.darpa.gov). 3701 North Fairfax Drive, Arlington, VA 22203-1714. Telephone: (571) 218-4219 e-mail: Web forms

Information Systems Security Association (http://www.issa.org) 9200 SW Barbour Blvd. #119-333 Portland, OR 97219. Telephone: (866) 349-5818 e-mail: Web forms

Institute for the Certification of Computing Professionals (http://www.iccp.org) 2350 East Devon Ave., Suite 115, Des Plaines, IL 60018-4610. Telephone: (847) 299-4227 e-mail: office@iccp.org

International Organization for Standardization (ISO) (http://www.iso.org) 1, ch de la Voie-Creuse, Case postale 56, CH-1211 Geneva 20, Switzerland. Telephone: +41 22 749 01 11 e-mail: Web forms

Internet Society (http://www.isoc.org/) 1775 Wiehle Ave., Suite 102, Reston, VA 20190-5108. Telephone: (703) 326-9880 e-mail: info@isoc.org

National Center for Supercomputing Applications (NCSA) (http://www.ncsa.uiuc.edu) University of Illinois at Urbana-Champaign, 1205 W. Clark St., Room 1008, Urbana, IL 61801. Telephone: (217) 244-0710 e-mail: tlbarker@ncsa.uiue.edu

National Telecommunications and Information Administration (http://www.ntia.doc.gov/) U.S. Dept. of Commerce, 1401 Constitution Ave. NW, Washington, DC 20230. Telephone: (202) 482-1840 e-mail: Web directory

Quality Assurance Institute Worldwide (http://www.qaiworldwide.com.qai.html) 2101 Park Center Drive, Suite 200, Orlando, FL 32835-7614. Telephone: (407) 363-1111 e-mail: Web directory

Urban and Regional Information Systems Association (http://www.urisa.org) 1460 Renaissance Drive, Suite 305, Park Ridge, IL 60068. Telephone: (847) 824-6300 e-mail: Web directory

World Wide Web Consortium (www.w3c.org) Massachusetts Institute of Technology, 32 Vassar St., Room 32-G515, Cambridge, MA 02139. Telephone: (617) 253-2613 e-mail: Web links

ADVOCACY GROUPS

Association for Women in Computing (http://www.awc-hq.org) 41 Sutter St., Suite 1006, San Francisco, CA 94104. Telephone: (415) 905-4663 e-mail: info@awc-hq.org

Black Data Processing Associates (http://www.bdpa.org) 6301 Ivy Lane, Suite 700, Greenbelt, MD 20770. Telephone: (800) 727-BDPA e-mail: Web forms

Center for Democracy and Technology (http://www.cdt.org) 1634 Eye St., NW, #100, Washington, DC 20006. Telephone: (202) 637-9800 e-mail: Web form

Computer Professionals for Social Responsibility (http://www.cpsr.org) 1370 Mission St., 4th Floor, San Francisco, CA 94103-2654. Telephone: (415) 839-9355 e-mail: cpsr@cpsr.org

Electronic Frontier Foundation (http://www.eff.org) 454 Shotwell St., San Francisco, CA 94110-1914. Telephone: (415) 436-9333 e-mail: information@eff.org

Electronic Privacy Information Center (http://www.epic.org) 1718 Connecticut Ave. NW, Suite 200, Washington, DC 20009. Telephone: (202) 483-1140 e-mail: Web form

Women in Technology (http://www.womenintechnology.com) 717 Princess St., Alexandria, VA 22314. Telephone: (703) 683-4033 e-mail: staff@womenintechnology.org

INDEX